P9-EMQ-148

Second Edition

EDITOR-IN-CHIEF

Renato Dulbecco
The Salk Institute

EDITORIAL ADVISORY BOARD

John Abelson
California Institute of Technology

Peter Andrews
Natural History Museum, London

John A. Barranger
University of Pittsburgh

R. J. Berry
University College, London

Konrad Bloch
Harvard University

Floyd Bloom
The Scripps Research Institute

Norman E. Borlaug
Texas A&M University

Charles L. Bowden
University of Texas Health Science Center at San Antonio

Ernesto Carafoli
ETH-Zentrum

Stephen K. Carter
Bristol-Myers Squibb Corporation

Edward C. H. Carterette
University of California, Los Angeles

David Carver
University of Medicine and Dentistry of New Jersey

Joel E. Cohen
Rockefeller University and Columbia University

Michael E. DeBakey
Baylor College of Medicine

Eric Delson
American Museum of Natural History

W. Richard Dukelow
Michigan State University

Myron Essex
Harvard University

Robert C. Gallo
Institute for Human Virology

Joseph L. Goldstein
University of Texas Southwestern Medical Center

I. C. Gunsalus
University of Illinois, Urbana–Champaign

Osamu Hayaishi
Osaka Bioscience Institute

Leonard A. Herzenberg
Stanford University Medical Center

Kazutomo Imahori
Mitsubishi-Kasei Institute of Life Science, Tokyo

Richard T. Johnson
Johns Hopkins University Medical School

Yuet Wai Kan
University of California, San Francisco

Bernard Katz
University College, London

Seymour Kaufman
National Institutes of Health

Ernst Knobil
University of Texas Health Science Center at Houston

Glenn Langer
University of California Medical Center, Los Angeles

Robert S. Lawrence
Johns Hopkins University

James McGaugh
University of California, Irvine

Henry M. McHenry
University of California, Davis

Philip W. Majerus
Washington University School of Medicine

W. Walter Menninger
Menninger Foundation

Terry M. Mikiten
University of Texas Health Science Center at San Antonio

Beatrice Mintz
Institute for Cancer Research, Fox Chase Cancer Center

Harold A. Mooney
Stanford University

Arno G. Motulsky
University of Washington School of Medicine

Marshall W. Nirenberg
National Institutes of Health

G. J. V. Nossal
Walter and Eliza Hall Institute of Medical Research

Mary Osborn
Max Planck Institute for Biophysical Chemistry

George E. Palade
University of California, San Diego

Mary Lou Pardue
Massachusetts Institute of Technology

Ira H. Pastan
National Institutes of Health

David Patterson
University of Colorado Medical Center

Philip Reilly
Shriver Center for Mental Retardation

Arthur W. Rowe
New York University Medical Center

Ruth Sager
Dana-Farber Cancer Institute

Alan C. Sartorelli
Yale University School of Medicine

Neena B. Schwartz
Northwestern University

Bernard A. Schwetz
National Center for Toxicological Research, FDA

Nevin S. Scrimshaw
United Nations University

Michael Sela
Weizmann Institute of Science, Israel

Satimaru Seno
Shigei Medical Research Institute, Japan

Phillip Sharp
Massachusetts Institute of Technology

E. R. Stadtman
National Institutes of Health

P. K. Stumpf
University of California, Davis (emeritus)

William Trager
Rockefeller University (emeritus)

Arthur C. Upton
University of Medicine and Dentistry of New Jersey, Robert Wood Johnson Center

Itaru Watanabe
Kansas City VA Medical Center

David John Weatherall
Oxford University, John Radcliffe Hospital

Klaus Weber
Max Planck Institute for Biophysical Chemistry

Thomas H. Weller
Harvard School of Public Health

Harry A. Whitaker
Université du Québec á Montréal

MONTGOMERY COLLEGE
TAKOMA PARK CAMPUS LIBRARY
TAKOMA PARK, MARYLAND

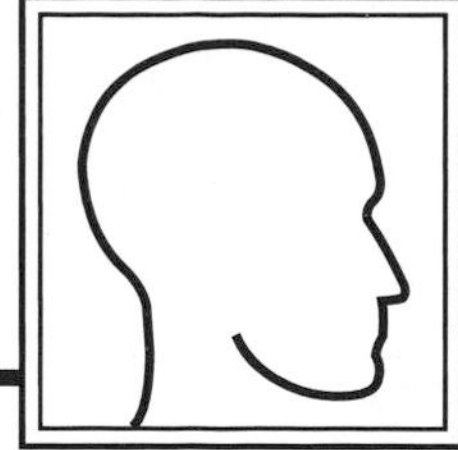

ENCYCLOPEDIA OF HUMAN BIOLOGY

Volume 7 **Po–Se**

Second Edition

Editor-in-Chief
RENATO DULBECCO
The Salk Institute
La Jolla, California

WITHDRAWN

ACADEMIC PRESS
San Diego London Boston New York Sydney Tokyo Toronto

NOV 2 2 1999

240955

This book is printed on acid-free paper. ♾

Copyright © 1997, 1991 by ACADEMIC PRESS

All Rights Reserved.
No part of this publication may be reproduced or transmitted in any form or by any means, electronic or mechanical, including photocopy, recording, or any information storage and retrieval system, without permission in writing from the publisher.

Academic Press
a division of Harcourt Brace & Company
525 B Street, Suite 1900, San Diego, California 92101-4495, USA
http://www.apnet.com

Academic Press Limited
24-28 Oval Road, London NW1 7DX, UK
http://www.hbuk.co.uk/ap/

Library of Congress Cataloging-in-Publication Data

Encyclopedia of human biology / edited by Renato Dulbecco. -- 2nd ed.
p. cm.
Includes bibliographical references and index.
ISBN 0-12-226970-5 (alk. paper: set). -- ISBN 0-12-226971-3 (alk. paper: v. 1). -- ISBN 0-12-226972-1 (alk. paper: v. 2). -- ISBN 0-12-226973-X (alk. paper: v. 3). -- ISBN 0-12-226974-8 (alk. paper: v. 4). -- ISBN 0-12-226975-6 (alk. paper: v. 5). -- ISBN 0-12-226976-4 (alk. paper: v. 6). -- ISBN 0-12-226977-2 (alk. paper : v. 7). -- ISBN 0-12-226978-0 (alk. paper: v. 8). -- ISBN 0-12-226979-9 (alk. paper: v. 9)
1. Human biology--Encyclopedias. I. Dulbecco, Renato, date.
[DNLM: 1. Biology--encyclopedias. 2. Physiology--encyclopedias.
QH 302.5 E56 1997]
QP11.E53 1997
612'.003-dc21
DNLM/DLC
for Library of Congress 97-8627
CIP

PRINTED IN THE UNITED STATES OF AMERICA
97 98 99 00 01 02 EB 9 8 7 6 5 4 3 2 1

CONTENTS OF VOLUME 7

Contents for each volume of the Encyclopedia appears in Volume 9.

S

PREFACE TO THE FIRST EDITION

We are in the midst of a period of tremendous progress in the field of human biology. New information appears daily at such an astounding rate that it is clearly impossible for any one person to absorb all this material. The *Encyclopedia of Human Biology* was conceived as a solution: an informative yet easy-to-use reference. The Encyclopedia strives to present a complete overview of the current state of knowledge of contemporary human biology, organized to serve as a solid base on which subsequent information can be readily integrated. The Encyclopedia is intended for a wide audience, from the general reader with a background in science to undergraduates, graduate students, practicing researchers, and scientists.

Why human biology? The study of biology began as a correlate of medicine with the human, therefore, as the object. During the Renaissance, the usefulness of studying the properties of simpler organisms began to be recognized and, in time, developed into the biology we know today, which is fundamentally experimental and mainly involves nonhuman subjects. In recent years, however, the identification of the human as an autonomous biological entity has emerged again—stronger than ever. Even in areas where humans and other animals share a certain number of characteristics, a large component is recognized only in humans. Such components include, for example, the complexity of the brain and its role in behavior or its pathology. Of course, even in these studies, humans and other animals share a certain number of characteristics. The biological properties shared with other species are reflected in the Encyclopedia in sections of articles where results obtained in nonhuman species are evaluated. Such experimentation with nonhuman organisms affords evidence that is much more difficult or impossible to obtain in humans but is clearly applicable to us.

Guidance in fields with which the reader has limited familiarity is supplied by the detailed index volume. The articles are written so as to make the material accessible to the uninitiated; special terminology either is avoided or, when used, is clearly explained in a glossary at the beginning of each article. Only a general knowledge of biology is expected of the reader; if specific information is needed, it is reviewed in the same section in simple terms. The amount of detail is kept within limits sufficient to convey background information. In many cases, the more sophisticated reader will want additional information; this will be found in the bibliography at the end of each article. To enhance the long-term validity of the material, untested issues have been avoided or are indicated as controversial.

The material presented in the Encyclopedia was produced by well-recognized specialists of experience and competence and chosen by a roster of outstanding scientists including ten Nobel laureates. The material was then carefully reviewed by outside experts. I have reviewed all the articles and evaluated their contents in my areas of competence, but my major effort has been to ensure uniformity in matters of presentation, organization of material, amount of detail, and degree of documentation, with the goal of presenting in each subject the most advanced information available in easily accessible form.

Renato Dulbecco

PREFACE TO THE SECOND EDITION

The first edition of the *Encyclopedia of Human Biology* has been very successful. It was well received and highly appreciated by those who used it. So one may ask: Why publish a second edition? In fact, the word "encyclopedia" conveys the meaning of an opus that contains immutable information, forever valid. But this depends on the subject. Information about historical subjects and about certain branches of science is essentially immutable. However, in a field such as human biology, great changes occur all the time. This is a field that progresses rapidly; what seemed to be true yesterday may not be true today. The new discoveries constantly being made open new horizons and have practical consequences that were not even considered previously. This change applies to all fields of human biology, from genetics to structural biology and from the intricate mechanisms that control the activation of genes to the biochemical and medical consequences of these processes.

These are the reasons for publishing a second edition. Although much of the first edition is still valid, it lacks the information gained in the six years since its preparation. This new edition updates the information to what we know today, so the reader can be confident of its full validity. All articles have been reread by their authors, who modified them when necessary to bring them up-to-date. Many new articles have also been added to include new information.

The principles followed in preparing the first edition also apply to the second edition. All new articles were contributed by specialists well known in their respective fields. Expositional clarity has been maintained without affecting the completeness of the information. I am convinced that anyone who needs the information presented in this encyclopedia will find it easily, will find it accessible, and, at the same time, will find it complete.

Renato Dulbecco

A GUIDE TO USING THE ENCYCLOPEDIA

The *Encyclopedia of Human Biology, Second Edition* is a complete source of information on the human organism, contained within the covers of a single unified work. It consists of nine volumes and includes 670 separate articles ranging from genetics and cell biology to public health, pediatrics, and gerontology. Each article provides a comprehensive overview of the selected topic to inform a broad spectrum of readers from research professionals to students to the interested general public.

In order that you, the reader, derive maximum benefit from your use of the Encyclopedia, we have provided this Guide. It explains how the Encyclopedia is organized and how the information within it can be located.

ORGANIZATION

The *Encyclopedia of Human Biology, Second Edition* is organized to provide the maximum ease of use for its readers. All of the articles are arranged in a single alphabetical sequence by title. Articles whose titles begin with the letters A to Bi are in volume 1, articles with titles from Bl to Com are in Volume 2, and so on through Volume 8, which contains the articles from Si to Z.

Volume 9 is a separate reference volume providing a Subject Index for the entire work. It also includes a complete Table of Contents for all nine volumes, an alphabetical list of contributors to the Encyclopedia, and an Index of Related Titles. Thus Volume 9 is the best starting point for a search for information on a given topic, via either the Subject Index or Table of Contents.

So that they can be easily located, article titles generally begin with the key word or phrase indicating the topic, with any descriptive terms following. For example, "Calcium, Biochemistry" is the article title rather than "Biochemistry of Calcium" because the specific term *calcium* is the key word rather than the more general term *biochemistry*. Similarly "Protein Targeting, Basic Concepts" is the article title rather than "Basic Concepts of Protein Targeting."

TABLE OF CONTENTS

A complete Table of Contents for the *Encyclopedia of Human Biology, Second Edition* appears in Volume 9. This list of article titles represents topics that have been carefully selected by the Editor-in-Chief, Dr. Renato Dulbecco, and the members of the Editorial Advisory Board (see p. ii for a list of the Board members). The Encyclopedia provides coverage of 35 specific subject areas within the overall field of human biology, ranging alphabetically from Behavior to Virology.

In addition to the complete Table of Contents found in Volume 9, the Encyclopedia also provides an individual table of contents at the front of each volume. This lists the articles included within that particular volume.

INDEX

The Subject Index in Volume 9 contains more than 4200 entries. The subjects are listed alphabetically and indicate the volume and page number where information on this topic can be found.

ARTICLE FORMAT

Articles in the *Encyclopedia of Human Biology, Second Edition* are arranged in a single alphabetical list by title. Each new article begins at the top of a right-hand page, so that it may be quickly located. The author's name and affiliation are displayed at the beginning of the article. The article is organized according to a standard format, as follows:

- Title and author
- Outline
- Glossary
- Defining statement
- Body of the article
- Bibliography

OUTLINE

Each article in the Encyclopedia begins with an outline that indicates the general content of the article. This outline serves two functions. First, it provides a brief preview of the article, so that the reader can get a sense of what is contained there without having to leaf through the pages. Second, it serves to highlight important subtopics that will be discussed within the article. For example, the article "Gene Mapping" includes the subtopic "DNA Sequence and the Human Genome Project."

The outline is intended as an overview and thus it lists only the major headings of the article. In addition, extensive second-level and third-level headings will be found within the article.

GLOSSARY

The Glossary contains terms that are important to an understanding of the article and that may be unfamiliar to the reader. Each term is defined in the context of the particular article in which it is used. Thus the same term may appear as a Glossary entry in two or more articles, with the details of the definition varying slightly from one article to another. The Encyclopedia includes approximately 5000 glossary entries.

DEFINING STATEMENT

The text of each article in the Encyclopedia begins with a single introductory paragraph that defines the topic under discussion and summarizes the content of the article. For example, the article "Free Radicals and Disease" begins with the following statement:

A FREE RADICAL is any species that has one or more unpaired electrons. The most important free radicals in a biological system are oxygen- and nitrogen-derived radicals. Free radicals are generally produced in cells by electron transfer reactions. The major sources of free radical production are inflammation, ischemia/reperfusion, and mitochondrial injury. These three sources constitute the basic components of a wide variety of diseases. . . .

CROSS-REFERENCES

Many of the articles in the Encyclopedia have cross-references to other articles. These cross-references appear within the text of the article, at the end of a paragraph containing relevant material. The cross-references indicate related articles that can be consulted for further information on the same topic, or for other information on a related topic. For example, the article "Brain Evolution" contains a cross reference to the article "Cerebral Specialization."

BIBLIOGRAPHY

The Bibliography appears as the last element in an article. It lists recent secondary sources to aid the reader in locating more detailed or technical information. Review articles and research papers that are important to an understanding of the topic are also listed.

The bibliographies in this Encyclopedia are for the benefit of the reader, to provide references for further reading or research on the given topic. Thus they typically consist of no more than ten to twelve entries. They are not intended to represent a complete listing of all materials consulted by the author or authors in preparing the article.

COMPANION WORKS

The *Encyclopedia of Human Biology, Second Edition* is one of an extensive series of multivolume reference works in the life sciences published by Academic Press. Other such works include the *Encyclopedia of Cancer, Encyclopedia of Virology, Encyclopedia of Immunology,* and *Encyclopedia of Microbiology,* as well as the forthcoming *Encyclopedia of Reproduction.*

Polarized Epithelial Cells

PRASAD DEVARAJAN
JON S. MORROW
Yale University School of Medicine

GLOSSARY

Cell junctions Regions, usually along the lateral membrane, where adjacent cells contact each other. Junctions may be impermeant (e.g., tight or occludens junctions) or may anchor two cells together (e.g., adherens junction or desmosomes)

Cell polarity A characteristic feature of epithelial cells, in which proteins and organelles are spatially and asymmetrically distributed into functionally distinct domains within the plasma membrane and cytoplasm

Cytoskeleton Structural and motile elements of the cell that provide a supportive scaffold, tracks for the transport of organelles and vesicles, and the framework by which many integral membrane proteins are organized

Endocytosis Process by which cells ingest macromolecules or other materials by a process of membrane invagination. Localized regions of the plasma infold and pinch off to form endocytic vesicles. Two types of endocytosis are recognized: (i) pinocytosis, the nonspecific uptake of water and dissolved molecules; and (ii) receptor-mediated endocytosis, in which specific molecules recognized by receptors on the plasma membrane are selectively taken up. The contents of endocytic vesicles are often either recycled or degraded in lysosomes

Epithelial cell Cells that line body cavities, ducts, blood vessels, or the external surfaces of the body. Collectively they form an epithelium, which is found wherever free surfaces exist in the body. Examples include the skin, the lining of the alimentary tract or airways, or the lining of kidney tubules or blood vessels. Epithelial cells play a protective role, as well as a critical role, in the vectorial movement of fluids, ions, nutrients, and waste products across cells and in the secretion of biologically active molecules into the appropriate extracellular compartment. These properties follow from their polarity

Transcytosis Process by which vesicles containing integral membrane proteins are moved from one polarized region of the cell's membrane to another within the same cell.

Vesicular transport Process by which newly synthesized integral membrane proteins and many secreted proteins are incorporated into vesicles and targeted to the appropriate membrane domain

EPITHELIAL CELLS FORM TIGHTLY APPOSED SUPERficial layers that line the body's surface (e.g., skin), all internal cavities (such as the digestive, respiratory, and genitourinary tracts), and the internal surfaces of both exocrine (e.g., pancreas) and endocrine (e.g., thyroid) glands. Such cells play a critical role in the vectorial movement of fluids, ions, nutrients, and waste products across cells and in the secretion of biologically active molecules into the appropriate extracellular compartment. This selective movement of molecules is achieved by a remarkable asymmetric distribution of the molecules responsible for transport into functionally and spatially distinct domains within the plasma membrane and cytoplasm, a phenomenon referred to as cell polarity. Although most eukaryotic cells require degrees of spatial and temporal organization for processes as diverse as morphogenesis, cell–cell interaction, migration, and vectorial transport, cell polarity has been most extensively studied in epithelial cells and is the focus of this article. This article will identify key epithelial cellular players that are polarized, examine the mechanisms by which they achieve and maintain their asymmetric form and function, and discuss examples of alterations in epithelial cell polarity that contribute to disease states.

ENCYCLOPEDIA OF HUMAN BIOLOGY, Second Edition, VOLUME 7. Copyright © 1997 by Academic Press. All rights of reproduction in any form reserved.

I. GENERAL ORGANIZATION

Polarized epithelial cells are characterized (Fig. 1) by (1) the asymmetric distribution of plasma membrane proteins and lipids into three functionally distinct domains, termed apical (facing the lumen), lateral (in contact with the adjacent cell), and basal (resting on the basement membrane); (2) occluding tight junctions that segregate the plasma membrane into apical and lateral domains; (3) adhesive cell–cell attachments along the lateral domain provided by the zonula adherens and desmosomes; (4) focal adhesions between the basal domain and the basement membrane; and (5) the asymmetric distribution of the cytoskeleton and cytoplasmic organelles.

A. The Apical Membrane: Specialized and Regulated

The apical surface contains most of the proteins required for organ-specific function and is therefore the most specialized domain. Indeed, even within a given tissue, the composition of apical membranes can differ markedly, as exemplified in the kidney, where each nephron segment is tailored to carry out specialized tasks. The apical membrane often contains numerous finger-like evaginations (microvilli) from the luminal surface, which are studded with mechanisms that facilitate the unidirectional movement of ions and nutrients. These include energy-driven pumps (e.g., Ca^{2+}-ATPase, H^+-ATPase, H^+/K^+-ATPase), ion chan-

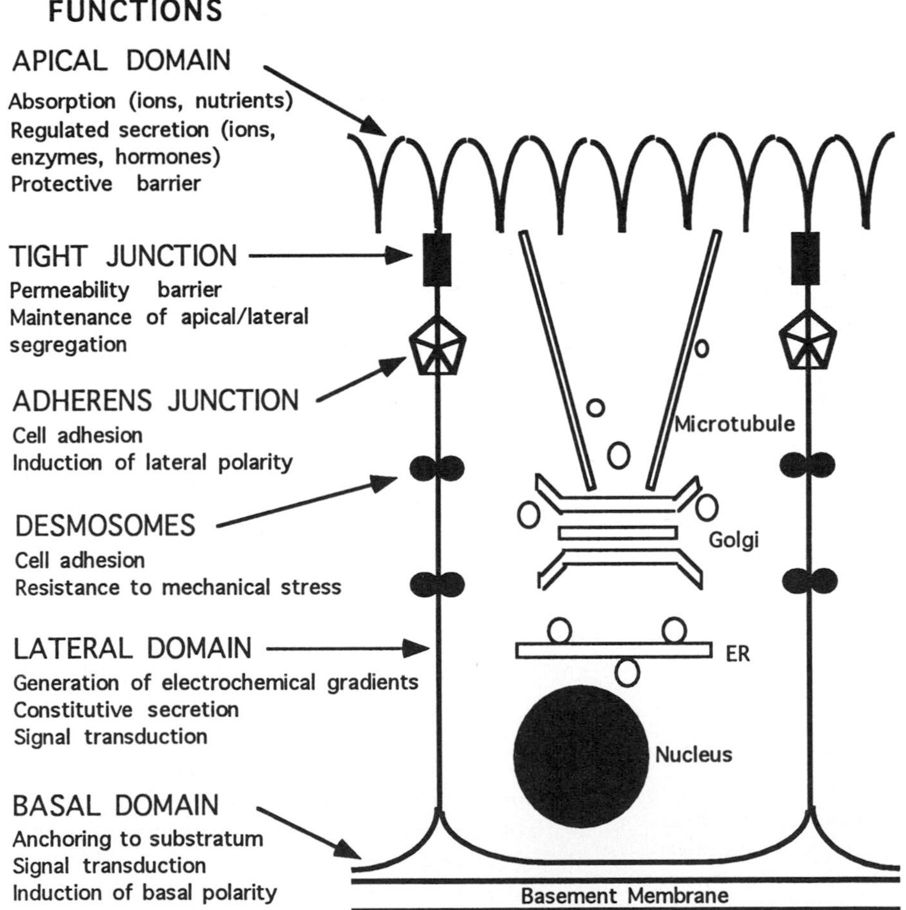

FIGURE 1 General organization of polarized epithelial cells. The apical domain faces the lumen and is specialized to carry out regulated vectorial transport and protective functions. The tight junction segregates the apical and lateral domains and acts as a permeability barrier. The lateral domain contains sites for cell adhesion and proteins for the generation of electrochemical gradients and constitutive secretion. The basal domain anchors the cell to the underlying basement membrane.

nels (e.g., Na^+, Cl^-, K^+, and water channels), cotransporters (e.g., sodium-dependent uptake of glucose and amino acids), and exchangers (e.g., Na^+/H^+ exchanger). Also abundant are luminal enzyme systems (e.g., hydrolases, alkaline phosphatase, aminopeptidases, dipeptidylpeptidases), glycolipids, and a large number of glycosyl phosphatidylinositol (GPI)-anchored proteins.

The characteristic structural and functional organization of the apical membrane is maintained and regulated by a distinct subapical, actin-based cytoskeleton (Fig. 2). Actin is a very abundant, highly conserved protein that can assemble and disassemble between a monomeric (G-actin) form and an oligomeric filamentous (F-actin) form. This interconversion is critical to the role of actin in cytokinesis, maintenance of cell shape, and protein trafficking, and is mediated by a host of actin-binding proteins. Within each microvillus, a central core of F-actin bundles exists with their plus (growing) ends toward the tip and their minus ends inserted into a subapical meshwork termed the terminal web. Microvillar actin filaments are cross-linked into a rigid structure by the bundling proteins villin and fimbrin. Other actin-binding proteins, such as myosin-I and ezrin, are thought to link the actin core to the apical plasma membrane. The terminal web at the base of the microvillus is a dense girdle of actin filaments bound together by bundling proteins (including spectrin and tropomyosin) and is linked laterally to the tight and adherens junctions. In addition to its role as a structural scaffold, the apical actin skeleton also serves as a regulated barrier for exocytosis, as a mechanism for rapid membrane recruitment by apical surface remodeling, as a track for movement of motor-driven secretory vesicles from microtubules to the apical surface, and as a tether for integral membrane proteins. [*See* Membranes, Biological.]

B. Cell Junctions: Sticking Together

Polarized epithelial cells characteristically adhere to each other at several points along the lateral membrane. The tight junctions or zonula occludens are formed just below the apical domain, by the "kissing" of the outer plasma membrane leaflets of adjacent cells. Several molecules that contribute to tight junction structure have been elucidated (Fig. 3). Occludin,

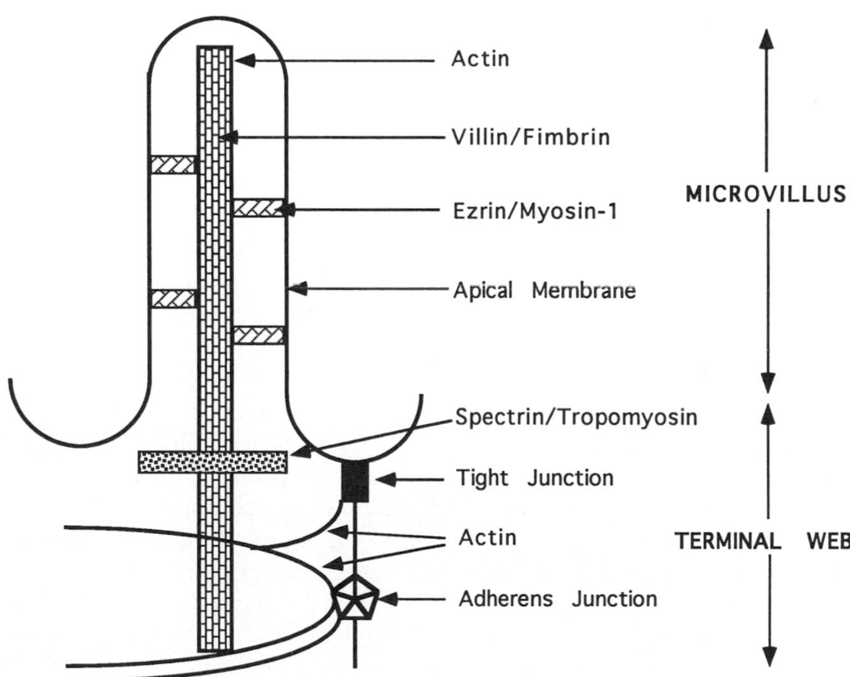

FIGURE 2 Apical and subapical actin-based skeleton. Within each microvillus is an actin core that inserts into the terminal web. Microvillar actin is bundled by villin/fimbrin and is linked to the plasma membrane via ezrin/myosin. Terminal web actin is bundled by spectrin/tropomyosin and is linked laterally to tight and adherens junctions.

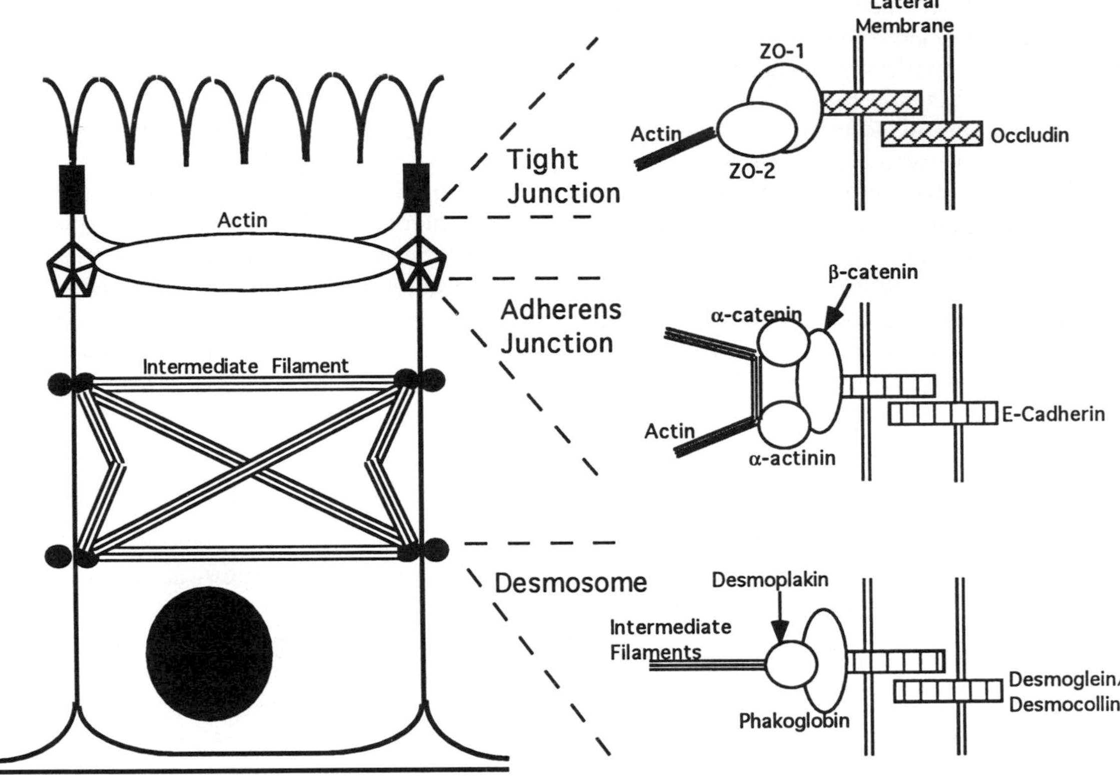

FIGURE 3 Cell junctions at the lateral membrane. Tight junctions form at the tip of the lateral domain. The core molecule is occludin, which is linked to actin via ZO-1 and ZO-2. The adherens junction, just below the tight junction, has E-cadherin in its core and interacts with terminal web actin through catenins and α-actinin. The desmosomes, scattered at multiple sites along the lateral membrane, contain desmogleins and desmocollins, which are linked to intermediate filaments via phakoglobin and desmoplakin.

a transmembrane protein located within the tight junctions, is thought to be the primary barrier-forming molecule utilized in homotypic contacts between adjacent cells. Occludin interacts with a cytoplasmic linker protein called ZO-1 (for zonula occludens protein-1), which in turn is linked to the subapical actin cytoskeleton via ZO-2 and other undefined attachments. Tight junctions are dynamic structures subject to multiple regulatory pathways, including intracellular calcium concentrations and phosphorylation by both ser/thr and tyrosine kinases. Both ZO-1 and ZO-2 belong to the membrane-associated guanylate kinase protein family, members of which are involved in organizing structural and functional links among transmembrane proteins, signaling pathways, and the cytoskeleton. The primary documented role of tight junctions is to seal the paracellular space by forming a selective permeability barrier across epithelial cell sheets. This is essential for the maintenance of electrochemical ionic and molecular gradients created by selective absorption (or vectorial excretion) via transcellular pathways. In addition, tight junctions prevent the diffusion of membrane components between apical and lateral membranes, thereby assisting in the maintenance of cell polarity. [*See* Cell Junctions.]

The adherens junction or zonula adherens in mature polarized epithelial cells is concentrated just below the tight junction along the lateral membrane, in the region of the subapical terminal web. The protein E-cadherin, a member of the classical cadherins family, is a transmembrane, calcium-dependent adhesion molecule that forms the core of the adherens junction by forming homotypic interactions between contacting cells (Fig. 3). Other components of the cadherin-based adhesion complex are at least two additional cytoplasmic proteins termed α- and β-catenin. The binding of the cytoplasmic domain of E-cadherin to β-catenin, which in turn binds α-catenin, forms a membrane-associated complex that in turn links cadherin to the submembraneous actin cytoskeleton. This linkage occurs by two mechanisms: a direct binding of α-catenin with F-actin and an indirect linkage in which actin-binding proteins such as α-actinin or β-spectrin also bind directly to α-catenin. E-cadherin function and assembly are regulated by intracellular calcium levels, the state of cell contact, and signal transduction events mediated by protein phosphorylation. E-cadherin is required for epithelial cells to remain attached, and in its absence the many other cell junction proteins cannot support intercellular adhesion. Because of this property, E-cadherin activity (or in activity) is one factor that determines the invasiveness of malignant epithelial cells and the ability of such cancers to metastasize. In addition, E-cadherin plays a crucial role in the induction of apical/lateral polarity, as it is one of the first molecules to assemble at sites of cell–cell contact. This initiates a cascade of events including recruitment of cytoskeletal and integral membrane proteins destined to form components of the lateral domain.

Desmosomes appear as pairs of electron-dense discs that sandwich the lateral membranes of adjacent cells at multiple contact sites. Analogous to adherens junctions, the core desmosomal adhesion molecules are proteins of the cadherin superfamily, termed desmocollins and desmogleins, which have the capability of homotypic interactions (Fig. 3). These proteins in turn are linked to an extensive intermediate filament network via desmoplakin and phakoglobin. The desmosome-intermediate filament system provides the cellular support required to withstand mechanical stress. This is best illustrated in the epidermis, where several blistering disorders are a result of autoantibodies to desmosomal components.

C. The Lateral Membrane: Business as Usual

The lateral domain contains highly conserved proteins that execute biological functions fundamental to polarized cells from all tissues, as well as nonpolarized cells. For example, Na,K-ATPase actively pumps sodium out of the cell (and potassium into the cell at a $3Na^+ : 2K^+$ ratio), thereby generating the electrochemical gradients required to drive several sodium-dependent absorptive processes at the apical domain. Other lateral membrane transporters are involved in translocation of absorbed nutrients into, and the exchange of ions with, the circulating extracellular compartment. Receptors for hormones and neurotransmitters participate in several constitutive cellular functions.

The lateral membrane is characterized by a highly developed spectrin-based cytoskeleton. Spectrin is a highly conserved, flexible, rod-shaped heterodimer that interacts with F-actin to form a submembraneous lattice-like scaffolding of five- and six-sided polygons. Spectrin–actin junctions are stabilized by actin-bundling proteins, including adducin and protein 4.1 (Fig. 4). The spectrin-based skeleton was previously thought to be merely a supporting infrastructure, providing mechanical support and elasticity to the plasma membrane. Recent observations suggest that it provides organizational stability by controlling the distribution of integral membrane proteins and that the control of such interactions is required for membrane stability as well as ordered function. One important way that spectrin interacts with the membrane is via the adapter molecule ankyrin, which links spectrin to a host of integral membrane proteins, including Na,K-ATPase, the anion exchanger AE1 or band 3, and H^+,K^+-ATPase. This interaction restricts the lateral mobility of membrane proteins, thereby maintaining their polarized distribution. In addition, at least two direct membrane-association domains on spectrin have been documented, called MAD1 and MAD2. MAD1 is responsible for the assembly of spectrin at the lateral membrane, and MAD2 possesses binding sites for putative regulatory molecules such as G-proteins and inositol-1,4,5-triphosphate.

D. The Basal Domain: Settling Down

The basal domain contains proteins that attach cells to the underlying extracellular matrix (ECM) and basement membrane. These adhesions are mediated primarily by integrins, a large family of heterodimeric transmembrane proteins that bind several components of the ECM (e.g., collagen, fibronectin). The cytoplasmic domains of integrins directly bind to at least two cytoskeletal proteins, α-actinin and talin, which in turn are linked to a cytoskeletal matrix containing actin, vinculin, and paxillin. Together, these form organized clusters called focal adhesions (Fig. 5), which are a protein scaffold for the assembly of an extensive intracellular signaling cascade, including the focal adhesion kinase, Src tyrosine kinase, MAP kinase, SH2/SH3-binding proteins (e.g., Grb2), and GTP-binding proteins (e.g., Ras). Consequences of this signaling network include the rearrangement of actin filaments, stabilization of focal adhesions,

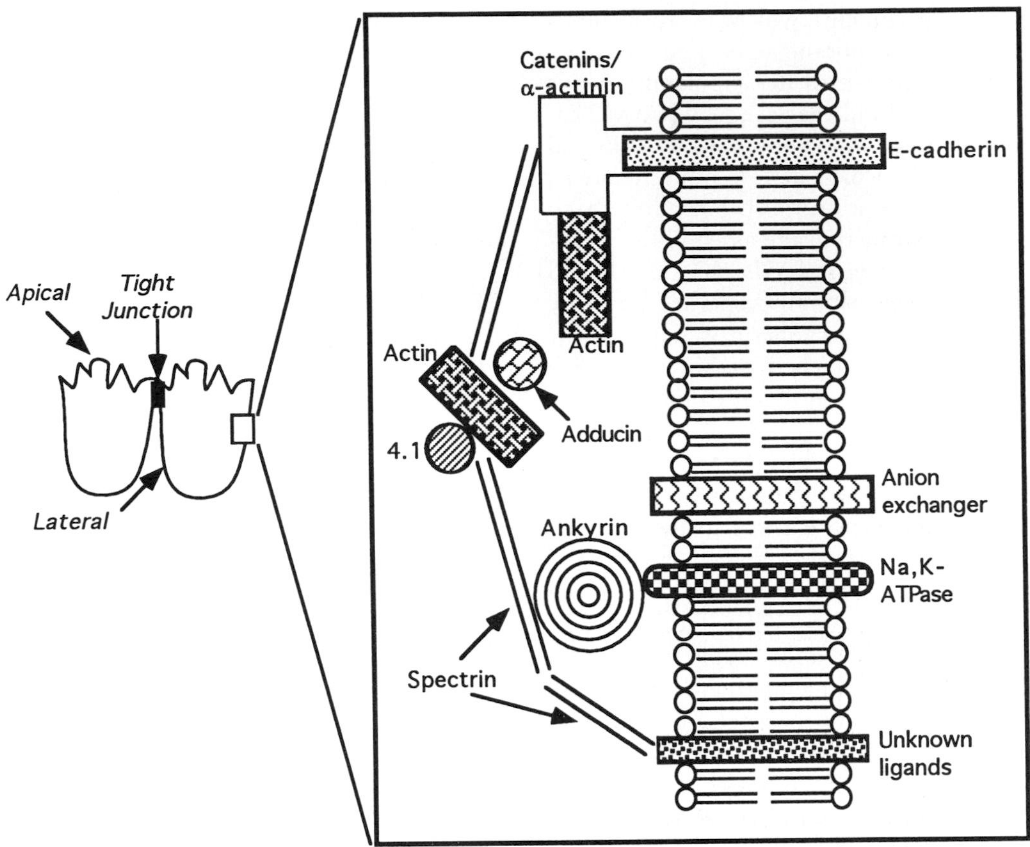

FIGURE 4 The spectrin-based skeleton at the lateral membrane. Spectrin is linked to Na,K-ATPase via ankyrin; other integral membrane proteins that bind ankyrin are not shown. The association of spectrin with E-cadherin via catenins has only been shown *in vitro*. Spectrin–actin bundling is promoted by adducin and protein 4.1. The direct membrane association of spectrin to unknown ligands is also shown. The anion exchanger does not bind ankyrin in epithelial cells.

modulation of cell adhesion (for changes in cell shape and spreading), and activation of various transcription factors. Thus, integrin-mediated interactions play critical roles in the induction and maintenance of the apicobasal phenotype, mechanical cellular adhesion and support, and regulation of cell migration, proliferation, and differentiation. Perturbations in these attachments are seen in muscular dystrophy, in several autoimmune blistering disorders of the skin, and in tumor cell growth and metastasis.

E. Cytoplasmic Organelles and Skeleton: Tracks, Engines, Coaches, and Stations

The cytoplasm of polarized epithelial cells is characterized by the microtubule-based skeleton and by an elaborate network of organelles [including the endoplasmic reticulum (ER), Golgi apparatus, endosomes, and lysosomes] that are in continuous bidirectional contact through a flow of small transport vesicles. Not only do these components play a fundamental role in the genesis of cell polarity, but they are also spatially and functionally polarized themselves. Microtubules are long tubular polymers composed of tubulin dimers and a heterogeneous group of microtubule-associated proteins. They are dynamic structures with a rapidly assembling plus end and a disassembling minus end. The bulk of microtubules in polarized epithelia is arranged with their minus ends directed toward the terminal web actin filaments at the apical pole and their plus ends toward the basolateral domain and *trans*-Golgi network (Fig. 6). These provides the "tracks" along which proteins destined for the apical domain are transported via direct or transcytotic pathways. In addition, microtubules

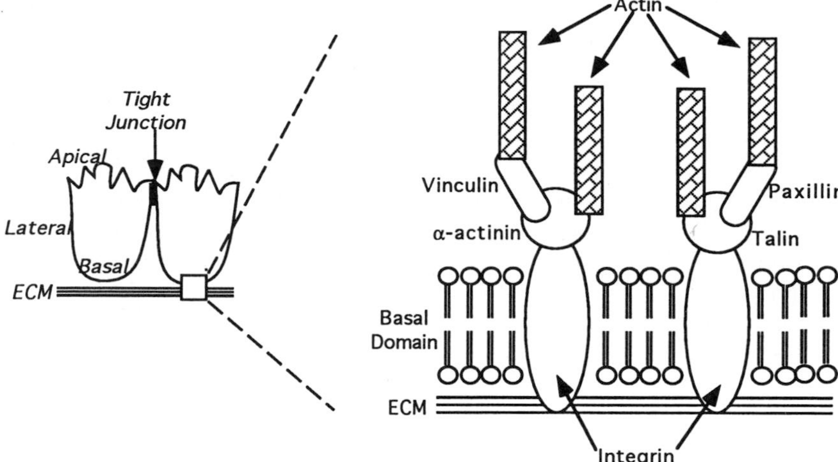

FIGURE 5 The integrin-based focal adhesion at the basal membrane. Integrins bind to components of the extracellular matrix (ECM) and to α-actinin and talin, which in turn are linked to a cytoskeletal matrix containing actin, vinculin, and paxillin to form a focal adhesion. This scaffold recruits an intracellular signaling cascade (including focal adhesion kinase, Src tyrosine kinase, GTP-binding proteins) that regulates actin filaments, stabilizes focal adhesions, and modulates cell adhesion to facilitate changes in cell shape and cell spreading.

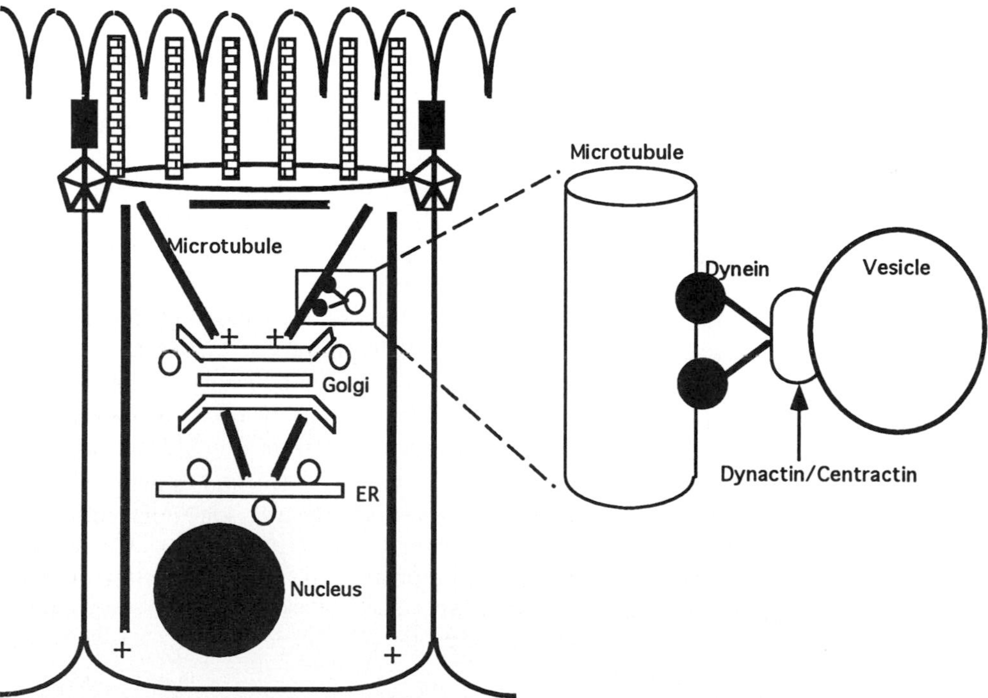

FIGURE 6 Microtubules and their motors. Microtubules are arranged predominantly with their minus ends toward the apical pole and plus ends toward the basal domain and *trans*-Golgi network. Other microtubules run parallel to the apical domain or between the ER and the Golgi apparatus. Microtubule kinesin motors (not shown) move vesicles to the plus end, whereas dyneins are retrograde motors. Dyneins link the microtubules to a dynactin/centractin complex, which in turn binds to a spectrin skeleton present in transport vesicles.

forming a subapical cap parallel to the apical membrane are involved in endocytosis and recycling of membrane proteins, and those associated with the ER and Golgi apparatus play a crucial role in organelle positioning as well as vesicular trafficking between and within these organelles. The "engines" that pull vesicles along microtubule "tracks" are called microtubule motors, of which there are two categories, called kinesins and dyneins. Both are large multi-subunit proteins that cross-link tubulin and accessory cytosolic proteins to a target vesicle and translocate it along the microtubule by an ATP-dependent process. Kinesins are anterograde motors that drive vesicles to the plus end. Dyneins are retrograde motors that require a group of recently described accessories (including dynactin, Arp1 or centractin, and actin capping protein) to move vesicles to the minus end (Fig. 6). Evidence suggests that the dynein/dynactin/centractin complexes are linked to transport vesicles by binding to a vesicular, spectrin-based skeleton.

The larger cytoplasmic organelles, namely the ER and Golgi apparatus, are the docking "stations" for vesicular transport. The ER is a labyrinth of branching, interconnected tubules and flattened sacs extending throughout the cytosol and enclosing a single convoluted space, the ER lumen. The ER synthesizes lipids, processes the majority of the cell's proteins, and sorts them to the Golgi apparatus via the process of vesicular transport. Organelle spatial and structural polarity is most dramatically exhibited by the Golgi apparatus. It consists of a series of flattened, curved cisternae with many small peripheral vesicles, asymmetrically situated in a perinuclear location. The Golgi apparatus itself displays a marked polarity in structure, cytochemistry, and function, resulting in the formation of three groups of cisternae. The *cis*-Golgi faces the ribosome-studded rough ER, the *trans*-Golgi network (TGN) faces the apical membrane, and the medial Golgi stacks are between. This arrangement allows the sequential modification of voyaging proteins and lipids and their ultimate targeting to the destined domain. [*See* Lipids; Microtubules.]

The trafficking of all protein and lipid "cargo" within cells is dependent on loading these cargo into vesicles ("coaches"), budding of these vesicles from the donor membrane, translocation along microtubules to a receiving membrane station, and delivery of cargo by vesicle fusion. Recent studies have identified several of the key players involved in this vesicular transport. The initial loading and budding of a vesicle from a membrane require the association of specialized cytosolic coat proteins, of which two major types have been identified, namely clathrin and coatomers. Clathrin-coated vesicles are involved in receptor-mediated endocytosis and in the traffic of lysosomal or vacuolar proteins from the TGN. This is achieved by the assembly of clathrin, cytosolic factors (including ADP-ribosylation factor or ARF, ATP, and GTP), and an adapter protein (AP) complex (including adaptins, AP50, AP17) at membrane sites determined by the receptor being endocytosed. Subsequent conformational changes in the clathrin structure impart curvature at the membrane site, which induces budding. The bud pinches off as a vesicle by a process termed periplasmic fusion, which requires the GTPase dynamin. At the receptor site, clathrin-coated vesicles are uncoated by the action of a 70-kDa heat shock protein.

Coatomer-containing vesicles are primarily involved in bidirectional ER–Golgi and intra-Golgi vesicular transport. The GTPase, ARF, is the first protein recruited to the donor membrane. Next, the cytosolic coatomer proteins (e.g., β-COP) and vesicular skeleton proteins (isoforms of spectrin and ankyrin) assemble at the donor site, resulting in the formation of a bud containing cargo derived from the lumen at the donor membrane. The bud pinches off as a vesicle and is targeted to the acceptor site, where GTP hydrolysis results in the dissociation of ARF and the release of vesicular contents.

The specific docking of clathrin- or coatomer-coated vesicles at the targeted site requires the presence of distinct "pilots" to encode the destination, as well as corresponding receptors at the targeted site to capture the vesicles. A novel class of cytoplasmically oriented integral membrane proteins, called SNAREs, fulfills these functions. Two families of SNAREs have been identified. The v-SNAREs are localized to transport vesicles and donor membranes and constitute the pilots, whereas the t-SNAREs are the vesicle receptors at the targeted station. Uncoating of targeted vesicles at the accepting site appears to expose the vesicles v-SNAREs to the t-SNAREs at the targeted site, allowing their interaction. The final steps in membrane fusion are unclear, but appear to require a specific ATPase termed *N*-ethylmaleimide-sensitive protein (NSF).

II. GENERATION OF CELL POLARITY

The previous section showed that polarized epithelial cells contain numerous membrane compartments (e.g., plasma membrane, Golgi apparatus, ER, endosomes, lysosomes), each with specialized functions bestowed by distinct biochemical compositions. All proteins originate in ribosomes and the ER, and are sorted to their appropriate destinations by vesicular transport pathways, guided by signals intrinsic to the targeted protein and extrinsic cues provided by cell–cell and cell–substratum contacts.

A. Pathways and Sorting Signals: All Roads Do Not Lead to Rome

Several pathways of flow are available to vesicles in polarized epithelial cells (Fig. 7). All proteins destined to be secreted or targeted to organelles are first imported into the ER, where they fold, oligomerize, and undergo partial N-linked glycosylation. They are then carried forward indiscriminately to the *cis*-Golgi network by coatomer-coated vesicles, a process that can be experimentally blocked by the drug brefeldin A, which disrupts the Golgi apparatus. Certain proteins contain an ER retention signal identified by the amino acids KDEL (Lys-Asp-Glu-Leu) at their COOH terminus; these proteins are selectively returned to the ER. This return pathway is microtubule dependent, and is blocked by nocodazole or taxol. All other proteins are transported through the Golgi stacks in the *cis*-to-medial-to-*trans* direction on coatomer-coated vesicles. Along the way, processing events such as N- and O-linked glycosylation are completed.

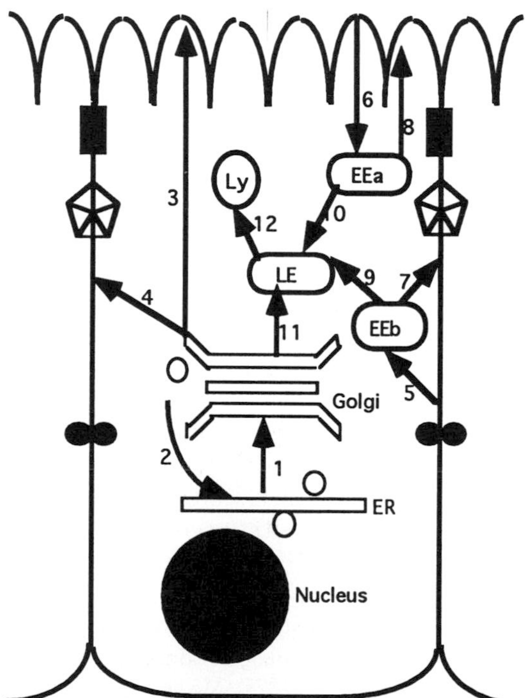

FIGURE 7 Principal pathways of vesicular transport. LE, late endosome; EEa, early apical endosome; EEb, early basal endosome; Ly, lysosome. 1, bulk transport from ER to Golgi; 2, return pathway for KDEL-containing ER resident proteins; 3, direct apical sorting (e.g., of GPI-linked proteins); 4, direct lateral sorting (e.g., for NPXY-containing proteins); 5, clathrin-mediated endocytosis from lateral domain; 6, clathrin-mediated endocytosis from apical domain; 7 and 8, recycling pathways; 9 and 10, transport from early to late endosome; 11, direct targeting from TGN to late endosomes; 12, targeting to lysosomes.

The *trans*-Golgi network is a major sorting station from which vesicles are targeted by a variety of pathways to the apical, basolateral, endosomal, or lysosomal membrane (Fig. 7). In the direct pathway, proteins are directly and vectorially delivered to the appropriate domain. In the indirect pathway, all newly synthesized proteins are delivered to the basolateral membrane; subsequently, those destined for the apical domain are removed by endocytosis and are transported to the apical membrane by transcytosis. In the random pathway, the proteins are randomly delivered to both apical and basolateral domains. Proteins are subsequently sorted to the correct domain by selective endocytosis and transcytosis. All three pathways are used, and the predominant pathway employed depends on the cell type. For example, kidney tubule cells (both native and cultured) appear to employ predominantly the direct pathway to polarize basolateral proteins such as the Na,K-ATPase, whereas intestinal epithelial cells and hepatocytes have well-developed transcytotic pathways.

Until recently, it was erroneously believed that only the apical targeting of proteins is signal mediated and that the basolateral targeting of proteins represented a "default" pathway taken in the absence of an apical targeting signal. This assumption was based on the observation that it is usually the apical domain that is specialized for characteristic epithelial cell functions such as absorption and secretion. Many recent studies have challenged this view and have redefined basolateral sorting as a specific signal-mediated event. Two classes of basolateral-sorting determinants can now be discerned. The first is related to clathrin-coated pit signals such as the NPXY motif (Asn-Pro-X-Tyr) in the cytoplasmic tails of proteins such as LDL receptor, vesicular stomatitis virus G protein, and Fc receptors. The second class of basolateral-targeting signals are found in proteins such as transferrin receptor and polymeric Ig receptor and contain a tyrosine residue that appears to be critical. Many proteins that are linked to GPI are found in the apical membrane, and splicing of a GPI anchor onto a normally basolateral protein makes it sort to the apical domain, suggesting that GPI anchors may serve as apical-targeting signals.

Proteins destined for the lumen of lysosomes, such as the acid hydrolases, are targeted from the TGN to late endosomes and lysosomes. These proteins contain a unique sorting signal in the form of mannose-6-phosphate (M6P) groups, which bind to M6P receptors at the TGN to form a receptor-dependent transport vesicle. When these vesicles reach late endosomes and lysosomes, the characteristic low intracellular pH in these organelles dissociates and releases the targeted proteins from the M6P receptor.

Endocytosis is a clathrin-mediated process by which localized areas of the plasma membrane invaginate and pinch off to form vesicles. Material endocytosed from either the apical or the basolateral domain first enters an early endosomal compartment that is domain specific. This allows endocytosed proteins to be recycled back to the original domain, unless they contain signals that target them to the other domain by transcytosis. Proteins that are not recycled are targeted to a common late endosomal compartment and then into lysosomes to be degraded. [*See* Protein Targeting, Basic Concepts.]

B. Spatial Cues: The Guiding Lights

Epithelial cells arise at multiple times and places during morphogenesis from apolar progenitors. In the absence of extracellular contacts, single epithelial cells do not exhibit structural and functional characteristics of polarized cells. Cell polarity starts with a spatial cue, which is cell–cell or cell–substrate adhesion. Although some controversy exists on the precise hierarchy of signaling, it is clear that both pathways of spatial signaling are important. Cell adhesion molecules, specifically E-cadherin, mediate cell–cell contact, and this alone may be sufficient to initiate a cascade of events that results in at least some forms of polarity. It is envisioned that these contact sites form "nucleating centers" for the recruitment of a spectrin-actin based cytoskeleton and associated signaling networks at the plasma membranes. Other spatial cues emanate from integrin-mediated cell–substratum contacts very early in tissue morphogenesis. These interactions establish the apico/basal axis of polarity and also recruit an actin-based cytoskeletal complex and signaling molecules to the basal domain. Randomly oriented microtubules organize themselves with respect to the polarized cytoskeletal complexes, and the process of polarized protein sorting begins.

III. ALTERED STATES

Many examples of disease states are associated with a disruption in epithelial cell polarity. The most frequent type of human cancer is the carcinoma, which is derived from epithelial tissues. Cancer cells are characterized by a loss of cell polarity and a decrease in cadherin-mediated cell–cell contact, leading to invasiveness and metastasis. The commonest cause for injury and failure of several organ systems is hypoxia or ischemia. This has been best studied in kidney cells, where hypoxic injury is characterized by a disruption of epithelial cell polarity, leading to the mislocation of crucial transporters and ion pumps. Recovery from ischemic kidney failure is characterized by the restitution of cell polarity. Polycystic kidney disease is one of the most common inherited disorders in humans; epithelial cells from polycystic kidneys also display several alterations in polarity.

BIBLIOGRAPHY

Brown, D., and Stow, J. L. (1996). Protein trafficking and polarity in kidney epithelium: from cell biology to physiology. *Physiol. Rev.* **76**, 245–297.

Clark, E. A., and Brugge, J. S. (1995). Integrins and signal transduction pathways: The road taken. *Science* **268**, 233–239.

Cole, N. B., and Lippincott-Schwartz, J. (1995). Organization of organelles and membrane traffic by microtubules. *Curr. Opin. Cell Biol.* **7**, 55–64.

Devarajan, P., and Morrow, J. S. (1996). The spectrin skeleton and organization of polarized epithelial cell membranes. *Curr. Top. Membr.* **43**, 97–128.

Drubin, D. G., and Nelson, W. J. (1996). Origins of cell polarity. *Cell* **84**, 335–344.

Gumbiner, B. M. (1996). Cell Adhesion: The molecular basis of tissue architecture and morphogenesis. *Cell* **84**, 345–357.

Matter, K., and Mellman, I. (1994). Mechanisms of cell polarity: Sorting and transport in epithelial cells. *Curr. Opin. Cell Biol.* **6**, 545–554.

Rothman, J. E., and Wieland, F. T. (1996). Protein sorting by transport vesicles. *Science* **272**, 227–234.

Schekman, R., and Orci, L. (1996). Coat proteins and vesicle budding. *Science* **271**, 1526–1533.

Polyamine Metabolism

PETER P. McCANN
British Biotech Inc.

ANTHONY E. PEGG
Milton S. Hershey Medical Center, Pennsylvania State University

GLOSSARY

α-Difluoromethylornithine Enzyme-activated, irreversible inhibitor of ornithine decarboxylase, which blocks the formation of putrescine

Ornithine decarboxylase First enzyme in the general polyamine biosynthetic pathway, converting ornithine to the diamine putrescine

Putrescine 1,4-Diaminobutane, the first amine in the biosynthetic pathway for polyamines

Spermidine Product of the biosynthetic reaction that adds an aminopropyl group to putrescine

Spermine Product of the addition of an aminopropyl group to spermidine

***S*-Adenosylmethionine decarboxylase** Enzyme that produces decarboxylated *S*-adenosylmethionine as an aminopropyl donor for both spermidine and spermine formation

THE BIOCHEMISTRY AND CELLULAR PHYSIOLOGY of polyamine metabolism (the synthesis and interconversion of putrescine, spermidine, and spermine from ornithine) have been a steadily growing field of research, and although the function of polyamines is still not well understood at the molecular level, a great deal of information attests to the importance of polyamines in cellular function. More recently, there has been a growing awareness that polyamine metabolism provides a useful target for the design of inhibitors, which have value as pharmacological agents. The availability of such inhibitors has led to a rapid increase in knowledge of the importance of polyamines in a wide variety of living organisms, and the clinical utility of these inhibitors in certain situations is now clearly evident. In view of the ubiquitous distribution of polyamines and their key role in a variety of cellular processes, further uses for these inhibitors as chemotherapeutic agents is expected.

I. GENERAL OUTLINE OF POLYAMINE METABOLISM

A. Biosynthesis and Retroconversion

The polyamine biosynthetic pathway (Fig. 1) has been well studied in mammalian and other eukaryotic cells. The precursor diamine putrescine is the result of the decarboxylation of ornithine by the pyridoxal phosphate-dependent enzyme ornithine decarboxylase (ODC). The polyamines spermidine and spermine are then sequentially formed by aminopropyl groups added to putrescine. Two quite distinct enzymes, spermidine synthase and spermine synthase, which both require an aminopropyl donor in the form of *S*-adenosylmethionine, catalyze the formation of these two polyamines. *S*-Adenosylmethionine decarboxylase

ENCYCLOPEDIA OF HUMAN BIOLOGY, Second Edition, VOLUME 7. Copyright © 1997 by Academic Press. All rights of reproduction in any form reserved.

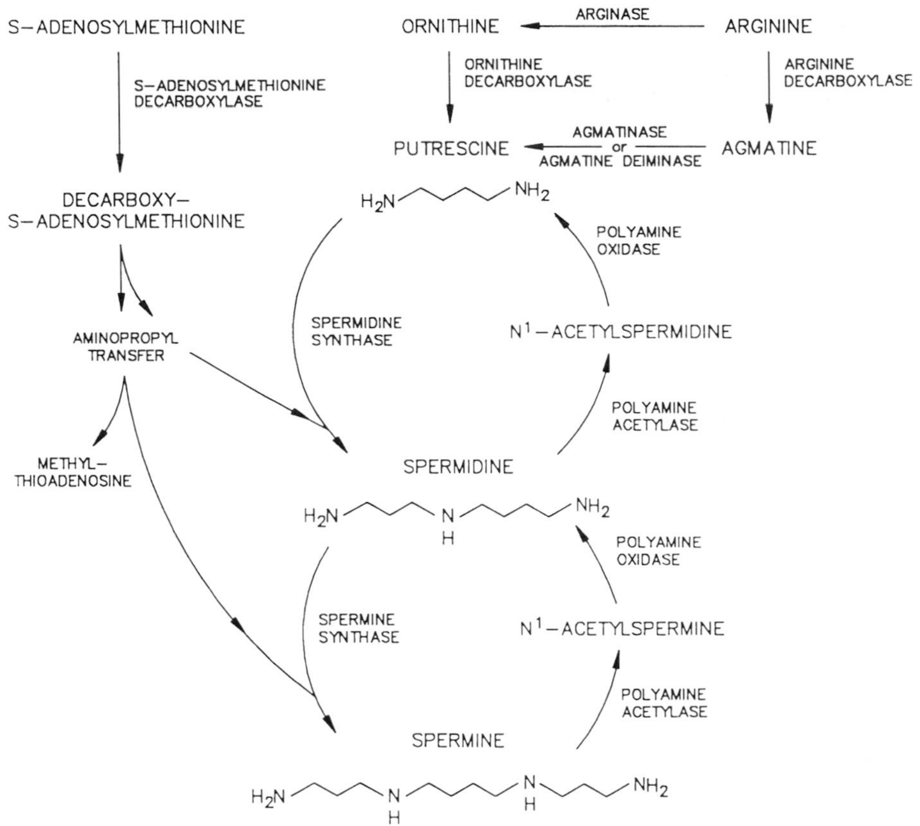

FIGURE 1 General overview of polyamine metabolism. It should be noted that not all of these reactions occur in all species.

(AdoMetDC), a pyruvoyl enzyme, is responsible for the formation of *S*-adenosylmethionine (AdoMet).

Although no other pathway is responsible for the *de novo* formation of polyamines in mammalian cells, a specific membrane transport system exists for putrescine, spermidine, and spermine; it is highly regulated and controls the uptake of exogenous sources of these amines. There is also a retroconversion of spermidine and spermine to putrescine controlled by both spermidine–spermine-N^1-acetyltransferase (SAT) (polyamine acetylase) and polyamine oxidase. SAT converts spermidine to its N^1-acetyl derivative, which in turn is cleaved by polyamine oxidase to putrescine and 3-acetamidopropanal (see Fig. 1). There is an analogous retroconversion of spermine into spermidine. Putrescine, the N^1-acetylspermidine derivative, and in some cases spermidine are excreted from various mammalian cells. Polyamines can also arise from exogenous sources such as diet. Serum oxidases may, in fact, degrade such extracellular amines. [*See* Cell Membrane Transport.]

B. Incorporation of Polyamines into Other Cellular Molecules

Posttranslational modification of specific proteins is one clearly defined and specific function of polyamines. For example, transglutaminase-catalyzed incorporation of spermidine and spermine into proteins occurs in higher eukaryotes, although the specific physiological function of such incorporation is not well understood.

The formation of hypusine in the initiation factor eIF-5A for protein synthesis is another example of polyamines serving as precursors of protein modification. Hypusine, N^{ε}-(4-amino-2-hydroxylbutyl)lysine, originates from the conversion of putrescine to spermidine and to deoxyhypusine, although only the 1,4-

diaminobutane fragment of spermidine is incorporated into hypusine itself.

Although not present in bacteria, hypusine is found in many mammalian cells and in yeast as part of the eIF-5A protein, suggesting its importance in some general, well-conserved biological process.

II. POLYAMINE METABOLISM IN MAMMALIAN CELLS

A. Ornithine Decarboxylase

ODC is a pyridoxal phosphate-dependent enzyme that is found only at very low levels in resting cells. Cellular levels of ODC can be significantly increased, sometimes by a 100-fold or more, by stimulation with growth factors, hormones, drugs, and other regulatory molecules; however, ODC still represents only a minor portion of the total cellular protein (e.g., from 0.01% of the cytosolic protein in androgen-stimulated mouse kidneys to 0.00012% in thioacetamide-stimulated rat liver). A macromolecular inhibitor of ODC, termed antizyme, has been isolated in a number of cells and may be another regulatory factor controlling ODC activity as well.

The use of the competitive inhibitor α-methylornithine conclusively demonstrated that formation of putrescine by ODC was essential for DNA replication and, consequently, mammalian cell growth. Later synthesis of the more potent inhibitor α-difluoromethylornithine (DFMO; eflornithine, Ornidyl®) alllowed extensive work to explore the relationship between polyamine depletion caused by ODC inhibition, its consequent effects on cell replication, and, in some cases, cell differentiation.

DFMO (an enzyme-activated, mechanism-based, irreversible inhibitor of ODC) and some other substrate and product-related inhibitors, which have somewhat improved biochemical properties, have been widely used as biochemical and pharmacological tools. In fact, DFMO has been extensively explored as therapy for several types of cancer and protozoan infections. A number of studies have shown that treatment of cells in culture with DFMO significantly decreases intracellular putrescine and spermidine with little or no effect on spermine, whereas an enormous increase occurs in the overall levels of decarboxylated AdoMet. Other more potent inhibitors such as (2*R*, 5*R*)-δ-methyl-α-acetylenic putrescine had greater effects on spermine levels but did not achieve complete depletion as was the case with spermidine.

B. S-Adenosylmethionine Decarboxylase

Mammalian AdoMetDC is the enzyme that provides decarboxylated AdoMet as an aminopropyl donor in the formation of spermidine from putrescine. Once it has been decarboxylated, AdoMet is committed to polyamine production, as no other reactions utilizing decarboxylated AdoMet at any physiologically significant rate are known. Therefore, the production of decarboxylated AdoMet is kept low and constitutes the rate-limiting factor in spermidine formation. Mammalian AdoMetDC is activated by putrescine and repressed by spermidine, linking the supply of decarboxylated AdoMet to the need for spermidine and the availability of the other substrate (putrescine) for spermidine synthesis. AdoMetDC has an enzyme-bound pyruvate as cofactor and, analogous to ODC, is present in cells only at very low levels (e.g., equal to 0.015% of the soluble protein in ventral prostate and to 0.0007% in liver). Similarly, AdoMetDC is also regulated by a number of external growth factors and stimuli.

Mammalian AdoMetDC is a dimer of two pairs of subunits with molecular weights of 30,621 and 7681. These subunits are formed by the cleavage of a proenzyme chain (M_r about 38,000) in a reaction that forms the pyruvate prosthetic group at the amino-terminal end of the larger subunit. The processing–cleavage step may be an autocatalytic reaction of the proenzyme chain and is accelerated in the presence of putrescine. Because this step is essential for the production of the active sites of the enzyme, it provides an attractive target for the future design of therapeutically useful inhibitors of AdoMetDC.

Methylglyoxal bis(guanylhydrazone) (MGBG) is a potent inhibitor of AdoMetDC and was used in early experiments to prevent formation of decarboxylated AdoMet and, thus, synthesis of spermidine. Although such experiments suggested the importance of polyamines for cell growth, they were not conclusive because MGBG is nonspecific and has a variety of effects on mitochondrial and other physiological functions unrelated to effects of inhibition of polyamine biosynthesis. MGBG can be considered as a structural analogue of spermidine and is taken up by cells via a specific polyamine transport mechanism. Thus, the reversal of the effects of MGBG by spermidine may be as much due to interference of its cellular uptake or binding as to its replacement of intracellular polyamines.

Even though other inhibitors of AdoMetDC have been utilized, only recently have newer specific and more potent compounds become available, such as 5′-deoxy-5′[*N*-methyl-*N*-(aminooxyethyl)]-aminoadenosine, 5′-deoxy-5′[*N*-methyl-*N*-(3-hydrazinopropyl)]aminoadenosine, and *S*-(5′-deoxy-5′-adenosyl)-methyl-thioethylhydroxylamine. These irreversible inhibitors of AdoMetDC apparently bind to the active site of the enzyme and form a covalent bond with the enzymatic pyruvate prosthetic group. An enzyme-activated inhibitor of AdoMetDC has also been described—5′-{[(*Z*)-4-amino-2-butenyl]methylamino}-5′-deoxyadenosine.

These inhibitors produce profound but expected changes in intracellular polyamine levels and consequently inhibit cell growth. For example, they reduce spermidine and spermine as well as decarboxylated AdoMet and 5′-methyl-thioadenosine and cause an enormous increase of putrescine in L1210 cells. Following depletion of spermidine and spermine, cell growth is arrested in spite of the large concentration of available putrescine; addition of either one depletes polyamine fully restored normal cell growth. Utilization of DFMO and an AdoMetDC inhibitor together completely blocks the increase of all three amines, demonstrating the potential utility of inhibition of AdoMetDC as a target for drug design.

C. Aminopropyltransferase

Spermidine synthase is responsible for the transfer of the aminopropyl group from decarboxylated AdoMet to putrescine. A second enzyme, spermine synthase, carries out a similar reaction and adds another aminopropyl group to spermidine. Although the mechanisms of action of these two aminopropyl transferases are analogous, both spermidine synthase and spermine synthase are specific and discrete enzymes, each having its own unique substrate. The cellular amounts of both of these aminopropyl transferases are normally much higher than either ODC or AdoMetDC and are apparently regulated by the levels of their substrates (e.g., decarboxylated AdoMet). However, apparently the disposition of available decarboxylated AdoMet toward spermidine or spermine is probably determined by the relative amounts of the two synthases, and marked changes in spermidine synthase activity have been observed in response to hormones, tissue regeneration, and cell growth factors.

The multisubstrate analogue *S*-adenosyl-1,8-diamino-3-thioctane (AdoDato), designed as an inhibitor of spermidine synthase, strongly inhibits this enzyme in mammalian cells. Another potent inhibitor of the same enzyme, cyclohexylamine, is competitive with respect to putrescine with a K_i of about 0.2 μM.

Although AdoDato is an effective *in situ* inhibitor of spermidine synthase in mammalian cells, its overall utility is limited because of the large increase in decarboxylated AdoMet resulting from the compensating rise in AdoMetDC after the inhibition of spermidine synthase. This decarboxylated AdoMet, in turn, can be used for the synthesis of spermine via spermine synthase, and thus the total polyamine pools are not significantly altered. Cyclohexylamine, however, does decrease overall polyamine levels in murine tumors.

Design and synthesis of specific spermine synthesis inhibitors have been of recent interest because of the relative ineffectiveness of ODC inhibitors in decreasing spermine levels and because the physiological significance of spermine synthase, present in mammalian cells but not found in many microorganisms, is not clearly understood at present.

The first inhibitor shown to appreciably block spermine synthase *in vitro* was *S*-methyl-5′-methylthioadenosine. Another compound, *S*-adenosyl-1,12-diamino-3-thio-9-azadodecane (AdoDatad), structurally related to AdoDato by addition of an aminopropyl moiety, was designed and found to be a potent, specific, multisubstrate analogue inhibitor of spermine synthase. This enzyme step is also blocked effectively by *n*-butyl-1,3-diaminopropane. However, neither AdoDatad nor *n*-butyl-1,3-diaminopropane had any effect on cell proliferation, although spermine was significantly depleted, suggesting either no requirement for spermine or that the increased amounts of spermidine found could compensate for the lack of spermine.

D. Polyamine Transport

Besides having the ability to synthesize polyamines, cells also possess a specific membrane transport system for the uptake of exogenous polyamines. Although its biochemical mechanism is not well understood, this uptake system is regulated by the intracellular polyamine content. Thus, when polyamines are depleted by use of specific inhibitors, the transport system responds to increased uptake and vice versa.

Of particular import is the fact that the transport system reduced the therapeutic effectiveness of polyamine biosynthetic inhibitors such as DFMO because extracellular polyamines (e.g., from diet, intestinal microorganisms, cell turnover) can be utilized via the transport system of polyamine depleted cells. For example, DFMO is significantly more active against a polyamine transport mutant L1210 cell line *in vivo* than against the parent cell line. DFMO is also more effective against tumors in rodents when intestinal polyamine oxidase is inhibited, thus preventing formation of polyamines from their N^1-acetyl derivatives.

Many polyamine analogues are actually substrates for the transport system and thus act as competitive inhibitors in relation to the natural polyamines. In fact, a number of carrier systems likely have some overlap in specificities. One system is Na^+-dependent and can be regulated by compounds that change Na^+ flux. Characterization of such systems should ultimately allow the design and synthesis of specific inhibitors to regulate entry and efflux of the polyamines.

MGBG toxicity is quite obviously related to its transport and ultimate accumulation within the cell. Such high levels of MGBG cause inactivation of mitochondria and finally inhibition of macromolecular biosynthesis. Although MGBG accumulation is via the polyamine transport system, it does not seem to mimic the natural polyamines and other polyamine analogues in down-regulating the polyamine transport system. As the system continues to transport MGBG into cells that already contain very high levels of the drug, MGBG's lack of repression of such transport may be a factor in its significant cytotoxicity.

Various *N*-alkyl polyamine analogues have been synthesized and are taken up by the polyamine transport system. For example, N^1,N^8-bis(ethyl)-spermidine reduced ODC levels and intracellular putrescine, spermidine, and also spermine in L1210 cells, whereas N^1,N^{12}-bis(ethyl)spermine was even more active in that it also reduced AdoMetDC levels and caused almost complete depletion of all the polyamines. Such effects are produced in part because these analogues are recognized by the polyamine regulatory systems for ODC and AdoMetDC biosynthesis. In addition, efflux of intracellular polyamines increases by some yet undefined mechanism, which normally regulates intracellular polyamine levels.

The bis(ethyl)polyamines are also potent inducers of SAT, and enzyme increases of several 100-fold are seen in cells within hours after exposure to the analogues. Such induced increases in SAT may also affect the rapid excretion of intracellular polyamines because such acetylation facilitates conversion of nonexcreted spermine to spermidine and putrescine, which are both excreted from the cell.

III. POLYAMINE METABOLISM IN DISEASE-CAUSING MICROORGANISMS AND VIRUSES

A. Protozoa

The polyamine metabolism of disease-causing protozoa has been of particular interest because of the pronounced sensitivity of several species to the effects of several polyamine inhibitors, primarily DFMO.

The original observations of the effects of DFMO on protozoa were made in African trypanosomes when the inhibitor was found to cure acute, lethal infections of *Trypanosoma brucei brucei* in mice. These initial results led to the rapid clinical use of DFMO in Africa against what would have been fatal cases of late-stage, arsenic-resistant West African sleeping sickness caused by *Trypanosoma brucei gambiense.*

Numerous experiments have been conducted concerning the role of polyamines in trypanosomal growth and differentiation. The African trypanosome appears to be unusually sensitive to DFMO, due in part to both its rapid doubling time and lack of spermine. ODC in the trypanosome, however, appears to be kinetically similar to the mammalian enzyme. Again, analogous to mammalian cells, DFMO enters the trypanosome by passive diffusion rather than facilitated transport. An intact host immune system is necessary for cures of trypanosome infections, and other evidence indicates that polyamines are required for the shift of antigenic determinants on the parasite membrane. Furthermore, DFMO and consequent polyamine depletion induces morphological and biochemical shifts of entire trypanosome populations from parasitic bloodstream forms to "short, stumpy" forms, which do not replicate. This differentiation event is independent of an earlier actual block of bloodstream form replication.

Trypanosomes also contain a novel spermidine-containing cofactor, which is necessary for glutathione reductase activity. This cofactor, trypanothione, uniquely present in trypanosomatids, is greatly reduced in *T. b. brucei* after DFMO treatment and is probably yet another target for polyamine depletion in these protozoa.

Recent experiments on molecular aspects of ODC from *T. b. brucei* have demonstrated that the trypanosomal ODC has a much longer half-life than the mammalian enzyme. Although the trypanosomal enzyme is very homologous to the mammalian enzyme, a likely explanation for this is that ODC from *T. b. brucei* lacks a specific 36-amino-acid carboxy-terminal peptide containing a "PEST" (proline–glutamate – aspartate – serine–threonine - containing) region, which promotes rapid protein degradation.

The other polyamine biosynthetic enzymes have been less studied in trypanosomes and, although the AdoMetDC of *T. b. brucei* is inhibited by drugs such as Berenil®, pentamidine, and MGBG, the effects of such inhibition contribute little or nothing to the antiprotozoal effects of these agents. As is the case with the mammalian enzyme, spermidine synthase from *T. b. brucei* is quite sensitive to both cyclohexylamine and AdoDato. Neither one has any *in vivo* effect on trypanosome infections as might be predicted, because neither one actually alters trypanosome polyamine levels.

DFMO will also inhibit the growth of promastigotes of the trypanosomatid *Leishmania donovani* but has no effect on its intracellular-infective cousin *Trypanosoma cruzi,* the cause of American trypanosomiasis (Chagas' disease). Unexpectedly, α-difluoromethylarginine, an irreversible inhibitor of arginine decarboxylase (an enzyme not present in mammalian cells), significantly inhibited *T. cruzi* infection of macrophages as well as actual replication of the parasite. These findings indicate the apparent presence of arginine decarboxylase in *T. cruzi* and a possible unique target for this now incurable disease.

A number of the sporozoea class of protozoa are quite sensitive to the effects of DFMO, including *Eimeria* spp. and *Plasmodia* spp., the latter causing various forms of malaria, the most common parasitic disease worldwide. DFMO inhibits the *in vitro* replication of *Plasmodium falciparum* and will cure exoerythrocytic infections of *P. berghei* in mice and also block the sporogonous cycle of *P. berghei* in mosquitoes. Although not curative of erythrocytic infection of *P. berghei,* DFMO will significantly lower parasitemia and, in combination with a bis(benzyl)polyamine analogue, will actually cure *P. berghei*-infected mice. [*See* Malaria.]

Pneumocystis carinii, the cause of the lethal opportunistic pneumonia in acquired immunodeficiency syndrome patients, has been thought to be related to the protozoal class Sporozoea, subclass Coccidia, although recently it has been shown to have ribosomal RNA more homologous to certain fungi. In any case, DFMO has been successfully used in numerous clinical instances as therapy for *P. carinii* pneumonia. Experiments in cell culture and immunosuppressed rats have shown that *P. carinii* is inhibited by DFMO, and its growth-inhibitory effects can be reversed by putrescine. [*See* Protozoal Infections.]

B. Bacteria

In bacteria and in plants, an alternate pathway exists for the synthesis of putrescine via the decarboxylation of arginine to form agmatine, which then yields putrescine (see Fig. 1). As bacteria have both ODC and arginine decarboxylase, inhibition of either enzyme alone induces the other to increase and thus maintain a specific intracellular concentration of putrescine.

Although bacteria, in general, contain only putrescine and spermidine and not spermine, it has been difficult to deplete intracellular polyamines and thus have specific effects on cell growth and replication.

Remarkably, several bacteria such as *Escherichia coli* and *Klebsiella pneumoniae* have ODC enzymes that are wholly unreactive to DFMO. Other analogues, however, such as α-difluoromethylputrescine and α-monofluoromethylornithine, do completely and irreversibly inhibit the enzymes from these organisms. α-Difluoromethylarginine, as well as even other more potent arginine and agmatine analogues, was found to irreversibly inhibit the arginine decarboxylases found in both bacteria and plants.

AdoMetDC from *E. coli* is inhibited by MGBG in a cell-free system, but this drug is not effective on intact cells. Some analogues of AdoMet as well as pentamidine and Berenil® are also *in vitro* inhibitors of cell-free bacterial AdoMetDC. With regard to spermidine synthase, both cyclohexylamine and AdoDato are potent inhibitors of the *E. coli* and *Pseudomonas aeruginosa* cell-free enzymes. However, cyclohexylamine is taken up by both organisms and will inhibit growth specifically due to depletion of spermidine, unlike AdoDato, which is not taken up.

The most significant growth inhibition of bacteria (e.g., *E. coli, P. aeruginosa,* and *Serratia marcescens*) was demonstrated with combinations of α-monofluoromethylornithine, α-difluoromethylarginine, and cyclohexylamine, wherein all three polyamine biosynthetic enzymes (ODC, arginine decarboxylase, and spermidine synthase) were simultaneously inhibited. Although appreciable growth inhibitory effects were noted, they were not as profound as in a number of protozoa and fungi. Therefore, the aforementioned inhibitors of polyamine biosynthesis will unlikely be useful as therapy for any bacterially caused infections.

C. Fungi

Filamentous fungi and most yeast synthesize putrescine solely via ODC, unlike bacteria. Although only limited experimental work has been done with zoophilic and disease-causing fungi and yeast, the ODC in *Saccharomyces cervisiae* and *S. uvarum* is sensitive to DFMO. Furthermore, polyamines are depleted by DFMO in several human pathogenic *Candida* spp., whereas *C. tropicalis* growth was also significantly inhibited by DFMO as well. α-Monofluoromethyldehydroornithine methylester was shown to be a potent inhibitor of two species of dermatophyte fungi, *Microsporum* and *Trichophyton*. In fact, it was 25 times more effective than DFMO itself, which also significantly arrested the growth of these organisms.

D. Viruses

Viruses as intracellular parasites depend completely on their metabolically active host cells for replication. However, in some cases, viral enzymes necessary for replication are encoded in the viral DNA or RNA. The poxviruses such as vaccinia apparently encode both ODC and AdoMetDC and have been shown to respond to polyamine inhibitors such as α-methylornithine, DFMO, and MGBG. DNA viruses in the herpes family, such as herpes simplex virus and human cytomegalovirus, have been shown to be sensitive to the effects of DFMO and MGBG as well. A correlation was also made between the ability of herpes simplex virus to induce the synthesis of polyamines and the antiviral effects of the inhibitors. It was found that depletion of polyamines was necessary in the host cells prior to infection with cytomegalovirus to have an effect on viral replication by DFMO. Overall, although polyamine biosynthesis is required for viral replication, more experimental work is needed to effectively determine any possible utility of inhibitors for viral diseases. [*See* Virology, Medical.]

BIBLIOGRAPHY

Marton, L. J., and Pegg, A. E. (1995). Polyamines as targets for therapeutic intervention. *Annu. Rev. Pharmacol.* **35**, 55–91.

McCann, P. P., and Pegg, A. E. (1992). Ornithine decarboxylase as an enzyme target for therapy. *Pharmacol. Ther.* **54**, 195–215.

McCann, P. P., Pegg, A. E., and Sjoerdsma, A. (eds.) (1987). "Inhibition of Polyamine Metabolism: Biological Significance and Basis for New Therapies." Academic Press, San Diego.

Pegg, A. E. (1986). Recent advances in the biochemistry of polyamines in eukaryotes. *Biochem. J.* **234**, 249–262.

Pegg, A. E. (1988). Polyamine metabolism and its importance in neoplastic growth and as a target of chemotherapy. *Cancer Res.* **48**, 759–774.

Pegg, A. E., and McCann, P. P. (1982). Polyamine metabolism and function. *Am. J. Physiol.* **243**, C212–C221.

Pegg, A. E., and McCann, P. P. (1992). *S*-Adenosylmethionine decarboxylase as an enzyme target for therapy. *Pharmacol. Ther.* **56**, 359–377.

Porter, C. W., and Bergeron, R. J. (1988). Enzyme regulation as an approach to interference with polyamine biosynthesis—An alternative to enzyme inhibition. *Adv. Enzyme Regul.* **27**, 57–82.

Sjoerdsma, A., and Schechter, P. J. (1984). Chemotherapeutic implications of polyamine biosynthesis inhibition. *Clin. Pharmacol. Ther.* **35**, 287–300.

Tabor, C. W., and Tabor, H. (1984). Polyamines. *Annu. Rev. Biochem.* **53**, 749–790.

Tyms, A. S., Williamson, J. D., and Bacchi, C. J. (1988). Polyamine inhibitors in antimicrobial chemotherapy. *J. Antimicrob. Chemother.* **22**, 403–427.

Polymerase Chain Reaction

JAMES R. THOMPSON
Cornell University Medical College

GLOSSARY

DNA polymerase An enzyme that is able to synthesize a new strand of DNA by extending an oligonucleotide primer in a 5′ to 3′ direction using free nucleotides and a preexisting DNA strand as a template

Major histocompatibility complex A complex of numerous genes that play a role in the generation of an immune response. In humans, these genes show sufficient variability as to be suitable for differentiation among individuals

Primer A short oligonucleotide chain or polymer of DNA that can hybridize to a DNA or RNA template and allow the initiation of polymerization of a new nucleic acid strand

Restriction enzyme An enzyme that is able to cleave DNA at a specific nucleic acid sequence

Restriction site A DNA sequence that is recognized and cleaved by a restriction enzyme

Reverse transcriptase An enzyme encoded by certain viruses that is capable of synthesizing a DNA strand using a DNA primer, free nucleotides, and RNA as a template

Taq A heat-stable DNA polymerase, isolated from a hot spring bacterium, that is used in PCR reactions. *Taq* lacks a DNA proofreading activity

Vent A heat-stable DNA polymerase, isolated from a submarine thermal vent bacterium, that is used in PCR reactions. *Vent* expresses a DNA proofreading activity

THE POLYMERASE CHAIN REACTION (PCR), MORE than any other discovery in recent history, has revolutionized both basic medical research and the clinical diagnosis of disease. The PCR reaction allows multiple rounds of amplification of a minimal amount of starting DNA material to obtain sufficient quantities of sample for analysis and use. The exponential amplification of DNA is facilitated by the use of a heat-stable DNA polymerase, isolated from a thermophilic bacterial organism. The use of this thermostable polymerase is essential to the reaction because each amplification cycle requires a high temperature DNA denaturing step. Since its first description in 1985 as a means to amplify DNA, PCR methodology has been modified and improved so that it is now utilized for many applications beyond this original purpose. In biomedical research, PCR is used to (1) introduce mutations into genes to study protein–structure function, (2) analyze gene transcriptional regulation, and (3) facilitate cloning. In paleontology, PCR is utilized to amplify and characterize fossil DNA. In forensic analysis, PCR is utilized to analyze minimal quantities of tissue evidence that can be used as a "fingerprint" in criminal identification.

I. INTRODUCTION

The PCR reaction is an extension of work originally done in defining the function of a class of enzymes called polymerases. Polymerases are enzymes that are able to catalyze a reaction in which a new strand of

ENCYCLOPEDIA OF HUMAN BIOLOGY, Second Edition, VOLUME 7. Copyright © 1997 by Academic Press. All rights of reproduction in any form reserved.

DNA is synthesized using a preexisting strand of DNA as a template. The steps of this reaction are seen in Fig. 1. In step 1, a double-stranded piece of DNA, the template, is mixed with another short oligonucleotide fragment of DNA, called a primer, that is complementary to a segment of the long template strand. The mixture is heated, typically at 94°C, in order to "melt" or dissociate the two DNA template strands. The mixture is then cooled to allow the primer sequence to anneal to the template (Fig. 1, step 2). The annealing temperature is typically 55°C; however, this can vary depending on the degree of homology between the primer and the template. The polymerase enzyme and a mixture of the four deoxynucleoside triphosphates (dNTPs: dATP, dCTP, dGTP, and dTTP) are added to the reaction (Fig. 1, step 3). Because typical polymerases are thermolabile, this addition is done following step 1 and step 2 as the high temperature of these previous steps can cause denaturation of the protein. The reaction is then incubated, allowing the polymerase to sequentially add free nucleotides in a 5′ to 3′ direction to the growing primer until a new DNA strand, complementary to the original template, is synthesized. Kary Mullis, who subsequently won the Nobel Prize for his work, first demonstrated that this reaction could be repeated or cycled multiple times in order to amplify a DNA target template over a millionfold.

II. POLYMERASE CHAIN REACTION (PCR): THEORY AND METHODOLOGY

The cycling or chaining of a single polymerase reaction to amplify a DNA template is driven by several

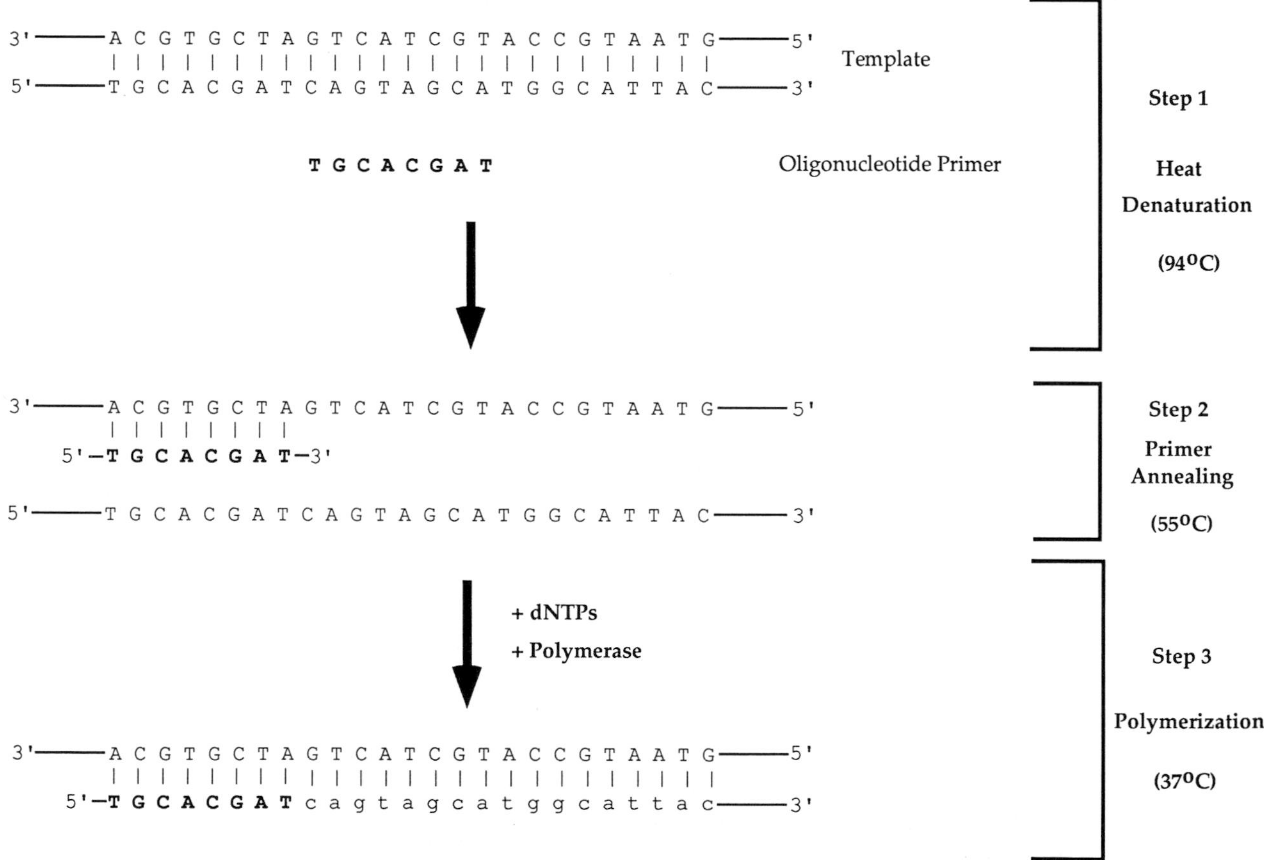

FIGURE 1 Steps of a standard polymerization reaction. The capital regular type letters indicate nucleotides of the double-stranded DNA template and the capital bold type letters indicate the primer. The reaction is heated to melt or denature the double-stranded DNA, and the oligonucleotide primer is then allowed to anneal to the now single-stranded template. Polymerase and free deoxynucleoside triphosphates (dNTPs) are added. The enzyme extends the primer by the addition of dNTPs to synthesize a new DNA strand complementary to the template. The newly synthesized DNA strand is indicated by lowercase regular type letters. The temperatures indicated are typical ones for a polymerization reaction.

features that are illustrated in Fig. 2. In the first step of a PCR reaction, two different oligonucleotide primers are used which flank the region of DNA to be amplified. The polymerase reaction is carried out as described earlier, and a single cycle will thus generate two new complementary DNA strands. These complementary strands are part of two new double-stranded DNA products that can be utilized in the next reaction cycle as template. Thus, each subsequent reaction step doubles the amount of the new double-stranded product generated. The ability to synthesize a new product at each step is driven by the addition of a large molar excess of the oligonucleotide primers added at the first cycle. Therefore, sufficient primer is available for use at each step to hybridize to the newly formed template and to initiate each subsequent reaction. PCR reactions are typically carried out for 25 cycles. Because the amount of product template generated is doubled at each step, the amount of amplification of the initial template is 2^{25} or 33.5 millionfold.

The practical use of PCR as a laboratory tool was initially hampered by two problems: (1) because different temperatures are used for the melting, annealing, and polymerization steps, samples had to be manually transferred to different water baths during the entire cycling amplification procedure and (2) the polymerase utilized for the reaction, Klenow enzyme derived from the bacteria *Escherichia coli,* was heat labile. Therefore, after each cycle, it was necessary to add new Klenow enzyme to the samples. Since 25 cycles of a PCR reaction take several hours to complete, the amplification process was a technically tedious procedure. The first problem was overcome by the development of a computerized cycling machine that could be programmed to alter the temperature of a heating block to the preset PCR parameters. The second problem was solved with the discovery and isolation of a heat-stable enzyme, called *Taq* polymerase, that could carry out the reaction.

Taq polymerase was cloned from a bacterium called *Thermus aquaticus,* which was first described in 1969. This thermophilic microorganism was isolated from a hot spring in Yellowstone National Park and is capable of growth at 70–75°C. Because of its adaptation for growth in this environment, the DNA polymerase expressed by *T. aquaticus* carries out its polymerization reaction at high temperatures (70°C) and is relatively stable and not denatured by exposure to even higher temperatures (90°C). The heat stability of *Taq* polymerase makes it highly suitable for use in

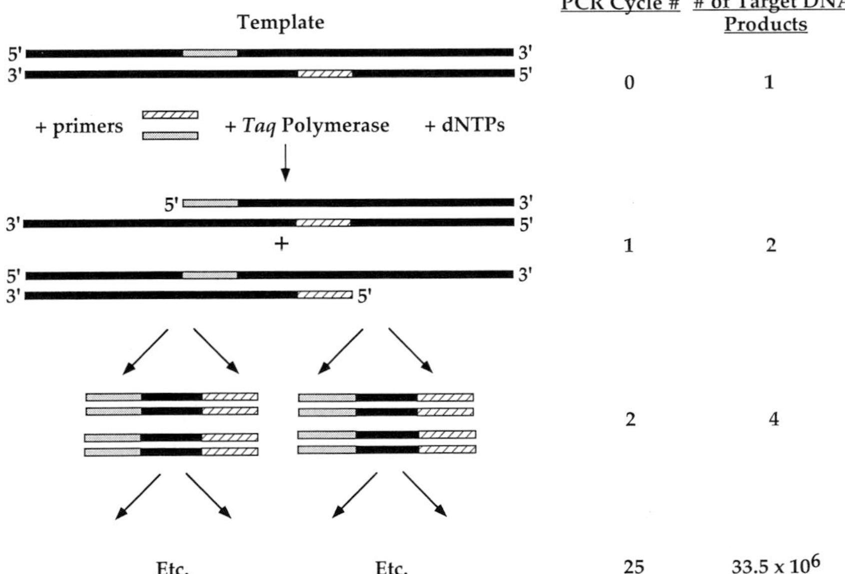

FIGURE 2 The polymerase chain reaction. The double-stranded DNA template is indicated in bold. Both primers and regions of the DNA template that are homologous to the primers are indicated by stippled or striped boxes. On the right, the PCR cycle number and the corresponding number of molecular DNA products that would be synthesized at that cycle number are indicated. At each step, the amount of product is effectively doubled. A PCR reaction is typically carried out for 25 cycles and can therefore give a 33.5-millionfold amplification of starting material.

PCR. *Taq* can be added to a PCR reaction at the first step and can be utilized without replenishing throughout all subsequent cycles, including each 90°C melting step. The use of *Taq* coupled with the advent of automatic robotic cyclers has made PCR an efficient routine process.

III. PCR PRACTICAL CONSIDERATIONS

Although PCR is a powerful tool, there are a number of technical considerations that are inherent in the method. These are important to note in designing experiments and in interpreting data. The three most important of these are the (1) introduction of mutations into DNA PCR products by *Taq* polymerase, (2) mishybridization of primers to related target sequences to generate incorrect products, and (3) contamination of starting material. The first of these relates to a feature of the *Taq* polymerase enzyme. When synthesizing a new strand of DNA from a template strand, it is essential that polymerases be able to correctly incorporate the proper nucleotide that is complementary to the nucleotide of the template strand being copied. The fidelity of replication is important because misincorporation of an incorrect nucleotide may introduce a mutation that can be lethal to an organism. Many polymerases contain an exonuclease enzymatic feature that allows them to overcome this problem. This exonuclease activity of a polymerase allows the enzyme to recognize that an incorrect base has been incorporated and to cleave the nucleotide, allowing reincorporation of the correct base. Polymerases that have this "proofreading" or exonuclease activity typically have error rates in the range of 1 in 10^5–10^6 nucleotides copied. *Taq* polymerase lacks this exonuclease activity and has an approximate error rate for misincorporation of a nucleotide of 1 in 10^4–10^5 nucleotides copied. This high error rate of *Taq* may present a problem in the interpretation of data. For example, if DNA samples from patients are being analyzed in order to determine if a gene mutation correlates with a clinical disease, it must be confirmed that nucleotide differences between the DNA of normal and that of affected patients did not arise artifactually from the PCR of samples. To address this problem, multiple individually prepared samples can be subjected to PCR reaction and characterized. Also, other polymerases have been described that have a much greater fidelity rate than *Taq*. One of these, called *Vent*, which was isolated from a submarine thermal vent, is heat stable and contains a "proofreading" exonuclease activity, giving it a 5- to 15-fold higher fidelity rate than *Taq*.

The second problem associated with PCR is the hybridization of primers to incorrect regions of target DNA. Primers that are not 100% matches to their specific targets must sometimes be used when the exact sequence of the region to be amplified is not known. Also, oligonucleotide primers, which are 100% homologous to their target regions, may also share some homology with other regions of a DNA sample being analyzed. This is especially true when the sample contains a large number of varied sequences such as that prepared from total cellular DNA. The mispriming caused by hybridization to incorrect target regions may cause the generation of false products. The ability of PCR to yield false product should be considered when interpreting data and when designing primers. Primers that form highly stable hybrids to the DNA target regions should preferentially be used.

The final problem associated with PCR is the contamination of samples being analyzed with other DNA. Because PCR is able to amplify a very small amount of material, a minimal amount of cross-contamination between samples can lead to the misinterpretation of data. An example of this is using PCR to examine patient samples to detect viral DNA sequences and to diagnose an infection. A negative sample can be contaminated by the aeration of droplets from the handling of another viral positive patient's sample in the same area. Because cross-contamination is such a problem, special precautions, such as aeration-filtered pipets, are often used in laboratories where specific qualitative analyses of samples are essential.

IV. APPLICATIONS IN MOLECULAR BIOLOGY AND GENE ANALYSIS

PCR has been adapted for a variety of uses in the laboratory in order to engineer DNA and to study gene expression. The three most common techniques, which will be discussed, are (1) the introduction of restriction sites into DNA, (2) the introduction of mutations into DNA, and (3) the analysis of gene expression using reverse transcriptase PCR (RT-PCR).

In order to amplify or clone a portion of DNA, the fragment must be introduced or ligated into a vector, such as a plasmid, for propagation. DNA products must be subcloned into sequences called restriction sites. Restriction sites are specific DNA sequences that are cleaved by enzymes. In order to subclone a DNA region of interest, the restriction site on the fragment of interest must match that of the vector. The lack of convenient compatible restriction sites is often a problem which can make cloning difficult. PCR can be used to introduce restriction sites on the end of a DNA fragment of interest. The method used to accomplish this is summarized in Fig. 3. Oligonucleotide primers are designed that have the desired restriction sites in their 5′ regions. Their 3′ regions contain sequences that are specifically homologous to the region of interest. The PCR reaction is carried out whereby the presence of nonhomologous sequences in the primers will not affect their ability to hybridize to their target sequences. The resulting PCR product contains newly introduced restriction sites that can be utilized for cloning.

A second useful technique is the use of PCR for the introduction of mutations into DNA. The method used to do this is summarized in Fig. 4. Two PCR reactions are carried out separately. Each reaction is performed with one outer completely homologous primer and an internal primer that contains the mutation of interest. This mutation can be a single or multiple nucleotide change as long as the primer contains enough completely homologous sequence to recognize its target region. Following the two PCR reactions, the samples are purified, allowed to anneal to each other, and the recessed 3′ ends filled in enzymatically. The resultant molecule can then be amplified with the original two outer primers to generate a DNA fragment with a specific mutation introduced. The introduction of mutations into DNA is useful for many reasons. One example is the study of protein function. A single amino acid of a protein can be examined by mutating it and analyzing the effect of this on protein activity. To mutate an amino acid, the three nucleotides encoding its corresponding codon in the DNA sequence are altered. The new protein sequence with the alteration can then be expressed and characterized. In addition to the introduction of mutations, PCR can also be used to insert and delete sequences from a DNA fragment of interest.

A third common use of PCR in the laboratory is

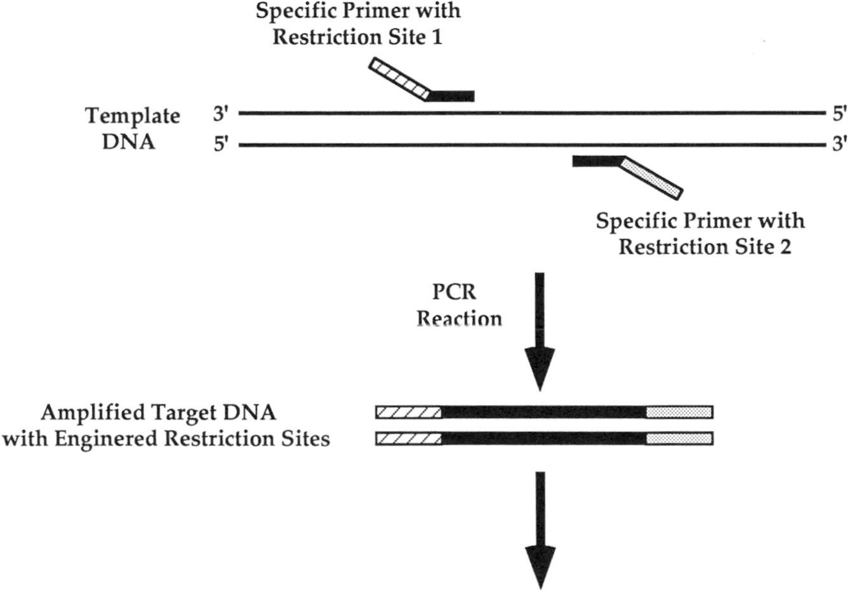

FIGURE 3 Molecular engineering of restriction sites onto a DNA fragment. The double-stranded DNA template is indicated. The primers shown contain two features. The regions indicated in black are homologous to the DNA target sequence of interest. The regions indicated by either the striped or the stippled boxes encode unique restriction sites that are not homologous to the DNA template. The PCR reaction is carried out, and an amplified product is obtained that contains the DNA region of interest flanked by restriction sites that can be utilized for cloning.

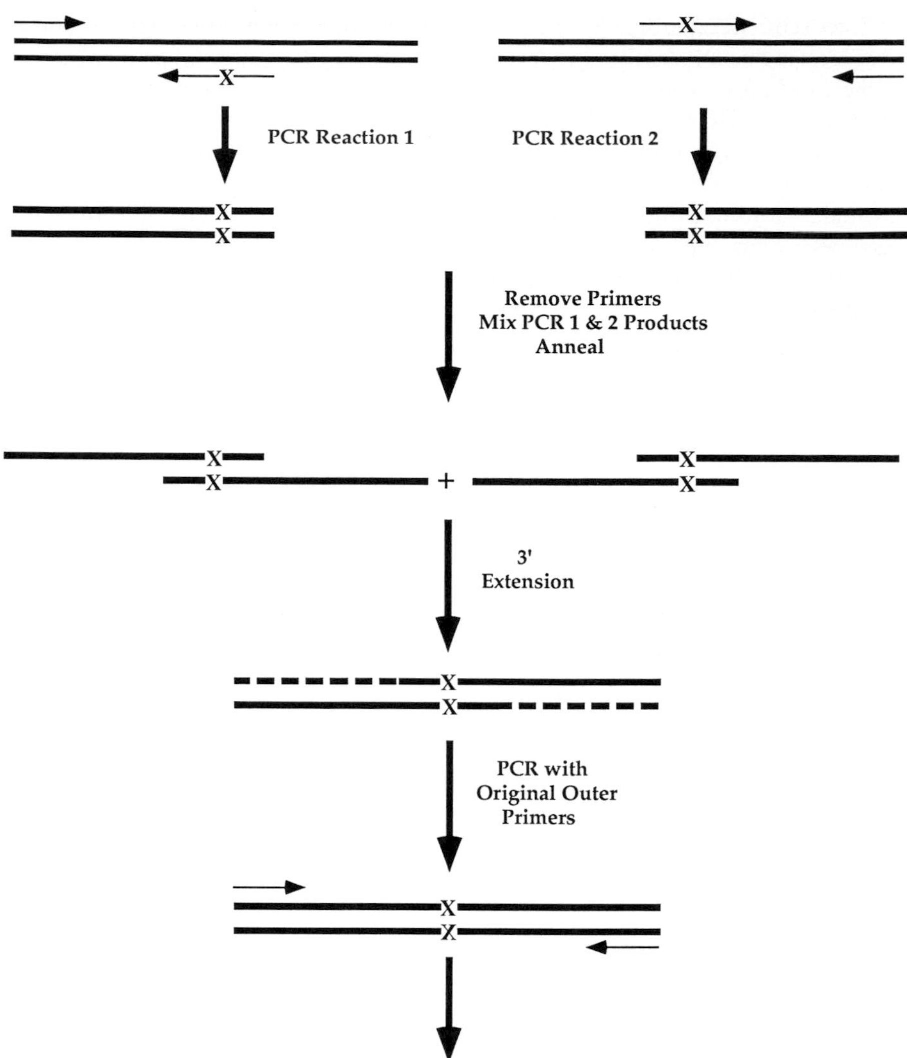

FIGURE 4 Introduction of mutations into a DNA target via PCR. The double-stranded DNA target in which a mutation will be introduced is indicated by the two heavy dark lines. The primers for amplification of the sequence are indicated by light lines with arrows. The nucleotide(s) that will be mutated is located in the internal primers and is indicated by an X. Initially, two PCR reactions are carried out to generate two smaller overlapping double-stranded DNA products that contain the mutation of interest. The products are purified, heat denatured, and allowed to reanneal. Overlapping DNA fragments are extended using polymerase to generate a double-stranded, full-length product containing the mutation. This product is subjected to PCR using the two original outer primers to amplify the target and yield a DNA fragment with a newly engineered mutation.

in the study of gene transcriptional regulation. It is often of interest to examine whether the level of RNA transcribed from a particular gene is upregulated or downregulated over time or in response to specific reagents. However, sometimes the amount of RNA being examined is insufficient to be detected by typical methods. Amplifying a sequence from a small amount of RNA is called reverse transcriptase PCR or RT-PCR and is summarized in Fig. 5. The RNA of interest is first subjected to a reverse transcription reaction. Reverse transcriptase is a polymerase encoded by viruses that is able to synthesize a DNA strand from a RNA strand by extension of a DNA primer. Several types of DNA primers can be used in the reverse transcriptase reac-

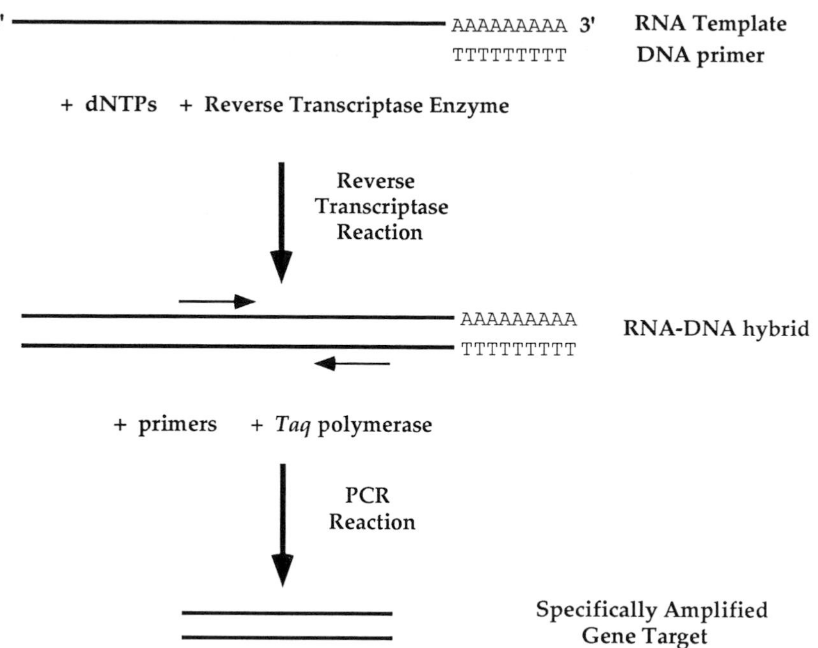

FIGURE 5 Reverse transcriptase polymerase chain reaction. An mRNA template is shown containing a polyadenylated 3′ tail. A DNA oligonucleotide primer is hybridized to the RNA. A polythymidine DNA primer is shown; however, a gene-specific primer or a collection of random hexamers can also be used. The reverse transcriptase reaction is carried out to generate a DNA template as part of a DNA–RNA hybrid. Gene-specific primers and *Taq* polymerase are then added and a PCR reaction is carried out. The product can be analyzed by electrophoresis and quantitated to determine the relative transcription levels of the original RNA.

tion. One type is a specific primer that is homologous to the gene of interest. A second type, illustrated here, is a polythymidine oligonucleotide that can hybridize to the polyadenine sequence that is present at the 3′ end of eukaryotic mRNAs. A third type is a mixture of random hexamer oligonucleotides that can hybridize and therefore, theoretically, prime at all RNA sequences present, thus converting all RNAs to DNAs. Following this first-strand DNA synthesis from RNA, gene-specific primers are utilized and a typical PCR reaction is carried out. Samples can be analyzed by electrophoresis and compared to determine the relative amounts of original RNA that was present.

V. APPLICATIONS IN THE CLINICAL DIAGNOSIS OF INFECTION AND DISEASE

PCR is now used routinely for clinical diagnosis of disease and infection. In the diagnosis of disease, PCR is most valuable as a tool in the prenatal diagnosis of genetic disorders. In order to be useful, the mutation in the human genome that causes a particular disease must be known. An example of such a disease is sickle cell anemia, which is caused by a mutation within the β-globin gene. In order for the disease to be evident, both allelic copies of the β-globin gene must carry this mutation. If only one copy of the gene is mutated, the patient is a carrier but will not express the severe sickle cell phenotype. The mutation in the DNA of the β-globin that causes sickle cell anemia also results in the alteration of a restriction enzyme site. Therefore, DNA isolated from a normal β-globin gene can be cleaved by the restriction enzyme whereas DNA isolated from a mutated gene that can give rise to sickle cell anemia cannot be cleaved by the restriction enzyme. For prenatal diagnosis, fetal cells are isolated by amniocentesis and DNA is prepared from this. The DNA is then subjected to PCR using primers that flank the region that is mutated in sickle cell anemia patients. The amplified DNA is cut with the restriction enzyme, and DNA fragments are separated by electrophoresis. The presence or absence of the restriction site in the amplified β-globin gene will indicate whether the fetus is a carrier of the disease or will develop sickle cell anemia. In addition to prenatal diagnosis, this technique can be utilized in the examination of parental DNA for genetic counseling and to determine the likelihood of any offspring developing the disease.

Although it is helpful for the specific mutation that causes a disease to be known, PCR can also be useful in the diagnosis of diseases that are caused by a variety of mutations. An example of such a disease is Duchenne muscular dystrophy (DMD). DMD is one of the most common human genetic disorders, affecting approximately 1 in 3500 male births. The dystrophin gene, which has been shown to be mutated in affected individuals with DMD, is enormous, spanning more than 2 million base pairs of DNA. One-third of all the alterations in this gene which result in DMD have been shown to have arisen by new, previously unreported mutations. Interestingly, 60% of all cases of the disease are caused by large intragenic deletions within the dystrophin gene. Because of the heterogeneity of the mutations that cause DMD, a single PCR reaction would not be useful in diagnosis of the disease. However, multiple primers can be used in a single PCR reaction to amplify nine separate regions of the dystrophin gene. Once these regions are amplified, the samples can be separated by electrophoresis and compared to PCR samples prepared from a normal dystrophin gene. In this way, a large region can be examined rapidly for gross deletions that may cause DMD. Because of the heterogeneity of mutations, this PCR test is not useful for all DMD diagnoses, but the coamplification method with multiple primers has permitted the rapid detection of 80–90% of all dystrophin gene deletions. [*See* Muscular Dystrophies.]

In addition to genetic disorders, PCR can be used to diagnose viral infections. This method is most useful for the detection of viruses that are able to remain latent in individuals and may cause no overt symptoms immediately. Early identification of these types of viral infections may be important for effective treatment. An example of this is the use of PCR to detect human immunodeficiency virus (HIV). Upon infection of an individual, HIV is able to integrate its DNA into that of the patient's genome and may remain latent for many years. The current method used for detecting the virus is identifying antibodies against HIV in the host. There are several drawbacks in using the indirect detection of host antibodies rather than the direct detection of virus. One is that it takes some time for antibodies to be produced following viral exposure. Therefore, if the HIV antibody test is performed prior to seroconversion, a false-negative diagnosis may occur. Second, in infants, maternal antibodies may persist in the newborn for up to 15 months. Detection of anti-HIV antibodies in an infant may therefore result in a false-positive result. Finally, antibody detection in individuals may sometime render inconclusive results. Therefore, a second detection method is required. The PCR test for HIV detection is performed on DNA prepared from an individual's blood samples. Although the HIV DNA sequence is highly diverse, the primers used are from regions shown to be highly conserved among all the different HIV isolates. Because of its level of sensitivity, PCR is now routinely used to diagnose HIV as well as a large variety of other viral infections. In addition to diagnosis, PCR is also used to differentiate between active and latent viral infections. In the case of a HIV active infection, RNA transcripts are generated from the integrated viral DNA and are used to synthesize more virus. In a latent infection, no such HIV-specific transcripts are present. RT-PCR can be used to detect the presence of HIV transcripts, indicating whether the infection is active or latent and allowing appropriate treatment. [*See* Required Immune Deficiency System, Virology.]

VI. AMPLIFICATION OF ANCIENT DNA

The application of PCR technology to amplify DNA from ancient sources has made PCR an invaluable tool in the fields of anthropology, archaeology, and evolutionary studies. Many important samples exist in museums, zoological collections, and various institutions that could be of great use if the DNA in these samples could be analyzed. Some examples of these are preserved ancient microorganisms, Egyptian mummy tissue, and animal skins from rare or extinct animal species. The direct analysis of these samples is typically hampered by an insufficient amount of tissue and the danger of destroying the samples. PCR can be used to amplify regions of DNA from these samples with a minimal amount of starting material required. Although PCR can generate sufficient quantities of material for extensive qualitative analysis, there are two problems with the amplification of ancient DNA. One is the problem of the quality of the DNA. Old DNA is heavily modified and typically degraded to a large degree so that only small fragments of several hundred base pairs are present in preserved samples. Therefore, the size of DNA that can be amplified and examined is limited. The second problem is the design of primers used in the PCR reaction. Because ancient samples are divergent in their DNA from modern organisms, guesses must be made for the sequence of the primers. Typically, primers are designed against highly conserved regions, such as that of mitochondrial DNA segments, to give the greatest probability of hybridization to target sequences. Despite these drawbacks, PCR has been used to examine the sequences of DNAs from rare sources that were previously inaccessible by routine methods.

VII. APPLICATIONS IN FORENSIC ANALYSIS

Because PCR enables the amplification of minute amounts of sample tissue, the technique has become an invaluable tool in forensic analysis. Prior to PCR, DNA was utilized as a fingerprint in criminal investigaitons and in paternity determination, but a significant amount of starting material was required. PCR allows the analysis of a single hair, a small number of cells, or even a single sperm to identify a suspect or confirm identity. In order to be of use, the DNA region of analysis must be sufficiently polymorphic in the population as to allow the differentiation of individuals. The common DNA region examined is that of the class II major histocompatibility complex or HLA-D genes. [*See* Major Histocompatibility Complex (MHC).]

The HLA-D genes are organized into three regions, HLA-DR, -DQ, and -DP, each of which encode an α and β gene. These genes encode proteins that are expressed on a number of cell types, including B lymphocytes, macrophages, and activated T lymphocytes. Their diversity arises from the fact that they are part of the immune response and are responsible for the recognition and binding of foreign antigens in the human body. The HLA gene routinely used in analysis is DQα. To date, eight unique alleles of DQα have been identified in the human population. To analyze evidence, forensic samples are subjected to PCR using primers that recognize conserved regions outside the highly polymorphic region of the DQα gene to generate a 242-bp DNA fragment. The DNA is then fixed to a substrate and hybridized to a series of radiolabeled or colorimetrically labeled allele-specific oligonucleotides (ASOs). These ASOs can differentiate among the variants of the DQα gene and reveal which two of the alleles are present in a forensic sample. These can then be compared to allelic samples obtained from a suspect. Allele frequency data have been accumulated for Caucasian, Black, and Asian populations and it therefore can be determined with what frequency the alleles will appear at random in a specific individual. The average frequency or occurrence of two specific DQα alleles in an individual in the population ranges from 0.5 to 15%. Although this is not a conclusive fingerprint, it may aid in the elimination of potential suspects in a criminal investigation. In addition, using the DQα as a fingerprint can be used in combination with other HLA-D regions. The DPβ gene has 21 allelic variants that can be characterized by similar methods used for the DQα gene. The use of this loci in combination with DQα and other loci can effectively multiply the discriminatory power of the genotyping and function essentially as a conclusive fingerprint. Presently, a variety of other loci are being examined for use in DNA fingerprinting, and population statistical data are being generated for these loci. Because of its ability to generate an individual identity from a minimal amount of evidence, PCR has become an invaluable tool in forensic analysis.

BIBLIOGRAPHY

Diffenbach, C. W., and Dveksler, G. S. (ed.) (1995). "PCR Primer: A Laboratory Manual." Cold Spring Harbor Laboratory Press, New York.

Erlich, H. A. (ed.) (1989). "PCR Technology: Principles and Applications for DNA Amplification." Stockton Press, New York.

Lenstra, J. A. (1995). The applications of the polymerase chain reaction in the life sciences. *Cell. Mol. Biol.* **41**(5), 603.

Mayr, W. R. (1995). DNA markers in forensic medicine. *Tranfus. Clin. Biol.* **2**(4), 325.

Rahtchian, A. (1995). Novel methods for cloning and engineering genes using the polymerase chain reaction. *Curr. Opin. Biotechnol.* **6**(1), 30.

Ronai, Z., and Yakubovskaya, M. (1995). PCR in clinical diagnosis. *J. Clin. Lab. Anal.* **9**(4), 269.

Polymorphism of the Aging Immune System

BERNHARD CINADER
University of Toronto

GLOSSARY

B cells Bone marrow-derived cells that differentiate, acquire receptors, and can finally interact with antigens and synthesize antibodies

Economic correction Designates a relation between youthful activity and middle-age decrease in that activity

LAK cells Lymphokine-activated killer cells

NK cells Nonspecific killer cells

T cells Thymus-derived cells, also coming from the bone marrow, that differentiate in the thymus and then leave

Tc cells Cytotoxic T cells that can kill target cells

AGING OF ANY PARTICULAR FUNCTION OR SYNthetic activity occurs at different rates in different individuals (i.e., the rate of progression is polymorphic, or variable). Functionally different tissues of the same individual age at different rates, that is, aging is not synchronized, it is compartmentalized. In many compartments, a direct relationship exists between the density of receptors, concentration or activity in youth, and rate of decrease in later life—an economic correction of youthful synthetic exuberance. Genetic programs, wear and tear, repair capacity, and feedback controls are involved in the progression of aging. In the immune system, involution of the thymus is an important factor in aging of the T-cell compartment; different T-cell types change at different rates in different individuals.

I. PROGRESSION OF AGING

In the influenza epidemics of 1957–1958 and 1962–1963, deaths occurred more frequently among the elderly than among the middle-aged. These deaths were due to many causes, including cardiovascular diseases and secondary bacterial pneumonias, but some evidence indicates that an age-related decline in immune response was a factor in the relatively high influenza infections among the elderly. In fact, this is a general feature of aging: the concentration of natural antibodies to flagella (bacterial surface appendages) is, on the average, twice as high in individuals in their thirties to forties than in those in their sixties. As the amount of antibody to foreign antigens decreases, the incidence of antibodies to self-antigens (i.e., of the individual's own body), such as DNA, increases. These changes in immune responsiveness occur at different ages in different individuals and in some individuals cannot be detected, even at a very advanced age. Thus, individual differences (i.e., polymorphism) are considerable in the rate and severity of age-related changes. In the same individual, age-related changes of different cell populations are initiated at different ages and progress at different rates. These differences in immune response depend on multiallelic genes and environmental influences.

II. AGING OF THE IMMUNE SYSTEM

The progression of age-related changes depends on the ability to repair biochemical damage and on the ability to generate precursors from stem cells and gen-

ENCYCLOPEDIA OF HUMAN BIOLOGY, Second Edition, VOLUME 7. Copyright © 1997 by Academic Press. All rights of reproduction in any form reserved.

erate end cells from precursors. Throughout life, the immune system is dependent on processes of differentiation from precursors to functionally specialized lymphocytes of two types: committed thymus-derived (T) and bone marrow-derived (B) cells. Because precursor cells replicate and their descendants differentiate to become mature cells, the immune system is profoundly affected by wear and tear of precursor cells and less by wear and tear of mature cells than are tissues such as muscle and brain. Therefore, to what extent do precursor cells lose their ability for self-renewal? There appears to be a correlation between the number of *in vitro* cell division of fibroblasts and age of donor. Argument for a biological link between this type of *in vitro* and *in vivo* aging can be found in the reported correlation between average life span of different species and the number of *in vitro* divisions that their fibroblasts can undergo. Nevertheless, the effect of age on the *in vivo* capacity for self-renewal remains a controversial subject. It may depend on the developmental distance from the primitive stem cell. The further the precursor cell differentiates toward a fully committed cell (end cell), the greater the probability that self-renewal declines. The polymorphism of these changes remains to be explored and may contribute to the resolution of contradictory results. [*See* Lymphocytes.]

In early postnatal life, T cells acquire their functional specialization in the thymus. Involution of the thymus is a programmed feature of postnatal development and reaches completion in middle age (in humans, at about 50 yr of age). In men it occures at an earlier age than in women; hormonal controls can affect the rate of thymus regression. The pituitary gland is implicated in some of the changes. Furthermore, the age-related decrease in the secretion of thymus hormones, such as thymulin, may be due to a decrease in synthesis of thyroxine, the thyroid hormone. Whether or not initiation of thymus regression is triggered by endocrine controls is not definitively established. Events in the thymus itself are multicentric, which results in different types of precursor cells, that is, precursors of suppressor cells for humoral and cell-mediated immunity age at different rates and even in different directions (one class decreases while another increases with age in some inbred mouse strains). [*See* Thymus.]

A decreasing supply of thymic peptides and cytokines may be pacemakers of the impact of involution on the immune response and contribute to the peripheral consequences of involution. The involution of the thymus initiates peripheral age-related changes in immune responsiveness during the second half of life. However, precursor cells, which leave the thymus, settle in spleen and lymph nodes and are sources for renewal of T-cell function in later life. Finally, the involution of the thymus affects peripheral function if or when the peripheral pools have exhausted their ability for self-renewal.

The general view on aging-progression, deduced from observations on humans, can be extended and confirmed in appropriate animal model systems, particularly with animals of different homogeneous genetic background (i.e., with strains of inbred mice). The validity of extension from animal to humans depends on the validity of the assumptions that the factors involved in aging are common to all mammals. In fact, there is indirect evidence for this view. A positive correlation exists between life span and concentration in the liver of carotenoids, α-tocopherol, ceruloplasmin, and ascorbate; an inverse correlation is evident between life span and concentration of cytochrome, gluthathione transferase, and catalase. It is, therefore, reasonable to conclude that aging of different mammals depends on similar processes and that it is legitimate to extrapolate from short-lived to long-lived mammals. In general, aging is apparently dramatic in some, but not all, compartments of the immune system. We shall briefly consider the polymorphism of some of the following cell populations:

1. T-suppressor capacity for antibody response;
2. T-suppressor capacity for cell-mediated immunity;
3. cytotoxic T cells (CTL);
4. B cells and T-cell helper capacity;
5. nonspecific killer (NK) and lymphocyte-activated killer (LAK) cells.

III. POLYMORPHISM OF AGING

In some inbred strains of mice, a marked decrease is noted in suppressor capacity for the humoral response with a simultaneous increase of suppressor capacity for cell-mediated immunity. In other strains of mice, these processes occur very slowly, and in yet others suppressor capacity for cell-mediated immunity either does not change with age or increases to middle age and then decreases.

A great diversity exists in age-related changes of CTL cells and in their inhibition by suppressor cells. The extent of LAK cell activity differs in young animals of different strains and, as animals age, the level decreases to a greater extent the greater the youthful activity (i.e., economic correction).

B cells undergo age-related changes in some individuals (e.g., inbred mouse strains), which apparently renders the ability to respond to antigens less dependent on T-cell help as the animal ages. Relatively few age-related changes are found in NK cells of most strains of mice; in some there even appears to be an increase with age in middle age followed by a decline in old age. In short, polymorphism occurs in the rate at which age-related changes occur and in the cell type that is predominantly affected by these changes. This can be easily established with inbred mice, because young and old animals of the same genetic background can be compared. This type of comparison cannot be made in the study of human aging; the heterogeneity observed in the rate at which the human immune system ages can be assumed to be a reflection of the differences in the type of aging that occurs in different individuals (i.e., with different genetic backgrounds).

IV. COMPARTMENTALIZED AGING AND POLYMORPHISM

Because individual differences of age-related changes will be illustrated in terms of suppressor capacity in various contexts, we will give a brief summary of how suppressor capacity of two classes (one affecting humoral immunity, the other cell-mediated immunity) can be assayed.

Suppressor capacity in the thymus, affecting humoral response (antibody production), can be measured by reconstitution experiments, in which lethally irradiated mice are given either lymphocytes of young donors or lymphocytes from the same young donors but mixed with thymus cells from donors of different ages. The reconstituted mice are then immunized, and antibody is measured. In some strains of mice, thymus cells from young donors reduce the humoral response of the lymphocytes, showing a suppressor effect, whereas thymus cells from old donors increase it, showing reduced suppression.

Changes in suppressor capacity for cell-mediated immunity can be measured by the ability of thymus cells to prevent lymphocytes to be sensitized by targets and to kill them. Results obtained by this method have been confirmed by a variety of other tests for cell-mediated immunity.

Differences within the same strain in age-related changes of these two classes of suppressor capacity are remarkable. The independence of the aging changes occurring in the two classes has been demonstrated by nutritional, hormonal, and pharmacological interventions, which can affect aging of one class without affecting aging of the other.

Reference to the rate of age-related changes modified by various external factors has already been made. These factors can be used to determine whether or not aging in different cellular subsets is under independent genetic controls (i.e., whether or not aging is compartmentalized). This distinction between independence and interdependence can be analyzed by interventions, which change the rate of one process without changing that of another, and by selecting different types of interventions, which act on fundamentally different metabolic or developmental processes. To this end, hormones, purine analogues, and different diets have been employed.

The rate of increase in precursors of suppressor cell for CTL can be accelerated by administration of β-estradiol and progesterone; this treatment does not bring about changes in humoral response. The rate of decrease in precursors for suppression of humoral response can be delayed with purine analogues, which are readily available as antiviral drugs. This type of drug may affect progression differentially because adenosine deaminase and purine nucleosidase deficiency have different effects on humoral and cell-mediated immunity. So far, two such antiviral compounds have been tested and found to modify age-related progression of the humoral response. A third strategy for modification of age-related progression is based on feeding diets, which differ in fatty acid composition. This is a strategy of choice, because lipid composition of membranes changes during differentiation and the ratio of polyunsaturated to saturated fatty acid (P/S) affects membrane fluidity. In fact, diets induce changes in suppressor capacity for humoral as well as for cell-mediated immunity: diets low in P/S prevent or delay resistance against down-regulation (i.e., against tolerance induction) and delay loss of thymus suppressor activity for antibody formation; diets high in P/S prevent or delay the age-related increase of suppressor activity for cell-mediated immunity.

It follows that: (1) suppressor capacity for humoral and cell-mediated immunity can age at different rates and even in different directions (i.e., one decreasing, the other increasing) and (2) the rate of aging can be changed in one of these types of suppressor capacity without changing it in the other. Therefore, it is reasonable to conclude that age-related progression in these two compartments is under the control of different genes, at least in individuals with certain types of genetic backgrounds.

Age-related changes in suppressor capacity are only one of several instances of aging, occurring at different rates in different compartments. Other instances of compartmentalized aging have been found in differential progression of aging of antibody responsiveness in the mucosal and peripheral immune system, in different isotypes, and in age-related progression of antigen-presenting cells. In short, aging within the immune system is compartmentalized; in the same individual, different compartments, as well as same compartments of different individuals, age at different rates. These conclusions are based on studies of aging in rodents but are clearly applicable to human aging. It is well known that a man may lose his ability to divide hair follicles (i.e., become bald), while retaining the ability to divide precursor cells of sperm (i.e., remain fertile). By extension, compartmentalization of the immune system, analyzed in rodents, can be assumed also to apply to humans though this remains to be rigorously demonstrated.

V. ECONOMIC CORRECTION: AN INTRACOMPARTMENTAL CONTROL PROCESS

Individual differences in the quantity of a given gene product at different ages are yet another aspect of the individuality of aging. The range of this polymorphism is great early in life, decreases in middle age, and then increases again in senility. Between early age and middle age, the synthesized quantity of various types of molecules and cell types decreases; the magnitude of this decrease is greater the greater the quantity of a given gene product produced early in life. This relationship between youthful quantity and magnitude of age-related decrease is referred to as economic correction. As a consequence of this process, the quantity of many macromolecules, produced by different individuals, varies over a much larger range in youth than in middle age. The level, attained in middle age, remains constant over a considerable portion of the second half of life and shows an individually variable decrease toward the end of life.

It is apparent from the foregoing survey of polymorphism and compartmentalization of aging processes that this type of data represents the fine structure of the heterogeneity of aging. Insight into the genetics of polymorphism is not only of biological but also of clinical interest. The initiation and progression of diseases of old age depend on a variety of controlling processes (including T-cell regulation), which prevent immune responses that might damage self; the effectiveness of this controlling process changes dramatically in some individuals, although not in others. Preventive medicine for degenerative diseases of old age can only be developed if the genetic individuality of the polymorphism is known. The ability to identify populations at risk would allow us to develop strategies that delay or prevent changes that predispose to a particular type of degenerative disease. The growing proportion of the elderly in the world's population makes this task an important target for improving the quality of life and preventing excessive pressure on our health services.

BIBLIOGRAPHY

Bergener, M., Ermini, M., and Stahelin, H. B. (eds.) (1988). "Crossroads in Aging," 1988 Sandoz Lecture in Gerontology. Academic Press, London.

Chandra, R. K. (ed.) (1985). "Nutrition, Immunity and Illness in the Elderly." Pergamon, New York/Oxford/Toronto/Sydney/Frankfurt.

Cinader, B. (ed.) (1989). "Symposium 2.4.2: Genetics of Aging, XVIth International Congress of Genetics." Genome, National Research Council, Ottawa, Ontario.

Cinader, B., and Kay, M. M. B. (1986). Differentiation and regulatory cell interactions in aging. *Gerontology* **32**, 340–348.

Courtois, Y., Faucheux, B., Forette, B., Knook, D. L., and Treton, J. A. (eds.) (1986). "Modern Trends in Aging Research." John Libby Eurotext, London/Paris.

de Weck, A. (ed.) (1984). Lymphoid cell functions in aging. *In* "Topics in Aging Research in Europe." Eurage, Rijswijk.

Goidl, E. A. (ed.) (1986). "Aging and the Immune Response." Marcel Dekker, New York.

Maddox, G. L., and Busse, E. W. (eds.) (1987). "Aging: The Universal Human Experience." Springer Publishing, New York.

Schneider, E. L., and Rowe, J. W. (1996). "Handbook of the Biology of Aging." Academic Press, San Diego.

Yamamura, Y., and Tomio, T. (eds.) (1983). "Progress in Immunology V." Academic Press, San Diego.

Population Differentiation and Racial Classification

MICHAEL D. LEVIN
University of Toronto

GLOSSARY

Cline Graded variation in gene frequency along a line of geographic transition

Geographical race Broad, geographically delimited collections of races (populations)

Local race Breeding population or population isolate

mtDNA Mitochondrial DNA inherited through females. Differences in individual mtDNA allow hypotheses on evolution

Population All the organisms of a species; within a species, a community of individuals where mates are usually found; a local or breeding group; the inhabitants of a place

Race Group of populations of a species distinct from other groups of the same species in at least a few characteristics; a subspecies

POPULATION DIFFERENTIATION IS THE RESULT OF genetic change. Populations of a species may be denoted as races, or subspecies if the differences in the gene frequencies in each are regarded as significant. The use of the term *race* to refer to human populations is, however, controversial scientifically and politically. The political use of the term, which is usually part of a racist ideology, has affected the scientific discussion about its use. It is possible to separate political and scientific questions to some extent, but not absolutely, as political arguments tend to build on scientific information and scientific questions are often stimulated by political statements. The distinction between scientific uses of the term and social and political uses of the term race is therefore important.

Two major questions depend on an understanding of the processes of population differentiation: the origin of modern humankind, *Homo sapiens sapiens*, and the maintenance of human genetic diversity. Both are connected to the use of the term race. Many scientists believe that answers to these questions do not depend on a division of humankind into races. The influence of language, distance, culture, and other nonbiological differences on evolutionary processes and the uses of genetic variation in the study of population histories are among the questions of current scientific interest.

The history of racial classification, a different problem, is marked by literary influences and scientific and political controversy and often appears to have been based on and to be motivated by notions of group superiority. Attempts to establish associations between various morphological and other biological differences and to correlate these differences with psychological and social characteristics are controversial not only politically but scientifically. Many scientists do not consider these associations legitimate. These attempts to extend the question of human variation beyond differences in genetic variation have generally been regarded as unscientific. The association of morphological characteristics with moral and behavioral characteristics, moreover, has been criticized because it supports racism, the ranking of subpopulations of

ENCYCLOPEDIA OF HUMAN BIOLOGY, Second Edition, VOLUME 7. Copyright © 1997 by Academic Press. All rights of reproduction in any form reserved.

the human species, and has been used to justify discrimination against individuals.

I. TERMINOLOGY AND THE HISTORY OF CLASSIFICATION

A. Terminology of Population Differentiation

The differentiation of population as races implies that the differences among the human beings are sufficient to subdivide the species into subspecies. The purpose of racial studies is the definition of relevant and significant differences that would allow this kind of classification. Racial studies have gradually been exhausted as criteria on which group differences rest are found to be inadequate for logical reasons (e.g., arbitrariness, failure to distinguish one group from another, and lack of correspondence to linguistic or cultural uniformities) and for statistical reasons (e.g., greater variance within the populations, defined as races, than between them, the inability to discover a single characteristic that can be associated exclusively with one racial population).

As different criteria were found inadequate, qualifications of the simple term race were developed. The terms *geographical race* and *local race* were suggested to associate distinct populations with specific places. These places of origin defined aboriginal populations retrospectively to a relatively fixed world population as of 1492. Clines were suggested as an alternate for race and for population as an object of study. Clines, although they did organize valuable data, have not replaced the concept of population, but have made understanding of its use more sophisticated. Use of clines has demonstrated that differences in single genes between populations were gradual along a spatial dimension. Isolation was suggested as a criteria of racial separation. With research revealing the clinal variation of genetic frequencies, the argument for isolation, or separation, of human populations was severely weakened. The failure to demonstrate isolation and the clear interdependence of environment and behavior in genetic changes are significant arguments against the use of the term race for humankind populations. More recent research, dependent on more sophisticated statistical multivariate techniques to be described later, indicates a range of boundaries of greater or lesser degrees of difference between populations. The scientists who have done this research recognize that clustering of populations into broad categories is possible but have not used race to describe these categories.

B. History of Classification

The discoveries of the New World and the expansion of contact with Africa and Asia had made Europeans aware of the great diversity of the human population. The growth of science led to attempts to understand and incorporate this diversity into existing ideas. The apparent isolation of and morphological differences among human populations seemed to be most significant. What had seemed, from a European perspective, to be a process of increasing differentiation of relatively isolated human populations came to an end in the fifteenth century with the voyages of discovery, which dramatically increased communication, migration, and change. The contrast makes the rate of change before 1492 seem slow and the populations seem isolated. The combination of the humanness and differences of New World and African and Asian peoples was a problem that demanded understanding. It is only recently that the Eurocentric dimension of this view has been recognized.

The first classification of humankind into races was probably that of François Bernier, a French physician, who in 1684 suggested four: Europeans, Far Easterners, "blacks," and Lapps. Despite criticisms, most notably of Leibniz in 1737, who wrote that "there is no reason why men who inhabit the earth should not be of the same race, which has been altered by different climates" (a view with truly modern echos), race classification was elaborated over the centuries.

In the eighteenth century, three classifications of humankind were suggested that have continued to influence forms of racial thinking. In 1737, as part of his classification of all living organisms, Linnaeus "divided human kind into four varieties: *Homo Europaeus, Homo Asiaticus, Homo Afer,* and *Homo Americanus.*" Linnaeus based his classification on the idea of the fixity of the species, that species were immutable. Varieties, however, were members of species that differed because of environmental conditioning. The immutability, or adaptability, of organisms is an underlying question in all race classifications as it raises questions of which factors are the basis of human differentiation. George Buffon in his "Natural History" (1749–1804) suggested five races: Lapland, Tartar, Southern Asiatic, European, and American. He believed the white race was the norm and others were exotic variations, but of one species. In 1775 in "The Natural Variety of Mankind," Johann Friedrich

Blumenbach, a professor of medicine at the University of Göttingen, who despite believing that classifications are arbitrary, subject to exceptions, and representations of continua, suggested five categories: Caucasian, Mongolian, Ethiopian, American, and Malay. This fivefold classification has had a long life in the literature on race in the more general form of white, yellow, black, red, and brown. Blumenbach coined the term Caucasian to describe the "white race." Buffon and Blumenbach are regarded as founders of physical anthropology, and Blumenbach was known as the "father of craniology," the study and use of skulls for scientific purposes. These fathers of the scientific study of human variation, especially Blumenbach, were highly critical of attempts to rank races. Among those who argued for the separate origin and for superiority and inferiority of races were Voltaire, Thomas Jefferson, Lord Kames, and an English physician, Dr. Charles White, who wrote "An Account of the Regular Gradation in Man" (1799). The questions of the unity or plurality of human origins (monogenesis or polygenesis) touched on questions of religion, politics (particularly the legality of slavery and, later, immigration policy), as well as science.

The nineteenth- and twentieth-century classifications were marked by debate on the formal or adaptive character of races, whether there were fixed characteristics that separated the races, as a result of separate origin or of adaptation, or whether human differentiation was a result of continuing adaptive processes affecting genetic and morphological characteristics differentially so that major taxonomic distinctions of *Homo sapiens* into subspecies or races were not useful for scientific thinking and work. The main expression of this debate was the question of types, their validity, and their use. The question of "pure" racial types arose as part of the debate on typology, and variations of the question of purity, miscegenation or mongrelization, race preservation, and dilution became scientific and popular issues. In 1899, Wm. Z. Ripley used cranial data in "The Races of Europe" in which he distinguished the Nordic, Alpine, and Mediterranean races, each characterized by a definite head shape, other morphological characteristics, and particular mental and temperamental traits. Modern classifications include that of J. Deniker (1900) of six categories, which, for example, defined Europeans as having fair, wavy, or straight hair and light eyes, and that of E. A. Hooton (1946), who suggested three primary races (White, Negroid, and Mongoloid) and four other general kinds of races (primary subrace, composite race, composite subrace, and residual mixed types), each comprising various numbers of races. The typologies of Deniker and Hooton indicate some of the problems of racial typologies, namely, that as classificatory systems they tend to be rigid. Accommodation of variant individuals in the schema is often difficult, resulting in increasingly numerous and complex sets of categories. The basic problems of establishing a typology become more difficult (e.g., which criteria are fundamental, which secondary, the number of criteria that should be used) and are all questions that cannot be determined in the absence of a theory. Cranial measurement and brain size had become important criteria of differentiation, yet whatever criteria were given prominence, none could be consistently applied and would differentiate all humans into categories. Neither could pure types be found, nor could a single or even a few characteristics be specified that would allow discrete categorization of all humans.

In 1971, S. M. Garn offered the most recent classification based on three categories: geographical races, which are defined as "a collection of populations whose similarities are due to long-continued confinement within set geographical limits"; local races, which "correspond more nearly to the breeding populations themselves . . . isolated by distance, by geographical barriers or by social prohibitions, . . . are totally or largely endogamous"; and micro races, which are related to "very real differences in the genetic makeup of cities and continual changes in the frequencies of genes," which create regional differences. Garn names nine geographical races: American, Polynesian, Micronesian, Melanesian, Australian, Asiatic, Indian, European, and African. He names 32 local races but does not name micro races, although he suggests they may be defined in terms of cities. This extension of the concept race to all magnitudes of populations, based on the concept of breeding, can be seen in retrospect as an attempt to preserve the concept by making it a synonym for population.

II. EVOLUTIONARY ASPECTS OF POPULATION DIFFERENTIATION

The failures of the attempts to establish a typology of human differentiation based on morphology were compounded as genetic and molecular information was brought to bear on the problems of human differentiation. It came to be recognized that what had been described as *racial characteristics* were adaptations to environments, or the result of other evolutionary phenomena, drift migration, and admixture. The black skin of Australians and Africans, for example, is a similar bodily feature that evolved independently, these populations being, in terms of their histories as shown by genetic research, as far apart as possible. Garn, in the same work in which he proposed his nine geographical races, acknowledged that "the distinctive characteristics of every race may now be understood in terms of the special environments in which they have lived." The confirmation of the variation of genes with environment weakened significantly any concept of absolute distinctiveness of populations within the human population, any notion of separateness between subpopulations in the human species. Moreover, to continue to define characteristics that were recognized to be adaptations as *racial* begs the question.

The study of variations of single characteristics, by use of clines, maps of the spatial variation in morphology, gene frequencies, or other biological traits, has demonstrated that these charcteristics vary more gradually than do linguistic, or cultural, or "racial" definitions of groups would lead us to believe. Although he did not use the term *cline*, G. M. Morant, in his "Races of Europe" (1939), demonstrated that morphological differences were gradual and less than linguistic and cultural differences on which the Nazis based their irredentist and racial supremacist claims. The study of variation by plotting gene frequencies as clines has opened the study of human variation to evolutionary theory. An interesting example of polymorphism is the presence of tawny hair among Australian aborigines as children, which is a maximum of 90% in the western Australian desert, declining from this center. Figure 1 illustrates the clinal distribution of tawny hair among Australian aborigine children.

Other genes are known to be indicative of local environments, notably those that determine the diseases thalassemia and sickle-cell anemia, and to vary in ways that are well represented in clinal distributions. Both thalassemia and sickle-cell disease, although strongly associated with certain regions (the Mediterranean for the former and Africa and India for the latter), vary independently of "racial" categories. Such variations are examples of microevolution and suggest emphatically that one characteristic cannot be used to define a subspecies or races. Afircans are more likely to have sickle-cell disease, but the disease is a result of a combination of cultural, environmental, and genetic factors. It has evolutionary advantage, namely, resistance to malaria, when an individual is heterozygous but is fatal for homozygous individuals.

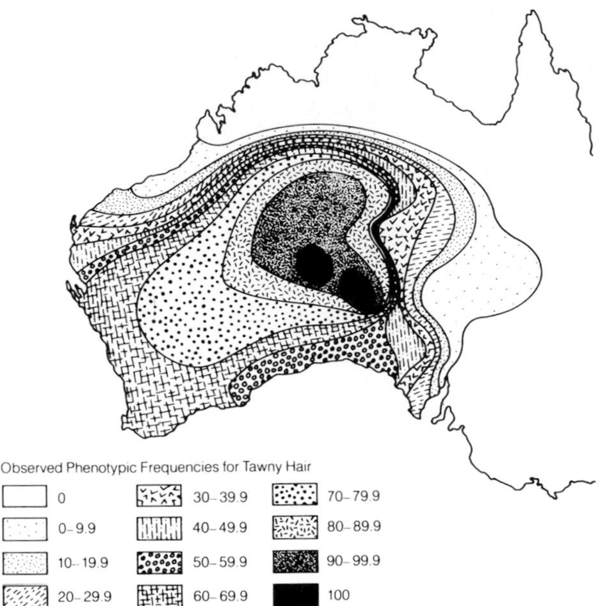

FIGURE 1 The gene distribution for tawny hair in Australia. Clines illustrate the decreasing gradient in distribution of the concentration of tawny hair (Birdsell, 1985).

African environments led to the evolution of certain morphological features; some African environments led to the evolution of the sickle-cell gene. The association is real, but the processes are separate.

Multivariate studies have been used to test directly differentiation due to race. R. C. Lewontin began with the classification of major geographical races and local races. Using categories suggested as races by previous authors, he examined the degree to which these categories explained human diversity. His conclusion was that the populations, described in the literature as "races," explain little of human diversity. "The mean proportion of the total species diversity that is contained within populations is 85.4%. . . . Less than 15% of all human diversity is accounted for by differences between human groups! Moreover, the difference between populations within a race accounts for an additional 8.3%, so that only 6.3% is accounted for by racial classification." The accumulation of evidence on genetic variation continues to indicate the continuity of variation in the human species, rather than discrete, separate groupings.

Multivariate studies have also demonstrated other adaptive effects and have been used to test racial classifications. Studies have shown that gene frequency distribution varies with the advance of agriculture in Europe and with climate. Language change and differences have been shown to be associated with genetic variations and population differentiation. These studies strongly support the possibility that evolutionary change and its interdependence with cultural or political factors (e.g., language) determined genetic boundaries and forms of migration. [*See* Population Genetics.]

III. POPULATION HISTORY

Recent research has been establishing increasingly sophisticated measures of population differentiation and suggesting hypotheses to explain differentiation. New techniques offer answers to particular aspects of the questions of the origin, or evolution, of modern humans and of the factors that created and maintain the differentiation of the present human population. The study of the population history of humankind is an area of active and extremely interesting and often controversial research.

A. Mitochondrial DNA and Paleoanthropology

In 1987, the proposal that genetic evidence demonstrated a common origin for humankind in an African "Eve" continued a debate in the study of human evolution that was at least five decades old. Figure 2, from that study, shows the divergence of mtDNA of 147 people from five geographical populations. The distribution of members of these populations across the dendogram indicates the biological diversity within these regional populations and the human species as a whole. This study also claimed that modern humans evolved relatively recently (about 200,000 years ago) in Africa. The debate has resolved into three positions: the recent Africa origin, or African "Eve," position taken by the authors of this study; an assimilationist position, which proposes that modern humans originated in Africa, migrated from there, and interbred to create local variation; and the multiregional model, which posits evolution of racial differences in different regions. The question under debate is whether humankind has a common origin in Africa, or whether modern humans evolved in more than one region. If the "Eve" hypothesis is correct, human variation and diversity, that is, modern "races" or ethnic groups, evolved from a common origin. If the multiregional evolutionary or the assimilationist arguments are correct, human differences are a product

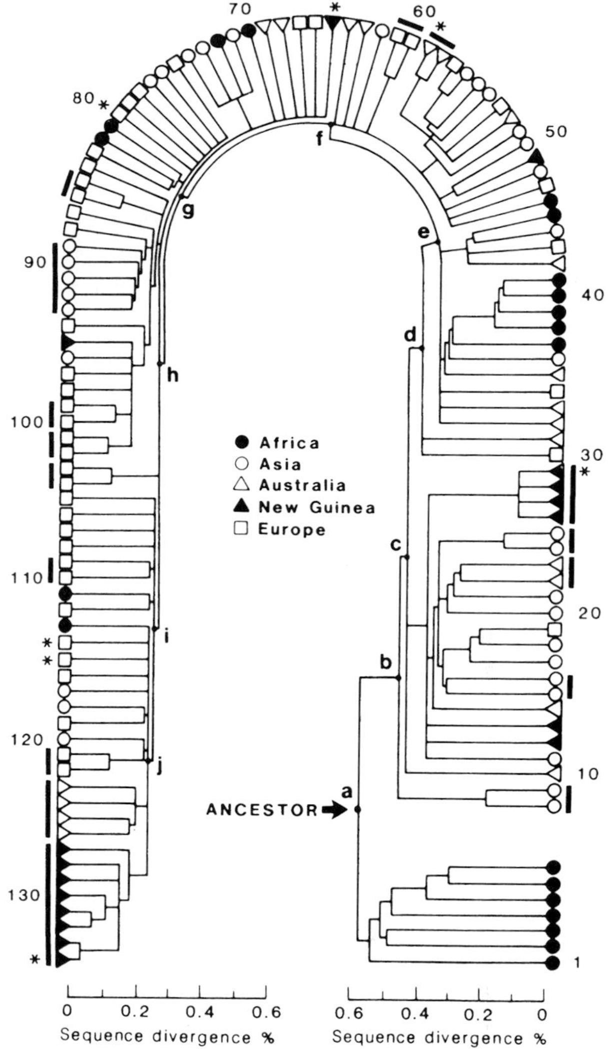

FIGURE 2 Genealogical tree for 134 types of human mtDNA (133 restriction maps plus reference sequence). The multiple origins of all populations represented except the African implies that each area was repeatedly colonized. On the basis of these data, *Homo sapiens* thus has a common African origin, and present-day populations were created by repeated migrations from Africa, replacing any earlier populations (Cann, R., *et. al.*, 1987).

of older divergences, perhaps among populations of *Homo erectus*, and the unity of the species is a consequence of contact and gene flow, not common origin. The common origin position is supported by genetic and paleoanthropological (fossil) evidence. The multiregional evolution and the assimilationist positions are supported by paleoanthropological evidence and a critique of the genetic evidence. Many issues, methodological and substantive, continue to be debated, but with fortunate fossils finds these could be resolved.

B. Geography, Language, and Culture

Spatial and cultural variables have been used to suggest answers for the question of origin and of population differentiation. Linguistic, archaeological, and genetic evidence has allowed a classification of human language phyla and genetic clusters that suggests six or seven main groups. This study suggests that genetic groupings and language families have similar patterns of divergence and form similar genealogical trees. This classification is compatible with a common origin of humankind.

The study of prehistoric migrations from Asia to Melanesia and the Americas, the spread of agriculture in Europe, and the migrations and separations of culturally distinct groups have been aided by genetic comparisons. For Europe and New Guinea, language boundaries have been shown to coincide with boundaries showing sharp genetic change. The emphasis in these studies is on the degree of morphological and genetic differentiation and the explanation for the maintenance of these differences, differences that depend on isolation allowing differentiation and reducing gene flow, which would have reduced variability. Both language differences and geography are factors in creating isolation and, thus, differentiation. The relative importance of each and the exact relation to genetic variation is still to be established. It should be emphasized that in these studies, linguistic and genetic divergence is not an exact match and boundaries do not always coincide, nor is change from group to group ordered in a particular way.

Variables affecting genetic change and differentiation in population include geography, gradual "demic" population movement and more rapid mass migration, language, and culture. Evidence of population separation includes DNA and non-DNA polymorphisms, dentition, and other paleoanthropological data. These studies can outline speculative histories of human evolution and of populations (e.g., estimated times of migration, population affinities), but the fundamental problems of analysis are as yet unresolved. Among these are questions of how to standardize measures of differences in frequencies of genetic distribution, linguistic differences and classification, and other characteristics (especially dentition, the rate of change of language and of DNA, rates of genetic drift and mutation, the effect of and rate of change of environmental change on gene frequencies). Empirical problems also exist: If, as is likely, small populations separated, genetic changes would have likely been more rapid than in large populations. Thus, the two populations would not only as expected diverge genetically, but would have done so at different rates. Broader issues are also in question: the shape and basis for deciding the form of the human genealogical diagram. Methodological problems caused by the wide range of possible error also limit accuracy of hypotheses of origin.

Studies attempting to trace population histories and origins will, nevertheless, continue to fascinate us as scientists and as human beings. Although the answers to these questions of origin are likely to remain speculative, the processes of population differentiation (i.e., differences in gene frequencies), it has become clear, are dependent not only on neutral processes, mutation, selection, gene flow, and genetic drift, but also on culturally determined marriage preferences, mass migrations, and gradual population expansions. The demonstration of cultural and spatial factors affecting genetic change and in creating and preserving population differences is extremely important in establishing the complexity and historical character of population differentiation.

IV. MEANING AND DIFFERENCE: TERMINOLOGY AND POLITICS

A. Changing Terminology in Anthropology

The use of the term race has carried with it the notion of fixed characteristics, much like the notion of fixity of species, which was problematic in the development of evolutionary theory. When morphological and anatomical characteristics were the variables of human differentiation and the data were discontinuous, the possibility of discrete human subspecies could be entertained. Before archaeological and molecular biological data indicated the unity of the human species, multiple origins of *Homo sapiens* could be argued. The main characteristics of human differentiation, the continuity of change, its gradualness, and the relative lack of difference between groups, which continue to be demonstrated with the advances of research, indicate that race as a category of evolution or a major division in the human species is unsupported. The repeated attempts to demonstrate the separation of races by reduction of morphological characteristics to genetic differences have largely failed. After this work is considered, only the differences in bodily features and aboriginal geographical status (i.e., as of the fifteenth century) remain as the basis of racial classification.

The language of discussion of human differentiation and diversity in textbooks in physical anthropology has changed significantly. Chapter headings have changed from "Human Races" or "Races of Man," to "Human Diversity" or "Human Variability." Terms such as "racial characteristic" and "racial admixture" have either disappeared in favor of "bodily features" or been abandoned in recognition that they implied an identification between a race and a specific characteristic. The remaining defenses of the concept of race are common sense (i.e., based on the evidence of our eyes) or simply that such distinctions and concepts are useful and interesting. The alternative to discussion of population differentiation sometimes takes the form of describing *Homo sapiens* as a polytypic species. The scientists studying the evolutionary impact of language and geography use the terms group, phyla, genetic entity, population cluster, and, the most general, population.

B. Use of Biology for Political Purpose

The use of the term race has often implied unbridgeable differences between individuals. The attempts to demonstrate differences between racial groups have been used as the basis of beliefs about the superiority and inferiority of racial groups (i.e., racism). Racism has led to the implementation of policies and practices of discrimination and genocide in the twentieth century. Racism depends on the belief that morphology or bodily features allow conclusions about behavior, intelligence, and morality, that in some way appearance and behavior derive from a common, in this case racial, source. Despite scientific evidence to the contrary, people cling to these prejudices. What is described as biological is often cultural, as is the belief of superiority.

Modern research makes it increasingly clear that human populations have had differing histories but that at the level that racial differences have been noted, the differences do not establish separate population groups. The arbitrary and general concept of population is useful for scientific purposes because its limits and definition can be specified, whereas race, not only because of its political implications, is not. Moreover, group membership and individual identification are not biological, but cultural.

BIBLIOGRAPHY

Ammerman, A. J., and Cavalli-Sforza, L. L. (1984). "The Neolithic Transition and the Genetics of Populations in Europe." Princeton Univ. Press, Princton, New Jersey.

Bateman, R., Goddard, I., O'Grady, R., Funk, V. A., Mooi, R., Kress, W. J., and Cannell, P. (1990). Speaking of forked tongues: The feasibility of reconciling human phylogeny and the history of language. *Curr. Anthropol.* **31**, 1–24.

Birdsell, J. B. (1985). "Human Evolution," 3rd Ed. Houghton Mifflin, Boston.

Cann, R. L., Stoneking, M., and Wilson, A. C. (1987). Mitochondrial DNA and human evolution. *Nature* **325**, 31–36.

Cavalli-Sfroza, L. L. (1991). Genes, peoples and languages. *Sci. Am.* **265**, 104–110.

Cavalli-Sfroza, L. L., Piazza, A., and Menozzi, P. (1988). Reconstruction of human evolution: Bringing together genetic and archaeological, and linguistic data. *Proc. Natl. Acad. Sci. USA* **85**, 6002–6006.

Garn, S. M. (1971). "Human Races," 3rd Ed. Charles C Thomas, Springfield, Illinois.

Gossett, T. F. (1963). "Race: The History of an Idea in America." Southern Methodist Univ. Press, Dallas.

Gould, S. J. (1979). "The Mismeasure of Man." Norton, New York.

Greenberg, J. H., Turner, C. G., II, and Zagura, S. L. (1986). The settlement of the Americas: A comparison of the linguistic, dental and genetic evidence. *Curr. Anthropol.* **27**, 477–496.

Kirk, R., and Szathmary, E. (eds.) (1985). "Out of Asia: Peopling the Americas and the Pacific." The Journal of Pacific History, Canberra.

Lewontin, R. C. (1972). The apportionment of human diversity. *Evolutionary Biol.* **6**, 381–398.

Menozzi, P., Piazza, A., and Cavalli-Sfroza, L. L. (1978). Synthetic maps of human gene frequencies in Europeans. *Science* **201**, 786–792.

Nelson, H., and Jumain, R. (1994). "Introduction to Physical Anthropology," 6th Ed. West Publishing, Minneapolis/St. Paul.

Piazza, A., Mennozzi, P., and Cavalli-Sfroza, L. L. (1981). Synthetic gene frequency maps of man and selective effects of climate. *Proc. Natl. Acad. Sci. USA* **78**, 2638–2642.

Ruvolo, M., Zehr, S., von Dornum, M., Pan, D., Chang, B., and Lin, J. (1993). Mitochondrial COII sequences and modern human origins. *Mol. Biol. Evol.* **10**, 1115–1135.

Sokal, R. R., Oden, N. L., Legendre, P., *et al.* (1990). Genetics and language in European populations. *Am. Naturalist* **135**, 157–175.

Spuhler, J. N. (1988). Evolution of mitochondrial DNA in monkeys, apes and humans. *Yearbook Phys. Anthropol.* **31**, 15–48.

Stoneking, M. (1994). In defense of "Eve." *Am. Anthropologist* **95**, 131–141.

Stringer, C. B. (1990). The emergence of modern humans. *Sci. Am.* **263**, 98–104.

Sussman, R. W. (1993). A current controversy in human evolution: Overview. *Am. Anthropologist* **95**, 9–13.

Thorne, A. G., and Wolpoff, M. H. (1992). The multiregional evolution of humans. *Sci. Am.* **266**, 76–83.

Washburn, S. L. (1963). The study of race. *Am. Anthropologist* **65**(3, Pt. 1), 521–531.

Wilson, A. C., and Cann, L. L. (1992). The recent African genesis of humans. *Sci. Am.* **266**, 68–73.

Population Genetics

RICHARD LEWONTIN
Harvard University

GLOSSARY

Allele frequency Relative proportion of all genomes in a population that carry a particular allele at a locus

Assortative mating Mating in which the choice of partner is based on some genetic, social, or geographical criterion. It may be positive, in which like mates with like, or negative, in which there is an avoidance of similar mates

Genetic drift Random fluctuation allele frequency at a locus arising from sampling error in breeding of a finite population

Natural selection Differential survival and reproduction of genotypes as a result of functional and anatomical differences in organisms that carry them

Polymorphism Presence in a population of more than one allele at a locus so that the frequency of the most common allele is not more than 99%

Reproductive fitness Expected number of offspring produced by individuals of a given genotype including the probability that the individuals survive to reproductive age

INDIVIDUAL HUMAN BEINGS DIFFER FROM EACH other genetically, as do geographical populations and ethnic and linguistic groups. Population genetics is concerned (1) with the description of that individual and group variation, (2) with the population forces that are acting to maintain or alter that variation at the present time, and (3) with the reconstruction of the forces in the past that have resulted in the current genetic composition of the human species. Because the human species is better known biologically and historically than any other species, human population genetics also provides models for the study of genetic variation in living organisms in general.

I. INDIVIDUAL VARIATION

Surveys of the genotypes of individuals within human populations have shown an immense variation at the level of individual nucleotide positions, genes, gene complexes, whole chromosomes, and genomes of cell organelles (e.g., mitochondria). No two individuals, even identical twins, are genetically identical because mutational errors in DNA replication are sufficiently frequent to guarantee that base pair differences will exist between the genomes of daughter cells of a single cell division. For example, if two genomes are surveyed for nucleotide differences by restriction endonucleases, which cut DNA at specific well-defined sequences, they will differ from each other at about one in every 200 base pairs. The functional significance of these differences is unknown because the restriction sites are not localized to particular genes or to particular base positions within codons. Much more revealing surveys of genetic variation come from studies of particular genes and gene complexes such as those specifying enzymes, blood group antigens, histocompatibility systems, and specific developmental features. [*See* Genes; Human Genome.]

Surveys of more than 100 enzyme-coding loci, using gel electrophoresis to detect amino acid substitutions caused by mutations, have shown that approximately one-quarter are polymorphic within a major racial group and one-third are polymorphic within the species as a whole. A locus is classified as polymorphic if the most common allele at the locus has a frequency

ENCYCLOPEDIA OF HUMAN BIOLOGY, Second Edition, VOLUME 7. Copyright © 1991 by Academic Press. All rights of reproduction in any form reserved.

of 0.99 or less. Another measure of genetic diversity of a population is the proportion of heterozygotes (i.e., in which the two genes of a pair are not identical) averaged over loci. For a typical human population, this proportion is roughly 0.065 for enzyme-coding loci. Both in the proportion of loci polymorphic and average heterozygosity, humans are typical of vertebrates in general.

In some cases one allele at a polymorphic locus is frequent, but in others there are two or more allelic forms with roughly equal frequencies. Table I shows the frequencies of the homozygotes (in which the two genes of the same pair are identical) and of the heterozygotes for the 15 most polymorphic enzymes in the English population. Polymorphisms of a similar degree are found for red cell antigens. Of 33 known red cell antigen genes, about one-third are polymorphic with an average heterozygosity of 0.16. Table II shows the allelic frequencies at the polymorphic blood group loci in the English population.

By far the most extraordinary polymorphism in human populations is for the HLA gene complex of tightly linked loci specifying histocompatability proteins. In Caucasians, for example, there are 14 alternate alleles at the A locus, 17 alleles at the B locus, 6 at the C, and 8 at the locus. The alleles at all the loci are generally in intermediate to low frequency, and none of the loci has a very common form (usually referred to as "wild type").

The consequence of the large amount of genic polymorphism is that the probability that two individuals chosen from a European population are genetically

TABLE I
Frequencies of Various Enzyme Variants Found in the English Population

	Homozygotes			Heterozygotes		
Enzyme	1	2	3	1/2	2/3	1/3
Red cell acid phosphatase	0.13	0.36	0	0.43	0.05	0.03
Phosphoglucomutase 1	0.59	0.06	—	0.35	—	—
Phosphoglucomutase 3	0.55	0.07	—	0.38	—	—
Placental alkaline phosphatase	0.41	0.07	0.01	0.35	0.05	0.12
Peptidase A	0.58	0.06	—	0.36	—	—
Adenylate kinase	0.90	0.01	—	0.09	—	—
Adenosine deaminase	0.88	0.01	—	0.09	—	—
Alcohol dehydrogenase 2	0.94	—	—	0.06	—	—
Alcohol dehydrogenase 3	0.36	0.16	—	0.48	—	—
Glutamate-pyruvate transaminase	0.25	0.25	—	0.50	—	—
Esterase D	0.82	0.01	—	0.17		
Malic enzyme	0.48	0.09	—	0.43	—	—
Phosphoglycolate phosphatase	0.68	0.03		0.29		
Glyoxylase 1	0.30	0.21		0.49		
Diaphorase 3	0.58	0.05		0.36	—	—

TABLE II
Blood Type Frequencies in the English White Population

System	Type	Frequency
ABO	A	0.447
	B	0.082
	AB	0.034
	O	0.437
MNS	MS	0.201
	Ms	0.093
	MNS	0.260
	MNs	0.236
	NS	0.060
	Ns	0.149
Rh	r	0.147
	R_1	0.535
	R_2	0.150
	R_1R_2	0.129
	R_0	0.022
	R'	0.011
	r''	0.006
P	P_1	0.266
	P_1P_2	0.499
	P_2	0.234
Secretor	Se^+	0.773
	Se^-	0.227
Duffy	Fy^a	0.177
	Fy^aFy^b	0.462
	Fy^b	0.301
Kidd	Jk^a	0.583
	Jk^aJk^b	0.361
	Jk^b	0.056
Dombrock	Do^a	0.664
	Do	0.336
Auberger	Au^a	0.857
	Au	0.143
Xg	Xg^a	0.894
	Xg	0.106
Sd	Sd^a	0.912
	Sd	0.088
Lewis	Le^a	0.224
	Le	0.776

identical, considering only the 35 most polymorphic loci, is about 8×10^{-14}. Nor do these highly polymorphic loci exhaust the known variation. About 2 per 1000 individuals carry a rare variant, recognizable electrophoretic allele for some enzyme-coding locus. "Inborn errors of metabolism," in which some enzymatic function is deficient or absent, are usually the consequence of homozygosity for a low-frequency mutant allele at a locus and thus are individually rare. Heterozygotes for these alleles are much more common, however. In a sample from the United States examined for 14 inherited disorders, it was estimated that 13% of newborns were heterozygous for a mutant allele. It is clear that all human individuals are heterozygous for scores of mutations that would be deleterious in the homozygous condition. [*See* Human Genetics.]

II. VARIATION BETWEEN GROUPS

Human geographical populations are obviously differentiated for some genetically influenced traits (e.g., skin color hair, form, facial characters, and height). These are not typical of all genes, however. For two-thirds of the human genome estimated to be monomorphic, all human individuals, irrespective of geographical population, are genetically identical with the exception of those individuals carrying rare mutant alleles. For the polymorphic one-third of the genome, the degree of differentiation between populations varies. Table III shows the three most and three least geographically differentiated protein-coding loci known. Even for the highly differentiated Duffy, Rhesus, and P loci, there is no allele that is 100% in one racial group but absent in another, although there are large frequency differences. Currently, 17 loci have been studied in a sufficiently wide geographical sample to make an estimate of the relative variation between and within groups. Table IV shows for each locus the proportion of the heterozygosity that occurs within local ethnically and linguistically homogenous populations (i.e., Kikuyu or French or Japanese), between such populations within major geographical race divisions (Black Africans, Caucasians, Asians, Oceanians, etc.), and between these races. Although there is some variation from gene to gene, on the average 85% of human genetic diversity occurs between individuals within a local population and the remaining 15% is split roughly evenly between population differentiation within major races and major racial differentiation.

TABLE III
Examples of Extreme Differentiation and Close Similarity in Blood Group Allele Frequencies in Three Racial Groups

Gene	Alleles	Caucasoid	Negroid	Mongoloid
Duffy	Fy	0.0300	0.9393	0.0985
	Fy^a	0.4208	0.0607	0.9015
	Fy^b	0.5492	—	—
Rhesus	R_0	0.0186	0.7395	0.0409
	R_1	0.4036	0.0256	0.7591
	R_2	0.1670	0.0427	0.1951
	r	0.3820	0.1184	0.0049
	r′	0.0049	0.0707	0
	Others	0.0239	0.0021	0
P	P_1	0.5161	0.8911	0.1677
	P_2	0.4839	0.1089	0.8323
Auberger	Au^a	0.6213	0.6419	
	Au	0.3787	0.3581	
Xg	Xg^1	0.67	0.55	0.54
	Xg	0.33	0.45	0.46
Secretor	Se	0.5233	0.5727	
	se	0.4767	0.4273	

TABLE IV
Proportion of Genetic Diversity Accounted for within and between Populations and Races

	Within races		
Gene	Within populations	Between populations	Between races
Hp	0.893	0.051	0.056
Ag	0.834	—	—
Lp	0.939	—	—
Xm	0.997	—	—
Ap	0.927	0.062	0.011
6PGD	0.875	0.058	0.067
PGM	0.942	0.033	0.025
Ak	0.848	0.021	0.131
Kidd	0.741	0.211	0.048
Duffy	0.636	0.105	0.259
Lewis	0.966	0.032	0.002
Kell	0.901	0.073	0.026
Lutheran	0.694	0.214	0.092
P	0.949	0.029	0.022
MNS	0.911	0.041	0.048
Rh	0.674	0.073	0.253
ABO	0.907	0.063	0.030
Mean	0.854	0.083	0.063

III. FORCES CONTROLLING VARIATION

Given the observed genetic variation within and between populations, how is the amount and nature of this variation to be explained? Genetic variation within a species is created, maintained, and altered by a combination of six forces: mutation and recombination, which create variation, and assortative mating, natural selection, random genetic drift, and migration, which modulate that variation.

A. Mutation

Ultimately all genetic variation comes from the random errors that occur in DNA replication either spontaneously or as a result of chemical and physical mutagens. There are no really good estimates of spontaneous mutation rates per gene in humans, but if they are typical of mammals, a rough estimate of 10^{-5} per gene (approximately 10^{-9} per nucleotide) per generation can be used. Because the rate of increase of an allele in a population from repeated mutations alone is equal to the mutations rate times the frequency of the allele *from* which it is mutating, mutation is a weak force in changing allele frequencies. The human species as a whole is probably not more than 250,000 years old, and present major geographical races probably diverged about 35,000 years ago. Given a human generation time of 25 years, only 10,000 generations have elapsed since the species was formed and 1500 generations since the divergence of the races. Then, by mutation alone, an allele could only have increased in frequency by a maximum of 0.10 since the founding of the species or 0.015 since the divergence of the races. Clearly other forces must operate on the mutant genes as they arise. [*See* Mutation Rates.]

B. Recombination and Sexual Reproduction

In the absence of sexual reproduction, genetic diversity among individuals would depend on the serial occurrence of different mutations in a given ancestral line. Sexual reproduction and the consequent recombination of genetic material from the maternal and paternal lines allow historically independent mutations to be combined in the same individuals, giving rise to vastly greater total variation and allowing variants that have arisen in different lines in the recent past to appear in a single genome. One consequence of nuclear genome recombination is that separate family lines do not preserve a trace of their ancestral history because they are constantly mixing. Because there is no recombination in mitochondria, however, which are entirely maternally inherited, it is possible to make reconstructions of their ancestral history and so trace the common ancestry of individuals and groups.

C. Natural Selection

Different genotypes may have different physiologies and morphogenesis as a consequence of different gene products. In turn, these physiological and anatomical differences may cause differences in the fertility and probability of survival of the different genotypes. In such a case, allele frequencies will change within the population. It is by no means certain, however, that genetic variation will have such effects on reproductive fitness. Many nucleotide changes occur in intergenic regions and in silent positions in codons and may have no effect at all on the organism. Changes in introns may be selectively neutral but, if they occur at splice junctions, may prevent normal protein synthesis. This is the case for the *thalassemias,* a collection of mutations that interfere with normal globin synthesis because they interfere with exon splicing. Even if nucleotide changes cause amino acid substitutions in proteins, these substitutions do not necessarily affect the biochemical activity or structural role of the protein. Moreover, even if physiological or anatomical changes occur, some may be irrelevant to differential reproduction or survivorship during reproductive ages. Genetic changes that influence the probability of survivorship during middle or old age have little or no effect on offspring production. Finally, genetic changes that do affect reproduction and survival may do so in a way that is sensitive to environment so that fluctuating environments may result in no clear trend in allelic frequency change. In sum, it cannot be assumed that a genetic variation will be subject to natural selection in the absence of clear experimental evidence on reproductive fitness. For example, attempts to find different reproductive rates among different blood group genotypes have failed, except for the fetal death associated with Rh incompatability. Various statistical associations have been claimed between blood groups or HLA genotypes and various diseases (e.g., the association between ABO blood types and peptic ulcer). But none of these associations, if real, have been shown to result in different reproductive rates. Do people who eventually get peptic ulcers leave fewer (or more) children? It has been claimed, for example, that darker-skinned populations occur where higher ultraviolet irradiation would cause skin cancer in light-skinned people, whereas very light skin in higher latitudes is an adaptation to making vitamin D with low ultraviolet irradiation. But it has never been shown that Finns would get a significant amount of skin cancer in Africa or that Africans would suffer from rickets in Finland or that, even if they did, a difference in reproductive schedules would follow. Natural selection has virtually never been demonstrated in the human species except for clearly deleterious genes, which are generally rare. Natural selective explanations for major polymorphisms are lacking. The only exception is the sickle cell polymorphism, in which the protection that heterozygotes have from malaria is established and explains the maintenance of the sickling allele in the population (discussed later).

There are essentially three forms of natural selection that may be relevant to human genetic variation. In the first form, the loss of reproductive fitness caused by an allele is either recessive or partly manifested in the heterozygote. If there are repeated mutations from the normal allele to the deleterious form, *a*, an equilibrium between the loss of the allele because of selection and the recruitment of the allele from new mutations will occur such that the frequency, q, of the deleterious allele will be

$$q = \sqrt{\mu/(1 - w_{aa})},$$

where μ is the rate of mutation to the deleterious allele and w_{aa} is the reproductive fitness of the deleterious homozygote. If there is any deleterious effect in heterozygotes, homozygotes are so rare as not to make any difference to the result and

$$q = \mu/(1 - w_{Aa}),$$

where w_{Aa} is the fitness of heterozygotes. In either case the equilibrium frequency of the allele will be low. For example, if homozygotes die before reproduction, $w_{aa} = 0$ and

$$q = \sqrt{\mu},$$

which for realistic mutation rates will be smaller than 1%. Because homozygotes only occur at a frequency q^2, the frequency of affected persons in the population will only be 1 in 10,000 for this lethal gene. Presumably the inborn errors of metabolism owe their characteristically low frequencies to this mutation–selection balance.

A second form of selection, in which heterozygotes between two alleles have a higher reproductive fitness than either homozygote, leads to a stable intermediate frequency of the alleles independent of the mutation rate. The frequency of the allele *a* will be

$$q = \frac{w_{AA} - w_{Aa}}{w_{AA} + w_{aa} - 2w_{Aa}},$$

where w is the reproductive fitness of the genotypes *aa, Aa,* and *AA*. Sickle cell anemia conforms to this prediction. Homozygotes for the sickling gene, *SS,* usually do not survive to adulthood ($w_{SS} = 0$) because of their severe anemia. In West Africa where there is a high frequency of falciparum malaria, homozygotes for the normal hemoglobin A die of malaria at a moderate rate, whereas heterozygotes seem to be totally protected against malaria ($w_{SA} = 1$). The observed incidence of the sickle cell allele is about 10%, which would be predicted from the formula for equilibrium if the malarial death rate was about 10% ($W_{AA} = 0.9$). A confirmation of this theory is that the frequency of the allele *S* has decreased among the descendents of West African slaves in North America at a rate more than twice that expected from the rate predicted simply from the interbreeding with Caucasians.

A third form of selection occurs when the heterozygote has a *lower* reproductive fitness than either homozygote, as in the case of the maternal–fetal incompatibility for Rh alleles. In such cases, an unstable equilibrium is expected and one or the other of the alleles should become fixed in the population, depending on the initial frequency of the alleles. This has not happened in human populations, all of which are polymorphic for Rh alleles in what appears to be a long-term polymorphism predating the present geographical divergence of human groups. The explanation of this contradiction remains unknown.

The failure to find satisfactory selective explanations for the widespread polymorphisms in the human species means that reproductive differentials between genotypes are small, about 1% or less. To detect such selection, age-specific mortality and fertility schedules would have to be compiled for extremely large samples of individuals of different genotypes, on the order of hundreds of thousands.

D. Genetic Drift

In a population of finite size, the genetic composition of the population will not be reproduced exactly from generation to generation but will be subject to sampling error because each generation is a sample of the gamete pool of the previous generation. The smaller the population size, N, the greater the sampling error. As a consequence, even in the absence of any force of selection or mutation, allele frequencies in a population will fluctuate from generation to generation, with the fluctuations being inversely proportional in size to the population number. This process of random change of allele frequencies is *genetic drift*. There are three consequences of genetic drift.

1. New mutations as they arise are almost always lost within a few generations. The probability of loss is $(2N - 1)/2N$. Even a slightly favorable new mutation, with selective advantage s, has a probability of only $2s$ of being retained. Thus, new favorable mutations must occur over and over again before they are incorporated.
2. Polymorphisms that are not being maintained by selection within populations will be lost as one of the alleles randomly drifts to fixation at a frequency of 100%.
3. Different populations will diverge from each other in genetic composition as one allele increases in frequency in some populations and decreases in others. The variation that is seen in allele frequencies of polymorphic loci in human populations may simply be the result of random divergence since the separation of the populations.

Eventually some alleles will be totally lost from some populations, and this process is accelerated if new populations are founded by only a few individuals or pass through a severe bottleneck in population size. This is the probable explanation, for example, for the absence or very low frequency of the B blood type in American Indians who presumably came over the Bering Strait in small numbers during the last glacial period and then were isolated from the ancestral populations in Asia when the ice melted. Eventually, if populations remain isolated from each other for a long enough period, they may become totally differentiated at a locus, with no alleles in common. This has not happened for any genes in the human species, presumably because human populations have not been totally isolated from each other for long periods. Again, it must be remembered that the major geographical races are only about 1500 generations old.

E. Migration

Populations exchange genes as a consequence of migrations, invasions, and conquests. Such gene exchanges reduce the genetic differences between populations that have built up as a result of isolation or different selection in different environments. If two populations with the frequencies q_1 and q_2 of an allele exchange a proportion m of their members in one generation, the difference in their allele frequencies in the next generation is simply

$$(q_1 - q_2)' = (1 - m)(q_1 - q_2).$$

This linear relation can be used to estimate how much gene flow has occurred between two populations over a few generations because the total migration will be roughly the sum of the one-generation rates. For example, the Duffy blood group allele Fy^a has a frequency of about 0.40 in Europeans and up to 70% in American Indians, but is absent in West Africans. The frequency is 0.081 among blacks in New York. Then, assuming that only Europeans had a significant genetic input into black populations, an estimate of the proportion of European ancestry in American blacks from New York is

$$m = 1 - \frac{(0.40 - 0.081)}{(0.40 - 0)} = 0.189.$$

The estimates for other northern cities are similarly high (0.26 for Detroit and 0.22 for Oakland), but southern samples show much lower values (0.037 for Charleston, SC, and 0.106 for rural Georgia). These figures are the total migration of European genes into the black population since slaves were first introduced in large numbers about 10 generations ago. The estimates would be more difficult to make if blacks have a significant amount of American Indian ancestry, which they do in some regions.

F. Human Population Structure

Human populations are not isolated islands of randomly mating individuals that occasionally receive migrants from other populations, nor is the human species one large interbreeding unit. There is a hierarchical mating structure. At the local level, even within small villages and tribes, there is assortative mating by clan, religion, caste, and appearance. This is sometimes a positive assortment, as in the tendency to mate within religions, castes, and social classes, or there may be negative assortment when there are taboos against mating within totem groups or families. Assortative mating by skin color is strong in the United States, with "cross-racial" marriages making up a fraction of 1% of all unions.

At the next level, there is selective mating by geographical locality. Despite a strong bias toward mating between individuals born within a small geographical radius, there are also traditions of marrying out to other villages and locales. There are also national and linguistic barriers that correspond to large geographical areas and finally major geographical separations by continent or widely spaced island groups. But none of these is absolute, and there is a continuous flow of migrants fluctuating in numbers depending on unique historical events. Because of asymmetrical social definitions of group membership, gene passage usually appears to be one way. Any person with known black ancestry is described as "black" in the United States, so it is impossible to estimate the flow of genes from Africans into the "white" population.

In summary, human populations have a structure that allows genetic drift to cause local differentiation among castes, religions, regions, ethnic groups, etc., but that structure is not one of rigidly isolated islands of population of a fixed size. Rather, there is partial isolation by geographical and social distance, differing in degree from region to region and historical epoch to historical epoch.

G. Synthesis of Forces

Mutation, migration, selection, and genetic drift all operate simultaneously, and the genetic diversity within and between populations is a consequence of opposing tendencies of the different forces. Mutation introduces variation into populations and in the short run causes differentiation between populations because the same mutations will not have occurred in all local groups. In the long run, however, recurrent mutation will give the same spectrum of changes for all populations. Migration increases the diversity within populations by the mixing of local differentiated groups but decreases the differentiation between populations. Selection has different effects depending on the nature of the fitness differences. Selection in favor of heterozygotes maintains variation within populations. Directional selection favoring one genotype causes populations to become homozygous and makes all local populations genetically similar unless different alleles are favored under different environmental conditions. There is no direct evidence in human populations for directional selection of different alleles in different populations, although that may be the case for skin color or body fat. Selection against heterozygotes, as in the case of maternal–fetal incompatability, causes populations to become homozygous and to differentiate from each other. Finally, random genetic drift causes a loss of genetic variation within populations and a divergence between populations.

The importance of random drift as opposed to the deterministic directional forces of selection, migration, and mutation depends on the product of population size, N, and the parameters of the deterministic forces of selection (w), migration (m), and mutation (μ). Generally, if the product Nw or Nm or $N\mu$ is greater than 1, the deterministic force predominates and little random differentiation between populations will occur by drift. If the product is smaller than 1, however, the deterministic force is insufficient to prevent random differentiation. So, if local populations of Brazilian Indians are approximately a few hundred in size, selection intensities of less than 1% will be insufficient to determine the frequencies of polymorphic genes, and the populations will drift in their allele frequencies in a random way. If, however, such local populations exchange even one migrant individual per generation with each other ($m = 1/N$), $Nm = 1$ and they will not diverge randomly from each other, although they may diverge as a whole from some other more distant population.

The pattern of genetic variation within and between human populations is then the result of some combination of these diversifying and homogenizing forces. Unfortunately, it has not been possible to measure with sufficient accuracy the rates of selection, mutation, migration, and population size to make a clear assignment of the forces for most human variation. This problem is immensely complicated by the rapid historical changes that have occurred in population sizes, migration patterns, environmental conditions, and social structures in relatively few generations of human history.

IV. RECONSTRUCTING HISTORY

Although we do not know what forces were responsible for establishing the widespread polymorphisms in

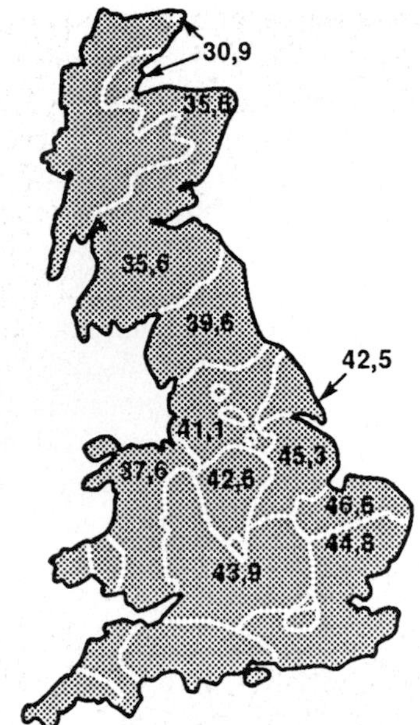

FIGURE 1 Cline of frequency of the blood type A in Britain. [Reproduced, with permission, from R. C. Lewontin, *Sci. Am.*, p. 19.]

human populations or the original differentiation of geographical groups, it is clear that the forces of selection and random drift have not been strong compared with the role of migration during the past 10,000 years. Geographical patterns of allele frequency distribution on national and continental scales show the clear traces of migrations and invasions whose histories are otherwise known from historical records. Figures 1 and 2 show examples of these genetic traces of migration. Figure 1 shows the percentage of blood type A in Britain. The decreasing cline northward and westward from the original invasion sites of the Danes on the east coast (so-called Danelaw) reflects the high frequency of A among Scandinavians and the low frequency among the Celts and Picts of Scotland. Although not shown on the map, there is a similar low frequency of A in Ireland and in Liverpool, which is the chief port of entry into Britain from Ireland. Reciprocally, there is an island of high A frequency in far western Ireland as a consequence of a long-time garrisoning of British troops there.

Figure 2 shows a cline in A-type frequencies from the northeast to the southwest of the Japanese archipelago with a concentration of high A in Honshu

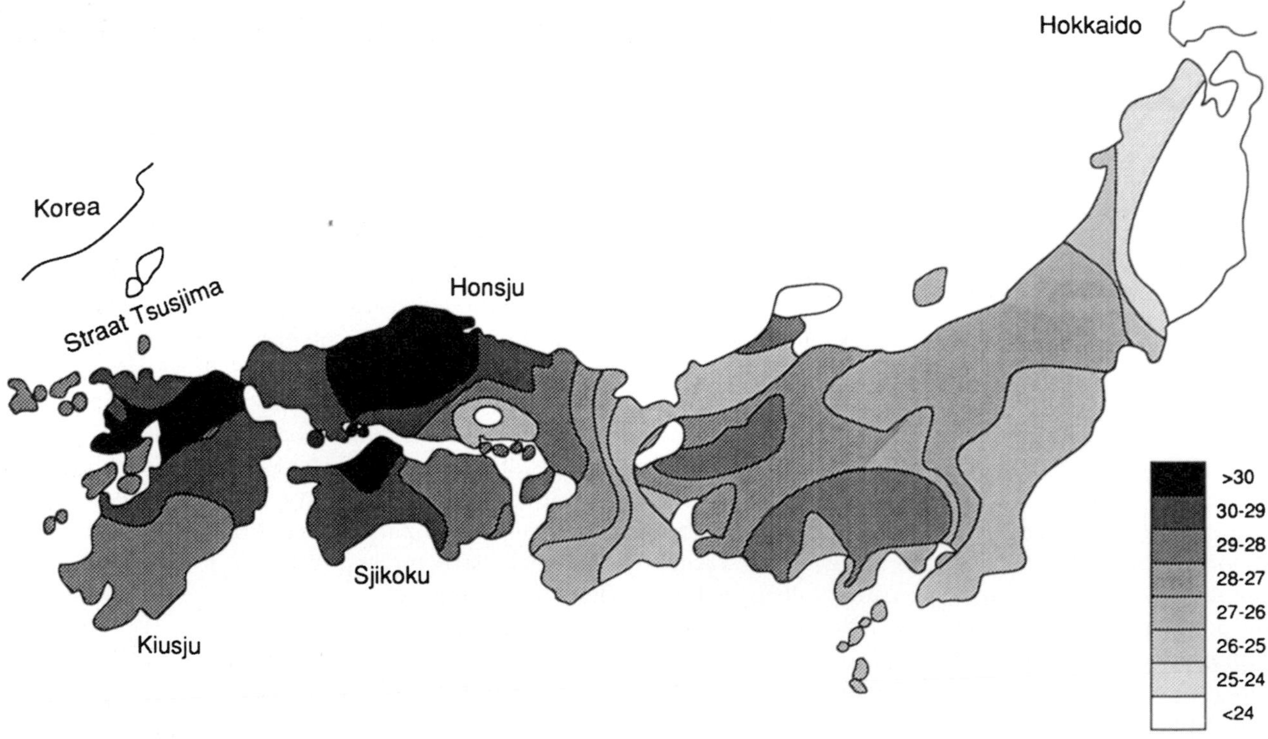

FIGURE 2 Cline of frequency of blood type A in the Japanese archipelago. [Reproduced, with permission, from R. C. Lewontin, *Sci. Am.*, p. 19.]

FIGURE 3 Clines of the first three principal components of allele frequencies in Eurasia, based on HLA and blood group polymorphisms. Each of the principle components is represented by a different color. The more intense the color, the higher the value of the principle component. [Reproduced, with permission, from Menozzi *et al.* (1978).]

and Kyushu bordering the straits of Tsushima. This corresponds to the known repeated invasions of Japan by mainland peoples of China through the closest point in Korea. The low frequency of A in the far north presumably is that of the aboriginal peoples (perhaps the modern-day Ainu) who mixed with the mainland invaders. What is extraordinary about these two cases is that the historically determined clines have persisted for more than 2500 years despite the high mobility and urbanization of Britain and Japan. In general, in Europe, differences in allele frequency between populations is well-correlated with geographical distance but not with linguistic family differences.

The parallel between allele frequency clines and known history for small regions makes it likely that large continental clines could be used to trace large-scale migrations. Information from a large number of different polymorphic systems can be combined into multivariate statistical indices, the *principal components,* and when these are mapped for Eurasia, as in Fig. 3, there is clear evidence of a cline from central Asia fanning out into Europe paralleling the postulated spread of agricultural people between 9000 and 5000 years ago. The possibility of reconstructing human migration patterns from gene frequency patterns is a consequence of the relative weakness of selective forces and of the maintenance of local breeding populations over short periods of time so that history is not obscured by mixing.

BIBLIOGRAPHY

Cappuccino, N., and Price, P. W. (1995). "Population Dynamics: New Approaches and Synthesis." Academic Press, San Diego.

Cavalli-Sforza, L. L., and Bodmer, W. F. (1971). "The Genetics of Human Populations." Freeman, San Francisco.

Cavalli-Sforza, L. L., Menozzi, P., and Piazza, A. (1994). "The History and Geography of Human Genes." Princeton Univ. Press, Princeton, NJ.

Daniels, G. (1996). "Genetics of the Human Blood Groups." Blackwell Science, Oxford.

Harris, H. (1980). "Human Biochemical Genetics," 3rd Ed. rev. Elsevier/North Holland, Amsterdam.

Lewontin, R. C. (1972). The apportionment of human diversity. *Evolution. Bio.* **6**, 381.

Menozzi, P., Piazza, A., and Cavalli-Sforza, L. L. (1978). Synthetic maps of human gene frequencies in Europeans. *Science* **201**, 786.

Race, R. R., and Sanger, R. (1975). "Blood Groups in Man," 6th Ed. Blackwell, Oxford.

Spiess, E. B. (1989). "Genes in Populations." Wiley, New York.

Porphyrins and Bile Pigments

DAVID SHEMIN[1]
Woods Hole Oceanographic Institution

HEINZ FALK
Johannes Kepler University, Linz, Austria

GLOSSARY

Isomer Compound having the same percentage of composition and molecular weight as another compound but differing in chemical structure

Isotope Any of two or more atoms that have the same number of protons but different atomic weights (e.g., ^{14}N has 7 protons and 7 neutrons whereas ^{15}N has 7 protons and 8 neutrons)

PORPHYRINS ARE HETEROCYCLIC COMPOUNDS composed of four pyrrole rings linked to each other by four methene bridge carbon atoms (Fig. 1). These cyclic compounds are extremely stable, having been found in petroleum shale oil and in fossilized excrements millions of years old. The basic structures of porphyrins are found in the heme molecules of hemoglobin, myoglobin, cytochromes, catalases, and the like, in the chlorophylls and bacteriochlorophylls of photosynthetic organisms, in siroheme and coenzyme F430, and in the corrin ring of vitamin B_{12} (Fig. 2). The porphyrin ring of heme is coordinated to iron, whereas chlorophyll contains magnesium, coenzyme F430 contains nickel, and the corrin ring is linked to cobalt. The heme proteins are concerned with oxygen transport (hemoglobin), oxygen storage (myoglobin), oxidation (cytochromes and peroxidases), and nitrogen fixation (leghemoglobin). Whereas heme is concerned primarily with the release of available energy, the role of chlorophyll is photosynthesis, by which solar energy builds up reducing and oxidizing potentials, so as to store chemical energy.

As discussed in Section I, the biosynthesis of porphyrins involves the formation of a "parent" porphyrin structure having an acetic acid (A) and a propionic acid (P) side chain in each of the β positions of each of the pyrrole rings. With an A and a P side chain in each of the pyrrole rings, one can construct four structural isomers. The parent structures of all biological functioning porphyrins or porphyrin-derived compounds have the following distribution of side chains, starting from ring *A* (see Fig. 1): AP, AP, AP, PA; they are designated isomer III. In particular pathological or metabolic altered states, isomer I (AP, AP, AP, AP) is formed as well. The unique structures of the side chains eventually found in the functioning porphyrin (e.g., protoporphyrin of heme) arise by subsequent alteration of the A and P side chains of the parent isomer III: for example, the methyl side chains of protoporphyrin are derived by decarboxylation of the four acetic acid (A) side chains, and the two vinyl side chains (rings *A* and *B*) are derived by decarboxylation and oxidation of the propionic acid (P) side chains (Figs. 1 and 3).

I. BIOSYNTHESIS OF PORPHYRINS

The first committed step in the synthesis of the porphyrin structure, and thereby in the synthesis of heme, chlorophyll, and the corrin ring, is the formation of

[1]Deceased.

ENCYCLOPEDIA OF HUMAN BIOLOGY, Second Edition, VOLUME 7. Copyright © 1997 by Academic Press. All rights of reproduction in any form reserved.

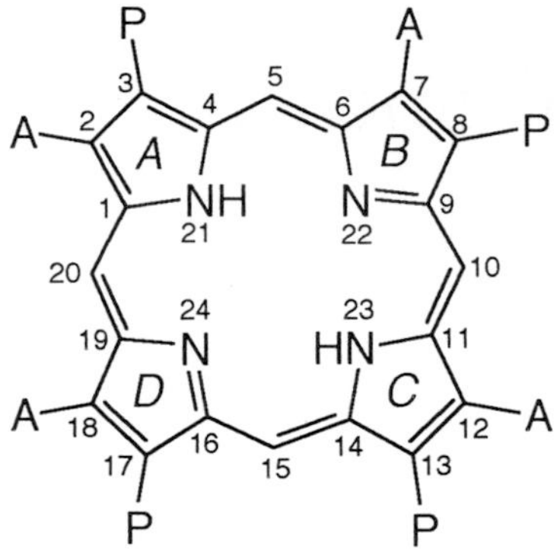

FIGURE 1 Uroporphyrin III. Side chain abbreviations: A, $-CH_2COOH$; P, $-CH_2-CH_2-COOH$.

δ-aminolevulinic acid (ALA). In humans as well as other members of the animal kingdom, and also in photosynthetic bacteria, ALA is enzymatically synthesized from the simple amino acid glycine and from succinic acid with the participation of two coenzymes—coenzyme A and pyridoxal phosphate—which are derivatives of two essential human dietary compounds: panthothenic acid and vitamin B_6, respectively (see Fig. 3). It was recently found that vertebrates have two separate genes that encode for the enzyme that catalyzes this reaction (i.e., δ-aminolevulinate synthase). The erythroid gene is expressed exclusively in the red blood cell, whereas the hepatic form of the enzyme is probably expressed ubiquitously. It is of further interest that plants, algae, and other microorganisms (e.g., *Cyanobacteria, Chromatium,* and *Methanobacterium*) synthesize ALA from glutamic acid, an amino acid.

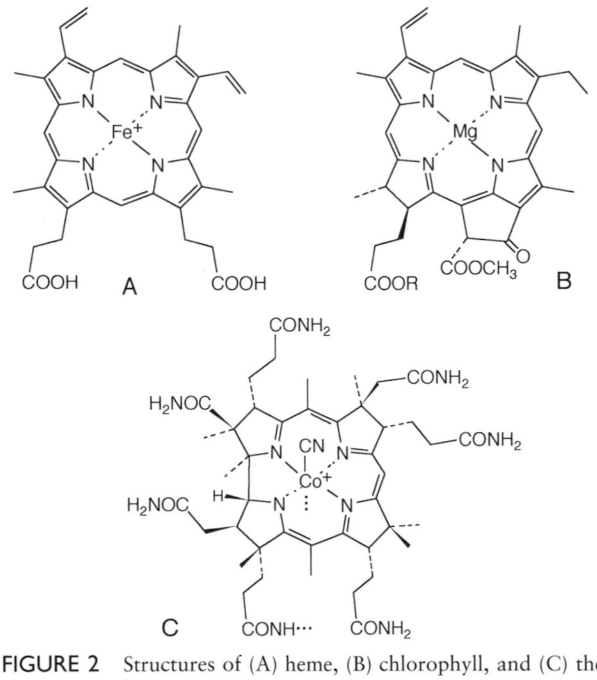

Subsequently, two molecules of ALA are enzymatically condensed to form a pyrrole derivative, porphobilinogen (PBG). The enzyme catalyzing this reaction (i.e., PBG synthase) is a zinc-containing enzyme markedly inhibited by the presence of lead. Four molecules of PBG are then condensed enzymatically with the participation of two enzymes (PBG deaminase and uroporphyrinogen III cosynthase) to form uroporphyrinogen III (reduced uroporphyrin III). (In the absence of cosynthase, the nonbiological functioning isomer I is formed.) The A side chains of uroporphyrinogen III are enzymatically decarboxylated to methyl groups to form coproporphyrinogen III and subsequently the P side chains of rings A and B are decarboxylated and dehydrogenated to vinyl groups to form protoporphyrinogen. The latter is oxidized to protoporphyrin, which is enzymatically converted to heme by the addition of iron. It is worth noting that the first step (in Fig. 3) and last three enzymatic steps occur in the mitochondria, whereas the intermediate enzymes are found in the cytosol. The rate-limiting step in the synthesis of porphyrins appears to be the synthesis of ALA, partially regulated by the concentration of heme. Chlorophyll and the corrin ring of vitamin B_{12} are synthesized from intermediates in the outlined scheme (see Fig. 3).

FIGURE 2 Structures of (A) heme, (B) chlorophyll, and (C) the corrin ring of vitamin B_{12}.

II. PORPHYRIAS

There are approximately seven well-defined human inborn errors in the metabolic process concerned with the synthesis of porphyrins. The genetic inheritance is usually autosomal dominant, except for congenital erythropoietic porphyria and ALA dehydratase deficiency porphyria, which are autosomal recessive disorders. Two of these disorders are briefly described here.

A. Congenital Erythropoietic Porphyria

Congenital erythropoietic porphyria is a rare disease in which the homozygote is characterized by an increase of porphyrins in the blood, urine, and feces. Since the abnormality lies in the deficiency of uroporphyrinogen III cosynthase, the excretion is that of

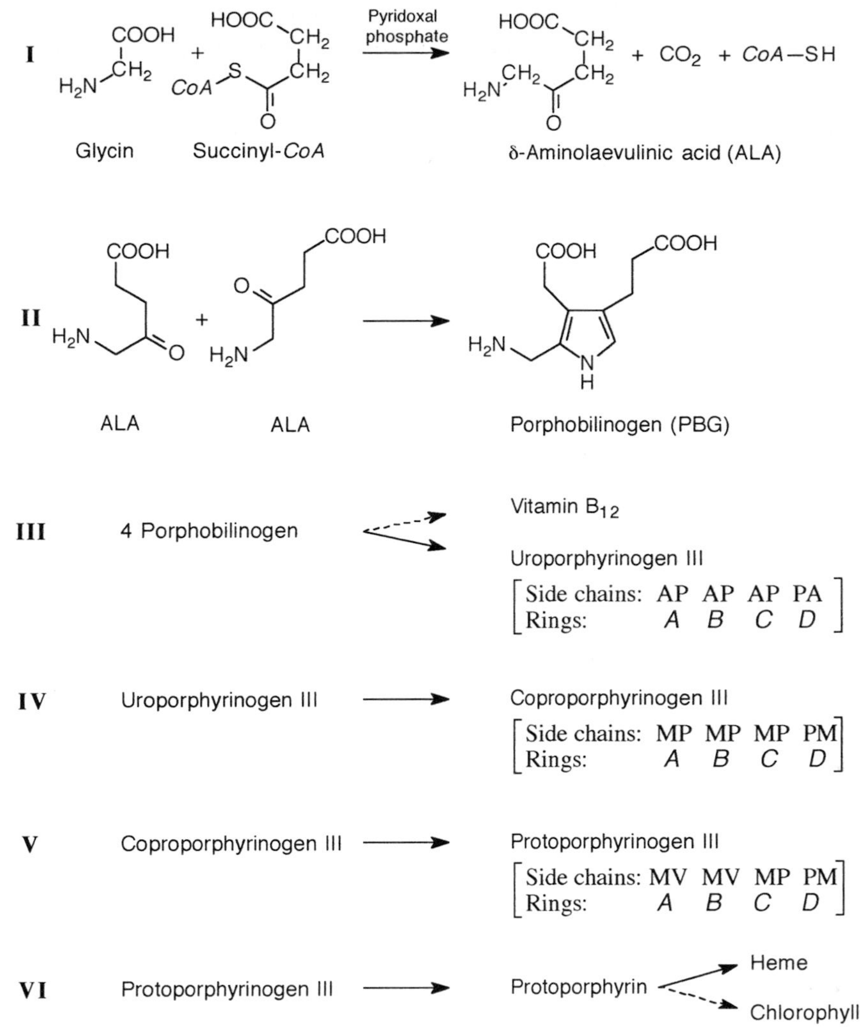

I $H_2N{-}CH_2{-}COOH + CoA{-}S{-}C(=O){-}CH_2{-}CH_2{-}COOH \xrightarrow{\text{Pyridoxal phosphate}} H_2N{-}CH_2{-}C(=O){-}CH_2{-}CH_2{-}COOH + CO_2 + CoA{-}SH$

Glycin Succinyl-*CoA* δ-Aminolaevulinic acid (ALA)

II ALA + ALA → Porphobilinogen (PBG)

III 4 Porphobilinogen ⇢ Vitamin B_{12}; → Uroporphyrinogen III

Side chains:	AP	AP	AP	PA
Rings:	*A*	*B*	*C*	*D*

IV Uroporphyrinogen III → Coproporphyrinogen III

Side chains:	MP	MP	MP	PM
Rings:	*A*	*B*	*C*	*D*

V Coproporphyrinogen III → Protoporphyrinogen III

Side chains:	MV	MV	MP	PM
Rings:	*A*	*B*	*C*	*D*

VI Protoporphyrinogen III → Protoporphyrin → Heme; ⇢ Chlorophyll

FIGURE 3 Outline of the biosynthesis of a porphyrin. *CoA*, Co-enzyme A. Side chain abbreviations: M, $-CH_3$; A, $-CH_2-COOH$; V, $-CH=CH_2$; P, $-CH_2-CH_2-COOH$.

uroporphyrin I and coproporphyrin I. The accumulation of the porphyrins in the tissues, bones, and teeth also causes the individual to be photosensitive, exposure to sunlight causing blisters that readily become infected. The bones and the teeth of individuals with this disorder are reddish. A similar disorder has been found in cattle and in the fox squirrel.

B. Acute Intermittent Porphyria

Acute intermittent porphyria is an autosomal dominant disorder in which the metabolic defect is in the deficiency of the enzyme porphobilinogen deaminase. Approximately 50% of the normal enzymatic activity is found in the subjects' tissues. Most of the subjects who inherit this genetic deficiency do not exhibit any clinical symptoms or biochemical abnormalities. However, when exposed to some drugs (e.g., barbiturates or sulfonamides) or some stresses, they accumulate PBG and ALA and express clinical symptoms. Along with marked increases in PBG and ALA in the urine, the subjects can have acute abdominal pains, neurological symptoms, paralysis, muscle weakness, and some abnormal behavioral patterns. It is of interest that, before being recognized as a metabolic disorder, many subjects, in the acute phase, have received appendectomies and even been assigned to mental institutions. The incidence of the disease is apparently more common in Scandinavia, South Africa, Great Britain, and Lapland. The occurrence in some instances has been traced to the original individual carrying the defective gene. The injection of hemin (heme–Fe^{3+}), which inhibits ALA synthesis, appears to lessen the symptoms occurring in the acute phase.

III. LIFE SPAN OF THE RED BLOOD CELL

Knowledge of the pathway by which porphyrins are synthesized enables one to specifically label the heme of hemoglobin by the ingestion, for example, of glycine tagged with the stable isotope of nitrogen (^{15}N) and to determine the life span of the red blood cell under normal physiological conditions. By plotting the ^{15}N concentration of the hemin samples isolated from blood samples against time, one can determine the average life span of the red blood cell, since the hemoglobin in the cell is not in a dynamic state of synthesis and degradation. The average life span is found to be about 120 days (Fig. 4). This survival time of the red blood cell in patients with congenital porphyria or polycythemia vera is similar to that of normal subjects, whereas in sickle-cell anemia or untreated pernicious anemia the red blood cell seems to be destroyed indiscriminately.

IV. METABOLISM OF PORPHYRINS

The heme moieties of hemoglobin and other heme proteins are eventually metabolized to bile pigments (Fig. 5) and eliminated in both the stool and the urine.

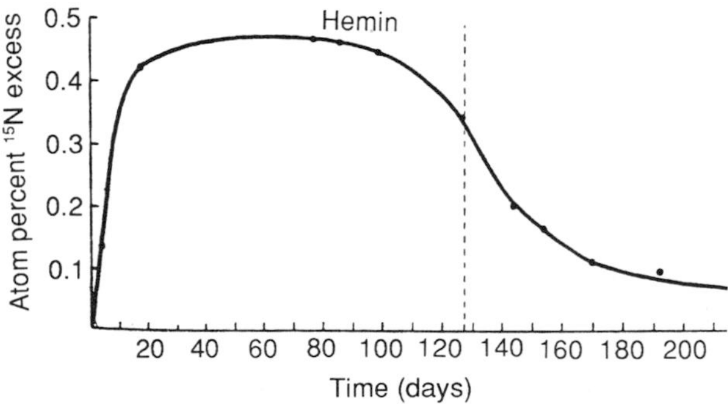

FIGURE 4 Life span of the human red blood cell. Plot of the ^{15}N concentration of the heme of hemoglobin versus time after the ingestion of ^{15}N-labeled glycine.

Approximately 80% of the bile pigment is derived from hemoglobin.

The heme molecule is metabolized by the reticuloendothelial system by microsomal heme oxygenases to yield the bile pigment biliverdin, which is then reduced by a reductase to the orange pigment, bilirubin (Fig. 5). In this process the 5-methine bridge carbon atom is split off as carbon monoxide. On the one hand, carbon monoxide could pose a problem as it is highly poisonous. This is due to its binding to hemoglobin and myoglobin, where it blocks the binding position of oxygen. The binding affinity of carbon monoxide to hemoglobin is about 200 times greater than that for oxygen. [*See* Reticuloendothelial System.] On the other hand, carbon monoxide released in the course of heme catabolism may function as a neurotransmitter comparable to nitric oxide. [*See* Neurotransmitter and Neuropeptide Receptors in the Brain; Synaptic Physiology of the Brain.]

It is interesting to note that heme binds carbon monoxide about 25,000 times stronger than oxygen. The reduction in affinity for carbon monoxide in hemoglobin is brought about by an energetically unfavorable bending of the Fe—C≡O angle, which is 180° in the more stable carbon monoxide–heme complex. This bending is enforced by the distal histidine of the protein.

The rather insoluble bilirubin is released into the circulation, where it binds to the serum albumin and subsequently is cleared by the liver. (The liberated iron is recycled for heme synthesis.) In the liver the bilirubin is esterified with glucuronic acid and thereby converted to a more soluble derivative. The esterified bilirubin and any unesterified pigment are secreted into the bile and subsequently into the intestines, where it is further acted on by bacteria to form urobilinogens. The urobilinogens are, in part, resorbed and reexcreted into the bile and, in part, excreted in the stool and the urine. Oxidized derivatives of urobilinogens are partially responsible for the color of these excretions. The absence of urobilinogens in the stool and in urine is indicative of obstruction of the bile ducts. It is of interest that it is advantageous to have bilirubin as the circulating bile pigment rather than biliverdin, which is the end product of heme catabolism in birds and reptiles, for bilirubin is a strong antioxidant. Bilirubin scavenges peroxyradicals and thereby protects fatty acids, transported by albumin, from oxidation. In this process the bilirubin is oxidized to biliverdin, which is then rapidly reduced to bilirubin.

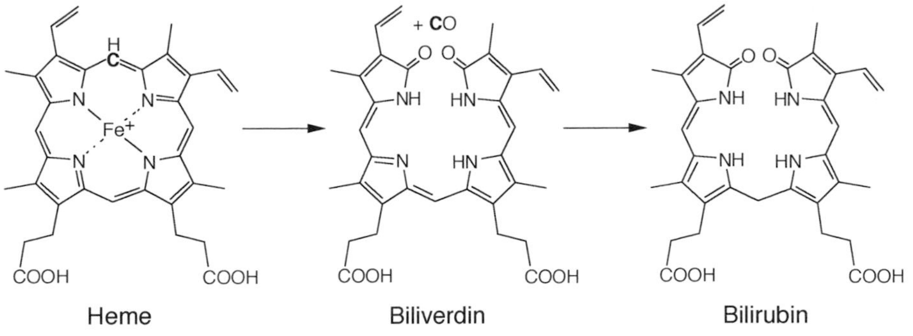

FIGURE 5 Structures of bile pigments in the conversion of heme to biliverdin and bilirubin.

V. DISORDERS IN BILIRUBIN METABOLISM

Bilirubin is a potentially toxic compound, having diverse effects, such as inhibition of protein synthesis, carbohydrate metabolism in the brain, ATPase activity, and many other enzymatic reactions; therefore, hyperbilirubinemia and the resulting jaundice are indeed a serious matter. Hyperbilirubinemia and jaundice can occur as a result of abnormalities in the process of formation, excretion, and metabolism of bile pigments. Obviously, liver disorders such as hepatitis, cirrhosis, and cancer play a role in causing an increased amount of unconjugated serum bilirubin, as does an abnormal degree of hemolysis of red blood cells. The lack of the functioning liver enzyme that catalyzes the esterification of bilirubin by glucuronic acid also results in an increased concentration of water-insoluble unconjugated bilirubin in the circulation. A rather significant percentage of newborns have hyperbilirubinemia, due to a combination of increased bilirubin production and insufficient glucuronyl transferase activity for bilirubin. Phototherapy provides an efficient means to clear the body of bilirubin. The molecular basis of phototherapy consists of a photo-induced isomerization of one or both exocyclic double bonds at the terminal rings of bilirubin, which renders the molecule more soluble and thus leads to promoted excretion.

BIBLIOGRAPHY

Baldwin, J. M. (1980). The structure of human carbonmonoxy hemoglobin at 2.7 Å resolution. *J. Mol. Biol.* **136,** 103–128.

Battersby, A. R. (1994). How nature builds the pigments of life: The conquest of vitamin B_{12}. *Science* **264,** 1551–1557.

Dolphin, D. (1978). "The Porphyrins." Academic Press, New York.

Falk, H. (1989). "The Chemistry of Linear Oligopyrroles and Bile Pigments." Springer, Vienna/New York.

Friedman, J. M. (1985). Structure, dynamics, and reactivity in hemoglobin. *Science* **228,** 1273–1280.

Kappas, A., Sassa, S., and Anderson, K. E. (1983). The porphyrians. *In* "The Metabolic Basis of Inherited Disease" (J. B. Stanbury, J. B. Wyngaarden, D. S. Fredrickson, J. L. Goldstein, and M. S. Brown, eds.), 5th Ed., pp. 1301–1384. McGraw–Hill, New York.

Ostrow, D. J. (1986). "Bile Pigments and Jaundice," Marcel Dekker, New York.

Shemin, D. (1982). From glycine to heme. *In* "From Cyclotrons to Cytochromes" (N. O. Kaplan and A. Robinson, eds.), pp. 117–129. Academic Press, New York.

Stocker, R., Glazer, A. N., and Ames, B. N. (1987). Antioxidant activity of albumin-bound bilirubin. *Proc. Natl. Acad. Sci. USA* **84,** 5918–5922.

Stocker, R., Yamamoto, Y., McDonagh, A. F., Glazer, A. N., and Ames, B. N. (1987). Bilirubin is an antioxidant of possible physiological importance. *Science* **235,** 1043–1046.

Verma, A., Hirsch, D. J., Glatt, C. E., Ronnett, G. V., and Snyder, S. H. (1993). Carbon monoxide: A putative neural messenger. *Science* **259,** 381–384.

Pregnancy, Dietary Cravings and Aversions

FORREST D. TIERSON
University of Colorado at Colorado Springs

GLOSSARY

Aversion A compelling distaste for a certain food item or items in the diet

Craving An urgent and imperative longing or intense, compulsive desire for one or more of a wide variety of articles of diet

Geophagy A special form of pica in which clay or earth is eaten

Pica Craving for a substance that is not normally considered edible

THE OCCURRENCE OF DIETARY CRAVINGS AND aversions during pregnancy is well known. Unfortunately, little is known of the etiology or epidemiology of this complex of symptoms. The anthropological and nutritional literatures contain numerous descriptions of cravings and aversions, food taboos, and foods for which there exist either prescriptions or proscriptions during pregnancy, in both traditional and modern populations. However, even though a wealth of qualitative information exists (usually in the form of narrative examples), little information of a quantitative nature has been published on the subject. In addition, most descriptive studies have been retrospective in nature, reporting observations made relatively late in pregnancy.

I. BACKGROUND

Special attention has been given to the diets of pregnant women over most of recorded history. Prohibitions or restrictions during pregnancy have long existed for some kinds of foods, while other foods have often been regarded as essential to successful pregnancy outcome.

Many factors are known to affect maternal diet during pregnancy. These factors include cultural practices, beliefs, food availability, economic conditions, and individual food preferences. Not surprisingly, women generally consume more food when they are pregnant than when they are not pregnant. Unexpectedly, we find increased consumption during the early stages of pregnancy when maternal caloric needs are not as great as during the second and third trimesters. This process of "excess" consumption in early pregnancy is thought to provide the fetus with necessary nutrients without putting undue physiological stress on the mother, although precisely how this process works in humans is not well known. The mechanisms that bring about changes in maternal appetite during pregnancy are also incompletely understood, but these influences on intake levels of nutrients can be substantial.

II. PICA OF PREGNANCY

Pica is probably the most notorious of the better known forms of cravings during pregnancy, especially the specific type of pica expressed as geophagy. However, the nutritional and obstetrical literatures contain many references to cravings during pregnancy for items other than clay, earth, and/or laundry starch, including cravings for such diverse items as baking soda, coal, soap, disinfectant, toothpaste, mothballs, gasoline, tar, paraffin, wood, chalk, and even pencil erasers. While certainly not exhaustive, this list does provide an idea of the wide range of items for which cravings during pregnancy have been reported.

ENCYCLOPEDIA OF HUMAN BIOLOGY, Second Edition, VOLUME 7. Copyright © 1997 by Academic Press. All rights of reproduction in any form reserved.

By far, cravings for clay, laundry starch, and soil (earth) outnumber cravings for other items in most areas of the world. Among black women of lower socioeconomic status in the rural southern United States, some studies have found that between 50 and 75% of pregnant women report pica for clay and/or laundry starch. Similar cravings for clay are found among aboriginal women in Australia and among several different groups of women in Africa. In general, explanations for the existence of pica of this type fall into four major categories, including (1) being caused by superstitious beliefs and/or as the result of folklore; (2) physiological explanations claiming that pica exists to compensate for some existing dietary deficiency; (3) psychological factors invoking explanations of insecurity, suggestion, attention seeking, and other more severely psychotic reasons on the part of the pregnant women; and (4) alterations in the sense of taste and smell during pregnancy. Most likely, the majority of pica can generally be explained by tradition and the folkways of the affected people. For example, black women in the rural south also experience pica for clay and starch when not pregnant. These items are considered to be especially important for successful pregnancy outcome and are supplied by friends and family members to the pregnant woman.

Generally speaking, pica of pregnancy is less frequent in Western nations and more recently in time. However, in more educated populations, pica often may not be reported due to the fact that many individuals consider cravings for odd, nonfood items to be aberrant and a perversion of sorts. Thus, the actual occurrence of pica among more educated, Western populations is not clear. During a recent prospective study of the cravings and aversions of 400 white pregnant women of high socioeconomic status in Albany, New York, no cases of pica were reported. [*See* Feeding Behavior.]

III. CRAVINGS AND AVERSIONS FOR FOOD ITEMS

Historically, most studies of dietary cravings and aversions during pregnancy have dealt mainly with pica, and secondarily with cravings for food items. Aversions during pregnancy were often not considered at all. However, it is clear that dietary aversions are at least as common as dietary cravings during pregnancy, and they are most certainly related.

The reasons why cravings and aversions for food items develop are essentially unexplained, although a number of explanations have been offered. Cravings may represent a physiological response to maternal and/or fetal nutritional needs. Aversions may represent a response to low levels of toxins present in the foods for which aversions have developed. Cravings or aversions may also be caused by mediating factors such as maternal metabolic changes and changes in olfactory and taste sensitivity during pregnancy. Cravings and aversions may not be biologically determined at all, but may simply be learned behaviors or they may be solely idiosyncratic in nature. No matter what the cause, it is significant that studies show that cravings and aversions for food items are not limited to a few isolated individuals, but that they affect a sizeable fraction of the pregnant population studied. [*See* Food Acceptance: Sensory, Somatic, and Social Influences.]

The widespread occurrence of dietary cravings and aversions during pregnancy among human populations is demonstrated by the high frequency with which this phenomenon is the focus of folk beliefs concerning pregnancy and health. Cravings and aversions are expressed for a number of common dietary constituents, not just idiosyncratic items such as pickles and ice cream, or clay and starch. Studies have demonstrated that at least 10% of women in specific populations report cravings for several different dietary items, with others also reporting high frequencies of aversions for some food items. Since some aversions are to substances known to be embryotoxic, such as tobacco smoke and alcohol, the occurrence of these aversions may result in reduced fetal exposure to such toxins.

If dietary cravings and aversions affect maternal nutrition by increasing the intake of nutrients or by decreasing the ingestion of embryotoxic agents, then their potential impact on maternal (and, subsequently, fetal) health in human populations might be considerable. Since dietary cravings and aversions during pregnancy exist among many populations, such an impact could be strong and widespread.

A. Occurrence of Dietary Cravings and Aversions

In a relatively recent prospective study of 400 white, well-nourished women in Albany, New York, whose pregnancies were ascertained by the 13th week, 76% of the women reported craving at least one item, while 85% reported at least one aversion during pregnancy.

The greatest changes in frequency of consumption of specific food and beverage items, according to information obtained from 7-day diet histories, occurred between the last menstrual period and the 12th week of pregnancy. Most cravings and aversions also occurred early in the pregnancy, with aversions occurring earlier, on the average, than cravings. Women reporting cravings increased their consumption of the items craved, and they decreased their consumption of those items for which they reported aversions. At least 15% of the women reported cravings for foods in the categories of ice cream, chocolate candy, cookies, citrus products, and fruits and/or fruit juices (other than citrus). A further 10% expressed cravings for foods with Italian sauce (essentially tomato sauce). At least 10% of the women reported aversions to foods categorized as fish, beef, foods with Italian sauce, and meat in general. Over 20% reported aversions to alcoholic beverages, and fully 34% reported aversions to coffee during their pregnancies. In addition, approximately 13% reported aversions to cigarette smoke.

Less comprehensive results reported in other studies substantiate many of these results—keeping in mind that most other studies have been retrospective in nature.

B. Cravings and Aversions and Fetal Outcome

Although associations between separate cravings and measures of fetal outcome were not demonstrated in the Albany study, several associations existed between the occurrence of specific dietary aversions and several measures of fetal outcome, especially fetal growth index (FGI), which is a measure of mean size for gestation length adjusted for population means in the specific geographic area. A significant positive association existed between the occurrence of any aversions during pregnancy and increased FGI. Several positive associations were noted between increased FGI and aversions to several different meats, implying that decreased consumption of meat resulted in increased FGI (greater birth weight). Other studies of cravings and aversions have not reported similar findings, although aversions to some meats have been routinely reported in the literature. Additionally, in most populations, there are always certain foods that are thought to be important in terms of their effect on pregnancy outcome—either foods necessary for successful outcome or foods to be carefully avoided during pregnancy.

C. Origins of Cravings and Aversions

The origins of the majority of dietary cravings and aversions reported during pregnancy are defined by the pregnant women as being endogenous in nature. The origins of aversions appear to be more closely related to endogenous reasons than the origins of cravings. It is possible that women might express cravings for items that they would prefer to eat anyway (e.g., chocolate), but which they would feel constrained against eating when not pregnant. Aversions, on the other hand, are usually more abrupt behaviors that are initiated in response to some specific stimulus (e.g., morning sickness). Most of the nausea and vomiting during pregnancy that can be attributed directly to the pregnancy also occurs early in pregnancy. Since the origin of most dietary aversions appears to be more directly related to physiological considerations, perhaps these two symptoms (cravings and aversions) should not be united into a single complex but should be considered as two separate patterns of behavior. [*See* Pregnancy, Nausea and Vomiting.]

BIBLIOGRAPHY

Dickens, G., and Trethowan, W. H. (1971). Cravings and aversions during pregnancy. *J. Psychosom. Res.* **15**, 259–268.

Fairburn, C. G., Stein, A., and Jones, R. (1992). Eating habits and eating disorders during pregnancy. *Psychosom. Med.* **54**, 665–672.

Hook, E. B. (1978). Dietary cravings and aversions during pregnancy. *Amer. J. Clin. Nutr.* **31**, 1355–1362.

O'Rourke, D. E., Quinn, J. G., Nicholson, J. O., and Gibson, H. H. (1967). Geophagia during pregnancy. *Obstet. Gynecol.* **29**, 581–584.

Profet, M. (1992). Pregnancy sickness as adaptation: A deterrent to maternal ingestion of teratogens. *In* "The Adapted Mind: Evolutionary Psychology and the Generation of Culture" (J. H. Barkow, L. Cosmides, and J. Tooby, eds.), Chap. 8, pp. 327–365. Oxford University Press, New York.

Tierson, F. D., and Hook, E. B. (1989). Dietary cravings and aversions during pregnancy and association with pregnancy outcome. *Amer. J. Phys. Anthropol.* **78**, 314–315. [Abstract]

Tierson, F. D., Olsen, C. L., and Hook, E. B. (1985). Influence of cravings and aversions on diet in pregnancy. *Ecol. Food Nutr.* **17**, 117–129.

Walker, A. R. P., Walker, B. F., Jones, J., Verardi, M., and Walker, C. (1985). Nausea and vomiting and dietary cravings and aversions during pregnancy in South African women. *Brit. J. Obstetr. Gynaecol.* **92**, 484–489.

Pregnancy, Gestation

THOMAS R. MOORE
University of California, San Diego

SEVERAL CRITICAL ASPECTS OF HUMAN PREGnancy are overviewed in this article. Beginning with information about birth and perinatal mortality rates in the United States, this article examines factors associated with suboptimal fetal, neonatal, and infant mortality statistics in this country. Facts concerning the duration of normal human pregnancy are presented, and normal adaptive physiological changes experienced by pregnant women are outlined. Finally, the issue of multifetal gestation is considered, with a review of current trends in twin and triplet births, expected survival and neonatal morbidity rates, and maternal complications.

I. NATALITY STATISTICS

A. Birthrates

The overall birthrate is typically defined as the number of births per 1000 population. Although the birthrate is rising worldwide, in developed countries the birthrate has been falling slowly since the mid-1970s. Since 1980 in the United States, despite a 5-year burst of births in the 1988–1992 period, the birthrate has declined by 1–2% per year to the current level of 15.2 births/1000 (Fig. 1). However, birthrates for specific age groups have redistributed, with the rate for teenagers falling, and the rate for women over 35 rising (see Fig. 2). This trend may be due to the increasing social emphasis on contraception and pregnancy avoidance among women under 18 years old and the availability of assisted reproductive technologies which augments fertility among women older than 35 years. Overall, a continued fall in birthrates in developed countries is projected over the next 20 years.

B. Delivery Mode

Cesarean section has become an increasingly common means of delivery worldwide possibly associated with wider availability of broad spectrum antibiotics and obstetrical facilities with surgical capability. In some U.S. hospitals, over 1 in 3 births are by cesarean section. However, the marked increase in cesarean sections in the United States, which rose most steeply from 1980 to 1985 at 6% per year, has finally abated. This upward trend from 4.5% in 1965 to approximately 24% of deliveries in 1989 provoked intense study and debate among health consumers, epidemiologists, and clinicians. Commonly cited justifications for increased use of cesarean section are hopes of improving neonatal outcome by avoiding traumatic vaginal birth, increased diagnosis of dystocia associated with larger babies, and routine cesarean section for breech presentation.

Nevertheless, the U.S. rate of cesarean delivery declined for the fifth consecutive year and was 7% lower in 1994 (21.2%) than in 1989 (22.8%). The primary cesarean rate was also 7% lower in 1994 than in 1989. The rate of vaginal birth following a previous cesarean delivery was 39% higher in 1994 than in 1989 (see Fig. 3).

The health and economic consequences of increased cesarean delivery are enormous. The hospital stay and costs incurred by cesarean patients are approximately twice that of vaginally delivered patients. The morbidity sustained by cesarean patients (wound and uterine

ENCYCLOPEDIA OF HUMAN BIOLOGY, Second Edition, VOLUME 7. Copyright © 1997 by Academic Press. All rights of reproduction in any form reserved.

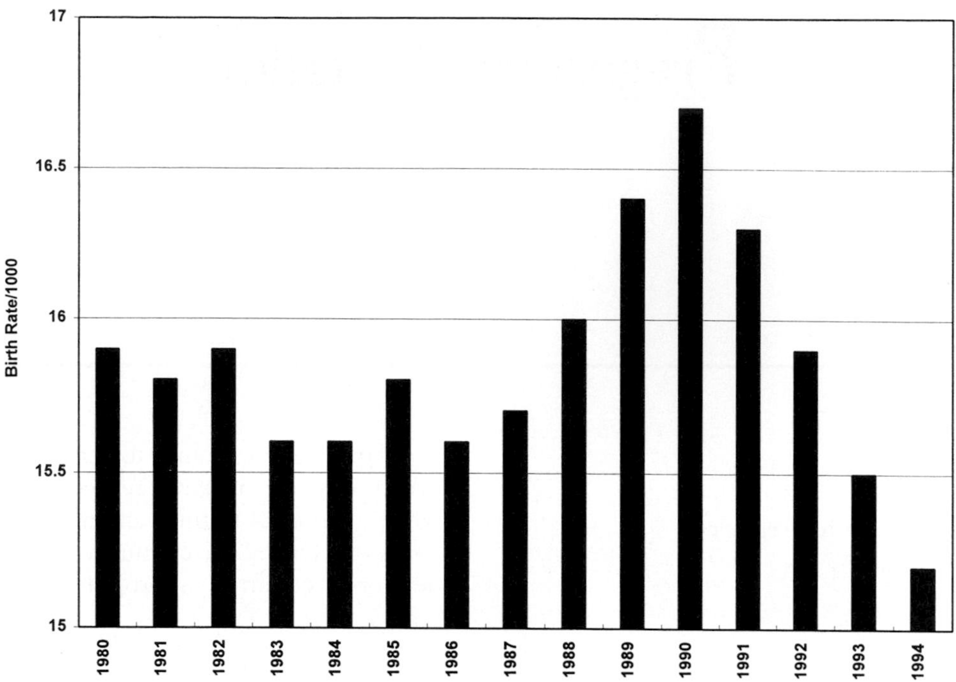

FIGURE 1 U.S. birthrates per 1000 population (1980–1994). Adapted from Ventura *et al.* (1996).

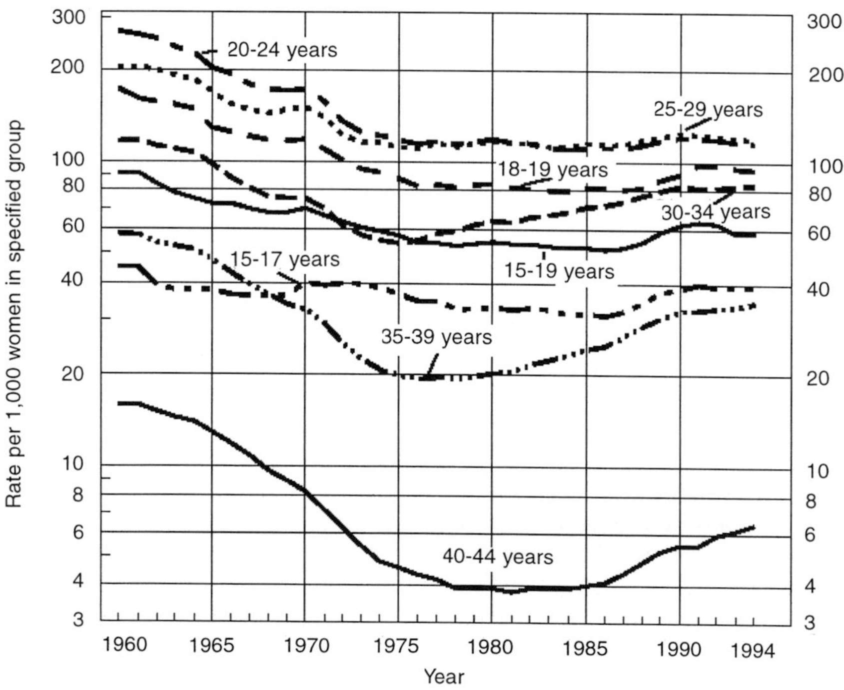

FIGURE 2 Birthrates by age of mother (1960–1994). From Ventura *et al.* (1996) with permission.

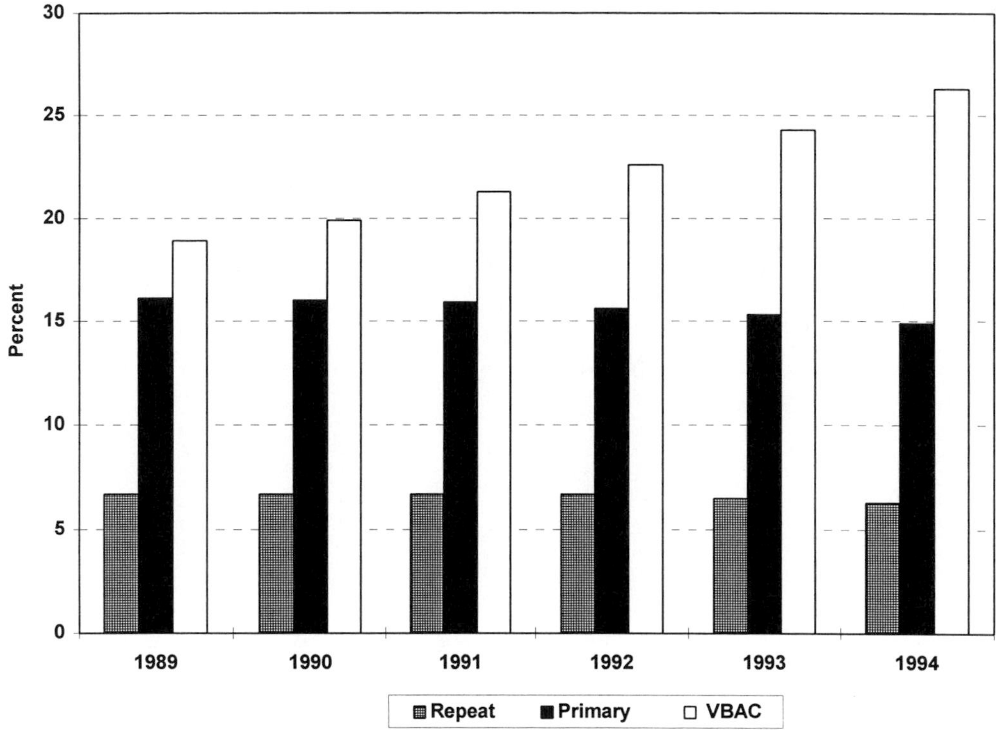

FIGURE 3 U.S. cesarean section and vaginal birth after cesarean (VBAC) rates (1989–1994). Adapted from Ventura *et al.* (1996).

infections, thromboses, and emboli) is increased severalfold over vaginally delivered patients. These differences are compounded by the fact that over 80% of patients with a prior cesarean section will be delivered abdominally in subsequent pregnancies. Although the cesarean rate has risen precipitously, there is little evidence that improved perinatal outcomes resulted. Several studies, comparing differing cesarean rates in similar populations, have failed to show a reduction in neonatal morbidity or mortality, despite markedly increased maternal morbidity.

C. Mortality Rates

The perinatal mortality rate (PMR) is a common method of comparing risks and outcomes of various health systems, economies, and ethnic groups. PMR is composed of fetal mortality and neonatal mortality. The fetal mortality rate (FMR) generally describes births after 20 weeks gestation in which no signs of life were present per 1000 births. The neonatal mortality rate (NMR) is typically determined from deaths of liveborns per 1000 live births up to 28 days of life.

1. U.S. Mortality Statistics

In 1991, U.S. PMR reached a record low of 8.7/1000, 19% lower than the rate in 1985. However encouraging this trend may be, in fact the U.S. PMR is not particularly favorable when compared with other developed countries. This is surprising, given the enormous expenditures on health care per capita in the United States. Despite a decline in the U.S. PMR by more than 70% between 1950 and 1990, the United States' international ranking deteriorated over the same period from 3rd lowest to 19th lowest primarily because the fall in the U.S. rate has not kept pace with the decline in other countries (Fig. 4). The reasons for this difference have been much debated, but the appropriate public health policies and corrective actions remain unresolved.

2. Factors Influencing Mortality

a. Low Birth Weight and Prematurity

Low birth weight (<2500 g) and premature birth (<37 gestational weeks) are the major influences on perinatal mortality, contributing over 75% of the deaths. In the United States the incidence of low birth

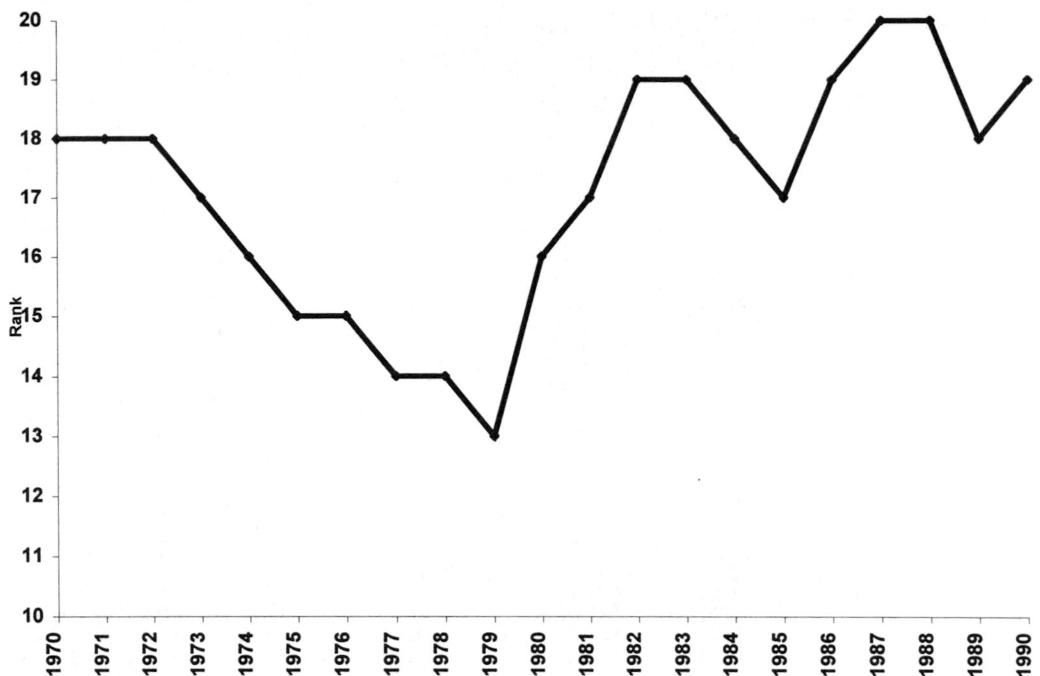

FIGURE 4 Rank of United States in perinatal mortality among world nations (1970–1990). Adapted from Ventura *et al.* (1996).

weight births has continued to climb overall, rising from 6.8% in the mid-1980s to a present 7.2%. Although the rate of low birth weight deliveries improved slightly among black mothers, dropping between 1993 and 1994 to 13.2%, this rate is twice that of white mothers, whose own low birth weight rate increased during this period.

Similarly, the proportion of preterm births (<37 weeks gestation) in the United States has increased steadily from 9 to 11% between 1981 and 1993. Although preterm births among black newborns fell slightly to 18%, the lowest proportion in almost a decade, the rate was more than twice that of other races (Fig. 5). Preterm birth rates were largely unchanged among American Indians, Hispanics, Asians, and Pacific Islanders.

Reasons for the increase in both low birth weight and prematurity over the past two decades are presently unclear and, as they are likely responsible for the majority of the poor rating in U.S. PMR will appropriately be the focus of considerable research effort in the future. The contributions of socioeconomic status, intrauterine infection, and ethnicity—all factors previously demonstrated to affect preterm delivery—must be further defined.

b. Race/Ethnicity

The PMR for the black population in 1991 (15.7/1000) was more than twice that of the white population (7.4/1000), a gap considerably wider than in 1985. This disparity in perinatal mortality between the two major races derives from a much greater drop in perinatal mortality for the white population compared with those of African-American descent (Fig. 6).

For the white population, the average annual decline in the perinatal mortality rate between 1985 and 1991 was 4.1% compared to 2.2% among blacks, with higher rates in both FMR and NMR contributing to the increase in the race differential. The PMR for the Hispanic population (7.9/1000) was 11% higher than that of the non-Hispanic white population (7.1/1000).

c. Prenatal Care and Educational Level

The PMR varies by maternal educational attainment, with higher rates among those with less schooling. This mortality differential by education persists when other socioeconomic and medical risk factors are taken into account. Utilization of prenatal care has a similar powerful effect on perinatal mortality. Although analyses of the effect of prenatal care on

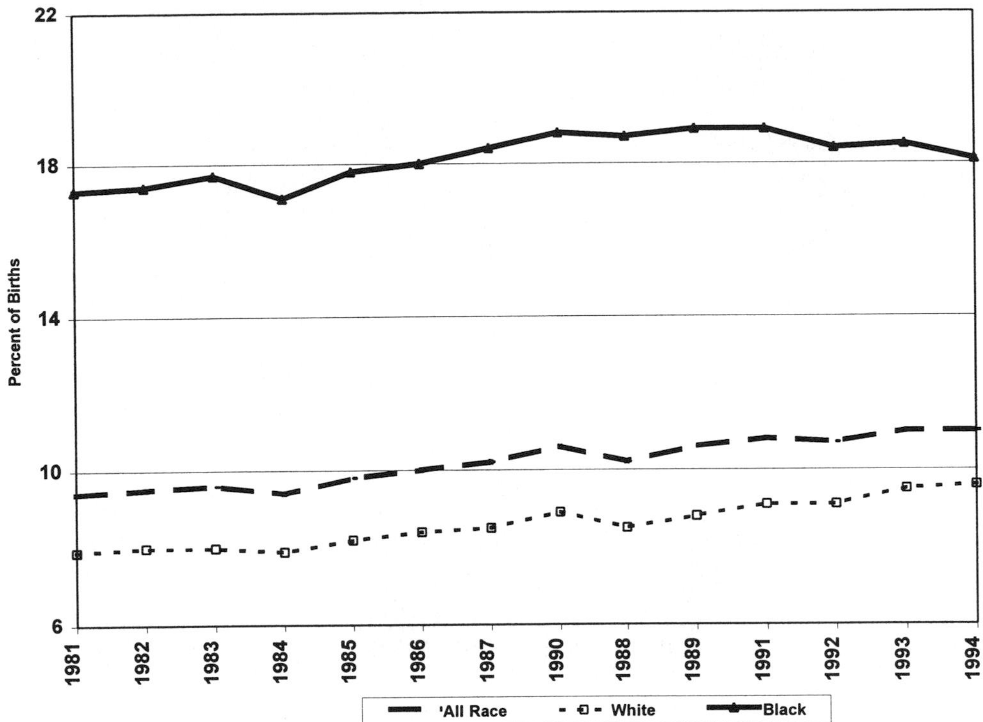

FIGURE 5 U.S. rate of premature births by race of mother (1981–1994). Adapted from Ventura *et al.* (1996).

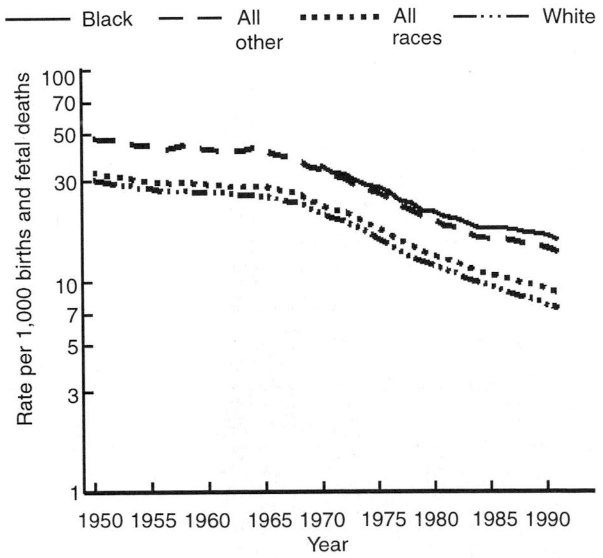

FIGURE 6 U.S. perinatal mortality rates by ethnicity. From Hoyert (1995).

PMR are complicated by factors such as differences in opportunities to obtain prenatal care, perinatal mortality rates are high in women who receive no prenatal care during pregnancy as compared to women who received some prenatal care. The association between prenatal care and fetal mortality holds for all ages, races and ethnic groups, birth order, length of gestation, and education (Fig. 7).

II. PREGNANCY DURATION

A. Statistics

1. Mean Length of Human Pregnancy

The expected length of pregnancy in the human is of intense interest to the gravida herself and her family and is of great importance to the clinician entrusted with her care, as delivery too early or too late may result in neonatal asphyxia or death.

Despite the fact that embryonic development does not begin until after fertilization (usually day 14 of the menstrual cycle), the predicted duration of pregnancy

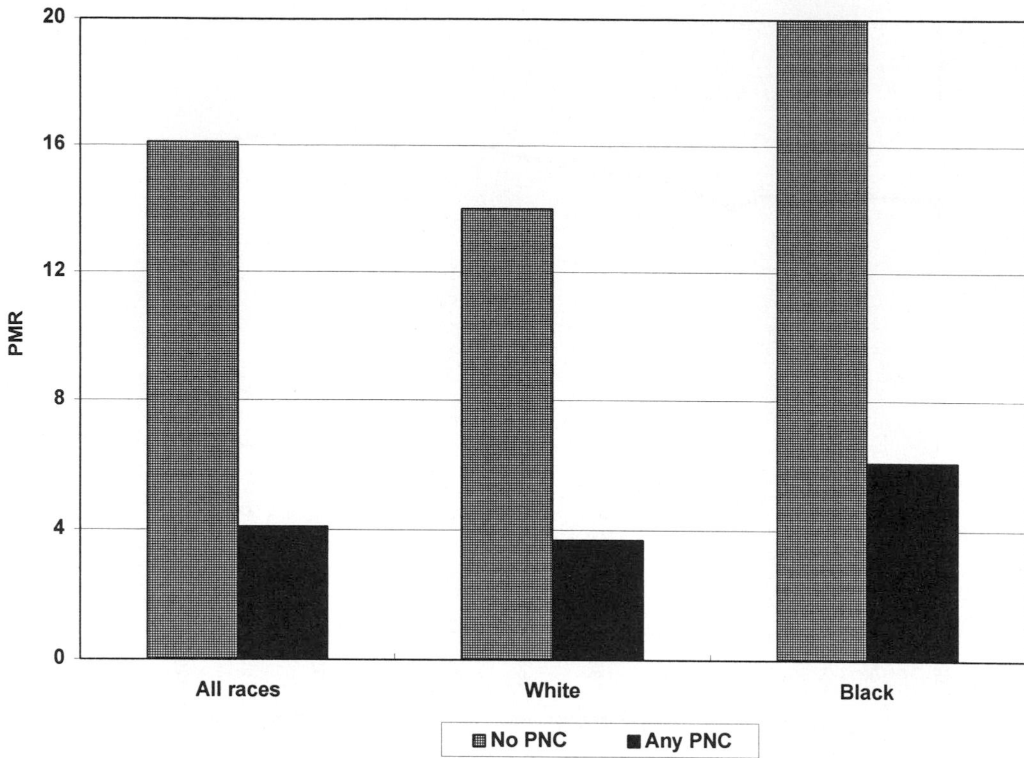

FIGURE 7 Perinatal mortality rate (PMR) by ethnicity and prenatal care (PNC). From Hoyert (1995).

has been traditionally calculated from the first day of the last menstrual period (LMP) to the most common day of spontaneous labor and delivery (280 days). However, a number of important factors may influence the calculation of the expected day of delivery.

1. Variation in the actual day of ovulation within the menstrual cycle. Although the majority of women with 28-day cycles ovulate on day 14, up to 90% of cycles may vary in a given year by ±7 days, more commonly 2–7 days later or earlier than predicted.
2. The precise timing of fertilization. The actual day of fertilization typically coincides with the day of ovulation (on cycle day 14 or 15), but may vary by up to ±3 days.
3. Variation in the onset of spontaneous labor. Numerous factors appear to influence the onset of spontaneous term labor, including changes in fetal, placental, or decidual hormone levels or bacterial invasion of the uterine cavity.

Although the length of gestation is clinically defined as 280 days, recent investigations suggest that the average length, based on the LMP, is actually 292 days. As might be predicted from the fact that ovulation tends to occur on or slightly later than the 14th day, the actual day of delivery predicted from ultrasound measurement of the fetal head in the second trimester is approximately 2 days earlier than predicted by LMP (Table I).

The increased precision in prediction of the day of delivery using ultrasound measurements of the fetus further refines the definitions of term and postterm gestation. As noted in Table II, the proportion of pregnancies delivering "at term" is 88% using ultra-

TABLE I
Pregnancy Duration in Normal Women with Regular Menstrual Cycles Determined from Ultrasound or Last Menstrual Period Prediction

	Ultrasound	Last menstrual period	*P*
Mean	281 days	283 days	<0.001
Median	279 days	282 days	<0.001
Mode	281 days	284 days	

TABLE II
Distribution of Spontaneous Births in Women with Regular Menses as Predicted by Ultrasound or Last Menstrual Period

Delivery	Ultrasound	Last menstrual period
±7 days	61%	56%
±14 days	88%	84%
<259 days (preterm)	4%	4%
≥296 days (postterm)	2%	9%

sound prediction versus 84% with LMP. Moreover, the number of pregnancies identified as "postdates"—more than 14 days past due—is increased by more than fourfold when the LMP is used instead of ultrasound. Differences in distribution of deliveries by the two methods are illustrated by Fig. 8.

B. Postdatism

The EDC is at the lowest point of perinatal mortality for the majority of pregnancies. There does not appear to be a significant change in perinatal morbidity across the 5-week time period recognized as "term" (37–42 weeks), but a clear rise in PMR begins just after 42 weeks of gestation, approximately doubling by 43 weeks, and is four to six times higher at 44 weeks than

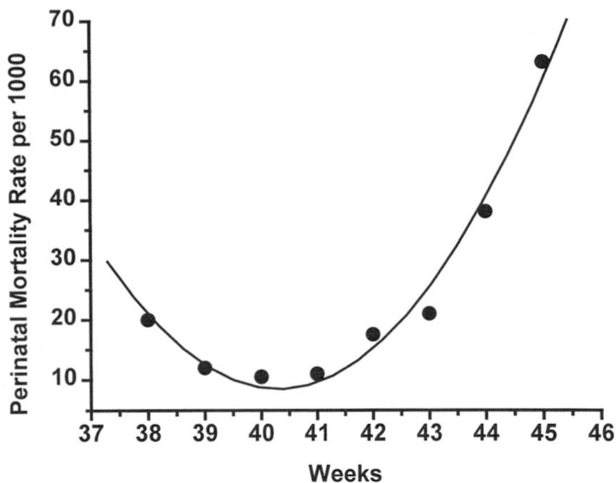

FIGURE 9 Perinatal mortality rate versus gestational age. Adapted from McClure-Browne (1963).

at term (Fig. 9). Consequently, a pregnancy should be considered prolonged when it has proceeded beyond 296 days or 42 weeks.

Although the frequency of true postdate pregnancy has been reported to range from 7 to 12%, many of these pregnancies appear prolonged because of problems in accurate obstetrical dating. However, for the approximately 4% of truly postterm gestations, the pathophysiology underlying prolongation of pregnancy is not clear. Some evidence indicates that bio-

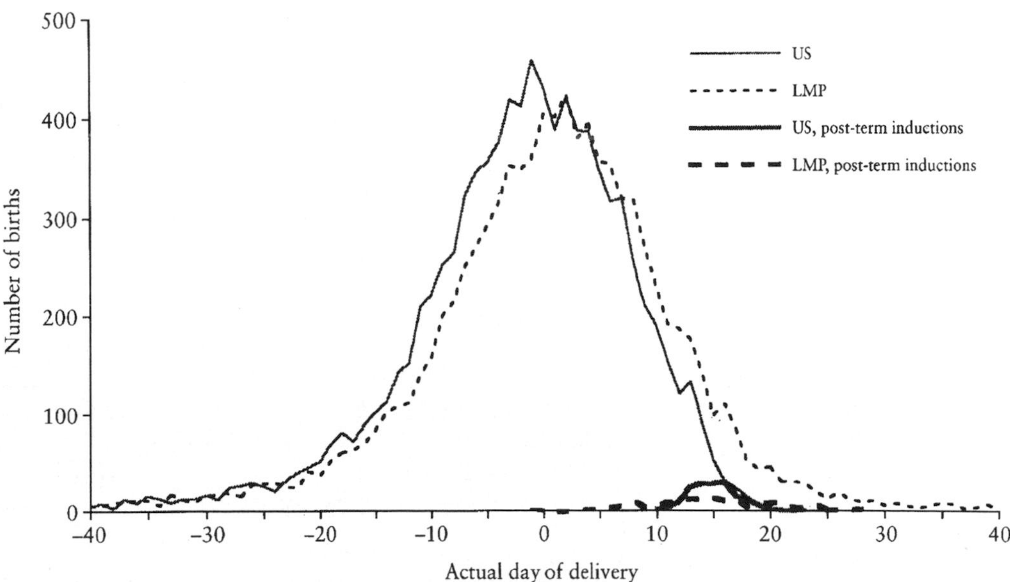

FIGURE 8 Distribution of spontaneous births around the day of delivery predicted by ultrasound (US; day 0) and the last menstrual period (LMP) in 9240 women with normal menstrual cycles. From Tunón *et al.* (1996).

chemical abnormalities in fetal or placental tissues may delay the onset of parturition. For example, anencephalic fetuses lacking pituitary tissue often are postdates whereas an anencephalic fetus with normal pituitary tissue usually delivers at term or even preterm. This suggests a central role for the fetal pituitary in initiating labor. The placenta probably also participates in the labor-initiation process as in the rare instance in which the placenta lacks a key enzyme, sulfatase, which results in abnormally low production of the hormone estriol, prolonged gestation frequently results.

Although the precise mechanisms leading to the increased risk of perinatal death in the postterm period remain unclear, the stigmata of prolonged pregnancy—a wasted, stillborn infant with little subcutaneous fat, stained with thick meconium accompanied by a calcified and infarcted placenta—supports the overall impression of placental failure, fetal dehydration, starvation, and asphyxiation. Modern obstetrical management focuses on avoiding this catastrophe. Typical management interventions include meticulous dating of the pregnancy, inducing labor as soon as possible after 41 weeks, and performing frequent testing of fetal heart rate and amniotic fluid volume.

C. Obstetrical Dating

The estimated delivery date (EDD) is frequently referred to by an older term, "estimated date of confinement" (EDC). The term EDC arose from the custom of requiring the gravida to closet herself during the last weeks of pregnancy (usually confined to her bed chamber). A more appropriate term today is the EDD, which is calculated by adding 280 days to the first day of the last menstrual period. Another technique, Naegele's rule, predicts EDD by subtracting 3 months from the LMP and adding 1 week. A number of other pregnancy events can be used to assess gestational age. Typical milestones and limitations in their accuracy are summarized in Table III.

III. MATERNAL ADAPTATION TO PREGNANCY

The changes in the structure and function occurring in the human female during pregnancy are remarkable for their variety and intensity, including a 40% increase in blood volume and cardiac output and a 10-fold increase in uterine blood flow. As a general theme, the physical and physiologic changes of pregnancy are mediated by rising steroid and peptide hormone levels produced by the placenta.

A. Hormones

Enormous quantities of peptide and steroid hormones enter the maternal circulation, with progesterone and estradiol levels increasing 7-fold during the first 30 weeks of gestation. Other steroid hormones, such as

TABLE III
Clinical Criteria for Dating Human Pregnancy

Dating criterion	Factors affecting certainty	Typical 90% confidence interval
Last menstrual period	Unusual menstrual cycle length, recent use of hormonal contraception prior to conception	±2 weeks
Positive urine or serum hCG	Continuously detectable from 12 to 20 days after fertilization	
Uterine size on physical palpation at first prenatal visit	Unreliable with obesity, after 12 or before 6 gestational weeks	±2 weeks
Embryonic crown–rump length on ultrasound	Most accurate from 6 to 10 weeks	±3 days
Doppler auscultation of fetal heart	Obtainable after 10–12 weeks	±2 weeks
Maternal perception of fetal movement	Nulliparas: 18–20 weeks Multiparas: 16–20 weeks	±2 weeks
Uterine fundus at maternal umbilicus	Typically at 20 weeks; unreliable with myomata or obesity	±2–3 weeks
Uterine fundal height (FH) in centimeters from symphysis pubis	FH (cm) = gestational weeks from 18 to 36 weeks; influenced by fetal size, amount of amniotic fluid	±2–3 weeks
Sonographic biometry at 15–22 weeks	Unreliable for dating after 26 weeks	±1.5 weeks

TABLE IV
Key Hormones Involved in Maternal Pregnancy Adaptation

Hormone	Source	End-organ effects
Progesterone	Placental biotransformation of maternal cholesterol	Relaxation of smooth muscle Reduced uterine motility Downregulation of maternal immune response to fetal antigens
Deoxycorticosterone	Progesterone transformation in maternal kidney	Promotes salt and water retention
Estrone, estradiol	Placental biosynthesis from maternal and fetal androgens	Increased uterine blood flow Increased renin–angiotensin–aldosterone activity Increased sodium and water retention Increased binding protein biosynthesis by the liver
Human chorionic somatomammotropin	Placenta	Mobilizes free fatty acids Increased insulin resistance Increased nitrogen retention

deoxycorticosterone (DOC), rise by as much as 20-fold. This increase in hormone quantity and activity promotes profound and overlapping effects on maternal physiology. The major hormones, synthesis site, and principal physiologic effects are shown in Table IV.

1. Steroid Hormones

The central role of progesterone in promoting maternal adaptation and maintenance of the pregnancy cannot be overemphasized. Withdrawal of progesterone in early pregnancy by failure of the corpus luteum, oophorectomy, or administration of the progesterone antagonist RU-486 results in abortion. Examples of progesterone mediation of maternal tolerance of pregnancy include maternal vasorelaxation, vascular volume expansion, and suppression of uterine contractions. A fall in the ratio of progesterone to estrogen has been proposed as the initiator of labor in human pregnancy.

More estrogens are produced during a single pregnancy than would be produced in 100 years of reproductive life in normal adult women. Pregnancy estrogens are derived from dehydroepiandrosterone (DHEA) and DHEA sulfate, primarily of fetal origin. Estrone and estradiol have central roles in maternal cardiovascular adaptation, significant cardiovascular effects (sodium retention and plasma volume expansion, decreased vascular resistance), promotion of liver protein synthesis (increased binding proteins and clotting factors), and on uterine circulation (increased uterine blood flow). [*See* Steroids.]

2. Peptide Hormones

Human chorionic somatomammotropin (hCS), exclusively of placental origin, is produced in enormous quantities by the end of pregnancy, with peak levels 40 times higher than those of progesterone (160 ng/ml). A main effect of hCS is to antagonize the effects of insulin, resulting in a progressive increase in "insulin resistance" during pregnancy, thus predisposing pregnant women to a gestational form of diabetes. hCS also promotes mobilization of free fatty acids, the substrate for gluconeogenesis, and functions as a fetal growth factor.

Relaxin is a small peptide hormone produced by the corpus luteum and the placenta. It may assist in maintaining uterine quietude as the conceptus grows and in promoting cervical softening as labor approaches. The placenta also synthesizes a number of peptides that are similar in structure and function to those produced by the adult pituitary and hypothalamus. Pituitary-like hormones include hCG (similar to luteinizing hormone), hCS (similar to growth hormone), growth hormone (GH), and chorionic thyrotropin and corticotropin. Hypothalamic-like CRH, TRH, GnRH, and somatostatin have been demonstrated to be of cytotrophoblastic origin. The placental production of homologs of all recognized pituitary and hypothalamic hormones indicates that the placenta has all elements required to function as an independent hypothalamic–pituitary target organ unit. However, the precise role of these substances in the regulation of fetal growth and maternal adaptation is still being clarified. [*See* Placenta.]

B. Cardiovascular Adaptation

The changes that occur in the cardiovascular system during pregnancy serve as a paradigm for the changes observed in other organ systems. The general principle is that placental hormones alter maternal end-organ function to maximize fetal development. Cardiovascular changes induced by placental hormones may superficially appear to be pathological, and yet failure of normal cardiovascular adaptation during pregnancy often becomes indeed pathological.

1. Vascular Volume

Figure 10 illustrates typical changes observed in cardiovascular parameters during normal singleton pregnancy. Blood and plasma volume increase linearly during gestation, peaking at 28–32 weeks, which is approximately 40% above nonpregnant values (the normal 5-liter blood volume expands to 7 liters during pregnancy). Accompanying this increase is an additional 4–6 liters of fluid distributed into the extravascular tissues, for a next fluid gain of 6–8 liters. In twin or triplet pregnancy, the vascular volume may increase by as much as 75–100%. After 28 weeks, the vascular volume remains relatively constant until delivery.

The progressive increase in vascular volume during the first 28 weeks of pregnancy appears to depend on rising levels of placental steroids and the concomitant fall in systemic vascular resistance associated with estrogen-induced vascular relaxation. Mean arterial pressure falls, establishing a relative state of hypovolemia that stimulates an increased renal production of aldosterone. Elevated aldosterone activity promotes renal tubular sodium and water retention.

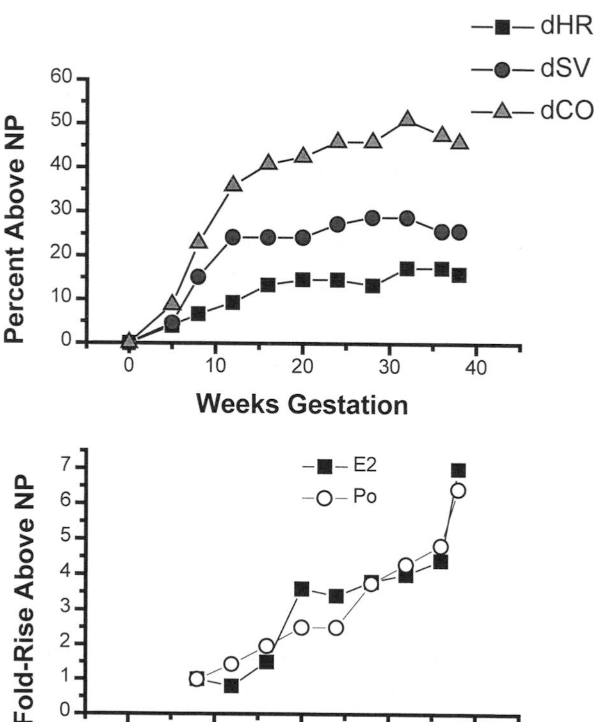

FIGURE 10 Changes in maternal cardiovascular parameters and hormone levels during gestation. NP, nonpregnant; E2, estradiol; Po, progesterone; dHR, change in heart rate; dCO, change in cardiac output; dSV, change in stroke volume. Adapted from Robson *et al.* (1989).

2. Erythropoiesis

The erythropoietic response to increasing plasma volume appears to follow rather than lead hypervolemia, such that the weekly increment in plasma volume at each point of gestation exceeds the rise in red cell mass, resulting in a progressive fall in hematocrit during the first and second trimesters, totaling 15–20% below prepregnant values. This progressive hemodilution, which occurs in iron-replete women, is a "physiologic anemia." Hemodilution, indicated by the decline in hematocrit, is important evidence of normal hemodynamic adaptation to pregnancy.

3. Maternal Hemodynamics

Figure 10 plots the progressive increase in cardiac parameters in normal pregancy derived from noninvasive Doppler measurements. These changes are driven by the hormone-induced increase in maternal vascular volume. Cardiac output augmentation is accomplished by a combination of increased heart rate and stroke volume, but most of the increase is mediated by stroke volume. The maternal heart rate increases by a maximum of 15–20% during pregnancy, and rarely exceeds 100 bpm in normal women.

The increase in cardiac output during pregnancy is not uniformly distributed. Blood flow to maternal brain, gastrointestinal organs, and musculoskeletal structures is unchanged. However, renal perfusion rises by 40%, uteroplacental flow increases 15-fold, and circulation to the skin and breasts also rises significantly.

Maternal hemodynamics are profoundly affected by maternal posture, particularly in the third trimester, especially when women are positioned on their backs. With the patient supine, the uterine fundus obstructs returning venous blood flow, reducing cardiac return and compromising cardiac output. Fainting may ensue. Similar but less dramatic effects are associated with maternal standing or sitting.

C. Pulmonary Adaptation

1. Chest Anatomy

Although most pregnant women feel somewhat breathless and compromised in their ventilatory status, in fact the opposite is true. Although in late pregnancy, the diaphragm is significantly elevated above nonpregnant positions; the chest diameter widens under the influence of the growing uterus and relaxation by estrogens. This results in a net increase in total lung capacity and vital capacity, although residual volume (end-expiratory volume) falls slightly.

2. Pulmonary Dynamics

Under the influence of progesterone, the resting respiratory volume—"tidal volume"—increases by 40%, resulting in a 40% increase in minute ventilation. The respiratory rate is unchanged. The functional effect of this pregnancy-induced "hyperventilation" results in a compensated respiratory alkalosis. Compared to nonpregnant values, the maternal oxygen level changes little because of essentially complete oxygen saturation, but blood carbon dioxide levels fall by 20%, which assists in extracting carbon dioxide from the fetal compartment. Additional changes in pulmonary function during pregnancy include decreased airway resistance (bronchial smooth muscle relaxation) and a fall in pulmonary vascular resistance.

D. Renal and Urinary Adaptation

1. Functional Changes

Urinary tract changes in pregnancy are mediated by placental hormones and the compensatory cardiovascular adjustments detailed previously: renal blood flow and renal plasma flow increase by 30–40% as vascular volume rises. The glomerular filtration rate increases from 122 ± 24 to 170 ± 23 ml/min in response to augmented renal circulation.

The increase in plasma and extracellular water during pregnancy (8 liters) requires the retention of 800–900 mEq of sodium, or approximately 20–30 mEq per week. As noted earlier, this process is mediated largely by the elevation in aldosterone. Within hours of delivery, maternal plasma levels of DOC and other steroid hormones fall to normal levels and diuresis is initiated with vigor.

2. Anatomical Changes

Smooth muscle relaxation induces anatomical changes that are apparent with ultrasound imaging as early as 8 weeks of gestation, including hydroureter and hydronephrosis. Decreased urethral sphincter and urinary bladder tone contribute to the urinary incontinence noted as pregnancy progresses. Relaxation of the urethral sphincter and bladder tone may also contribute to increased frequency of cystitis.

E. Gastrointestinal Adaptation

The predominant effects of pregnancy on the gastrointestinal organs are related to progesterone-induced smooth muscle relaxation. Esophageal reflux and heartburn, common complaints, are due to hormonal relaxation of the gastroesophageal sphincter more than to upward pressure by the uterus. Gastric emptying is significantly delayed during pregnancy because of smooth muscle relaxation. Clinically, this places the pregnant woman at increased risk of aspiration of gastric contents after vomiting. Most anesthesiologists consider pregnant patients to be at high risk in regurgitation of retained gastric contents during the induction of general anesthesia. In the large bowel, colonic transit times are prolonged, resulting in the common complaint of constipation. [*See* Pregnancy, Nausea and Vomiting.]

Progesterone action on the biliary system results in sluggish emptying of the gall bladder, predisposing to stone formation. The sluggish flow of bile through the hepatic canaliculi may contribute to the occurrence of cholestatic jaundice of pregnancy.

Changes in liver function during pregnancy are complex. Estrogen stimulation of hepatic protein synthesis increases serum-binding proteins and clotting factors, promoting an increased risk of thrombosis during pregnancy. Cholesterol and triglyceride concentrations are elevated, presumably associated with increased free fatty acid mobilization to augment the enhanced gluconeogenesis noted in pregnancy.

F. Integumentary Changes

Pregnancy-associated changes in the skin and its appendages are very evident to the pregnant woman and her family. As in other systems, placental hormones play a key role. Striae gravidarum, commonly called "stretch marks," are typically distributed over the abdomen and lower breasts, and form in response to the high plasma steroid levels in the pregnant woman rather than being caused by stretching of the skin.

1. Pigmentation

In general, melanocyte activity is increased during pregnancy. Darkening of the aerolae, skin in the line from the umbilicus to the pubis ("linea nigra"), areas of the face and neck ("melasma"), and the vulva occurs to some extent in 90% of women. Nevi may become more pigmented, but the effect on the incidence or progression of melanoma is probably unchanged.

2. Vascular Changes

Superficial varicosities arise in 40% of women primarily in the lower extremities but may also involve the vulva and rectum (hemorrhoids). Spider angiomata, characterized by a dilated central artery with smaller radiating branches, are distributed over the upper chest, extensor surfaces of the hands and arms, and face. As similar lesions are observed in hyperestrogenic males, it is postulated that the vascular lesions of pregnancy are mediated by estrogens. All of these changes resolve after pregnancy.

3. Hair

Hair loss occurs after hair has been in the resting, or telogen, phase. Ordinarily, telogen hair is a small proportion of the total follicles present. During pregnancy, many hairs are stimulated into the growth, or anagen, phase. Many women note increased hair growth during pregnancy. After delivery, these hairs enter the telogen phase and, 4–6 weeks postpartum, the telogen hairs are shed ("telogen effluvium"). Although the amount of hair lost in a short period of time can be disturbing, the condition is self-correcting.

IV. MULTIFETAL PREGNANCY

A. Current Trends

Babies born in multifetal pregnancies are more vulnerable to early death and disability than are babies born in singleton deliveries. More than half of all multiple births are low birth weight (less than 2500 g) compared with 6% of singleton births. Multiple births are seven times more likely than singletons to die in the first week of life. Multiples also require more health care dollars; each birth in a twin or triplet delivery is reported to cost two to three times that of a birth in a singleton delivery.

The incidence of twin gestation in the United States is approximately 1% of all pregnancies, but this includes twins arising from two eggs (dizygotic twins) and those resulting from splitting of a single developing embryo (monozygotic twins). The incidence of monozygotic ("identical") twinning is fairly constant throughout the world at approximately 1 : 250 births and is not inherited in families. The etiology of "identical" twinning is unclear, but is thought to be a random event of embryo splitting that may occur at any point from day 1–15 postconception. Embryo division after day 13 can result in conjoined twinning. [*See* Embryology and Placentation of Twins.]

Dizygotic twins result from the fertilization of two ova. Production of multiple ova during a menstrual cycle may occur through the use of fertility drugs, because of genetic or familial predisposition, or under the influence of advancing maternal age. In the United States, wide ethnic differences exist in the incidence of dizygotic twinning, varying from a low of 1 : 140 pregnancies among Asians to a high of 1 : 70 for African-Americans. The familial association of dizygotic twins is apparently passed from mother to daughter, and the tendency to multiple ovulation increases with maternal age and parity. Higher-order multiples (triplets, quadruplets) are much more rare in nature, with an expected frequency for triplets of only 1 in 6400 pregnancies and 1 in 512,000 for quadruplets.

In recent years the number of multiple births has grown markedly, partly because of the natural increase in multiple ovulation in older women, but mostly because of the increased frequency of multifetal pregnancies created with assisted reproductive technologies ("test-tube pregnancies") (Fig. 11). In 1994, the total of 101,658 multiple births included 97,064 twin (96%), 4233 triplet (4%), 315 quadruplet (.3%), and 46 quintuplet (0.05%) or greater multiples. Over the last decade the number of twin births has risen by 33%, and the number of higher-order multiples by 178%. The overall ratio of deliveries of multiples-to-singletons has increased 33% since 1980 (19.3 per 1000) and the higher-order multiple birth ratio (primarily triplet births) jumped 12%.

B. Outcomes and Complications

The iatrogenic increase in multifetal pregnancies is of more than academic interest as the risk of premature birth and its attendant complications rises precipitously with each additional fetus in the uterus. The major expected outcomes are outlined in Table V.

1. Prematurity, Neonatal Mortality, and Morbidity

The major negative effect of multifetal gestation is on gestational age at delivery. Interestingly, the mean gestational age and birth weight at delivery of triplets today are little different than reported in the mid-1960s (33.5 weeks, 1600 g), suggesting that modern techniques of inhibiting preterm labor are not very effective. This problem is magnified in quadruplets, with a median gestational length of 30.5 weeks and a mean birth weight of approximately 1250 g. Clearly the outcome prognosis for the early half of quadruplet

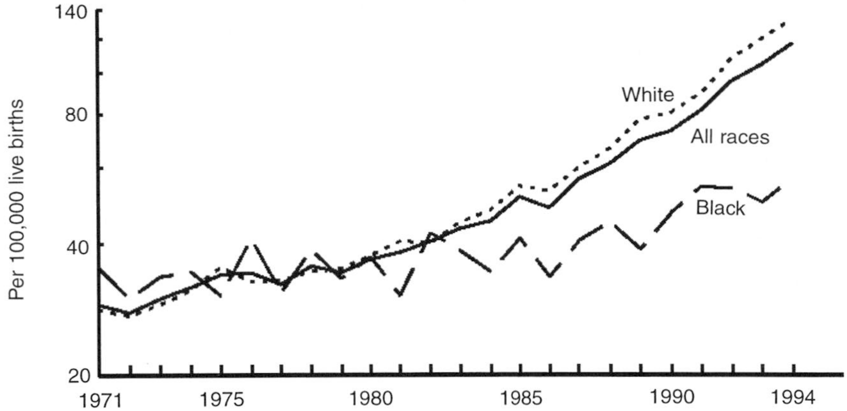

FIGURE 11 Incidence of multifetal pregnancy in the United States (1971–1994). From Ventura *et al.* (1996).

pregnancies, delivered more than 10 weeks early and at less than 2.5 lbs each, is extremely guarded, with cerebral palsy risk approaching 50% per pregnancy.

However, recent comparisons of outcomes of very low birth weight infants with prior eras suggest that substantial progress has been made in improving survival and reducing handicap among very premature neonates. As shown in Fig. 12, substantial improvement in survival was achieved in the 24- to 26-week gestational age period when the 1990 cohort of neonates was compared so a similar group delivered in 1988. Further inspection of Fig. 12 demonstrates the importance of achieving 28 weeks gestation where 90% survival can be expected.

2. Maternal Morbidity

Multifetal pregnancy also carries a significant risk of maternal morbidity, primarily because of the high rate of cesarean section and the increased incidence of preeclampsia. Cesarean section rates are 50–75% for twin pregnancy and are almost 100% for higher-order multiples. Regardless of delivery mode, uterine overdistention with multiples may precipitate abruptio placentae or postpartum uterine atony along with severe maternal hemorrhage.

The increase in preeclampsia among mothers of multiples is most remarkable. This condition, which seems to afflict only humans, occurs in 8–10% of all gestations. It is characterized by severe hypertension, edema, and heavy renal protein excretion and typically occurs in first pregnancies. However, in its full-blown form, pulmonary edema, seizures, brain hemorrhage, and death may ensue. Risk factors that increase the risk of preeclampsia include diabetes (2×), preexisting hypertension (2–3×), prior preeclampsia (2×), and multifetal pregnancy (2–3×). As noted in Table V, the incidence of preeclampsia roughly doubles as each additional fetus is added to the pregnancy (8% singletons, 17% twins, 38% triplets, 55% quadruplets). Also, preeclampsia is as frequent a cause

TABLE V
Outcome and Morbidity in Triplet and Quadruplet Pregnancy

Outcome parameter	Twins	Triplets	Quadruplets
Gestational age at birth	36 weeks	32.5 weeks	30.5 weeks
Birthweight	2400 g	1600 g	1250 g
Perinatal mortality	4.6%	8.9%	12.5%
Cerebral palsy (risk per pregnancy)	3%	15%	48%
Preeclampsia	17%	39%	55%
Premature labor	42%	92%	74%

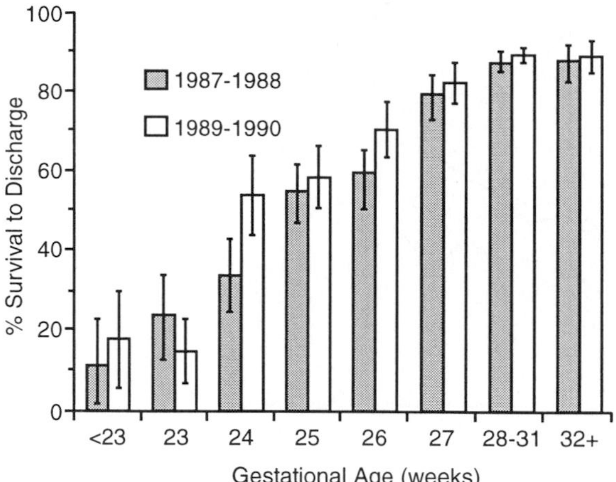

FIGURE 12 Changing survival rates by gestational age for very low birth weight infants. Error bars indicate a 95% confidence interval. From Hack *et al.* (1995).

of preterm birth as is premature labor and ruptured membranes in women with multifetal gestations. [*See* Maternal Mortality.]

V. SUMMARY

Human pregnancy is an enormously complex process, comprising the better part of a year to complete fetal development and delivery, typically, to a singleton offspring. Although maternal physical adaptations and labor timing have been refined through evolutionary millennia to optimize outcome, significant morbidity afflicts pregnancies in which the normal gestational processes are not followed precisely.

BIBLIOGRAPHY

Arlettaz, R., and Duc, G. (1994). Triplets and quadruplets in Switzerland, 1985–1988. Schweizerische Medizinische Wochenschrift. *Journal Suisse de Medecine* **122**(14), 511–516.

Callahan, T. L., Hall, J. E., Ettner, S. L., *et al.* (1994). The economic impact of multiple-gestation pregnancies and the contribution of assisted-reproduction techniques to their incidence. *N. Engl. J. Med.* **331**(4), 244–249.

Chelmow, D., Penzias, A. S., Kaufman, G., and Cetrulo, C. (1995). Costs of triplet pregnancy. *Am. J. Obstet. Gynecol.* **172**(2), 677–682.

Davison, J. M., Gilmore, E. A., Durr, J., Robertson, G., and Lindheimer, M. D. (1984). Altered hemodynamics and tubular function in normal human pregnancy. *Am. J. Physiol.* **246,** F105.

Hack, M., Wright, L. L., Shankaran, S., Tyson, J. E., Horbar, J. D., Bauer, C. R., *et al.* (1995). Very low birthweight outcomes of the National Institute of Child Health and Human Development Neonatal Network, November 1989 to October 1990. *Am. J. Obstet. Gynecol.* **172,** 457–464.

Hardardottir, H., Kelly, K., Bork, M. D., Cusick, W., Campbell, W. A., and Rodis, J. F. (1996). Atypical presentation of preeclampsia in high-order multifetal gestations. *Obstetr. Gynecol.* **87**(3), 370–374.

Ho, S. K., and Wu, P. Y. K. (1975). Perinatal factors and neonatal morbidity in twin pregnancy. *Am. J. Obstet. Gynecol.* **122,** 979–987.

Hoyert, D. L. (1995). Perinatal mortality in the United States, 1985–91: National Center for Health Statistics. *Vital Health Stat* **20**(26).

Hytten, F. E., and Leitch, I. (1964). "Cardiovascular Dynamics in the Physiology of Human Pregnancy," p. 50. Blackwell, Oxford.

Imaizumi, Y. (1994). Perinatal mortality in single and multiple births in Japan, 1980–1991. *Paediatr. Perinat. Epidemiol.* **8**(2), 205–215.

Keith, L. G., Papiernik, E., Keith, D. M., and Luke, B. (eds.) (1995). "Multiple Pregnancy: Epidemiology, Gestation, and Perinatal Outcome." Parthenon, New York/London.

McClure-Browne, J. C. (1963). Postmaturity. *Am. J. Obstet. Gynecol.* **85,** 573.

Moore, T. R. (1993). Maternal adaptation to pregnancy. *In* "Gynecology and Obstetrics, a Longitudinal Approach" (T. R. Moore, R. C. Reiter, R. W. Rebar, and V. V. Baker, eds.), pp. 223–234. Churchill Livingstone, New York.

Pritchard, J. A., and Hunt, C. F. (1958). A comparison of the hematologic responses following the routine prenatal administration of intramuscular and oral iron. *Surg. Gynecol. Obstet.* **106,** 516.

Robson, S. C., Hunter, S., Boys, R. J., and Dunlop, W. (1989). Serial study of factors influencing changes in cardiac output during human pregnancy. *Am. J. Physiol.* **256,** H1060–H1065.

Ron-El, R., Mor, Z., Weinraub, Z., Schreyer, P., Bukovsky, I., Dolphin, Z., Goldberg, M., and Caspi, E. (1992). Triplet, quadruplet and quintuplet pregnancies: Management and outcome. *Acta Obstet. Gynecol. Scand.* **71**(5), 347–350.

Rosen, M. G., Dickinson, J. C., and Westhoff, C. L. (1991). Vaginal birth after cesarean: A meta-analysis of morbidity and mortality. *Obstet. Gynecol.* **77,** 465–470.

Seoud, M. A., Toner, J. P., Kruithoff, C., and Muasher, S. J. (1992). Outcome of twin, triplet, and quadruplet in vitro fertilization pregnancies: The Norfolk experience. *Fertil. Steril.* **57**(4), 825–834.

Taffel, S. M., Placek, P. J., Moien, M., and Kosary, C. L. (1991). 1989 U.S. cesarean section rate studies: VBAC rate rises to nearly one in five. *Birth* **18,** 73–77.

Tunón, K., Eik-Nes, S. H., and Grøttum, P. (1996). A comparison between ultrasound and a reliable last menstrual period as predictors of the day of delivery in 15,000 examinations. *Ultrasound Obstet. Gynecol.* **8,** 178–185.

Ventura, S. J., Martin, J. A., Mathews, T. J., and Clarke, S. C. (1996). "Advance Report of Final Natality Statistics, 1994: Monthly Vital Statistics Report," Vol. 44, No. 11. National Center for Health Statistics, Hyattsville, MD.

Weinberger, S. E., Weiss, S. T., Cohen, W. R., *et al.* (1980). Pregnancy and the lung. *Am. Rev. Respir. Dis.* **121,** 559.

Yokoyama, Y., Shimizu, T., and Hayakawa, K. (1995). Incidence of handicaps in multiple births and associated factors. *Acta Genet. Med. Gemellol.* **44**(2), 81–91.

Pregnancy, Nausea and Vomiting

FORREST D. TIERSON
University of Colorado at Colorado Springs

GLOSSARY

Emesis gravidarum Nausea and vomiting of pregnancy (NVP) primarily in the first trimester (morning sickness)

Hyperemesis gravidarum Severe NVP extending beyond the 16th week, interfering with maternal nutrition and fluid balance

NAUSEA AND VOMITING OF PREGNANCY (NVP) IS a ubiquitous complex of symptoms. Although the relationship between NVP and pregnancy is well known (and is often mentioned in the popular literature), relatively little is known about the cause of NVP or its epidemiology. NVP is often used as an indicator of pregnancy well-being. Many studies have demonstrated a positive association between the presence of early NVP (morning sickness) and favorable pregnancy outcome. In addition, epidemiological studies have reported associations between the presence of early NVP and smoking, between the presence of early NVP and increased infant birthweight, and between the absence of NVP early in pregnancy and increased risk of spontaneous abortion (usually early in pregnancy as well).

I. HISTORICAL BACKGROUND

Nausea and vomiting during early pregnancy is so common that it is considered to be a normal and expected part of pregnancy. Indeed, nausea of pregnancy often provides the first indication that a woman is actually pregnant. NVP has been of interest over most of recorded history, with the first known description of vomiting in pregnancy coming from a papyrus source dated around 2000 B.C. The earliest explanations for the cause of NVP relied mostly on excessive food intake. Supposedly, this excessive intake resulted in nerve compression, which in turn activated NVP. By the end of the nineteenth century, psychological causes were advanced to explain NVP. It was thought that vomiting of pregnancy was a manifestation of neurosis and that, therefore, vomiting was readily amenable to suggestive treatment to affect a cure. Women with NVP were said to exhibit significant psychopathology of emotion—especially anxiety and tension, depression, and resentment. Positive correlations were found between the severity and duration of symptoms of NVP and the degree of emotional disturbance experienced by the pregnant woman. In addition, it was claimed that NVP was used by pregnant women to avoid work and to avoid or refuse intercourse. NVP was also thought to be the manifestation of a wish that the pregnancy be aborted.

Although a wealth of evidence now exists that implies organic causes for NVP, these psychologically based explanations have remained in vogue until quite recently. An overreliance in the past on psychogenic explanations still detrimentally influences our understanding and perception of NVP today.

Despite popular interest in NVP, little scientific inquiry has been expended in developing an understanding of this complex of symptoms. Until about 1990, no more than approximately 45 papers had been published on NVP, and the majority of these papers were concerned almost solely with hyperemesis. Although several studies have reported some data on prevalence of symptoms of NVP, little information has been presented on time of onset or duration of symptoms.

ENCYCLOPEDIA OF HUMAN BIOLOGY, Second Edition, VOLUME 7. Copyright © 1997 by Academic Press. All rights of reproduction in any form reserved.

In addition, partly because of the nature of when pregnancies are ascertained, most of the descriptive studies have been retrospective, reporting observations late in pregnancy or even well after pregnancy termination.

Over approximately the past five years, a mild renewal of interest in NVP has occurred, especially regarding how NVP may be adaptive in nature by serving as a mechanism for reducing fetal exposure to toxins contained in some foods. Although this idea is not exactly new, its clearest expression is in a 1992 comprehensive article by M. Profet. In addition, meta-analysis of previously published data on NVP has solidified thought regarding the relationship between NVP and pregnancy outcome.

II. PREVALENCE OF NVP IN MODERN POPULATIONS

On the basis of retrospectively obtained data, many investigators have reported that nausea of pregnancy affects between 50 and 70% of pregnant women. In a relatively recent prospective study of 414 pregnant women of high socioeconomic status in Albany, New York, 89.4% reported having nausea and/or vomiting of pregnancy. The total number of women reporting vomiting of pregnancy represented 56.5% of the study population, while 32.9% had nausea alone. Only 10.6% of the women had no symptoms of NVP.

Among women who had nausea, the mean and median week of onset was the sixth week of gestation. Women who had nausea with vomiting did not differ in the mean week of onset of symptoms from women having only nausea. Figure 1 displays the pattern of nausea and vomiting over the course of pregnancy by showing the percentage of women in the entire sample who were experiencing nausea with or without vomiting (dashed line) and vomiting (solid line) at any specific period during gestation.

As seen in Fig. 1, only about 20% of all women had developed symptoms of nausea with or without vomiting by the fourth week of gestation. Thereafter, the number of women developing symptoms of nausea increased sharply, so that by the eighth week nearly 80% of the women in the sample had developed nausea, and 70% were still actually experiencing NVP. By Week 16, 98% of all women who ever developed nausea (about 86% of the total sample) had begun having symptoms. A total of 52% of all women in the study (91.5% of the women who ever developed

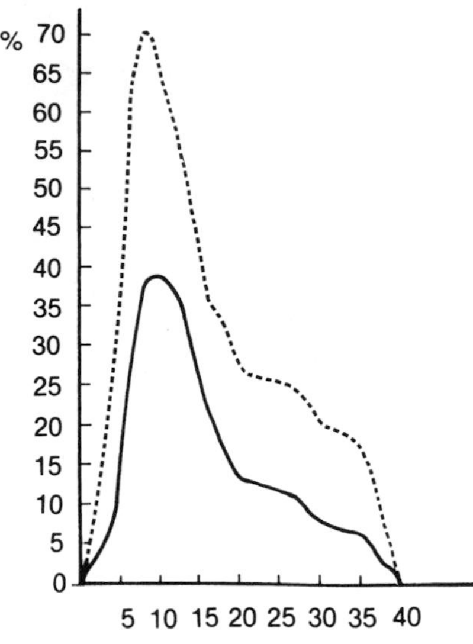

FIGURE 1 Pattern of nausea (dashed line) and vomiting (solid line) of pregnancy. Displayed is the percentage of women in the entire sample of 414 women experiencing symptoms of nausea and vomiting of pregnancy by time of pregnancy when symptoms were experienced. [From F. D. Tierson, C. L. Olsen, and E. B. Hook (1986). Nausea and vomiting of pregnancy and association with pregnancy outcome. *Am. J. Obstet. Gynecol.* **155**(5), 1019. Reproduced with permission of the C. V. Mosby Company.]

vomiting) had developed vomiting by the 12th week—at which time about 35% were still experiencing vomiting of pregnancy.

Of those women who had nausea, 30% had stopped having symptoms by the 12th week. Fifty percent of the women had stopped having symptoms by the 15th week. Even so, by the 20th week of gestation, 25% of the women ever having nausea were still experiencing symptoms. Of those women expressing vomiting of pregnancy, 50% had stopped having symptoms by the 15th week. Nausea persisted longer among women whose nausea was accompanied by vomiting than among women with nausea alone.

The incidence of NVP determined in the Albany study is somewhat higher than that reported by others, although nausea and vomiting are rarely treated as separate symptoms. Much of the difference between this study and previous ones is most likely due to the greater ascertainment of women with only minor symptoms of nausea in this prospective study (where pregnancies were ascertained by the 13th week of gestation). Again, several retrospective studies have reported that about 70% of the women in the study population experienced nausea and vomiting of pregnancy.

III. NVP AND PREGNANCY OUTCOME

Hyperemesis gravidarum, which is usually accompanied by maternal weight loss, has been associated with low-birthweight (small for gestational age) infants owing to the fact that hyperemesis interferes with maternal nutrition and fluid balance. Particularly severe cases can be life-threatening—to both mother and fetus. The effects of "normal" NVP are not quite as dramatic. Initially, women experiencing vomiting of pregnancy gain weight more slowly than women with no NVP, or women with only nausea. However, after their symptoms of vomiting decrease in severity, women who previously experienced vomiting of pregnancy gain weight faster than women who had only nausea and faster than women who had no symptoms of NVP. As a consequence, by about the 25th week of pregnancy, no significant difference exists in maternal weight gain between women with symptoms of NVP and women with no symptoms.

In the Albany study, women without any symptoms of NVP accounted for a significantly larger proportion of the fetal deaths (due to spontaneous abortion or miscarriage). This finding is in line with results obtained from other studies. The generally accepted reason for this association is that a low level of steroidal hormones (progesterone and estrogens) is not enough to induce NVP—but at the same time the low level is not enough to maintain pregnancy and, as a consequence, the pregnancy is spontaneously aborted.

The effects of NVP on pregnancy outcome may also vary with maternal age. In a study of teen-age pregnant women in New Jersey, early NVP alone was not found to be significantly associated with lower infant birthweight, although late NVP was correlated with significantly lower infant birthweight, especially when maternal weight gain was inadequate. A meta-analytical analysis that examined the data available from 11 previous studies confirmed the decreased risk of miscarriage associated with gestational nausea and vomiting. In addition, the analysis indicated that the association of NVP with decreased fetal mortality was restricted to the first 20 weeks of gestation. The fail-safe n for this meta-analysis was greater than 150 (meaning that over 150 additional possibly unreported studies with contradictory evidence would be required to refute the observed association). [*See* Abortion, Spontaneous; Steroids.]

IV. ORIGINS OF NVP

Little information is available that would allow comparison of hormone levels with NVP in early pregnancy since most studies of nausea and vomiting of pregnancy have been concerned with hyperemesis gravidarum. The relationship between levels of steroidal hormones in the third trimester (obtained from studies of hyperemesis) and hormone levels early in pregnancy (especially as they influence NVP) remains poorly understood.

Although the etiology of NVP is still not well defined, a hormonal influence is suspected. Since some women experience nausea similar to NVP with the use of oral contraceptives or when they are on estrogen medication, increased levels of sexual steroids (progesterone and estrogens) during pregnancy are thought to be related to the development of NVP. In addition, NVP and nausea during the use of oral contraceptives are both usually confined to the first few months—either after conception or after first starting to take oral contraceptives. In both cases there is a rapid increase in hormone levels at this time. As mentioned earlier, low levels of steroidal hormones during early pregnancy would not trigger the development of NVP, but the same low levels would also be insufficient for proper pregnancy development and the pregnancy would be aborted.

It has been suggested that women who experience early NVP may have lower functional liver capacities. As a result of the lower capacities, these women would be overly sensitive to estrogens or their metabolites and would be more likely to develop NVP. The functional load on the liver would be similar during early pregnancy and when oral contraceptives were first administered—both conditions when the production of estrogens increases rapidly.

Because there is often a consistent pattern to the daily time, frequency and duration of episodes of NVP, some researchers have suggested that prostaglandin E_2 released from decidual cells and macrophages of the decidera basalis may be the main cause of symptoms of NVP. The symptoms may be suppressed by rising maternal serum levels of progesterone and cortisol, owing to their immunosuppressant properties. Again observing the diurnal and episodic nature of the release of steroid and peptide hormones, other research suggests that digestive dysfunctions during pregnancy are associated with elevated steroid and peptide [β-endorphin, neuropeptide Y (NPY)] hormone interaction with innate biological rhythms in controlling the gastrointestinal tract.

A correlation between increased levels of human chorionic gonadotropin (HCG) and NVP has been reported. Such a correlation would certainly be expected, since levels of HCG increase rapidly in early pregnancy. However, several studies fail to demonstrate any causative connection. There are also suggestions that immunological factors may possibly be involved in the origin of nausea and vomiting of pregnancy, but again the actual nature of this relationship is not clear.

A multitude of changes occur in the maternal system during pregnancy, many of which are intimately interrelated. One problem with establishing the immediate *cause* of NVP is that, although correlations are relatively easy to demonstrate, causal connections are much more difficult owing to the myriad changes taking place. In addition, a solitary cause is unlikely because of the high degree of interconnectedness among changes.

A further problem in establishing causal relationships relates to the confusion of different levels of explanation. An explanation of the origin of NVP at the proximate level (hormonal changes) may be viewed as being in opposition to an explanation at the level of evolutionary significance or purpose (often called the ultimate level of explanation), when in fact they explain the exact same event. In addition, we are often driven to explain the origin of NVP by placing it in an adaptive context. This proclivity for defining behaviors in the context of a selectionist universe fails to take into account that many biological features are the result of stochastic processes. Furthermore, though selection may be operating, it may not be operating in the manner that we suspect to be the case. For example, although NVP may reduce fetal exposure to certain toxins, selection may actually be selecting for earlier fetal maturation, and NVP may be a simple by-product of this selection.

BIBLIOGRAPHY

Behrman, C. A., Hediger, M. L., Scholl, T. O., and Arkangel, C. M. (1990). Nausea and vomiting during teen-age pregnancy: Effects on birthweight. *J. Adolescent Health Care* **11,** 418–422.

Gadsby, R., Barnie-Adshead, A. M., and Jagger, C. (1993). A prospective study of nausea and vomiting during pregnancy. *Br. J. General Practice* **43,** 245–248.

Hill, P. (1990). Pickles, peptide hormones and pregnancy: A hypothesis. *Med. Hypotheses* **32,** 255–259.

Jarnfelt-Samsioe, A. (1987). Nausea and vomiting in pregnancy: A review. *Obstet. Gynecol. Surv.* **42**(7), 422–427.

Jarnfelt-Samsioe, A., Samsioe, G., and Velinder, G-M. (1983). Nausea and vomiting in pregnancy—A contribution to its epidemiology. *Gynecol. Obstet. Invest.* **16,** 221–229.

Kasper, A. S. (1980). Nausea of pregnancy: An historical medical prejudice. *Women and Health* **5**(1), 35–44.

Profet, M. (1992). Pregnancy sickness as adaptation: A deterrent to maternal ingestion of teratogens. *In* "The Adapted Mind: Evolutionary Psychology and the Generation of Culture" (J. H. Barhow, L. Cosmides, and J. Tooby, eds.), Chap. 8, pp. 327–365. Oxford Univ. Press, New York.

Tierson, F. D., Olsen, C. L., and Hook, E. B. (1986/1989). Nausea and vomiting of pregnancy and association with pregnancy outcome. *Am. J. Obstet. Gynecol.* **155,** 1017–1022; **160,** 518–519.

Weigel, M. M., and Weigel, R. M. (1989). Nausea and vomiting of early pregnancy and pregnancy outcome. An epidemiological study. *Br. J. Obstet. Gynaecol.* **96,** 1304–1311.

Weigel, R. M., and Weigel, M. M. (1989). Nausea and vomiting of early pregnancy and pregnancy outcome. A meta-analytical review. *Br. J. Obstet. Gynaecol.* **96,** 1312–1318.

Prenatal Diagnosis

STUART K. SHAPIRA
Baylor College of Medicine

GLOSSARY

Aminocytes Desquamated fetal cells within the amniotic fluid

Amniotic fluid Fluid contained within the amniotic sac that surrounds the fetus, predominantly consisting of fetal urine

Chorion Outer layers of cells enclosing the developing embryo, of which a portion will develop into the fetal contribution of the placenta

Congenital anomaly A structural feature present from birth that is different from "normal"

Dysmorphic A description of a body part that has not followed the normal pattern of growth or formation, and is often disproportionate when compared to the normal body part

Intrauterine Within the uterus

Prenatal Before birth

Preterm Before the usual delivery time, as in preterm labor.

Teratogen An agent which acts during pregnancy to cause or influence the development of anomalies in the fetus; these can include viruses, drugs, chemicals, occupational or environmental exposures, and maternal conditions

Transabdominal Through the abdominal wall

Transcervical Through the cervical canal (lower most portion of the uterus)

PRENATAL DIAGNOSIS ENCOMPASSES MORE THAN just the technology to identify congenital developmental abnormalities and serious and/or lethal genetic conditions in a fetus during a pregnancy. It is also utilized to identify "normal" pregnancies in couples that are at risk for having a child with a serious congenital anomaly or genetic condition so that they can be reassured about the outcome of a pregnancy. Prenatal diagnostic investigations and screening cannot be separated from the sensitive issue of reproductive choices. The goal of prenatal diagnostic services is to provide nondirective information to couples about their risk of having an abnormal fetus, the tests that would be available to diagnose an abnormality, the limitations of the testing, and the risks associated with each test. Prenatal diagnosis also includes supportive and genetic counseling for couples faced with difficult reproductive decisions.

I. GENETIC DISORDERS AND CONGENITAL ANOMALIES

Genetic diseases and congenital abnormalities are not rare occurrences. Approximately 3% of all infants are born with a major congenital anomaly, the majority of which have a genetic basis. Additionally, approximately 7% of all individuals will manifest symptoms of a genetic disorder during childhood or adolescence. A large proportion of genetic diseases, specifically those in which congenital anomalies are present, can be recognized during fetal life by prenatal ultrasonography or can be identified in the newborn period. The causes of various types of congenital anomalies are summarized in Table I.

Although many congenital anomalies have a primary genetic basis, it is important to differentiate these conditions from anomalies caused by fetal exposure to maternal factors. Such factors which cause congenital anomalies and/or dysmorphic features via teratogenic effects include several types of viral infec-

ENCYCLOPEDIA OF HUMAN BIOLOGY, Second Edition, VOLUME 7. Copyright © 1997 by Academic Press. All rights of reproduction in any form reserved.

TABLE I
Causes of Congenital Anomalies

Chromosomal abnormalities	6–10%
Single gene defects	3–7.5%
Multifactorial inheritance (polygenic and environmental factors)	20–30%
Environmental factors (maternal conditions and teratogens)	4–5%
Unknown causes	50%

TABLE II
Common Prenatally Diagnosed Disorders

Chromosomal abnormalities
Submicroscopic chromosomal deletions or duplications
Congenital abnormalities
Spina bifida, other neural tube defects
Many congenital heart defects
Abnormal kidney development
Body wall defects
Limb reduction defects
Abnormal head size and/or brain development
Skeletal development disorders
Metabolic disorders
Galactosemia
Amino acid and organic acid disorders
Storage disorders
Mitochondrial and peroxisomal disorders
Single gene defects
Cystic fibrosis
Fragile X syndrome
Duchenne/Becker muscular dystrophy and myotonic dystrophy
Sickle cell disease and other hemoglobin disorders
Hemophilia
Congenital infections

tion during pregnancy, some maternal medical conditions, and maternal exposure to alcohol, drugs, and certain medications. These causes of congenital anomalies would not be considered genetic, but rather "environmental." Based on numerous observations and studies, there is no doubt that fetal development can be severely altered by many factors which perturb the normal intrauterine environment.

In contrast to environmental and teratogenic effects, genetic causes of congenital anomalies are far more prevalent and include numerous chromosomal abnormalities (abnormal number or structure of chromosomes) and deletions or duplications of small chromosomal segments not discernible by routine chromosome analysis (submicroscopic deletions or duplications). Additionally, many genetic conditions are caused by mutations within single genes or by mutations in several genes (polygenic inheritance). Therefore, a fetus or newborn with congenital anomalies likely occurred as a result of genetic factors, which warrants appropriate genetic counseling for the family. [*See* Genetic Diseases; Genetic Counseling.]

II. PRENATAL DIAGNOSIS: UTILIZATION

When there is a genetic problem in a family or if a child is born with a congenital anomaly, three questions are generally expressed: "What is the problem?" "Why did it occur?" "What is the chance of it happening again?" In addition, couples often ask what can be done to avoid a recurrence or to test for the possibility. Couples that have had one child with a genetic condition and couples who are "at risk" for having children with genetic problems by virtue of a family history or an abnormal screening test result often consider utilizing prenatal diagnosis. Prenatal diagnosis of congenital anomalies or genetic conditions has become aroutine part of comprehensive prenatal care. It is also recognized that physicians need to discuss prenatal diagnostic options with couples because failure to do so can have medicolegal implications. Several hundred fetal conditions may be identified with prenatal diagnostic techniques; a few examples are listed in Table II.

III. TECHNOLOGY

A. Ultrasonography and Visual Examination

Ultrasonography is a valuable imaging technique that is available in almost every maternity hospital and obstetrics clinic in developed countries. However, the highest quality or high-resolution ultrasound machine can be quite complicated, requiring operators to be very experienced with the function of the machine in order to provide the best possible diagnostic services. Ultrasonography is a safe procedure and has not caused any documented physical danger to the fetus or its mother. [*See* Ultrasound Propagation in Tissue.]

Ultrasound is defined as sound waves that are reflected off internal structures back to the machine, which reconstructs an image of the objects being viewed. Ultrasound utilizes sound frequencies above the audible limit of 0.02 megahertz (MHz). Obstetric ultrasound equipment utilizes frequencies between 2.0 and 7.5 MHz, with lower frequencies allowing greater penetration through pelvic, uterine, and fetal structures, whereas higher frequencies permit better resolution of the scanned structures. An experienced ultrasound operator chooses the frequency that gives optimal views of the fetus, given each particular situation. A good ultrasound machine should be able to show detail in structures that are as small as 5 mm.

Congenital anomalies occur in 3–4% of all infants, and lethal anomalies account for 25% of all perinatal deaths. Prenatal ultrasonography can detect many fetal anomalies, of which 90% occur in couples that have no recognizable risk factors. Examination of the fetus for congenital anomalies with high-resolution ultrasound equipment by an experienced fetal ultrasonographer can often provide valuable prenatal diagnostic information about the well-being of the fetus and the fetal environment. In contrast, prenatal diagnosis of fetal anomalies by ultrasonography can permit informed decisions to be made regarding management of the pregnancy, delivery, and neonatal care. Other fetal imaging studies, such as fetal skeletal X-rays, fetal magnetic resonance imaging, and transcervical fetoscopy (direct visual examination of the fetus utilizing specialized endoscopic equipment) have also been utilized to provide prenatal evaluation of congenital anomalies.

B. Fetal Cell and Fluid Testing

Fetal tissue sampling for chromosome analysis, metabolic testing, gene mutation testing, assessment of congenital infection, or testing for blood disorders has become routine for providing prenatal diagnostic information. Fetal tissue sampling is an invasive technique that carries certain risks to the fetus, as well as to the mother. However, fetal tissue sampling techniques have become common obstetrical practice, particularly when the risk to the pregnancy is outweighed by the benefit of detecting an abnormality so that informed management decisions can be made. The practice of fetal tissue sampling involves (1) transabdominal amniocentesis beyond 12 weeks of pregnancy, (2) obtaining placental tissue by chorionic villus sampling (CVS) at 9–10 weeks of pregnancy, (3) aspiration of fetal blood by percutaneous umbilical blood sampling, or (4) fetal tissue biopsy. [*See* Genetic Testing.]

1. Amniocentesis

Amnioentesis was initially applied in the 19th century for the management of pregnancies with polyhydramnios (too much amniotic fluid). The technology has now become routine for obtaining samples of amniotic fluid for biochemical testing, as well as for obtaining amniocytes that can be grown and analyzed for their chromosomal complement, specific gene mutations, and/or metabolic aberrations. Most often the procedure is performed using ultrasound guidance, usually during the second trimester of pregnancy or beyond. Under aseptic conditions, a needle, usually no larger than 20 gauge, is inserted through the mother's abdominal wall and the uterus into an accessible pocket of amniotic fluid. No more than 10% of the total amniotic fluid volume is removed, and the first few milliliters are discarded in order to minimize the risk that the sample is contaminated with maternal cells.

2. Chorionic Villus Sampling

The technique in widest practice today for obtaining chorionic villus specimens for prenatal diagnosis was pioneered in the early 1980s. Because the placental and fetal tissues are both derived from cell divisions of the fertilized egg, sampling the chorionic villi of the future placenta can give prenatal diagnostic information about the fetus. During the latter part of the first trimester of pregnancy (9–12 weeks), the amniotic cavity does not yet fill the uterine cavity, making it possible to insert an instrument through the cervix into the developing placenta for obtaining a chorionic sample. The chorion has begun to differentiate into the chorion frondosum, which will become the placental site, and the frondosum contains the actively dividing villus cells which are ideal for tissue biopsy. An instrument (plastic catheter threaded over a blunt malleable metal obturator) is inserted through the cervix into the lower border of the chorion frondosum under ultrasound guidance. The metal obturator is withdrawn and a syringe is placed at the operator's end of the catheter; suction is applied to obtain villi. Transabdominal ultrasound-guided needle insertions for CVS procedures have also been performed since the mid-1980s and may be the more common approach today, but the method chosen for the CVS procedure is largely influenced by the location of the placenta in the uterus. Although the tissue samples obtained transabdominally are, on the average, slightly smaller than those obtained transcervically, the samples are uniformly adequate for routine prenatal testing.

3. Fetal Blood Sampling

The technique used to obtain a sample of pure fetal blood has allowed the prenatal diagnosis of hemoglobin disorders, bleeding disorders (coagulopathies), platelet and metabolic disorders, immunodeficiencies, and fetal infections. In addition, the technique has allowed for direct fetal blood transfusion therapy to correct fetal anemia. The ultrasound-guided transabdominal approach for fetal blood sampling, developed in the early 1980s, is now widely practiced, with the umbilical cord being the commonest site of sampling. However, because maintaining the target (usually the site that the umbilical cord inserts into the placenta) while advancing the needle in the same ultrasound view can be a tricky task, only experienced physicians, often with a well-trained assistant or ultrasonographer, should perform the procedure. No serious maternal complications have been reported, but risks of complications for the fetus [infection, leaking of amniotic fluid, bleeding, fetal blood clotting (thrombosis), fetal death, and miscarriage] can be as high as 2%.

4. Fetal Tissue Biopsy

Various fetal tissues have been sampled by the transcervical or transabdominal insertion of instruments in order to obtain biopsy material for making a specific diagnosis of a genetic or developmental abnormality. Fetal skin has been sampled in order to diagnose congenital skin disorders by histological and ultrastructural studies. Fetal liver samples have been obtained to diagnose some rare inborn errors of metabolism that cannot be diagnosed from chorionic villi, amniocytes, or fetal blood. Fetal muscle biopsies have been helpful in diagnosing certain muscular dystrophies. Fetal lung or kidney biopsies have been utilized to clarify genetic or developmental abnormalities in these structures. Techniques and indications for prenatal diagnosis by fetal tissue sampling have changed rapidly, and this evolution will likely continue.

C. Maternal Serum Testing

Measurement of maternal serum α-fetoprotein (MSAFP) has previously been recommended as a routine prenatal test. An elevated MSAFP level can indicate an open neural tube defect (such as spina bifida or anencephaly), an abdominal wall defect, or other serious birth defects. In the absence of fetal anomalies, an elevated MSAFP level may be due to placental anomalies, such as cysts, fetal bleeding into the maternal circulation via the placenta, fetal growth problems, or a fetal death. A low MSAFP level indicates an increased risk for Down syndrome or perhaps other chromosomal abnormalities.

Advances in maternal serum testing have added human chorionic gonadotrophin (hCG) and unconjugated estriol (uE3) to MSAFP testing, resulting in a "triple screen" for levels of all three analytes. Women carrying Down syndrome pregnancies more likely exhibit low levels of MSAFP, but will more likely also have low uE3 and elevated hCG levels. The levels of the three analytes are used to calculate a relative risk that the fetus will be affected with Down syndrome. Levels of these analytes may also show specific elevations or depressions in pregnancies with other chromosomal abnormalities, although there is not the same degree of sensitivity and specificity as in Down syndrome pregnancies. Multiple factors can alter the analyte levels and affect the relative risk calculation. Such factors include the age of the pregnancy, maternal smoking, obesity, racial origins, and multiple gestations (twins, triplets, etc.); in fact, the test is considered unreliable when multiple gestations exist. This technology is best designed to screen between 16 and 18 weeks of pregnancy for neural tube defects, abdominal wall defects, and Down syndrome. [*See* Down Syndrome, Molecular Genetics.]

D. Complications and Limitations

Common indications for prenatal diagnostic evaluation are listed in Table III. Of course the use of amniocentesis or CVS poses risks to the pregnancy; the complication rate (infection, hemorrhage, cramping, preterm labor, fetal trauma, miscarriage) with these procedures is 1 : 100 to 1 : 400. Other problems include failure of the fetal cell culture to grow or con-

TABLE III
Indications for Prenatal Diagnosis

Advanced maternal age (≥35 years)
Previous child with a chromosomal abnormality
Previous child with a major congenital abnormality
Previous child with certain skeletal disorders
Previous child with a biochemical genetic disroder
Previous child with certain single gene disorders
High risk for certain genetic or biochemical disorders
Balanced chromosomal rearrangement in a parent
Abnormal maternal serum α-fetoprotein level (high or low)
Exposure to teratogens

tamination of the fetal cells with maternal cells; in either case, a repeat amniocentesis procedure may be required. Unfortunately, normal results on fetal ultrasonography, a fetal chromosome analysis, or a "triple screen" do not completely eliminate the possibility that the infant will have a genetic condition. However, normal screening test results can be quite reassuring to the couple and the physician. If a genetic condition or particular congenital anomalies are diagnosed prenatally, the family and physician have the option to modify the care of the fetus or newborn, to be psychologically and medically prepared for the birth of the affected child, and to discuss and possibly implement the option of pregnancy termination.

IV. ALTERNATIVES AND CHOICES

Couples at high risk for having infants with genetic conditions or congenital anomalies may decline prenatal diagnosis on moral, individual, or financial grounds. However, they may desire to decrease their risk of having an affected child by utilizing other available reproductive options. Although some couples decide to "take their chances," all couples need to be informed of appropriate alternatives, including artificial insemination by donor sperm, surrogate motherhood, and *in vitro* fertilization with a donor egg, as well as the options of adoption, contraception, and sterilization. [*See* In Vitro Fertilization.]

Advances have also been made in diagnosing certain genetic conditions before implantation. With this technique, a woman's eggs are fertilized *in vitro* (in a culture dish), and once reaching the eight-cell blastocyst stage, a single blastomere is removed and studied for the presence or absence of certain genetic conditions. Only unaffected blastocysts are subsequently implanted in the mother. This technology has been successfully used for the preimplantation diagnosis of embryos that are unaffected with cystic fibrosis, Tay-Sachs disease, and Lesch-Nyhan syndrome, and could potentially be used for prenatal testing for numerous single gene disorders as well as certain chromosomal abnormalities. However, this technology is new and expensive, and only a low proportion of women undergoing the procedure actually achieve a viable pregnancy.

BIBLIOGRAPHY

Brock, D. J. H., Rodeck, C. H., and Ferguson-Smith, M. A. (eds.) (1992). "Prenatal Diagnosis and Screening." Churchill Livingstone, New York.

Rimoin, D. L., Connor, J. M., and Pyeritz, R. E. (eds.) (1996). "Emery and Rimoin's Principles and Practice of Medical Genetics," 3rd Ed. Churchill Livingstone, New York.

Stevenson, R. E., Hall, J. G., and Goodman, R. M. (eds.) (1993). "Human Malformations and Related Anomalies." Oxford University Press, New York.

Trent, R. J., and Trent, R. J. (eds.) (1995). "Handbook of Prenatal Diagnosis." Cambridge University Press, Cambridge, UK.

Primate Behavioral Ecology

PAUL A. GARBER
University of Illinois

GLOSSARY

Behavioral ecology Study of relations among ecological factors, social organization, and mating systems in animal species

Behavioral strategies Alternative patterns of behavior that influence reproductive success, access to resources, and/or social position

Female-bonded group Social groups composed mainly of female matrilines and unrelated males. In these species, females generally remain in the group in which they are born, males emigrate from their natal group, and kinship appears to play a significant role in female social interactions

Food patch The spatial and temporal clumping of food items; generally defined as the crown of a tree or an area in which food resources are distributed such that an animal can continue to feed without having to switch to a new food type. The size of the food patch sets some upper limit on the number of animals that can feed simultaneously

Mating system Patterns of sexual interactions, mate preferences, and sexual competition among reproductively active individuals residing in a social group. The mating system differs from the social system in that not all adults within a group may be reproductively active

Mental map An internal geometric representation of the angles and distances between a large number of feeding sites or landmarks in an animal's home range. Animals that encode spatial information in the form of an internal geometric representation can compute novel and direct routes of travel between known locations

Phylogeny Evolutionary history of a species or taxonomic group

Primate community Species that live in the same environment, overlap in their use of resources, and interact such that the activities of each species are likely to have some effect on the activities of the other species

Social group Aggregation of individuals that are spatially cohesive, interact regularly, form social bonds, and travel together. Individuals within a social group tend to have more frequent interactions with each other than individuals from different social groups

THE ORDER PRIMATES REPRESENTS A DIVERSIFIED group of some 200 species of prosimians, monkeys, apes, and humans that are generally characterized by a single infant at birth, a relatively long period of infant dependence, and complex behavioral and social interactions. With few exceptions, primates live in social groups composed of individuals of all ages and both sexes. The structure and form of these groups vary considerably, however, and recent studies in the behavior and ecology of nonhuman primates have identified a number of primary factors that influence the size, composition, and cohesiveness of social groups, including (1) the threat of predation, (2) food distribution and availability, (3) ranging patterns and use of spatial information, (4) kinship, (5) infant care, and (6) intrasexual aggression and mating patterns.

The study of primate *behavioral ecology* has centered on the comparative approach. Comparisons of the *social organization* and feeding ecology of closely related species living in the same ecological community indicate that although phylogeny and body size constrain the ways in which a species responds to changes in the environment, interspecific differences

ENCYCLOPEDIA OF HUMAN BIOLOGY, Second Edition, VOLUME 7. Copyright © 1997 by Academic Press. All rights of reproduction in any form reserved.

in social organization are often explained in terms of subtle differences in diet, food quality, and the distribution of preferred food resources. It has generally been assumed that even among members of the same social group, individual competition for access to important feeding sites tends to limit the size of the social unit. When resources are scarce or distributed in small scattered feeding sites, within-group *feeding competition* can result in a decrease in group size, a decrease in group cohesion, the formation of foraging subgroups, or a reduction in the health and nutritional status of low-ranking group members. In contrast, larger groups may be favored under conditions of high food density or in areas where increased vigilance and opportunities for predator detection play a significant role in reducing mortality.

Factors that are important in understanding the behavioral ecology of nonhuman primates are also important in understanding the behavior of human primates. Relations among behavioral, ecological, and social patterns found in our closest living relatives provide explanatory models for identifying ancestral and derived character traits and adaptations that have shaped human biology and evolution. This article examines and compares the behavioral ecology of a select group of primate species in an attempt to focus on the ways in which group size and composition influence individual behavioral strategies and reproductive success. Table I lists the scientific names of primate species discussed in this article.

TABLE I
Common and Scientific Names of Primate Species Discussed in this Article

Common name	Scientific name
Black spider monkey	*Ateles paniscus*
Bonnet macaque	*Macaca radiata*
Brown capuchin monkey	*Cebus apella*
Chacma baboon	*Papio ursinus*
Chimpanzee	*Pan troglodytes*
Costa Rican spider monkey	*Ateles geoffroyi*
Dusky titi monkey	*Callicebus moloch*
Gelada baboon	*Theropithecus gelada*
Hamadryas baboon	*Papio hamadryas*
Hanuman langur	*Presbytis entellus*
Howler monkey	*Alouatta palliata*
Humans	*Homo sapiens*
Indri	*Indri indri*
Japanese macaque	*Macaca fuscata*
Lar gibbon	*Hylobates lar*
Long-tailed macaque	*Macaca fascicularis*
Mountain gorilla	*Gorilla gorilla*
Moustached tamarin	*Saguinus mystax*
Night monkey	*Aotus trivirgatus*
Olive baboon	*Papio cynocephalus anubis*
Orangutan	*Pongo pygmaeus*
Pigtailed macaque	*Macaca nemistrina*
Red colobus	*Colobus badius*
Red howler monkey	*Alouatta seniculus*
Rhesus macaque	*Macaca mulatta*
Ring-tailed lemur	*Lemur catta*
Saddle-back tamarin	*Saguinus fuscicollis*
Siamang	*Hylobates syndactylus*
Squirrel monkey	*Saimiri sciureus*
Talapoin monkey	*Miopithecus talapoin*
Vervet monkey	*Cercopithecus aethiops*
Woolly spider monkey	*Brachyteles arachnoides*
Yellow baboon	*Papio cynocephalus cynocephalus*
Yellow-handed titi monkey	*Callicebus torquatus*

I. PREDATION

In addressing questions concerning the advantages of social group living, it has been suggested that the threat of predation is a major factor promoting social cohesion, group stability, and increased group size in diurnal primates. For small-bodied arboreal monkeys, raptors such as hawks and eagles are likely to pose the greatest predatory threat. Defense against aerial predators commonly involves concealment, vigilance, a cessation of movement, or flight. In the case of larger-bodied and more terrestrial primates, mammalian carnivores present the greatest danger. Antipredator tactics against these threats include vigilance, retreat to an arboreal haven, frequent alarm calls, and a variety of branch-shaking and arboreal threat displays. If group size is a direct response to predator activity, associated patterns of behavior (both protective and avoidance) and grouping should exist in areas of high predator density that are not commonly observed in areas of lower predator pressure. For example, although the night monkey exhibits a noctural activity pattern throughout most of its range, this primate has been observed to travel and forage frequently during the day in areas where large predatory owls are common. Similarly, in areas free of predators, some populations of long-tailed macaques live in smaller social groups than populations in areas where the threat of predation is greater. However, in the absence of systematic information on predation or predatory attempts on these species, conclusions regarding the relation between group size and group protection must be viewed with caution.

Many species of primates exhibit characteristic alarm calls and behavioral responses that are specific to particular classes of predators. The existence of these calls suggests that predators have exerted a strong influence on at least some aspects of primate behavior. In a series of taped playback experiments, for example, it was determined that vervet monkeys display acoustically distinct alarm calls for six different types of predators. In this species there is evidence that females give alarm calls more frequently in the presence of kin than in the presence of nonkin. However, because high-ranking animals call more frequently than low-ranking animals, it is difficult to separate the effects of kinship, social status, and spatial position on calling behavior.

Data on a population of ground-living yellow baboons in Amboseli, Kenya, indicate that at least 25% of all the known or presumed deaths occurring during a 14-month period were the result of predatory attacks. Although there have been reports that adult male baboons may defend the group from predators, such behavior rarely occurs and its effect on individual survivorship is unknown. In contrast, red colobus males have been observed to defensively attack and chase chimpanzees attempting to prey on group members. In many cases these males were effective in reducing the hunting success of the chimpanzees. Although other factors such as injury, disease, and starvation may contribute more to mortality than predation, the threat of predation appears to have a significant effect on behavior and the development of antipredator tactics.

If group size is a primary response to predator pressure, primate species living in the same ecological community are expected to have groups of similar overall size. This, however, is often not the case. For example, in a community of 10 species of neotropical primates in Peru, group sizes appear to reflect more closely the breeding system and feeding behavior of each species than the degree of predator pressure. Species exhibiting a *monogamous mating pattern* had the smallest group size, followed by *polyandrous* and *polygynous* species. Although body size is an important factor in vulnerability to predators, small-bodied species were characterized by both the smallest and the largest group sizes. In this and many other primate communities, the availability of resources during food-limited periods of the year may have a more direct effect on group size than does predation.

II. FOOD DISTRIBUTION AND AVAILABILITY

The abundance, availability, and distribution of food resources exert a strong influence on the structure and size of primate foraging groups. A number of hypotheses have been proposed to explain the adaptive basis of group size and *foraging behavior*. In species for which the costs of foraging increase rapidly with additional group size, smaller groups are likely to exploit their environment more efficiently than larger groups. When resources are distributed in small, scattered patches, individuals living in larger groups may be forced to travel greater distances, visit more feeding sites, and compete more aggressively over limited food rewards than members of the same species living in smaller groups. However, when exploiting highly productive, clumped, or rapidly renewing feeding sites, individuals foraging together in a larger social unit may obtain feeding advantages by cooperative hunting (e.g., chimpanzees and humans) or collectively disturbing concealed prey (squirrel monkey), traveling together to minimize return times to exploited food patches (e.g., mixed species troops of tamarins), or by successfully defending major feeding sites from smaller neighboring groups (e.g., female-bonded groups of Old World monkeys). Although many researchers have assumed that individuals in smaller groups have lower net *foraging costs* than individuals in larger groups, this is likely to depend on the number and distribution of both large and small food patches as well as on the costs of *subgroup formation*. In species characterized by flexible subgrouping patterns, individuals may avoid competition by temporarily separating from the main group and exploiting smaller resource patches. This foraging pattern requires high levels of social communication, an ability to quickly reestablish social bonds, and an exchange of information about resource availability in the environment. Among nonhuman primates, such extreme flexibility in foraging behavior is reported in chimpanzees, black spider monkeys, woolly spider monkeys, and some populations of long-tailed macaques.

If subgroup formation is a general response to resource scarcity and patchiness, intragroup feeding competiton can have an important influence on group stability and individual reproductive success. Among chacma baboons, within-group feeding competition for limited resources can result in loss of body weight, inadequate nutrition, and death for adult females and infants. This reflects the ability of larger and more aggressive adult and subadult males to displace females and young from feeding sites. Similarly, in the brown capuchin monkey the aggressive behavior of the dominant male is the most important factor affecting the feeding success of other group members. Subordinates avoid competing directly with higher ranking individuals by foraging independently and at lower-quality feeding sites. When forced to feed apart from the main group, low-ranking group members suffer a decrease in energy intake as well as a higher risk of predation. In this brown capuchin population, the high cost of subgroup formation and the ability of dominant animals to control access to high-quality resources limited the size of the social group to between 5 and 14 individuals.

Given their large body size, substantial daily food requirements, and absence of major predators (except humans), the great apes (gorilla, chimpanzee, and orangutan) are expected to forage in small groups. Although small foraging parties (1–5 animals) are common in orangutans and chimpanzees, the heaviest of all living primates, the mountain gorilla, lives and forages in stable, cohesive social groups of 10–20 animals. This apparent anomaly is explained by the fact that these African apes exploit superabundant patches of terrestrial herbaceous vegetation. The plant species eaten by gorillas are among the most common in their habitat. These large patches can accommodate larger feeding groups, and aggressive feeding competition between group members is infrequent.

III. USE OF SPATIAL INFORMATION IN FORAGING

Free-ranging primates face complex problems associated with locating and traveling to feeding sites that are ephemeral and exhibit a scattered and patchy distribution within their home range. In many species, foraging efficiency may be more dependent on behaviors and morphologies that aid in prey location (perceptual abilities, complex spatial learning skills, memory of temporal information associated with production schedules, and rates of resource renewal in preferred feeding patches) than adaptations associated with prey capture (strength, speed, large body size). For primates exploiting relatively open habitats, sighting to distant landmarks may serve a primary function in spatial orientation and navigation. Anthropoid primates are characterized by excellent visual acuity, stereopsis, and color discrimination. Evolutionary changes in the primate visual system are likely to have had an important effect on perceptual cues used in foraging, social communication, predator detection, and navigation. For example, field observations of yellow baboons and hamadryas baboons indicate that troops frequently move along habitual and preferred routes of travel and appear to use major topographical features of their habitat as landmarks for orientation in travel. Features of the habitat such as sleeping cliffs, water sources, groves of shade trees, and large patches of grasses and corms serve as a focus of troop activity.

In the case of rain forest primates, however, physical constraints imposed by the denseness of the forest canopy may limit the effectiveness of landmarks as sighting cues to a distance of some 10–15 m. Given the patchy distribution of many tropical tree and liana species, an ability to retain detailed information on the location and availability of seasonal food sources is likely to be a significant factor in foraging success.

Reports from several field studies offer evidence of a high degree of spatial learning in rain forest primates. In most instances, however, the evidence is correlational and based on observations of goal-directed behavior in which an animal travels in a relatively straight line to feeding sites that lie distant or far outside its field of view. Although goal-directed behavior and straight-line travel suggest that a forager has specific expectations regarding where it is heading, a memory and sense of direction from where it has traveled, and temporal information of when it was there last, there is little empirical information on how primates encode spatial information and use perceptual cues to navigate between feeding sites.

A review of patterns of habitat exploitation in rain forest primates indicates that some arboreal species are best described as "central place foragers," i.e., they exhibit a ranging pattern in which travel to feeding sites emanates from either a centrally located sleeping area or a small set of frequently revisited trees. Over the course of several days or weeks the same feeding sites and travel paths are reused consistently. As the location and availability of resources change over time, a new set of target trees and arboreal pathways become the focus of daily activity. A central place foraging pattern has been reported in howler monkeys and indris. Other primate species exhibit a more flexible ranging pattern in which travel and foraging are oriented to multiple sites located throughout the home range (multiple central place foraging). This type of foraging pattern may reduce overall travel costs by allowing a forager to switch from one to several previously visited refuge, sleeping, and feeding sites depending on its present location in the forest. In the case of the Costa Rican spider monkey, a multiple central place foraging pattern eliminates long return trips to a single refuge site and is likely to be of greatest advantage during the period of the year when resources are scarce. Although it may be hypothesized that primates that use a wider range of feeding sites and travel routes maintain more detailed knowledge of the spatial location of rare or ephemeral feeding sites, it remains unclear whether different primate species encode spatial information in the form of route based (memory of a small set of reused and interconnecting pathways) or geometrically based (knowledge of angles and distances between out-of-view feeding sites, and an ability to calculate novel paths of travel) memory. It has been found that moustached and saddle-back tamarins exhibited a foraging pattern in which preferred feeding sites were reached using a variety of novel travel routes and from a variety of different directions. Based on a sample of the top 15 fruiting trees in their diet, it was determined that 78% of travel routes used by the tamarins to reach these sites were novel or nonrepetitive. These data offer indirect evidence that these primates maintain an internal geometric representation (mental map) of the angles and distances between a large number of feeding sites in their range.

Field observations of chimpanzee transporting stone and wooden tools to nutting sites offer additional correlational evidence of use of geometric spatial information. Chimps in the Tai National Park, Ivory Coast, crack open hard nuts using clubs and stone hammers. Transport of hammer stones to *Panda oleosa* trees occurs frequently. These trees were rare and widely scattered along a 5-km^2 area of the forest. Forty percent of the hammer stones used by the chimps were transported a distance of between 20 and 500 meters. In these cases, hammer stones retrieved by chimps were not visible from the location of the feeding tree. An analysis of the stones transported greater than 20 m to a nutting tree indicates that in 63% of cases the nearest stone to the nutting tree was selected. Based on the distribution of stones in the area, the probability of choosing the nearest stone by chance alone was only 26%. The chimps performance in selecting nutting stones involved an ability to (1) measure and conserve distance, (2) compare distances for several alternative stone selections, (3) remember the locations of newly placed stones relative to other stones, and (4) mentally rotate the relationship between the distances and the locations of individual stones to individual trees and the present location of the animal. Although additional testing and field experimentation are required to determine the manner in which nonhuman primates store, represent, and use spatial information, it is possible that many species maintain a geometric representation of large numbers of landmarks in the environment and can compute novel and direct travel routes to reach feeding sites.

IV. KINSHIP

Evidence among several species of primates suggests that patterns of cooperative behavior such as coalition formation, hunting, food sharing, resource defense, and assistance in caring for young occur mainly between family members and close kin. However, because precise information on the genetic relatedness of group members is generally unknown or restricted to individuals related through the female line, attempts to examine patterns of kin recognition and kin-associated behavior in primates have primarily focused on matrilineal relations. In Japanese macaques, for example, female relatives are found to groom, rest, feed, and travel together more frequently than expected based on the number of potential social partners present in the group. Similar mother–mature-offspring associations and sister associations have been documented among female rhesus macaques, bonnet macaques, pig-tailed macaques, olive baboons, gelada baboons, Hanuman langurs, vervet monkeys, squirrel monkeys, ring-tailed lemurs, and chimpanzees. In many cases, females of one matriline socially dominate females of a second matriline. During agonistic encounters or at limited feeding sites, support provided by matrilineal kin may play a key role in determining priority access to resources.

Among male primates, there is only limited evidence of patrilineal-based kin affiliations. This may reflect the general mammalian pattern of intense intrasexual competition among reproductively active males or parental uncertainty through polygynous matings, as well as the fact that dispersal patterns result in social groups that are largely composed of a single or unrelated set of adult males in many primate species. In species in which related males are reported to reside in the same group (e.g., the gorilla, red howler monkey, and chimpanzee), male coalitions appear to play a critical role in preventing aggressive and infanticidal attacks by nonresident males, as well as maintaining the integrity of the breeding unit.

V. INFANT CARE

Considerable evidence shows that matrilineal relatives directly contribute to the care and protection of infants in female-bonded primate societies. However, with few exceptions, information on male–infant interactions in primate species has failed to identify a direct link between probable paternity and caregiving behavior. In the case of olive baboons, for example, protective and affiliative interactions between adult males and infants are often better explained in terms of the maintenance of a social bond between the male and the infant's mother than by kin-related behavior. In this species, recent immigrant males spend more time grooming and interacting with infants than do resident males who are probable fathers.

Among Old World primates in which paternity is more certain, extreme variation exists in the nature and frequency of male–infant interactions. In species exhibiting a harem-like mating pattern (harem polygyny), fathers tend to interact minimally with their offspring. Although it is difficult to quantify the importance of vigilance behavior in protecting young, at present there is no evidence that the single breeding male in a harem group invests more in infant care than do reproductively active adult males in multimale polygynous breeding groups.

In New World species characterized by a nuclear family social group, a single infant at birth, and a pair-bonded mating pattern (e.g., dusky titi monkey, yellow-handed titi monkey, night monkey), fathers substantially contribute to infant care. In these primates, the adult male is the primary infant caretaker and devotes considerable time and effort in carrying and sharing food with the young. Among monogamous lesser apes, however, patterns of male caregiving are more viable. Whereas siamang males may spend up to 78% of the day transporting and grooming a single young, lar gibbon males provide little direct infant care. Thus, despite the fact that paternal certainty can be an important factor influencing infant care, that alone is not a sufficient condition from which to predict male caregiving behavior. It is likely that additional factors such as the relation between nonmaternal care and *infant survivorship,* the cost of reproduction to the female, and the role of male–infant alliances in establishing a *social–sexual bond* between the male and the infants mother have contributed importantly to the evolution of male caregiving behavior.

Recent information on the social and mating patterns of tamarin and marmoset monkeys provides additional support for a critical relation between the high cost of reproduction to the female and patterns of male caregiving behavior. Tamarins and marmosets are unusual among higher primates in their production of twin offspring, a polyandrous mating system in some species, the fact that females can produce two litters per year, and the degree to which adult males cooperatively participate in infant care. Studies on both captive and free-ranging groups indicate that although virtually all group members assist in caring for the young, adult males are the principal caretakers. In the case of moustached and saddle-back tamarins, groups commonly contain two or three adult males. Evidence shows that the number of adult male helpers residing in groups is positively correlated with infant survivorship. Although it remains uncertain whether adult helpers are related to the breeders and act to increase the kinship component of their reproductive success or are unrelated and act to increase the individual component of their reproductive success, the size and structure of tamarin social groups are strongly influenced by the high costs of reproduction to the breeding female and the need for nonmaternal infant care. In the absence of at least two adult nonmaternal helpers, it is unlikely that a female tamarin or marmoset could successfully rear twin infants.

It has been proposed that a monogamous mating pattern and adult male provisioning of females and young were an important and early component of the behavioral ecology of human ancestors. However, based on a broader understanding of primate mating and social patterns, it is now apparent that the bases of male–infant interactions are extremely complex and may reflect factors other than *paternity.* Moreover, among modern humans there exists extreme variation in the degree to which adult males provide direct infant care. In some societies fathers may spend up to 14% of their day interacting with young (e.g., !Kung San Bushman). However, in most cases, male–infant interactions are infrequent and limited to small amounts of play behavior. Although fathers do provide food for their offspring, it remains uncertain whether this behavior developed as a mechanism to maintain a social bond between an adult male and an adult female or whether it evolved as a mechanism to increase the survivorship of a male's offspring.

VI. INTRASEXUAL AGGRESSION AND MATING COMPETITION

Among mammals, females invest more heavily in the production of offspring than do males. This begins with the initial expenditure of energy required to produce an egg and continues through the period of gestation and lactation. Given that the cost of reproduction to the female is significantly greater than it is for a male, female reproductive success is constrained by the number of offspring that survive (not produced), whereas male reproductive success is limited by availability and access to fertile females. Males often compete aggressively for access to reproductive partners, resulting in extreme variance in male reproductive opportunities and reproductive success. In contrast, although females can exercise some level of mate choice and select male partners based on particular physical or behavioral qualities, at present there is little direct evidence that a female's mating preferences have a significant impact on her lifetime reproductive success.

In primates characterized by a polygynous mating system, intrasexual competition among males has a direct effect on patterns of social tolerance, aggression, and migration. In the case of unimale breeding groups, intense competition for *reproductive sovereignty* between the harem leader and individuals or bands of nonresident males often exists. Depending on the size of the harem, the age of the harem leader, and the number of extragroup males in the population, attempted takeovers of the group can occur rarely or every few months. These takeovers are extremely aggressive and result in severe injuries and sometimes death.

Social interactions among males residing in multimale polygynous groups are highly complex and include elements of social cooperation, alliances, and physical aggression. In many baboon and macaque species, males compete with each other for positions in a dominance hierarchy. These hierarchies are unstable, however; as resident males leave the group, new males enter, and changing alliances alter dominance relations. In these groups a single male rarely is able to monopolize access to all reproductive females. Rather, a number of males may mate with a given female, although some males may mate more frequently and at a more optimal time for conception. In these primates, a variety of tactics or alternative behavioral strategies exist by which males are able to gain access to reproductive partners (e.g., immigration into a new social group, coalition and alliance formation, development of a social–sexual bond with particular females, formation of consort relationships).

Females can also compete for reproductive partners. This can occur through direct solicitation of copulations, transfer into another social unit, or directed aggression against other females. In the case of tamarins and marmosets discussed in the previous section, it has been proposed that females mate preferentially with males that provide direct care for their offspring. Alternatively, dominant female rhesus macaques, yellow baboons, and red howler monkeys have been observed to attack subordinate females who were in the process of copulating. In addition, female primates engage in a variety of behavioral and reproductive tactics that serve to confuse paternity, reduce the opportunities for *infanticide,* and limit the amount of aggression directed against their offspring.

Increasing evidence shows that dominant females in some primate species can impair or suppress the reproductive cycle of subordinate group members. Although the physiological mechanisms through which this occurs are not well understood, both stress and *pheromone* cues can disrupt normal endocrine function. In captive female talapoin monkeys, for example, subordinates that receive high levels of aggression exhibit a decrease in sexual activity and show steroid hormone levels indicative of neuroendocrine dysfunction. Similarly, social stress can result in low levels of conception and increased rates of abortion in vervet monkeys. Finally, among free-ranging tamarins and marmosets, regardless of the number of adult females in a group, only a single female produces offspring. Data from captive tamarin and marmoset groups indicate that subordinate females are frequently anovulatory. Ovulation inhibition may be related to scent marking and the resultant pheromone signals given by the dominant female. Thus, factors such as social dominance, stress, hormone production, and nutritional condition all appear to influence individual reproductive success and the behavioral interactions of group-living primates.

The form and structure of a primate group represents a compromise between the threat of predation and predator protection, foraging costs and access to high-quality feeding sites, reproductive opportunities and mating exclusivity, and the importance of group members in providing care for young. These factors interact in complex ways and produce the variety of behavioral, mating, and social patterns found in living primates. Although certain qualitative relations exist among diet, feeding ecology, group size, and group cohesion, detailed information on the distribution, patch size, seasonality, renewal rate, and nutritional quality of foods exploited by primates is needed before more quantitative and predictive models of behavioral ecology can be tested. In particular, studies must focus on differences in the costs and benefits of particular foraging patterns for males and females, the manner in which female foraging patterns change during pregnancy and lactation, and whether subordinates, migrants, and extragroup individuals suffer higher foraging costs and greater risks of predation. In those species for which the advantages of social group living are not directly related to feeding and predation, access to mating partners and assistance in caring for young may play a more formidable role in the structure and form of the social group.

BIBLIOGRAPHY

Boesch, C., and Boesch, H. (1984). Mental map in chimpanzees: An analysis of hammer transports for nut cracking. *Primates* **25**, 160–170.

Fleagle, J. G. (1988). "Primate Adaptation and Evolution." Academic Press, New York.

Garber, P. A. (1987). Foraging strategies among living primates. *Annu. Rev. Anthropol.* **16**, 339–364.

Garber, P. A. (1989). The role of spatial memory in primate foraging patterns: *Saguinus mystax* and *Saguinus fuscicollis*. *Am. J. Primatol* **19**, 203–216.

Jolly, A. (1985). "The Evolution of Primate Behavior," 2nd Ed. MacMillan, New York.

Richard, A. F. (1985). "Primates in Nature." Freeman, New York.

Smuts, B. B., Cheney, D. L., Seyfarth, R. M., Wrangham, R. W., and Struhsaker, T. T. (1986). "Primate Societies." University of Chicago Press, Chicago.

Terborgh, J., and Janzen, C. H. (1986). The socioecology of primate groups. *Annu. Rev. Ecol. Systemat.* **17**, 111–136.

Primates

ERIC DELSON
Lehman College and Graduate School, City University of New York, and American Museum of Natural History

IAN TATTERSALL
American Museum of Natural History

GLOSSARY

Cathemeral Activity pattern of an animal wherein activity is distributed over the entire 24-hr cycle, rather than limited to the daytime (diurnal), night-time (nocturnal), or the times around dawn and dusk (crepuscular)

Morphology Visible attributes; anatomy

Phylogeny Evolutionary history; a statement of evolutionary relationships

Systematics Study of the diversity of life and of the relationships among taxa

Taxonomy Theory and practice of classifying organisms; derived from the noun "taxon" (pl. taxa), which denotes a named unit at any rank of the Linnean hierarchy

PRIMATES IS THE ORDER OF MAMMALS CONtaining the lemurs, lorises, tarsier, monkeys of the Old and New Worlds, apes, and humans, totaling approximately 200 living species. Also usually regarded as primates, besides the fossil relatives of extant species, are the members of an early Tertiary group of "archaic" forms known as Paromomyiformes.

I. DEFINITION AND CLASSIFICATION

Attempts to define the order Primates in terms of morphological characteristics uniquely shared by all of its members have routinely been frustrated by the fact that phylogeny, rather than morphology, gives the group its unity. However, if we disregard the archaic forms of the Paleocene and early Eocene, we can follow W. E. Le Gros Clark in identifying several "progressive trends" that have some value in demarcating developments in primate evolution from those in other mammalian lineages. These include a decreased significance of olfaction compared with stereoscopic vision; enhancement of grasping and manipulative capabilities, associated with the replacement of claws by nails on all or nearly all digits; and a tendency toward relative enlargement of the "higher" centers of the brain, notably the association areas of the cortex.

A diagram of primate evolutionary relationships at the family level is shown in Fig. 1. Many of its details, perhaps most importantly the affinities of the tarsier, are actively debated, and we do not regard this statement as definitive. However, it does serve as the basis for the outline classification presented in Table I. Geological ranges of the families identified in Fig. 1 are shown in Fig. 2.

II. "ARCHAIC" PRIMATES

The primate order originated prior to about 65 Myr (million years) ago, and early primates flourished in the northern continents during the Paleocene epoch, around 65–56 Myr ago. The most successful of these archaic forms were *Plesiadapis* (Figs. 3 and 4) and its relatives, which were characterized by small braincases, long faces, clawed feet, and an enlarged, specialized anterior dentition. "Archaic" primates lacked the postorbital bar, a strut that delineates the orbit laterally in all living primates, and are included in the

ENCYCLOPEDIA OF HUMAN BIOLOGY, Second Edition, VOLUME 7. Copyright © 1997 by Academic Press. All rights of reproduction in any form reserved.

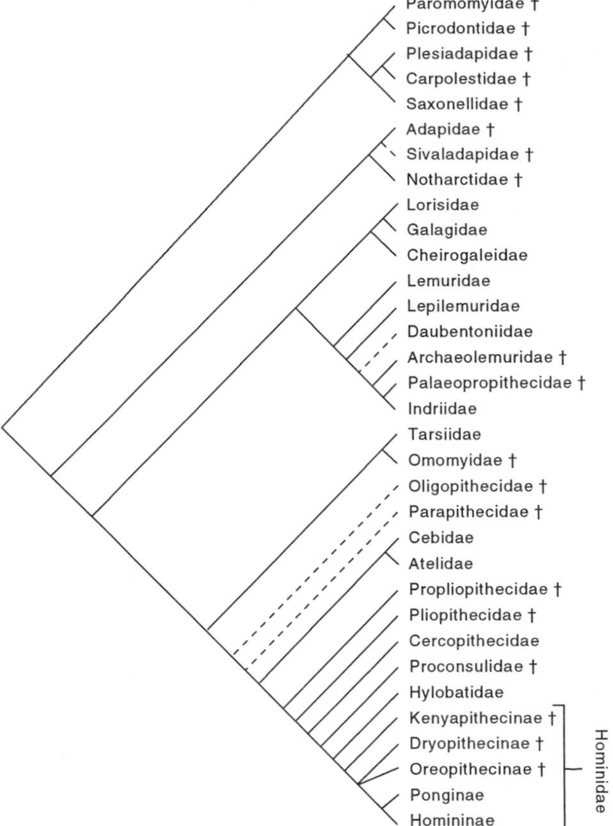

FIGURE 1 A provisional phylogeny of the primates. † denotes extinct taxon. Dotted lines indicate alternative placement of disputed taxa.

order principally on the bases of the structure of their bony ear and postcranial bones and their molar tooth morphology. Generally, in early primates, the molars show some lowering of cusp relief, which has been taken to suggest that the origin of the primate order lay in a dietary shift away from insects and toward plant foods. Postcranial remains are rare but indicate that *Plesiadapis* itself, though heavy-boned, was arboreal. Although some studies suggest that the paromomyiforms were more closely related to other mammalian groups and should be removed from the primates, we retain them within this order at present.

III. "LOWER" PRIMATES OF MODERN ASPECT

A. The Living "Lower" Primates

The extant strepsirhine (or "lower") primates (suborder Strepsirhini; Fig. 3) are most abundantly represented on the island of Madagascar, where higher primates were absent before the recent arrival of humans. All living strepsirhines are united in possessing a "grooming claw" on the second pedal digit and a "tooth comb," in which the slender, elongated front lower teeth lie horizontally and are closely pressed together. They also retain a primitive form of the nasal apparatus, with a "tethered" upper lip (fixed to the upper jaw in the midline, rather than fully mobile) and a moist rhinarium (or "wet nose," as in dogs). In comparison with the "higher" primates, the sense of smell, as expressed by these anatomical retentions and the presence of olfactory marking behaviors using feces, urine, or the exudations of specialized glands, is relatively important. On the other hand, the diurnal visual sense is less developed, as reflected in weak or absent color vision. Notably, however, in contrast to the higher primates, many strepsirhines are nocturnal.

Surviving strepsirhines in Africa and Asia are all classified as lorisoids, belonging to the families Lorisidae (the quadrumanous lorises and pottos) and Galagidae (the leaping bushbabies); so also are the dwarf lemurs of Madagascar (family Cheirogaleidae). All lorisoids are nocturnal primates, which, although social, are relatively nongregarious; as noted, scent-marking provides an important channel for intraspecific communication among these forms.

The Malagasy families Lemuridae (the lemurs, ruffed lemurs, and bamboo lemurs), Indriidae (the sifakas, babakotos, and woolly lemurs), Daubentoniidae (the aye-aye), and Lepilemuridae (sportive lemurs) are more diverse. Of these families, the latter two are exclusively nocturnal, and the last probably lost a diurnal genus only recently; the other two (the principally quadrupedal Lemuridae and the vertical-clinging and leaping Indriidae) are chiefly composed of diurnal or cathemeral species and show a great deal of variation is social organization. Groups range from bonded pairs (permanent associations of an adult male with an adult female) with immature offspring through small multimale, multifemale units to large heterosexual groups of 30 individuals or more. These groups, as in most higher primates, appear to be based around female nuclei, and male exchange is common, especially during the brief annual breeding season. Diets appear to vary widely by season and locality, and despite their generally lower ratios of brain to body size, these diurnal strepsirhines seem to be every bit as opportunistic in the exploitation of their habitats as are the higher primates.

TABLE I
Partial Classification of the Order Primates

Rank	Taxon	"Common" name
Semiorder	Paromomyiformes†	"Archaic" forms
Suborder	Plesiadapiformes†	"Archaic" forms
Superfamily	Paromomyoidea†	"Archaic" forms
Superfamily	Plesiadapoidea†	"Archaic" forms
Genus	*Plesiadapis*†	
Semiorder	Euprimates	"Primates of modern aspect"
Suborder	Strepsirhini	"Lower" primates, strepsirhines
Infraorder	Adapiformes†	Extinct "lemur-like" forms
Genus	*Notharctus*†	
Genus	*Smilodectes*†	
Genus	*Leptadapis*†	
Infraorder	Lemuriformes	Lemurs, lorises
Superfamily	Lemuroidea	"True" lemurs
Family	Lemuridae	
Superfamily	Indrioidea	Sifakas, aye-aye, etc.
Family	Indriidae	Sifakas
Family	Daubentoniidae	Aye-aye
Family	Lepilemuridae	Sportive lemurs
Superfamily	Lorisoidea	Lorises, galagos, mouse lemurs
Family	Lorisidae	Lorises
Family	Galagidae	Galagos, bushbabies
Family	Cheirogaleidae	Mouse and dwarf lemurs
Suborder	Haplorhini	"Higher" primates and tarsiers
Hyporder	Tarsiiformes	Tarsier, fossil relatives
Family	Tarsiidae	Tarsiers
Family	Omomyidae†	Extinct "tarsier-like" forms
Genus	*Necrolemur*†	
Hyporder	Anthropoidea	"Higher" primates
Infraorder	Paracatarrhini†	"Archaic" anthropoids
Family	Oligopithecidae†	"Archaic" protoanthropoids
Genus	*Catopithecus*†	
Family	Parapithecidae†	Extinct Oligocene monkeys
Genus	*Apidium*†	
Infraorder	Platyrrhini	New World anthropoids
Superfamily	Ateloidea	New World monkeys
Family	Atelidae	Howler, spider, saki
Subfamily	Atelinae	Howler, spider monkeys
Subfamily	Pitheciinae	Saki, titis, owl monkeys
Family	Cebidae	Marmosets, capuchins, squirrel monkeys
Subfamily	Cebinae	Squirrel, capuchin monkeys
Subfamily	Callitrichinae	Marmosets, tamarins
Subfamily	Branisellinae†	Extinct early platyrrhines
Infraorder	Catarrhini	Old World anthropoids
Parvorder	Eocatarrhini†	"Archaic" catarrhines
Family	Propliopithecidae†	Extinct common ancestors of eucatarrhines
Genus	*Propliopithecus*†	
Family	Pliopithecidae†	Extinct early catarrhines
Genus	*Pliopithecus*†	
Parvorder	Eucatarrhini	Advanced catarrhines
Superfamily	Cercopithecoidea	Old World monkeys
Family	Cercopithecidae	Old World monkeys
Subfamily	Cercopithecinae	Cheek-pouched Old World monkeys
Subfamily	Colobinae	Leaf-eating Old World monkeys
Genus	*Mesopithecus*†	
Subfamily	Victoriapithecinae	Extinct early cercopithecids
Superfamily	Hominoidea	Apes and humans
Family	Proconsulidae	Extinct early apes
Family	Hylobatidae	Lesser apes, gibbons
Genus	*Hylobates*	

(*continues*)

TABLE I (*Continued*)

Rank	Taxon	"Common" name
Family	Hominidae	Great apes, humans and extinct relatives
Subfamily	Kenyapithecinae	Extinct "archaic" great apes
Subfamily	Dryopithecinae	Extinct early great apes
Subfamily	Oreopithecinae	*Oreopithecus*
Subfamily	Ponginae	Orangutans and extinct relatives
Genus	*Pongo*	
Genus	*Sivapithecus*†	
Subfamily	Homininae	African apes, humans, and extinct relatives
Genus	*Ardipithecus*†	
Genus	*Australopithecus*†	
Genus	*Paranthropus*†	
Genus	*Homo*	
Genus	*Graecopithecus*†	
Genus	*Gorilla*	
Genus	*Pan*	

†Denotes extinct taxon; genera listed are those illustrated or mentioned in the text.

B. The Fossil Record of Lower Primates

By the end of the Paleocene, new kinds of primates were beginning to appear alongside the archaic forms, and in the following Eocene epoch, between about 56 and 34 Myr ago, these more modern-looking forms rapidly replaced the archaic types. The fossil record, as currently known, provides no clear ancestry within the Paleocene primate radiation for that of the Eocene; it is possible that the African record, still poorly known, holds the key to this relationship. Eocene primates are generally classified into the lemur-like Adapiformes and the tarsier-like Omomyidae, although this simple division probably obscures a situation of considerably greater complexity, and some of today's lemurs may bear more than general affinities to elements of the Eocene primate fauna. Whatever the exact geometry of the relationships between the lower primates of today and those of the Eocene may ultimately prove to be, however, as functioning organisms, the latter were already clearly similar to today's lower primates. Undoubtedly arboreal in habit, Eocene primates had grasping hands and feet, with an opposed hallux and sensitive digital pads backed by flat nails; braincases were relatively large whereas the face was reduced, presumably in correlation with a deemphasis of the sense of smell. Conversely, the eyes faced forward to enhance overlap of the visual fields, thus promoting stereoscopic vision, and were ringed by bone (Figs. 4 and 5).

As climates cooled subsequent to the Eocene, primates began to disappear from the northern continents, and later fossil evidence, principally from the Miocene of Africa and Asia, reveals strepsirhine primates closely related to those that now live in those areas. Late Holocene deposits in Madagascar indicate the survival until recently, on that island, of a much more varied lemur fauna than exists there today; extinct genera include several large-bodied and specialized forms.

IV. "HIGHER" PRIMATES AND TARSIERS

The living primates have been subdivided alternatively into the Prosimii (=Plesiadapiformes, Strepsirhini, and Tarsiiformes here) versus Anthropoidea or into the Strepsirhini versus Haplorhini, with the main difference involving the position of the tarsier. Most current research, both genetic and morphological, supports the phylogenetic linkage of tarsiers to the other "dry-nosed" or haplorhine primates (monkeys, apes, and humans), as opposed to the "wet-nosed" strepsirhines characterized, as noted earlier, by a tethered upper lip and greater sensory emphasis on olfaction. The resulting division is usually recognized at the taxonomic level of the suborder, as in our classification. However, as universally accepted today, the platyrrhines (New World monkeys) and the catarrhines (Old World monkeys, apes, humans, and their extinct relatives) shared a common ancestry distinct from all other primates, thus forming a natural

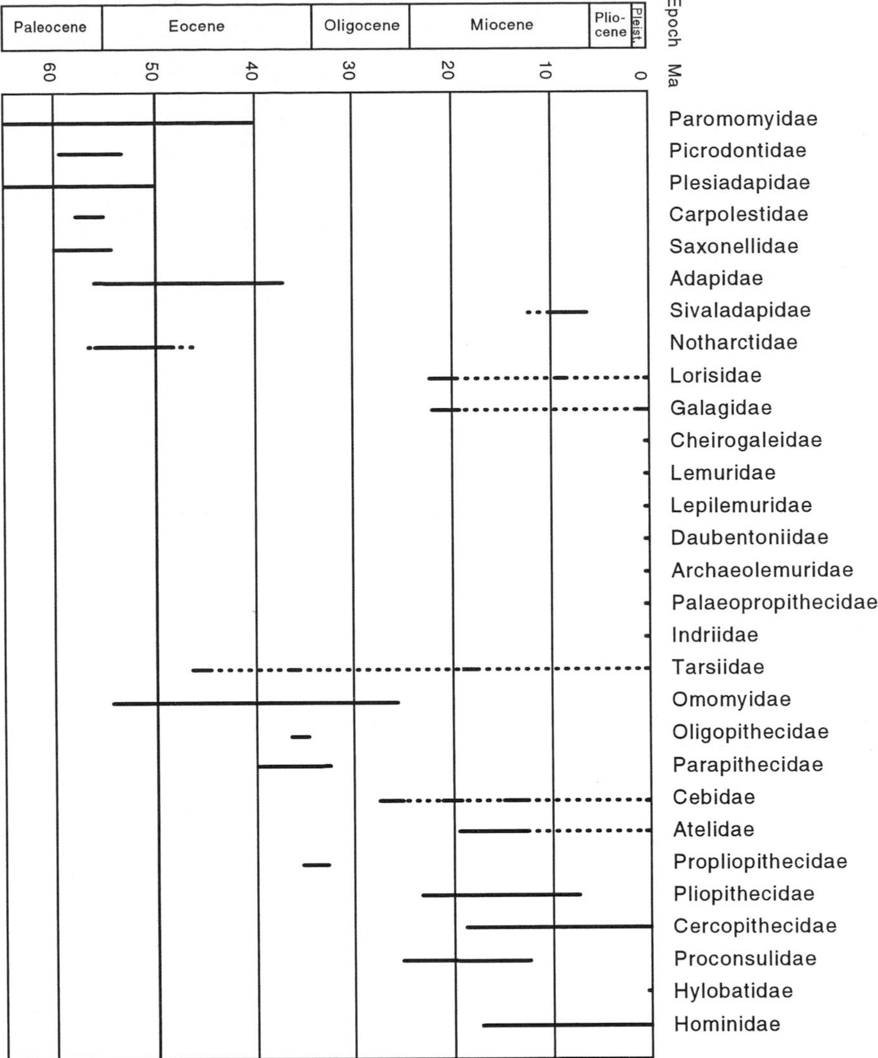

FIGURE 2 Temporal distribution of the primate families shown in Fig. 1. Dotted lines represent presumed range continuity between known occurrences.

group—the "higher" primates, or anthropoids. This group is formally recognized as the Anthropoidea and was previously ranked as a suborder, when tarsiers were considered prosimians. To recognize both of these important systematic evolutionary concepts, we include the Anthropoidea as a subunit within the Haplorhini, giving it the rank of hyporder.

A. The Living Haplorhines

Modern tarsiers (Tarsiidae) occur as four species of the genus *Tarsius*. They are among the smallest primates, restricted to islands of Southeast Asia and the Philippines, where they live in pair-bonded groups. Their huge eyes are an adaptation to their principally nocturnal activity pattern, and their elongated leg and foot bones allow a rapid-leaping locomotion (see Fig. 3).

The South American platyrrhines are considered by many to be the most diverse primate superfamily, in terms of diet, locomotion, and morphology. The family Cebidae includes the subfamilies Callitrichinae (marmosets and tamarins) and Cebinae, both characterized by a short face and lightly built masticatory system designed for a diet of insects, fruits, and gums. The marmosets are the smallest living anthropoids;

FIGURE 3 Representative living lower primates. Clockwise from top left: forkmarked lemur (Cheirogaleidae), bamboo lemur (Lemuridae), sportive lemur (Lepilemuridae), avahy (Indriidae), potto (Lorisidae), bushbaby (Galagidae), tarsier (Tarsiidae). Center: aye-aye (Daubentoniidae). Not to scale. (Illustration by Don McGranaghan.)

they are generally pair-bonded, and the male takes an active role in the care of the young, often twins. The cebines include both large- (capuchins) and small-bodied (squirrel monkey) species, which live in large groups and have relatively large brains for platyrrhines. The family Atelidae includes mostly larger platyrrhines with a heavily built masticatory apparatus (jaws, teeth, projecting face, chewing muscles) adapted to a frugivorous (fruit-eating) or folivorous (leaf-eating) diet. The uniformly large atelines (including the spider and howler monkeys) are characterized by prehensile (grasping) tails and suspensory locomotion and feeding postures, although some species are acrobatic, whereas others move cautiously; most atelines live in

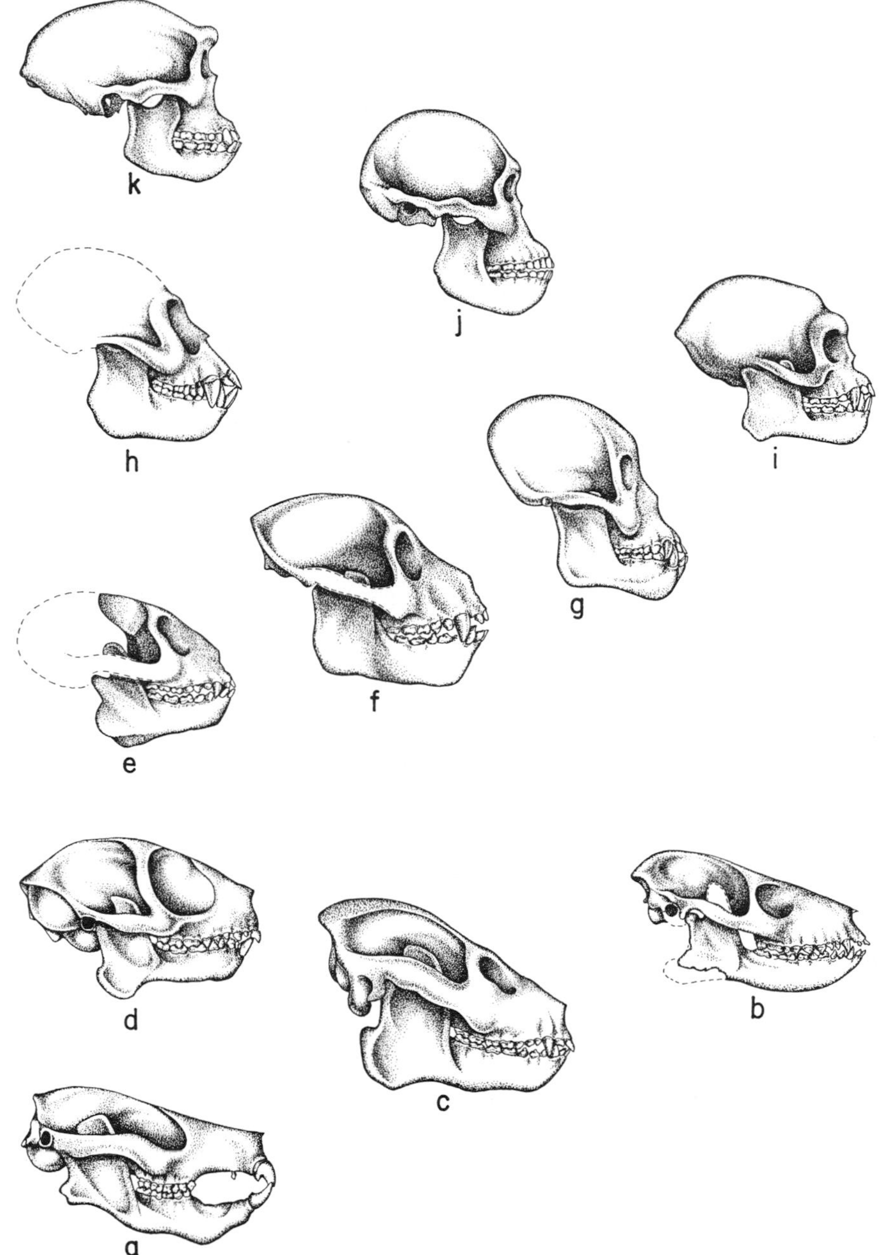

FIGURE 4 Skulls in lateral view of various extinct primates. a, *Plesiadapis tricuspidens;* b, *Notharctus tenebrosus;* c, *Leptadapis magnus;* d, *Necrolemur antiquus;* e, *Apidium phiomense;* f, *Propliopithecus* (=*Aegyptopithecus*) *zeuxis;* g, *Proconsul heseloni;* h, *Cebupithecia sarmientoi;* i, *Pliopithecus vindobonensis;* j, *Australopithecus africanus;* k, *Homo erectus.* Not to scale. (Illustration by Don McGranaghan.)

moderately large social units. By contrast, the pitheciines may form pair-bonded social groups (the titi and the nocturnal–cathemeral owl monkey) or small heterosexual units (the somewhat larger-bodied sakis and uakaris). The latter forms have adapted to a diet of hard-shelled or hard-seeded fruits, for which they have developed a specialized dentition (Fig. 6).

The Old World monkeys (superfamily Cercopithecoidea) are the most numerous living primate group, with some 75 species in about 15 genera,

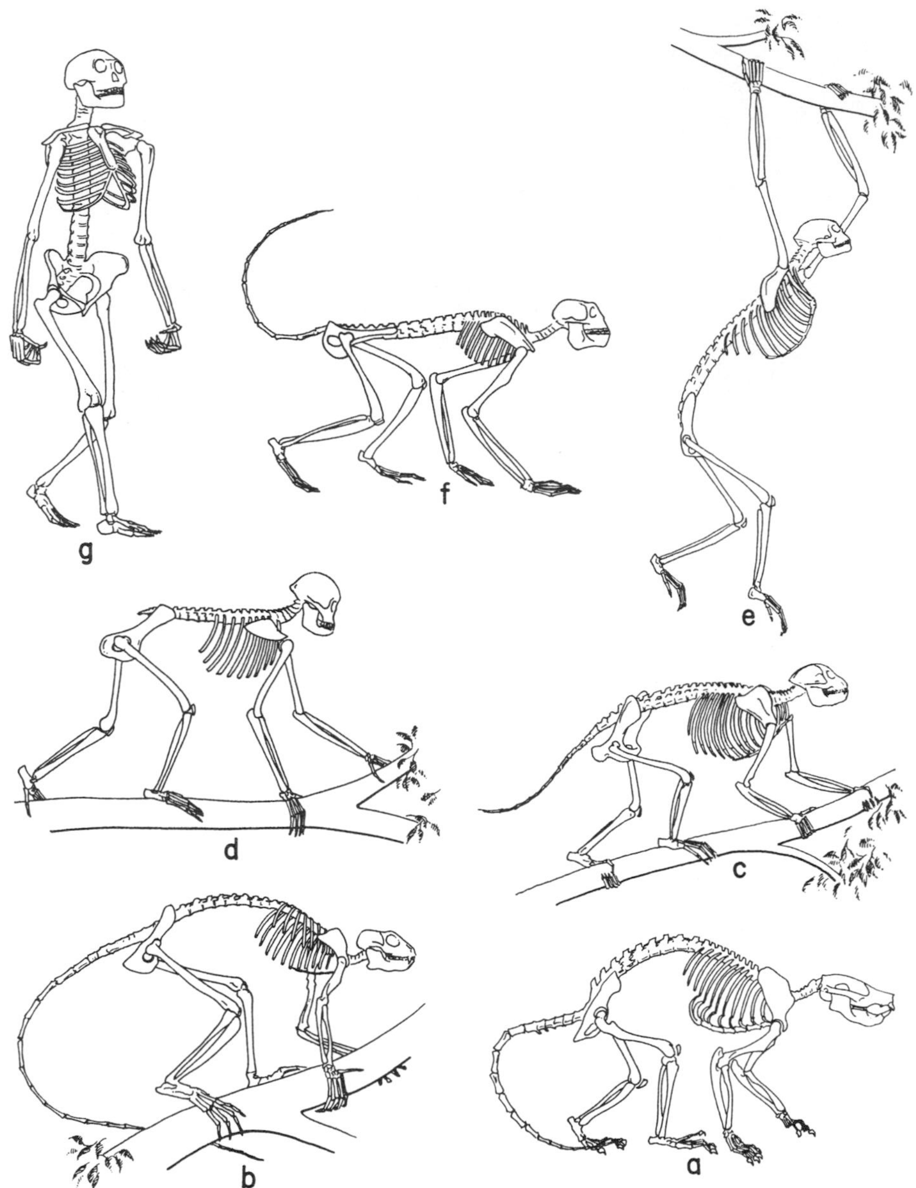

FIGURE 5 Skeletons of various extinct primates. a, *Plesiadapis tricuspidens;* b, *Smilodectes gracilis;* c, *Propliopithecus* (=*Aegyptopithecus*) *zeuxis;* d, *Proconsul heseloni;* e, *Pliopithecus vindobonensis;* f, *Mesopithecus pentelicus;* g, *Australopithecus afarensis.* Not to scale. (Illustration by Don McGranaghan.)

and among the most widely distributed. They share a distinctive dental pattern and the presence of ischial callosities, tough rump patches on which they sit in trees (also found in gibbons and possibly an ancient catarrhine feature). One family, Cercopithecidae, is divided into two subfamilies, Cercopithecinae and Colobinae. The former is characterized by cheek pouches for temporary food storage, a simple stomach, and generally a somewhat to greatly elongated snout; the diet is eclectic, including fruits, some leaves and shoots, nuts, insects, and even rare vertebrate prey. A partly terrestrial quadrupedal pattern of locomotion probably is typical for the cercopithecines, with several groups independently either

FIGURE 6 Representative living higher primates. Clockwise from upper left: tamarin (Callitrichinae), spider monkey (Atelinae), orangutan (Ponginae), colobus (Colobinae), chimpanzee, human (Homininae), saki (Pitheciinae), macaque (Cercopithecinae). Not to scale. (Illustration by Don McGranaghan.)

strongly terrestrial or more fully arboreal. The colobines, on the other hand, are more acrobatic arboreal runners and leapers, with only one living species spending much time on the ground. Their diet is concentrated on leaves and shoots, as mirrored in their enlarged and sacculated stomach. Social organization varies widely among the cercopithecids, including bonded pairs, unimale "harems" with several females, and groups with roughly equal numbers of both sexes and between 15 and 200 members. African cercopithecines include mangabeys, baboons, mandrills, guenons, and patas monkeys, whereas the macaques occur in Northwest Africa and range from Pakistan to Japan and Indonesia; they live in environments from desert fringe through savannah, rain forest, deciduous forest to high mountains. Colobines are less tolerant, with the African colobus varieties and the Asian langurs and doucs found mainly in deciduous or evergreen forest, although "golden monkeys" range to high elevations in China and proboscis monkeys live in mangrove swamps.

To humans, the most intriguing primates are the apes, grouped (along with people) in the superfamily Hominoidea. The lesser apes, or gibbons (*Hylobates*), form one family found only in Southeast Asia. These pair-bonded fruit-eaters are most famous for their brachiating pattern of locomotion, involving ricochetal arm-swinging and below-branch suspension, as contrasted with the above-branch running of most monkeys. Gibbon males and females both defend their home ranges or territories by means of "song" calls of several types, which indicate their position to neighbors and cement the pair bond.

In the past, great apes (Asian orangutan, African chimpanzee and gorilla) were often classified together in the family Pongidae, as opposed to humans in Hominidae, but a variety of mainly genetic studies have demonstrated that humans and great apes are extremely close to each other in evolutionary terms and that African apes and humans form a phyletic unit or evolutionary subgroup as contrasted with orangutans. Thus, by analogy to other primates (and mammals in general), only a single family Hominidae is recognized to reflect the low diversity of this group, and it is in turn divided into the subfamilies Ponginae for the orangutan and its extinct relatives and Homininae for humans, chimpanzees, and gorillas. The orangutan is, almost like lorisoids, relatively nongregarious: the large adult males patrol broad ranges overlapping those of females and their offspring, but the two sexes seldom travel together. Gorillas live in social groups with one or two adult males and several females, whereas chimpanzees usually form small groups of males, females, and young or consort pairs, which fluidly "fuse" into larger bisexual groups and divide again every few days. Diets of most hominoids consist mainly of fruits, although gorillas also eat other plant parts, and chimpanzees have been seen to actively hunt various animals including monkeys. Chimpanzees also use twigs to obtain termites from their mounds, stones to break open nutshells, and leaves to get water from tree cavities; thus, they are considered to make and use tools on a regular basis, and they teach this behavior to their offspring. Great apes do not truly brachiate (like gibbons do) as adults: orangutans use hands and feet interchangeably when climbing and walk quadrupedally on the ground; chimpanzees and gorillas "knuckle-walk" on the ground but also climb actively and swing manually in trees (except for adult gorillas).

B. The Fossil Record

The living tarsiers represent only a small fraction of the past diversity of the Tarsiiformes. The earliest tarsier-like primates, known as the Omomyidae, first appeared in the early Eocene, along with the lemur-like adapiforms (see earlier). This group was diverse throughout the Eocene in North America, with several species often found together. Additional species, including several in a distinct subfamily, occurred in Europe, and rare fossils possibly belonging to the Omomyidae have been recognized in Asia and northern Africa. Recently, fragmentary remains of possible close relatives of *Tarsius* itself have been recovered from the Eocene of China, Oligocene of Egypt, and Miocene of Thailand. Most likely, an animal broadly similar to the Eocene omomyids was ancestral to the anthropoids, but no known form is a good candidate for such an ancestor. A variety of poorly known species from eastern Asia (45–40 Myr ago) and northern Africa (55–40 Myr ago) have been suggested as documenting a previously unknown major radiation of protoanthropoids. The fragmentary nature of these fossils makes it difficult to test this hypothesis.

The 36- to 33-Myr-old Fayum fossil beds in northern Egypt provide an important "window" onto both anthropoid and catarrhine origins. Of the three families known from there, two may belong to an archaic anthropoid infraorder, the Paracatarrhini; some authors consider these animals to be members of the catarrhine radiation, but they appear too conservative to be retained in that group. The Oligopithecidae includes the recently discovered *Catopithecus*, for which a skull and jaws are known. These reveal a mosaic pattern of derived and conservative features: for example, the orbits appear to have been closed off as in anthropoids, whereas the mandibular symphysis was probably unfused and the teeth are like those of lower primates. The oligopithecids may represent a late-surviving member of the protoanthropoid radiation. Somewhat more derived are the Parapithecidae, which were morphologically monkey-like, sharing general features of locomotor ability, relative brain size, and probably lifeways with living monkeys of both the Old and New Worlds. Although once proposed as ancestors of the cercopithecoids, they are today viewed as an early side-branch radiation of either the catarrhines or the anthropoids as a whole. They were mainly arboreal fruit-eaters; at least one species was unique among primates in having lost all lower incisor (front) teeth (see Figs. 4 and 5).

The earliest New World monkeys appeared in South America late in the Oligocene, about 27 Myr ago, with no trace of earlier primates south of Texas. During the early Cenozoic, South America was an island continent, isolated from Africa by the expanding South Atlantic Ocean and from North and Central America by a broad Caribbean–Pacific connection, and the source area of the platyrrhines is, thus, a continuing problem in primate evolution. Both North America (with Eocene omomyids) and Africa (with somewhat earlier anthropoids; see the following) have been championed, with the latter view more widely but not universally accepted. In one of the most fascinating aspects of New World monkey evolution, fossils that can be readily identified as belonging to each of the two living families occur by 20 Myr ago, and by 15 Myr ago, fossils that belong to each of the four living subfamilies have been identified; some of those fossils can even be identified as close evolutionary relatives or ancestors of certain living genera. This very early diversification of living lineages is uncommon among primates (see Figs. 1, 4, and 5).

By contrast, the history of catarrhine primates is best seen as a series of radiations, each replacing its antecedents after a brief overlap interval. The third group of anthropoids represented in the Egyptian Fayum, contemporary with the parapithecids, includes several species of the undoubted archaic catarrhine *Propliopithecus* (one is often called *Aegyptopithecus*). In tooth and skull morphology, these animals are more similar to the living catarrhines and may be similar to the common ancestor of all later Old World anthropoids. Additional genera of "primitive" catarrhines, not closely related to either monkeys or apes, are known from the Miocene of Africa, Europe, and Asia. These forms, along with the older *Propliopithecus,* offer insight into the morphology and diversity of early catarrhines.

The oldest definite Old World monkeys (cercopithecoids), recognized by their distinctive tooth morphology, are known from the earlier Miocene (15–20 Myr) of eastern and northeastern Africa. The two modern subfamilies possibly differentiated in Africa by 12 Myr ago, but they are not documented until about 10 Myr ago. Today, the leaf-eating colobines are almost entirely arboreal and generally of rather small size, but large, terrestrially active forms lived in Europe and Africa between 5 and 2 Myr ago and briefly extended into Mongolia. The same interval (the Pliocene) saw a great radiation of cercopithecine monkeys in Africa, the spread of macaques across Eurasia, and the development of a number of apparently independent terrestrial lineages.

The ape and human superfamily Hominoidea is first represented in the late Oligocene of Africa, where *Proconsul* and related forms occur between 26 and 14 Myr ago. This group is now usually thought to be similar to the common ancestor of the two living hominoid families: the gibbons (Hylobatidae) and the humans and great apes (Hominidae). No gibbon ancestors have been discerned in the fossil record, although previous researchers suggested certain pliopithecoids for that role; a major problem is that modern gibbons combine very conservative craniodental morphology with (secondarily?) small body size and unique locomotor adaptations, so that recognizing a "proto-gibbon" that lacks these unique features is operationally difficult.

The earliest members of Hominidae (the kenyapithecines) appeared in Africa and Arabia between about 20 and 14 Myr ago, and closely related species inhabited Turkey and central Europe slightly later; none of them can be clearly linked with either modern subfamily. The dryopithecines of Europe and eastern Asia lived between 13 and 8 Myr ago; they are somewhat more like modern hominids in their faces and limb bones especially, but they still do not show any characters linking them unequivocally to either living subfamily. The orangutan lineage is the first to be documented in the fossil record, with members of the genus *Sivapithecus* known from Turkey, Pakistan, and India between 12 and 7 Myr ago. *Sivapithecus* shares with orangutans several specialized features of the face and teeth [such as frontal sinuses derived from the maxillary sinus, narrow interorbital region, ovoid orbits, outwardly rotated upper canines, and constricted incisive canal (the passageway linking the floor of the nose to the roof of the mouth)], which can be readily recognized in even fragmentary fossils.

The hominine lineage is far less well documented in the 10- to 4-Myr interval, with only a few teeth or jaws in eastern Africa and perhaps the European *Graecopithecus* (10–8 Myr) representing potential early relatives of the African apes and humans. No later fossils are linked to the chimpanzee or gorilla, but human evolution is rather well known after 4

Myr ago. *Australopithecus, Paranthropus,* and *Ardipithecus* include eight species characterized by brains slightly expanded relative to body size as compared with those of apes, small canine teeth incorporated into the incisor complex, and a variety of features related to their bipedal locomotor adaptation. By about 2.5 Myr ago, the earliest members of the genus *Homo* can be diagnosed by an increase in brain size, an apparently more modern postcranium and locomotor system, and the presence of stone tools. Humans occupied Eurasia by at least 1 Myr ago, but anatomically modern fossils are not known until around 100,000 years ago, in Africa and then the Near East. By about 30,000 years ago, modern humans had spread over the Old World including Australia, replacing more archaic varieties such as the European Neanderthals, and reached the New World by at least 12,000 years ago. [*See* Comparative Anatomy; Evolving Hominid Strategies; Human Evolution.]

BIBLIOGRAPHY

Delson, E., Tattersall, I., Van Couvering, J. A., and Brooks, A. S. (eds.) (1997). "Encyclopedia of Human Evolution and Prehistory," 2nd Ed. Garland, New York.

Fleagle, J. G. (1988). "Primate Adaptation and Evolution." Academic Press, New York.

Fleagle, J. G., and Kay, R. F. (eds.) (1994). "Anthropoid Origins." Plenum, New York.

Gebo, D. L. (ed.) (1993). "Postcranial Adaptation in Nonhuman Primates." Northern Illinois Univ. Press, De Kalb.

Martin, R. D. (1990). "Primate Origins and Evolution." Princeton Univ. Press, Princeton, New Jersey.

Mittermeier, R. A., Tattersall, I., Konstant, W. R., Meyers D. M., and Mast, R. B. (1994). "Lemurs of Madagascar." Conservation International, Washington, D.C.

Napier, J. R., and Napier, P. H. (1985). "The Natural History of the Primates." MIT Press, Cambridge, Massachusetts.

Richard, A. (1985). "Primates in Nature." W. H. Freeman, New York.

Szalay, F. S., and Delson, E. (1979). "Evolutionary History of the Primates." Academic Press, New York.

Prions

STANLEY B. PRUSINER
University of California, San Francisco

GLOSSARY

Creutzfeldt–Jakob disease Fatal prion disease of humans that frequently presents with progressive dementia

Gerstmann–Sträussler–Scheinker disease Fatal, inherited prion disease of humans that most often presents with ataxia and incoordination

Kuru Fatal, infectious prion disease of humans that usually presents with ataxia and incoordination

Mad cow disease (bovine spongiform encephalopathy) Prion disease of cattle

Prion Proteinaceous infectious particle that resists inactivation to procedures that modify nucleic acids. Prions are composed largely, if not exclusively, of PrP^{Sc}, an abnormal, disease-causing form of prion protein

Scrapie Naturally occurring prion disease of sheep and goats

PRIONS ARE A NOVEL CLASS OF BIOLOGICALLY active particles that appear to be composed only of protein; yet, they are infectious and mimic viruses in some respects. Prions cause scrapie and mad cow disease in animals and a quartet of neurodegenerative diseases in humans: Creutzfeldt–Jakob disease (CJD), Gerstmann–Sträussler–Scheinker disease (GSS), fatal familial insomnia (FFI), and kuru. Unlike any other group of illnesses, the human prion diseases manifest as infectious, inherited, and sporadic diseases. All prion diseases are characterized by degeneration of the central nervous system (CNS) and the absence of an inflammatory response. A protein designated prion protein or PrP is responsible for all prion diseases. PrP is encoded by a host gene found in all mammals examined to date, and it exists in at least two forms: the cellular, normal form (PrP^C) and the disease-causing abnormal form (PrP^{Sc}). The conversion of PrP^C into PrP^{Sc} is a posttranslational process that involves a profound conformational change. Prions are composed largely, if not entirely, of PrP^{Sc}. The prion diseases, once classified as slow virus infections, are disorders of protein conformation. [*See* Proteins.]

I. PRION CONCEPT

As knowledge accumulated about the infectious agent causing scrapie, the unusual properties of these particles began to be appreciated. Radiobiological studies offered the most provocative data suggesting that the scrapie agent is fundamentally different from viruses.

Once an effective protocol for partial purification of the scrapie agent was developed, convincing data showing that a protein molecule is necessary for infectivity were obtained. Attempts to use the same protocol to demonstrate that the scrapie agent contains a nucleic acid were unsuccessful. Five procedures that modify or hydrolyze nucleic acids failed to inactivate the scrapie agent, yet these same procedures were capable of inactivating numerous viruses as well as small infectious nucleic acids called viroids. On the basis of these studies, the term "prion" was introduced to distinguish the class of infectious particles causing scrapie from those responsible for viral illnesses.

To avoid prejudging the structure of these infectious particles, three hypothetical structures for the

ENCYCLOPEDIA OF HUMAN BIOLOGY, Second Edition, VOLUME 7. Copyright © 1997 by Academic Press. All rights of reproduction in any form reserved.

prion were proposed: (1) proteins surrounding a nucleic acid that encodes them (a virus), (2) proteins surrounding a small noncoding polynucleotide, and (3) a proteinaceous particle devoid of nucleic acid. Over the past 15 years, a large body of experimental data argues persuasively for the third postulate.

Although both prions and viruses multiply, their properties, structure, and mode of replication exhibit many fundamental differences. Viruses contain a nucleic acid genome that encodes progeny viruses that include most or all of the proteins in their protective shells. Although prions contain an abnormal form of the prion protein, designated PrP^{Sc}, and PrP is encoded by a cellular gene, no evidence for a nucleic acid within the prion particle exists. Although viruses evoke an immune response during infection, prions do not. [*See* Virology, Medical.]

II. PRION PROTEINS

Development of a more rapid and economical bioassay greatly facilitated purification of the hamster scrapie agent and led to the discovery of a unique glycoprotein initially designated PrP 27–30. This protein migrates during electrophoresis in polyacrylamide gels containing sodium dodecyl sulfate as a broad band with apparent molecular weight of 27,000 to 30,000. Subsequent studies showed that PrP 27–30 is derived from the larger protein PrP^{Sc}. Other investigations demonstrated that PrP^{Sc} is derived from PrP^{C} by a posttranslational process involving a profound conformational change.

A. Prions Contain PrP^{Sc}

Many different lines of evidence argue that the protein PrP^{Sc} is a component of the infectious prion particle: (1) PrP^{Sc} and PrP 27–30 and scrapie infectivity copurify using biochemical methods and the concentration of PrP 27–30 is proportional to prion titer. (2) The kinetics of proteolytic digestion of PrP 27–30 and infectivity are similar. (3) PrP^{Sc} and infectivity copurify by immunoaffinity chromatography and α-PrP antisera neutralize infectivity. (4) PrP^{Sc} is detected only in clones of cultured cells producing infectivity. (5) PrP amyloid plaques are specific for prion diseases of animals and humans. Deposition of PrP amyloid is controlled, at least in part, by the PrP sequence. (6) PrP^{Sc} (or PrP^{CJD}) is specific for prion diseases of animals and humans. Deposition of PrP^{Sc} precedes spongiform degeneration and reactive gliosis. (7) Genetic linkage was established between the mouse (Mo) PrP gene and scrapie incubation times in mice, with short and long incubation times encoding PrP molecules differing at residues 108 and 189, respectively. The PrP^{C} molecules of mice with short and long incubation times are designated PrP-A and PrP-B, respectively. The length of the incubation time is determined by the level of PrP expression and the PrP sequence. (8) Expression of Syrian hamster (SHa) PrP in transgenic (Tg) SHaPrP mice renders them susceptible to SHa prions. The primary structure of PrP^{Sc} in the inoculum governs the neuropathology and prion synthesis. (9) Although expression of human (Hu) PrP did not render Tg mice susceptible to human prions, Tg(MHu2M) mice expressing a chimeric Hu/Mo PrP transgene are susceptible to Hu prions. (10) Genetic linkage between PrP gene point mutations at codons 102, 178, 198, or 200 and the development of inherited prion diseases in humans has been demonstrated. Genetic linkage was also established between the mutation insert of six additional octarepeats and familial CJD. (11) Mice expressing MoPrP or MHu2M transgenes with the P102L point mutation of GSS spontaneously develop neurological dysfunction, spongiform brain degeneration, and astrocytic gliosis. Serial transmission of neurodegeneration was initiated with brain extracts from these Tg mice. (12) Ablation of the PrP gene in mice prevents scrapie and propagation of prions after intracerebral inoculation of prions. (13) Mice expressing chimeric Mo/SHaPrP transgenes produce "artificial" prions with novel properties. (14) Overexpression of MoPrP-B and SHaPrP produces spongiform degeneration, myopathy, and peripheral neuropathy in older transgenic mice; serial transmission of neurodegeneration was initiated with brain extracts from the mice overexpressing foreign PrP transgenes.

B. Amino Acid Sequence of PrP

Purification of PrP 27–30 to homogeneity allowed determination of its NH_2-terminal amino acid sequence. This sequence permitted the synthesis of an isocoding mixture of oligonucleotides that was subsequently used to identify incomplete PrP cDNA clones from hamster and mouse. cDNA clones encoding the entire open reading frames (ORF) of SHa and Mo PrP were subsequently recovered.

III. PRION PROTEIN GENE

PrP is encoded by a chromosomal gene and not by a nucleic acid within the infectious scrapie prion particle. Levels of PrP mRNA remain unchanged throughout the course of scrapie infection—an observation that led to the identification of the normal PrP gene product PrP^{C}. The entire open reading frame or protein coding region of all known mammalian and avian PrP genes is contained within a single exon (Fig. 1). This feature of the PrP gene eliminates the possibility that PrP^{Sc} arises from alternative RNA splicing; however, mechanisms such as RNA editing or protein splicing remain a possibility. The two exons of the SHaPrP gene are separated by a 10-kb intron: exon 1 encodes a portion of the 5′ untranslated leader sequence, whereas exon 2 encodes the ORF and 3′ untranslated region. The Mo and sheep PrP genes comprise three exons with exon 3 analogous to exon 2 of the hamster. The promoters of both the SHa and Mo PrP genes contain multiple copies of G-C-rich repeats and are devoid of TATA boxes. These G-C nonamers represent a motif that may function as a canonical binding site for the transcription factor Sp1.

Mapping PrP genes to the short arm of Hu chromosome 20 and the homologous region of Mo chromosome 2 argues for the existence of PrP genes prior to the speciation of mammals. Hybridization studies demonstrated <0.002 PrP gene sequences per ID_{50} unit in purified prion fractions, indicating that a gene encoding PrP^{Sc} is not a component of the infectious prion particle. This is a major feature that distinguishes prions from viruses, including those retroviruses that carry cellular oncogenes, and from satellite viruses, which derive their coat proteins from other viruses previously infecting plant cells.

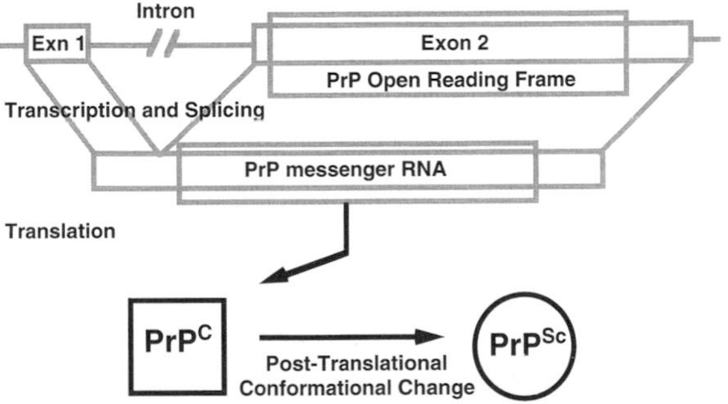

FIGURE 1 Structure and organization of the chromosomal PrP gene. In all mammals examined the entire ORF is contained within a single exon. The 5′ untranslated region of the PrP mRNA is derived from either one or two additional exons. Only one PrP mRNA has been detected. PrP^{Sc} is thought to be derived from PrP^{C} by a posttranslational process. The amino acid sequence of PrP^{Sc} is identical to that predicted from the translated sequence of the DNA encoding the PrP gene, and no unique posttranslational chemical modifications have been identified that might distinguish PrP^{Sc} from PrP^{C}. Thus, it seemed likely that PrP^{C} undergoes a conformational change as it is converted to PrP^{Sc}.

Although PrP mRNA is constitutively expressed in the brains of adult animals, it is highly regulated during development. In the septum, levels of PrP mRNA and choline acetyltransferase were found to increase in parallel during development. In other brain regions, PrP gene expression occurred at an earlier age. *In situ* hybridization studies show that the highest levels of PrP mRNA are found in neurons. PrP^{C} expression in the brains of Syrian hamsters and mice brain was most intense in the stratum radiatum and stratum oriens of the CA1 region of the hippocampus and was virtually absent from the granule cell layer of the dentate gyrus and the pyramidal cell layer throughout Ammon's horn.

IV. PRION PROTEIN STRUCTURE AND FORMATION OF PrP^{Sc}

PrP^{C} molecules destined to become PrP^{Sc} exit to the cell surface prior to their conversion into PrP^{Sc}. Like other glycophosphoinositol (GPI)-anchored proteins, PrP^{C} appears to reenter the cell through a subcellular compartment bounded by cholesterol-rich, detergent-insoluble membranes that might be caveolae or early endosomes (Fig. 2). Within this nonacidic compartment, GPI-anchored PrP^{C} seems to be either converted into PrP^{Sc} or partially degraded. The partially degraded fragment of PrP^{C} appears to be the same as the protein previously designated PrP^{C}-II in partially purified fractions prepared from Syrian hamster brain. After denaturation, PrP^{Sc}, like PrP^{C}, can be released from the cell membranes by digestion with phosphatidylinositol-specific phospholipase C, suggesting that PrP^{Sc} is tethered only by the GPI anchor. In scrapie-infected cultured cells, PrP^{Sc} is trimmed at the N terminus to form PrP 27–30 in an acidic compartment. Whether this acidic compartment is endosomal or lysosomal where PrP 27–30 accumulates remains to be determined. In contrast to cultured cells, the N-terminal trimming of PrP^{Sc} is minimal in brain, where little PrP 27–30 is found. Deleting the GPI addition signal resulted in greatly diminished synthesis of PrP^{Sc}. In contrast to PrP^{C}, PrP^{Sc} accumulates primarily within cells, where it is deposited in lysosomes.

A. Structures of Purified PrP^{C} and PrP^{Sc}

PrP^{Sc} was analyzed by mass spectrometry and gas-phase sequencing in order to identify any amino acid

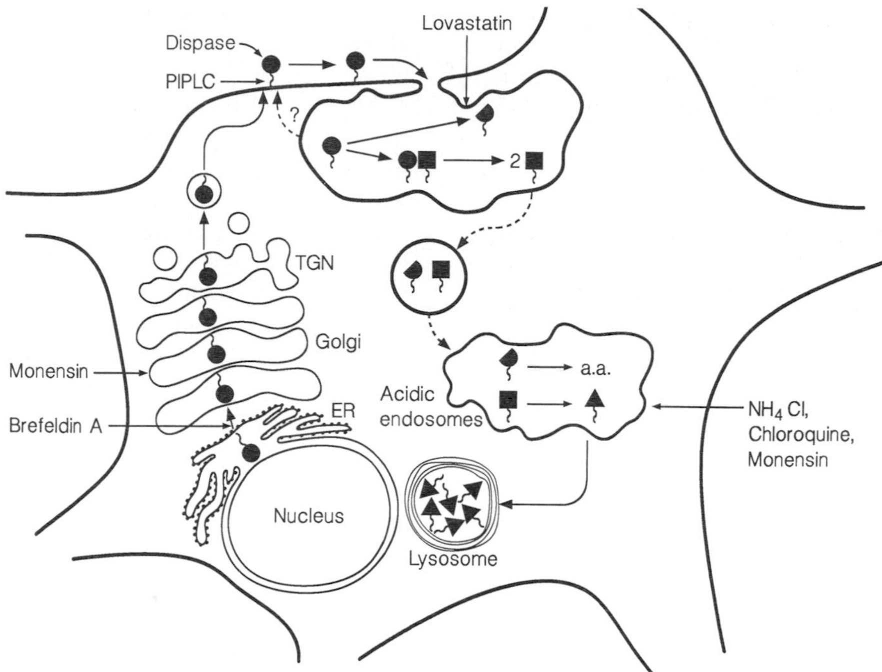

FIGURE 2 Pathways of prion protein synthesis and degradation in cultured cells. PrP^{Sc} is denoted by solid squares and solid circles designate PrP^{C}. Prior to becoming protease resistant, the PrP^{Sc} precursor transits through the plasma membrane and is sensitive to dispase or phosphatidylinositol-specific phospholipase C (PIPLC) added to the medium. PrP^{Sc} formation probably occurs in a compartment accessible from the plasma membrane, such as caveolae or early endosomes, both of which are nonacidic compartments. The synthesis of nascent PrP^{Sc} seems to require the interaction of PrP^{C} with existing PrP^{Sc}. In cultured cells, but not brain, the N terminus of PrP^{Sc} is trimmed to form PrP 27–30; PrP^{Sc} then accumulates primarily in secondary lysosomes. The inhibition of PrP^{Sc} synthesis by brefeldin A demonstrates that the endoplasmic reticulum (ER)–Golgi is not competent for its synthesis and that transport of PrP down the secretory pathway is required for the formation of PrP^{Sc}.

substitutions or posttranslational chemical modifications. The amino acid sequence was the same as that deduced from the translated ORF of the PrP gene, and no candidate posttranslational chemical modifications that might differentiate PrP^{C} from PrP^{Sc} were found. These findings forced consideration of the possibility that conformation distinguishes the two PrP isoforms.

To gather evidence for or against the hypothesis that a conformational change features in PrP^{Sc} synthesis, PrP^{C} and PrP^{Sc} were purified using nondenaturing procedures that determined the secondary structure of each. Fourier transform infrared (FTIR) spectroscopy demonstrated that PrP^{C} has a high α-helix content (42%) and no β-sheet content (3%), findings that were confirmed by circular dichroism measurements. In contrast, the β-sheet content of PrP^{Sc} was 43% and the α-helix content was 30% as measured by FTIR. As determined in earlier studies, N-terminally truncated PrP^{Sc} derived by limited proteolysis and designated PrP 27–30 has an even higher β-sheet content (54%) and a lower α-helix content (21%).

Neither purified PrP^{C} nor PrP^{Sc} formed aggregates detectable by electron microscopy, whereas PrP 27–30 polymerized into rod-shaped amyloids (Fig. 3). Although these findings argue that the conversion of α helices into β sheets underlies the formation of PrP^{Sc}, we cannot eliminate the possibility that an undetected chemical modification of a small fraction of PrP^{Sc} initiates this process. Since PrP^{Sc} seems to be the only component of the "infectious" prion particle, it is likely that this conformational transition is a fundamental event in the propagation of prions. In support of the foregoing statement is the finding that denaturation of PrP 27–30 under conditions that reduced scrapie infectivity resulted in a concomitant diminution of β-sheet content.

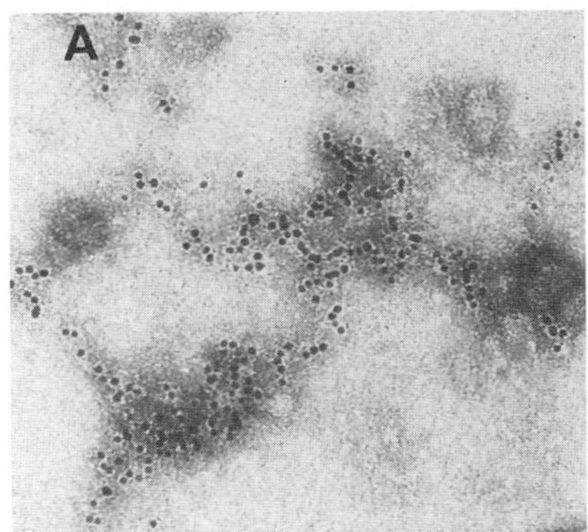

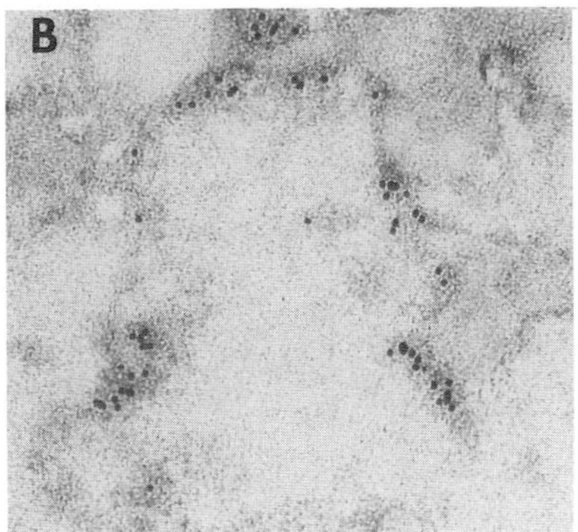

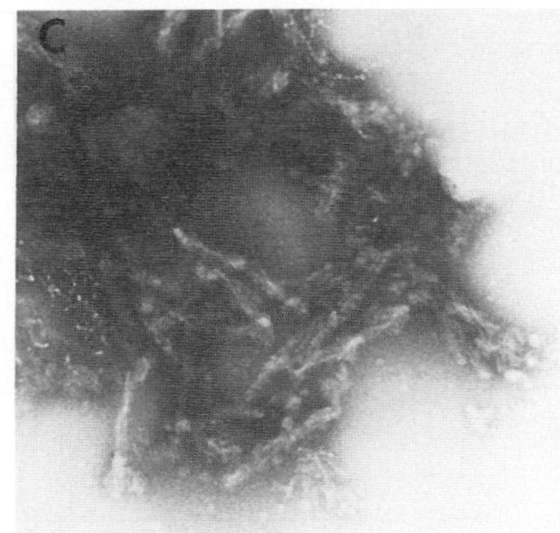

FIGURE 3 Electron micrographs of negatively stained and immunogold-labeled prion proteins: (**A**) PrP^C and (**B**) PrP^{Sc}. (**C**) Prion rods composed of PrP 27–30 were negatively stained with uranyl acetate. Bar = 100 nm.

B. On the Mechanism of Conversion of PrP^C into PrP^{Sc}

As noted, the conversion of PrP^C into PrP^{Sc} involves an increase in the β-sheet content, diminished solubility, and resistance to proteolytic digestion. Transgenetic studies argue that PrP^C and PrP^{Sc} form a complex during PrP^{Sc} formation; thus, synthetic PrP peptides, which mimic the conformational pluralism of PrP, were mixed with PrP^C to determine if its properties were altered. Peptides encompassing two α-helical domains of PrP when mixed with PrP^C produced a complex that displayed many properties of PrP^{Sc}. The PrP^C/peptide complex formed fibrous aggregates and up to 95% of complexed PrP^C sedimented at 100,000 *g* for 1 hr, whereas PrP^C alone did not. These complexes were resistant to proteolytic digestion and displayed a high β-sheet content. Unexpectedly, the peptide in a β-sheet conformation did not form the complex, whereas the random coil did. Addition of 2% Sarkosyl disrupted the complex and rendered PrP^C sensitive to protease digestion. Although the pathogenic A117V mutation increased the efficacy of complex formation, α-PrP monoclonal antibody (mAb) prevented interaction between PrP^C and peptides. These findings, in concert with transgenetic investigations, argue that PrP^C interacts with PrP^{Sc} through a domain that contains the first two putative α helices. Whether PrP^C/peptide complexes possess prion infectivity as determined by bioassays remains to be established.

Experiments with immunoprecipitated [^{35}S]Met-labeled $SHaPrP^C$ and a >50-fold excess of purified $SHaPrP^{Sc}$ have been interpreted as showing that PrP^C becomes protease resistant under these conditions. Whether the protease resistance is due to the conversion of PrP^C into PrP^{Sc} under these conditions, or it results from the binding of PrP^C to PrP 27–30, which is quite hydrophobic, is unknown. When the ratios of PrP^C and PrP^{Sc} approximate those found *in vivo*, no evidence for the conversion of PrP^C into PrP^{Sc} could be found in cell-free systems.

Transgenic mice expressing human and chimeric PrP genes were inoculated with brain extracts from humans with inherited or sporadic prion disease to investigate the mechanism by which PrP^C is transformed into PrP^{Sc}. Although Tg(HuPrP) mice expressed high levels of $HuPrP^C$, they were resistant to Hu prions. They became susceptible to Hu prions upon ablation of the mouse PrP gene. In contrast, mice expressing low levels of the chimeric transgene were susceptible to Hu prions and registered only a modest decrease in incubation times upon MoPrP gene disruption. These and other findings argue that a species-specific macromolecule, provisionally designated protein X, participates in prion formation. Though the results demonstrate that PrP^{Sc} binds to PrP^C in a region delimited by codons 96 to 167, they also suggest that PrP^C binds protein X through residues near the C terminus. Protein X might function as a molecular chaperone in the formation of PrP^{Sc}.

To investigate proteins that might feature in the conversion of the cellular prion protein (PrP^C) into the scrapie isoform (PrP^{Sc}), neuroblastoma (N2a) cells were examined for the expression and cellular distribution of heat-shock proteins (Hsp's), some of which function as molecular chaperones. In scrapie-infected (ScN2a) cells, Hsp 72 and Hsp 28 were not induced by heat shock, sodium arsenite, or an amino acid

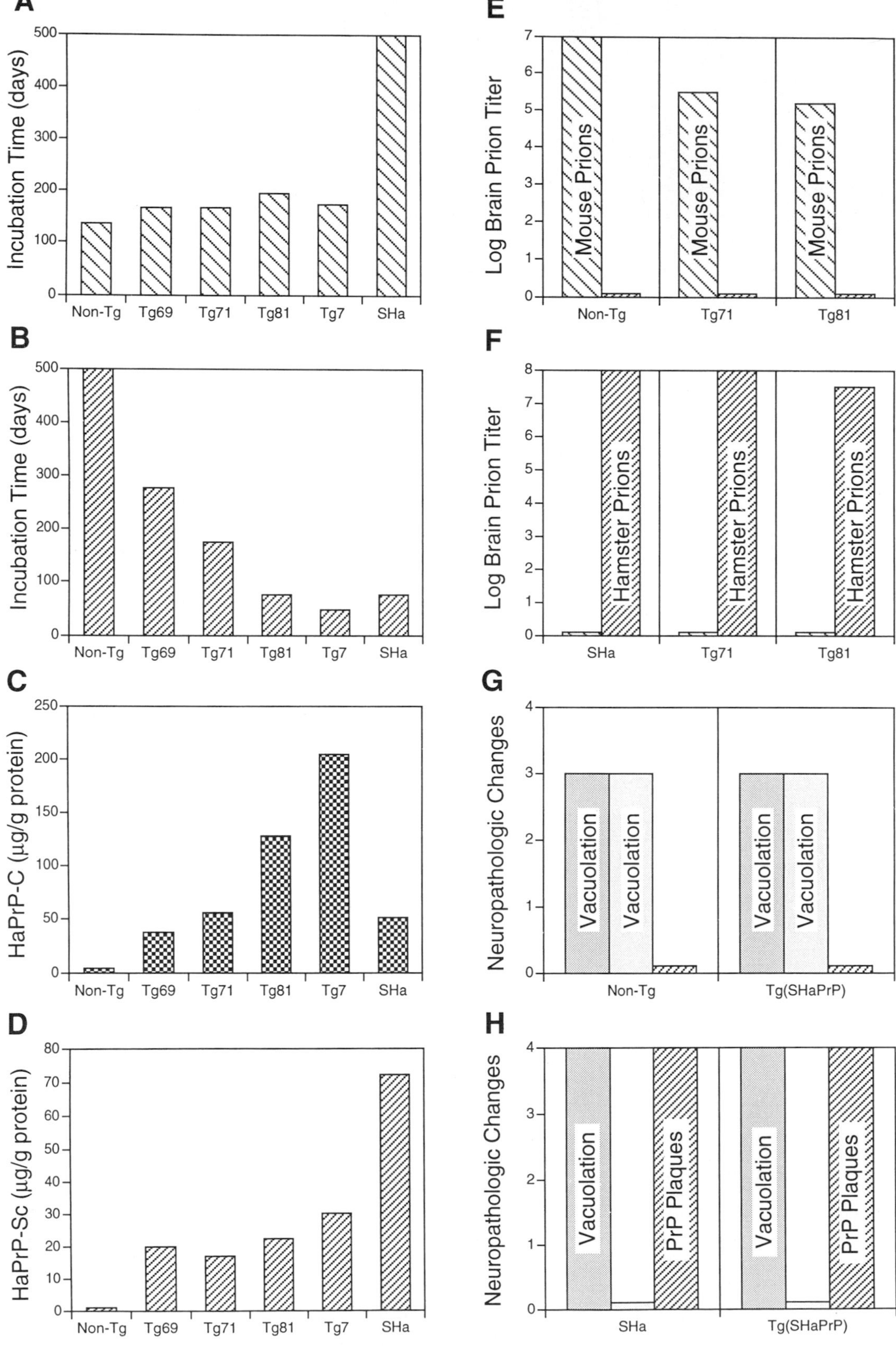
A
Incubation Time (days)
500
400
300
200
100
0
Non-Tg
Tg69
Tg71
Tg81
Tg7
SHa
B
Incubation Time (days)
C
HaPrP-C (µg/g protein)
250
200
150
100
50
D
HaPrP-Sc (µg/g protein)
80
70
60
50
40
30
20
10
E
Log Brain Prion Titer
Mouse Prions
F
Hamster Prions
G
Neuropathologic Changes
Vacuolation
Tg(SHaPrP)
H
PrP Plaques

analog in contrast to uninfected control N2a cells, whereas other inducible Hsp's were increased by these treatments. Following heat shock of the N2a cells, constitutively expressed Hsp 73 was translocated from the cytoplasm into the nucleus and nucleolus. In contrast, the distribution of Hsp 73 in ScN2a cells was not altered by heat shock; the discrete cytoplasmic structures containing Hsp 73 were largely resistant to detergent extraction. These alterations in the expression and subcellular translocation of specific heat-shock proteins in ScN2a cells may reflect the cellular response to the accumulation of PrP^{Sc}. Whether any of these Hsp's feature in the conversion of PrP^{C} into PrP^{Sc} or the pathogenesis of prion diseases remains to be established.

C. Species Barriers for Transmission of Prion Diseases

The passage of prions between species is a stochastic process characterized by prolonged incubation times. Prions synthesized *de novo* reflect the sequence of the host PrP gene and not that of the PrP^{Sc} molecules in the inoculum. On subsequent passage in a homologous host, the incubation time shortens to that recorded for all subsequent passages and it becomes a nonstochastic process. The species barrier concept is of practical importance in assessing the risk for humans of developing CJD after consumption of scrapie-infected lamb or bovine spongiform encephalopathy (BSE) infected beef.

To test the hypothesis that differences in PrP gene sequences might be responsible for the species barrier, Tg mice expressing SHaPrP were constructed. The PrP genes of Syrian hamsters and mice encode proteins differing at 16 positions. Incubation times in four lines of Tg(SHaPrP) mice inoculated with Mo prions were prolonged compared to those observed for non-Tg, control mice (Fig. 4A). Inoculation of Tg(SHaPrP) mice with SHa prions demonstrated abrogation of the species barrier, resulting in abbreviated incubation times due to a nonstochastic process (Fig. 4B). The length of the incubation time after inoculation with SHa prions was inversely proportional to the level of $SHaPrP^{C}$ in the brains of Tg(SHaPrP) mice (Figs. 4B and 4C). $SHaPrP^{Sc}$ levels in the brains of clinically ill mice were similar in all four Tg(SHaPrP) lines inoculated with SHa prions (Fig. 4D). Bioassays of brain extracts from clinically ill Tg(SHaPrP) mice inoculated with Mo prions revealed that Mo prions but no SHa prions were produced (Fig. 4E). Conversely, inoculation of Tg(SHaPrP) mice with SHa prions led to the synthesis of only SHa prions (Fig. 4F). Thus, the *de novo* synthesis of prions is species specific and reflects the genetic origin of the inoculated prions. Similarly, the neuropathology of Tg(SHaPrP) mice is determined by the genetic origin of prion inoculum. Mo prions injected into Tg(SHaPrP) mice produced a neuropathology characteristic of mice with scrapie. A moderate degree of vacuolation in both the gray and white matter was found whereas amyloid plaques were rarely detected (Fig. 4G). Inoculation of Tg(SHaPrP) mice with SHa prions produced intense vacuolation of the gray matter, sparing of the white matter, and numerous SHaPrP amyloid plaques, characteristic of Syrian hamsters with scrapie (Fig. 4H).

D. Ablation of the PrP Gene

Ablation of the PrP gene in $Tg(Prnp^{0/0})$ mice has, unexpectedly, not affected the development of these animals. In fact, they are healthy at almost 2 years of

FIGURE 4 Transgenic (Tg) mice expressing Syrian hamster (SHa) prion protein exhibit species-specific scrapie incubation times, infectious prion synthesis, and neuropathology. (**A**) Scrapie incubation times in nontransgenic mice (Non-Tg) and four lines of Tg mice expressing SHaPrP and Syrian hamsters inoculated intracerebrally with $\sim 10^6$ ID_{50} units of Chandler Mo prions serially passaged in Swiss mice. The four lines of Tg mice have different numbers of transgene copies: Tg69 and 71 mice have two to four copies of the SHaPrP transgene, whereas Tg81 have 30 to 50 and Tg7 mice have >60. Incubation times are number of days from inoculation to onset of neurological dysfunction. (**B**) Scrapie incubation times in mice and hamsters inoculated with $\sim 10^7$ ID_{50} units of Sc237 prions serially passaged in Syrian hamsters and as described in (A). (**C**) Brain $SHaPrP^{C}$ in Tg mice and hamsters. $SHaPrP^{C}$ levels were quantitated by an enzyme-linked immunoassay. (**D**) Brain $SHaPrP^{Sc}$ in Tg mice and hamsters. Animals were killed after exhibiting clinical signs of scrapie. $SHaPrP^{Sc}$ levels were determined by immunoassay. (**E**) Prion titers in brains of clinically ill animals after inoculation with Mo prions. Brain extracts from Non-Tg, Tg71, and Tg81 mice were bioassayed for prions in mice (left) and hamsters (right). (**F**) Prion titers in brains of clinically ill animals after inoculation with SHa prions. Brain extracts from Syrian hamsters as well as Tg71 and Tg81 mice were bioassayed for prions in mice (left) and hamsters (right). (**G**) Neuropathology in Non-Tg mice and Tg(SHaPrP) mice with clinical signs of scrapie after inoculation with Mo prions. Vacuolation in gray (left) and white matter (center), and PrP amyloid plaques (right). Vacuolation score: 0 = none, 1 = rare, 2 = modest, 3 = moderate, 4 = intense. (**H**) Neuropathology in Syrian hamsters and transgenic mice inoculated with SHa prions. Degree of vacuolation and frequency of PrP amyloid plaques as described in (G).

age. $Prnp^{0/0}$ mice are resistant to prions and do not propagate scrapie infectivity. $Prnp^{0/0}$ mice were sacrificed 5, 60, 120, and 315 days after inoculation with Rocky Mountain Laboratory (RML) prions passaged in CD-1 Swiss mice. Except for residual infectivity from the inoculum detected at 5 days after inoculation, no infectivity was detected in the brains of $Prnp^{0/0}$ mice.

$Prnp^{0/0}$ mice crossed with Tg(SHaPrP) mice were rendered susceptible to SHa prions but remained resistant to Mo prions. Since the absence of PrP^C expression does not provoke disease, it is likely that scrapie and other prion diseases are a consequence of PrP^{Sc} accumulation rather than an inhibition of PrP^C function.

Mice heterozygous ($Prnp^{0/+}$) for ablation of the PrP gene had prolonged incubation times when inoculated with Mo prions. The $Prnp^{0/+}$ mice developed signs of neurological dysfunction at 400–460 days after inoculation. These findings are in accord with studies on Tg(SHaPrP) mice in which increased SHaPrP expression was accompanied by diminished incubation times (see Fig. 4B).

Since $Prnp^{0/0}$ mice do not express PrP^C, we reasoned that they might more readily produce α-PrP antibodies. $Prnp^{0/0}$ mice immunized with Mo or SHa prion rods produced α-PrP antisera that bound Mo, SHa, and Hu PrP. These findings contrast with earlier studies in which α-MoPrP antibodies could not be produced in mice, presumably because the mice had been rendered tolerant by the presence of $MoPrP^C$. That $Prnp^{0/0}$ mice readily produce α-PrP antibodies is consistent with the hypothesis that the lack of an immune response in prion diseases is due to the fact that PrP^C and PrP^{Sc} share many epitopes. Whether $Prnp^{0/0}$ mice produce α-PrP antibodies that specifically recognize conformational-dependent epitopes present on PrP^{Sc} but absent from PrP^C remains to be determined.

E. Modeling of GSS in Tg(MoPrP-P101L) Mice

The codon 102 point mutation found in GSS patients was introduced into the MoPrP gene and Tg(MoPrP-P101L)H mice were created expressing high (H) levels of the mutant transgene product. The two lines of Tg(MoPrP-P101L)H mice designated 174 and 87 spontaneously developed CNS degeneration, characterized by clinical signs indistinguishable from experimental murine scrapie and neuropathology consisting of widespread spongiform morphology and astrocytic gliosis and PrP amyloid plaques. By inference, these results contend that PrP gene mutations cause GSS, familial CJD, and FFI.

Brain extracts prepared from spontaneously ill Tg(MoPrP-P101L)H mice transmitted CNS degeneration to Tg196 mice expressing low levels of the mutant transgene product and some Syrian hamsters. Many Tg196 mice and some Syrian hamsters developed CNS degeneration between 200 and 700 days after inoculation, whereas inoculated CD-1 Swiss mice remained well. Serial transmission of CNS degeneration in Tg196 mice required about 1 year and serial transmission in Syrian hamsters occurred after ~75 days. Although brain extracts prepared from Tg(MoPrP-P101L)H mice transmitted CNS degeneration to some inoculated recipients, little or no PrP^{Sc} was detected by immunoassays after limited proteolysis. Undetectable or low levels of PrP^{Sc} in the brains of these Tg(MoPrP-P101L)H mice are consistent with the results of these transmission experiments, which suggest low titers of infectious prions. Though no PrP^{Sc} was detected in the brains of inoculated Tg196 mice exhibiting neurological dysfunction by immunoassays after limited proteolysis, PrP amyloid plaques as well as spongiform degeneration were frequently found. The neurodegeneration found in inoculated Tg196 mice seems likely to result from a modification of mutant PrP^C that is initiated by mutant PrP^{Sc} present in the brain extracts prepared from ill Tg(MoPrP-P101L)H mice. In support of this explanation are the findings in some of the inherited human prion diseases, as described earlier, where neither protease-resistant PrP nor transmission to experimental rodents could be demonstrated. Furthermore, transmission of disease from Tg(MoPrP-P101L)H mice to Tg196 mice but not to Swiss mice is consistent with earlier findings demonstrating that homotypic interactions between PrP^C and PrP^{Sc} feature in the formation of PrP^{Sc}.

In other studies, modifying the expression of mutant and wild-type (wt) PrP genes in transgenic mice permitted experimental manipulation of the pathogenesis of both inherited and infectious prion diseases. Although overexpression of the wtPrP-A transgene by ~8-fold was not deleterious to the mice, it did shorten scrapie incubation times from ~145 days to ~45 days after inoculation with Mo scrapie prions. In contrast, overexpression at the same level of a PrP-A transgene mutated at codon 101 produced spontaneous, fatal neurodegeneration between 150 and 300 days of age in two new lines of Tg(MoPrP-P101L) mice designated 2866 and 2247. Genetic crosses of Tg(MoPrP-P101L)2866 mice with gene-targeted mice lacking both PrP alleles ($Prnp^{0/0}$) produced animals with a highly synchronous onset of illness between 150 and 160 days of age. The

Tg(MoPrP-P101L)2866/Prnp$^{0/0}$ mice had numerous PrP plaques and widespread spongiform degeneration in contrast to the Tg2866 and 2247 mice, which exhibited spongiform degeneration but only a few PrP amyloid plaques. Another line of mice designated Tg2862 overexpresses the mutant transgene by ~32-fold and develops fatal neurodegeneration between 200 and 400 days of age. Tg2862 mice exhibited the most severe spongiform degeneration and had numerous, large PrP amyloid plaques. Although mutant PrP^{C}(P101L) clearly produces neurodegeneration, wtPrP^{C} profoundly modifies both the age of onset of illness and the neuropathology for a given level of transgene expression. These findings and those from other studies suggest that mutant and wtPrP interact, perhaps through a chaperone-like protein, to modify the pathogenesis of the dominantly inherited prion diseases.

V. HUMAN PRION DISEASES

A. Clinical Manifestations

The human prion diseases are manifest as infectious, inherited, and sporadic disorders and are often referred to as kuru, CJD, GSS, and FFI depending on the clinical and neuropathological findings.

Infectious forms of prion diseases result from the horizontal transmission of infectious prions, as occurs in iatrogenic CJD and kuru. Inherited forms, notably GSS, familial CJD, and FFI, comprise 10–15% of all cases of prion disease. A mutation in the ORF or protein coding region of the PrP gene has been found in all reported kindreds with inherited human prion disease. Sporadic forms of prion disease comprise most cases of CJD and possibly some cases of GSS. How prions arise in patients with sporadic forms is unknown; hypotheses include horizontal transmission from humans or animals, somatic mutation of the PrP gene ORF, and spontaneous conversion of PrP^{C} into PrP^{Sc}. Numerous attempts to establish an infectious link between sporadic CJD and a preexisting prion disease in animals or humans have been unrewarding.

B. Diagnosis of Human Prion Diseases

Human prion disease should be considered in any patient who develops a progressive subacute or chronic decline in cognitive or motor function. Patients are typically adults between 40 and 70 years of age and often exhibit clinical features helpful in providing a premorbid diagnosis of prion disease, particularly sporadic CJD. There is as yet no specific diagnostic test for prion disease in the cerebrospinal fluid. A definitive diagnosis of human prion disease, which is invariably fatal, can often be made from the examination of brain tissue. Over the past four years, knowledge of the molecular genetics of prion diseases has made it possible to diagnose inherited prion disease in living patients using peripheral tissues.

A broad spectrum of neuropathological features in human prion diseases precludes a precise neuropathological definition. The classic neuropathological features of human prion disease include spongiform degeneration, gliosis, and neuronal loss in the absence of an inflammatory reaction. When present, amyloid plaques that stain with α-PrP antibodies are diagnostic.

The presence of protease-resistant PrP (PrP^{Sc} or PrP^{CJD}) in the infectious and sporadic forms and most of the inherited forms of these diseases implicates prions in their pathogenesis. The absence of PrP^{CJD} in a biopsy specimen may simply reflect regional variations in the concentration of the protein. In some patients with inherited prion disease, PrP^{Sc} is barely detectable or undetectable; this situation seems to be mimicked in transgenic mice that express a mutant PrP gene and spontaneously develop neurological illness that is indistinguishable from experimental murine scrapie.

In humans and Tg mice that have no detectable protease-resistant PrP but express mutant PrP, neurodegeneration may, at least in part, be caused by abnormal metabolism of mutant PrP. Because molecular genetic analyses of PrP genes in patients with unusual dementing illnesses are readily performed, the diagnosis of inherited prion disease can often be established where there was either little or no neuropathology, atypical neurodegenerative disease, or misdiagnosed neurodegenerative disease, including Alzheimer's disease.

Although horizontal transmission of neurodegeneration to experimental hosts was for a time the "gold standard" of prion disease, it can no longer be used as such. Some investigators have reported that transmission of the inherited prion diseases from humans to experimental animals is frequently negative when using rodents, despite the presence of a pathogenic mutation in the PrP gene, whereas others state that this is not the case with apes and monkeys as hosts. The discovery that Tg(MHu2M) mice are susceptible to Hu prions promises to make many transmission studies that were not possible in apes and monkeys amenable to experimental investigation.

The hallmark common to all of the prion diseases whether sporadic, dominantly inherited, or acquired by infection is that they involve the aberrant metabolism of the prion protein. Making a definitive diagnosis of human prion disease can be rapidly accomplished if PrP^{Sc} can be detected immunologically. Frequently PrP^{Sc} can be detected by either the dot blot method or Western immunoblot analysis of brain homogenates in which samples are subjected to limited proteolysis to remove PrP^{C} prior to immunostaining. The dot blot method exploits enhancement of PrP^{Sc} immunoreactivity following denaturation in the chaotropic salt guanidinium chloride. Because of regional variations in PrP^{Sc} concentration, methods using homogenates prepared from small brain regions can give false negative results. Alternatively, PrP^{Sc} may be detected *in situ* in cryostat sections bound to nitrocellulose membranes followed by limited proteolysis to remove PrP^{C} and guanidinium treatment to denature PrP^{Sc}, and thus enhance its avidity for α-PrP antibodies. Denaturation of PrP^{Sc} *in situ* prior to immunostaining has also been accomplished by autoclaving fixed tissue sections.

In the familial forms of the prion diseases, molecular genetic analyses of PrP can be diagnostic and can be performed on DNA extracted from blood leukocytes *ante mortem*. Unfortunately, such testing is of little value in the diagnosis of the sporadic or infectious forms of prion disease. Although the first missense PrP mutation was discovered when the two PrP alleles of a patient with GSS were cloned from a genomic library and sequenced, all subsequent novel missense and insertional mutations have been identified in PrP ORFs amplified by polymerase chain reaction (PCR) and sequenced. The 759 base pairs encoding the 253 amino acids of PrP reside in a single exon of the PrP gene, providing an ideal situation for the use of PCR. Amplified PrP ORFs can be screened for known mutations using one of several methods, the most reliable of which is allele-specific oligonucleotide hybridization. If known mutations are absent, novel mutations may be found when the PrP ORF is sequenced. [*See* DNA Markers as Diagnostic Tools.]

When PrP amyloid plaques in brain are present, they are diagnostic for prion disease as noted earlier. Unfortunately, they are thought to be present in only ~10% of CJD cases, and by definition all cases of GSS. The amyloid plaques in CJD are compact (kuru plaques). Those in GSS are either multicentric (diffuse) or compact. The amyloid plaques in prion diseases contain PrP. The multicentric amyloid plaques that are pathognomonic for GSS may be difficult to distinguish from the neuritic plaques of Alzheimer's disease except by immunohistology. In these kindreds the diagnosis of Alzheimer's disease was excluded because the amyloid plaques failed to stain with β-amyloid antiserum but stained with PrP antiserum. In subsequent studies, missense mutations were found in the PrP genes of these kindreds. [*See* Alzheimer's Disease.]

In summary, the diagnosis of prion or prion protein disease may be made in patients on the basis of (1) the presence of PrP^{Sc}, (2) mutant PrP genotype, or (3) appropriate immunohistology and should not be excluded in patients with atypical neurodegenerative diseases until one or preferably two of these examinations have been performed.

C. Iatrogenic Creutzfeldt–Jakob Disease

Accidental transmission of CJD to humans appears to have occurred by corneal transplantation, contaminated electroencephalograph (EEG) electrode implantation, and surgical operations using contaminated instruments or apparatus. A cornea unknowingly removed from a donor with CJD was transplanted to an apparently healthy recipient who developed CJD after a prolonged incubation period. Corneas of animals have significant levels of prions, making this scenario seem quite probable. The same improperly decontaminated EEG electrodes that caused CJD in two young patients with intractable epilepsy were found to cause CJD in a chimpanzee 18 months after their experimental implantation.

Surgical procedures may have resulted in accidental inoculation of patients with prions during their operations, presumably because some instrument or apparatus in the operating theater became contaminated when a CJD patient underwent surgery. Although the epidemiology of these studies is highly suggestive, no proof for such episodes exists.

Since 1988, >60 cases of CJD after implantation of dura mater grafts have been recorded. All of the grafts were thought to have been acquired from a single manufacturer whose preparative procedures were inadequate to inactivate human prions. One case of CJD occurred after repair of an eardrum perforation with a pericardium graft.

Thirty cases of CJD in physicians and health care workers have been reported; however, no occupational link has been established. Whether any of these cases represent infectious prion diseases contracted during care of patients with CJD or processing specimens from these patients remains uncertain.

D. Human Growth Hormone Therapy

The possibility of transmission of CJD from contaminated human growth hormone (HGH) preparations derived from human pituitaries has been raised by the occurrence of fatal cerebellar disorders with dementia in >90 patients ranging in age from 10 to 41 years. Though one case of spontaneous CJD in a 20-year-old woman has been reported, CJD in patients under 40 years of age is very rare. These patients received injections of HGH every 2 to 4 days for 4 to 12 years. Interestingly, most of the patients presented with cerebellar syndromes that progressed over periods varying from 6 to 18 months. Some patients became demented during the terminal phases of their illnesses. The clinical courses of some patients with dementia occurring late resemble kuru more than ataxic CJD in some respects. Assuming these patients developed CJD from injections of prion-contaminated HGH preparations, the possible incubation periods range from 4 to 30 years. Incubation periods of two to three decades have been suggested to explain cases of kuru in recent years. Many patients received several common lots of HGH at various times during their prolonged therapies, but no single lot was administered to all the American patients. An aliquot of one lot of HGH has been reported to transmit CNS disease to a squirrel monkey after a prolonged incubation period. How many lots of the HGH might have been contaminated with prions is unknown.

Five cases of CJD have occurred in women receiving human pituitary gonadotropin.

E. PrP Mutations and Genetic Linkage

The discovery of the PrP gene and its linkage to scrapie incubation times in mice raised the possibility that mutation might feature in the hereditary human prion diseases. A proline (P) → leucine (L) mutation at codon 102 was shown to be linked genetically to development of GSS with a logarithm of odds (LOD) score exceeding 3. This mutation may be due to the deamination of a methylated CpG in a germline PrP gene, resulting in the substitution of a thymine (T) for cytosine (C). The P102L mutation has been found in ten different families in nine different countries, including the original GSS family.

An insert of 144 bp at codon 53 containing six octarepeats has been described in patients with CJD from four families all residing in southern England (Fig. 5). This mutation must have arisen through a complex series of events since the human PrP gene contains only five octarepeats, indicating that a single recombination event could not have created the insert. Genealogical investigations have shown that all four families are related, arguing for a single founder born more than two centuries ago. The LOD score for this extended pedigree exceeds 11. Studies from several laboratories have demonstrated that two, four, five, six, seven, eight, or nine octarepeats in addition to the normal five are found in individuals with inherited CJD, whereas deletion of one octarepeat has been identified without the neurological disease.

For many years the unusually high incidence of CJD among Israeli Jews of Libyan origin was thought to be due to the consumption of lightly cooked sheep brain or eyeballs. Recent studies have shown that some Libyan and Tunisian Jews in families with CJD have a PrP gene point mutation at codon 200, resulting in a glutamate (E) → lysine (K) substitution. One patient was homozygous for the E200K mutation, but her clinical presentation was similar to that of heterozygotes, arguing that familial prion diseases are true autosomal dominant disorders. The E200K mutation has also been found in Slovaks originating from Orava in the former Czechoslovakia, in a cluster of familial cases in Chile, and in a large German family living in the United States. Some investigators have argued that the E200K mutation originated in a Sephardic Jew whose descendants migrated from Spain and Portugal at the time of the Inquisition. It is more likely that the E200K mutation has arisen independently multiple times by the deamidation of a methylated CpG as described earlier in the codon 102 mutation. In support of this hypothesis are historical records of Libyan and Tunisian Jews indicating that they are descended from Jews living on the island of Jerba, where Jews first settled around 500 BC, and not from Sephardim.

Many families with CJD have been found to have a point mutation at codon 178, resulting in an aspartate (D) → asparagine (N) substitution. In these patients as well as those with the E200K mutation, PrP amyloid plaques are rare; the neuropathological changes generally consist of widespread spongiform degeneration. Recently a new prion disease that presents with insomnia has been described in three Italian families with the D178N mutation. The neuropathology in these patients with FFI is restricted to selected nuclei of the thalamus. It is unclear whether all pa-

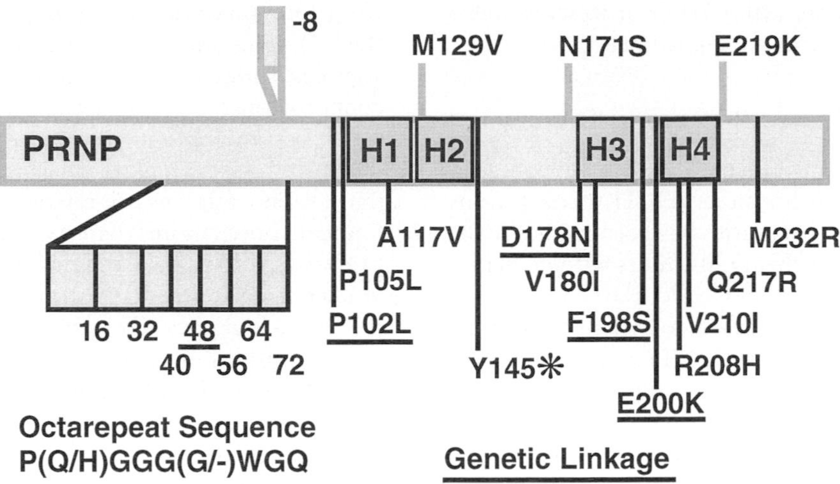

FIGURE 5 Human prion protein gene (PRNP). The open reading frame (ORF) is denoted by the large gray rectangle. Human PRNP wild-type polymorphisms are shown above the rectangle and mutations that segregate with the inherited prion diseases are depicted below. The wild-type human PrP gene contains five octarepeats [P(Q/H)GGG(G/-)WGQ] from codons 51 to 91. Deletion of a single octarepeat at codon 81 or 82 is not associated with prion disease; whether this deletion alters the phenotypic characteristics of a prion disease is unknown. There are common polymorphisms at codons 117 (Ala → Ala) and 129 (Met → Val); homozygosity for Met or Val at codon 129 appears to increase susceptibility to sporadic CJD. Octarepeat inserts of 16, 32, 40, 48, 56, 64, and 72 amino acids at codon 67, 75, or 83 are designated by the small rectangles below the ORF. These inserts segregate with familial CJD, and genetic linkage has been demonstrated where sufficient specimens from family members are available. Point mutations are designated by the wild-type amino acid preceding the codon number and the mutant residue follows (e.g., P102L). These point mutations segregate with the inherited prion diseases and significant genetic linkage (underlined mutations) has been demonstrated where sufficient specimens from family members are available. Mutations at codons 102 (Pro → Leu), 117 (Ala → Val), 198 (Phe → Ser), and 217 (Gln → Arg) are found in patients with GSS. Point mutations at codons 178 (Asp → Asn), 200 (Glu → Lys), and 210 (Val → Ile) are found in patients with familial CJD. Point mutations at codons 198 (Phe → Ser) and 217 (Gln → Arg) are found in patients with GSS who have PrP amyloid plaques and neurofibrillary tangles. Additional point mutations at codons 145 (Tyr → Stop), 105 (Pro → Leu), 180 (Val → Ile), and 232 (Met → Arg) have been recently reported. Single-letter code for amino acids is as follows: A, Ala; D, Asp; E, Glu; F, Phe; I, Ile; K, Lys; L, Leu; M, Met; N, Asn; P, Pro; Q, Gln; R, Arg; S, Ser; T, Thr; V, Val; Y, Tyr.

tients with the D178N mutation or only a subset present with sleep disturbances. It has been proposed that the allele with the D178N mutation encodes an M at position 129 in FFI, whereas a V is encoded at position 129 in familial CJD. The discovery that FFI is an inherited prion disease clearly widens the clinical spectrum of these disorders and raises the possibility that many other degenerative diseases of unknown etiology may be caused by prions. The D178N mutation has been linked to the development of prion diseases with a LOD score exceeding 5. Studies of PrP^{Sc} in FFI and familial CJD caused by the D178N mutation show that after limited proteolysis the M_r of the FFI PrP^{Sc} is ~2 kD smaller. Whether this difference in protease resistance reflects distinct conformations of PrP^{Sc} that give rise to the different clinical and neuropathological manifestations of these inherited prion diseases remains to be established.

Like the E200K and D178N(V129) mutations, a valine (V) → isoleucine (I) mutation at PrP codon 210 produces CJD with classic symptoms and signs. It appears that this V2101 mutation is also incompletely penetrant.

Other point mutations at codons 105, 117, 145, 198, 217, and possibly 232 also segregate with inherited prion diseases. Patients with a dementing or telencephalic form of GSS have a mutation at codon 117. These patients as well as some in other families were once thought to have familial Alzheimer's disease, but are now known to have prion diseases on the basis of PrP immunostaining of amyloid plaques and PrP gene mutations. Patients with the codon 198 mutation have numerous neurofibrillary tangles that stain with antibodies to τ and have amyloid plaques composed largely of a PrP fragment extending from residues 58 to 150. A genetic linkage study of this family produced a LOD score exceeding 6. The neuropathology of two patients of Swedish ancestry with the codon 217 mutation was similar to that of patients with the codon 198 mutation.

Patients with GSS who have a proline (P) $\rightarrow$ leucine (L) substitution at PrP codon 105 have been reported. One patient with a prolonged neurological illness spanning almost two decades who had PrP amyloid plaques was found to have an amber mutation of the PrP gene, resulting in a stop codon at residue 145. Staining of the plaques with α-PrP peptide antisera suggested that they might be composed exclusively of the truncated PrP molecules. That a PrP peptide ending at residue 145 polymerizes in amyloid filaments is to be expected, since an earlier study already noted showed that the major PrP peptide in plaques from patients with the F198S mutation was an 11-kDa PrP peptide beginning at codon 58 and ending at ~150. Furthermore, synthetic PrP peptides adjacent to and including residues 109 to 122 readily polymerize into rod-shaped structures with the tinctorial properties of amyloid.

One view of the PrP gene mutations has been that they render individuals susceptible to a common "virus." In this scenario, the putative scrapie virus is thought to persist within a worldwide reservoir of humans, animals, or insects without causing detectable illness. Yet 1 in 10^6 individuals develop sporadic CJD and die from a lethal "infection" while ~100% of people with PrP point mutations or inserts appear to eventually develop neurological dysfunction. That germline mutations found in the PrP genes of patients and at-risk individuals are the cause of familial prion diseases is supported by experiments with Tg(MoPrP-P101L) mice described earlier. These and other Tg mouse studies also argue that sporadic CJD might arise from either a somatic mutation of the PrP gene or the spontaneous conversion of PrP^C to PrP^{CJD}. [*See* Genetic Diseases.]

VI. A PERSPECTIVE ON PRIONS

A. Prions Are Not Viruses

The study of prions has taken several unexpected directions over the past few years. The discovery that prion diseases in humans are uniquely both genetic and infectious has greatly strengthened and extended the prion concept. To date, 18 different mutations in the human PrP gene all resulting in nonconservative substitutions have been found to be either linked genetically to or to segregate with the inherited prion diseases (see Fig. 5). Yet the transmissible prion particle is composed largely, if not entirely, of an abnormal isoform of the prion protein designated PrP^{Sc}. These findings argue that prion diseases should be considered pseudoinfections, since the particles transmitting disease appear to be devoid of a foreign nucleic acid and thus differ from all known microorganisms as well as viruses and viroids. Because much information, especially about scrapie of rodents, has been derived using experimental protocols adapted from virology, we continue to use terms such as infection, incubation period, transmissibility, and end point titration in studies of prion diseases.

B. Do Prions Exist in Lower Organisms?

In *Saccharomyces cerevisiae,* ure2 and [URE3] mutants were described that can grow on ureidosuccinate under conditions of nitrogen repression such as glutamic acid and ammonia. Mutants of ure2 exhibit Mendelian inheritance, whereas [URE3] is cytoplasmically inherited. The [URE3] phenotype can be induced by ultraviolet irradiation and by overexpression of ure2p, the gene product of ure2; deletion of ure2 abolishes [URE3]. The function of ure2p is unknown but it has substantial homology with glutathione-*S*-transferase; attempts to demonstrate this enzymatic activity with purified ure2p have not been unsuccessful. Whether the [URE3] protein is a posttranslationally modified form of ure2p that acts upon unmodified ure2p to produce more of itself remains to be established.

Another possible yeast prion is the [PSI] phenotype. [PSI] is a non-Mendelian inherited trait that can be induced by expression of the PNM2 gene. Both [PSI] and [URE3] can be cured by exposure of the yeast to 3 m*M* guanidine HCl. The mechanism responsible for abolishing [PSI] and [URE3] with a low concentration of guanidine HCl is unknown. In the filamentous fungus *Podospora anserina,* the het-s locus controls the vegetative incompatibility; conversion from the S^s to the s state seems to be a posttranslational, autocatalytic process.

If any of these cited examples can be shown to function in a manner similar to prions in animals, then many new, more rapid, and economical approaches to prion diseases should be forthcoming.

C. Common Neurodegenerative Diseases

The knowledge accrued from the study of prion diseases may provide an effective strategy for defining the etiologies and dissecting the molecular pathogenesis of the more common neurodegenerative disorders such as Alzheimer's disease, Parkinson's disease, and amyotrophic lateral sclerosis (ALS). Advances in the molecular genetics of Alzheimer's disease and ALS suggest that, like the prion diseases, an important subset is caused by mutations that result in nonconservative amino acid substitutions in proteins expressed in the CNS. Since people at risk for inherited prion diseases can now be identified decades before neurological dysfunction is evident, the development of an effective therapy is imperative.

D. Future Studies

Tg mice expressing foreign or mutant PrP genes now permit virtually all facets of prion diseases to be studied and have created a framework for future investigations. Furthermore, the structure and organization of the PrP gene suggested that PrP^{Sc} is derived from PrP^{C} or a precursor by a posttranslational process. Studies with scrapie-infected cultured cells have provided much evidence that the conversion of PrP^{C} to PrP^{Sc} is a posttranslational process that probably occurs within a subcellular compartment bounded by cholesterol-rich membranes (see Fig. 2). The molecular mechanism of PrP^{Sc} formation remains to be elucidated, but chemical and physical studies have shown that the conformations of PrP^{C} and PrP^{Sc} are profoundly different.

The study of prion biology and diseases is a new and emerging area of biomedical investigation. Though prion biology has its roots in virology, neurology, and neuropathology, its relationships to the disciplines of molecular and cell biology, as well as protein chemistry, have become evident only recently. Certainly, learning how prions multiply and cause disease will open up new vistas in biochemistry and genetics.

BIBLIOGRAPHY

Gajdusek, D. C. (1977). Unconventional viruses and the origin and disappearance of kuru. *Science* **197**, 943–960.

Guilleminault, C., Lugaresi, E., Montagna, P., and Gambetti, P. (1994). "Fatal Familial Insomnia: Inherited Prion Diseases, Sleep, and the Thalamus." Raven, New York.

Parry, H. B. (1983). "Scrapie Disease in Sheep" (D. R. Oppenheimer, ed.). Academic Press, New York.

Prusiner, S. B. (1982). Novel proteinaceous infectious particles cause scrapie. *Science* **216**, 136–144.

Prusiner, S. B. (1991). Molecular biology of prion diseases. *Science* **252**, 1515–1522.

Prusiner, S. B. (1994). Inherited prion diseases. *Proc. Natl. Acad. Sci. USA* **91**, 4611–4614.

Prusiner, S. B. (1995). The prion diseases. *Sci. Am.* **272**, 48–57.

Prusiner, S. B., Collinge, J., Powell, J., and Anderton, B. (1992). "Prion Diseases of Humans and Animals." Ellis Horwood, London.

Problem Solving

K. J. GILHOOLY
Aberdeen University

GLOSSARY

Problem reduction representation Representation of a problem as a set of goals, subgoals, sub-subgoals, and so on, which need to be accomplished before solution

State space representation Representation of a problem as a set of possible states into which the problem material can be manipulated

I. GENERAL CHARACTERIZATION

A. Defining Problem Solving

What is a problem? The definition offered by the Gestalt psychologist Karl Duncker many years ago is still serviceable. He wrote that "a problem arises when a living organism has a goal but does not know how this goal is to be reached."

This is a useful initial formulation that signals a number of points. First, a "task" set by an experimenter is not necessarily a problem for a given individual. Whether a task is a problem depends on the person's knowledge and on his or her ability to locate relevant knowledge, should he or she have it. Second, a problem may vanish or be dissolved if the person changes his or her goals. A third point is that a problem does not effectively exist until the person detects some discrepancy between his or her goals and the current situation.

B. Dimensions of Problems

Most psychological studies of problem solving have dealt with well-defined problems. If we accept the useful proposal that problems in general can be viewed as having three components (viz., a starting state, a goal state, and a set of processes that may be used to reach the goal from the starting state), then a problem is well-defined if all three components are completely specified. Problems in mathematics, in logic, and in board games tend to be well-defined. Although well-defined, such problems can be difficult, and the psychologist is faced with the task of explaining how we humans, with our various limitations, manage to solve geometry, chess, and similar scale problems in reasonable times. Of course, it will be still more difficult to explain how we tackle those ill-defined problems that are more typical of real life than the well-defined variety.

Ill-defined problems leave one or more components of the problem statement vague and unspecified. Problems can vary in degree of definedness, ranging, for example, from "make a silk purse out of a sow's ear" (which leaves possible methods unspecified) to "make something useful from a sow's ear" (which has a vague goal and unspecified methods) to "make something useful" (which leaves vague the goal, the starting state, and the methods available).

It seems a reasonable strategy for psychologists to start by studying people's ways of handling apparently well-defined problems and then to move on to consider ill-defined tasks. Research suggests that people tackle ill-defined tasks by seeking a well-defined version of the problem, which they then work within until the problem is solved or a new definition is tried. Thus studies of how well-defined problems are solved will be relevant to part of the process of solving ill-defined problems.

Another useful distinction is that between adversary and nonadversary problems. In an adversary problem, the person is competing with a thinking opponent, whereas in nonadversary problems, the struggle is with inert problem materials (real or symbolic) that

ENCYCLOPEDIA OF HUMAN BIOLOGY, Second Edition, VOLUME 7. Copyright © 1997 by Academic Press. All rights of reproduction in any form reserved.

are not reacting to what the person does with a view to defeating him or her (despite what the problem solver may feel).

A third distinction that has become increasingly important in problem-solving research is that between knowledge-rich and knowledge-lean problems. To a large extent, this distinction refers to the solver's view of the problem. A problem is knowledge-rich for solvers who bring considerable relevant knowledge to the problem. For example, if someone has just been told the basic rules of a game that they have never encountered before, a problem in that game would be knowledge-lean for that person, but the same problem would be knowledge-rich for an expert player of the game. Many artificial puzzles used in studies of problem solving are knowledge-lean for most people. Until relatively recently, most studies of problem solving focused on knowledge-lean puzzles, particularly in the case of nonadversary problem solving. More recently there has been considerable interest in the study of knowledge-rich nonadversary tasks (e.g., computer programming and physics problem solving). Such studies frequently involve contrasts between the behavior of people for whom the problems concerned are knowledge-rich (experts in the area) with people for whom the problems are knowledge-lean (novices in the area).

C. Outline of Main Stages in Problem Solving

A few steps are generally discerned in problem solving. These are the following:

1. Detecting that a problem exists; realizing that there is a discrepancy between the current situation and a goal and that no way is known to reach solution without search. In the course of working toward the major goal, numerous specific subgoals may arise and provide further targets for problem solving.

2. Formulating the problem more completely. How are the starting conditions to be defined and represented? How are the goals to be defined and represented? What relevant actions are available? A problem formulation is an internal representation (or mental model) of the problem components, and solution attempts are made within the representation.

Two broad ways of representing problems are (a) in state space terms and (b) in problem reduction terms. In the state space approach, the solver typically works forward from the starting state and explores alternative paths of development as these branch out from the starting state. Within the state space approach, the solver may also work backward by inverse actions from goal states toward the starting state. A resulting successful sequence discovered by mental exploration could then be run forward to actually reach a goal state from the starting state. Indeed, both forward and backward exploration may be tried in various mixtures within the same problem attempt. In the problem reduction approach, the overall problem is split up into subproblems. For example, if the goal state involves a conjunction of conditions (achieve X *and* Y *and* Z), problem reduction would seek to achieve each condition separately (either by further problem reduction or by state space search). A key method within problem reduction is means–ends analysis. In this method, subgoals are set up for achievement such that those subgoals would lead to achievement of higher level goals. These subgoals correspond to subproblems, which can themselves be reduced. Problem and subproblem reduction continue until subgoals are reached that can be immediately solved because a method is already known for them.

3. Given a representation and a choice of approach (state space or problem reduction), solution attempts can begin. Although in the case of well-defined problems, computers could in theory continue until solution or until all possible move sequences have been exhausted, humans usually give up—or at least set problems aside temporarily—long before all possible state action or problem reduction sequences have been explored. That is to say, stop rules are invoked. Different types of stop rules can be readily imagined. For example, one type of rule would be to stop exploring within a particular representation if solution has not been reached by a certain time and return to the problem formulation stage to produce a revised formulation for further exploration. A second type of rule would be to accept a "good enough" problem state rather than continue exploration for an optimal state (sometimes called *satisficing*). A third type of rule would be to only stop if an optimal solution is reached (i.e., a "maximizing" rule). A further type of stop rule would be an "abandon" rule (i.e., a judgment that further reformulation and/or exploration would not be worthwhile and the problem should be left unsolved). Intuitively, this would seem to involve rather complex assessments of the chances of succeeding or failing and of the cognitive and other costs of continuing work or abandoning the problem. Again, this type of stop rule appears to have been little researched.

II. COMMON BLOCKS TO PROBLEM SOLVING

A. Set

"Sets" are fixed approaches to solving certain classes of problems, and these fixed approaches can be powerful blocks to solution. A robust demonstration of such effects can be obtained using water jar problems. In these tasks, the subjects are asked to say how they could get exactly a specified amount of water using jars of fixed capacities and an unlimited source of water [e.g., given three jars (A, B, C) of capacities 18, 43, and 10 units, respectively, how could you obtain exactly 5 units of water?]. The solution may be expressed as B − A − 2C. After a series of problems with that same general solution, subjects usually have great difficulty with the following problem: Given three jars (A, B, and C) of capacities 28, 76, and 3 units, respectively, how could you obtain exactly 25 units of water? In fact, the solution to this problem is very simple (i.e., A − C), but when it comes after a series of problems involving the long solution (B − A − 2C), many people fail to solve or are greatly slowed down compared with control subjects.

Similar results have also been found with other types of problems such as anagrams. If subjects have a run of anagrams with the same scrambling pattern, they will be slowed down when an anagram with a different scrambling pattern appears in the sequence. Again, if there is a run of "animal" solutions to a sequence anagram, solving will be slowed down when a "nonanimal" solution is required.

As well as training effects, the layout of a problem can induce strong sets. For example, if subjects are given a pattern of nine dots as below

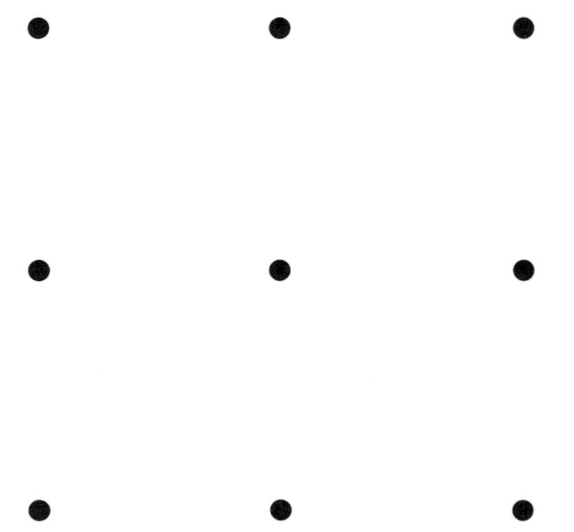

and asked to draw four straight lines to connect up all the dots without raising the writing tool from the paper, few solve it within 20 min. Most subjects are mislead by the square layout into keeping their efforts within the square (although no such restriction was imposed in posing the problem). The problem cannot be solved while this "set" is dominant. Solution

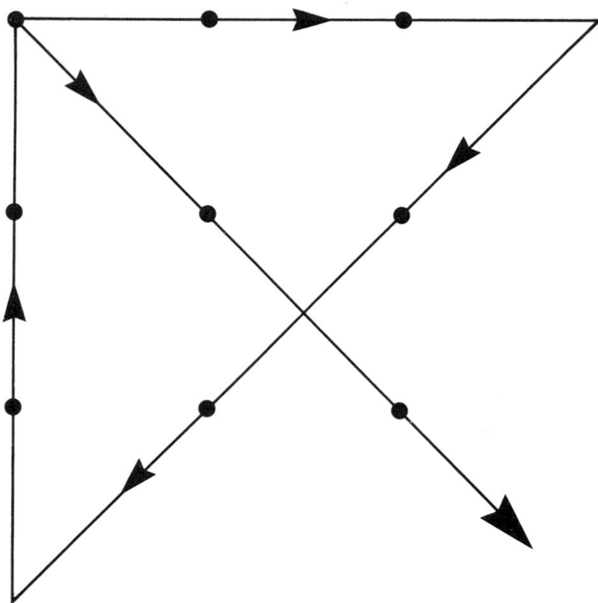

requires the lines to go outside the square shape.

B. Functional Fixity

A related block to effective problem solving, known as *functional* fixity, has also been identified. Functional fixity tends to arise when an object has to be used in a new way to solve some problem. The classic study of functional fixity was carried out by Duncker in 1945 using the "box" (or "candle" problem). In this task, subjects were presented with tacks, matches, three small boxes, and three candles. The goal was to put the candles side by side on a door, in such a way that they could burn in a stable fashion. For one group of subjects the boxes were empty, but for the other group (experimental group) the boxes were used as containers and held the matches, tacks, and candles. The solution is to use the boxes as platforms and fix them to the door using the tacks. It was found that the solving rate was much higher in the control group than in the experimental group. Duncker explained this result in terms of a failure to perceive the possible platform function of the boxes when they were presented as containers. Functional fixity has been independently demonstrated and further investigated in a large number of later studies. The phenomenon of functional fixity is a robust one and, along with "set," is doubtless a major source of difficulty in real-life problem solving.

III. EXPERTISE IN PROBLEM SOLVING

It is a striking fact that an expert in a given area can come up with much better and often faster solutions than can a beginner. How is this done? Does the expert, for instance, mentally run through a much larger number of possible actions more rapidly than the beginner?

Studies of chess skills have been informative about the nature of expertise; chess is a good area for this type of research because differences in skill level are easily measured. Early studies by the Dutch psychologist De Groot reported that experts and less-skilled players differed relatively little in the amount of mental searching they did about possible moves and countermoves. De Groot was able to keep track of the mental search patterns of the players by having them think aloud as they chose moves. Both his expert and less-expert groups of players looked ahead a maximum of about six steps and considered totals of about 35–50 moves and countermoves before deciding on their own best move. However, the more-skilled players always chose much better moves than did the less-skilled. A clue as to how the chess masters differed from the less-skilled players came from studies in which the subjects were shown chess positions for short times (5 sec). After such short exposures, master level players could recall the board with greater than 90% accuracy, whereas less-skilled players performed with around 40% accuracy. When the chess positions were made up by placing pieces at random on the board, however, both the master level and less-skilled players scored about 40% correct recall of the boards after short exposures. So it was inferred that the master level players had built up in their permanent or long-term memories a large number of familiar chess patterns that they could then recognize when a new but realistic position appeared. If pieces were arranged at random, few if any familiar patterns would appear, and performance would fall back to amateur levels. Similarly, in playing the game and choosing moves, familiar patterns will be recognized and will guide the search more effectively. More recent research, using larger numbers of subjects and a wider range of chess skill, has also shown that more skilled players do in fact tend to search somewhat more extensively and more rapidly than do less-skilled players and are better at evaluating the potential of intermediate positions. The amount of searching is, however, but a fraction of that carried out by the best computer chess programs.

Evidence from biographies of leading chess players indicates that about 10 years of concentrated study is required to reach expert levels. An early start also seems beneficial. Studies of other areas of mental skills likewise indicate about a 10-year period of intensive practice before high-level performance is reached. Thus experts' apparently effortless solving of new problems in their fields is purchased with many years of prior effort. Prior experience has built up the extensive background knowledge that experts use to describe new problems effectively, to fill in implicit details, and to anticipate likely directions of solution. Results similar to those in chess have been reported now in a wide range of problem areas, including physics, political science, mathematics, and programming. However, when experts encounter problems that exceed their expertise, they too must revert to the more general methods of search outlined in Section I of this article.

BIBLIOGRAPHY

Ericsson, K. A., and Smith, J. (1991). "Toward a General Theory of Expertise: Prospects and Limits." Cambridge Univ. Press, Cambridge, England.

Garnham, A., and Oakhill, J. (1994). "Thinking and Reasoning." Routledge, London.

Gilhooly, K. J. (1996). "Thinking: Directed, Undirected and Creative," 3rd Ed. Academic Press, London.

Sternberg, R. J. (ed.) (1994). "Thinking and Problem Solving." Academic Press, New York.

Programmed Cell Death, Apoptosis[1]

JOHN F. BOYLAN
Dupont Merck Pharmaceutical Company

GLOSSARY

Apoptosis A descriptive term used to describe a series of specific cellular events leading to the systematic death and destruction of the cell. Such a process is part of a signal transduction pathway initiated as a result of an environmental stress or intracellular damage

Cell cycle Highly ordered process of cellular duplication and division resulting in the production of two identical daughter cells; often defined by the amount of replicated DNA present in the cell: G1 phase being the normal amount of DNA in a single cell, S phase being the period of new DNA synthesis copied from the original cellular DNA, and G2 and M phase being twice the amount of cellular DNA present in the cell, followed by mitosis and cytokinesis

DNA ladder Cleavage of cellular DNA between the nucleosomal sections of DNA resulting in a "ladder" of increasing sized DNA fragments at 100-bp intervals. Often used as a marker of apoptosis; however, not all cells produce a DNA ladder

Necrosis Unregulated destruction of the cell in response to environmental stress or large-scale intracellular damage

Programmed cell death A type of highly regulated cell death that occurs as part of normal development and tissue remodeling in response to developmental cues. Apoptosis and programmed cell death share some of the same characteristics

AN OVERALL INCREASE OR DECREASE IN CELLS OF a complex organism is a fine balance between the rate of growth and death. The process of regulated cellular self-destruction has gained increased notoriety as a cellular signaling system as important and complex as cellular proliferation.

I. INTRODUCTION

Programmed cell death (PCD) describes the events surrounding the specific and often times systematic removal of proliferating cells through a signal transduction pathway whose endpoint is self-elimination. PCD was described earlier this century when during development selected cells appeared to shrink in size and the chromatin condensed into a large mass. Many years later it became apparent that PCD is a normal physiological and developmental mechanism present in many cell types regulating the overall cellular population within an organ system. Such a system plays a central role in maintaining tissue homeostasis as well as coordinating the schedule destruction of cells during normal development. The term "apoptosis" was derived from the Greek word describing leaves falling from the tree and was used to first describe a histological phenotype observed in tissue sections that had no obvious disease. Sometimes the terms PCD and apoptosis are used synonymously. PCD relates to events during development and tissue remodeling whereas apoptosis relates more to events induced by cellular damage. Although PCD and apoptosis are both highly regulated and require *de novo* gene ex-

[1]For any commercial use of the information obtained in this article, or to reproduce or otherwise use this article in whole or in part, permission must be obtained from the DuPont Merck Pharmaceutical Company, the author, and Academic Press.

ENCYCLOPEDIA OF HUMAN BIOLOGY, Second Edition, VOLUME 7. Copyright © 1997 by Academic Press. All rights of reproduction in any form reserved.

pression, the two events are different. For example, a study comparing cell death in T cells and intersegmental muscles of a developing moth demonstrated differences in cell surface morphology, nuclear, and DNA fragmentation. It remains to be determined exactly which characteristics of PCD and apoptosis overlap.

II. TYPES OF CELL DEATH

A. Necrosis

Necrosis and apoptosis are descriptive terms used to describe the morphological characteristics of cell death (Table I). Necrosis is very different from apoptosis. Necrosis is the rapid and complete breakdown of all cellular systems in response to irreversible damage. This process is characterized by random clumping and distribution of nuclear chromatin; swelling of organelles, particularly the mitochondria; and the rupture of membranes and release of contents. At later stages the cell membrane ruptures and a host immune response follows with mononuclear phagocytic cells arriving to scavenge the debris.

B. Apoptosis/PCD

Apoptosis and PCD are systematic, energy-dependent, controlled events regulating the death and destruction of a cell (Table I). Apoptosis has been technically easier to study than PCD and so a better understanding of apoptosis exists. The rest of this article focuses on events common to both apoptosis and PCD. The regulation of apoptosis is an active process involving new gene transcription. Early apoptosis is observed when nuclear chromatin condense and segregate into groups along the inner nuclear membrane. In contrast to necrosis, apoptotic cells begin to shrink. Folding along the nuclear membrane and condensation of the cytoplasm follow, with organelles remaining intact. In fact, it has been shown that cells undergoing early stages of apoptosis are still able to carry out several normal metabolic activities. Nuclear fragmentation and continued condensing of the cytoplasm signal the last stage of apoptosis, leaving encapsulated cellular debris termed "apoptotic bodies" that is phagocytized by neighboring cells and degraded by lysosomal enzymes. Apoptosis is characterized by the absence of a localized immune response.

III. OVERVIEW OF SIGNALING EVENTS CONTROLLING PCD/APOPTOSIS

This is an area of active research that remains incomplete. A number of important regulatory molecules have been identified, but their precise role and interrelationship remain to be uncovered. Clearly the events surrounding PCD are highly conserved between a wide range of species from bacteria to plants to mammals. For example, the apoptosis regulatory gene bcl-2 first isolated in mammalian cells is highly homologous to the worm *Caenorhabditis elagans* gene, ced-9. Ectopic expression of bcl-2 in *C. elagans* produces a phenotype similar to a ced-9 gain of function mutation. Such conservation of gene function suggests that once early signaling events evolved they became so important that their function was conserved during

TABLE I
Morphological Comparison of Apoptosis and Necrosis

	Apoptosis	Necrosis
Nucleus	Chromatin condenses	Irregular clumping of chromatin
	Membrane involution and budding	Membrane rupture
	Nucleosomal DNA fragmentation	Random DNA degradation
Cytoplasm	Condensation of cytoplasm	Swelling of endoplasmic reticulum and mitochondria
	Microvilli disappear	Rupture of organelle membranes
	Increase in presence of vacuoles	Destruction of all cellular structures
	Normal metabolism	No metabolic activity
	Organelles remain intact	
Cell membrane	Membrane blebbing	Cell swelling
	Formation of apoptotic bodies	Total lyses of cell
	Shrinkage of cell	

mammalian evolution. Such information allows selected gene function to be accurately studied in primitive *in vivo* model systems that are much more amenable to experimental manipulation. The apoptotic function of four gene families is discussed next. [*See* Cell Signaling.]

A. Bcl-2 Family

Members of the Bcl-2 gene family play a central role in the regulation of apoptosis. Two groups of genes are represented: the "death promoting" composed of the Bax, Bak, and Bcl-Xs genes and the "death inhibiting" composed of the Bcl-2, Bcl-Xl, and Bag-1 genes. The exact function of the encoded proteins remains unclear. Bcl-2 has been shown to block apoptosis following genotoxic insult. This protection is blocked by Bax, allowing apoptosis to occur. It has been suggested that these proteins interact as homo- and heterodimers and that the ratio between the expression level of the "death-promoting" group and the "death-inhibiting" group signals apoptosis in response to genotoxic damage, serum starvation, or activation of the Fas receptor. Bcl-2 related proteins are localized within the outer mitochondria, nuclear, and endoplasmic reticulum membranes. Recent research supports the idea that these proteins function as pore-forming channels regulating membrane permeability. [*See* Mitochondria; Nuclear Pore, Structure and Function.]

B. p53

Monitoring the cell's health as it traverses the cell cycle is an important mechanism by which a cell can ensure correct growth and division. G1-S and G2/M are two important checkpoints during the cell cycle that allow the cell to survey itself for improper cell division. Cellular p53 exists as a multicomponent transcriptional regulatory complex that plays a central role in either growth arrest or apoptosis at these checkpoints in response to improper growth signals or DNA damage. In general, the function of p53 is to protect the cell and its progeny from improper cell division and mutations through the regulation of gene transcription. The exact p53 signaling mechanism remains elusive. Expression of the p53 protein is increased following DNA damage or removal of serum from cultured cells correlating with a G1 cell cycle arrest. Additionally, p53 can act as a transcriptional regulator in controlling the expression of selected growth and apoptosis-related genes. For example, evidence shows that p53 may regulate the expression of Bax and Bcl-2. [*See* Cell Cycle; Tumor Suppressor Genes, p53.]

C. c-*myc*

The protooncogene c-*myc* encodes a nuclear transcription protein. Like p53, c-*myc* functions to monitor the quality of the cellular environment and can act as an inducer of either apoptosis or proliferation under certain circumstances. c-*myc* was first identified for its role in regulating proliferation. Overexpression of c-*myc* stimulates proliferation. Interestingly, the same cells grown in the absence of certain growth factors undergo apoptosis.

D. ICE Protease

This group of proteases consists of specific cysteine proteases, the prototype of which is the ICE protease or interluekin (IL)-1β-converting enzyme. The ICE protease derives its name from the cleavable substrate of the enzyme: IL-1β. Other members of this enzyme family include Nedd2/Ich-1, TX/ICE-II/Ich-2, ICE-III, CPP32/Yama, and Mch2. Any member of this group of proteases, when transfected into a variety of cells, can induce apoptosis, presumably due to the inherent enzyme activity. Not all members share the same substrate specificity, suggesting separate functions. The potential variety of substrates and the precise role of these proteases in apoptosis remain unclear. It has been suggested that lamin B1, poly(ADP-ribose)polymerase, topoisomerase I, retinoblastoma protein, and β-actin may serve as protease substrates during the activation of apoptosis.

E. Tumor Necrosis Factor Superfamily

The tumor necrosis factor (TNF) gene family is composed of membrane receptors that share specific domain homology. Members of this group include TNF, CD95 (Fas/Apo-1), NGF, CD40, and CD27 receptors. These receptors span the plasma membrane of specific cell types. Once bound with their respective ligands, they transmit the apoptotic signal within the cell. The events leading up to DNA degradation and completion of the apoptosis signal are intense areas of investigation.

IV. ROLE OF PCD/APOPTOSIS IN DISEASE

The inappropriate activation or inactivation of apoptosis relates to several different disease states.

Several degenerative diseases produce substantial cell loss through apoptosis. For example, Alzheimer's disease (AD) is characterized by neuronal cell death. *In vitro* and *in vivo* studies support the role of the secreted B-amyloid protein in the degenerative nature of AD, although the exact sequence of events is still unknown. [*See* Alzheimer's Disease.]

The absence or reduction of the apoptotic mechanism has a profound effect on the development and progression of many different forms of cancer. It is thought that one of the early events in cellular transformation is the deregulation of apoptosis. Such a change in cellular proliferation would help serve as a source for additional gene mutations. A normal cell carrying such mutations would presumably destroy itself through apoptosis. In cancer, the reduction of apoptosis creates a way for a cell to generate mutations that are then selected for as a growth or survival advantage driving progression toward a malignant phenotype. Such mutations may provide increased resistance to DNA-damaging agents, secretion of activated proteases for metastasis, growth factors for promoting vascularization, or simply the elimination of checkpoint controls in the cell cycle. Mutations of bcl-2, p53, or c-*myc* have been suggested to be early events in the development of cancer.

Diseases associated with the immune system often stem from improper apoptosis. The immune system normally undergoes a high level of apoptosis. Autoreactive cells are destroyed as lymphocytes proliferate and differentiate in various parts of the body, preventing immune cells from reacting against ones own self. The breakdown of this control may lead to various autoimmune diseases such as arthritis and systemic lupus erythematosis. Both the suppression and the activation in the rate of T-cell apoptosis have been observed following HIV infection. In an effort to prolong T-cell survival and increase viral replication, HIV delays the initial apoptosis event in infected T cells. Following viral replication and packaging, apoptosis is initiated, helping to release mature infectious virions. [*See* Lymphocytes; Mutation Rates; T-Cell Activation.]

V. EXPERIMENTAL METHODS USED TO STUDY PCD/APOPTOSIS

Current techniques used to evaluate and quantitate apoptosis center around DNA degradation. These methods range in difficulty. The simplest is staining cells with either a DNA-specific dye (DAP I, Hoechst 33258) or hematoxylin and eosin (H&E) staining. The breakdown of the nuclear membrane and DNA aggregates within the cell are scored as apoptotic and are often expressed as an apoptotic index. In comparison, necrotic cell DNA stains diffusely. A more difficult method is the enzymatic labeling of degraded DNA ends using the TUNNEL procedure or direct visualization of degraded DNA via gel electrophoresis. Other apoptotic assays include measuring endonuclease activity, changes in intracellular pH, intracellular Ca^+ alterations, or changes in phospholipid membrane content. It is important to be aware that normal physiological events can produce similar changes and each assay must be validated with the proper controls.

VI. SUMMARY

Many unanswered questions remain, despite the ever-increasing research into the mechanism of programmed cell death and apoptosis. Although several key regulatory proteins have already been identified, the exact mode of action is unclear. It is not known how such proteins signal PCD or substrate specificity. For example, little is known surrounding the identity of various protease targets. How does c-*myc* or bcl-2 signaling play a role in PCD signal transduction versus cell proliferation? How are cell proliferation and PCD related and regulated? How are apoptosis and PCD related and how do they differ? As often is the case, as more information is revealed, the actual complexity and highly regulated nature of PCD becomes appreciated.

Apoptosis-related WEB sites:

1. http://www.access.digex.net/~regulate/apolist.html
2. http://www.celldeath-apoptosis.org/

BIBLIOGRAPHY

Kumar, S. (1995). ICE-like proteases in apoptosis. *Trends Biochem. Sci.* **20**, 198–202.

Schwartz, L. M., and Osborne, B. A. (eds.) (1995). "Methods in Cell Biology: Cell Death." Academic Press, San Diego.

Uren, A. G., and Vaux, D. L. (1996). Molecular and clinical aspects of apoptosis. *Pharmacol. Ther.* **72**, 37–50.

Wyllie, A. H. (1994). Apoptosis: Death gets a brake. *Nature* **369**, 272–273.

Zakeri, Z., Bursch, W., Tenniswood M., and Lockshin R. A. (1995). Cell death: Programmed, apoptosis, necrosis, or other. *Cell Death Differ.* **2**, 87–96.

Proprioceptive Reflexes

S. C. GANDEVIA
DAVID BURKE
Prince of Wales Medical Research Institute, University of New South Wales

GLOSSARY

Antagonistic muscles Muscles with actions that are anatomically opposite (e.g., flexion and extension of a joint)

Corticopontine Pertaining to fibers that arise in (or close to) the motor cortex and are destined for the brain stem

Corticospinal Pertaining to fibers that arise in (or close to) the motor cortex and reach the spinal cord

Disynaptic Involving two synapses

Homonymous Homonymous motoneurons (or receptors) are those innervating (or in) the same muscle

Interneuron Relay cell within the spinal cord

Latency Time taken between a stimulus and a response

Monosynaptic Involving only one synapse

Motoneuron Cell within the brain stem or spinal cord that sends an axon to a skeletal muscle and innervates a number of muscles fibers

Oligosynaptic Involving a few synapses (perhaps three to five)

Postsynaptic inhibition Mechanism for reducing the excitability of a neuron via a neurotransmitter, which hyperpolarizes the neuron

Presynaptic inhibition Mechanism for reducing the efficacy of a synapse by reducing the amount of neurotransmitter released at the synapse

Proprioceptive input Input to the nervous system from receptors in muscles, joints, and skin

Proprioceptive reflex Relatively stereotyped automatic motor response to a proprioceptive input

Reticulospinal pathways Descending pathways from the brain stem to the spinal cord

Synapse Specialized connection between one neuron and a second neuron

Synergistic muscles Muscles that act together to produce a particular action

ALL HUMAN MOVEMENTS, EXCEPT PERHAPS EXtremely rapid ones, can be modified during their performance. These modifications may be deliberately initiated by the subject once they are aware that the initial movement is no longer appropriate or has run into an unanticipated obstacle. In addition, however, movement is continuously modulated by activity in reflex pathways, and the subject may be quite unaware that these modifications are occurring. Reflex activity forms an integral part of normal voluntary movement: only rarely does one occur without the other. Apart from voluntary motor acts and their reflex modulation, some acts can be programmed voluntarily to occur only in response to a particular stimulus (triggered reactions). Some skilled motor acts (e.g., signing one's name, driving a car) can become so learned and automatic that the intrusion of consciousness rarely occurs. By convention, these are not considered reflex acts. The following discussion considers some of the circuits responsible for the reflex modulation of movement.

Reflexes depend on information detected by specific receptors and transmitted to the central nervous system to modify either the motor program (including the motor command reaching neurons in the motor cortex) or the effects of the descending corticospinal drive onto the motoneuron pools in the brain stem and spinal cord. The feedback that acts to modify movement may involve the special senses (e.g., vision or balance mechanisms in the inner ear), or may come from the periphery as a consequence of the movement

ENCYCLOPEDIA OF HUMAN BIOLOGY, Second Edition, VOLUME 7. Copyright © 1997 by Academic Press. All rights of reproduction in any form reserved.

itself (e.g., from receptors in the skin, joint, and muscles of the moving limb). Only the latter (i.e., proprioceptive reflexes) will be considered in this article.

I. REFLEX CIRCUITS

A. Spinal Circuitry

The spinal cord, particularly at the cervical and lumbar enlargements (which contain the innervation for the limbs), has complex circuitry, the function of which is only now being fully dissected in experimental animals such as the cat. In human subjects, a number of simple circuits have been defined physiologically and their role in the control of movement studied. Only those circuits that have been defined for human subjects will be discussed in detail: monosynaptic Ia excitation, olgiosynaptic Ia excitation, reciprocal inhibition, group Ib inhibition, recurrent inhibition, and presynaptic inhibition of the Ia afferent terminals. Ia and Ib afferents innervate primary muscle spindle endings and Golgi tendon organs, respectively. [*See* Proprioceptors and Proprioception.]

In addition, although the precise circuitry has not been determined for human subjects, cutaneous and joint afferents exert reflex effects on motoneurons, and this circuit is subject to supraspinal control. [*See* Spinal Cord.]

1. Reflexes Involving Muscle Afferents

The simplest spinal circuit is the direct (monosynaptic) connection between group Ia afferents from muscle spindles and the motoneurons of their own (homonymous) muscle (Fig. 1). There are also monosynaptic connections to the motoneurons innervating synergistic muscles and a disynaptic inhibitory pathway mediating reciprocal inhibition to motoneurons innervating antagonistic muscles. The monosynaptic influence on synergists is usually weaker than the homonymous connection and can even affect muscles that act at nearby joints. Experimental evidence indicates that each primary muscle spindle afferent makes synaptic connections with virtually all motoneurons in the homonymous muscle. However, the absolute strength of the synaptic connection between a single spindle afferent and motoneuron is relatively weak compared with the synaptic drive required to discharge a motoneuron. Any movement or disturbance that synchronizes the muscle spindle input will provide a proportionately greater effect on the motoneuron pool. The reciprocal inhibition is mediated via an interneuron termed the Ia inhibitory interneuron. Each such interneuron projects to a small fraction of the antagonist motoneuron pool.

Apparently, as in the cat, another excitatory connection exists between group Ia afferents and their homonymous motoneurons, exerted through an oligosynaptic pathway. This has been documented for some upper and lower limb muscles in humans. The relevant interneurons (and the Ia inhibitory interneuron) are the site of convergent input from cutaneous afferents and from descending pathways activated during a voluntary contraction (presumably including the corticospinal pathway). Group II muscle afferents innervating secondary muscle spindle endings probably also exert an excitatory effect on homonymous motoneurons in humans, although the synaptic details are not yet clear.

The monosynaptic Ia pathway contains no interneurons, but the level of activity in this circuit can still be modulated through the mechanism of presynaptic inhibition (Fig. 2). This is mediated by interneurons that synapse on the terminals of the Ia afferent fibers before they reach the motoneuron, thus forming an axo–axonic synapse that is presynaptic to the Ia–

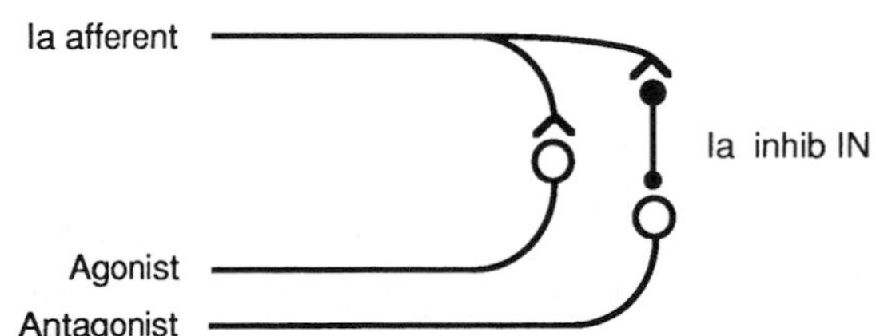

FIGURE 1 The monosynaptic excitatory connection between muscle spindle afferents (Ia afferents) and the agonist (homonymous) motoneuron pool, and the disynaptic inhibitory connection to the antagonist muscle via the Ia inhibitory interneurons (Ia inhib IN). Solid symbols indicate inhibitory interneurons.

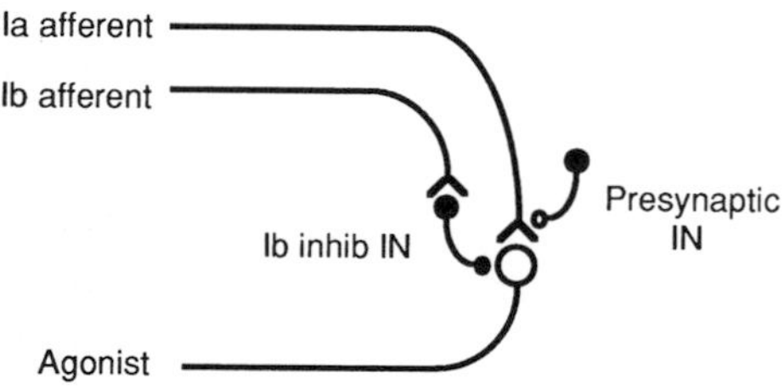

FIGURE 2 The inhibitory connection between Golgi tendon organ afferents (Ib afferent) and the agonist motoneuron pool via the Ib inhibitory interneuron (Ib inhib IN). Also shown is a presynaptic interneuron which can reduce the effectiveness of synaptic transmission.

motoneuron synapse. Activity of this interneuron diminishes the amount of excitatory transmitter released by an impulse traveling along the Ia afferent fiber and so decreases the excitatory effect of that impulse on the motoneuron. The presynaptic inhibitory interneuron receives powerful inputs from cutaneous and muscle afferents from the periphery and is also subjected to supraspinal control, its activity being depressed during contraction of the homonymous muscle. Although the full extent and functional importance of presynaptic inhibition are not established, it provides the central nervous system with a potent means of controlling the input to spinal and, possibly, supraspinal sites. Thus, when standing unsupported, presynaptic inhibition may be increased for some muscles (soleus) and decreased for others (quadriceps).

Because fusimotor neurons (both gamma and beta) are activated in a deliberate voluntary contraction of a muscle, the activity of group Ia afferents is increased, and this activity enters spinal reflex circuits that are already "primed" by the actions on interneurons described in the foregoing. Hence, voluntary contraction of a muscle is associated with enhancement of the reflex influence of group Ia afferents from the contracting muscle on its homonymous alpha motoneurons and on the alpha motoneurons of synergists. There is also parallel inhibition of antagonists via the disynaptic reciprocal inhibitory pathway. By contrast with their powerful reflex effects on alpha motoneurons, group Ia afferents have relatively weak reflex effects on gamma motoneurons, which may be just as well because otherwise the Ia input could drive gamma motoneurons to produce more Ia activity, and this might create an unstable oscillation. [*See* Fusimotor System.]

Group Ib afferents from Golgi tendon organs convey a message related to the force of contraction and have an inhibitory effect on homonymous and synergistic motoneurons through a disynaptic pathway (see Fig. 2). The group Ib effect via the Ib inhibitory interneuron has a widespread influence within the spinal cord, possibly more diffuse than the group Ia effect. The Ib inhibitory interneuron receives other peripheral inputs, for example, from cutaneous and joint afferents (see the next section). In the cat, it also receives some excitatory input from homonymous group Ia afferents, with the result that group Ia activity can theoretically both excite and inhibit homonymous motoneurons. This apparently paradoxical arrangement is one of several means by which the central nervous system may modulate the excitatory effect of muscle spindle inputs to motoneurons. It is not known whether or not this connection exists in humans.

2. Reflexes Involving Cutaneous and Joint Afferents

The afferent fibers from mechanoreceptors in the skin have complex actions on both alpha and gamma motoneurons, the effect being dependent on the skin region stimulated. The spinal reflex effects are mediated by oligosynaptic pathways, containing at least two interneurons, both of which receive inputs from other sources, both peripheral (e.g., group Ib and group II muscle afferents) and supraspinal (e.g., corticospinal and reticulospinal pathways).

Relatively little is known about the specific reflex effects of the different classes of cutaneous mechanoreceptors. Clearly, however, afferents from specialized cutaneous receptors can exert complex oligosynaptic reflex effects within a single motoneuron pool, as illustrated by the following examples. First, cutaneous afferents may exert differential effects onto the different types of motoneurons within the motoneuron pool. Thus, low-threshold, early-recruited motoneurons of the first dorsal interosseous muscle (an intrinsic muscle of the hand) may be inhibited by cutaneous afferents from the index finger (the digit the muscle moves), whereas higher-threshold, later-recruited motoneurons are excited. This reflex arrangement favors the activation of motoneurons that are capable of rapid, forceful contractions. Cutaneous inputs from functionally related digits (e.g., the thumb) probably have a similar reflex effect. Second, there is a highly synchronized afferent input from cutaneous afferents within the fingers and hand when a grasped object slips fractionally against the skin surface. This input leads to a rapid reflex increase in force, which is probably mediated through an oligosynaptic spinal pathway. [*See* Skin and Touch.]

Similarly, specialized mechanoreceptors in joints have oligosynaptic pathways to motoneurons, in part mediated by convergence on the Ib inhibitory interneuron. The overall pattern of activity from joint afferents can cause a marked change in the activity of muscles operating across a joint. As a joint moves into full flexion, the extensor muscles are facilitated at a spinal level; conversely, as the joint moves into full extension, the flexor muscles are facilitated. Such a reciprocal arrangement, documented for the knee and elbow joints in the cat, serves to protect a joint from the potentially damaging effects of excessive flexion or extension. The detailed neuronal basis for this arrangement has not been defined, although the

Ib inhibitory interneuron is probably involved. There is evidence for reflex inhibition of muscles acting about the human knee when the joint capsule is distended by an effusion. [*See* Proprioceptors and Proprioception.]

In summary, cutaneous and joint afferents almost certainly have excitatory and inhibitory connections to the same motoneuron pool, the reflex operating being the one "chosen" by descending commands to be appropriate for the particular movement. In spinal animals and in humans with spinal cord injury, this supraspinal control is lost, and the dominant reflex effect appears to be inhibition of motoneurons innervating extensor muscles and facilitation of those innervating flexor muscles, with the result that even innocuous stimuli can provoke "flexor spasms" in such patients.

3. Recurrent Inhibition

The output of the motoneuron can be modulated by an intraspinal feedback circuit using the Renshaw cell to produce recurrent inhibition (Fig. 3). A collateral banch coming off the axon of the alpha motoneuron excites nearby Renshaw cells, which directly inhibit the homonymous motoneuron pool. Renshaw cells inhibit both alpha and gamma motoneurons and, in addition, the Ia inhibitory interneuron directed to antagonistic motoneurons. Hence, activity in the Renshaw circuit tends to curtail a contraction and decrease the inhibition of the antagonist. In addition, the Renshaw cell circuit may help to focus the motor output on a restricted group of muscles. Like all interneurons, the Renshaw cell is subject to peripheral and descending control. During strong voluntary contractions, its excitability is turned down, so that there is less recurrent inhibition of the contracting muscle and the reciprocal inhibitory pathway to its antagonist is open. During weak voluntary contractions, Renshaw cells are excited to provide graded inhibition of the agonist motoneuron pools and indirect facilitation of the antagonist.

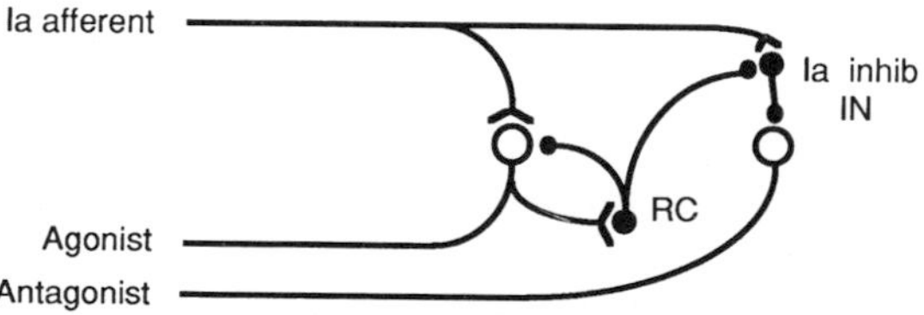

FIGURE 3 The Renshaw cell (RC) is excited by a collateral of the motor axon to provide recurrent inhibition of the motoneuron pool. The Renshaw cell also inhibits the Ia inhibitory interneuron (Ia inhib IN). These actions curtail activation of the agonist and favor activation of the antagonist. Solid symbols indicate inhibitory interneurons.

B. Transcortical (Long-Loop) Circuits

Rapidly conducting afferents from mechanoreceptors in skin, joint, and muscle (particularly the muscle spindle and the Golgi tendon organ) have oligosynaptic projections to the primary motor and somatosensory areas of the cerebral cortex. This afferent information reaches the cortex by many pathways. The major ones traverse the posterior columns and, for muscle afferents from the lower limbs, the dorsal spinocerebellar tract to the dorsal column nuclei (and nucleus Z), where, after a relay, the input enters the medial lemniscus, only to relay again in the thalamus on the way to the cortex. At each level—spinal, thalamic, and cortical—there may be intramodality convergence (within the muscle afferent or cutaneous afferent types) and intermodality convergence (between the cutaneous and muscle afferents). In addition, the ascending volley in the sensory pathways can be modified at these three levels by descending motor pathways or local influences.

Within the primary motor cortex, the ascending sensory information has excitatory synapses on the upper motor neuron, which projects via corticospinal pathways to spinal motoneuron pools. The corticospinal fibers pass through the medullary pyramids, forming the pyramidal tract. These descending pathways make direct (monosynaptic) connections with their target motoneurons. Individual corticospinal axons may synapse with many motoneurons within a pool and even with many functionally related motoneuron pools. In addition, evidence in cats and monkeys shows indirect connections via interneurons at a local segmental level or via propriospinal neurons, which are located in the high-cervical spinal cord. The relative importance of these indirect connections in humans is not known, although recent evidence supports the presence of analogous propriospinal interneurons, which are activated in voluntary contractions and on which afferents from skin and muscle also converge. The presence of interneurons in the pathways from motor cortex to motoneuron provides an additional means by which sensory feedback from the periphery or the activity of other descending pathways can influence the resulting muscle contraction.

II. Proprioceptive Reflexes

A number of reflexes from mechanoreceptors in muscle and skin can be readily demonstrated in humans, the most easily elicited being the tendon jerk. However, the majority of these reflexes depend on subjecting the nervous system to an intense afferent volley generated by an artificial stimulus and, though clinically useful, such reflexes reveal little about how reflex activity is integrated into the control of movement. Nevertheless, reflexes such as the tendon jerk are useful because they enable physicians to check the integrity of spinal circuits and to quantify the changes in reflex intensity in different neurological diseases.

In normal human subjects, any reflex that is elicited repeatedly will habituate (i.e., the same stimulus will produce a smaller response). The habituation can be minimized by using irregularly repeated stimuli and keeping the repetition rate low, approximately once every 5–10 sec. The degree of habituation is greater for cutaneous reflexes such as the glabellar tap (the reflex blinking that occurs when the skin of the forehead is tapped), possibly because cutaneous reflexes contain more interneurons. However, if the test muscle is contracted voluntarily, the degree of habituation is much less, whether the reflex is primarily of cutaneous or muscle origin, and stimulus rates of >1/sec are then feasible (Fig. 4).

A. Tendon Jerk

If the tendons of a number of muscles are tapped briskly, a brief twitch contraction will occur in the relevant muscle. The time interval (latency) between the tendon tap and the onset of the reflex contraction depends on the distance of the muscle from the spinal cord, but it is short [e.g., 30–40 msec for the calf muscles (Fig. 5) and 20–25 msec for the thigh muscles]. An abrupt tap on the tendon excites not only mechanoreceptors in the appropriate muscle but also those in nearby muscles and skin and produces a number of impulses from the more sensitive receptors. It is generally accepted that the major afferent input produced by a tendon tap is group Ia activity from the percussed muscle, although group Ib and cutaneous afferents are also excited. The major excitatory pathway for the tendon jerk is the group Ia monosynaptic pathway. However, the excitation produced in the motoneurons lasts so long that other spinal pathways and other afferent fibers can also affect the reflex discharge. The duration of this excitation is much longer for the tendon jerk than for the H reflex (see Section II,C) in human motoneurons (see Fig. 5).

The tendon jerk can be potentiated in normal subjects by a number of different maneuvers; this phenomenon is known as reflex reinforcement. Perhaps the simplest method of reinforcement is to ask the subject to contract the relevant muscle weakly because, as discussed earlier, this potentiates transmission in group Ia excitatory pathways and suppresses relevant inhibitory interneurons. However, if the contraction is too strong, the muscle stiffens and it can be difficult to see the enhanced reflex. In practice, most physicians use a technique called the Jendrassik maneuver, in which the subject contracts a remote muscle rather than the one being tested. The potentiation produced by this maneuver is less than is produced by contraction of the test muscle and occurs through activation of descending pathways to alter reflex transmission and the excitability of motoneurons.

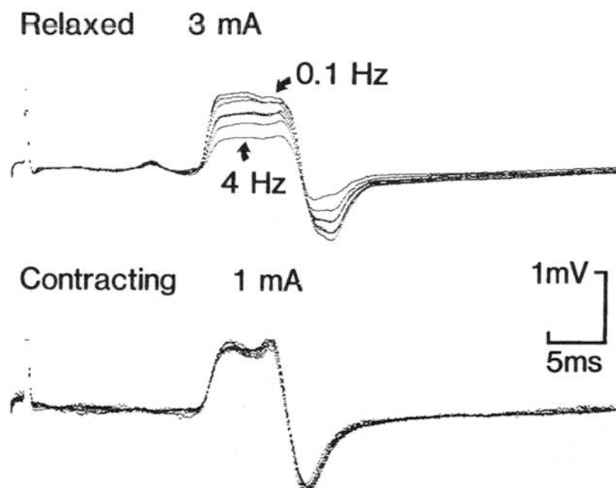

FIGURE 4 H reflex of a flexor muscle of the wrist produced by electrical stimulation at a range of frequencies (0.1, 0.2, 0.5, 1, 2, 3, and 4 Hz). When the muscle is relaxed (upper records), the reflex response attenuates as stimulus rate is increased; however, when the muscle contracts, the reflex can be obtained at a lower stimulus intensity (1 mA rather than 3 mA), and its amplitude is not affected by the stimulus rate. [Reproduced, with permission, from D. Burke, R. W. Adams, and N. F. Skuse (1989). *Brain* **112**, 417–433.]

B. Unloading Reflex

If a subject contracts a muscle steadily against a resistance that is suddenly removed, the contracting muscle abruptly shortens. Recordings of the electromyogram of the contracting muscle reveal that the muscle contraction is transiently interrupted at short latency after the removal of the load. The unloading reflex is produced by the abrupt removal of ongoing reflex support to the contracting motoneurons, particularly from group Ia afferents from spindle endings in the contracting muscle. The latency of the silent period in the contracting muscle is similar to the latency of the tendon jerk of the same muscle; the unloading reflex can be considered the inverse reflex to the tendon jerk.

The physiological significance of the unloading reflex is that it reveals the presence of supportive excita-

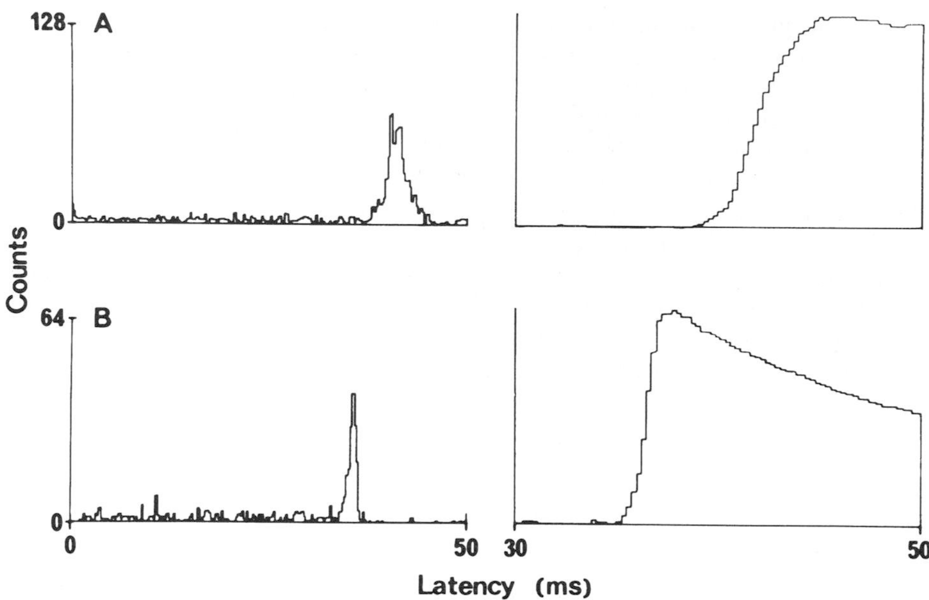

FIGURE 5 A histogram of the discharge of a voluntarily activated motor unit in soleus in response to percussion on the Achilles tendon (A) and to electrical stimulation of the tibial nerve to produce the H reflex (B). The occurrence of a discharge in the motor unit produces a single count in the histogram. The tendon tap and H-reflex stimuli are delivered at time 0. On the left, the increase in probability of discharge probably reflects the differential of the rising phase of the responsible excitatory event within the motoneuron (excitatory postsynaptic potential, or EPSP). On the right, the cumulative increase in probabilities of discharge is therefore presumed to represent the rising phases of the EPSPs. In these plots, latency has not been corrected for the trigger delay (2.5 msec). Note that with the tendon tap (A), the reflex response occurs approximately 5 msec later than with H-reflex stimuli (B). Also note that the duration of the excitation is much longer with the tendon tap. Numbers of triggers used in the histograms were about 2500. [Reproduced, with permission, from D. Burke, S. C. Gandevia, and B. McKeon (1984). *J. Neurophysiol.* **51**, 185–194.]

tion to the contraction from peripheral reflex sources. Clearly, the ongoing activity in reflex pathways must have a significant influence on motoneuron firing for its removal to be capable of silencing motoneurons, even if only transiently.

C. H Reflex

This reflex is named after Paul Hoffmann, who first described that a reflex muscle contraction can be produced by electrical stimulation of the afferent fibers coming from the muscle (See Fig. 4). Group I muscle afferents are activated by low-intensity stimulation of the relevant nerve. The latency of the reflex contraction is approximately 5 msec less than the latency of the tendon jerk for the same muscle, the difference being attributed to the fact that the electrical stimulus does not require activation of the muscle spindle receptor and conduction along the distal part of the afferent fibers (see Fig. 5). The H reflex has been considered the electrically induced analgoue of the mechanically induced tendon jerk, and differences between the two reflexes in different experimental situations have been attributed to receptor (muscle spindle) mechanisms. Thus, comparisons of these two reflexes have been used as measures of the degree of sensitization of spindle endings by fusimotor drive. [*See* Fusimotor System.] However, the afferent volleys for the two reflexes differ in a number of respects, all of which could alter the resulting reflex. Perhaps the major difference is that the electrically evoked volley will contain significant Ib activity, whereas the mechanically induced volley will not.

Group Ia afferents are responsible for the H reflex largely via a monosynaptic pathway. However, the group Ib contamination of the afferent volley produces disynaptic inhibition in the homonymous motoneurons, and this curtails the duration of the Ia excitation. Changes in the H reflex can be produced by affecting the Ib inhibitory interneuron. The H reflex provides a probe with which the function of the human spinal cord can be investigated. Virtually all the studies of reflex circuitry within the human spinal cord have depended on demonstrating changes in the H reflex. As a final example, recent studies of the H reflex of soleus during walking suggest that the reflex excitability may not parallel that of the motoneuron pool. The reflex responsiveness is high during the stance phase of walking, and even higher during standing, both circumstances in which active maintenance of body position against gravity is required.

It is often (erroneously) stated that the H reflex is more restricted in distribution than the tendon jerk. However, if the subject contracts the test muscle, both the H reflex and tendon jerk can be elicited from many limb muscles.

D. Stretch Reflexes

Stretch reflexes refer to the responses observed following stretch of a muscle (Fig. 6). The stretch is usually applied rapidly and produces a burst of discharges from dynamically sensitive receptors (particularly muscle spindle endings) within the muscle. However, the disturbance is not highly localized so that intramuscular receptors from adjacent muscles and nearby cutaneous and joint receptors are also activated. For most human muscles, the response to abrupt stretch of a contracting muscle shows three components. The first occurs at a latency consistent with a monosynaptic reflex and is probably equivalent to the tendon jerk (see earlier). The second occurs at a longer latency and is thus consistent either with a more slowly conducting afferent volley reaching the spinal cord or with a longer neuronal pathway within the central nervous system. The latencies for the tendon jerk and long-latency response are about 25 to 50 msec, respectively, for the long flexor muscle of the thumb. The third, and often blending temporally with the second component, is the deliberate voluntary reaction to the stretch.

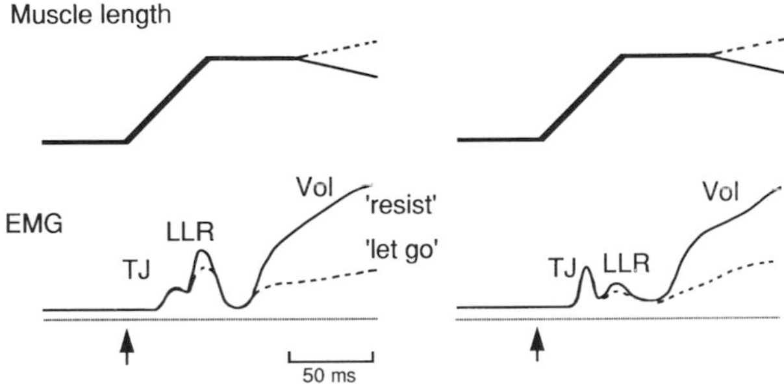

FIGURE 6 Diagrammatic representation of the responses of human muscles to stretch. Upper panels show muscle lengthening (with stretch as an upward deflection), and lower panels show the electromyographic responses (EMG) when the stretches are given during a steady contraction. On the left, the patterns obtained for some distal muscles are depicted, and on the right, for some proximal muscles. Following the muscle stretch, there are three phases of the response: a short-latency spinal response, equivalent to the tendon jerk (TJ), the long-latency response (LLR), and the voluntary response (Vol). The long-latency response is bigger when the subject is instructed to resist the applied stretch than when instructed to let go immediately after the stretch. In the former case, the muscle length begins to return toward the initial value.

The second component of the stretch reflex has aroused considerable debate because of the suggestion that it may represent the operation of a long-loop, possibly transcortical reflex (see earlier). Evidence supporting this conclusion includes the relationship between the latency of the second component and the distance between the stretched muscle and the brain, and observations that the component is reduced or absent in patients with lesions interrupting the pathways to and from the motor cortex. Furthermore, in patients in whom the motor cortex projects muscles of both hands, muscle stretch on one side produces a bilateral long-latency response. However, not all findings have supported this explanation, and it is clear now that other factors must also be considered, including the oligosynaptic activation produced by more slowly conducting group II afferents from the secondary endings of muscle spindles, and the continuation of excitation due to multiple discharges of primary muscle spindle endings and even from cutaneous receptors. Different mechanisms acting together probably are responsible for the second component of the stretch reflex, and the relative importance of any one mechanism will differ for different muscle groups (see Fig. 6).

Not only does rapid muscle stretch (or shortening) produce excitation (or inhibition) of the relevant muscles, but extremely slow perturbations may have a similar effect. Much as described earlier, the reflex responses produced by slow stretch can be guided according to the subject's intent to resist or to relax during the perturbation. It is as if reflex circuits can be present prior to a voluntary contraction or prior to an externally induced adjustment to movement. In other words, the overall gains of the stretch reflex response, and thus the effective stiffness of the muscle, can be controlled. It must be emphasized that the resting (background) discharge of muscle and cutaneous mechanoreceptors provides "tonic" input to the central nervous system. Small changes in the discharge rates of muscle spindle afferents from a muscle group will affect motoneuron excitability. Thus, while phasic or transient stimuli (e.g., sudden stretch, H-reflex inputs) are often used to assess reflex pathways, the slowly changing tonic input accompanying normal movement will continuously modulate motoneuronal output. Furthermore, many proprioceptors, including cutaneous receptors, respond disproportionately to small mechanical disturbances, such as occur during natural movement, rather than to large mechanical disturbances, such as a tendon tap.

E. Reflex Effects of Muscle Vibration

Vibration activates many mechanoreceptors, in muscle and skin, but receptors in the muscle spindle are particularly sensitive and, when vibration is applied directly to a muscle or tendon, the reflex effects are generally attributed to group Ia afferents from the muscle spindle. If a muscle or its tendon is vibrated, two phenomena may occur: the tendon jerk and the H reflex will be inhibited, and a slowly increasing contraction develops in the vibrated muscle. The former phenomenon (vibratory inhibition) is largely due to presynaptic inhibition of the monosynaptic group Ia pathway due to the vibration-induced activity in muscle (and cutaneous) afferents. The slow-developing contraction (tonic vibration reflex) is probably due to group Ia afferent activity transmitted to the homonymous and synergistic motoneurons along nonmonosynaptic pathways. At least in some muscles, however, cutaneous afferents contribute to this nonmonosynaptic excitation. The tonic vibration reflex is subject to activity in descending motor pathways. On the one hand, it can be completely overridden voluntarily and, on the other, it can be potentiated by a weak voluntary contraction of the test muscle.

Muscle vibration also results in inhibition of the motoneurons of antagonistic muscles, presumably a reflection of Ia reciprocal inhibition. In addition, vibration of the Achilles tendon can disturb postural equilibrium in standing human subjects. When the Achilles tendons are vibrated, the subject sways backward as if to compensate for apparent stretch of the calf muscles, and this may be sufficient to provoke a fall.

III. CONCLUSIONS

This article focused initially on some of the spinal and suprapinal circuitry that is available to mediate reflex contributions to muscle performance, and then on some of the proprioceptive reflexes that can be evoked in human subjects. The latter emphasis reflects the need for neurologists and neurophysiologists to have means of monitoring particular neuroanatomical pathways. Clinically, the proprioceptive reflexes are often evoked by stimuli that are more intense and synchronized than those occurring naturally. Thus, perhaps not surprisingly, much current research is shifting toward examining the coordinated responses of many muscles within a limb, or in muscles acting over many body segments, to smaller and more natural stimuli. Examples include the responses to upper limb movement, which can be recorded throughout the lower limb and trunk, and the response of neck, trunk, and leg muscles to small disturbances of stance. Analyses of these more complex situations are providing a picture of a less stereotyped and more modifiable reflex organization. Such an organization would fit with the increasing number of classes of interneurons that have been characterized within the spinal cord of experimental animals.

BIBLIOGRAPHY

Baldissera, F., Hultborn, H., and Illert, M. (1991). Integration in spinal neuronal systems. *In* "Handbook of Physiology" (V. B. Brooks ed.), Sect. I. Vol. II, Part 1, pp. 509–597. American Physiological Society, Bethesda, Maryland.

Burke, D., Adams, R. W., and Skuse, N. F. (1989). The effects of voluntary contraction on the H reflex of human limb muscles. *Brain* **112**, 417–433.

Burke, D., Gandevia, S. C., and McKeon, B. (1983). Afferent volleys responsible for spinal proprioceptive reflexes in man. *J. Physiol. (London)* **339**, 535–552.

Colebatch, J. G., and McCloskey, D. I. (1987). Maintenance of constant arm position or force: Reflex and volitional components in man. *J. Physiol. (London)* **386**, 247–261.

Fournier, E., and Pierrot-Deseilligny, E. (1989). Changes in transmission in some reflex pathways during movement in humans. *News Physiol. Sci.* **4**, 29–32.

Hultborn, H., Meunier, S., and Pierrot-Deseilligny, E. (1987). Changes in presynaptic inhibition of Ia fibres at the onset of voluntary contraction in man. *J. Physiol. (London)* **389**, 757–772.

Jankowska, E., and Lundberg, A. (1981). Interneurones in the spinal cord. *Trends Neurosci.* **4**, 230–233.

Marsden, C. D., Rothwell, J. C., and Day, B. L. (1983). Long-latency automatic responses to muscle stretch in man: Origin and function. *In* "Motor Control Mechanisms in Health and Disease" (J. E. Desmedt, ed.), pp. 509–539. Raven, New York.

Neilson, P. D., and Lance, J. W. (1978). Reflex transmission characteristics during voluntary activity in normal man and in patients with movement disorders. *In* "Cerebral Motor Control in Man: Long Loop Mechanisms" (J. E. Desmedt, ed.), Vol. 4, pp. 263–299. Karger, Basel.

Rack, P. M. H. (1981). Limitations of somatosensory feedback in control of posture and movement. *In* "Handbook of Physiology" (V. B. Brooks, ed.), Sect. I. Vol. I, Part 1, pp. 229–256. American Physiological Society, Bethesda, Maryland.

Proprioceptors and Proprioception

S. C. GANDEVIA
Prince of Wales Medical Research Institute, University of New South Wales

GLOSSARY

Golgi tendon organs Sensory receptors within musculotendinous junctions that are capable of signaling the force of muscle contraction and its rate

Joint receptors Sensory receptors within joint capsules and ligaments that respond to local stress or pressure changes within the joint

Muscle spindles Complex sensory and motor structures located within mucsles; afferent fibers from muscle spindle endings can signal the length of the muscle and its velocity of movement

Nerve fibers Afferent and efferent nerve fibers that convey information to and from the central nervous system, respectively

Perceived motor commands Signals of motor command generated within the central nervous system, which can directly influence proprioceptive sensation

Proprioception Global term including sensations of the position and movement of the limbs and trunk, together with sensations of heaviness and force; loosely used synonyms include kinesthesia and joint position sense

Somatosensory cortex Area(s) of the cerebral cortex that receive ordered inputs from muscle, joint, cutaneous, and other afferent fibers

THE PHYSIOLOGICAL PROPERTIES OF THE MAJOR classes of proprioceptor located in muscles, tendons, joints, and skin are the focus of this article. The proprioceptive sensations evoked by these afferents are examined, together with the roles for perceived motor commands. It is emphasized that proprioception encompasses a group of sensations, including limb position and movement, sensations of force and heaviness, and sensation of the timing of muscle contraction.

I. BACKGROUND

This article is concerned mainly with the physiological properties of proprioceptors (i.e., the peripheral sensory receptors, which provide signals about proprioception) and, secondly, with their contributions to specific aspects of proprioception. Emphasis has been placed deliberately on the properties of proprioceptors in human subjects, partly because there are differences in proprioceptors among species, but also because, based on neurophysiological studies in humans, the principles governing proprioception have been established in the last two decades. In addition, it is necessary to consider briefly the role of motor signals generated internally within the central nervous system in proprioception; these are loosely termed signals of motor command, corollary discharges, or perceived motor commands. Unlike some other senses, such as smell, vision, and hearing, voluntary movement and muscle contraction are usually a prerequisite for generation of proprioceptive signals. For this reason alone, not surprisingly signals of central motor command are important potential contributors to proprioception.

Since early in this century, the terms proprioception and kinesthesia have coexisted with little attempt to define them explicitly. The role of the relevant peripheral receptors and the central neural mechanisms have, at times, been hotly debated for different aspects of proprioception. However, in the last two decades, a number of psychophysical and electrophysiological

ENCYCLOPEDIA OF HUMAN BIOLOGY, Second Edition, VOLUME 7. Copyright © 1997 by Academic Press. All rights of reproduction in any form reserved.

studies have led to greater agreement about the principal neural mechanisms. The terms proprioception and kinesthesia are commonly used synonymously; they encompass a group of sensations that *includes* (1) sensations of limb (or joint) movement and position, (2) sensations of force and heaviness, and (3) sensations of the timing of muscular contraction.

Specialized sensory receptors within muscles and tendons, and in joint capsules and ligaments, can provide useful information about the "movement status" of any body part. In addition to their role in cutaneous sensation, even receptors within the skin can assist proprioceptive sensation, partly because some are inevitably excited as a result of joint movement. On this basis, no major class of specialized peripheral receptor receptor in skin, joint, or muscle can be denied a role in proprioception. To determine their potential role requires a knowledge of their response properties, that is, the specific peripheral events, which they can transmit to the central nervous system. It is then necessary to confirm that role by appropriate psychophysical experiment.

II. PROPERTIES OF PROPRIOCEPTORS

A. Muscle Spindle Endings

The muscle spindle is an intricate apparatus consisting of both sensory and motor components and is found in almost every skeletal muscle (Fig. 1). Each human muscle spindle usually contains 3–15 specialized muscle fibers, termed intrafusal fibers. The spindles lie parallel to the main (extrafusal) muscle fibers but do not produce significant contribution to total output of muscle force. The intrafusal fibers may be more numerous in human muscles than in the muscles of cats, the species in which much of the physiological work has been undertaken. There are at least three types of intrafusal fiber within each spindle: two types of nuclear bag fiber [dynamic (bag 1) and static (bag 2)] and the nuclear chain fiber. [*See* Skeletal Muscle.]

There are two types of specialized sensory endings associated with muscle spindles: the primary spindle ending, which has spiral sensory terminals on the central portion of all the intrafusal fibers from a spindle (particularly the dynamic bag fiber in human spindles), and the secondary spindle, ending with terminals on all but the dynamic bag intrafusal fiber. The density of spindles is highest for small muscles (of the trunk and distal extremities) when measured as the number of spindles per unit of muscle weight, but it is more uniform if measured per number of motoneurons innervating the muscle. The detailed anatomy of muscle spindles may vary within and between muscles.

Afferent fibers innervate sensory receptors and conduct signals to the spinal cord. They are often classified according to their diameter: the largest and most

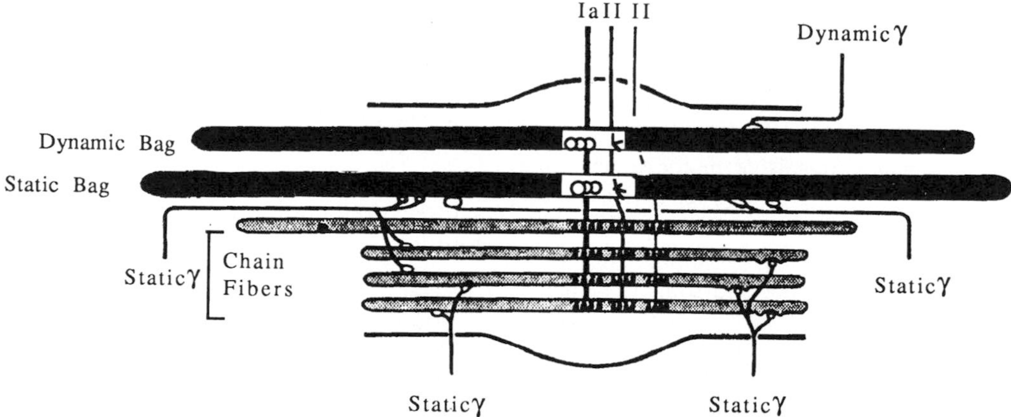

FIGURE 1 Simplified diagram of the structure and innervation of a typical cat muscle spindle. Two nuclear bag fibers and four nuclear chain fibers are shown. There is one dynamic nuclear bag intrafusal fiber (dynamic bag) and one static nuclear bag fiber (static bag). The diagram shows one primary spindle ending (Ia), which has sensory spiral terminals of a group Ia axon around every intrafusal fiber. It shows two secondary spindle endings (II); one secondary ending (left) has terminals on both chain fibers and nuclear bag fibers, and the other secondary ending (right) has sensory spirals on chain fibers only. Note the selective innervation of dynamic bag fiber by dynamic γ axon, and both selective and nonselective innervation of the static bag fiber and the chain fibers by static γ axons. [Simplified from I. A. Boyd (1985). Muscle spindles and stretch reflexes. *In* "Scientific Basis of Clinical Neurology" (M. Swash and C. Kennard, eds.). Churchill Livingston, London.]

rapidly conducting afferents innervate primary spindle endings (group Ia afferents) and Golgi tendon organs (Ib afferents; see Section II,B). Secondary endings are innervated by smaller-diameter myelinated fibers (group II afferents). Although the basic anatomy of muscle spindles is similar in different species, some differences do exist. For instance, in humans the majority of group I muscle afferents in the upper and lower limbs have a conduction velocity similar to that of the specialized cutaneous and joint afferents (60–70 m/sec). In contrast, in the cat and monkey, the fastest cutaneous afferents conduct more slowly than group I muscle afferents. Human spindles have greater attachments to the adjacent extrafusal muscle fibers than those in the cat, and this may render their afferents more sensitive to forces transmitted "in series" from adjacent muscle fibers. [*See* Muscle, Anatomy.]

The primary and secondary spindle endings signal the length, and changes in length, of the muscle in which they reside. The discharge frequency of the primary spindle ending encodes mostly the velocity of movement, although the length and, to a lesser extent, the acceleration are also encoded (Fig. 2A; see also Fig. 3). The discharge of the secondary ending encodes preferentially the static muscle length or joint position. Properties of the different proprioceptors are summarized in Table I. The traditional way to quantify the responses of spindle endings is to stretch the muscle, increasing its length over about 1 sec; the primary ending shows an abrupt increase in firing frequency and then a slower increase throughout the stretch. The firing frequency declines abruptly once the new length is reached. In contrast, the discharge frequency of the secondary ending follows the profile of the length change more closely. Thus, the primary spindle ending shows a greater dynamic response to the change in length than does the secondary. Both have a static, or steady, discharge, which increases as

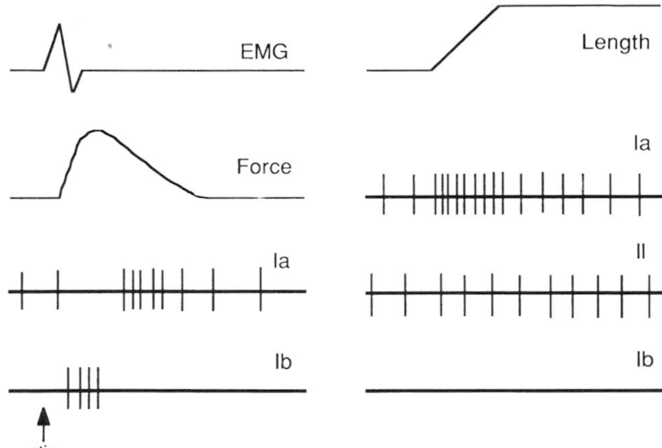

FIGURE 2 (Left) Diagrammatic representation of the procedures used for physiological identification of muscle spindle afferent (Ia) and Golgi tendon organ afferent (Ib). Following electrical stimulation to produce a large muscle twitch, the Ia afferent fires during the falling phase of the twitch force, whereas the Ib afferent fires during the rising phase. The electromyographic response (EMG) and force profiles are shown. The discharges of the single afferent fibers are shown by the action potentials (represented as vertical lines). (Right) Diagrammatic representation of the response of muscle spindle endings [group Ia (primary) and group II (secondary)] and tendon organ afferent (Ib) to a passive increase in length. There is a dynamic response in the Ia afferent to the onset of stretch. The discharge of the group II spindle afferent has a higher discharge frequency at the longer muscle length. The Ib afferent shows no response to the passive length change, although it discharges to dynamic active changes in muscle force.

TABLE I
Summary of Properties of Human Joint, Muscle Spindles, and Cutaneous Receptors in the Hand to Passive Movement[a]

	Response across middle of range	Unidirectional response	Multiaxial response	Proprioceptive signal
Muscle spindle endings	100%	100% unidirectional (i.e., increases with muscle lengthening)	73%	Signals muscle lengthening and rate of lengthening
Joint receptors (80% of population)	0%	60% unidirectional	70%	Detection of limit to movement
Joint receptors (20% of population)	100%	34% unidirectional	66%	Minor signal of joint angle and movement
Slow-adapting cutaneous receptors	0%	67% unidirectional	24%	Minor signal of joint angle and movement
Rapid-adapting cutaneous receptors	0%	18% unidirectional	27%	Event detector

[a]The term "response" denotes a change in discharge frequency of the afferent.

the muscle length increases, but the secondary ending has a greater sensitivity (i.e., greater discharge frequency per unit length change).

Over the full physiological range of joint excursion, the discharge frequency of primary spindle endings is not linearly related to the length change. The primary ending has a small "linear range," being exquisitely sensitive to very small increments in length (up to 100–200 μm in the cat), but is much less sensitive to large increments. In contrast, the secondary ending responds in a more linear way over a much wider range of lengths. This means that the primary spindle ending responds more readily than other intramuscular receptors when vibration at high frequency and low amplitude is applied to the tendon. The discharge may be locked to the phase of vibration with one discharge per vibratory cycle. This susceptibility to vibratory stimuli has been exploited in psychophysical experiments (see Section IV) as a means to stimulate (preferentially) one class of proprioceptor. Because of mechanical properties of the intrafusal (dynamic bag) fiber, the primary ending's sensitivity to small length changes can be "reset" once a new operating length is achieved, so that its sensitivity is preserved over the physiological range of joint positions. The sensitivity of spindle endings is such that they respond to local events within the muscle, including the twitch contraction of nearby extrafusal muscles fibers and the mechanical disturbance set up by the arterial pulse. Responsiveness also varies depending on the spindle's exact location in the muscle, presumably because the stretch "seen" by each spindle varies slightly according to its specific location.

Output from the central nervous system to intrafusal muscle fibers on whch the spindle's sensory endings lie can alter both the resting discharge of the ending and its responsiveness to changes in length. This output consists mainly of small-diameter γ motoneurons, which synapse toward the poles of the dynamic nuclear bag fibers (γ dynamic) or the static nuclear bag and the nuclear chain fibers (γ static). In addition, some spindles receive a second motor output from β motoneurons, which have a shared output to both the spindle and extrafusal fibers. This form of innervation is the sole intrafusal innervation in amphibia. It is also present in primates and has been visualized in human muscles. It may be especially common in muscles of small size, such as the intrinsic muscles of the hand.

Activation of different elements of the fusimotor system (through γ and β motoneurons) can produce changes in the background discharge rate and in the response of the ending to imposed length changes. Dynamic fusimotor inputs (via the dynamic nuclear bag fiber) may produce only a small change in background discharge of primary spindle endings but a marked increase in the response to length changes (increase in the dynamic response). Static fusimotor inputs via the static nuclear bag fiber markedly increase the background discharge rate of primary endings (and less so for the secondary ending), with little effect on the primary ending's response to length changes. Static fusimotor input via the nuclear chain fiber increases the background discharge and the static response to muscle stretch, and it can cause the primary (but not the secondary) endings to fire in one-to-one relation with the fusimotor-induced contraction of the chain fiber (a phenomenon termed driving). When both dynamic and static inputs to a primary spindle ending are activated, the effect of the latter may be dominant. [*See* Fusimotor System.]

B. Golgi Tendon Organs

Golgi tendon organs have been studied in less detail than muscle spindle endings, although their role in the control of movement may be equally important. Golgi tendon organs lie at the musculotendinous junctions throughout the muscle, in series with the extrafusal muscle fibers. They are innervated by large-diameter (group Ib) afferent fibers. They encode the contractile force developed by the extrafusal muscle (and its rate of change) with little adaptation over time and are relatively insensitive to passive lengthening of the muscle (see Fig. 2). A number of muscle fibers from different motor units insert into the capsule of each tendon organ. The high sensitivity of the tendon organ to active muscle force is indicated by the ability of individual muscle fibers innervated by a single motoneuron (i.e., a motor unit) to increase the discharge of the tendon organ. Indeed, this response to contraction forms the critical basis for distinction between the spindle and tendon organs during experimental studies (see Fig. 2). Thus, the tendon organ discharges during the rising phase of contractile force, whereas the muscle spindle falls silent (i.e., is "unloaded") during this phase and discharges preferentially during the falling phase of the contraction as the muscle lengthens slightly. Motor units of different size may produce a similar discharge from a single Golgi tendon organ, suggesting that these receptors are designed to sample the force from different regions of the muscle. They retain an ability to signal fluctuations in force when many motor units are recruited or even during shortening contractions. The ultimate contractile force "seen" by the tendon organ is a complex mixture of the direct in-series and in-parallel component of force.

Both individual tendon organs and muscle spindle endings are extremely sensitive transducers. They can respond with high discharge rates to increments in local active force of about 10 mg and length changes of about 10 μm, respectively. However, both tend to saturate and reach a maximal response with the much larger static force and length changes that occur during normal behavior. This has led to the suggestion that these proprioceptors are especially important in transducing the local events according to their microenvironment within the muscle. The overall "average" of the activity in these endings may provide an accurate picture of the "average" force or length of the parent muscle. However, the sensitivity of the intramuscular afferents is such that the central nervous system also receives information about the dynamic performance of subvolumes of muscle. The discharge of these receptors in response to relatively large force and length changes contributes to conscious proprioception, and it is also likely that the microsensitivity of the endings to small mechanical changes has a perceptual role.

C. Joint Receptors

Receptors within joints were once thought to provide the only signals required for sensation of limb position and movement; indeed, this sensation was often termed *joint* position sense. One basis for this view is the presence of specialized receptors within joints and ligaments. The most important joint receptors for proprioception are probably the Ruffini endings, located within the joint capsule. Golgi endings, histologically similar to those at musculotendinous junctions, are found within the capsule and ligaments. Paciniform endings are also associated with joints. Controversy has surrounded the capacity of specialized joint receptors to signal the angle at a joint to the central nervous system. Classic studies in the 1950s suggested that individual (slowly adapting) joint afferents were "tuned" to signal a particular region of the joint angular range and that they responded to both the static angle of the joint and the velocity of joint movement. Such properties would clearly make joint receptors powerful candidates to provide proprioceptive information. However, although each joint probably contains some receptors with these properties, the large majority of human joint afferents respond preferentially as a joint is moved toward and beyond the limits of its usual range (see Fig. 3). Some joint receptors discharge at both ends of the movement range, for example, at the extremes of flexion *and* extension (i.e., bidirectional response), and thus would signal joint position ambiguously. Furthermore, their discharge may increase at the extremes of movement in more than one axis, for example, flexion and extension, and abduction and adduction (i.e., multiaxial response). These properties would make the receptors unambiguous detectors of movements, which are going outside the usual range of joint movement and, therefore, may harm the joint. Coupled with this, muscles whose contractions would continue to move the joint out of its physiological range are probably reflexly inhibited by these "limit detectors" within the joints, thus preventing damage to the joint.

D. Cutaneous Receptors

There are at least four specialized types of cutaneous receptors that respond to mechanical events involving the glabrous or nonhairy skin surface. The afferent fibers associated with two of these specialized receptors respond with a rapidly adapting discharge to mechanical indentation of the skin; they innervate the Pacinian and Meissner's corpuscles. Afferents innervating the Merkel–cell neurite complex and Ruffini endings are slowly adapting in their response to mechanical stimuli. Details about these receptors and the specific contribution of these afferent fibers to specialized aspects of cutaneous sensitivity are addressed elsewhere. Theoretically, these afferents could have an important role in sensation of limb position and movement. First, the rapidly adapting afferents may signal in a nonspecific way that a disturbance has occurred at or near a joint (i.e., event detector); second, slowly adapting cutaneous afferents may signal that the skin overlying a particular joint has been stretched or distorted; and third, the generalized facilitation of central nervous system pathways produced by the continually changing input from cutaneous afferents has an indirect role in proprioception. This is particularly so for the hand with its high density of cutaneous innervation in the tips of the fingers. Thus, anesthesia of the fingertip impairs proprioceptive judgments from proximal finger joints and even adjacent fingers. This is less evident for skin overlying large joints in the lower limb. [*See* Skin and Touch.]

Direct recordings of the discharge of cutaneous afferents innervating the human hand show that all receptor classes discharge during active and passive movements of the fingers. The background discharge

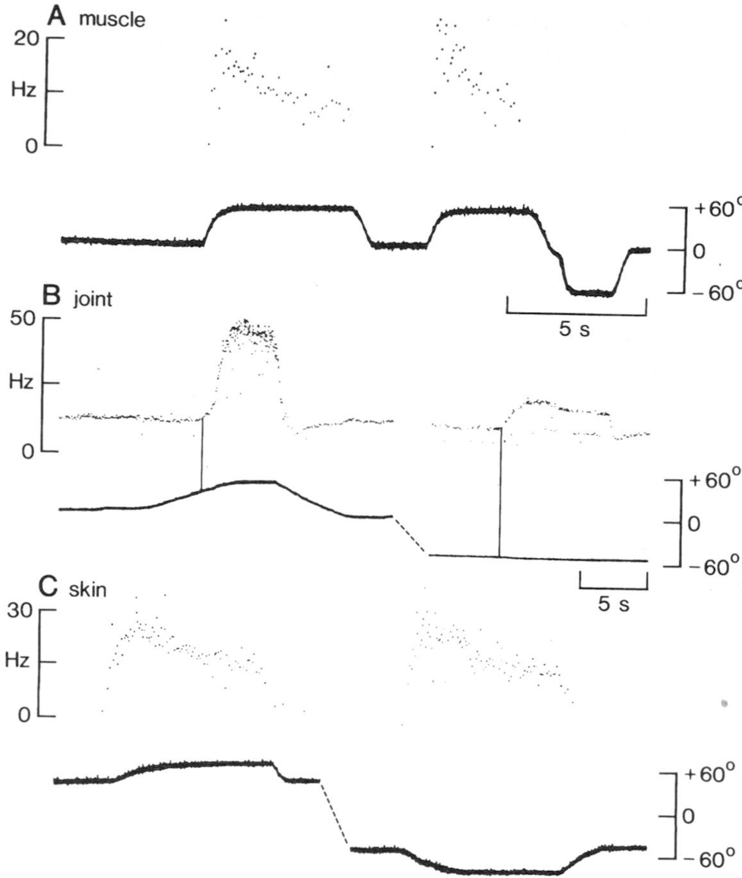

FIGURE 3 Typical responses of human proprioceptive afferents to passive movements. In each of the three panels, the upper record shows the discharge frequency of the afferent (plotted in Hertz or impulses per second), and the lower record shows the goniometer signal or joint angle. The time calibration is shown for each panel. (A) Response of a muscle spindle afferent from the adductor of the ring finger to passive stretch and shortening applied by passive flexion and extension of the proximal interphalangeal joint. In all panels, the discharge frequency is plotted against time. This goniometer record shows stretch as an upward deflection. The afferent shows a slowly adapting response to the lengthening movements away from the rest position. (B) Response of a typical joint afferent from the proximal interphalangeal joint of the index finger. The afferent had a background discharge across the full angular range (120°). This discharge increased only with hyperextension (left) *and* hyperflexion (right) of the joint (see goniometer record below). This is termed a bidirectional response. (C) Response of a slowly adapting cutaneous afferent innervating skin near the proximal interphalangeal joint. The afferent discharged toward both ends of the range of passive movement. [Panel B is modified from D. Burke, S. C. Gandevia, and G. Macefield (1988). Responses to passive movement of receptors in joint, skin and muscle of the human hand. *J. Physiol. (London)* **402**, 347–361.]

of these afferents may be minimal when the hand has been in a rested position for some minutes. Single, slowly adapting cutaneous afferents innervating the skin of the fingers, commonly discharge, like typical joint receptors, at both ends of a normal angular range of movement (see Fig. 3). Despite this potential ambiguity, signals from cutaneous afferents may help to resolve difficulties that may exist in the central nervous system when interpreting the discharge from muscle and joint proprioceptors. These may arise for signals of muscle length and velocity from spindle endings in muscles that cross several joints.

III. PROPRIOCEPTIVE MECHANISMS

A. Sensations from Proprioceptors

Given the response properties of the proprioceptors, it is possible to state which of them have the potential to contribute to a particular sensation. For example, when a joint is passively moved across its normal angular range, muscles will be stretched (exciting muscle spindle endings), the joint capsule will be stressed (exciting some joint receptors), and the skin near the joint will be disturbed (exciting slowly and rapidly adapting cutaneous receptors). Thus, theoretically, muscle spindle, joint, and cutaneous receptors could provide the sensorium with information about what happened at the joint. However, this is dependent on central perceptual mechanisms having access to the neural activity generated by the different classes of proprioceptor. If the movement is made actively, then Golgi tendon organs will also be activated, and there will also be signals related to the central commands for movement.

For some decades, it was denied that the discharge of specialized intramuscular receptors, particularly muscle spindle endings, could be perceived. It was argued that signals in these endings would be hard to decode because of the interaction between the fusimotor input and the changes in muscle length, both of which affect the muscle spindle discharge. Furthermore, it proved more difficult to demonstrate a projection to the sensorimotor cortex from primary muscle spindle endings than from cutaneous and joint receptors. However, now both psychophysical and electrophysiological evidence indicates that muscle spindle endings in primate project to the sensorimotor cortex and play a major role in sensation of limb position and movement sense. This evidence is the following.

(1) Subjects experience illusory movements (and changes in limb position) when a muscle is vibrated in a way that excites the muscle spindle endings within it. The direction of these illusory movements is consistent with a perceived elongation or stretch of the vibrated muscle. These illusory movements can also be detected during a voluntary movement. Furthermore, electrical stimulation of muscle afferents can also produce similar illusory movements. These illusions are best explained by the perception of the discharge of muscle spindle afferents. They are quantitatively influenced by the number of spindle afferents activated and their discharge frequency. By contrast with sensations derived from some single cutaneous afferents innervating the finger tip, the discharge of many muscle receptors may be required to generate perceived proprioceptive signals. The illusions produced by muscle spindles are powerful, but labile, being less obvious if vision of the relevant limb is permitted. This presumably indicates that there is a dynamic (as well as a static) representation of the limb within the central nervous system, which can be influenced by many cues.

(2) The ability to detect changes in limb position and movement remains when the contribution from joint and cutaneous receptors is abolished by anesthesia. Although this procedure does not impair proprioceptive acuity for large joints in the leg, acuity is impaired for joints in the hand. Clearly, for the former joints, proprioceptive acuity can be entirely normal (across the usual angular range) when reliance is placed solely on specialized intramuscular receptors. Consistent with this view is the observation that surgical removal of a joint and associated tissues and its replacement with a prosthesis does not produce an overt deficit in detection of changes in joint position and movement in the hand as well as in the hip and knee.

(3) Some joints can be positioned so that the muscles that normally move them are unable to do so. Movement at these joints can no longer stretch these muscles, and the ability to detect joint rotation is then severely impaired.

(4) Muscle spindle endings (both primary and secondary) and tendon organs have anatomically documented projections to the sensorimotor cortex in experimental animals. Based on several techniques, the projections to the somatosensory cortex have been determined in human subjects for rapidly conducting muscle afferents from distal and proximal muscles (Fig. 4), and even from muscles of the trunk. It is likely that the signals from intramuscular receptors undergo significant central processing at or before reaching a cortical level. This may be required (a) to determine useful signals of angular position, velocity, and other derivatives of muscle length; (b) to determine which particular joint is moving, given that some muscles span several joints; and (c) to decode the component of muscle spindle discharge that results from peripheral events rather than fusimotor activity.

Though available experimental evidence suggests that joint receptors have only a minor role to play

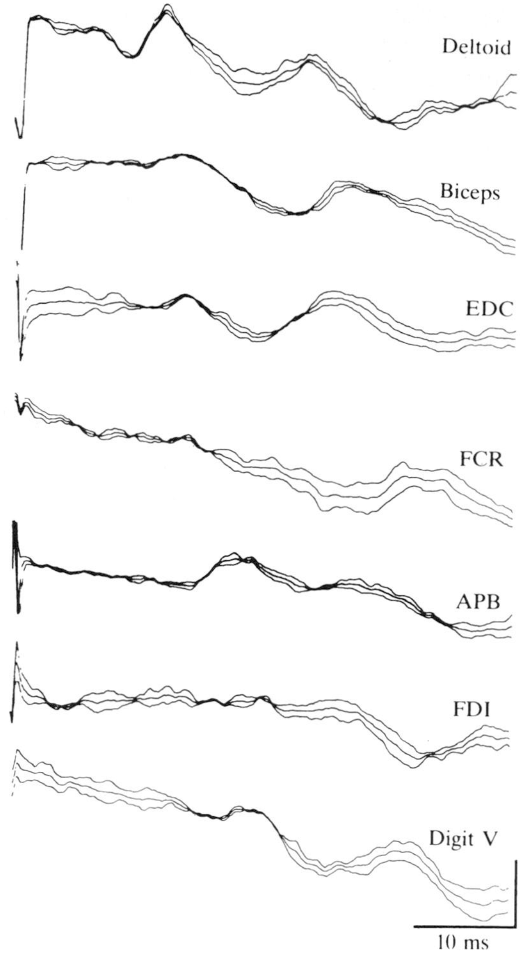

FIGURE 4 Cerebral potentials recorded from the parietal scalp overlying the somatosensory cortex in one subject. Potentials were produced by intramuscular stimulation of the motor point of anterior deltoid (top traces), short head of biceps brachii, extensor digitorum communis (EDC), flexor carpi radialis (FCR), abductor pollicis brevis (APB), and first dorsal interosseous (FDI), and by stimulation of the digital nerves of the little finger (bottom traces). Stimuli were delivered at three times the relevant motor or sensory threshold. The number of responses in each average and vertical calibration were as follows: deltoid, $n = 1024$, 1.4 μV; EDC, 1024, 1 μV; FCR, 3000, 1 μV; APB, 2048, 1 μV; FDI, 2048, 1 μV; little finger–hand area electrode, 1024, 2 μV. Potentials were somatotopically organized with those from distal muscles (e.g., APB, FDI) more lateral on the scalp than those from proximal muscles (e.g., deltoid). [From S. C. Gandevia and D. Burke (1988). Projection to the cerebral cortex from proximal and distal muscles in the human upper limb. *Brain* **111,** 389–403.]

in proprioceptive sensations when the joint is in its midrange, they likely play a greater role when the joint approaches the end of its usual angular range. Thus, both the proprioceptive acuity and reflexes evoked by joint afferents are enhanced when a joint is distended following a traumatic effusion or affected by some forms of arthritis.

Golgi tendon organs respond to active muscle force, and signals from them project to the somatosensory cortex in the cat and presumably also in human subjects. However, though it is likely that their signals can directly influence force sensations, definitive evidence for their role in the perception of muscle force has not been easy to obtain. As argued next, under many circumstances, subjects appear to use signals of centrally generated motor command rather than peripheral signals of tension in the estimation of forces and tensions.

B. Signals of Perceived Motor Commands

Additional components of proprioceptive sensation depend, in a less direct way, on signals derived simply from peripheral afferents. The best-known example of this is the sensation of muscular force (or heaviness). Whereas signals from skin, joint, and tendon organ afferents would register the force used to support or move an object, when subjects are formally required to signal the perceived force (or heaviness) they do so by reference to a centrally generated signal of the effort required in the task. Thus, as muscle fibers fatigue, a greater effort is required, and so the perceived force increases. This phenomenon has been studied in many situations in which the relationship between the central motor command or effort and the achieved force is altered (Fig. 5). This preferred reliance on a partly centrally generated signal for estimates of force, effort, and heaviness may have a teleological basis. For example, it means that the sensorium becomes aware of the increased central motor command required for a fatiguing task rather than the overt tension failure that necessitated it. This would ensure that associated changes in proprioceptive reflexes are generated in proportion to the increase in central command. Indeed, it is as if subjects can check on the status of a particular motoneuron pool and its muscle fibers by reference to a central signal. Accuracy of judgment about force, as for movement sensations, may be relatively uniform for different joints within a limb. Certainly, the marked distal acuity for cutaneous sensation is not paralleled by marked acuity for some proprioceptive sensations.

Some limited evidence suggests that the neural mechanisms underlying the generation of the relevant signal of motor command for force perception probably involve the primary motor cortex. However, the

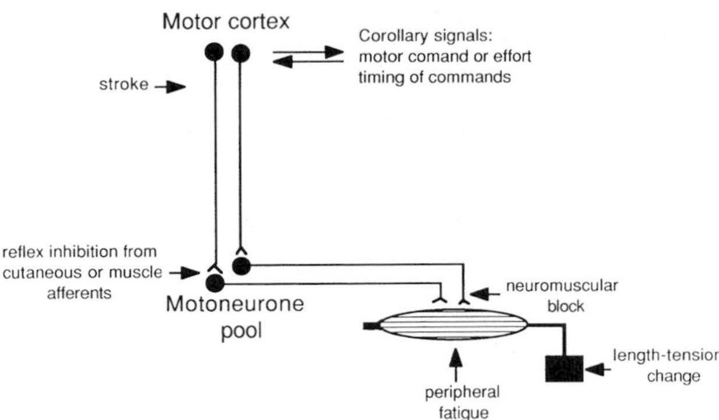

FIGURE 5 Simplified presentation of descending motor pathways, the motoneuron pool, and muscle. The modifications that produce an increase in perceived force and heaviness are marked. Signals related to the motor outflow are involved in the perception of motor command or effort and in perception of the time at which motor commands are dispatched. They do not give rise to the sensation that a limb has moved.

interaction of such internal signals with those from peripheral receptors warrants further study. It is, nonetheless, now well established that signals of motor commands do not provide directly a sense that the relevant body part has moved: attempts to move a paralyzed limb are associated with a feeling of great effort or weight, not with a sense that the limb actually moved.

Signals related to centrally generated motor commands probably have a number of other proprioceptive functions apart from simple movement production. They act principally to allow evaluation and interpretation of ongoing signals from proprioceptors. Some additional functions follow. First, they influence the transmission of somatosensory inputs to the sensorimotor cortex just prior to and during movement. Second, they influence the potency of many reflexes that affect motoneurons during movement. Third, they provide signals that can be used to decode the signal from muscle spindles (and presumably other proprioceptors) into that component due to an external perturbation to a movement and that expected based on the descending voluntary drive to the fusimotor system. Such decoding is especially relevant given that the fusimotor system can alter selectively the bias and the gain of muscle spindle signals. Fourth, some signals that rely on force *and* length (e.g., compliance and stiffness) may be biased by the level of perceived motor command. Finally, the perceived *time* at which a muscle contraction is commanded occurs before movement onset and must also represent perceptual access to a centrally generated motor signal. [*See* Fusimotor System.]

IV. CONCLUSIONS

Many of the proprioceptive mechanisms discussed here have related to simple movements or perceived responses to passive movements. Although neither of these situations occurs routinely in daily life, the relevant natural mechanisms are likely to be based on those revealed under simpler experimental conditions. Proprioception covers a range of sensations associated with the central commands needed to produce force and movement, and the monitoring of the peripheral consequences of the movement. The importance of peripheral proprioceptors in motor control is revealed by the deficits following loss of proprioceptive input in the ability to execute complex movements, to learn movement sequences, and to sustain a constant level of muscle contraction.

ACKNOWLEDGMENTS

The author's work is supported by the National Health and Medical Research Council of Australia. Comments on the manuscript by D. Burke and G. Macefield are gratefully acknowledged.

BIBLIOGRAPHY

Boyd, I. A., and Gladden, N. H. (eds.) (1985). "The Muscle Spindle." Stockton Press, New York.

Burgess, P. R., Wei, J. Y., Clark, F., and Simon, J. (1982). Signalling of kinesthetic information by peripheral sensory receptors. *Annu. Rev. Neurosci.* 5, 171–187.

Burke, D., Gandevia, S. C., and Macefield, G. (1988). Responses to passive movement of receptors in joint, skin and muscle of the human hand. *J. Physiol. (London)* **402**, 347–361.

Gandevia, S. C., and Burke, D. (1992). Does the nervous system depend on kinesthetic information to control natural limb movements? *Behav. Brain Sci.* **15**, 614–632, 815–819.

Goodwin, G. M., McCloskey, D. I., and Matthews, P. B. C. (1972). The contribution of muscle afferents to kinaesthesia shown by vibration induced illusions of movement and by the affects of paralysing joint afferents. *Brain* **95**, 705–748.

Hulliger, M. (1984). The mammalian muscle spindle and its central control. *Rev. Physiol. Biochem. Pharmacol.* **101**, 1–110.

Matthews, P. B. C. (1972). "Mammalian Muscle Receptors and Their Central Actions." Arnold, London.

Matthews, P. B. C. (1988). Proprioceptors and their contribution to somatosensory mapping: Complex messages require complex processing. *Can. J. Physiol. Pharmacol.* **66**, 430–438.

McCloskey, D. I. (1981). Corollary discharges: Motor commands and perception. *In* "Handbook of Physiology, Sect. I, The Nervous System," Vol. 2, Part 2, pp. 1415–1447. American Physiological Society, Bethesda, Maryland.

Proske, U., Schaible, H.-G., and Schmidt, R. F. (1988). Joint receptors and kinaesthesia. *Exp. Brain Res.* **72**, 219–224.

Prostate Cancer

MAARTEN C. BOSLAND
New York University Medical Center

GLOSSARY

Adenocarcinoma Neoplasm that (1) is derived from epithelial cells, (2) grows essentially in a glandular fashion, and (3) is malignant, that is, it invades surrounding tissues and disseminates to form metastases

Androgen receptor Intracellular (probably intranuclear) protein that binds the active androgenic hormone 5α-dihydrotestosterone with high affinity and low capacity; the hormone–receptor complex can interact with specific DNA sequences (androgen response elements) to regulate expression of genes that are involved in androgen-specific responses of the target cell

Androgens Natural androgens are a group of steroid hormones that, either directly or after metabolic conversion in target cells to the active androgenic hormone 5α-dihydrotestosterone, bind to the androgen receptor and thereby have trophic effects on target cells; they are produced by the testes and the adrenal cortex

Antiandrogens Compounds that bind to the androgen receptor without having biological activity, and thereby competitively inhibit the action of 5α-dihydrotestosterone on the target (cancer) cell

Benign prostatic hyperplasia (BPH) Prostatic enlargement caused by a joint proliferation of the glandular epithelium and the fibromuscular stroma of the prostate that does not display malignant properties (such as invasive growth and metastases); BPH is very common in aging men, in whom it usually causes obstruction of the urinary flow as the first and most significant clinical complaint

Hormonal therapy Therapy of hormone-sensitive cancers that involves either (1) reduction or elimination of the production of the hormone in question by drugs or by removal of its source or (2) inhibition of the action of the hormone on the target (cancer) cell by a compound that blocks the receptor for that hormone (e.g., antiandrogens) or that inhibits the formation of the active form of the hormone (e.g., 5α-reductase inhibitors)

Latent prostate cancer Early stage of prostate cancer that is very common in aging men, but can be detected only by histological examination of the prostate; it does not progress to a cancer that causes symptoms within the lifetime of most of these men

LH–RH analogs Peptides that are analogs of luteinizing hormone-releasing hormone (LH-RH), which is produced in the hypothalamus; they bind to the LH-RH receptor in the pituitary and either have LH-RH activity and exhaust the production of LH by the pituitary or have no biological activity and thereby competitively inhibit the action of LH-RH

Male accessory sex glands Exocrine, androgen-dependent glands that are part of the male genital tract and produce secretion that is part of the ejaculate; besides the prostate, they include the seminal vesicles and bulbourethral glands

Prostate-specific antigen (PSA) PSA is proteolytic enzyme that is secreted by the prostate to the glandular lumens, where it is normally contained and excreted with ejaculation of the prostatic fluid. Disruption of the normal structure of the prostate by disease, such as cancer, BPH, and prostatitis, can lead to leakage of PSA outside the glands, leading to detectable levels of this substance in the blood. Measurement of blood concentrations of PSA is clinically used as a marker to detect prostate cancer and to monitor the response to treatment

Tumor heterogeneity Presence of tumor cells within the same neoplasm that have different biological and phenotypical characteristics, for example, differ in hormone sensitivity

THIS ARTICLE DESCRIBES KEY ASPECTS OF THE BIOLogy and pathology of cancer of the prostate gland,

ENCYCLOPEDIA OF HUMAN BIOLOGY, Second Edition, VOLUME 7. Copyright © 1997 by Academic Press. All rights of reproduction in any form reserved.

and it briefly summarizes important clinical aspects of this disease. A more detailed description is given of research into the causes of prostate cancer, including epidemiology and studies in animal and tissue culture systems.

I. INTRODUCTION

Prostate cancer is the most frequently occurring cancer in men in the United States and many western European countries. This disease is the second most frequent cause of death due to cancer in males in these countries, after lung cancer and before cancer of the large bowel and rectum. It is estimated that in the United States, approximately 10 out of every 100 men who reach the age of 50 will develop prostate cancer in their lifetime, and that 3 to 4 of every 100 men will die from this disease. In 1995, there were approximately 244,000 newly diagnosed cases of prostate cancer in the United States and approximately 40,400 deaths from prostate cancer. Because prostate cancer is such a widespread disease, there is a large amount of information about its diagnosis, treatment, and prognosis. Furthermore, the prostate is the major target organ for male sex hormones, and therefore much research has been conducted on aspects of endocrinology and biochemistry of this gland. In contrast, surprisingly little is known about the causes (etiology) of prostate cancer, unlike other major human cancers, such as lung cancer (known to be predominantly caused by smoking).

II. ANATOMY AND BIOLOGY OF THE PROSTATE GLAND

The prostate is one of the accessory sex glands in male mammals; the other major male accessory sex glands are the seminal vesicle and the bulbourethral gland. The human prostate is a single glandular structure that completely surrounds the urethra at the base of the urinary bladder; the larger part of the gland is dorsally located between the urethra and the rectum. The gland is approximately 2.5–3 cm long and 3–4 cm in diameter, and it weighs 18 to 25 g in young adult males. It is composed of 30–50 compound tubuloalveolar glands that share 16–32 excretory ducts that open out into the part of the urethra that is surrounded by the prostate. The ducts of the paired seminal vesicles merge with the paired deferent ducts to form the ejaculatory ducts that project dorsally through the prostate and open out in a structure that projects into the lumen of the prostatic part of the urethra, termed the colliculus seminalis or verumontanum. The glands comprising the prostate are embedded in firm connective tissue (stroma) consisting mainly of fibroblasts and smooth muscle cells. A fibroelastic capsule surrounds the prostate.

In the mammalian fetus, the prostate develops as four paired groups of endodermal epithelial buds that grow from the urogenital sinus (which develops from the primitive hind-gut) into the surrounding endodermal mesenchyme. This mesenchyme plays a decisive role in determining the future development of the bud epithelium. In the male, the Müllerian duct regresses almost completely, and the Wolffian (mesonephric) duct gives rise to, for example, the deferent ducts and seminal vesicles. One of the paired groups of prostatic buds does not further develop in the human fetus, but in rodents this group forms into the ventral lobe of the prostate. The other three groups of buds develop into separate prostate lobes in rodents (dorsal, lateral, and anterior lobes; the latter is synonymous with the coagulating gland). In the human and several other species, however, these three groups merge into a single glandular mass. Various subdivisions of the human prostate have been proposed based on the embryology of the gland and on morphological and pathological considerations. From a clinical and pathological point of view, it is most useful to think of the prostate as composed of three glandular zones: the glands that directly surround the urethra close to the bladder neck (transition zone; occupies 5% of the total prostatic volume in a young adult man); the glands that surround the deferent ducts as they run through the prostate to open into the urethra at the verumontanum (central zone; 25% of volume); and the glands that are located in the periphery (peripheral zone; 70% of volume).

The prostatic ducts are lined by simple or pseudostratified columnar epithelium, and the prostatic glands contain two different cell types: the glandular cells, which are also referred to as luminal or secretory cells, and a more or less continuous lining of nonsecretory basal cells. The secretory cell contains the intricate apparatus for producing the prostatic secretion: rough endoplasmic reticulum, mitochondria, Golgi structures, and secretory granules that are surrounded by a single membrane. The mechanism of secretion is of the merocrine type and, possibly, also the apocrine type. The basal cells possibly function as reserve cells that proliferate to make up for losses of secretory cells. The latter cell type, however, can also divide.

The major function of the prostate and seminal vesicle is to produce the seminal fluid during ejaculation (approximately 30% of this fluid is produced by the prostate and 70% by the seminal vesicles). The function of the bulbourethral gland is unclear at present. The seminal fluid plays a critical role in maintaining sperm viability and sperm motility after ejaculation. The secretion of the prostate contains many constituents. Quantitatively the most important constituents are citric acid, zinc, and spermine (a polyamine), which all probably protect the sperm cells in some manner, as well as the enzymes acid phosphatase, the function of which is unclear, and prostate-specific antigen (PSA), which is a proteolytic enzyme that is important in the normal function of the prostatic fluid. The major ion in prostatic secretion is sodium. Some of the major constituents of the seminal vesicle secretion are fructose (provides energy for the sperm cells) and protective substances such as protease inhibitors and uric acid. The synthesis and secretion of the prostatic fluid are under nervous as well as hormonal control (predominantly androgens).

The considerable fibromuscular stroma of the human prostate also acts as a sphincter of the urethra that controls micturition in conjunction with the pelvic musculature. Furthermore, this fibromuscular tissue plays a major role in the process of the ejaculation of the seminal fluid and, thereby, of the semen. These functions are neuronally controlled.

Although the prostate is not an endocrine gland in the real sense (it does not synthesize hormones), the gland forms the active androgen metabolite 5α-dihydrotestosterone (DHT) from the male sex hormone, testosterone. Testosterone is produced predominantly in the testes by the Leydig cells and is to a much lesser extent (approximately 5%) derived from other sources, particularly the adrenal cortex. The conversion from testosterone to DHT is irreversibly effectuated by the enzyme 5α-reductase in the prostate epithelial and stromal cells. The formation of DHT by this enzyme is under tight control, as is the further metabolism of DHT in prostate cells by oxidoreductase enzymes (reversibly) to metabolites that are further enzymatically hydroxylated (irreversibly) to inactive steroids. This regulation is necessary because DHT is the active hormone that binds to a DHT-specific androgen receptor protein in the prostate epithelial and stromal cell. This binding is currently thought to occur in the nucleus, and not in the cytoplasm. The androgen receptor molecule contains a ligand (DHT)-binding domain and a DNA-binding domain. The binding of DHT alters the structure of the DNA-binding domain in such a way that zinc-containing DNA folds (so-called zinc fingers) are exposed that can reversibly bind to specific DNA sequences called androgen-response elements (AREs); AREs are comparable to estrogen-responsive elements in the DNA, which have been demonstrated to play a role in the response to estrogens of estrogen-responsive tissues. [*See* DNA and Gene Transcription.] When bound to these AREs, the DHT–receptor complex acts as a transcription factor that, in conjunction with other transcription factors, activates transcription of genes downstream from the ARE. Through this mechanism, DHT regulates the expression of genes that are involved in the response of the prostate cells to androgens. This response can be either stimulation of cell division and inhibition of programmed cell death (apoptosis) or cellular differentiation (production and secretion of the prostatic fluid).

Cell proliferation and secretory function of the prostate epithelium are also regulated by the stromal cells that surround it via secretion of humoral factors, which locally diffuse to the epithelial cells. The epithelial and stromal cells, particularly the smooth muscle cells, both contain androgen receptors and form DHT from testosterone. Under normal conditions, the primary action of androgen on prostatic epithelial cells is thought to be mediated by the stromal cells, which secrete diffusible molecules (called growth factors) that transmit the androgenic signal to the epithelial cells; this cell–cell interaction mechanism, mediated by growth factors that locally diffuse from one cell type to another, is called paracrine action. In prostate cancer, this mechanism is probably disrupted in a manner that leads to a shift in the primary response to androgen from cellular differentiation to increased cell proliferation and decreased programmed cell death (with increased numbers of cells as a result).

Control of cell proliferation in the prostate is under primary control of androgenic steroids, but estrogens probably also play a role in this respect. The pituitary peptide hormone prolactin likely facilitates the effect of androgens on the prostate. The human and rodent prostate contains estrogen receptors and possibly prolactin receptors. Synthetic and secretory activities of the prostate epithelium are not only under the control of all of these hormones, but are also regulated by the autonomic nervous system. The secretion of the steroid hormones that regulate prostate function is under control of the hypothalamo–pituitary system. The secretion of testosterone by the testes, for example, is regulated by the pituitary peptide "luteinizing hormone" (LH), and the secretion of LH is regulated by the hypothalamic substance "LH-releasing hormone" (LH-RH). The secretion of LH-RH, in turn, is controlled by feedback influence of the blood concentration of testosterone and by the central nervous system; circulating testosterone also feeds back directly on the pituitary to regulate LH secretion. This delicate regulating system maintains testosterone in the blood at a constant level, and thereby maintains the prostate gland. An additional regulating system is the serum protein sex-hormone-binding globulin (a serum protein) that reversibly binds most testosterone in the blood, and thereby determines the fraction of free testosterone that is available to enter tissues and cells.

III. PATHOLOGY AND TUMOR BIOLOGY OF PROSTATE CANCER

The great majority of prostate tumors in men are of the adenocarcinoma type, that is, a malignant neoplasm derived from epithelial cells that grow in a glandular fashion. Prostate cancer should be distinguished from benign hyperplasia of the prostate (BPH), which is very common in aging men. BPH is a joint proliferation of the glandular epithelium and the fibromuscular stroma and does not display malignant properties (such as invasive growth and metastases). BPH is predominantly localized in the transition zone of the prostate surrounding the urethra and usually causes obstruction of the urinary flow as the first and most significant clinical complaint. Prostate cancer, on the other hand, is a proliferation of only prostatic epithelial cells, and it is predominantly located in the peripheral zone of the prostate but can also occur in the transition zone. The normal balance between cell proliferation and programmed cell death (apoptosis) that regulates the size of the tissues of the body is lost in prostate (and other) cancer cells. In prostate cancer cells, the rate of apoptosis is more profoundly decreased than the rate of cell proliferation is increased, more so than in most other cancers. Prostate cancer is a malignant disease: the tumor cells invade surrounding tissues, they penetrate the prostatic capsule and blood and lymph vessel walls, and, once in the blood or lymph fluid, they will disseminate and lodge in other tissues, forming metastases. These metastases most frequently occur in the pelvic, spinal, and other bones, in the pelvic lymph nodes, and in the lungs. The bone metastases are a major cause of pain in patients with advanced prostate cancer. Although the reason for the predilection of prostate cancer to form skeletal metastases is not fully understood, there is some evidence to suggest that growth factors secreted by bone cells create a microenvironment that is particularly receptive for prostate cancer cells.

Prostate carcinomas cause the release of a number of substances into the bloodstream. Of these, the enzyme prostate-specific acid phosphatase and prostate-specific antigen (PSA) are clinically most important, because they can be used as markers to detect prostate cancer and to monitor the response to treatment of prostate cancers. The proteolytic enzyme PSA is secreted by the prostate to the glandular lumens, where it is normally contained and excreted with ejaculation of the prostatic fluid. Disruption of normal glandular structure of the prostate by disease, such as cancer, BPH, and prostatitis, can lead to leakage of PSA outside the glands, leading to measurable levels of this substance in the blood. Blood levels of prostate-specific acid phosphatase are elevated only in advanced stages of the disease because it is associated with bone metastases. A newly discovered marker of prostate cells and of prostate cancer is prostate-specific membrane antigen (PSMA), a cell-surface antigen, but its value as a clinical marker is not yet clear. Another recent development is the use of the reverse transcriptase polymerase chain reaction (RT-PCR) method to detect metastatic prostate cancer cells that are in the blood circulation and produce PSA and PSMA in small amounts. The RT-PCR method can measure the presence of PSA and PSMA mRNA from these circulating cells in the blood. Another development was the finding that PSA circulates in the blood either as such (free) or bound to the serum proteins α_1-antichymotrypsin (the predominant form of PSA in the serum) and α_2-macroglobulin, and that the amount of free PSA and the relative binding of PSA to these proteins may differ in men with prostate cancer, with BPH, or without prostate disease. In addition, binding to these proteins affects recognition of the molecule in the immunoassays that are used for PSA measurement.

Prostate cancer develops very slowly, probably over a period spanning several decades. Clinically evident prostate cancer is frequent only in men over 60–65 years of age, and its occurrence increases rapidly with increasing age. Very small prostate cancers, which cannot be detected clinically and are found only by microscopic examination of prostate tissue (histological prostate cancer), are common in aged men and even occur in a significant fraction of younger men (in 25–35% of men aged 30–40). These cancers are often multifocal and do not penetrate the prostate capsule or metastasize; they are detected only by histological examination of prostate tissue that is removed surgically by a transurethral (via the urethra) resection procedure (for treatment of BPH), or in tissue that is collected at autopsy after death. It is estimated that at least 50% of all men will develop histological prostate cancer, and some investigators estimate that it occurs in as much as 80% of men over 80 years of age. Even more common is a putative precursor of histological cancer, termed prostatic intraepithelial neoplasia (PIN), which is not a real neoplasm but can progress to become cancer. All evidence available today indicates that many, but not all, prostate cancers develop from PIN, and that the small histological prostate cancers can progress (i.e., continue to grow and become more malignant) to clinically evident prostate cancer later in life. This stepwise progression toward clinically evident, metastasizing prostate cancer is viewed as an expression of the multistage nature of human cancer, that is, the occurrence of discrete steps in development of cancer that are each associated with additional critical alterations in the genetic material of the cells.

Two important characteristics of prostate cancer are the very slow growth rate of this tumor and the large variability in biological properties, such as degree of hormone sensitivity, production of PSA and acid phosphatase, and degree of histological differentiation, the latter categorized as well, moderately, or poorly differentiated based on morphological characteristics of the cancer cells and their nuclei, and on the growth pattern of the carcinoma (see later description of grading). Both properties create a problem for the treatment of prostate cancer because slow-growing tumors are relatively resistant to radiotherapy and chemotherapy, and the variability obviously has a negative effect on the reliability of diagnostic procedures and prognosis.

Most prostate cancers are androgen-dependent, that is, they will regress when the production of androgens in the body is stopped by some means or when the action of androgens on the prostate (cancer) cell is blocked in some way. Eventually, however, almost every prostate carcinoma that initially regressed upon hormonal treatment will relapse to a hormone-insensitive state and grow in the absence of androgens. Why this happens is not completely clear—the cancer cells could adapt to the absence of androgen; the tumor could be the result of a confluence of multiple small tumors, some of which are hormone-sensitive and others are not; or small numbers of hormone-insensitive cells could develop in a tumor that was initially 100% androgen-sensitive, and the hormonal treatment may select for these cells. The last possibility is most likely the predominant mechanism of the relapse to a so-called hormone-refractory tumor. The development of tumor heterogeneity (a mix of androgen-sensitive and -insensitive cells) is thought to be caused by a genetic instability of the cancer cells.

A number of genetic alterations have recently been identified in the DNA of prostate carcinoma cells, such as changes in the level of expression of some genes (i.e., changes in the amount of their protein products produced), the loss of pieces of certain genes (such as so-called tumor suppressor genes), and loss or gain of (pieces of) chromosomes (aneuploidy). The increasing genetic instability of prostate cancer cells as they progress probably causes some of these changes. Some of these genetic alterations can be viewed as a consequence rather than a cause of the cancer, but many genetic changes are probably involved in the actual causation and development of prostate cancer. Some genetic changes appear to be related to the loss of specific normal cell–cell interactions. Others seem to be associated with the gain of the ability to enzymatically dissolve surrounding connective tissue (needed for invasive growth) and induce the growth of new blood vessels (angiogenesis) that are needed to provide blood supply to the growing tumor. Changes in the expression of certain genes are probably related to the earlier-mentioned loss of androgen responsiveness, although this is not completely clear at present. The loss of function of certain tumor suppressor genes (such as *p53*) appears to be related to the acquisition of metastatic properties. However, none of these changes has yet been definitively linked to a specific step in the development of prostate cancer, or to a specific risk factor for prostate cancer, nor have any prostate cancer-specific genetic alterations or prostate cancer susceptibility genes been identified at present. [*See* Tumor Suppressor Genes, p53.]

IV. CLINICAL ASPECTS OF PROSTATE CANCER

Until measurement of serum PSA levels was widely used as a method to detect prostate cancer, the disease was usually diagnosed when a patient encountered symptoms related to obstruction of urinary flow or pain from metastatic processes, or when a digital rectal examination (palpation of the dorsal aspect of the prostate) was carried out by a physician. This rectal exam can detect a large percentage of prostate carcinomas, but unfortunately only some 20% of these are in an early stage of their development, at which time they are potentially curable by total (or radical) prostatectomy (surgical removal of the prostate). Small, early cancers can also be detected by transrectal (via the rectum) ultrasound techniques, but this method is not suitable for screening purposes. Measurement of serum PSA, however, has been viewed as a potentially effective way to detect prostate cancer at an early stage and it is easy to apply in a screening setting. The use of serum PSA measurements as a screening technique has not yet been proven to be of value in reducing prostate cancer mortality (a large clinical trial is currently testing this assumption), but it is nevertheless widely used as such. The advent of this approach since 1988 has resulted in an explosion of newly diagnosed prostate cancers per year from approximately 80/100,000 men in 1982 to over 180/100,000 in 1992, a rise of 225%.

Serum PSA values of 4 ng/ml and above are indicative of prostatic disease, either cancer or BPH. Values of 10 ng/ml and higher are usually not associated with BPH, but almost always with cancer, and PSA values of 40 ng/ml and higher are frequently associated with local extension of the tumor outside the prostate and/or metastases. Some urologists have proposed to use lower cutoff values, for example, 2.5 ng/ml for younger men (40–49 years with little or no BPH expected) and higher values (6.5 ng/ml) for older men (70 years and older) who often have considerable BPH. Confirmation or rejection of the suspicion of prostate cancer because of elevated PSA levels and/or a positive rectal exam is usually done by histological examination of one or more needle biopsies, which are sometimes taken with the assistance of an ultrasound examination.

The most widely used and reliable pathological method to provide information about the prognosis that can help to guide the therapy of prostate cancer is the so-called grading of the cancer in biopsy material or surgical specimens by a pathologist using the Gleason grading system. This method assigns a numerical score from 1 to 5 to the morphological (light microscopic) pattern of the cancer according to the degree to which this pattern deviates slightly (1) to markedly (5) from normal. It is common to assign two such scores to the cancer, one for the predominant pattern and one for the second most common pattern observed in the tumor. This results in a two-digit score (or grade), for example, 2,4, and can also be expressed as a so-called Gleason sum (e.g., 6 = 2 + 4), which can range from 2 to 10. This Gleason sum correlates well with prognosis for low scores (2–4), which indicate a good prognosis, and high scores (8–10), which point to a poor prognosis. Unfortunately most cancers (75%) fall in the intermediate range (5–7), which is a poor predictor of how well or poorly the patient will fare.

Prostate cancers are clinically divided into groups according to their size and extent and the presence of metastases in regional (pelvic) lymph nodes or at distant sites in the body. This procedure is called staging and will determine the treatment that is proposed. Staging usually includes methods to detect bone metastases (e.g., a bone scan examination) and may involve the surgical removal of the regional pelvic lymph nodes that drain the prostate to examine them for the presence of metastases. A summary of the method of staging for prostate cancer that is most widely used in the United States is given in Table I. Many stage A cancers are found incidentally in transurethral resection material, but some are found when they caused elevation of serum PSA levels. Stages A and B are without clinical complaints.

Approximately 35–40% of new prostate cancer cases are at an advanced stage at the time of detection, that is, they are locally extended and/or have metastasized, and they cannot be cured. About 50% of new cases are treated with a surgical procedure in which the entire prostate is removed (radical prostatectomy). Unfortunately, in 25–45% of the cases that undergo radical prostatectomy there are indications that not all tumor tissue is removed. Thus, more than 50% of prostate cancer patients have a poor prognosis, despite the recent advances in early detection of the disease.

Some stage A cancers are not treated but followed ("watchful waiting"), particularly in older patients (≥70 years). Stage A and B cancers are usually treated with either prostatectomy or radiation. Radiation can be applied by implantation of needles containing a source of radiation, most often iodine-125, or more commonly by external radiation methods that can be accurately focused at the prostate. External radiation is also applied to stages C and D1. Hormonal therapy is used for all stages beyond stage A, and chemotherapy is usually reserved for patients who have stage D2 disease that has become resistant to hormonal therapy. Combinations of different modes of therapy are common. Since the disease is essentially incurable

TABLE I
Staging of Prostate Cancer

Stage[a]	Local lesion	Elevation of PSA (>4 ng/ml)	Metastases: Pelvic lymph node	Metastases: Distant	Clinically important	Prognosis	Percentage of patients[b]
A1	Not palpable (T1a)[c]	Sometimes	No	No	Probably no	Treatment probably unnecessary	
A2	Not palpable (T1b)	Frequently	No	No	Possibly yes	Curable, but treatment possibly unnecessary	~15
B0	Not palpable, localized (T1c)	Frequently	No	No	Probably yes	Often "curable"[d]	
B1	Palpable, localized, small (T2a & b)	Usually	No	No	Yes	Often "curable"	~40
B2	Palpable, localized, large (T2c)	Usually	No	No	Yes	Often "curable"	
C	Local extension beyond capsule (T3a–c & T4)	Almost always[e]	No	No	Yes	Occasionally "curable"	~15
D1	Any of above (T1–T4)	Almost always[e]	Yes (N1–N3)	No	Yes	Rarely "curable"	~15
D2	Any of above (T1–T4)	Yes[e]	Yes/No	Yes (M1)	Yes	Incurable	~15

[a]According to the American Urologic Association, *clinical* stage is the stage based on clinical findings and *pathological* stage is the stage based on pathological examination of prostate and lymph nodes (after surgery) in combination with clinical findings.

[b]Percentage of patients that have that stage of prostate cancer at the time of first diagnosis.

[c]The TNM system (American Joint Commission on Cancer) indicates the size/extent of the tumor (from T1–T3), absence (N0) or presence of metastases in regional lymph nodes (N1–N3, depending on size and extent of the metastasis), and the presence (M1) or absence (M0) of distant metastases; this is a pathological staging system and is currently most widely used.

[d]"Curable" is defined here as removal of the primary tumor (by surgery or radiation) without clinical evidence of recurrence or development of detectable metastases in the lifetime of the patient, but it does not necessarily mean that all cancer has absolutely been removed. Thus, "curability" depends heavily on the age of the patient: the same stage and grade of tumor may be curable in a 70-year-old man but not in a 50-year-old man.

[e]PSA levels are often considerably higher than 4 ng/ml in cinical stages C (mean ~100 ng/ml), D1 (mean ~100 ng/ml), and D2 (mean ~500 ng/ml), with PSA ranging on the order of 10–200, 20–200, and 50–1000 ng/ml, respectively. For pathological stages C and D, the mean PSA levels are lower, but the ranges of PSA values are even larger.

beyond (pathological) stage B, most patients receive one type of treatment (e.g., hormonal treatment) until it is not effective anymore; then, other therapy modes will be applied (e.g., chemotherapy).

Hormonal treatment consists of some form of lowering the blood concentration of testosterone to very low levels or of blocking the effect of testosterone on the androgen-sensitive cancer cell. Lowering the testosterone concentration can be achieved by castration (or orchiectomy), by treatment with estrogens (particularly diethylstilbestrol, DES), or by treatment with LH-RH analogs. The latter treatment either competitively inhibits the effects of natural LH-RH on the pituitary (analogs with no biological activity) or exhausts the LH production by the pituitary following an initial increase due to constant hyperstimulation of LH secretion (analogs with agonistic activity). These treatments will take away or inhibit testicular androgen production, but they will not suppress androgen production by the adrenal gland. Therefore, drugs have been developed that inhibit adrenal androgen production or have antiandrogenic properties because they block the binding of DHT to the androgen receptor in the prostate epithelial (cancer) cells. Combination of inhibition of testicular androgen production and application of inhibitors of adrenal androgen production or antiandrogens to take away the effect of adrenal androgens, so-called total androgen blockade, is thought to be more effective than either treatment alone. Indeed, there is evidence that such combination treatment is slightly more effective than only inhibition of testicular androgen production. The androgen dependence of most prostate cancers and the effectiveness of castration and estrogen treatment were first demonstrated in the early 1940s by Dr. Charles Huggins (and co-workers), who received the Nobel Prize for his work.

Only surgery, and sometimes radiation therapy, can be definitively curative in cases of early stage (A and B) prostate cancer. However, because of the old age of most patients, any treatment that will prolong life is of great value, particularly when symptoms are suppressed and side effects are minimal or absent (palliative therapy). In cases of locally extended or advanced prostate cancer, radiation and/or hormonal therapy are commonly applied. Radiation therapy improves survival somewhat in many cases. Hormonal treatment will effectively suppress symptoms in 75–85% of patients for variable periods of time, ranging from a few months to up to many years, but the stage and grade of the disease are not good predictors of the response. The median response duration is approximately 12–18 months. However, overall survival is not markedly improved in most cases, and survival after failure to respond to hormonal therapy is predictably on the order of only 12 months. The results of chemotherapy trials of hormone-insensitive stage D patients have been largely disappointing. There is clearly great need for better therapeutic approaches to control hormone-refractory prostate cancers.

V. CAUSES (ETIOLOGY) OF PROSTATIC CANCER

Although prostate cancer is a very common cancer in men, little is known with certainty about the causes of this disease. Studies in animal models and in tissue culture and, in particular, epidemiological investigations have provided some insights in the possible causes of prostate cancer.

Epidemiological methods can be distinguished into two broad categories, *descriptive epidemiology* and *analytical epidemiology*. Descriptive epidemiological studies are concerned with prostate cancer occurrence in relation to time and characteristics of the study population such as age, ethnicity, geographic location and degree of urbanization of site of residence, marital status, and socioeconomic status. Analytical epidemiology examines specific hypotheses and can be subdivided into four subcategories: (1) *Correlation* (*or ecological*) *studies* determine the correlation between, for example, the estimated consumption of a particular food in different countries (international) or in different areas within one country and the mortality or incidence of a type of cancer in those countries/areas. (2) *Case–control* (*or case–referent*) *studies* compare a group of cancer patients (usually 50–200 individuals) with one or more appropriate control groups and determine whether certain factors occur more or less frequently among cases than in controls; for example, whether occupational exposure to certain chemicals or the consumption of certain foods is more or less frequent. These studies can be based on information on death certificates (retrospectively) or on information gathered by interview or questionnaire. In the latter cases, not only frequency of occurrence of factors can be determined but sometimes more exact quantitative information, such as amounts of certain foods consumed. (3) *Prospective cohort studies* collect information from a large group (cohort) of individuals (usually many thousands) by interview, questionnaire, or some other method. Then the cohort is followed until cancer occurs in a sufficient number of persons to allow a comparison between them and a group of persons from the cohort that did not develop this cancer. (4) A special type of epidemiology, sometimes called *metabolic or molecular epidemiology,* determines the correlation between the presence of cancer and parameters that are related to the causation of the cancer, such as the concentration of certain hormones, indicators of nutritional status, markers of exposure to a specific chemical carcinogen, or genetic changes. Factors such as consistency and degree of the associations found, as well as "dose–effect" relationships, coherence of associations between studies of different types, and biological plausibility, all play a role in the evaluation of the results of epidemiological studies.

A. Descriptive Epidemiology of Prostate Cancer

Prostate cancer is not diagnosed before the age of 65–75 years in most men, and it is thereby typically a disease of old age, more so than any other type of cancer. However, the occurrence of prostate cancer rises more steeply with increasing age than any other cancer and is the most frequent cancer in men of over 80 years of age in many countries.

There is considerable variation in the occurrence of prostate cancer between different countries. In general, prostate cancer incidence and mortality rates are high in North American and western European countries; lower rates are found in eastern and southern Europe and in some Asiatic and most South American countries; and prostate cancer is infrequent in some Central and South American countries, in southeastern Asiatic countries, and in Japan. In fact, the ratio of occurrence in the highest-risk population (U.S. Blacks) to that in the lowest-risk populations (e.g., Chinese living in Shanghai) is larger for prostate cancer than for any other type of cancer. Some examples of this variability are given in Table II. The actual numbers of new cases (incidence) and deaths (mortality) from prostate cancer are given in Table III. These figures and most data that epidemiologists use to compare (prostate) cancer rates in different groups of people are age-adjusted, that is, mathematically corrected for differences in the age distribution (and life expectancy) of the populations studied. This is necessary because as the proportion of older individuals increases, so does the number of cancer cases.

Larger-scale migration to the United States occurred at the end of the last century and the first half of this century, predominantly from Asiatic, Central American, and European countries. Many of the countries from which these migrants originated belong to the areas that have an intermediate or low risk for prostate cancer (see Table II). These migrant populations in the United States appear to have a risk for prostate cancer that is higher than the risk that prevails in their home country and that approaches the risk in the general U.S. population. This change

TABLE II
Examples of the Geographic Variation in Prostate Cancer Occurrence

High-risk countries	Intermediate-risk countries	Low-risk countries
Sweden	Chile	Mexico
United States of America	Spain	Ecuador
Norway	Poland	Singapore
Federal Republic Germany	Paraguay	China
France	India	Japan
Hungary	Greece	Thailand

TABLE III
Examples of Age-Adjusted Incidence and Mortality Rates for Prostate Cancer[a]

Incidence (annual number of new cases per 100,000) of prostate cancer in:	1970	1985[b]
U.S.A. (Blacks, San Francisco Bay area)	93.7	151.6
U.S.A. (Whites, San Francisco Bay area)	66.5	88.5
U.S.A. (Blacks, Detroit)	114.8	158.2
U.S.A. (Whites, Detroit)	52.3	62.7
U.S.A. (Japanese in Hawaii)	30.7	44.3
U.S.A. (Whites in Hawaii)	76.2	102.8
Japan (Osaka)	3.0	8.0
China (Shanghai)	0.3	2.5
Incidence (annual number of prostate cancer deaths per 100,000) in:	**1965**	**1985[b]**
Norway	17.3	22.2
U.S.A.	17.0	17.5
Puerto Rico	12.1	16.6
Poland	9.1	11.9
Japan	2.3	3.3

[a]The actual rate figures depend on which population is used as standard for the age adjustment; the data presented here were age-standardized to the "Standard World Population."

[b]Before PSA was widely used to detect prostate cancer.

in prostate cancer risk has been best studied among Japanese living in Hawaii (see Table III) and the San Francisco Bay area. Clearly, the environment and not some genetic factor determines the risk for prostate cancer in the first place.

Nevertheless, prostate cancer is less frequent in most of these migrant populations in the United States, including Mexicans, Japanese, Chinese, and Filipinos, than among U.S. White men. The indigenous people in the United States, American Indians and Alaskan Inuit, also have a lower prostate cancer risk than White men. These observations might suggest that there are racial (genetic) differences in susceptibility to the factors that cause prostate cancer. On the other hand, not all American men share the same environment; for example, their life-style (which is an environmental factor) may differ considerably, and this may be related to these differences in prostate cancer risk. Furthermore, prostate cancer is very frequent among U.S. Blacks (see Table III), whereas this disease is not very frequent among the predominantly Black populations of some Caribbean islands, and it is probably also infrequent in sub-Saharan Africa. In fact, American Blacks have the highest risk for prostate cancer in the world. The differences in prostate cancer occurrence among Black men living in different parts of the world also indicate that environment is a more important determinant of prostate cancer risk than is genetic makeup.

Genetic factors are nevertheless important in approximately 6–8% of prostate cancer cases, who have blood relatives with prostate cancer. Men with such a family history of prostate cancer have a higher risk for developing this disease than men who do not; their excess risk is on the order of 2- to 3-fold for men with one blood relative with prostate cancer and increases to 5- to 10-fold for men with more than two such relatives. These findings indicate that genetic factors can contribute to a person's risk for developing prostate cancer, but they do not explain the differences found, for example, between U.S. Blacks and Whites and between Japanesse and western European men.

The occurrence of prostate cancer, both mortality and incidence, has increased, for the most part slowly, over the past several decades in almost every country that has been studied (see Table III). Since the method of age adjustment has been used in these studies to correct for differences in life expectancy, this increase in prostate cancer is real and not related to the increasing length of life that has occurred in many countries. Increases in the quality of medical diagnostic procedures has not resulted in increased detection of prostate cancer as cause of death. The increased use of PSA measurements is certainly responsible for the explosion of new prostate cancer cases in the United States in recent years, but cannot explain the continuing slow rise in prostate cancer death rates. The increased occurrence of the disease over time suggests that the presence and influence of environmental determinants of prostate cancer have increased. Prostate cancer occurrence has increased much more rapidly in Japan during the last two decades than in, for example, the United States (see Table III); this has been ascribed to the rapid westernization of the Japanese life-style. The increase in U.S. prostate cancer frequency has been much larger among Black men than among Whites until 1988, but this trend has since reversed due to the disproportionally infrequent use of PSA and rectal exams to detect prostate cancer in African-American men as compared to Caucasian men. In fact, the Black-to-White ratio in prostate cancer death rates has changed from 0.6 in 1930 to approximately 2 in the past two decades. Interestingly, this change coincided with a large-scale migration of Blacks from rural to urban areas, and this change in environment may well have contributed in some way to the change in prostate cancer occurrence.

Old men very frequently have a form of prostate cancer that is not clinically evident and only histologically detectable, as pointed out earlier. The occurrence of this form of prostate cancer (often referred to as latent prostate cancer) does not vary much in different countries. Even in populations in which the occurrence of prostate cancer as a disease differs widely, for example, Japanese living in Japan and Hawaii, and U.S. White and Black men, this latent type of prostate cancer is about equally frequent. However, there is a type of latent cancer that is larger and grows more aggressively than average, and this type does show the same geographic variation as clinical prostate cancer. Based on these observations, it is believed that the environmental factors that appear to determine prostate cancer risk influence the progression from small, not very invasively growing prostate cancer to larger, more invasive cancer (which ultimately develops into a tumor that causes clinical symptoms).

Cancer patterns in some populations that differ in their life-style environment from the general population have been studied to determine the influence of life-style on cancer development. Best studied in this regard are groups of Mormons in Utah and Seventh Day Adventists in California. Because of religious reasons, these groups usually do not smoke or drink alcoholic beverages, tea, and coffee, have rather strict sexual mores, and the Adventists in particular refrain from a rich diet and many are vegetarians. Cancer of the breast (in females) and colon (in both sexes) occurs less frequently in both groups, and this has been ascribed to their moderate life-style. Prostate cancer, however, is equally frequent in these two religious groups and in the rest of the U.S. population. The differences in prostate cancer occurrence among countries and, with several exceptions, among ethnic groups in the United States are rather similar to the differences in occurrence found for breast and colon cancer. Also, the changes in cancer patterns in migrants in the United States are very comparable for breast, colon, and prostate cancer. Thus, there are probably some similarities in the causes of these three important human cancers, but there are likely also major differences in the etiology of prostate cancer on the one side and of breast and colon cancer on the other. Both colon and breast cancer are thought to be related to dietary risk factors, and breast cancer also to hormonal factors. Prostate cancer, however, probably has a more complex etiology. [*See* Breast Cancer Biology.]

To estimate whether factors related to social and economic status are associated with cancer risk, cancer patterns of groups that differ in income, occupation, or level of education are often compared. No consistent relation has been found between prostate cancer risk and such indicators of socioeconomic status. A consistent finding, however, is that in many European countries prostate cancer is more frequent in urbanized than in rural areas. In the United States, this pattern has been found only among Blacks, but not Whites. This excess risk for prostate cancer in urban areas may be related to pollution or an increased likelihood of occupational exposure to carcinogens, but it may also be related to urban–rural differences in, for example, life-style.

Prostate cancer and benign hyperplasia of the prostate (see earlier) are most likely unrelated processes. The epidemiology of the two diseases is rather different, and the morphology and site of occurrence in the prostate gland are very dissimilar, as pointed out earlier.

In general, married men have a higher risk for prostate cancer than single men, and widowed and divorced men have a higher risk than married men. Among U.S. Blacks, however, single men are at the highest risk, followed by married and then widowed/divorced men. In fact, single U.S. Black men are probably the population at highest risk for prostate cancer worldwide at the present time. These associations between risk and marital status might be related to sexual behavior, as discussed later.

In summary, the descriptive epidemiology has provided the following information on the probable causes of prostate cancer:

1. Environment rather than genetic factors seems to determine a person's risk of developing prostate cancer (geographic variation and migrant studies).
2. These environmental determinants probably influence the progression from very slow growing, very little malignant latent cancers to more malignant cancers that will eventually metastasize (lack of geographic variation for latent cancers).
3. The presence and influence of these environmental factors have probably increased over the past several decades (increasing occurrence of prostate cancer).
4. Aspects of life-style, in particular, dietary habits (similarity with the epidemiology of breast and colon cancer), exposure to environmental or occupational carcinogens (higher urban than rural occurrence of prostate cancer), and aspects of sexual behavior (relation with marital status), may be related to the development of prostate cancer.
5. Notwithstanding the probably great importance of environmental determinants, genetic factors may influence prostate cancer risk on an individual basis in a small fraction of cases (familial aggregation of risk).
6. For some unknown reason, African-American men have a higher risk for developing prostate cancer than any other population in the world.

B. Environmental Factors

Environment is thought to play a major role in the causation of prostate cancer. Aspects of life-style, such as diet and nutrition and sexual behavior, and occupational and environmental exposure to carcinogenic factors, all probably play a role in this respect. Smoking, which is a major factor in the causation of several human cancers, is not related to prostate cancer risk.

1. Diet and Nutrition

Most indications and evidence that dietary factors are involved in the etiology of prostate cancer are derived from epidemiological studies, but there are also some animal and *in vitro* investigations that have addressed this question. Rather than providing details of these various studies, a summary of their results is presented in Table IV. Notwithstanding the many difficulties in interpreting epidemiological results, there are a few associations that are consistent and coherent: (1) a strong positive association (more consumption is associated with higher risk) has been found for consumption of fats and oils, and total fat consumption; (2) a moderate positive association has been established for eggs, animal (saturated) fats, total and, possibly, animal protein, and dietary fiber; and (3) there is no association between risk and the consumption of alcoholic beverages. It is often not clear which of these factors are causally related with prostate cancer risk, because they are often interrelated, for example, the consumption of animal fat and animal protein usually go hand in hand. Animal model studies have, for the most part, not provided support for the asser-

TABLE IV

Summary of the Results of Epidemiological, Animal Model, and Tissue Culture Studies on the Relation between Dietary Factors and Prostate Cancer Risk[a]

Dietary factor	Correlation studies	Case–control studies	Prospective studies	Animal studies	*In vitro* studies
Food					
Meat, all/cattle	++	+?	0		
Pork	+	0?			
Poultry	0	0?			
Edible fats/oils	+++	++			
Eggs	+?	++	+		
Milk		+?			
Fish	0	−	0		
Cereals	−−				
Sugars	++				
Vegetables, all	0	−?			
Green-yellow (Japanese)	−	?			
Carrots	−−				
Pulses/nuts/seeds	−−	−			
Nutrients					
Fat, total	+++	++	0?	0/++?	
Animal/saturated	++	++		0	
Vegetable/unsaturated	0	?		0/++?	
Protein, total	++	+?			
Animal	++				
Carbohydrates					
Simple	++				
Complex	0				
Dietary fiber		−?			
Caloric intake (energy)	+				
Vitamin A/carotenes	0	???	?		−−−
Vitamin C	0	0?	0		
Vitamin D					−
Selenium	−	+?			
Zinc	+?	+?			
Beverages					
Coffee	++				
Tea	0				
Alcohol, total	?	0	0		
Beer	+?	0			
Wine	0	?			
Hard liquor	0	0			

[a]Associations are indicated as: +++, strongly positive; ++, moderately positive; +, weakly positive; 0, no association; −, weakly negative; −−, moderately negative; −−−, strongly negative; ?, inconclusive results; ? after a + or − sign, limited evidence for the association; ???, very contradictory results.

tion that dietary fat enhances prostate cancer development. However, a recent study with human prostate cancer cells (LNCaP cells) transplanted into immunodeficient mice (so-called nude mice that lack a thymus and T cells) showed that a high-fat (corn oil) diet (~40% of calories from fat) enhanced the growth of tumors that developed from these cells as compared with diets with lower amounts of fat (~31, 21, 12, or 2.5% of calories from fat); the lower the fat content, the slower was the tumor growth. [*See* Nutrition, Fats and Oils.]

The findings for vitamin A and β-carotene are very contradictory. *In vitro* studies have shown that retinol and other retinoids counteract the effects of chemical carcinogens on the morphology of rodent prostate explants. The results of some epidemiological studies, however, indicate that vitamin A and/or carotenes may increase risk for prostate cancer, whereas others suggest a protective effect. Most studies that showed increased risk found this only for patients 70 years and older, but not for younger patients, and they estimated total dietary intake of vitamin A and carotene. Most studies that found a negative or no association determined only intake of one or a few vitamin A/β-carotene-rich foods and they often did not distinguish between younger and older cases. It is possible that the observed effects of retinoids and carotenes may be related to the presence or absence of another factor, the exposure to which is associated with the consumption of these micronutrients. There are as yet no reports of animal studies examining vitamin A or carotenes, but there is evidence that a synthetic vitamin A analog, fenretinide [or *N*-(4-hydroxyphenyl)-retinamide], does not influence prostate carcinogenesis in laboratory rats. Thus, the possible influence of retinoids and carotenes on prostatic carcinogenesis is presently unresolved. However, vitamin A has been shown to be essential for normal function of the prostate and normal differentiation of the prostate epithelium, both in studies on human vitamin A deficiency cases and in animal and organ culture studies. In addition, it should be noted that vitamin A has been shown to protect against the development of cancer in a number of animal models other than prostate cancer. [*See* Vitamin A.]

Zinc is important in the functional activity of the prostate gland, and it is one of the major constituents of the prostatic fluid as mentioned earlier. Whether this trace mineral plays a role in the development of prostate cancer is, however, not known.

The evidence in favor and against factors related to energy expenditure (such as exercise) is mixed. It is also unclear at present whether obesity, which is in part related to the diet and physical activity, is associated with an increased risk for prostate cancer. Interestingly, however, obesity has been shown to influence the production and metabolism of sex steroid hormones in men.

One possible mechanism by which dietary factors can influence the development of prostate cancer is via effects on the endocrine system. Indeed, a number of studies in animals and humans have shown that changes in the composition of the diet can affect the concentrations of circulating testosterone, estradiol-17β, and prolactin in males. However, the data are too fragmentary at present to draw firm conclusions, but they do indicate that both the production of hormones and their metabolic clearance from the blood can be influenced by diet. [*See* Endocrine System; Obesity.]

2. Sexual Factors, Venereal Disease, Vasectomy, and Prostatitis

The results of some case–control studies have suggested that prostate cancer risk is related to aspects of sexual behavior, such as sexual drive and ejaculatory activity. Although it is plausible to assume that such relations exist, it is far from clear exactly which sexual factors are in fact related with increased risk. A consistent positive association has been found in case–control studies between prostate cancer risk and a history of venereal disease. It is not clear, however, which venereal diseases are more important in this respect, and by what mechanism they would influence prostatic carcinogenesis. Prostatitis (inflammation of the prostate) is a very common condition, and venereal diseases are frequently associated with prostatitis. There is some suggestive, but not conclusive, evidence that a history of prostatitis is associated with increased prostate cancer risk.

Vasectomy has been identified as a possible, but weak, risk factor in several epidemiological studies. However, there are also some studies that indicate that vasectomy does not increase prostate cancer risk. There are no known mechanisms by which vasectomy can influence the development of prostate cancer. More research is needed to clarify this issue.

3. Occupational Factors, Environmental Pollution, and Radiation

Increased occurrence of prostate cancer has been found among workers in a number of occupations. These are summarized in Table V. There are no indications that exposure to specific chemicals is related to

TABLE V
Summary of Associations between Occupations and Prostate Cancer Risk

Occupational group	Association
Armed forces personnel	Increased risk has consistently been observed in this very diverse group in England and the United States.
Farmers and farm workers	Increased prostate cancer risk has been found in several case–control and cohort studies from several countries, but this finding is not consistent; pesticide exposure is possibly related to this finding.
Rubber industry workers	Increased risk has been found in workers in the mixing and batch preparation division in some plants, but not in others; some animal studies found prostate cancer after exposure to chemicals used in these divisions.
Nuclear industry workers	Increased risk has been observed in several studies, but not in other studies.
Iron/steel foundry workers	Increased prostate cancer risk was found in one study, but not in several other studies.

increased prostate cancer risk in men, although animal studies have shown that a variety of chemical carcinogens can produce prostate cancer under specific favorable conditions. Cadmium has often been reported to be a potential prostate carcinogen, however, at present there is absolutely no epidemiological evidence to support this. On the other hand, there are some indications from animal and *in vitro* studies that cadmium can produce neoplastic changes in the rodent prostate.

There is some evidence that prostate cancer is more frequent in areas with a high level of air pollution. This is consistent with the earlier mentioned higher prostate cancer rates in urban compared to rural areas.

The rat ventral prostate, and probably also the human prostate (studied *in vitro*), can accumulate, secrete, metabolize, and bind a variety of chemical carcinogens. Some of these processes can produce active carcinogens in the prostate cells and, thereby, contribute to the formation of prostate cancer.

There are a number of reports of increased prostate cancer risk in workers exposed to radiation on the job, particularly in the nuclear industry, but there also studies that did not find this. Recent studies indicate no increase in the frequency of prostate cancer among men who had been exposed to heavy radiation from the Hiroshima and Nagasaki atomic bomb explosions. However, X irradiation of the pelvis can cause prostate carcinomas in rats. Thus, exposure to gamma and alpha radiation is possibly a risk factor for prostate cancer.

C. Endogenous Factors

As pointed out earlier, hormones, particularly androgenic steroids, regulate normal function of the prostate gland, and the majority of human prostate cancers are initially dependent on the presence of androgens. Therefore, it is plausible that hormones, and particularly androgens, are involved in the development of prostate cancer.

Differences in hormonal parameters between prostate cancer patients and controls have not yielded insights in this respect, because such differences are related to the presence of the disease, but not necessarily to its development. A relation between changes in hormonal status with aging and the development of prostate cancer has been suggested, particularly a decline in testosterone and an increase in estrogens in the blood.

A number of studies have compared hormonal parameters in groups of men who differed in risk for prostate cancer, for example, South African Blacks, U.S. Blacks, Japanese men, and U.S. Whites. One somewhat consistent finding was that the concentrations of testosterone and LH were often slightly higher, and never lower, in the high-risk compared to the low-risk populations. In addition, evidence from prospective hormonal studies suggests that higher concentrations of precursors of testosterone or higher activity of the enzyme 5α-reductase are related to increased risk for prostate cancer. Finally, some studies suggest that concentrations of estrogenic hormones in the blood are higher in high-risk individuals or populations than in low-risk groups.

As mentioned earlier, men with blood relatives that had prostate cancer are at increased risk for the disease, suggesting that genetic predisposition may play a role in the development of some prostate cancers. The blood concentration of testosterone has been shown to be lower in such men than in appropriate controls in one study. This finding raises the possibility that genetic predisposition to prostate cancer is mediated via mechanisms other than those that are responsible for the differences in prostate cancer risk between, for example, U.S. White and Black men or between U.S. Blacks and African Blacks (testosterone, see previous paragraph).

A number of animal model studies demonstrate that testosterone very effectively enhances the formation of prostate cancer, both by increasing the sensitivity of the prostate tissue to chemical carcinogens and particularly by a tumor promotor-like action when administered chronically after treatment with carcinogens. Testosterone stimulates cell proliferation in the prostate and thereby increases the probability that lesions created in the DNA by a carcinogen become permanent inheritable genetic DNA alterations in a cell (mutations). It probably also acts by enhancing the further proliferation of cells that carry such mutations, which allows them to form clones of neoplastically transformed cells that will grow out to become a cancer. The joint administration of testosterone and estrogenic hormones (without any carcinogen treatment) also very effectively produces prostate cancer in rats; this finding provides support for the earlier mentioned assertion that estrogens are also involved in prostate carcinogenesis.

Since the development of prostate cancers probably depends on the presence of DHT in the prostate epithelial cells, inhibition of the formation of DHT from testosterone by 5α-reductase has been suggested as a possible way to prevent this disease from developing. A large clinical trial is currently under way testing this hypothesis, in which an inhibitor of 5α-reductase (finasteride or Proscar) is given to men at high risk for developing prostate cancer.

VI. SUMMARY AND PERSPECTIVE

Prostate cancer is a very frequently occurring cancer in men in Western countries. The hallmark features of prostate cancer are its dependence on androgenic steroids to grow, its sensitivity to hormonal treatment, and its relapse to a hormone-refractory state after initial response to hormonal treatment. Early detection by measuring blood levels of PSA has led to a marked recent increase in the incidence of this cancer, but has also probably contributed to improved surgical intervention in early stages of the disease. Descriptive epidemiology indicates that environmental factors determine the risk for developing prostate cancer. Life-style factors, particularly Western dietary habits (high-fat and high-protein foods, obesity), are probably related to the development of prostate cancer. Further environmental risk factors are a history of venereal disease and, as yet unspecified, exposures to chemicals and ionizing radiation in certain occupations. Hormones are most likely involved in the causation of prostate cancer, in particular, high circulating levels of androgens are associated with increased risk for cancer of the prostate. Thus, although little is known about the etiology of prostate cancer, the following list includes risk factors that have been identified with certainty:

1. A Western life-style, particularly Western dietary habits and high fat intake;
2. A family history of prostate cancer, with risk increasing with increasing number of affected blood relatives;
3. African-American descent or African-American life-style;
4. History of venereal disease or, possibly, prostatitis and vasectomy;
5. Employment in the armed services or, possibly, farm work.

More epidemiological research is clearly needed to better define risk factors, and more animal model experiments and metabolic and molecular studies in humans are needed to substantiate epidemiological findings. These studies will have to concentrate on the role of hormones and life-style factors, such as diet and sexual behavior, and they need to focus on reasons for the extremely high prostate cancer rates among African-American men. Very little is known about the mechanisms of prostate cancer development and progression, for example, how androgens may stimulate prostate carcinogenesis and how specific genetic changes are involved. The application of *in vitro* models and molecular biological methods is required to resolve such questions. Major challenges for the future are (1) to find better methods for the early detection of prostate cancer that will identify clinically relevant tumors when they can still be cured by surgery; (2) to establish effective prevention strategies; and (3) to develop therapies that will kill the androgen-insensitive cell populations in human prostate cancers.

ACKNOWLEDGMENT

This work was supported in part by Grants 43151, 48084, and 58088 from the National Cancer Institute.

BIBLIOGRAPHY

Bosland, M. C. (1988). The etiopathogenesis of prostate cancer with special reference to environmental factors. *Adv. Cancer Res.* **51,** 1.

Bosland, M. C. (1994). Male reproductive system. *In* "Carcinogenesis" (M. P. Waalkes and J. W. Ward, eds.), p. 339. Raven, New York.

Brawer, M. K. (1995). How to use prostate specific antigen in the early detection or screening for prostate carcinoma. *CA Cancer J. Clin.* **45,** 148.

Coffey, D. S. (1992). The molecular biology, endocrinology, and physiology of the prostate and seminal vesicles. *In* "Campbell's Urology" (P. M. Walsh, A. B. Retik, T. A. Stamey, and E. D. Vaughan, eds.), 6th Ed., p. 221. Saunders, Philadelphia.

Coffey, D. S., Bruchovsky, N., Gardner, W. A. Resnick, M. I., and Karr, J. P. (eds.) (1987). "Current Concepts and Approaches to the Study of Prostate Cancer," Vol. 239, *Progress in Clinical and Biological Research.* Liss, New York.

Nomura, A. M. Y., and Kolonel, L. N. (1991). Prostate cancer: A current perspective. *Am. J. Epidemiol.* **13,** 200.

Rinker-Schaeffer, C. W., Partin, A. W., Isaacs, W. B., Coffey, D. S., and Isaacs, J. T. (1994). Molecular and cellular changes associated with the acquisition of metastatic ability by prostate cancer cells. *The Prostate* **25,** 249.

Slawin, K. M., Ohori, M., Dillioglugil, O., and Scardino, P. T. (1995). Screening for prostate cancer: An analysis of the early experience. *CA Cancer J. Clin.* **45,** 134.

Spring-Mills, E., and Hafez, E. S. E. (eds.) (1980). "Male Accessory Sex Glands: Biology and Pathology." Elsevier/North-Holland Biomedical Press, Amsterdam.

Stamey, T. A., and McNeal, J. E. (1992). Adenocarcinoma of the prostate. *In* "Campbell's Urology" (P. M. Walsh, A. B. Retik, T. A. Stamey, and E. D.Vaughan, eds.), 6th Ed., p. 1159. Saunders, Philadelphia.

Protein Detection

CARL R. MERRIL
National Institute of Mental Health

GLOSSARY

Autoradiography Method of using emitted rays from radioactively labeled material, such as labeled proteins, to expose photographic film to detect the labeled substance

Coomassie Blue Organic stain that binds to proteins in a noncovalent manner, providing detection of proteins in the microgram range

Fluorography Detection method that uses fluors to convert the energy from radioactive decay into photons of light

Lowry procedure One of the most popular methods developed for the quantitative detection of proteins, it can detect as little as 0.1 μg protein

Operationally constitutive proteins Subset of proteins that have constant intragel density ratios in all of the gels used in an experiment; this subset can be used as a set of endogenous reference standards for density normalization for intragel protein comparisons

Protein silver stains Procedures developed for the detection of proteins by the reduction of ionic silver to metallic silver in the presence of proteins

Schlieren Patterns or shadows caused by variations in the refractive index in nonhomogeneous protein solutions; these patterns may be used to monitor protein separations

SOME PROTEINS, SUCH AS HEMOGLOBIN, MAY BE detected by their color; however, most proteins do not absorb light in the visible range of the spectrum, and other methods have been required to detect their presence. Development of these protein-detection methods has progressed in parallel with the ability to resolve proteins from tissues and body fluids. Although most proteins can be visualized by their absorption of ultraviolet light, greater sensitivity has been achieved with organic and inorganic stains and color-producing reactions. Currently, the most commonly used stains are the organic stains, such as Coomassie Blue, and the silver stains, which depend on the reduction of ionic silver to metallic silver in the presence of proteins. In some cases, proteins may be monitored by labeling them with radioactive isotopes. However, to achieve a significant signal from a radioactive protein, it must be labeled to a high specific activity. This requirement limits the use of radioactive labeling to *in vitro* studies for the detection of the less-abundant proteins, as prohibitively large amounts of expensive radioactive precursors would be needed to label human proteins *in vivo*. In addition, the radiation hazards of such studies would be unacceptable.

I. UNDERLYING PRINCIPLES OF PROTEIN DETECTION

A. General Perspective

Certain proteins, such as those that make up human hair, are in sufficient abundance that their detection can often be made by a glance. Others, such as hemoglobin, are colored and can be detected by eye when they are present in milligram quantities. However, most proteins are not colored. Furthermore, proteins are generally present in complex mixtures in tissues and body fluids. The ability to detect and study the

ENCYCLOPEDIA OF HUMAN BIOLOGY, Second Edition, VOLUME 7. Copyright © 1997 by Academic Press. All rights of reproduction in any form reserved.

proteins contained in these complex mixtures depends on the ability to separate and visualize them. Currently, the most powerful separation methods are based on high-resolution, two-dimensional electrophoresis, a technique that separates the proteins by charge in the first dimension and by mass in the second dimension. Application of this technology has revealed that the concentrations of individual protein species, in most tissues and body fluids, varies by more than a millionfold. The capacity to detect the less-abundant, or trace, proteins, which may include proteins that function as vital enzymes and receptor molecules, has required the development of more sensitive detection methods, such as fluorescent dyes and the silver stains. [*See* Proteins.]

B. Historical Perspective

Proteins were first recognized in 1838 by the Dutch chemist Gerardus J. Mulder as "albuminoid substances" of biological origin, which often contain nitrogen, sulfur, and phosphorus. Many of the earliest protein-detection methods, such as the Kjeldahl and Folin methods, made use of the presence of these elements or other reactive groups generally present in proteins. The Kjeldahl method is based on the determination of the organic nitrogen contained in the protein. This is accomplished by boiling the protein in concentrated sulfuric acid in the presence of a catalyst such as cupric sulfate. The resulting carbon dioxide and water are discarded, and the amount of protein is determined by measuring released ammonia. In the Folin method, copper and phenol are reacted with the phenolic side groups of the proteins, such as tyrosine, to produce a blue color. One of the micro modifications of the procedure, known as the Lowry method, is capable of detecting as little as 0.1 μg of protein. This method is still one of the most commonly used for the determination of protein concentration. Although this method is about 10-fold more sensitive than detection of protein by the ultraviolet absorption of light, it still cannot detect <20,000 billion protein molecules (MW 30,000). A similar sensitivity can be achieved with the organic stain Coomassie Blue, which binds noncovalently to proteins. This stain is currently used in a general protein assay and for the visualization of proteins separated electrophoretically on polyacrylamide gel. More recently, silver stains have been demonstrated to have the capacity to detect as little as 0.01 ng protein, or as few as 200 million protein molecules. Although attaining even greater sensitivities would be useful, our ability to detect proteins has increased >100 millionfold since they were first recognized in the middle of the nineteenth century.

II. DETECTION WITH LIGHT

The earliest studies of proteins relied on direct observations utilizing either the naturally colored proteins, such as myoglobin, hemoglobin, ferritin, and cytochrome *c*, or chemical methods, such as by precipitating relatively large amounts of proteins. Although most protein molecules cannot be observed directly with visible light, early investigators were able to study their electrophoretic properties by observing the migration of quartz microspheres with the proteins adsorbed to their surface in electric fields.

Arne Tiselius demonstrated the use of ultraviolet light for the detection of noncolored proteins in the 1930s. Most of the proteins, with the exception of the protamines, contain aromatic amino acids, which provide them with a 280-nm absorption band. Though it is possible to use this ultraviolet absorption to estimate the concentration of proteins, the absorption is not uniform for all proteins, as the content of aromatic amino acids varies considerably. The use of ultraviolet absorption for the detection of proteins during and after separation by electrophoresis requires a special light source, filters, and optical components that are transparent to ultraviolet light.

As the addition of protein to a solution will cause a change in the refractive index of the solution, determining protein concentration with a refractometer is possible. However, the solvent without the protein must be available as a blank. Tiselius demonstrated that the shadows, or schlieren, created by boundaries between regions with different refractive indices, due to the varying concentrations of proteins in electrophoretic systems, could be used to monitor the electrophoretic separation of proteins. The complex optical systems required for the use of schlieren systems has limited its application.

III. DETECTION WITH ORGANIC STAINS

A. Background

The use of solid support media, ranging from moist filter paper to polyacrylamide gels, stimulated the introduction and use of organic stains, particularly those that had already proven their use in histological staining applications for the detection of proteins. These stains eliminated many of the complications inherent in the ultraviolet and schlieren detection systems. The organic stains also provided increased sensitivities. These stains are often bound to the proteins with noncovalent bonds, which requires that they be used to detect proteins after they are separated by procedures such as electrophoresis. Some of the first organic proteins employed were Bromophenol Blue and Amido Black. Some of these stains have proven to be useful because of their ability to differentially stain proteins; for example, Oil Red O preferentially stains lipoproteins.

B. Coomassie Blue Stains

The most sensitive of the organic stains are the Coomassie Blue stains, with their capacity to detect as little as 0.1 μg protein. These organic stains were originally developed in the middle of the nineteenth century as acid wool dyes. They were named Coomassie dyes to commemorate the 1896 British victory in the battle for the occupation of the Ashanti capital, Kumasi, or Coomassie, in Africa. These triphenylmethane Coomassie stains have superior protein-staining abilities.

The first of triphenylmethane Coomassie Brilliant Blue stains introduced was Coomassie Brilliant Blue R250 (the letter R stands for a reddish hue, and the number 250 is a dye strength indicator). It can detect as little as 0.1 μg protein and gives a linear response up to 20 μg. However, the relationship between stain density and protein concentrations varies for each protein. Another Coomassie stain, Coomassie Brilliant Blue G250 (G indicates a greenish hue), is used for rapidly staining gel. Its capability as a rapid stain is based on its limited solubility in trichloroacetic acid, which permits its use as a colloidally dispersed dye. This dye does not penetrate gels and stains the protein only on or near the surface. Another Coomassie stain, Coomassie Violet R150, has gained some favor by virtue of its ability to rapidly stain proteins separated by isoelectric focusing (a procedure that separates proteins by their charge) on polyacrylamide gels, while not staining the carrier ampholytes (which are used to create the pH gradient in these gels).

Though the mechanism of Coomassie Brilliant Blue staining is not fully understood, it is known that these dyes require an acidic medium for electrostatic attraction to be exerted between the dye molecules and the amino groups of the proteins. The binding is fully reversible, as has been demonstrated by the loss of the stain following dilution under appropriate conditions. The relatively high staining intensity of Coomassie Blue stains, compared with other organic dyes, appears to be due to their ability to form secondary bonds between the dye molecules, permitting the accumulation of additional dye molecules. The basic amino acids have been proposed to supply the major binding sites for the Coomassie dyes. This proposal is based on the following observations: polypeptides, which are rich in lysine and arginine, are aggregated by Coomassie G stain, suggesting that the dye interacts with the basic groups in the polypeptides and a significant correlation exists between the intensity of Coomassie Blue staining and the number of lysine, histidine, and arginine residues in the protein.

C. Other Common Organic Protein Stains

Amido Black (Acid Black 1) and Fast Green (Food Green 3) are also commonly utilized for protein detection; however, Coomassie Blue staining exhibits three times the intensity of Fast Green and six times the intensity of Amido Black. The staining intensities of these dyes are approximately proportional to their relative molar adsorption coefficients.

D. Organic Stains that Provide Fluorescent Protein-Detection Methods

Anilinonaphthalene sulfonate was the first fluorescent stain used to visualize proteins in gels. It is used as a postelectrophoretic stain and appears to form a fluorescent complex with the protein's hydrophobic sites. It has a sensitivity limit of about 20 μg protein. Higher sensitivity has been achieved with preelectrophoretic fluorescent stains, such as dansyl chloride. This stain reacts with proteins to form fluorescent derivatives in 1–2 min at 100°C, with a sensitivity limit of 8–10 nm.

A number of fluorescent stains were designed to increase the detection limits of amino acid analyzers. One such compound, fluorescamine, is nonfluorescent prior to its reaction with a protein. This compound reacts with the primary amines of the basic amino acids within each protein and the N-terminal amino acids to yield a fluorescent derivative. It has proven capable of detecting as little as 6 ng protein. The protein derivative of a related compound, 2-methoxy-2,4-diphenyl-3(2*H*)-furanone (MDPF), has the same speed and simplicity of reaction as fluorescamine, but it is 2.5 times as fluorescent as a fluorescamine-labeled protein. As little as 1 ng protein has been detected with MDPF, and it has a linear response from 1 to 500 ng. As with most other protein stains, the relationship between the relative fluorescence and the protein concentration varies for each protein, depending on the number of reactive groups in the protein.

Most of the fluorescent stains form covalent bonds with the proteins. Reaction conditions are generally best provided prior to the separation of the protein. Preelectrophoretic staining has certain advantages: the possibility of performing stoichiometric reactions with proteins without the diffusion limitations imposed by staining within a gel matrix, the feasibility of following the process of electrophoresis visually with prestained proteins, and the absence of background problems due to dye-trapping or reaction of the dye with the gel. However, the covalent bonds formed between the proteins and these stains may alter the charge of the protein and affect its resolution in methods that separate proteins by their charge, such as with isoelectric focusing. Such charge alterations are generally not of significant consequence for electrophoretic techniques that separate proteins on the basis of molecular weight. The addition of a few relatively small dye molecules will generally have an insignificant effect on the mass of most proteins. Furthermore, even in purification techniques that may be affected by these stains, as long as the stains react with the proteins in a stoichiometric manner, the shifts in protein patterns should be highly reproducible, permitting the construction of valid protein maps and protein identifications.

Although fluorescent stains may achieve a greater sensitivity than most other organic stains, they require ultraviolet light for visualization, and direct quantitation requires fairly sophisticated equipment. These problems, coupled with the altered mobility of the proteins during isoelectric focusing, have inhibited the utilization of the fluorescent stains.

IV. DETECTION WITH METALLIC STAINS

A. Background

In the twelfth century, Count Albert von Bollstädt reported that silver nitrate could blacken human skin. However, silver was not used as a stain in modern scientific applications until C. Krause adapted it to stain fresh tissues for histological examination in 1844. By 1873, Camillo Golgi, followed by Ramon S. Cajal, utilized silver stains to revolutionize the understanding of the anatomy of the central nervous system. Silver was introduced as a general stain for proteins separated by polyacrylamide gel electrophoresis by C. R. Merril and Robert Switzer in 1979. This stain was adapted from a histological silver stain. The silver stains have permitted the detection of as little as 0.01 ng protein. The gain in sensitivity afforded by silver staining has stimulated the recent widespread use of this method for the detection of proteins separated by polyacrylamide gel electrophoresis. The first silver stains used for the detection of proteins were adapted from histological stains, and they were often tedious, requiring hours of manipulation and numerous solutions. However, during the 10 years since the introduction of silver staining as a general method for the detection of proteins, simplified staining protocols have been developed. These silver staining protocols can generally be divided into three main categories: the diamine, or ammoniacal silver stains; the nondiamine chemical stains, based on photographic chemistry; and stains based on the photodevelopment or photoreduction of silver ions to form a metallic silver image.

B. Chemistry of the Silver Stains

Diamine silver stains were first developed for the visualization of nerve fibers by Cajal. They rely on ammonium hydroxide to form silver diamine complexes to stabilize the silver ion concentrations. Image production is initiated by acidifying the stain solution, usually with citric acid in the presence of formaldehyde, to lower the concentration of free ammonium ions, thereby permitting the liberation of silver ions. The silver ions are reduced with the formaldehyde to form a metallic silver image of the protein pattern. An optimal concentration of citric acid is necessary to provide a controlled rate of silver ion reduction and thus prevent the nonselective deposition of silver. Diamine stains may become selectively sensitive for glycoproteins if the concentration of silver ions is too low. This specificity can be minimized by maintaining a sufficient sodium–ammonium ion ratio in the diamine solution. However, some investigators have enhanced this tendency of the diamine stains to develop a protein stain that selectively stains neurofilament polypeptides.

The nondiamine chemical-development silver stains were developed by adapting photographic photochemical protocols. These use silver nitrate to supply silver ions. Image formation is initiated by the selective reduction of the silver ions to metallic silver by formaldehyde under alkaline conditions. Sodium carbonate and/or hydroxide and other bases are used to maintain an alkaline pH during development. The formic acid, produced by the oxidation of formaldehyde, is buffered by these bases.

The photodevelopment stains utilize energy from photons of light to reduce ionic to metallic silver. Proteins enhance the photoreduction of ionic to metallic silver in gels impregnated with silver chloride. The use of photoreduction provides for a rapid, simple silver stain method for detecting proteins.

In some cases, combining the photodevelopment and chemical-development methods has been advantageous. One such stain permits the detection of proteins in the nanogram range in <15 min. This stain uses silver halide to provide a light-sensitive detection medium and to prevent the loss of silver ions from membranes or thin-layer gels; photoreduction to initiate the formation of silver nucleation centers; and chemical development to provide a high degree of sensitivity by depositing additional silver on the silver nucleation centers. The stain's rapidity of action and ability to stain samples spotted on membranes, such as cellulose nitrate, have permitted its use as a quantitative protein assay.

The basic mechanism underlying all of these protein-detection protocols with silver staining involves reduction of ionic to metallic silver. Detection of proteins in gels or on membranes requires a difference in the oxidation–redution potential between the sites occupied by proteins and the adjacent sites in the gels or on the membranes. If a protein site has a higher reducing potential than the surrounding gel or matrix, then the protein will be positively stained. Conversely, if the protein site has a lower reducing potential than the surrounding gel or matrix, the protein will appear to be negatively stained. These relative oxidation–reduction potentials can be altered by the chemistry of the staining procedure.

The reactive groups that are necessary for the silver stains are the sulfur and amino groups. Evidence also indicates that the silver stains require the cooperative effects of several intramolecular functional groups to form complexes with the silver prior to the reduction of ionic to metallic silver. The importance of the basic and the sulfur-containing amino acids has been corroborated by observations with purified peptides and proteins of known amino acid sequence. For example, leucine enkephalin, which has neither sulfur-containing nor basic amino acids, does not stain with silver, whereas neurotensin, which also has no sulfur-containing amino acids but does have three basic amino acid residues (one lysine and two arginines), does stain. The importance of the basic amino acids has been further substantiated by evaluations of the relationship between the amino acid mole percentages of proteins and their ability to stain with silver.

Some proteins have proven to be difficult to stain with silver, for as yet undetermined reasons. Calmodulin and troponin C require pretreatment with glutaraldehyde, whereas histones require pretreatment with formaldehyde and Coomassie Blue. However, even with this pretreatment the sensitivity for histones is decreased 10-fold as compared with detection of neutral proteins.

C. Color Effects with Silver Stains

Whereas most proteins stain with monochromatic brown or black colors, certain lipoproteins tend to stain blue and a number of the glycoproteins appear yellow, brown, or red. This color effect has been shown to be analogous to a photographic phenomenon first described by Thomas Seebeck in 1810, when he noted that if the spectrum of visible light obtained by passing sunlight through a prism was projected onto a silver chloride-impregnated paper, some of the colors of the spectrum appeared on the paper. The production of color by both the photographic experiment conducted by Seebeck and certain proteins separated on polyacrylamide gels is caused by the size of the metallic silver particles, produced by the photoreduction effects of the light; the refractive index of the photographic emulsion or electrophoretic gel; and the distribution of the silver particles. The smaller silver grains (<0.2 μm in diameter) transmit reddish or yellow red light, whereas grains >0.3 μm give bluish colors, and larger grains produce black images.

The production of color by proteins separated on polyacrylamide gels may be enhanced by lowering the concentration of reducing agent in the image development solution, prolonging the development time, adding alkali, or elevating the temperature during staining. These colors may aid in identification of certain proteins.

D. Sensitivity of Protein Detection with Silver Stains

Currently, silver stains offer one of the most sensitive nonradioactive methods for detecting proteins, particularly for proteins separated by gel electrophoresis. They are generally 100-fold more sensitive than the

Coomassie stains, and they can often detect as little as 0.01 ng protein. The chemical-development silver stains are, in general, more sensitive than photodevelopment silver stains. However, the loss in sensitivity by photodevelopment silver stains may be compensated for by their ability to produce a protein image within 10–15 min after gel electrophoresis. High sensitivities with silver stains require care in the selection of reagents, including the water that is used to make up the solutions. The water should have a conductivity of <1 mho/cm. Contaminants may cause a loss of sensitivity and result in staining artifacts. One common contaminant, a keratin-like protein, produces artifactual bands with molecular weights ranging from 50,000 to 68,000.

V. DETECTION OF SPECIFIC PROTEINS

A. Detection with Stains

Silver stains have been developed that demonstrate considerable specificity for nucleolar proteins and neurofilament polypeptides. Other silver staining protocols produce colors with certain types of proteins: sialoglycoproteins may be stained yellow, whereas lipoproteins tend to stain blue. Lipoproteins are also preferentially stained with the stain Oil Red O. With other protocols, the glycoproteins appear yellow, brown, or red, whereas other proteins are stained brownish black. Glycoproteins may also be detected by the red color they produce following oxidation with periodic acid and treatment with fuchsin sulfurous acid (Schiff's reagent).

B. Detection of Specific Enzymes

Enzymatically active proteins may be detected and quantified by monitoring their enzymatic activity. A number of enzymatic reactions can be manipulated to produce a color. In some cases, secondary reactions must be coupled to the initial enzymatic reaction. These colors can be used to detect the proteins in solution and in some cases directly in the electrophoretic support media. For example, the enzyme acid phosphatase can catalyze the modification of phenophthalein to produce a pink color, whereas amylase can be detected by placing a gel containing this enzyme in contact with a starch plate. Subsequent treatment of the starch plate with iodine will produce a purple color, except in regions that were in contact with amylase. These regions will appear as clear areas. A number of these assays use electron transfer dyes, such as methyl thiazolyl tetrazolium, to detect enzymes involved in electron-transfer reactions. Over 100 specific enzymes can now be detected by specific color reactions in a manner similar to those mentioned here. For this type of detection to be successful, electrophoretic parameters, buffers, and temperature must be optimized so that the proteins maintain their enzymatic activities. However, when pancreatic proteins were separated by the relatively harsh conditions of high-resolution, two-dimensional electrophoresis, which involves denaturation and treatment with detergent, 15 pancreatic proteins still displayed enzymatic activity.

C. Detection of Protein by Specific Antibody Binding

Directly detecting proteins separated in gels with antibodies is difficult, due mainly to the inability of the antibodies to diffuse into the gel matrix. However, antibody detection can be enhanced by transferring the proteins to a thinner matrix, such as nitrocellulose paper, diazo-modified cellulose, cyanogen bromide-activated paper, or nylon membranes. This transfer is usually facilitated by electroblotting or Western blotting. By carefully selecting the transfer buffer and electrical parameters, >90% of the protein from the gel on the membrane matrix can be captured. Many of the protein stains developed to visualize protein in gels can be applied to proteins bound to membranes. Detection of specific proteins with immunostains have been achieved with both polyclonal and monoclonal antibodies. It has been possible to detect subnanogram quantities of a specific protein by reacting it with antibodies that are complexed with peroxidase. The antigen–antibody is visualized by immunoperoxidase staining. It is also possible to use antibodies that have been complexed with enzymes, fluorescein, or rhodamine. Some investigators have been able to detect as little as 1 pg of a specific protein with these techniques.

VI. DETECTION OF RADIOACTIVE PROTEINS

A. Background

Radioactively labeled proteins may be visualized without staining by autoradiographic methods. These autoradiographic methods were first introduced by Abel Niépce de Saint-Victor in 1867 and Antoine-Henri Becquerel and Marie Curie in 1896, when they used photographic plates to search for elements and mineral crystals that emit rays that could penetrate the paper and expose and darken the photographic plates in a manner similar to those discovered by Wilhelm Roentgen. Becquerel discovered that uranium crystals emitted such rays, whereas Curie found that similar rays could be emitted by crystals containing other elements. She named this phenomenon radioactivity. While autoradiography has been used successfully to detect ^{14}C-labeled proteins, fluorographic techniques were introduced by Wilson in 1958 to study tritium (^{3}H)-labeled compounds.

Proteins need to be labeled to a high specific activity with radioactive isotopes to provide for the detection of the trace proteins. For example, 0.1 ng of a 40,000-MW protein labeled to a high specific activity of 1000 Curies per millimole produces only 5 disintegrations per minute. Proteins labeled to a high specific activity may be detected with sensitivities equal to, and often better, than those obtained by most stains. However, the use of such radioactively labeled proteins is limited, as it is difficult to achieve these high specific activities in animal studies and unethical to utilize such high specific activities in research involving humans.

Proteins are generally labeled during their synthesis in cells with either ^{3}H, ^{14}C, ^{35}S, ^{32}P, or ^{125}I. These isotopes may be incorporated in a ubiquitous amino acid, such as methionine or leucine, to label most of the proteins in a cell culture or tissue slice, or they may be incorporated in a specific precursor, such as glucosamine, to label only the glycoproteins. Labeling the proteins *in vitro* is also possible with reagents such as iodinated *p*-hydroxyl-phenylpropionic acid or *N*-hydroxysuccinimide ester (Bolton-Hunter reagent). Although these labeling methods often affect the electrophoretic migration of the protein, other procedures such as reductive methylation with boro-^{3}H-hydride do not appear to affect the migration.

B. Detection with Autoradiography

Autoradiographic detection depends on the formation of a latent image in the film following exposure of photographic film to ionizing radiation from beta, gamma, or X rays. The image is made visible by normal photographic development. In general, a film's sensitivity to ionizing radiation increases linearly with temperature until 60°C. Dehydrating gels prior to autoradiography are generally useful to reduce the path length from the emitting isotope to the film. Certain isotopes such as ^{3}H emit such weak beta rays that they provide inefficient labels for autoradiographic detection, even when the gels are dehydrated. It is possible to enhance the sensitivity of the film by bathing the preexposed film in a dilute solution of silver nitrate or by baking the film for 2 hr in a hydrogen atmosphere at 66°C.

C. Detection with Fluorography

It is possible to improve the detection of proteins labeled with ^{3}H, ^{14}C, and ^{35}S by converting some of the energy from the radioactive decay to light by the addition of a fluor or scintillation agent to the gel or membrane containing the labeled protein. Dispersal of the fluor throughout the gels is necessary when the gel contains proteins labeled with ^{3}H, as the path length of the weak beta particles emitted by this isotope is often too short to penetrate the gel to reach the photographic film. The sensitivity of this method will be enhanced if the film's spectral sensitivity matches the emission spectra of the fluor used. In contrast to autoradiography, low-temperature exposure (−70° to −80°C) during fluorography increases the sensitivity. This low-temperature enhancement may be due to reduced energy loss of the fluor molecules through intramolecular vibrations, thereby making more energy available for light production. The low temperature may also prolong the lifetime of unstable latent images, until they can be stabilized by reduction of four or more silver ions by multiple photon hits. By maintaining fluorographs at low temperatures, sensitivity can increase about 10-fold. A further 3-fold increase in sensitivity may be achieved if the film is prefogged to a uniform optical density of 0.15. This prefogging reduces the threshold for the creation of stable latent images. Fluorography may also enhance the detection of isotopes, which are high-energy emitters, such as ^{32}P or ^{125}I. These isotopes have emissions that generally pass through the film with energy to spare. By placing the film between the gel and an image intensification screen (containing phosphors or fluors), any radiation passing through the film will strike the screen, releasing photons of light, which will provide for a fluorographic image in addition to the autoradiographic image produced by a particle's original traverse of the film.

D. Detection by Photostimulated Luminescence

In addition to the use of autoradiography and fluorography for the detection of radioactive proteins, it is also possible to use a sensor containing a photostimulatable phosphor. Such sensors are available as "imaging plates." These imaging plates generally consist of a photostimulatable phosphor, such as barium fluorobromide containing a trace of europium, for luminescence, on a polyester film backing. The activated image, caused by exposing the plate to radiation, is read by illuminating the plate with a red laser. The resulting photostimulated luminescence is detected by a photomultiplier. Detection by photostimulated luminescence provides a wide linear dynamic range and high sensitivity. The sensitivity for ^{14}C has been reported to be 82 times that of X-ray film. The major drawback with photostimulated luminescence is a lower resolution than that which can be obtained by X-ray film. This loss of resolution is caused by the use of a plastic protective layer on the imaging plates and resolution limitations of the image readers. This limitation may be partially overcome through the use of nonprotected imaging plates.

VII. QUANTITATIVE DETECTION OF PROTEINS

A. Background

Many protein-detection techniques may be employed quantitatively, provided that their methodological limitations are respected. The major limitation for almost all of the methods, including stain and autoradiographic methods, is that most proteins exhibit protein-specific quantitative responses. These protein-specific variations are indicative of a dependence of these methods on the content of specific groups within each type of protein that permits detection. In the case of radioactively labeled proteins, detection is determined by the mole percent of the labeled amino acids; for the detection of protein with ultraviolet light, it depends on the number of aromatic amino acids; whereas for silver stain detection, it depends primarily on the mole percent of the basic amino acids. Because each type of protein has a specific amino acid arrangement, they will each respond to a detection method in a unique manner. Although these protein-specific detection variations may prove troublesome, they may also be utilized to differentiate proteins. In intergel quantitative studies, they limit quantitative comparisons to homologous proteins.

B. Quantitation with Specific Protein-Detection Methods

One of the most common methods used for the quantitative detection of proteins, the Lowry method, produces specific staining curves when the density of the color reaction is plotted against the concentration of the protein. Because this method is based on the Folin reaction, in which copper ions and phenol are reacted with the phenolic side groups of the proteins, such as tyrosine, to produce the blue color, the slope of the staining curve will depend mainly on the number of phenolic side groups in the protein.

Proteins detected with Coomassie Blue also display protein-specific staining curves. The variation in Coomassie Blue staining has been shown to be related to the mole percent of the basic amino acids in the protein. These curves are generally linear over a protein range of 0.5–20 μg.

Proteins detected with silver staining display a similar correlation between the intensity of the staining reaction and the number of basic amino acids in each protein. The linear portion of the silver-staining curve generally begins at 0.02 ng and extends over a 40-fold range. It is possible to extend the useful quantitative range of the protein-detection methods beyond the linear range by using curve-fitting techniques.

Protein concentrations of $>$2 ng usually cause a saturation in the silver image, which results in nonlinearity above that concentration. On analysis of gel images, saturated bands or spots can be recognized by either a plateau of the staining density in the center of the bands or spots or the presence of a center that is less intensely stained than the regions near the edges. This effect is similar to the "ring-dyeing" noted with some of the organic stains. In ring-dyeing, the stain concentration is less in the center of a band or spot than at the edge—an artifact due to an insufficient diffusion of dye molecules into the protein band or spot.

C. Quantitation Standards

Because most of the protein-detection methods depend on specific reactive groups within each protein, the standards chosen should reflect this bias. For example, because there is a correlation between the intensity of Coomassie Blue staining and the number of basic amino acids in each protein, it is critical that the standard chosen should have an equivalent number of basic amino acids. If a general protein standard is needed, it should contain about 13 mole percent basic amino acids, as the basic amino acid content of proteins ranges between 10 and 17 mole percent, with a modal content of 13 mole percent. The common choice of bovine serum albumin as a protein standard may be flawed in this respect, as it has a basic amino acid content of 16.5 mole percent. Similar consideration must also be made for silver-staining standards, as the silver stains also have a dependence on the presence of basic amino acids.

D. Quantitative Intergel Protein Comparisons

The occurrence of protein-specific staining curves with most staining protocols and the specific emissions of radiolabeled proteins labeled with specific radioactive amino acids require the limitation of quantitative intergel comparisons to homologous protein bands or spots on each gel. For example, the actin spot on one gel can be compared with an actin spot on another gel, but not with a transferrin spot. Furthermore, gels to be compared must have been run under similar conditions (percentage acrylamide, stacking gel specifications, etc.) because migration distance affects band or spot compression, which in turn may influence the dyeing reactions. Quantitative intergel variations may influence the dyeing reactions; quantitative intergel comparisons also require the presence of reference proteins for the normalization of spot or band densities. Endogenous reference proteins may be found by searching each gel for a subset of proteins that have constant intragel density ratios. This subset of spots can be defined as a set of "operationally constitutive proteins." These proteins can be used to calculate specific normalization factors for each of the gels. This scheme corrects for variations in the protein-detection procedure, in image digitization, and in initial protein-loading.

BIBLIOGRAPHY

Gershoni, J. M., and Palade, G. E. (1983). Protein blotting: Principles and applications. *Anal. Biochem.* **131**, 1.

Merril, C. R. (1986). Development and mechanisms of silver stains for electrophoresis. *Acta Histochem. Cytochem.* **19**, 655–667.

Merril, C. R. (1990). Silver staining of proteins and DNA. *Nature* **343**, 779–780.

Merril, C. R., Harasewych, M. G., and Harrington, M. G. (1986). Protein staining and detection methods. *In* "Gel Electrophoresis of Proteins" (M. J. Dunn, ed.). Wright, Bristol, England.

Wilm, M., Shevchenko, A., Houthaeve, T., Breit, S., Schweigerer, L., Fotsis, T., and Mann, M. (1996). Femtomole sequencing of proteins from polyacrylamide gels by nanoelectrospray mass spectrometry. *Nature* **379**(1), 466–469.

Young, E. G. (1963). Occurrence, classification, preparation and analysis of proteins. *In* "Comprehensive Biochemistry" (M. Florkin and E. H. Stotz, eds.). Elsevier, New York.

Protein-Energy Malnutrition

J. C. WATERLOW
London School of Hygiene and Tropical Medicine

GLOSSARY

DNA Deoxyribose nucleic acid is a component of the nucleus in all cells and is the carrier of the genetic code. The amount of DNA per cell is fixed in each species and is not affected by environmental influences such as malnutrition. DNA therefore provides a constant baseline against which losses or gains of other cell components can be assessed

Edema Accumulation of fluid in the subcutaneous tissues, usually starting in the feet and legs and then spreading over the whole body. The term does not include extravasation of fluid into the pleural or abdominal cavities, which is not a feature of PEM

Kwashiorkor Condition whose outstanding characteristic, and the defining factor, is edema, usually accompanied by an enlarged liver, which on biopsy or autopsy proves to be loaded with fat. Sometimes but not always there are extensive changes in the skin and mucosae, with hypopigmentation of normally black hair. The condition was originally described in children, but may occur in adults in times of famine

Marasmus Synonymous with athrepsia. The characteristics of this condition are those of starvation, that is, total lack of food. The body weight is between 60 and 30% of normal, with gross loss of muscle and subcutaneous fat

Protein-energy malnutrition Generic term for malnutrition resulting from a diet deficient in protein or energy or both, to be distinguished from specific deficiencies of vitamins or minerals, but often accompanied by them. It is most commonly seen in children in developing countries, but may also occur secondary to diseases that produce malabsorption, such as celiac disease, inflammatory bowel disease, or cancer

THE COMMONEST FORM OF MALNUTRITION worldwide is that produced by diets deficient in energy and protein, so-called protein-energy malnutrition (PEM). In developing countries, PEM is a disease of poverty, aggravated by recurrent infections, and is responsible for the deaths of many millions of children below the age of 5. In industrialized countries, PEM is mainly secondary to disease and is seen, for example, in children with cystic fibrosis or untreated celiac disease, and in adults with conditions that interfere with the absorption of food, in anorexia nervosa, in neoplastic disease, and after severe trauma. Most of our knowledge of the pathophysiology of PEM has been derived from research on children; only more recently has there been increasing awareness, particularly among surgeons, of its importance in adults.

I. MODERATE PROTEIN-ENERGY MALNUTRITION IN THE COMMUNITY

In children, PEM of mild or moderate severity is diagnosed by growth failure, which can take two different forms: a deficit, by international growth standards, in weight for height ("wasting") and a deficit in linear growth ("stunting"). These conditions, although often combined, represent two different biological processes. The prevalence of wasting peaks in the second year of life, after weaning, and thereafter declines. In some populations up to 20% of children between 1

ENCYCLOPEDIA OF HUMAN BIOLOGY, Second Edition, VOLUME 7. Copyright © 1997 by Academic Press. All rights of reproduction in any form reserved.

and 2 years may be diagnosed as wasted (2 SD or more below normal weight for height). Stunting manifests differently; linear growth retardation is often apparent by 3 months of age and continues throughout the preschool years, so that by the age of 5 the child may be 15–20 cm below normal height, but with normal weight for height. Wasting results simply from lack of food and tends to disappear when the child gets better access to family meals. The cause of stunting in linear growth is not yet clear; recent evidence suggests that it may result from a diet deficient in quality, in terms of indispensable amino acids and micronutrients, rather than in quantity. Infections may also play an important part. Stunting is reversible if the child moves to a normal environment, but catch-up to normal height takes a long time.

In adults it has been proposed that the body mass index (BMI or Quetelet's index, i.e., weight/height^2) should be used for the assessment of nutritional state. A BMI below 18 kg/m^2 may be regarded as the lower limit of normal or acceptable weight. From 18.5 to 16 is a zone of risk, and below 16 subjects are considered to be definitely malnourished. Patients with anorexia nervosa typically have a BMI of about 14; at levels below 12 death is likely.

II. SEVERE PROTEIN-ENERGY MALNUTRITION: CLINICAL CHARACTERISTICS

Almost all our knowledge of the pathophysiology of severe PEM has been derived from work on children. In adults the pathology is probably the same, but the conditions of war and famine under which severe PEM occurs make scientific studies very difficult. The description that follows is therefore based largely on children.

The clinical characteristics of severe PEM cover a spectrum, ranging from marasmus at one end to the clinical picture commonly called "kwashiorkor" at the other. In marasmus (synonyms: atrophy, athrepsia, starvation) there is a gross loss of body weight, mainly at the expense of muscle and subcutaneous fat. There is no edema and the skin shows no remarkable changes, but if there is dehydration resulting from diarrhea, the loss of turgor makes the diagnosis difficult. Sometimes there is atrophy of the mucosa of the tongue and cracks at the corners of the mouth, which are probably caused by riboflavin deficiency. Xerophthalmia may be a complication in countries where vitamin A deficiency is common. Marasmus has been defined for purposes of classification as a condition in which body weight is less than 60% of normal, with no edema (Table I). The weight may be as low as 30% before death occurs. However, this classification is not entirely satisfactory, because marasmic children are often extremely retarded in linear growth, indicating that the condition has been of long standing.

The term kwashiorkor (from the Kwa language of coastal Ghana) has been widely used to describe a child who is moderately underweight, with edema, enlarged fatty liver, often a severe desquamating dermatosis ("flaky paint" skin), lesions of the mucosae, and sparse depigmented hair. These latter lesions are by no means always present and it is now accepted that the essential diagnostic feature of kwashiorkor is the presence of edema, so that it might be more appropriate to name the condition "edematous malnutrition." Intermediate forms occur in which the wasting characteristic of marasmus is combined with edema (see Table I). This is the type of PEM that is most difficult to treat and has the highest fatality rate.

III. PATHOPHYSIOLOGY OF SEVERE PEM

In spite of the striking clinical differences between the extremes of marasmus and kwashiorkor, these differences should not be overstressed because most of the changes described here are found in greater or lesser degree throughout the spectrum.

The pattern of organs is altered; the visceral organs, particularly the brain, are relatively well preserved in terms of mass and protein content, whereas muscle and skin are greatly reduced. This distorts any comparison between normal and malnourished on the basis of body weight. In the liver, the ratio of nitrogen to DNA is reduced, which gives an index of the extent of protein depletion of the tissue. Particularly in edem-

TABLE I
Classification of Severe Protein-Energy Malnutrition

	Edema	
Weight	0	+
80–60%	Undernutrition	Kwashiorkor
Less than 60%	Marasmus	Marasmic kwashiorkor

atous cases there is fatty infiltration of the liver, which starts in the periphery and eventually fills the whole lobule. The fat content may amount to 50% of the wet weight. This condition is completely different from the small droplet centrilobular fatty degeneration that results from toxins or infections. The most probable cause of the immense accumulation of fat is reduced hepatic synthesis of the apolipoprotein responsible for transporting fat out of the liver.

Cardiac output and stroke volume are reduced and central venous pressure is raised. The peripheral circulation is sluggish and the skin cold. Renal function is impaired, with a reduction in glomerular filtration rate and renal plasma flow and decreased ability to concentrate the urine and to excrete acid. Important changes are found in the gastrointestinal tract. There is atrophy of the exocrine pancreas and of the gut mucosa, with reduced production of the digestive enzymes, particularly of the disaccharidases. This may lead to osmotic diarrhea. Nevertheless, there appears to be enough reserve capacity for digestion and absorption of protein to be well maintained. Almost always, when it has been looked for, bacterial colonization of the small intestine has been found, and suppression of it with metranidazole is thought to have a significant effect on mortality. There is usually a normocytic anemia of moderate degree unless aggravated by folic acid deficiency. The anemia is remarkably well tolerated, presumably because of the low metabolic demand for oxygen.

Edema, the hallmark of kwashiorkor, starts in the feet and then becomes generalized. Ascites is not found in children, but has been described in severely malnourished adults in refugee camps. The edema is accompanied by hypoalbuminemia, and an albumin level below 1 g/dl is a serious prognostic sign. The extent of the edema may be almost double the normal body water content. It is probable that most of the excess water is extracellular, but, given the imprecision of the methods, it is not possible to rule out some increase in intracellular water. Although sodium is retained along with water and the absolute amount of sodium in the body is increased, some cases, particularly of marasmic kwashiorkor, exhibit hypotonicity, with serum sodium concentrations below 120 mmol/liter. In this situation, cell water may well be increased. There is also evidence of impairment of membrane permeability, allowing sodium to leak in and potassium to leak out.

Children with advanced PEM are often severely depleted of potassium. This has been shown by potassium balances and by whole-body ^{40}K counting. The finding of a low ratio of potassium to DNA in muscle biopsies is evidence that the reduction of potassium per unit body weight is a true depletion and not merely a reflection of a loss of cell mass. Historically, when potassium depletion was recognized and corrected, clinical signs of magnesium deficiency—muscular rigidity and convulsions—sometimes appeared. Thus in severe PEM there are major disturbances in electrolyte and water balance, which may be disguised by dehydration resulting from diarrhea. Signs of dehydration in the upper part of the body may coexist with edema in the lower part—a condition that is very difficult to treat.

There are endocrine changes that give some pointers to etiology because they differ somewhat in kwashiorkor and marasmus. Insulin levels are low and corticosteroids are high in marasmus, whereas the opposite changes tend to be found in kwashiorkor, as might be expected if the energy intake has been relatively well maintained. The activity of the renin–angiotensin system may be increased in kwashiorkor. In both, growth hormone levels are raised, whereas concentrations of insulin-like growth factor-1 and of thyroid hormones are reduced.

Basal metabolism and protein turnover per kilogram body weight are both low in the initial stages, when the child is acutely ill and in danger of death. With treatment, both rates rise to supranormal levels, presumably as a result of the distortion of body composition referred to earlier, in which actively metabolizing organs are spared at the expense of those with a low metabolic rate. In parallel with the low basal metabolic rate, the child with severe PEM tends to be hypothermic and is unable to produce a normal febrile response to infection, which greatly complicates diagnosis.

Serum antibody levels are normal or raised, but the antibody response to vaccines may be decreased. There is reduced secretory immunity, shown by low secretory immunoglobulin levels in nasal washings and duodenal fluids and diminished response to oral measles and polio vaccines. The mortality rate after measles is particularly high. Complement factors are reduced. Circulating neutrophils retain the ability to phagocytose bacteria but not to kill them. There is a profound atrophy of the thymus and impairment of cell-mediated immunity, as shown by a reduced response to tuberculin and other skin tests. It is small wonder that these children are extremely susceptible to infections and that their presence is often missed.

This susceptibility may be aggravated by deficiencies of trace elements, such as zinc, copper, and selenium. Zinc deficiency plays an important role in changes in the immune system and possibly also in skin and gut. Selenium and copper are components of enzymes responsible for scavenging free radicals. The plasma concentration of vitamin E, an important antioxidant, is reduced. The concentration of glutathione in red cells is reduced in kwashiorkor, but is apparently normal in marasmus. Ferritin levels are increased, and high ferritin is a bad prognostic sign. Damage by free radicals could cause lipid peroxidation of cell membranes and so be responsible for some of the changes described here. [*See* Food Antioxidants; Nutrition, Trace Elements.]

IV. TREATMENT OF SEVERE PEM

Treatment falls into three stages. In the first, which may last 3 days or so, the main objective is to counter dehydration and infections. Fluids, hypotonic in sodium but with added glucose, potassium, and magnesium, should be given wherever possible by nasogastric tube; with intravenous therapy, death may easily occur from pulmonary edema caused by overloading the weakened heart. It may be assumed that infections are present, even if not diagnosed, and most experienced pediatricians give broad-spectrum antibiotics prophylactically. In the second stage, which may last 7–10 days, the aim is to restore the child's capacity to assimilate energy and protein. Many enzymes, notably those of the urea cycle, are reduced and overenergetic feeding can be fatal. At this stage it is best to give no more than 1 g protein and 100 kcal/kg per day. In the third stage, which is marked by a return of the child's appetite, a diet providing 4–5 g protein and 150–200 kcal/kg per day will promote rapid weight gain and an early discharge from hospital. Supplementary potassium and magnesium should be given throughout, but iron should not be given until stage 3, because of the danger of oxidative damage.

With careful treatment along these lines, the fatality rate in severe childhood PEM has been reduced to about 5%, although it still remains in the region of 25–30% in many hospitals in the developing world. Recent work shows that even very severe cases can be successfully treated at home, provided that mothers are suitably instructed.

It is probable that the same principles apply in adults. Studies in refugee camps in Somalia have shown that edema, which was common, was associated with a high fatality rate. There was a significant reduction in deaths when the edematous subjects were given moderate amounts of protein, rather than the large amounts that are customary, confirming the idea that depleted enzymes need time to be restored.

V. ETIOLOGY

The classic view is that marasmus, often seen in breastfed babies after weaning, is caused by an overall deficiency of food, whereas in kwashiorkor there has been a relative deficiency of protein in relation to energy. It has been said that the marasmic lives on his or her own meat. The hormone profile suggests that the child is in a catabolic state, in which the breakdown of muscle protein provides amino acids for the maintenance of protein synthesis by the liver. In kwashiorkor, with an unbalanced and inadequate protein intake, there is a reduction in hepatic synthesis of albumin and apolipoprotein, hence the combination of edema and fatty liver. Edema is aggravated by potassium deficiency and consequent sodium retention. [*See* Nutrition, Protein.]

This view is still controversial. It has been suggested that there is no difference in the diets that lead to kwashiorkor and marasmus, although the epidemiological evidence indicates clearly that kwashiorkor is most common in those countries where the staple of the diet is low in good-quality protein, as in diets based on maize or cassava. Others stress the importance of free radical damage. At present we are unable to quantify the role of infections, which are an important source of free radicals. These may well be responsible for some of the life-threatening changes that occur in the terminal stages of PEM.

Much of the work on the pathophysiology of PEM in children was done in the 1960s and 1970s. More recently, physicians and surgeons have become increasingly aware that PEM may occur in patients with a variety of diseases and are rediscovering knowledge that was gained earlier from the study of children.

BIBLIOGRAPHY

Brooke, O. G., and Cocks, T. (1974). Resting metabolic rate in malnourished babies in relation to total body potassium. *Acta Paediatr. Scand.* **63**, 817–825.

Golden, M. H. N., and Ramdath, D. (1987). Free radicals in the pathogenesis of kwashiorkor. *Proc. Nutr. Soc.* **46**, 53–68.

James, W. P. T. (1971). Effects of protein-calorie malnutrition on intestinal absorption. *Ann. N.Y. Acad. Sci.* **176**, 244–261.

Keusch, G. T., and Farthing, M. J. G. (1986). Nutrition and infection. *Annu. Rev. Nutr.* **6**, 131–154.

Shetty, P. S., and James, W. P. T. (1994). "Body Mass Index: A Measure of Chronic Energy Deficiency in Adults," FAO Food and Nutrition Paper No. 56. Food and Agriculture Organization, Rome.

Tomkins, A. M. (1986). Protein-energy malnutrition and risk of infection. *Proc. Nutr. Soc.* **45,** 289–304.

Viart, P. (1978). Hemodynamic findings during treatment of protein-calorie malnutrition. *Am. J. Clin. Nutr.* **31,** 911–926.

Waterlow, J. C. (1992). "Protein-Energy Malnutrition." Edward Arnold, London.

Waterlow, J. C. (1994). Childhood malnutrition in developing nations: Looking back and looking forward. *Annu. Rev. Nutr.* **14,** 1–20.

Waterlow, J. C., and Schürch, B. (eds.) (1994). Causes and mechanisms of linear growth retardation. *Eur. J. Clin. Nutr.* (Suppl. 1).

Protein Phosphorylation

CLAY W SCOTT
THEODORA W. SALCEDO
Zeneca Pharmaceuticals

GLOSSARY

Calmodulin Intracellular calcium-binding protein that mediates many functions of calcium in mammalian cells

Protein kinases Enzymes that catalyze the transfer of a phosphate group from ATP to a specific amino acid residue of a substrate protein

Protein phosphatases Enzymes that remove the phosphate group from phosphoproteins

Second messenger Intracellular molecule generated in response to an extracellular signal, which modulates the behavior of the cell

SH2 domain Motif present in many signal-transducing proteins that binds to phosphotyrosine domains on activated receptors

Signal transduction Intracellular biochemical cascade activated by an extracellular signal that leads to a change in the behavior of the cell

PROTEIN PHOSPHORYLATION IS A COMMON mechanism used to modulate the biological activity of intracellular proteins. This process is a reversible modification regulated by protein kinases and phosphoprotein phosphatases. Protein kinases catalyze the transfer of the γ-phosphate from a donor adenosine triphosphate (ATP) molecule onto a specific amino acid residue in the protein (Fig. 1). Either an increase or decrease in the function of a protein may occur, depending on the particular protein and the actual site being phosphorylated. The covalent attachment of a phosphoryl group (PO_4^{2-}) with its high charge density onto a protein induces a conformational change in the protein. The consequences of phosphorylation of an enzyme can be an increase or decrease in kinetic properties (K_m or V_{max}) or a response to allosteric effectors. Phosphorylation of a cell-surface receptor can reduce its ability to respond to agonist, a process termed desensitization. Phosphorylation of structural proteins can either stimulate or inhibit their interaction with the cytoskeleton. Many proteins are phosphorylated at multiple sites by distinct protein kinases with differing effects on biological activity (discussed in the following). Phosphoprotein phosphatases catalyze the removal of phosphate groups from phosphoproteins, allowing the protein to return to its former functional state. Both protein kinases and phosphatases are under regulatory control, and therefore the relative activities of these enzymes will generally determine the phosphorylation state of a given protein and, hence, its functional activity. In some circumstances, however, the availability or accessibility of the substrate protein will dictate its phosphorylation state.

The profound effect of phosphorylation on protein function allows a cell to coordinately regulate a number of diverse biological processes, including metabolism, secretion, contractility, electrical excitability, and gene expression. Protein phosphorylation plays an especially prominent role in the transduction and amplification of extracellular signals into intracellular responses. Hormones, neurotransmitters, and growth factors bind to cell-surface receptors, initiating a cascade of biochemical events that generates an intracellular second messenger (Fig. 2). These second messen-

ENCYCLOPEDIA OF HUMAN BIOLOGY, Second Edition, VOLUME 7. Copyright © 1997 by Academic Press. All rights of reproduction in any form reserved.

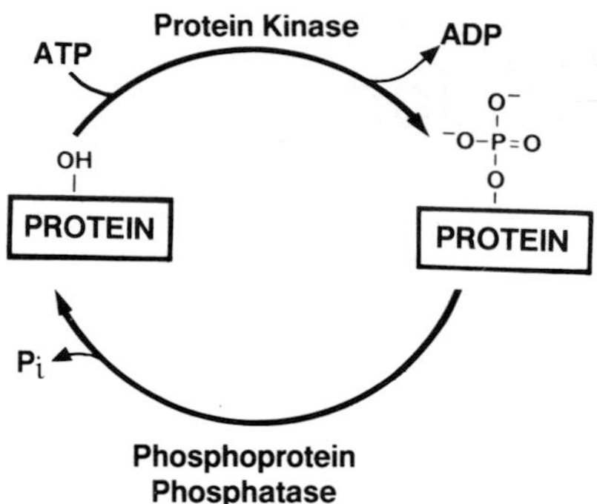

FIGURE 1 Protein phosphorylation is the enzymatic transfer of a phosphoryl group onto a specific amino acid residue (usually serine, threonine, or tyrosine). ATP serves as the primary phosphate donor molecule, although some protein kinases can utilize GTP. P_i, inorganic phosphate.

gers activate specific protein kinases, which then amplify the biochemical cascade by phosphorylating various cellular proteins. These phosphorylated proteins ultimately cause a change in the behavior of the cell, that is, the response to the extracellular signal. Some cell-surface receptors have protein kinase or phosphatase domains in their intracellular segments. Agonist binding to these receptors stimulates the enzyme activity inside the cell, resulting in propagation of the external stimuli. In these circumstances, the requirement for a second messenger may be abrogated.

I. PROTEIN KINASES

Approximately 200 protein kinases have been identified thus far. It is estimated that over 1000 protein

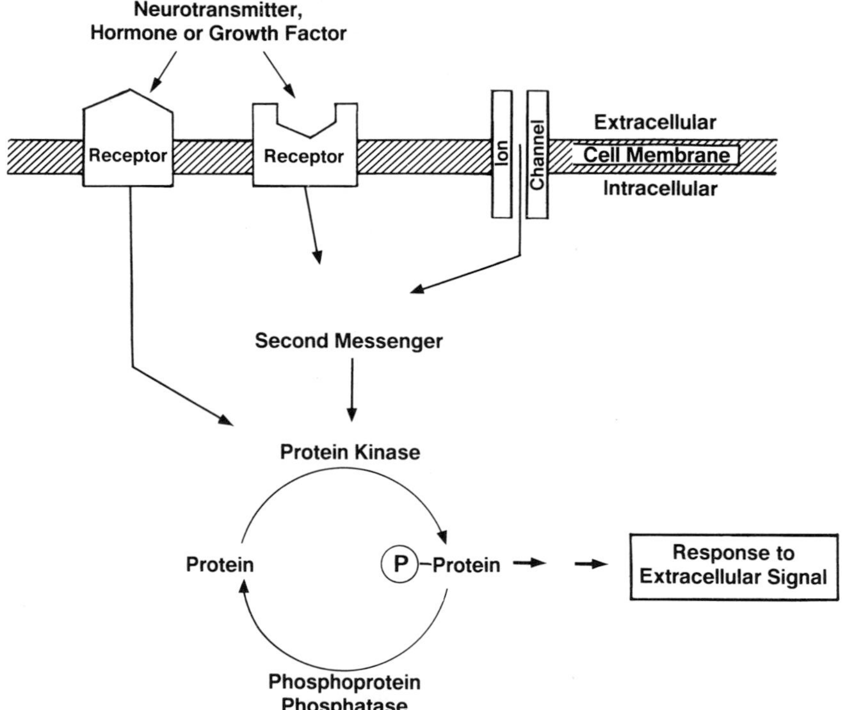

FIGURE 2 Protein phosphorylation is a common mechanism used to transduce extracellular stimuli into intracellular responses. Various extracellular molecules bind to their cell-surface receptors and regulate protein phosphorylation either directly (when the receptor itself is a kinase or phosphatase) or indirectly (through changes in levels of second messengers). Activation of a kinase or phosphatase produces a change in the phosphorylation state of selective proteins, which leads to a change in the behavior of the cell (e.g., contraction, secretion, gene expression, or metabolism). The second messengers can be generated by activating membrane-bound enzymes (in the case of cAMP, cGMP, diacylglycerol, and inositol triphosphate) or by passage through receptor- or voltage-activated ion channels (Ca^{2+}).

kinases exist in mammalian cells. Historically, these protein kinases have been categorized by the amino acid residue they phosphorylate; protein serine/threonine kinases, protein tyrosine kinases, and protein histidine kinases (in bacteria). More recently, some protein kinases have been shown to phosphorylate serine, threonine, and tyrosine residues. These kinases have been called dual specificity kinases. Protein kinases can be further divided on the basis of the effector system that controls their activity, for example, cAMP-dependent protein kinase and calcium/calmodulin-dependent protein kinases. As noted earlier, some protein kinases are not directly regulated by second messengers, but instead are coupled to cell-surface receptors and are activated upon ligand binding. Other protein kinases are members of kinase cascades and are themselves activated by phosphorylation. Some effector-independent protein kinases may be modulated by substrate availability, although they might be modulated by other, as yet unidentified second messengers. The following sections will describe particular protein kinases, emphasizing their mechanism of regulation.

A. cAMP-Dependent Protein Kinase

Many different hormones and neurotransmitters bind to their cell-surface receptors and stimulate the enzyme adenylate cyclase, resulting in the synthesis of the second messenger cyclic adenosine monophosphate (cAMP). [*See* Cell Signaling.] cAMP activates the enzyme cAMP-dependent protein kinase, resulting in the phosphorylation of numerous cellular proteins. cAMP-dependent protein kinase exists as an inactive holoenzyme composed of a dimer of regulatory subunits (R) and two catalytic subunits (C). Both R (monomer) and C have relative molecular weights (M_r) in the 40,000–45,000 range. There are two cAMP-binding sites on each R subunit or four binding sites per dimer. Activation of the holoenzyme occurs as

$$R_2C_{2(\text{inactive})} + 4\text{cAMP} = R_2(\text{cAMP})_4 + 2C_{(\text{active})}$$

Four molecules of cAMP bind to the regulatory dimer and induce dissociation of the holoenzyme complex. The free catalytic subunits are active and phosphorylate substrate proteins. Hydrolysis of cAMP by phosphodiesterases shifts the equilibrium toward the inactive holoenzyme, thereby terminating the actions of the catalytic subunit. The two cAMP-binding sites in each R monomer have extensive amino acid homology, suggesting they were formed via a gene duplication event. Molecular cloning has revealed several forms of R and C. The biological significance of these different isoforms is not yet understood.

Numerous substrates for cAMP-dependent protein kinase have been identified. A few examples include enzymes involved in glycogen metabolism in liver and muscle, enzymes involved in lipid metabolism in adipocytes, and proteins that control the contractile activity of various types of muscle. This kinase also plays an important role in regulating the transcription of certain genes. The active catalytic subunit migrates into the nucleus, where it phosphorylates nuclear proteins such as CREB (cAMP response element-binding protein). CREB binds to a specific DNA sequence on cAMP-inducible genes and promotes their transcription. Phosphorylation of CREB by the kinase results in enhanced transcription of cAMP-responsive genes.

B. cGMP-Dependent Protein Kinase

cGMP-dependent protein kinase exists as a dimer with each monomer (M_r 80,000) containing two cyclic nucleotide-binding sites and a catalytic domain. These domains show extensive homologies to the corresponding regions in cAMP-dependent protein kinase, although for cGMP-dependent protein kinase the cyclic nucleotide-binding sites and catalytic domain are linked by peptide bond and do not separate upon binding cGMP. Activation of the holoenzyme occurs by binding four molecules of cGMP:

$$4\text{cGMP} + E_{2(\text{inactive})} = 4\text{cGMP} \cdot E_{2(\text{active})}$$

As is the case with cAMP-dependent protein kinase, dissociation and hydrolysis of cyclic nucleotide cause a shift in the equilibrium toward the inactive holoenzyme. The two cGMP-binding sites on each monomer share homology in terms of amino acid sequence, yet are distinct in that synthetic cGMP-like molecules bind with different affinities to the two sites.

Although cGMP-dependent protein kinase has been extensively characterized in terms of physicochemical properties, little is known about its substrates or biological function. A role for cGMP-dependent protein kinase in cellular metabolism has been most clearly defined in the regulation of vascular smooth muscle contractility. In this tissue, substances that relax the muscle such as nitrovasodilators and atriopeptins stimulate guanylate cyclase, resulting in increased synthesis of cGMP and activation of cGMP-dependent protein kinase. The mechanism by which cGMP-dependent protein kinase relaxes the muscle, however, is not understood.

C. Calcium/Calmodulin-Dependent Protein Kinases

The family of protein kinases that are regulated by calcium plus calmodulin include calcium/calmodulin-dependent protein kinases I, II, and III, myosin light-chain kinases, and phosphorylase kinase. With the exception of calcium/calmodulin-dependent protein kinase II (CaM-kinase II), these enzymes have few physiological substrates. CaM-kinase II will be described as an example of this class of protein kinases with the understanding that the other enzymes, although similar in terms of second-messenger requirements, display distinct physicochemical properties.

CaM-kinase II is present in most mammalian tissues and is abundant in brain, where it can account for up to 2% of total protein in some regions. Its substrates include proteins involved in neurotransmitter synthesis and release, cytoskeleton formation, and carbohydrate metabolism. Studies with mutant mice that lack CaM-kinase II have indicated that this enzyme is necessary for the development of long-term potentiation, a form of memory. In the nervous system, CaM-kinase II is an oligomer with an apparent molecular weight of 460,000–650,000. It is composed of varying ratios of two or three distinct subunits each of M_r 50,000–60,000. Each subunit contains all of the functional domains that comprise the oligomer, namely, an ATP-binding domain, a calmodulin-binding domain, and a catalytic domain. Other domains identified within each monomer include autophosphorylation domains and an inhibitory domain.

Although the regulation of CaM-kinase II activity is not completely understood, a number of points have been recently clarified. In the absence of calcium and calmodulin, the dephosphorylated enzyme remains in an inactive state by its inhibitory domain. An increase in intracellular calcium, due to an influx through activated calcium channels or release from intracellular storage sites, results in the formation of Ca^{2+}/calmodulin complexes that can then bind to CaM-kinase II. This eliminates the action of the inhibitory domain and allows the enzyme to phosphorylate its substrates, which includes itself (autophosphorylation). The autophosphorylated enzyme loses its dependency on calcium–calmodulin and remains active until it is dephosphorylated. This calcium-independent state represents a potential amplification step in the signal transduction process, that is, small changes in calcium concentrations could produce sustained activation of CaM-kinase II and substantial changes in cellular function.

D. Protein Kinase C

Protein kinase C (PKC) is a serine/threonine kinase that is activated by Ca^{2+}/phospholipid or phorbol esters. It exists as a monomer with a molecular weight of approximately 85,000. Although abundant in brain, it is present in most tissues. It is now clear that PKC is actually a family of isoenzymes or isoforms. Ten different isoenzymes have been identified using molecular cloning techniques. Nine isoenzymes (α, γ, δ, ε, ζ, η, θ, λ, ι) are single gene products. Isoenzymes β_{I} and β_{II} are produced by alternative splicing of a single gene.

The different isoenzymes are often grouped into three broad categories depending on mode of activation. Classical or conventional PKCs (cPKC) are calcium-dependent PKCs that are activated via the classical phosphatidylinositol pathway requiring calcium, phospholipid, and 1,2-diacylglycerol (DAG). The α, β_{I}, β_{II}, and γ isoenzymes are members of the cPKC group. Calcium-independent PKCs, also known as novel or nonclassical PKCs (nPKCs), are activated by DAG and phospholipid alone and include the δ, ε, η, and θ isoenzymes. Members of the third group are called atypical PKCs (aPKCs) and require neither calcium nor DAG for activation. Isoenzymes ζ, λ, and ι are examples of aPKCs. The biological significance of PKC diversity is not fully understood. Subtle differences in various biochemical properties, tissue distribution, and subcellular localization suggest that different isoforms have distinct functions in the modulation of or responses to various physiological stimuli. Increased understanding of isoenzyme function will no doubt result from the development of isozyme selective inhibitors as well as genetic knockout and mutation experiments.

Of the three PKC groups described here, the activation of the cPKCs is best understood. Many receptors for extracellular ligands are coupled to phospholipase C, which upon activation catalyzes the hydrolysis of phosphatidylinositol-4,5-bisphosphate (PIP_2) within the plasma membrane. Two second messengers, inositol 1,4,5-trisphosphate (IP_3) and DAG, are produced. IP_3 initiates the mobilization of calcium from intracellular stores, whereas DAG stimulates cPKC by increasing its affinity for calcium. Activation of the kinase can now occur with calcium concentrations found in resting cells. The localization of both DAG and phospholipid to the plasma membrane suggests that PKC must interact with this membrane for activation. Indeed, many studies have shown that cPKC moves from the cytoplasm to the plasma membrane after cells are treated with agents that stimulate PIP_2 hydrolysis. Inactivation of PKC occurs as a result of the conversion of DAG to phosphatidic acid. However, under some circumstances inactivation may occur by proteolytic degradation—initially to the active fragment termed protein kinase M and subsequently to inactive fragments.

PKC is also a receptor for the tumor-promoting phorbol esters (present in croton oil). Phorbol esters appear to elicit their deleterious effects owing to inappropriate activation of PKC. The phorbol ester 12-*O*-tetradecanoyl-phorbol-13-acetate (TPA) has structural similarity to DAG and can stimulate PKC activity in a manner similar to DAG. Both the cPKCs and nPKCs are phorbol ester-sensitive, whereas the aPKCs are phorbol ester-insensitive. This ability of phorbol esters to bind and activate PKC has resulted in their frequent use as pharmacological tools to study the role of PKC (and PKC isoforms) in cellular responses.

E. Growth Factor–Receptor Tyrosine Kinases

Receptors for certain serum growth factors contain intrinsic protein kinase activities within their cytoplasmic domains. These receptor kinases specifically phosphorylate tyrosine residues and can be grouped into the following protein–tyrosine kinase receptor families on the basis of structural similarities: platelet-derived growth factor (PDGF), epidermal growth factor (EGF), fibroblast growth factor (FGF), insulin-like growth factor (IGF), hepatocyte growth factor (HGF), and neurotrophin.

As an example, members of the IGF receptor family are structurally homologous oligomers composed of two α subunits, located extracellularly and linked by disulfide bonds to each other, and two β subunits. The β subunits transverse the plasma membrane and contain the intracellular tyrosine kinase domain. The α subunits compose the ligand-binding domain. In contrast to the IGF receptor family, members of the EGF and PDGF receptor families contain both the extracellular ligand-binding domain and the cytoplasmic tyrosine kinase domain in a monomeric protein.

Stimulation of receptor tyrosine kinase activities requires binding of the growth factor (the agonist). The current hypothesis to explain subsequent activation of signal transduction pathways is ligand-induced aggregation of receptor subunits. Receptor dimerization appears to be an intrinsic function of the EGF receptor and may occur with all of the tyrosine receptor kinases. It was recently suggested that monomeric ligands such as EGF bind simultaneously to two receptors, favoring the model of receptor activation following dimerization. It is interesting to note that antibodies to the insulin receptor, which cause receptor cross-linking, also stimulate the receptor tyrosine kinase activity. In addition, both PDGF and colony-stimulating factor (CSF) exist as disulfide-linked dimers and may be capable of interacting with the agonist-binding domain of two monomers, thereby promoting dimerization.

The dimerization of tyrosine kinase receptors usually results in receptor autophosphorylation on distinct tyrosine residues. Phosphorylation of residues located in the kinase domain itself can result in regulation of the kinase catalytic activity. Both the insulin and hepatocyte growth factor receptors undergo rapid intramolecular autophosphorylation reactions that further enhance their tyrosine kinase activity. The autophosphorylated receptor kinase phosphorylates substrate proteins, thereby propagating the response to the agonist. In contrast to these receptors, the kinase activity of other receptors (EGF) does not appear to be regulated by autophosphorylation. In addition to tyrosines in the kinase catalytic domain, tyrosine sites located outside the catalytic domain are autophosphorylated. These phosphotyrosines function as binding sites for proteins containing src homology domains (SH2) that connect the receptor to downstream signal transduction pathways. Proteins with SH2 domains contain a phosphotyrosine-binding pocket. The actual SH2 domains are stretches of 100 amino acid residues found in different key intracellular signal transduction molecules, including Ras GTPase-activating protein (GAP), phospholipase C-γ1 (PLCγ1), src kinase, and phosphatidyl 3-kinase (PI3K), as well as adaptor proteins such as Grb2. The binding affinity between an SH2 domain and a specific sequence is of high affinity (10–100 n*M*), though there is lack of measurable affinity to unphosphorylated sequences. Amino acids that surround the phosphotyrosine act to increase the affinity and specificity of the interaction. As a result of SH2 domain binding to a phosphotyrosine, the enzymatic activity of the protein containing the SH2 domain may be stimulated, the protein may relocalize within the cell, or the protein may itself become phosphorylated.

A large body of evidence supports the hypothesis that the receptor kinase activities play crucial roles in mediating the actions of various growth factors. Some of the substrates for these receptor kinases have been defined. In the case of the PDGF-β receptor (PDGF-βR), progress has been made in mapping the autophosphorylation of specific tyrosines. The PDGF-βR has at least nine tyrosine residues that undergo autophosphorylation. Experiments performed with PDGF-βR mutants lacking specific autophorylation sites were used to map the binding sites for proteins associated with key signal transduction pathways. Tyr-857 was found to be involved in regulating the kinase activity of the catalytic domain, whereas Tyr-579 and Tyr-581 were identified as binding sites for src kinases (Fig. 3). These experiments further emphasize the specificity of the interaction between various SH2 domain-containing proteins and receptor phosphotyrosines.

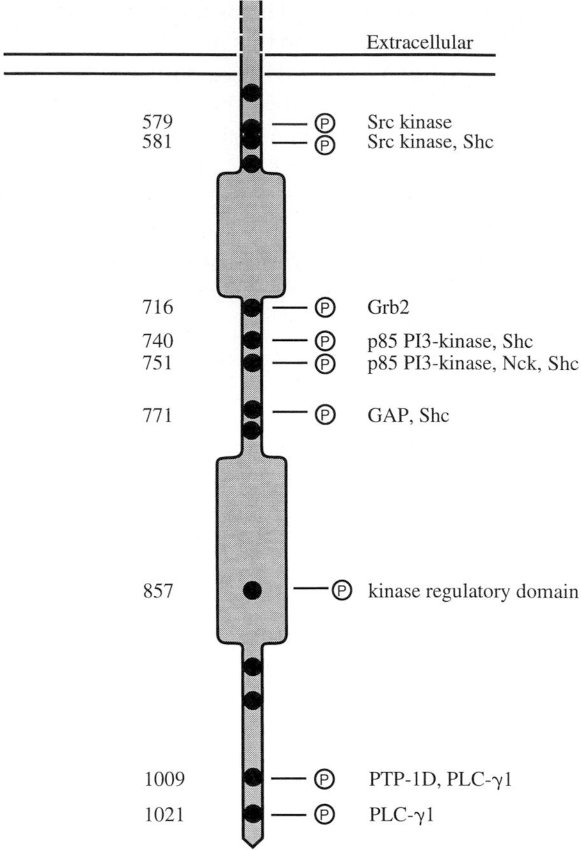

FIGURE 3 Autophosphorylation sites in the PDGF-β receptor and binding of signal transduction molecules via SH2 domains. Solid circles indicate phosphorylated tyrosine residues and numbers indicate amino acid position. Binding of signal-transducing proteins containing SH2 (src homology 2) domains to particular phosphotyrosine residues are indicated. PDGF, platelet-derived growth factor receptor; Src kinase, src kinase family members; Shc, Grb2, Nck, various adaptor molecules; P85 P13-kinase, phosphatidylinositol 3′-kinase p85 subunit; GAP, GTPase-activating protein; PTP-1D, protein tyrosine phosphatase 1D; PLC-γ1, phospholipase C-γ1. The location of the phosphorylation site in the second part of the kinase domain, important for kinase catalytic activity, is also indicated.

F. Nonreceptor Tyrosine Kinases

Some receptors lack an intrinsic receptor tyrosine kinase domain and activate signaling pathways by stimulating tyrosine kinases that reside within the cell. These tyrosine kinases are not activated exclusively by receptors that lack intrinsic kinase domains, but can also be activated by some receptor tyrosine kinases. Examples of cytoplasmic tyrosine kinases include the src-like kinases, syk/ZAP-70 kinases, and Janus (Jak) kinases.

v-Src is an example of a viral oncogene. Certain retroviruses induce the malignant transformation of cells through the expression of viral oncogenes. For several of these retroviruses, the oncogene product (such as v-src) is a tyrosine-specific protein kinase. The transforming proteins are structurally altered variants of normal cellular protein tyrosine kinases. Because the viral variants induce neoplastic transformation, their cellular homologues are thought to be important regulatory enzymes that control normal cell growth and metabolism. Members of the src-like kinases are intracellular enzymes that attach to the inner plasma membrane via a fatty acid moiety. pp60src was the first member of the src family identified, which now has seven additional well-characterized members (p59fyn, p59hck, p62yes, p56lck, p56lyn, p55fgr, and p55blk). Src kinases share a conserved kinase domain, SH2 and SH3 domains, but have unique N termini. Src family kinases function in signaling cascades mediated by T-cell receptors, B-cell receptor, interleukin-2 (IL-2) receptor, and Fc receptors. [*See* Interleukin-2 and the IL-2 Receptor; T-Cell Receptors.]

Members of the Syk family of tyrosine kinases include p72syk and ZAP-70. These kinases possess two SH2 domains, but lack any SH3 domain or fatty acid attachment site. p72syk is implicated in B-cell signaling, and ZAP-70 is believed to be more important in T-cell signaling.

Members of the Jak kinase family include Jak1, Jak2, Jak3, and Tyk2. Structurally, the members of this family contain two kinase domains, only one of which is active. In addition, they share five homologous domains whose functions are not completely understood. Jak kinases have been implicated in the signaling cascades of numerous cytokines.

At least in part, cytokine receptors signal by activating Jak kinases. In most cases, a Jak kinase is constitutively associated with the membrane-proximal domain of the cytokine receptor. Ligand binding induces receptor oligomerization and subsequent activation of the Jaks, apparently by cross-phosphorylation of two Jaks. The active Jaks phosphorylate the receptor on tyrosine residues. These phosphotyrosine residues become docking sites for various cellular proteins via their SH2 domains. Some of these proteins become phosphorylated by the Jaks, resulting in a change in their biological activity. One example are the STATs (signal transducers and activators of transcription). These proteins are recruited to the phosphorylated receptor, where they become phosphorylated by Jaks. The phosphorylated STATs are able to translocate to the nucleus, where they bind DNA and ultimately alter gene expression. This current model of Jak and STATs is likely to undergo refinement with continued progress in this field.

G. Consensus Sequences for Protein Kinases

Protein kinases recognize substrates on the basis of their amino acid sequence. Studies with synthetic peptides as substrates have shown that changing amino acids within the vicinity of a phosphorylation site can dramatically affect the ability of the peptide to undergo phosphorylation. Cyclic AMP-dependent protein kinase, for example, recognizes the sequence -Arg-Arg-X-Ser(Thr)-X-, where X represents any amino acid. Peptides, which lack either of the two basic arginines, or in which the arginines are displaced from the phosphorylatable serine, exhibit significant increases in K_m and/or decreases in V_{max}. Although the primary sequence of a protein is an essential determinant for recognition by protein kinases, higher orders of structure resulting from the folding of the polypeptide chain can influence substrate–enzyme interaction. Small peptides, for example, which contain identical primary sequence to natural substrate proteins, typically display poorer phosphorylation kinetics. In addition, some proteins are not substrates for protein kinases but can be phosphorylated after denaturation. Higher orders of structure can, therefore, have either positive or negative effects on substrate specificity.

II. PROTEIN PHOSPHATASES

In contrast to protein kinases, of which over 200 distinct enzymes have been identified, protein phosphatases are fewer in number and promiscuous in activity. Indeed, the broad substrate specificity of these enzymes can create difficulty in defining distinct enzyme forms. The protein phosphatases are currently classified according to their specificity for phosphoamino acids: protein serine/threonine and protein tyrosine phosphatases. Just as with the protein kinases, a few protein phosphatases show dual specificity. The protein serine/threonine phosphatases are subdivided based on their preference of substrate (α or β subunit of phosphorylase kinase) and susceptibility to inhibition by two thermostable proteins (inhibitor-1 and inhibitor-2). The identification of enzymes within these classes that have differing structural or regulatory properties has led to further subcategories. Molecular cloning has revealed multiple isoforms for most all of the catalytic and regulatory subunits of the phosphatases. A central question in the study of protein phosphorylation is how a relatively few number of phosphatases can modulate the activities of a large and diverse number of protein kinases. Increasing evidence suggests that the function, activity, and cellular location of protein phosphatases may be modulated by their regulatory subunits.

A. Protein Phosphatase-1

Protein phosphatase-1 (PP1) preferentially dephosphorylates the β subunit of phosphorylase kinase and is inhibited by inhibitor-1 and inhibitor-2. PP1 exists in a number of heteromeric structures, each composed of a common catalytic subunit complexed to various regulatory or targeting subunits. The targeting subunits appear to direct the catalytic subunit to particular locations within the cell and thereby restrict access to substrates. Four different forms of PP1 have been identified: (1) a MgATP-dependent phosphatase, (2) a myosin-associated phosphatase, (3) a glycogen-bound phosphatase, and (4) a nuclear phosphatase. The MgATP-dependent complex is composed of an inactive catalytic subunit bound to inhibitor-2. Activation of the complex requires phosphorylation of inhibitor-2 by protein kinase GSK-3 (hence the requirement for Mg^{2+} and ATP). The phosphorylated, active heterodimer undergoes autodephosphorylation and subsequently returns to the inactive conformation.

The glycogen-bound form of PP1 consists of the catalytic subunit complexed to a glycogen-binding protein termed the G subunit, or R_{GL}. The PP1 catalytic subunit is active in this configuration, although physically constrained to glycogen particles. The glycogen-bound catalytic subunit is also afforded protection from the actions of inhibitor-1 and inhibitor-2. Phosphorylation of R_{GL} by cAMP-dependent protein kinase promotes release of the active catalytic subunit, thereby limiting the dephosphorylation of glycogen phosphorylase. cAMP-dependent protein kinase can also phosphorylate and activate inhibitor-1, which then binds to the free catalytic subunit of PP1 and inhibits its activity. Thus, activation of cAMP-dependent protein kinase by epinephrine results in a decrease in the dephosphorylation of enzymes involved in glycogen metabolism, thereby promoting glycogen breakdown. [*See* Glycogen.] R_{GL} also appears to target PP1 to the sarcoplasmic reticulum in skeletal and cardiac muscle. At this location, PP1 may dephosphorylate phospholamban, a protein that regulates the uptake of calcium into the sarcoplasmic reticulum. [*See* Muscle, Physiology and Biochemistry.]

PP1 binds to myosin as part of a heteromeric structure, the composition of which depends on the type of cell being evaluated. At least one subunit of the enzyme complex directs the catalytic subunit to the myofibril, where it can dephosphorylate myosin light chains. The catalytic subunit of PP1 has also been identified in the nuclei of some eukaryotic cells, where it is complexed to chromatin via an inhibitory subunit. The function of nuclear PP1 has not yet been defined.

B. Protein Phosphatase-2A

Protein phosphatase-2A consists of a catalytic subunit (M_r 36,000) complexed to an A subunit (M_r 61,000) and one of several different B subunits. The type 2A phosphatases have a broad substrate specificity but show a preference for the α rather than the β subunit of phosphorylase kinase. This class of phosphatases is stimulated by basic proteins and polyamines. The biological functions of the A and B subunits are not clear, although reconstitution experiments have shown that they alter the enzymatic properties and substrate specificity of the catalytic subunit. Recently, certain DNA tumor viruses have been shown to encode proteins that replace the B subunit and thereby alter the activity of the catalytic activity. This change in PP2A activity presumably contributes to the transformation of the cells by these viruses.

C. Protein Phosphatase-2B

PP2B, or calcineurin, is a calcium- and calmodulin-binding protein found primarily in the brain. It exists as an inactive heterodimer of A (M_r 61,000) and B (M_r 16,000) subunits. The A subunit contains the catalytic domain as well as a calmodulin-binding domain. The B subunit has amino acid homology with calmodulin and can bind four molecules of calcium. Thus, the A subunit interacts with two related but distinct calcium-binding proteins: the B subunit, which does not dissociate from the A subunit, and calmodulin, which binds to the A subunit only transiently. Understanding the activation of phosphatase-2B has been difficult because the enzyme can exist in a number of different conformational states with different activities. Current evidence suggests that the B subunit binds calcium and undergoes a conformational change that stimulates the catalytic activity of the A subunit. This also enhances the binding of Ca^{2+}/calmodulin to the A subunit, which results in further stimulation of its catalytic activity.

These calcium-regulated interactions suggest that PP2B functions in calcium-dependent signal transduction pathways. Indeed, most of the known substrates for PP2B function to regulate the activity of protein kinases or phosphatases. For example, PP2B dephosphorylates both inhibitor-1 and DARPP-32, two closely related proteins that are inhibitors of PP1 when phosphorylated by cAMP-dependent protein kinase. Dephosphorylation of these inhibitors abolishes their inhibitory actions on PP1. Thus, an increase in intracellular calcium could initiate a phosphatase cascade by activating PP2B, which then activates PP1. This response represents a pathway to antagonize the actions of cAMP and is another example of cross-communication among different second-messenger systems.

D. Protein Phosphatase-2C

PP2C is present in tissues at much lower levels than other serine/threonine phosphatases. It is unique among serine/threonine phosphatases in that it exists as a monomer (M_r 43,000). It requires Mg^{2+} or Mn^{2+} for activity and has a broad substrate specificity. Like the other type 2 protein phosphatases, phosphatase-2C prefers the α subunit of phosphorylase kinase and is not regulated by inhibitor-1 or inhibitor-2. Its biological substrates have not been defined, and its biochemical function is unknown.

E. Phosphotyrosyl-Protein Phosphatases

Phosphotyrosyl-protein phosphatase (PTPases) are implicated in a number of cellular processes, including T-cell activation, tumor suppression, cytoskeletal organization, cell proliferation, differentiation, and growth factor receptor signaling. The field of PTPases has undergone rapid growth since 1988 when the first PTPase was isolated. Recent evidence indicates that the PTPase family now consists of over 40 members all with the common "active site" signature sequence of (I/V)HCXAGXGR(S/T)G. Based on additional structural features, PTPases are commonly classified into two groups. The first group includes receptor-like PTPases that have an extracellular domain, a single membrane-spanning domain, and an intracellular domain composed of either one or two catalytic units. The second group includes the intracellular PTPases with the common feature of a single catalytic domain flanked by additional sequences that are important for localization of the protein.

The hematopoietic cell-surface protein CD45 is an example of a receptor-like PTPase. CD45 is involved in T-cell receptor (TCR)-mediated signal transduction and may activate protein kinases important for TCR signaling (p56lck or p59fyn) by dephosphorylating a regulatory phosphotyrosine in these proteins. Other receptor-like PTPases have cell adhesion molecule-like extracellular domains, which may be involved in cell–cell interactions or carbonic anhydrase-like domains of unknown function.

Examples of intracellular PTPases include PTP-1B, PTP-1C (HCP), and PTP-2C (Syp). PTP-1B was the first PTPase purified and a crystal structure was recently solved. PTP-1C and PTP-2C are interesting in that these cytoplasmic PTPases each contain two SH2 domains in addition to the PTPase catalytic domain. It is thought that the SH2 domains direct the association of these PTPases to phosphotyrosines on activated growth factor receptors. Experiments *in vitro* have confirmed that PTP-1C binds via SH2 domain to activated EGF receptors, as well as receptors for hematopoietic factors such as interleukin-3 and Steel factor. Likewise, PTP-2C binds to PDGF receptors, EGF receptors, and the insulin-receptor substrate (IRS-1) of insulin receptors. Although it is clear that the SH2-containing intracellular PTPases bind to growth factor receptors, the actual function of the bound PTPase is much less clear.

A key regulatory function for at least one SH2-containing PTPase was revealed in studies with *moth-eaten* mutant mice. The genetic mutation of the *moth-eaten* mouse was mapped to the gene that encodes PTP-1C. Mice with this mutation have severe hematopoietic abnormalities with elevated numbers of autoreactive B cells and erythroid and myeloid cells, and they die within 9 weeks of birth as a result of autoimmune disease. Thus, a normal function of PTP-1C appears to be as a regulator of hematopoietic cell proliferation, probably in response to growth factors.

III. SUBSTRATES FOR PROTEIN PHOSPHORYLATION

Because protein phosphorylation has been implicated in regulating a vast number of cellular functions, it is not surprising that the number of proteins undergoing reversible phosphorylation is quite large. It has been estimated that over one-sixth of the proteins in mammalian cells are phosphorylated. The proteins described in this section represent some well-characterized examples and demonstrate the breadth of involvement of protein phosphorylation in cell function.

A. Cell-Surface Receptors

Cell-surface receptors function as bridging molecules between extracellular signals and intracellular machinery. Any mechanism that modulates receptor function will have dramatic effects on the response of the cell to extracellular stimuli. Protein phosphorylation represents one such mechanism. A number of receptors for hormones, neurotransmitters, growth factors, and essential nutrients are subject to phosphorylation. Examples include the α_1-adrenergic and β-adrenergic receptors, the glucocorticoid receptor, the receptor tyrosine kinases, and the transferrin receptor. Phosphorylation at intracellular sites can regulate both receptor activity and subcellular distribution.

The β-adrenergic receptor binds catecholamines and activates adenylate cyclase via the stimulatory GTP-binding protein G_S. Sequential activation of the β-receptor produces a diminished response, a phenomenon termed homologous or agonist-specific desensitization. The biochemical basis for this effect appears to be due to the phosphorylation of the β

receptor by an enzyme termed the β-adrenergic receptor kinase (βARK). βARK phosphorylates the agonist-bound form of the receptor causing a functional uncoupling with G_S. Following phosphorylation, the receptor is sequestered within an intracellular compartment and is unable to interact with extracellular agonists. Presumably, dephosphorylation and redistribution to the plasma membrane complete the resensitization of the β-receptor. Recent data demonstrate that βARK phosphorylates other receptors *in vitro* and may play a more general role in modulating the activity of a family of cell-surface receptors.

Heterologous desensitization, due to the exposure of a cell to a specific agonist, can also result in a diminution in subsequent responses to agonists acting through other receptor types. Cyclic AMP-dependent protein kinase can phosphorylate the β-receptor *in vitro* and is implicated in the heterologous desensitization of this receptor. As is the case with βARK, phosphorylating the β-receptor by cAMP-dependent protein kinase reduces the coupling between receptor and G_S. In contrast to βARK, however, cAMP-dependent protein kinase can phosphorylate the β-receptor independent of agonist occupation. This suggests a negative feedback loop, wherein stimulation of adenylate cyclase by any positively coupled receptor causes increased cAMP production and activation of cAMP-dependent protein kinase with subsequent phosphorylation of the β-receptor. In effect, this would dampen any potential stimulation of adenylate cyclase by the β-receptor pathway.

The tyrosine kinase activities of growth factor receptors are often regulated by serine/threonine phosphorylation. Protein kinase C phosphorylates both the insulin and EGF receptors *in vitro,* inhibiting ligand binding and decreasing tyrosine kinase activity. Similar results are observed in cells treated with phorbol ester, suggesting that the protein kinase C effects are physiologically significant. Although cAMP-dependent protein kinase does not appreciably phosphorylate the insulin receptor *in vitro,* activating cAMP-dependent protein kinase in intact cells results in phosphorylation of the receptor along with decreased insulin binding and diminished insulin receptor tyrosine kinase activity. Whether these effects are due to phosphorylation of the receptor by the cAMP-dependent protein kinase itself or they occur via another kinase activated by this enzyme is unclear. In either case, these observations provide direct evidence of cross-talk between different effector systems and demonstrate the diverse pathways that can be utilized to modulate a particular biochemical response.

B. Structural Proteins

A wide variety of structural proteins undergo reversible phosphorylation, from contractile proteins in muscle cells to cytoskeletal proteins in virtually all mammalian cell types. MAP2 and tau are neuronal proteins that bind to and stabilize microtubule structures. MAP2 resides in the cell body and dendrites of neurons and binds to both microtubules and neurofilaments in these locations. MAP2 appears to act as a coupling agent to form cross-bridges between these cytoskeletal elements.

Tau binds primarily to microtubules within the axon of neurons. However, in Alzheimer's disease, tau accumulates in a structure termed the neurofibrillary tangle, which are located within the cell bodies of certain neuron populations. Neurofibrillary tangles are pathological hallmarks of this disease; their location and density correlate with the dementia that is characteristic of Alzheimer's patients. Although the pathophysiology of this disease is not clear, one hypothesis suggests that the accumulation of tau within neurofibrillary tangles in the cell body disrupts cell function, resulting in neuronal death and development of the clinical signs of Alzheimer's disease. An alternative hypothesis suggests that tangles develop as a consequence of neuronal degeneration, but do not play a causal role in this disease. Immunocytochemistry studies with antibodies that recognize different phosphorylation sites on tau have shown that the neurofibrillary tangle form of tau and axonal tau differ in their phosphorylation states. It is not yet clear whether the abnormal phosphorylation of tau promotes its accumulation into neurofibrillary tangles, or whether altered phosphorylation is merely a consequence of its abnormal accumulation as tangles in the cell bodies of diseased neurons.

Purified MAP2 and tau contain varying amounts of covalently bound phosphate. Dephosphorylating MAP2 and tau enhances their ability to promote microtubule assembly. Both proteins can be phosphorylated *in vitro* by cAMP-dependent protein kinase and CaM-kinase II; in each case, phosphorylation inhibits their microtubule-forming activity. Many other protein kinases can phosphorylate MAP2 or tau *in vitro,* some of which reduce the microtubule-forming activity of these proteins. It is not yet clear which kinases are responsible for phosphorylating MAP2 and tau *in vivo.* Nonetheless, these observations support the hypothesis that phosphorylation–dephosphorylation of cytoskeletal-associated proteins may be one mechanism by which changes in cell shape, structure, and function can be regulated. [*See* Phosphorylation of Microtubule Protein.]

C. Ion Channels

Protein phosphorylation has been indirectly linked to modulation of both voltage-regulated and ligand-regulated ion channels. In some cases, most notably the nicotinic acetylcholine receptor and voltage-sensitive calcium channels, a direct link between channel activity and phosphorylation of the channel (or of channel-associated regulator proteins) has been made. All four of the major second messenger-dependent protein kinases modulate the electrical excitability of different cells. Phosphorylation can regulate different characteristics of ion channels, from altering the time for opening or closing of a channel to regulating the number of functional channels capable of opening upon membrane depolarization.

The nicotinic acetylcholine receptor (nAchR) is a neurotransmitter-regulated ion channel located on the postsynaptic membrane of neuromuscular junctions. The biochemical properties of this receptor channel have been established using receptor purified from the electric organ of electric rays and electric eels. The purified receptor consists of four subunits in a $\alpha_2\beta\gamma\delta$ complex. The receptor exists as a phosphoprotein *in vivo* and can be phosphorylated under physiological conditions by three protein kinases that are present on the postsynaptic membrane. Each of these kinases, cAMP-dependent protein kinase, protein kinase C, and a protein tyrosine kinase, phosphorylates multiple subunits of the receptor. Phosphorylation of the purified receptor by cAMP-dependent protein kinase or the tyrosine kinase increases the rate of receptor desensitization. This results in diminished ionic conductance through the postsynaptic membrane, that is, a decrease in cholinergic synaptic transmission. Protein kinase C has not been shown to have a direct effect on purified receptors, although phorbol esters increase nAchR desensitization in cultured cells. The extracellular signal that regulates tyrosine-specific phosphorylation of the nAchR is not yet known. These data indicate that different extracellular signals can induce the phosphorylation of the nAchR by different protein kinases, resulting in decreased responsiveness to acetylcholine.

D. Metabolic Enzymes

Glycogen synthase, the rate-limiting enzyme in glycogen synthesis, was one of the first phosphorylatable enzymes identified, yet the phosphorylation mechanisms that affect this enzyme are not completely understood. Glycogen synthase is a classic example of regulation through multisite phosphorylation and demonstrates the complexity involved in deciphering the effects of multiple phosphorylation events on enzyme activity. This complexity is due to the number of residues phosphorylated on glycogen synthase *in vivo* (minimally seven) and the fact that almost all serine-threonine protein kinases can phosphorylate glycogen synthase *in vitro*. Obviously, identifying the protein kinase(s) activated by an extracellular signal and responsible for *in vivo* phosphorylation is a difficult challenge.

Figure 4 depicts the current understanding of the multisite regulation of glycogen synthase. Phosphorylation of glycogen synthase at site 1, 2, or 3 inhibits

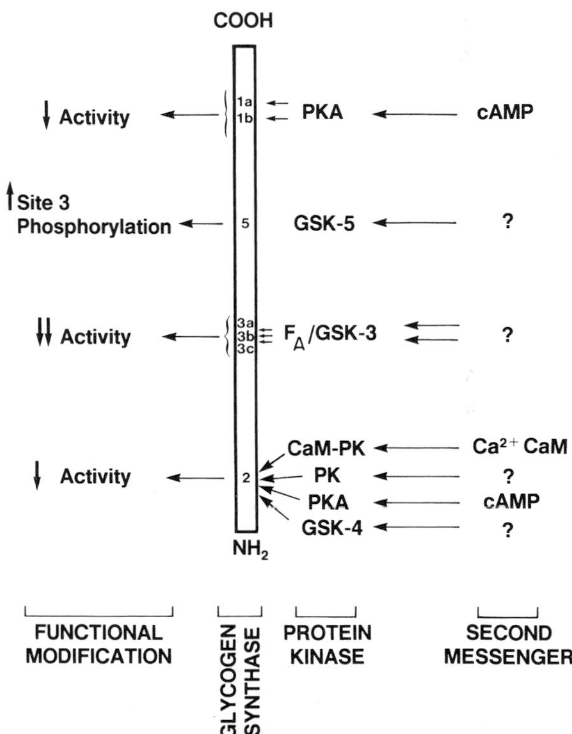

FIGURE 4 Multisite regulation of glycogen synthase activity. Phosphorylation of various sites is catalyzed by cAMP-dependent protein kinase (PKA), CaM-kinase II (CaM-PK), phosphorylase kinase (PK), and several glycogen synthase kinases (F_A/GSK-3, GSK-4, GSK-5). The consequence of each phosphorylation event is indicated under Functional Modification. [Adapted from J. Patel (1988). Protein phosphorylation: A convergence site for multiple effector pathways. *In* "Neuronal and Glial Proteins: Structure, Function and Clinical Application" (P. J. Marangos, I. Campbell, and B. M. Cohen, eds.). Academic Press, New York.]

enzyme activity, although site 3 produces a greater inhibitory effect. Phosphorylation of site 3 by F_A/GSK-3 requires prephosphorylation at site 5 by GSK-5 (also known as casein kinase II). The type 1 and 2A protein phosphatases are the predominant phosphatases that act on glycogen synthase.

Epinephrine induces the activation of cAMP-dependent protein kinase, resulting in phosphorylation of sites 1a and 1b. Site 2 is phosphorylated, either by cAMP-dependent protein kinase or phosphorylase kinase (which itself is activated by cAMP-dependent protein kinase). The enhanced phosphorylation of site 2 is possibly due to inhibition of protein phosphatase activity via phosphorylation of inhibitor-1 by cAMP-dependent protein kinase. Site 3 is also phosphorylated in response to epinephrine, presumably by F_A/GSK-3, although this has not been demonstrated directly. The net effect of these phosphorylations is inhibition of glycogen synthase activity.

Insulin causes a selective dephosphorylation of site 3, either by inhibiting F_A/GSK-3 activity or by activating a protein phosphatase. Insulin also may alter the conformation of glycogen synthase such that it is a poorer substrate for F_A/GSK-3 (or better substrate for phosphatase). Insulin-stimulated dephosphorylation of glycogen synthase results in enhanced activity of the enzyme. Thus far, a direct connection between the tyrosine kinase activity of the insulin receptor and phosphorylation–dephosphorylation of glycogen synthase has not been established.

IV. MULTISITE PHOSPHORYLATION AS AN INTEGRATION MECHANISM

An increasing number of proteins appear to be multiply phosphorylated with varying consequences on biological activity. Multiple site phosphorylation represents one approach by which different extracellular signals, activating distinct pathways and generating various second messengers, converge to regulate the function of a single protein. In this respect, multisite phosphorylation represents a convergence mechanism for coordinated regulation of cellular function. The consequences of multisite phosphorylation can be minimally separated into the following four categories.

(1) Phosphorylation of a common site by multiple effector-dependent protein kinases. Different extracellular signals act in an additive fashion to regulate the function of a single focal protein, thus producing a concerted biochemical response. Examples include site 5 of glycogen synthase and serine-40 of tyrosine hydroxylase.

(2) Phosphorylation of distinct sites by multiple effector-dependent protein kinases producing similar functional responses. This phenomenon is observed with the nicotinic acetylcholine receptor as well as a number of cytoskeletal proteins.

(3) Phosphorylation of distinct sites by multiple effector-dependent protein kinases with opposing functional responses. The receptor tyrosine protein kinases are prototypic examples of this type of regulation.

(4) Phosphorylation of one site generates a recognition site for a second protein kinase. Although phosphorylation by the first kinase may or may not produce a change in protein function, it is a prerequisite for secondary phosphorylation, which does alter protein function. This mechanism is exemplified by phosphorylation of sites 5 and 3 on glycogen synthase (see earlier). Sequential phosphorylation may be a prerequisite for F_A/GSK-3 activity; in addition to its actions on glycogen synthase, sequential phosphorylation of phosphatase-1 by F_A/GSK-3 has been demonstrated using cAMP-dependent protein kinase as the initiating protein kinase. Sequential phosphorylation of glycogen synthase by cAMP-dependent protein kinase and casein kinase I has also been described. Another example where phosphorylation of one site generates a recognition site for a second protein (often with SH2 domains) was illustrated for the PDGF-Rβ in Fig. 3.

BIBLIOGRAPHY

Boyer, P. D., and Krebs, E. G. (eds.) (1986). "The Enzymes," Vol. 17. Academic Press, Orlando, Florida.

Boyer, P. D., and Krebs, E. G. (eds.) (1987). "The Enzymes," Vol. 18. Academic Press, Orlando, Florida.

Cohen, P. (1988). Protein phosphorylation and hormone action. *Proc. Roy. Soc. London, Ser. B* **234**, 115–144.

Edelman, A. M., Blumenthal, D. K., and Krebs, E. G. (1987). Protein serine/threonine kinases. *Annu. Rev. Biochem.* **56**, 567–613.

Mumby, M. C., and Walter, G. (1993). Protein serine/threonine phosphatases: Structure, regulation and functions in cell growth. *Physiol. Rev.* **73**, 673–699.

Yarden, Y., and Ulrich, A. (1988). Growth factor receptor tyrosine kinases. *Annu. Rev. Biochem* **57**, 443–478.

Proteins

THOMAS E. CREIGHTON
European Molecular Biology Laboratory

GLOSSARY

Polypeptide chain Structural unit of a protein; a linear chain of amino acids linked together by peptide bonds

Primary structure Sequence of amino acids in the polypeptide chain of a protein

Quaternary structure Assembly of individual polypeptide chains in large multisubunit proteins

Secondary structure Regular local conformations of the polypeptide chain, especially α helices, β strands, and β turns

Tertiary structure Three-dimensional structure of the polypeptide chain in a functional protein

PROTEINS ARE VITAL CONSTITUENTS OF THE human body and are essential for life as we know it. In a human, there are about 10^5 different proteins that perform these varied functions. Yet they all share the same fundamental architecture, which is adapted in many different ways. The structures of proteins are the key to understanding how they work and why human beings and other living organisms are possible.

I. FUNCTIONS

A. Structural

Many proteins are simply architectural, particularly the *fibrous proteins. Keratins* are the proteins that largely make up hair, nails, and skin. *Collagen* constitutes almost a quarter of the dry weight of the human body and plays a major role in holding it together. Some collagens form ropes and straps in tendons and ligaments, whereas others make woven sheets in skin, filtration membranes in kidneys, and calcium-reinforced skeletal frameworks in bone and teeth. [*See* Collagens, Structure and Function.]

Another fibrous structural protein is *elastin,* a rubber-like protein that occurs in most connective tissues of the body, especially in the walls of blood vessels and in ligaments. It is the major component of elastic fibers, which can stretch several times in length and then rapidly return to their original size and shape when the tension is released. [*See* Elastin.]

Proteoglycans are proteins with carbohydrates attached that form much of the matrix between cells in connective tissue. They are important in determining the viscoelastic properties of joints and other parts of the body subject to mechanical deformation. [*See* Proteoglycans.]

B. Informational

Proteins are involved in all steps in transmitting and propagating nerve impulses. They do this by sensing the presence of nerve transmitters and by controlling the permeabilities of nerve membranes to different ions. The photoreceptor protein of the eye, rhodopsin, absorbs light, undergoes a change in its shape, and then triggers a nerve impulse, all so efficiently that a single photon is sufficient to activate a rod cell of the eye. Proteins are also involved in our other senses—touch, hearing, smell, and taste—but less is known about the molecular details of these processes.

Many hormones are proteins, including insulin, glucagon, growth hormone, and corticotrophin. The receptors that recognize the presence of all hormones,

ENCYCLOPEDIA OF HUMAN BIOLOGY, Second Edition, VOLUME 7. Copyright © 1997 by Academic Press. All rights of reproduction in any form reserved.

and trigger the appropriate response, are also proteins. Finally, proteins control all the steps in expression of the genetic information of a cell.

C. Immunochemical

The body's defenses against microorganisms and all other extraneous agents depend on proteins known as *antibodies* or *immunoglobulins,* which specifically recognize and bind foreign molecules. A huge number of different antibody molecules, with different specificities, are possible, yet they all have very similar structures.

D. Transport

Many molecules are transported through the body by binding to specific proteins with an appropriate affinity. The O_2 required for metabolism is transported through the blood from the lungs by binding to the protein hemoglobin in the red blood cells and is stored in muscle cells bound to myoglobin. Serum albumin carries a variety of small molecules. Iron is a necessary nutrient, but is also toxic in its free form, so it is transported in the blood by transferrin and stored in tissues inside another protein, ferritin, which has a cavity capable of holding up to 4500 ferric ions.

E. Catalysis

All the myriad of chemical reactions that occur within the human body are catalyzed by proteins, which in this case are known as *enzymes*. Even simple chemical reactions that occur spontaneously are catalyzed by enzymes, such as the hydration of CO_2 by carbonic anhydrase.

F. Pumping across Membranes

Within membranes are many proteins, some of which act as highly selective pumps or gates for various ions and other small molecules. Protein pumps concentrate small molecules on one side of a membrane, against a concentration gradient. They function by binding the molecules tightly on one side of the membrane, then using chemical energy to release them on the other side. In contrast, gates simply regulate the flow of specific molecules through membranes in both directions, depending on the concentration difference across the membrane. [*See* Ion Pumps; Membranes, Biological.]

G. Movement

Some proteins, especially actin and myosin of muscle, convert chemical energy into mechanical energy. Actin and myosin accomplish this by binding to each other, changing shape so that one moves relative to the other, then releasing and repeating the cycle.

II. STRUCTURES

The diverse functional properties of proteins can be understood in terms of their structures. At the simplest level, proteins are simply linear, unbranched polymers of amino acids. All proteins are made up of the same backbone and the same 20 amino acids. It is the sequence of amino acids linked together that distinguishes one protein from another, and this level of structure is called the *primary structure*.

A. Primary Structure

1. Amino Acids

Of the 20 amino acids normally used to build proteins, 19 have the general structure

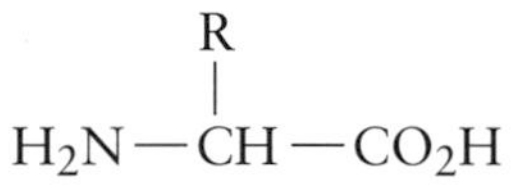

$$\begin{array}{c} \mathrm{R} \\ | \\ \mathrm{H_2N-CH-CO_2H} \end{array}$$

and differ only in the chemical structure of the side chain, R (Fig. 1). Proline, the twentieth natural amino acid, is similar but has the side chain bonded to the nitrogen atom:

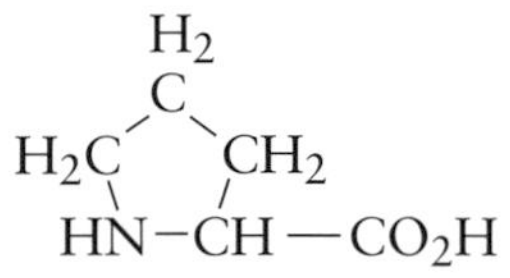

$$\begin{array}{c} \mathrm{H_2} \\ \mathrm{C} \\ \mathrm{H_2C} \quad \mathrm{CH_2} \\ \mathrm{HN-CH-CO_2H} \end{array}$$

Except in the amino acid glycine, where the side chain is simply a hydrogen atom, the central carbon atom (C^{α}) is asymmetric and always of the L-isomer:

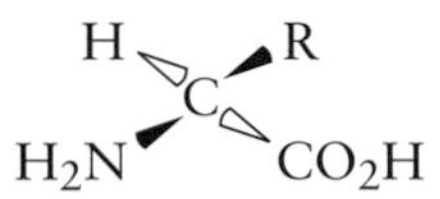

$$\begin{array}{c} \mathrm{H} \quad \mathrm{R} \\ \mathrm{C} \\ \mathrm{H_2N} \quad \mathrm{CO_2H} \end{array}$$

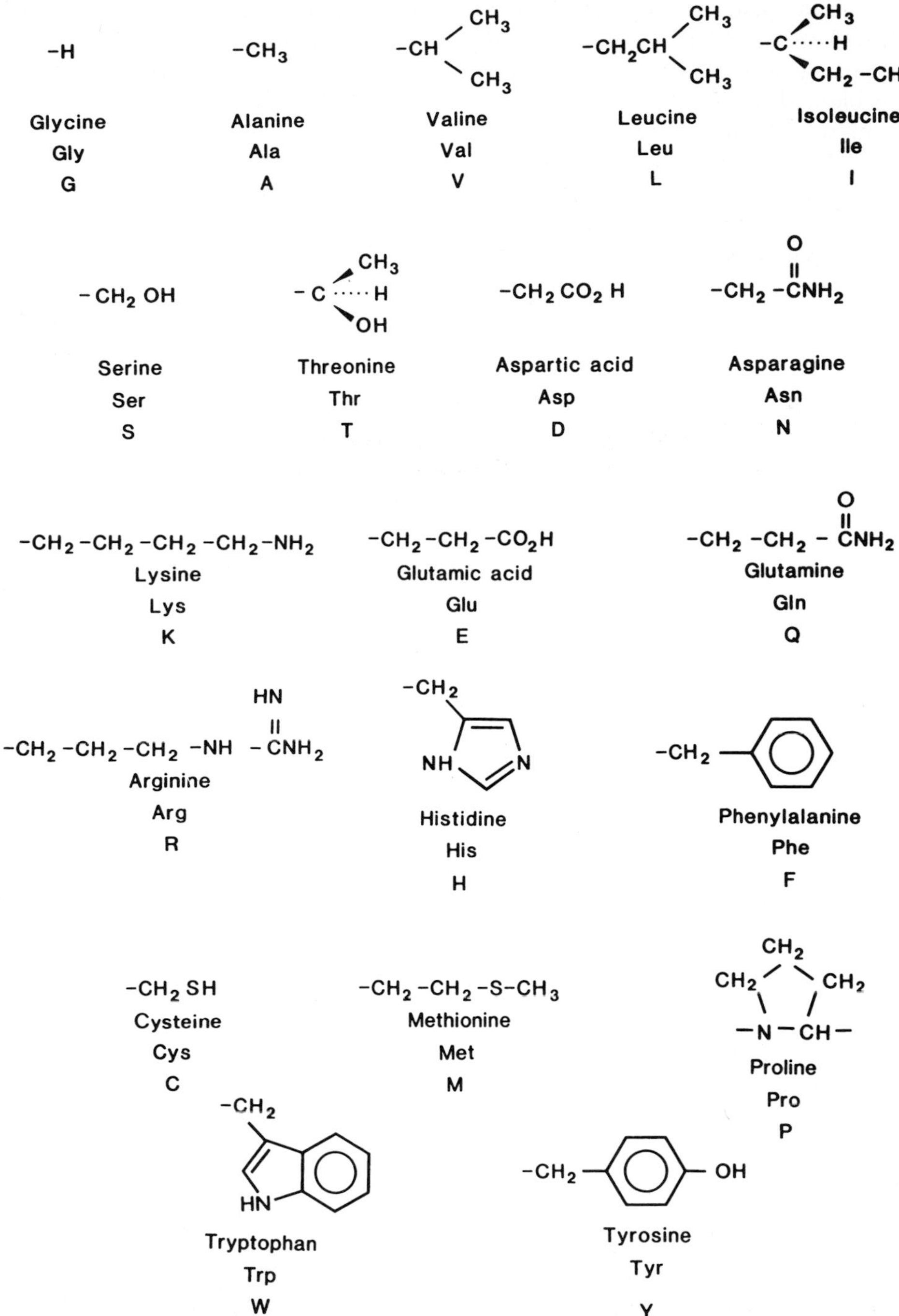

FIGURE 1 Chemical structures of the side chains of the 20 amino acids used to synthesize proteins. Below each side chain is the full name of the corresponding amino acid, plus the three- and one-letter abbreviations generally used for residues in proteins. The side chains of Ile and Thr have asymmetric centers, and only the isomer illustrated occurs naturally. The C^{α} and N atoms of the backbone are also included in the unique case of Pro.

The 20 different amino acid side chains possess a variety of chemical properties that, when combined on a single molecule, give a protein properties far beyond those possible with smaller molecules or simpler polymers. Some side chains are simple nonpolar hydrocarbons, with no functional groups, but they have the important property of not interacting favorably with water, that is, are *hydrophobic.* The other side chains have polar functional groups, particularly amino, carboxyl, hydroxyl, thiol, and amide groups, which can participate in a variety of chemical reactions.

2. Polypeptide Backbone

Amino acids are assembled into proteins by linking them together via *peptide bonds,* expelling one water molecule per peptide bond in the process:

$$-NH-\underset{}{\overset{R_n}{\overset{|}{C}}}-CO_2H + H_2N-\overset{R_{n+1}}{\overset{|}{CH}}-CO_2H$$

$$\xrightarrow{\nearrow H_2O} -NH-\overset{R_n}{\overset{|}{C}}-\overset{O}{\overset{\|}{C}}-HN-\overset{R_{n+1}}{\overset{|}{CH}}-CO_2H$$

Many amino acids—usually between 50 and 1000, on average 300, but as many as 5000—are linked together in this way to form a linear polypeptide chain.

The polypeptide backbone is simply a repetition of the basic amino acid unit, comprising three atoms (the C^α and the amide N and carbonyl $\acute{C}$ of the peptide bond). The polypeptide chain has a free amino group at one end, the N terminus, and a free carboxyl at the other, the C terminus. The group of atoms originating from the same amino acid is designated as a *residue,* and individual residues are generally numbered sequentially, starting at the N terminus.

The number and sequence of amino acids in the polypeptide chain distinguish one protein from another. The amino acid sequences of approximately 10^4 proteins are known at the present time, containing some 2 million residues. There are no striking patterns in most protein sequences, and the distribution of residues is almost random. Only in certain structural proteins, with repetitive structures, are the amino acid sequences remarkable.

With any of 20 amino acids at each position, an immense number of different protein sequences are possible. For example, an average polypeptide chain of 300 residues could exist with any of 20^{300}, or 10^{390}, different sequences. One molecule of each such sequence would fill the entire universe 10^{287} times over. Obviously, not every protein sequence can exist today, nor could it have existed during the lifetime of the universe.

3. Biosynthesis

Natural proteins are synthesized within the cytoplasm of cells, using the genetic information and the individual amino acids. Of the 20 amino acids incorporated into proteins, only 11 are synthesized by humans; the remainder (histidine, isoleucine, leucine, lysine, methionine, phenylalanine, threonine, tryptophan, and valine) must be supplied by the diet, generally by the breakdown of proteins consumed.

Biosynthesis of proteins is a complex process in which the sequence of amino acids assembled into the polypeptide chain is determined by the nucleotide sequence of the corresponding gene (Fig. 2). The nucleotide sequence of the gene is first transcribed into a complementary sequence of nucleotides in the *messenger RNA* (mRNA). The individual amino acids are attached to specific adaptor molecules, known as *transfer RNA* (tRNA). On a very large molecular apparatus of proteins and RNA known as the *ribosome,* the tRNA molecules recognize the sequence of nucleotides in the mRNA, three at a time (one *codon*), to insert the appropriate amino acid into the growing polypeptide chain. Synthesis starts at the amino end of the polypeptide chain, always with the residue Met, and progresses stepwise toward the C terminus. The genetic code that is used to translate the nucleotide gene sequence into amino acid sequence of the protein is presented in Fig. 3. [*See* DNA Synthesis.]

Protein synthesis is usually regulated at the initial stage of transcription of the gene into mRNA. Regulatory proteins bind near the 5′ end of the gene, or at more distant sites, and determine the level of its transcription. In this way, proteins are synthesized only in the appropriate cell type and under the appropriate circumstances. [*See* DNA and Gene Transcription.]

By the technology of *protein engineering,* virtually any protein can be made in large quantities by expressing its gene in appropriate cells under artificial circumstances. Short polypeptide chains can also be synthesized chemically.

Most genetic diseases of humans are caused by the absence or alteration of a particular protein, due to mutation of its gene. For example, the well-known mutation that produces sickle-cell anemia causes the sixth residue of the β chain of hemoglobin to be changed from the normal Glu to Val. This causes

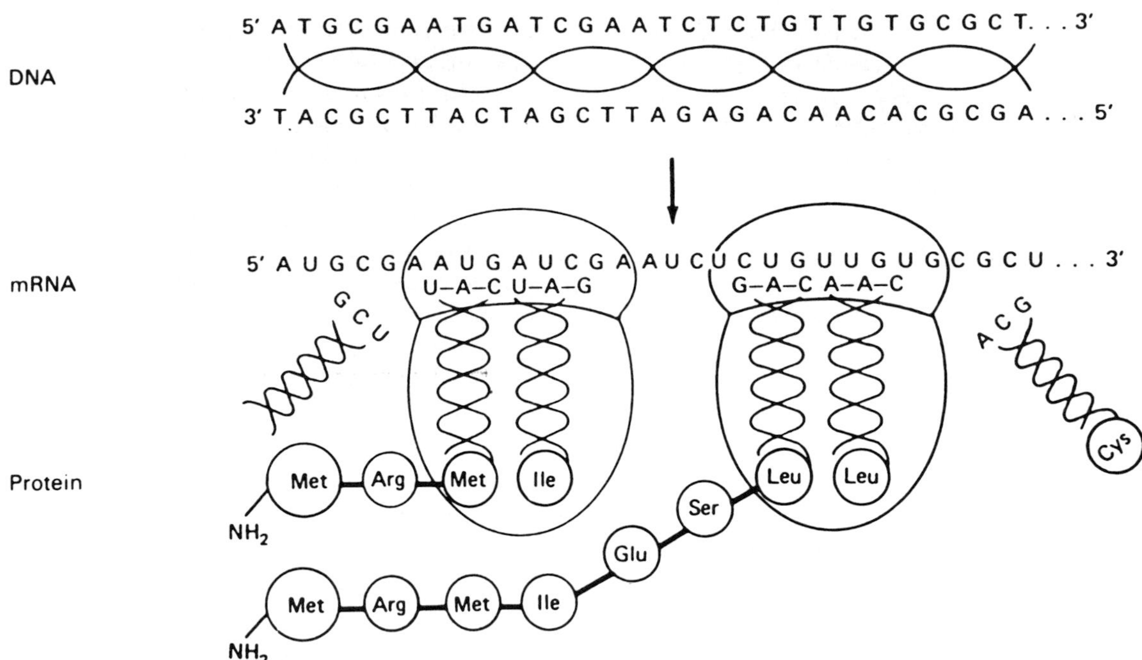

FIGURE 2 Biosynthesis of a polypeptide chain. At the top is shown the double-stranded DNA, with the complementary nucleotide sequences of the two antiparallel strands containing the information for the protein primary structure. The nucleotide A is always paired with T, C with G. One of the DNA strands is transcribed into a complementary mRNA strand, with the nucleotide U replacing T. The mRNA is translated by ribosomes, the two large "acorn"-shaped structures. Starting at the first AUG codon from the 5′ end, the triplet codons are recognized by "hairpin"-shaped tRNA molecules with the complementary anticodon, to which the appropriate amino acid is attached. On the ribosome, the amino acids are linked together by peptide bonds. The ribosome on the left is adding the fourth residue, Leu, to the nascent chain, whereas that on the right is further along the mRNA and adding the eighth residue. Upon reaching a termination codon, the completed polypeptide chain is released from the ribosome.

the protein to aggregate under certain conditions to produce the "sickling" of the red blood cells that results in the physiological symptoms. The hundreds of genetic diseases of humans known to result from absent or abnormal proteins include the hemophilias, the thalassemias, phenylketonuria, Duchenne muscular dystrophy, and color blindness. [*See* Genetic Diseases.]

4. Posttranslational Modifications

Only the 20 amino acids of Fig. 1 are included in the genetic code and are incorporated into the initial polypeptide chain synthesized. In some proteins, the polypeptide chain undergoes alterations to the primary structure, usually in a very specific fashion, in which all the molecules of a particular protein are modified in the same way. Residues may be removed from either end of the polypeptide chain, the chain may be cleaved by proteolytic enzymes, and the terminal amino and carboxyl groups may be modified. A wide variety of specific alterations of amino acid side chains may also occur. The most common are addition of carbohydrate units to Asn, Thr, and Ser residues, phosphorylation of Ser, Thr, His, Lys, and Tyr residues, and attachment of various prosthetic groups that are involved in function, such as heme groups. Cys residues can also be cross-linked by oxidative formation of disulfide bonds between their thiol groups.

After biosynthesis on ribosomes in the cell cytoplasm, it is important that a protein be directed to its appropriate place in the cell. Proteins destined to be secreted from the cell, or incorporated into membranes, have special sequences at their amino terminus (signal peptides) that serve this purpose. The signal peptides are usually cleaved off subsequently. The signals that send proteins to other compartments such as the nucleus, mitochondria, lysosomes, and peroxisomes are also special sequences, some of which are still being sought.

5. Evolutionary Origins of Proteins

The vast number of amino acid sequences possible with a polypeptide chain implies that whenever two

First position \ Second position	U	C	A	G	Third position
U	Phe	Ser	Tyr	Cys	U
	Phe	Ser	Tyr	Cys	C
	Leu	Ser	Terminate	Terminate	A
	Leu	Ser	Terminate	Trp	G
C	Leu	Pro	His	Arg	U
	Leu	Pro	His	Arg	C
	Leu	Pro	Gln	Arg	A
	Leu	Pro	Gln	Arg	G
A	Ile	Thr	Asn	Ser	U
	Ile	Thr	Asn	Ser	C
	Ile	Thr	Lys	Arg	A
	Met	Thr	Lys	Arg	G
G	Val	Ala	Asp	Gly	U
	Val	Ala	Asp	Gly	C
	Val	Ala	Glu	Gly	A
	Val	Ala	Glu	Gly	G

FIGURE 3 The genetic code used to specify the amino acid residues incorporated during biosynthesis of proteins. Each codon of the mRNA consists of three of the nucleotides A, U, C, or G. The first nucleotide of each codon is given on the *left,* the second at the *top,* and the third on the *right.* Of the 64 possible codons, three normally code for no amino acid, but terminate the polypeptide chain.

protein sequences are substantially more similar than expected from chance, they must be related and must have arisen evolutionarily from a common ancestor, that is, are *homologous.* Two random sequences are expected to have about 6% of their residues identical, just by chance. Residues can be inserted and deleted by genetic mutations, so gaps may be introduced in either sequence when they are compared to maximize their similarities, but this decreases the significance of any similarities. Consequently, two proteins are generally taken to be homologous if more than 20% of their residues are identical when aligned with a reasonable number of gaps.

The sequences of proteins provide a great deal of information about the evolutionary relatedness of the species from which they come. Generally, the more closely related the species, the more similar the amino acid sequences of homologous proteins. For example, the sequences of many proteins, including cytochrome *c*, the hemoglobin α, β, and γ chains, and the fibrinopeptides, from humans are identical to those of chimpanzees. Myoglobin and the hemoglobin δ chain from the two species each differ at a single residue.

Differences between proteins presumably have arisen during evolution by mutation of their genes after separation from their last common ancestor. The extent of the differences depends on both the time since the genes diverged and the rate at which mutations are permitted to occur. Some proteins change very slowly, some very rapidly. Most of the changes that have occurred appear to be neutral and to have no significant effect on the function of the protein. Most mutations that alter the function presumably are selected against during evolution, so the rate of change is usually greater with fewer functional constraints of the protein.

After the evolutionary appearance of life, new proteins are believed to have arisen primarily by duplication and subsequent mutation of an existing gene. For example, the polypeptide chains of myoglobin and the α, β, γ, δ, and ε chains of hemoglobin are all homologous and undoubtedly arose by duplications of an ancestral globin gene. Whether all present proteins ultimately arose by duplication of one or a very few common ancestors is not clear, for over long periods of evolutionary time homology becomes undetectable owing to the large number of mutations that have accumulated.

Some proteins have been elongated by duplication of all or part of the gene, so different segments of the protein sequence are homologous to each other. In other cases, parts of different genes have been duplicated and rearranged to generate new mosaic proteins.

In most cases, similarity of protein amino acid sequences implies similarity of function, although there are exceptions. The biological function of a new gene or protein is often inferred from any homology to genes or proteins of known function.

B. Folding of the Polypeptide Chain

The polypeptide chains of natural proteins differ from unnatural polypeptides and other polymers in that they fold back upon themselves to adopt compact, well-ordered, three-dimensional structures, in which the positions of most atoms are fixed in space by interactions with neighboring atoms. These folded conformations are essential for the biological functions of proteins.

1. Polypeptide Conformation

The three-dimensional conformations of proteins result from the ability of single covalent bonds to undergo rotation. For each residue (except Pro), rotations are possible around two single bonds of the backbone; the angles of rotation are given the designations ϕ and ψ (Fig. 4). In the case of Pro, its cyclic side chain restricts the value of ϕ to be about $-60°$. The peptide bond has partial double-bond character and is constrained to be planar, but two different forms are possible. The trans form of the planar peptide bond (with the C^α atoms on opposite sides of the bond) is intrinsically favored, unless the following residue is Pro, when the cis isomer (with the C^α atoms on the same side of the bond) has a similar intrinsic stability. In addition to the backbone, the side chains of most of the amino acids have intrinsic flexibility comparable to that of small molecules.

In a fully unfolded polypeptide chain, rotations about single bonds are expected to be random, but not all combinations are possible, for clashes between atoms would occur in some cases. Local restrictions on flexibility of the backbone vary with the nature of the side chain; for example, 61% of the combinations of ϕ and ψ are possible with Gly residues, which have no side chain, but only about 5% are feasible with the most restrictive side chains of Val and Ile residues. For most amino acid residues, about 30% of the possible combinations of ϕ and ψ are possible, so the polypeptide backbone has substantial flexibility.

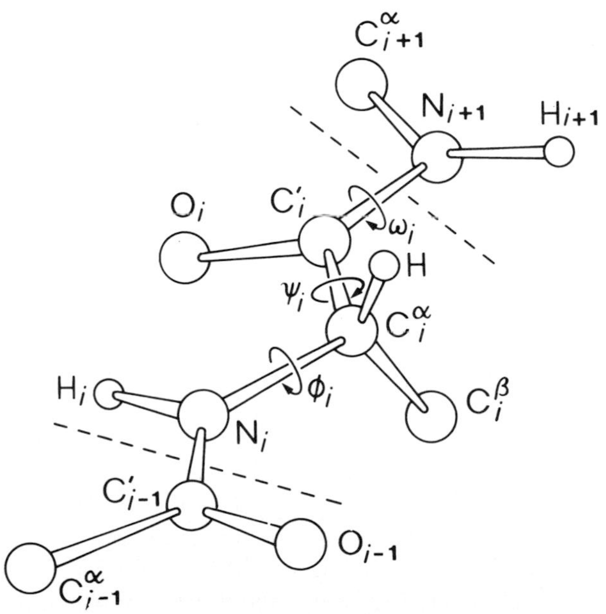

FIGURE 4 A short segment of polypeptide backbone, with the limits of one residue indicated by the dashed lines. Rotations about the bonds of the backbone produce different three-dimensional conformations. The angles of rotation ϕ and ψ have considerable flexibility, but ω must be close to either 0° or 180°, because the peptide bond has partial double-bonded character and must be planar. The polypeptide chain is illustrated in its fully extended conformation, where $\phi = \psi = \omega = 180°$.

Each residue in a flexible polypeptide chain can adopt a number of different conformations, perhaps eight on average. In this case, an average-sized protein chain of 300 residues might be expected to adopt 8^{300} ($=10^{270}$) different conformations. Not all of these would be feasible, because atoms distant in the primary structure would clash in some, but even if only one in 10 billion of these conformations were possible, the total number would be reduced only to 10^{260}.

The enormous number of conformations possible with an unfolded protein could not be encountered on a practical time scale. Rotations about single bonds to produce a different conformation occur about every 10^{-11} seconds. Therefore, to sample 10^{260} conformations would require, on average, 10^{249} seconds. This is vastly greater than the age of the earth, which is about 10^{17} seconds. Proteins obviously cannot fold to their tertiary strutures by a random search, since they are observed to fold within a few seconds after completion of biosynthesis.

Completed proteins also can be unfolded and will usually refold on a similar time scale. The mechanism of folding is not known for certain, but refolding probably occurs so quickly because the unfolded protein tends to adopt a limited number of nonrandom conformations after being placed under refolding conditions; most of these species are too unstable to be detectable. The slowest step is at a late stage of folding, when the protein gets very close to the final folded conformation.

Another way in which large polypeptide chains manage to fold so quickly is by folding initially into smaller structural units of some 50 to 200 residues, called *domains,* that are independently stable folding units. This type of folding of independent domains occurs sequentially during biosynthesis, as the polypeptide chain is being extended from the N terminus.

Folding *in vivo* after or during biosynthesis is generally assisted by *chaperones,* which inhibit aggregation of the unfolded polypeptide chain. In addition, there are catalysts of other intrinsically slow steps: formation of disulfide bonds and cis–trans isomerization of peptide bonds.

2. Protein Folded Conformations

The overall three-dimensional conformation of a protein is known as its *tertiary structure*. It can be determined in great detail by X-ray crystallography or by nuclear magnetic resonance analysis. The former gives the structure in the solid, crystalline state, the latter in solution. Fortunately, it is now clear that there is a single folded conformation for most proteins that is not altered substantially by crystallization. Although nearly 300 different protein tertiary structures are known, they have a number of characteristics in common; one folded protein is illustrated in Color Plate 1. Actual proteins differ from such models in having varying degrees of flexibility and in being very dynamic.

Small folded proteins consist of single structural domains, and their overall shapes are roughly spherical. Individual domains rarely contain more than 200 amino acid residues, and larger proteins are folded into multiple domains.

The interior of a folded protein domain is remarkably compact, with very few cavities the size of a water molecule. In some proteins, a few water molecules are present within the protein interior, but they are usually involved in hydrogen bonding to the protein and are essentially part of the protein structure. The close-packing of atoms within a protein is such that about 75% of the interior volume is occupied by atoms, as in crystals of small molecules, but considerably greater than the 44 to 58% packing densities of liquids. Yet this close-packing in proteins is accomplished within the constraints of the covalent structure of the polypeptide chain, with relatively little strain due to unfavorable bond lengths, angles, or rotations.

Within a single domain, the polypeptide backbone generally pursues a moderately straight course across the domain, turns on the surface, and then returns across the domain in a more-or-less direct path. The impression is of segments of somewhat stiff polypeptide chain interspersed with relatively tight turns or bends, which are almost always on the surface of the protein. On the other hand, an increasing number of exceptional structures are being determined.

There are marked differences between the surfaces and interiors of proteins, depending on whether the protein occurs in aqueous solution or is embedded within membranes. With water-soluble proteins, any amino acid can occur on the surface, but there are restrictions on which occur in the interior. Virtually all the side chains that are normally ionized (Asp, Glu, Lys, and Arg residues) occur only on the surface, where they interact favorably with water, and not in the interior. Gly and Pro residues also tend to be on the surface, because they are commonly used in reverse turns of the polypeptide chain. The interior is composed almost entirely of nonpolar side chains (Ala, Val, Leu, Ile, Met, Phe, Tyr, Trp); virtually all the buried polar groups, principally the NH and CO groups of the backbone, are paired in hydrogen bonds. With proteins that are embedded in the hydrophobic interiors of membranes, nonpolar side chains are on the protein exterior and interacting with the membrane. Polar side chains are less frequent than in water-soluble proteins and tend to be in the interior, where they are believed to have functional roles, such as guiding ions through channels in the membrane.

a. Secondary Structures

The relatively straight segments of polypeptide backbone in a folded protein often adopt relatively regular conformations, known as *secondary structures.* The *α helix* is one and the *β strand* the other (Fig. 5). Their primary common characteristic is that all polar groups of the polypeptide backbone are paired in hydrogen bonds.

The α helix is a right-handed helix with 3.6 residues per turn and a translation along the axis of 1.50 Å (0.15 nm) per residue. Most conspicuous is the hydrogen bond between the CO oxygen of the backbone of each residue and the backbone NH of the fourth residue along the chain. The polypeptide main chain forms the inner part of the rod-like helix, and the side chains radiate outward in a helical array. The side chains of residues three and four apart in the primary structure are near each other spatially. A local variation is the 3_{10} *helix,* with one less residue per turn; it generally occurs only locally at the ends of α helices.

In the β-strand conformation, the polypeptide chain is nearly fully extended, with the CO and NH groups of each residue pointing to one side, those of the adjacent residue to the other. The side chains alternately point up and down, so those of residues two apart in the sequence are near each other. Multiple β strands aggregate side-by-side, forming hydrogen bonds between the CO of one strand and the NH of another, to form a *β sheet.* Adjacent strands may be either parallel or antiparallel, but the geometry of the chain is slightly different in the two cases, so sheets are often all of one type or the other. β sheets in proteins are not flat, but are almost always twisted, with each strand having a right-handed twist. A recently discovered variant is the *β helix,* in which one long β strand is distorted into a helix, with the usual hydrogen bonding between adjacent turns of this helix.

Approximately one-third of residues in proteins tend to be in helices and another third in β sheets. Most of the interiors of proteins are composed of β sheets, α helices, or both, packed together. Some

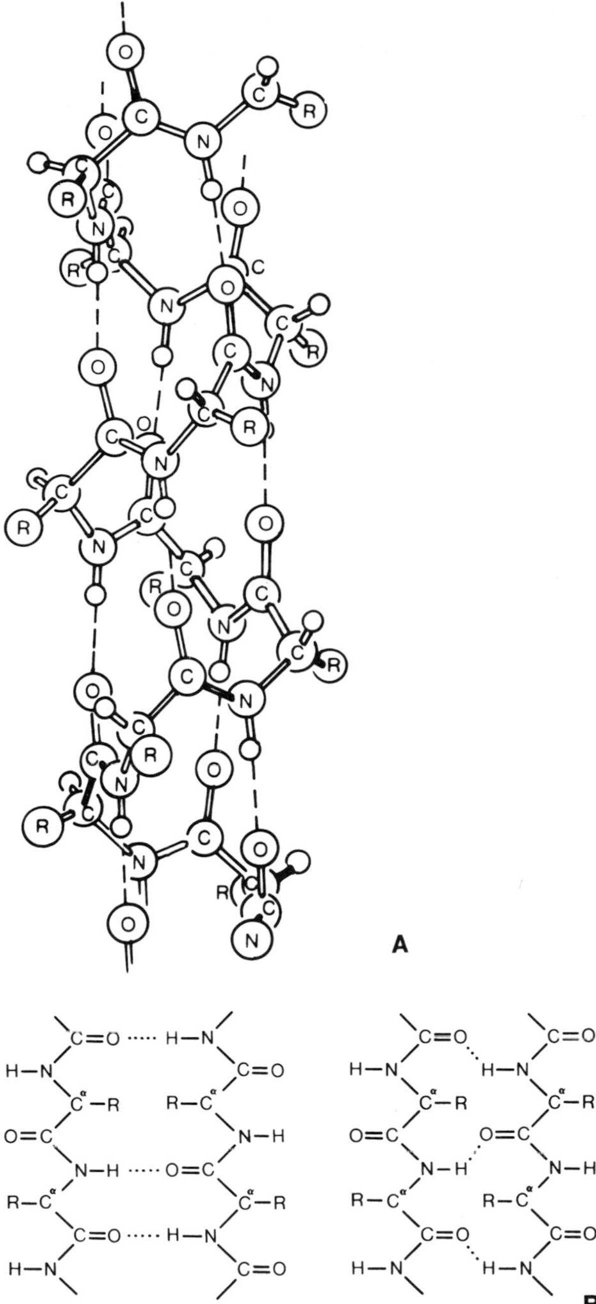

FIGURE 5 Secondary structures found commonly in proteins. (A) An α helix. (B) β sheets, antiparallel (left) and parallel (right). The side chains of the residues in β strands alternately project above and below the plane of the figure.

proteins, such as myoglobin and hemoglobin (Fig. 6), are composed only of α helices and turns; others consist entirely of β sheets and turns, whereas many contain both β sheeets and α helices. Simplified repre-

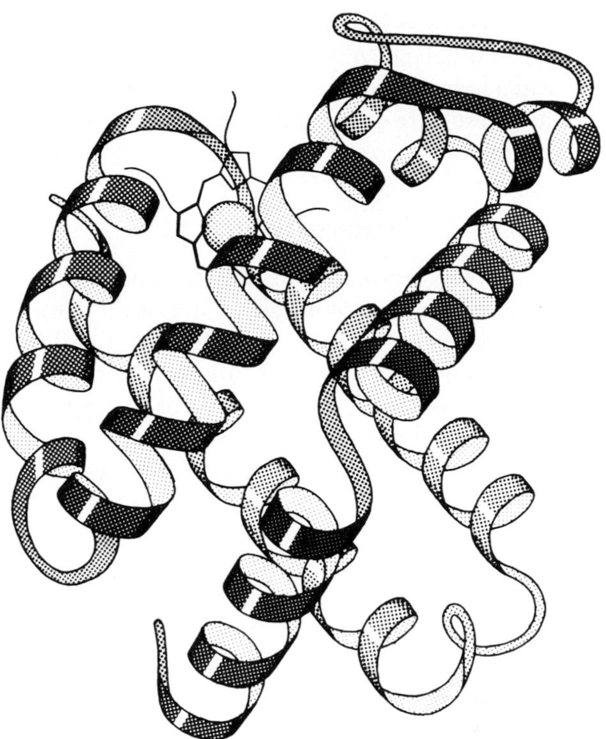

FIGURE 6 Schematic drawing of the topology of the polypeptide chain of the β subunit of human hemoglobin. Only the polypeptide backbone is depicted, as a ribbon, plus the heme prosthetic group. The coiled ribbon depicts the α-helical conformation. (Drawing kindly supplied by Jane Richardson.)

sentations of proteins often depict α helices as cylinders or coiled ribbons and β strands as broad arrows pointing from the amino to the carboxyl end of the polypeptide chain (Fig. 6).

b. Fibrous Proteins

Fibrous proteins, such as collagen and keratins, differ from the globular types of proteins described earlier, in that they have repetitive, extended structures, without the reverse turns to give a compact structure. In *collagen,* three polypeptide chains of about 1000 residues each are in helical conformations and wound about each other, to give a molecule 14 Å in diameter and 3000 Å in length (0.14 × 300 nm). This unusual architecture results from the repetitive sequence of collagen, in which every third residue in the sequence is Gly and many of the remainder are Pro. The absence of a side chain on the Gly residues permits the chains to coil about each other closely, and the restricted flexibility possible with Pro residues imparts rigidity to the structure.

Keratins consist of two or three polypeptide chains, each in an α-helical conformation, twisted about each other in a long extended structure. This regular structure also results from repetitive features of the amino acid sequence.

C. Quaternary Structure

Many proteins normally function as part of larger structures made up of multiple copies of the same or different polypeptide chains. A wide variety of different macromolecular structures are found, up to large complexes containing hundreds of polypeptide chains. Spherical viruses are made with an outer shell of many identical polypeptide chains interacting specifically to generate a regular icosahedron structure; a typical number is 180. This higher level of structure is designated as *quaternary structure*. When part of a larger aggregate, an individual polypeptide chain is referred to as a *subunit* or *protomer* and is often designated by a Greek letter.

Proteins interact with each other specifically by having surfaces on their folded conformations that are complementary, both in shape and in physical properties. These surfaces generally fit together so that most of the surface atoms are in close, energetically favorable contact; polar groups are usually paired in hydrogen bonds. Less frequently, ionized groups (which are electrically charged) are paired, one positive and one negative, to avoid buying any net charge. The interfaces between associated proteins are very like the interiors of the individual folded subunits. The strengths of the interactions vary enormously, depending on both the degree of complementarity and the area of the interacting surfaces.

Aggregation of identical polypeptide chains requires that the interacting surfaces of each protein subunit be self-complementary. When there are two complementary surfaces and they are not contiguous on the surface of the individual monomers, the association is said to be *heterologous*. An indefinite head-to-tail association will result, and many filamentous protein structures are generated by linear head-to-tail aggregation of globular subunits. Stable trimers, pentamers, hexamers, and heptamers of identical subunits are also made in this way, by fixing the geometry of interaction such that closed rings of the appropriate size result.

A more common type of association, *isologous*, is observed between identical molecules when a strictly limited aggregation is required. In this case, two molecules associate using the same surface on each. This requires that the single interacting surface on each subunit have a mirror plane of complementarity, so that the two identical surfaces are complementary. The result is a stable dimer in which the two monomers are equivalent. The dimer has a twofold symmetry axis through the junction of the two subunits, such that rotating the dimer by 180° about the axis reproduces the same molecule as the original (rotational symmetry). No further aggregation is possible, unless additional interacting sites are present. Tetramers are often formed by two different types of isologous association, so they are dimers of dimers.

D. Relationship between Structural Levels

Which secondary, tertiary, and quaternary structures are adopted by a protein depends entirely on its primary structure. Proteins with related amino acid sequences invariably have essentially the same secondary and tertiary structures, so a particular structure does not require a unique primary structure. In particular, the relative importance of the primary and tertiary structures in determining the secondary structure is not certain. Whereas the various amino acids tend to occur with different frequencies in α helices, β sheets, reverse turns, and irregular conformations, the stabilities of the various conformations depend on the presence of the tertiary structure. It should not be simply assumed that the primary structure determines the secondary structure, which then defines the tertiary structure.

The quaternary structure is determined only by the residues within the surfaces that interact. Therefore, related proteins need not have the same quaternary structure, because changing only a few amino acid residues within the interacting surface can alter dramatically the interaction with other proteins. For example, myoglobin functions as a monomer, whereas the homologous α, β, γ, δ, and ε chains of hemoglobin exist as $\alpha_2\beta_2$, $\alpha_2\gamma_2$, $\alpha_2\delta_2$, and $\alpha_2\varepsilon_2$ complexes.

E. Stability of Protein Folded Conformations

Whether or not the folded conformation of a protein is stable depends on both the protein's primary structure and its environment, that is, the nature of the solvent, the pH, salt concentration (ionic strength), temperature, pressure, and so on. Proteins can be unfolded, often reversibly, by high or low temperatures, extremes of pH (either very acidic or very basic), or addition of substances that promote unfolding (denaturants), such as urea or guanidinium chloride. Even under optimal conditions, the folded state of a protein is only marginally stable relative to the unfolded conformation. The many interactions holding together the folded structure of a single domain are only barely able to compensate for the energetic cost of fixing the flexible polypeptide chain in a single conformation. Unfolding of a single domain is usually cooperative, in that once unfolding starts, it proceeds to completion; partially folded conformations are relatively unstable. This implies that the interactions stabilizing the folded state are also cooperative, in that the stabilizing contribution of each interaction depends on the simultaneous presence of the others. Only by having a multiplicity of intrinsically weak interactions present simultaneously can folded conformations be stable.

The major contributions to the stability of the folded state are the hydrogen bonds between polar groups and the van der Waals interactions between nonpolar atoms that occur within the protein. Hydrogen bond interactions also occur between the unfolded polypeptide chain and the solvent, but those within the folded protein are intramolecular and, consequently, usually more favorable energetically. The complexity and cooperativity of these interactions, however, have thus far precluded a rigorous accounting for protein stability.

III. INTERACTIONS WITH OTHER MOLECULES

The functional properties of proteins become apparent only when they interact with other molecules.

A. Binding of Specific Ligands

1. Binding Sites

Every protein has a binding site for at least one other molecule, or *ligand,* with which it interacts during its function. Such binding sites are sufficiently specific that the protein can distinguish between the appropriate ligand and all other molecules that it is likely to encounter. For most ligands, the binding site is a specific, relatively small area on the surface of the protein that is complementary, sterically and physically, to the ligand. Such interactions have been described here already in terms of quaternary structure (see Section II,C). For very small ligands, such as oxygen atoms and electrons, which would be difficult to recognize specifically on such a basis, additional groups (prosthetic groups) are incorporated into the protein to add to the specificity. For example, heme groups are used for binding electrons and oxygen molecules in cytochromes and hemoglobins, respectively. The protein first binds the appropriate prosthetic group, and together they can recognize specifically the correct small ligand. Nevertheless, many proteins are able to distinguish between many small ligands without prosthetic groups. For example, calcium-binding proteins can bind Ca^{2+} selectively in the presence of 1000-fold higher concentrations of Mg^{2+}, a very similar ion.

Specific binding is accomplished by having within the binding site a constellation of several protein groups of intrinsic affinity for the ligand, arranged in the appropriate stereochemistry so that the ligand interacts with all the groups simultaneously. The binding sites on a protein usually preexist, formed by the folding of the protein. Binding of the ligand usually has only minor effects on the structure of the protein binding site. Exceptions usually are of functional importance.

Some proteins have only a single binding site at which they bind a single ligand, whereas others have binding sites for two or three ligands. Only in very large, complex proteins are there likely to be a greater number of binding sites. Each polypeptide chain generally has no more than one binding site for any one ligand. In a multimeric protein, with n identical polypeptide chains, there should be n equivalent binding sites for each ligand. In some cases, however, the binding sites bridge the symmetry axes relating different polypeptide chains, so only a fraction of the n sites can be occupied simultaneously. [*See* DNA Binding Sites.]

2. Binding Affinities

The binding affinity is usually expressed as a *dissociation constant,* which is the concentration of free ligand at which the protein binding site is occupied half the time at equilibrium; greater binding requires higher ligand concentrations. The affinities of ligands for different binding sites vary enormously. Some may be so great that binding is essentially irreversible. At the other extreme are many examples of binding so weak as to be of questionable significance. When binding is of biological significance, the dissociation constant is generally similar to the concentration of the ligand likely to be encountered naturally.

B. Interactions between Binding Sites: Allostery

Many proteins bind multiple ligands, and very often the binding of one ligand does not affect significantly the binding of another at a distant binding site. If the binding sites overlap, of course, binding of one ligand will inhibit binding of the other. If the binding sites are adjacent, binding of one ligand can increase or decrease the affinity for the other by directly affecting its binding site.

Much biological regulation occurs by the binding of one ligand altering the binding of another at a totally separate site; such proteins are said to be *allosteric.* The structural basis of the interaction between binding sites is best understood in the protein hemoglobin, where O_2 binding by each of the four subunits, $\alpha_2\beta_2$, is cooperative, in that binding at each site increases the affinity of the other sites. In addition, other ligands bind at other sites and also affect the O_2 affinity.

The effects on O_2 affinity occur primarily because the four subunits of the hemoglobin tetramer can exist in two somewhat different quaternary structures. One, designated T for "tense," is characterized by low affinity for O_2; the other, designated R for "relaxed," has high affinity. T is the predominant structure in the absence of O_2, but binding of O_2 shifts the equilibrium toward the R state, and thereby increases the O_2 affinity of the remaining sites. Other ligands affect O_2 affinity by binding preferentially to either the R or T state.

By this simple mechanism, the binding of O_2 can be regulated very closely. Similar mechanisms probably occur with other allosteric proteins.

C. Immunoglobulins

Related proteins need not bind the same ligands, for changing only a few residues in a binding site can dramatically alter its specificity. Excellent examples are the immunoglobulins of the immune system, which can recognize virtually any molecules foreign to the body. Yet all immunoglobulins have remarkably similar three-dimensional structures. Their binding sites are composed of only 10 to 15 residues in six loops on their surface, out of a total of nearly 700 for an antibody molecule, so changing only these few residues can drastically alter the antigen specificity.

IV. CHEMICAL CATALYSIS BY ENZYMES

A. Enzyme Function

Enzymes are proteins that bind ligands but then cause them to undergo a specific chemical reaction. In this case, the ligands are known as *substrates,* and the region of the binding site is known as the *active site.* Only a single reaction generally occurs on any enzyme, even though its substrate in solution in the absence of the enzyme might undergo a number of different reactions at similar rates. An enzyme generally causes unimolecular reactions to occur at rates some 10^6 to 10^{14} times faster than would occur under the same conditions in the absence of the enzyme. The enzyme is not altered by the reaction and is only a catalyst. Therefore, the enzyme does not alter the normal equilibrium constant for the reaction and must also increase the rate of the reverse reaction to exactly the same extent under the same conditions. The products of the reaction can also be used as substrates in the reverse reaction, if the equilibrium permits.

A characteristic of enzyme-catalyzed reactions is that the rate of the reaction is proportional to the concentration of substrate only at low concentrations. With increasing concentration, the rate plateaus and approaches a maximum value (V_{max}) at high substrate concentrations. This kinetic behavior is a consequence of the enzyme binding the substrate molecules at a finite number of active sites. Maximal activity is obtained when the enzyme is saturated with substrate. The substrate concentration at which the rate is half-maximal, the Michaelis constant or K_m, is a function of both the affinity of the substrate for the enzyme and the rate of the reaction on the enzyme.

Enzymes increase the rates of chemical reactions in a variety of ways. The most fundamental is that the enzyme does not have optimal affinity for either the substrate or product, but for the *transition state* for the reaction. The transition state for any chemical reaction is that species along the reaction pathway that is least stable, with the highest free energy. The rate of the reaction is determined by the free energy of the transition state, relative to that of the substrate; the lower the free energy of the transition state, the more rapid the reaction. By binding the transition state for a particular reaction most tightly, an enzyme lowers its free energy relative to the substrate and product and thereby increases the rate of the reaction on the enzyme. Enzymes catalyze specific reactions by binding tightly only specific substrates and transition states. In contrast, immunoglobulins are simply complementary to their ligands and do not catalyze reactions. But if the antibody is raised against a transition state analog, that antibody will catalyze the appropriate reaction.

Enzymes are especially effective in catalyzing reactions between two or more reactants, because they can be bound simultaneously on the enzyme, at neighboring positions with the active site. The reaction between multiple substrates on the enzyme is effectively unimolecular, whereas the same reaction in the absence of the enzyme would require that the multiple reactants encounter each other simultaneously in solution, which is very improbable at low concentrations. The effective concentration between reactants bound to a protein can be very high. In some reactions, in which a chemical group is transferred from one substrate to another, the enzyme can be used as an intermediary, accepting temporarily the group from one substrate and subsequently transferring it to the second substrate.

Although no enzymatic reaction is understood entirely, the basic principles of enzyme catalysis appear to be relatively straightforward. The importance of enzymes for life cannot be overstated, because they cause the chemical reactions of metabolism to occur rapidly under physiological conditions at 37°C. [*See* Enzymes, Coenzymes, and the Control of Cellular Chemical Reactions.]

B. Regulation of Enzyme Activity

Enzymes do not function in isolation, and it is crucial that each enzyme be an integral part of the chemical reaction pathways that take place. Its substrate must be available at the appropriate time and concentration, and its reaction must be catalyzed only when there is a need for its product. Accordingly, the activities of enzymes are linked together in a complex network of chemical reactions, by which metabolism occurs, and they must be regulated.

One common method of integrating the activities of enzymes is to make them responsive to the availability of substances other than just their substrates. This is accomplished by having a binding site for the regulatory metabolites on the appropriate enzymes. Just as binding of a ligand at one site can affect binding at another by allostery (see Section III,B), so can catalysis at the active site be affected. Many enzymes are allosteric, and this is a common way of controlling the activity of an enzyme. For example, the end product of a biosynthetic pathway usually inhibits the first enzyme of that pathway, a phenomenon known as *feedback inhibition*. In this way, the end product is produced by the enzymes of the pathway only when its level falls and biosynthesis is required.

Another very common method of regulating the activity of an enzyme is to alter its functional properties by modifying it covalently, most often by phosphorylation of Ser, Thr, or Tyr residues. A variety of *protein kinases* that catalyze such reactions are known, with varying specificities for the proteins that they phosphorylate. It is important that such covalent modifications be reversible, so there are a variety of *protein phosphatases* that remove the phosphoryl groups from the target enzymes. The activities of the protein kinases and phosphatases are usually regulated by various physiological stimuli, such as hormones and growth factors. Often there are complex cascades of phosphorylation of several kinases and phosphatases, so there can be a wide range of such control, and very small signals can be amplified to have substantial physiological effects. [*See* Protein Phosphorylation.]

C. Energy Transduction

Proteins catalyze many processes other than chemical reactions in solution; in particular, they can convert one form of energy to another. The ubiquitous adenosine triphosphate (ATP) is the usual currency of chemical energy and is involved in most energy interconversions, being either utilized or generated. For example, the chemical energy of ATP is converted to mechanical energy by moving one protein assembly (myosin thick filaments) relative to another (actin thin filaments) in muscle. This movement of one protein assembly relative to another is caused by a transient interaction between them, a conformational change in one (myosin) to move it relative to the other, followed by dissociation of the two and reassociation at a different point. In the process, the two proteins have moved relative to each other by about 100 Å (10 nm). Muscle contraction is simply the result of numerous such cycles, in each of which one filament slides past the other within the muscle.

The chemical energy for the movement comes from the hydrolysis of ATP, one molecule of which is consumed per myosin unit in each cycle. The details of the crucial conformational change in myosin are not known, but the complete cycle requires that the affinities of the myosin and actin for each other be closely controlled.

In general, transduction of the chemical energy of ATP requires only that there be two different forms of the enzyme (E and E*) that can be interconverted only by hydrolyzing ATP to ADP (adenosine diphosphate) and P_i (inorganic phosphate) or by carrying out the second reaction (A → B) to which the first is to be coupled:

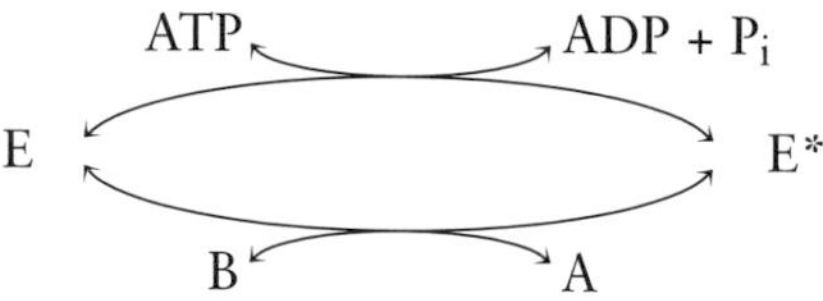

E* can differ from E only in having the phosphate group covalently attached to it. So long as E and E* can be interconverted only by these two reactions, the chemical energy of ATP hydrolysis will drive the reaction A → B. That reaction can be anything capable of being carried out by a protein: mechanical movement by a conformational change, different affinities for ligands (to pump molecules across membranes against a concentration gradient), or emission or absorption of light.

Pumping ions and other metabolites across membranes, against concentration gradients, is accomplished by protein pumps embedded in the membrane, with a suitable channel for permitting passage of the appropriate small molecule through the membrane. In the best-understood ion pump, that for pumping Na^+ and K^+ ions, the chemical energy of ATP is used to phosphorylate the protein. Both phosphorylated and nonphosphorylated forms of the protein must exist in two conformational states, in which the channel is open to opposite sides of the membrane. By permitting only certain interconversions of the various forms of the protein pump, the chemical energy of ATP is used to transport Na^+ and K^+ ions against concentration gradients.

Light detection in the eye occurs upon its absorption by the pigment *retinal,* attached to the protein *opsin;* the complex is known as *rhodopsin.* Absorption of light causes the retinal to alter its shape, which affects the structure of rhodopsin. The rhodopsin then binds to another protein, transducin, and alters its affinity for its ligands. By a complex cascade of subsequent events, the absorption of a single photon of light by rhodopsin causes the closure of hundreds of Na^+ and K^+ channels in the membrane of the rod cell of the eye, which produces an electrochemical signal that is transmitted to other neurons of the retina.

V. DEGRADATION

Proteins have only limited life spans, which vary markedly from one protein to another. Their ultimate fate is to be degraded to the constituent amino acids, and there is extensive turnover of proteins within most cells. Cells need mechanisms for surviving temporary conditions of starvation, by degrading "luxury" proteins in order to utilize their amino acids for biosynthesis of those more essential for survival. Maturation of reticulocytes into oxygen-carrying erythrocytes of the blood is accompanied by the selective degradation of mitochondria, ribosomes, and many proteins no longer required once the cell has synthesized its complement of hemoglobin. There is also a need for degrading abnormal proteins, either those that have succumbed to old age by chemical modifications or those that were synthesized in an incorrect form because of mistakes in the biosynthetic process. Abnormal proteins are readily recognized by most cells and are rapidly degraded. Degradation of some proteins is also necessary to permit regulation of their function. A major control over protein function in humans is at the stage of protein biosynthesis, and the level of any protein can be decreased only if there is a means by which it is degraded.

The proteins that turn over most rapidly are often those that catalyze rate-determining metabolic reactions. For example, the most rapidly degraded protein in the liver is ornithine decarboxylase, with a half-life of only 11 minutes. It catalyzes an important step in polyamine biosynthesis and is not subject to any known regulation by allosteric control or by covalent modification. Instead, its rate of biosynthesis can vary more than a 1000-fold, so decreases in its activity can occur only if there is a mechanism for degrading the protein. The eukaryotic cell cycle is controlled by the rapid, but controlled, degradation of the regulatory protein cyclin.

In contrast, many proteins are long-lived, especially structural proteins. Proteins of the eye lens are not degraded, so those synthesized in the embryo are still present in the human eye lens 70 or more years later. During that time, chemical modifications of the proteins are inevitable and result in pigmentation, decreased solubility, and formation of cataracts. Other proteins exist for the life of the cell they are in, such as hemoglobin, which is stable throughout the 3-month life span of an erythrocyte. At the end of that time, the entire erythrocyte is degraded.

The factors that determine the rate of degradation of different proteins are still being sought, as is elucidation of the mechanism. Degradation of individual molecules of a protein is random, and newly synthesized molecules are just as likely to be degraded as old, unless chemically modified significantly. A single event is also sufficient to cause degradation. In some systems, covalent attachment of another protein, ubiquitin, is believed to be the event that marks a protein for degradation; the identity of the N-terminal residue of the polypeptide chain is important for determining its rate of ubiquitination.

Many enzymes are known that degrade proteins; they are known as *proteases*. They have varying degrees of specificity, usually for the amino acid sequence. The cytosol of human cells contains a very large complex, the proteasome, which degrades misfolded or ubiquitinated proteins. Such degradation is regulated and ATP-dependent, which distinguishes it from most other proteolytic systems, and the proteasome is believed to be the catalyst of most regulatory protein degradation.

BIBLIOGRAPHY

Creighton, T. E. (1993). "Proteins: Structures and Molecular Properties," 2nd Ed. Freeman, New York.

Fersht, A. (1985). "Enzyme Structure and Mechanism." Freeman, New York.

Kyte, J. (1995). "Structure in Protein Chemistry." Garland, New York.

Lesk, A. (1991). "Protein Architecture: A Practical Approach." IRL Press, Oxford, England.

Lodish, H., Baltimore, D., Berk, A., Zipursky, S. L., Matsudaira, P., and Darnell, J. (1995). "Molecular Cell Biology," 3rd Ed. Scientific American Books, New York.

Perutz, M. F. (1991). "Protein Structure and Function." Freeman, New York.

Protein Targeting, Basic Concepts

GUNNAR VON HEIJNE
Stockholm University

GLOSSARY

Endocytosis Mechanism by which particles and molecules from the extracellular medium are internalized by the cell

Endosome Vesicular structure that serves as a relay station for protein transport to lysosomes

Golgi apparatus Vesicular structure that receives proteins from the endoplasmic reticulum and moves them toward lysosomes or the plasma membrane

Lysosome Vesicular structure that contains hydrolytic enzymes responsible for the digestion of macromolecules and other endocytosed materials

Mitochondrion Vesicular structure that contains the enzymes of the citric acid cycle and the respiratory chain where fatty acids and pyruvate are oxidized to carbon dioxide and energy-containing ATP

Peroxisome Vesicular structure that contains enzymes to carry out various oxidative reactions; also contains catalase, an enzyme that converts potentially toxic hydrogen peroxide to water and free oxygen

Plasma membrane Outer membrane of a eukaryotic cell

THE EUKARYOTIC CELL CONTAINS A NUMBER OF distinct membrane-delimited compartments or organelles (e.g., the nucleus, the endoplasmic reticular network, the Golgi apparatus, lysosomes, mitochondria, and peroxisomes). These organelles import most or all of their constituent proteins from the cytosol. This cell can also secrete proteins to the surrounding medium, constitutively as well as in a regulated fashion. In addition to soluble proteins, integral membrane proteins can also be directed to their correct locations, be it the plasma membrane or one of the organellar membranes. These sorting processes depend on specific targeting sequences present in the proteins and on transport machineries in the target membranes.

I. THE SORTING PROBLEM

Except for a handful of proteins encoded in mitochondrial DNA and made inside this organelle, all proteins of the human cell are synthesized in the cytosol. Nevertheless, many of these proteins have to be transported to other locations inside or outside the cell to perform their function (Fig. 1). [*See* Proteins.]

Histones and other DNA-binding proteins (e.g., transcription factors, DNA and RNA polymerases, and splicing enzymes) must go into the nucleus. Most of the polypeptides involved in respiration must find their way to mitochondria, as must the polymerases and ribosomal proteins required for organelle-specific protein synthesis and DNA replication. [*See* Histones.]

The secretory pathway starts in the endoplasmic reticulum (ER), continues through the Golgi stacks (where a subsidiary route leading to the lysosomes branches off), and ends at the plasma membrane. Peroxisomal proteins are imported directly from the cytosol into the organelle. Proteins and other molecules can also be ingested by the eukaryotic cell through endocytosis, a pathway that leads from the plasma membrane via endosomes to the lysosomes.

Over the past 10–15 years our knowledge of the basic mechanisms responsible for these phenomena has grown substantially, mainly as a result of rapid

ENCYCLOPEDIA OF HUMAN BIOLOGY, Second Edition, VOLUME 7. Copyright © 1997 by Academic Press. All rights of reproduction in any form reserved.

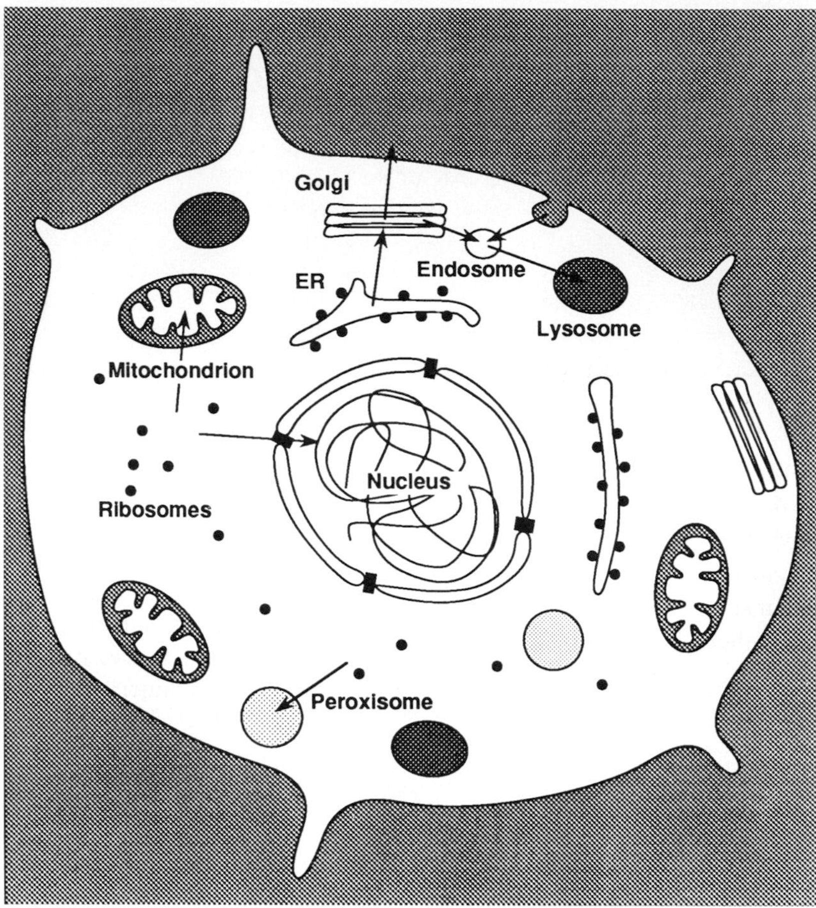

FIGURE 1 Protein-sorting pathways in a eukaryotic cell. ER, endoplasmic reticulum.

developments in recombinant DNA technology, *in vitro* reconstitution of organellar protein import and subcellular vesicular transport, and genetic and biochemical analyses of the import apparatuses. It has been discovered that a sharply delimited segment(s) of a polypeptide chain can serve as a targeting signal, routing the molecule to the appropriate organelle, where membrane-bound protein complexes recognize the targeting signal and translocate the protein through one or more membranes to its final location. The basic designs of many of the classes of targeting signals are also known.

II. THE SECRETORY PATHWAY

The secretory pathway is arguably the most important and the most complicated sorting pathway in the eukaryotic cell. During its passage from the ER to the plasma membrane, a protein travels through many organelles with different internal environments and is subject to the actions of a number of modifying enzymes. Among the most common posttranslational modifications are the removal of the targeting, or signal, peptide, the addition and trimming of oligosaccharides, and the addition of lipid moieties to the protein chain.

At the level of the Golgi apparatus, some proteins are shunted from the main pathway to the lysosomes. Of those that continue along the main branch, some are stored in secretory granules that release their contents only in response to an extracellular signal, whereas others are secreted directly into the surrounding medium in a constitutive fashion. [*See* Golgi Apparatus.]

The transit from one compartment to the next (i.e., from the ER to the cis Golgi, from one Golgi compartment to the next, and from the trans Golgi network to the plasma membrane) takes place in small "shuttle" vesicles that bud off from one membrane only to fuse with the next one in the series. Cytosolic proteins are apparently involved in both budding and fusion processes, thus directing the shuttle vesicles to their target membranes.

A. Events at the Endoplasmic Reticulum

1. Signal Peptides and the Signal Recognition Particle Cycle

The steps involved in the initial targeting to and translocation across the ER membrane are shown in Fig. 2. The amino-terminal 15–20 residues of a secretory protein contain the targeting signal. As soon as this signal peptide (SP) emerges from the ribosome, it is recognized by the so-called signal recognition particle (SRP), a complex consisting of six protein chains and one RNA molecule. The SRP binds to the ribosome, and the whole ribosome–SRP complex then binds to the SRP receptor, an integral component of the ER membrane. The interaction with the SRP receptor breaks the ribosome–SRP association. The nascent chain is then translocated across the membrane through an aqueous channel composed of the Sec61/TRAP/TRAM protein complex. Shortly after translocation has been initiated, the signal peptide is removed by the action of signal peptidase, another integral membrane protein. Finally, the completed chain falls off the ribosome and folds into its native structure in the lumen of the ER in a process catalyzed by one or more chaperones such as BiP, a protein belonging to the hsp70 family.

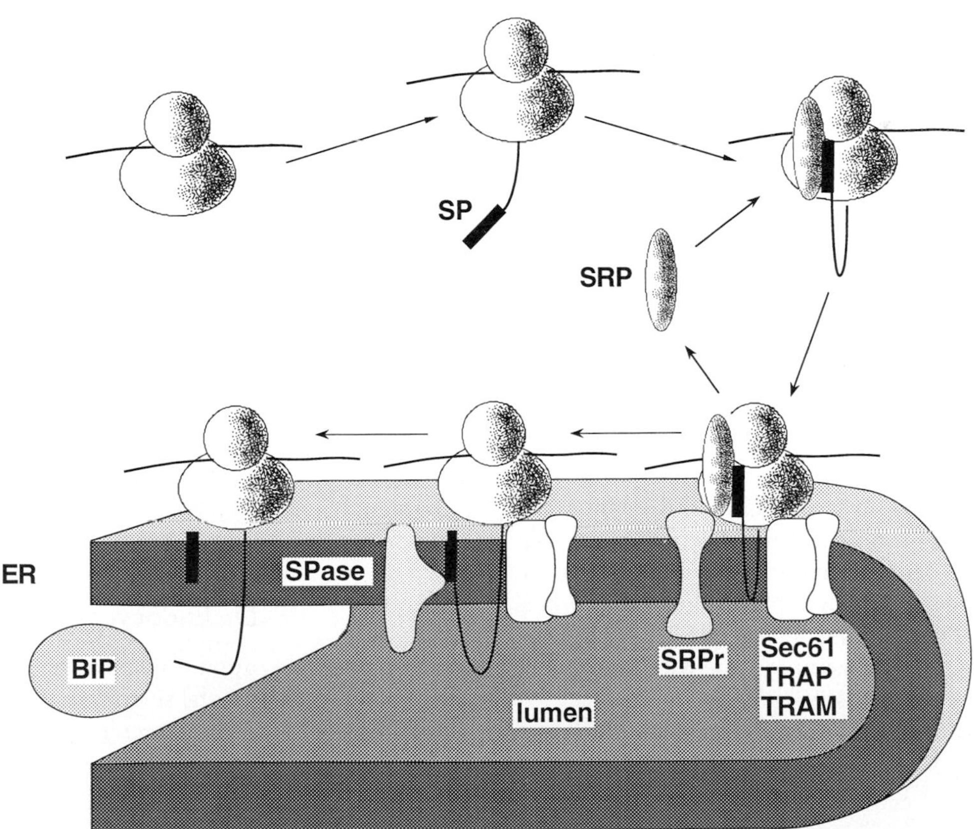

FIGURE 2 The signal recognition particle cycle. As the ribosome starts translating the mRNA, an amino-terminal signal peptide (SP) on the nascent chain emerges. The signal recognition particle (SRP) binds to the SP and the ribosome. This complex is recognized by the SRP receptor (SRPr) in the endoplasmic reticulum (ER). The SRP recycles and the nascent chain is translocated through the translocase complex (composed of the Sec61, TRAP, and TRAM proteins) into the lumen of the ER, where it is bound by the chaperone BiP and ultimately folds into its native conformation. The signal peptide is removed by the signal peptidase (SPase).

Amino acid sequences are known for many signal peptides from different organisms. They lack strong sequence homology, but they all conform to the same basic pattern (Fig. 3): a positively charged amino-terminal region, a central apolar region, and a carboxy-terminal, more polar, region, including the cleavage site.

Gene fusion experiments have shown that the presence of a signal peptide is necessary, but not always sufficient, for translocation into the ER. In addition to having an SP, it seems that the nascent chain must be prevented from folding into a stable tertiary structure prior to translocation, or else the process will be aborted. Recent data indicate that chaperones found in the cytosol help keep the nascent chain in an unfolded state.

The early steps of glycosylation of the secretory proteins also take place in the ER. Core oligosaccharides are added to acceptor sites in the nascent chain during or immediately after translocation into the luminal compartment.

2. Membrane Protein Assembly

Proteins that span a membrane at least once are called integral membrane proteins. Most integral membrane proteins use the secretory machinery described earlier in the initial phases of the membrane insertion process, but additional elements in their sequences serve as stop–transfer signals that prevent the more distal parts of the chain from being translocated. This results in a transmembrane topology.

The four classes of integral membrane proteins are shown in Fig. 4. Class I proteins are made with a standard cleavable SP and have a stretch of 15–25 apolar amino acids that constitutes the stop–transfer signal (ST) and ends up spanning the ER membrane, with its carboxyl-terminal end facing the cytosol. Class II proteins can be thought of as originating from secreted proteins, in which the cleavage of the SP has been blocked. The resulting uncleaved SP thus anchors a class II protein in the membrane, leaving the amino terminus facing the cytosol.

Class III proteins also lack a cleavable SP. They have a long apolar region at their amino terminus, but they do not have the typical cluster of positive charges found on the amino terminus of normal signal peptides. This allows them to insert into the membrane in the orientation opposite that of the class II proteins.

Class IV proteins, finally, have multiple apolar regions and span the membrane many times. Many transport proteins and ion pumps belong to this group.

B. Golgi Apparatus

The Golgi stacks serve a number of important functions in the secretory pathway: they contain enzymes involved in the trimming and modification of the oligosaccharide chains attached to secretory and membrane proteins in the ER, they sort apical from basolateral proteins (see Section II,E), and they shunt lysosomal proteins away from the main pathway.

Lysosomal targeting is signaled by a specific oligosaccharide modification: mannose 6-phosphate. A receptor with affinity for this moiety is located in the Golgi apparatus and shuttles appropriately modified proteins to endosomes and, ultimately, to lysosomes.

C. Constitutive versus Regulated Secretion

Many cells store some of their secretory proteins in dense secretory granules. When stimulated by the appropriate signal, these granules fuse with the plasma membrane and release their content into the medium. Other proteins are secreted directly from the trans Golgi network in an unregulated constitutive fashion. The signals for these two alternative pathways of secretion are not known.

D. Endocytosis

Endocytosis, the uptake of particles, nutrients, and other molecules from the surrounding medium, is intimately coupled with the secretory pathway. Solutes

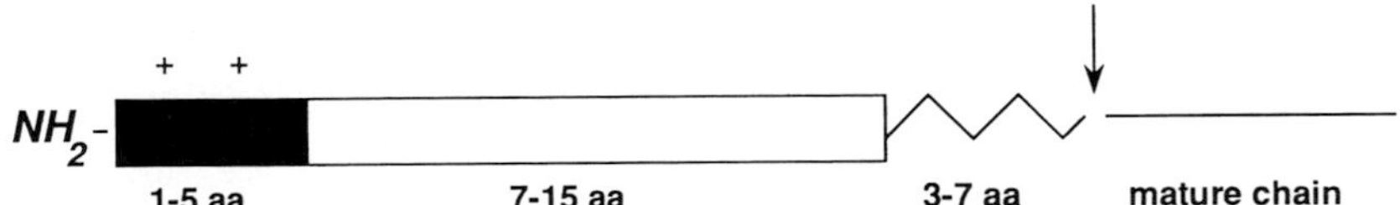

FIGURE 3 A secretory signal peptide. Typical lengths [i.e., the numbers of amino acids (aa)] of the three regions that make up the signal peptide are indicated.

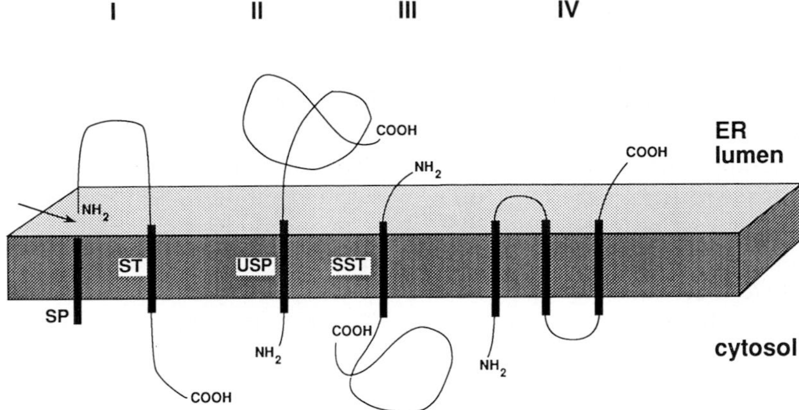

FIGURE 4 Class I membrane proteins are made with a cleavable signal peptide (SP) and an apolar stop–transfer sequence (ST). Class II proteins are anchored by an uncleaved SP (USP). Class III proteins are anchored by an uncleaved amino-terminal start–stop translocation sequence (SST). Class IV proteins have multiple membrane-spanning regions and might be oriented with their amino and carboxy termini on either side of the membrane.

can be internalized nonspecifically through bulk-phase uptake, but many peptides and other compounds are first bound to specific receptors on the cell surface. These receptors are transmembrane proteins with both extracellular and intracellular domains. After binding the ligand they cluster in invaginations of the plasma membrane, called coated pits. The coated pits pinch off from the membrane to form coated vesicles, which in turn can fuse with endosomal vesicles. In the low pH milieu of the endosome the ligand dissociates from the receptor, and the latter in many cases recycles to the plasma membrane.

Many enveloped viruses also enter the cell via this route. In the acidic environment of the endosome, proteins in the virus envelope undergo a conformational change that exposes fusinogenic regions, and the viral membrane fuses with the endosomal membrane, expelling the viral genome into the cytosol.

E. Sorting in Polarized Cells

Many cells are polarized in the sense of having two sides, with distinct cell-surface properties. Thus, epithelial cells have an apical surface facing the external milieu and a basolateral surface facing the internal milieu of the organism. Many soluble proteins are known to be secreted preferentially from one side or the other, and many integral membrane proteins also display preferential sorting. It is not known what the signals for this kind of intracellular sorting are, but the two pathways are thought to diverge in the trans Golgi network or shortly after passage through the Golgi compartment. [*See* Polarized Epithelial Cell.]

III. IMPORT INTO THE NUCLEUS

The nucleus is surrounded by a double membrane studded with nuclear pores that allow the diffusion of small molecules of less than about 60,000 Da. Thus, small proteins can enter and exit the nucleus by passive diffusion, and their distribution between the nucleus and the cytosol depends on whether they bind to other molecules specifically localized to either compartment.

Larger proteins can be imported into the nucleus only by active transport through the nuclear pores. The details of this mechanism have not been elucidated, but it has been shown for a number of nuclear proteins that the signals for nuclear localization typically consist of short stretches of positively charged amino acids. Unlike most other targeting signals, the nuclear localization signals are not cleaved after import, which could be related to the fact that nuclear proteins have to reenter the newly formed nuclei after mitosis. Nuclear localization signals also seem to work cooperatively, such that multiple signals enhance the import efficiency when present in the same protein. [*See* Nuclear Pore, Structure and Function.]

IV. IMPORT INTO MITOCHONDRIA

Mitochondria are surrounded by two membranes: outer and inner. Proteins are imported into the matrix compartment through contact sites, where the two membranes are in close apposition (Fig. 5). Many precursor proteins are made as full-length molecules in the cytosol and are imported posttranslationally; others may be imported cotranslationally.

A number of distinct steps on the import pathway have been identified. First, the precursor binds to a receptor on the mitochondrial surface. The amino-terminal part of the precursor, including the targeting peptide, is then translocated across a contact site in a process that requires the chain to be in a more or less unfolded conformation. This initial translocation step also requires an electrical transmembrane potential across the inner membrane. When the amino terminus has penetrated into the matrix space, the remaining parts of the molecule can enter the organelle, even in the absence of a membrane potential. Finally, the targeting peptide is removed by soluble proteases located in the matrix.

Cytosolic chaperones are thought to be important for maintaining the precursor in an unfolded conformation, whereas chaperones located in the mitochondrial matrix may pull the nascent chain into the organelle and catalyze its subsequent folding.

Mitochondrial targeting peptides are rich in positively charged residues, arginine in particular, and lack negatively charged acidic residues. They can fold into so-called amphiphilic α helices (Fig. 6), with one charged and one apolar face. Such structures bind tightly to the surface of lipid bilayers, and it is conceivable that a direct protein–lipid interaction of this kind is an important step on the import pathway.

Proteins targeted to the intermembrane space have composite targeting signals with an amino-terminal matrix-targeting part, followed by a stretch with many of the properties normally found in secretory SPs. The SP-like part may function either as a stop–transfer sequence to interrupt translocation

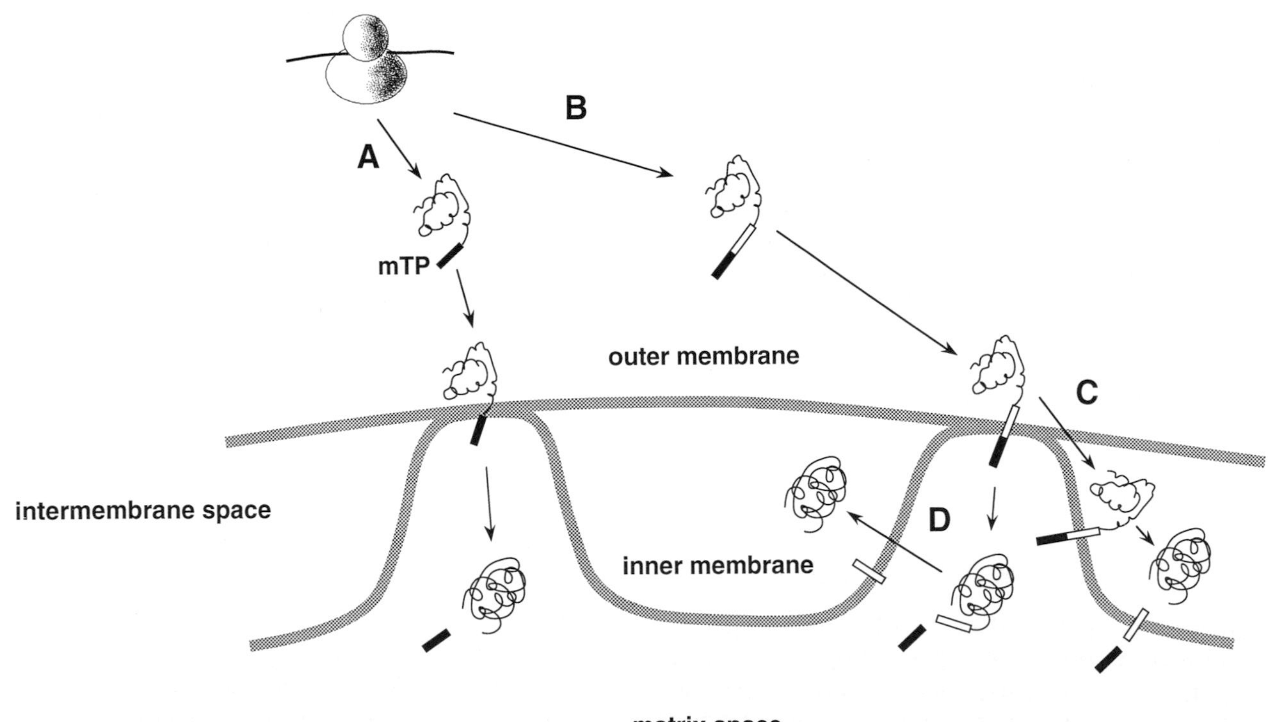

FIGURE 5 Mitochondrial protein import. (A) The unfolded precursor protein, with its mitochondrial targeting peptide (mTP), binds to a receptor in the outer membrane and is imported through a contact site where the outer and inner membrane are in close juxtaposition. (B) Proteins destined for the intermembrane space are made with a composite targeting peptide and are either imported only through the outer membrane ("stop–transfer" model or (C) first imported into the matrix, only to be immediately reexported back through the inner membrane ("conservative sorting" model, D).

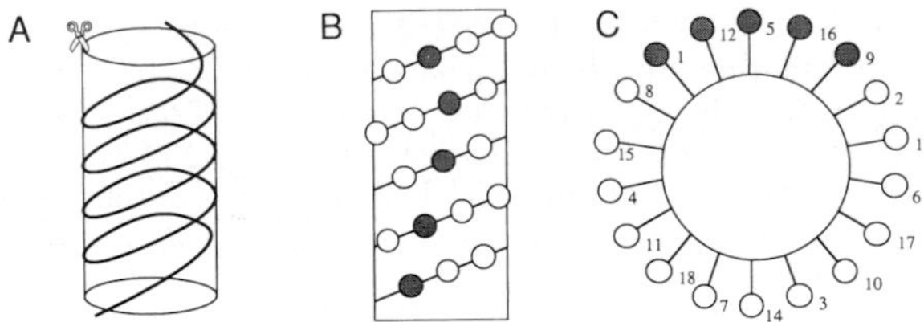

FIGURE 6 A typical mitochondrial matrix-targetig peptide, the amphiphilic helical structure of which is shown in three representations. (A) The backbone of an idealized α helix. (B) A helical net diagram in which the cylindrical helix has been cut open and folded flat. The positions of the side chains are indicated. (C) A helical wheel diagram looking down the axis of the helix. Numbering refers to the position in the linear sequence. Note that the positively charged residues (shaded areas) line up on one face of the helix in B and in C.

across the inner membrane or as a reexport signal if the protein is first fully imported all the way into the matrix. The matrix-targeting signal is cleaved upon entry into the matrix, exposing the SP-like structure that guides the protein back through the inner membrane.

V. IMPORT INTO PEROXISOMES

Peroxisomes are small organelles that contain enzymes involved in various oxidative reactions. Protein import into peroxisomes is posttranslational and requires ATP. Gene fusion studies indicate that both amino- and carboxy-terminal regions in the nascent protein may be active in the import process, and a highly conserved carboxy-terminal targeting signal with the sequence Ser–Lys–Leu–COOH is found in many, but not all, of the peroxisomal proteins.

VI. EVOLUTION OF SORTING PATHWAYS

In bacteria, SPs direct proteins into or across the inner membrane. The SPs, as well as the mechanisms of targeting and translocation, are similar to what are found in the secretory pathway of eukaryotic cells. Indeed, bacterial proteins can often be secreted when made in eukaryotic cells and vice versa. Thus, it seems that protein secretion is a highly conserved process in evolution.

Sorting to the intermembrane space of mitochondria provides a nice example of this evolutionary conservation. Mitochondria, being descendants of bacteria, seem to have retained the capacity to translocate proteins from the matrix across their inner membrane into the intermembrane space in response to an SP-like structure. Since most of the genes originally present in the genome of the protomitochondrion have been transferred to the nuclear genome during evolution, the specific import pathway (i.e., cytosol to matrix) must have been added at an early stage. Intermembrane space proteins that presumably were first made inside the mitochondrion with a secretory SP to ensure correct targeting are now made in the cytosol with a matrix-targeting signal, followed by a secretion signal. Once inside the organelle, the ancient secretory pathway is engaged, and the protein is routed to its final location.

The early eukaryotic cell must also have faced the general problem of mistargeting of its proteins. This could have been avoided in at least two ways: by increasing the specificity of targeting and by preferential proteolytic degradation of missorted proteins. So far, specific targeting, rather than specific degradation, seems to be the rule, although quantitative data are lacking.

The problem of mistargeting is intimately connected with the origin of the targeting signals. Targeting to both the secretory pathway and mitochondria involves sorting signals based on structures with one positively charged and one apolar part, and both kinds of peptides are known to interact strongly, though in different ways, with lipid bilayers. Thus, we might look for the origins of protein sorting in the simple physical chemistry of protein–lipid interactions.

BIBLIOGRAPHY

Gething, M.-J., and Sambrook, J. (1992). Protein folding in the cell. *Nature* **355**, 33–45.

Gilmore, R. (1993). Protein translocation across the endoplasmic reticulum—A tunnel with toll booths at entry and exit. *Cell* **75**, 589–592.

Glick, B. S., Beasley, E. M., and Schatz, G. (1992). Protein sorting in mitochondria. *Trends Biochem. Sci.* **17**, 453–459.

Izard, J. W., and Kendall, D. A. (1994). Signal peptides: Exquisitely designed transport promoters. *Mol. Microbiol.* **13**, 765–773.

Jungnickel, B., Rapoport, T. A., and Hartmann, E. (1994). Protein translocation: Common themes from bacteria to man. *FEBS Lett.* **346**, 73–77.

Kornfeld, S., and Mellman, I. (1989). The biogenesis of lysosomes. *Annu. Rev. Cell Biol.* **5**, 483–525.

Pfanner, N., and Neupert, W. (1990). The mitochondrial protein import apparatus. *Annu. Rev. Biochem.* **59**, 331–353.

Rothman, J. E. (1994). Mechanism of intracellular protein transport. *Nature* **372,** 55–63.

Schwarz, E., and Neupert, W. (1994). Mitochondrial protein import: Mechanisms, components and energetics. *Biochim. Biophys. Acta* **1187,** 270–274.

Silver, P. A. (1991). How proteins enter the nucleus. *Cell* **64,** 489–497.

von Heijne, G. (1994). Membrane proteins: From sequence to structure. *Anu. Rev. Biophys. Biomol. Struct.* **23,** 167–192.

Wachter, C., Schatz, G., and Glick, B. S. (1994). Protein import into mitocondria: The requirement for external ATP is precursor-specific whereas intramitochondrial ATP is universally needed for translocation into the matrix. *Mol. Biol. Cell* **5,** 465–474.

Protein Targeting, Molecular Mechanisms

JACK A. VALENTIJN
Yale University School of Medicine

GLOSSARY

Molecular chaperones Regulatory proteins (e.g., BiP, Hsp70) that bind to preproteins and assist during various stages of the translocation process

Organelles Intracellular membrane compartments (e.g., endoplasmic reticulum, mitochondrion) in eukaryotic cells

Signal recognition particle Cytosolic receptor protein that binds to preproteins destined for translocation across the endoplasmic reticulum membrane

Signal sequence A stretch of amino acids in the peptide sequence of nascent or newly synthesized proteins that serves as a "zip code" for protein targeting; the size of a signal sequence can vary as much as from 3 to 70 amino acids

Topogenic sequence A sequence in proteins that encodes a transmembrane domain

Translocation Active (i.e., energy-dependent) transport of nascent or newly synthesized preproteins across organellar membranes

EXCEPT FOR A FEW MITOCHONDRIAL PROTEINS, eukaryotic cells synthesize all their proteins in the cytosol. Therefore, protein targeting mechanisms are required in order to deliver newly synthesized proteins to their proper organelles. Cells utilize several distinct strategies for the correct targeting of proteins. These strategies include translocation across organellar membranes via aqueous channels, sorting and retention within organelles, and posttranslational modifications such as the covalent addition of lipid anchors.

I. INTRODUCTION

Human cells, like all eukaryotic cells, are equipped with highly specialized membrane compartments called organelles, including the nucleus, peroxisomes, mitochondria, the endoplasmic reticulum (ER), and the Golgi apparatus. This compartmentalization allows cells to establish distinct intracellular microenvironments, each offering optimal conditions for a specific subset of metabolic reactions. For instance, lysosomes—organelles that degrade aged or defective macromolecules and membranes—are filled with degradative enzymes whose action is favored by an acidic pH maintained by proton pumps in the lysosomal membranes.

In order to supply their organelles with the correct proteins, cells utilize a number of different protein targeting strategies. Given the complexity of the cellular architecture and the large variety of proteins found in cells, it is remarkable how rarely proteins are mistargeted. Improper protein targeting does occur, however, in a number of hereditary diseases such as cystic fibrosis which is caused by a mutation in the CFTR gene encoding for a chloride channel protein. Because of this mutation, the chloride channel protein misfolds and remains stuck in the ER, and can therefore not insert into the plasma membrane to exert its normal function.

As illustrated in Fig. 1, four main targeting routes can be distinguished in eukaryotic cells: (1) targeting to various components of the secretory pathway, (2) mitochondrial targeting, (3) peroxisomal targeting,

ENCYCLOPEDIA OF HUMAN BIOLOGY, Second Edition, VOLUME 7. Copyright © 1997 by Academic Press. All rights of reproduction in any form reserved.

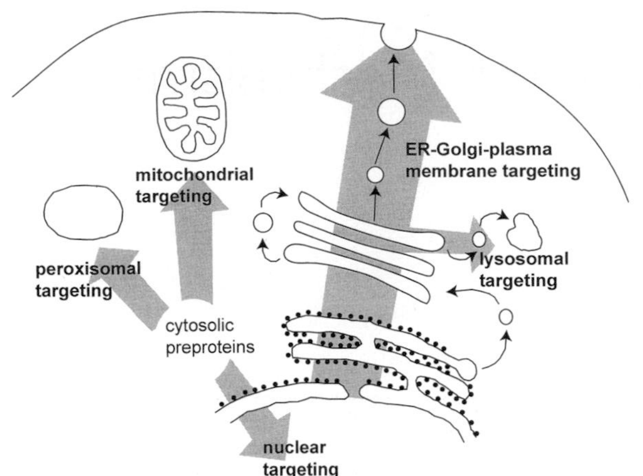

FIGURE 1 Schematic representation of the main protein targeting pathways in eukaryotic cells.

and (4) nuclear targeting. Signal sequences on nascent or newly synthesized proteins determine the preproteins' itineraries and final destinations. Some examples of signal sequences are given in Table I.

II. PROTEIN TARGETING TO THE SECRETORY PATHWAY

A. Translocation to the ER

Proteins that are destined to enter the secretory pathway usually start their itinerary by translocating across the ER membrane. In mammalian cells, this process appears to be mediated exclusively by the signal recognition particle (SRP)-dependent protein translocation pathway. This pathway is cotranslational, which means that proteins are translocated while they are still being synthesized, i.e., translated from mRNA, on ribosomes. SRP is a cytosolic protein that binds to a hydrophobic signal sequence—16 to 30 amino acids in size—on the amino terminus of a nascent protein, temporarily halts translation in the cytosol, and directs the ribosome from which the protein emerges toward the ER by binding to a SRP receptor in the ER membrane (Fig. 2). Both SRP and the SRP receptor are GTPases. Upon the binding of the SRP–polypeptide–ribosome complex to the SRP receptor, SRP is displaced and the ribosome is enabled to associate with a translocon, a transmembrane protein that forms an aqueous pore in the ER membrane allowing the nascent protein to move from the cytosolic to the lumenal side of the ER membrane. During the translocation process, the signal peptide of many nascent polypeptide chains binds to an ER membrane protein called translocating chain associating membrane protein (TRAM), whose function is still unclear. So-called topogenic sequences in nascent transmembrane proteins determine which parts of the protein remain inserted into the membrane and also the orientation of the protein. Stop-transfer sequences in transmembrane proteins cause the translocon–ribosome complex to disassemble during protein translation; as a result, the ribosome will continue protein translation in the cytosol and the remainder of the translated protein will protrude into the cytosol.

The SRP-mediated protein translocation is guided at various stages by molecular chaperones. One such chaperone, called the nascent polypeptide-associated complex, binds to nascent polypeptide chains in the cytosol in order to confer specificity to the interaction between the signal sequence and SRP. Another chaperone, called the binding protein (BiP), binds to short hydrophobic regions on the elongating polypeptide in the ER lumen, thereby protecting the nascent protein against denaturing, incorrect folding, and nonspecific aggregation. BiP is an ATPase that uses the energy produced by the hydrolysis of ATP to dissociate from the preprotein it was bound to. The dissociation of BiP from the preprotein will cause the polypeptide chain to fold correctly, in which case the hydrophobic-binding regions for BiP will be burried in the protein's interior. But if the protein is folded incorrectly, BiP-binding regions will be exposed on the surface of the protein and will be able to associate with BiP again.

TABLE I
Examples of Signal Sequences[a]

Signal sequence	Type of signal
-KDEL-COO^-	ER retention
-CAC-COO^-	Geranylgeranylation motif (membrane anchor)
-SKL-	Peroxisomal import
-PPKKKRKV-	Nuclear import
H_3^+N-MKWVTFLLLLFISGSAFSR-	ER import
H_3^+N-MLSLRQSIRFFKPATRTLCSSRYLL	Mitochondrial import

[a]Amino acid sequences of signal peptides are represented using the one letter code. The carboxy terminus of a protein is indicated by COO^-, and the amino terminus by H_3^+N.

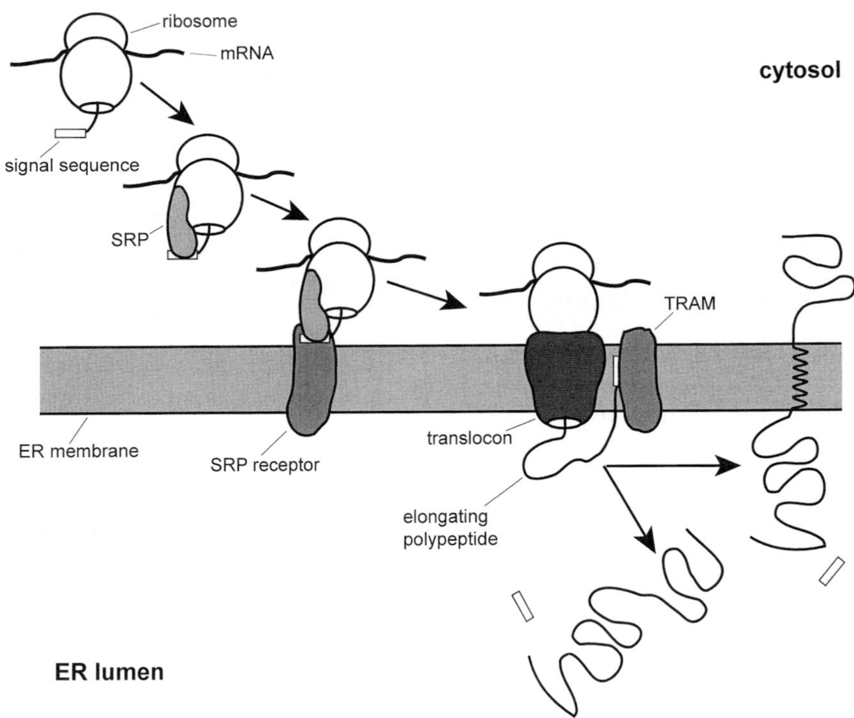

FIGURE 2 Model of protein translocation across the endoplasmic reticulum pathway. See text for explanation.

Subsequently, the misfolded proteins are degraded. Thus BiP also plays an important role in the quality control of protein biosynthesis. Many more chaperones are involved in ensuring proper translocation and conformation of nascent proteins. To name but a few, protein disulfide isomerase catalyzes the rearrangement of disulfide bonds in nascent proteins, whereas calnexin and calreticulin are involved in the folding of glycoproteins and the quality control thereof.

B. ER to Golgi Transport

Correctly folded and assembled proteins are transported to the Golgi apparatus in carrier vesicles that bud off from the ER membrane and fuse with the membrane of the *cis*-Golgi network. Even proteins that are part of the ER lumenal protein machinery are transported to the *cis*-Golgi cisternae, but are subsequently returned to the ER. This is enabled by a four amino acid sequence (KDEL) that many ER resident proteins display on their carboxy terminus and which is recognized by receptors present in the *cis*-Golgi membrane. Specialized carrier vesicles transport the proteins displaying the ER retention signal back to the ER.

C. Targeting to Lysosomes

Lysosomal enzymes (acid hydrolases) are phosphorylated in the Golgi on carbon atom 6 of one or more mannose residues in the glycosyl moiety that was added in the ER. One of the enzymes responsible for this phosphorylation, *N*-acetylglucosamine phosphotransferase, recognizes a specific signal sequence carried only by lysosomal enzymes. Once phosphorylated, the acid hydrolases can bind to mannose-6-phosphate receptors in the membrane of the *trans*-Golgi network. Special carrier vesicles containing the lysosomal enzymes still bound to mannose-6-phosphate receptors then bud from the *trans*-Golgi membrane and fuse with lysosomes. Because of the acidic pH in the lysosomes, the newly arrived hydrolases will dissociate from the mannose-6-phosphate receptors, and following removal of the phosphate(s), mature lysosomal enzymes are formed.

I-cell disease or mucolipidosis II is a lysosomal storage disease characterized by severe psychomotor retardation and skeletal deformities. It is caused by the mistargeting of several lysosomal enzymes due to their lack of mannose-6-phosphate. Instead of being targeted to lysosomes, these enzymes are secreted and found at high levels in the patients' blood and urine.

D. Targeting to the Plasmalemma

Secretory proteins and plasma membrane proteins are sorted and packed by the Golgi apparatus in secretory vesicles that bud off from the *trans*-Golgi cisternae and fuse with the plasma membrane (exocytosis). The sorting mechanisms employed are thought to be analogous to those used in the lysosomal targeting pathway. Two types of secretory vesicles can be distinguished: constitutive and regulated exocytotic vesicles. Constitutive secretion occurs in all cell types mainly to supply the plasma membrane with newly synthesized proteins and phospholipids. The regulated secretory pathway is used by cells that specialize in secreting hormones, digestive enzymes, or neurotransmitters in response to an appropriate signal such as the activation of membrane receptors coupled to intracellular signaling routes. In polarized cells, the constitutive and secretory vesicles target to distinct specialized membrane domains (basolateral and apical membrane, respectively) and supply these domains with different subsets of membrane-associated proteins. [*See* Membranes, Biological.]

III. PROTEIN TARGETING TO MITOCHONDRIA

Although mitochondria possess their own DNA, the vast majority of mitochondrial proteins (over 95%) are encoded by nuclear DNA. The precursors to these mitochondrial proteins are synthesized in the cytosol on free ribosomes and are subsequently translocated into the mitochondria. In contrast to ER translocation, transport across mitochondrial membranes occurs posttranslationally, i.e., after ribosomal translation is completed. Cytosolic chaperones belonging to the Hsp70 family of heat shock proteins (of which BiP is a member) bind to the newly synthesized mitochondrial precursor proteins to prevent them from folding, thus keeping them translocation competent.

Because mitochondria possess an outer membrane (OM) and an inner membrane (IM), proteins can be targeted to four different mitochondrial subcompartements: OM, IM, intermembrane space, and mitochondrial matrix. At the ultrastructural level, the mitochondrial OM and IM appear to be joined together at numerous focal areas. It is at these so-called contact sites that protein translocation is thought to occur. Many mitochondrial precursor proteins carry an amino-terminal signal sequence that binds to a specific receptor in the OM. The signal sequence is between 15 and 70 amino acids in size and its composition is highly variable between different mitochondrial preproteins. However, all signal sequences are hydrophilic and contain mostly positively charged residues. After binding of the signal sequence to its receptor, a translocation channel forms in both the OM and IM, allowing the preprotein to translocate directly into the mitochondrial matrix (Fig. 3). The driving force for the translocation process is provided by the membrane potential across the IM. In addition, evidence suggests that mitochondrial matrix proteins belonging to the Hsp70 family, which bind to precursor proteins, facilitate the import of preproteins by acting as "molecular motors" fueled by the energy released from ATP hydrolysis. These Hsp70 proteins, together with Hsp60 proteins, also play a role in the correct folding and assembly of mitochondrial preproteins. The preprotein signal sequences are cleaved off in the mitochondrial matrix by signal peptidases. The mechanisms of protein targeting to the mitochondrial OM, IM, and intermembrane space are still poorly understood, but may involve stop-transfer sequences (as encountered in ER translocation) and additional signal peptides that are exposed after cleavage of the primary signal sequence. [*See* Mitochondria.]

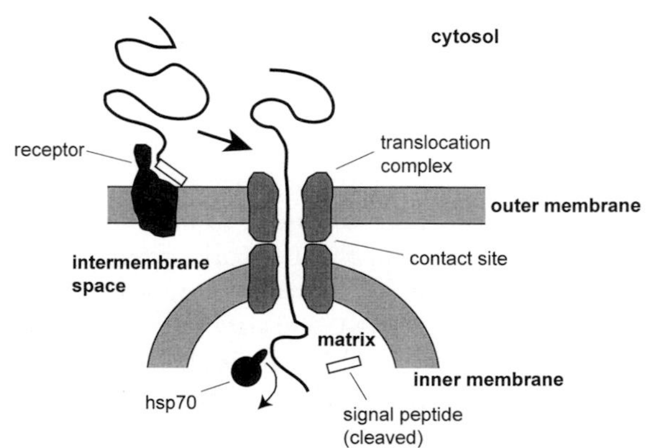

FIGURE 3 Model of protein translocation across the mitochondrial inner and outer membranes. See text for explanation.

IV. PROTEIN TARGETING TO PEROXISOMES

Most peroxisomal precursor proteins carry a noncleavable, carboxy-terminal signal sequence composed of only three amino acids, SKL, which is essential for the correct targeting to peroxisomes. A few proteins, however, including peroxisomal thiolase and sterol carrier protein 2, employ a cleavable, amino-terminal signal sequence. It follows that peroxisomes possess more than one type of signal sequence receptor. Although at present little is known about the peroxisomal translocation mechanism, it is assumed to be similar to that of the ER and mitochondria.

Several human diseases have been linked to defects in peroxisomal protein targeting. In particular, primary hyperoxaluria type 1 is caused by the mistargeting of alanine : glyoxylate aminotransferase (AGT), a peroxisomal enzyme responsible for the detoxification of glyoxylate, to mitochondria. This peroxisome-to-mitochondrion mistargeting is the consequence of only two single amino acid substitutions in the AGT precursor. First, the substitution of a leucine for a proline on position 11 generates a mitochondrial targeting sequence, and second, the substitution of an arginine for a glycine on position 170 inhibits peroxisomal import.

V. TARGETING OF NUCLEAR PROTEINS

Nuclear proteins, although encoded by DNA inside the nucleus, are synthesized in the cytosol and then transported into the nucleus through nuclear pores. These nuclear pores are very large protein complexes—they can be seen by electron microscopy—that perforate the double-membrane nuclear envelope to form an aqueous channel with a diameter of about 9 nm. Small water-soluble molecules can permeate these channels by passive diffusion, but the larger nuclear proteins are imported by an active transport mechanism.

To ensure their proper targeting, nuclear proteins carry one or more nuclear localization signals consisting of a sequence of four to eight amino acids that is rich in positively charged amino acids (lysine and arginine) and often contains proline. Cytosolic factors bind to the nuclear localization signals and direct the nuclear protein to a nuclear pore complex. After binding to a component of the nuclear pore complex, the nuclear protein is transported across the nuclear envelope through an ATP-dependent mechanism. During this translocation process, the nuclear pore can actually enlarge up to a diameter of 26 nm. [*See* Nuclear Pore, Structure and Function.]

VI. PROTEIN TARGETING BY MEANS OF POSTTRANSLATIONAL MODIFICATIONS

Numerous proteins insert into organelle or plasma membranes after the covalent addition of a lipid anchor, which can consist of isoprenoid, fatty acid, or glycosylphosphatidylinositol. The proteins that undergo these posttranslational modifications display specific sequences that are recognized by the enzymes responsible for adding the lipid anchor. For instance, ras GTPases and the γ subunit of heterotrimeric G proteins possess a carboxy-terminal CAAX motif in which the cysteine residue becomes isoprenylated after a series of enzymatic reactions. Thus, these motifs can be regarded as specific targeting signals.

Phosphorylation is a posttranslational modification that can dynamically control the subcellular localization of a protein. An interesting example is the small GTP-binding protein rab4, which inserts into membranes of early endosomes by means of a carboxy-terminal geranylgeranyl moiety added posttranslationally. Rab4 is thought to control a discrete step in endocytosis. During mitosis, endocytic activity is arrested, rab4 becomes phosphorylated by a cdc2 kinase, and, as a direct consequence, redistributes to the cytosol. After mitosis, rab4 is dephosphorylated and becomes endosome associated again.

BIBLIOGRAPHY

Gorelick, F. S., and Jamieson, J. D. (1994). The pancreatic acinar cell: Structure–function relationships. *In* "Physiology of the Gastrointestinal Tract" (L. R. Johnson, D. H. Alpers, E. D. Jacobson, J. Christensen, and J. H. Walsh, eds.), pp. 1353–1376. Raven Press, New York.

Hammond, C., and Helenius, A. (1995). Quality control in the secretory pathway. *Curr. Opin. Cell Biol.* **7**, 523–529.

Neupert, W. (1994). Transport of proteins across mitochondrial membranes. *Clin. Invest.* **72**, 251–261.

Palade, G. E. (1975). Intracellular aspects of the process of protein synthesis. *Science* **189**, 347–356.

Rapoport, T. A., Jungnickel, B., and Kutay, U. (1996). Protein transport across the eukaryotic endoplasmic reticulum and bacterial inner membranes. *Annu. Rev. Biochem.* **65**, 271–303.

Rassow, J., and Pfanner, N. (1996). Protein biogenesis: Chaperones for nascent polypeptides. *Curr. Biol.* **6**, 115–118.

Schatz, G., and Dobberstein, B. (1996). Common principles of protein translocation across membranes. *Science* **271**, 1519–1526.

Tager, J. M., Aerts, J. M. F. G., van den Bogert, C., and Wanders, R. J. A. (1994). Signals on proteins, intracellular targeting and inborn errors of organellar metabolism. *J. Inherit. Metab. Dis.* **17**, 459–469.

Walter, P., and Johnson, A. (1994). Signal sequence recognition and protein targeting to the endoplasmic reticulum membrane. *Annu. Rev. Cell Biol. 10,* 87–120.

Proteoglycans

THOMAS N. WIGHT
University of Washington

GLOSSARY

Aggregation Phenomenon that occurs when specific proteoglycans bind to a single molecule of hyaluronic acid via noncovalent bonds

Basement membrane Specialized form of extracellular matrix that contains type IV collagen, laminin, and heparan sulfate proteoglycan; it is present beneath several types of epithelial cells and surrounds other cells such as muscle and nerve

Biglycan Form of small interstitial proteoglycan (160–240 kDa) containing two chains, either chondroitin sulfate or dermatan sulfate chains, that is similar but not identical to decorin

cDNA Refers to the cloning of DNA: therefore, a cDNA molecule made by copying a specific mRNA molecule

Decorin Small interstitial proteoglycan (90–140 kDa) containing either chondroitin sulfate or dermatan sulfate chains that is intimately associated with collagen fibrils

Glycosaminoglycans Unbranched chains of repeating disaccharide units in which one of the monosaccharides is an amino sugar and the other is usually a hexuronic acid

Hyaluronic acid High-molecular-weight glycosaminoglycan consisting of a repeat disaccharide pattern of D-glucuronic acid and D-glucosamine

Proteoglycans Family of charged molecules containing a core protein to which are covalently attached one or more glycosaminoglycan chains

Syndecan Proteoglycan that is present on the surface of epithelial cells and interacts with collagen

PROTEOGLYCANS (PGs) ARE A GROUP OF COMplex and diverse macromolecules that are present in almost all tissues and are synthesized by a variety of cell types. They can be found in four distinct locations: (1) throughout the extracellular matrix (ECM), where they are referred to as interstitial PGs; (2) associated with specialized structures of the ECM such as basement membranes and basal laminae; (3) part of or associated with the plasma membrane of cells; and (4) in intracellular structures such as secretory storage granules and synaptic vesicles. These macromolecules consist of a core protein to which at least one glycosaminoglycan chain is covalently bound. Glycosaminoglycans are unbranched chains of repeating disaccharide units in which one of the monosaccharides is an amino sugar (hexosamine) and the other is usually a hexuronic acid. These chains vary in length and may contain 100 or more repeating disaccharide units, except hyaluronic acid, which can have tens of thousands of repeats. All glycosaminoglycans are highly negatively charged and with the exception of hyaluronic acid all are O-sulfated, and in the case of heparan sulfate and heparin N-sulfated as well. Four basic types of glycosaminoglycans are recognized and defined according to the type of hexosamine (either *N*-acetylglucosamine or *N*-acetylgalactosamine) and/or the conformation of the uronic acid residues (either glucuronic or iduronic). The four basic glycosaminoglycan types include hyaluronic acid, chondroitin/dermatan sulfate (CS/DS), heparin/heparan sulfate (Hep/HS), and keratan sulfate (KS). The chemical structure of the disaccharide patterns of the different glycosaminoglycans is diagrammed in

ENCYCLOPEDIA OF HUMAN BIOLOGY, Second Edition, VOLUME 7. Copyright © 1997 by Academic Press. All rights of reproduction in any form reserved.

Fig. 1. In addition to glycosaminoglycans, both N- and O-linked oligosaccharides may be covalently linked to the protein core.

I. STRUCTURE

The complexity and diversity of these macromolecules are derived largely from the number of different protein cores within specific proteoglycan families and from the variety in the complex carbohydrate structures produced by a large number of posttranslational modifications involved in the synthesis of the completed molecule. Usually, one type of glycosaminoglycan predominates on a single core protein, giving rise to four major families: chondroitin sulfate proteoglycan (CS-PG), dermatan sulfate proteoglycan (DS-PG), heparan sulfate proteoglycan (HS-PG), and keratan sulfate proteoglycan (KS-PG). More than one type of glycosaminoglycan can, however, be inserted on the same core, such as is found in cartilage CS/KS and in some epithelial-derived PGs (HS/CS). The presence of sulfate and/or carboxyl groups on each of the disaccharide units makes the chains strong polyanions and this characteristic contributes significantly to their physical properties as well as governs many of their interactions with other molecules. Hyaluronic acid differs from the other glycosaminoglycans in that it is nonsulfated, and current available evidence indicates that hyaluronic acid is not covalently linked to protein.

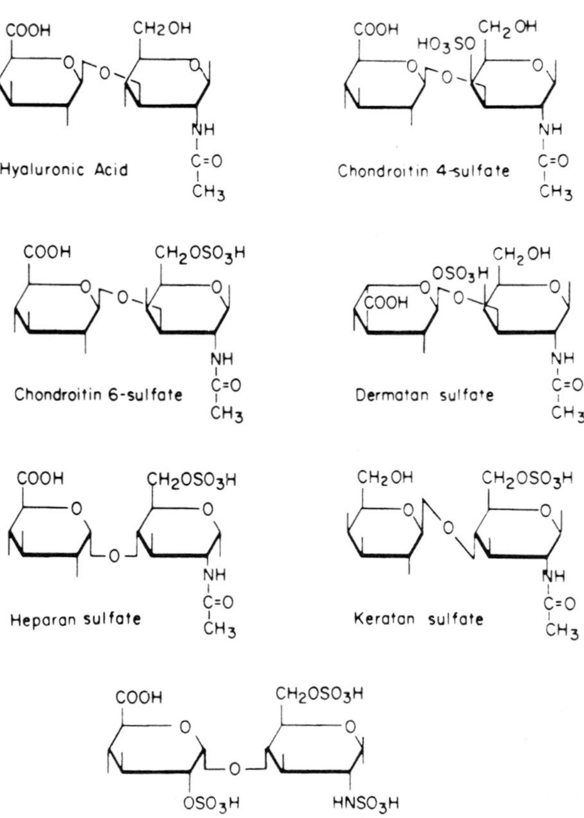

FIGURE 1 Chemical structure of the repeating disaccharide units of the glycosaminoglycan.

Chondroitin sulfate has a disaccharide repeat pattern similar to that of hyaluronic acid, but it contains galactosamine instead of glucosamine, and the galactosamine usually has a sulfate ester attached at the 4 or 6 position. The degree of sulfation can vary within a single preparation and from one tissue to another. The isomeric chondroitin 4- and 6-sulfates occur in tissues independently of each other, although in some instances, chondroitin 4-sulfate and chondroitin 6-sulfate disaccharide units may be present in the same chain.

Dermatan sulfate is an isomer of the chondroitin sulfates in which some (varies between 10 to 80%) of the D-glucuronic acid is replaced by L-iduronic acid; however, not all the glucuronic acid is replaced in this polysaccharide and, therefore, dermatan sulfate may be viewed as a copolymer of disaccharides containing both iduronic and glucuronic acid. The formation of iduronic acid residues occurs by the conversion of glucuronic acid already incorporated into the growing polymer by an epimerization reaction that is tightly coupled to the sulfation process.

Keratan sulfate has a repeating disaccharide residue containing *N*-acetylglucosamine and galactose but no uronic acid. It is usually of relatively low molecular weight and has limited distribution, being present in cornea, cartilage, and nucleus pulposus. Keratan sulfate differs from other glycosaminoglycans not only by the fact that it lacks uronic acid, but it also is not formed on the typical xylose–serine linkage between the core protein and the GAG chain, which is characteristic of all the other glycosaminoglycans.

Heparin and heparan sulfate are glycosaminoglycans of closely related structure. Heparin has a repeating disaccharide unit composed of glucosamine and either L-iduronic acid or D-glucuronic acid. A large portion of the glucosamine residues contain *N*-sulfate groups instead of *N*-acetyl groups, although a small proportion of the glucosamine residues are N-acetylated. Most glucosamine residues also carry a sulfate ester group in the C-6 position, and in addition most of the iduronic acids are sulfated at C-2. Heparin exists in tissue as single polysaccharide chains with molecular weights ranging from 5000 to 15,000 but it is synthesized as a proteoglycan with a molecular weight of 1,000,000 containing a small core protein and several long (~70-kDa) chains.

The structure of heparan sulfate is based on the same disaccharide repeating unit as that of heparin but it differs markedly in its sulfate content. It contains as much *N*-acetyl as *N*-sulfate substitution and a lower degree of O-sulfation than heparin such that on the average there is only one sulfate group per disaccharide unit. The chain appears to contain a block structure in which some regions contain no *N*-sulfate and very little *O*-sulfate. In addition, glucuronic acid is the predominant uronic acid component in heparan sulfate rather than iduronic acid, as is the case for heparin. The L-iduronic acid: D-glucuronic acid ratio generally increases with increasing sulfate content. In broad terms, low-sulfated, D-glucuronic acid-rich polysaccharides are classified as heparan sulfate, whereas high-sulfated, L-iduronic acid-rich species are designated as heparin. Since samples of each class do possess intermediate properties from a structural standpoint, it is most likely appropriate to consider heparin, heparan sulfate, and the intermediates as members of the same family of heparin-like polysaccharides.

Proteoglycans from cartilage have been the most widely studied primarily because cartilage contains large amounts of proteoglycans. The principal PG in cartilage is a large macromolecule (~2000 kDa) that contains both CS and KS chains. The core protein is approximately 250 kDa and about 300 nm in length. The protein portion constitutes only 5–10% of the entire molecule, and the rest of the molecule consists of complex carbohydrates. For example, there are approximately 80–100 CS chains (~20 kDa) on a typical cartilage CS-PG. The sulfated glycosaminoglycan chains are involved in strong intramolecular electrostatic repulsion, which give the PG molecule a very large solvent volume and contribute to the ability of PGs to absorb compressive loads. Each chain is attached to the core protein through a glycosidic bond between the hydroxyl group of serine and a xylose residue at the reducing end of the chain. Two galactose residues and one glucuronic acid residue complete the specialized linkage region. This linkage region is typical for all other PGs with the exception of KS-PG. Keratan sulfate is linked to protein via an *N*-acetylgalactosamine residue linked to either serine or threonine (in cartilage) or via an *N*-acetylglucosamine residue attached directly to asparagine (in cornea). In addition, both N- and O-linked oligosaccharides are covalently attached to the core protein of the cartilage CS-PG. Some of the serines and xyloses appear to be phosphorylated. A diagram showing the complex carbohydrate substitutions on the core protein is presented in Fig. 2. [*See* Cartilage.]

The large CS/KS proteoglycan in cartilage also possesses the capacity to noncovalently interact with hyaluronic acid and form large multimolecular aggregates. A specific region located in the N-terminal portion of the core protein (approximately 65 kDa) is responsible for this binding activity and is localized to a globular domain (G1) of the core protein. This globular domain contains a tandemly repeated double-loop structure enriched in cysteine residues. This portion of the molecule is termed the hyaluronic acid binding region. A set of one to three small molecules (~45 kDa), called link glycoproteins (LP), interacts with both hyaluronic acid and the core protein to stabilize the aggregate structure. Interestingly, the LPs also have a double-loop structure near the C-terminal end that is highly homologous to the same structure in the core protein. Both the hyaluronic acid binding region and the link glycoprotein–hyaluronic acid binding site require five consecutive repeating hyaluronic acid disaccharides for tight binding.

Other domains within the CS/KS core protein have been recognized using monospecific probes such as monoclonal antibodies and by examining amino acid sequences derived from cDNA clones to the core protein. For example, a specific region of the molecule (approximately one-half of the protein core) contains 117 Ser–Gly sequences and these serines are the sites for the attachment of the bulk of the CS chains. The distribution of Ser–Gly in this domain leads to clustering of the CS chains. A specific region of the core protein near the N terminus is the site for the attachment of KS chains. A region at the C-terminal end of the molecule is globular and deduced amino acid sequences derived from cDNA clones show that this region has a high homology with a hepatic lectin and other vertebrate carbohydrate binding proteins. Thus, this region may have an organizational function involving interaction with other molecules in the extracellular matrix. A diagram of the different domains of a PG molecule is shown in Fig. 3.

The strategic placement of the component complex carbohydrates and protein domains within the cartilage CS/KS proteoglycan confers an important biological property to this class of molecule. The high charge density achieved when PGs aggregate with hy-

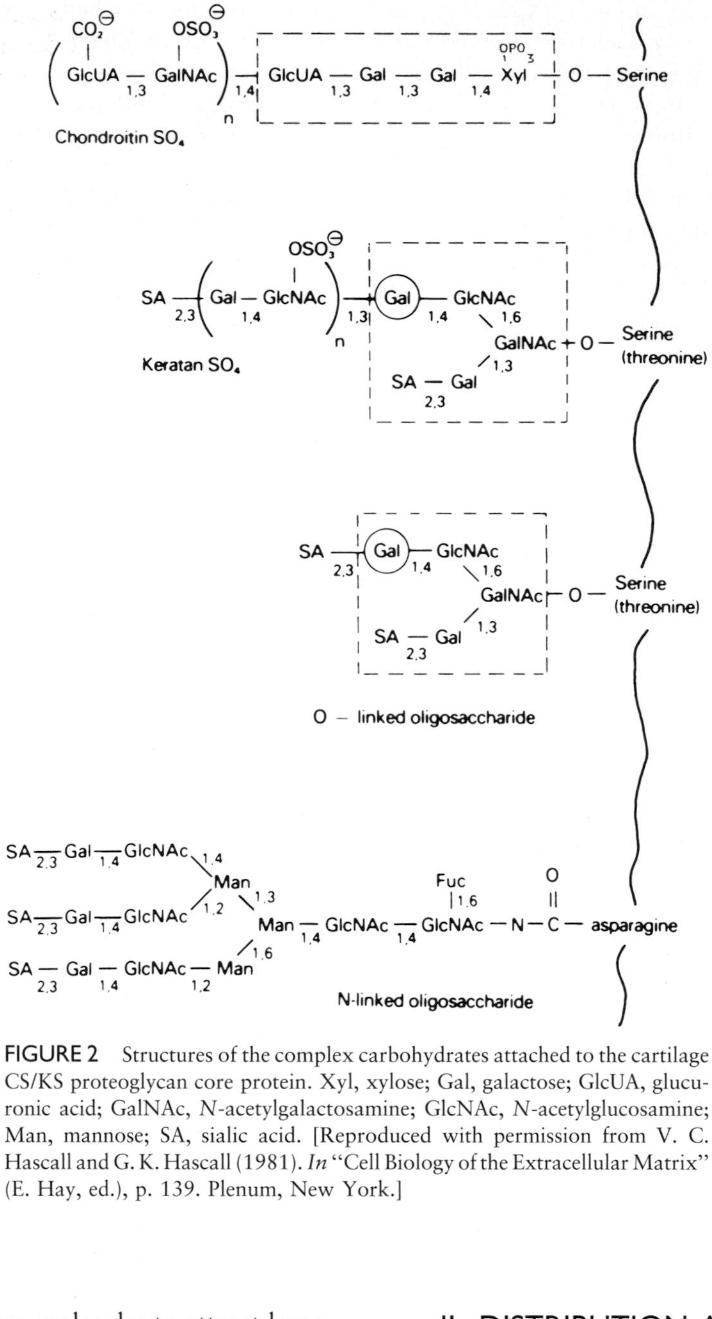

FIGURE 2 Structures of the complex carbohydrates attached to the cartilage CS/KS proteoglycan core protein. Xyl, xylose; Gal, galactose; GlcUA, glucuronic acid; GalNAc, *N*-acetylgalactosamine; GlcNAc, *N*-acetylglucosamine; Man, mannose; SA, sialic acid. [Reproduced with permission from V. C. Hascall and G. K. Hascall (1981). *In* "Cell Biology of the Extracellular Matrix" (E. Hay, ed.), p. 139. Plenum, New York.]

aluronic acid permits these molecules to attract large volumes of water. These molecules can occupy a solution volume of 30–50 times their dry weight. In cartilage, these large hydrodynamic volumes are limited to much smaller volumes by the network of collagen fibrils. The resulting swelling pressure provides cartilage with its resiliency and its ability to absorb compressive forces with minimal deformation. Any alteration in the charge density of this molecule severely compromises its ability to absorb compressive loads.

II. DISTRIBUTION AND FUNCTION

Large aggregating PGs are not restricted to cartilage but are found in other tissues that are subjected to compressive forces. For example, large aggregating CS-PGs have been found in aorta and tendon and they appear similar but not identical to the CS/KS proteoglycan from cartilage. Some tissues such as skin contain *large nonaggregating CS-PGs.* The amino acid composition of skin CS-PG is similar to that of the large aggregating cartilage PG but it appears to

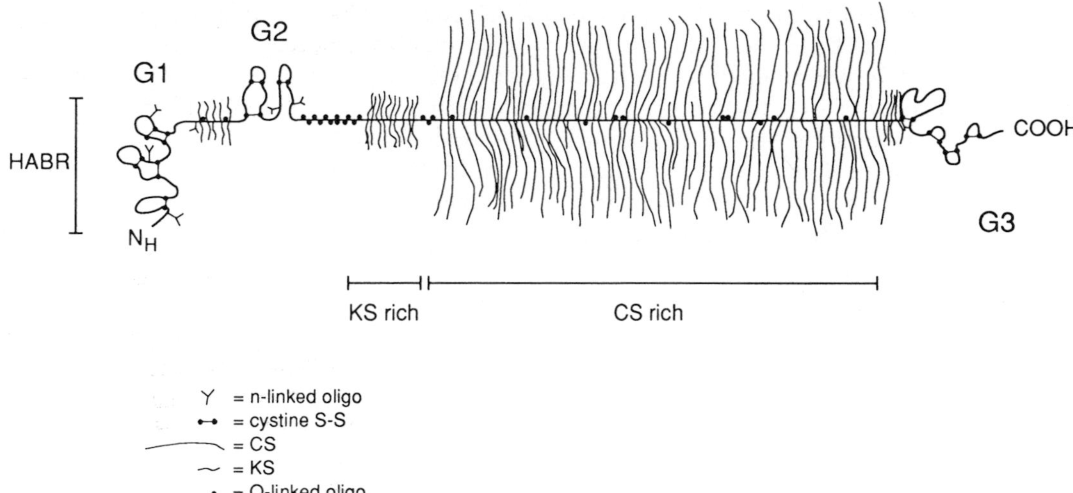

FIGURE 3 Schematic model of the core protein of the large CS/KS proteoglycan from cartilage. The molecule is divided into a number of domains. G1, G2, andG3 indicate the globular domains. The hyaluronic acid binding region (HABR) is followed by a keratan sulfate-rich region. The bulk of the protein core has CS chains attached, forming the CS-rich region. [Adapted from V. C. Hascall (1988). *ISI Atlas of Science: Biochemistry* **1**, 189.]

lack cysteine and methionine. These two amino acids are present in the hyaluronic acid binding region of the cartilage molecule, where cysteine disulfide linkages are essential for the functional integrity of this region of the core protein.

There is a class of *small interstitial proteoglycans* (~100–300 kDa) that contain either CS or DS chains that are present in the ECM of most tissues. Two major forms of DS-PGs have been described that vary in size and iduronic acid content. One form has an overall size of 90 to 140 kDa with a core protein of ~40 kDa. It is thought to contain only one DS chain. This molecule has been named *decorin* because it appears to "decorate" collagen fibrils in a regular array. The iduronate content of the chains varies from 45 to 90%. Peptide mapping and cloning studies reveal similarities in the core protein of this DS-PG class among various tissues, such as skin, cartilage, sclera, aorta, and tendon. Deduced amino acid sequences from cDNA clones reveal that decorin contains a repeat of a 24-amino-acid unit that has an arrangement of conserved leucine sequences, characteristics of many proteins that bind to other proteins. This small DS-PG binds to type I collagen through its core protein and is capable of influencing collagen fibrillogenesis and may regulate the organization of collagen in some tissues. In addition to DS-PG, a small KS-PG binds to collagen in some tissues such as cornea. [*See* Collagen, Structure and Function.]

A larger form of DS-PG (~160–240 kDa) has been isolated and characterized from bone and cartilage, although it appears to be present in other tissues. This form, called *biglycan,* has (nearly) an identical core protein size to that of decorin but the two core proteins are only 55% homologous. The precise location and function of biglycan are not yet known. It may be that these two DS-PG forms arose as a result of gene duplication from a short gene that may have been used many times in diverse organisms for generating protein domains with the capacity to bind to other proteins.

Other PGs are confined to specialized structures of the ECM such as *basement membranes.* Basement membranes, which lie beneath epithelial and endothelial cells, consist of a sheet of ECM containing a number of different proteins that provide a framework for the attachment of these cells as well as a filtration barrier for the exchange of oxygen and plasma constituents. Although basement membranes from different sources contain a number of different proteins, HS-PG as well as type IV collagen and laminin is believed to occur in all basement membranes. The basement membrane HS-PGs appear to be quite heterogeneous but the major one is large, containing a core glycoprotein that varies between 200 and 400 kDa depending on the tissue source. This large HS-PG may act as a precursor, generating smaller HS-PGs that have been identified in some basement membranes. This particular PG confers a selective charge filtration barrier to basement membranes by virtue of the strategic placement of charged glycosaminoglycan chains attached to the core protein. Disruption of this charge barrier leads to marked alterations in vascular permeability, which occurs in some diseases such as diabetes and glomerular nephritis. In addition, HS-PGs interact with other basement membrane components such as laminin, type IV collagen, and fibronectin through specific binding sites within these molecules. Thus, these PGs contribute to the structural stability of this ECM structure. A specialized form of the basement membrane is found at the nerve muscle synapse, where both HS-PGs and CS-PGs are concentrated. These PGs may function in the transport of synaptic proteins, regulate diffusion of ions, and/or provide proper spacing for the pre- and postsynaptic membranes.

III. PLASMA MEMBRANE PROTEOGLYCANS

Proteoglycans appear to be associated with the plasma membrane of cells in at least three ways: (1) intercalated within the membrane, (2) associated with the membrane via a phosphatidylinositol linkage, and (3) bound to particular membrane receptors. Heparan sulfate PGs, which appear distinct from those present in basement membrane, are common as part of the surfaces of most cells. There are a number of different HS-PGs associated with cell membranes. For example, an 80-kDa species that is lipophilic is believed to be intercalated in the plasma membrane of rat liver hepatocytes. Another similarly sized, nonlipophilic species is associated with the membrane via a putative receptor. A larger species ($\sim$350 kDa) is present on the surface of a variety of cells. For example, a number of different HS-PGs have been identified on the surface of cultured human fibroblasts. One of them has two core proteins of about 90 kDa linked together by disulfide bonds. This molecule closely resembles the transferrin receptor and may play a role in cellular growth control.

Other HS-containing PGs are associated with endothelial cell surfaces, such as the HS-PG that binds antithrombin III, facilitating the inactivation of thrombin at the endothelial cell surface. An HS-PG that specifically binds lipoprotein lipase has also been identified on the endothelial cell surface and therefore participates in the metabolism of lipids at that site. One of the better-known HS-containing plasma membrane PGs is a 250-kDa PG containing both HS and CS chains. The molecule is present and restricted to the surface epithelia in a variety of tissues. This PG interacts with a variety of ECM components such as collagen types I, III, and V, fibronectin, and thrombospondin and therefore is thought to be one form of an ECM receptor. Recent success at achieving cDNA clones to this molecule reveals specific domains within the 33-kDa protein core. The sequence shows discrete cytoplasmic, transmembrane, and NH_2-terminal ECM domains. The amino-terminal domain contains about 235 amino acids and the Ser–Gly glycosaminoglycan attachment sites, whereas the transmembrane domain contains a hydrophobic stretch of about 25 amino acids. The cytoplasmic domain is short, containing only 34 amino acids. This molecule has been named syndecan (from the Greek *syndein,* "to bind together").

Although HS-PGs appear to be associated with most cell surfaces, CS-PGs are also associated with plasma membranes. For example, a large CS-PG ($\sim$400 kDa) has been identified on the surface of cultured cells from human malignant melanoma tissue. This PG contains a core protein of 250 kDa and two to three large (60-kDa) CS chains. Although the core protein has not been sequenced, this PG fulfills most of the criteria for an intercalated plasma membrane component. The PG is hydrophobic, it can be inserted into liposomes, and it can be localized to specific microspike structures as part of the cell surface. Functional studies suggest that this PG is important in cell adhesion and may be a factor in regulating the growth of these malignant cells. A specific ligand for this membrane-associated PG has not yet been identified.

There are some PGs that are present within the membrane that bind to a specific external protein in a highly selective manner. One example is the high-molecular-weight component of the TGF-β receptor. This is a 200- to 350-kDa PG (core protein $\sim$100–200 kDa) that binds to transforming growth factors $\beta 1$ and $\beta 2$ in some cells. TGF-β is part of a large family of factors that influence differentiation and growth of a variety of cell types.

IV. INTRACELLULAR PROTEOGLYCANS

A number of cells that participate in immune and inflammatory reactions, including mucosal mast cells, basophils, monocytes, eosinophils, neutrophils, and natural killer cells, contain PGs within their storage granules. Perhaps the best example of an intracellular PG is the heparin PG, which is synthesized and stored within mast cells as 1000-kDa PG. The heparin PG consists of a small core glycoprotein (~20 kDa) containing multiple Ser–Gly repeats that are attachment sites for heparin chains of 80–100 kDa. These chains are subsequently cleaved by a mast cell-derived endoglycosidase to yield partially degraded chains of 7–25 kDa. These chains represent active heparin, which is secreted by mast cells upon activation. Heparin is a powerful anticoagulant in that it potentiates the inactivation of thrombin by antithrombin III. Heparin binds to lysyl residues in antithrombin and accelerates thrombin–antithrombin formation. This accelerating activity is believed to reside in a specific pentasaccharide sequence within the heparin molecule. Some HS-PGs contain HS chains with "heparin-like" sequences. Other cells that participate in the immune response also synthesize and store similar but not identical PGs. For example, basophils synthesize and store a highly sulfated CS-PG (~150 kDa). This PG differs from other mammalian CS-containing PGs in that it is highly resistant to proteolysis and thought to be important in stabilizing, concentrating, and regulating the activity of specific storage granule enzymes such as proteases important in immune and inflammatory reactions.

Other cell types contain PGs within their secretory vesicles. For example, a pituitary tumor cell line At-T-20 stores a CS-PG in its secretory vesicle and secretes it coordinately with pituitary adrenocorticotropic hormone. Proteoglycans are also present in cholinergic synaptic vesicles and may facilitate delivery of certain components of the synapse such as acetylcholinesterase, which is known to interact with a HS-PG. Examples of the structural diversity of different PGs are illustrated in Fig. 4, and the structural characteristics of some of the well-characterized PG types are presented in Table I.

V. BIOSYNTHESIS

The synthesis of PGs occurs in a number of well-defined steps. The synthesis of the core protein and the assembly of N-linked high-mannose oligosaccharides takes place in the rough endoplasmic reticulum of the cell. The majority of posttranslational modifications occur when the core protein (sometimes referred to as precursor protein) is processed in the Golgi complex. This processing includes: (1) stepwise addition of monosaccharides to form the GAG chains on appropriate serine residues, (2) addition of O-linked oligosaccharides onto appropriate serine and threonine residues, (3) conversion of high-mannose N-linked oligosaccharides to complex forms, (4) possible processing of protein to remove a portion of the peptide, and (5) O-sulfation, N-sulfation, and glucuronic to iduronic acid epimerization. In some cases, phosphate may be added onto serine residues in the core protein and/or onto many of the xylose residues that link the GAG chains to the core protein. Sulfation of the GAGs is mediated by 3′-phosphoadenosine-5′-phosphosulfate, either simultaneously with chain polymerization or as a final step after chain completion.

Once synthesized, the PGs can have a variety of fates. They may enter a storage granule such as observed for the heparin PG in mast cells and the

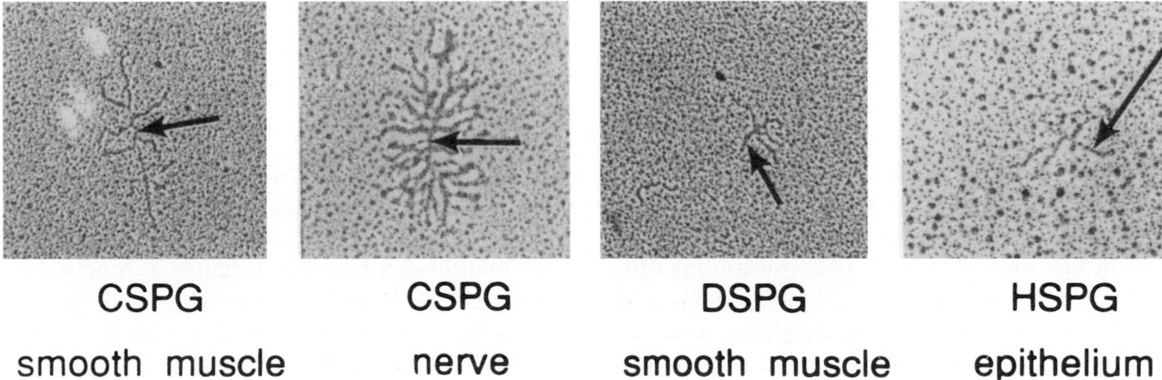

FIGURE 4 Electron micrographs of purified PGs prepared from different sources. Micrographs are printed at the same magnification (×84,750). Arrows indicate the core protein with varying numbers of glycosaminoglycans attached. [Reproduced with permission from T. N. Wight (1989). *Arteriosclerosis* 9, 1.]

TABLE I
Diversity of Proteoglycan Families

Type (suggested names)	Average size[a] (kDa)	No. of GAG chains[b]	GAG (kDa)	Core protein[c] (kDa)	Source
CS/KS-PG (Aggrecan)	2500	100 CS 30–60 KS	20 6	250	Cartilage
CS-PG	1000–2000	12–20	40–60	300–400	Aorta
CS-PG	1000	50	20	50	Skin
CS-PG (Tap-1) (PG-1000)	1000	20	40	200	Nerve muscle synapse
CS-PG (Verisican)	1000	ND[d]	ND	350	Skin fibroblasts
CS-PG	400–500	4	60	250	Melanoma surface
CS-PG (PG-19) (Serglycin)	500	14	40	20	Parietal yolk sac
DS-PG	2000–3000	20	56	500	Ovarian granulosa cells
DS-PG (DS-PG-1) (Biglycan)	160–285	2	20	40	Bone, cartilage
DS-PG (DS-PG-II) (Decorin)	70–120	1	20	40	Tendon, skin, aorta, cartilage
Heparin-PG	750–1000	10–15	60–100	4	Mast cells
HS-PG	750	3–4	70	350–400	Basement membrane tumor
HS-PG	350	8–12	20	2 × 90	Fibroblast cell surface
HS/CS-PG (Syndecan)	200–260	4 HS 2 CS	36 17	56	Epithelial cell surface
HS-PG	130–200	4	14–30	18–250	Kidney glomerular basement membrane
HS-PG	80	4	15	30	Hepatocyte cell surface
KS-PG	80	1–3	5–7	40–55	Cornea

[a]Most of these values are estimates based on molecular sieve properties of intact molecules and therefore should be considered as approximations.
[b]GAG, glycosaminoglycans.
[c]Most of these values are derived from size estimates after removing the GAG chains. Since all of these proteins also contain N- and O-linked oligosaccharides that are not removed by the enzymes, the size estimates will include these carbohydrates.
[d]ND, no data.

HS-PG in the synaptic vesicles. They may insert into the membrane of secretory vesicles and be deposited on the surface as an intercalated, integral membrane component like the CS-PG of human melanoma cells and hybrid CS/HS proteoglycans of mouse mammary epithelial cells, or they may be packaged into secretory vesicles and secreted into the ECM as a structural component such as the large CS-PG from both cartilage and blood vessels and the small DS-PGs in cartilage skin and tendon.

VI. INVOLVEMENT IN DEVELOPMENT AND DISEASE

Formation of a precisely organized, functional tissue or organ is the culmination of a complex series of specific cellular events. These events involve several common types of cell behavior, notably movement, proliferation, shape change, recognition, and adhesion. Molecules present in the ECM and associated with the surface of cells are important for providing the proper environment for these events to take place and for contributing appropriate signals to the cells involved. There are a number of developing systems where changes in PGs and hyaluronic acid have been correlated with different stages of morphogenesis. For example, in the developing cornea, hyaluronic acid is prominent at early epithelial stages. This environment is conducive for mesenchymal migration into the hydrated cornea stroma. This phase is followed by removal of hyaluronic acid by a secreted form of the hyaluronidase enzyme, which causes the tissue to lose water and condense. At this time, mesenchymal cells differentiate into corneal keratinocytes and synthesize CS-PGs, KS-PGs, and collagen to form the differentiated tissue.

Some other examples of PG changes during development include changes during hematopoiesis, branching morphogenesis in kidney, lung, and salivary gland, and the development of the vascular system. The manner by which PGs influence development in all of these systems appear to be mediated through their ability to modify cellular events such as cell adhesion, proliferation, and migration. For example, HS-PGs have been found in points of cell contact with their ECM and it is thought that these molecules stabilize adhesion by interacting with other adhesion proteins. Whereas HS-PGs appear to promote cell adhesion, both CS-PG and DS-PG inhibit the attachment of cells to a variety of substrata and therefore may facilitate cell detachment and movement. In fact, increased amounts of CS-PG have been observed during cell movement in a number of systems. Certain PGs also appear to influence cell proliferation. The glycosaminoglycan heparin and some forms of HS-PG inhibit the proliferation of some types of cells such as vascular smooth muscle cells. The precise mechanism by which these molecules influence growth is not understood. Proteoglycans can interact with growth factors. For example, HS-PGs have been shown to bind growth factors such as fibroblast growth factor, preventing its proteolysis and possibly making the growth factor available for maximal activity within tissue.

Proteoglycans have also been implicated in a number of human diseases. Usually, any change in the structure and/or content of PGs within tissue creates an abnormal ECM that severely compromises the normal functioning of the tissue. For example, PGs that are undersulfated in embryonic cartilage contribute to a reduction in the size of the forming growth plate, which consequently results in stunted and shortened growth—a condition known as chondrodystrophy. A group of genetic diseases, the mucopolysaccharidoses, is characterized by the abnormal accumulation of specific types of glycosaminoglycans in lysosomes of different cell types. The accumulation is due to a deficiency in specific lysosomal enzymes involved in the degradation of glycosaminoglycans. These diseases often lead to severe joint and corneal abnormalities and early death.

Proteoglycans also accumulate in the intimal layer of blood vessels during the early phases of atherosclerosis and this accumulation is thought to predispose the blood vessels to further complications of atherosclerosis, such as lipid accumulation, calcification, and thrombosis, by virtue of their ability to interact with component molecules involved in these processes. In cancer, changes in PG in the stroma surrounding malignant cells are often observed and thought to contribute to the proliferative and invasive properties of the malignant cells themselves. In some diseases of the nervous system, such as Alzheimer's disease, PGs have been observed as components of amyloid deposits and may play a role in amyloid deposition in this disease. [*See* Alzheimer's Disease; Atherosclerosis.]

BIBLIOGRAPHY

Evered, D., and Whelan, J. (eds.) (1986). "Functions of Proteoglycans," CIBA Symposium No. 124. John Wiley & Sons, New York.

Gallagher, J. T., Lyon, M., and Steward, W. P. (1986). Structure and function of heparan sulfate proteoglycans. *Biochem J.* **236,** 313.

Hascall, V. C. (1989). Proteoglycans: The chondroitin sulfate/keratan sulfate proteoglycan of cartilage. *ISI Atlas of Science: Biochemistry* **1,** 189.

Hascall, V. C., Heinegard, D. K., and Wight, T. N. (1991). Proteoglycans: Metabolism and pathology. *In* "Cell Biology of the Extracellular Matrix" (E. Hay, ed.), 2nd Ed. Plenum, New York.

Hay, E. (ed.) (1991). "Cell Biology of the Extracellular Matrix," 2nd Ed. Plenum, New York.

Höök, M., Woods, A., Johanson, K., Kjellen, L., and Couchman, J. R. (1986). Function of proteoglycans at the cell surface. *In* "Functions of Proteoglycans" (D. Evered and J. Whelan, eds.), CIBA Symposium No. 124. John Wiley & Sons, New York.

Muir, H. (1983). Proteoglycans as organizers of the intercellular matrix. *Biochem. Trans.* **11,** 613.

Poole, A. R. (1986). Proteoglycans in health and disease: Structure and functions. *Biochem. J.* **236,** 1.

Ruoslahti, E. (1988). Structure and biology of proteoglycans. *Annu. Rev. Cell Biol.* **4,** 229.

Toole, B. P. (1981). Glycosaminoglycans in morphogenesis. *In* "Cell Biology of the Extracellular Matrix" (E. Hay, ed.). Plenum, New York.

Wight, T. N. (1989). Cell biology of arterial proteoglycans. *Arteriosclerosis* **9,** 1.

Wight, T. N., and Mecham, R. (eds.) (1987). Biology of proteoglycans. *In* "Cell Biology of the Extracellular Matrix" (E. Hay, ed.). Plenum, New York.

Wight, T. N., Heinegard, D. K., and Hascall, V. C. (1991). Proteoglycans: Structure and function. *In* "Cell Biology of the Extracellular Matrix" (E. Hay, ed.), 2nd ed. Plenum, New York.

Protozoal Infections

J. EL-ON
Ben Gurion University of the Negev, Beer Sheva, Israel

Revised by
F. KIERSZENBAUM
Michigan State University

GLOSSARY

Commensalism Two organisms living together without causing injury to each other

Cyst Nonactive, resting form that is responsible for disease transmission

Prophylaxis Treatment given to people at risk for protection against disease

Protozoan parasite Microorganism that lives within or on another organism (the host)

Pseudopods, cilia, flagella, undulating membrane Locomotion organelles

Reservoir Animal harboring a disease in addition to and while humans are not available

Trophozoite Active, mobile form that is generally responsible for pathological manifestations

Vector Arthropod that transmits parasitic disease from one host to another

Zoonosis Disease affecting both animals and humans

PROTOZOAN PARASITES ARE MICROSCOPIC UNIcellular organisms with worldwide distribution affecting both animals and humans. These parasites generally live either within or on the host in one of two ways: as commensals, causing no injury to the host, or as parasites responsible for various damages to host tissues that may lead to severe disease or even death. In the past, the high prevalence of tropical diseases has been limited to certain areas, especially tropical and subtropical climates. In recent decades, these diseases have expanded their geographical distribution, mainly as a result of the increased frequency of travelers to the tropics. This article has been compiled as a source of information covering the important protozoal diseases of humans.

I. INTRODUCTION

Among the infectious diseases affecting humans, those caused by protozoa are still highly important. These unicellular, eukaryotic cells are microscopic and are capable of performing all the vital functions within a single cell. They have a worldwide distribution and cause morbidity and mortality in millions of people, particularly in temperate and tropical areas. The distribution and prevalence of these infectious diseases are considerably affected by socioeconomic conditions, malnutrition, and genetic factors. Acquired and

ENCYCLOPEDIA OF HUMAN BIOLOGY, Second Edition, VOLUME 7. Copyright © 1997 by Academic Press. All rights of reproduction in any form reserved.

natural immune responses are also significantly important and can regulate infections and affect their expression. Protozoan parasites have generally complex life cycles that may involve vertebrate and invertebrate hosts and might be associated with morphological, biological, and biochemical changes. They replicate within the host, either asexually or sexually, to produce overwhelming infections. The development of resistant protozoa to available drugs and the complicated mechanisms that they have developed to invade the host immune response have raised serious problems regarding their control and eradication.

II. AMOEBIOSIS

Amoebas are cosmopolitan parasites with variable infective rates and are highly common in populations with poor hygienic conditions. Apart from *Entamoeba gingivalis*, which inhabits the mouth, all the others—*Entamoeba histolytica, E. hartmani, E. coli, E. polecki, Endolimax nana, Iodoamoeba butchlii*, and *Dientamoeba fragilis*—are intestinal parasites. Amoebas are variable in size (6–50 μm), they move by protoplasmic extensions called pseudopodia, and they multiply by binary fission. Amoebas appear as a motile active trophozoite or as a nonmotile round cyst with a thick wall. The trophozoite form is involved in the pathogenicity, whereas the cyst transfers the disease. Only two amoebas are pathogenic to humans, *E. histolytica* and *D. fragilis*, of which the former may cause fatal infection. Invasive amoebiosis is a major health and social problem in Africa, southeast Asia, China, and Latin America and represents one of the fatal parasitic intestinal diseases. Improving environmental sanitation may reduce infectivity and mortality.

A. *Entamoeba histolytica*

Entamoeba histolytica is a parasite generally limited to adults, although occasionally young children up to 2 years old may also be infected. It infects almost 10% of the world's population and results in about 40,000 deaths annually. [*See* Amoebiosis, Infection with *Entamoeba histolytica*.]

Humans become infected by ingesting the cyst in contaminated food or water. The cyst (see Color Plate 2), 8–20 μm in diameter, may have up to four nuclei and a bar-shaped condensed chromatoid material containing ribosomes and glycogen. Excystation occurs in the ileocecal region of the intestine, forming one to four trophozoites from each cyst. The trophozoite is 15–30 μm in diameter with one nucleus containing a central karyosome (see Color Plate 3). Most infections are asymptomatic. Invasion of the intestinal mucosa of the large intestine generally starts 10 days to several weeks after infection, leading to flash-shaped lesions accompanied by diarrhea or amoebic dysentery. The rectosigmoid area is generally less affected with minimal tissue reaction. In severe cases, the disease is accompanied by bloody diarrhea, sharp abdominal pain, vomiting, constipation, fever, nausea, and leukocytosis.

Amoebic granuloma (amoeboma), an inflammatory lesion resembling carcinoma, may develop in the intestinal wall mainly due to mixed bacterial and amoebic infection. The amoeboma is generally loaded with eosinophils, lymphocytes, and fibrous tissue. It usually occurs in the cecum and disappears within several weeks after successful treatment.

Penetration into circulation through the intestinal wall may lead to extraintestinal infection. The upper part of the right lobe of the liver is the most commonly infected region, although the parasites may spread to the lung, the brain, or other organs. The development of liver abscess may be associated with fever, chills, sweating, and a low erythrocyte sedimentation rate. Pain in the right side, the right lower chest, and the stomach area may also develop. Approximately 50% of the patients with amoebic liver abscess have no history of intestinal infection. It is interesting to note that the amoebic liver abscess is generally not associated with inflammatory reaction of the surrounding tissue, therefore, no scar tissue forms after successful treatment.

For diagnostic purposes, *E. histolytica* should be distinguished from other nonpathogenic amoebas harboring in the intestine. Amoebic intestinal infection is confirmed by detecting the trophozoite or the cyst in fresh stools. The presence of red blood cells in the *Amoeba* cytoplasm is highly indicative of invasive amoebiosis. The presence of Charcol–Leyden crystals (coalescence of eosinophil granules in the stool) may also suggest amoebic infection. A selective medium supplemented with starch is generally used for cultivating the parasite. Negative examination results of three stools obtained on alternate days is highly indicative for lack of amoebic infection. Sigmoidoscopy may reveal typical amoebic ulcers in about 85% of moderate cases. Tests for circulating antibodies [including indirect hemagglutination (IHA) test, indirect fluorescent antibody test (IFA), enzyme-linked immunosorbent assay (ELISA), and gel diffusion] are indicative in severe intestinal (60%) and extra intestinal (95%) amoebiosis. In cases of liver abscesses, amoebic cysts may be found in approximately 15% of the patients. Computerized tomography scanning of the liver using X rays will demonstrate an altered area, and ultrasound examination is important for both diagnosis and abscess aspiration.

Because nonpathogenic *E. histolytica* may be transformed into a virulent invasive parasite, treatment should be given for all cases of amoebiosis, both symptomatic and asymptomatic. The latter is treated with diiodohydroxyquin (D iodoquin), diloxanide furoate (Furamide), or paromomycin sulfate (Humatin). Mild intestinal disease is treated with a combination of metronidazole (Flagyl) or tinidazole (Fasigyn) and diiodohydroxyquin or paromomycin sulfate, and severe intestinal amoebiosis is treated with a combination of metronidazole or dehydroemetine plus diiodohydroxyquin. A hepatic abscess is treated with metronidazole plus diiodohydroxyquin or dehydroemetine followed by chloroquine phosphate (Aralen) plus diiodohydroxyquin. Dehydroemetine may be highly toxic. It reduces dysentery but does not terminate the infection if given alone. Occasionally, aspiration of the abscess is required in addition to chemotherapy.

It has been shown that carbohydrate recognition plays an important role in the interaction of the amoeba with both bacterial and mammalian cells. Recognition is based on the interaction of specific sugars on the surface of the cell and sugar-recognizing proteins (lectins) on the surface of the parasite. The interaction induces the conversion of nonpathogenic parasites to the virulent state, characterized by changes of amoebic enzymes recognizable by their electrophoretic profile (zymodeme).

Virulency of invasive amoebiosis after contact with the target cells is initiated by the release of proteins that create holes in the host cell membrane (amoebapore) and toxins with protease activity. Patients with invasive amoebiosis may suffer from immune dysfunction caused by deficiencies of macrophages and other cells.

Dientamoeba fragilis is an additional pathogenic intestinal amoeba. This parasite occurs only as a trophozoite having either one (40%) or two (60%) nuclei with a karyosome composed of four to eight granules. It is limited to the intestine, in which it may cause abdominal pain associated with diarrhea. Diiodohydroxyquin, furamide, or tetracycline is considered an effective drug against the disease.

III. TRYPANOSOMIOSIS

The trypanosomes are elongated hemoflagellates, 15–40 μm long, belonging to the family Trypanosomatidae. Two major diseases are caused by the genus *Trypanosoma:* sleeping sickness in tropical Africa and Chagas' disease in Central and South America. Parasites appear in the blood of the vertebrate host as a trypomastigote, an elongated flagellate with lateral undulating membrane that emerges as free flagellum with a single nucleus and an additional rod-like organelle, the kinetoplast, containing mitochondrial DNA. Beneath the parasite's outer membrane, there are a number of microtubules. When blood forms of African trypanosomes are ingested by the invertebrate host, they transform into procyclic forms with a single, well-developed mitochondrion; the American trypanosomes transform into epimastigotes. These forms multiply in the midgut of the host and after several multiplication cycles they develop into metacyclic infective forms in either the salivary gland (African trypanosomes) or the rectal area (American trypanosomes) of the insect. Once they have reached the mammalian host by either inoculation during the insect feeding time or contamination with insect feces containing trypanosomes, they initiate infection. [*See* Trypanosomiosis.]

A. African Trypanosomiosis

African trypanosomes are the causative agents of sleeping sickness in humans and animals in Africa. The disease is considered endemic in tropical Africa between the latitudes of 20° north and south of the equator. The distribution of the disease corresponds to that of the tsetse fly (genus *Glossina*), which transmits it during the feeding process. Although there are several species of trypanosomes, only two affect humans: *T gambiense* causes chronic disease in west and central Africa and *T. rhodesiense* causes acute disease in east Africa with a 100% mortaility rate in untreated patients. Other species, e.g., *T. brucei* (see Color Plate 4), *T. evansi*, and *T. congolense*, are animal parasites. *Trypanosoma gambiense* is considered mainly a human parasite, whereas *T. rhodesiense* is a human and game animal parasite. Both parasites are transmitted by several species of *Glossina* flies, which are either highly associated with water (i.e., *G. palpalis* and *G. tachinoides*, which transmit *T. gambiense*) or associated with drier areas (i.e., *G. morsitans*, which transmits *T. rhodesiense*).

An estimated 50 million people living in Africa are at risk; there are about 20,000 new cases annually. Inoculation of the parasites into the skin by the tsetse fly may lead to variable induration and swelling at the site of inoculation, which may then develop into a trypanosomal chancre within 2–3 days postinfection. After several additional days, the parasites spread into the blood circulation via the lymph nodes, where they undergo extensive multiplication by binary fission. Their invasion of the circulation is associated with pathological manifestations, including lymphadenopathy, hepatosplenomegaly, hypochromic anemia, increased erythrocyte sedimentation rate, and a high elevation of IgM titer. Headaches, joint pains, cramps, weakness, and erythematous rash also develop. In advanced infection, the nervous system may be involved, leading to mental disturbances, apathy, and paralysis, followed by diffuse meningoencephalitis associated with coma, convulsions, and death. The invasion of the nervous system is more common in *T. gambiense* infections, whereas *T. rhodesiense* generally kills the host before brain damage occurs.

African trypanosomes in the blood have wave appearances with 1- to 8-day intervals. The survival of trypanosomes within the circulation is due to their ability to vary their surface membrane glycoproteins, a phenomenon known as "antigenic variation." Most of the parasites are killed by antibodies produced by the host, but several resist by expressing different variant surface glycoproteins which cannot be recognized by those antibodies. Each trypanosome might show more than 100 variable antigens during infection, and the variability causes fatal infection by avoiding host defenses. Antigenic variation, which occurs in either the presence or the absence of an antibody response, represents a major obstacle regarding the development of effective vaccination.

Several methods have been used for the detection of trypanosomal infection. Active motile trypanosomes may be detected in either the lymph node aspirate or by a microscopic examination of a wet drop or stained smear of blood. Negative daily examinations for 12 consecutive days are required to exclude infection. Cerebral spinal fluid aspirate should also be examined to confirm dissemination to the nervous system. Invasion of the nervous system may be associated with raised white cell counts and up to five times the protein concentration, particularly IgM in the spinal fluid. Parasites may also be cultivated in special culture media and inoculated into rats, but this technique is useful only in *T. rhodesiense* infection. Because the African trypanosomes are morphologically similar, electrophoretic enzyme variants (isoenzymes) and DNA profile analysis have to be done for species identification. Pentamidine isethionate (Lomidine) and sodium suramin (Naphuride, Antrypol) are the drugs of choice for *T. gambiense* for *T. rhodesiense* infections, respectively. Arsenic compounds, e.g., melarsen oxide (Mel B, Arsobal), are used in the meningoencephalitic stages when the nervous system is involved. Pentamidine has a marked prophylactic activity. All these drugs are toxic. Both pentamidine and suramine are nephrotoxic, and pentamidine may also cause a form of reversible diabetes. Melarsen oxide occasionally causes encephalopathy. α-Difluoromethylornithine (DFMO), an inhibitor of polyamine biosynthesis, has shown antitrypanosomal activity in laboratory animals and humans.

B. American Trypanosomiosis

American trypanosomiosis (Chagas' disease) is a zoonotic disease caused by the hemoflagellate *T. cruzi*. This disease is transmitted to humans by domestic species of blood-sucking triatomine bugs that have adapted to life in the cracks of walls of poor rural housing. The parasite ingested with infected blood develops in the insect gut as an epimastigote, which can multiply and transform into the metacyclic infective form in the posterior part of the gut within 10 to 14 days after infection. Humans become infected when the metacyclic trypanosomes are deposited with the insect feces at the time of blood feeding and the feces are rubbed into the bite site, breaks of the skin caused by scratching or mucous membranes such as those of the eyes. In the mammalian host, the parasite generally grows and multiplies as intracellular amastigotes of 1.5–4 μm length in various cells (Fig. 1), including those of the reticuloendothelial system and heart muscle. In acute infections, the parasite can reach the peripheral blood, in which they appear as elongated trypomastigotes, 20 μm long with a large posterior organelle, the kinetoplast (see Color Plate 5). The amastigotes replicate intracellularly, leading to host cell bursting, while the trypomastigotes, being invasive, can infect other cells, be carried in the blood to other organs, spreading the infection, or ingested by insects seeking a blood meal.

Parasites cause indurated swelling at the site of infection called chagomas. If the conjunctiva is the site

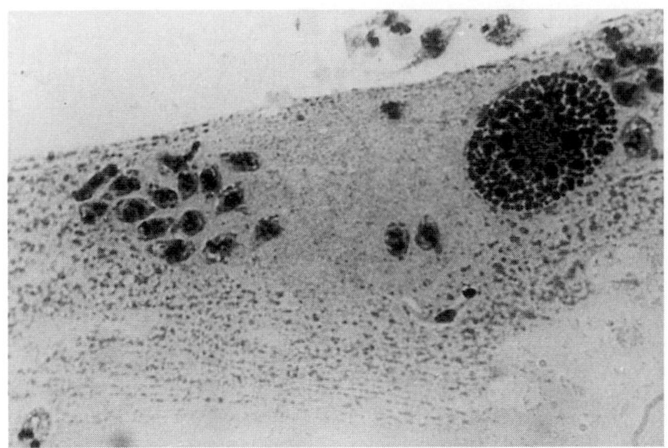

FIGURE 1 *Trypanosoma cruzi* amastiogotes in a fibroblast.

Typical trypomastigotes with big kinetoplasts can be detected in the blood within 1–2 weeks after infection. The disease can also be confirmed serologically using a complement fixation (CF) test. Xenodiagnosis is an additional test for *T. cruzi;* 4 to 10 uninfected bugs are fed on a suspected patient and within several weeks the bugs and their feces are examined for the presence of trypanosomes.

Therapy is limited. Nifurtimox (Lampit), benznidazole, and ketoconazole have been only partially effective and cause undesirable side effects. In endemic areas, 1% gentian violet is frequently added to blood used for transfusion to kill the trypanosomes and to prevent infection of the blood recipient.

of trypanosome entry, a unilateral painless periorbital edema (Romañas' sign) appears. From the initial port of entry, the parasites are disseminated to the internal organs, where they can invade cells of many tissues.

Trypanosoma cruzi infection is characterized by an early acute phase during which amastigotes multiply intracellularly and trypomastigotes may appear in the blood. Infected patients, particularly children, may die during the acute phase, but most patients survive and generally develop a chronic, slow-progressing infection leading to severe heart or gastrointestinal disease. The mortality rate among individuals with these manifestations tends to be relatively high.

Patients suffering from Chagas' disease develop lymphadenitis and hepatosplenomegaly accompanied by irregular fever. It is believed that in advanced cases the parasite causes damage to parasympathetic ganglia and, consequently, to the organs that such ganglia control. Invasion of nervous plexuses associated with the gastrointestinal tract may produce massive dilatation and alteration of muscular movement and the development of megacolon and megaesophagus.

In mammalian hosts, *T. cruzi* develops mainly as intracellular amastigotes which, eventually, transform into trypomastigotes that are released to the surrounding body fluids. It has been suggested that the parasite shares antigens with the host's tissue, leading to autoimmunologic reactivity causing pathology. Although controversial, this notion has posed theoretical difficulties to vaccine development. Further evidence is required to demonstrate that autoimmunity is responsible for cardiac damage or other damage occurring during Chagas' disease. It is not clear whether autoimmunity is elicited by the parasite or by altered tissue antigens resulting from damage caused by the parasite.

IV. LEISHMANIOSIS

Leishmaniosis is an infectious disease caused by protozoa of the genus *Leishmania.* Depending on the species, the disease may be either generalized (systemic), destructive, affecting skin and internal mucosas (mucocutaneous), or chronic, being purely cutaneous. *Leishmania* are transmitted by several species of sand flies belonging to the genus *Phlebotomus* (Old World) or the genus *Lutzomiya* (New World). The disease is widely distributed in many parts of the New and Old World, including the Far East, the Middle East, India, southern Russia, the Mediterranean, Africa, and Central and South America. So far, no accurate estimate of the number of people suffering from leishmaniosis is available. A figure of 12 million has been suggested, with 5000 deaths and 400,000 new cases annually. Leishmaniosis is a zoonotic disease with various animals, including rodents, small animals, and canine species serving as reservoirs. This disease is of clinical importance due to its chronicity, local tissue destruction, resistance to chemotherapy, and potentially high mortality. The leishmanias are obligate intracellular parasites of the mononuclear phagocyte, in which they grow and multiply within the digestive vacuole. It appears that *Leishmania* amastigotes avoid digestion by interfering with the activity of lysosomal enzymes and the oxidative burst activity, probably through secreted glycoproteins and glycolipids. [*See* Leishmaniosis.]

In the sand fly, the parasite is developed into an elongated (20 μm) free-living flagellate—the promastigote (Fig. 2)—with a single nucleus and a rod-like kinetoplast containing mitochondrial enzymes. After 5–7 days of several biological cycles in the midgut of the insect, it travels to the insect's mouth, from which

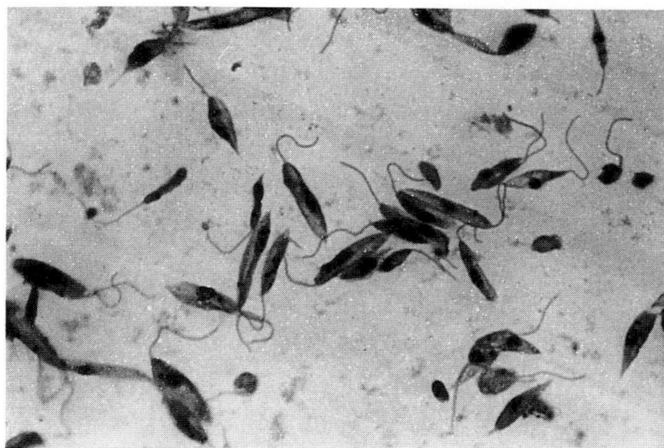

FIGURE 2 *Leishmania major* promastigotes.

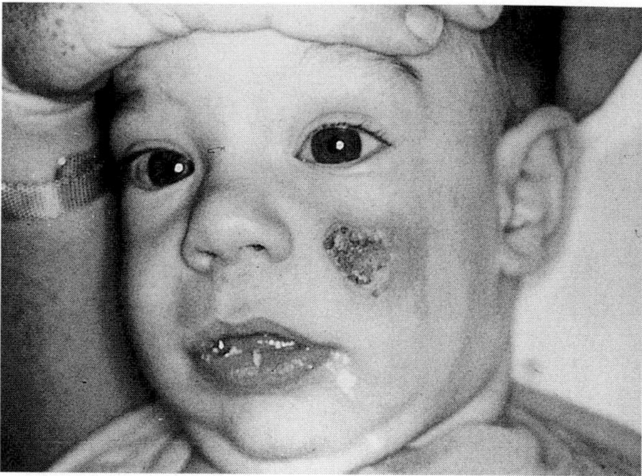

FIGURE 4 Ulcerative skin lesion caused by *L. major*.

it is transferred to the mammalian host through bites. In the mammalian host, including humans, the parasite is engulfed by macrophages, in which it transforms into a smaller (2–5 μm) form: the amastigote (Leishman Donovan body (Fig. 3). All *Leishmania* are morphologically similar and their classification is generally based on serological and biochemical characterization (isoenzyme profiles and DNA pattern).

A. Cutaneous Leishmaniosis

Cutaneous leishmaniosis (CL) is one of the most important cases of chronic ulcerative skin lesions (Fig. 4). The disease is endemic to several tropical and subtropical parts of the world and occurs in several clinical forms: acute CL, chronic CL, recurrent CL, and diffuse CL. Introduction of the parasite into the skin by fly vector bite is followed by the development of a papular lesion at the site of the bite, which may enlarge, ulcerate, and persist for up to 18 months. The ulcer tends to be painless unless secondarily infected. Spontaneous resolution of a large lesion often leaves an unsightly scar. The acute form of CL, caused by *Leishmania major*, can develop into chronic CL, which follows a course of years, whereas that caused by *L. tropica* may develop into recurrent CL or leishmaniosis recidivans. The diffuse form of the disease (DCL), which occurs in a small number of cases infected with *L. aethiopica*, *L. mexicana amazonensis*, or *L. mexicana pifanoi*, is characterized by multiple lesions in the form of nodules and plaques widely disseminated on the entire body surface. In humans, a self-healing lesion is followed by a long-lasting immunity to reinfection.

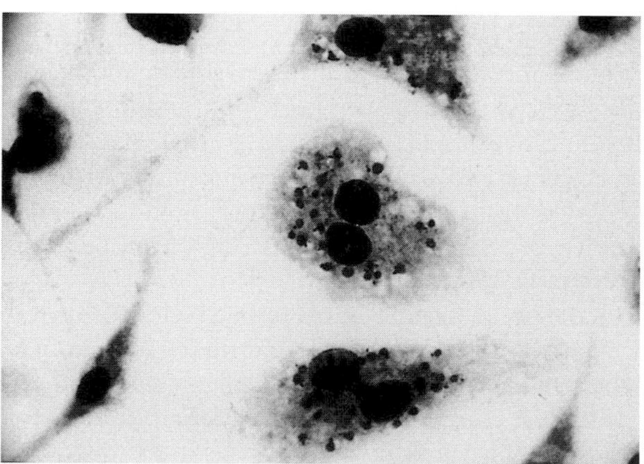

FIGURE 3 *Leishmania major* amastigotes in mononuclear phagocytes.

B. Mucocutaneous Leishmaniosis

The most destructive lesion of CL is the mucocutaneous form (MCL) caused by *L. braziliensis braziliensis* and *L. braziliensis panamensis*. The disease starts as a small nodule that ulcerates within several weeks and tends to metastasize to the mucocutaneous junction either spontaneously or years after the original lesion has healed, causing destruction of the adjacent tissue over a period of years.

C. Visceral Leishmaniosis

Visceral leishmaniosis (kala azar), caused by *L. donovani*, is characterized by systemic infection of the reticuloendothelial system. This disease is associated with enlargement of the spleen and the liver, accompanied by massive involvement of the bone marrow and the lymph nodes. Irregular fever, anemia, leukopenia, thrombocytopenia, and an elevated IgM level may also develop. Advanced stages of the disease are generally associated with generalized immunosuppression, leading to secondary bacterial and viral infections. The visceral disease is the most lethal, producing almost 100% mortality in untreated cases.

Diagnosis of all forms of leishmaniosis is made by demonstrating the amastigotes in aspirates taken from either the edge of the lesion (CL and MCL) or lymph nodes (VL). The parasite is detected microscopically after Giemsa staining. The parasite could also be easily cultivated in blood agar medium at 28°C, in which free-living promastigotes develop. Serological tests, including the IFA test, hemagglutination, the CF test, and ELISA, are also available and are mainly used for the diagnosis of visceral leishmaniosis. The leishmanin test is a skin test that is particularly suited for the cutaneous and mucocutaneous disease. This test measures delayed-type hypersensitivity reactivity and becomes positive during the acute phase of the foregoing diseases, but remains negative in the diffuse and visceral forms. The pentavalent antimonials, pentostam and glucantime, are the drugs of choice against all the clinical forms of the disease. Infections resistant to antimonials are treated with either pentamidine or amphotericin B. These drugs are highly toxic and are generally associated with severe side effects. Rifampicin and ketoconazole, given orally, have been found to be effective against cutaneous and mucocutaneous forms. Topical treatment using paromomycin sulfate and methylbenzethonium chloride in soft white paraffin is also available against the cutaneous forms.

Immunization against CL with living, fully virulent parasites has been used in Iran, Israel, and the former Soviet Union. This is the only protozoan "vaccine" presently available for humans. However, it entails the inoculation of living promastigotes, leading to normal infection, but at a selected site on the body and at a suitable time. The disadvantage of this vaccination procedure is the possible development, over many years, of microbial contamination, allergic reactions, and chronic disease, in addition to lesion formation. This vaccination process, which mimics long-term natural infection, is limited to CL and cannot be applied against MCL or VL.

V. TRICHOMONIOSIS

Trichomoniosis is caused by flagellated protozoan parasites in animals and humans. The parasites inhabit the gastrointestinal and the urogenital systems, living on bacteria and cell debris. The trophozoite, the only form that exists, is pear shaped, measures 5–12 μm, and has three to five anterior flagella, an axostyle, and an undulating membrane. The parasite multiplies by binary fission and is transmitted from person to person through intimate contact.

Three *Trichomonas* species infest humans: *T. tenax* (6–10 μm) inhabits the mouth and is transmitted by mouth-to-mouth contact; *T. hominis* (8–12 μm) inhabits the intestine and is transmitted by the fecal–oral route; and *T. vaginalis* (7–10 μm), a parasite of the urogenital tract, is generally transmitted during sexual intercourse. *Trichomonas tenax* and *T. hominis* are considered commensal, harmless flagellates and are highly prevalent where hygiene conditions are poor. Of these three parasites, *T. vaginalis* is the only one that can cause disease in humans, being the most common pathogen in the urogenital tract. In females, *T. vaginalis* inhabits the vagina, urethra, paraurethral glands, and the bladder. In male, the urethra and prostate gland are the most affected organs, although the bladder, the paraurethral gland, seminal vesicles, testes, and epidydimis may also be infected. The disease is generally transmitted during sexual intercourse, although nonvenereal routes of transmission through contaminated towels cannot be excluded. The frequency of urogenital trichomoniosis is relatively lower in men than in women, and thus it is generally agreed that the infected female is the reservoir of the parasite and that the male is usually the carrier.

Trichomoniosis is an international health problem, and the distribution of the disease in Western countries is relatively high as a result of more liberal attitudes and practices. Reports of infection rates in women vary widely. According to various sources, the prevalence of trichomoniosis in the United States is 3–15% among healthy women, 10% in private gynecological practices, 30% in gynecological clinics, and 38–56% in women attending venereal disease clinics. Data from the 1978 survey of clinics in the United Kingdom showed a prevalence of about 5%. Similar findings have been reported recently in other countries. Trichomoniosis in women is generally associated with vaginal, vulva, and cervix inflammation, accompanied by a discharge that ranges from scanty and diluted to malodorous and cream colored. Vaginal tenderness, vulval pruritis, and burning are the most common symptoms. Men harboring the parasites are generally asymptomatic, although urethral and other urological manifestations might develop.

Trichomoniosis elicits cellular and humoral responses. Serological tests, such as hemagglutination, complement fixation, immunofluorescence, radioimmunoassay, and ELISA, have been used to detect specific antibodies. In patients infected with *T. vaginalis,* there is an antitrichomonal response that includes the production of IgG, IgM, and IgA antibodies. Sera from infected patients have also been shown to exhibit complement-mediated lytic activity on *Trichomonas* and to protect rodents against parasite infection. Antibodies of the IgG and IgA classes have also been found in the vaginal secretions of infected women. However, there is very little correlation between the level of specific antibodies and the severity of the vaginal inflammation.

The most accurate diagnostic method involves detection of the mobile trophozoite in fresh vaginal or urethroprostatic secretions under a light microscope. Because of the rapid death of the parasite outside the body, the microscopic examination of the sample should be done as soon as possible after it has been obtained. The parasite may also be cultivated in a variety of liquid media at pH 5.5–6.0 and 37°C. Cultivation of vaginal exudate samples is a more sensitive diagnostic procedure that may detect twice as many *Trichomonas* infections.

Metronidazole (Flagyl) and other 5-nitroimidazoles (tinidazole and secnidazole) are the drugs of choice against the disease. Alternative treatments include a mixture containing diiodohydroxyquinoline, dextrose, lactose, and boric acid (Floraquin) or a combination of furazolidone and nifuroxime (Tricofuron). Both partners should be treated simultaneously.

VI. GIARDIOSIS

Giardia lamblia occurs worldwide and is a flagellated parasite of humans, particularly children, and probably domestic animals. Up to 30% of the world population is infected. Adults generally develop resistance to the local strain, but exposure to a new strain in a new geographical area may initiate new infections. Infection occurs through food or water contaminated by fecal cysts. It has been estimated that approximately 100 cysts are required to infect a person. Person-to-person contact may serve as an additional route of transmission, particularly among children in nursery school, mentally handicapped people, and homosexuals. Most infections are endemic, but contamination of water supplies with the parasite may cause epidemics.

The parasite has a simple asexual life involving motile trophozoites that divide and develop into infective cysts that are shed in the stool. The trophozoite, the form causing the disease in humans, is pear shaped with bilateral symmetry (see Color Plate 6). It is 10–18 μm long with four pairs of flagella, two nuclei, and two central karyosomes. it has an axostyle, parabasal bodies, and a sucking disc that occupies about half of the ventral surface of the parasite and serves to attach the parasite to the intestinal wall. The cysts (Fig. 5), 10–14 μm long with two to four nuclei, are produced when conditions are unfavorable. They are protected by a thick wall and are hardly affected by the normal level of water chlorination. The cysts remain infective even after several weeks in fresh water.

Giardiosis is asymptomatic in most cases, despite the presence of cysts in the feces. In invasive cases, the parasite may cause irritation to duodenal and jejunal mucosa associated with acute or chronic diarrhea and impaired absorption of fat (steatorrhea), D-xylose, and vitamin B_{12}. In severe cases, anorexia, watery diarrhea with yellow stools, abdominal cramps, urticaria, weakness, and general malaise may develop. Symptom development is due partly to host characteristics and partly to parasite strain differences. Most of the pathology mediated by *Giardia* is due to mechanical injury of the epithelial cells of the intestinal villi to which the parasite is attached via its sucking disc. Several other components, e.g., parasite toxin, and other microbial pathogens may also be responsible for the pathological effect.

The role of immunological response in giardiosis has been reviewed by den Hollander *et al.* The fact that adults develop resistance to infection, whereas immunologically compromised individuals, particularly those with hypogammaglobulinemia, show increased incidence of infection, suggests that the immu-

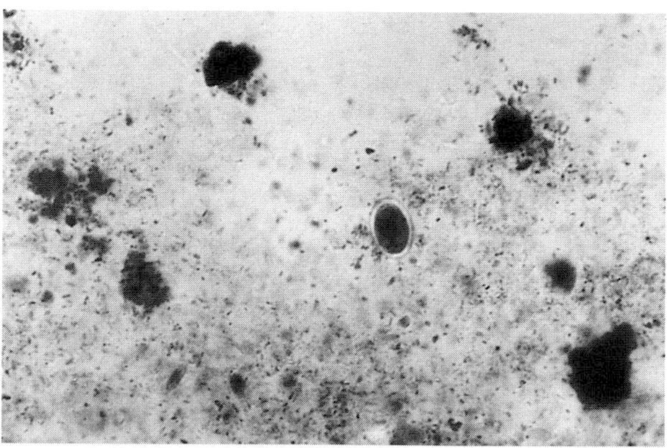

FIGURE 5 *Giardia lamblia* cyst in stool.

nological status of the host may play an important role in the outcome. However, although circulating antibodies against *Giardia* have been demonstrated in 18–77% of infected individuals, very little is known about their function in the disease. It was also suggested that prolonged infection may segregate the human leukocyte antigens (HLA) markers. No correlation with ABO blood groups was demonstrated.

Diagnosis is based on microscopical detection of the trophozoite and/or the cyst in the stool. The formol-ether concentration technique may increase the chances of finding the cyst. In cases suspected for giardiosis but with negative stool examinations, the hairy string test (Enterotest) is required to detect the parasite. This test uses a small capsule containing a coiled thread. The capsule is swallowed by the patient while the free end remains attached to the patient's cheek. Two hours later, the released thread is withdrawn and examined for attached trophozoites. Cultivation of the parasite in artificial medium is possible, although this procedure has not yet been applied for routine examinations. An ELISA has been developed for detecting parasite antigens in the stool. Several negative stool examinations over a period of 1 to 2 weeks are required to exclude infection.

The nitroimidazole compounds metronidazole (Flagyl) and tinidazole (Fasigen), as well as quinacrine and chloroquine, are the drugs of choice against this disease. Flagyl and Fasigen may be associated with nausea and a metallic taste in the mouth. Flagyl was found to be mutagenic *in vitro* and teratogenic in experimental animals and therefore should be avoided during pregnancy.

VII. BALANTIDIOSIS

Balantidium coli is a ciliated protozoan of humans, primates, pigs, and swines of worldwide distribution. The parasite, which has a cyst and trophozoite form, is the largest protozoan of humans. The trophozoite is pear shaped, 30–150 μm long, with an elongated, kidney-shaped macronucleus, a single micronucleus, and a subterminal cytostome. The body is covered with fine cilia that are arranged in rows and are responsible for the parasite's movement. The cyst, 40–60 μm, has a spherical appearance and is involved in the disease transmission. The parasite is reproduced by either simple binary fission or conjugation. *Balantidium coli* is a parasite of the large intestine and, in most cases, it is commensal, living on starch, bacteria, and cell debris in the lumen of the large intestine. Invasive balantidiosis is associated with the formation of flask-shaped ulcers, similar to those caused by *E. histolytica*, accompanied by diarrhea and dysentery. In severe cases, urethritis, cystitis, and myeloneplevitis may also develop. Penetration of the parasite into the peritoneum through intestinal perforation is highly dangerous, causing about 30% mortality.

Microscopical detection of either the cyst or the trophozoite in the stool is the most reliable diagnostic method available. The parasite may also be cultivated in various media, although a large inoculum is required for successful results. Tetracycline, diiodohydroxyquine, and metronidazole are the drugs of choice against the disease.

VIII. MALARIA

Malaria, a sporozoan infection, is still one of the most important protozoal diseases affecting many millions of people in tropical and subtropical areas. The disease has a global distribution and kills about 1 million children in Africa every year. It is a threat to 200 million additional people all over the world. The disease is caused by intracellular parasites of the genus *Plasmodium*, and four species affect humans. *Plasmodium falciparum* and *P. vivax* cause malignant and benign tertian malaria, respectively, and account for more than 95% of all malarial infections. The other two species, *P ovale* and *P. malariae*, are of regional importance, causing benign tertian and benign quartan malaria. [*See* Malaria.]

The disease is initiated by the bite of an infected female anopheline mosquito, which releases small (2–3 μm) spindle-shaped sporozoites into the circulation. The inoculated sporozoites rapidly penetrate the liver parenchymal cells in which they multiply asexually by binary fission to produce preerythrocytic schizonts (see Color Plate 7). During this stage, the patient remains free of symptoms. Within several days, the infected hepatocytes rupture and mature merozoites are released into circulation, where they invade the red blood cells and start the erythrocytic phase. *Plasmodium falciparum* and *P. malariae* have only one preerythrocytic cycle, after which they are passed from the liver cells into circulation, whereas *P. vivax* and *P. ovale* are reproduced simultaneously in both erythrocytes and liver cells to form the exoerythrocytic (EE) phase over a period of years. This may account for the relapses of the disease observed long after the original blood forms have been cleared by chemotherapy. In addition, merozoites that have emerged from ruptured erythrocytes are incapable of infecting hepatocytes. Therefore, patients infected by blood transfusions will suffer only from blood-phase parasites no matter what species it is and should be treated with anti-blood form drugs only.

In the erythrocyte, the merozoite grows into a trophozoite (see Color Plate 8) and multiplies by asexual schizogony, forming a blood form schizont (see Color Plate 9). After 48 to 72 hr, depending on the *Plasmodium* species, the erythrocyte is ruptured and the releasing merozoites invade new red blood cells and start the cycle again. Within several cycles, a synchronous maturation and liberation of parasites and their metabolites into circulation is established, leading to typical malaria fever attacks every 48 hr (*P. falciparum*, *P. vivax*, and *P. ovale*) or 72 hr (*P. malariae*). During this period, the trophozoites may develop into sexual forms, the gametocytes (see Color Plate 10). These forms are able to further their development only in the insect gut, where fertilization occurs to form the ookinete, which penetrates the intestinal walls and develops into an oocyst. The mature oocyst is then ruptured and thousands of sporozoites are released. These initiate a new infection following the bite of an infective mosquito. Malaria infection may also be acquired congenitally during pregnancy, by blood transfusions, and by contaminated syringes among drug users.

The clinical manifestations of malaria are caused by the erythrocytic phase of the parasite. Merozoite invasions of red blood cells and rupture of the mature schizont-infected erythrocyte are followed by new invasions of other red blood cells by the released merozoites. Once inside the erythrocyte, *Plasmodium* ingests 75% of the hemoglobin content of the cell. Consequently, toxic metabolites and malaria pigment are accumulated in the red blood cells and are released into the circulation when the cell ruptures, leading to a malarial fever attack. The attack generally lasts 8 hr and includes a cold stage (0.3–1 hr) and a hot stage (3–8 hr), followed by sweating (2–3 hr), and is the main symptom of malaria. As a result of parasite development, many erythrocytes are destroyed and anemia develops. The severity of anemia is affected not only by the level of the parasitemia, but also by immunological mechanisms and depression of the hematopoietic system. Clinical signs include headaches, malaise, nausea, vomiting, abdominal discomfort, aching muscles, and joint pain. Hepatosplenomegaly is commonly observed in malaria patients, and jaundice as well as renal failure associated with edema and hemoglobinuria may also occur.

In the *P. falciparum* infection, severe symptoms that could be fatal may develop suddenly. It was found that erythrocytes infected with late trophozoites and schizont presented electron-dense excrescences (knobs) on their surface membrane. These knobs used to be attached to receptors on venous endothelium, thus leading to massive sequestration of parasites in deep capillaries. As a result, brain capillaries are blocked and damaged and cerebral malaria develops and is associated with convulsions and comas. In addition, damage to the fetus with a high mortality rate may be caused from the sequestration of infected cells in the placenta.

Conditioned protective immunity (premunition) to malaria develops very slowly in residents living in endemic areas. Immunological studies clearly indicate roles for both cellular and humoral responses in the defense mechanism against malaria. However, malarial infection may also induce immunosuppression, leading to increased frequency and high susceptibility to viral and bacterial secondary infection.

Genetic and epidemiological studies indicate that patients with certain red blood cell abnormalities, i.e., gluclose-6-phosphate dehydrogenase (G-6PD) deficiency, sickle cell anemia, and β-thalassemia, are protected from lethal infections of *P. falciparum*. Oxidative damage was shown to limit the parasite multiplication within these cells, allowing a better immunological response and recovery from infection.

Malaria is diagnosed by detecting the parasite in blood smears and thick blood drops. Several blood films taken every 6 hr over a period of 2 to 3 days should be examined in order to exclude malaria. Serological tests, including IFA, ELISA, and RIA, are also available. Parasite DNA has been used as a target molecule for the development of diagnostic tests based on hybridization technology. This technique was shown to be highly sensitive and could detect 5–25 pg of parasite DNA and a parasitemia level of 0.001%.

Chloroquine (Aralen) and other 4-amino quinolines, i.e., amodlaquin and camoquin, are the drugs of choice against the disease. These compounds are effective only against blood forms. For a radical cure of *P. vivax* and *P. ovale* infections, an additional treatment with primaquine is required. The drugs are usually well tolerated with only minor side effects of nausea and dysphoria. In cases of long-term administration, myopathy and retinopathy may develop. Drugs useful for resistant malaria are quinine, quinidine, or mefloquine, given either alone or in combination with sulfa-antifolate. Quinine treatment may be associated with massive intravascular hemolysis leading to dark urine (black water fever) and renal failure. Chemoprophylaxis includes chloroquine or other 4-amino quinolines, either alone or in combination with Fansidar (pyrimethamine and sulfadoxine), in areas having chloroquine-resistant parasites. Proguanil has been suggested as a substitute for Fansidar, as the combination of chloroquine and Fansidar could be fatal. Prophylaxis should start 1 week before exposure and be continued for 4 weeks after exposure.

IX. BABESIOSIS

Babesiosis is a tick-borne, highly fatal, economically important disease of animals and livestock. The causative agents are several members of the genus *Babesia* (i.e., *B. microti* and *B. divergens*) that also infect humans through hard ticks of the genus *Ixodes*. The parasite invades the reproductive organs of the tick and is then transmitted to the tick's eggs. The developing tick embryo becomes infected, and once the parasite reaches the salivary gland, the tick becomes infective and may transmit the disease to the vertebrate host during the feeding process. Adult ticks generally do not transmit the disease.

In the vertebrate host the 3-μm-long parasite invades the erythrocyte, in which it grows and multiplies by either binary fission or schizogony, causing infection resembling that of *P. falciparum* malaria. Hemolytic anemia and hemoglobinuria are the major pathological manifestations, accompanied by fevers, chills, sweating, and malaise. In splenectomized patients, the disease may be fatal from damage caused to the liver and the kidneys. Microscopic examination of Giemsa-stained blood smears is used to detect the parasites. An IFA test is also available, although it is generally used for epidemiological studies. Inoculation of the patient's blood into splenectomized hamsters may be useful in cases with negative smear results. Chloroquine phosphate, although not curative, is the drug of choice against the disease. Pentamidine isethionate is used mainly in severe infections.

X. TOXOPLASMOSIS

Toxoplasmosis is a common sporozoan disease of animals and humans. The disease is economically important in veterinary medicine, as it is a major cause of ovine abortion. Toxoplasma, the causative agent of toxoplasmosis, is an obligatory intracellular parasite, 3–5 μm long, with a crescent shape, and generally appears in pairs or groups in various nucleated cells. Very limited information is available regarding the timing of initial infection in either humans or animals.

Humans become infected either by ingesting the oocyst that is shed in cat feces or by eating raw meat contaminated with *Toxoplasma* cysts. Infection may also be acquired through organ transplantation and blood transfusion. Parasites released from either an ingested oocyst or cyst-contaminated food invade the epithelial intestinal cells, reproduce by endodyogeny, and destroy the cells. The released parasites may invade additional intestinal cells or, alternatively, reach the internal organs through the lymphatic and blood vessels. The parasite invades various nucleated cells, particularly those of the reticuloendothelial system and epithelial cells. Within these cells, the parasite reproduces very fast, producing pseudocysts containing many tachyzoites. The mature pseudocyst is then reputured and the released parasites invade new cells. Most of the parasites are killed by the host's immune response 8–10 days after infection, leaving only a few parasites (bradyzoites) that multiply very slowly to form 5- to 100-μm cysts in various organs, in striated muscles, and particularly in the brain. The cysts, which may contain more than 100 bradyzoites, remain in an active resting state with no inflammatory reaction over a period of years and may be long-lived. Both humoral and cellular responses are involved in this conditioned immunity (premunition), and immunosuppression of the host may cause a relapse of the disease followed by invasive infection.

In cats, the parasite invades only the intestinal epithelial cells. After ingesting oocysts or meat containing cysts, parasites are released in the intestine and invade the epithelial cells, in which they reproduce asexually. After several cycles of reproduction, microgametes (male) and macrogametes (female) are formed to start the sexual cycle, producing zygotes that develop into oocysts containing two spores, each with four sporozoites. The oocysts are shed in the cat's feces and become infective within several days. The oocysts may remain infective in the ground over a period of a year.

In humans, toxoplasmosis is generally asymptomatic. In symptomatic infants the disease is considered neurotropic whereas in adults it is viscerotropic. Lymphadenopathy is the major clinical manifestation of acute, acquired toxoplasmosis. Malaise, fever, headache, sore throat, and hepatosplenomegaly with or without chorioretinitis may develop. In severe infection, pathological effects may also appear in the heart, kidneys, lungs, and other organs. An immunosuppressed host may develop fatal infection if not treated. The fetus carried by an infected mother may be infected through the placenta; such congenital infection may cause severe eye injury, hydrocephalus, and mental disturbances associated with convulsions in the newborn. In the central nervous system of these infants, the parasites are present in the neurological cells, the capillary endothelium, and the mononuclear phagocyte, causing small lesions and abscesses.

In mononuclear phagocytes, the parasite enclosed in endocytic vesicles can resist lysosomal digestion by interfering with phagosome–lysosome fusion. This is thought to be the effect of a protein released into the host cell that sticks to the surface of the vesicle.

Several serological tests, including hemagglutination, the IFA test, and ELISA, are available to detect infection. The Sabin–Feldman dye test, which measures IgG antibodies, is considered highly specific and sensitive. This test, introduced in 1948, is based on the lytic effects of live parasites mediated by serum containing anti-*Toxoplasma* antibodies. These antibodies are detectable 1 to 2 weeks after infection and reach a titer of ≥1 : 1000 within 6 to 8 weeks. A low titer of 1 : 4 to 1 : 64 commonly persists for the patient's entire lifetime. Approximately 5 to 30% of young individuals and 10 to 67% of individuals over 50 years old may have positive serological reactions. The detection of IgM anti-*Toxoplasma* antibodies is important for the diagnosis of acute and infant infections. IgM antibodies appear as early as 5 days after infection, rise rapidly to 1 : 80–1 : 1000, and fall to 1 : 10–1 : 20 within several weeks.

A combination of pyrimethamine and sulfonamide in conjunction with folinic acid is used for treating *Toxoplasma* infection. These drugs should be avoided during pregnancy because of the possible teratogenic effect of pyrimethamine. Spiramycin is an alternative, safer drug.

XI. SARCOSPORIDIOSIS

Sarcosporidiosis is caused by several members of the genus *Sarcocystis*, which includes a group of parasites of various domestic and wild animals with worldwide distribution. Most species of *Sarcocystis* are nonpathogenic, but several of them may cause severe symptomatic infection, particularly in animals. The parasites are mainly located in the fibers of striated muscles, in which they appear as elongated cyst-like bodies (sarcocysts) from a few microns to 5 cm long and containing many mononucleated crescent-like spores 10 to 15 μm in length.

Several species, e.g., *S. bouinominis*, *S. suihominis*, and *S. linemani*, have been found to infect humans. The disease is acquired by ingesting uncooked meat contaminated with sarcocysts. In the intestine, the released parasites invade the intestinal epithelial cells, within which they reproduce by binary fission. They are then carried out by the lymph and blood vessels to the muscles, in which they reproduce and form sarcocysts. Infection in humans is very rare and generally asymptomatic, causing neither damage nor inflammatory reactions. Fever, diarrhea, abdominal pain, and weight loss occasionally develop. The disease is generally detected at histological examinations. No treatment is available.

XII. CRYPTOSPORIDIOSIS

Cryptosporidium is a coccidian protozoan parasite of humans. Since 1976, when *Cryptosporidium* was first demonstrated as a human pathogen, a considerable number of biological and epidemiological studies have described the zoonotic characteristics of this disease, which consists of gastrointestinal manifestations in both humans and livestock. So far, very little information is available regarding host specificity, although more than 30 host species have been described as susceptible to infection. The disease may be transmitted directly from human to human or animal to human, as well as by waterborne routes. The parasite appears to be an extremely widespread organism of global distribution. People of all ages are susceptible to infection, although young children are considered the most susceptible. The parasite is now accepted to be a major cause (7%) of diarrhea among children in Third World countries.

Cryptosporidium is a small protozoan with a life cycle that includes six major stages: release of infective sporozoites (excystation), asexual reproduction (merogony), gamete formation (gametogony), fertilization, oocyst production, and sporozoite formation (sporogony). The parasite is considered an obligatory intracellular parasite of epithelial cells confined to the microvillus region of the intestine. The oocyst, up to 6 μm in size, sporulates within the host cell and is infective when passed in the feces.

Humans generally become infected by ingesting food and water contaminated with oocysts. The disease in humans is usually limited to between 10 and 20 days and is associated with watery diarrhea. In malnourished children and in immunodeficient hosts, such as patients with AIDS, the disease may cause a prolonged life-threatening, cholera-like syndrome associated with abdominal discomfort, anorexia, fever, nausea, and weight loss. Staining of fecal smears with carbol fuchsin and methylene blue is generally used for detecting the parasite oocyst by light microscopy (see Color Plate 11). If needed, the formaline-ether technique may be used to concentrate the oocysts. A commercial, indirect fluorescent antibody test for stool examination is also available.

Almost no data are available regarding the role of the immune response in human cryptosporidiosis, although specific anti-parasite IgM and IgG antibodies are detected in the sera of most patients. The major therapy in humans is supportive dehydration care. Spiramycin, a macrolide antibiotic and the only drug available against the disease, is only occasionally effective.

XIII. ISOSPORA INFECTION

Ispospora is a common, worldwide intestinal coccidian protozoan that causes intestinal manifestations in animals and humans. Although most of the *Isospora* species are animal parasites, three—*I. hominis, I. belli, I. natalensis*—may also infect humans, particularly children in warm climates. Similar to other coccidian parasites, *Isospora* is an obligatory intracellular parasite of the epithelial cells of the small intestine, which reproduces both sexually and asexually within the same host. Humans may acquire the disease by ingesting undercooked meat containing oocysts. Each oocyst (20–33 μm) contains two sporocysts (12–14 μm), each of which contains four sporozoites. In the intestine, the crescent-shaped sporozoites are liberated from the ingested oocysts and penetrate the epithelial cells of the mucosa. After extensive reproduction, the cells are ruptured, releasing merozoites that invade adjacent epithelial cells. After several asexual cycles, gametocytes are formed, followed by fertilization and oocyst formation. The oocyst may be mature by the time it passes into the feces (*I. hominis*) or it may require 1 (*I. natalensis*) to 4 days (*I. belli*) outside the body for complete maturation. The disease in humans is generally asymptomatic, causing no harm to the patient. In a few cases, fever, headache, and abdominal cramps associated with mild diarrhea, lasting a few days, may develop within about a week after infection. In immunocompromised and immunosuppressed patients, severe disease accompanied by water diarrhea, anorexia, and weight loss may develop over a long period.

Diagnosis of the disease is made by the microscopic demonstration of the oocyst in the stool. The formalin-ether concentration technique may also be employed. No effective treatment is available and only symptomatic treatment is possible. *Isospora belli* is partially sensitive to a combination of either pyrimethamine/sulfonamide or trimethoprime/sulfamethoxazole.

XIV. PNEUMOCYSTOSIS

Pneumocystis carinii is a cosmopolitan organism found in the lungs of malnourished children and immunocompromised hosts. Both sexes and all ages are susceptible to infection. No final agreement has been made regarding the classification of this parasite and some authors have classified it as a yeast. Four stages of the parasite have been described in the infected lungs of humans and animals: trophozoites, intermediate immature cysts, mature cysts, and intracyst bodies. The trophozoites, 2–8 μm long, develop from the ruptured cyst and multiply by binary fission, budding, or endodyogeny. The trophozoite may appear either as a single cell with one or two nuclei or as clumps. The cyst is spherical (4–6 μm) and generally contains eight crescent-shaped intracyst bodies, 1–1.5 μm each, which multiply by sporogony. All forms occur extracellularly in the alveoli, and although intracellular forms may be detected, it is not clear whether they can survive intracellularly.

Both the cyst and the intracyst bodies are involved in the transmission of the disease. Because more than 75% of all children present anti-*Pneumocystis* antibodies by the age of 4 years, it was suggested that infection with *Pneumocystis* occurs in very early childhood, probably through inhalation of contaminated air. Intrauterine infection has also been reported. Most initial infections are asymptomatic, leading to a latent phase that may flare up in immunocompromised patients. Children with congenital immune deficiency, organ transplant recipients receiving immunosuppressive agents, patients with malignancies receiving antineoplastic therapy, and those with AIDS are considered highly susceptible to infection, with 100% mortality if not treated.

The parasite lives in the alveolar lining layer attached to the alveolar type I epithelial cells. It causes an acute, diffuse alveolar pneuomonitis with marked impairment of the pulmonary functions. The lungs fill up with thick and tenacious material containing masses of plasma cells and parasites that act as a mechanical barrier to oxygen. Hypoxia, cyanosis, dyspnea, and tachypnea associated with a nonproductive cough may develop.

Diagnosis of the disease is made by detecting the trophozoite and/or the cyst in sputum, bronchopulmonary lavage, or lung biopsy. Samples are examined microscopically after staining with Giemsa or Gomori's methenamine silver stain. Serological tests for detecting both anti-*Pneumocystis* antibodies and parasite antigen are available, but have only minor use, as antigens and antibodies are also demonstrated in asymptomatic patients.

Trimethoprim/sulfamethoxazole or pyrimethamine/sulfadiazine are the drugs of choice against the disease. In resistant cases, pentamidine is introduced. This drug is highly toxic and may cause hyper- or hypoglycemia, hypotension, renal damage, and sterile abscesses at the site of injection. Trimethoprim/sulfamethoxazole is also used as prophylactic therapy to individuals at high risk, i.e., before transplantation or prior to toxic therapy administration.

BIBLIOGRAPHY

Benenson, A. S. (1985). "Control of Communicable Diseases in Man," 14th Ed. Am. Public Health Assoc., Washington, D.C.

Bunyaratvej, S., Bunyawongwiroj, P., and Nitiyanant, P. (1982). Human intestinal sarcosporidiosis: Report of six cases. *Am. J. Trop. Med. Hyg.* **31**, 36.

Current, W. L., Reise, N. C., Errest, J. V., Baily, W. S., Heyman, M. B., and de Raadt, P. (1985). African trypanosomiasis. *Med. Int.* **4**, 146.

den Hollander, N., Riley, D., and Befus, D. (1988). Immunology of giardiasis. *Parasitol. Today* **4**, 124.

Denis, M., and Chadee, K. (1988). Immunopathology of *Entamoeba histolytica* infections. *Parasitol. Today* **48**, 247.

Despommier, D. D., Gwadz, R. W., and Hotez, P. J. (1994). "Parasitic Diseases," 3rd Ed. Springer-Verlag, New York.

Gilles, H. M. (1985). Malaria. Med. *Int.* **4**, 141.

Greenblatt, C. L. (1980). The present and future of vaccination for cutaneous leishmaniasis. *In* "New Developments with Human and Veterinary Vaccines" (A. Mizrahi and I. Hertman, eds.), pp. 259–285. A. R. Liss, New York.

Kierszenbaum, F. (1984). The chemotherapy of *Trypanosoma cruzi* infections (Chagas' disease). *In* "Parasitic Diseases" (J. M. Mansfield, ed.), Vol. 2. Dekker, New York.

Kierszenbaum, F. (1986). Autoimmunity in Chagas' disease. *J. Parasitol.* **72**, 201.

Kierszenbaum, F., ed. (1994). " Parasitic Infections and the Immune System." Academic Press, San Diego.

Kierszenbaum, F., and Hudson, L. (1985). Autoimmunity in Chagas disease: Cause or symptom? *Parasitol. Today* **1**, 4.

Marsden, P. D. (1985). Chagas' disease: American trypanosomiasis. Med. *Int.* **4**, 151.

Matsumoto, Y., and Yoshida, Y. (1986). Advances in *Pneumocystis* biology. *Parasitol. Today* **3**, 137.

Mirelman, D., Feingoid, C., Wexler, A., and Bracha, R. (1983). Interactions between *Entamoeba histolytica,* bacteria and intestinal cells. *In* "Cytopathology of Parasitic Disease," pp. 2–18. Ciba Foundation Symposium No. 99, Pitman, Bath, England.

Tzipori, S., Smith, M., Brich, C., Barnes, G., and Bishop, R. (1983). Cryptosporidiosis in hospital patients with gastroenteritis. *Am. J. Trop. Med. Hyg.* **32**, 931.

Walsh, J. A. (1983). Problems in recognition and diagnosis of amoebiasis: Estimation of the global magnitude of morbidity and mortality. *Rev. Infect. Dis.* **8**, 228.

Yoshida, Y., Matsumoto, Y., Yamada, M., Okabayashi, K., Yoshikawa, H., and Nakazawa, M. (1984). *Pneumonocystis carinii:* Electron microscopic investigation on the interaction of trophozoite and alveolar living cell. *Abl. Bakt. Hyg. A* **256**, 390.

Pseudoautosomal Region of the Sex Chromosomes

FRANÇOIS ROUYER
INSERM, Institut Pasteur

GUDRUN A. RAPPOLD
Institute of Human Genetics, University of Heidelberg

GLOSSARY

Alu sequences Repetitive deoxyribonucleic acid (DNA) sequence family; about 3×10^5 Alu sequences are interspersed in the human genome

Crossing-over Reciprocal exchange of DNA segments between homologous chromatids during meiosis

Genetic linkage Cosegregation of two genetic markers at meiosis according to their proximity along the chromosome

Positive interference Effect that decreases the probability of a second crossing-over when a first crossing-over has occurred on the same chromatid

Pseudoautosomal Genetic behavior of a locus located on the sex chromosomes but displaying partial or no genetic linkage to the sex locus

Recombination fraction (or recombination frequency) Summed frequency of recombinant chromatids among the total number; it varies from 0 to 0.5 (0–50%) and can be expressed as a genetic distance in centiMorgans (0–50 cM) between the two markers

Restriction fragment-length polymorphism DNA polymorphism that alters the length of a restriction fragment; it can be generated by mutations suppressing or creating restriction sites and by deletion or insertion events within the fragment

Synaptonemal complex Structure observed between paired homologous chromosomes at meiosis, bearing some nodules supposed to be the sites of crossing-over and named recombination nodules

Telomere End of linear chromosome; human chromosomes are divided into short arms (p) and long arms (q), separated by the centromere; a distal position lies toward the telomere, and a proximal one toward the centromere

VNTRs DNA polymorphism that alters the length of a restriction fragment, based on variable number of tandem repeats

IN MAMMALS, SEX DETERMINATION IS DIRECTED by a pair of heterologous chromosomes, X and Y, and the presence of a Y chromosome triggers male differentiation of the organism. This pair of sex chromosomes is thought to derive evolutionarily from a common ancestor. The divergence between the X and Y chromosomes during evolution would have started with the acquisition by the Y chromosome of a male differentiating function, the testis-determining factor (TDF). Nevertheless, the X and Y chromosomes of mammals retain two common regions in which there is X–Y crossing-over, possibly ensuring their proper segregation at male meiosis. Because they are not strictly sex-linked, these portions of the sex chromosomes are named the pseudoautosomal regions, PAR1 and PAR2. They are located at the tip of the short and long arms of X and Y. In humans, the occurrence of an X–Y genetic interchange at each male meiosis

ENCYCLOPEDIA OF HUMAN BIOLOGY, Second Edition, VOLUME 7. Copyright © 1997 by Academic Press. All rights of reproduction in any form reserved.

confers to the pseudoautosomal segments some properties unique in the genome. This makes it an interesting model for studying normal and abnormal meiotic recombination.

I. DEFINITION

A. Meiotic Pairing Between the X and Y Chromosomes

In 1934, P. Koller and C. Darlington observed that the sex chromosomes of *Rattus norvegicus* paired at male meiosis. Since then, this observation has been enlarged to numerous mammals including humans. Pairing is a well-known behavior of homologous autosomes at meiotic prophase and is supposed to allow the exchange of genetic material and proper segregation of homologues to the haploid gametes. The molecular basis of such a mechanism is unknown but is supposed to be directed by DNA sequence homology. In the case of the human sex chromosomes, the synapsis encompasses the entire short arm of the Y chromosome and the tip of the short arm of the X chromosome (Fig. 1). Occasionally pairing of the sex chromosomes has been noticed not only between Xp and Yp regions but also between Xq and Yq, reflecting sequence homologies

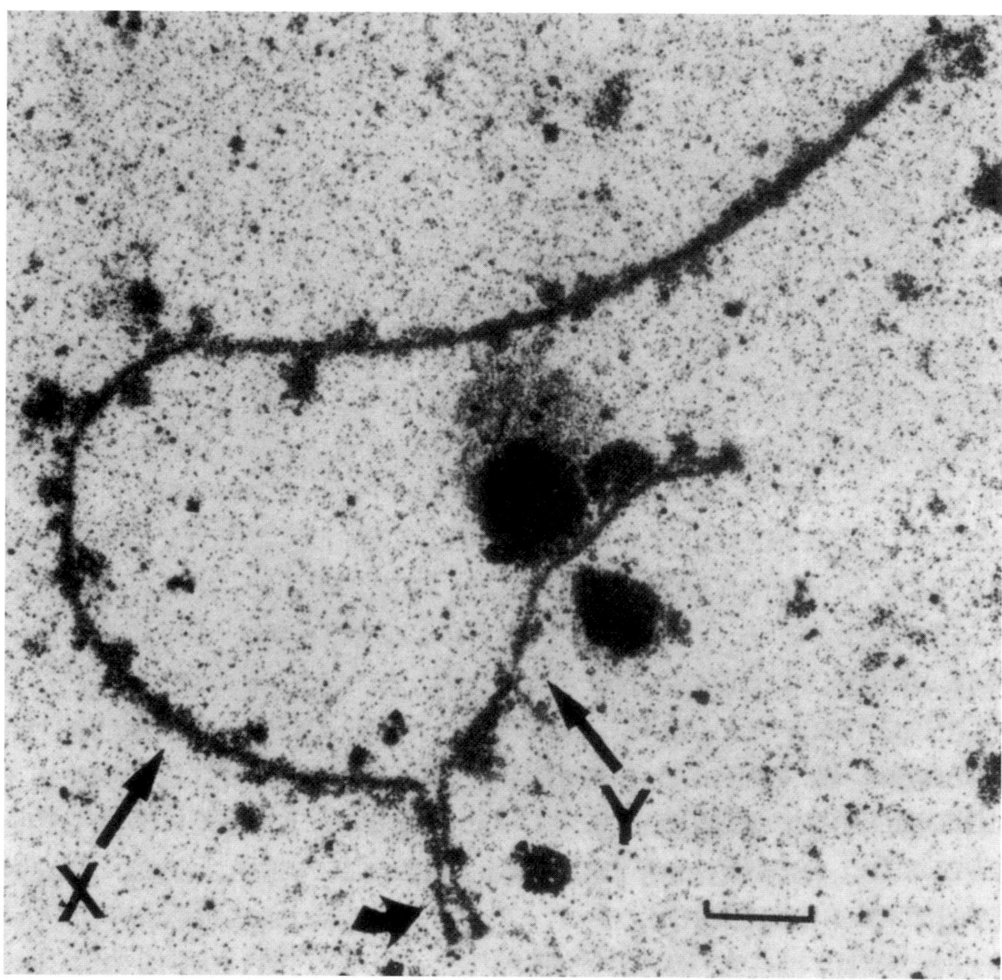

FIGURE 1 The X–Y pair. Electron microscopy microspread preparation showing the X–Y pairing at the pachytene stage of meiosis. Arrow indicates the location of the synaptonemal complex between the short arms of the chromosomes. Bar-1 μm. [Reproduced, with permission, from A. C. Chandley (1984). On the nature and extent of X–Y pairing at meiotic prophase in man. *Cytogenet. Cell Genet.* **38**, 241–247, S. Karger AG, Basel.]

of the pseudoautosomal region on the long arm of the sex chromosomes.

B. Pseudoautosomal Sequences at the Tip of the Short and Long Arms of the Sex Chromosomes

The mammalian X and Y chromosomes undergo a meiotic crossing-over in either one of their pairing regions and this crossing-over is limited to two small regions at the tip of their short and long arms, which are strictly homologous between X and Y. Such DNA sequences show partial or no sex linkage and are thus pseudoautosomal sequences.

In humans, the existence of the pseudoautosomal region of the sex chromosomes has been demonstrated by the finding of DNA fragments, which recognize loci unlinked to the sex on the Y chromosome. Probes detecting restriction fragment-length polymorphisms (RFLPs) or variable number of tandem repeats (VNTRs) represent useful genetic markers along the chromosome. The allelic forms of such polymorphic markers are represented by DNA fragments of different sizes, as observed on a Southern blot. Figure 2 shows the segregation of a pseudoautosomal locus in a five-child family and demonstrates the inheritance of the X-linked allele of the father by two of his sons. This is interpreted as an X–Y interchange, which occurred in these two meioses, between the locus detected by the probe and the sex-determining locus TDF. Results obtained from numerous family studies can be cumulated to determine the recombination fraction between two such loci.

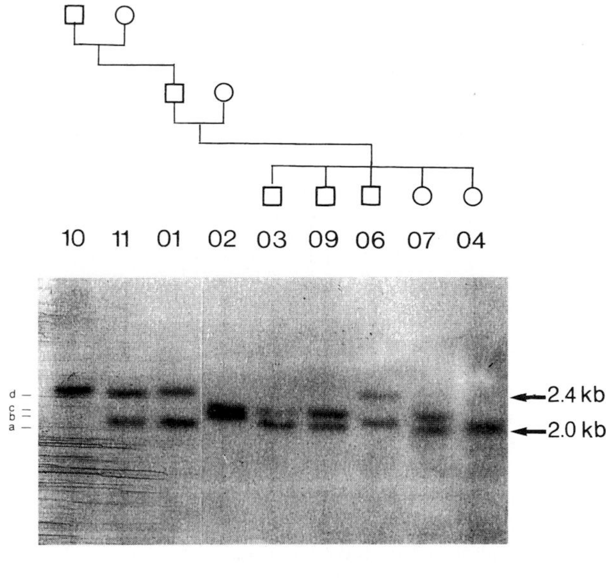

FIGURE 2 Inheritance of the pseudoautosomal locus DXYS15 in a three-generation family. Individuals are two paternal grandparents (10 and 11), two parents (01 and 02), and five children (3, 9, 6, 7, and 4). Squares and circles represent male and female individuals, respectively. DNA samples are digested with the restriction enzyme TaqI, electrophoresed on agarose gel, and transferred to a nylon membrane by the Southern blot technique. The membrane is hybridized to the radiolabeled probe, washed, and then autoradiographed. Four different allelic DNA fragments (a, b, c, and d) with sizes ranging from 2 to 2.4 kb are observed in this family. The genotypes of each individual are shown at the bottom and indicate the alleles of the DXYS15 locus borne by the sex chromosomes. The X-linked allele (a) of the father, 01 (inherited from the grandmother, 11), segregates with the Y chromosome in the two sons 03 and 09. Therefore, two X–Y recombination events can be detected between DXYS15 and TDF among the five paternal meioses analyzed in this progeny. [Reprinted, with permission, from M. C. Simmler, F. Rouyer, G. Vergnaud, M. Nyström-Lahti, K. Y. Ngo, A. de la Chapelle, and J. Weissenbach (1985). Pseudoautosomal DNA sequences in the pairing region of the human sex chromosomes. *Nature* **317**, 692–697.]

II. GENETIC PROPERTIES

A. A Gradient of Sex Linkage in Male Meiosis

The use of several DNA probes allowed an analysis of genetic recombination in the pseudoautosomal region and genetic maps from PAR1 have been established by family haplotyping and single-sperm haplotyping. These studies have established a linear gradient of recombination with frequencies between 50% at the Xp/Yp telomeres and 0% at the pseudoautosomal boundary (PAB). The crossing-over can take place at varying positions in the region, and the different pseudoautosomal loci show a hierarchy of partial sex linkage, which reflects their order on the map (Fig. 3). For example, the proximal locus MIC2 is exchanged between X and Y chromosomes in only 2% of meiosis, whereas the pseudoautosomal boundary (locus PAB) is totally sex-linked and displays a 0% sex recombination rate. [*See* Telomeres and Telomerase.]

After a large number of meiosis analyses, double recombination events have been rarely observed in the pseudoautosomal region (PAR1), using family and single-sperm haplotyping. Positive interference thus seems not to be complete.

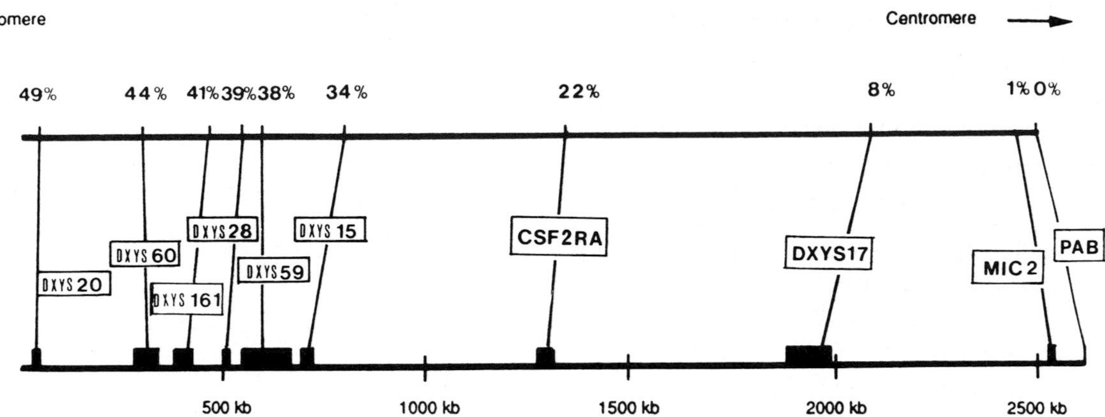

FIGURE 3 Genetic and physical maps of the pseudoautosomal region PAR1. (Top) The complete genetic map stretches over 50% recombination or 50 cM in male meiosis. Sex recombination frequencies of ten loci, including the pseudoautosomal boundary (PAB), are indicated. (Bottom) Physical map of the Y pseudoautosomal region. Shaded boxes above the map reflect the localization intervals of the corresponding loci as defined by pulse-field gel electrophoresis. Distances (in kilobases) are measured from the telomere.

B. Comparison of Male and Female Recombination Frequencies

The recombination rates measured between pseudoautosomal loci in female meiosis strongly contrast with the rates in male meiosis. For the same interval, comprising almost the entire pseudoautosomal region, 2–3% recombination is observed in X–X meiosis compared with nearly 50% recombination in X–Y meiosis. This considerable excess of male recombination can be interpreted as a consequence of the short length of the pseudoautosomal segment compared with the size of the X chromosome. In male meiosis, the X–Y exchange is limited to the pseudoautosomal regions, whereas the X–X crossing-over can occur along the entire X chromosome in female meiosis.

C. Comparison of Physical and Genetic Maps

Physical maps of both human pseudoautosomal regions have been established by pulse-field gel electrophoresis. The PAR1 spans approximately 2600 kb and PAR2 about 320 kb and are delimited on their distal side by the telomere and on their proximal side by sequences specific to the X or Y chromosomes, which define the pseudoautosomal boundary. The physical map of PAR1 e.g. can be compared to the genetic maps obtained in male and female meioses and is shown in Fig. 3. On average, around 53 kb of DNA correspond to one centiMorgan in the PAR1. This represents a recombination excess of about 10-fold compared to both the genome average and that measured in the pseudoautosomal region in female meiosis.

In PAR2, the recombination rate between two specific loci has been estimated to be 2% in male and 0.4% in female meiosis. Thus, there is an approximately 5-fold difference in PAR2 between the frequencies of recombination in male and female germ cells.

If crossing-over is necessary to ensure a proper segregation of the X–Y pair, the recombination excess observed in male meiosis could be a direct consequence of an interchange having to take place between X and Y chromosomes in the short or long arm pseudoautosomal segment. This very high rate of recombination measured in the pseudoautosomal segment could simply be due to elimination of the nonrecombined X–Y pairs during meiosis by a selection mechanism (this represents 90–95% meiosis according to the recombination excess in the male). Alternatively, a particular recombination process could act in the pseudoautosomal region at male meiosis to ensure one crossing-over in each X–Y pair. Electron microscopy studies show that a recombination nodule is present on each X–Y pair at meiosis, as it is for other chromosomes. This suggests the existence of a control mechanism that could promote the occurrence of crossing-overs in chromosome pairs including X and Y, whatever the extent of their pairing.

III. ABNORMAL INTERCHANGES

Two major types of sex reversal syndromes are known in the human: XX males and XY females. XX males (about 1/20,000 newborn males) have a 46,XX karyotype and show a sterile male phenotype, whereas XY females (about 1/50,000 newborn girls) have a 46,XY karotype and are sterile females. The majority of XX males and a small proportion of XY females are due to abnormal genetic interchanges between X and Y chromosomes during paternal meiosis.

At each meiosis, a normal crossing-over takes place between the X and Y chromosomes within the pseudoautosomal region. Therefore, this portion of the Y chromosome constitutes a hot spot of recombination located just distal to the sex-determining region. Exceptionally, some accidental events occur and promote X–Y recombination outside the pseudoautosomal segment. A normal interchange between the pseudoautosomal regions of the X and Y chromosomes is shown in Fig. 4A. The crossing-over can take place proximally to the TDF locus (SRY) on the Y chromosome, as shown in Figs. 4B and 4C. In such cases, the sex-determining region is transferred from the Y to the X chromosome, and the region distal to the X breakpoint is transferred to the Y chromosome. The spermatozoa bearing such abnormal paternal sex chromosomes will give birth to XX males if the rearranged X is inherited (chromosomes I and III in Fig. 5) or XY females if the rearranged Y is inherited (chromosomes II and IV in Fig. 5).

Notably, the majority of XY females show the presence of the sex-determining region on their Y chromosome and, therefore do not result from this etiology.

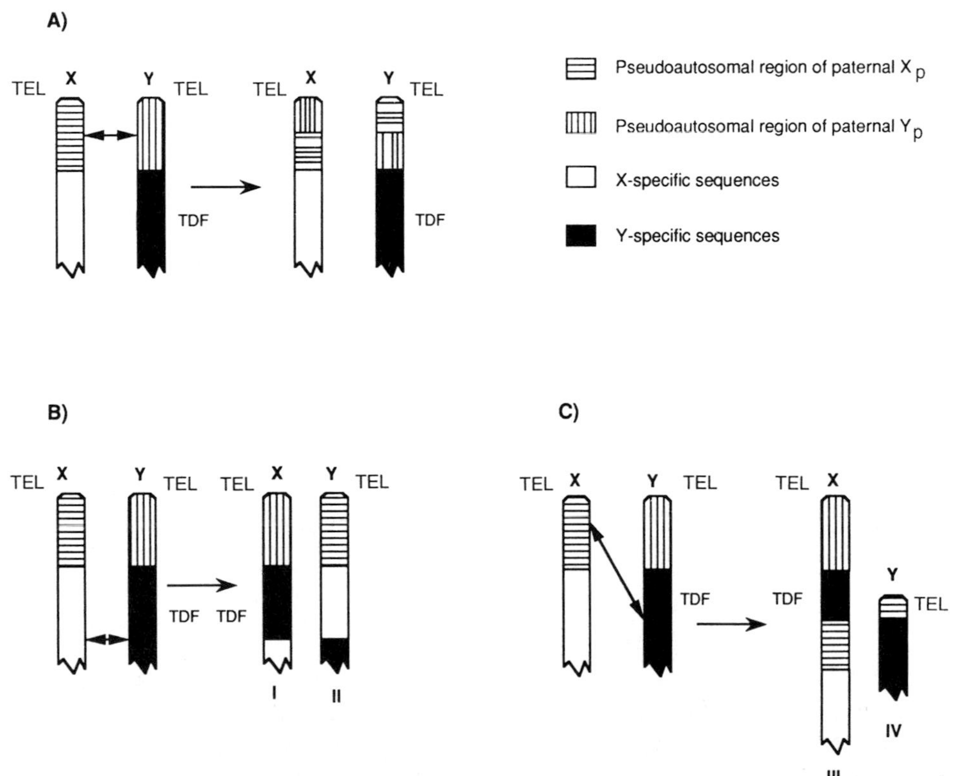

FIGURE 4 Normal and abnormal X–Y interchanges at male meiosis. Paternal chromosomes and the reciprocal meiotic products are drawn on the left and right sides of the single-head arrows, respectively. Double-head arrows indicate the localization of the recombination sites. The terminal part of the chromosomes is indicated by TEL. *TDF is the testis-determining factor SRY. Roman numerals below the chromosome drawings refer to the text. (A). Normal crossing-over within the pseudoautosomal region. (B) Abnormal crossing-over with a breakpoint located in X-specific sequences. (C) Abnormal crossing-over with a breakpoint located in the pseudoautosomal region of the X chromosome.

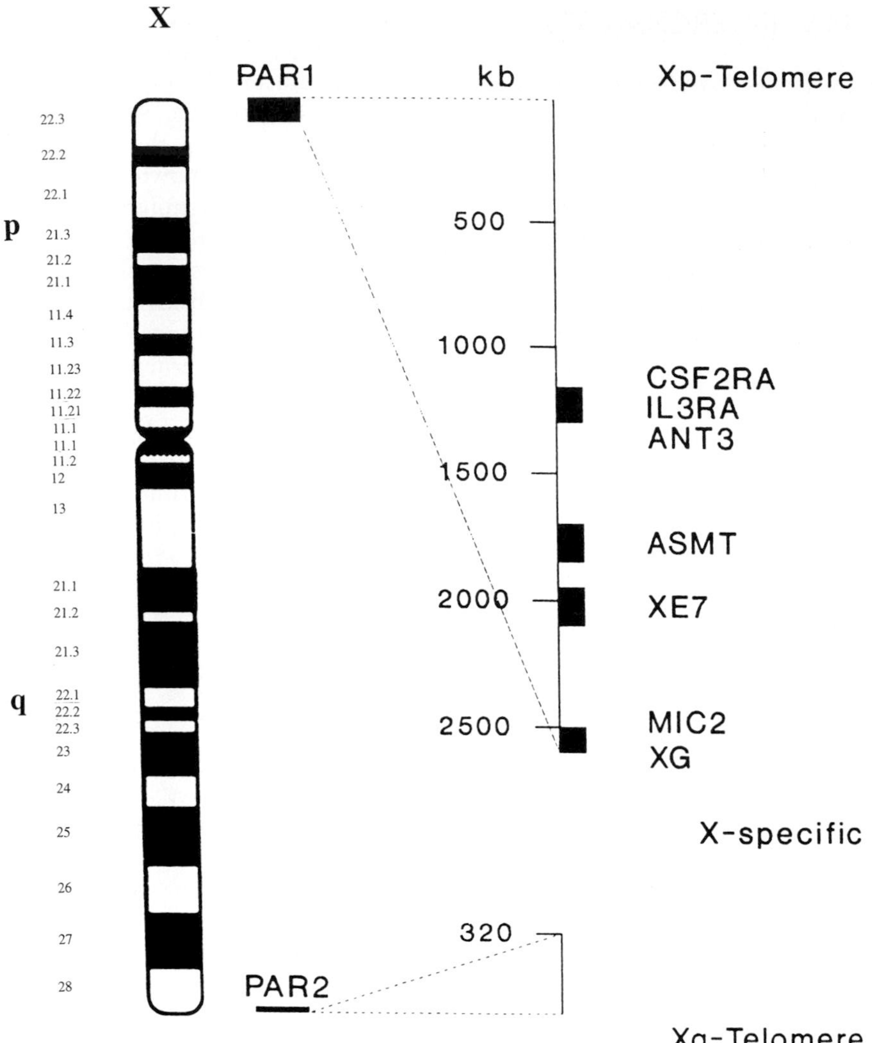

FIGURE 5 Map of the genes of the pseudoautosomal region PAR1. Distances (in kilobases) are measured from the telomere. The PAR1 extends for approximately 2.6 Mb and PAR2 for 0.32 Mb. Seven different genes have been assigned to the PAR1. They are represented by black boxes according to their respective distance from the telomere. Genes in the PAR2 have not been identified to date.

Mutations in SRY or related autosomal genes could account for the female phenotype of such patients. Conversely, abnormal X–Y interchange represents the major cause of XX maleness and has essentially been studied through this disorder. The position of the recombination breakpoint on the short arm of the Y chromosome is variable among XX males but obviously proximal to the SRY locus. On the short arm of the X chromosome, the breakage occurs principally within X-specific sequences (chromosome I in Fig. 4B), but in some cases it is located inside the pseudoautosomal region (chromosome III in Fig. 4C). These patients exhibit a partial trisomy of the pseudoautosomal region but do not seem to differ phenotypically from the other XX males. Analysis of DNA sequences surrounding the breakpoints of rearranged chromosomes provides clues about the mechanism involved in these X–Y interchanges. The breakpoint was analyzed at the molecular level in a single case of XX maleness, with an X chromosome of type III according to Fig. 4C. This study revealed that the abnormal X–Y interchange had been promoted by homologous recombination between two Alu repetitive sequences from two nonhomologous loci of the X and Y chromosomes.

IV. EXPRESSION

A. Pseudoautosomal Genes

In the era of systematic gene cloning and region-specific gene identification, molecular access to virtually all genes in the pseudoautosomal regions is now made possible. At present seven pseudoautosomal genes of PAR1 are cloned in humans (Fig. 5).

MIC2 was the first pseudoautosomal gene described. The MIC2 gene encodes a ubiquitous cell-surface antigen, defined by the monoclonal antibody CD99, a glycoprotein involved in T-cell adhesion processes. Elevated levels of expression have been observed in Ewings sarcoma and primitive neuroectodermal tumors. Further studies on the genomic locus of MIC2 have established that sequences related to several exons of MIC2 exist and that these form part of a pseudogene called MIC2R. MIC2R has been suggested to represent the result of a duplication event in this region.

Another gene related to MIC2, *XG,* the blood group gene, is close to and coregulated with MIC2. XG represents the most proximal pseudoautosomal gene, spanning the pseudoautosomal boundary of the X chromosome. Three of its exons are inherited truly pseudoautosomal, whereas the other seven exons reside in the X-specific region.

More distal, a pseudoautosomal gene with unknown function, *XE7,* has been isolated from a cDNA library, which was made from a cell line containing the inactive X as its sole human component.

ASMT (acetylserotonin methyltransferase or hydroxy-indole-O-methyltransferase, HIOMT) catalyzes the final reaction in the synthesis of the hormone melatonin, which is secreted from the pineal gland. This enzyme was cloned by sequence homology to the bovine ASMT gene and genetically mapped to the PAR1.

The most distal gene cloned so far in the PAR1 is *CSF2RA* (colony-stimulating factor receptor α), which encodes the α subunit of the GM-CSF receptor. It maps about 1200–1300 kbp from the telomere. GM-CSF (granulocyte-macrophage colony-stimulating factor) is a growth and differentiation factor that acts on the monocyte/macrophage lineage. Another cytokine receptor gene, *IL3RA* (interleukin receptor subunit α) resides in close proximity to CSF2RA, between CSF2RA and ANT3. Interestingly, both genes IL3RA and CSF2RA share the same β subunit, raising the hypothesis that a cytokine receptor gene cluster containing growth factor receptors may reside in PAR1.

Further proximal on the chromosome, *ANT3* (adenine nucleotide translocase) has been mapped, which plays a fundamental role in the energy metabolism of the cell, catalyzing the exchange of ATP and ADP across the mitochondrial membrane. Of all genes of the PAR1 isolated to date, it represents the most highly conserved gene, ideally suited for comparative mapping studies in other species. Interestingly, a homolog of ANT3, ANT2, maps on the long arm of the X chromosome and undergoes X-inactivation, whereas ANT3 escapes X-inactivation. The two genes ANT3 and ANT2, therefore, despite being closely related, show striking differences in their X-inactivation behavior.

B. Pseudoautosomal Genes Escape X-Inactivation

In mammals, one of the two X chromosomes is inactivated in females. The inactivation process allows the dosage compensation of X-linked genes between females with two X chromosomes and males with a single X chromosome; therefore, genes shared by both sex chromosomes, such as pseudoautosomal genes, should be noninactivated.

Of the approximately 120 genes cloned and mapped to the X chromosome, only a few have been shown to escape X-inactivation. Among those that are transcribed from both active and inactive X chromosomes are genes such as STS, KAL, GS1, ZFX, UBE1, SMCX, and RPS4X and genes from within the pseudoautosomal region (CSF2RA, IL3RA, ANT3, ASMT, XE7, MIC2, XG). Most of the genes that have been shown to escape X-inactivation map to the most distal band on the short arm of the X chromosomes, which was already previously suggested by studies on chromatin structure and DNA replication. At the molecular level, ANT3, XE7, MIC2, and XG have been investigated and shown to be noninactivated by studying their expression in human–rodent hybrid cells containing an inactive X chromosome. The molecular basis of X-inactivation is unknown, but studies with X-autosome translocations indicate that inactivation can spread over several million base pairs within autosomal sequences, although it is not as stable as it is in X-linked sequences. The noninactivation of the pseudoautosomal segment could, therefore, be promoted by specific sequences of this region.

V. BOUNDARIES

The distal and proximal limits of the pseudoautosomal region display some original features (Fig. 6). At the distal end (see Fig. 6A), the telomere is constituted of tandem arrays of the sequence TTAGGG, which has been identified as the component of all vertebrate telomeres as well as telomeres of some lower eukaryotes. A telomeric restriction fragment containing the last kilobases of the pseudoautosomal region can be used as a functional telomere in the yeast *Saccharomyces cerevisiae*, showing the very high conservation of these structures during evolution. In humans, the $(TTAGGG)_n$ array is variable in length according to the cell lineage by occupies a few kilobases at the end of the chromosome, and has a particularly large size in the germ line. Several kilobases proximal to the telomere, two stretches of hypervariable sequences occur (hypervariable regions I and II in Fig. 6A), separated by an invariant region. Hypervariable sequences are due to copy number variations of small repeated nucleotide (or minisatellite) sequences. They provide very polymorphic markers (i.e., highly variable in different individuals, but inherited in a Mendelian fashion) for the end of the pseudoautosomal region, which are particularly useful for genetic analysis. Moreover, other minisatellite sequences are present in other parts of the pseudoautosomal segment.

At the proximal end (see Fig. 6B), a discrete boundary separates pseudoautosomal sequences regularly exchanged between sex chromosomes, and proximal X- or Y-specific unrelated sequences. These X- or Y-specific DNA sequences could have been brought into the vicinity of the pseudoautosomal region by a major rearrangement on the Y chromosome during evolution (see Section VI). On the Y chromosome, but not on the X, an Alu repetitive sequence has been inserted 225 bp distal to the former pseudoautosomal border, defining a new boundary. The accumulation of mutations in this short 225-bp interval led to a sequence

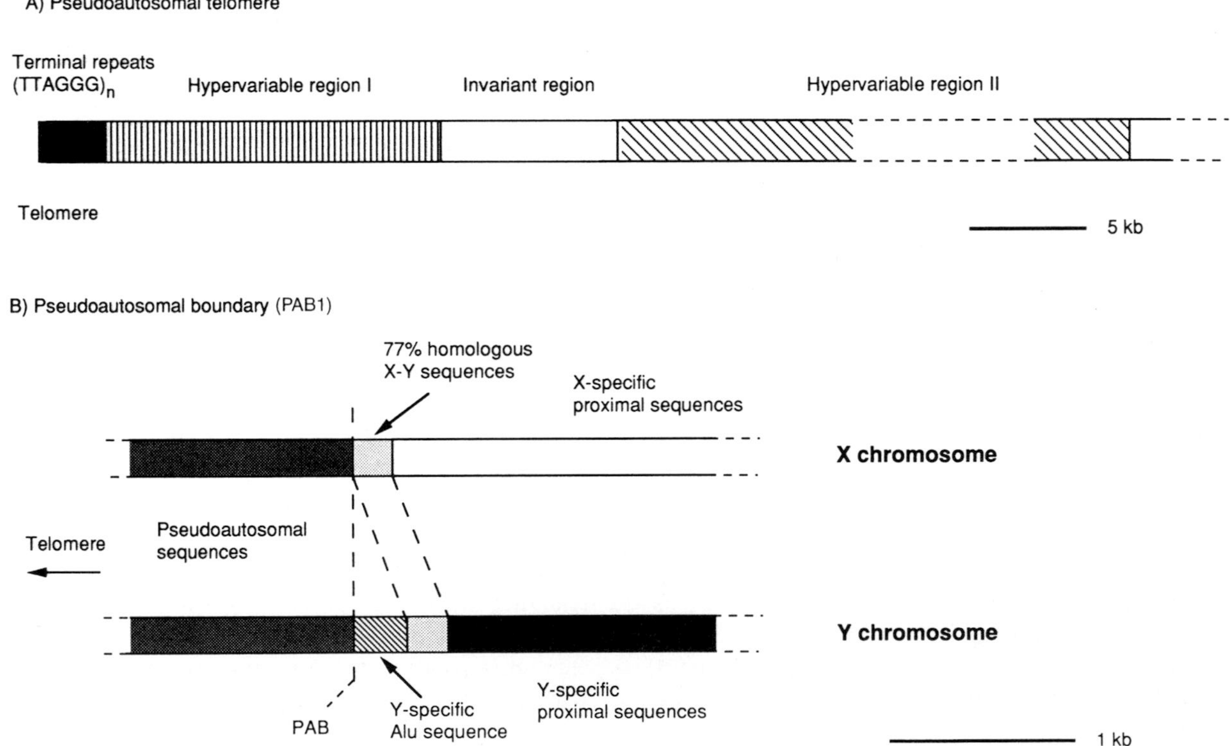

FIGURE 6 Boundaries of the pseudoautosomal region PAR1. (A) Telomere. The length of the terminal repeats at the end of the chromosome is arbitrarily defined, as it is variable according to the cell type. The size of the hypervariable region II is in the range of a few tens of kilobases. (B) Boundary region. The Y-specific Alu sequence and the partially X–Y homologous region are 300 and 225 bp long, respectively.

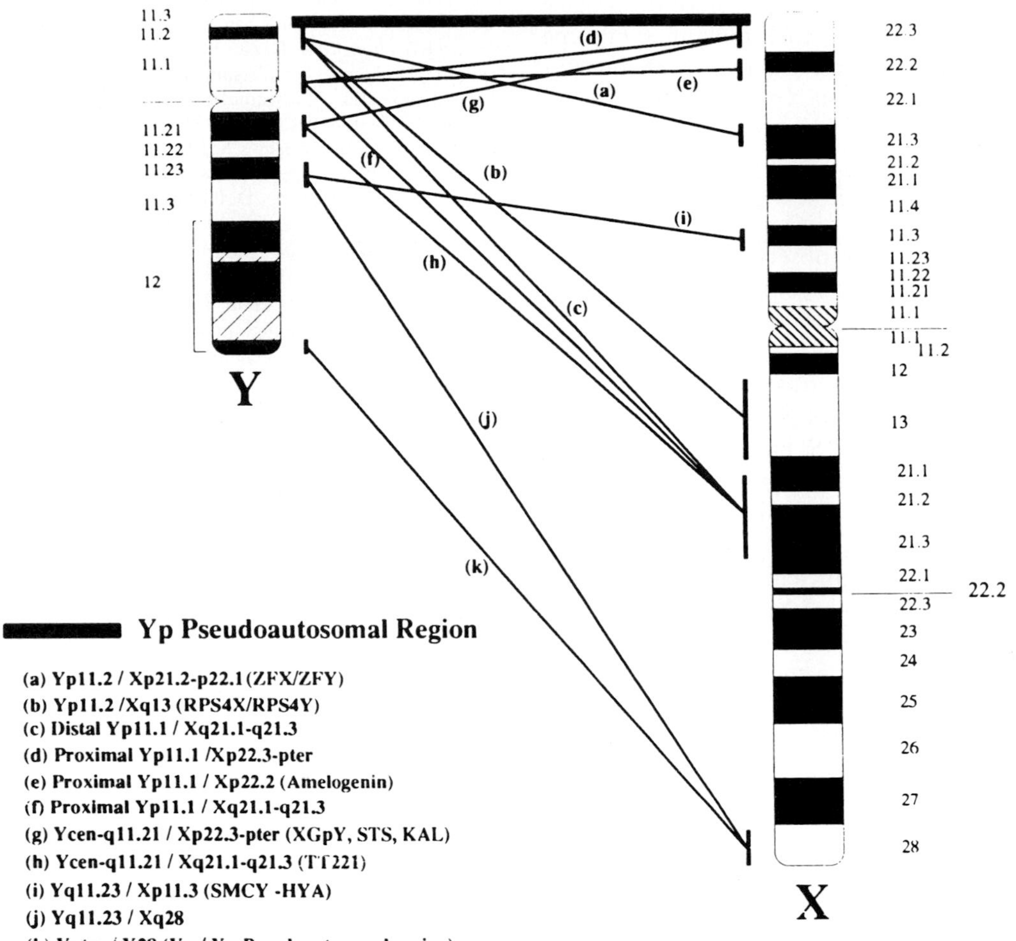

FIGURE 7 DNA sequence homologies between X and Y chromosomes. Chromosomes are drawn according to the conventional representation of their respective banding patterns, including numbering of the bands. Vertical bars joined by arrows show regions displaying DNA sequence homology between X and Y chromosomes. [Reproduced with permission from N. A. Affara and C. Lau (1994). Report of the First International Workshop on Human Y Chromosome Mapping. *Cytogenet. Cell. Genet.* **67**, 359–402, S. Karger AG, Basel.]

divergence of 23% between X and Y chromosomes, showing that the insertion of the Alu element on the Y chromosome has prevented subsequent proximal X–Y recombination. Among 150 sex chromosomes tested for the occurrence of this Alu element, neither gain of the sequence by an X chromosome nor its loss by a Y chromosome has been observed. The presence of the Alu sequence on the Y chromosome of great apes and its absence in Old and New World monkeys suggest this new boundary to be about 50 million years (Myr) old.

The pseudoautosomal boundary on Xq/Yq (PABY2 and PABX2) is defined by a LINE repeat sequence. It was suggested that the boundary arose relatively recently as a result of an ectopic recombination event mediated by the LINE sequences which were originally present on nonhomologous stretches of X and Y chromosomes DNA.

VI. EVOLUTIONARY CONSIDERATIONS

A. Conservation of Pseudoautosomal Sequences in Mammals

In the absence of polymorphic DNA markers, the occurrence of pseudoautosomal sequences has not been established in most species except for human, chimpanzee, sheep, and mouse.

It has been previously noted that steroid sulfatase (STS) and the gene leading to Kallmann syndrome (KAL) escape X-inactivation and map near, but not within, the human PAR1 on Xp22.3. Both STS and KAL have been found to have pseudogenes on the long arm of the X chromosome. In mouse, the steroid sulfatase gene (Sts) is pseudoautosomal and has functional X- and Y-linked alleles that also appear to escape X-inactivation. This observation led to the hypothesis that the present-day human PAR1 represents only a part of a previous considerably larger segment of homology that has been disrupted by a pericentric inversion. However, in contrast to all other genes from the X-specific portion of Xp22.3, PRKX, which resides between XG and STS, has a homolog on Yp rather than on Yq. This is intriguing, as it indicates that the single pericentric inversion event on the Y chromosome hypothesized to have occurred during primate evolution is not sufficient to explain the present X/Y homology pattern of Xp22.3.

Two pseudoautosomal genes, CSF2RA and IL3RA, that map within 100 kbp in the human PAR1 have been shown to map to two different mouse autosomes, MMU19 and MMU14, respectively. Comparative *in situ* hybridization in primate and lemur species has also revealed that ANT3 maps to an autosome in two different lemur species, suggesting that the PAR1, too, may be of relatively recent origin.

Sequences from the small PAR2 on Xq/Xq have been transposed from the X onto the Y chromosome very recently in human evolution and are therefore considered to be human-specific. All of these results strongly suggest that the PAR1 and PAR2 are of relatively recent origin on the mammalian sex chromosomes.

BIBLIOGRAPHY

Burgoyne, P. S. (1982). Genetic homology and crossing over in the X and Y chromosomes of mammals. *Hum. Genet.* **61**, 85.

Cooke, H. J., Brown R. A., and Rappold, G. A. (1985). Hypervariable telomeric sequences from the human sex chromosomes are pseudoautosomal. *Nature* **317**, 687.

Disteche, M. C., Brannan, C. J., Larsen, A., Adler, A., Schorderet, D. F., Gearing, D., Copeland, N. G., Jenkins, N. A., and Park, L. S. (1992). The human pseudoautosomal GM-CSF receptor α subunit gene is autosomal in mouse. *Nature Genet.* **1**, 333.

Ellis, N. A., Goodfellow, P. J., Pym, B., Smith, M., Palmer, M., Frischauf, A.-M., and Goodfellow, P. N. (1989). The pseudoautosomal boundary in man is defined by an Alu repeat sequence on the Y chromosome. *Nature* **337**, 81.

Ellis, N. A., Ye, T.-Z., Patton, S., German, J., Goodfellow, P. N., and Weller, P. A. (1994). Cloning of *PBDX*, an *MIC2*-related gene that spans the pseudoautosomal boundary on chromosome Xp. *Nature Genet.* **6**, 394.

Ellison, J. W., Ramos, C., Yen, P. H., and Shapiro, L. J. (1993). Structure and expression of the human pseudoautosomal gene XE7. *Hum. Mol. Genet.* **9**, 691.

Freije, D., Helms, C., Watson, M. S., and Donis-Keller, H. (1992). Identification of a second pseudoautosomal region near the Xq and Yq telomeres. *Science* **258**, 1784.

Goodfellow, P. J., Darling, S. M., Thomas, N. S., and Goodfellow, P. N. (1986). A pseudoautosomal gene in man. *Science* **234**, 740.

Gouch, N. M., Gearing, D. P., Nicola, N. A., Baker, E., Pritchard, M., Callen, D. F., and Sutherland, G. R. (1990). Localisation of the human GM-CSF receptor gene to the X-Y pseudoautosomal region. *Nature* **345**, 734.

Henke, A., Fischer, C., and Rappold, G. A. (1993). Genetic map of the human pseudoautosomal region reveals a high rate of recombination in female meiosis at the Xp telomere. *Genomics* **18**, 478.

Klink, A., Schiebel, K., Winkelmann, M., Rao, E., Horsthemke, B., Claussen, U., Lüdecke, H. J., Scherer, G., and Rappold, G. A. (1995). The human protein kinase gene PKX1 on Xp22.3 displays Xp/Yp homology and is a site for chromosomal instability. *Hum. Mol. Genet.* **4**, 869–878.

Kvaloy, K., Falvagni, F., and Brown W. R. A. (1994). The sequence organization of the long arm pseudoautosomal region of the human sex chromosomes. *Hum. Mol. Genet.* **3**, 771.

Milatovich, A., Kitamura, T., Miyajima, A., and Francke, U. (1993). Gene for the alpha-subunit of the human interleukin-3 receptor (IL3RA) localized to the X-Y pseudoautosomal region. *Am. J. Hum. Genet.* **53**, 1146.

Page, D. C., Bieker, K., Brown, L. G., Hinton, S., Leppert, M., Lalouel, J.-M., Lathrop, M., Nyström-Lahti, M., de la Chapelle, A., and White, R. (1987). Linkage, physical mapping and DNA sequence analysis of pseudoautosomal loci on the human X and Y chromosomes. *Genomics* **1**, 243.

Petit, C., Levilliers, J., and Weissenbach, J. (1988). Physical mapping of the human pseudoautosomal region; Comparison with genetic linkage map. *EMBO J* **7**, 2369.

Rappold, G. A., Klink, A., Weiss, B., and Fischer, C. (1994). Double crossover in the human Xp/Yp pseudoautosomal region and its bearing on interference. *Hum. Mol. Genet.* **3**, 1337.

Ried, K., Mertz, A., Nagaraja, R., Trusgnich, M., Riley, J. H., Anand, R., Lehrach, H., Page, D., Ellison, J. W., and Rappold, G. (1995). Characterization of a YAC contig spanning the pseudoautosomal region. *Genomics* **29**, 787.

Rouyer, F., Simmler, M.-C., Johnsson, C., Vergnaud, G., Cooke, H. J., and Weissenbach, J. (1986). A gradient of sex linkage in the pseudoautosomal region of the human sex chromosomes. *Nature* **319**, 291.

Schiebel, K., Weiss, B., Wöhrle, D., and Rappold, G. (1993). A human pseudoautosomal gene, ADP/ATP translocase, escapes X-inactivation whereas a homologue on Xq is subject to X-inactivation. *Nature Genet.* **3**, 82.

Schmitt, K., Lazzeroni, L. C., Foote, S., Vollrath, D., Fisher, E. M. C., Goradia, T. M., Lange, K., Page, D. C., and Arnheim, N. (1994). Multipoint linkage map of the human pseudoautosomal region based on single-sperm typing: Do double crossovers occur during meiosis? *Am. J. Hum. Genet.* **55**, 423.

Toder, R., Rappold, G., Schiebel, K., and Schempp, W. (1995). ANT3 and STS are autosomal in prosimian lemurs: Implications for the evolution of the pseudoautosomal region. *Hum. Genet.* **95**, 22.

Yi, H., Donohue, S. J., Klein, D. C., and McBride, O. W. (1993). Localization of the hydroxyindole-O-methyltransferase gene to the pseudoautosomal region: Implications for mapping of psychiatric disorders. *Hum. Mol. Genet.* **2**, 127.

Psychoanalytic Theory

JACOB A. ARLOW
New York University College of Medicine

ARNOLD D. RICHARDS
New York Psychoanalytic Institute

GLOSSARY

Borderline state Descriptive term referring to a group of conditions that manifest both neurotic and psychotic phenomena without fitting unequivocally into either diagnostic category. Some ego functions are fairly well preserved but other ego functions show impairment, resulting in reduced flexibility, adaptability, and interference with the overall evaluation of reality

Identification Psychological process whereby the subject assimilates an aspect, property, or attribute of the other and is transformed wholly or partially after the model the other provides. It is by means of a series of identifications that personality and character of the individual are constituted

Narcissism Concentration of psychological interest upon the self. This may range from healthy self-esteem and pride and pleasure in one's own body and mind, relationships, and achievements to pathological forms of brooding, painful self-consciousness, and an increased propensity for shame

Psychic apparatus Hypothetical division of the mind into various systems, agencies, or groups of functions to aid in the understanding of the psychological development, experience, and behavior of human beings. It does not imply specific anatomical or neurophysiological nervous system structures

Transference Displacement of patterns of feelings and behavior originally experienced with significant figures of one's childhood to individuals in one's current relationships

PSYCHOANALYSIS HAS A THREEFOLD CHARACTER. It is a method for studying the function of the human mind, a means of treating certain psychological disorders, and a body of knowledge derived from psychoanalytic investigation. Mental activity reflects the function of the brain. It constitutes the result of the dynamic interaction of conflicting psychological forces operating within and beyond the scope of consciousness. The nature of an individual's intrapsychic conflicts and the various compromise formations instituted in an attempt to resolve conflicts are deeply influenced by the vicissitudes of individual development.

I. PSYCHOANALYSIS AS A BIOLOGICAL SCIENCE

Psychoanalysis takes its place among the biological sciences as a naturalistic discipline deriving conclusions from observations within a standard setting. Psychoanalysis views the functioning of the mind as a direct expression of the activity of the brain. This activity reflects the experience of the total organism and operates according to certain inherent biological principles. A primary principle is the tendency of the human organism to seek pleasure and to avoid pain or unpleasure. Clearly, this principle must have had survival value in the course of evolution since painful sensations are likely to be noxious in nature (i.e.,

ENCYCLOPEDIA OF HUMAN BIOLOGY, Second Edition, VOLUME 7. Copyright © 1997 by Academic Press. All rights of reproduction in any form reserved.

threatening the integrity of the organism) whereas pleasurable sensations are usually associated with gratification of biological needs, security, and safety.

This last point is especially pertinent because of the helpless, immature state of the human infant at birth. The newborn child is totally incapable of fending for itself. Without the nurturing and protective care of adults, it would soon perish. Furthermore, this state of dependency continues for a long time, longer than in the case of other mammals. The consequences of this fact for the development of the human psyche are profound. It underlies the importance of the "others" upon whom the individual depends. Perforce, humans are destined to become social animals, keenly aware of the distinction between self and others, as well as relations to others. Recent studies of infant development have demonstrated certain features of preadaptation to socialization in the form of inherent patterns of behavior, patterns that serve to stimulate pleasurable reactions in the mother and dispose the infant in turn to respond to the mother's expressions. Human communication begins in the context of the *pleasure–unpleasure principle.*

At the beginning of life, it would appear that the quest for complete and instantaneous gratification is the paramount principle of mental activity. The exigencies of human existence, however, are such that attaining endless, unalloyed pleasure is impossible. Pleasurable sensations disappear as new needs arise. Tensions rise when needs are not immediately met and new experiences do not correspond fully with memories of lost pleasures. Thus, in pursuit of pleasure, gratification of needs must be postponed and some activity relating to the external world must be instituted. Help must be enlisted and substitute gratifications accepted.

The capacity for pleasure is biologically rooted. It is related to the physiology of the body and ordinarily seems to follow a consistent course of maturation and development. During the earliest months of life, the pleasurable sensations connected with feeding seem to be the most important ones. These include not only the alleviation of hunger but also the concomitant sensations associated with the experience—bodily contact, warmth, the mother's gaze, the sound of her voice. The central need of the child is to be nourished and the mother is the object that satisfies the need. Wishes emanating from this phase are called the *oral phase* and remain active throughout life in one form or another, even though augmented and modified by subsequent wishes and events. During the second year of life, other interests come to the fore. The child becomes aware of him- or herself as an independent entity. He begins to assert his independence from his mother and begins to appreciate developing capacities and a sense of mastery. The biological sources of pleasure now center about digestive and excretory functions, the mastery of one's body and its content. The child tests his ability to manipulate his body contents in fact and in fantasy and observes how these activities influence others. This period, roughly covering the age span from two to four, is referred to as the *anal phase.*

Perhaps the most crucial period for psychological development begins about the age of three to three-and-a-half and culminates between the ages of five-and-a-half to six. During this period the child is intensely preoccupied with the activity of his genitals and their potential for pleasure. He begins to contemplate such fundamental issues as the differences between the sexes, the mysteries of conception and birth, the powers and privileges of the adult, and the puzzle of death. Sexual urges become very strong and become manifest in speech and play or covertly in fantasies and dreams. Since, in both sexes, the possession of a phallus becomes a central issue and because of the fact that, at the same time, sexual wishes are often directed toward the parents, this phase has been called the *phallic-oedipal period.* Sexual wishes appear in the content of fantasies accompanying masturbation, a practice common during this phase.

From the study of sexual perversions, observations of children, the psychology of dreams, and the structure of neurotic symptoms, as well as many aspects of normal sexual experiences, Sigmund Freud concluded that these biologically based pleasure-seeking activities of childhood represented constituent elements of the sexual impulse, which attains its final genitally dominated character only toward the end of adolescence.

II. FUNDAMENTALS OF PSYCHOANALYTIC PSYCHOLOGY

Several other principles are fundamental to psychoanalytic theory. Foremost among these is the concept of *determinism.* Mental life is not random or chaotic. Psychological experiences demonstrate the persistent effects of significant antecedent events in the life of the individual. What happens in mental life is part of an ongoing historical process that gives form and meaning to what the individual thinks and feels. Both nature and nurture contribute to developmental sequences of mental functioning, resulting in a patterning of experience that is clearly motivated and in which events are causally related. Psychoanalysis, furthermore, is a *dynamic* psychology. It conceptualizes mental functioning in terms of an apparatus that performs work, an apparatus that is propelled into motion, driven to action by inner urges consciously experienced as wishes. Some of these urges, as indicated earlier, are clearly biological in nature. They stem from the physiological functioning of the organs of the body. This is clearly true in the case of the sexual drives. Other dynamic forces in the mind are less clearly linked to specific physical zones or bodily function, but they are equally important. Foremost among these are the propelling drives toward aggression, hatred, and destructiveness. Unlike the sexual drive, evidence suggesting the operation of a persistent urge toward aggression has no clear base in biology. The evidence for this concept is psychological in nature. Psychoanalysts differ as to whether other motivational forces, such as safety, security, self-esteem, integrity, and mastery, should be considered primary drives or should be subsumed in some way under the broader categories of the *sexual and aggressive drives*. [*See* Aggression.]

A note on the history of terminology must be inserted here. Freud used the German word *trieb* to indicate the dynamic, driving forces acting upon the mind. Unfortunately, the term was mistranslated into English as "instinct" and it is in this form that the term for the concept has persisted in the literature, sometimes expressed as instinctual drive. The term "drive" conveys the precise meaning of the concept.

The drive concept, of course, is an abstraction. In practice, what one observes are mental representations of the drives. They take the form of specific, concrete wishes and are designated drive derivatives. As the individual develops and matures, the manifest forms of the *drive derivatives* undergo change and transformation. This occurs in keeping with the pleasure principle, with the tendency to avoid or mitigate the potential experiencing of unpleasant or painful affects.

Certain specific fears play a leading role in the transformation of the drive derivatives. The first is the danger that the organismic distress the infant experiences in situations of unfulfilled needs might assume devastating proportions. At a later stage of cognitive functioning, with a beginning appreciation of the perception that the mother's face signals impending relief, the failure of the mother to reappear melds into a fear of separation from her. This combination of events becomes a danger situation, fraught with the possible evolution of intense unpleasure or pain. At a still more advanced level of interaction with the mother and a fuller appreciation of her as an independent object, the child comes to face another set of potential dangers. He feels threatened by situations that may lead to the loss of the mother's love. Somewhat later, the child perceives another set of threatening situations, the all-pervasive fear of punishment, specifically by means of physical mutilation of the genitals. Subsequently, as the individual conscience begins to develop, a new source of danger comes into being, namely, the painful affects connected with self-condemnation, the sense of guilt, and the need for punishment. The latter constitute an example of self-directed aggression. It is important to note at this juncture that the drives, sexual and aggressive, operate toward the self and one's own body just as they do toward other objects. An individual may feel love and hate for himself even at the same time, very much as he may experience these feelings toward others. The urge to punish one's self evokes unpleasant affects, which the individual perceives as a signal of impending unpleasure or pain.

These considerations lead us to another fundamental principle of psychoanalysis. Psychoanalysis is a psychology of *conflict,* a psychology of dynamic forces in opposition to each other. Intrapsychic conflict is ubiquitous and never-ending. The forces in conflict are multiple and diverse. They bespeak the contradiction among wishes and fears, threats and warnings, hopes and anticipations for the future, regrets of the past. Freud formulated his final theory of the structure of the psychic apparatus according to the role that each mental element played in intrapsychic conflict. Thus, the persistent wishes of the past, operating as continuous stimuli to the mind and giving rise to innumerable, repetitive, relatively predictable patterns of mental representations, collectively constitute a structure of the mind, a system of the psychic apparatus designated as the *id.* The term id derives from the Latin word for *it*, and the choice of the term reflects how alien and unacceptable some of its derivative manifestations are when presented to consciousness. The id is the vast reservoir of motivational dynamic. It consists of sexual and aggressive wishes, primitive in nature, self-centered and often antisocial.

Another source of motivational dynamic consists of the ideal aspirations, the moral and behavioral imperatives, the judgment of right and wrong. This group of relatively stable functions of the mind Freud called the *superego,* in recognition of the fact, as he thought, that it developed later in the life of the mind, but also from its function as an observer and critic, seeming to stand above and beyond the self, passing judgment on it. Frequently, but by no means always, impulses emanating from the superego oppose the demands of the id. This applies particularly to cases of hysteria. Moral considerations are repetitive and relatively predictable but, like the id, the superego is full of internal contradictions.

The third structural component of the psychic apparatus is made up of those functions that serve to integrate and to mediate the complementary or contradictory aims of the other agencies of the mind. At the same time it takes into account the nature of the objective, realistic situation in which the individual finds himself. Freud called this set of functions the *ego*. It comprises activities that identify it as the executant for all the agencies of the mind. It is the mediator between the internal and the external world, between the world of thoughts and feelings on one hand, and the world of perception and objects on the other.

The concept of psychic structure should not be taken too rigidly. The essential criterion that applies to any mental representation is the role it plays in intrapsychic conflict. Id, ego, and superego constitute only abstract conceptualizations of the patterning of the forces in conflict within the mind. In the clinical setting, situations of intense conflict delineate most clearly the boundaries of the constituent structures of the psychic apparatus. Under more harmonious psychological circumstances, the contributions of the component systems tend to fuse. It is the function of the ego to integrate and to resolve intrapsychic conflicts in order to avoid the danger of unpleasant affects, particularly depressive affect and anxiety. The end products of the ego's efforts, so to speak, represent compromise formations, which is to say that all the various participants in the internal conflict find at least some representation of their dynamism in the final mental product. The compromise formations effected by the ego may be successful ones in the sense that they ward off or circumvent the appearance of pain or unpleasure or they may fail in the sense that the final product of the process is fraught with a greater or lesser component of pain or brings the individual into conflict with the environment and actual danger to his person. Various affects, anxiety in particular, serve as signals, warning that a danger situation of one of the several types mentioned earlier may be developing. The potential danger may be actual or imagined, real or fantasy. The functioning of the mind is an endless exercise in adaptation, reconciliation, and integration. Needless to say, this effort is not always successful. Failure may eventuate in inhibitions, symptoms, or maladaptive character traits.

The ego has at its disposal many different methods for dealing with danger situations. For example, certain mental processes signaling danger, from whatever source, may be rendered nonexistent, which is to say they are promptly forgotten and cannot be recalled to mind. They no longer are available to consciousness. The individual is not aware of them and, as far as he is concerned, they never happened. Such mental processes are said to have been *repressed*. In a definitive act of repression, the repressed element leaves no traces, which is to say it ceases to function as a driving force stimulating the psychic apparatus into action. Psychically it has become nonexistent. It is only when the process of repression is incomplete, when the element excluded from consciousness continues to exert some dynamic impetus, as evidenced by the appearance of derivative representations, that one may infer the presence of a repressed element, of an unconscious mental process. Unconscious mental processes are not apprehended directly. They are inferred from an examination of the data of observation, utilizing criteria of interpretation applicable to any form of communication as will be discussed later.

The ubiquitous influence of unconscious processes on conscious mental functioning is a major principle of psychoanalytic theory. The role of unconscious processes in mental functioning follows inexorably from the dynamic principle in psychoanalysis, from the theory of forces activating the mental apparatus. Deriving from the metaphorical use of surface for conscious and depth for unconscious, psychoanalysis has come to be considered a "depth psychology." Though the concept of surfaces or layering is hardly appropriate, it has nonetheless served as the basis for a term applied to this aspect of psychoanalytic theory, namely, the *topographic principle,* the relationship between conscious and unconscious mental processes.

One of the empirical findings of psychoanalysis is the persistent and powerful influence of early childhood experience. Although the events and wishes of the years before the age of six seem to fade from memory and very little of them can be recalled in later years, these events nonetheless affect psychological development and personality structure in the most profound ways. For ages, educators have appreciated this principle intuitively. Freud reached the conclusion empirically. Ethologists have confirmed the principle experimentally.

The human mind develops as the interplay between the maturation of inherent, biologically determined capacities and the vicissitudes of experience. In the transformation of the newborn infant into a mature human being, every stage of development presents the individual with a specific set of problems, with fresh goals to be achieved. How problems are solved at one stage will influence the ability of the individual to negotiate the next set of developmental challenges. The specific needs, achievements, and conflicts of one phase are superseded but not displaced or eliminated in the next phase. As a rule, but not always, the successful resolution of developmental challenges of earlier phases seems to facilitate successful resolution of later developmental challenges. On the other hand, accidents of fate, such as severe illness, inborn physical or psychological deficits, cruel treatment, negligent care, or abandonment, may all have adverse effects upon individual development, causing pathological development immediately or rendering the individual incapable of mastering subsequent developmental challenges. Such events or influences constitute psychological trauma when they overwhelm the ego's capacity to master the terrifying and painful dangers that may occur at all periods of life. This is the beginning of the process of pathogenesis, which may lead to inhibitions, symptoms, perversions, and character deformations. Emphasis on the importance of childhood events for normal and pathological development constitutes the *genetic principle* in psychoanalysis. It is an empirical finding, not a derivative hypothesis.

Before considering the form that the persistent wishes of childhood take, it is necessary to appreciate the nature of children's thinking. The child's wishes are urgent, imperious, and uncompromising. His interests are self-centered. The distinction between objective perception and inner wishful thinking is not firmly established. Fantasies of the magical power of thought, of omnipotence and destructiveness are taken very seriously. The child's grasp of reality and causality take a long time to develop. In many ways, infantile notions of magic persist in the minds of many adults. Evidence for this fact is readily available in the prevalence of superstitious beliefs.

It is against this background that the child tries to formulate answers to fundamental existential problems. He must confront the inevitability of frustrations of his needs and wants, the limitations of his control over his own body, and the discrepancy between himself and adults. Issues of procreation, life and death, sex and violence, the different roles of men and women—the challenges are universal, true for children all over the world, no matter the cultural level of the environment in which they are being raised. Each child attempts to answer these questions with the limited intellectual resources at his command. The assistance that he gets from grownups is not always cognitively useful or affectively satisfying.

The child creates his own fantasy solutions to these problems and these fantasies serve as vehicles for the powerful driving wishes, fears, and self-punitive notions typical for that period of development. Originally, it may be presumed, such fantasies are exclusively or primarily imagined representations of wishes fulfilled but, as the child becomes aware of the dangers connected with the emergence or the expression of such wishes, even their fantasy expressions are modified by the process of compromise formation. Because the more primitive expressions of these wishes prove to be dangerous, they are repressed and forgotten. As an adult, the individual is not at all aware that he ever harbored such notions. Such repressions, however, are rarely definitive. Although the wishes remain unconscious, they continue to exert a dynamic role, making their effects discernible in derivative forms. Some of these derivative forms emerge as fantasies, watered-down versions, symbolically altered, less threatening representations of the original wishes. Derivative, acted-out forms of the same wishes may take the shape of habits, character traits, special interests, choice of profession, and a wide range of psychopathological formations. In summary, the unacceptable wishes of childhood take the form of persistent unconscious fantasies, exerting a continuous stimulus to the mind, eventuating in compromise formations. Some of these compromise formations are adaptive and are considered normal. Others are maladaptive—"abnormal."

Just how early in life the ability to create fantasy emerges is difficult to say. The process is certainly facilitated by the acquisition of language, but the ability to fantasize seems to antedate the appearance of language to some degree. This occurs probably in the second year of life. The complex fantasy life that certain observers believe to characterize the psychological activity of infants even as far back as the first six months of life does not seem to be a tenable proposition. Fantasy thinking is metaphoric in nature and makes extensive use of symbolism. This is an inevitable outcome of the fact that human thought is inherently metaphoric. Fundamentally, perceptual experience is processed according to the criteria of pleasant or unpleasant, familiar or unfamiliar. Memories of perceptual experiences are stored, organized, and patterned in keeping with these principles. The facile transfer of meaning from one mental element to another on the basis of similarity or difference, of association in memory or experience, leads quite directly to metaphoric thinking and symbolism.

Although the nature of the unconscious fantasies remains constant, their derivative manifestations evolve and are transformed in the course of time. They change with the advancing cognitive capabilities, with the appreciation of the real environment, and with the consolidation of moral values. The fundamental plot of the fantasies, however, remains the same. The characters and the settings change and "grow up" with time. There is good evidence to suppose that unconscious fantasying goes on all the time we are awake and a good deal of the time we are asleep. Every individual harbors a set of unconscious fantasies typical for him or her. They represent the special way in which that individual integrated the major experiences and relationships, the important traumata and drive conflicts of his childhood years. The persistent unconscious fantasies serve as a mental set against which the sensory data are perceived, interpreted, and responded to. Furthermore, specific perceptions of events in the external world may resonate with elements in the individual's unconscious fantasy system and may evoke conscious representations of unconscious fantasy wishes. The derivatives consist of compromise formations that may be adaptive or maladaptive. The emergence of maladaptive compromise formations of unconscious fantasy wishes marks the beginning of the process of pathogenesis.

III. PSYCHOANALYTIC METHODOLOGY AND THERAPY

The fundamental operational principles of psychoanalysis—determinism, dynamic conflict, and the role of unconscious mental elements—all enter into the organization of the standard mode of psychoanalytic investigation. This is known as the psychoanalytic situation. The patient reclines on the couch, looking away from the analyst. He is asked, as far as possible, to report with complete candor whatever thoughts or feelings present themselves to consciousness. In effect, he is asked to function as a nonjudgmental reporter of his own mental functioning. No consideration justifies the exclusion of any element that occurs to the analysand's mind. This technique of reporting is called *free association*. Its aim is to obtain a dynamic record of the analysand's mode of mental functioning, reflecting an endogenously determined flow of the individual's thought. External influences are reduced to a minimum. When external influences do intrude upon the analysand's awareness, they are examined from the point of view of the dynamic, evocative power they exert upon the stream of the analysand's associations, on his mode of mental functioning. During the analytic session, external intrusions may take many forms. The siren of a passing fire engine, the perception of a change in the decor or furniture arrangement in the analyst's office, the odor of flowers in the room, and, most important of all, whatever the analyst does or says. In any event, the approach to each of these perceptual experiences remains the same. What the analyst studies is the dynamic, evocative effect of the perceptual experience upon the nature of the analysand's thoughts.

The analyst has a dual role in this procedure. He is an observer of how the patient's mind works. At the same time, however, through the things that he says and does, he becomes a participant in the process. Functioning as a participant observer, the analyst pays special attention to the effects his interventions produce in the stream of the analysand's associations. It is for this reason that the analyst must pursue rigorously a stance of nonjudgmental neutrality regarding the analysand's realistic decisions, moral dilemmas, or partisan conflicts. The fundamental concern of the analyst is to observe and to understand how the patient's mind works. The analyst's opinions and prejudices are irrelevant to this work and, in fact, introducing them may be counterproductive. Such efforts would be suggestive and educational in nature, subtly directing the analysand away from trends in his own thinking that he would consider to be counter to the analyst's point of view or interests. The technical aim of the analyst is to supply understanding and insight, not to furnish a set of directives or to act as a model for the analysand to emulate. This conjoint investigation of the workings of a person's mind has no parallel in any other form of human communication.

From the point of view of theory, the results of this inquiry may be examined from several hierarchically related levels of abstraction. From the fundamental, experiential level of *clinical observation*, the analyst gets to know things not available to other observers. More than that, he becomes aware of the form and content in which thoughts appear, the patterning and configurations they assume, the repetition of certain themes, the irrelevant or unexpected intrusions of ideas and actions, the struggle of the patient to hide some elements or to minimize or repudiate their significance.

The next level is that of *clinical interpretation.* As in any form of communication or dialogue, more meaning is conveyed than is contained in the explicit verbal or motor expressions alone. The context in which ideas occur, the relationship to contiguous elements, the position of the idea in a sequential series of thoughts, the similarity to antecedent mental presentations or their persistence and repetition—all of these, augmented, to be sure, by the nature of the quality of the analysand's speech, the affective mood that is projected, and the motor concomitants of communication, enable the analyst to make connections that are not immediately apparent in the manifest text of the analysand's productions. In this respect, he is aided by the analysand's use of figurative speech, especially metaphor and symbol. Their use enhances and extends the communicative significance of the analysand's productions. At this level of the analytic work, what the analyst does is to make connections among the analysand's thoughts, connections that the analysand has been unable or unwilling to acknowledge on his own. From the stream of the analysand's associations, the analyst infers meanings and motives unknown to the analysand, which he communicates to him.

How is it possible for the analyst to learn something about another person's thinking, something of which the latter is himself not at all aware? In the genesis of interpretation, a number of important processes take place in the mind of the analyst. First is the experience of *empathy.* The analyst identifies with the patient, that is, he puts himself in the patient's position psychologically but is aware at the same time that the moods and thoughts that occur to him represent his reflections and reactions to what the patient has been telling him. The second process is *intuition,* according to which the analyst organizes the patient's productions and integrates them with what he has learned about the analysand previously, but it is a process that takes place outside the scope of the analyst's awareness. The end result of this integrative working over of the patient's productions into a meaningful hypothesis presents itself to consciousness through the process of *introspection.* The thought thus formed may be incomplete, incorrect, or usually only a step in the direction of the proper apprehension of the meaning of the analysand's associations. A more accurate interpretation comes about when the results of this intrapsychic communication to the analyst are consciously and cognitively examined in the light of the criteria already mentioned. Meaning derives from context, contiguity, sequence, similarity, figurative language, especially metaphor and symbolism, and other elements that, in general, hold sway in communication.

An interpretation is actually a hypothesis offered to the analysand. Interpretations vary in the amount of data they attempt to comprehend. There are interpretations concerning minute sequences of mental processes, ranging to comprehensive formulations concerning the meaning, origin, and purpose of lifelong patterns of behavior, thought, and feeling. Like any hypothesis, an interpretation must be consistent with the data of observation and it must be coherent. Furthermore, a psychoanalytic interpretation is not proferred in the course of analytic investigation merely as a summary of the relationship among observable data. The dynamic impact of a particular intervention called interpretation is what is important. An immediate acceptance or rejection of the interpretation by the analysand can be totally misleading and is actually beside the point. The effect of the intervention on the subject's flow of thought, the new material that it brings to light—these are of paramount significance. In effect, each interpretation the analyst makes constitutes a sort of experimental intervention.

From the experience of clinical observation and interpretation, certain *clinical generalizations* become possible. Such generalizations may apply to the meaning of repetitive, diverse, but related patterns of mental activity or syndromes articulating similar compromise formations. They may be observed repetitively in the individual; they may be recapitulated from patient to patient. In a compulsion neurosis, for example, the compulsive symptom generally protects the individual against unpleasant affects resulting from conflicts over murderous impulses. For the fetishist, the presence of the fetish is an essential condition for making sexual pleasure possible by presenting the individual with an actualization of an unconscious fantasy of a female phallus, a concept necessary in the case of the fetishist, to deny the possibility of genital mutilation.

From clinical interpretations and clinical generalizations, it becomes possible to formulate certain theoretical concepts that flow logically from the interpretations and to which the interpretations may lead. Most of the basic, operational, theoretical concepts of psychoanalysis belong to this category. Among these, for example, are the concepts of repression, defense, unconscious fantasy, and compromise formation. This is the level of *clinical theory*.

Finally, there are those abstract theoretical concepts not directly derived from clinical observational experience. Accordingly, this level of abstraction is referred to as *metapsychology*, that is, beyond psychology. Metapsychology concerns issues such as the compulsion to repeat, the nature and origin of mental energy, the relationship of quantitative changes in drive energy to the experience of pleasure and pain, or whether repression takes place because a mental element is divested of its drive energy or is opposed to countervailing drive force supplied by the ego. Most of these concepts Freud borrowed from physics and biology. Robert Waelder, who delineated the levels of psychoanalytic propositions just mentioned, said that these levels are not of equal importance for psychoanalysis. The data of observation and clinical interpretation are entirely indispensable, not only for the practice of psychoanalysis but for an appreciation of the empirical basis upon which psychoanalytic propositions are founded. Clinical generalizations and clinical theory are necessary too, though perhaps not to the same degree. Studying the record of free associations in context, a person may be able to understand a situation, a symptom, or a dream with little knowledge of clinical theory. Metapsychological abstractions, however, bear little relevance to the interpretations of the observational data and to the generalizations drawn from them. In actual therapeutic experience, metapsychological theories play hardly any role in the formulation of psychoanalytic conclusions.

IV. PSYCHIC STRUCTURE

Psychoanalytic theory is neither static nor unchanging. It has been the subject of continuing revision, sometimes radical in nature, in keeping with new insights and fresh discoveries. The current division of mental organization according to the specific function that each psychological element plays in intrapsychic conflict has been designated the *structural hypothesis*. Persistent patterns of functioning, repetitive in nature, more or less predictable in function, are grouped together as a component structure of the mind. This theory replaces an earlier concept of mental organization that Freud enunciated, the topographic theory. According to that theory, accessibility to consciousness was the paramount criterion of mental organization. Repressed elements collectively constituted the system *unconscious*, usually designated *ucs*, and the elements it contained shared certain characteristic modes of functioning. They were primarily instinctual wishes, driving impulsively toward gratification. The wishes were basically primitive in nature and operated in complete disregard for logic or the fixed categories of mental concepts, functioning indifferently to realistic considerations. Freud abandoned this theory when his observational data indicated that forces opposed to the drives, fixed mental concepts, defensive activities, and adaptive mechanisms also operated outside of the individual's awareness. According to structural theory, portions of the ego, the id, and the superego, that is to say, parts of all of the psychic structures, function outside of consciousness and exert a significant influence on conscious mental activity.

The delimitations of the structural components of the psyche are not as sharp as one might think. In a manner reminiscent of behavior in political organizations, there are shifting alliances among the component psychic systems and there are even contradictions within the component parts of each system. Severe intrapsychic conflict lays bare most clearly the outlines of the different psychic agencies. Typically, in the hysterias, the ego must deal with a sharp conflict between the wishes of the id for forbidden gratifications and the countervailing condemnation of the superego. In severe depressions, the id and superego impulses join forces in a murderous assault upon the self. In certain forms of psychopathy, an alliance is made between the id and the ego. In asceticism, superego and ego combine against the pleasure-seeking impulses of the id. To be sure, these formulations represent extreme simplifications but they serve to illustrate the clinical conceptualization of intrapsychic conflict within the structural theory.

Pursuing its function of attempting to reconcile the conflicting demands made upon it, the ego has available to it a wide range of mental mechanisms. The mechanism of repression has already been mentioned, but there are many others in addition. A drive impulse may be diverted from its primary object onto other objects, objects less important and less threatening. Or the impulse may be transformed into or displaced by a different impulse, even an opposite one, for example, hatred into love, or vice versa. The existence of the disturbing impulse may be acknowledged, but it may be mistakenly perceived as being present in someone else. These mechanisms rarely function in isolation in regard to a particular conflict. Because they were first described as modes of ego functioning, serving the common purpose of fending off an impending danger, these operations were referred to as mechanisms of defense. Closer examination, however, reveals that the ego makes use of such operations for many purposes other than defense. Fundamentally, these are mechanisms of the mind, serving the process of adaptation in the broadest sense of the word.

The unconscious defensive displacement of fantasy wishes from the original or primary object onto others is a constant feature of psychological life. Unconsciously, the individual may transfer these persistent wishes onto other objects who in some way become associated with the original object of the individual's drives. It is as if the individual foists a preconceived scenario onto people and events, a process that endows a unifying pattern upon the course of the individual's life. This process of transferring fantasy wishes onto other persons, called *transference,* plays an important role in psychoanalytic treatment. In contrast to other people in the course of everyday life, the analyst does not respond to the transferred wishes the patient directs toward him. Accordingly, derivatives of the patient's fantasy life and experiences of the past emerge with special clarity in the course of the analytic relationship. Analyzing the transference enables the analyst to demonstrate to the patient how the unconscious wishes of childhood have persisted in the patient's life, causing him to confuse fantasy with reality, past and present. Analysis of the transference is a particularly effective instrument to demonstrate to the analysand how the past is embedded in the present.

In addition to the formative role played by gratification and/or frustration of the drives, there are other aspects of the interrelationship with the "others" that play a crucial role in developing psychic structure. The term used for the interaction with other individuals is *object relations.* The term distinguishes between self and object, but it probably owes its origin to earlier concepts in the history of psychoanalysis. Freud used the term originally to apply to the mental representation that was the object of an instinctual drive. Strictly speaking, the term "object" could refer to the self, to a portion of one's body, or to mental representations of other individuals. In actual practice, it has come to apply almost exclusively to other persons. The term *narcissism* applies when the self is taken as the object of the erotic drive.

From his interaction with objects in his immediate world, the individual acquires a wide catalog of methods for coping with difficult situations, preferred solutions to conflicts, skills to master, ideals to aspire to, and so on. Basically, these acquisitions result from an identification that the individual effects with certain objects, that is, the individual remodels himself after the object and takes over some aspect of the personality of the other. Identification, however, is never complete; one person is never a psychological clone of another. Nor is it possible to predict in advance with which aspect of an object an individual will identify. Affective considerations play an important role in this process. Nor is it essential for one to have direct experience with an object in order to identify with him. Fantasy objects or individuals from literature, history, or religious teaching may become models from whom the individual acquires modes of thinking and acting. Although the personality of the individual is shaped greatly by the identifications effected with primary objects during childhood, it is a fact that identification is possible later in life, especially during adolescence. In many instances, such late identifications help to give the final stamp of character upon the personality.

Identifications that the individual makes in childhood tend to be primitive, highly idealized, and invested with grandiose illusions. Furthermore they are self-centered, that is to say, narcissistic. Part of the process of attaining maturity consists of replacing the exaggerated, grandiose, ideal aspirations of childhood with more realistic, obtainable goals, as well as developing a more objective evaluation of the primary object and of one's objects and of one's self. Some residues of early ideal formation persist in everyone and they serve as a standard against which the individual measures self-worth. The feeling of self-esteem depends, to a large extent, upon how one judges one's self in terms of the distance between the actual situation and the ideals and standards one sets for one's self. According to the conclusions reached, one will either like or dislike one's self. Some individuals, however, continue to harbor the grandiose ego ideals of childhood. They are impelled toward impossible goals and exaggerated expectations, all of which are doomed to failure and disillusionment. Such issues become central in the psychology of narcissistic persons. They require endless supplies of narcissistic gratification in order to feel worthwhile. When successful in the pursuit of their goals, they are elated; when unsuccessful, they become depressed and even suicidal.

V. COMPROMISE FORMATION

The turbulent conflicts of childhood come to a head during the phallic-oedipal phase. Powerful though his wishes are, the child ultimately recognizes that they can never be fulfilled. He dare not risk the loss of his parents nor can he master the fear of retaliation. In a sense, his own wishes come to represent a source of great danger to him. Either he renounces his wishes or masters them by modifying them. Injury to the genitals, the so-called castration anxiety, is the danger typical for this period. How the child resolves the conflicts of this period has fateful consequences for the development of his personality and the course of his life. The manner in which such conflicts are resolved varies from individual to individual. The inherent drive endowment of the individual, the nature of the relationships with his parents and other significant figures, the specific events in his life, and other experiential and maturational factors all play a role in shaping the final outcome of the conflicts of the oedipal phase. Certain forms of the wishes may be renounced entirely; others may be repressed but only incompletely. Taking the form of persistent unconscious fantasies, such wishes may exert a continuing influence upon mental functioning for the rest of the individual's life.

The turmoil of the oedipal phase is followed by a period of relative quiescence. This is known as the latency period and it lasts until the onset of adolescence. The intensity of conflicts abates and the individual becomes socialized and educable. But, though the conflicts have become relatively quiescent, they are hardly extinct. The turmoil of adolescence represents a secondary, more sophisticated attempt to resolve the conflicts of the oedipal phase at a higher level. A certain degree of restructuralization of the psychic apparatus occurs, out of which emerge the individual's sexual identity, his social and professional roles, and his inner moral commitments. Whether the secondary recapitulation of the conflicts of the oedipal phase during adolescence is primarily the result of biological or sociological factors is impossible to ascertain.

How the individual resolves the inevitable conflicts of life decides the issues of normality or pathology, of health or illness. Largely, this is a question of degree, since it is impossible for all compromise formations instituted by the ego to be equally effective under all circumstances. Some individuals, for example, seem never to have been able to master their childhood fears. They continue to suffer from irrational inhibitions, fears, and compulsions throughout latency and adolescence and into adult life. In other cases, pathology originates when some event in life upsets the balance of compromises that the ego has managed to maintain. This situation activates a latent conflict from childhood, and the individual responds to fantasied dangers articulating the effects of persistent unconscious childhood wishes. The reactivation of such latent unconscious conflicts is called *regression*. The responses to the inherent dangers are automatic, that is to say, outside of the individual's control. In addition, they are maladaptive and may be painful. The goal of psychoanalytic therapy is to give the patient insight into the nature of his fears and to demonstrate to him the automatic, inappropriate, largely unconscious measures he has undertaken to cope with his irrational anxieties. As the patient comes to recognize how the past is operative in his functioning in the here and now, he becomes able to organize more effective, less conflictual compromise formations to manage the effects of his persistent conflicts.

VI. SOCIAL APPLICATIONS

By applying some of the principles derived from the knowledge of the psychology of individuals, psychoanalysis has been able to afford some measure of insight into social and group phenomena. Certain group phenomena appear to replicate *en masse* some of the themes and psychological mechanisms observable in individuals. Themes and modes of expression common to myths, fairy tales, literary works, and religious traditions and rituals often repeat, sometimes in minute detail, the derivative expressions of unconscious conflicts observed in individuals. They prove appealing to the individual because, like fantasies and dreams, they represent disguised representations of repressed wishes. In large measure, the fantasy life of each individual represents a secret rebellion against the need to grow up and to renounce the gratification of his drives. In a very specific sense, it constitutes rebellion against society and certain of the strictures that civilization imposes upon the individual. In the process of mythopoesis and literary creation, the poet and the mythmaker transform their private daydreams into creations compatible with the ideals of the community, capable of giving pleasure and conveying at the same time disguised, transformed expressions of forbidden wishes that other members of the community share in common with them.

In the myth or the work of art, an individual may find in the external world a projected representation of his own unformed and unexpressed wishes. By identifying with the principal characters in a work of fiction or with the hero of the historical or religious myth, the individual attains some measure of gratification of his repressed, unconscious wishes, impulses that are ordinarily forbidden. Those who participate in this process, the audience or the members of a cult, constitute a group by virtue of sharing with each other disguised indulgence in forbidden wishes. In doing so, they exculpate each other by the knowledge that they have all participated, however transiently, in the pleasure of an unconscious fantasy shared in common. As Hanns Sachs said, everybody's guilt is nobody's guilt.

An unconscious fantasy shared in common serves as the bridge from individual conflict to mass participation. The capacity of the political or religious leader to evoke unconscious fantasies shared by large numbers of the population has played a significant role in many of the mass movements of this century. Mass movements are often organized around a central charismatic figure, who is intuitively perceived as the hero of a latent or manifest group mythology and who comes to represent the ego ideal of the members of the group. Because of the inherent methodological problems, however, validation of psychoanalytic insights into mass phenomena remains problematic.

VII. MORE RECENT THEORETICAL FORMULATIONS

Many new developments have occurred in psychoanalytic theory during the past few decades. Some may be considered extensions or elaborations of the basic concepts of the drive–conflict–compromise formulation with renewed emphasis on certain vicissitudes of development. Other approaches are clearly revisionary in nature. Foremost among these are Kleinian theory, British and American object relations theory, self-psychology, relational psychoanalysis, and intersubjectivity theory. Some of these have coalesced into identifiable orientations or schools of thought with their own organizations, conferences, and publications. Whether these newer theoretical formulations expand the internal consistency or explanatory power of the more traditional Freudian theory remains to be seen.

These theoretical approaches, which are sometimes subsumed under the general category of object relations theory, challenge the motivational importance of biologically grounded drives in human experience, including the role of the drives in establishing the infant–mother bond. Instead, they emphasize that drive-related strivings gain expression and are really understandable only in terms of the early experiential interactions with gratifying or nongratifying objects. The early patterns of self–object relations are internalized and repressed but they exercise a dynamic power upon the mind, leading the individual to repeat earlier patterns of interaction with significant objects. Thus, from these perspectives, it is not the derivatives of conflictual childhood wishes per se that are repudiated but the direct relational constellations through which these wishes came into being. It is the tendency to repeat earlier pathological patterns of interaction with significant objects that brings about pathology. Proponents of this approach believe that object relations formulations are especially relevant to the understanding of more disturbed patients, patients often referred to as borderline or narcissistic, as well as to neurotic patients. Characteristic of more disturbed patients is a tendency to split the mental representation of the primary objects into two parts, one entirely good and the other entirely bad. This primitive defense of splitting, it is maintained, is more germane to the psychopathology of such patients and their regressive reactivation of unconscious fantasy wishes typical for neurotics.

A variant of the object relations approach, one that is somewhat closer to the conflict concept in psychoanalytic theory, is the separation-individuation paradigm proposed by Margaret Mahler. On the basis of observations of infants and toddlers, she conceptualized a series of stages through which the individual achieves psychological as well as physical separation from the mother. Successful separation culminates in subjective feelings of autonomy and the emergence of the individuality that become the foundations of the self, of psychological personhood. From this standpoint, pathology results from an inability to negotiate successfully the psychological challenges of the separation-individuation process. In such instances, the individual finds it difficult to engage and resolve the conflicts of the oedipal period. Psychotic pathology reflects the failure to emerge from the state of symbiotic fusion with the mother; neurotic pathology would point to an inability to reach that stage of autonomy in which the object is firmly conceptualized as an independent and constant one. Though Mahler's research concentrated on the neonate's dependence on the mother, more recent studies by infant researchers have stressed the infant's inborn perceptual preferences, experiences of perceptual unity, and programmed capacity to interact with the environment. Evidence of the infant's ability to order and differentiate a variety of stimuli and, by inference, to experience the process of emerging organization has led some researchers to postulate a series of different senses of selfhood present from birth on. The findings of such investigations present a direct challenge to some of Freud's earlier metapsychological theories, such as the notion of "primary narcissism," a theory that assumed that, at the beginning of life, all pleasurable drive tendencies are vested in the self.

Theories that delineate successive senses of self-experience during the first two years of life, based on direct observations of neonates, parallel certain psychological theories of the self introduced by other psychoanalysts in recent decades. These theories elevate the concept of the self to a superordinate position in explaining mental development, psychopathology, and psychoanalytic treatment. Such theories hope to replace explanations framed in terms of Freud's structural hypothesis and the concepts of intrapsychic conflict. Theories of the self claim to address two types of deficiencies they discern in traditional psychoanalytic therapy. They assert their explanations of clinical phenomena to be closer to actual experience than those explanations framed in terms of the traditional model, which they regard as mechanistic, experience-distant, and wed to a metapsychology based on dated neurobiological and energic concepts. Second, they claim superior insight into the more disturbed patients, particularly those with severe character pathology. According to them, the pathology of the more disturbed patients is based not on the conflicts associated with the oedipal phase of development but on what they call "archaic," "preoedipal," "narcissistic," or "self object" transferences that characterize the object relations of such patients. Conflicts concerning mothering and nurturing become central in treatment. The classic conflicts of the oedipal phase, conflicts concerning envy, rivalry, hostility, and fear, are held to be of secondary importance. Self psychologists see their patients relating to them in primitive ways that correspond to the modes in which infants and young children use and depend on their parents, especially their mothers. For such analysands, the analyst becomes an object who promotes their development by mirroring and fostering emergent feelings of adequacy and self-worth. The analyst becomes an object for idealization, someone who calms and soothes the patient, thus aiding the patient in the regulation of inner tensions.

Unlike most object relations theories, these theories of the self tend to understand early development in terms of the programmed unfolding of a constellation of functions that collectively constitute the self. The self in its regulatory, mediating, integrating, and initiative-taking activities thus supplants the ego in Freud's structural theory. Whether in the guise of a "self schema," a "self organization," or a "bipolar nuclear self," such theories understand mental health as the adequate realization of certain maturational potentials inherent in the self. Correlatively, pathology is viewed as a derailment of this biopsychological program at some point in early life. Thus, some self theories, construing early development in terms of successive stages or phases to be transversed, are led to view psychopathology from the standpoint of a deficit psychology, whereby pathology seems to develop because the child's caretakers have not supplied the necessary crucial self-affirming interactions. Because of these deficiencies in the parent–child interaction, developmental arrests develop, nodal points in early life from which the self was unable to proceed with its programmed agenda. This position stands in contrast to the conflict psychology through which psychoanalysis has traditionally approached psychopathology and psychotherapy. The relationship between deficit and conflict psychologies, as is also the question of whether the self has explanatory or heuristic power greater than that of the Freudian ego, is a subject of ongoing debate within contemporary psychoanalysis. According to some self psychologists, pathology is a consequence of a failure to achieve a vital, cohesive, "nuclear" self, a failure generally rooted in unempathic, unresponsive mothering that has not mirrored or reinforced the early feelings of grandiosity that, under conditions nearer optimal, can blossom into healthy feelings of vitality and self-worth. Although self-psychology is offered as a rival theory to the conflict–danger–compromise approach, it may be that actually it only adds a further danger situation to the four typical dangers mentioned previously, namely, the danger of loss of self.

The relational/intersubjective schools conceive of mind as consisting of "transactional patterns and internal structures derived from an interactive and interpersonal field" (see Mitchell, 1988). A central human concern is to achieve authentic personal meeting. Sexuality is a response rather than an internal pressure. Patterns of relating derived from a tendency to preserve continuity are repeated in different forms throughout development and the impact of early experience is less crucial than in Freudian theory. These schools stress that the analyst must acknowledge the analysand's privileged understanding of her own experience and the analyst's own subjectivity as a codeterminant of the treatment process. The analyst is advised to work against the asymmetry inherent in any doctor–patient relationship. Technique, these new theories hold, should foster a sense of mutuality and further the construction of new relational and intersubjective meanings that focus on the analyst–analysand interaction as seen through the eyes of the latter. This is very different from the classic analytic view of the analyst who listens neutrally but with an expectation borne of experience that the analyst will be able to infer unconscious meanings from the patient's symptoms, free associations, interactions with people, and other behavioral descriptions.

Among recent theoretical developments, biological formulations of psychoanalytic concepts deserve special mention, but with a proviso. Descriptions of neurobiological mechanisms have not as yet been shown capable, and indeed may never supplant psychological propositions. The most productive work in neurobiology involves work on the neuroanatomical structures and the neurophysiological processes that subtend psychoanalytic concepts. Rather than modifying its understanding of mental functioning on the basis of neurobiological findings, psychoanalysis has pointed neurobiology in the direction of particularly promising conceptual and experimental endeavors. Thus we now have a body of literature that addresses topics of analytic concern: neurophysiological demonstrations of unconscious mental processes, for example, between right brain and left brain, that relate to Freud's notions of primary and secondary processes; research on the neurophysiological pathways of affective expression; and research on the impact of perceptual environmental experiences on neurological development. Notions of "neuroplasticity" (the notion of critical periods during which certain experiences are necessary for optimal brain development) and of a "neurorepresentational system" have been proposed as bridge concepts linking neurobiological functioning to mental activity, including unconscious mental processes. A recent study equates the Freudian unconscious with biogenetically ancient mechanisms that involve rapid eye movement (REM) sleep and are located in the prefrontal cortex and associated structures. Research that seeks to elaborate a neurobiological substrate to psychoanalytic concepts provides the same useful function as research into the psychological processes associated with neurobiological events. In the case of neurobiology and psychoanalysis, benefit is derived from reporting research in coordinate rather than in casual terms so that, in the words of J. H. Smith and J. C. Ballinger, "For the neurobiologist, psychological events are markers of neurobiological processes and for the psychologist, neurobiological events as markers of psychological processes."

BIBLIOGRAPHY

Blum, H., Kramer, Y., Richards, A. K., and Richards, A. D. (eds.) (1988). "Fantasy, Myth, and Reality: Essays in Honor of Jacob A. Arlow." International Universities Press, Madison, Connecticut.

Eagle, M. N. (1984). "Recent Developments in Psychoanalysis: A Critical Evaluation." McGraw–Hill, New York.

Edelson, M. (1988). "Psychoanalysis: A Theory in Crisis." Univ. of Chicago Press, Chicago/London.

Kernberg, O. F. (1984). "Severe Personality Disorders: Psychotherapeutic Strategies." Yale Univ. Press, New Haven, Connecticut.

Kohut, H. (1984). "How Does Analysis Cure?" Univ. of Chicago Press, Chicago.

Mitchell, S. (1988). "Relational Concepts in Psychoanalysis." Harvard Univ. Press, Cambridge, Massachusetts.

Pine, F. (1985). "Developmental Theory and Clinical Process." Yale Univ. Press, New Haven, Connecticut.

Reiser, M. F. (1984). "Mind, Brain, Body: Toward a Convergence of Psychoanalysis and Neurobiology." Basic Books, New York.

Richards, A. K., and Richards, A. D. (1995). Notes on psychoanalytic theory and its consequences for technique. *J. Clin. Psychoanal.* 4(4).

Richards, A. D., and Willick, M. (eds.) (1986). "Psychoanalysis, the Science of Mental Conflict: Essays in Honor of Charles Brenner." The Analytic Press, Hillsdale, New Jersey.

Smith, J. H., and Ballinger, J. C. (1981). Psychology and neurobiology. *Psychoanal. Contemp. Thought* 4(3), 407–421.

Stern, D. N. (1985). "The Interpersonal World of the Infant." Basic Books, New York.

Wallerstein, R. S. (1986). "Forty-two Lives in Treatment: A Study of Psychoanalysis and Psychotherapy." Guilford Press, New York.

Psychoneuroimmunology

ROBERT ADER
NICHOLAS COHEN
University of Rochester School of Medicine and Dentistry

GLOSSARY

Autoimmunity Immune reactions generated against an individual's own tissue or cellular antigens that can lead to diseases such as systemic lupus erythematosus and rheumatoid arthritis

Cellular immunity Immunity mediated by antigen-stimulated, thymus-derived effector T lymphocytes (e.g., cytotoxic T cells) that does not involve antibody

Conditioned response Response to a previously neutral (conditioned) stimulus after the neutral stimulus has been paired with a stimulus that unconditionally elicits the particular response being studied

Humoral immunity Effector mechanism or immunity that is mediated by circulating antibodies produced by bone marrow-derived B lymphocytes. For many antigens, B cells produce antibody only after the occurrence of complex interactions among helper and suppressor T lymphocytes, certain accessory cells, and B lymphocytes

PSYCHONEUROIMMUNOLOGY IS, PERHAPS, THE most recent convergence of disciplines (i.e., the behavioral sciences, the neurosciences, and immunology) that has evolved to achieve a more complete understanding of how the interactions among systems serve homeostatic ends and influence health and disease. The hypothesis that the immune system constitutes an underlying mechanism mediating the effects of psychosocial factors on the susceptibility to and/or the progression of some disease processes is tenable, however, only if it can be shown that the brain is capable of exerting some regulation or modulation of immune responses. Several laboratories are now exploring such relationships at levels of organization ranging from the molecular to the organismic. Psychoneuroimmunology, then, refers to the study of the interaction among behavioral, neural, endocrine, and immune processes of adaptation. The present synopsis concentrates on the behavioral component of psychoneuroimmunological research.

I. INTRODUCTION

In practice, research in immunology has proceeded on the implicit assumption that the immune system is autonomous—an agency of defense that operates independently of other psychophysiological processes. Indeed, the self-regulating capacities of the immune system are remarkable and the interactions among subpopulations of lymphocytes occur and can be studied *in vitro*. The same can be said for other physiological processes. The endocrine system, too, was once thought to be autonomously regulated, and it is only in modern times that neural influences were identified. It may well be that immune responses can be made to occur in a test tube, devoid (presumably) of the (confounding or modulating) influences of the natural environment, but it is the immune system functioning in the natural environment that is of ultimate concern. Without denying the idiosyncratic properties of the immune system, it is becoming clear that behavioral factors and neuroendocrine factors play a critical role in the immunoregulatory processes that contribute to homeostasis.

ENCYCLOPEDIA OF HUMAN BIOLOGY, Second Edition, VOLUME 7. Copyright © 1997 by Academic Press. All rights of reproduction in any form reserved.

It has been known for some time that manipulations of brain function can influence immune function. Lesions or electrical stimulation of the hypothalamus, for example, alters humoral and cell-mediated immune responses. Conversely, elicitation of an immune response (exposing an animal to an antigen) can influence hypothalamic activity. Following stimulation with different antigens, there is an increase in the firing rate of neurons within the ventromedial hypothalamus, and this increase corresponds to the time of the peak antibody response. Some of the most recent data documenting the potential for neural–endocrine–immune system interactions come from neuroanatomical studies of the innervation of lymphoid tissues (Fig. 1), which provides a foundation for functional relationships among these systems.

The observation that neurochemical and endocrine signals influence immune responses is supported by the identification of receptors on lymphocytes for a variety of hormones and neurotransmitters. Also, recent data suggest that lymphocytes themselves are capable of producing neuropeptides and hormones.

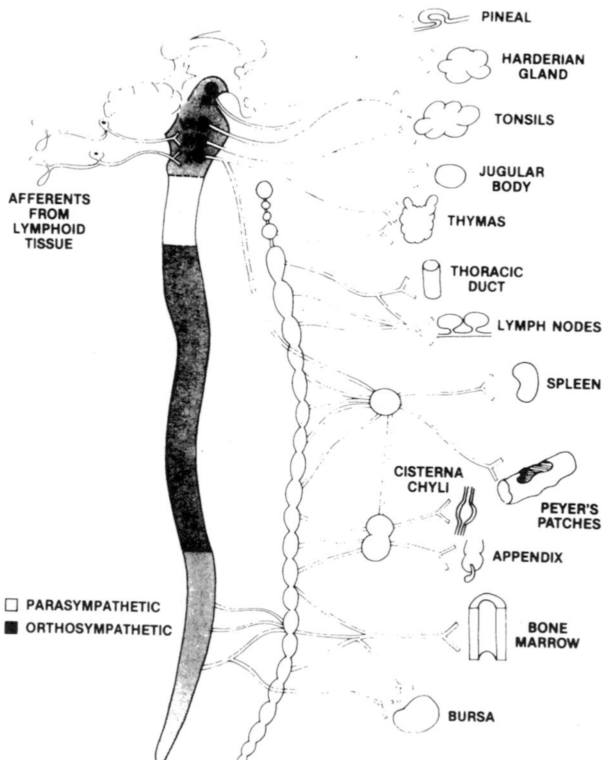

FIGURE 1 Recently collected data provide evidence for the innervation of lymphoid organs. [Reproduced, with permission, from J. B. Martin (1988). *Prog. Neuroendocrinimmunol.* **1**, 5–8.]

Again, not only do variations of hormonal state or neurotransmitter levels influence immune responses, but the immune response to antigenic stimulation induces neuroendocrine changes. At the neural and endocrine levels, then, there is abundant evidence of the potential for interactions between the brain and the immune system. [*See* Endocrine System; Lymphocytes.]

II. STRESS AND IMMUNE FUNCTION

A. Studies in Humans

Observations suggesting a link between behavior and immune function go back to the earliest observations of a relationship between psychosocial factors, including "stress," and susceptibility to those disease processes that we now recognize as involving immunological mechanisms. There are now abundant clinical data documenting an association between psychosocial factors and disease.

The death of a family member, for example, is rated highly on scales of stressful life events and has been associated with depression and an increased morbidity and mortality in the case of a variety of diseases, many of which are presumed to involve immune defense mechanisms. Bereavement and/or depression has also been associated with changes in some features of immunological reactivity such as a reduced lymphoproliferative response to mitogenic stimulation and impaired natural killer cell activity. Other studies have documented changes in immune reactivity associated with the affective responses to other "losses" such as marital separation and divorce. These are provocative and important observations. It should be emphasized, however, that the association between the response to "losses" and increased morbidity and the association between the response to "losses" and alterations in immune function do not, in themselves, establish a causal link between psychosocial factors, immune function, and health or disease. [*See* Depression.]

The death of a spouse is intuitively stressful, but changes in immune function are observed in humans exposed to less severe, but nonetheless effective, naturally occurring "stressful" circumstances. The level of distress during student examination periods, for example, is invariably greater than that during control periods and, in a series of experiments conducted at Ohio State University, transient impairments in several parameters of immune function were observed in medical students at such times. Relative to a non-stressful baseline measurement, examination periods are associated with a decrease in mitogen responsiveness, natural killer cell activity, percentage of helper T lymphocytes, and interferon production by stimulated lymphocytes. In students who are seropositive for Epstein–Barr virus (EBV), exam periods are associated with elevated EBV titers, interpreted as a poorer cellular immune response control over the latent virus. The incidence of self-reported symptoms of infectious illness is also increased during examination periods. Personality tests have not revealed differences between the student volunteers and their classmates, and other life changes that could have influenced immune function during examination periods were minor and unrelated to changes in immunological reactivity.

An old (and continuing) experimental and clinical literature suggests that immune function can be altered by psychological means (e.g., hypnosis), and an increasing number of studies are attempting to relate different personality characteristics and affective states with alterations in immune function. These correlational studies must be considered preliminary at this time. They are difficult studies to implement, so it is hardly surprising that unequivocal evidence for causal relationships does not yet exist.

B. Studies in Animals

Most of the evidence for stress-induced alterations in immunity comes from basic research on animals. A variety of stressors (avoidance conditioning, restraint, noise) can, under appropriate experimental circumstances, influence the susceptibility of mice to a variety of infectious diseases (Coxsackie B, herpes simplex, polyoma, and vesicular stomatitis viruses). It thus appears that stressful circumstances can alter the host's defense mechanisms, allowing an otherwise inconsequential exposure to a pathogen to develop into clinical disease. Adult mice are generally resistant to Coxsackie virus and show no manifestations of disease when exposed to either virus or "stress" alone. Symptoms of disease are observed, however, when mice are inoculated with virus and exposed to stressful environmental conditions. These results parallel clinical observations. For example, the presence of pollen alone may not be sufficient to elicit symptoms in a subject with hay fever, but the combination of pollen and a threatening life situation is.

Separation experiences (i.e., "losses") have also been studied in animals. Periodic interruptions of mother–litter interactions and/or early weaning decrease lymphocyte proliferation to mitogenic stimulation and reduce the plaque-forming response to subsequent challenge with sheep red blood cells (SRBC) in rodents. Monkey infants and their mothers respond to separation with a transient depression of *in vitro* mitogen resonsiveness. Separation of squirrel monkeys from their mothers results in several changes in immunological reactivity, including a decline in complement protein levels (an effector mechanism in humoral immune responses), macrophage function, and immunoglobulin G antibody responses to immunization with a benign bacteriophage, the magnitude of one or another of the effects being a function of the psychosocial environment in which the animals were housed following separation.

In adult animals, a variety of behavioral manipulations are capable of influencing a variety of immune responses in a variety of species—and in a variety of ways. Heat, cold, and restraint (each of which is commonly referred to as a stressor and each of which is commonly thought to elicit a representative "stress" response) have different effects on the same immune response, and the same stressor (e.g., restraint) has different effects on different immune responses. The intensity and "chronicity" of the stressor are among the several parameters of stimulation that are likely to influence immune responses. The initial response to auditory stimulation, for example, is a depression of mitogenic responsivity. Repeated exposure to the loud noise (i.e., chronic stress), however, is associated with an enhancement of this same response. Moreover, the initial association of elevated adrenocortical steroid levels with decreased immunological reactivity is not necessarily observed under conditions of chronic stress. In another series of experiments, there was a graded suppression of mitogen responses with increasing intensities of electric shock stimulation, a relationship that persisted in adrenalectomized animals. As has been observed for other psychophysiological responses, the data suggest that it may be the organism's capacity to cope with stressful environmental circumstances that determines the extent to which immune processes are affected.

Neuroendocrine states provide the internal milieu within which immune responses occur. Stimulating animals during prenatal and early life, varying social interactions among adult animals, and exposing animals to environmental circumstances over which they have no control are among the psychological manipulations that induce neuroendocrine changes that are now implicated in the modulation of immune responses. Our knowledge of interactions between neuroendocrine and immune function under normal and stressful conditions, however, is incomplete. For example, glucocorticoids are, in general, immunosuppressive. It is therefore assumed that adrenocortical steroid elevations, the most common manifestation of the stress response, are responsible for the frequently observed suppression of immune function. There are numerous examples of stress-induced, adrenocortically mediated alterations of immune responses, particularly *in vitro*. However, there are numerous other observations of stress-induced alterations in immune function that are independent of adrenocortical activation. The subtleties that can characterize the involvement of hormones and neuropeptides in the mediation of stress-induced alterations of immunity are illustrated by the results of different regimens of electric shock stimulation. Intermittent and continuous schedules of inescapable electric footshocks result in an analgesia to subsequent footshock. Only the intermittent shock, however, was found to be an opioid-mediated analgesia and only the intermittent footshock resulted in a suppression of natural killer cell activity. These results apparently reflect the immunomodulating potential of endogenously released opioids. Thus, even a cursory review of the literature makes it evident that the *in vivo* immunological consequences of stress involve extremely complex neural, endocrine, and immune response interactions. Considering that immune responses are themselves capable of altering levels of circulating hormones and neurotransmitters, these interactions are likely to include complex feedback and feedforward mechanisms as well. [*See* Neuroendocrinology.]

Stressful life experiences affect immunity, but the data yield an inconsistent picture of the direction, magnitude, and duration of the effects. At the very least, the effects of stress appear to be determined by: (1) the quality and quantity of stressful stimulation; (2) the capacity of the organism to cope effectively with stressful circumstances; (3) the quality and quantity of immunogenic stimulation; (4) the temporal relationship between stressful stimulation and immunogenic stimulation; (5) the parameters of immune function and the times chosen for measurement; (6) the experiential history of the organism and the prevailing social and environmental conditions upon which stressful stimulation and immunogenic stimulation are superimposed; (7) a variety of host factors such as species, strain, age, sex, and nutritional state; and (8) interactions among these variables. Although we are far from a definitive analysis of the effects of stress and the means by which perceived events are translated into altered physiological states capable of modulating immune functions, the available data do provide a body of evidence that the immune system is sensitive to the modulating effects of psychobiological processes ultimately regulated by the brain.

III. EFFECTS OF CONDITIONING ON IMMUNE RESPONSES

A. Background

Behaviorally conditioned alterations in immune function provide one of the more dramatic lines of research implicating the brain in the modulation of immune responses. These studies derived from serendipitous observations of mortality among animals in which a saccharin-flavored drinking solution had been paired with an injection of cyclophosphamide (CY), a powerful immunosuppressive drug. Pairing a novel taste stimulus, a conditioned stimulus (CS), with an agent that induces temporary gastrointestinal upset, the unconditioned stimulus (UCS), will, after a single pairing, result in an aversion to that taste stimulus. Repeated reexposures to the CS in the absence of the UCS will result in extinction of the avoidance response. During the course of repeated extinction trials, conditioned animals began to die and, more critically, mortality rate, like the magnitude of the avoidance response, varied directly with the volume of saccharin consumed on the single conditioning trial. In an attempt to explain these results, it was hypothesized that, at the same time that a behavioral response was conditioned, an immunosuppressive response was conditioned and being elicited in response to reexposure to the CS (Fig. 2). If so, this might have increased the susceptibility of these animals to any latent pathogens in the laboratory environment.

B. Conditioned Changes in Immune Function

To test this hypothesis, the authors designed a study in which rats were conditioned by a single pairing of saccharin-flavored water and an injection of cyclophosphamide. When the animals were subsequently immunized with SRBC, one subgroup of conditioned animals remained untreated (to assess the effects of conditioning per se), another subgroup of conditioned animals was injected with CY (to establish the uncon-

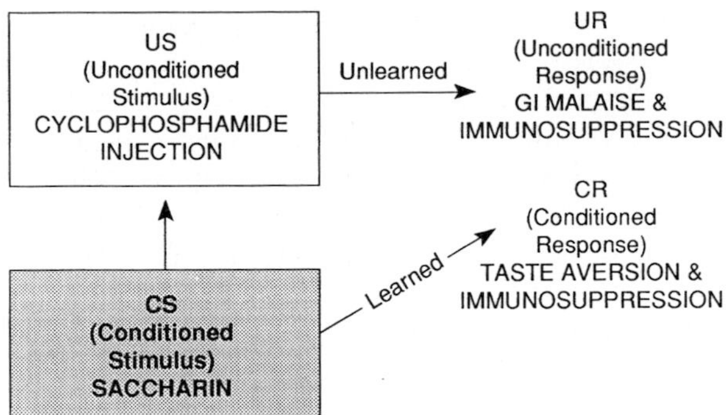

FIGURE 2 Schematic representation of the relationship between conditioned and unconditioned stimuli and conditioned and unconditioned responses. [Reproduced, with permission, from R. Ader (1987). *Immunopathol. Immunother. Lett.* 2, 6–7.]

ditioned immunosuppressive effects of the drug on the response to antigen), and an experimental subgroup was reexposed to the saccharin-flavored water (the CS). A nonconditioned group was also provided with saccharin-flavored water that had not previously been paired with their injection of CY. As hypothesized, conditioned animals reexposed to the CS at the time of antigenic stimulation showed an attenuated antibody response to SRBC in relation to nonconditioned animals and conditioned animals that were not reexposed to the CS. These results were taken as evidence of behaviorally conditioned immunosuppression and have now been independently verified and extended by several investigators.

The effects of conditioning have been quite consistent under a variety of experimental conditions. The magnitude and/or the kinetics of the unconditioned response vary as a function of the dose of CY, but one can still observe conditioned changes in immunological reactivity using different doses of CY. Also, there are several experiments in which the immunomodulating effects of "stress" have been conditioned. In addition to the effects of conditioning on antibody responses described here, conditioning is capable of influencing different parameters of cell-mediated immunity as well as a variety of nonspecific host defense responses. Not only has the release of histamine by sensitized animals been conditioned, but conditioned increases in a specific mediator of mucosal mast cell function have also been demonstrated.

The notion that conditioned immune responses are inextricably linked with conditioned avoidance responses or that there is some direct relationship between conditioned behavioral responses and conditioned immunological responses receives no support from the available literature. Taste aversions can be expressed without concomitant changes in immune function, and conditioned changes in immune function can be obtained without observable conditioned avoidance responses. Consistent with the relationship between conditioned behavioral and autonomic or endocrine responses, the available data suggest that different (multiple) conditioning processes and mechanisms are involved in the conditioning of behavioral responses and in the conditioning of different immune responses.

Both the acquisition and extinction of the conditioned enhancement of immunological reactivity have been observed using an antigen rather than pharmacological agents as the UCS. In one study, CBA mice were anesthetized, shaved, grafted with skin from C57BL/6J mice, and left bandaged for 9 days. In reponse to the grafting of allogeneic tissue, there was an increase in the number of precursors of cytotoxic T lymphocytes (CTLp) that could react against alloantigens of the foreign tissue. Since CTLp numbers did not return to baseline levels for 40 days, the conditioning manipulations were repeated three times at 40-day intervals. On the fourth trial, the procedures were repeated again, except that the experimental animals did not receive the tissue graft. Approximately half the animals, however, showed an increase in CTLp in response to the grafting procedures. When these "responders" were divided into groups that were exposed to additional conditioning trials or extinction trials (sham grafting), all of those that experienced additional conditioning trials showed a conditioned increase in CTLp, whereas none of the previous "responders" that were given unreinforced trials showed a conditioned response.

In another study, mice were repeatedly immunized with low doses of a common laboratory antigen following exposure to a gustatory conditioned stimulus. Three weeks after the last conditioning trial, half the animals in each of several experimental groups were reexposed to the CS alone, or they were reexposed to the CS in the context of a low-dose booster injection of the same antigen. Since it seemed unlikely that a CS could initiate antibody production by itself, this booster dose of antigen, insufficient to elicit more than a small antibody response, was administered to assure that a salient stimulus was available to activate the immune system. Reexposure to either the CS or the booster dose of antigen alone was not sufficient to elicit a robust antibody response. However, relative to several control groups, an enhancement of antibody titers was observed when conditioned mice were reexposed to the CS in the context of a minimally immunogenic dose of antigen.

The physiological mediation of conditioned alterations in immune function is not yet known. Some investigators have hypothesized that the conditioned suppression of immunological reactivity was the direct result of stress-induced responses since glucocorticoid elevations (taken as an index of "stress") frequently suppress some immune responses. The existing data, however, provide no support for stress-induced elevations in "stress hormones," notably adrenocortical steroids, as the mediator of conditioned alterations in immune function. In fact, most of the data stand in direct contradiction to such a hypothesis.

Although there are problems in interpreting extirpation experiments, it should be noted that, in one study, immunosuppression was not observed in adrenalectomized mice. On the other hand, no conditioned suppression of antibody production is observed when a drug, lithium chloride (LiCl), that elevates steroid levels but is immunologically neutral under defined experimental conditions is used as the UCS—or when steroid levels are elevated by injections of LiCl or corticosterone at the time of immunization. Conditioned suppression and/or enhancement of antibody- and/or cell-mediated responses occurs in the presumed absence of or with equivalent elevations in corticosterone and, presumably, other "stress hormones." In a two-bottle, preference testing procedure, fluid consumption is equal in experimental and control groups, and the thirsty animal is not faced with the conflict of choosing between drinking and the noxious effects that are associated with the CS solution. Under these experimental circumstances, conditioned immunosuppression is still observed in the antibody response to T-dependent and T-independent antigens, a graft-versus-host reaction, and in the white blood cell response to CY. Finally, in a discriminative conditioning paradigm, both the stimulus that signaled presentation of the UCS (the CS+) and a signal that was not associated with the UCS (the CS−) induce an elevation in adrenocortical steroid levels, but only the CS+ induces a conditioned release of histamine. It is reasonable to hypothesize that conditioned alterations of immunological reactivity may be mediated by conditioned neuroendocrine responses, but the data that have been collected thus far are inconsistent with the hypothesis that such effects are mediated simply by nonspecific, stress-induced changes in hormone levels.

Although the mechanisms underlying conditioned suppression and enhancement of immunological reactivity are not known, there is no shortage of potential mediators of such effects. Multiple processes are probably involved. Conditioned immunosuppressive responses, for example, occur when conditioned animals are reexposed to the CS before as well as after immunization. This observation could imply that the mechanisms do not involve antigen-induced immunological or neuroendocrine changes; they could also indicate that different mechanisms are involved when conditioning is superimposed on a resting or on an antigen-activated system. Also, different immunomodulating agents have different sites of action and the same immunomodulating drug may have different effects on activated or nonactivated lymphocytes. We now know that the immune system is innervated, that leukocytes and neurons share certain neuropeptide/neurotransmitter receptors, that lymphocytes can produce several neuroendocrine factors, and that cells of the immune system and the nervous system can produce and respond to the same cytokines. Thus, conditioned changes in neural and/or endocrine activity that can be recognized by activated lymphocytes or, conversely, the effects of conditioning on the release of immune products capable of being recognized by the nervous system constitute potential pathways for the conditioned modulation of immune functions. Indeed, in attempting to account for the conditioned enhancement of antibody production when antigen was used as the UCS, it was hypothesized that two signals were involved: one from the immune system in response to the booster injection of antigen, and one from the nervous and neuroendocrine systems in response to the conditioned stimulus.

Current research provides compelling evidence for the acquisition and extinction of conditioned suppression and enhancement of immunological reactivity. These studies dramatically illustrate the role of behavior in the modulation of immune responses, a modulation that is, presumably, ultimately regulated by the brain.

C. Clinical Implications of Conditioning Studies

There are strains of mice that spontaneously develop an autoimmune disease that is strikingly similar to systemic lupus erythematosus in humans. In this disorder, a suppression of immunological reactivity is in the biological interests of the organism. The (NZB× NZW)F1 female mouse, for example, develops a lethal glomerulonephritis that can be delayed by weekly injections of CY. Therefore, in an effort to determine the biological significance of conditioned alterations in immunological reactivity, the effects of conditioning were assessed in this animal model of autoimmune disease. If, as indicated earlier, the immunosuppressive effects of CY can be conditioned, could conditioned immunopharmacological effects be applied to a pharmacotherapeutic regimen, that is, could conditioned stimuli be substituted for some proportion of the active drug treatments received by these animals to delay the development of disease?

One group of mice (Group C100), treated under a traditional pharmacotherapeutic regimen, was given a saccharin-flavored solution to drink and, after each exposure to saccharin, the animals were injected with CY. As expected, this protocol delayed the development of proteinuria and mortality. An experimental (conditioned) group (Group C50) was injected with CY following saccharin on only half of the weekly occasions when the CS was presented. A nonconditioned control group (Group NC50) received the same number of saccharin and CY presentations as Group C50, but for this group the saccharin and CY were never paired.

Nonconditioned and placebo-treated animals did not differ. This indicated that half the total dose of CY administered to mice that were treated under the standard pharmacotherapeutic protocol was ineffective in modifying the course of autoimmune disease. Group C50 was also treated with half the cumulative dose of CY given to Group C100, but these animals developed proteinuria significantly more slowly than placebo-treated mice and significantly more slowly than nonconditioned mice treated with the same cumulative amount of drug.

The mortality data yielded the same results. There was no difference between nonconditioned animals and untreated controls, but conditioned animals treated with the same amount of CY as nonconditioned animals survived significantly longer than untreated controls and nonconditioned animals. The mortality rate of conditioned mice did not differ significantly from animals in Group C100, which received twice as much drug. These results indicate that, within the context of a pharmacotherapeutic regimen, conditioning effects were capable of influencing the onset of autoimmune disease using a cumulative dose of active drug that was not, by itself, sufficient to alter the course of disease.

After the period of pharmacotherapy, groups were divided into thirds that (1) continued to receive saccharin and CY on whatever schedule existed during therapy, (2) continued to receive saccharin and intraperitoneal injections of saline but no CY, or (3) received neither saccharin nor CY. Consistent with the interpretation that the effects on the development of lupus were conditioning effects, unreinforced presentations of the CS influenced the development of autoimmune disease in conditioned animals but not in nonconditioned animals. Among the animals conditioned under a traditional or continuous schedule of reinforcement (Group C100), mice that continued to receive CS exposures after the termination of active drug therapy survived significantly longer than similarly treated mice that were deprived of both the CS and the drug. In fact, animals that continued to be exposed to saccharin plus intraperitoneal injections of saline did not differ from animals that continued to be treated with active drug. These data, too, are consistent with the conditioned immunopharmacological effects described earlier. Consensual validity for these results is provided by studies in which repeated exposures to a CS previously associated with CY accelerated tumor growth and mortality in response to a transplanted syngeneic plasmacytoma and attenuated an experimentally induced arthritic inflammation. These findings document the biological impact of conditioned immunopharmacological responses. They also suggest that there may be some heuristic value in conceptualizing pharmacotherapeutic protocols as a conditioning (learning) process.

IV. IMMUNOLOGICAL EFFECTS ON BEHAVIOR

In the same way that there are reciprocal relationships between neural and immune functions and endocrine and immune functions, data are accumulating to suggest that there are immunological influences on behavior in addition to behavioral influences on immune function. Cytokines, a variety of chemical substances released by activated lymphocytes, facilitate intercellular communication within the immune system and, in addition, constitute a channel of communication from the immune system to the nervous system. Interleukin-1, in particular, is a cytokine released by activated macrophages that is capable of altering electrical activity in the brain, neurotransmitter functions, and a variety of behavioral and affective states. Several investigators have described the behavioral effects of (early) viral infections, the cognitive and emotional sequelae of autoimmune diseases, and behavioral differences between normal mice and those with a genetic susceptibility to autoimmune disease. [*See* Autoimmune Disease.]

Furthermore, recent data suggest that behavioral changes associated with immunological dysfunctions may actually be adaptive with respect to the maintenance or restoration of homeostasis within the immune system. Lupus-prone (NZB×NZW)F1 mice do not acquire conditioned taste aversions in response to immunosuppressive doses of CY that are effective in inducing conditioned avoidance responses in healthy control (C57BL/6) mice. Also, when tested after the development of signs of autoimmune disease (lymphadenopathy and elevated autoantibody titers), Mrl-lpr/lpr mice, another strain of animals that spontaneously develop a lupus-like disorder, do not avoid flavored solutions paired with doses of CY that are effective in inducing taste aversions in congenic Mrl+/+) control mice. These differences in behavior do not result from a learning deficit in the lupus-prone mice, since there are no substrain differences prior to the development of symptoms of disease nor when a nonimmunosuppressive drug is used as the UCS. Phenomenologically, it would appear that lupus-prone mice "recognize" the existence of their immunological deficit and/or the ameliorating effects of the immunosuppressive drug, despite its noxious gastrointestinal effects.

Mrl-lpr/lpr mice with symptoms of autoimmune disease also voluntarily consume more of a flavored drinking solution containing CY than do asymptomatic controls. Moreover, they drink sufficient amounts of the CY-laced solution to attenuate lymphadenopathy and autoantibody titers. Although not previously described with respect to the immune system, these data are consistent with a large literature indicating that behavioral responses are a primary means by which animals maintain and regulate some physiological states. Whether, in the case of a dysregulated immune system, the animal is responding to nonspecific, immunologically induced pathophysiological changes in one or another target organ or, consistent with the bidirectional pathways that link the central nervous system and immune system, whether the brain is capable of receiving and processing information emanating from the (dysregulated) immune system directly remains to be determined. To the extent that the brain is capable of acting on information provided by the immune system, it would appear that behavioral processes have the potential to serve an *in vivo* immunoregulatory function.

V. SUMMARY

The observations and research described in this article derive from a nontraditional view of the "immune system." It has become abundantly clear that there are probably no organ systems or homeostatic defense mechanisms that are not, *in vivo,* subject to the influence of interactions between behavioral and physiological events. The complex mechanisms underlying these interactions and their relationship to health and illness, however, are imperfectly understood. The most imperfectly understood, perhaps, are the interrelationships among brain, behavior, and immune processes.

Without attempting to be exhaustive, we have, using "stress" effects and conditioning phenomena as illustrations, pointed out that behavior is capable of influencing immune function. We have also noted that the immune system is capable of receiving and responding to neural and endocrine signals. Conversely, it would seem that behavioral, neural, and endocrine responses are influenced by an activated immune system. Thus, a traditional view of immune function that is confined to cellular interactions occurring within lymphoid tissues is insufficient to account for changes in immunity observed in subhuman animals and humans under conditions that prevail in the real world. The clinical significance of these interactions will not be fully appreciated until we understand more completely the extent of the interrelationships among brain, behavior, and immune functions. Behavioral research represents a new dimension in the study of immunity and immunopharmacology, but it has already yielded basic data that suggest new integrative approaches to an analysis of clinically relevant issues.

BIBLIOGRAPHY

Ader, R., and Cohen, N. (1993). Psychoneuroimmunology: Conditioning and stress. *Annu. Rev. Psychol.* **44,** 53–85.

Ader, R., Cohen, N., and Felten, D. L. (eds.) (1990). "Psychoneuroimmunology," 2nd Ed. Academic Press, New York.

Felten, D. L., Felten, S. Y., Bellinger, D. L., Carlson, S., Ackerman, K. D., Madden, K. S., Olschowka, J. A., and Livnat, S. (1987). Noradrenergic sympathetic neural interactions with the immune system: Structure and function. *Immunol. Rev.* **100,** 225–260.

Fricchione, G. (ed.) (1994). Stress mechanisms. *Adv. Neuroimmunol.* **4,** 1–56.

Glaser, R., and Kiecolt-Glaser, J. (eds.) (1994). "Handbook of Human Stress and Immunity." Academic Press, New York.

Locke, S., Ader, R., Besedovsky, H., Hall, N., Solomon G., and Strom, T. (eds.) (1985). "Foundations of Psychoneuroimmunology." Aldine, New York.

Scharrer, B., Smith, E. M., and Stefano, G. B. (eds.) (1994). "Neuropeptides and Immunoregulation." Springer-Verlag, Berlin.

Sheridan, J. F., Dobbs, C., Brown, D., and Zwilling, B. (1994). Psychoneuroimmunology: Stress effects on pathogenesis and immunity during infection. *Clin. Microbiol. Rev.* 7, 200–212.

Psychophysiological Disorders

LAWRENCE WARWICK-EVANS
University of Southampton, United Kingdom

GLOSSARY

Approach–avoidance Situation in which the competing drives to approach for a reward and to avoid for fear of punishment are simultaneously present

Asthma Recurrent respiratory disorder involving reversible narrowing of the airways due to autonomic and/or immune responses

Atherosclerosis Relatively permanent reduction in the cross-sectional area of (particularly coronary) arteries due to thickening of the arterial wall and deposition of mainly fatty substances

Biofeedback Procedure whereby physiological activity (heart rate, muscle tension, eccrine sweat gland activity, etc.) is recorded and, with humans, is continuously displayed to the subject while he or she attempts to control the activity; with animals, reward and/or punishment is contingent on increasing or decreasing the activity

Conditioning There are two types of conditioning: (1) Pavlovian or classical conditioning in which a neutral stimulus (e.g., a bell) is presented at the same time as an adequate stimulus (e.g., food) for a response (e.g., salivation); after one or more pairings the previously neutral stimulus evokes the response; (2) operant or instrumental conditioning in which any behavior followed by reinforcement tends to be repeated, while nonreinforced behavior becomes less frequent

Coronary heart disease Occlusion of the coronary arteries that is serious enough to produce symptoms ranging from minor electrical abnormalities to infarction and death

Hypertension Resting blood pressure that is elevated with respect to a particular criterion such as 120 mm Hg systolic or 80 diastolic. The degree of elevation is relatively arbitrary: it may be either systolic or diastolic or both and is often age related, for example, over 160 systolic or over 90 diastolic at age less than 40 years

Neuropeptides Molecules of two or more amino acids linked by peptide bonds and capable of either facilitating or inhibiting neural transmission

Raynaud's syndrome Intermittent but excessive vasoconstriction of the peripheral circulation, precipitated by causes such as a drop in temperature, smoking, or psychological factors

Repression Hypothesized mental process in which disturbing ideas, feelings, or memories are kept out of conscious awareness, but may be brought into consciousness via techniques of word association or dream analysis

Risk factors Characteristics that indicate an increased probability of the presence or future occurrence of a disease (sometimes but not always causal)

Stress Pressure defined subjectively in terms of the psychological demands (real or imaginary) on a person relative to their coping resources and generally accompanied by emotional arousal

Ulcers Local lesions of the skin or mucous membranes exposing deeper unprotected tissue

IT IS CONSPICUOUSLY CLEAR THAT THE MIND (psyche) can influence many aspects of bodily (somatic) activity, as when fear quickens the pulse, or even that the mind can completely control bodily functions, such as the skilled sensorimotor coordination of reading aloud or driving an automobile. Moreover it is widely believed that this influence or control may adversely effect a wide range of physiological functions. The resulting dysfunctions are popularly known as psychosomatic disorders, but this term has been so extensively abused, in particular to refer to condi-

ENCYCLOPEDIA OF HUMAN BIOLOGY, Second Edition, VOLUME 7. Copyright © 1997 by Academic Press. All rights of reproduction in any form reserved.

tions in which patients believe that they are ill even in the absence of any physiological or organic pathology, that an alternative description became necessary. Consequently the American Psychiatric Association introduced into its Diagnostic and Statistical Manual (DSM) of Mental Disorders (1952) the category of "Psychophysiologic Autonomic and Visceral Disorders." These were defined as disorders due to disturbance of innervation or of psychic control and were subdivided into the following ten categories of reaction: musculoskeletal, respiratory, cardiovascular, gastrointestinal, genitourinary, nervous system, organs of special sense, endocrine, skin, and hemic and lymphatic. In the second edition of the DSM (1968) and in the eighth edition of the World Health Organization's (WHO) International Classification of Disease, the description "Psychophysiologic Disorders" was retained but redefined in terms of "physical symptoms that are caused by emotional factors and involve a single organ system, usually under the control of the autonomic nervous system."

However, a radically new diagnostic structure was introduced in the third edition of the DSM (1980) and retained in its revised version (1987). The innovation was the introduction of five independent axes of diagnostic classification. The third of these axes involves all physical symptoms or states of the patient. A new and much broader category of "psychological factors affecting physical condition" was introduced and defined as follows: "[Symptoms] can apply to any physical condition to which psychological factors are judged to be contributory. It can be used to describe disorders that in the past have been referred to as either 'psychosomatic' or 'psychophysiological.'" They then cite the following as common but not exhaustive examples: obesity, tension headache, migraine headache, angina pectoris, painful menstruation, sacroiliac pain, neurodermatitis, acne, rheumatoid arthritis, asthma, tachycardia, arrhythmia, gastric ulcer, duodenal ulcer, cardiospasm, pylorospasm, nausea and vomiting, regional enteritis, ulcerative colitis, and frequency of micturation. No reason was given for this change of emphasis, but it may well have resulted from the following two considerations. First, there was a growing appreciation that psychiatric illness was not a necessary prerequisite for the development of a psychophysiological disorder and, second, that the "normal" behavior of a patient was a contributory cause to most illnesses. Nevertheless, a contemporary redefinition of the original term is still required, and the following is proposed: "Psychophysiological disorders are real, not imaginary, disorders of physiological systems, the onset, severity or duration of whose dysfunction is at least partly due to psychological factors."

Traditionally, psychological factors have been interpreted to refer only to psychodynamic processes, conditioning (both classical and operant), emotional experience, and other aspects of private mental life. If, however, psychological factors are understood to include an organism's behavior, then the implications broaden. Since very little pathophysiology is of purely genetic origin, it must follow that environmental factors are the major influence. But since the organism has the choice of an infinitude of alternative behaviors ranging from those with the highest risk (e.g., substance abuse) to those with the lowest risk (e.g., the choice of a healthy diet and adequate exercise), it must be behavior that mainly determines the extent of the environmental contribution to physiological dysfunction. On this broader interpretation of psychological factors it is clear that not just a limited set of dysfunctions invites the description psychophysiological but that the description applies to the majority of illness and disease states. An appreciation of the inevitability of this conclusion is essential for the prevention and management of illness. This article, however, will be restricted to psychophysiological disorders as defined by the traditional interpretation of "psychological factors."

I. THE CONCEPTS OF A PHYSIOLOGICAL SYSTEM AND OF CAUSATION

A. The Systems Analysis of the Human Organism

It is an almost universal convention in the teaching of medicine and physiology to subdivide the body into a set of systems; to a great extent the practice of and specialization within medicine follow a similar convention and, to a lesser extent, research tends to focus within a particular system. Thus people learn about, specialize in, or research into the cardiovascular, respiratory, gastrointestinal, genitourinary, musculoskeletal, endocrine, nervous, or immune systems much as if they were *separate* systems, which they are not. They are, of course, *separable* systems in theory and to treat each as a closed system has many advantages. For example, one's area of interest or responsibility is thereby precisely delimited; quantifiable analytic concepts such as negative or positive feedback, system variable, set point, gain, delay, transfer function, damping, oscillation, and homeostasis may be used to model the system's behavior; and above all we are provided with a feeling of potential completeness of understanding. But the cost of these advantages is that we can lose sight of the more complex realities that (1) the dividing line between systems is to some extent arbitrary and (2) there is a hierarchy of systems with each system open for dynamic interaction with other systems at the same level in the hierarchy, open to control from above, and also open to disturbance from the external world.

The first point may be illustrated by the following example: a major function of the cardiovascular system is to perfuse the tissues with oxygen, but this is also a function of the respiratory system, so any understanding of how the body achieves this necessitates the idea of a cardiorespiratory system. The second point, which is the more fundamental with respect to psychophysiological disorders, merits further expansion. The fundamental point is that it is the *brain* that has the capacity to regulate or disrupt the efficient functioning of all other systems and that has the potential for control of each of the systems in which psychophysiological disorders may arise. This is not to deny that a system is capable of autonomy or self-regulation or that a peripheral system does not exercise feedback control (e.g., via neuropeptides) on the brain itself. Dynamic interaction between systems is well illustrated by the interaction of visual, vestibular, and other proprioceptive systems, the striate musculature, and the autonomic nervous system in the normal control of balance and movement and in the production of motion sickness. Finally, the direct influence of the external world over each system should not be forgotten: the immune system reacts dramatically to bacteria, foreign protein, and viruses (and catastrophically to human immunodeficiency virus), gastrointestinal efficiency depends on the quality and quantity of food, and effective sexual functioning depends on the presence of appropriate stimulation. In conclusion, the systems-analytic approach not only offers a useful classificatory framework for psychophysiological disorders but because of the overarching influence of the brain this approach suggests how psychological factors might exercise an influence on most body systems.

B. Causation

Whereas many illnesses are defined in terms of a description of their symptoms, psychophysiological disorders are defined in terms of their causes. This immediately requires us to consider what is meant by a cause and what criteria must be satisfied before something can be accepted as a cause. But contemporary medicine has tended to relinquish the concept of cause in favor of the idea of a risk factor. This refers to any characteristic (genetic, biochemical, physiological, psychological, or sociological) the degree of whose presence increases the probability that a person has or will develop a disease. Epidemiological research, particularly prospective studies, has established that for each disease there are many risk factors. But the relationship between causes and risk factors is not straightforward, since any definition of cause must acknowledge two ideas: (1) that causes precede their effects and (2) to the extent that a cause is present then the probability of its effects will be increased. Additionally, if we can describe, preferably quantitatively, the intervening processes whereby a hypothesized cause produces its effect then we feel more justified in accepting that there is a causal relation between the two events. But since this third requirement merely reflects what happens at a particular moment to be the state of our knowledge, it cannot be used as a necessary criterion of causality. Just as there are many risk factors for each disease, the current consensus is that causation is usually multifactorial.

The main conclusion from these considerations is that whereas all causes of diseases must be risk factors, it is not necessarily the case that all risk factors have a causal relationship to a disease. For example, a risk factor may be associated with a disease either because the risk factor causes the disease or because the disease caused the risk factor or because a third variable was responsible for both the disease and the risk factor. In theory these three criteria would only be fully met by the following type of research study. Human subjects would be randomly allocated into either a control or experimental group. The latter would be exposed to severe and protracted psychological stress in a standardized laboratory environment (which should be both objectively described and subjectively rated), while continuous invasive measurements were made across an extensive range of variables, and the experiment would continue until severe physiological disturbances arose, at which point the stress would be discontinued but the subjects would continue to receive intensive and invasive study so as to establish whether the effects were acute and reversible or chronic and irreversible. Even if severe pathology did result, one could conclude only that psychological factors are *capable* of causing the disorder, not that whenever a disorder arises it is due to psychological factors. But since such an experiment is both impractical and unethical, the demands for conclusive proof that psychological factors cause physiological disorders have been set impossibly high. Nevertheless, there is an abundance of alternative lines of evidence in support of the hypothesis that psychophysiological disorders are both widespread and severe in contemporary societies. The evidence is presented and critically reviewed in Sections II and III.

II. EVIDENCE FOR THE PREVALENCE OF PSYCHOPHYSIOLOGICAL DISORDERS

A. Theoretical Approaches

The central requirement of all theories is that they attempt to explain how mental activity can exercise an adverse effect on physiological functions. In the case of disorders of the striate muscles such as tics, torticollis, tension headache, and muscle spasms, this presents no problem. The cortex (the locus of our essential humanness, of personality, of learning, and of psychodynamic conflicts) is hardwired via motor neurons through to the relevant muscle beds and can produce its effect directly, though not necessarily voluntarily. A similar argument can be deployed that the cortex via the limbic system, hypothalamus, and pre- and postganglionic fibers can dysregulate all effectors of the autonomic nervous system (ANS). Similarly, cortical efference to the hypothalamus and thence to the pituitary can effect hormonal activity, and therefore to some extent could disrupt the neurohormonal environment that is required for the optimal functioning of the immune system.

The main theoretical approaches to explain the occurrence of psychophysiological disorders are psychodynamic, evolutionary, conditioning, and personality or dispositional. Each will be described in turn. The evolutionary approach emphasizes the very protracted time scale of natural selection of those best adapted to survive and reproduce in a competitive world where predators abounded and prey was scarce. The chances of survival were greatly increased by the development of bodily capacities such as the fight and flight reactions for dealing with immediate demands for dynamic activity and a variety of catabolic processes that mobilize energy resources on a longer time scale. This emphasis on the adaptation of early man the hunter and the hunted has been invoked to explain both the metabolically inappropriate high cardiac output often seen in early hypertension and the excess free fatty acids and cholesterol associated with atherosclerosis. But the evolutionary approach also emphasizes the steady exercise requirements of man the nomadic gatherer, the amount of time spent in leisure and relaxation after basic needs for survival have been satisfied, and an individual's interdependence on other members of a relatively small group. By contrast, contemporary humans take little exercise, enjoy less leisure, and tend to live relatively independently of others in very large cities. Essentially the argument is that humans evolved via natural selection to cope successfully with the demands of a world that has virtually ceased to exist, and those very reflexes and dispositions that once served so well have become maladaptive and counterproductive in terms of physical (and mental) health.

Psychodynamic explanations derive from Freudian conceptions of the development and structure of the mind (ego, superego, id, subconscious, and unconscious), the dynamic power of repressed intrapsychic conflicts to be converted into physical symptoms, and the importance of symbolism. More specifically, a psychosomatic disorder was believed to result from a combination of the following three characteristics: (1) a personality with an important intrapsychic conflict pattern that evoked a defense mechanism such as repression, (2) an emotionally important precipitating life crisis, and (3) a constitutional weakness in an organ system. The conflict was said to be resolved or alleviated by being converted into bodily symptoms (hysterical conversion reaction). The main themes of this analysis have received so little experimental support and achieved so few therapeutic successes that they have been largely abandoned, but the idea of a constitutional (possibly genetic) predisposition to develop a psychophysiological disorder in a particular system may be incorporated into other theoretical approaches. [*See* Psychoanalytic Theory.]

Not all people are equally susceptible to psychophysiological disorders and there have been alternatives to the psychodynamic approach in explaining this differential susceptibility. It has been proposed that some people have a genetically determined tendency to overreact to psychologically disturbing situations in a particular way, for example, by oversecretion of digestive acids or by excessive vasoconstriction. (This is clearly similar to the organ weakness theory of the preceding paragraph.) Although there is little evidence that this tendency is either widespread or strong in humans, laboratory animals may be selectively bred for a predisposition to develop ulcers in response to psychological stress. An alternative explanation of individual differences in susceptibility proposes that, particularly with increasing age, some people develop a fixed or "stereotyped" tendency to react in one specific system to a wide range of emotional experiences such as fear, anger, or anxiety, each of which would in most people evoke different patterns of physiological responses. This repeated disturbance will then, it is argued, lead to a breakdown of normal homeostatic restraints and ultimately to illness. There is little or no systematic evidence that this occurs.

Personality theory also attempts to explain not only why certain people succumb to psychophysiological disorders but also which particular disorders they will develop. It has been suggested that some people have personalities characterized by the frequent or protracted presence of particular attitudes or emotions. Each attitude or emotion is said to be accompanied by a disturbance of a particular physiological system, for example, the experience of helpless suffering and frustration is accompanied by hives or eczema, whereas the feeling of being in a state of constant readiness to deal with psychological demands is associated with hypertension. There is little evidence for the truth of these suggestions except that feelings of anger, hostility, and preparedness to meet real or imagined demands have quite frequently been reported in people with mildly elevated blood pressure.

The final theoretical viewpoint asserts that psychophysiological disorders, since by definition they are not purely genetic, are acquired or learned at the hands of the environment. This learning may take the form of either classical (Pavlovian) conditioning or instrumental (operant) conditioning. In classical conditioning a previously neutral or unconditional stimulus (e.g., the sound of a bell) is presented along with a potent stimulus (e.g., food) that evokes a response, usually within the ANS (e.g., salivation). Usually, after several pairings the conditional stimulus when presented *without* the unconditional stimulus evokes the response, although the effect soon disappears. Sometimes, particularly if fear or avoidance of an aversive situation is involved, only a single pairing of the two stimuli is required for the association to be learned, moreover the conditional stimulus may retain its power to evoke the response for thousands of test trials or over many years. By contrast, in instrumental conditioning a response (e.g., bar pressing) that is rewarded or reinforced (e.g., by food or avoidance of shock) comes to be repeated with a probability that is a function of the frequency and intensity of the reinforcement. As with classical conditioning, the response usually disappears soon after the reinforcement is discontinued, but if the original reinforcer was avoidance of an aversive stimulus then the response may remain for years. Learning theorists have certainly demonstrated classical conditioning in which asthmatic attacks occur in response to a previously neutral visual stimulus. However, the phenomenon appears to occur only in a very few individual animals or humans. More surprisingly by pairing sodium saccharine (conditional stimulus) with cyclophosphamide (unconditional stimulus that depresses immune function) it has recently been shown that saccharine subsequently acquires the capacity to depress both humoral and cell-mediated immunity. This example of a classically aquired response is relatively easy to establish in several species, though there is as yet no evidence that the effect is long-lasting. By contrast, several animal studies have successfully used the instrumental paradigm of shock avoidance either alone or with food as the reward for the response of increases in blood pressure. The majority of animals learned to elevate their mean blood pressures by 30–60 mm Hg, but when the reinforcers were removed, the blood pressure returned to baseline. [*See* Conditioning.]

B. Empirical Research Paradigms

In theory, relevant research may be classified according to the following schema: human versus animal experiments (each subdivided into acute or chronic) and experimental versus epidemiological studies (the latter subdivided into retrospective or prospective).

1. Human Studies

Human experiments are limited by ethical considerations to brief and relatively mild manipulations of stressful stimuli (e.g., white noise, IQ tests, competitive reaction time tasks, mental arithmetic, penetrating interviews about anxiety-provoking incidents in the subject's life, or threat of electric shock), while the effects are recorded over a wide range of physiological and biochemical variables (e.g., blood pressure, cardiac output, regional blood flow, muscle tension, airway resistance, digestive acids, blood or urinary catecholamines, or corticosteroids). Even if only slight and temporary disturbances are found, it is usually implied that if the stimulation had been more severe and protracted then the disturbance would have been correspondingly more serious and chronic. Epidemiological studies of the retrospective type select patients with functional or even organic disorders (e.g., asthma, hypertension, myocardial infarction, dermatitis, migraine, impotence, or ulceration of the digestive tract), then using interviews or questionnaires they establish a link with possible psychological aspects of the patient's personality or history. These include intensive single-patient studies and surveys extending to many thousands of subjects. A large proportion of retrospective studies have concentrated not merely on stressful experiences but on any events that require major psychological changes in life-style.

The original inventory was the Social Readjustment Rating Scale, which included not only negative items such as death of a spouse, divorce, or jail term but also more positive events such as marriage, pregnancy, and vacations. Each event counted on a scale of life change units (LCUs) and a high overall score was expected to be associated with a high incidence of illness. Modest negative correlations have often been found between high scores and general measures of well-being (depression, anxiety, and minor infections), but the correlation between LCUs and real physiological disorders is often not significantly different from zero. In a series of attempts to improve these results, the rating scale has undergone a number of revisions and amendments, as in the Psychiatric Epidemiology Research Interview (PERI). Also it is suggested that the effect of stress and change on health are buffered in some people by the presence of moderating variables such as physical fitness, social support, or characteristics like hardiness of personality. Even after statistical allowance is made for these factors, most correlations remain in the region of $r = 0.1$ or less. And even if clinically and statistically significant correlations were found, they would still be open to the criticisms (1) that there was bias on the part of the interviewer or the respondent to the questionnaire and (2) that the correlation did not indicate causality but was due to the influence of confounding variables such as diet, alcohol, smoking, poverty, or many others. The first of these criticisms may be wholly and the latter partly countered by the use of prospective studies. In these, initially healthy subjects are recruited; they are assessed along many dimensions of biochemical, physiological, psychological, and sociological characteristics; these assessments are repeated over many years; and the records of those who eventually become ill are then searched for the characteristics that distinguish them from those who remain healthy. Such studies are rare.

2. Animal Studies

With respect to animal studies there is always the question of whether they provide a realistic physiological model, but in connection with psychosomatic disorders one must consider also whether their mental processes are sufficiently comparable to our own to justify extrapolating to humans. Epidemiological studies of free-living animals report that the incidence of any of the illnesses mentioned so far as possibly being psychologically caused has an extremely low prevalence.

In acute animal studies, a hypothesized and often very intense stressor is continuously or repeatedly applied, usually for just sufficient time for a physiological effect to be found. In better experiments, careful control is maintained over confounding physical variables like fighting and nutrition. For example, in the yoked control design, pairs of animals are restrained and wired electrically in series; each animal therefore receives the same number of shocks of the same intensity at the same time; each animal has access to an identical bar or lever; and one of these is connected to the circuitry and, if pressed at the correct time, will delay or cancel the shock, whereas the other is not connected. The only differences between the two conditions are the psychological experience of being in control rather than helpless or being able or not able to predict whether a shock will occur. Significantly more stomach ulcers are reliably found in the helpless animals for whom the shock is unpredictable.

In other designs a wide variety of psychosocial stimuli (exposure to the sight, sound, and smell of a predator, overcrowding, or disruption of established social networks) have been employed and the effects are frequently severe and extensive (elevated catecholamine and corticosteroid levels, increased blood pressure, reduced estrous and spermatogenesis, ulceration, increased rates of infection and cancer, adrenocortical enlargement, thymic involution, atherosclerosis, and premature death). Chronic experiments in which all animals are studied until their deaths are rarely reported; this is a serious gap in our knowledge since without them we cannot be certain whether some of the physiological effects (notably hypertension) are reversible when the stressor is removed. In short, animal work has established incontrovertibly that psychological factors can cause not only functional physiological disorders but eventually organic damage. What is more contentious is whether animals provide a reasonable model for human psychophysiological disorders.

If research over the last few decades had been systematically coordinated, then we would have from each of these paradigms evidence for or against the validity of each of the preceding theories with respect to each of the types of disorder discussed in the next section. In fact, research has been disproportionately concentrated on acute animal experiments, on retrospective epidemiological studies, and on only a few disorders.

III. CURRENTLY AVAILABLE EVIDENCE FOR PARTICULAR DISORDERS

In the 1950s there would have been no difficulty about which disorders to include in this section: they would have been the seven that were then the focus of attention for psychoanalysts, that is, peptic ulcer, bronchial asthma, essential hypertension, ulcerative colitis, thyrotoxicosis, rheumatoid arthritis, and neurodermatitis. By the 1970s the list would have been classified according to the body system analysis proposed in DSM-II and would have excluded thyrotoxicosis and rheumatoid arthritis, but would have been extended to include most disorders of each system, for example, musculoskeletal—tension headache, tics, torticollis, and muscle spasms; cardiovascular—essential hypertension, tachycardia, vasovagal fainting, arrhythmias, migraine headache, and Raynaud's syndrome; and respiratory—psychogenic breathlessness. But with a growing belief that psychological factors (even if behavior is excluded and in the absence of psychiatric disorder) can precipitate, exacerbate, or prolong a wide range of disorders, the list is greatly extended so that only a very small proportion can be selected for discussion. The following criteria are proposed for the selection process: (1) the disorder must be serious, (2) it must have a high prevalence, and (3) evidence of a causal role for psychological factors must come from at least two of the paradigms described in Section II,B. The following three illustrative areas meet all of these criteria—ulcers of the alimentary tract, cardiovascular disorders, and asthma.

A. Ulcers of the Alimentary Tract

Ulcers form a heterogeneous set of disorders, even if they are subdivided by locus or hypothesized cause. Mechanisms may involve over- or undersecretion of acid or digestive hormones, alterations in smooth muscle motility or regional blood supply, reduced mucus production, or impaired neutralization. As usual, both genetic and environmental factors influence these functions. But mere introspection, unfashionable though this is, reveals how susceptible the delicate interrelationships of the system are to psychological upsets. Invasive studies of patients with gastric fistulae confirm their functional vulnerability to psychological events such as anger, frustration, and distressing interviews. Epidemiological work is inconclusive since, although sustained and/or severe exposure to stress frequently precedes ulceration, this may be mediated by behavioral responses to the stress such as increased smoking or alcohol intake. But the most consistent and dramatic evidence that psychological stress causes ulcers comes from laboratory work on rats and monkeys performed by a variety of experimenters over the last three decades. Stresses have included physical restraint, approach–avoidance conflict, and the use of controllable versus uncontrollable and predictable versus unpredictable electric shock; experimental durations have varied from 6 hours to 30 days.

Experimental designs have become increasingly sophisticated and typically employ multiple genetically matched control groups, often in a yoked control paradigm. By sacrificing animals at different times during and after exposure to the stressful situation it is possible to study the development of the pathology. Perhaps the most interesting result is that the ulcers seem to result not so much for sympathetic overarousal but from poststress parasympathetic rebound, although corticosteroid activity also appears to be implicated. One frequently overlooked aspect of these studies is that there is always disruption of the animals' normal eating patterns. Animals are either food-deprived for 6 to 72 hours before the experiment or the experimental procedures themselves disrupt normal patterns of eating. This disruption per se is not usually enough to cause ulceration since the control animals experience equal disruption without adverse effects. The essential point is that the *combination* of the psychological stress and food deprivation is required to provoke the ulcers—the effect is interactive.

B. Disorders of the Cardiovascular System

Cardiovascular disorders range from minor dysfunctions of vasoconstriction and vasodilation such as Raynaud's syndrome or migraine through arrhythmias and hypertension to coronary heart disease. The latter two conditions each meet the criteria of seriousness, prevalence, and evidence, and will be considered separately.

All definitions of hypertension are arbitrary, but all refer to a relatively permanent elevation of systolic and/or diastolic blood pressure with respect to a theoretical "normal" level that is often age related (e.g., current North American practice but not the WHO recommendation). Secondary hypertension is defined as being due to an organic cause like coarction of the aorta or pheochromocytoma and therefore cannot be of psychological origin. But primary or essential hypertension is defined in terms of the absence of organic cause and thereby becomes a candidate for a psychophysiological disorder. Estimates of the prevalence of the latter vary from 10 to 30% in Europe and the United States. It is serious both in its own right as a cause of death but more because of its causal role for stroke and coronary heart disease. Consequently its etiology has been the subject of extensive research for many decades. Nevertheless, our understanding of its causes is still very limited. Some form of polygenic inheritance is estimated to account for 20 to 65% of the population variance in blood pressure, but even the genetic effect may act only by increasing one's susceptibility to as yet unconfirmed environmental causes. Suggested risk factors include dietary aspects (sodium or sodium/potassium intake, obesity, low fiber, high protein, excess trace elements, and high alcohol intake), smoking, caffeine, and psychological factors such as personality and stress. Evidence for a causal role of the latter two will be reviewed next. [*See* Hypertension.]

Acute experimental studies of animals and humans are quite conclusive; psychological factors such as classical conditioning, operant conditioning, and stress reliably evoke large increases in blood pressure on the order of 10 to 50% of pretreatment levels. In explanation of this effect there are numerous regulatory mechanisms that may mediate between the cortically based psychological processes and the response of elevated blood pressure, for example, cardiac and vagal regulation of heart rate and stroke volume, vasoconstriction, renal blood flow, or hormonal changes in catecholamines, renin, or aldosterone. But the central and hitherto unresolved issue is whether psychological factors can produce relatively permanent (on the order of years) effects.

Chronic human experimental work does not exist at all and very few animal studies follow the animals for more than a few weeks after the end of the stress sessions. The general picture is that animals recover back to baseline levels either overnight or a few weeks or months after the last stress sessions. However, a few individual animals do develop left ventricular hypertrophy or interstitial nephritis, and this may sustain irreversible hypertension. Cross-sectional epidemiological surveys frequently report an association between personality variables (anger or anxiety) and level of pressure, or between stressful employment (air traffic control, bus driving), stressful environments (inner city housing), and elevated blood pressure, but the effect is usually slight (<10 mm Hg).

True prospective studies over more than a decade and focusing on psychological factors do not appear to have been carried out, although people with large short-term blood pressure increases to a variety of stimuli have been found to have a significantly higher incidence of hypertension at follow-ups of 5–45 years. Nevertheless, the role of psychological factors in provoking acute blood pressure increases is indisputable and, in theory, repeated or protracted stress could lead from elevated cardiac output to elevated peripheral resistance via baroreceptor resetting, hypertrophy of the ventricular or arterial wall, or whole-body autoregulation—but this is currently unproven.

There are three arguments why coronary heart disease (CHD) could be regarded as a psychophysiological disorder. First, hypertension is a contributory cause of CHD; therefore, to the extent that hypertension is psychogenic, so also is CHD. Second, large-scale prospective studies have shown that individuals with a type A personality (time urgent, competitive, and ambitious) have double the normal probability of developing CHD over the subsequent 8 years, though not over the subsequent 24 years. Finally, numerous animal experiments and human epidemiological studies have shown that laboratory and real-life stresses are associated with increased atherosclerosis, although there have also been some negative findings. But it should be added that the effects of stress, even when present, are relatively small compared with major risk factors such as age, smoking, diet, and serum cholesterol. By contrast, the acute effects of stress in people with preexisting CHD may be calamitous, involving angina, arrhythmias, or myocardial infarctions. [*See* Atherosclerosis.]

C. The Respiratory System

The major psychologically influenced disturbance of the respiratory system is asthma, which may be defined as a reversible, recurrent breathing disorder associated with hyperreactivity of the bronchial tree. It has a prevalence of roughly 10% and is potentially lethal if an attack cannot be controlled. There are two different types of reactivity; the first involves the parasympathetic branch of the ANS, and the second comprises an allergic/immune response. The relationship between these two types of response is very poorly understood and it is easiest to illustrate psychological influences on the ANS component. The initial events involve constriction of the smooth muscles of the bronchioles and an increase in the rate and viscosity of neurosecretion, both being under parasympathetic control and possibly leading to local edema and hyperemia. There is little animal evidence for the relevance of psychological factors, but epidemiological work confirms the relevance of emotional disturbance; acute experimental work clearly demonstrates that the onset and course of an attack may be influenced by the expectation and beliefs of the sufferer and suggests that conditioning may be involved. [*See* Pulmonary Pathophysiology.]

Numerous experimenters have explored the effects of suggestion on asthmatics and controls who are either healthy or suffer from other lung diseases. It has been frequently shown that if it is suggested to an asthmatic that he or she is inhaling a substance that will precipitate an attack then the respiratory efficiency will decline. Measures of breathing efficiency have ranged from self-report to whole-body plethysmography and forced oscillation techniques. The strength of the effect has varied from slight but statistically significant increases in airway resistance to full-blown dyspnea and wheezing. Such experiments also tend to confirm the direct dependency of the bodily response on the mental attitude because the effect can be reversed by providing a therapeutic placebo such as nebulized saline along with suitable instructions. These effects cannot always be obtained and in partial explanation of this it is suggested that asthmatics can be divided into three groups. One group's asthma is said to derive essentially from severe infectious disease of the lung, another group is said to suffer from an allergic form of asthma, and only in the last group are psychological factors a cause, and it is these sufferers who show the dramatic effects of suggestion. Case histories and surveys provide some support for this classification, and there is a relatively consistent subgroup who often identify emotional disturbance as the precipitating factor for their asthmatic episodes. However, it should be noted that some subjects show evidence of all three aspects.

IV. CONCLUDING COMMENTS

Although the idea that the mind can produce adverse effects on the body is still central to any conception of psychosomatic/psychophysiological disorders, there have been several changes of emphasis. First, such effects are no longer seen as necessarily symptomatic of mental illness. Second, whatever arbitrary analysis of the body systems is used, it is clear that psychological factors can either directly or indirectly influence them. Third, although a disorder may at first be "merely" functional, it may develop into structural pathology; consequently it may not necessarily be reversible. Finally, because of the multifactorial causation of illness, it is clearly a logical error to classify certain groups of disorders as psychological in origin and others as due to physical causes. The point is that on one occasion an ulcer or asthmatic episode may be due to excess smoking or animal danders while on another occasion the same event in the same patient may be due to psychological factors.

The logic of the relationship between etiology and treatment is frequently misunderstood, and nowhere more so than with respect to psychophysiological disorders. The logic of this error is as follows: "A entails B," therefore "not-A entails not-B." Stripped to its logical bones, this argument lacks all conviction, but embedded in the form "stress causes psychophysiological disorders so suitable stress management or avoidance will cure the disorder," its appeal is more insidious. The conclusion about cure may well be correct but it is not logically entailed by the premise, particularly if the disorder has progressed to an organic form. In this case it may require surgical or pharmacological intervention. Nevertheless, psychotherapy, including psychoanalysis, meditation, relaxation, behavior modification, biofeedback, or family therapy, may assist the process of recovery from psychosomatic disorders.

BIBLIOGRAPHY

Dutevall, G. (1985). "Stress and Common Gastrointestinal Disorders." Praeger, New York.

Fava, G. A., and Wise, T. N. (eds.) (1987). "Research Paradigms in Psychosomatic Medicine." Karger, Basel.

Friedman, H. S. (ed.) (1990). "Personality and Disease." John Wiley & Sons, New York.

Husband, A. H. (ed.) (1992). "Behavior and Immunity." CRC Press, Boca Raton, Florida.

Serafino, E. P. (1990). "Health Psychology: Biopsychosocial Interactions." John Wiley & Sons, New York.

Steptoe, A. (1983). Psychological aspects of bronchial asthma. *In* "Contributions to Medical Psychology" (S. Rachman, ed.). Pergamon, Oxford.

Psychotherapy

MICHAEL J. LAMBERT
Brigham Young University

JOHN M. LAMBERT
University of Utah

GLOSSARY

Classical conditioning A form of learning in which a previously neutral stimulus comes to elicit a response that was previously a biological reflex

Defense mechanism Psychic mechanism that acts to maintain a person's feelings of adequacy and worth rather than to cope directly with an anxiety-provoking situation

Operant conditioning A form of learning in which a particular response is reinforced and therefore becomes more likely to occur

Organismic valuing process Process by which a person's self-actualizing tendency evaluates experiences as maintaining or enhancing the person

Self-actualization A basic inborn tendency for human beings to move in the direction of achieving their full potential

Transference An unconscious process whereby a client projects attitudes and emotions applicable to another significant person onto the therapist

I. INTRODUCTION

Psychotherapy, defined within the broader context of the field of psychology, is a skilled and intentional process whereby the behavior of a person or group of persons is modified using clinical methods and interpersonal experience derived from well-established principles. Depending on the theoretical orientation of the therapist, psychotherapy can be conceptualized in many different ways.

The human past holds many interesting examples of attempts to alleviate the misery and discomfort of individuals perceived to be suffering from psychological disorders. A given culture's frame of reference, which entails complete information and beliefs exhibited by that culture, reveals the strategy that would be taken to alleviate or cope with behavior unexpectedly exhibited by a member of the society. By posing a threat to the societal structure and going contrary to the way one ought to act to be a functional, contributing organism, an aberration in conventional behavior may be perceived by other members of the culture as nonadaptive and therefore undesirable. Attempts to assimilate individuals perceived as "abnormal" have taken several forms that are contingent on culture-bound factors. The antecedents of modern psychotherapy can be found in traditional medico-religious practices, which have been shown to have significant positive effects in times of cultural distortion as well as allowing individual aberrant behavior to be addressed in a healthy manner. Traditional societies generally show an absence of "privacy" and "individuality" in the sense that Western peoples perceive these entities. Thus, traditional healing is community based and more reliant on allowing a person perceived as abnormal to be assimilated by finding a role for the person which heightens his or her capacity for participation and sense of belonging.

Since the industrial revolution, the Western world has undergone several changes which make living in a modern, industrialized country a fast-paced, time-

ENCYCLOPEDIA OF HUMAN BIOLOGY, Second Edition, VOLUME 7. Copyright © 1997 by Academic Press. All rights of reproduction in any form reserved.

limiting, and stressful endeavor. As our society changes, so must our definitions of mental health. The "private" sphere, which is an invention of the industrial world, is divided as the household, nuclear family world from the workplace institutions and more "public" sphere. The public sphere does not appear to be a place where meaning is found and is not a place where individuals concentrate a large amount of their emotional energy. The private sphere of family and friend relationships is fragile and susceptible to recurrent stress. With the absence of a meaningful community to provide support for family structure, people are experiencing a high amount of stress and are more and more apt to seek help. Should we continue to adapt as individuals to the exceedingly busy world in a way which society deems necessary and normal, or is it perhaps a "natural" phenomena to become depressed or maladjusted in today's world? These questions hinge on our cultural expectations of "what" needs to be reorganized within an individual and "who" has the professional authority to guide or control this process.

Throughout prescientific Europe several measures were taken by "doctors" or "practitioners" who were believed to be competent which appear today as unacceptable ways to treat persons in need of therapy. As Western culture has learned from the past, the ability to refine knowledge of the relationship between mental disturbance and causation has been significantly narrowed. We have been able to systematically approach the mind in a way that takes the entire body into consideration and does not separate psychological processes from the entire physiology. People in North America and Europe have slowly given authority to scientific explanation as the best explanation over former traditional authorities. This is not true of every world culture. Many populations are keeping established traditions of medicine and religious practices relative to their own cultural identities.

As observation and experimentation through the use of the scientific method became more widespread and mandatory in the 1800s and as discoveries and theories, such as the theory of evolution put forth by Charles Darwin (1809–1882), became more popular, the modern field of psychology was born. Modern founders William James (1842–1910) and Wilhelm Wundt (1832–1920) began to systematize knowledge before the turn of the 20th century. From these beginnings came the birth of modern psychotherapy as a treatment method grounded in theory. Attempts to classify and understand the similarities and differences of persons suffering from aberrative symptoms led to several patterns which Kraepelin (1856–1926) noted in a classification system. This system was adapted and is currently accepted as the "Diagnostic and Statistical Manual of Mental Disorders" (DSM). Sigmund Freud (1856–1939), a medical doctor interested in the relationship among culture, biology, and psychological processes, espoused a detailed theory of human nature and function which sought to explain how events in the past and our physiology determine our behavior in the present. Many psychotherapies have been developed since Freud's time, yet his influence is unquestioned. [*See* Mental Disorders.]

Providers of treatment today are goal oriented toward the alleviation and reduction of disorders diagnosed through the criteria agreed upon in the currently updated DSM. The methods chosen by the therapist are provisional upon the preferences and theoretical leanings of the therapist as well as client preferences with regard to time and financial considerations. There are hundreds of different therapies currently employed in variation throughout the therapeutic community, yet most are variations of four general categories: psychodynamic, humanistic–phenomenological, cognitive–behavioral, and biological–drug therapy. Recipients of treatment are categorized either as an inpatient, who are treated in a hospital or residential institution, or as an outpatient, which designates more highly functioning individuals receiving psychotherapy while maintaining function in the world. Providers of treatment are psychiatrists, psychologists, and clinical social workers. A psychiatrist is a medical doctor trained in the treatment of mental disorders who has the authorization to prescribe medications. A psychologist specializing in psychotherapy is trained in a clinical or counseling Ph.D. program. Currently, psychologists are unable to prescribe drugs, but there is ongoing debate to extend this practice to clinical psychologists who have specific training in this area. Finally, clinical social workers trained in a masters' degree program as well as a host of paraprofessionals are involved in the treatment process.

Psychotherapy is most often performed in a one-on-one or individual therapy format. This typically takes the form of one 50-min session per week or daily sessions in inpatient settings. A common mode of treatment delivery is group psychotherapy. This modality usually involves 8 to 12 patients in sessions that typically last one and one half or more hours. Patients in these groups often have similar problems and are able to focus on and practice methods of dealing with their failing coping strategies. Psychotherapy offered in a group format is often as effective as that offered to individuals. Another mode of treatment delivery involves intervening with naturally occurring groups such as married couples and families. These treatment modalities are used when problems are conceptualized as directly growing out of the interactions of partners and family members and when change in these interactions becomes a primary goal of treatment.

II. PSYCHODYNAMIC PSYCHOTHERAPY

Formally, the field of psychotherapy was begun when Freud began treating patients in Vienna, Austria, in the late 1800s. Psychoanalysis, the method espoused and performed by Freud and the most influential therapy during the 1930s and through the 1950s, is theory, investigative method, and therapy combined in consideration of human behavior. For over 100 years Freud's views of personality have been the source of comparison as new theories have departed from his and as new knowledge has been gathered in psychology. [*See* Psychoanalytic Theory.]

Briefly, psychoanalytic techniques are based on allowing unconscious processes to be expressed consciously. Freud's personality theory has been controversial and misunderstood, but it is through careful observation that it was derived and altered throughout his lifetime.

The libido, or sexual drive, and the ego, which is essentially a drive for self-preservation, are the fundamental drives motivating outward human behavior. The structure of the personality is based on expressions of psychological energy: id (physiological forces, hunger, thirst, elimination, and basic drives), ego (rational thinking), and superego (societal standards). Conflicts arise among these expressions, resulting in different types of anxiety based on the external stimuli, namely neurotic, objective, or moral anxiety. To cope with the anxiety, the ego denies or distorts reality in a defensive way to reduce stress. The "defense mechanisms" include repression, denial, reaction formation, projection, displacement, sublimation, rationalization, regression, identification, and intellectualization. These mechanisms deal with unconscious conflict that comes about in psychosexual development during the childhood years.

The psychosexual stages posited by Freud occur within the first 5 or 6 years of life (oral, anal, and phallic), although the final stage (genital), if reached, continues throughout life. Freud considered these stages to be biologically determined, but evidence from anthropological studies has shown child-rearing practices to be widely varied. The oral stage lasts from birth up to 18 months and is characterized by a focus on the mouth for pleasure. Functions such as eating and holding are related directly to the mother as the care giver and main object of dependency. Overstimulation in this stage may cause overdependence in childhood and in later life whereas understimulation may cause insecurity.

The anal stage lasts from 18 months to 3 years and is characterized by a focus on the anal area as a main source of pleasure. At this time, the child is being toilet trained and develops conflicts centered around the success or failure of this process. Conflicts obtained during this stage can be seen in later life as personality characteristics (anal retentive: orderly and clean vs anal expulsive: destructive and disorderly).

The phallic stage last from age 3 until age 5 or 6 and is characterized by a focus on the genitals as the main source of gratification. The attention payed to the genital area and stimulation for sensual pleasure takes place in an environment where there is castration anxiety for boys and penis envy (fear that they have already been castrated) for girls. These fears cause anxiety which is accentuated by oedipal and Electra complexes. The Oedipus complex is a reference to a boy's sexual interest in the opposite sex parent and hostility and anger toward the same sex parent. A child must learn to express love for the opposite sex parent in a nonsexual way by identifying with the same sex parent. Difficulties in this stage may cause later problems in relations with the opposite sex.

Latency is the period lasting from 6 to 12 years old where there is no real psychosexual development as children repress their libido into active sports, school, and friends. The genital stage begins at puberty and continues throughout the life cycle, focusing on members of the opposite sex as sexual objects instead of the self-love emphasis of the previous stages.

The concepts of the unconscious, regression, transference, and resistance are important in understanding the logic of psychoanalytic techniques. There are mental processes that are outside of the awareness of an individual that have powerful influences on conscious experience. These unconscious processes are manifest in slips of the tongue and come forth as contradictory ideas or thoughts. Threatening thought content is often repressed and denied expression in the conscious mind. Conflicts arising between the conscious and the unconscious mind are defended using mechanisms mentioned previously. Therapist attempts to work through the threatening content of the unconscious are often met with resistance as the patient will not allow the repression to be undone. The mechanisms of defense are normal developments, but usually represent limitations to coping with reality as gauged in their severity.

Particularly important for successful therapy is the defense mechanism known as transference. Patients relive memories and feelings from the past in the context of therapy. Unaware that present feelings are actually the product of past experience, the patient unconsciously transfers feelings onto the therapist which enable them to tolerate treatment. The therapist is able to interpret the transference as a reexperiencing of past repression in the immediate and intense context of the concrete present, thereby providing the patient with a reassurance and needed "truth" about his or her childhood past.

The basic techniques used to evoke transference and facilitate bringing the unconscious to consciousness are analysis of dreams, analysis of defense mechanisms, and free association. The patient lies on a couch and the analyst sits behind or away from the patient. In free association, the patient is asked to relate everything of which they are aware in verbal fashion. The content is analyzed in hopes of finding meaning in omitted material, disruptions, and slips of the tongue. The analyst interprets and shares the interpretation with the client in hopes of allowing the achievement of insight into the unconscious. The analyst must be neutral yet empathic to reduce extraneous interference of the therapeutic process. Dreams, as expression of the unconscious, are also analyzed to reveal wishes, fears, and needs which appear in symbolic form. Depending on the patient's disorder and the point of view of the analyst, a dream may be seen in many different ways.

The legacy of psychoanalysis has taken many different forms, but has become increasingly less and less viable as a treatment option due to its high cost, long length, and the existence of new and different approaches to therapy. There are now numerous forms of analytically based psychotherapies that are offered over short durations (4–6 months) that respond to the limitations of psychoanalysis.

III. HUMANISTIC/ PHENOMENOLOGICAL PSYCHOTHERAPY

Phenomenologists, also referred to as humanistic psychologists, place an emphasis on subjective feelings relating to growth and self-actualization in reality as it is experienced. Clients treated under this paradigm are required to take responsibility for their choices in terms of how they will think, behave, and perceive the world. The approach stresses how one feels in the "here and now" as opposed to the approach of psychodynamic therapy which stresses unconscious childhood conflicts. As a result the client is expected to improve on his or her own. This is facilitated by allowing the relationship between client and therapist to be as equal as possible and to help the client start a natural growth process with current feelings.

A. Person-Centered Therapy

Originally trained as a psychodynamic therapist, Carl Rogers, is responsible for pioneering the person-centered approach to the treatment of psychological disorders over his lifetime (1902–1987). His is a nondirective therapy that is free of judgement and therapist interpretation. His theory of personality maintains that to understand another person's world it is important to place yourself within their internal frame of reference to the best of your ability. Each person actively differentiates between self and other and begins to validate the special experiences that make up one's consciousness through relations to the world and others, and through the formation of values based on these perceptions. The self is in need of positive regard through acceptance and love from others. Individuals acquire conditions of worth when they act according, not to their internal organismic valuing process, but to the conditional love and internalized values of others. These conditions of worth are set points which individuals feel they must live in accordance with in order to see themselves positively.

In reality, it is impossible to avoid incongruencies between the "ideal" self and "real" self. Highly functioning individuals have small differences between how they perceive themselves and how they actually experience life, but for many people the ideal self is far different than the real self or total experience of the organism. Rogers views this incongruity as the core conflict of psychological maladjustment and as a tragedy since a basic birthright of humans is to live as a unified whole. When a person experiences threats to his or her self concept there are processes of defense which allow the person to "subceive" stimuli and distort reality in a way which lessens anxiety. Of course the defense is made at a cost to the person as perceptions of the world become inadequate and sometimes rigid.

The main goal of therapy thus would be self-integration/self-actualization by way of reducing the gap between the self-concept and real experience. The therapeutic relationship, which is based on the following six factors, is indispensable in effecting change in the personality of the person in therapy.

1. The relationship between the two people involved must be of some perceived importance to both parties.

2. The vulnerability to anxiety by the person in therapy is based on the aforementioned use of defenses to defer reality in a way which furthers a rift between ideal and real self. This incongruence, realized by the client, is the motivating force for a successful outcome in therapy.

3. Genuineness is a necessary characteristic of the therapist. This means maintaining integrity in the relationship with the client and in expressing immediate feelings and self-disclosing as honestly and appropriately as possible.

4. Unconditional positive regard for the client must be experienced by the therapist. Because a client's incongruence has been influenced in part by conditions set by others as to how and why to behave, the therapist must be unconditional in his regard of the client to reduce the distortions to self-awareness and to increase positive self-regard.

5. Accurate empathy is needed in the therapeutic relationship as a therapist must put all confusions, fears, and other personal feelings aside in an attempt to fully sense the inner world of the client in a way which allows complete awareness and acceptance of the client as a person. Therapist empathy and unconditional positive regard are complementary and needed for the client to be open and honest.

6. The client needs to perceive the therapist as genuine in order for trust to build. This perception of genuineness must occur for the therapeutic relationship to be successful.

These factors have been recognized as helpful if integrated into other therapies as the therapist–client relationship is seen as the most important aspect of treatment.

B. Gestalt Therapy

Developed by Frederick S. Perls and influenced by Gestalt psychology, this method espouses a phenomenological philosophy which hinges on the idea that people tend to actively systematize their view of the world by creating their own versions of reality and contends that people must enact their "true" feelings for psychological growth to continue. The therapy is based on helping clients to become aware of themselves and accepting of themselves through dramatic and direct methods.

IV. COGNITIVE/BEHAVIORAL THERAPY

Behavior therapy is centered around trying to help clients view their psychological aberrations as learned behavior which can be changed independent of underlying problems or insight into self-awareness. The roots of this paradigm are bedrocked in the work of Ivan Pavlov, J. B. Watson, and B. F. Skinner. The term behavior therapy is grounded in classical conditioning techniques whereas behavior modification is focused on operant conditioning techniques. [*See* Conditioning.]

Cognitive/behavior therapy focuses on changing thinking patterns and outward behavior. There is continuous debate over conceptual assumptions, and although there is a common methodological orientation, behavior therapy entails a huge array of techniques grounded in learning theory. This paradigm views abnormal behavior as learned and acquired in much the same fashion as normal behavior and views modification as applying social learning principles to current behavioral determinants. Treatment is derived from scientific principles brought forth through experiments that are specific and replicable. The goals of therapy are mutually contracted with the client, and although there may be a general method for a specific problem, each therapy is individually tailored to the client based on his or her thoughts, feelings, and actions. Disordered behavior can be modified as maladaptive thoughts and actions are replaced by systematic new learning experiences. Treatment is characterized by the following characteristics.

1. Good therapist–client relationship.
2. Listing of thoughts and behaviors to be changed; establishment of goals based on this list would facilitate focus on specific thoughts and behaviors.
3. Therapist acts as a teacher by assigning tasks and specific plans for time spent away from the therapist.
4. Treatment is continuously monitored and modified in response to patient progress in order to bring about the best outcome.

A. Behavioral Techniques

Systematic desensitization was developed by Joseph Wolpe in the late 1950s. It is a treatment for phobias and other types of irrational anxiety in which the client is hierarchically and progressively taught relaxation techniques in an attempt to weaken the association between anxiety and the feared object. Whether or not anxiety is extinguished or simply replaced with a calmer response is unclear.

Modeling is a technique where clients watch other people perform the correct and desired behaviors. This vicarious treatment is used in teaching assertiveness and appropriate social skills by role playing and graded practice.

Extinction is the process of trying to make undesirable behaviors less likely by removing the rewards which reinforce maladaptive behavior. Exposure techniques, such as flooding, attempt to extinguish a given behavior by exposing the client to an inordinate amount of fear-eliciting stimuli in hopes that the response to the fear will diminish.

Punishment is employed as a method to disrupt behavior and apply a negative stimulus so as to discourage further behavior after it has occurred.

Aversive conditioning, which may include punishment, attempts to extinguish habits by associating them with discomfort either directly or covertly.

Positive reinforcement is employed systematically through the setting up of contingencies which reward positive behavior and change behavior patterns for the better.

B. Rational Emotive Behavior Therapy

Started by Albert Ellis in the 1950s, rational emotive behavior therapy is an approach that is prominent today. The basis of the therapy is the principle that depression, anxiety, and feelings of guilt, as well as other psychological problems, are a result of *how* people think about the world. The therapist aids the client in learning to recognize cognitions that are self-defeating or irrational and attempts to replace faulty cognitions with advantageous and realistic ones through modeling techniques and logic. There is a direct challenge of the clients belief system set against external reality. Altering thinking patterns, otherwise known as cognitive restructuring, allows clients to behave in appropriate ways.

C. Beck's Cognitive Therapy

Although developed independently from Ellis, Aaron Beck's therapy is also focused on cognitive restructuring. His therapy has been especially useful in treating depression and anxiety disorders. The therapy is grounded on the assumption that people tend to be unrealistic in their goal setting and that their efforts become self-defeating which reinforces their negative cognitions. Although the clients' beliefs about the world are not disputed directly, as in rational emotive behavior therapy, the therapist helps the client realize that his or her thinking errors are what lead to low self-esteem, anxiety, or depression. For instance, such specific depressogenic assumptions as overgeneralizing, selective abstraction, dichotomous thinking, self-referencing, and excessive responsibility lead to non-adaptive behaviors. An example is the cognitive triad, which is the ideational theme for depression. A person (1) interprets events negatively, (2) dislikes himself or herself because of this interpretation, and (3) appraises the future in a negative fashion. The therapist encourages the client to test automatic thoughts empirically and logically. A client can successfully correct maladaptive cognitions by becoming aware of what he or she is thinking, recognizing which thoughts are astray, substituting accurate judgements where the automatic thoughts occurred, and accepting feedback from a therapist who will inform him or her as to whether or not the changes are effective.

V. BIOLOGICAL/DRUG THERAPY

Anxiety disorders, depression, obsessive compulsive disorder, forms of mania, and schizophrenia are treated with psychoactive drugs that alter neurotransmitters in the brain. Antipsychotics (neuroleptics) reduce the disordered thinking, paranoia, hallucinations, delusions, and other psychotic symptoms in those with schizophrenia and related disorders. Drugs such as chlorpromazine and haloperidol are effective in 60–70% of patients, enabling a better quality of life and a better responsiveness to their environment. Side effects include blurred vision, dizziness, retention of urine, dry mouth, and movement disorders increasing in severity over time.

Drugs for the treatment of depression, subdivided into monoamine oxidase inhibitors and tricyclic antidepressants (TCAs) have been shown to have significant effects on depressive symptoms. TCAs have been prescribed more as they have fewer side effects and better effectiveness. A new generation of drugs, including fluoxetine (prozac), which is the most widely prescribed antidepressant today, have been found useful in the treatment of panic disorder and obsessive–compulsive disorder as well as depression.

Of all legal drugs, the most prescribed are the anxiolytics, such as diazepam, chlordiazepoxide, and other benzodiazepines. These drugs are somewhat helpful for the treatment of generalized anxiety disorder and have an immediate calming effect on the patient. These drugs promote physical dependence and may have side effects such as light-headedness and impaired psychomotor functioning. Research has shown that there are ethnic differences in the effectiveness of drugs on psychological disorders. The differences are believed to be related to dietary practices as well as genetically influenced factors of drug metabolism. All of these drugs are typically offered in the context of a human relationship that is supportive and encouraging, often in conjunction with psychotherapy.

VI. EFFECTIVENESS

In 1952, Hans J. Eysenck published a paper that challenged the effectiveness of psychotherapy as a viable treatment for psychological disorders over naturally occurring life experience alone. This study encouraged research which employed traditional therapies investigating several questions: Is psychotherapy effective? Do the effects last over time? How does psychotherapy compare with drug therapy? Is one type of therapy better than another? Overall, the research has demonstrated quite conclusively that those people entering therapy can expect substantial beneficial effects greater than time alone can provide. Measures, both diverse and comprehensive, from the perspective of patients, therapists, and the gauging of social role functioning, have reflected that most patients receiving therapy are better off than those who do not. Improvement has been shown even when the symptoms presented suggest serious psychopathology. Positive effects have also been shown in therapy where the goals are narrowly defined or when the goals apply to broad areas, such as work performance and social functioning.

The lastingness of psychotherapeutic intervention is somewhat dependent on the severity of the disorder and the length of treatment. Many patients who undergo treatment achieve healthy adjustment for long periods of time, and some research suggests that improvement can continue even after the termination of therapy. Despite the fact that many patients have a long history of recurrent problems, improvement can and does occur, yet clear evidence shows that some patients relapse and certain problems have been shown to be particularly susceptible to relapse, including drug addictions, alcohol abuse, smoking, obesity, and possibly depression. A major factor in maintaining positive results seems to be the degree to which patients recognize that changes are partially the result of their own effective efforts.

Influenced by psychoanalytic views, psychotherapy has traditionally been a long process, often lasting several years. In today's current health care environment, where much of the payment for psychotherapy is from third-party sources, there is an increasing emphasis on cost effectiveness. Brief therapy, which typically lasts from 1 to 20 sessions, has become the standard form of treatment. Research suggests that a wide variety of therapies of this duration are highly effective, with about 70% of patients significantly improving. Unfortunately, there are cases when a patient can get worse as a result of therapy. This worsening is often due to the difficulty of treating more severely disturbed clients, such as those diagnosed with schizophrenia, but also to wide value differences between therapist and client, therapist errors such as poorly timed or inappropriate interventions, and a poor therapist/client relationship.

Psychotherapeutic medication prescribed by a psychiatrist or other physician is utilized as often as psychotherapy. Research has shown that many psychotherapies are just as effective or more effective than antidepressant medication in the treatment of depression, panic disorder, and agoraphobia, whereas pharmacotherapy for schizophrenia and similar disorders generally indicates the greater value of medication over psychotherapy for alleviating symptoms. Although it has been shown that psychotherapy can produce effects equal to or above that of medication, the success rates of one psychotherapy over another are still debatable.

Of the aforementioned traditional therapies, research has found little difference in success, although behavioral and cognitive therapies have been shown to have significant success with a number of specific problems such as panic, phobias, health-related problems, and a wide variety of childhood disorders. Continued research on the most effective treatment for specific disorders will enhance the ability of practitioners to bring about beneficial outcomes in an efficient manner.

The lack of consistent differences in the overall success rates of varied therapies and the growing trend toward an eclectic approach have spurred research on factors found to be common across all therapies. The factors most researched are those associated with the person-centered or humanist school. Reviews of factors necessary for change cite empathy, positive regard, nonpossessive warmth, and unconditional positive regard as vital to the therapist–patient relationship. Psychodynamically oriented therapists call attention to the therapeutic alliance, which is based on such variables as the patient's affective relationship to the therapist, therapist empathic understanding of the patient, the congruence of patient–therapist goals of therapy, and the ability of the patient to work purposefully in therapy. These variables, if evoked positively, have a positive effect on outcome. Also, a client's perceptions of a therapist's personal qualities, such as sensitivity, gentleness, and honesty, as well as the degree to which the therapist was perceived as understanding and accepting, seem to be highly correlated with a beneficial outcome across all psychotherapies. Research devoted to common therapist variables has shed light on the subtlety and depth of the processes of psychotherapy as something which cannot be reduced to a list of techniques, but must be assessed as a meaningful process whose effectiveness hinges on the interaction of the individuals involved. This is not to suggest that technical proficiency has no unique contributions to make. In fact, the use of manuals to guide the practitioner in implementing psychotherapies is an important development that will enable more reliable research and practice to be carried out. Most of today's psychotherapists do not feel a need to secure themselves exclusively within the confines of one theoretical paradigm, but are more likely to procure their treatment options based on all of the available techniques and procedures. Indeed, most therapists in practice today describe themselves as "eclectic" or mixed in their approach.

BIBLIOGRAPHY

Bergin, A. E., and Garfield, A. E. (eds.) (1994). "Handbook of Psychotherapy and Behavior Change, 4th Ed. Wiley, New York.

Bonger, B., and Beutler, L. E. (eds.) (1995). "Comprehensive Textbook of Psychotherapy: Theory and Practice." Oxford University Press, New York.

Freedman, D. K. (ed.) (1992). "History of Psychotherapy: A Century of Change." American Psychological Association, Washington, D.C.

Prochaska, J. O., and Norcross, J. C. (1994). "Systems of Psychotherapy: A Transtheoretical Approach, 3rd Ed. Brooks/Cole, Pacific Cove, CA.

Puberty

ZVI LARON
Schneider Children's Medical Center, Petach-Tiqva

GLOSSARY

Bone age Important maturational index determined with the aid of an atlas using the number, size, and shape of the small bones of the hand and epiphyseal centers of the finger phalanges

Estrogens Female sex hormones

GnRH Gonadotropin-releasing hormone, a hormone secreted in pulsatile manner (every 90 min) by the hypothalamus and stimulating the synthesis and secretion of the gonadotropins: follicle-stimulating hormone (FSH), also developing the male sperm cells, and luteotrophic hormone (LH), also stimulating the testosterone-secreting cells in the male testicles

Gonadostat Hypothetical center of the central nervous system that governs the secretion of GnRH, the gonadotrophic secretion–stimulating hormone in the hypothalamus

Gonads Sex glands: the ovaries in the female, which produce the female sex hormone (estrogen) and ova, and the testicles in the male, which produce testosterone and spermatozoa

Hypothalamus Part of the central nervous system that manufactures the stimulatory and inhibitory hormones of the pituitary

Menarche First menstrual bleeding

Neuroendocrine Pertaining to the interrelation between nervous and endocrine tissues

Pituitary Hormone-secreting gland situated in the bony part of the skull called the sella turcica, the anterior part of which, originating from the pharynx, secrete a series of trophic hormones including the gonadotropins

Pubertal growth spurt Accelerated growth velocity caused by an increased secretion of sex hormones during puberty

Puberty Maturational stage of childhood that comprises the stages of sexual development

Testosterone Male sex hormone

PUBERTY IS A MATURATIONAL STAGE BETWEEN childhood and adulthood. It comprises the stages of sexual development, the end of which is the capacity to reproduce. It is a complex process governed by neuroendocrine factors modulated by hereditary traits.

During this period, boys and girls undergo profound physical changes (e.g., the appearance and progression of the secondary sexual characteristics and a growth spurt) (Fig. 1). In addition, profound psychological changes take place. The onset of puberty is governed by a hypothetical center in the hypothalamus, the "gonadostat," which is latent in the prepubertal period and, for reasons so far not understood, is activated at around age 8 years in girls and 10 years in boys. The relatively dormant period of the gonadostat is paralleled by a relative state of "lowered sensitivity" of the gonads and their target tissues. An imbalance of either gonadostat or the sensitivity of the peripheral tissues may cause an abnormal sequence of puberty.

ENCYCLOPEDIA OF HUMAN BIOLOGY, Second Edition, VOLUME 7. Copyright © 1997 by Academic Press. All rights of reproduction in any form reserved.

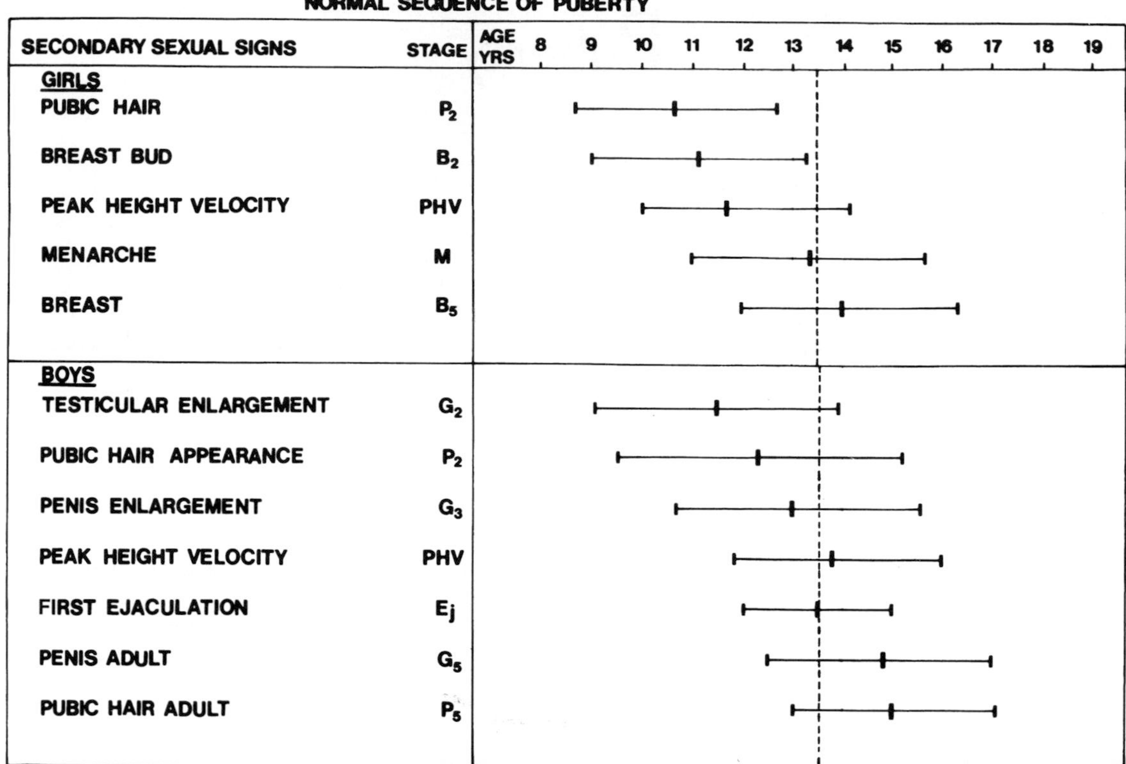

FIGURE 1 Sequence of pubertal development.

I. NOMENCLATURE OF PUBERTAL GRADING

The usual rating of prepuberty and pubertal stages is done according to J. M. Tanner, who graded the prepubertal stage P_1 and rates the pubertal stages from P_2 to P_5 (the adult stage). Although four ratings are insufficient to grade the developmental stages, Tanner's staging is the most frequently employed, and we shall refer to it in this article. It is also advised for means of accuracy to rate each sign separately [e.g., breasts (B_{1-5}), axillary hair (A_{1-5}), pubic hair (P_{1-5}), genitalia (G_{1-5})].

II. EVALUATION OF THE PROGRESSION OF PUBERTY

Ascertainment of normal progression of puberty (i.e., the age at onset) and normal rate necessitates periodic complete physical examinations, including standing height (and/or supine length), ascertainment of appearance of secondary sexual characteristics such as preacne or acne, moustache, beard, axillary and pubic hair, and a change in the voice in boys, and, in females, development of breast tissue and redness and presence of white discharge from the vagina. In addition, the size of the testes and penis in boys and small and large labia in girls should be evaluated. An important indicator of the degree of maturation is the bone (skeletal) age, determined from a hand and wrist X ray.

III. SEQUENCE OF PUBERTAL DEVELOPMENT IN GIRLS

A relation exists between the age of onset of breast development, height spurt, and age at menarche (see Fig. 1). The first pubertal event may be the onset of the growth spurt and/or pubic hair. The growth spurt may precede the onset of breast development by as much as 1 year. Peak height velocity coincides with midpuberty (i.e., Tanner P_3). Menarche occurs after the time of peak height velocity has passed, at a bone age of 13–13.5 years. Growth after menarche ranges between 5 to 9 cm. In most girls, menarche occurs within 4–5 years after the onset of breast growth.

IV. SEQUENCE OF PUBERTAL DEVELOPMENT IN BOYS

The initial manifestation is enlargement of the testicular volume and appearance of creases (rugae) of the scrotum to be followed by the appearance of pubic hair (see Fig. 1). The penis growth usually parallels that of pubic hair. The onset of the pubertal growth spurt occurs approximately 1 year after the start of increase in gonadal growth. The peak height velocity occurs late in puberty, approximately at P_4. At this stage, there occurs also a change in voice and an increase in muscle mass. The milestone of male puberty is the age at first conscious ejaculation, which occurs at a bone age of 13–13.5 years. The growth after the first ejaculation to final height ranges between 16 and 22 cm.

V. ENVIRONMENTAL FACTORS INFLUENCING PUBERTY

Normal, early, or delayed puberty is influenced by hereditary characteristics. The normal age of onset of puberty in girls is around age 8 and in boys 10 years, but slight variations occur according to geographical areas, with populations in warm climates often having earlier onset of puberty than nordic people. It has also been claimed by R. E. Frisch that there is an invariant mean weight of approximately 48 kg associated with the age of menarche, but this axioma is not universally accepted. Malnutrition or insufficient nutrition in absolute terms (e.g., in anorexia) or in relative terms (e.g., in youngsters participating in competitive sports) may delay onset or progress in puberty. Growth hormone deficiency or primary IGF-I deficiency (Laron Syndrome) may also delay puberty. Obesity, however, may enhance the appearance of puberty.

The average onset of puberty during the past 200 years or more shows a secular declining trend. This has been explained to result from a deterioration in socioeconomic, nutritional, and health conditions in the Middle Ages and in improvement in these conditions during the past two centuries. In developing countries, this trend seems to have stopped.

VI. DEVELOPMENT OF SECONDARY SEXUAL SIGNS

In many children, true gonadal-induced puberty is preceded by an arousal of the adrenal and sex hormones (adrenarche), the androgens playing the major role. These hormones account for the appearance of pimples (preacne), above or on the side of the nose, and axillary or pubic hair (mainly in females).

Whereas in females pubic hair seems to be mainly under the control of adrenal androgens, in males both the gonadal and adrenal androgens appear to play a role. The appearance of pubic hair is gradual from soft "fuzzy" to harder "curly" hair. The initial hair growth is at the base of the scrotum or along the labia majora and extends toward the pubis. In the adult stage (P_5), the female hair is distributed in an inverse triangular fashion, whereas in men there may be further spread up the pubic area along the linea alba.

Breast growth in females is controlled by the ovarian estrogen secretion. The first stage is characterized by appearance of subareolar tissue growth "breast buds," to be followed by areolar growth and bulging, nipple growth, and subsequent breast enlargement to full size. The size of the breasts at stage B_5 is variable among individuals and is influenced by hereditary characteristics. The growth is best determined by repeated measurements of the half circumference and grading of the nipples, areolae, and whole breasts.

During full-fledged puberty in males, the massive testosterone to estrogen conversion leads to swelling (sometimes painful) of the breasts. This is called pubertal gynecomastia. In most instances, this is transitory.

The initial manifestation of the male pubertal development is the enlargement of the testes (Fig. 2). The changing size of the testes can be appraised by the comparative palpation using an orchidometer or by measurement with a caliper. As there are individual variations, the early onset of testicular growth can be determined only if prepubertal measurements have been performed. Testicular growth is followed by enlargement and coarsening of the scrotal skin, which also reddens and becomes more pigmented.

The phallic growth starts in stage G_2 and is progressive in both length and width. It can be measured with a caliper, applying light streching. Knowing that mean penis length at birth is 3.5 ± 0.5 cm and that mean prepubertal length is 5.5 ± 0.5 cm, early pubertal growth can be determined by sequential measurements (Fig. 3).

The first observed signs in females are redness and whitish discharge from the vagina and increase in size of the labia minora and majora, followed by progressive growth until adult dimensions.

Pubertal milestones are, in the female, the "menarche" and, in the male, the "first conscious ejacula-

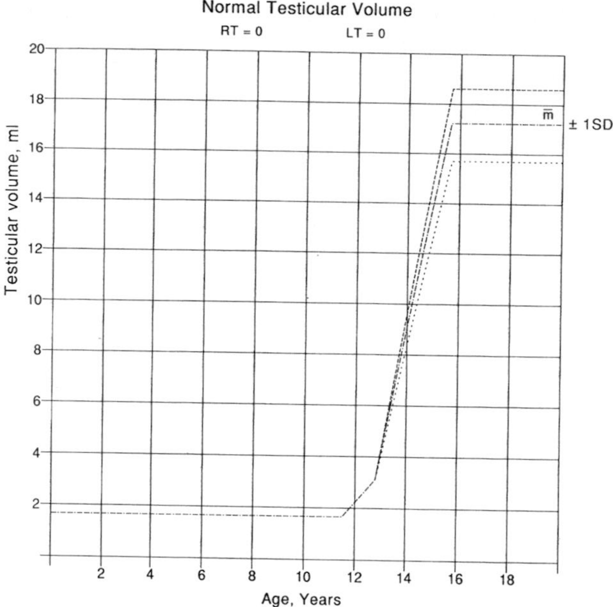

FIGURE 2 Normal testicular development. [From E. Zilka and Z. Laron (1969). *Harefuah* 77, 511–513, with permission.]

tion." Both occur at a bone age of 13.5 years, which concurs with pubertal stages of P_{3-4}.

VII. HORMONAL CHANGES AT PUBERTY

There are marked changes of gonadotropins and gonadal and adrenal sex hormones before and during puberty; the progressive changes are clearly evident when we examine large groups, but not always when following single individuals.

After a period of increased secretion postnatally and during infancy, the levels of basal gonadotropins (LH and FSH), as well as the sex hormones, decrease to constant prepubertal levels.

During the peripubertal period, FSH levels rise in girls, whereas the LH rise in boys is later; however, there is a progressively increasing response of LH in boys with each pubertal stage and of FSH in girls. Yet in some girls the FSH response may be strongest in peripuberty.

As a consequence of the rise of the gonadotropins, the sex hormones [estrogens (mainly estradiol) in females (Table I) and testosterone in boys (Table II)] rise progressively.

There is also a rise in the adrenal androgens delta-4-androstenedione and dehydroepiandrosterone sulfate (DHEA-S), which may precede the rise of gonadal steroids. [*See* Steroids.]

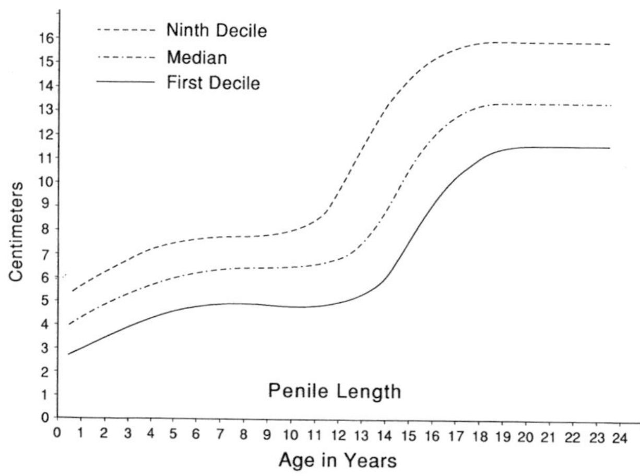

FIGURE 3 Normal penile growth.

VIII. INDUCTION OF PUBERTY

The increased secretion of gonadotropins before onset of puberty is a consequence of a change in neural and hormonal modulation of synthesis and secretion of

TABLE I
Normal Values for Plasma Estrone (E1) and Estradiol (E2) during Sexual Development (pg/ml) in Females[a,b]

Age (yr)	*n*	E1	E2
Prepubertal (>7 years)	32	14 7–29	10 <7–20
Stage 2	30	16 <7–37	14 17–35
Stage 3	24	29 8–53	30 7–60
Stage 4	31	38 10–77	45 12–93
Stage 5	36	53 12–142	63 12–250
Adults	47	69 20–182	111 17–290

[a]Based on data from F. Bidlingmaier and D. Knorr (eds.) (1978). "Oestrogens, Pediatric Adolescent Endocrinology," Vol. 4. p. 80. Karger, Basel, with permission.
[b]E1 and E2 given as median and range. *n* = number of determinations.

TABLE II
Mean (±S.D.) Plasma Concentration of Testosterone in Males According to Pubertal Stages[a]

Pubertal stage	Age (yr)	Testosterone (nM/liter)
P_1	9–11	0.6 ± 0.3
P_2	11.5	1.8 ± 1.0
P_3	12.5	3.8 ± 2.0
P_3	13.5	7.3 ± 5.5
P_4	14.5	12.8 ± 4.3
P_5	15.7	15.2 ± 4.1

[a]Based on data from M. G. Forest (ed.) (1989). "Androgens in Childhood, Pediatric Adolescent Endocrinology," Vol. 19. p. 114. Karger, Basel, with permission.

GnRH. GnRH secretion changes from a tonic to a pulsatile form—the type of stimulation needed to induce and maintain LH and FSH release and subsequently the female menstrual cycle. The area in the hypothalamus responsible for the change to pulsatility is called the gonadostat. Hereditary differences in the period of maturation of the gonadostat account for individual variations in the onset of puberty. The limits of what is considered within physiological range and abnormal are not clear and not uniformly accepted.

IX. DELAYED PUBERTY

The upper limit to distinguish between constitutional delay and abnormal delay seems to be age 13 in girls and 15 in boys (e.g., when no signs of puberty are present at those ages an abnormality should be suspected and investigations done). However, it is practical to perform these even earlier, especially when there is no familial history of delayed puberty. Constitutional delay of puberty is more frequent in boys, and these are more concerned with the accompanying slowing in the growth rate than the slow sexual maturation. The differential diagnosis is between constitutional delay and true hormonal deficiency, which includes isolated gonadotopin deficiency [normal prepubertal growth sometimes associated with anosmia (lack of smell), i.e., Kallman syndrome, Prader Willi syndrome (marked obesity, short stature, and some mental retardation), and multiple pituitary hormone deficiencies including gonadotropins]. Possible other diagnoses are primary gonadal failure (high basal or stimulated LH and/or FSH), such as in gonadal dysgenesis, Turner syndrome, anorchia, or acquired damage to the ovaries or testes (e.g., torsion, mumps, accidents).

Hyperprolactinemia is a frequent cause of delayed puberty in girls but rarely in boys. It also causes secondary amenorrhea in girls.

Constitutionally delayed puberty can be treated by either short courses of small doses of androgens or anabolic steroids or by pulsatile GnRH infusions. The latter form is more physiological, the former simpler and more acceptable by the individuals. As longer prepubertal height gain leads to a higher final height, it is advised to treat delayed puberty only if psychosocial factors demand it. Sex hormone replacement therapy in states of permanent hormone deficiency should be initiated as late as possible for the same reasons. Hyperprolactinemia is treated medically in cases of pituitary microadenoma or surgically in macroadenoma and selective cases of microadenoma. If delayed puberty is accompanied by growth arrest, a cerebral tumor such as craniopharyngioma should be excluded.

X. PRECOCIOUS PUBERTY

Precocious puberty is defined as the appearance of sexual characteristics before age 8 in girls and age 10 in boys. When it is due to a premature activation of the gonadostat it is defined as "true" or "central" in contradistinction to precocious "pseudo" puberty when the pubertal signs are caused by sex hormones not stimulated by the pituitary (e.g., congenital adrenal hyperplasia, gonadal or adrenal tumors, or exogenous sex hormones). There is also a form of "incomplete" sexual precocity [e.g., precocious thelarche (breast development) or precocious adrenarche (pubic and/or axillary hair development)]. These forms may be transitory (thelarche only) or stationary, or may develop into true precocious puberty.

True precocious puberty is four to five times more frequent in girls than in boys. In girls, in most instances, the cause is unidentified. In approximately half of the affected boys, an organic lesion or anomaly of the central nervous system [e.g., brain tumors (hamartoma, pinealoma) or hydrocephalus] is found. Precocious puberty should be suspected when an acceleration of the growth spurt and an advancement of bone age occur before the expected age.

Organic lesions are treated whenever possible by surgery. Idiopathic true precocious puberty is presently treated successfully with slow-release long-acting depot preparations of GnRH analogues. [*See* Endocrinology, Developmental.]

It has been documented that some of the infants who present with early breast enlargement (premature thelarche) may subsequently develop early puberty. It has also been observed that growth hormone or IGF-I treatment may advance the age of onset of puberty and accelerate the pubertal process, therefore close follow-up of these children is indicated.

XI. CONCLUSIONS

Puberty is a stormy period in the life of children. The sexual maturation leads to marked external and internal changes of the body and a spurt in linear growth and profoundly affects the behavior and emotions.

Any change in the orchestration of these complex processes may have long-lasting effects. Thus growth and development during prepuberty should be carefully followed and explored, if indicated, to ensure a normal pubertal process.

BIBLIOGRAPHY

Bidlingmaier, F., and Knorr, D. (eds.) (1978). "Oestrogens, Physiological and Clinical Aspects, Pediatric Adolescent Endocrinology," vol. 4. Karger, Basel.

Bronson, P. H., and Rissman, E. F. (1986). The biology of puberty. *Biol. Rev.* **61**, 157–195.

Forest, M. G. (ed.) (1989). "Androgenes in Childhood, Pediatric Adolescent Endocrinology," Vol. 19. Karger, Basel.

Frisch, R. E. (1985). Fatness, menarche and female fertility. *Perspect. Biol. Med.* **28**, 611–633.

Frisch, H., and Laron, Z. (eds.) (1988). "Induction of Puberty in Hypopituitarism," Serono Symposia Review No. 16. Ares-Serono, Rome.

Grave, G. D., and Cutler, G. B., Jr. (eds.) (1993). "Sexual Precocity. Etiology, Diagnosis and Management." Raven, New York.

Kauli, R., Galatzer, A., Kornreich, L., Lazar, L., Pertzelan, A., and Laron, Z. (1997). Final height of girls with central precocious puberty, untreated versus treated with cyproterone acetate or GnRH analogue. *Hormone Res.* **47**, 54–61.

Laron, Z., Arad, J., Gurewitz, R., Grunebaum, M., and Dickerman, Z. (1980). Age at first conscious ejaculation: A milestone in male puberty. *Helv. Paediatr. Acta* **35**, 13–30.

Laron, Z., Sarel, R., and Pertzelan, A. (1980). Puberty in Laron-type dwarfism. *Europ. J. Pediatr.* **134**, 79–83.

Laron, Z., Dickerman, Z., and Kauli, R. (1987). Treatment of premature and delayed puberty with LH-RH and its analogs. *In* "Seminars in Reproductive Endocrinology" (R. H. Asch. ed.), pp. 421–429. Thieme Med. Publ., New York.

Stanhope, R., Abdulwahid, N. A., Adams, J., Jacobs, H. S., and Brook, C. G. D. (1985). Problems in the use of pulsatile gonadotropin releasing hormone for the induction of puberty. *Hormone Res.* **22**, 74–77.

Styne, D. M. (ed.) (1991). Puberty and its disorders. *Endocrinol. Metab. Clinics of North America* **20**(1), 245.

Tanner, J. M. (1962). "Growth at Adolescence," 2nd Ed. Blackwell Scientific Publ., Oxford, England.

Pulmonary Circulation, Pharmacology

RAJAMMA MATHEW
New York Medical College

BURTON M. ALTURA
State University of New York Health Science Center at Brooklyn

GLOSSARY

Eicosanoids Twenty-carbon fatty acids, which include prostaglandins, leukotrienes, and epoxy-, mono-, and dihydroeicosanoic acids

Endothelial-derived relaxing factor Labile relaxing agent produced by endothelium in response to certain stimuli

Oxygen radicals Partially reduced species of O_2 free radicals, which are reactive and cytotoxic by-products of normal aerobic metabolism

Pulmonary hypertension Mean pulmonary artery pressure of more than 18 mm Hg

Vascular tone Partial contractile state of the vessel

MOST OF OUR UNDERSTANDING OF HUMAN pulmonary circulation is based on animal experiments. The pulmonary circulation is a low-resistance system and is influenced by ventricular compliance and pressure, and thoracic pressure. Not only does it serve the vital function of gaseous exchange, but it is also important in numerous metabolic functions that affect pulmonary arterial and bronchial smooth muscle cells, pulmonary microvascular permeability, and the systemic circulation. The pulmonary vasculature will react to external stimuli such as an increase in pulmonary artery pressure, flow secondary to cardiopulmonary defects, and various drugs. It is also influenced by local changes (e.g., oxygen tension, acid–base balance, injury, and the maturation of the system). The pulmonary vasculature has a large surface area lined by endothelial cells (EC), which provide a dynamic barrier between the underlying smooth muscle cells and the various vasoactive substances in the circulating blood. Various vasoactive substances are synthesized and metabolized in the lungs, and the end products are distributed throughout the systemic circulation to act on peripheral tissues. Existing vascular tone in the arteries dictates the responses to various stimuli. *In vivo* and *in vitro* studies, including isolated perfused lungs, isolated vessel, and isolation and culture of endothelial cells, have been designed to understand the physiology and pathophysiology of the pulmonary circulation. In this article, an attempt will be made to review some of these issues.

I. MATURATIONAL ASPECTS

During fetal life, lungs receive only 8–10% of the cardiac output. The rest is diverted to the systemic circulation via a patent ductus arteriosus. The fetal lungs do not participate in gas exchange but are involved in various metabolic functions that, in fact, prepare the fetus to adapt to an independent life at birth. Phospholipid, glucocorticoid, and antioxidant

ENCYCLOPEDIA OF HUMAN BIOLOGY, Second Edition, VOLUME 7. Copyright © 1997 by Academic Press. All rights of reproduction in any form reserved.

synthesis increases as the gestational age advances. Prostaglandins (PGs) of the E and F series are present in both the fetal and maternal circulations. The functional activity of endocrine cells and peptide (i.e., bombesin-like) immunoreactivity increase during late gestation and the neonatal period, which may indicate that they have a role in regulation of the neonatal circulation; however, their exact roles are not understood.

The fetal circulation has a high pulmonary vascular resistance and tone. The changes in such resistance play a critical role in the overall adaptation to extrauterine life. After birth, with the onset of air breathing, pulmonary vascular resistance falls and it is accompanied by an increase in pulmonary flow. Several factors are known to promote pulmonary vasodilation at birth. Part of the fall may be related to mechanical factors and/or to prostaglandin release associated with ventilation and local changes in pO_2 and pCO_2. Recent studies in animals have shown that endogenous endothelium-derived relaxing factor (EDRF), identified as nitric oxide, modulates basal vascular tone and contributes to the normal reduction in pulmonary vascular resistance.

Fetal pulmonary circulation has a greater vascular reactivity compared with the adult pulmonary circulation. Acetylcholine and bradykinin, which have little effect on normal adult pulmonary circulation, produce dramatic drops in pulmonary arterial pressure and resistance. Endothelium-dependent vasodilation increases with gestational age. In the fetal circulation, small doses of epinephrine and norepinephrine cause intense vasoconstriction. These profound effects seen in fetal pulmonary circulation are diminished after the pulmonary vascular resistance has fallen. Thus, the effects of these compounds depend on the initial vascular tone in the fetus as well as in the adult. Bradykinin, a potent vasoactive peptide, is thought to have a physiological role in the perinatal circulation. It produces vasodilation by PG-dependent and -independent mechanisms in fetal lambs. The magnitude of the acetylcholine response progressively increases with gestational age. This is thought to be caused by maturation of the effector system rather than the vascular receptors. In the rabbit pulmonary artery, from Day 1 to 5 months, there is a progressive increase in the beta-adrenergic receptor-dependent relaxation response to isoproterenol, and subsequently there is reduction in the relaxation response to isoproterenol with advancing age. However, relaxation to sodium nitrite, which is unrelated to beta-adrenergic receptors, is independent of age.

The fetal carotid arteries contract in response to norepinephrine and serotonin earlier than they do to a neurogenic stimulus, suggesting that the alpha-adrenergic receptors on smooth muscle cells develop before the adrenergic nerves function. Whether it holds true for pulmonary arteries or arterioles is not known. In piglets, the receptors for norepinephrine appear within 1–2 days after birth in intrapulmonary arteries, but receptors for serotonin, angiotensin II, and vasopressin are not present, although these receptors are present in intrapulmonary veins. Beta-adrenergic-related relaxation is seen within the first day of birth, but histamine H_2-receptors are not present even after 10 days of birth. Thus, there is a wide variation in development of various receptors. The contractile response to histamine increases with age in dogs as well as in lambs. It is thought to be caused by an increase in quantity of H_1-receptors and also by an increase in intrinsic contractile responsiveness.

Some of the PGs, too, show age-related changes in their action on pulmonary vasculature. PGD_2 causes marked decreases in pulmonary artery pressure and resistance in the first 3 days of life, and after 15 days it acts as a vasoconstrictor. Likewise PGH_2 (endoperoxide, a metabolite of arachidonic acid) causes decreases in pulmonary artery pressure in the unventilated fetal lamb, but in ventilated fetal lambs it causes an increase in pulmonary artery pressure. Vasoconstricting effects of endothelin-1, an endothelium-derived contracting factor, augment with increasing gestational age. Some of the metabolic functions also appear late in fetal life and do not reach the adult level at birth. The changes related to maturation also consist of changes in the ratio of subtypes of receptor populations. Thus, there are distinct differences in responses to various vasoactive substances in the neonatal and the adult pulmonary circulation. These differences will have effects on drug metabolism, clearance, and treatment of pulmonary disease processes.

II. NEUROHUMORAL ASPECTS

In fetal lungs, the nervous tissue and neuroepithelial bodies appear early. Lungs are innervated from the anterior and posterior pulmonary plexi. The latter are formed by fibers from the sympathetic trunk and the vagi. From this network, the major pulmonary arteries receive a plexus of large nerve trunks coursing within the adventitial layer (myelinated and nonmyelinated). The terminal twigs pass to the region of smooth muscle cells (SMC) in both large and small elastic, as well as muscular, arteries down to 30 μm in diameter. Fewer fibers are seen in veins. In small arteries, the fibers remain external to the medial coat. In arteries less than 30 μm in diameter, the fibers are not present. Both sympathetic and parasympathetic fibers are found in close association with pulmonary vessels. In adult dogs, the major effect of sympathetic activation is to increase stiffness of large arteries. Adrenergic receptors in the pulmonary vessels are involved in local responses to hypoxia and hypercapnia. In the cat pulmonary vascular bed, both subtypes of postjunctional alpha-adrenoreceptors have been demonstrated.

Forebrain stimulation of the cat causes an active change in pulmonary vascular resistance, suggesting the presence of a descending pathway that, when stimulated, causes vasodilation. It has been suggested that adrenal–medullary mechanisms may play a role in cardiovascular responses elicited by stimulation of central nervous system (CNS) structures. The role of the hypothalamus in regulation of vasoconstrictor tone in the pulmonary vascular bed still remains to be established. Electrical stimulation of the left vagus causes vasodilation of the ipsilateral unexpanded lung, an effect abolished by atropine. The pulmonary vasculature of the unexpanded fetal lung is sensitive to both parasympathetic and sympathetic stimulation. It is not yet clear if neurotransmitters are released in the unstimulated state. Thus, it is not certain if the CNS has any role in modulating pulmonary circulation under physiological conditions.

Pulmonary smooth muscle is equipped with functioning alpha- and beta-adrenergic systems. Alpha receptors predominate over beta receptors, thus a vasoconstriction response is favored. These receptors are functionally independent of the available autonomic nerve endings. Propranolol, a mixed beta-adrenergic blocker (acting on β_1 and β_2 receptors), increases the vasoconstrictor response to epinephrine and norepinephrine in perfused lungs and in pulmonary vascular smooth muscle (VSM). Histamine is contained in relatively large concentrations in the lungs of rats, dogs, and other mammals. Two subtypes of histamine receptors exist, H_1-receptors, which mediate vasoconstriction, and H_2-receptors, which subserve vasodilation. A significant population of H_2-inhibitory receptors exist, which may play a role in hypoxic pulmonary vasoconstriction. Thus, histamine and adrenergic receptors may have a role in modulating vascular tone and may play a role in partly mediating hypoxic vasoconstriction.

Serotonin, a circulating hormone, is a powerful vasoconstrictor in the pulmonary circulation. It is contained in mast cells and platelets and is metabolized by ECs. In pulmonary arteries, infusion of serotonin causes pulmonary vasoconstriction but no platelet aggregation. Catecholamines (e.g., norepinephrine and epinephrine) are potent vasoconstrictors and are endogenous in origin. A large number of peptides have been identified in the lungs. Many of these peptides influence pulmonary smooth muscle tone, permeability, and inflammatory responses. Angiotensin II, spasmogenic lung peptide, substance P, and bombesin are vasoconstrictors. Substance P can affect pulmonary vessels directly, but it is also thought to be involved in pulmonary reflexes and in inflammation. Vasoactive intestinal peptide (VIP) is the only peptide so far shown to relax pulmonary vessels, and it is more potent than prostacyclin (PGI_2). Activated complement C_5a increases pulmonary vascular permeability, as do the peptides containing lipids such as leukotrienes (LTC_4, LTD_4). Bradykinin, together with hypoxia or PGE_2, can induce moderate degrees of a high-permeability type of pulmonary edema. A physiological role for these peptides, however, is not established as yet.

III. FACTORS INFLUENCING VASOMOTOR TONE

A number of physiological, pharmacological, and experimental conditions can affect vasomotor tone. VSMs differ in their ability to respond to various vasoactive substances depending on their location, structure, sensitivity and density, local temperature, pH, integrity of ECs, cell metabolism, and tissue injury, to name a few. There are also species- and strain-related changes. The role of aging and neurohumoral factors are covered elsewhere in this article. Improper handling of the tissue has been found to alter ionic equilibrium, with the result that Ca^{2+} and K^+ are lost, and intracellular Na^+ level is increased, which can alter vascular responsiveness to various stimuli. Some of the amines and zwitterion buffers used in various *in vitro* studies (e.g., Tris, HEPES, and MOPS) exert significant inhibitory effects on exchangeability and transmembrane movement of Ca^{2+}in VSM. General anesthetics commonly used can attenuate contractile responses to agonists in a dose-dependent fashion, which are not inhibited by known pharmacological antagonsits or PG synthase inhibitors. These anesthetic agents, in a dose-dependent manner, prevent uptake of radiolabeled Ca^{2+} in VSM.

For excitation–contraction coupling in VSM, Ca^{2+} is required. The process of contraction can involve influx of extracellular Ca^{2+} or release of Ca^{2+} from intracellular stores. Depending on the agonists used and the type of blood vessel, VSMs may vary in their mechanism for accumulation and recruitment of Ca^{2+}. It has been shown that calcium channel blockers (e.g., nitrendipine and verapamil) attenuate K^+-induced contractions but are ineffective in attenuating vasoactive agonists such as phenylephrine, norepinephrine, serotonin, angiotensin II, and prostanoids. These results are consistent with the concept that K^+-induced contractions of pulmonary VSM are primarily mediated via influx of extracellular Ca^{2+}, whereas most vasoactive agonists recruit Ca^{2+} from intracellular stores. Recently, two vasoactive peptides (VIP and substance P) have been identified within the nerve fibers and have terminals innervating pulmonary and systemic vessels. It is possible that these neuropeptides have a physiological influence on VSM tone.

IV. ISOLATED VESSEL STUDIES

Isolated vessel studies can be performed with relative ease, and this method has been used extensively with various vessel segments from various sites and species to study the effects of different vasoactive substances. The pulmonary arteries have low resting tone. Therefore, in all such *in vitro* studies, basal tone has to be increased with a vasoconstrictor agent such as phenylephrine, 5-hydroxytryptamine, $PGF_{2\alpha}$, or their analogues. The vascular response varies with species and the segment of the vessel studies. An obligatory role for endothelium was described for acetylcholine-induced vascular relaxation. It was thought to be secondary to synthesis and release of EDRF. The work done with isolated pulmonary arteries is too numerous to summarize here and is beyond the scope of this article. The segment of a vessel can be used either in a ring form or spirally cut to a narrow strip. Care is taken not to injure ECs, and in some cases endothelium is removed by gently rubbing the luminal surface with a pared wooden stick or a filter paper. The arterial segment is mounted in a tissue bath containing physiological buffer solution, kept at 37°C and at pH 7.4, with an oxygen–carbon dioxide mixture. The contractile responses to various vasoactive agents are usually monitored isometrically by a force displacement transducer. The resting tone required to obtain the optimal response varies with the origin of the vessel and the species.

Pathological destruction of endothelium can alter the delicate balance between the vasodilating and vasoconstricting agents in the circulation. In atherosclerosis and pulmonary hypertension, EC damage has been demonstrated, and this method provides an opportunity to study EC cell function in experimentally induced diseases. The effects of hypoxia have been extensively studied. In all mammalian species studied so far, contractile responses of pulmonary arteries to acute hypoxia have been demonstrated. It has been suggested that during hypoxia, endothelium may release a contractile agent(s) or there may be inhibition of a cyclic 3′5′-guanosine monophosphate (cGMP)-associated relaxation mechanism. The role of Ca^{2+} has also been evoked.

V. ISOLATED LUNG PERFUSION STUDIES

Isolated perfused lung preparations ahve been used to evaluate the effects of various noxious substances, such as hypoxia, and other chemicals (e.g., bleomycin, α-naphthyl thiourea), and also to evaluate the uptake and metabolism of various amines and drugs and how they are affected in states of injury. The metabolism of amines reflects the functional state of ECs. The response to pharmacological agents varies depending on the perfusate used. In physiological salt-perfused lungs, more PGs are released compared with blood-perfused preparations. The vascular tone of the preparation is important, as the elevated tone would magnify the effects of vasodilators. It has been shown, that blood-perfused isolated rat lungs develop a large hypoxic pressor response compared with salt-perfused lungs, but this difference might be dependent on differences in glucocorticoid activity in the two preparations. Effects of commonly used anesthetic agents on lung metabolism of various biogenic amines have been studied. Ketamine, halothane, and nitrous oxide inhibit 5-hydroxytryptamine metabolism, whereas fentanyl, an opioid, does not have any effect. Thus, isolated perfusion studies of lungs allow us to examine the functions of endothelium in normal and abnormal states as well as in the presence of various pharmacological agents. However, as explained elsewhere, not all the results obtained can be extrapolated to the living organism.

VI. PHARMACOLOGY OF ENDOTHELIAL CELLS

Tremendous progress has been made in the understanding of ECs, primarily because of the advent of the electron microscope and techniques for the isolation and culture of the ECs. The pulmonary EC lining comes in contact with the entire cardiac output—bloodborne hormones, vasoactive substances, and formed elements of the blood itself. The ECs not only provide a physical barrier between the blood and VSM but are actively involved in synthesis, uptake, and metabolism of various vasoactive substances, antithrombolytic activities, and respiratory function. ECs from different sites differ in morphology and biochemical functions.

A. Barrier Functions

The EC lining provides a physical barrier between the circulating blood elements and the underlying smooth muscle cells. Various substances such as lipoproteins and nucleotides are degraded on the luminal surface. Thus, these substances are prevented from reaching the extravascular space. Other compounds (e.g., adenosine) are taken up intracellularly and metabolized.

B. Modulation of Vascular Tone

Pulmonary ECs modulate pulmonary vascular tone and structure by releasing vasoactive substances in response to numerous stimuli.

1. Endothelium-Derived Relaxing Factor

Since the discovery of endothelium-derived relaxing factor in 1980, it has been established that the presence of endothelium is required for the relaxation of smooth muscle by agents such as acetylcholine, adenine nucleotides, thrombin, substance P, calcium ionophore (A23187), and bradykinin. EDRF, now identified to be nitric oxide (NO), is a potent vasodilator in pulmonary circulation and is continuously released under basal conditions. It has a short half-life, and superoxide ions contribute to its instability. In the vascular compartment it is rapidly inactivated by hemoglobin, therefore it acts locally at the site of its release. NO is formed in the presence of oxygen from the guanidino group of L-arginine by the action of a constitutive isoform of nitric oxide synthase (cNOS), an endothelial cell membrane-bound enzyme. This reaction is NADPH and Ca^{2+}/calmodulin dependent. Endogenous NO diffuses from EC to neighboring smooth muscle cells, where it activates soluble guanylate cyclase through a reaction with the heme moiety of this enzyme. Activated guanylate cyclase increases cyclic guanosine monophosphate (cGMP) in smooth muscle, leading to relaxation.

In addition to its vasodilator role, endogenous NO functions as an inhibitor of smooth muscle mitogenesis and proliferation by a cGMP-mediated mechanism. Thus, endogenous NO not only maintains the pulmonary vasculature in a relaxed state, but also contributes to vascular remodeling. NO also inhibits platelet aggregation and regulates Ca^{2+} mobilization by a cGMP mechanism. Atropine and methylene blue (inhibitor of soluble guanylate cyclase), carbonyl group reagents, K^+-borohydride, diethiothreitol, phenylhydrazine, phemidone, and nordihydroguaretic acid all inhibit SMC relaxation as well as cGMP accumulation, whereas the relaxation is potentiated by MB22948, a type II cGMP-specific phosphodiesterase inhibitor. L-Arginine analogs such as nitro L-arginine, nitro L-monomethyl arginine, nitro L-arginine methyl ester, and so on, inhibit NO synthesis. These agents are used as experimental tools to study the biology of NO.

In immunostimulated macrophages, the formation of nitric oxide is dependent on NADPH and is induced by Mg^{2+}. Activated macrophages produce NO by an inducible form of nitric oxide synthase (iNOS) that is Ca^{2+} independent. NO thus synthesized is considered to be a key mediator in the host immune response to infection. Cytokines and endotoxin can induce iNOS in EC and SMC to produce a large amount of NO, which results in negative inotropy and cardiovascular collapse.

Another isoenzyme of NOS is present in the neuronal cells of the brain in the peripheral nonadrenergic and noncholinergic neurons (NANC). NO, via NANC, acts as an atypical neurotransmitter for various tissues, including SMC, causing relaxation.

2. Endothelium-Derived Hyperpolarizing Factor

EDRF-independent and prostanoid-independent relaxation is produced by an as yet unidentified factor, endothelium-derived hyperpolarizing factor (EDHF). EDHF hyperpolarizes SMC by opening K^+ channels and closing voltage-dependent Ca^{2+} channels, resulting in SMC relaxation. Production and release are regulated by cytosolic concentration of Ca^{2+} ions. In small blood vessels, EDHF works in cooperation with NO.

3. Endothelium-Derived Contracting Factor

One of the most potent contractile factors produced by EC is endothelin (ET), a 21-amino-acid peptide. Sarafotoxins, derived from snake venoms, are found to be analogous to ET. Three isoforms of ET (ET1,

ET2, and ET3) are widely distributed. ET1 plays an important role in the pulmonary vasculature by constricting underlying smooth muscle via ETA receptors. It produces a long-lasting contractile effect. The pressor effects of ET1 are limited by its removal within the pulmonary circulation and also by the production of EDRFs such as NO, PGI_2, and EDHF. A vasodilatory effect of ET1 via ETB receptors on EC, mediated by NO generation, is a transient phenomenon. Thus, the net response to ET1 remains vasoconstriction. However, in the newborn the vasodilatory effect of ET1 predominates. ETB receptors on SMC cause contraction. Various chemical and mechanical factors, including epinephrine, thrombin, Ca^{2+} ionophore (A23187), arachidonic acid, stretch, and increased transmural pressure, can increase ET1 production. An influx of Ca^{2+} via dihydropyridine-sensitive channels is required for ET1 activity. This process may involve protein kinase C activation. The time course of contraction is accompanied by a similar time-dependent influx of Ca^{2+}. The contractile responses are abolished by Ca^{2+} antagonists.

Pulmonary veins are more sensitive to ET1 compared with pulmonary arteries. In addition to vasocontractile properties, ET1 has a wide range of biological actions. ET1 stimulates SMC and fibroblasts, and it acts synergistically with other growth factors, such as platelet-derived growth factor (PDGF) and epidermal growth factor (EGF). Thus, ET1 may modulate vascular tone and structure in health and disease.

Other EC-derived contractile factors include angiotensin II, histamine, 5-hydroxytryptamine, superoxides, and the products of cyclooxygenase and lipoxygenases.

C. Metabolic Activity

Pulmonary EC surfaces are important sites of pharmacokinetic functions. These include removal, biosynthesis, and release of vasoactive substances that affect cardiovascular regulation in health and disease states. Biogenic amines (e.g., norepinephrine and 5-hydroxytryptamine) are removed by a carrier-mediated and drug-sensitive transport process and are degraded intracellularly by catechol-O-methyl transferase and monoamine oxidase, respectively. A substantial amount of these amines is removed during a single passage through the pulmonary circulation. However, other biogenic amines (e.g., histamine, dopamine, and epinephrine) generally are not removed. Altered biogenic amine metabolism may thus be related to alteration in cardiovascular regulation. PGs of the E and F series are removed by the lungs via carrier-mediated and energy-requiring processes, and also rapid enzymatic degradation occurs. However, PGI_2 and thromboxane are not removed by the lungs. In response to shear stress and increased flow, the ECs increase the release of PGI_2, thus affecting lung function.

Angiotensin-converting enzyme (ACE) hydrolyzes peptidyl dipeptide bonds from the terminal group carboxyl of bradykinin and angiotensin I, resulting in inactivation of the former and the conversion of the latter to a circulating vasoactive hormone, angiotensin II. Several biologically active peptides (e.g., VIP, bombesin, and oxytocin) escape degradation. Substance P is hydrolyzed by cultured EC but not intact lung. Other substances (e.g., insulin and atrial natriuretic factor) are also removed from pulmonary circulation. ECs are also thought to play a role in modifying interstitial concentration of adenosine, which is an important link between tissue metabolism and VSM activity.

A number of drugs are metabolized and cleared by the lung. Several pharmacological agents (e.g., antihistamines, antimalarials, morphine-like analgesics, anorectics, tricyclic antidepressants, and anesthetics) are concentrated in the lung as basic amines. The mechanism of accumulation of several xenobiotic amines has been extensively studied by the isolated lung perfusion method. Persistence of some of the amines in the lung is important in drug-induced pulmonary phospholipidosis. Phospholipidosis decreases uptake and metabolism of serotonin, but it enhances uptake and accumulation of chlorpromazine, chlorphentermine, and imipramine, which may limit their access to metabolic sites. Cocaine inhibits serotonin and norepinephrine clearance in rabbit perfused lungs. However, in *in vitro* studies, norepinephrine clearance is not affected. Therfore, not all the results obtained from isolated perfused lung can be extrapolated to the entire organism.

D. Homeostasis and Antithrombogenic Function

Normal ECs do not activate platelets, leukocytes, blood coagulation, or fibrinolytic factors. Endothelial proteoglycans (mainly heparin sulfate) provide a nonthrombogenic surface. The ECs shield the blood elements and coagulation factors from the underlying basement membrane, which is extremely toxic. Active antithrombogenic mechanisms involve synthesis and release of prostacyclin, secretion of plasminogen activators, degradation of proaggregatory ADP by membrane-associated ADPase, uptake, inactivation, and clearance of thrombin, and contribution of thrombomodulin in thrombin-dependent activation of protein C. The thrombin activity is limited to the site of injury. Hemostasis is a complex interaction among endothelial cells, platelets, and coagulation factors to form a mechanical seal, which subsequently gets removed by fibrinolysis.

E. Immune System Responses

Endothelial cells, when stimulated with appropriate agents, can be induced to express class II histocompatibility antigens and Fc receptors, engage in phagocytosis, and produce toxic oxygen radicals. In this respect, ECs resemble macrophages.

F. Respiratory Function

The ECs also participate in CO_2 release from the lungs. Carboanhydrase is a cytosolic enzyme and a membrane-bound ectoenzyme and adds significantly to the conversion of CO_2 from plasma bicarbonate.

VII. ROLE OF EICOSANOIDS

The term eicosanoid refers to oxygenated 20-carbon fatty acids, including PGs, leukotrienes, and epoxy-, mono-, and dihydroxy eicosanoic acids. Eicosanoids are released by enzymatic oxygenation of fatty acids [e.g., 8,11,14-eicosatrienoic acid (di-homo-γ-linoleic acid), 5,8,11,14-eicosatetraenoic acid (arachidonic acid), and 5,8,11,14,17-eicosapentaenoic acid]. Other cells (e.g., leukocytes and platelets) are also capable of generating arachidonate products. Arachidonic acid is found as an esterified complex in the phospholipid component of the cell membrane. PGs are not stored in the cell but are synthesized and released on demand or with stimuli. These stimuli include mechanical stimulation, embolization with air, endotoxin, and chemical stimulation with agents (e.g., acetylcholine, bradykinin, histamine, angiotensin II, serotonin, leukotrienes, and antigen). They have effects locally and, by and large, are not circulating hormones. Most of the eicosanoids are vasoactive substances, thus they are capable of modulating vascular tone and membrane permeability. They are also involved in events that occur concomitantly with lung injury.

Free fatty acids from membrane lipids are released by oxygenation, which is preceded by activation of phospholipase A_2 or diacylglyceride lipase. Once arachidonic acid is released, it is metabolized by different pathways.

A. Cyclooxygenase Pathway

A cyclooxygenase enzyme, PG endoperoxide synthetase, converts arachidonic acid to biologically active and unstable endoxyperoxides (PGG_2 and PGH_2). These endoxyperoxides are further converted to PGE_2, prostacyclin (PGI_2), PGD_2, and thromboxane A_2 (TxA_2). PGI_2 and TxA_2 are unstable products, and they are converted to 6-keto PGF_α by nonenzymatic degradation and to TxB_2 by spontaneous hydrolysis, respectively. The presence of these stable compounds is taken as an indicator of the parent compounds. PGs of E and F series both are released during anaphylactic shock, and they play significant roles in fetal pulmonary circulation and childbirth. PGD_2 is a potent vasoconstrictor (a vasodilator during the first 3 days of life), and it is released during anaphylaxis. TxA_2 is a potent vasoconstrictor and a powerful platelet-aggregating agent. It comprises one-third of the cyclooxygenase products released from human lung cells challenged with immunoglobulin or a calcium ionophore. There is species variation in TxA_2 production. Histamine (acting via the H_1 receptor), bradykinin, a slow-reacting substance of anaphylaxis (SRS-A), or mechanical stimulation all can release TxA_2. There is experimental evidence that TxA_2 plays a major role in mediating bronchoconstriction. A large amount of PGI_2 is released by the lungs. ECs produce much more PGI_2 compared with smooth muscle cells. It is a potent vasodilator and has platelet antiaggregatory properties. In large vessels, the ability to produce PGI_2 is of physiological importance for the prevention of platelet aggregation and injury to the EC surface.

B. Lipoxygenase Pathway

Arachidonic acid metabolized via the lipoxygenase pathway results in peptidoleukotrienes, which make up the slow-reacting substances of anaphylaxis. Peroxidation of arachidonic acid results in formation of unstable hydroperoxyeicosatetraenoic acids (HPETEs) and the unstable intermediate LTA_4, which is converted to LTB_4, LTC_4, LTD_4, and LTE_4. Peptidyltransferases in the tissue cleave the terminal amino acids from LTC_4 to yield LTD_4 and LTE_4. The main effects of leukotrienes are on the immunological system and smooth muscle in respiratory, cardiovascular, and gastrointestinal systems. LTB_4 is produced mainly by leukocytes and has chemotactic and cytokinetic properties. It induces adhesion of leukocytes to ECs and has a weak contractile property. Thus, it is involved mainly in inflammation. LTC_4, LTD_4, and LTE_4 are potent vasoconstrictors, and they can induce increased mucus secretion and bronchoconstriction in various species. Leukotriene release has been shown to be increased in various pathological states.

C. Cytochrome Pathway

The products from the cytochrome P450 pathway have effects on sodium transport and the vasculature. These products have been identified by their oxygenated peak formation (P_1 and P_2) in the presence of NADPH. Doses of P_1 relax isolated rabbit pulmonary arterial rings. A major component of P_1 appears to be a 5,6-epoxyeicosatrienoic acid derivative of arachidonic acid. P_2 has no effect on the vasculature.

VIII. ROLE OF MAGNESIUM

Mg^{2+} is found in abundance in the living organism, and it plays a critical role in numerous biological processes. As early as 1938, it was shown that $MgCl_2$ antagonized histamine- and $BaCl_2$-induced constriction in rabbit pulmonary arteries. Since then it has been established that Mg^{2+} is a natural Ca^{2+} blocker and extracellular Mg^{2+} modulates membrane permeability to Ca^{2+}. Mg^{2+} content in pulmonary arteries is lower compared with pulmonary veins and heart. Under normal conditions, Mg^{2+} has no effect on pulmonary arteries, because of their low basal tone. However, when the vascular tone is increased with various contractile agents, deficiency of Mg^{2+} results in potentiation of contraction. In low-Mg^{2+} states, relaxation effects of isoproterenol and prostacyclin are also attenuated. Mg^{2+} deficiency in rats induces coronary artery damage, edema, fragmentation of elastic laminae, and smooth muscle hyperplasia. In bovine coronary arteries, guanylate cyclase is sensitive to activation by NO and nitrosoguanidine in the presence of Mg^{2+}, and these activations markedly increase tissue cGMP, whereas Ca^{2+} markedly inhibits activation of guanylate cyclase in the presence of Mg^{2+} but not Mn^{2+}. This inhibition is dependent on the Ca^{2+}/Mg^{2+} ratio. Mg^{2+} has also been shown to relax the bronchial smooth muscle. Mg^{2+} therapy attenuates bronchoconstriction induced by histamine and methacholine. The common pathway of asthma is thought to be related to translocation of Ca^{2+} in activation of the smooth muscle contractile system, mast cells to produce histamine, the mucus secretory system, and nerve impulse initiation and conductance of vagal fibers. Mg^{2+}, on the other hand, facilitates the uptake of Ca^{2+} by sarcoplasmic reticulum and it also inhibits slow inward Ca^{2+} currents and Ca^{2+}-induced Ca^{2+} release. Therefore, it is not surprising that Mg^{2+} therapy has been useful in many cases of asthma.

Mg^{2+} is also involved in protein synthesis. It activates RNA polymerase, which allows RNA to be copied by the messenger. Most of the enzymes involved in DNA repair are more or less Mg^{2+} dependent. Mg^{2+} decreases the number of DNA replication errors, and it is essential for both nuclear transcription and ribosomal translation of gene expression.

Mg^{2+} has an important role in angiogenesis and atherogenesis, since Mg^{2+} influences EC migration and proliferation as well as lipid metabolism and VSM tone. Therefore, chronic Mg^{2+} deficiency may adversely affect EC migration and proliferation, resulting in delay or prevention of wound healing and reendothelialization of vascular injuries. This could facilitate thrombosis and excessive subintimal SMC proliferation, resulting in atherosclerotic changes. Whether attempts to increase extracellular Mg^{2+} will influence cell proliferation and migration *in vivo* is not known.

The host Mg^{2+} levels determine the result of injury. Mg^{2+}-deficient young rats develop hypertension, which is exacerbated by noise stress, and when subjected to audiogenic seizure shock they show significant pathological changes in the pulmonary vasculature and parenchymal components of the lungs. Low Mg^{2+} levels are found to magnify hyperoxic lung injury and Mg^{2+} deficiency is associated with a prolonged recovery from myocardial stunning. Mg^{2+} deficiency *in vitro* has been shown to enhance free radical-induced intracellular oxidation and cytotoxicity in EC and cardiac myocytes. Thus, Mg^{2+} appears to protect a variety of cells during stress. More Mg^{2+} may be required to attenuate the ensuing pathological changes.

IX. INFLUENCE OF INJURY

Endothelial cells play an important role in lung injury. The abnormal EC function occurs concomitant with enhanced pulmonary vascular reactivity.

A. Endotoxin Injury/Sepsis

Lung injury is often the first symptom of sepsis in one-third of all trauma patients. In burn patients, sepsis-induced adult respiratory distress syndrome (ARDS) has a high mortality. Alterations in pulmonary circulation can result in pulmonary hypertension (PH) and increased pulmonary vascular permeability. In addition, these patients develop ventilatory abnormalities such as decreased lung compliance and ventilation perfusion mismatch. These abnormalities have been duplicated in animal models by administering endotoxin or bacteria. Thromboxane A_2 is considered to be responsible for early PH and bronchoconstriction seen after bacterial infusion, and thromboxane inhibitors can markedly attenuate these symptoms. Furthermore, an increase in plasma TxA_2 has been demonstrated in humans during sepsis-induced ARDS. However, TxA_2 is not responsible for the later stage of PH. Lipoxygenase products (e.g., LTC_4 and LTD_4) are released during inflammation, and they are thought to play a role in accentuating the lung injury. However, neutrophils play an important role in increasing lung vascular permeability in sepsis and oxygen toxicity. The activated neutrophils release proteases and oxygen radicals, which cause lung injury. Arachidonate products may modulate endotoxin-induced injury indirectly through interactions with the complement system and platelet or leukocyte functions. Endotoxemia releases both constrictors and dilator cyclooxygenase products as well as lipoxygenase products. Arachidonate products of leukocytic origin may affect airways. However, such products of platelet origin are not considered important in the pathogenesis of lung injury.

Antibody binding with ECs can result in platelet-activating factor (PAF) release. The released PAF will recruit polymorphonuclear cells (PMN) and induce their aggregation and degranulation, leading to occlusion of small vessels, tissue injury, and pulmonary edema. PAF-stimulated neutrophils can induce *de novo* synthesis of LTB_4, which will contribute to, amplify, and perpetuate the inflammatory reaction. The activated neutrophils can release H_2O_2 and O_2^-, which have cytotoxic effects. It is suggested that H_2O_2 is converted to OH·, which may lead to lipid peroxidation injury.

B. Oxygen Toxicity

Prolonged exposure to high oxygen invariably leads to oxygen-mediated lung injury. The mechanism of oxygen toxicity is believed to be related to the intracellular production of free radicals, which are cytotoxic by-products of a variety of normal processes of aerobic tissue metabolism. These free radicals are superoxide radicals ($O_2\cdot$), hydrogen peroxide (H_2O_2), hydroxyl radicals (OH·), singlet oxygen ($^1O_2$), and peroxide radicals (ROO·). Under normal circumstances, they are detoxified by enzymatic antioxidants (e.g., superoxide dismutase, catalase, and glutathione peroxidase). There are nonenzymatic antioxidants such as vitamin E, beta-carotene, nonessential polyunsaturated fatty acids, and other thiols.

ECs are more susceptible to oxygen toxicity than are epithelial cells. Within 48 hr of hyperoxic exposure, the biochemical and morphological changes in ECs can be detected, which lead to increases in microvascular permeability. Normally in ECs the rate of cyanide-resistant O_2 consumption is high, which increases markedly in hyperoxia, thus creating intracellular imbalance and making ECs more vulnerable to hyperoxic insult. Bleomycin and radiation therapy injury are also related to the generation of oxygen radicals. If the capillary leak induced by oxygen toxicity is not restored, the lung damage becomes progressive and ventilation–perfusion mismatch develops, eventually leading to low respiratory reserve. A classic example of oxygen toxicity is seen in premature babies with hyaline membrane disease who go on to develop bronchopulmonary dysplasia after oxygen therapy. Increased O_2 concentration inhibits DNA and protein synthesis, therefore lung maturation is also affected in these newborns, because the maturation continues postnatally.

C. Pulmonary Hypertension

Pulmonary hypertension, a common sequel of many cardiopulmonary diseases, is associated with high morbidity and mortality rate. Persistent PH in newborn infants is believed to be due to a lack of regression of high pulmonary vascular resistance of the fetal lung at the onset of air breathing at birth. Regardless of the etiology, the majority of the cases are accompanied by medial hypertrophy and widespread histological changes of the pulmonary arteries. The pathogenesis of PH, however, is not fully understood. Pulmonary endothelial dysfunction in pediatric patients with congenital heart defect and pulmonary hypertension is associated with high morbidity and mortality despite favorable pathological changes in lung vessels. This suggests that EC dysfunction is an early event preceding the morphological changes. Chronic hypoxia, which is associated with pulmonary diseases such as chronic bronchitis, interstitial pneumonitis, and chronic obstructive pulmonary disease, also causes vascular EC injury PH. Thus, endothelial dysfunction appears to underlie most forms of PH. Impaired endothelial function also plays an important role in diseases like atherosclerosis, diabetes, reperfusion injury, and other forms of vasculopathy.

The injured ECs can activate cellular and humoral inflammatory pathways, arachidonic acid products, coagulation factors, and leukocytes, resulting in generation of oxygen radicals. The injured ECs can also increase the release of procoagulant factor VIII and reduce plasminogen activator and prostacyclin, thus favoring thrombosis (blood clotting). EC dysfunction also leads to decreased metabolism of vasoactive peptides. These cells can release various growth factors, including ET1, which could then act on underlying smooth muscle.

Histological changes that occur in the pulmonary vasculature are swelling, hypertrophy, and hyperplasia of the cells in the intima and the medial coat of the pulmonary arteries, leading to narrowing and occlusion of the lumen. There is also extension of the smooth muscle cells into nonmuscular pulmonary arteries. In some small arterial lumens, platelet or leukocyte aggregates or thrombosis completely occlude the lumen. These changes lead to pH and right ventricular hypertrophy and, in late stages, right heart failure. Some of the changes seen in various clinical forms of PH can be reproduced in experimental models. Acute exposure to hypoxia leads to increased PH, which is reversible. However, subacute and chronic hypoxia lead to persistent vasoconstriction, PH, and structural remodeling. On return to room air, some of the changes start to regress. Chronic hypoxia and monocrotaline (MCT) in rats induce progressive pulmonary hypertension. MCT is a pyrrolizidine alkaloid of plant origin that causes PH 10–14 days after a single subcutaneous injection. It is metabolized in the liver to an active pyrrol form, which causes pulmonary EC damage. EC dysfunction, such as decreased uptake of serotonin and norepinephrine, and diminished ACE levels are present.

Impaired NO-related EC-dependent relaxation has been observed in pulmonary arteries from humans with pulmonary hypertension and also from rats with chronic hypoxia- and MCT-induced PH. In a number of species, including the human, a continuous release of NO contributes to low basal pulmonary vascular resistance. ECs of pulmonary resistance vessels are thought to generate NO. Increased ET1 levels have been found in the lungs of patients and rats with PH. A daily infusion of BQ123, an ETA blocker, has been found to partially inhibit MCT-induced PH. This is not surprising since NO regulates ET1 production in a reciprocal manner. Ca^{2+} antagonists and Mg^{2+} attenuate cardiopulmonary changes in MCT-induced PH without preventing endothelial dysfunction. The effects of Ca^{2+} antagonists may be the result of inhibition of the increase in vascular tone and/or inhibition of activity of growth factors, including ET1. In this context, it is worth noting that an increased intracellular Ca^{2+} level is required for DNA synthesis and cell proliferation and, in addition, ET1 requires Ca^{2+} for its activity. It is conceivable that an ETA blocker and Ca^{2+} antagonists inhibit DNA synthesis and cell proliferation by blocking ET1 activity, thus attenuating MCT-induced PH. Taking these observations together, it appears that endothelial injury resulting in an inhibition of NO generation may be the key event in the pathogenesis of PH. In the absence of inhibitory effects of NO, ET1 and the other mitogens have the potential to exert vasoconstrictive and vasoproliferative activities, thus contributing to the pathogenesis of PH.

X. POTENTIAL CLINICAL APPLICATION

Knowledge of the pharmacology of eicosanoids has already led to a greater understanding of perinatal pulmonary circulation and has become useful in clinical practice. In a number of premature infants, the patent ductus arteriosus can be successfully closed by administering cyclooxygenase inhibitors (e.g., indomethacin). In infants with ductus arteriosus-dependent cardiac defects, the ductus can be kept patent temporarily by administering PGE_1. PGI_2 has been used in various cardiac catheterization laboratories to assess pulmonary vascular reactivity and, to a limited extent, to predict the outcome of vasodilator therapy in cases of primary PH. The role of arachidonic acid cascade and oxygen radicals is well known in lung injury, and thus manipulation of PGs and administration of antioxidants will become a part of therapy.

Inhaled NO as a therapy has been developed because of its unique property of acting locally. Thus, inhaled NO (20–80 ppm) causes selective pulmonary vasodilation and systemic circulation is not affected. It has been successfully used in treating persistent PH of the newborn and PH occurring during the postoperative period. Research is continuing to produce drugs that will release NO to the desired site.

The adverse reaction of endotoxin/septic shock is the result of excessive NO generation. Recent attempts to treat humans and experimental animals with inhibition of NOS activity have met with considerable success. However, caution needs to be exercised here. The NOS blocker may lower the cardiac output, which would interfere with tissue perfusion and make the condition worse. In the presence of PH this treatment is not likely to be successful. More research is required in this area.

Considerable interest has been generated in the possible important role of Mg^{2+} in modulating vascular tone and integrity of excitable cells and in atherogenesis. Mg^{2+} deficiency has been shown to be associated with a higher incidence of sudden death, ischemic heart disease, stroke, and preeclampsia. Dietary Mg^{2+} deficiency is known to cause systemic hypertension, structural changes in the peripheral and cerebral microvasculature, and pulmonary lesions that resemble those found in adult respiratory disease. Loss of cellular Mg^{2+} results in loss of critically important phosphagens (e.g., Mg-ATP and creatine phosphate). Therefore, under certain pathophysiological conditions (e.g., hypoxia, anoxia, ischemia, and cellular injury) in which cellular Mg^{2+} is depleted, the Na^+/K^+ pump, phosphagen stores, and membrane structure will be compromised, leading to alterations in resting membrane potentials. Increased synthesis of some eicosanoids may be linked to enhanced influx and translocation of Ca^{2+}. In Mg^{2+} deficiency, the lowered cyclic AMP can permit a high cyclooxygenase activity and a drastic increase in thromboxane. Thus, Mg^{2+} deficiency can affect the arachidonic acid cascade, which has been considered extremely important in pathogenesis of lung injury. There is evidence that Mg^{2+} acts primarily at the VSM cell membrane on specific Mg^{2+}/Ca^{2+} and Na^+/Ca^{2+} exchange sites to regulate finely the entry of Ca^{2+} in smooth muscle cells, vascular tone, blood pressure, and local flow. During stress, Mg^{2+} leaves the cell, thus making it more vulnerable to toxicity from Ca^{2+}. External Mg^{2+} plays an important role in modulating PG-induced contractions and relaxations and also enhances prostacyclin synthesis in cultured cells. It is a powerful inhibitor of blood coagulation and is a requirement for intracellular cement and endothelial–endothelial cell junctions. There have been few reports of Mg^{2+} treatment in PH. In acute hypoxia-induced PH in dogs, pulmonary artery pressure can be successively lowered with infusion of $MgCl_2$. Mg aspartate HCl has been demonstrated to attenuate MCT-induced pulmonary hypertensive changes in rats. Recently Mg^{2+} therapy has been successfully used in treating newborns with PH.

Although magnesium deficiency per se is not implicated as an etiological factor in any of the diseases mentioned here, the deficiency certainly can make the cellular injury and response much worse through the various mechanisms mentioned. Thus, by increasing intracellular Mg^{2+} by dietary or therapeutic means, it is possible that the cells might be able to withstand injury much better.

XI. SUMMARY

Our understanding of pulmonary circulation has undergone a considerable expansion during the past decade. Recent studies have shed a great deal of light on the delicate balance between the vasocosntrictors and vasodilators and on the multifaceted activities of ECs that are needed to maintain homeostasis. Further insight into these areas will lead to a better understanding of the pathophysiology of various disease states of the lung and the prevention and treatment of pulmonary diseases.

BIBLIOGRAPHY

Abu-Osba, Y. K., Gala, O., Manasra, K., and Rejjal, A. (1992). Treatment of severe persistent pulmonary hypertension of the newborn with magnesium sulphate. *Arch. Dis. Child.* **87**, 31–35.

Altura, B. M. (1987). Pharmacology of the pulmonary circulation. *In* "The Pulmonary Circulation in Health and Disease" (J. A. Will, C. A. Dawson, K. E. Weir, and C. K. Bruckner, eds.), pp. 79–95. Academic Press, New York.

Altura, B. M., and Altura, B. T. (1990). Role of magnesium in pathogenesis of high blood pressure: Relationship to its actions on cardiac and vascular smooth muscle. *In* "Hypertension: Pathophysiology, Diagnosis and Management" (J. H. Laragh and B. M. Brenner, eds.), pp. 1003–1025. Raven, New York.

Altura, B. M., and Altura, B. T. (1995). Magnesium in cardiovascular biology. *Sci. Am. Sci. Med.* **2**(3), 28–37.

Banai, S., Haggroth, L., Epstein, S. E., and Casscells, W. (1990). Influence of extracellular magnesium on capillary endothelial cell proliferation and migration. *Circ. Res.* **67**, 645–650.

Barnes, P. J. (1994). Endothelins and pulmonary diseases. *J. Appl. Physiol.* **77**, 1051–1059.

Carrol, M. A., Schwartzman, M., Baba, M., Abraham, N. G., and McGiff, J. C. (1987). Formation of biologically active cytochrome p450–arachidonate metabolites in renomedullary cells. *In* "Advances in Prostaglandins, Thromboxane and Leukotrienes Research." (B. Samuelssomn, R. Paoletti, and P. W. Ramwell, eds.), Vol. 17, pp. 714–718. Raven, New York.

Chand, N., and Altura, B. M. (1980). Occurrence of inhibitory histamine H_2-receptors in isolated pulmonary blood vessels of dogs and rats. *Experientia* **36**, 1186–1187.

Chand, N., and Altura, B. M. (1981). Acetylcholine and bradykinin relax intrapulmonary arteries by acting on endothelial cells: Role in lung vascular diseases. *Science* **213**, 1376–1379.

Dickens, B. F., Weglicki, W. B., Li, Y. S., and Mak, T. I. (1992). Magnesium deficiency *in vitro* enhances free radical-induced intracellular oxidation and cytotoxicity in endothelial cells. *FEBS Lett.* **311,** 187–191.

Fishman, A. P. (1985). Pulmonary circulation. *In* "Handbook of Physiology. Section 3: Respiratory System" (A. P. Fishman and A. B. Fisher, eds.), pp. 93–165. *Am. Physiol. Soc.,* Washington, D.C.

Gillis, N. C. (1986). Pharmacological aspects of metabolic processes in the pulmonary microcirculation. *Annu. Rev. Pharmacol. Toxicol.* **26,** 183–200.

Harker, L. A. (1988). Endothelium and hemostasis. *In* "Endothelial Cells" (U. S. Ryan, ed.), Vol. 1, pp. 167–177. CRC Press, Boca Raton, Florida.

Hyman, A. L., Lippton, H. L., and Kadowitz, P. J. (1988). Neurohumoral regulation of the pulmonary circulation. *In* "Vasodilators, Vascular Smooth Muscle, Peptides, Autonomic Nerves and Endothelium" (P. M. Vanhoutte, ed.), pp. 311–319. Raven, New York.

Kinsella, J. P., Ivy, D. D., and Abman, S. H. (1994). Ontogeny of NO activity and response to inhaled NO in the developing ovine pulmonary circulation. *Am. J. Physiol.* **36,** H1955–H1961.

Komori, I. K., and Vanhoutte, P. M. (1990). Endothelium-derived hyperpolarizing factor. *Blood Vessels* **27,** 238–245.

Martin, D. C., Carr, A. M., Livingston, R. R., and Watkins, C. A. (1989). Effects of ketamine and fentanyl on lung metabolism in perfused rat lungs. *Am. J. Physiol.* **257,** E379–E384.

Mathew, R. (1991). Fetal and neonatal circulation. II. Metabolic aspects. *In* "Neonatal and Fetal Medicine: Physiology and Pathophysiology" (R. A. Polin and W. W. Cox, eds.), Vol. 1, pp. 678–682. Saunders, Philadelphia.

Mathew, R., and Altura, B. M. (1990). Physiology and pathophysiology of pulmonary circulation. *Microcirc. Endothelium Lymphatics* **6,** 211–252.

Mathew, R., and Altura, B. M. (1991). The role of magnesium in lung diseases: Asthma, allergy and pulmonary hypertension. *Magnesium Trace Elem.* **92,** 220–228.

Mathew, R., Colt, S., and Gewitz, M. H. (1994). Nitric oxide production is impaired in rat pulmonary artery during pulmonary hypertension. *In* "Biology of Nitric Oxide" (M. Feelisch, R. Busse, and S. Moncada, eds.). Portland Press, London.

Miyauchi, T., Yorikane, R., Sakal, S., Sakurai, T., Okada, M., Nishikibe, M., Yano, M., Yamaguchi, I., Sugishita, Y., and Goto, K. (1993). Contribution of endothelin 1 to the progression of cardiopulmonary alterations in rats with monocrotaline-induced pulmonary hypertension. *Circ. Res.* **73,** 887–897.

Moncada, S., and Higgs, A. (1993). The L-arginine–nitric oxide pathway. *N. Engl. J. Med.* **329,** 2002–2012.

Nowak, J. (1984). Eicosanoids and the lungs. *Ann. Clin. Res.* **16,** 269–286.

Petros, A., Lamb, G., Leone, A., Moncada, S., Bennett, D., and Vallance, P. (1994). Effects of nitric oxide synthase inhibitor in humans with septic shock. *Cardiovasc. Res.* **28,** 34–39.

Ryan, U. S. (1988). Phagocytic properties of endothelial cells. *In* "Endothelial Cells" (U. S. Ryan, ed.), Vol. 3, pp. 33–49. CRC Press, Boca Raton, Florida.

Said, S. I. (ed.) (1985). "The Pulmonary Circulation and Acute Lung Injury." Futura Publishing, Mt. Kisko, New York.

Wu, F., Altura, B. T., Gao, J., Barbour, R. L., and Altura, B. M. (1994). Ferrylmyoglobin induced by acute Mg^{2+} deficiency. *Biochim. Biophys. Acta Molec. Basis Dis.* **1225,** 158–164.

Pulmonary Pathophysiology

A. VERSPRILLE
Erasmus University, Rotterdam

GLOSSARY

Airway obstruction Increased airflow resistance in the airways

Compliance Change in lung volume for a given change in lung recoil pressure, measured as intraesophageal pressure

Equal pressure point Sites in the bronchial tree where, during expiration, the pressure in the bronchi is equal to the pressure around the bronchi

FEV_1 Forced expiratory volume in 1 sec, that is, the maximal lung volume to be expired in 1 sec after maximal inspiration

FIV_1 Forced inspiratory volume in 1 sec, that is, the maximal lung volume to be inspired in 1 sec after maximal expiration

FRC Functional residual capacity, that is, lung volume at the end of a normal expiration

Hyperinflation Increase in FRC

Hyperpnea Increase in ventilation corresponding to an increase in metabolic rate, for example, during exercise, when the normal arterial P_{CO_2} is maintained

Hyperventilation Ventilation that is too large for a certain metabolic state, leading to a decrease in the CO_2 tension of the arterial blood below 30 mm Hg

Hypoventilation Insufficient ventilation to maintain a normal arterial P_{CO_2}, which increases above 49 mm Hg

Hypoxemia Decrease in oxygen saturation of the arterial blood below 90%

$MEF_{75,50,\text{or }25}$ Maximal expiratory flow when 75, 50, and 25% of the vital capacity is in the lungs, respectively

MVV_{30} Maximal voluntary volume that can be breathed in and out in 30 breaths/min

PEFR Peak expiratory flow rate, that is, maximal air flow shortly after the start of a forced expiration

Pneumotachometer Tube with a resistance to measure airflow based on the measurement of a pressure difference

Pulmonary edema Condition in which an increased amount of fluid is present in the lung interstitium and in the alveoli

Restriction Decrease in alveolar volume at TLC (total lung capacity), FRC, and RV (residual volume) levels

RV Residual volume, that is, lung volume after a maximum expiration

Specific compliance Lung compliance per liter of lung volume

Spirometer Container submerged in a water bath, used to measure lung volume and volume displacement during breathing maneuvers

Spirometry Recording of lung volumes with use of a spirometer

TLC Total lung capacity, that is, lung volume after a maximal inspiration

PATHOPHYSIOLOGY IS A FIELD IN MEDICAL BIOLogy in which fundamental mechanisms of disorders in physiological functions are studied. The main goal is directed at defining the character of the relationships between the disordered processes. These relationships are expressed in quantitative terms to assess the gravity of an illness. The description of disorders in founded theories, that is, the explanation of empirical knowledge in logic models, will contribute to medical science in two ways: (1) by detecting the essential variables for use as indices in an adequate diagnostic procedure, and (2) by predicting the results of medical interventions at a higher level of confidence than before.

ENCYCLOPEDIA OF HUMAN BIOLOGY, Second Edition, VOLUME 7. Copyright © 1997 by Academic Press. All rights of reproduction in any form reserved.

In pulmonary pathophysiology the functional disorders of airways, alveoli, and gas exchange are the subjects of study.

A logical consequence of fundamental knowledge in this field is the application of physiological measurements in clinical routine. This type of applied physiology can be called clinical physiology, which deals especially with the methodology and interpretation of functional tests in patients.

In this article, the pulmonary disorders will be considered in three categories:

- disorders of the airways that increase airflow resistance;
- disorders of the alveolar volume that impair lung expansion; and
- disorders of gas exchange between alveolar capillary blood and ambient air.

Usually the functional disorders will have a structural basis.

I. AIRWAY OBSTRUCTION

A. The Nose

Three important functional structures in the nose must be considered.

1. The ciliated, pseudostratified columnar epithelium contains mucus cells, that is, goblet cells, and mucus glands, which produce a fluid mixture of water, salts, and mucus. The water evaporates to saturate the inspiratory air with water vapor at 37°C. The mucus collects particles from the inspiratory air in order to clean the air before it enters the lungs.

A second important structural part of the epithelium is formed by the cilia, which move the mucus and dust particles to the pharynx, where it is accumulated and swallowed. The cilia have a fast beat in the direction of the nasal pharynx, and a much slower movement back. During the fast beat they are stiffened, and the viscous mucus moves in the direction of the pharynx. During the backward movement the cilia relax, more or less, and bend and pass through the mucus more easily without moving it backward. The movement of the cilia is coordinated in such a way that a wave moves from the front of the nose to the nasal pharynx. The mechanisms of coordination are unknown.

The cleaning process is usually supported by blowing one's nose. This forceful action causes a high pressure in the nose, which might press some mucus and dust, including bacteria, into the *paranasal sinuses,* causing infections. Sniffing repeatedly, or *sniffling,* to prevent mucus from running of the nose is also used as a cleaning process. During sniffling a negative pressure develops in the nasal cavity.

The motility of the cilia is affected by tobacco smoke, other chemical substances, and infectious materials.

2. Below the epithelium a rich plexus of blood vessels is present, supplying heat for both warming the air and evaporating water.

3. The nasal conchae, shell-like bony projections into the nasal cavity and covered with epithelium, enlarge the surface area of the nose and intensify all the processes mentioned above. These conchae decrease the width of the nasal passage, which allows obstruction to occur more easily when epithelial swelling occurs.

B. The Nasal and Oral Pharynx

The nasal pharynx and the oral pharynx are conducting parts of the airway without special function. At this level the alimentary tract and the respiratory tract cross each other. During swallowing the larynx is moved upward and the epiglottic cartilage is pushed against the root of the tongue and bent over the entrance of the larynx. This closure function of the epiglottis seems redundant: after removal of the epiglottis, the entrance of the larynx is directly pushed against the base of the tongue and sufficiently closed.

C. The Larynx

The main structures in the larynx cavity are the vocal folds, between which is a cleft, the rima glottidis. The vocal folds are a double layer of membraneous and mucosal tissue around the vocal ligaments. These ligaments are fixed in front to the thyroid cartilage and dorsally to the arytenoid cartilage. Movement of the latter can stretch the ligaments. [*See* Larynx.]

Coughing is a function necessary to evacuate mucus mixed with foreign material from the large bronchi and trachea into the pharynx. The material will be either expectorated or swallowed; if the latter occurs, bacteria are destroyed by gastric acid in the stomach. Coughing is a deep inspiration followed by a forced expiration, initially with a closed vocal cleft, resulting in an elevated subglottic pressure that opens the vocal cleft shortly and repeatedly. Closure occurs each time by the pressure drop when air escapes. During such a short opening of the vocal cleft, the forceful expiratory flow blows out the mucus. It even carries mucus away from the large bronchi all along the trachea. Coughing is elicited by mechanical or chemical stimulation of the tracheal and bronchial epithelium.

Reflexive closing off of the entrance of the trachea occurs in the presence of irritating and noxious gases, which stimulate the nasal mucosa. Water also stimulates such a reflex, causing a spasm of the larynx muscles, which stretch the vocal ligaments to close off the trachea. This phenomenon explains death by "dry drowning" during immersion.

Pressing is a voluntary action during defecation, micturition, and during labor in women, which is achieved by an expiratory action against closed vocal folds. The increased pressure in the lungs supports the diaphragm, when forceful contractions of the abdominal muscles are exerted to increase intraabdominal pressure.

The speech function of the vocal folds occurs in a coordinated action of the laryngeal muscles with the respiratory muscles, and the muscles of the pharynx and mouth. The respiratory muscles deliver the energy for the airflow, which causes the vocal folds to vibrate. The laryngeal muscles stretch the vocal cords, the degree of which more or less determines the pitch of a sound. The muscles of the throat, tongue, and mouth determine the shape of the sounding board, which acts as a filter on the higher harmonics of a tone. The higher harmonics render the tone of a sound. This accounts for the vowels. The consonants are produced in the mouth by closed or narrowed passages. However, they can be extended with a sound containing a vowel, for example, when the alphabet is said.

Paralysis of the larynx muscle, for example, after thyroid surgery during which the recurrent nerves are injured, causes an extreme increase in airway resistance, especially during inspiration. The vocal ligaments are not stretched and the folds behave like half-moon-shaped sacs, which bulge out downward during inspiration by the lower pressure below the sacs. The ligaments are moved in a medial direction and cause a narrowing of the vocal cleft, causing a serious obstruction. During expiration the sacs are pressed in a lateral direction and the cleft is less narrowed. The obstruction can be so serious that formerly extirpation of the vocal folds was necessary and speech function had to be sacrificed. At present, the vocal folds are fixed laterally in the larynx, giving enough cleft for air passage and maintaining the possibility of speech, although in a hoarse, throaty voice.

D. The Trachea and Bronchial Tree

1. Structure

The trachea, or windpipe, which extends from the larynx into the thorax, consists of open, horseshoe-shaped rings of cartilage. The open part of the horseshoe is located dorsally. The ends of the cartilage are connected to each other with fibrous tissue. This shape is important for the passage of food through the esophagus, which is located between the trachea and the cervical vertebrae. During the passage of food the esophagus bulges into the trachea.

The cartilage rings are joined by connective tissue. This structure of cartilage with connective tissue in between containing elastic fibers makes the trachea rigid transversely and flexible longitudinally. It keeps the trachea open for air passage and allows the trachea to bend with neck movements. The elastic fibers recover the length of the trachea after stretch during both breathing and swallowing.

The pseudocolumnar ciliated epithelium, located in the trachea, functions like its counterpart in the nose. The cilia, however, beat in the opposite direction and move the mucus upward to the larynx. The same occurs in the bronchi. This cleaning function is inhibited by the same agents affecting the cilia of the nose.

Two main bronchi are formed at the bifurcation of the trachea. Their angle forms inside a keel, the carina, in the direction of the trachea. By further branching a bronchial tree is formed. The main function of the trachea and bronchial tree is conduction of air, but the functions mentioned for the nose are also still exhibited: warming of air, humidifying air by evaporation of water, and cleaning the air of dust. However, this cleaning process often fails. When the respiratory air is full of dust, these particles enter the alveoli, in the long run causing alveolitis, followed by fibrosis. A common cold and, more seriously, a flu are signs of insufficient protection of the epithelium by the mucus and cilia, and by our immune system as a secondary barrier.

The bronchial tree is subdivided into generations. The trachea is called generation zero. The two main bronchi are generation 1. The bronchial tree is composed of 16 generations.

The main bronchi also contain horseshoe cartilage in their walls. In generations 2 to 11 the cartilage is irregularly spiralized around the bronchi. These generations are surrounded by connective tissue and run in parallel with the pulmonary arteries. In these connective tissue layers the pressure is equal to pleural pressure. From generation 12 up to 16, the very small bronchi are called bronchioles. Bands of smooth muscle cells are formed into spirals around these bronchioles. The inner layer of the bronchioles consists of cubic epithelium without cilia. Connective tissue fibers run radially from the bronchioles into the septa of the surrounding alveoli, like the guy ropes of a tent. Contraction of the smooth muscle cells constricts the bronchioles and stretches the radial fibers. Relaxation of the muscles allows the fibers to dilate the bronchioles.

During inspiration, airway resistance decreases by two mechanisms: (1) the more negative pleural and intrathoracic pressure, which dilates the larger bronchi, and (2) the radial fibers of the bronchioles, which stretch more and exert more traction on the bronchioles to dilate them. This latter mechanism is supported by the more negative intrathoracic pressure, which also surrounds the bronchioles.

The smooth muscles of the bronchioles have an important control function on distribution of the respiratory air in the alveoli. We will consider the effects of this control function when smooth muscles contract in one part of the respiratory tract and relax in another part. Next, we will analyze the effects of a general bronchiolar constriction, as could occur in asthmatic patients.

2. Airway Resistance and Airway Obstruction

Airflow needs a pressure difference (P) between the alveoli and the mouth. During normal inspiration, alveolar pressure is 1 to 2 mm Hg lower than ambient air pressure at the entrance of the nose. During expiration, alveolar pressure is higher than ambient. When no turbulence occurs, the relationship between the pressure difference and air flow obeys the aerodynamic equivalent of Ohm's law: $P = V'R$, where V' is volume flow of air (in ml/sec) and R is flow resistance in mm Hg/(ml/sec). R is a variable by definition and cannot be measured independently from V' and P; it is the proportionality constant between the two. When turbulence occurs, the linear relationship between V' and P is lost. Then, the relationship is described by $P = k_1V' + k_2V'^2$. Because of turbulence a much larger P value is needed for the same volume flow.

During normal breathing, turbulence occurs at the bifurcations of the bronchi. During the forced breathing action, turbulence also occurs in the larger bronchi themselves. Therefore, flow resistance is usually higher than can be predicted from the geometry of the bronchi assuming a nonturbulent flow. Such a prediction can be made from the equation experimentally found by the French physician J. L. M. Poiseuille: $R = 8\eta l/\pi r_4$, where η is the viscosity of the respiratory air, and l and r are the length and radius of each generation. When using this formula for R and the geometric data of the trachea and bronchial tree as published by Weibel (1964) at the lung volume level after normal expiration, the resistance of generations 0 to 11 is about 87% of total airway resistance. Thus, the remaining part of flow resistance in generations 12–16, including generations 17–23 (or 26), is about 13%. The approximation of airway resistance is given in Fig. 1. Because of turbulence, the calculated value of R in the larger bronchi is slightly underestimated. For our modeling we therefore assume a ratio of 90 : 10 between generations 0–11 and 12–23, respectively. In the electric analogy, this ratio is used to demonstrate the effects of distribution and extensive bronchial constriction in an asthmatic attack.

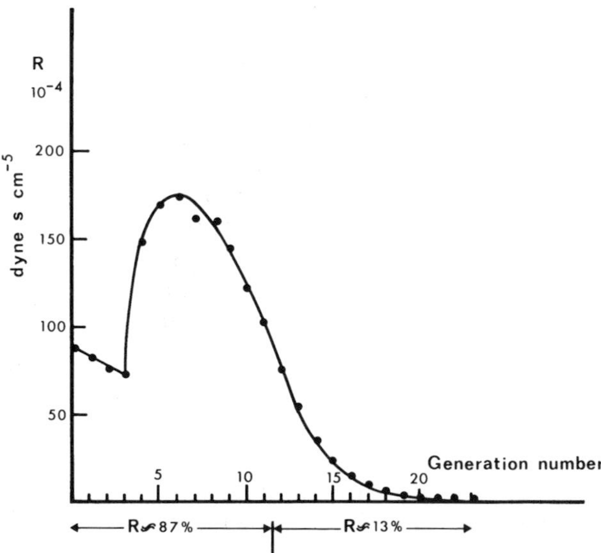

FIGURE 1 Airway resistance per generation. The calculation of airway resistance per generation is based on the flow resistance (R_i) of an individual average bronchus of a generation, which is determined according to the Poiseuille equation (see text) and the mean data of r (radius) and l (length) of each generation. The effective flow resistance (R_g) in a generation is found from $1/R_g = n(1/R_i)$, where n is the number of bronchi per generation.

Figure 2a presents the normal model. Total airway resistance (100%) has been set at 100 Ω. Then, trachea and bronchi, that is, generations 0–11, are represented by resistance R_1 of 90 Ω and the bronchioles and alveolar ducts by 10 Ω. The bronchioles are subdivided into two compartments to demonstrate the effects of mutual changes in flow resistance on the distribution of airflow, using electric current as its equivalent. Both parts have a resistance of 20 Ω and exhibit an effective resistance of 10 Ω according to the formula $1/R_{eff} = 1/R_{2.1} + 1/R_{2.2}$. A charge of these resistances with 110 V results in a current of 1.1 A. A current of 0.55 A runs through both $R_{2.1}$ and $R_{2.2}$.

In Fig. 2b, $R_{2.1} = 25$ Ω and $R_{2.2} = 16.67$ Ω, giving again an effective resistance $R_{eff} = 10$ Ω and a current of 1.1 A. However, the current is differently distributed between $R_{2.1}$ and $R_{2.2}$. The voltage on $R_{2.1}$ and $R_{2.2}$ is 10% of 110 V. Thus, $R_{2.1}$ gets a current of 11 V/25 Ω = 0.44 A, and through $R_{2.2}$ runs a current of 11 V/16.67 Ω = 0.66 A. To increase flow resistance of a tube with 25% (from 20 to 25 Ω), a diameter decrease of approximately 6% according to the Poiseuille equation is necessary. Such a decrease in diameter causes a decrease in flow of 20% (from 0.55 to 0.44 A). The decrease in resistance of $R_{2.2}$ is 16.67%, corresponding to a diameter increase of approximately 4.5%, causing an increase in flow of 20%. This example of modeling demonstrates a large shift in airflow by a small change in diameter of the bronchioles.

In Fig. 2c we modeled a general constriction of the bronchioles elicited during an asthmatic attack. We assume a constriction to one-half the normal diameter. That implies a 2^4 or 16-fold increase in resistance. Total airway resistance then is 90 + 160 = 250 Ω. The decrease in current is from 1.1 to 0.44 A, which is 60%. To compensate for this decrease in airflow, an effort 2.5 times larger than normal is needed during breathing. A constriction to one-quarter of the bronchiolar diameter increases bronchial resistance 256 times and total flow resistance 26.5 times, which means very large efforts are needed to breathe.

The ability of the bronchioles to constrict and dilate by means of their smooth muscle cells layer is a favorable function to control the distribution of air. However, this ability is paid for by asthmatic patients, who easily develop high flow resistances in the bronchioles from a variety of stimuli.

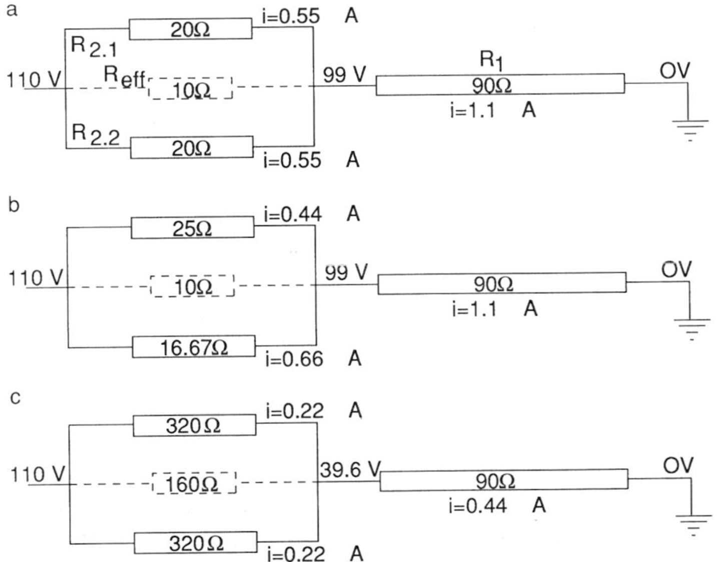

FIGURE 2 Effects of changes in bronchiolar flow resistance. The baseline resistance of each of the two bronchiolar compartments is 20 Ω, giving an effective resistance (R_{ef}) of 10 Ω. In between the bronchiolar resistances at the right side and the resistance of the large bronchi at the left, a voltage of 11 V exists in (a) and (b). The figures in the model give the changes as discussed in the text. In (c) the voltage is 39.6 V between the "bronchioles" and the "bronchi." For further explanation see text.

3. Control of Bronchiolar Diameter

The bronchial muscular tone is dependent on the activity of the autonomic system. Sympathetic activity is increased during physical work, causing a dilatation in favor of better ventilation without too much effort. The sympathomimetic hormones norepinephrine and epinephrine support this bronchial dilatory effect.

Isoprenaline, a sympathomimetic drug, is usually used to test the origin of increased airway resistance. When airway resistance diminishes after administration of this drug with use of a nebulizer during inhalation, increased muscular tone is assumed to be the reason for bronchial obstruction. Nowadays another sympathomimetic drug, terbutaline, with fewer cardiac side effects, is used for this test.

In normal subjects a dilating effect of isoprenaline and terbutaline is undetectable. This can be easily understood from Fig. 2a, when we assume a double diameter of the bronchioles in the normal model. The resistance will decrease 16 times ($R = 8\eta l/\pi r^4$), which is a fall to less than 1 Ω in the electric analogy. Then the total flow resistance is decreased to 90%, which results in a 10% increase in airflow. This difference, which is the maximal one, is barely significant.

The vagus nerve and the parasympathomimetic agents acetylcholine, pilocarpine, and carbachol have a bronchial dilatory effect. The effect of the vagus is small.

An important stimulus is the alveolar CO_2 tension (P_{A,CO_2}). A decrease in P_{A,CO_2} causes a bronchial constriction. In some way the vagus nerve is involved in this reaction, because after the nerve is blocked by atropine the constrictor effect of a decrease in P_{A,CO_2} is smaller. A decrease in P_{A,CO_2} can be caused by either hyperventilation or a decrease in perfusion of blood to some pulmonary area. By this mechanism, regional ventilation is adapted to regional perfusion. Bronchial constriction occurs where blood perfusion is small, causing a smaller ventilation, matched to the perfusion.

4. Pathological Stimuli of the Bronchial Smooth Muscle

During exercise, spasmodic contractions of the bronchial smooth muscles can occur, causing a serious asthmatic attack. The hypothesis of a decrease in P_{A,CO_2} is not very popular anymore. Another hypothesis to explain the effect is a stimulation of the bronchial mucosa with cold air or cooling of the epithelium by evaporation of water. The bronchoconstrictive response on exercise is increased by cooling of the inspiratory air. Heating the inspirate to body temperature and saturation with water vapor abolish the exercise-induced bronchoconstriction.

Allergens are external substances that when inhaled, provoke an asthmatic attack of a type called extrinsic asthma. When an asthmatic attack cannot be traced to such external allergens, by use of a skin test, the spasmodic smooth muscular contractions are called intrinsic asthma. [*See* Allergy.]

Psychological factors by themselves will not be sufficient to cause a spasmodic contraction of the bronchial muscles. Emotions might facilitate an attack of asthma in patients who are constitutionally predisposed to hyperreactivity of the bronchial smooth muscles.

5. Other Types of Airway Obstruction

Other reasons for airway obstruction can be swelling of the bronchial walls, as in bronchitis, edema, caused by many disorders, and loss of elasticity of the connective tissue, as in emphysema.

In emphysema, two main mechanisms cause an increased obstruction of the airways, especially during expiration. The loss of elasticity causes two changes to occur: (1) negative thoracic pressure is less negative than normal and (2) the walls of the bronchioles are insufficiently connected by connective tissue fibers to the alveolar interstitial tissue. The effect of both mechanisms is to decrease the diameters of the airways, especially during expiration.

During inspiration this effect does not have much influence, because the walls of the bronchi are flaccid. Thus a negative alveolar pressure is needed to cause the flow of ambient air (which pressure is reset to zero) to the alveoli. This negative pressure causes a more negative intrathoracic pressure. Then a positive pressure exists between pressure in the airways and pressure surrounding the airways. The flaccid airways will be kept open easily and airflow will hardly be obstructed.

During expiration a positive alveolar pressure is needed with respect to ambient air pressure. As a consequence, intrathoracic pressure surrounding the airways will be increased above ambient air pressure. As considered for normal breathing, this increased intrathoracic pressure causes a larger airway resistance during expiration than during inspiration. In emphysema patients, this mechanism is worsened owing to the higher intrathoracic pressure and the less firm bronchi. Emphysema patients can inspire well but have great difficulties in expiration.

During expiration a pressure drop occurs from the

alveoli through the airways to the nose or mouth due to the flow resistance of the bronchi (Figs. 3a and 3b). This pressure gradient will have a value at a certain point in the airways equal to intrathoracic pressure: this is the equal pressure point. Downstream in the upper airways, pressure in the lumen of the airways is even smaller than the surrounding intrathoracic pressure, resulting in a positive pressure difference from outside to inside. The bronchi will be easily compressed, resulting in a large airway obstruction.

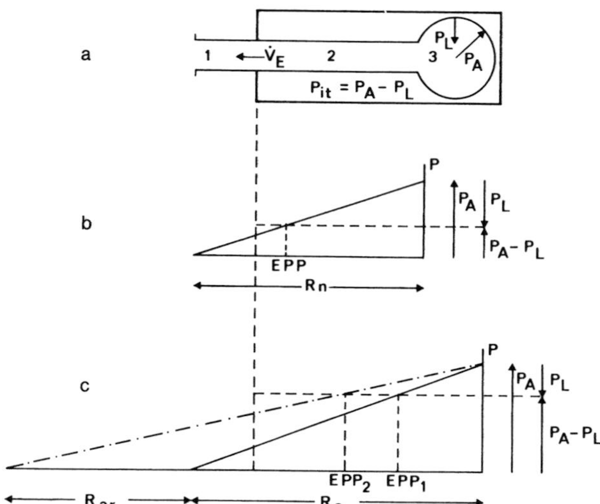

FIGURE 3 Pressure fall during expiration in the intra- and extrathoracic airways. (a) A schematic conceptual model of extra- (1) and intrathoracic (2) airways as a tube of constant resistance between alveoli (3) and mouth or nostrils. This implies a nonlinear projection of anatomical length of the airways on this tube. Parts with a relatively large resistance, such as the upper airways, will take a longer part of the tube than peripheral bronchioli, where the resistance is low (see Fig. 1). P_A is alveolar pressure during expiration; V_E' is expiratory flow; P_L is recoil pressure; P_{it} (intrathoracic pressure) $= P_A - P_L$; P_{it} surrounds the airways. (b) Pressure fall from alveoli to nostrils over a normal airway resistance (R_n). At the site where pressure in the airways is decreased to $P_A - P_L$, transmural pressure is zero because airway pressure is equal to P_{it}. This is the equal pressure point (EPP). Downstream (toward the left) the pressure outside the airways is higher than the intralumenal pressure, causing some compression of the airways. (c) Conceptual diagram for a patient with emphysema. Airway resistance is increased, therefore the abscissa is lengthened to the right (R_e) with respect to R_n. Extrathoracic airway resistance is constant. A higher P_A is needed for expiration. Because of loss of elasticity P_L is decreased. Therefore, the difference between P_{it} and P_A is smaller than in normal conditions. The point of zero transmural pressure (EPP_1) is shifted to the periphery with respect to normals. An additional external resistance (P_{ar}) decreases the pressure gradient, giving a shift of EPP_1 to proximal airways (EPP_2) and a smaller positive pressure on the airways downstream of EPP_2.

In an advanced state of expiratory airway resistance, many patients expire with pursed lips. Pursed lip breathing increases airway resistance even more (Fig. 3c). But the additional resistance is not within, but downstream from the intrapulmonary airways. As a consequence of the increased total expiratory resistance, airflow is decreased and, therefore, the pressure gradient from the alveoli through the airways is decreased. This results in a shift of the equal pressure point to the upper airways, which are more rigid than the smaller bronchi. This shift diminishes the possibility of airway collapse. These emphysematous patients try to control a more gradual expiration.

E. Tests of Airway Obstruction

1. Spirometry

Airway resistance has been defined as a variable calculated from the pressure difference between alveoli and the ambient air divided by the expiratory and inspiratory airflow, respectively. Alveolar pressure cannot be measured directly. Therefore, several indirect measures have been applied as substitutes for airway resistance. The first substitutes were based on the capability to breathe out or to breathe in a maximal volume of air in a limited time, for example, a second. Such a volume is called a dynamic lung volume. FEV_1 is the forced expiratory volume that can be maximally expired in 1 sec after maximal inspiration. FIV_1 is the forced inspiratory volume that can be maximally inspired in 1 sec after maximal expiration. MVV_{30} is the maximal voluntary volume that can be breathed in and out at a rate of 30/min. The rate is given by a metronome. These volumes can be obtained by means of spirometry, which is presented in Fig. 4 and explained in its legend. Figure 5 shows a recording of FEV_1, FIV_1, and MVV_{30} of a normal person. These values are diminished when airway obstruction is present.

2. Pneumotachography

Another method is pneumotachography, which is the recording of the forced expiratory and inspiratory flow with use of a pneumotachometer (Fig. 6). The flow signal is not recorded as a function of time but plotted against the volume that is breathed out or in, respectively (Fig. 7). This volume is obtained from the integration of the flow signal. From this flow–volume curve several indices of obstruction are obtained.

Peak expiratory flow rate (PEFR) is the maximal flow obtained shortly after the start of forced expiration. The maneuver is performed after maximal inspi-

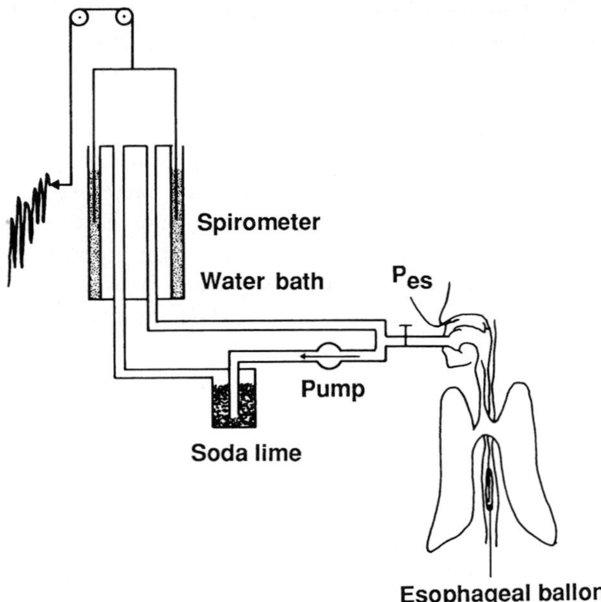

FIGURE 4 A spirometer is a container in a waterbath. The patient's lung volume changes are reflected in this container by volume displacement. CO_2 production is trapped in soda lime. O_2 from the spirometer is used. Therefore, the volume of the spirometer decreases gradually, as seen by the gradual increase of the level on the recording paper. With use of an esophageal balloon, the change in pleural (i.e., intrathoracic) pressure can be measured during normal breathing (P_{es}).

ration. This variable is often used in follow-up studies of a patient at home. The device is simple and is based on a measurement that indicates only the maximal flow expired. Several methods of measurement have been developed. This variable is not an accurate measure of airway obstruction because the value is very sensitive to the efforts of the patient, hence this procedure requires maximal cooperation.

MEF_{75} is the maximal expiratory flow when 75% of the volume that will be expired during this maneuver is not yet expired. MEF_{50} and MEF_{25} are the corresponding flows when 50 and 25%, respectively, are not yet expired.

These indices have relatively large standard variations when compared in a group of normal volunteers to be used as a reference group for comparison with a patient's value.

The shape of the curve can also give us important information. In obstruction of the peripheral airways, that is, the smaller bronchi and bronchioles, usually the down-slope of the flow–volume curve from PEFR to the end is concave.

II. ABNORMAL CHANGES IN VOLUME

Changes in volume are mainly due to changes in alveolar volume. The main features of the alveoli will be

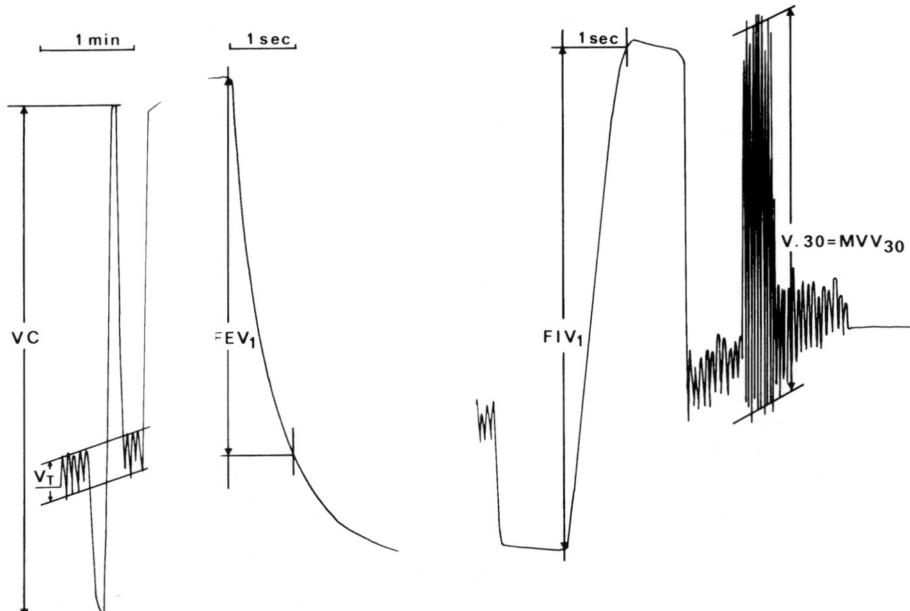

FIGURE 5 Spirometric recordings. The procedures used to determine several static and dynamic volumes are defined in the text. Note: During the recording of the dynamic volumes, the recording paper is transported faster.

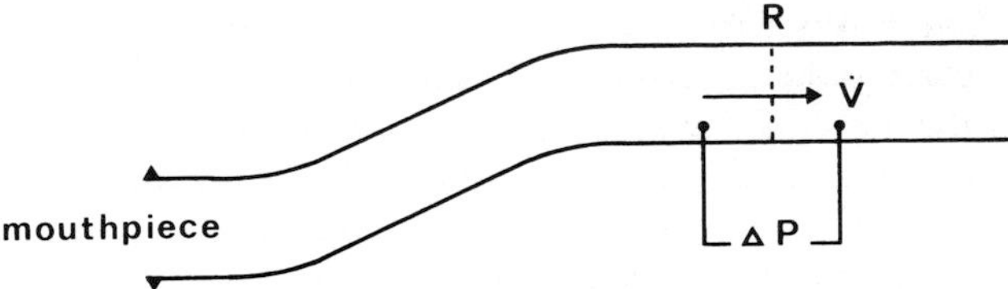

FIGURE 6 Pneumotachometer. This design gives only the principles of the pneumotachometer and is not a longitudinal section of an industrial type. In a tube, a grid causes a slight flow resistance (R), which has a mathematical relationship to P and V', according to the aerodynamic law $P = RV'$. The tube near the mouth is bent to trap the sputum droplets.

considered first to better understand their functional disorders.

A. The Primary Lobule

Generation 16, the terminal bronchioles, gives entrance to the alveolar region of the lungs. Each tubule of generation 17 with all its branches is called a primary lobule. It also could be called a functional unit, because it represents the principal function of the lung, as the nephron does in the kidney. The first three generations of the primary lobule, 17–19, are partly conducting tubules and partly gas-exchanging alveoli, because alveoli bulge from the walls of the tubules. Figure 8 shows a schematic drawing of generations

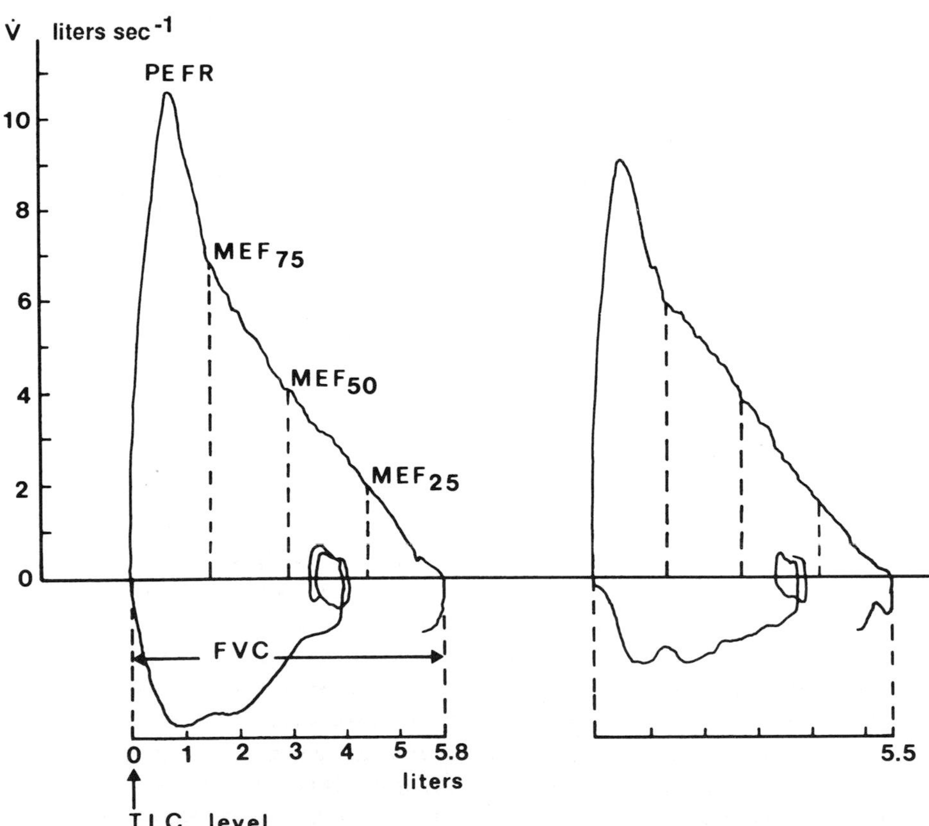

FIGURE 7 Two flow–volume curves. These two curves represent flow velocity, plotted on the y axis, against expired volume, plotted on the x axis, during a forced and maximal expiration from maximal inspiratory level (TLC). In normal lungs, this forced vital capacity (FVC) is equal to the inspiratory VC (Fig. 5). In airway obstruction, FVC is usually smaller than inspiratory VC. These two curves were made consecutively. They demonstrate the much larger variation in PEFR than in MEF_{50} and MEF_{25}. These latter two values are almost independent of the amount of the voluntary muscle forces on the expiratory air. The reason is that a larger pressure inside the alveoli increases intrathoracic pressure with the same amount, causing a higher flow resistance, which compensates for the higher alveolar expiratory driving pressure.

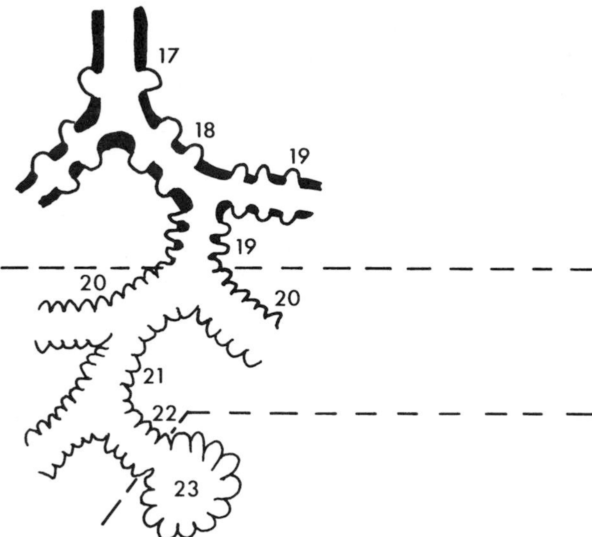

FIGURE 8 A schematic model of the primary lobule and the distribution of alveoli. More detailed information is given in the text and Table I.

17–23. Three generations, 20–22, of alveolar ductules have walls of alveoli. The last generation, 23, contains alveolar saccules. Table I presents data about the number of alveoli in a primary lobule. The total number of alveoli in the lungs is $2277 \times 2^{17} \approx 300{,}000{,}000$. Other studies of the total number of alveoli estimated the number to be between 200×10^6 and 600×10^6.

The diameter of the alveoli is approximately 0.2 mm (200 μm). However, the diameter is different for different lung regions. At the top of the lungs the diameter is larger and at the bottom it is smaller. At the bottom, the lung tissue is less stretched than at the top, because the top is stretched by the weight of the lung. The specific gravity of the lung tissue is approximately 0.3. This vertical gradient from top to bottom changes in a vertical gradient from sternum to spine when the vertical position is changed for the supine position.

TABLE I
The Number of Alveoli in a Primary Lobule[a]

Generation number	Number of alveoli per unit	Units per generation	Number of alveoli
17	5	1	5
18	8	2	16
19	12	4	48
20	20	8	160
21	20	16	320
22	20	32	640
23	17	64	1088
		Total number of alveoli:	2277

[a]Data according to Weibel (1964).

B. The Structure of the Alveoli

The alveoli contain different types of cells that have different functions.

1. Pneumocytes Type I

Pneumocytes type I are simple squamous cells that compose the thin epithelial layer of the alveoli. The diffusion membrane between alveolar air and blood is composed of this epithelium, its basement membrane, a thin layer of connective tissue, and the endothelial layer plus its basement membrane of the blood capillaries. On the epithelium a thin fluid layer is present, containing at its surface substances that decrease the surface tension of the fluid–air interface.

2. Pneumocytes Type II

Pneumocytes type II are secretory cells connected to the squamous cells and are located at the surface of the alveolar tissue. They contain lamellar bodies in which the secretions are collected. These secretions contain surfactants that decrease the surface tension. Phospholipids (68%), especially diplamitoyl lecithin (47%), are the main constituents. Proteins (9%) and cholesterol (8%) are other components of the surfactant.

Surface tension is the result of the mutual attractive forces exerted by the molecules in fluid phase compared with the mutual attractive forces of the molecules in gas phase. A drop of water, without molecular attraction, would immediately fall apart into molecules. The attractive forces keep the molecules together in a drop and cause lateral tension at the surface: surface tension. When the surface of the fluid–air interface is curved, as in a water drop, the surface tension exerts a pressure on the fluid. In the negative image of the drop, the water bubble, pressure is exerted on the air in the bubble. This pressure (P) depends on the surface tension (γ) and the radius (r) of the bubble according to Laplace's law: $P = 2\gamma/r$. But when the pressure is kept at zero in the bubble the surface tension exerts a negative tension on the surrounding water.

In pure water $\gamma = 72$ mN/m. Lung surfactant decreases this value to $\gamma = 20$ mN/m. The alveoli with a diameter of 0.2 mm will exert a negative pressure on the surrounding tissue of $-P = -2.20$ mN m^{-1}/ 0.0001 m $= -400$ $Nm^{-2} = -0.4$ kPa ≈ -4 cm H_2O. This is a mean value based on the assumption that the alveoli are balloon shaped. Actually the alveoli are irregular polyhedra, with flat interalveolar membranes and strongly curved parts at the edges of the membranes, where two or more alveoli are connected. Therefore we must assume different values of negative pressure in the alveolar interstitium. Overall, the surface tension of the alveolar lining fluid contributes for about half the value to the negative pleural pressure.

3. Alveolar Macrophages

Alveolar macrophages are phagocytes that originate from monocytes in the blood. They are present in the interstitium and the alveolar lumen. Their phagocytic function depends on the destructive enzymes from the lysosomes, one of which is elastase. These enzymes are set free in the surrounding tissue when macrophages die, then autodigestion of the alveolar septa is a real danger. Under normal conditions, two mechanisms protect the lung against this threat: (1) the elastase is transported with mucus to the larynx to be coughed out and (2) the elastase is inactivated by the formation of a conjugate with the plasma protein α_1-antitrypsin. In the blood the conjugate is changed to α_2-macroglobulin and next destroyed in the liver. Smoking has a negative effect on both protection mechanisms. It is generally accepted that a deficiency in α_1-antitrypsin is one of the conditions leading to the development of emphysema, although not a *sine qua non*. [*See* Macrophages.]

C. The Interstitium

The interstitium is in between the basement membranes of the alveolar epithelium and the vascular endothelium. On one side of the alveolar septum it is a very thin layer. There the blood–air barrier is about 0.4 μm. At this side, the active side, the respiratory gases are exchanged. At the other side, the service side, the connective tissue layer is much thicker. This layer contains collagen, elastic and reticular fibers. The fibers provide the lungs with a fragile framework. Active and service sides alternate to both sides of the alveolar septa, and the fibers are more or less plaited with the blood vessels.

The framework can be subdivided into two parts: (1) a basic rete of thick elastic fibers, in a relatively small amount, which are the main constituents of the framework, and (2) a more delicate network of reticular and collagenic fibers, which supports the capillaries and the epithelium. The fibers are also connected to the muscular bronchioles.

In the elastic fibers, recoil forces are present due to the stretch of the lung tissue. Together with the surface tension these recoil forces cause the negative pleural, and thus intrathoracic, pressure. When pulmonary volume increases, the surface area of the alveoli increases. As a consequence, the concentration of the surfactant at the surface of the alveolar lining decreases, which increases the surface tension γ. Moreover, the elastic fibers are more stretched, which increases their recoil forces. Thus, during inspiration, pleural and intrathoracic pressure will become more negative by both mechanisms.

The distinction between active and service sides could make sense from a pathophysiological point of view. When pulmonary edema develops, which is an excess of extravascular fluid in the interstitium and secondarily in the alveoli, the fluid will first accumulate at the service side before entering between the basement membranes at the active side. Therefore, pulmonary edema does not coincide necessarily with a decrease in the diffusion of oxygen from alveoli to blood.

In patients suffering from pulmonary fibrosis, which is an excessive accumulation of connective tissue, some are characterized by a decreased diffusion capacity whereas others are not. We might conclude from such observations that different types of fibrosis exist with respect to the distribution of the increased amount of connective tissue. When diffusion capacity is not changed, the amount of connective tissue will not (yet?) be located at the active side.

D. Stability and Atelectasis

According to Laplace's law, $P = 2\gamma/r$, the negative pressure developed by a small alveolus on the surrounding tissue is larger than that by a large alveolus, when both have the same surface tension in their alveolar lining fluid. This could imply that in a normal lung, where alveoli with different diameters exist, the smaller alveoli empty into the larger alveoli. This does not occur because of two self-controlling mechanisms: (1) when the alveoli enlarge the elastic fibers are more stretched and a larger counterforce is developed, and (2) the concentration of surfactant in the superficial layer decreases per unit of surface area and γ increases; when the alveoli become smaller the opposite happens.

When insufficient surfactant is available, as in premature infants, this control of stability fails more or less. Small alveoli empty into the larger ones and collapse, which is called atelectasis. The adult respiratory distress syndrome (ARDS) is among other features characterized by a deficiency of surfactant, caused by edema and destruction of the type II pneumocytes. Theoretically γ could increase maximally to 72 mN/m. For the averaged alveolus this should need a pressure of $(72 - 20) \times 2/0.001 \approx 10$ cm H_2O. To keep open the smaller alveoli, this pressure must be even higher. Moreover, the stability mechanism of the surfactant is lost and the smaller alveoli will empty into the larger ones, often causing excessive atelectasis.

E. Alveolar Volumes and Lung Volumes

Alveolar volume is estimated to obtain specific information of pulmonary pathology. Usually the volume of the airways, that is, the anatomical dead space of about 150 ml, is included in the value, giving lung volume. Lung volume is routinely determined at three specific levels.

1. Functional residual capacity (FRC) is the volume of alveoli and airways after a normal expiration (Fig. 9).
2. Total lung capacity (TLC) is the volume in the lungs and airways after maximal inspiration.
3. Residual volume (RV) is the lung volume after maximal expiration.

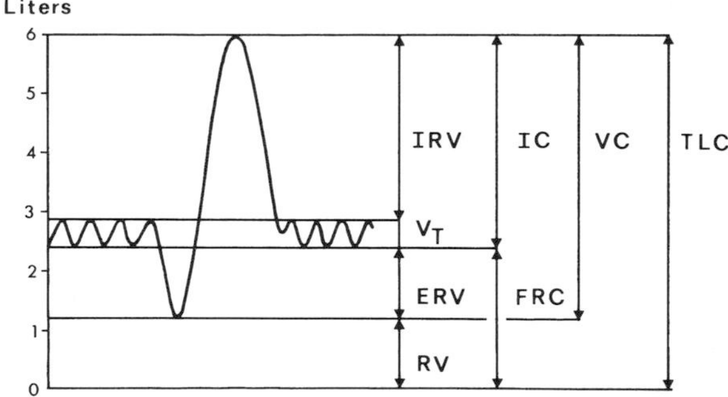

FIGURE 9 An idealized diagram giving all static volumes. The RV is determined by the helium-dilution method. Inspiratory reserve volume (IRV) is the maximal volume to be inspired after a normal inspiration. Expiratory reserve volume (ERV) is the maximal volume that can be expired after a normal expiration, and inspiratory capacity (IC) is that volume that can be maximally inspired from the normal expiratory lung volume level (FRC). All other symbols are given in the text.

The difference between TLC and RV is the vital capacity (VC), which is the maximal volume that a person can inspire after maximal expiration. A normal person can expire as much as he or she can inspire in one maneuver. Emphysema patients cannot easily expire VC after maximal inspiration. They mostly interrupt expiration by an inspiration before the level of RV is attained. Therefore, routinely inspiratory VC (see Fig. 5) must be determined.

The FRC is determined by use of the spirometer. First an amount, for example, z ml, of pure helium is injected in the spirometer. This causes a helium concentration of y% after equilibration, giving the volume of the spirometer and additional tubes according to $V_1 = z \times 100/y$. The patient is breathing ambient air through a three-way stopcock of the spirometer by way of a mouthpiece and a clamp on the nose. The patient is connected to the spirometer at the end of a normal expiration, when the patient's lung volume is FRC, by turning the three-way stopcock. The patient must breathe quietly for a few minutes to distribute the helium concentration equally in the spirometer and his lungs. It is not necessary to maintain V_1 in the spirometer during this test. Thus, when oxygen consumption decreases the volume and carbon dioxide production does not compensate for it (because it is trapped by the soda lime), the spirometer volume will be decreased. The recording will indicate accurately the volume decrease. When the helium is distributed, the test is stopped again after a normal expiration. Then the total volume of spirometer and lungs is V_2 + FRC. The helium concentration is x% now, thus the following mass balance can be formulated: $V_1 y = (V_2 + \text{FRC})x$, from which FRC can be solved.

When FRC is known, RV can be found easily by maximal expiration from the FRC level. TLC is found when VC is added to RV.

The volumes TLC, FRC, RV, and VC are called static volumes, because the patient can take his time for their determination, in contrast to the dynamic volumes FEV_1, FIV_1, and the flow-volume indices.

Another volume indicated in Fig. 5 but not yet mentioned is V_T, tidal volume, which is the normal volume breathed in and out with each ventilation. Expiratory volume is smaller than inspiratory volume, because oxygen uptake from the alveolar air into the blood is larger than carbon dioxide output from the blood into the alveolar air. An accurate indication of the differences is given by the higher N_2 concentration in the expiratory air with respect to the inspiratory air, whereas no N_2 exchange between blood and alveolar air will occur.

F. Pathophysiological Aspects of Lung Volume Changes

1. Obstruction

Lung volume changes in patients are usually due to changes in alveolar volume. They can be typical for certain diseases. In obstructive diseases such as emphysema, where extensive expiratory obstruction coincides with hardly any increase in inspiratory obstruction, patients accumulate alveolar volume and breathe at a much larger lung volume. The FRC is increased, which is called hyperinflation. The TLC is also increased. Because of the decreased elasticity the lung recoil forces are diminished and inspiration can reach a higher level. The RV is increased because of a collapse of the flaccid bronchi during maximal expiration. Alveolar air remains trapped behind these collapsed bronchi. The increase in RV is usually larger than the increase in TLC, giving a decrease of VC. A decrease in VC will also decrease FEV_1. However, in such an obstructive disease, FEV_1 is much more decreased than VC, which is indicated with the ratio FEV_1/VC as an additional but important indication for obstruction (see Section II,F,2).

These volume changes are characteristic for an airway obstruction due to a loss in elasticity. However, they can also be seen in chronic obstructive diseases by swellings of mucosa and mucus accumulations in the bronchi and bronchioles. Thus, the pattern of volume changes is characteristic but not specific for emphysema.

It happens regularly that patients with an extensive obstruction do not show any increase in these static volumes. This can be due to the fact that part of the alveolar regions is trapped behind closed bronchi during normal breathing when FRC is determined by the helium dilution method as described earlier. These closed regions will not be involved in the dilution of helium. Thus, the FRC value is observed as too small, as are the RV and TLC values.

Using body plethysmography, a description of which is beyond the scope of this article, total thoracic gas volume can be determined, including the trapped air. An additional test with this device will reveal the differences when trapped air is present.

In asthma, when an expiratory obstruction coincides with an increased inspiratory obstruction, the static volumes are not necessarily increased.

2. Restriction

Restriction of lung volume means a decrease in alveolar volume. Pneumonectomy and lobectomy, that is, resection of one lung and one of the lung lobes, respectively, are obvious reasons. Total lung capacity and VC will be decreased, and therefore also FEV_1, but FEV_1/VC will not be decreased. Thus, the decrease in FEV_1 is not indicative of an obstruction; only when FEV_1/VC is decreased do we suspect the patient to have an obstructive disorder.

Restriction is found in a variety of interstitial diseases. These disorders cause restriction via two mechanisms: (1) a replacement of air volume by tissue expansion and (2) a decrease in alveolar volume by increased elasticity (stiffness) of the lung tissue.

In both types of disorder, TLC, RV, FRC, and VC will be decreased. FEV_1 will also be smaller, but FEV_1/VC is usually normal, or even higher than normal. When the lung tissue is more elastic, that is, when it develops larger recoil forces than normal at the same lung volume level, negative intrathoracic pressure is more negative and the bronchioles are more stretched by the radial fibers running into the alveolar interstitium. Then, the airway resistance is decreased and expiration is facilitated. Although FEV_1 is decreased due to the decrease in VC, it is increased with respect to VC and thus FEV_1/VC is larger than normal. An increase in this ratio suggests increased elasticity or stiffness of the pulmonary tissue. By itself it is not a measure for it.

G. Compliance

As a measure for lung elasticity, compliance or, even better, specific compliance is determined. Compliance is the reverse of elasticity: it indicates how much the lungs yield when a certain pressure is established. During normal breathing, pressure in the lungs is zero at end-inspiration and at end-expiration. Only during these actions is alveolar pressure slightly negative and positive, respectively. The established pressure is the pleural pressure outside the lungs, which is negative at end-expiration and becomes more negative at end-inspiration. These changes in negative pleural pressure can be measured in the esophagus by use of an air-filled balloon about 0.5–1 cm in diameter and about

10 cm long, fixed on a sufficiently rigid but flexible tube.

Lung volume changes are measured by use of the spirometer. The changes in lung volume and intrathoracic pressure are plotted in an x–y diagram (Fig. 10).

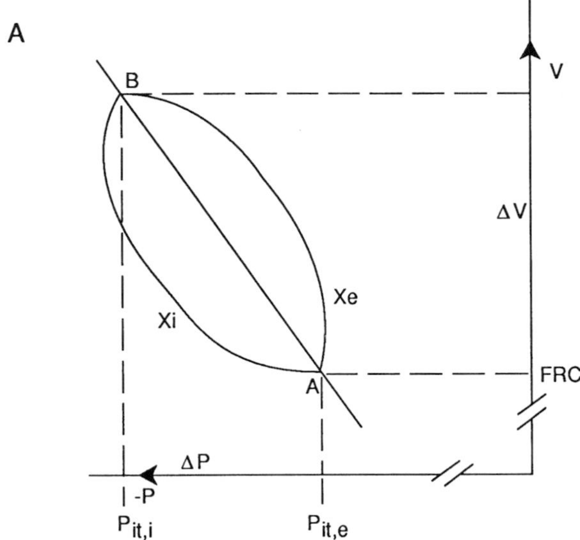

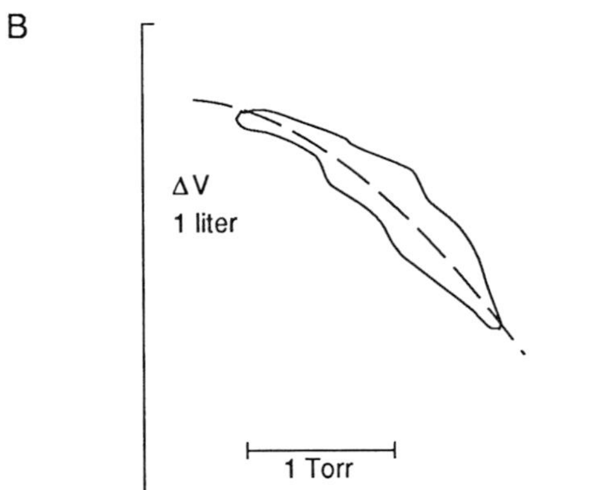

FIGURE 10 Volume–pressure diagrams. (A) Lung volume (V) is plotted on the x axis against esophageal pressure on the y axis ($-P$). $P_{it,i}$ and $P_{it,e}$ are the esophageal pressures as a substitute for intrathoracic pressure at inspiratory and expiratory volume levels, respectively. Starting inspiration at point A, an extra negative intrathoracic pressure is developed, following curve X_i owing to the negative alveolar pressure. The reverse occurs during expiration, following curve X_e starting at point B. The slope of the straight line from point A to point B is the value of compliance. (B) A curve of a patient suffering from fibrosis. The slope is smaller, indicating a lower compliance. Moreover, the slope is not straight, because the curve has a typical banana shape. The waves in the curve are due to cardiac oscillations.

Then, compliance $C = V_T/(P_{it,e} - P_{it,i})$, where $P_{it,e}$ and $P_{it,i}$ are expiratory and inspiratory pressures, respectively. C is the slope of the straight line in Fig. 10. For one-half the lung volume, the same change in intrathoracic pressure (P) is caused by only $\frac{1}{2}V_T$. Then a compliance value will be found that is one-half the value of the total lung, whereas no change in elasticity occurred. Thus, compliance depends on lung volume, which is not a good indication of the elasticity of its tissue. Therefore, compliance is expressed per liter of lung volume, giving specific compliance (C_{sp}). Specific compliance is calculated according to $C_{sp} = V_T[(P_{it,e} - P_{it,i})\cdot(\text{FRC} + \frac{1}{2}V_T)]$.

III. DISORDERS IN GAS TRANSFER

A. Ventilation–Perfusion Ratio

A schematic model of ventilation and perfusion is given in Fig. 11 for the total lung and pulmonary circulation. V_T is tidal volume, which is the volume breathed in and out. Because inspiratory volume (V_I) is larger than expiratory volume (V_E), due to a higher oxygen uptake from the alveoli into the blood than carbon dioxide output from the blood into the alveoli, V_T is replaced by the two mentioned volumes: $V'_E = fV_E$ and $V'_I = fV_I$, where f is the ventilatory rate per minute. V_D is airway dead space, that is, anatomical

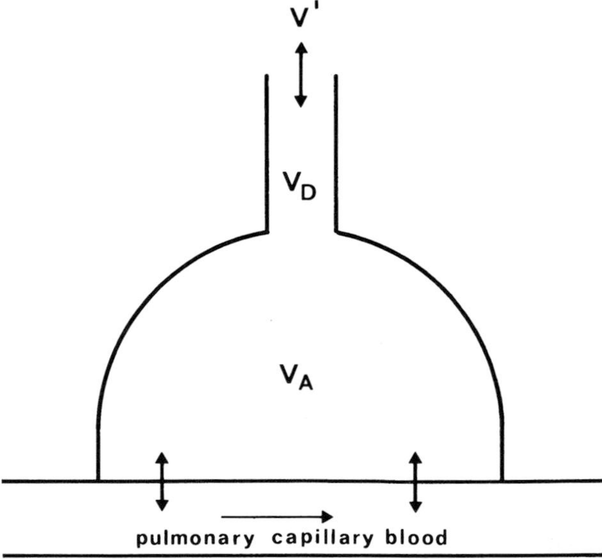

FIGURE 11 Three-compartment model of airways, alveoli, and pulmonary capillaries. V_D is the volume of the airways (anatomical dead space), V_A is the volume of the alveoli, and V is lung ventilation.

dead space. During each breath a part of the tidal volume is left behind in V_D, which is (per minute) $V'_D = fV_D$. Thus, alveolar ventilation V'_A is either equal to $V'_I - V'_D$ for inspiration or equal to $V'_E - V'_D$ for expiration.

Normal ventilation is about 7 liters/min. When ventilatory rate is 14 per min and the volume of the airways is about 150 ml, $V'_D \approx 2$ liters/min. Thus, total alveolar volume exchange is 5 liters/min of fresh air and 2 liters/min of alveolar air, returning from the airways. V'_A is used only for ventilation with fresh air.

Normal alveolar perfusion, that is, cardiac output, is about 6 liters/min. Thus, the overall ratio between alveolar ventilation (with fresh air) and perfusion (V'/Q' ratio) is 5/6 = 0.8. Normal values are usually given between 0.8 and 1.

B. Disorders in CO_2 and O_2 Exchange

Perfusion delivers CO_2 from the tissues into the alveoli and takes up O_2 from the alveoli into the blood. Ventilation transports CO_2 from the alveoli to the ambient air and refreshes the alveoli with ambient air to maintain alveolar O_2 concentration. Therefore, alveolar CO_2 and O_2 concentrations depend on both ventilation and perfusion of the alveoli.

1. Carbon Dioxide

CO_2 output (V'_{CO_2}) in both lungs is equal to the product of alveolar ventilation and CO_2 concentration in the alveolar air, according to

$$V'_{CO_2} = (V'_E - V'_D)F_{A,CO_2} - (V'_I - V'_D)F_{I,CO_2} \quad (1)$$

where F_{A,CO_2} and F_{I,CO_2} are the CO_2 fractions of alveolar air and inspiratory air, respectively. For accurate calculations, all volumes must be corrected to standard temperature (273K) and pressure (760 Torr), which we have neglected here for reasons of simplicity. F_{I,CO_2} is very low and can be neglected for the same reason. Then, Eq. (1) can be rewritten as

$$V'_{CO_2} = V'_A F_{A,CO_2} \quad (2)$$

$$F_{A,CO_2} = P_{A,CO_2}/(P_B - P_{s,H_2O}) \quad (3)$$

where P_B is ambient air pressure and P_{s,H_2O} is the saturated water vapor tension in the alveolar air. Thus, the fraction of CO_2 is part of the dry alveolar air.

From Eqs. (2) and (3) it follows that

$$V'_{CO_2} = V'_A P_{A,CO_2}/(P_B - P_{s,H_2O}) \quad (4)$$

When perfusion decreases to an alveolar area, where ventilation is normal, total CO_2 output (V'_{CO_2}) from the blood will be decreased, but CO_2 output per milliliter of blood will be increased. As a consequence, F_{A,CO_2} will be decreased. Thus, in circumstances of high V'/Q' ratios a decrease in F_{A,CO_2} and thus in P_{A,CO_2} will occur. When P_{A,CO_2} decreases, arterial P_{CO_2} (P_{a,CO_2}) will also decrease.

Under normal V'/Q' conditions, P_{a,CO_2} is about 40 Torr. When CO_2 production in the tissues is also normal, CO_2 output into the lungs and from the lungs into the ambient air will be normal, otherwise an accumulation or a depletion of CO_2 should occur.

A depletion will occur when V'_A increases. As a consequence, P_{A,CO_2} and thus P_{a,CO_2} will decrease until a new equilibrium in Eq. (4) is established. When P_{a,CO_2} falls below 30 Torr, the condition is called hyperventilation.

An accumulation occurs when V'_A decreases. P_{a,CO_2} will rise until the multiplication between P_{A,CO_2} and V'_A in Eq. (4) is equal to V'_{CO_2}. When P_{a,CO_2} rises above 49 Torr, the condition is called hypoventilation.

In lung diseases accompanied with V'/Q' disorders, regions with a variety of V'/Q' ratios will exist, from high V'/Q' ratios to low V'/Q' ratios. Because breathing is controlled, among other factors, by P_{a,CO_2}, overall ventilation will be controlled as long as possible to maintain the normal P_{a,CO_2}. Then the effects on P_{a,CO_2} by the regions with low V'/Q' ratios will be compensated by the effects on P_{a,CO_2} in the regions with the high V'/Q' ratios.

When the production of CO_2 is increased during exercise and the ventilation is also increased to maintain normal P_{a,CO_2} at 40 Torr, we call this type of ventilation hyperpnea.

When considering ventilation at high altitude, Eq. (4) can be easily misinterpreted. At high altitude P_B is lowered, thus we could be tempted to conclude that P_{A,CO_2} will be decreased to maintain the same V'_{CO_2}. However, that is a hasty conclusion. To understand the changes at high altitude we cannot neglect the corrections to standard conditions, STPD (i.e., standard temperature and pressure and dry air). Assuming a constant body temperature we need to apply the law of Boyle: PV = constant.

At high altitude the total pressure in the ventilatory volume V'_A is decreased as much as P_B. If V'_A is constant in terms of volume at lower pressure (we regard ventilatory rate as constant), it implies a decrease in V'_A if corrected to STPD. This decrease is proportional to the decrease in $P_B - P_{s,H_2O}$. Thus, according to Eq. (4), P_{A,CO_2} will remain constant if V'_{CO_2} is constant, but F_{A,CO_2} will increase proportionally to the decrease in $P_B - P_{s,H_2O}$ [Eq. (3)]. As a consequence, the product $V'_A F_{A,CO_2}$ in Eq. (2) is constant, where V'_A is at STPD. When a hypoxic drive of the ventilation is not yet present, there is no reason to assume a change in the V'_A in the terms of ventilatory volume at lower pressure. [*See* Respiratory System, Physiology and Biochemistry.]

2. Oxygen

Arterial oxygen tension (P_{a,O_2}) is approximately equal to alveolar oxygen tension (P_{A,O_2}) under normal circumstances. The difference between both variables ($D_{(A-a)O_2}$) is a few Torr. Nevertheless, in normal conditions our considerations of alveolar P_{O_2} count also for arterial P_{O_2}.

Oxygen consumption V'_{O_2} is equal to oxygen entering the lungs during inspiration $[f(V_I - V_D)F_{I,O_2}]$ minus oxygen leaving the lungs during expiration $[f(V_E - V_D)F_{E,O_2}]$:

$$V'_{O_2} = f(V_I - V_D)F_{I,O_2} - f(V_E - V_D)F_{A,O_2} \quad (5)$$

Although we are aware of the fact that V_I is slightly larger than V_E, they are assumed equal for the benefit of showing the effects on alveolar and arterial P_{O_2}. Then these equations follow:

$$V'_{O_2} = V'_A(F_{I,O_2} - F_{A,O_2}) \quad (6)$$

$$F_{A,O_2} = F_{I,O_2} - V'_{O_2}/V'_A \quad (7)$$

$$P_{A,O_2} = F_{A,O_2}(P_B - P_{s,H_2O}) \quad (8)$$

From Eqs. (7) and (8) come the following:

$$P_{A,O_2} = F_{I,O_2}(P_B - P_{s,H_2O}) - (V'_{O_2}/V'_A)(P_B - P_{s,H_2O}) \quad (9)$$

$$P_{A,O_2} = [F_{I,O_2} - (V'_{O_2}/V'_A)](P_B - P_{s,H_2O}) \quad (10)$$

When oxygen consumption is normal, the reasons for a fall in P_{A,O_2}, and thus in P_{a,O_2}, which is called hypoxemia, can be easily deduced from Eq. (9): (1) a decrease in F_{I,O_2} by oxygen consumption in an enclosed room (e.g., a bathroom with a gas water heater), (2) a decrease in P_B at high altitude in the mountains, and (3) a decrease in V'_A.

When we regard F_{I,O_2} in the term $F_{I,O_2}(P_B - P_{s,H_2O})$ of Eq. (9) as the fraction of the air coming into the alveoli after having passed the airways, we can rewrite Eq. (9) as

$$P_{A,O_2} = P_{I,O_2} - (V'_{O_2}/V'_A)(P_B - P_{s,H_2O})$$

Three other reasons leading to arterial hypoxemia will now be considered.

(4) Impaired diffusion: When the active side of the alveolar membrane between air and blood is thickened (e.g., as in fibrosis), diffusion is inhibited and $D_{(A-a)O_2}$ can increase. This does not usually occur in resting conditions, but as soon as some work is done the equilibrium between alveolar air and capillary blood is no longer attained and arterial P_{O_2} is decreased. Therefore, an exercise test could be useful to detect diffusion disorders. However, a direct measurement of diffusion with carbon monoxide as indicator gas is more specific.

(5) Right-to-left shunt: When blood passes atelectatic regions it will not be oxygenated. In this part of the blood, P_{O_2} and oxygen saturation of hemoglobin (S_{O_2}) will remain at the level of the P_{O_2} and S_{O_2} of the mixed venous blood. This blood mixes with normal arterialized blood, causing a lower S_{a,O_2} and also P_{a,O_2}. The phenomenon of no saturation of a part of the blood during lung passage is called right-to-left shunting. Such a shunt occurs in the nonventilated regions. The phenomenon of admixture of mixed venous blood with normal oxygenated blood is called venous admixture. A right-to-left shunt also causes an arterial hypoxemia. The amount of shunt or venous admixture can be calculated from the amount of oxygen transported in the shunt blood, which is $Q'_{sh}C_{v,O_2}$, and the amount transported in the ideal saturated blood, which is $(Q'_p - Q'_{sh})C_{pc,O_2}$, where Q'_p is the total lung perfusion (= cardiac output) and C_{pc,O_2} is the oxygen content in the pulmonary capillary blood entering the veins. The sum of the oxygen transport in the shunt blood and that in the blood effectively perfusing the lungs is equal to the amount transported in the arterial blood, which is $Q'_p C_{a,O_2}$. Thus

$$Q'_{sh}C_{v,O_2} + (Q'_p - Q'_{sh})C_{pc,O_2} = Q'_p C_{a,O_2}$$

which can be resolved into

$$Q'_{sh}/Q'_p = (C_{pc,O_2} - C_{a,O_2})/(C_{pc,O_2} - C_{v,O_2})$$

Q'_{sh}/Q'_{p} is called the shunt fraction of total cardiac output. To determine the shunt (in ml/min), cardiac output must also be determined.

(6) V'/Q' disorders: When pulmonary disorders are characterized by disorders in the V'/Q' ratios of the different regions, hypoxemia will occur also. The regions with high V'/Q' will be characterized by a high ventilation rate with respect to perfusion, that is, oxygen uptake (local V'_{O_2}). In the blood perfusing this region, P_{A,O_2} and P_{a,O_2} will increase according to Eq. (10). However, because of the almost full saturation of the hemoglobin in this blood, hardly any extra oxygen will be taken up.

The regions with low V'/Q' will cause a decrease of P_{A,O_2} and P_{a,O_2} [see Eq. (10)]. When P_{a,O_2} decreases, S_{a,O_2} also decreases. The arterial blood is a mixture of blood with a normal oxygen content and blood with a decreased oxygen content. It will be characterized by hypoxemia.

V'/Q' disorders and a shunt have the same effect on arterial carbon dioxide content. When the regions with low V'/Q' or no V' at all in the shunting are compensated by some extra ventilation in the normal regions, CO_2 output will be normal and P_{a,CO_2} will be normal (see Section III,B,1).

A differentiation between the low V'/Q' and a shunt is possible by application of oxygen, that is, by an increase of F_{I,O_2}. An increase in F_{I,O_2} in the regions with low V'/Q' will cause an increase in P_{A,O_2} according to Eq. (10). We assume V'/Q' is half its normal value, which implies a relatively large uptake of oxygen (V'_{O_2}) with respect to V'_A [see second term at the right side of Eq. (10)]. P_{A,O_2} will be decreased. When F_{I,O_2} is increased, P_{A,O_2} in that region becomes normal, and P_{a,O_2} in the blood leaving that region will be normal again.

In the case of a shunt, an increase in F_{I,O_2} is not effective, because when blood passes an atelectatic region without any ventilation, no exchange with oxygen is possible.

3. Pulmonary Disorders and Gas Exchange

In obstructive disorders the amount of obstruction will usually not be distributed homogeneously in both lungs. Consequently a disturbance in the distribution of the inspired air will occur. The alveolar regions, which are drained by bronchi with a low air flow resistance, will be emptied easier and earlier during expiration than the regions with a high airflow resistance. The same is assumed to occur during inspiration. The alveolar regions of bronchi with a high airflow resistance will be ventilated less than the other regions. There will be a high P_{A,CO_2} in the regions with low V'/Q' and a low P_{A,CO_2} in the regions with high V'/Q'. When we register the expiratory CO_2 concentration, an indication of the existence of V'/Q' disorders can be obtained. In Fig. 12a, a normal curve of the expiratory CO_2 concentration is given. Phase I is the phase without CO_2, when fresh air from the airways passes the CO_2 cuvette. Phase II is a transient phase when air passes coming partly from the alveoli and partly from the most distant airways. It is a mixture of fresh air and alveolar air in which the contribution of the first one decreases and that of the second increases. This transient phase also depends on the gradient in the CO_2 concentration from alveolar concentration to fresh air concentration in the region of the respiratory bronchioles. Phase III is the alveolar plateau, which increases slightly. In emphysema (Fig. 12b), phases II and III are one continuous curve, often ending at a CO_2 concentration that has a corresponding partial pressure above 49 Torr. This indicates regions with hypoventilation, or low V'/Q' ratios.

In adult respiratory distress syndrome, V'/Q' disor-

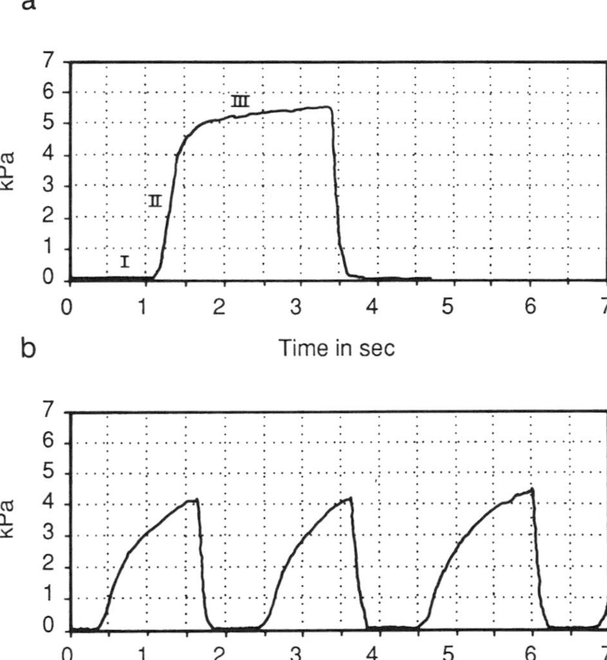

FIGURE 12 A normal and a pathologic capnogram. (a) A capnogram is the expiratory CO_2 concentration (%) or CO_2 tension (kPa as in this figure, 1 kPa = 7.5 mm Hg) plotted against time (sec). A normal capnogram is composed of three phases (see text). (b) Phases II and III coincide in the curve of a patient suffering from emphysema.

ders and extensive shunting due to edema causing atelectic regions are often present. Oxygen therapy does not help satisfactorily. The best therapy is to reopen alveoli. This needs mechanical ventilation with an increased (positive) end-expiratory pressure and superimposed insufflations.

In fibrosis, the active side of the alveolar membrane can be thickened, which will impair gas exchange. The measurement of O_2 transfer is not possible in clinical routine. Diffusion depends linearly on the difference in gas tension between both sides of the membrane. The determination of the difference in oxygen tension between the alveolar air and the pulmonary blood not only needs invasive techniques, such as catheterization of the pulmonary artery, but also demands complicated calculations, because the oxygen tension in the capillary blood increases nonlinearly during passage through the capillaries, and the oxygen tension in the alveoli changes continuously during breathing. Therefore, carbon monoxide (CO) is used. The diffusion capacity (D_{CO}) is defined as the amount of CO that passes the diffusion membrane per second and per CO pressure difference in Torr.

A well-known volume of air, approximately equal to VC, containing a low CO and helium concentration, is inspired after maximal expiration. The dilution of the CO fraction in the alveolar air is calculated from the dilution of the helium, assuming that no helium disappeared from the alveolar air between the moment of inspiration and that of expiration, giving the CO fraction in the alveolar air immediately after inspiration. The air is kept in the lungs for about 10 sec. This time of breath holding is accurately recorded. Then the air is breathed out fast and CO and helium concentrations are measured. The difference in CO concentration between the start and the end of the breath-holding period gives (1) the amount that diffused into the blood and (2) the two principal values of the monoexponential decay during that period. D_{CO} can be calculated from these data. D_{CO} depends on the surface area and thickness of the alveolar–capillary membrane and the amount of blood, specifically hemoglobin, in the pulmonary capillaries able to take up the CO. In large lungs the diffusion area is larger than in small lungs, coinciding with a larger D_{CO}. To avoid the effect of volume on D_{CO}, total D_{CO} is corrected for alveolar volume and expressed per liter lung volume from D_{CO}/TLC.

In fibrosis, usually a restriction of alveolar volume is present. Diffusion capacity (D_{CO}) will be impaired not only by a thicker air–blood barrier but also by the decrease in alveolar volume (i.e., the decrease in alveolar surface area). Therefore, D_{CO}/TLC is determined. A decrease in this value will not only be caused by a thick membrane at the active side, but also might result from a decrease in alveolar capillary blood. It is not possible to discriminate between both factors.

IV. PULMONARY CIRCULATION

A. Pressure Fall

The pressure gradient in the pulmonary circulation from about 12 mm Hg at the arterial side to about 2 mm Hg at the venous side is not only smaller than that in the systemic circulation (100–0 mm Hg) but it is also different in percentages of the total pressure fall in corresponding vessels compared with the systemic circulation. In the systemic circulation, about 60% of the pressure fall is located in the arterioles and 10% in the capillaries. In the pulmonary circulation, these figures are approximately 10 and 50%, respectively. Although about half of the pressure gradient is lost in the pulmonary capillaries, the total fall from 10 to 5 mm Hg is smaller than that in the capillaries of the systemic circulation, from 30 to 20 mm Hg. These figures indicate that the level of pressure in the lung capillaries is also lower than that in the capillaries of the systemic circulation.

We will see in Section IV,D that low pressure in the pulmonary circulation is important to avoid filtration of fluid from the blood vessels into the pulmonary interstitium and alveoli.

B. Pulmonary Vascular Flow Resistance

The pulmonary circulation is more complicated with respect to pressure fall and perfusion than might be concluded from the general data presented here. The pulmonary vessels, especially the capillaries, are collapsible tubes. When the pressure at the venous side of the capillaries is about 5 mm Hg ($\approx$6.8 cm H_2O), measured in the horizontal plane at the level of the entrance of the pulmonary artery into the lungs, this capillary pressure will be negative at a height more than 6.6 cm above the pulmonary artery (6.8/1.03; 1.03 is the specific gravity of blood), implying a collapse of the capillaries at the venous side by the alveolar pressure. Analogies of this type of perfusion are the Starling resistor and the waterfall concept, which are illustrated in Figs. 13a and 13b.

As in aerodynamics (Section I,D,2), flow resistance in the circulation obeys the Poiseuille resistance when

a

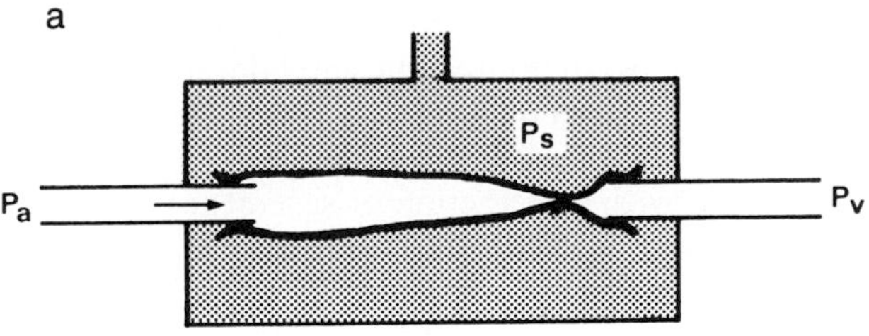

b

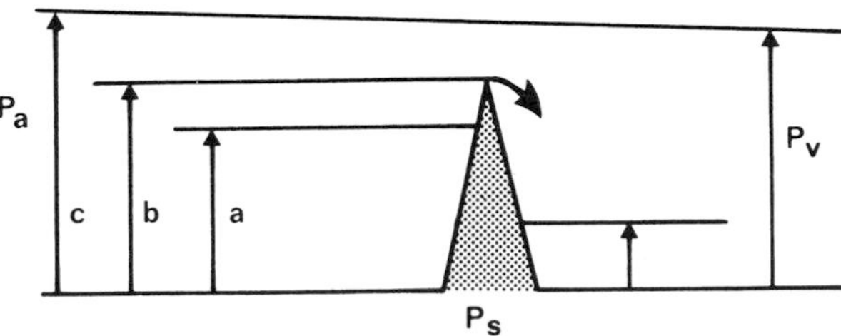

FIGURE 13 Starling resistor and waterfall model. (a) A Starling resistor composed of a latex tube in a closed chamber. When a flow from the left meets a closed Starling resistor due to P_s and flow is kept constant, pressure before the resistor will increase until it exceeds P_s. Then the "valve" will open and fluid passes the resistor. The pressure before the resistor will be maintained at the level of P_s independent of the amount of flow and independent of the pressure behind the resistor. (b) The waterfall concept represents the same mechanism. When the water before the rim of the fall is lower (situation a) than the height of the rim, no water will pass the fall. When water is continuously supplied before the rim, the water level will rise until it reaches the rim and starts to fall down, maintaining the level before the rim at the same height (situation b) independent of the amount of flow. The level downstream of the rim does not affect the flow over the rim. Only when the downstream level is higher than the rim (situation c) will we see a normal streaming river. The same is true for the Starling resistor; when the downstream pressure exceeds P_s, the latex tube remains fully open and a continuous flow dependent on the upstream-to-downstream pressure difference will be established.

flow is not turbulent, blood viscosity is constant, and the vessels are continuously open. Then the hemodynamic law $\Delta P = Q'R$, where ΔP is the pressure fall from pulmonary artery to left atrium, can be applied. However, when Starling resistors exist in part of the branched circuit of the pulmonary circulation, this hemodynamic equation, equivalent to Ohm's law, cannot be applied, because in those parts the capillaries are not continuously open. As illustrated by the waterfall model (see Fig. 13b), in these vessels no continuous pressure gradient exists, and a sudden pressure drop occurs after the rim (in the Starling resistor after the collapse in the capillaries). Then the downstream pressure has no effect and flow is not governed by the arterial-to-venous pressure difference. Under these circumstances, flow depends on the arterial-to-capillary (i.e., alveolar) pressure difference. As a consequence, calculations of the pulmonary blood flow resistance (R_p), based on the hemodynamic law mentioned earlier, must lead to unreliable data. This certainly occurs when alveolar pressure is increased, as during mechanical ventilation.

C. Control of Pulmonary Perfusion

1. Sympathetic Control

When the activity of the sympathetic part of the autonomic nervous system is increased, the small muscular arteries are dilated. This dilation increases pulmonary flow for the same driving pressure. A corresponding effect is produced by the sympathomimetic hormones. During exercise, sympathetic activity is increased, which supports a large flow through the lungs without an increase in pulmonary artery pressure. This effect is beneficial for the right ventricle, when it delivers a large cardiac output during exercise.

2. Alveolar O_2 Concentration

Another important stimulus on the small muscular arteries in the lungs is the alveolar O_2 concentration. This stimulus is important for the maintenance of normal V'/Q' ratios in the different lung regions.

When in an alveolar region the O_2 concentration decreases, the muscular arteries that supply this region with blood will constrict, causing a smaller flow. This is a useful reaction to avoid an insufficient oxygen transfer to this part of the blood. When for some reason (e.g., a clot of mucus in a bronchus) ventilation of the corresponding alveolar area is decreased, the O_2 concentration in this area will decrease, as explained in Section III,B [Eq. (10)]. Then perfusion of the region is adapted by the hypoxic vasoconstriction and the V'/Q' ratio is restored. In Section I,D,3, we considered the adaptation of ventilation to alveolar CO_2 concentration. This stimulus is also a control mechanism to maintain normal V'/Q' ratios in the different lung regions. To attain this, it is much more efficient to control V' as well as Q' than to adapt only one to the other.

Another important function of the hypoxic vasoconstriction is its effect on the pulmonary circulation before birth. A high flow resistance in the pulmonary arteries allows only a small blood flow through the lungs. Therefore, the majority of the blood from the inferior vena cava is shunted to the left side of the heart through the foramen ovale. Thus, the oxygenated blood coming from the placenta is transmitted primarily to the aortic arch, supplying the brain and heart.

The hypoxic vasoconstriction does not have only beneficial functions. Under pathological conditions, coinciding with atelectasis and low V'/Q' ratios, as in ARDS and emphysema, the flow resistance in the pulmonary muscular arteries can be increased to such an extent that a considerable increase in pulmonary arterial pressure results. This leads to a hypertrophy of the right ventricle, known as *cor pulmonale.*

At high altitude the decrease in P_B will lead to a decrease in P_{A,O_2} [see Eq. (10)], causing hypoxic vasoconstriction. Under these circumstances, edema could develop, which will contribute to a further development of vasoconstriction, causing a severe illness called mountain sickness. The reasons leading to edema under these high-altitude conditions have been attributed to many different mechanisms, which usually occurs if evidence is not strong. Recently, an attractive hypothesis has been postulated, implying stress failure of the lung capillaries. Because of an uneven constriction of the small muscular pulmonary vessels, a high pressure will occur in the capillaries downstream of the less constricted vessels. The walls of these capillaries will be affected by a high stress causing lesions and a high permeability edema.

D. Extravascular Lung Water and Pulmonary Edema

The extravascular fluid in the lung or extravascular lung water (EVLW) is continuously produced from the blood capillaries and is removed by the system of lymph vessels. The lymph vessels drain the lungs in the septa along the bronchi up to the respiratory bronchioles, but are not present in the alveolar septa. The amount of lymph transported by the lymph vessels in humans is unknown. About 5 ml/hr has been reported for dogs and sheep. The transport of lymph is insufficiently increased when relatively large amounts of EVLW develop in the interstitium. This is a reason for the supposition of insufficiently developed lymph valves in these lymph vessels.

Production of lymph from the blood capillaries is governed by the interaction of several forces, which are combined into the Starling equation:

$$Q_f' = K_f[(P_c - P_t) - \sigma(\pi_p - \pi_t)]$$

where

- Q_f' is the amount of fluid (ml/sec) that is filtered through the vascular wall of the pulmonary capillaries.
- K_f is the fluid filtration coefficient. This term indicates the amount of fluid filtration through a membrane for a given pressure difference. The term between the brackets implies the effective pressure gradient over the vascular wall. This term can be simplified to ΔP, giving the equation $Q_f' = K_f \Delta P$. Compared with the hemodynamic law $\Delta P = Q'R$, we conclude that $K_f \approx 1/R$. K_f is a reciprocal value of flow resistance, which is conductance. Thus, K_f indicates the fluid permeability of the vascular wall.
- P_c is the hydrostatic pressure in the lung capillaries and perhaps also in the lung arterioles and venules.
- P_t is the hydrostatic pressure of the interstitial fluid.
- π_p is the oncotic pressure of the plasma proteins.
- π_t is the oncotic pressure of the interstitial fluid.
- σ is the osmotic reflection coefficient, which corrects the effectivity of the proteins, because the capillary wall is not perfectly impermeable for these proteins. When the wall should be perfectly impermeable, $\sigma = 1$. For substances that pass the wall freely, such as the small molecular minerals in the plasma, $\sigma = 0$.

Most of these factors are not very well known. π_p is the only one that can be measured, because it is easy to obtain blood plasma from a patient. P_c can be approximated only by measuring the pulmonary capillary wedge pressure, P_{pcw}, with use of a Swan–Ganz catheter in the pulmonary artery. When the balloon at its tip is filled up with air, the catheter will close the artery, and no flow will exist in this artery and its branches. In these vessels the pressure will become equal to the pressure at the venous side of the system, where the veins coming from the branches of the closed artery connect with vessels where flow was maintained.

The other terms are unknown.

P_t will be different in different parts of the alveolar septa, as explained in Section III,B,2. P_t will be increased when surfactant is insufficiently active. Edema decreases surfactant activity and causes a higher surface tension, which causes a more negative interstitial pressure, leading to a vicious circle of EVLW formation.

Destruction of the capillary wall by noxious substances will decrease σ and k_f. The decrease in σ will diminish $\pi_p - \pi_t$ and the decrease in K_f will increase the passage of fluid. Both will contribute to the production of edema.

An increase in P_c by left ventricular failure or by mitral valve stenosis will cause a cardiogenic pulmonary edema.

An increase in EVLW has two counteracting effects on its own production, suggesting some self-limiting mechanism. An increase in EVLW increases P_t and decreases π_t. These two changes have a negative effect on Q_f', according to the Starling equation.

BIBLIOGRAPHY

Clark, J. T. H., and Godfrey, S. (1977). "Asthma." Chapman & Hall, London.

Clausen, J. L. (ed.) (1984). "Pulmonary Function Testing. Guidelines and Controversies." Grune & Stratton, New York.

Cotes, J. E. (1979). "Lung Function. Assessment and Application in Medicine," 4th Ed. Blackwell Scientific Publications, Oxford, England.

Crystal, R. G., and West, J. B. (eds.) (1991). "The Lung: Scientific Foundations," Vols. 1 and 2. Raven, New York.

Harris, P., and Heath, D. (1986). "The Human Pulmonary Circulation," 3rd Ed. Churchill Livingstone, Edinburgh.

Staub, N. C. (ed.) (1978). "Lung Water and Solute Exchange." Marcel Dekker, New York.

Weibel, E. R. (1964). Morphometrics of the lung. *In* "Handbook of Physiology, Section 3: Respiration" (W. O. Fenn and H. Rahn, eds.), Vol. 1. American Physiological Society, Washington, D.C.

Weir, E. K., and Reeves, J. T. (eds.) (1989). "Pulmonary Vascular Physiology and Pathophysiology." Marcel Dekker, New York.

West, J. B. (ed.) (1977). "Regional Differences in the Lung." Academic Press, New York.

West, J. B., and Mathieu-Costello, O. (1992). High altitude pulmonary edema is caused by stress failure of pulmonary capillaries. *Int. J. Sports Med.* **13**, Suppl. 1, S54–S58.

Wilson, A. F. (ed.) (1985). "Pulmonary Function Testing. Indications and Interpretations." Grune & Stratton, New York.

Pulmonary Toxicology

ROGER O. McCLELLAN
Chemical Industry Institute of Toxicology

GLOSSARY

Aerodynamic equivalent diameter Diameter of a unit density sphere with the same settling velocity as the particle of interest

Aerosol Relatively stable suspension of particles, fibers, or droplets in a gaseous medium

Alveolar macrophages Cells in the lungs that ingest dead or dying cells, bacteria, and other debris

Alveoli Thin-walled air sacs in the deep lung that are the location for exchange of oxygen and carbon dioxide between the air space and the blood capillaries

Fibers Solid forms whose length exceeds diameter by more than a factor of 3

Mucociliary escalator Ciliated epithelium and associated layer of mucus found in the upper respiratory tract that transports macrophages and debris from the lungs to the mouth

Ozone Oxidant gas formed as a secondary pollutant from interactions of oxides of nitrogen, hydrocarbons, and sunlight

Pneumoconiosis Chronic disease caused by inhaled dust and characterized by fibrosis of the lungs

PULMONARY TOXICOLOGY IS DEFINED AS THE study of inhaled poisons or toxicants and how these materials adversely affect the body. The root of pulmonary is *pulmo,* Latin for lung. However, for this article, the pulmonary system will be considered to have a broader definition and consist of the gas-conducting airways beginning at the nose or nares and extending into the lung, which consists largely of alveoli or air sacs, where gas exchange occurs. Toxicants are materials that at a specified level of intake over a specified time period cause damage to structures or disturbances of function of the body. Pulmonary toxicology is concerned with (1) the chemical and physical characteristics of toxicants in the air, (2) the basic biology of the respiratory tract, (3) the deposition and retention of inhaled materials in the body and their interaction with critical biological units, and (4) how such interactions with the respiratory tract and other systems produce disease (Fig. 1). In an even broader sense, pulmonary toxicology is also concerned with toxicants that may enter the body by routes other than inhalation, such as ingestion, and affect the respiratory tract.

I. CONCERN ABOUT AIR POLLUTION

There has been concern for centuries over airborne materials producing disease. Undoubtedly, this concern precedes recorded history and coincided with the human use of fire for cooking and heating and later for manufacturing. Concern for air pollution in the Western world accelerated after the introduction of coal as an energy source. As early as the thirteenth century, concern was recorded about coal smoke and odor in London. It is said that the Queen moved from London to Nottingham to escape the insufferable smoke. In 1661, a publication by John Evelyn entitled "Fumifugium: Or the Inconvenience of the Air and Smoke of London Dissipated" drew attention to air pollution.

ENCYCLOPEDIA OF HUMAN BIOLOGY, Second Edition, VOLUME 7. Copyright © 1997 by Academic Press. All rights of reproduction in any form reserved.

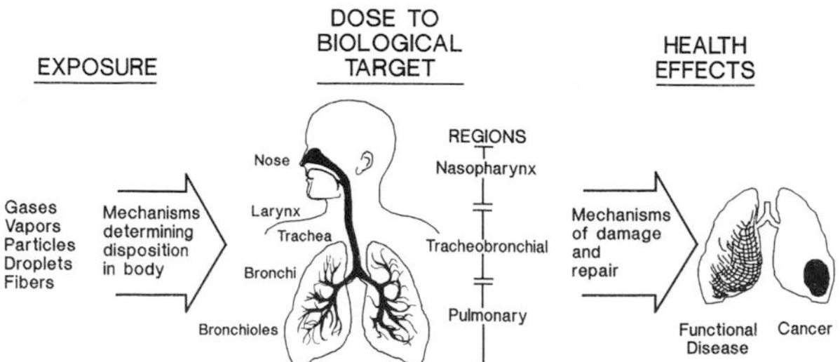

FIGURE 1 Interrelationships in assessing the health effects of airborne toxicants.

At the beginning of the industrial revolution, industry was located where wood was available for fuel. The development of steam engines in the eighteenth century provided a major impetus for the use of coal as a fuel. With the use of coal, industrialization accelerated and factories developed near coal fields and along waterways that transported coal. Increasingly, the smoke and ash from power plants, factories, and locomotives drew attention. Although air pollution was recognized in many countries, nowhere was the problem more apparent than in London. The high concentrations of pollutants, combined with the city's notorious fog, created a serious problem. The term smog, from a contraction of smoke and fog, is reputed to have originated in London. Serious pollution incidents in the early 1900s in Belgium, London, and Pennsylvania resulted in increased morbidity and mortality from respiratory effects, heightening public concern for air pollution.

Concern for the toxic effects of airborne materials is also traceable to World War I, when airborne chemicals, most particularly nitrogen mustard, were used. The tragic death of thousands of soldiers in World War I from exposure to mustard gas gave rise to concern during World War II that this experience not be repeated with other agents. This stimulated much research on the toxicity of various airborne chemical and biological agents and how their effects could be prevented or treated.

With development of nuclear weapons at the end of World War II, concern arose for possible inhalation hazards from exposure to airborne uranium, plutonium, and other radionuclides arising from the fission of uranium and plutonium. This stimulated major research efforts on the inhalation toxicity of these materials. This research, which benefited from the ability to trace materials by virtue of their radioactivity, has contributed much to our understanding of the basic biology of the respiratory tract and how it handles and responds to inhaled materials, especially particles. Unfortunately, during World War II and in the years after, when the nuclear industry was developing in the United States, the hazards of inhaling radon were not adequately recognized. Radon is a radioactive noble gas produced by decay of uranium. Radon emanates from soil, and its decay gives rise to radioactive decay product daughters. Exposures to high levels of radon in uranium mines, particularly in combination with cigarette smoking, resulted in many miners developing fatal lung cancers. This unfortunate experience is in part the basis for concern about radon exposure in homes, although most homes have radon levels substantially lower than found in uranium mines.

After World War II, air pollution arising from automobiles received increased attention. The Los Angeles basin stands out as an example of significant automobile-related air pollution. A marked increase in population, industry, and vehicles resulted in increased air pollution. Yet the potential for problems in this region was apparent much earlier. In 1542, the explorer Juan Rodrigues Cabrillo used the term "Bay of Smokers" for San Pedro Bay. In the 1950s, Professor A. J. Haagen-Smit of the California Institute of Technology discovered that Los Angeles smog originated from critical reactions between oxides of nitrogen and hydrocarbons from vehicle exhaust and sunlight producing ozone and other photochemical oxidants. This discovery helped stimulate the development of exhaust control technology to reduce vehicle emissions of hydrocarbons and oxides of nitrogen.

Finally, in considering the historical aspects of pulmonary toxicology, one must address the issue of cigarette smoking. About 150,000 new cases of lung cancer are diagnosed each year in the United States. Lung cancer is the leading cause of cancer deaths in men by a wide margin, and in women it has recently surpassed breast cancer as the leading cancer causing death. In both men and women, it is recognized that the vast majority of lung cancers, in excess of 80% of the cases, are due to cigarette smoking. The lung cancer death rate in women lags behind that of men because smoking became popular at a later date for women compared with men. Although the lung cancer rates appear to have peaked in men, they are still increasing in women. Thousands of other cigarette smoking-related deaths occur each year due to chronic respiratory disease. Smoking is also a causative factor in thousands of deaths each year due to cardiovascular disease. Recently, data have mounted suggesting that passive exposure of nonsmokers to "second-hand" smoke results in health effects in nonsmokers. [*See* Tobacco Smoking, Impact on Health.]

Cigarette smoking is also thought to influence the incidence of lung cancer produced by other air pollutants such as radon and asbestos fibers. Data from individuals who smoke and were exposed to asbestos can be used to illustrate the interactive effects. Individuals who do not smoke and are not exposed to asbestos will experience about 10 lung cancer deaths/100,000 persons. This is increased about 5-fold with exposure to asbestos (60 lung cancers/100,000 persons) and 10-fold in smokers (100–150 lung cancers/100,000 persons) and 50-fold in individuals who both smoked and were exposed to asbestos (600 lung cancers/100,000 persons). [*See* Asbestos.]

II. AIRBORNE TOXICANTS

Airborne toxicants may be subdivided into three categories based on their physical characteristics: (1) gases and vapors, (2) particles or droplets, and (3) fibers. These categories will be briefly described and examples of each given. Later the biological effects of some of the examples will be described.

A. Gases and Vapors

Gases and vapors exist in the air as dispersed molecules. A wide range of gases are toxic, including carbon monoxide, benzene, vinyl chloride, formaldehyde, ozone, oxides of nitrogen, and oxygen. Some may be surprised that oxygen has been identified as toxic. Obviously, oxygen at normal concentrations is essential for life. However, at very high concentrations and with a sufficiently long period of exposure, it can be injurious to the lungs. This illustrates an important concept in toxicology emphasized over four centuries ago by the Swiss-born alchemist and physician Paracelsus: "All substances are poisons; there is none which is not a poison. The right dose differentiates the poison and a remedy."

B. Particles and Droplets

Particles or droplets are aggregations of molecules of materials in either a solid or liquid form. They may range from a small fraction of a micrometer up to tens of hundreds of micrometers in diameter. The term aerosol is used to describe a relatively stable suspension of solid particles or liquid droplets in a gaseous medium. Gases or vapors are frequently adsorbed onto the surface or dissolved in particles or droplets. Some aerosols are dusts arising from grinding, milling, or blasting, with particles having the same chemical composition as the parent material. Coal dust or silica dust are examples. Fumes are formed by combustion, condensation, or sublimation and usually differ in chemical composition from the parent material. Examples are fumes from welding or lead from a smelter or in the exhaust from cars using lead-containing gasoline. Smokes are formed by incomplete combustion of organic materials. Examples are wood smoke and soot from diesel engines. Mists and fogs are liquid aerosols typically formed by condensation on minute particles or the uptake of liquid by hygroscopic, that is, water-adsorbing, particles. An example is sea mist.

C. Fibers

Fibers in a sense are a special kind of particle in which the aspect ratio, that is, length divided by diameter, is greater than 3. Fibers may be considered as microscopic javelins. Some fibers are naturally occurring and others are man-made. Perhaps the best-known mineral fibers are several kinds of asbestos that occur in nature and that were used extensively because of their insulating properties (e.g., brake linings and covering for high-temperature vessels and pipes). Cotton and silk are examples of organic fibers. Because fibers have many desirable commercial properties, a variety of man-made fibers have been produced. For example, refractory ceramic fibers and glass wool have been developed as substitutes for asbestos. Rayon and acrylics are substitutes for silk and cotton.

III. PULMONARY SYSTEM AS A TARGET FOR TOXICANTS

If toxicants are viewed as bullets, then it is appropriate to view the pulmonary or, more appropriately, the respiratory system as the target, in order to understand how the target handles and responds to inhaled toxicants. An appropriate starting place is to consider the anatomical characteristics of the respiratory tract. The respiratory system can be subdivided into three regions: nasopharyngeal, tracheobronchial, and pulmonary (see Fig. 1).

A. Nasopharyngeal Region

The nasopharyngeal region extends from the nose to the larynx. Most of the nasal passages are lined with well-vascularized epithelium containing cells that secrete mucus. Except at the entrance and in the olfactory region, the epithelium cells have cilia. In this region, large particles or droplets are removed and the air is warmed and becomes moist. Odor perception, that is, olfaction, also occurs in this region.

B. Tracheobronchial Region

The tracheobronchial region extends from the larynx to the end of the ciliated bronchioles that open into respiratory bronchioles. This region starts as a single conducting airway, the trachea, and then divides successively in a bifurcating manner. In humans, each airway divides into two equal-sized smaller airways. With each division the diameters decrease and the cross-sectional area increases. All of the airways are lined by ciliated epithelium coated with a thin layer of mucus produced by goblet cells and mucus-secreting cells interspersed with the ciliated epithelial cells. The airways conduct gases from the nasopharynx to the pulmonary region. The mucus layer, propelled by the beating of underlying cilia, moves particles and cellular debris from the lower airways and pulmonary region to the mouth, where the mucus and debris may be swallowed or expectorated.

C. Pulmonary Region

The pulmonary region consists of the last several divisions of respiratory bronchioles and more than 10 million alveolar ducts leading to 100 to 500 million alveoli. The walls of alveoli are covered with very thin pulmonary epithelial cells overlying a dense bed of blood capillaries. It is in this area where gas exchange takes place and is facilitated by the minimal distance between the air space of the alveoli and red blood cells in the capillaries. The alveolar surface area is on the order of 100 m^2. The alveolar surfaces are covered with a thin layer of surfactant that decreases surface tension, thus playing a key role in maintaining the dimensions of alveoli. Occasional macrophages, cells that ingest (i.e., clean up) dead or dying cells, bacteria, or other debris, are found in the alveoli. [*See* Macrophages.]

D. Respiratory Tract Cell Composition and Function

It is estimated that the respiratory tract contains over 40 different cell types. This includes many different kinds of epithelial cells that line this airway and alveoli, supporting connective tissue cells and endothelial cells lining the tract's abundant blood vessels. Each of these cell types may be affected in different ways by various toxicants. The interrelationships among these cells under normal conditions are only beginning to be understood with even less known about how they respond to toxicants. [*See* Pathophysiology of the Upper Respiratory Tract.]

The respiratory tract is functionally very complex. Typically, when considering the functional role of the respiratory tract, prime consideration is given to the movement of gases from the nose or mouth to the alveoli and the exchange of gases with a net movement of oxygen into and carbon dioxide out of the body. Beyond this essential role, it is important to recognize that the respiratory tract has other important capabilities. This includes an elaborate biochemical system that aids in olfaction and in the detoxification (decreasing toxic activity) or activation (increasing toxic activity) of inhaled chemical toxicants. It also has immunological capabilities and plays a key role in acid–base balance. Each of these various functional capabilities may be altered by toxic agents.

IV. DISPOSITION OF INHALED TOXICANTS

Large quantities of air enter the body to provide for adequate gas exchange. Typically, at rest, an individual takes about 15 breaths/min and with each breath inspires about 500 ml of air. This amounts to about 10 m^3 of air being inhaled each day. Individuals engaged in hard work have increased breathing rate and volumes. Unless special precautions are taken, for example, wearing a mask with a special chemical or particle filter, any toxicants in the air enter the body. The disposition of these toxicants will be governed by their physical and chemical characteristics and the nature of the target—the respiratory tract. The ultimate effect on the individual will be determined by both the deposition and retention of the toxicant.

A. Gases and Vapors

Uptake of toxic gases and vapors may occur throughout the respiratory tract. Because toxicants are typically present in very low concentrations in the air compared to the tissue, blood, or mucus, the domi-

TABLE I
Disposition and Effect of Inhaled Gases

Toxicant	Solubility	Reactivity	Metabolism	Site of effect	Health effect
Formaldehyde	High	High	Local	Nasal cavity and to a lesser extent tracheobronchial region	Irritation, nasal cancer in rats with prolonged high exposures
Sulfur dioxide	High	High	None	Tracheobronchial region	Acute respiratory effects, bronchitis
Ozone	Low	High	None	Distal bronchioles	Acute respiratory effects
Carbon monoxide	Low	Low	Binding to hemoglobin in red blood cells	High-oxygen-consuming tissues	Impaired oxygen transport, effect on brain and heart
Benzene	Low	Low	Detoxification and activation at site remote from lungs	Bone marrow	Bone marrow injury, leukemia

nant driving force for deposition is diffusion. The solubility and reactivity of the gas are critical determinants for retention in the respiratory tract and translocation to other tissues via the bloodstream. In recent years, mathematical models that take into account physiological characteristics of the subject as well as chemical properties of the toxicant have been used to describe the movement or kinetics of inhaled toxicants within the body. These physiologically based pharmacokinetic models have proved especially useful in extrapolations among laboratory animal species and humans. These models were originally developed to describe the kinetics of drugs, hence the term pharmaco (i.e., drug) kinetic. In this application, they might more appropriately be called toxico (i.e., poison) kinetic.

Consider the disposition of five gases with differing characteristics regarding solubility, reactivity, and metabolism by the body (Table I). Formaldehyde deposits primarily in the nasopharyngeal region and to a lesser extent in the tracheobronchial region. In both regions it remains near where it deposits with very little reaching the bloodstream. Sulfur dioxide deposits primarily in the lower tracheobronchial region with some reaching the bloodstream. It may also be carried on particles to the pulmonary region. Ozone deposits in the highest concentrations in the lower portion of the tracheobronchial region. Both carbon monoxide and benzene reach the pulmonary region, where they diffuse into the bloodstream. The carbon monoxide is bound to hemoglobin in red blood cells and the benzene is transported via the blood to other tissues, including the bone marrow.

As an aside, gases present in higher concentration in the bloodstream may diffuse from the blood to alveolar air and be exhaled. This is the basis of using the analysis of exhaled ethanol as a basis for estimating an individual's blood ethanol concentration. It should also be noted that by using sophisticated analytical instruments, numerous individual chemicals can be identified in the exhaled breath of individuals. These chemicals may serve as a fingerprint of the individual's dietary and other personal habits, including the environment in which they live and work.

B. Particles, Droplets, and Fibers

1. Deposition

Inhaled particles and droplets are deposited throughout the respiratory tract by mechanisms that are most strongly influenced by their physical dimensions and density. The principal mechanisms are impaction, sedimentation, interception, and diffusion (Fig. 2).

Impaction, sedimentation, and interception are all processes strongly influenced by the aerodynamic characteristics of particles, droplets, and fibers. Impaction is the process by which particles suspended in air, due to their inertia, travel along their original path and impact on a surface, rather than following the airflow around a bend, as might occur in an airway near a bifurcation of the airways. Sedimentation or a settling out of particles is dominant when airflow velocity is low. The rate of sedimentation is determined by the terminal velocity of particles. For particles 0.5 μm in diameter, it is about 0.001 μm/sec or the same as the diffusional displacement. For particles of smaller size, the terminal velocity will be less than diffusional displacement. As the particle increases in

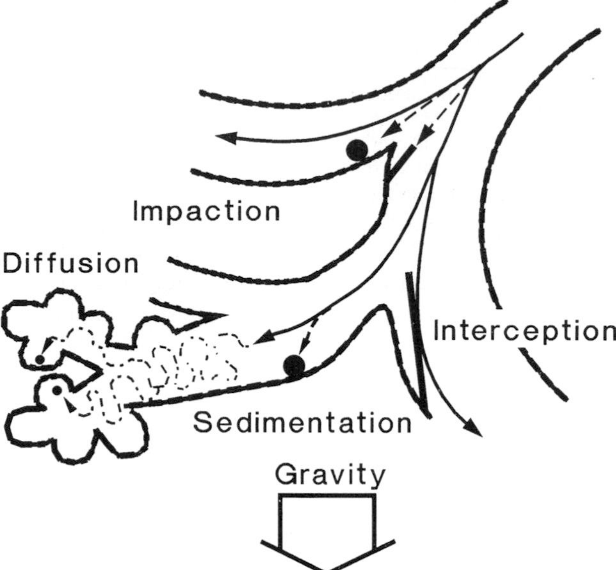

FIGURE 2 Mechanisms of deposition of particles, droplets, and fibers. —, flow streamline; ---, particle/fiber trajectory.

aerodynamic size, the terminal velocity increases markedly such that it is more than 0.1 μm/sec for a particle slightly more than 5 μm in aerodynamic diameter.

Diffusion is of greatest importance for particles less than about 0.5 μm in diameter. A particle of this size has a diffusional displacement of about 0.001 μm/sec. As particle size decreases, the diffusional displacement motion increases, thereby increasing the probability that particles will come into contact with the surface of an airway. Diffusional deposition appears to be highest when particles have just entered the airways, that is, in the proximal portion of the nasopharynx, and then in the pulmonary region when airflow is low. Small particles, formed by condensation and agglomeration, of radioactive polonium and lead from the radioactive decay of radon gas appear to deposit by diffusion.

The importance of aerodynamic characteristics in governing deposition of inhaled aerosols has given rise to the development of techniques that characterize aerosols as to this parameter. Aerodynamic equivalent diameter is defined as the diameter of a unit density sphere with the same settling velocity as the particle of interest. The aerodynamic equivalent diameter of a particle that is roughly spherical is most heavily influenced by particle diameter and density. For example, three particles, one with a real diameter of 2.0 μm and a density of 1.0 g/cm^3, a second with a real diameter of 0.63 μm and a density of 10 g/cm^3, and a third with a real diameter of 1.4 μm and a density of 2.0 g/cm^3, all have an aerodynamic equivalent diameter of 2.0 μm.

Real diameter and density are also important in influencing the aerodynamic equivalent diameter of fibers. The length of the fiber is of lesser influence. For example, three fibers all with a density of 2 g/cm^3 and differing slightly in real diameter (0.86, 0.68, and 0.55 μm) and markedly in real length (3, 10, and 30 μm) all have an aerodynamic equivalent diameter of 2 μm. The basis for this is apparent when one considers the analogy of a fiber as a microjavelin. The roles of the several processes influencing deposition of inhaled particles as related to various features of the respiratory tract are summarized in Fig. 3. The integrated impact of these processes on deposition is illustrated in Fig. 4.

Our recently acquired knowledge of the influence of particle size on deposition of inhaled particles has

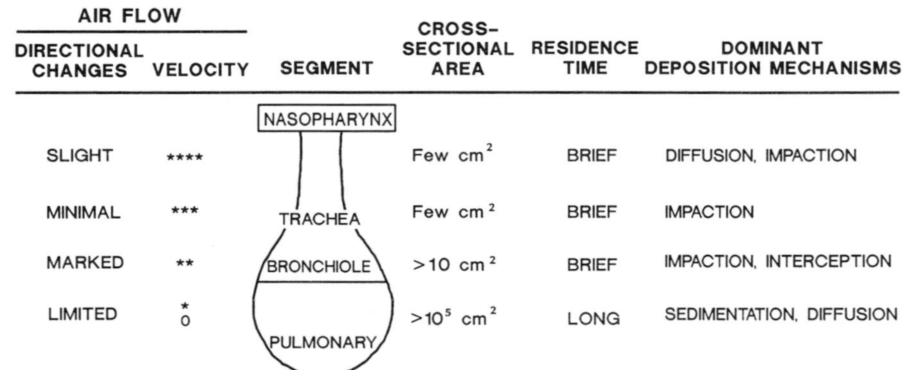

AIR FLOW
DIRECTIONAL CHANGES
VELOCITY
SEGMENT
CROSS-SECTIONAL AREA
RESIDENCE TIME
DOMINANT DEPOSITION MECHANISMS
NASOPHARYNX
SLIGHT **** Few cm² BRIEF DIFFUSION, IMPACTION
TRACHEA
MINIMAL *** Few cm² BRIEF IMPACTION
BRONCHIOLE
MARKED ** >10 cm² BRIEF IMPACTION, INTERCEPTION
PULMONARY
LIMITED * 0 >10⁵ cm² LONG SEDIMENTATION, DIFFUSION

FIGURE 3 Respiratory tract characteristics and mechanisms influencing deposition of particles.

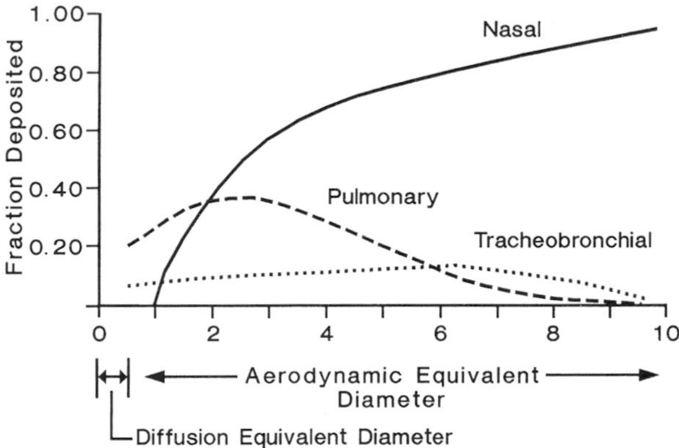

FIGURE 4 Fractional deposition of inhaled particles in various regions of the respiratory tract.

recently been considered in the setting of standards for airborne particles. The current National Ambient Air Quality Standard for particulate material is based on particles less than 10 μm in aerodynamic equivalent diameter, those that are readily inhaled and deposited in the respiratory tract. In 1996, the U.S. Environmental Protection Agency proposed revising the particulate material standard to also include samples of particles less than 2.5 μm in aerodynamic diameter since recent epidemiological studies suggest that such particles may be especially harmful. Prior to 1987, the standard was based on total suspended particles, measurements of which included many particles that did not have potential health consequences because they were not readily inhaled.

2. Clearance and Retention

For inhaled particles, droplets, and fibers, deposition of the respiratory tract is only the first step in influencing their potential for causing damage. Clearance, or the converse, retention, of the material markedly influences its toxicity. The clearance of liquid droplets is primarily influenced by the chemical characteristics of the liquid and materials contained in the liquid. Water from droplets is rapidly absorbed, leaving dissolved or suspended materials to be absorbed or taken into macrophages or other cells.

The processes by which particles and fibers are cleared are common for all regions of the respiratory tract, although their relative importance differs markedly for the various regions. Particles can (1) be cleared directly via the mucociliary apparatus, (2) be cleared via macrophage ingestion and removal via the mucociliary apparatus, (3) be ingested by macrophage and dissolved, (4) be ingested by macrophages and carried into the interstitial tissue of the lung, or (5) be ingested by macrophages and carried via the lymphatic vessels to regional lymph nodes, or (6) dissolve directly into the surrounding fluid or tissue. Materials deposited in the nasopharynx are clearly rapidly, being moved forward on the mucociliary blanket, by sneezing and nose blowing, or to the rear by mucociliary action to the mouth, where the material is swallowed or expectorated. Materials deposited in the tracheobronchial region are also cleared rapidly, in a matter of hours. Movement up the bronchioles and trachea on the mucociliary escalator occurs with material frequently contained within macrophages. The clearance mechanisms that remove particles also serve to remove bacteria and viruses, helping to protect the respiratory tract from infectious agents.

The rate of removal of materials deposited in the pulmonary region is quite variable and influenced greatly by dissolution of the material. The rate of dissolution is influenced primarily by the solubility of the material in body fluids and the surface area of the particle or fiber. Recall that aerodynamic diameter was a key factor determining deposition of particles. Once deposited, this factor is no longer of concern and the key consideration becomes the surface area of the particles. The smaller the particles, the greater the surface area relative to their mass, that is, 1 mg in small particles will dissolve more rapidly than if contained within large particles. Once dissolved, the fate of the material is determined by its chemical characteristics and how it is metabolized by the cells of the respiratory tract. In some cases, materials may be transported via the bloodstream or lymphatics to organs where they are metabolized.

Consider the clearance and retention of several different particles or fibers. Silica or coal dust particles encountered in mining or carbon particles from incomplete combustion, such as diesel engine exhaust, are relatively insoluble. A portion of the inhaled particles is clearly rapidly; witness what one can observe in the handkerchief or in the expectorate when working in a dusty environment. A small portion has a long residence time in the lung and adjacent lymph nodes, which are very discolored if the tissues are observed at death. The overall effectiveness of the clearance mechanisms is apparent when it is recognized that a cigarette smoker or coal miner may inhale and deposit kilograms of materials of which only a small fraction remains.

In contrast, inhaled lead particles leave the lungs relatively rapidly and the lead deposits in various tissues throughout the body, including the skeleton. The content of lead in urine can be used as a good index of lead exposure. This is a type of bioassay.

Many aspects of the retention of fibers are similar to that of particles. However, one aspect appears to be different. Certain of the highly insoluble asbestos fibers move over time from the airways, where they are initially deposited, to the periphery of the lung. In some cases, high concentrations of fibers can be found in the pleura, the exterior surface lining the lung and the chest wall. This may be a factor in the induction of mesotheliomas, cancers of the pleura.

V. RESPONSES TO INHALED TOXICANTS

The respiratory tract is a complex organ system and, not surprisingly, can respond in many ways to inhaled toxicants. In this section, the responses of the respiratory tract will be briefly reviewed, first from the aspect of functional alterations and then from a structural view considering both noncancer and cancer diseases. However, before considering effects in the respiratory system, it is appropriate to emphasize that in some cases the respiratory tract serves only as the portal of entry; the diseases of concern involve other organs or systems. For example, inhaled benzene causes leukemia, lead affects the central nervous system and bone marrow, vinyl chloride causes liver cancer, and carbon monoxide impairs oxygen transport.

A. Acute Functional Alterations

The parameters most typically measured in assessing the functional status of the respiratory tract and the extent of any disturbances related to toxicant exposure include measurements of respiratory frequency, lung volumes (tidal and minute volume), flow rates, airway resistance, and airway reactivity. Alterations in function may result from tissue injury or be mediated by reflex alterations of respiratory control and airway smooth muscle tone.

Some chemicals, examples being acrolein and formaldehyde, present in automobile exhaust and cigarette smoke, are capable of evoking an acute decrease in both respiratory frequency and tidal volume. The response is very quickly elicited, and normal function is rapidly restored after the irritant is removed. This action has been termed "sensory irritation" and is thought to be mediated by stimulation of the trigeminal nerve in the nasal passages and cornea. It is the nasal equivalent to irritant bronchial asthma.

Other chemicals, an example being sulfuric acid, cause a transitory increase in resistance to airflow. The effect is thought to be mediated through constriction of the muscles in the lower airways or bronchioles. Effects of exposure to some agents, such as sulfuric acid mists, are more pronounced in individuals with asthma and are especially noted in exercising individuals. At least in part, the exercising effect is due to increased quantities of the toxicant reaching the target tissue with increased minute volume, that is, an increase in dose. For some chemicals such as sulfuric acid mists, the individual's response to low doses of a drug producing airway reactivity, that is, bronchoconstriction, has been shown to be a good indicator of responsiveness to sulfuric acid exposures.

With some chemicals, an example being some isocyanates, after repeated exposures, the individual may become sensitized to the specific chemical and respond to very low concentrations of the chemical. Asthma, a condition characterized by an acute narrowing of the airways and difficulty breathing, has been increasing in incidence in recent years to on the order of 5% of the population. The basis for this increase in incidence is not known. Ironically, it is occurring even when outdoor air pollution is decreasing.

Other functional responses to inhaled toxicants require longer exposures for functional changes to develop and typically result from tissue injury. Exposure of human subjects under controlled conditions to ozone at concentrations similar to those observed in the Los Angeles basin result in reduced pulmonary function, especially as evaluated on the basis of the volume of air the individual can force out of their lungs in 1 sec (i.e., forced expiratory volume, 1 sec). Changes can be observed at lower concentrations of ozone when the individuals are exercising during exposure, that is, a greater "dose" may be delivered to the lungs. The effects are most severe when the exposures are extended from an hour up to 6 hr. It is interesting to note that, unlike with sulfuric acid, asthmatics do not respond differently than nonasthmatics. Individuals repeatedly exposed to ozone "adapt" and respond to a lesser degree compared to the response to their initial exposure.

Increasingly, controlled studies with human subjects exposed to low levels of pollutants are using new investigative techniques, such as bronchioalveolar lavage (BAL), to aid in understanding the mechanisms by which toxicant exposure alters function and structure. The BAL procedure involves introducing a small quantity of normal saline fluid into the lung and then removing the fluid. In addition to recovering a large portion of the fluid, cells and other lung constituents are recovered for examination. Use of this technique has revealed increased numbers of neutrophils (inflammatory white cells) from the lungs of individuals exposed to low levels of ozone for several hours. The presence of the neutrophils suggests the presence of an inflammatory process that may be the precursor of serious disease occurring later. [*See* Neutrophils.]

Ethical considerations sharply constrain the nature of the clinical studies that can be done with human subjects. The kind of test materials studied and the level of duration of the exposures must not pose a risk of producing irreversible effects.

For these reasons, our human experience with regard to development of chronic, irreversible disease has been gleaned from studies of persons exposed at high concentrations and, typically, for long periods of time in occupational or environmental settings. Most such exposures occurred some time ago, and it is difficult to retrospectively establish the characteristics of exposures to try to link exposure and disease as is the objective of epidemiological studies. And it is unlikely that most environments today are producing measurable chronic disease. As a result, it is becoming increasingly difficult to obtain human data on chronic diseases produced by airborne toxicants. Inevitable when such epidemiological studies are conducted, the substantial effects of cigarette smoking make it difficult to detect small effects due to toxicants.

In view of the inadequacies of human data, it is necessary to complement human data with information obtained from studies using laboratory animals and cellular and molecular model systems. In these studies, a broader range of variables can be evaluated under controlled conditions. The use of data from laboratory animal, cell, and molecular level studies is essential when evaluating new chemicals to which people may be exposed.

B. Chronic Disease

1. Diseases Other Than Cancer

Respiratory tract diseases are a major cause of morbidity and mortality in the United States. Over 5% of the population is afflicted with chronic bronchitis and emphysema resulting in substantial medical costs, restrictions in activity, and lost work days. Over 60,000 individuals die of these diseases each year. The overwhelming factor associated with these diseases and deaths is cigarette smoking. Only a very small portion of these diseases and deaths may be attributed to exposure to air toxicants other than cigarette smoke.

Analysis of data from earlier times, when pollutant levels were high, suggests that high-level exposures to airborne particulate material may influence the prevalence of these diseases.

Of the various chronic lung diseases, diffuse pulmonary fibrosis is the one most clearly linked to exposure to airborne toxicants. Most of this information has been obtained from studies on individuals exposed at earlier times before adequate control measures were instituted in the workplace. Examples of toxicants causing pulmonary fibrosis are coal dust, silica, asbestos, and cotton dust. When the disease is caused by dust it is termed "pneumoconiosis." Pulmonary fibrosis results in a thickening of the space between the alveoli and the blood capillaries. This thickened barrier slows down gas exchange.

Emphysema is also a disease of the respiratory tract. It results from inhaling tobacco smoke or oxidant gases, such as oxides of nitrogen, over a period of years. It is often seen in conjunction with chronic bronchitis, and together the two are called chronic obstructive pulmonary disease. This is a progressive and debilitating disease. Emphysema causes abnormal enlargement of the alveoli and destruction of alveolar walls. This results in a decreased surface area for gas exchange—which is the primary function of the lung. Many people who have emphysema must inhale oxygen-rich air in order to obtain sufficient oxygen to support cellular processes.

2. Respiratory Tract Cancers

Cancers of the respiratory tract are a major cause of death in the United States. The age-adjusted lung cancer death rate in males has steadily increased from 1930 to the mid-1990s, when it leveled off and appears to be decreasing. The rate for women began to rise in the 1960s and continues to increase. In California, the rate for women approaches that for men. It is generally acknowledged that well over 80% of the lung cancers are attributable to cigarette smoking.

Because of the overwhelming impact of cigarette smoking it has been difficult to identify from epidemiological studies other toxicants that cause lung cancer. An increased risk of lung cancer has been observed among people exposed to external γ or X irradiation, radon and its radioactive daughters, arsenic, beryllium, coke oven emissions, nickel, chromates, bischloromethyl ether, and mustard gas. Typically, the cancers are diagnosed as being bronchogenic carcinomas. They are usually malignant and cause death within 6–12 months. Studies with laboratory animals have suggested that materials such as plutonium, fission product radionuclides, and diesel exhaust are capable of causing lung cancer.

Cancers other than bronchogenic carcinomas are also of concern. Of particular note are mesotheliomas, which are cancers consisting of cells found in the pleural lining of the thoracic cavity. These have been observed 25 to 45 years after exposure of humans to certain types of asbestos.

Concern also exists for toxicants causing nasal cancers. For example, prolonged high levels of formaldehyde exposure cause nasal cancers in rats. To date, studies of human populations that have typically been exposed at lower levels have yielded negative or equivocal results. However, nasal cancers have been noted in workers exposed to chemicals such as nickel, wood particles, or chromium in industrial settings. The occurrence of nasal tumors in the general public is very rare.

Numerous studies have been conducted with a view to determining if environmental air pollution contributes to the overall incidence of lung cancer. One approach has been to compare the lung cancer rates for urban versus rural residents. After correcting for differences in smoking, the difference due to the urban factor is small

VI. ASSESSING RISKS AND CONTROLLING EXPOSURES

The foregoing discussion of necessity has been largely qualitative and focused on whether particular materials are toxic, that is, on hazard identification. This is the first step in the risk assessment process (Fig. 5). As difficult as this step is, the second step, risk characterization, is even more difficult—the establishment of quantitative relationships between exposure and health response. There is a high degree of uncertainty in estimating risk for low-level, short-duration exposures. This requires extrapolation downward from observations of people exposed at high levels or, in the absence of human data, a double extrapolation from laboratory animals exposed at high levels to people at low levels. The problem is formidable, recognizing that epidemiological and animal studies are typically able to detect an increase of only 20% or more in the incidence of a specific disease, whereas society is concerned with avoiding risk on the order of one in ten thousand or one in a million individuals. There is much debate over the appropriate methods for extrapolating to low levels of risks and the associated low levels of exposure. The debate is likely to continue until we have a much improved knowledge of the mechanisms by which cancers are caused.

The third step in the risk assessment process, exposure assessment, is also difficult. How do we estimate

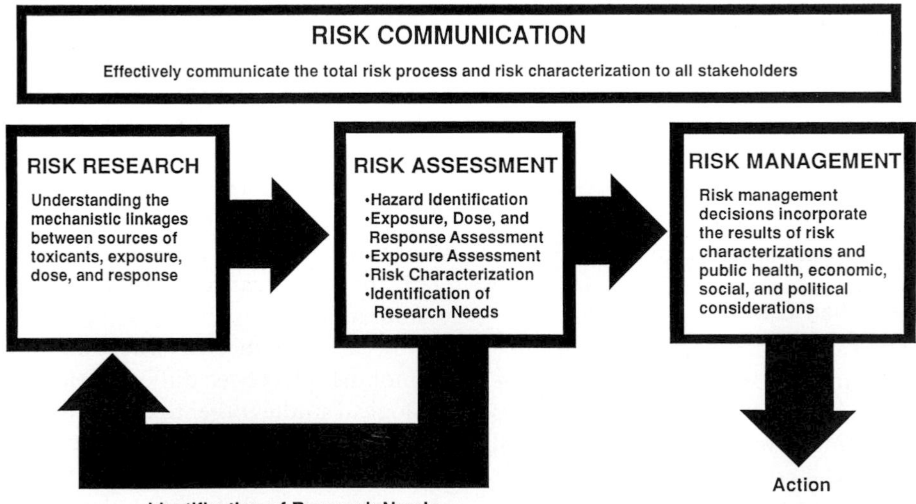

FIGURE 5 The interrelated process of risk research, risk assessment, risk management, and risk communication.

TABLE II
Basis for National Ambient Air Quality Standards

Pollutant	Susceptible population	Principal health effects
Particles	Individuals with preexisting respiratory diseases	Changes in mortality in sensitive populations, increase in respiratory symptoms, reduced pulmonary function
Sulfur dioxide	Asthmatics	Increased respiratory symptoms, reduced pulmonary function
Carbon monoxide	Individuals with heart disease	Aggravation of angina pectoris
Nitrogen dioxide	Young children and asthmatics, individuals with preexisting respiratory disease	Increased respiratory symptoms, reduced pulmonary function
Ozone	None identified	Increased respiratory symptoms, reduced pulmonary function
Lead	Fetus and young children	Neurobehavioral development, impaired heme synthesis

specific toxicant exposure (both concentration and duration) for an individual or a small population when the relevant exposure time for producing disease, such as lung cancer, may be 20 or more years? Significant progress has been made in this area in recent years, but much remains to be done. It should be emphasized that despite the inadequacies of exposure assessments, our ability to identify and measure hundreds of compounds in the air greatly exceeds our ability to understand and assess the effects of such exposures.

The final step of the risk assessment process is to bring estimates of the exposure together with estimates of exposure–response relationships to provide an overall estimate of the likely health consequences of exposure to a specific toxicant. Or alternatively, having established an acceptable level of health risk, it is possible to calculate the corresponding level to which exposures should be limited to achieve the health risk limit. This kind of approach has been taken by regulatory agencies with materials such as coke oven emissions. This qualitative approach has increasingly been taken when dealing with cancer risks. The risk assessment process can also aid in identifying uncertainties that may be reduced through future research.

An alternative approach has been taken in establishing occupational and environmental exposure limits for agents for which the concern is primarily health effects other than cancer. Examples are workplace threshold limit values established by the American Conference of Governmental Industrial Hygienists and National Ambient Air Quality Standards established by the United States Environmental Protection Agency (Table II). This approach is more qualitative and involves establishing from human or laboratory animal studies or both a "no observed effect" level. Typically, safety factor(s) are then used to establish an exposure limit that is lower. The safety factor may

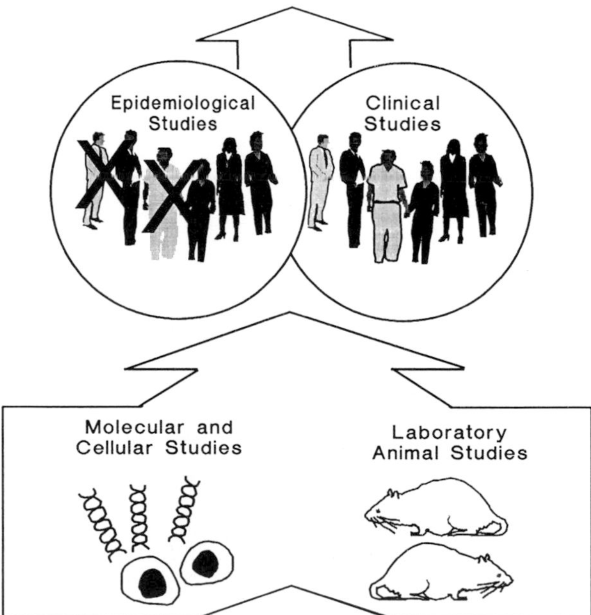

FIGURE 6 Integration of data obtained with complementary approaches.

be used to account for uncertainties in extrapolation from laboratory animals to humans or the presence of sensitive individuals in the population to be protected.

Ultimately, an understanding of the health effects of inhaled toxicants in humans will be achieved by integrating information obtained in epidemiological, human clinical, laboratory animal, cellular, and molecular studies (Fig. 6). Each particular kind of study has value, which is enhanced when used to complement information gained with other approaches.

BIBLIOGRAPHY

Gardner, D. E., Crapo, J. D., and McClellan, R. O. (eds.) (1993). "Toxicology of the Lung," 2nd Ed. Raven, New York.

Lippmann, M. (ed.) (1992). "Environmental Toxicants: Human Exposures and Their Health Effects." Van Nostrand–Reinhold, New York.

McClellan, R. O., and Henderson, R. F. (eds.) (1995). "Concepts in Inhalation Toxicology," 2nd Ed. Taylor & Francis, Washington, D.C.

Mohr, U., Dungworth, D., Kimmerle, G., Lewkowski, J., McClellan, R., and Stoeber, W. (eds.) (1988). "Inhalation Toxicology: The Design and Interpretation of Inhalation Studies and Their Use in Risk Assessment." Springer-Verlag, New York.

Mohr, U., Bates, D. V., Dungworth, D. C., Lee, P. N., McClellan, R. O., and Roe, F. J. C. (eds.) (1989). "Assessment of Inhalation Hazards: Integration and Extrapolation Using Diverse Data." Springer-Verlag, New York.

National Research Council, Committee on Risk Assessment of Hazardous Air Pollutants (1995). "Science and Judgment in Risk Assessment." National Academy Press, Washington, D.C.

Parkes, W. R. (ed.) (1994). "Occupational Lung Disorders," 3rd Ed. Butterworth–Heinemann, Oxford, England.

Warheit, D. B. (ed.) (1993). "Fiber Toxicology: Contemporary Issues." Academic Press, San Diego.

Watson, A. Y., Bates, R. R., and Kennedy, D. (eds.) (1988). "Air Pollution, the Automobile and Public Health." National Academy Press, Washington, D.C.

Purine and Pyrimidine Metabolism

THOMAS D. PALELLA
Rheumatic Disease Center

IRVING H. FOX
Biogen Incorporated

MICHAEL A. BECKER
University of Chicago

GLOSSARY

Base Generic term for the heterocyclic (i.e., containing two kinds of atoms—carbon and nitrogen) ring structures of purines and pyrimidines

***De novo* Biosynthesis** Metabolic pathways through which nonpurine and nonpyrimidine precursors are combined to form purines and pyrimidines, respectively

Nucleoside Compound of a sugar (e.g., ribose or deoxyribose) with a purine or pyrimidine base by way of an *N*-glycosyl link

Nucleotide Combination of a purine or pyrimidine base, sugar, and phosphate group or groups, comprising the fundamental units of nucleic acids; the base sugar linkage is *N*-glycosyl; the sugar–phosphate combination is via an ester linkage, usually to the 5′ hydroxyl group; mononucleotides contain a single phosphate group, dinucleotides contain two phosphate groups linked by a phosphodiester bond, and trinucleotides contain three phosphate groups

Purine Heterocyclic nine-member ring ($C_5H_4N_4$) that is the parent compound of adenine, guanine, and hypoxanthine

Pyrimidine Heterocyclic six-member ring ($C_4H_4N_2$) that is the parent compound of thymine, cytidine, and uracil

PURINES AND PYRIMIDINES ARE PRESENT IN ALL forms of plant and animal life. They constitute the building blocks of deoxyribonucleic acid (DNA) and ribonucleic acid (RNA) and, thus, are fundamental to reproduction and inheritance. In addition, purine nucleotides serve as molecular energy sources. Both purine and pyrimidine nucleotides are donors of high-energy phosphate in a plethora of enzymatic reactions. Finally, they also serve as coenzymes, neurotransmitters, and second messengers. In humans, several important diseases result from disordered purine or pyrimidine metabolism. For all of these reasons, purine and pyrimidine metabolism has been extensively studied. This article summarizes only the most important aspects of these expansive topics.

I. PURINES

The overall scheme of human purine metabolism is depicted in Fig. 1. This representation is conveniently divided into interfacing compartments: (1) *de novo* purine synthesis, (2) salvage of preformed purine bases, (3) nucleotide interconversions, and (4) degradation. The parent compound, the purine ring, is the starting point. Because the final degradative product in humans is uric acid, disorders of purine metabolism are often reflected in abnormal concentrations of uric acid or urate in the blood and other body fluids.

A. Metabolic Pathways

1. *De Novo* Purine Synthesis

The purine nucleus, consisting of fused pyrimidine and imidazole rings, is the parent compound for all purines. The origins of its individual atoms have been defined in a variety of prokaryotic and eukaryotic

ENCYCLOPEDIA OF HUMAN BIOLOGY, Second Edition, VOLUME 7. Copyright © 1997 by Academic Press. All rights of reproduction in any form reserved.

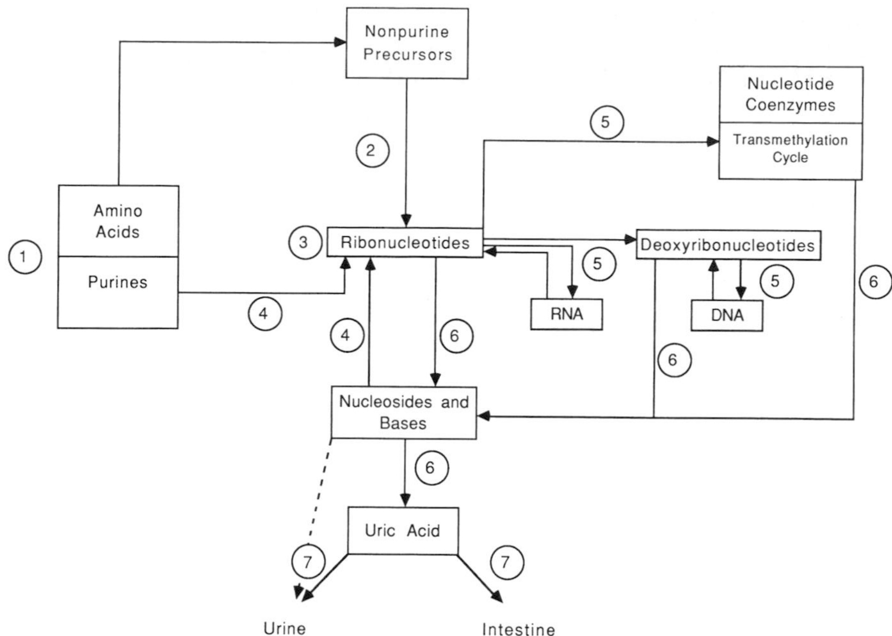

FIGURE 1 Overview of human purine metabolism. Human purine metabolism is focused on the synthesis of purine ribonucleotides by two pathways. In the *de novo* pathway, dietary intake of amino acids (1) provides nonpurine precursors for purine biosynthesis (2) leading to inosine-5′-monophosphate (IMP) formation. IMP is converted to other purine ribonucleotides by the interconversion pathways (3). Alternatively, preformed purine bases and nucleosides may be reutilized by conversion to ribonucleotides by the salvage pathways (4). The substrates for the salvage reactions derive from dietary purines (1) or breakdown of nucleotides (6). Purine ribonucleotides are converted from the monophosphate form to diphosphate and triphosphate forms, which serve as essential substrates for a variety of pathways (5). Purine nucleoside monophosphates are also the main substrates for the degradative pathways of purine metabolism (6). The final step in the degradation of purine nucleotides in humans is the formation of uric acid. Uric acid is the major excretory product of human purine metabolism, although small amounts of nucleosides and bases are excreted into the urine (7). Uric acid is excreted primarily in urine but smaller amounts are excreted into the intestine in which bacterial uricolysis occurs (7). [Modified, with permission, from I. H. Fox (1982). *In* "Clinical Medicine" (J. A. Spittell, Jr., ed.). Harper & Row, New York.]

systems. The biosynthetic pathway through which nonpurine precursors are combined to form the purine ring is referred to as *de novo* purine synthesis.

The first reaction in this synthesis is the formation of an important regulatory intermediate, 5-phosphoribosyl-α-1-pyrophosphate (PP-ribose-P). It has two main roles: it condenses with L-glutamine in the first committed step of the *de novo* pathway, and it serves as the phosphoribosyl donor in salvage reactions whereby purine bases are converted into ribonucleotides.

In a reaction catalyzed by PP-ribose-P synthetase, in the presence of magnesium (Mg^{2+}) and inorganic phosphate (P_i), the terminal pyrophosphate (PP_i) of adenosine-5′-triphosphate (ATP) is transferred to carbon 1 of ribose-5-phosphate yielding adenosine-5′-monophosphate (AMP). This allosteric enzyme is regulated by negative feedback inhibition by purine and pyrimidine nucleotides. [*See* Adenosine Triphosphate (ATP).]

$$\text{ribose-5-phosphate} + \text{ATP} \xrightarrow{Mg^{2+}} \alpha\text{-PP-ribose-P} + \text{AMP} \quad \text{(Reaction 1)}$$

The first committed step in *de novo* synthesis generates 5-β-phosphoribosyl-1-amine. In this irreversible reaction, which is catalyzed by the enzyme amidophosphoribosyltransferase (AmidoPRT), the amide group of glutamine displaces pyrophosphate from PP-ribose-P. The activity of this enzyme is regulated by PP-ribose-P and nucleotide concentrations. In the process, the substituents are inverted such that a β-linkage is generated:

$$\alpha\text{-PP-ribose-P} + \text{glutamine} + H_2O \xrightarrow{Mg^{2+}} 5\text{-}\beta\text{-phosphoribosylamine} + \text{glutamic acid} + PP_i \quad \text{(Reaction 2)}$$

Subsequent reactions in the *de novo* pathway form the purine ring and lead to inosine-5′-monophosphate (IMP) formation (Fig. 2).

2. Nucleotide Interconversions and Catabolism

All purine ribonucleotides and deoxyribonucleotides derive from IMP. These reactions are summarized in Fig. 3 (Reactions 12–15).

Mononucleotides are converted to nucleosides by a variety of enzymes falling into two broad classes: (1) nonspecific phosphatases and (2) 5′-nucleotidases (Reaction 16). Purine nucleosides may be further catabolized by phosphorolysis via purine nucleoside phosphorylase (Reaction 17). AMP deaminase converts AMP to IMP with the liberation of ammonia (Reaction 18).

Adenosine is converted to inosine via adenosine deaminase (Reaction 19), whereas adenosine kinase converts adenosine to AMP, consuming ATP in the process (Reaction 20).

3. Salvage Pathways

De novo purine synthesis is metabolically expensive. A minimum of six ATP molecules are consumed per purine nucleotide formed. Thus, salvage of purine moieties is energetically economical. Catabolized nucleotides are salvaged in two distinct ways. As noted

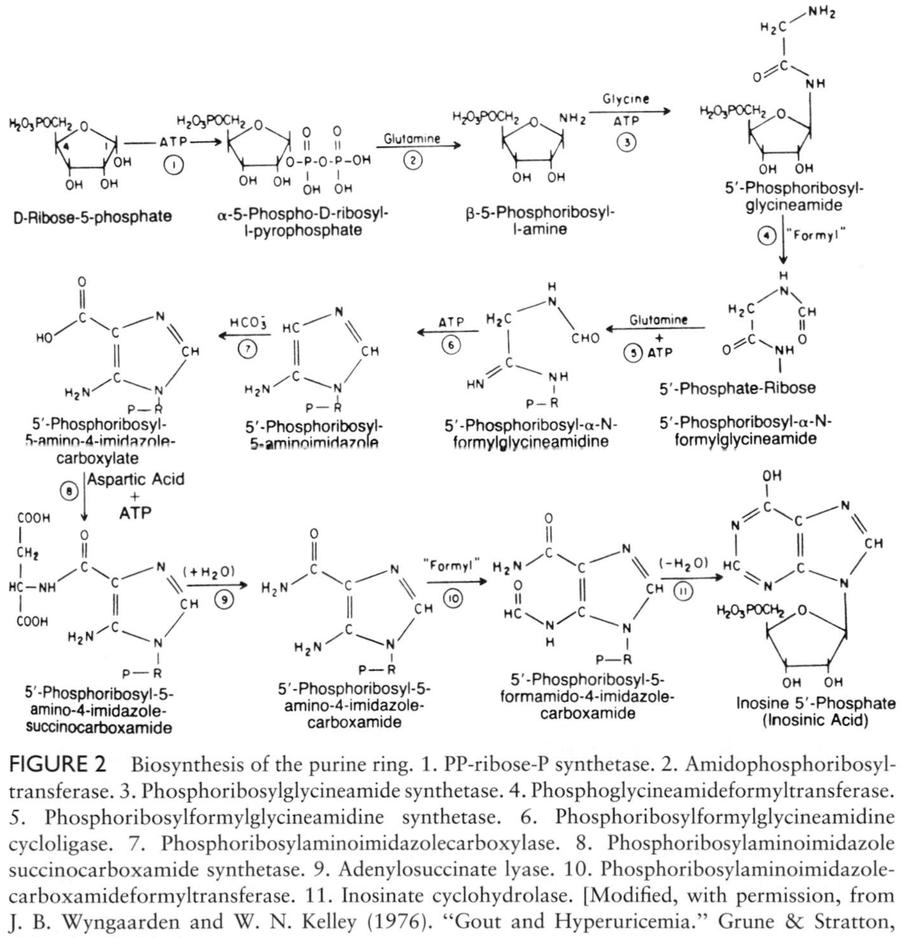

FIGURE 2 Biosynthesis of the purine ring. 1. PP-ribose-P synthetase. 2. Amidophosphoribosyltransferase. 3. Phosphoribosylglycineamide synthetase. 4. Phosphoglycineamideformyltransferase. 5. Phosphoribosylformylglycineamidine synthetase. 6. Phosphoribosylformylglycineamidine cycloligase. 7. Phosphoribosylaminoimidazolecarboxylase. 8. Phosphoribosylaminoimidazole succinocarboxamide synthetase. 9. Adenylosuccinate lyase. 10. Phosphoribosylaminoimidazolecarboxamideformyltransferase. 11. Inosinate cyclohydrolase. [Modified, with permission, from J. B. Wyngaarden and W. N. Kelley (1976). "Gout and Hyperuricemia." Grune & Stratton, Orlando, Florida.]

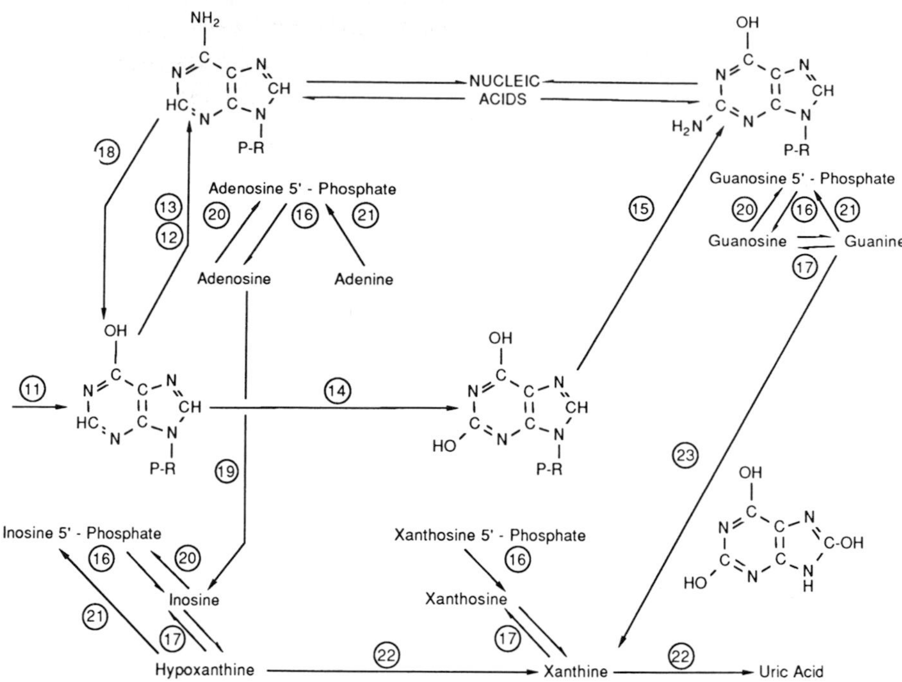

FIGURE 3 Biosynthesis and degradation of purine ribonucleotides, ribonucleosides, and bases. 12. Adenylosuccinate synthetase. 13. Adenylosuccinate lyase. 14. IMP oxidase. 15. XMP deaminase. 16. 5′-Nucleotidase. 17. Purine nucleoside phosphorylase. 18. AMP deaminase. 19. Adenosine deaminase. 20. Adenosine kinase. 21. Adenine phosphoribosyltransferase and hypoxanthine–guanine phosphoribosyltransferase. 22. Xanthine oxidase. [Modified, with permission, from J. B. Wyngaarden and W. N. Kelley (1976) "Gout and Hyperuricemia." Grune & Stratton, Orlando, Florida.]

earlier, purine nucleoside phosphorylase is a freely reversible reaction in which nucleosides are converted to the corresponding base and ribose-1-phosphate. The characteristic of this reaction favors the synthetic direction.

Furthermore, the purine nucleus is salvaged via the actions of phosphoribosyltransferases. In these reactions, the free bases condense with PP-ribose-P, forming ribonucleotides in a single step. Phosphoribosyltransferase reactions have the following general form:

$$\text{purine base} + \text{PP-ribose-P} \xrightarrow{Mg^{2+}} \text{purine mononucleotide} + PP_i \quad \text{(Reaction 21)}$$

In humans, two different purine phosphoribosyltransferases exist. Adenine phosphoribosyltransferase (APRT) converts adenine to AMP. Hypoxanthine–guanine phosphoribosyltransferase (HPRT) converts hypoxanthine and guanine to IMP and guanosine-5′-monophosphate (GMP), respectively. Physiologically, the phosphoribosyltransferase pathways are the most active in the salvage of preformed purine bases. The origin of these bases is from endogenous catabolism of purines, nucleic acid breakdown, and the exogenous intake or administration of purine compounds.

4. Synthesis of Uric Acid

Xanthine oxidase is a flavoprotein containing both iron and molybdenum, which oxidizes a wide variety of purines and pteridines. A soluble form of the enzyme having dehydrogenase activity (xanthine dehydrogenase, or D-form) has been isolated from a variety of sources and is likely the normal form of the enzyme. The D-form is converted to the oxidase form (O-form) by the oxidation of thiol groups in the protein. The O-form uses oxygen directly as a substrate. This enzyme converts hypoxanthine to xanthine and xanthine to uric acid:

$$\text{hypoxanthine} + H_2O + 2O_2 \rightarrow \text{xanthine} + 2O_2^- + 2H^+ \text{ (Reaction 22a)}$$

$$\text{hypoxanthine} + H_2O + O_2 \rightarrow \text{xanthine} + H_2O_2 \text{ (Reaction 22b)}$$

$$\text{xanthine} + H_2O + 2O_2 \rightarrow \text{uric acid} + 2O_2^- + 2H^+ \text{ (Reaction 22c)}$$

or

$$\text{xanthine} + H_2O + O_2 \rightarrow \text{uric acid} + H_2O_2 \text{ (Reaction 22d)}$$

Superoxide anion or hydrogen peroxide is thus generated. Hydrogen peroxide may then be converted to free hydroxyl radicals:

$$Fe^{2+} + H_2O_2 \rightarrow Fe^{3+} + OH^- + OH^\cdot$$

These compounds (O_2^-, H_2O_2, OH^-) are important mediators of inflammation and tissue destruction. Xanthine oxidase may thus have a significant pathophysiologic role when the flow of blood is blocked (ischemia) or there is tissue injury, which both lead to accentuated adenine nucleotide breakdown.

B. Regulation of Purine Biosynthesis

1. *De Novo* Purine Synthesis

A major site of regulation of purine biosynthesis involves the domain defined by the PP-ribose-P synthetase and AmidoPRT reactions (Reactions 1 and 2). PP-ribose-P synthetase activity is subject to inhibition by purine and pyrimidine products of the respective nucleotide biosynthetic pathways. Thus, the availability of PP-ribose-P to exert its role as an allosteric activator of *de novo* purine synthesis is determined by a feedback mechanism responsive to concentrations of both purine and pyrimidine nucleotides. Specificity in the purine regulatory process is introduced, however, at the level of the kinetic properties of AmidoPRT. This latter reaction is the irreversible first step committed to purine nucleotide synthesis and is the rate-limiting reaction in the pathway leading to the production of inosinic acid, from which all other purines are ultimately derived. Purine nucleotide pathway products are more potent inhibitors of AmidoPRT than of PP-ribose-P synthetase, and inhibition of the former is antagonized by the rate-limiting substrate, PP-ribose-P. These observations are consistent with a dual regulatory system in which fine control of purine nucleotide synthesis is exerted mainly at a site specifically committed to the pathway (AmidoPRT reaction), but more severe and generalized alterations in nucleotide levels can be modulated by control over the production of the key nucleotide regulatory substrate, PP-ribose-P.

2. Purine Ribonucleotide Interconversions

The next major site of regulation of purine ribonucleotide synthesis is at the level of interconversions. As noted before, a variety of reactions occur by which the various nucleotides are interconverted. Each of these reactions is regulated by its nucleotide end product, which typically inhibits the first enzyme in the pathway. The conversion of IMP to AMP is inhibited by the products of the reaction, AMP and guanosine-5′-diphosphate (GDP). Correspondingly, the synthesis of GMP from IMP is inhibited by guanosine-5′-triphosphate (GTP) in a two-step process, which requires ATP as an energy donor. Thus, availability of GTP and ATP, each of which controls the synthesis of the other nucleotide, as well as inhibition by products of the reactions with respect to their own biosynthesis, regulates the steady-state levels of adenyl and guanyl nucleotides.

3. Ribonucleotide Degradation

In vivo and *in vitro* evidence indicates that nucleotide breakdown is regulated in a complex manner. When cultured ascites tumor cells are incubated with 2-deoxyglucose or glucose, these compounds are rapidly phosphorylated by ATP. The abrupt decrease of ATP concentrations from its rapid utilization results in elevations of AMP and IMP. These nucleotides are dephosphorylated to purine nucleosides by 5′-nucleotidases and are subsequently either catabolized or reutilized. An important product of 5′-nucleotidase activity of AMP is adenosine, which is a biologically active compound. Adenosine acts by stimulating its own cell-surface receptors.

Regulation of nucleotide degradation is critically controlled by AMP deaminase (Reaction 18) and 5′-nucleotidases (Reaction 16). AMP deaminase is an allosteric enzyme that is activated by ATP and adenosine-5′-diphosphate (ADP) and inhibited by GTP and

P_i. Release of inhibition of AMP deaminase results in accelerated production of uric acid. Regulation at the level of dephosphorylation is more complex and is the focus of intensive investigation. At least three soluble 5′-nucleotidase activities have been described, differing in substrate specificities and responsiveness to effectors.

C. Production and Excretion of Uric Acid

The major determinants of purine synthesis are (1) rate of *de novo* purine synthesis, (2) rates of purine salvage, and (3) rate of exogenous supply of purines. The elimination of purines is controlled by the rates of (1) nucleotide breakdown, (2) synthesis of uric acid, and (3) purine excretion in urine and by extrarenal routes. Thus, the overall rate of purine synthesis reflects the difference between purine intake and *de novo* synthesis and excretion of purine metabolic products, particularly uric acid. In practice, these variables are complex and interrelated, making it very difficult to assess synthetic rates in intact organisms. Estimates of the minimal level of purine production have been obtained by severe dietary restriction of purine intake in humans. A very wide distribution of these rates in normal adults is seen. Correspondingly, urinary uric acid excretion rates also vary widely, because they are a reflection of overall purine production. Values for urinary uric acid excretion in 24 hr are defined on a statistical basis under controlled conditions of dietary intake on the basis of the mean, plus or minus two standard deviations. On isocaloric purine-free diets, the value in normal adult men was 418 ± 70 mg in one study and 426 ± 81 mg in another.

Sustained overexcretion of uric acid indicates increased synthesis of purines *de novo*. Normal excretion of uric acid in a gouty subject does not exclude the possibility of overproduction, however, because renal impairment may alter the evident excretor status of the patient. In such subjects, extrarenal disposal of urate is increased and may account for the majority of urate turnover. Probably $<10\%$ of subjects with primary gout overproduce urate.

1. Renal Handling of Uric Acid

Uric acid is fully excreted by the glomeruli. The relative contributions of reabsorption and secretion of urate in the human kidney are difficult to estimate. A four-component model of the renal handling of urate has been proposed, which agrees closely with a variety of observations regarding the renal handling of uric acid. In this model, there are two sites for urate reabsorption in the proximal nephron, separated by the urate secretory site. This model represents the best explanation to date for the accumulated observations on renal urate handling in humans. The relative fluxes of urate through each component remain undetermined and authentication of their existence awaits direct demonstration.

2. Extrarenal Excretion of Uric Acid

When ^{15}N-labeled uric acid is injected into normal subjects, only 75% of the radiolabel is recovered in the urine. Approximately 25% ^{15}N is recovered in urinary allantoin, urea and ammonia, and fecal nitrogen. These compounds originate from bacterial uricolysis within the intestines.

D. Important Disorders of Purine Metabolism

Many acquired and inherited disorders are associated with the purine metabolic pathways (Table I). In addition, these pathways are the target for drugs that have anticancer, antiviral or immunosuppressive properties. Hyperuricemia and gout are the most common abnormalities.

1. Hyperuricemia and Gout

Gout may be defined as a heterogenous group of diseases found exclusively in man, which in their full development are manifest by (a) an increase in the serum urate concentration; (b) recurrent attacks of a characteristic type of acute arthritis, in which crystals of monosodium urate monohydrate are demonstrable in leukocytes of synovial fluid; (c) aggregated deposits of monosodium urate monohydrate (tophi) occurring chiefly in and around the joints of the extremities and sometimes leading to severe crippling and deformity; (d) renal disease involving glomerular, tubular, and interstitial tissues and blood vessels; and (e) uric acid urolithiasis (kidney stones). These manifestations can occur in different combinations. (Wyngaarden and Kelley, 1976)

Hyperuricemia needs to be distinguished from gout. Although only a minority of hyperuricemic patients ever become gouty, all patients with gout have hyperuricemia at some stage in their clinical course. In one study, only 12% of 200 hyperuricemic patients had gout.

Using the uricase differential spectrophotometric method for measuring urate valves, the serum urate

TABLE I
Disorders Associated with Altered Purine Metabolism

Disorder	Basis for disorder
Hyperuricemia and gout	Idiopathic
	Inherited
	Hypoxanthine–guanine phosphoribosyltransferase deficiency
	PP-ribose-P synthetase overactivity
	Glucose-6-phosphatase deficiency
	Decrease renal urate clearance
	Secondary to drug or disease
Renal calculi	
Uric acid	Hyperuricemia and gout
Xanthine	Xanthine oxidase deficiency
	Inherited
	Acquired
	Allopurinol therapy
2,8-Dehydroxyadenine	Complete adenine phosphoribosyltransferase deficiency
Oxypurinol	Allopurinol therapy
Immune deficiency	Adenosine deaminase deficiency
	Inherited
	Acquired
	Purine nucleoside phosphorylase deficiency
	Ecto-5′-nucleotidase deficiency
	Inherited
	Acquired
Anemia	
Hemolytic	Adenylate kinase deficiency
	Adenosine deaminase increased activity
	Purine nucleoside phosphorylase deficiency
	Acquired adenosine deaminase deficiency
Megaloblastic	Hypoxanthine–guanine phosphoribosyltransferase deficiency
	Purine nucleoside phosphorylase deficiency
	PP-ribose-P synthetase deficiency
Central nervous system disease	Hypoxanthine–guanine phosphoribosyltransferase deficiency
	PP-ribose-P synthetase overactivity
	Adenylosuccinate lyase deficiency
	Guanine deaminase deficiency
	Purine nucleoside phosphorylase deficiency
Myopathy	Myoadenylate deaminase deficiency
	Xanthine oxidase deficiency
	Metabolic myopathies
No disease	Partial deficiency of adenine phosphoribosyltransferase

value is theoretically elevated when it is >7.0 mg/100 ml, the limit of solubility of monosodium urate in serum at 37°C. An elevated serum urate concentration is defined as a value exceeding the upper limit of the mean serum urate value plus two standard deviations in a sex- and age-matched healthy population. In most epidemiological studies, the upper limit has been rounded off at 7.0 mg/100 ml in men and 6.0 mg/100 ml in women. Finally, a serum urate value >7.0 mg/100 ml begins to carry an increased risk of gouty arthritis or renal stones. When methods are used for measuring uric acid that are not specific for urate, the upper limit of normal will be >7.0 mg/100 ml.

a. Clinical Features

Symptomatic hyperuricemia includes acute gouty arthritis, chronic gout, and renal disease. These complications of hyperuricemia result from the precipitation of uric acid in body fluids and tissues. The occurrence of this phenomenon depends on the physical properties of uric acid. It is a weak acid that is ionized at position 9 with pK of 5.75. The pH determines the form of uric acid that is deposited. In normal body tissues at physiologic pH, sodium urate is deposited, whereas in urine, which generally has a lower pH, uric acid is precipitated.

The most common complication of hyperuricemia is the development of gouty arthritis. Untreated gout often passes through three stages, after a period of asymptomatic hyperuricemia: acute gouty arthritis, intercritical gout, and chronic tophaceous gout. Asymptomatic hyperuricemia most often ends with an attack of gouty arthritis or of urolithiasis. Although only a minority of patients with hyperuricemia will ever develop gout, in most instances when gout does develop, it is after 20–30 yr of sustained hyperuricemia.

After gouty arthritis, renal disease appears to be the most frequent complication of hyperuricemia, with several types being associated with it. The first type, urate nephropathy, is attributed to the deposition of monosodium urate crystals in the renal interstitial tissue and is thought to be associated with chronic hyperuricemia. In contrast, uric acid nephropathy is related to the formation of uric acid crystals in the collecting tubules, pelvis, or ureter, with subsequent impairment of urine flow. This disorder is caused by elevated concentrations of uric acid in the urine and can appear as either acute uric acid nephropathy or uric acid calculi.

Hyperuricemia and gout are well-recognized complications of chronic lead intoxication, but its occurrence varies considerably. The basis for this association may be related to the ability of lead urate to undergo nucleation (nucleus around which a crystal forms) at a lower concentration than is required for the nucleation of monosodium urate.

Acute renal failure can result from the precipitation of uric acid crystals in the collecting ducts and ureters. This complication most commonly occurs in patients with leukemias and lymphomas as a result of rapid malignant cell turnover, often during chemotherapy.

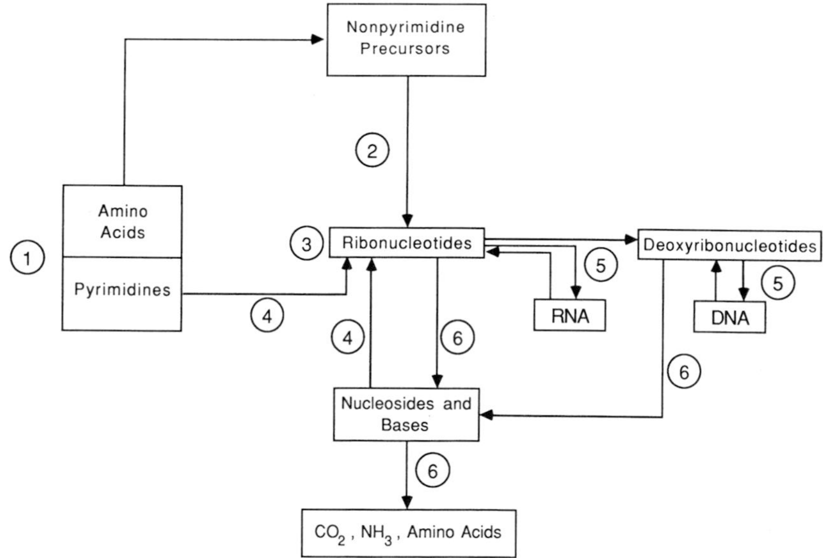

FIGURE 4 Overview of pyrimidine metabolism. Human pyrimidine metabolism is focused on the synthesis of the pyrimidine ribonucleotides in a manner analogous to purines (see Fig. 1). Dietary intake of amino acids (1) provides nonpurine precursors for *de novo* biosynthesis (2) leading to uridine-5′-monophosphate (UMP) formation. UMP is converted to other pyrimidine nucleotides by interconversion pathways (3). Preformed pyrimidine nucleosides and bases from the diet (1) and nucleotide catabolism (5) may be reutilized via salvage pathways (4). Pyrimidine nucleoside monophosphates are converted to diphosphate and triphosphate forms (3), which are used in the synthesis of nucleic acids (5). Degradation proceeds through the ribonucleotides to nucleosides and bases (6). Free pyrimidine bases are further degraded to amino acids, CO_2, and NH_3, the final products of pyrimidine catabolism. [From T. D. Palella and I. H. Fox (1989). *In* "The Molecular and Metabolic Basis of Acquired Disease" (R. D. Cohen, K. G. M. M. Alberti, B. Lewis, and A. M. Denman, eds.). Balliere Tindall, London. Used with permission.]

b. Treatment

The natural history of hyperuricemia and gout may be modified by lowering the serum urate levels. This is accomplished by drug therapy, which either inhibits uric acid synthesis at xanthine oxidase (allopurinol) or increases the renal excretion of uric acid (sulfinpyrazone or probenecid). The acute attacks of gout are treated by nonsteroidal anti-inflammatory drugs, such as indomethacin.

II. PYRIMIDINES

Pyrimidine metabolism is analogous to purine metabolism (Fig. 4). In pyrimidine metabolism, all pyrimidine nucleotides derive from uridine-5′-monophosphate (UMP). Pyrimidine nucleotides may be synthesized *de novo* or by reutilization of preformed bases and nucleosides. Interconversion of pyrimidines occurs at the monophosphate level. Degradation of pyrimidines results in formation of an amino acid, ammonia, and carbon dioxide. The overall economy of pyrimidine metabolism is regulated to some extent at each level. [*See* Nucleotide and Nucleic Acid Synthesis, Cellular Organization.]

A. Metabolic Pathways

1. Pyrimidine Biosynthesis

The *de novo* synthesis of the parent pyrimidine compound UMP is depicted in Fig. 5. The initial step is the formation of the unstable high-energy compound carbamylphosphate (CAP). Although two different enzymatic activities catalyzing the formation of CAP have been described in eukaryotes, CAP-synthetase II (CPS II) appears to be the activity associated with *de novo* pyrimidine metabolism.

CPS II is a cytoplasmic enzyme that exists in mammalian tissues as part of a three-enzyme complex. The other two enzymes, aspartate transcarbamylase and dihydroorotase, catalyze the second and third steps of *de novo* pyrimidine biosynthesis, respectively. This complex is designated *pyr* 1–3 (see Fig. 5). The product of these reactions, dihydroorotic acid, is reversibly oxidized to orotic acid to form the pyrimidine ring in a reaction catalyzed by dihydroorotic acid dehydrogenase. This enzyme is associated with the outer surface of the inner mitochondrial membrane in contrast to the cytosolic location of the other five enzymes of the pyrimidine *de novo* pathway.

In the penultimate steps of the *de novo* pathway (*pyr* 5,6), orotidine-5′-monophosphate (OMP)

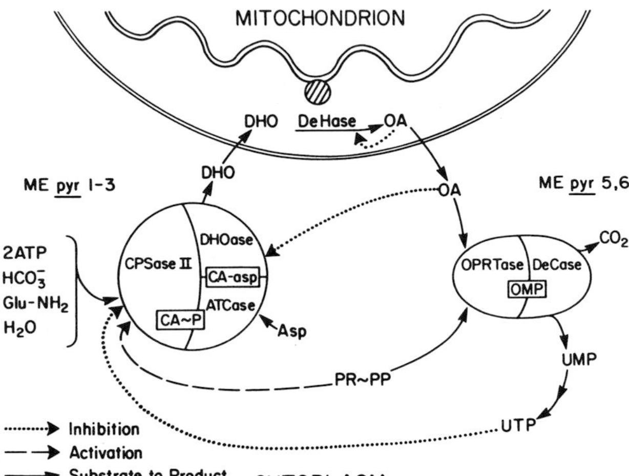

FIGURE 5 *De novo* pyrimidine biosynthesis. Multienzyme (ME) *pyr* 1–3 and *pyr* 5,6 are cytosolic proteins. ME *pyr* 1–3 has the enzyme centers for carbamylphosphate synthetase (CPSase II), aspartate transcarbamylase (ATCase), and dihydro-orotase (DHOase). ME *pyr* 5,6 has the active centers for orotate phosphoribosyltransferase (OPRTase) and orotidylate decarboxylase (OMP DeCase). Dihydro-orotate dehydrogenase (DHO DeHase) is on the outer surface of the inner membrane of the mitochondrion. The substrates are ATP, L-glutamine (Glu-NH_2), HCO_3^-, L-aspartate (Asp), and 5-phosphoribosyl-1-pyrophosphate (PR-PP). The intermediates of the pathway are carbamyl phosphate (CA-P), carbamyl-L-aspartate (CA-asp), dihydro-orotate (DHO), orotate (OA), orotidylate (OMP), and uridylate (UMP). Solid lines with arrows indicate substrate-to-product reactions. CA-P, CA-asp, and OMPase were placed in boxes because these intermediates are not readily released from the multienzymatic proteins into the cytoplasmic solution. [Reproduced, with permission, from M. E. Jones (1980). *Annu. Rev. Biochem.* **49.** © 1980 by Annual Reviews, Inc.]

is formed by orotate phosphoribosyltransferase (OPRT), a reaction similar to the salvage reactions of purine metabolism catalyzed by HPRT and APRT. OPRT activity is closely associated with the catalytic activity of the next enzyme in the pathway, orotidine-5′-monophosphate decarboxylase (ODC). Whether OPRT and ODC activities are associated as a single multifunctional protein or as a complex whose aggregation is dependent on the presence of two enzymes is not clear.

Pyrimidine nucleotides may also be formed by reutilization of preformed bases and nucleosides from dietary sources or nucleotide catabolism (Fig. 6). Uracil may be salvaged by two mechanisms: directly converted to UMP by OPRT (Reaction 1) or converted to uridine by the ubiquitous enzyme uridine phosphorylase (Reaction 2), which has specificity only for uracil and its analogues. Uridine is then converted to UMP by the action of uridine

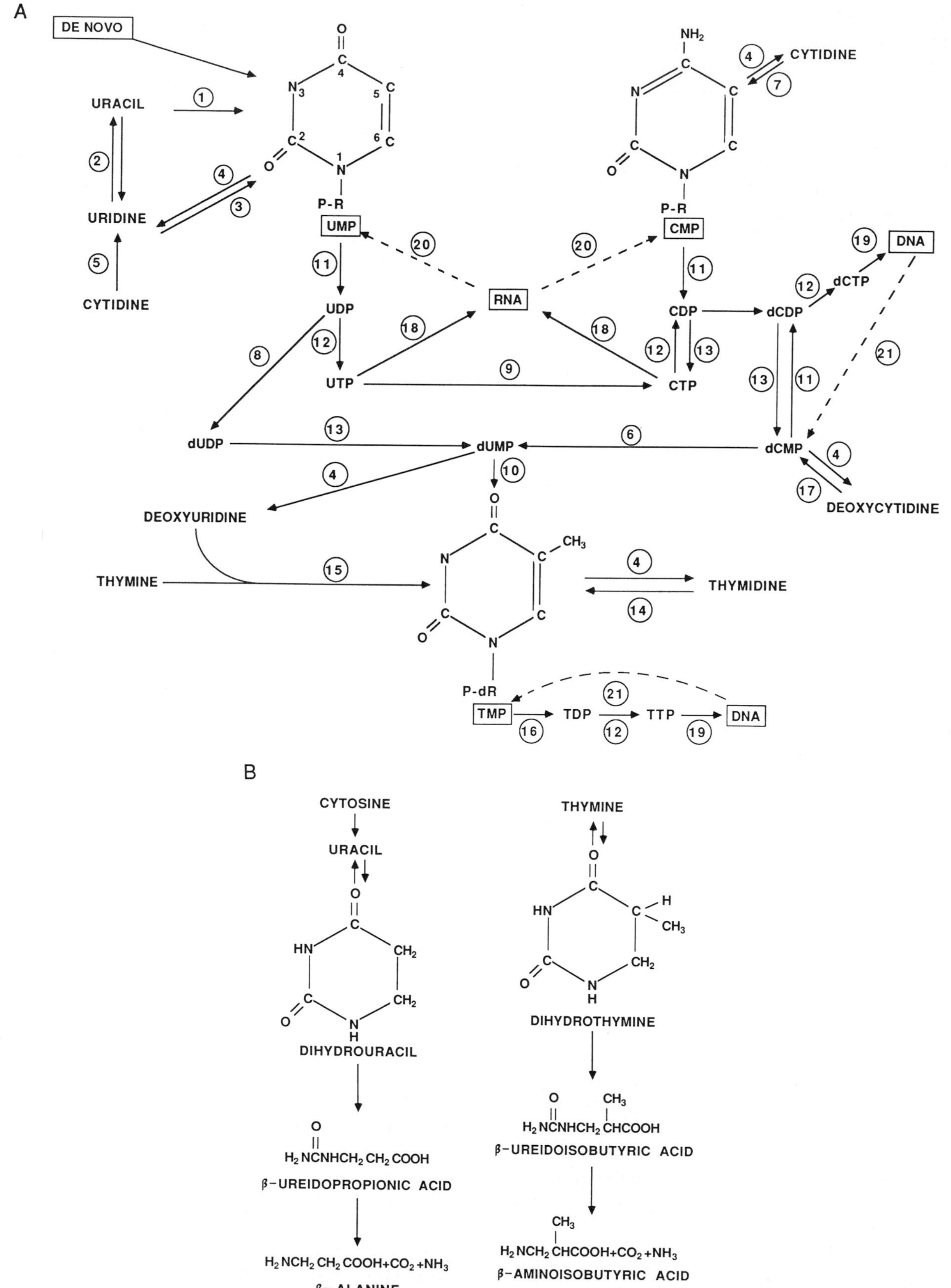

A
DE NOVO
URACIL
URIDINE
CYTIDINE
UMP
P-R
UDP
UTP
dUDP
RNA
CYTIDINE
CMP
CDP
CTP
dCDP
dCTP
DNA
dCMP
DEOXYCYTIDINE
dUMP
DEOXYURIDINE
THYMINE
THYMIDINE
P-dR
TMP
TDP
TTP
DNA
B
CYTOSINE
URACIL
DIHYDROURACIL
β-UREIDOPROPIONIC ACID
β- ALANINE
THYMINE
DIHYDROTHYMINE
β-UREIDOISOBUTYRIC ACID
β-AMINOISOBUTYRIC ACID

kinase (Reaction 3), which phosphorylates uridine and cytidine.

Uracil and thymine are ultimately converted to deoxyribonucleosides by a distinct enzyme, deoxythymidine phosphorylase, which requires deoxyribose-1-phosphate as a substrate. A separate reaction catalyzed by deoxythymidine synthetase (Reaction 15) converts thymine to deoxythymidine. Thymidine (i.e., 2-deoxythymidine) is converted to its monophosphate form (dTMP) by thymidine kinase (TK):

$$\text{deoxythymidine} + \text{ATP} \rightarrow \text{dTMP} + \text{ADP} \quad \text{(Reaction 14)}$$

TK is widely distributed in human tissues. Like uridine kinase, TK is present in fetal and adult liver in distinct forms. Deoxycytidine is converted to 2′-deoxycytidine-5′-monophosphate (dCMP) by a different enzyme, deoxycytidine kinase (Reaction 17), which also phosphorylates deoxyguanosine and adenosine.

2. Interconversions

Pyrimidine interconversions are also summarized in Figure 6. Ribonucleoside monophosphates are phosphorylated by kinases to di- and triphosphate forms. One kinase subserves phosphorylation of cytidine-5′-monophosphate (CMP), UMP, and dCMP (Reaction 11). A second activity phosphorylates deoxythymidine-5′-triphosphate (dTMP) and 2′-deoxyuridine-5′-monophosphate (dUMP) (Reaction 16). Either ATP or 2′-dTMP (dATP) can donate the phosphate moiety. Deoxycytidylate deaminase (Reaction 6) catalyzes the deamination of dCMP to dUMP. TMP (dTMP) synthesis from dUMP is controlled by the availability of folate derivatives by a reaction catalyzed by thymidylate synthetase:

$$\text{dUMP} + \text{5,20-methylene-5,6,7,8-tetrahydrofolate} \rightarrow \text{dTMP} + \text{7,8-dihydrofolate} \quad \text{(Reaction 10)}$$

Pyrimidine nucleoside diphosphates are converted to triphosphates via nucleoside diphosphokinase. The reaction has the general form

$$\text{NDP} + \text{N}_1\text{TP} \rightarrow \text{NTP} + \text{N}_1\text{DP}$$

$$\text{dNDP} + \text{N}_1\text{TP} \rightarrow \text{dNTP} + \text{N}_1\text{DP} \quad \text{(Reaction 12)}$$

In this reaction, N and N_1 may be any purine or pyrimidine base. (Deoxynucleotides are indicated as dNDP and dNTP.)

Nucleoside diphosphates are reduced to deoxy forms by ribonucleotide reductase:

$$\text{NDP} + \text{thioredoxin-(SH)}_2 \xrightarrow{\text{Mg}^{2+}} \text{dNDP} + \text{thioredoxin} - \text{S}_2 \quad \text{(Reaction 8)}$$

Thioredoxin-$(SH)_2$ is regenerated by the action of thioredoxin reductase. Ribonucleotide reductase is subject to complex feedback regulation.

3. Degradation

Pyrimidine mononucleotides are catabolized to nucleosides by the action of both nonspecific phosphatases and 5′-nucleotidase (Reaction 4). Cytidine is deaminated to uridine by cytidine deaminase (Reaction 5). Uridine and thymidine are catabolized to uracil and thymine by pyrimidine nucleoside phosphorylase (Reaction 2). Uracil and thymine are converted to dihydro forms by specific reactions catalyzed by dihydrouracil dehydrogenase and dihydrothymine dehydrogenase, respectively (see Fig. 6). The resultant dihydropyrimidines are hydrolyzed to their respective carbamyl-β-amino acids by dihydropyrimidinase. Products of these reactions are degraded to corresponding amino acids, CO_2, and ammonia by specific enzymes (e.g., β-ureidoproionase). Thus, the end products of pyrimidine metabolism are common metabolites that are not unique to a single metabolic pathway.

B. Regulation of Pyrimidine Metabolism

The first reaction in *de novo* pyrimidine biosynthesis, which is catalyzed by CPS II, is regulated by several mechanisms. CPS II is saturated with respect to

FIGURE 6 (A) Biosynthesis of pyrimidine nucleosides, nucleotides, and bases. 1. Orotate phosphoribosyltransferase. 2. Uridine phosphorylase. 3. Uridine kinase. 4. 5′-Nucleotidase. 5. Cytidine deaminase. 6. Cytidylate deaminase. 7. Cytidine kinase. 8. Ribonucleotide reductase. 9. CTP synthetase. 10. Thymidylate synthetase. 11. CMP–UMP–dCMP kinase. 12. Nucleoside dephosphate kinase. 13. Phosphatase. 14. Thymidine kinase. 15. Thymidine synthetase. 16. TMP–dUMP kinase. 17. Deoxycytidine kinase. 18. RNA polymerase. 19. DNA polymerase. 20. Ribonucleases. 21. Deoxynucleases. (B) Degradation of pyrimidines. [From T. D. Palella and I. H. Fox (1989). *In* "The Molecular and Metabolic Basis of Acquired Disease" (R. D. Cohen, K. G. M. M. Alberti, B. Lewis, A. M. Denman, eds.). Balliere Tindall, London. Used with permission.]

HCO_3^- and glutamine but not to ATP. CPS II activity increases with elevated PP-ribose-P or ATP concentrations as well as with reduced UTP concentrations.

The association of CPS II, aspartate transcarbamylase, and dihydroorotase activities in the multifunctional protein *pyr* 1–3 results in coordinated regulation of the initial three steps of *de novo* pyrimidine synthesis. This reflects changes in the amount of enyzme protein, which has been demonstrated in a variety of systems, including rapidly growing tissues, fetal tissues, neoplastic tissues, and blast-transformed lymphocytes. In one experimental model, gene amplification resulting in increased amounts of a single mRNA and higher production of the enzyme species has been suggested. Evidence also suggests that PP-ribose-P may increase *pyr* 1–3 levels by direct action on the gene encoding the multifunctional complex.

Other enzymes may also serve as regulatory points within the pyrimidine synthetic pathway. Evidence suggests that OPRT may be rate-limiting for the entire pathway. Because OPRT works at 10% saturation with PP-ribose-P, in cells, levels of this substrate may regulate the rate of the entire pathway through this enzyme. Finally, the rate of pyrimidine catabolism may influence the synthetic rate through modulation of uridine-5′-triphosphate (UTP) concentrations. The terminal enzyme in pyrimidine catabolism, *N*-carbamyl-β-alanine (NCβA) aminohydrolase, is regulated in opposing fashion by its substrate, NCβA, and its product, β-alanine. This effect is mediated by ligand-induced changes in enzyme polymerization.

In summary, the weight of evidence suggests that CPS II predominates as the major regulatory point in pyrimidine biosynthesis. When ATP concentrations are increased and both PP-ribose-P and uridine nucleotide levels are low, OPRT may become rate-limiting and thus an important regulatory site.

C. Interrelationship of Purine and Pyrimidine Metabolism

PP-ribose-P is an essential substrate and important regulatory intermediate in both purine and pyrimidine metabolism. In purine biosynthesis, PP-ribose-P is rate-limiting in three reactions—that is, those catalyzed by amidophosphoribosyltransferase, hypoxanthine–guanine phosphoribosyltransferase, and adenine phosphoribosyltransferase. In pyrimidine metabolism, PP-ribose-P activates CPS II, may induce *pyr* 1–3, and is a rate-limiting substrate for OPRT. Thus, increases in intracellular PP-ribose-P levels accelerate both purine and pyrimidine biosynthesis, whereas decreases in PP-ribose-P levels decrease the synthetic rates. Therefore, a potential mechanism for the coordinated synthesis of nucleotides for nucleic acid synthesis exists.

D. Important Disorders of Pyrimidine Metabolism

Only a few clinical disorders exist in the pyrimidine pathway. Orotic aciduria refers to the excessive excretion of orotic acid. It may be a rare inherited disorder in babies with severe anemia or occur secondarily to other diseases or drug intake.

Like the purine pathways, the pyrimidine pathways are the target for antiviral and anticancer drugs. A relatively common defect is pyrimidine 5′-nucleotidase deficiency, which has been observed in the erythrocytes of patients with an inherited hemolytic anemia or an acquired disorder related to lead intoxication.

Patients with lead intoxication have pyrimidine 5′-nucleotidase deficiency that worsens with increasing lead concentrations in the blood, a mild anemia, and large numbers of red cells with basophilic stippling. The inherited disorder associated with erythrocyte pyrimidine 5′-nucleotidase deficiency is characterized by congenital hemolytic anemia, enlargement of the spleen, and conspicuous basophilic stippling of peripheral erythrocytes. The hemoglobin ranges from 6 to 12 g/dl and the hematocrit (red cell volume) from 28 to 34%. There is a prominent reticulocytosis (presence of red cell precursors in circulations) with values ranging as high as 50%, poikilocytosis (red cell shape irregularities), moderate polychromasia, and mild anisocytosis; there frequently is hyperbilirubinemia.

The diagnosis of hemolytic anemia due to 5′-nucleotidase deficiency may be proven by demonstrating the enzyme deficiency in peripheral circulating erythrocytes using either UMP or CMP as substrate:

$$\text{UMP} + H_2O \rightarrow P_i + \text{uridine}$$

$$\text{CMP} + H_2O \rightarrow P_i + \text{cytidine}$$

The enzyme deficiency is a sensitive indicator of lead intoxication and is found to be decreased even when most other biological indicators of lead intoxication are negative.

The enzyme deficiency leads to an inability to degrade pyrimidine nucleotides, causing elevated cytidine nucleotide diphosphodiesters. Cytidine diphosphate choline and cytidine diphosphate ethanolamine account for 55% of the abnormal red cell pyrimidine nucleotides accumulated in this disorder. This leads ultimately to the basophilic reticulum of normal reticulocytes, and basophilic granules or stippling in erythrocytes are composed of RNA. The granules result from the retarded RNA degradation secondary to inability to dephosphorylate and render diffusible pyrimidine degradation products of RNA. The hemolysis results from reversible inhibition of the pentose phosphate shunt by elevated levels of lead and the pyrimidine 5′-nucleotides.

Treatment of lead-induced 5′-nucleotidase deficiency can be performed with calcium disodium edetate intramuscularly in addition to stopping toxic lead intake. Blood lead levels show a linear decrease whereas erythrocyte pyrimidine 5′-nucleotidase levels show a linear increase in value.

BIBLIOGRAPHY

Becker, M. A., and Roessler, B. J. (1995). Gout and hyperuricemia. *In* "The Metabolic and Molecular Bases of Inherited Disease" (C. R. Scriver, A. L. Beaudet, W. S. Sly, and D. Valle, eds.), pp. 1655–1677. McGraw–Hill, New York.

Dieppe, P. A., Doherty, M., and MacFarlane, D. (1983). Crystal-related arthropathies. *Ann. Rheumatic Dis.* **42,** 1.

Fox, I. H. (1981). Metabolic basis for disorders of purine nucleotide degradation. *Metabolism* **30,** 616.

Fox, I. H. (1982). Disorders of purine and pyrimidine metabolism. *In* "Clinical Medicine" (J. A. Spittel, ed.). Harper & Row, Philadelphia.

Kelley, W. N., Fox, I. H., and Palella, T. D. (1989). Gout and related disorders of purine metabolism. *In* "Textbook of Rheumatology" (W. N. Kelley, E. D. Harris, Jr., S. Ruddy, and C. Sledge, eds.). Saunders, Philadelphia.

Palella, T. D., and Fox, I. H. (1989). Disorders of purine and pyrimidine metabolism. *In* "The Metabolic and Molecular Basis of Acquired Disease" (R. D. Cohen, K. G. M. M. Alberti, B. Lewis, and A. M. Denman, eds.). Bailliere Tindall, London.

Wyngaarden, J. B., and Kelley, W. N. (1976). "Gout and Hyperuricemia." Grune & Stratton, New York.

Quinolones

VINCENT T. ANDRIOLE
Yale University School of Medicine

GLOSSARY

Cinnolones Subgroup of the quinolone class of antimicrobial agents, identified by an additional nitrogen in the 2 position of the nucleus and also referred to as 2-aza-4-quinolones

DNA gyrase Bacterial enzyme that nicks double-stranded chromosomal DNA, introduces supercoils, and then seals the nicked DNA; also referred to as DNA topoisomerase and nicking–closing enzyme

Naphthyridines Subgroup of the quinolone class of antimicrobial agents, identified by an additional nitrogen in the 8 position of the nucleus and also referred to as 8-aza-4-quinolones

Pyrido-pyrimidines Subgroup of the quinolone class of antimicrobial agents, identified by additional nitrogens in the 6 and 8 positions of the nucleus and also referred to as 6,8-diaza-4-quinolones

Quinolones Completely synthetic class of highly effective antimicrobial agents

QUINOLONES ARE A CLASS OF ANTIMICROBIAL agents that are completely synthetic and highly effective in the treatment of many different types of infectious diseases, primarily those caused by bacteria. These agents also have variable activity against *Mycoplasma, Chlamydia, Rickettsia,* and some *Plasmodium* species, but are not known to be active against viruses and fungi. The quinolone antimicrobial agents are analogs of nalidixic acid, which was developed and introduced in 1962. The older analogs–pipemidic acid, oxolinic acid, and cinoxacin—were developed later. Shortly thereafter the development of the newer quinolones progressed rapidly and was spearheaded by the insertion of a fluorine at the 6 position in the basic nucleus. This chemical modification was observed to enhance and broaden the antibacterial activity of these agents and led to the discovery of newer 4-quinolones with antibacterial activities 1000 times that of nalidixic acid. More than 1000 quinolones and their analogs have been synthesized and evaluated for their activity against bacteria. The clinical importance of the newer quinolones is based on their extremely broad antibacterial spectrum, unique mechanism of action, good absorption from the gastrointestinal tract after oral administration, excellent tissue distribution, and low incidence of adverse reactions.

I. CHEMISTRY AND CLASSIFICATION

The quinolone antibacterial agents are all structurally similar compounds. Yet there are some differences in the basic nucleus so that they can be divided into four general groups: naphthyridines, cinnolines, pyrido-pyrimidines, and quinolines (Fig. 1). A common skeleton, 4-oxo-1,4-dihydroquinolone, more commonly

ENCYCLOPEDIA OF HUMAN BIOLOGY, Second Edition, VOLUME 7. Copyright © 1997 by Academic Press. All rights of reproduction in any form reserved.

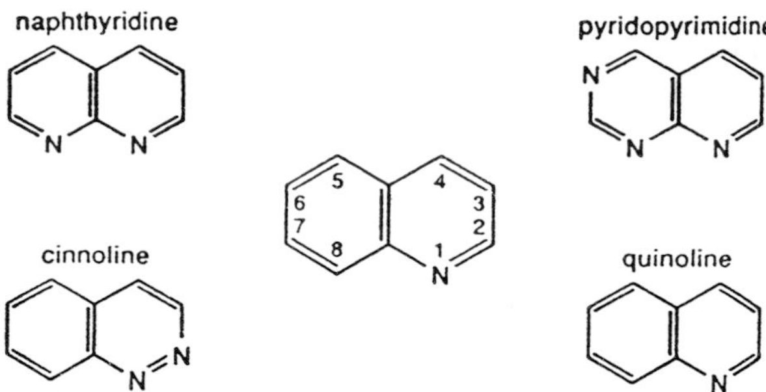

FIGURE 1 Chemical structures of the four general groups of the 4-quinolones and its system of ring numbering. [Reprinted with permission from V. T. Andriole (1989). Quinolones. *In* "Principles and Practice of Infectious Diseases" (G. L. Mandell, R. G. Douglas, and J. E. Bennett, eds.), 3rd Ed. Churchill Livingstone, New York.]

called 4-quinolone, is produced by the addition of an oxygen at the 4 position in the basic nucleus. The naphthyridines (nalidixic acid, enoxacin, tosufloxacin, and trovafloxacin), with an additional nitrogen in the 8 position, are 8-aza-4-quinolones. The cinnoline cinoxacin, with an additional nitrogen in the 2 position, is a 2-aza-4-quinolone. The pyrido-pyrimidines (pipemidic and piromidic acids), with additional nitrogens in the 6 and 8 positions, are 6,8-diaza-4-quinolones (Fig. 2). All of the other highly active agents (amifloxacin, ciprofloxacin, clinafloxacin, fleroxacin, levofloxacin, lomefloxacin, norfloxacin, ofloxacin, oxolinic acid, pefloxacin, sparfloxacin, and temafloxacin) are classified as 4-quinolones (Fig. 3). Numerous additional compounds have been synthesized and are undergoing development.

The 4-quinolones have certain common structural features that include a carboxyl group at the 3 position and a piperazine ring (except oxolinic acid) at the 7 position of the quinolone nucleus. Substitutions at the 1 and 8 positions of the quinolone or the naphthyridine nucleus and the para position of the piperazine ring, plus the introduction of a fluorine at the 6 position, are responsible for differences in the *in vitro* activity and pharmacologic properties of the newer 4-quinolone antimicrobial agents (see Figs. 2 and 3).

The 1,8-naphthyridine derivatives include: nalidixic acid (1-ethyl-7-methyl-1,8-naphthyridine-4-one-3-carboxylic acid), which is only slightly soluble in water but is soluble in dilute alkali and is stable in urine, marketed as Negram (Winthrop, New York); and enoxacin [1-ethyl-6-fluoro-1,4-dihydro-4-oxo-7-(1-piperazinyl)-1,8-naphthyridine-3-carboxylic acid], marketed as Penetrex (Rhone Poulenc Rorer, Fort Washington, Penn.). Oxolinic acid (5-ethyl-5,8-dihydro-8-oxo-1,3-dioxolo [4,5-*g*]quinoline-7-carboxylic acid), a quinolone derivative, is a crystalline substance that is a weak organic acid. It is no longer available for clinical use. Cinoxacin (1-ethyl-1,4-dihydro-4-oxo-1,3-dioxolo[4,5-*g*]cinnoline-3-carboxylic acid) is a yellow-white crystalline solid with a pK_a of 4.7. It is insoluble in water, poorly soluble in lipids, but is soluble in alkaline solution. It is marketed as Cinobac (Oclassen, San Rafael, Calif., and Biocraft, Fairlawn, N.J.). Norfloxacin [1-ethyl-6-fluoro-1,4-dihydro-4-oxo-7-(1-piperazinyl)-3-quinolone carboxylic acid] is a yellow-white crystalline solid only slightly soluble in water and is marketed as Noroxin (Merck, West Point, Penn., and Roberts, Eatontown, N.J.). Ciprofloxacin [1-cyclopropyl-6-fluoro-1,4-dihydro-4-oxo-7-(1-piperazinyl)-3-quinolone carboxylic acid hydrochloride] is a light yellow crystalline substance slightly soluble in water and is marketed as Cipro (Bayer, West Haven, Conn.). Other newer quinolones include: ofloxacin [9-fluoro-2,3-dihydro-3-methyl-10-(4-methyl-1-piperazinyl)-7-oxo-7*H*-pyrido[1,2,3-*de*]-1,4-benzoxacinc-6-carboxylic acid], which is marketed as Floxin (McNeil, Springhouse, Penn.); fleroxacin [6,8-difluoro-1-(2-fluoro-ethyl)-1,4-dihydro-7-(4-methyl-1-piperazinyl)-4-oxo-3-quinoline carboxylic acid] (Roche, Nutley, N.J.); lomefloxacin [1-ethyl-6,8-difluoro-1,4-dihydro-7-(3-methyl-1-piperazinyl)-4-oxo-3-quinoline carboxylic acid], which is marketed as Maxaquin (Searle, Skokie, Ill.); and pefloxacin [1 - ethyl - 6 - fluoro - 1,4 - dihydro - 7 - (4-methyl-1-piperazinyl)-4-oxo-3-quinoline carboxylic acid] (Bellon/Dianippon).

Many other newer quinolone compounds are currently under various stages of development. Some of these include: trovafloxacin [7-(1*a*,5*a*,6*a*)-6 - amino - 3 - azabicyclo(3.1.0)hex - 3 - yl - 6 - fluoro - 1 - (2,4 - difluorophenyl) - 1,4 - dihydro - 4 - oxo - 1,8 - naphthyridine-3-carboxylic acid] (Pfizer, New York); sparfloxacin [5-amino-1-cyclopropyl-7-(*cis*-3,5-dimethyl-1-piperazinyl)-6,8-difluoro-1,4-dihydro-4-oxo-3-quinoline carboxylic acid] (Rhone Poulenc Rorer); clinafloxacin [7-(3-amino-1-pyrrolidinyl)-8-chloro-1-cyclopropyl-6-fluoro-1,4-dihydro-4-oxo-3-quinoline carboxylic acid (Park Davis, Ann Arbor, Mich.); and levofloxacin [(−)-(*S*)-9-fluoro-2,3-dihydro-3-methyl-10-(4-methyl-1-piperazinyl)-7-oxo-7*H*-pyrido[1,2,3-*de*][1,4]-benzoxazine-6-carboxylic acid] (Ortho, Raritan, N.J.). [*See* Antimicrobial Drugs.]

II. MECHANISM OF ACTION

The molecular basis for the potent antibacterial effects of the newer quinolones has not yet been determined

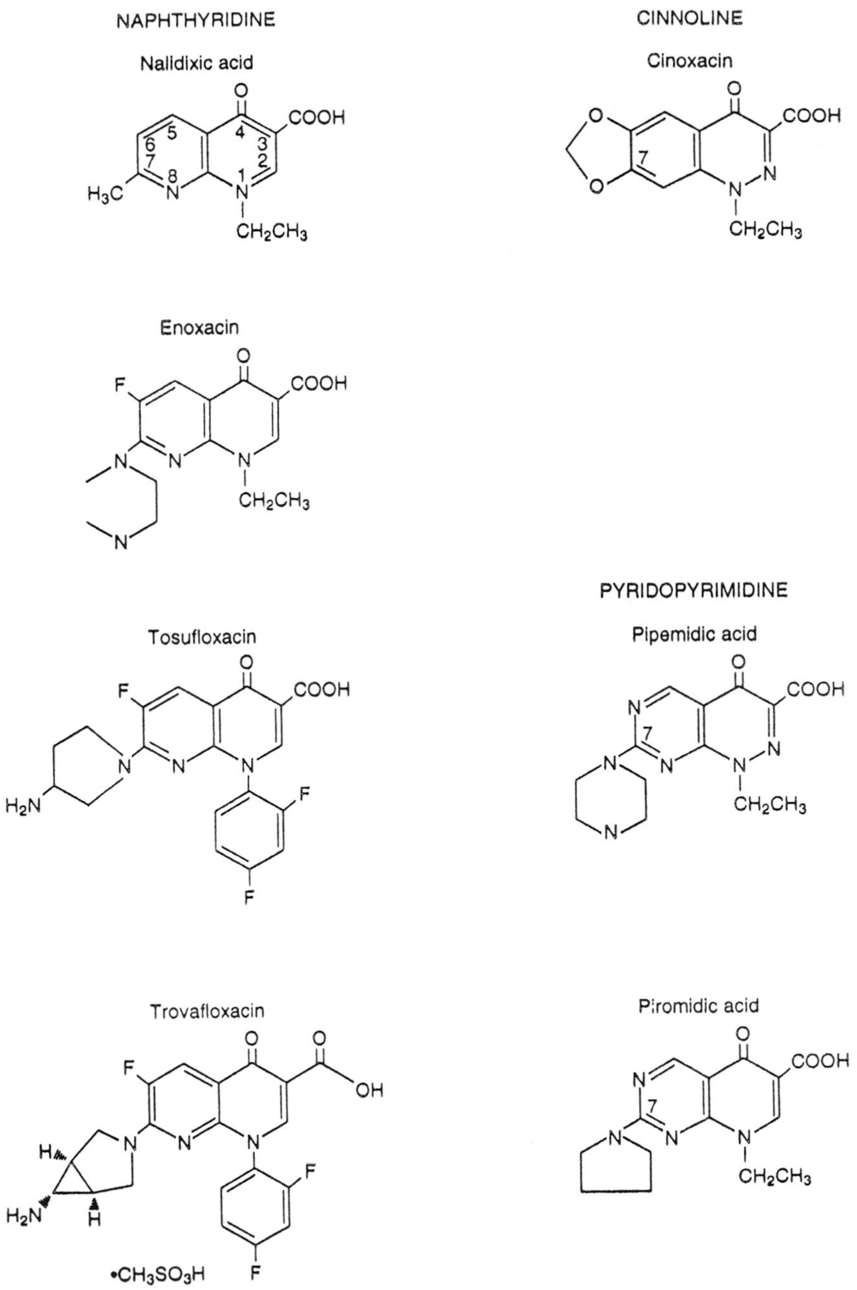

NAPHTHYRIDINE

Nalidixic acid

CINNOLINE

Cinoxacin

Enoxacin

PYRIDOPYRIMIDINE

Tosufloxacin

Pipemidic acid

Trovafloxacin

Piromidic acid

FIGURE 2 Chemical structures of naphthyridine, cinnoline, and pyrido-pyrimidine derivatives. [Reprinted with permission from V. T. Andriole (1996). Quinolones. *In* "Infectious Diseases" (S. L. Gorbach, J. G. Bartlett, and N. R. Blacklow, eds.), 2nd Ed. W.B. Saunders Co., Philadelphia.]

definitively. However, DNA synthesis is rapidly inhibited in susceptible bacterial cells, resulting in cell death, but not in mammalian cells. Inhibition of DNA synthesis by the quinolones is reversible. Also, the bactericidal activity of the quinolones is reduced significantly if RNA or protein synthesis is inhibited. All 4-quinolones are bactericidal. However, these drugs have a single concentration that is most bactericidal, so that higher or lower concentrations produce less bacterial death. This paradoxical effect of decreased killing at greater concentrations is most likely caused by a dose-dependent inhibition of RNA synthesis.

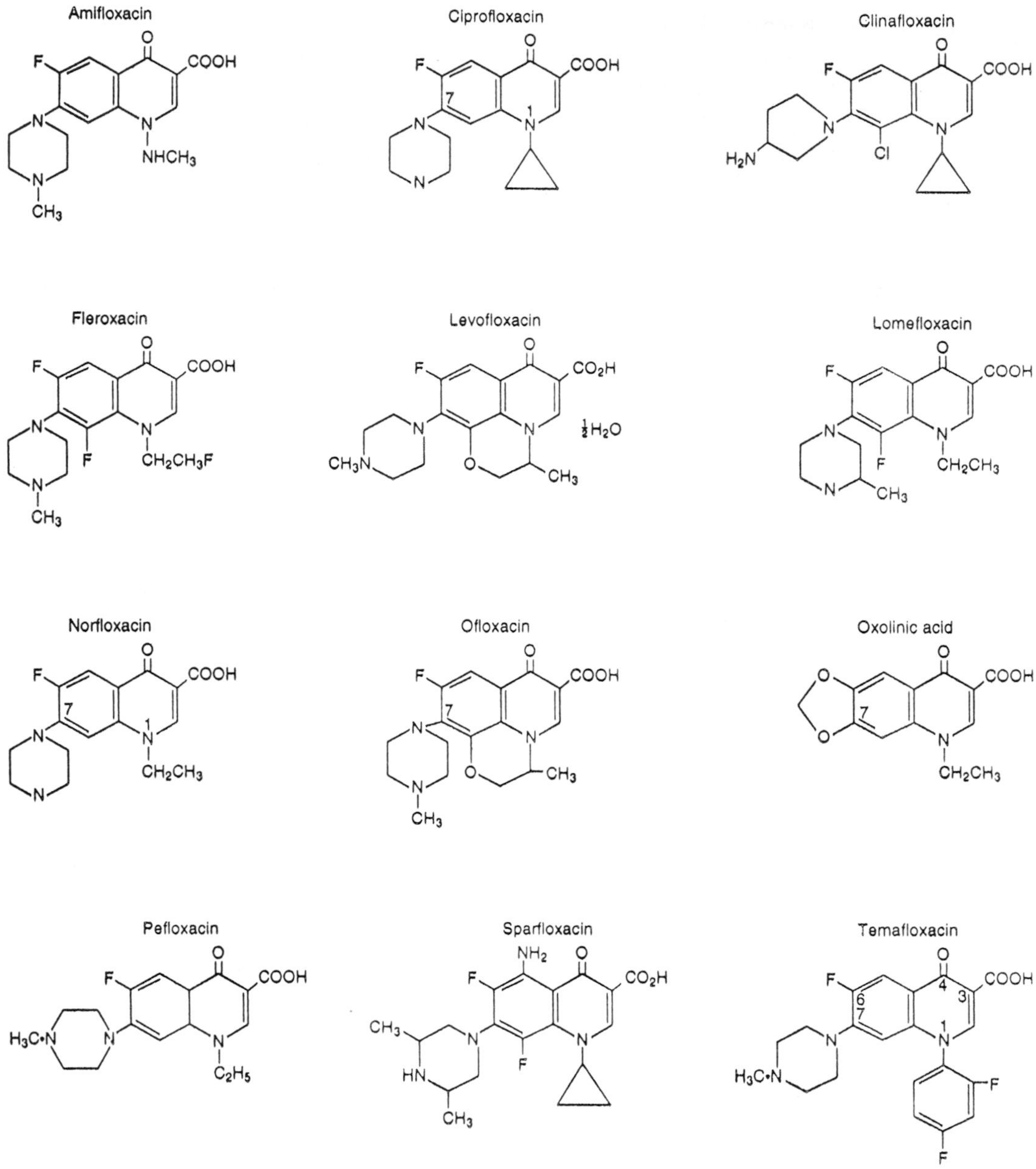

FIGURE 3 Chemical structures of the 4-quinolones that are derivatives of the quinolone nucleus. [Reprinted with permission from V. T. Andriole (1996). Quinolones. *In* "Infectious Diseases" (S. L. Gorbach, J. G. Bartlett, and N. R. Blacklow, eds.), 2nd Ed. W. B. Saunders Co., Philadelphia.]

The mechanism of action of the quinolones is by inhibition of DNA topoisomerases (gyrases), of which four subunits (two A monomers and two B monomers) have been defined. The topoisomerases, which have been found in every organism examined, supercoil strands of bacterial DNA in the bacterial cell. Each chromosomal domain is transiently nicked during supercoiling, which results in single-stranded DNA. When supercoiling is completed, the single-stranded DNA state is abolished by an enzyme that seals the nicked DNA. The sealing action of this enzyme is inhibited specifically by the quinolones. Thus the enzyme, termed DNA gyrase or topoisomerase II (nicking–closing enzyme), nicks double-stranded chromosomal DNA, introduces supercoils, and seals the nicked DNA. The A subunits are thought to introduce the nicks, the B subunits to supercoil, and then the A subunits seal the nick that they produced initially. Quinolones bind to the DNA–DNA gyrase complex and also inhibit topoisomerase IV. Some quinolones may kill by more than one mechanism. Although nalidixic acid prevents the A subunits from sealing the nicks in chromosomal DNA, newer quinolones may affect both the A and B subunits of DNA gyrase since mutations that affect the B subunit change the bacterial sensitivity to the 4-quinolones. The identification of DNA gyrase has provided the opportunity to develop new quinolone compounds that may have increased activity against DNA gyrase.

III. ANTIMICROBIAL ACTIVITY

Nalidixic acid has greater antimicrobial activity against gram-negative rods than against gram-positive bacteria (Table I). It is active against most Enterobacteriaceae, including strains of *Escherichia coli, Proteus mirabilis,* other *Proteus* species, *Klebsiella* species, *Enterobacter* species, and other coliform bacteria at concentrations (16 mg/ml or lower) that are easily achieved in the urine. Some strains of *Salmonella, Shigella,* and *Brucella* may be sensitive, whereas *Pseudomonas* and *Serratia* species are resistant. Bacterial resistance to nalidixic acid may develop by exposure to increasing concentrations of the drug *in vitro* and in patients during treament of bacteriuria. Gram-positive bacteria (e.g., *Staphylococcus aureus, Streptococcus pneumoniae,* and *Streptococcus faecalis*) are resistant to nalidixic acid.

Cinoxacin has an antibacterial spectrum of activity similar to that of nalidixic acid so that it is also active against most strains of gram-negative bacteria that cause urinary tract infections, but has negligible activity against *Pseudomonas aeruginosa* and gram-positive cocci such as *S. aureus* and *Staphylococcus saphrophyticus,* and streptococci and enterococci species. Cinoxacin also has *in vitro* activity against *Alcaligenes, Acinetobacter, Moraxella,* and *Hemophilus* species, as well as against *Clostridium perfringens, C. tetani, Neisseria meningitides,* and *Pseudomonas pseudomallei.* Cross-resistance among bacterial isolates develops against cinoxacin and nalidixic acid.

The *in vitro* antimicrobial activity of selected 4-quinolones is shown in Table I. The newer quinolones are very active against aerobic enteric gram-negative rods and against aerobic gram-negative cocci. Most of the newer quinolones also have activity against staphylococci. They have moderate activity against *P. aeruginosa.* The currently available quinolones have variable activity against streptococci and poor activity against anaerobes. However, a few agents under development, for example, trovafloxacin, clinafloxacin, and sparfloxacin, do have significant activity against streptococci, and trovafloxacin has excellent activity against gram-negative anaerobes.

The newer currently available 4-quinolones (ciprofloxacin, enoxacin, fleroxacin, lomefloxacin, and ofloxacin) and others under development (levofloxacin, sparfloxacin, and trovafloxacin) have excellent activity against *Legionella pneumophila.* Although ciprofloxacin and ofloxacin are active against *Mycoplasma pneumoniae* and *Chlamydia pneumoniae,* levofloxacin, sparfloxacin, and trovafloxacin are more active. The newer quinolones are also active against *Chlamydia trachomatis, Mycoplasma hominis,* and *Ureaplasma urealyticum.*

The 4-quinolones have variable activity against *Mycobacteria.* Although they are not very active against *M. avium-intracellulare,* some of the newer compounds are active against *M. tuberculosis, M. kansasii,* and *M. fortuitum,* but have lesser activity against *M. chelonae.* Also, sparfloxacin and ofloxacin have activity against *M. leprae* and have been used successfully in patients with lepromatous leprosy.

IV. MECHANISMS OF BACTERIAL RESISTANCE

Serial exposure of both gram-positive and gram-negative bacteria to subinhibitory concentrations of the quinolones *in vitro* leads to the selection of bacterial variants with reduced susceptibility to the drug. The resulting strains may exhibit cross-resistance to other quinolones. The mechanism of resistance usually involves either (1) mutations in the gene coding for DNA gyrase so that there is reduced quinolone affinity for the A or B subunit or (2) mutations that change the outer membrane porins. Relative resistance to antibiotics unrelated to the quinolones has been observed when reduced susceptibility to the quinolones is caused by reduced outer membrane porin F activity.

Although the exact mechanism for the development of bacterial resistance to the quinolones has not been determined, recent work suggests that high-level resistance occurs when normally occurring amino acids in specific positions of subunit A are replaced by other amino acids, for example, when serine in the 83 position is replaced by tryptophan.

TABLE I
In Vitro Antimicrobial Activity of Selected 4-Quinolones

Organism	MIC$_{90}$ (range) (ug/ml)			
	Nalidixic acid	Ciprofloxacin	Enoxacin	Norfloxacin
Gram-negative aerobes				
Escherichia coli	8 (4–128)	0.03 (0.015–0.06)	0.5 (0.25–1)	0.125 (0.06–0.5)
Klebsiella pneumoniae	8 (1–128)	0.125 (0.06–0.25)	0.5	0.25 (0.125–1)
Enterobacter spp.	32 (4–128)	0.125 (0.03–0.5)	0.5 (0.25–4)	0.5 (0.125–2)
Citrobacter spp.	8 (4–>100)	0.03 (0.03–0.06)	0.5	0.25 (0.125–0.5)
Serratia marcescens	>128 (16–>256)	0.25 (0.06–0.5)	2 (0.5–4)	1 (0.5–8)
Shigella spp.	4	0.03 (0.015–0.06)	0.125	0.06 (0.03–0.125)
Salmonella spp.	8 (4–8)	0.015 (≤0.015–0.03)	0.25 (0.125–0.25)	0.125 (0.06–0.125)
Proteus mirabilis	16 (4–32)	0.06 (0.03–0.125)	0.5 (0.25–1)	0.25 (0.125–0.5)
P. vulgaris	8 (4–16)	0.12 (0.03–0.25)	0.25 (0.25–0.5)	0.125 (0.06–0.125)
Morganella morganii	8 (2–8)	0.015 (0.015–0.03)	0.125 (0.03–0.25)	0.125 (0.03–0.25)
Brucella melitensis	—	0.5 (0.25–0.5)	—	—
Legionella spp.	0.25 (0.12–0.25)	(0.03–0.125)	0.2	(0.125–0.5)
Pseudomonas cepacia	16	0.5 (0.12–2)	25 (16–25)	8 (8–50)
P. aeruginosa	≥128	0.5 (0.25–1)	4 (2–8)	2 (0.06–8)
Haemophilus influenza	1 (1–2)	0.03 (0.007–0.06)	0.125 (0.06–0.25)	0.06 (0.03–0.125)
Neisseria gonorrhoeae	1 (1–2)	≤0.015 (≤0.015)	0.03 (0.015–0.06)	0.06 (0.015–0.125)
N. meningitidis	2	0.004	0.06	0.03
Acinetobacter spp.	32–256	0.5 (0.015–1)	1–2	8–64
Aeromonas spp.	0.5	≤0.008 (≤0.008–0.05)		0.06
Campylobacter jejuni	8	(0.12–0.78)	(1–32)	(0.25–2)
Moraxella catarrhalis	2	0.06 (0.007–0.06)	0.06	0.4
Providencia rettgeri	16	1 (≤0.008–2)	1 (0.5–6.25)	2 (0.25–3.1)
P. stuartii	32	0.25 (≤0.008–1)	1–2	2 (≤0.25–2)
Xanthomonas maltophilia	16	4 (1–4)	8 (3–16)	4 (4–>64)
Yersinia enterocolitica	4 (1–8)	0.06 (0.015–0.06)	0.12–0.25	≤0.12
Hafnia alvei	—	0.03 (0.015–0.06)	—	—
Helicobacter pylori	—	0.31 (0.039–0.31)	—	—
Gram-positive aerobes				
Enterococcus faecalis	>128 (>128)	4 (0.5–4)	8 (8–16)	8 (4–32)
E. faecium	>64	4 (2–8)	32	≥12.5
Staphylocci (coag neg)	128 (32–256)	0.5 (0.06–0.5)	1 (0.5–1)	1 (0.25–2)
Staphylocci (coag neg/meth res)	>128 (16–>128)	64 (0.12–>64)	—	—
S. aureus	64 (32–128)	0.5 (0.12–2)	2 (0.5–2)	2 (0.5–4)
S. aureus (meth susc)	64 (8–64)	1 (0.25–2)	—	—
S. aureus (meth res)	>128 (32–>128)	64 (0.25–>64)	—	—
Streptococcus (viridans group)	>128 (128–>128)	4 (0.12–8)	—	—
S. pyogenes	>100	1 (0.25–2)	>8	4 (2–16)
S. agalactiae	>128	2 (0.5–4)	>8	16 (4–16)
S. pneumoniae	>128 (128–>128)	4 (0.5–4)	16 (8–16)	16 (4–>16)
Listeria monocytogenes	>64	2 (0.5–4)	8–16	8 (4–16)
Corynebacterium spp.	0.06 (0.03–0.12)	1 (0.05–128)	8 (4–>128)	4 (4–>128)
Anaerobic bacteria and other organisms				
Peptostreptococcus spp.	>128 (64–128)	4 (0.25–16)	—	—
Clostridium spp.	>128 (4–>128)	0.5 (0.25–2)	(16–32)	(2–8)
C. difficile	>128 (64–>128)	12.5 (8–25)	128	128
Prevotella spp.	128 (64–128)	16 (0.25–16)	—	—
Mycobacterium fortuitum	—	0.3	—	2
M. tuberculosis	128 (16–128)	1	>5	8
M. avium complex	>256 (128–>256)	16	>256	≥16
M. chelonae	—	8	—	>16
M. kansasii	—	1	≥5	8
Bacteriodes spp.	512	16 (4–>16)	16	128
B. fragilis	128 (128–>128)	16 (4–>16)	32	>128
Fusobacterium spp.	128 (32–>128)	8 (2–8)	32	16
Anaerobic gram-positive cocci	(256–512)	(2–6.25)	(2–8)	(16–64)
Chlamydia pneumoniae	(>64)	(1–2)	—	—
C. psittaci	—	1	—	—
C. trachomatis	(>64)	(1–1.56)	6.3	≥16
Mycoplasma hominis	>256	(0.5–2)	8	(8–16)
M. pneumoniae	—	(0.78–2)	8	12
Ureaplasma urealyticum	—	4	—	—

TABLE I (*Continued*)

Organism	MIC$_{90}$ (range) (ug/ml)			
	Ofloxacin	Pefloxacin	Lomefloxacin	Fleroxacin
Gram-negative aerobes				
Escherichia coli	0.125 (0.06–0.25)	0.125 (0.125–0.25)	0.2 (0.06–1.0)	0.1 (0.03–2)
Klebsiella pneumoniae	0.25 (0.03–0.25)	0.5	1 (0.2–6.25)	0.5 (0.12–6.25)
Enterobacter spp.	0.5 (0.125–1)	0.5 (0.25–1)	0.5 (≤0.25–1)	0.12–0.25
Citrobacter spp.	0.5 (0.03–2)	0.5	0.5 (0.12–25)	0.12 (≤.06–25)
Serratia marcescens	1 (0.25–2)	1 (0.25–2)	2 (0.25–25)	0.5 (0.25–25)
Shigella spp.	0.125 (0.06–0.125)	0.125	(0.06–0.25)	≤0.125
Salmonella spp.	0.125 (0.06–0.125)	0.125 (0.06–0.25)	0.25	≤0.12–0.25
Proteus mirabilis	0.25 (0.25–0.5)	0.5 (0.25–1)	0.5–1	0.5 (≤0.12–0.5)
P. vulgaris	0.12 (0.03–1)	0.25	0.5 (0.25–1)	0.12 (≤0.12–0.25)
Morganella morganii	0.125 (0.125–0.25)	0.25 (0.25–0.5)	0.25 (0.25–12.5)	0.12 (<0.06–12.5)
Brucella melitensis	2	—	—	—
Legionella spp.	0.015 (0.008–0.15)	—	≤0.06	≤0.06
Pseudomonas cepacia	3.1 (3–32)	4 (2–8)	4 (4–>50)	2 (2–>50)
P. aeruginosa	4 (0.5–4)	—	16 (16)	4 (4–16)
Haemophilus influenza	0.03 (0.03–0.06)	0.06 (0.03–0.06)	≤0.06–0.12	(≤0.06–1)
Neisseria gonorrhoeae	0.03 (0.015–0.06)	0.06 (0.03–0.06)	0.12	0.2
N. meningitidis	0.015	0.03	≤0.06–0.42	(0.03–0.25)
Acinetobacter spp.	1 (0.12–2)	(1–8)	4	0.5–4 (0.5–32)
Aeromonas spp.	0.03–0.5 (0.03–10)	0.03	0.12	0.12–0.25 (0.12–1)
Campylobacter jejuni	(0.12–2)	0.5	(0.125–1)	0.5
Moraxella catarrhalis	(0.06–0.50)	0.25	≤0.1–1	(0.25–2)
Providencia rettgeri	1 (0.25–2)	0.5	4 (1.6–6.2)	0.5 (0.12–1)
P. stuartii	1 (0.6–4)	4	1 (1–4)	1 (0.5–2)
Xanthomonas maltophilia	4 (0.5–4)	4.0	8 (8–25)	3 (3–25)
Yersinia enterocolitica	(0.06–0.25)	0.25	(≤0.06–0.25)	(≤0.06–2)
Hafnia alvei	0.12 (0.15–0.25)	—	—	—
Helicobacter pylori	—	—	—	—
Gram-positive aerobes				
Enterococcus faecalis	4 (2–6.2)	4–8	8 (4–16)	8 (8–>16)
E. faecium	6.2 (2–16)	—	8	8
Staphylocci (coag neg)	0.5 (0.25–0.5)	0.5 (0.25–1)	1 (0.5–1)	1 (0.25–8)
Staphylocci (coag neg/meth res)	64 (0.12–>64)	—	—	4 (0.5–64)
S. aureus	0.5 (0.12–1)	0.5 (0.25–1)	1 (0.5–2)	1 (0.25–1)
S. aureus (meth susc)	0.5 (0.25–32)	—	—	32 (0.2–64)
S. aureus (meth res)	32 (0.25–>128)	—	—	64 (32–64)
Streptococcus (viridans group)	4 (1–8)	—	—	—
S. pyogenes	2 (0.5–4)	8 (8–16)	8 (4–12.5)	8 (4–12.5)
S. agalactiae	2 (1–4)	32	16 (8–32)	8 (≥8)
S. pneumoniae	4 (0.5–4)	12 (8–16)	8 (2–16)	8 (8–25)
Listeria monocytogenes	2 (1–4)	6–8	(6.2–8)	8 (4–>16)
Corynebacterium spp.	1 (0.5–64)	8 (8–>128)	>12.5	2 (1–32)
Anaerobic bacteria and other organisms				
Peptostreptococcus spp.	8 (0.25–32)	—	—	—
Clostridium spp.	8 (0.25–16)	(1–8)	(2–16)	(2–32)
C. difficile	8 (8–16)	64	≥32	16–32
Prevotella spp.	16 (0.5–16)	—	—	—
Mycobacterium fortuitum	(1–3.1)	2	—	≤0.05
M. tuberculosis	1 (0.12–4)	8	4	≤0.05
M. avium complex	32 (4–64)	>64	—	16
M. chelonae	>20	>64	—	>32
M. kansasii	(1–3.1)	4	—	≥0.5
Bacteriodes spp.	(2–32)	—	(8–32)	(2–64)
B. fragilis	(2–12.5)	16	(8–64)	≥16
Fusobacterium spp.	4 (0.5–8)	32	16	16
Anaerobic gram-positive cocci	(2–8)	16	(4–25)	(8–12.5)
Chlamydia pneumoniae	0.25	—	4	2
C. psittaci	0.5	—	—	—
C. trachomatis	(1–1.56)	—	(2–3.1)	(1.5–6.3)
Mycoplasma hominis	(1–2)	4	2	2
M. pneumoniae	(0.78–2)	4	(4–8)	4
Ureaplasma urealyticum	(2–4)	—	—	—

continues

TABLE I *(Continued)*

Organism	MIC$_{90}$ (range) (ug/ml)			
	Sparfloxacin	Trovafloxacin	Clinafloxacin	Levofloxacin
Gram-negative aerobes				
Escherichia coli	0.03 (≤0.015–0.12)	0.06 (≤0.008–0.12)	≤0.03 (≤0.03–0.06)	0.10
Klebsiella pneumoniae	0.012 (≤0.008–0.25)	0.05 (0.05–1)	0.13 (≤0.03–0.13)	0.25
Enterobacter spp.	0.06–.25 (0.015–1)	0.05	0.13 (≤0.03–0.25)	(0.18–0.39)
Citrobacter spp.	0.06–0.5 (≤0.015–1)	0.1 (0.06–0.25)	—	(0.03–0.78)
Serratia marcescens	1 (0.25–1)	0.5 (0.05–2)	—	3.13
Shigella spp.	0.06 (0.015–0.12)	0.06 (0.03–0.05)	—	—
Salmonella spp.	0.06 (0.015–0.12)	0.05	≤0.03 (≤0.03)	0.12
Proteus mirabilis	0.5 (0.06–.5)	(0.2–2)	≤0.03 (≤0.03)	0.19
P. vulgaris	0.5 (0.12–1)	0.5	—	0.20
Morganella morganii	0.25 (≤0.015–0.5)	0.25	≤0.03 (≤0.03)	0.12
Brucella melitensis	0.25	—	0.06 (0.03–0.06)	—
Legionella spp.	≤0.06	—	—	(0.05–0.12)
Pseudomonas cepacia	(1–8)	3 (0.5–32)	0.25 (≤0.03–0.25)	5.12
P. aeruginosa	1 (0.12–2)	0.25	0.25 (≤0.03–0.5)	3.13
Haemophilus influenza	≤0.015 (≤0.015–0.03)	0.03 (0.03–0.12)	≤0.03 (≤0.03)	0.02
Neisseria gonorrhoeae	≤0.015 (≤0.015–0.03)	0.25 (≤0.001–0.25)	—	0.02
N. meningitidis	≤0.06	0.008 (0.004–0.015)	—	—
Acinetobacter spp.	0.25 (0.015–0.5)	8	—	—
Aeromonas spp.	0.12 (≤0.015–0.25)	0.03	—	—
Campylobacter jejuni	(0.1–0.12)	—	—	—
Moraxella catarrhalis	(0.01–0.12)	0.03 (0.007–0.03)	(≤0.03)	0.10
Providencia rettgeri	0.5 (0.12–1)	0.5	—	—
P. stuartii	2 (0.12–8)	2	—	0.39
Xanthomonas maltophilia	0.5 (0.03–1)	2 (0.2–2)	—	3.13
Yersinia enterocolitica	0.06 (≤0.015–0.25)	(0.03–0.05)	≤0.03 (≤0.03)	—
Hafnia alvei	0.25 (0.015–0.25)	0.06 (0.008–0.06)	—	—
Helicobacter pylori	(0.25–4)	0.25 (0.31–0.25)	—	—
Gram-positive aerobes				
Enterococcus faecalis	0.5 (0.25–0.5)	2	0.25 (0.01–1)	1.56
E. faecium	1 (0.5–1)	2	0.5 (0.015–0.5)	3.13
Staphylocci (coag neg)	0.125 (0.03–8)	1–4 (0.015–16)	0.03 (≤0.007–0.25)	0.78
Staphylocci (coag neg/meth res)	4 (0.06–>16)	4 (0.015–16)	—	2.12
S. aureus	0.06 (0.03–2)		0.03 (≤0.007–0.125)	
S. aureus (meth susc)	0.25 (0.03–0.25)	0.06 (0.015–4)	—	0.39
S. aureus (meth res)	0.25 (0.03–1)	2 (0.015–8)	—	0.78
Streptococcus (viridans group)	0.25 (0.03–0.5)	0.25 (0.015–0.25)	0.06 (0.06–0.13)	(≤0.007–0.06)
S. pyogenes	0.5 (0.12–0.5)	0.12	(≤0.03–0.06)	1.56
S. agalactiae	0.5 (0.12–0.5)	0.25	(≤0.06–0.13)	2 (0.5–2)
S. pneumoniae	0.5 (0.064–0.5)	0.25 (0.064–0.5)	0.06 (≤0.008–0.06)	1.56
Listeria monocytogenes	2 (0.12–2)	0.25 (0.12–0.25)	—	—
Corynebacterium spp.	0.25 (0.25–64)	(0.03–>32)	—	—
Anaerobic bacteria and other organisms				
Peptostreptococcus spp.	4 (0.25–4)	1 (0.06–2)	0.5 (0.015–0.5)	5.56
Clostridium spp.	0.25 (0.015–2)	(<0.06–4)	0.12 (0.06–0.12)	0.78
C. difficile	6.25	4 (0.5–4)	0.5 (0.03–1)	6.25
Prevotella spp.	8–16	2 (0.06–4)	—	8.0
Mycobacterium fortuitum	1.56	—	—	2.78
M. tuberculosis	0.2	32 (2–64)	—	0.32
M. avium complex	12.5	64 (4–64)	—	16.50
M. chelonae	6.25–>100	—	—	—
M. kansasii	—	—	—	2.39
Bacteriodes spp.	2	0.5 (0.5–2)	—	5.12
B. fragilis	2 (1–2)	0.5 (0.12–4)	—	4.0
Fusobacterium spp.	2	2 (0.12–2)	—	—
Anaerobic gram-positive cocci	—	—	—	—
Chlamydia pneumoniae	(0.01–0.25)	0.12	—	0.5
C. psittaci	0.03	—	—	—
C. trachomatis	(0.05–0.063)	0.06	—	—
Mycoplasma hominis	≤0.06 (0.01–0.06)	0.03	—	—
M. pneumoniae	(0.1–0.25)	0.18	—	0.50
Ureaplasma urealyticum	0.5	0.5	—	2.0

Since quinolones interfere with DNA gyrase activity, which is necessary for plasmid replication, quinolones were expected to promote loss of plasmids and to inhibit transfer of R factor-mediated resistance. However, recent work suggests that plasmid-mediated resistance may be possible, though rare.

V. PHARMACOLOGY

All of the quinolones are well absorbed from the gastrointestinal tract after oral administration and most are excreted by the kidney into the urine. Some are metabolized in the liver.

After oral administration and absorption from the gastrointestinal tract, nalidixic acid is rapidly metabolized in the liver to biologically active hydroxynalidixic acid and inactive monoglucuronide conjugates, which, along with the parent compound, are rapidly excreted by the kidney into the urine. Plasma levels of 20–50 mg/ml may be attained 2 hr after a single oral dose of 1 g of the drug. Nalidixic acid does not accumulate in tissues even after prolonged administration, and the kidney is the only organ in which tissue concentrations may exceed plasma levels. The drug does not diffuse into prostatic fluid, but does appear in mother's milk and may be harmful to the newborn. Excretion is almost completely via the kidney into the urine in concentrations of active drug, after a 0.5- to 1-g oral dose in adults, in the range of 25–250 mg/ml. Bactericidal levels of the drug are also attained in the urine of patients with moderate or advanced renal failure. Although increased toxicity has not been observed in these patients, nalidixic acid should be used cautiously in patients with advanced renal failure and in patients with liver disease, since conjugation of the drug may be impaired in this latter group.

Cinoxacin is rapidly and almost completely absorbed from the gastrointestinal tract after oral administration and has a serum half-life of approximately 1 hr, which may increase threefold in patients with creatinine clearances of less than 30 ml/min/1.73 m^2. Peak plasma levels occur in 2–3 hr and range from <4 to 14.8 mg/ml and from 2.8 to 28 mg/ml after an oral dose of 250 and 500 mg, respectively. Urine concentrations are high and bactericidal for susceptible microorganisms but are decreased in patients with impaired renal function. Cinoxacin concentrations in human prostatic tissue range from 0.6 to 6.3 mg/g, and concentrations in renal tissue exceed those in serum. Cinoxacin is excreted by both glomerular filtration and tubular secretion and probenecid inhibits cinoxacin excretion by the kidney. Although it is not known whether cinoxacin is excreted into human milk, the drug should not be used by nursing women because of the potential for serious adverse reactions in nursing infants.

Norfloxacin, after oral administration, is also well absorbed from the gastrointestinal tract, producing low plasma levels and high concentrations in the urine. Norfloxacin crystals are occasionally observed during microscopic examination of freshly voided urine collected after large doses of 1200 and 1600 mg, but crystalluria is not encountered at lower doses. Norfloxacin is not excreted into human milk in concentrations detectable by bioassay.

Ciprofloxacin, after oral administration, is also rapidly absorbed from the gastrointestinal tract and is the most thoroughly studied quinolone in all respects. Peak plasma levels occur in 1–1.5 hr and are approximately 2–3 mg/ml after an oral dose of 500 mg. Its extravascular penetration into tissues and other body compartments is better than or comparable to other, newer quinolones. Ciprofloxacin penetrates blister fluid well, with 57% of the serum concentration recoverable. It is not known whether ciprofloxacin is excreted into human milk. A parenteral preparation of ciprofloxacin is also available.

The pharmacokinetic properties of some of the newer quinolones are summarized in Table II. In general, the newer quinolones exhibit linear pharmacokinetics. Peak serum concentrations occur 13 hr after oral administration. Food and histamine H_2-receptor antagonists (ranitidine) delay absorption so that serum peaks appear later and are moderately lower. Absorption is also reduced by concomitant administration of magnesium or aluminum hydroxide antacids, and by other drugs that decrease peristalsis or delay gastric emptying time. The newer quinolones are not extensively bound to serum proteins. Their long serum half-life allows dosing once or twice daily. The newer quinolones undergo renal and hepatic metabolism. Renal elimination is by glomerular filtration and active tubular secretion, which is blocked by probenicid (except fleroxacin). The antibacterial activity of the quinolones is reduced at lower urinary pH values (5.5–6.0 versus 7.4). Pefloxacin undergoes extensive hepatic metabolism, followed by enoxacin and, to a lesser degree, norfloxacin, ciprofloxacin, and fleroxacin. Hepatic metabolism is least with lomefloxacin and ofloxacin. Biliary concentrations of ciprofloxacin, enoxacin, ofloxacin, and pefloxacin are two to eight times the simultaneous serum concentrations. Ciprofloxacin, ofloxacin, nalidixic acid, and pefloxacin are excreted into breast milk

TABLE II
Pharmacokinetic Properties of Selected Newer Quinolones

	Dose (mg)	Peak serum concentration (mg/liter)	Half-life (hr)	Protein binding (%)	Bioavailability (%)	Volume of distribution (liters)	Urinary excretion	
							Unchanged (%)	Metabolites (%)
Ciprofloxacin	500	2–3	3–4.5	35	85	250	30–60	10
Enoxacin	400	2–3	4–6	43	90	190	50–55	15
Fleroxacin	400	4–6	10	23	96	100	60–70	10
Levofloxacin	500	5.7	7.6	—	—	102	—	—
Lomefloxacin	400	3	8	14–25	>95	190	70	10
Norfloxacin	400	1.5	3–4.5	15	80	225	20–40	20
Ofloxacin	400	3.5–5	5–6	8–30	85–95	100	70–90	5–10
Pefloxacin	400	4–5	10–11	25	90	110	5–15	55
Sparfloxacin	400	1.2–1.6	15.2–20.6	45	>60%	322	<15	25
Trovafloxacin	100–300	1.4–4.3	7.1–9.6	70	60–85	90–110	8.8	—

of lactating women. Ciprofloxacin and norfloxacin have high concentrations inside human neutrophils, whereas pefloxacin penetrates poorly into neutrophils and alveolar macrophages. The tissue penetration of some of the newer quinolones is summarized in Table III. [*See* Pharmacokinetics.]

VI. TOXICITY AND ADVERSE REACTIONS

Oral quinolones are usually well tolerated. Adverse reactions include nausea, vomiting, diarrhea, and abdominal pain (gastrointestinal); pruritus, nonspecific

TABLE III
Penetration of Selected Quinolones into Body Fluids and Tissues[a]

	Ciprofloxacin	Norfloxacin	Ofloxacin	Enoxacin	Pefloxacin	Fleroxacin	Sparfloxacin	Trovafloxacin
Blister fluid	++++	++++	++++	++++	+++	++++	++++	+++
Saliva	++	++	+++	+++	+++	+++	+++	—
Bronchial secretions	++	—	+++	++++	++++	—	++++	—
Pleural fluid	+++	—	—	—	—	—	+++	—
Nasal secretions	+++	+++	++++	++++	+++	++++	++++	—
Tears	++	++	+++	++	+++	+++	+++	—
Sweat	+	+	++	++	++	++	+++	—
Cerebrospinal fluid	+	—	++	—	+++	—	++	++
Prostatic fluid	+++	++	++++	++	—	++	—	—
Ejaculate	+++++	—	++++	++++	—	++++	—	—
Vaginal secretions	—	—	—	—	—	—	+++++	—
Lung	++++	—	++++	+++++	++++	++++	+++++	+++++
Kidney	+++++	+++++	+++++	++++	—	—	—	—
Bone	++++	—	++	++	++	++++	—	—
Skin	++++	—	—	++++	—	—	—	—
Muscle	++++	—	—	++++	—	—	—	—
Fat	++++	—	—	+++	—	++	—	—

[a]Key: + = AUC ratios or concentration ratios <0.1; ++ = AUC ratios or concentration ratios 0.1–0.5; +++ = AUC ratios or concentration ratios 0.5–1; ++++ = AUC ratios or concentration ratios 1–4; +++++ = AUC ratios of concentration ratios >4. Adapted from F. Sorge, U. Jaehde, K. Naber, U. Stephan (1989). *Clin. Pharmacokinet.* **16**, S5. Reprinted by permission.

rashes, urticaria associated with eosinohilia, and edema of the extremities (dermatologic); photosensitivity reactions involving skin surfaces exposed to sunlight manifested as a sunburn or rarely as a bullous eruption (especially nalidixic acid, lomefloxacin, pefloxacin, and fleroxacin); blurred vision, diplopia, photophobia, abnormal accommodation, and changes in color perception, all of which disappear with cessation of therapy (ophthalmologic); and headaches, drowsiness, asthenia, giddiness, vertigo, syncope, restlessness, insomnia, tinnitus, sensory changes, grand mal seizures, and acute reversible toxic psychosis, as well as pseudotumor cerebri with intracranial hypertension, papilledema, and bulging fontanelles in infants and young children, which reverse after cessation of therapy, especially nalidixic acid (neurologic). Rarely, abnormal liver function tests, renal function values, and reduced hematocrit, hemoglobin, and leukocyte counts have been observed. Nalidixic acid has been associated, rarely, with cholestatic jaundice and with hemolytic anemia that sometimes is associated with glucose-6-phosphate dehydrogenase (G6PD)-deficient red blood cells. The use of nalidixic acid and cinoxacin in prepubertal children and during pregnancy is not recommended.

The newer fluoroquinolones—norfloxacin, ciprofloxacin, ofloxacin, enoxacin, and pefloxacin—are considered relatively safe agents. They have very similar side effects but in low incidences. Gastrointestinal side effects are the most frequent (0.8–6.8% of patients) and include nausea, vomiting, dyspepsia, epigastric/abdominal pain, anorexia, diarrhea, flatulence, and dry mouth. Antibiotic-associated colitis has been seen rarely.

Central nervous system side effects occur in 0.9–1.8% of patients. Mild reactions include headache, dizziness, tiredness, insomnia, faintness, agitation, listlessness, restlessness, abnormal vision, and bad dreams. Hallucinations, depressions, psychotic reactions, and grand mal convulsions (severe reactions) are rare and disappear when therapy is discontinued.

Skin and allergic reactions occur in 0.6–2.4% of patients and include erythema, urticaria, rash, pruritis, and photosensitivity reactions of skin surfaces exposed to sunlight. Hypotension, tachycardia, nephrotoxicity, thrombocytopenia, leukopenia, anemia, and transient elevations in liver enzymes have been observed rarely. Anthropathy, gait abnormalities, and articular cartilage lesions in weight-bearing joints in juvenile animals have been observed. Although anthropathy has not been observed in children who have been treated with quinolones, the newer quinolones have not been approved for use in pediatric patients in the United States. [*See* Antimicrobial Agents, Impact on Newborn Infants.]

VII. DRUG INTERACTIONS

The quinolones are known to interact with a variety of other compounds. Bioavailability of some quinolones is reduced after oral administration with antacids containing calcium, aluminum, or magnesium, with sucralfate, and with iron or multivitamins containing zinc, resulting in low systemic levels of the quinolones. These agents should not be taken within the 2-hr period before or after quinolone administration.

Nalidixic acid–glucuronide conjugates may produce a false-positive reaction for urine glucose when tested with Benedict's solution, but not with glucose oxidase test strips. Nitrofurantoin interferes with the therapeutic action of nalidixic acid.

Some of the newer quinolones increase theophylline plasma concentrations, for example, enoxacin (111%), ciprofloxacin (23%), pefloxacin (20%), and ofloxacin (12%). Theophylline doses probably should be halved in patients receiving enoxacin. No routine reduction in theophylline dose is recommended for patients receiving ciprofloxacin, ofloxacin, or pefloxacin, but theophylline levels should be monitored.

Caffeine clearance is interfered with by the newer quinolones. Enoxacin increases the plasma concentration of caffeine by 41% and reduces the clearance by 78%. Ciprofloxacin increases the half-life of caffeine only modestly (15%), and ofloxacin only minimally.

The quinolones may interact to varying degrees with other drugs, including warfarin, H_2-receptor antagonists, cyclosporine, rifampin, and nonsteroidal anti-inflammatory drugs (NSAIDs). The concomitant administration of an NSAID with a quinolone may increase the risk of central nervous system stimulation and convulsive seizures.

Disturbances of blood glucose, including symptomatic hyper- and hypoglycemia, have been reported, usually in diabetic patients receiving concomitant treatment with an oral hypoglycemia agent or insulin. In these patients, careful monitoring of blood glucose is recommended, and the quinolone should be discontinued if a hypoglycemic reaction occurs.

VIII. CLINICAL USES

The newer quinolones have proved to be effective therapies for infection of the urinary tract, respiratory tree, gastrointestinal tract, skin, soft tissue, and bone; some sexually transmitted bacterial diseases; and some pelvic infections.

Adequate evidence of efficacy with the newer quinolones, particularly ciprofloxacin and ofloxacin, has been demonstrated in most of these infections. Enoxacin, fleroxacin, and lomefloxacin have been efficacious in some of these infections; norfloxacin in urinary tract infections and some sexually transmitted diseases; and cinoxacin and nalidixic acid in urinary tract infections only. The availability of intravenous preparations of ciprofloxacin, fleroxacin, ofloxacin, and pefloxacin has broadened the use of quinolones for these and other types of infections.

A. Urinary Tract Infections

Nalidixic acid, cinoxacin, norfloxacin, ciprofloxacin, ofloxacin, enoxacin, fleroxacin, lomefloxacin, and pefloxacin have established roles in treating urinary tract infections. For acute and recurrent uncomplicated urinary infections due to susceptible organisms, doses are as follow:

1. Nalidixic acid: adults, 1 g qid (four times daily) for 1–2 weeks, then 0.5 g qid if needed; children, 55 mg/kg/day in four divided doses for 1–2 weeks, then 33 mg/kg/day if needed.
2. Cinoxacin: adults, 250 mg qid or 500 mg bid for 1–2 weeks; children, not recommended.

Long-term therapy with nalidixic acid for frequently recurrent bacteriuria has resulted in poor follow-up cure rates and resistance has commonly emerged during treatment.

The newer quinolones are as effective as other well-established agents for the treatment of uncomplicated urinary infections. Single doses of norfloxacin (800 mg), ciprofloxacin (100 or 250 mg), fleroxacin (400 mg), and ofloxacin (200 mg) are highly effective in women with simple cystitis caused by Enterobacteriaceae, but may be less effective against *Staphylococcus saprophyticus*. Although norfloxacin, ciprofloxacin, ofloxacin, fleroxacin, lomefloxacin, or enoxacin given for 3–10 days has resulted in excellent bacteriologic cure rates in uncomplicated urinary infections, recent studies indicate that 3 days of treatment is sufficient and provides a very high cure rate for uncomplicated urinary tract infections in women.

Norfloxacin, ciprofloxacin, ofloxacin, fleroxacin, lomefloxacin, and enoxacin, given for 5–10 days to patients with nosocomial or complicated urinary infections, have resulted in excellent cure rates.

Ciprofloxacin (1000 mg/day), ofloxacin (300–600 mg/day), pefloxacin (800 mg/day), and norfloxacin (800 mg/day), given to patients with either acute or chronic prostatitis for 28 (range 5–84) days, cured 63–92% of patients.

B. Respiratory Tract Infections

Patients with purulent bronchitis, acute bacterial exacerbations of chronic bronchitis, or pneumonia who were treated for 10 (range 7–15) days with ciprofloxacin, ofloxacin, enoxacin, fleroxacin, lomefloxacin, or pefloxacin experienced clinical cure or improvement (76–91%) and bacteriologic cure (68–83%). However, in 49% of patients with *Pseudomonas aeruginosa* infections, 39% with *Streptococcus pneumoniae,* and 33% with *Staphylococcus aureus* infections, bacteriologic persistence, relapse, or treatment failure occurred. Most physicians are reluctant to use the newer quinolones to treat either community-acquired or aspiration pneumonia because of their reduced activity against *Streptococcus pneumoniae* and against those microaerophilic and anaerobic bacteria associated with aspiration pneumonia. However, intravenous ciprofloxacin, ofloxacin, pefloxacin, and fleroxacin have been used successfully in hospital-acquired pneumonia caused by aerobic gram-negative bacteria, although the bacteriologic eradication rate for *P. aeruginosa* is lower. Ciprofloxacin (750 mg twice daily) has been effective in cystic fibrosis patients with exacerbations of acute pulmonary infections, although resistant organisms may emerge. Malignant external otitis caused by *P. aeruginosa* may also respond to ciprofloxicin therapy.

The newer quinolones should not be used for acute sinusitis because of the possible presence of pneumococci and anaerobic streptococci, but may be useful in specific cases of chronic sinusitis caused by susceptible aerobic gram-negative bacteria. These agents should not be used for otitis media in pediatric patients. Norfloxacin has not been approved and should not be used for any type of respiratory tract infection.

C. Gastrointestinal Infections

The newer quinolones are highly active against those bacterial pathogens causing diarrheal disease, including toxigenic *E. coli, Salmonella, Shigella, Campylobacter,* and *Vibrio* species. The quinolones provide high drug concentrations in the lumen of the gut and the mucosa, which contribute to the eradication of these pathogens from the intestine within 48 hr of initiating therapy. Ciprofloxacin (500 mg bid) and norfloxacin (400 mg bid) for 3 to 5 days, or a single oral 400-mg dose of fleroxacin, cure greater than 90% of patients with either acute bacterial diarrhea or acute traveler's diarrhea and are comparable to trimethoprim-sulfamethoxazole. Single-dose fleroxacin was as effective as 2 or 3 days of therapy in patients with cholera, shigellosis, and *Vibrio parahemolyticus* infections. Single-dose therapy with 800 mg of norfloxacin or 1 g of ciprofloxacin is also effective in treating shigellosis except for patients infected with *Shigella dysenteriae* type 1. Bacterial resistance may develop more rapidly with the indiscriminate use of the newer quinolones. Thus, these newer agents should not be used as prophylactic agents to prevent acute traveler's diarrhea because this disease responds promptly to treatment once symptoms develop. However, prophylaxis is recommended for patients with impaired health and daily doses of norfloxacin (400 mg), ciprofloxacin (500 mg), ofloxacin (300 mg), or fleroxacin (400 mg) have been highly effective in preventing traveler's diarrhea. Most patients with typhoid fever treated with ciprofloxacin [500 mg twice daily for 2–15 (mean 13) days], ofloxacin (200 mg twice daily for 6–30 days), or fleroxacin (400 mg daily for 7 days) were cured. None of these patients relapsed or became chronic carriers. Also, the chronic salmonella carrier state was eliminated in 86% of patients treated with ciprofloxacin at 500–750 mg twice daily for 4 weeks, and followed for 10–12 months.

Helicobacter pylori, which has been associated with antral gastritis, is inhibited by the newer quinolones. However, these agents have not been effective in the treatment of *H. pylori*-associated gastritis. Although a preliminary report suggests that ciprofloxacin may have some value in *Clostridium difficile* enterocolitis, the newer quinolones have also been associated with this disease. Some relapses have been reported in patients with *Brucella* infections who have been treated with the newer quinolones.

D. Skin and Soft Tissue Infections

Ciprofloxacin, ofloxacin, fleroxacin, and enoxacin, given orally, effectively treat a variety of bacterial skin and skin structure infections in patients with cellulitis, subcutaneous abscesses, wound infections, and infected ulcers in diabetic patients. Clinical cure or improvement was observed in 95% of patients treated with oral ciprofloxacin (750 mg twice daily for 14 days). Bacteriologic cures were lower in patients infected with gram-positive organisms than were observed for infections caused by gram-negative aerobic bacteria. Quinolone therapy failed in 25% of anaerobic infections. Colonization with methicillin-resistant *Staphylococcus aureus* (MRSA) was eradicated in 50–79% of evaluable patients treated orally with ciprofloxacin (750 mg twice daily for 7–28 days). When rifampin was combined with ciprofloxacin, the eradication rate was 100% when the isolates were susceptible to both agents. These patients remained free of MRSA for at least 1 month. Although ciprofloxacin may eradicate MRSA colonization, or cure MRSA infection, the development of ciprofloxacin-resistant strains may occur.

E. Osteomyelitis

The newer oral quinolones, primarily ciprofloxacin, have been effective as monotherapy for osteomyelitis, particularly when caused by gram-negative aerobic pathogens. Oral ciprofloxacin in a dose of 750 mg twice daily for 8 weeks (range 4 days to 6 months) has been used in most patients with either acute or chronic osteomyelitis, in either native bone or complicated by a foreign body. Clinical cure or improvement occurred in approximately 80% of patients with follow-up of at least 6 months to more than 1 year. Treatment failures occurred in 15 to 20% and recurrent infection occurred in a few patients. The development of resistant strains occurred in some patients, primarily those with *P. aeruginosa* infections. Thus, ciprofloxacin, and possibly other newer quinolones, has an established efficacy in the treatment of osteomyelitis.

F. Sexually Transmitted Diseases

The newer quinolones have been used to treat a variety of sexually transmitted diseases. Ciprofloxacin, and other newer quinolones, are extremely active *in vitro* against *Neisseria gonorrhoeae,* including penicillinase-producing strains (PPNG), and against *Hemophilus ducreyi. Chlamydia trachomatis* isolates are most susceptible to ciprofloxacin, ofloxacin, sparfloxacin, and trovafloxacin but are resistant to enoxacin and norfloxacin. Although *Gardnerella vaginalis* and *Ureaplasma urealyticum* are relatively resistant to these agents, both sparfloxacin and trovafloxacin are active against *U. urealyticum,* and ciprofloxacin is active about 50% of the time. [*See* Sexually Transmitted Diseases.]

1. Gonococcal Infections

Single oral doses of 250 mg of ciprofloxacin, 400 mg of enoxacin, 400 mg of pefloxacin,400 mg of ofloxacin, 400 mg of fleroxacin, 200 mg of sparfloxacin, 100 mg of trovafloxacin, or 800 mg of norfloxacin cured 95–100% of uncomplicated gonococcal infections in both men and women, including patients infected with PPNG. Thus, the lowest effective oral single dose of the newer quinolones (100 mg of trovafloxacin) has cured almost 100% of patients with urethral as well as rectal gonorrhea, and is probably effective for pharyngeal gonococcal infections. However, there is little experience with these newer agents in the treatment of *disseminated gonococcal* infections. Also, although the quinolones are highly effective in curing uncomplicated gonorrhea, strains of gonococci resistant to the quinolones have begun to emerge.

2. *Chlamydia* Urethritis, Postgonococcal Urethritis, and Nongonococcal Urethritis

The current quinolones are not effective as single-dose therapy for *C. trachomatis* urethritis, nor do they prevent postgonococcal urethritis (PGU) when used as single-dose therapy in gonococcal infections. Although ciprofloxacin in a dose of 750 mg orally twice daily for 4 days eradicated *C. trachomatis* in 60% of coinfected patients, and reduced the incidence of PGU from 35 to 13%, the response rate was not optimal. However, trovafloxacin at 200 mg once daily for 5 days was as effective as 7 days of doxycycline for documented *C. trachomatis* cervicitis. Also, ofloxacin (300 mg twice daily) for 7 days was as effective as doxycycline in patients with nongonococcal urethritis (NGU)/*C. trachomatis* cervicitis. However, for NGU, the quinolones are less effective than doxycycline or azithromycin regardless of the presence or absence of *Chlamydia* or *Ureaplasma* infection.

3. Chancroid

A single oral dose of 500 mg of ciprofloxacin or 500 mg of ciprofloxacin twice daily for 3 days has cured 95 and 100% of patients with chancroid and *H. ducreyi* infections, respectively. Fleroxacin, 400 mg single dose in HIV-negative men, or 400 mg daily for 5 days in HIV-positive men, cured more than 90% of these patients with chancroid.

G. Other Infections

1. Immunocompromised Host

In granulocytopenic patients, oral ciprofloxacin (500 mg twice daily), norfloxacin (400 mg twice or three times daily), ofloxacin (300 mg twice daily), and pefloxacin (400 mg twice daily) have been used successfully for prophylaxis. The incidence of bacteremia and colonization with gram-negative bacilli was significantly reduced in febrile neutropenic patients, and was most effectively accomplished with ciprofloxacin. The emergence of quinolone-resistant gram-negative organisms has been observed recently in neutropenic cancer patients who received quinolone prophylaxis. Limited clinical experience with intravenous ciprofloxacin, ofloxacin, and pefloxacin in the *treatment* of severe infections in immunocompromised patients suggests a potential role for these agents when used in combination with other agents, for example, aminoglycosides. Monotherapy with the quinolones in febrile neutropenic patients is not recommended.

2. Central Nervous System Infections

Ciprofloxacin, ofloxacin, and pefloxacin do penetrate into the cerebrospinal fluid and brain tissue but there is little clinical experience with these agents for central nervous system bacterial infections. These quinolones should be reserved for special cases caused by multiantibiotic-resistant aerobic gram-negative bacteria. Recently, oral trovafloxacin (3 mg/kg for 5 days) was as highly successful as parenteral ceftriaxone and cured 95% of patients (18 years of age or younger) with acute meningococcal meningitis during an epidemic outbreak in Nigeria. Also, ciprofloxacin has been shown to be effective in eradicating the meningococcal nasopharyngeal carrier state.

Ciprofloxacin and ofloxacin have been used with some success, *in combination* with other drugs, to treat multidrug-resistant pulmonary tuberculosis (ciprofloxacin and ofloxacin) and *M. avium-intracellulare* bacteremia in AIDS patients (ciprofloxacin). In general, the current quinolones are, at best, second-line agents to be used in combination for treating mycobacterial disease, except for *M. leprae.* Sparfloxacin, pefloxacin, and ofloxacin have shown efficacy in patients with lepromatous leprosy.

Oral ciprofloxacin (500 mg twice daily for 2 days) cured all patients with Mediterranean spotted fever caused by *Rickettsia conorii.* Ciprofloxacin in combination with rifampicin has also been used to treat patients with right-sided *Staphylococcus aureus* endocarditis in recalcitrant patients. A small number of patients with cat-scratch disease experienced rapid improvment when treated with 500 mg twice daily of oral ciprofloxacin.

In the future, further modifications of the chemical structure of the newer 4-quinolones should lead to newer agents that may have the potential to treat infections in addition to those mentioned here.

BIBLIOGRAPHY

Andriole, V. T. (ed.) (1988). "The Quinolones." Academic Press, New York.

Andriole, V. T. (1989). The quinolones. *In* "Principles and Practices of Infectious Diseases" (G. L. Mandell, R. G. Douglas, and J. E. Bennett, eds.), 3rd Ed. Churchill Livingstone, New York.

Andriole, V. T. (1992). Quinolones. *In* "Infectious Diseases in Medicine and Surgery" (S. L. Gorbach, J. G. Bartlett, and N. R. Blacklow, eds.). Saunders, Philadelpia.

Chu, D. T. W., and Fernandes, P. B. (1989). Structure activity relationships of the fluoroquinolones. *Antimicrob. Agents Chemother.* **33**, 131.

Hooper, D. C., and Wolfson, J. S. (eds.) (1993). "Quinolone Antimicrobial Agents," 2nd Ed. American Society for Microbiology, Washington, D.C.

New, H. C. (1992). Quinolone antimicrobial agents. *Annu. Rev. Med.* **41**, 465.

Sorgel, F., Jaehde, U., Naber, K., and Stephan, U. (1989). Pharmacokinetics disposition of quinolones in human body fluids and tissues. *Clin. Pharmacokin.* **16S**, 5.

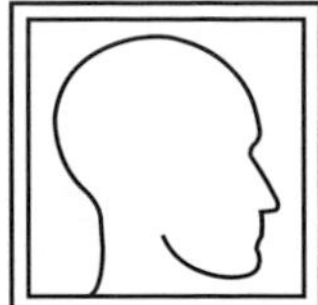

Radiation, Biological Effects

JOSEPH H. GAINER
Potomac, Maryland

GLOSSARY

Antibody affinity Binding energy of antibody sites for specific antigen epitopes. It determines the stability of antigen–antibody complexes, influencing the ability of antibody to neutralize microbial agents and toxins

Aplastic anemia Disease in which there is a persistent reduction in the number of circulating erythrocytes caused by a failure of adequate stem cell production

Gray (Gy) ISU unit for the absorbed dose of ionizing radiation. 1 Gy = 1 joule of absorbed energy per kilogram of irradiated material. 1 cGy = 10^{-2} Gy

Inflammation Response of a tissue or organ to a physical, biological, or chemical irritant with swelling (edema), redness (erythema), and pain, accompanied by cellular infiltration

Phagocytosis Engulfment of bacteria, fungi, and other foreign particles by white blood cell elements

Rad Absorption of 100 ergs per gram of tissue. One rad is approximately one roentgen (R)

Sievert (Sv) SI unit of dose equivalent. This is calculated by multiplying the dose in Gy by a quality factor, which adjusts for differences in relative biological effectiveness of different radiations. 1 cSv = 10^{-2} Sv

RADIATION SICKNESS IS THE ILLNESS PRODUCED after exposure to ionizing radiation. With heavy exposure it is characterized by malaise, nausea, emesis, diarrhea, and leukopenia. The mean survival time of mammals, including human beings, is inversely a function of radiation dose. At doses to 10,000 R or greater, the exposed subject dies in a matter of minutes to a couple of days, with symptoms consistent with pathology of the central nervous system. From 1000 R to several thousand R, there is a constant mean survival time. Below 1000 R, individuals die with symptoms of bone marrow pathology.

The radiation dose producing an LD_{50}, or lethal dose 50% (at which 50% of exposed subjects die), varies among the animal species and human beings, computed during a 30- to 42-day period (Table I).

Radiation sickness was reported soon after the discovery of the X ray. Antoine-Henri Becquerel and Marie Curie suffered from acute radiation dermatitis. The earliest known case of radiation-induced cancer was reported in 1902. Madame Curie herself died of aplastic anemia, probably caused by her prolonged exposure to radiation. Nine deaths caused by bone cancer were recorded in 1924 among watch industry workers who painted watch dials with radium, which emits radiation. Leukemia was induced in mice by a single whole-body exposure of 400 R or by a closely spaced, fractionated dose of 900 R. A high incidence of skin cancer and leukemia was observed among radiologists exposed to X rays. Human data on the effects of ionizing radiation come primarily from the people of Hiroshima and Nagasaki exposed to the atomic bombs in World War II. The explosions at Three Mile Island and at Chernobyl, especially with injuries and deaths at Chernobyl, and the potential

ENCYCLOPEDIA OF HUMAN BIOLOGY, Second Edition, VOLUME 7. Copyright © 1997 by Academic Press. All rights of reproduction in any form reserved.

TABLE I
Lethal Dose 50% in Animals and Humans

Species	Roentgens
Sheep and burros	155
Swine	195
Monkeys	395
Marmosets, goats, dogs, and humans	300–400
Gerbils	1059
Wild mice	1100–1200

for increased incidence of cancer and leukemia have heightened the concerns over the safety of nuclear power.

Living with radiation from "natural exposure" presents the following risks of cancer (Table II). Smoking tobacco products adds significantly to lung cancer deaths, and certainly a substantial portion of these deaths is caused by radiation.

I. THEORIES ON THE BIOLOGICAL EFFECTS OF RADIATION

French scientists in 1906 proposed that less-differentiated cells are more radioresistant than are highly differentiated ones and that proliferating tissues are more radiosensitive than are the nonproliferating tissues. An exception to this is that lymphocytes are sensitive, although they neither differentiate nor divide. Target theory predicts that the proportion of unaltered biological molecules decreases as a negative exponential function of dose; however, it was found inadequate to explain cellular radiation injuries. The indirect action of radiation was then proposed; it stated that biological molecules in aqueous solution are inactivated by free radicals formed when radiation interacts with water. Oxygenated tissues are more sensitive to radiation than anoxic ones owing to formation of oxygen free radicals. Another hypothesis is that radiosensitivity of a cell is directly proportional to its interphase chromosomal volume. This is consistent with observations on plant tissues, but for mammalian species its validity is not established.

TABLE II
Natural Exposure to Radiation

Condition	Radiation dose (mrem/yr)
Living within 20 miles of a nuclear power plant for 1 yr	0.02
One round-trip transcontinental flight	5.0 (135 cancers if entire U.S. population was exposed)
Normal body radioactive potassium	30 (1000 cancers)
Cosmic radiation at sea level and medical X rays	40 each (1100 cancers)
Radon gas (1.5 pCu/Liter of air)	500 (13,500 cancers if entire population was exposed)
Persons smoking tobacco products that contain radioactive elements	106,000 lung cancer deaths in 1985, only 3800 in nonsmokers

The concept of RBE (relative biological effectiveness) evolved because different types of radiation produce different degrees of damage with the same dose. This is due to the fact that the linear energy transfer (LET)—the amount of energy released per unit length along the radiation path—is different for each type of radiation. For the same total dose, the radiation of high LET (i.e., alpha particles, protons, or neutrons) produces greater damage than that of low LET (X rays and gamma rays). Furthermore, the oxygen effect, so marked with low-LET radiation, is negligible with high LET.

A. Determining Person-Years of Life Lost Using the BEIR V Method

The BEIR V method permits the calculation of the estimated person-years of life lost when groups of males and females are exposed to low-LET ionizing radiation; this determination is informative in rating risks to populations from radiation exposures. For example, when 1,000,000 females have received an acute dose of 10 mSv, the total number of radiation-induced fatal cancers is predicted to be 835. The total number of person-years of life lost is 1.4×10^4 or an average of about 5 days. However, the 22 who die of leukemia lose an average of about 34 yr; those who die of other forms of cancer lose an average of 16 yr.

The discovery of several radioprotective and radiosensitizing agents increased the knowledge of radiation injuries. A therapeutic regimen for exposed individuals involves a "functional replacement therapy" requiring transfusions of fresh platelets and whole blood and the administration of antibiotics. Spleen, splenic cells, and bone marrow transplantation are also beneficial. Current research in immunology, including studies on radiation and its effects on immunity, and the development of products such as biological response modifiers, including interferons, interleukins, cytokines, and monokines, provide additional chemotherapeutic treatment modalities for radiation sickness.

II. EFFECTS OF RADIATION ON ORGAN SYSTEMS

A. Central Nervous System

Neuroblasts and neuroglia are radiosensitive, whereas neurons are radioresistant. The acute response of the central nervous system to radiation is an acute inflammation. Oligodendrocytes show swelling and degenerative changes soon after exposure and early demyelination occurs. In the subacute period, the patient may completely recover or may have convulsions, ataxia, and incoordination. In the chronic period, delayed necrosis and demyelination are observed. Malignant intracranial neoplasms after radiation therapy have been observed, as has been spinal cord myelitis. Peripheral nerves are highly radioresistant.

Learning behavior in animals is not adversely affected by ionizing radiation, but nonintellectual behavior is affected. Depression of total body activity in monkeys and depression of running-wheel activity in rats have been observed. Decreased mating activity in boars and rats has been seen after fetal radiation; the learning capacity of animals radiated *in utero* markedly decreases.

Radiation injury to the central nervous system is a rare event in clinical medicine, but it is catastrophic for the patient in whom it occurs. By the mid-1960s, 57 patients with documented clinical cerebral necrosis had been reported. Cerebral necrosis can be diagnosed with the computerized tomogram scan; such damage does not self-repair. Necrosis has been reported as less than 1% for doses less than 5200 rads, as high as 4% for doses up to 6000 rads, and 16% for doses greater than 6000 rads.

1. Mental Retardation and Radiation

Analysis of the occurrence of mental retardation and/or decline in intelligence test scores for individuals exposed *in utero* to the atom bombs in Japan revealed the following. (1) The data did not suggest a comingling of different distributions, such as might arise from inclusion of a qualitatively different group of individuals (i.e., unrecognized cases of mental retardation). The cumulative distribution suggests a general phenomenon, a shift in the distribution of scores describing mental intellection with exposure. (2) There is no evidence of a radiation-related effect on mental retardation or intellection generally for those individuals exposed in the first 8 weeks of life. (3) The mean test scores are consistently significantly heterogeneous among exposure categories for those individuals exposed at 8–15 weeks after conception and less heterogeneous for those groups exposed at 16–25 or 26 or more weeks of gestational age. (4) Among the latter two groups of individuals there is a significant decrease in intelligence test scores with increasing exposure. (5) Within the most sensitive group, individuals exposed 8–15 weeks after conception, and with the better-fitting linear-quadratic model, the diminution in intelligence score is 21–26 points per Gy of exposure, or about 0.2 points per cGy.

2. Radiation Myelopathy

Because of the morbidity and mortality associated with radiation myelopathy, radiation injury to the spinal cord has been the subject of more clinical reports than any other normal-tissue injury. Its response has been investigated in several species of laboratory animals, yet few animal studies serve the needs of human medicine well. The trend of limiting the cord dose to 45 Gy is a reasonable action as long as the cancer is adequately treated.

B. Bone Marrow

The bone marrow, with its various cellular elements, is especially sensitive to radiation (Table III). [*See* Hemopoietic System.]

C. Skin and Mucous Membranes

The degree of radiation response of the skin and mucous membranes depends on (1) radiation dose, (2) quality of radiation, (3) time over which the dose is administered, (4) size of the field, and (5) anatomical location. Damage may range from minimal degenerative changes in germinal cells to total necrosis. Dermal changes are an initial erythema, desquamation, and then necrosis. Hyperpigmentation or depigmentation may occur, and neoplasms of the skin may develop. grafted skin radiosensitivity depends on the age of the graft; a 3-month-old graft shows responses similar to normal skin, and an old graft, more than 1 year, shows little reaction to radiation. [*See* Skin, Effects of Ultraviolet Radiation.]

The mucous membrane effect is radiation mu-

TABLE III
Effects of Radiation on the Bone Marrow

Radiation dose (R)	Effects
Red blood cells	
200	Depression of red cell precursors, regeneration occurs early
400–800	Red cell count falls slowly for 7–8 days then more quickly
700	Red cell precursors are depleted in 1 day and are totally absent by Day 3
Hematopoietic elements	
<100	No effects (no changes)
175	Granulocytes fall markedly initially
200–500	Granulocytosis at first, then a fall with recovery beginning at Day 36; lymphocytes are low at 3–4 days, remaining low for 5 weeks; platelets fall to a minimum in 8–10 days and return to normal in 7 weeks

cositis; it initially appears patchy and later spreads over the entire irradiated area. Varying degrees of squamous cell metaplasia and hyperplasia occur, and neoplasms of the laryngopharynx have been seen. [*See* Radiosensitivity of the Integumentary System.]

D. Urogenital System

The kidneys are only moderately sensitive to radiation. Acute radiation nephritis occurs 6–12 months after radiation therapy. Five adults died of acute radiation nephritis 5–6 months after 2300 R to the kidneys.

Concerning the genital system, the ovary and the testis are radiosensitive, whereas spermatids and spermatozoa are radioresistant. In the male, radiation-induced sterility is never immediate because of the high resistance of the spermatids and the spermatozoa. Male accessory organs are highly radioresistant. In animals, the lower the dose rate, within a certain range, the higher the sterility. Short-time fractionated doses produce a higher incidence of sterility in males than does a single dose.

In the female, radiation destroys not only the radiosensitive gametogenic epithelium but also much of the production of the sex hormones. In growing follicles, postradiation damage of the granulosa cells is seen before changes in the oocytes. Hence the female is more sensitive than the male; sterility in the female is immediate if the dose level is sufficiently intense. The uterus is radioresistant, and the vaginal mucosa is similar to that of other mucous membranes. The mammary glands, vulva, labia, and clitoris are relatively radiosensitive.

Oocyte sensitivity to radiation varies as a function of the stage of development. In mice the radiation of oocytes varies with age, species, strain of mouse, stage of growth of the follicles about the oocytes, and chromosomal configuration.

E. Human Fetus

Recent data suggest that a dose of 1–5 rads may be harmful to the human fetus. The gestation period between 18 and 38 days is highly radiosensitive for the fetus and gives a greater number of organ anomalies. Nervous tissue and optic tissue are especially sensitive. Diagnostic radiation (i.e., 0.7–5 rads) of young pregnant women increases the probability of a Down's syndrome baby by a factor of 10.

F. Liver

The liver is a highly radioresistant organ based on morphological changes. Functionally hepatic enzyme changes have been reported after radiation.

G. The Cardiovascular System

1. Heart

Myocardial cells are very radioresistant based on morphological changes, whereas small vessels of the heart are moderately radioresistant. Direct irradiation of the heart involves primarily the fine vasculature, with secondary effects on the connective tissue and indirect effects on the myocardium.

Radiation induces swelling of the myocardial fiber, loss of striations, homogenization and granulation of the sarcoplasm, disappearance of protoplasm, and persistence of hollow sarcolemma.

Nuclei show pyknosis or hyperchromasia around

the nuclear membrane, fragmentation, and lysis. Arterioles show thickened intima and degeneration and hyalinization of the media. The structural changes in the myocardium are observed only after high doses of radiation (i.e., 3000–10,000 R); however, damage to the pericardium is produced at lower doses.

Electrocardiographic changes have been reported after irradiation of the thorax. Myocardial infarction, although rare, may occur within 6 months of radiation therapy. Progressive deleterious changes in the fine vasculature of the pericardium or myocardium may lead to secondary focal degeneration of the pericardial or cardiac tissue and focal replacement fibrosis many months or even years after the irradiation. The degree and the extent of the damage depend on the dose and the contribution of other factors. Pericarditis, pericardial effusions, pericardial adhesions, and pericardial fibrosis also occur.

2. Aorta and Major Vessels

Hemorrhage is the main manifestation of damage to the major vessels after radiation therapy. Doses in the range of 5000 R in 41–47 days have produced rupture of the aorta. The intima was thickened and granular. In some areas, fibrin covered the intimal surface, and elsewhere there were mural thrombi.

Medium-sized muscular arteries show degeneration similar to that in large arteries, but they are not as severely damaged as are the arterioles and capillaries. The large- and medium-sized veins show similar changes, as do the lymphatics. Repair of acute injury to the blood vessel walls by fibrosis causes advanced sclerosis, sometimes completely blocking the vessel lumen.

Thus irradiation can induce vascular changes such as degeneration, necrosis, and proliferation of the endothelium, with edema, fibrosis, thickening, and infiltration of the vascular wall. Subsequent thrombosis and/or occlusion may occur in the severely damaged blood vessels. Vascular alterations can produce marked changes in tissues dependent on them for their blood supply.

H. Respiratory System

The richly ramified vascular system and lymphatic tissues of the respiratory system are radiosensitive, but cartilage is radioresistant. Infection renders lungs more radiosensitive. In human beings, radiation pneumonia has been observed 4–6 months after a fractionated exposure of 3000–5000 R.

The accumulation of fibrin-rich exudate within alveoli and the thickening of alveolar septa by fibrillar material, cellular proliferation, or fibrous tissue are seen following therapeutic irradiation of the thorax. Outpouring of alveolar edematous fluid rich in fibrin, associated with initial congestion and edema, in the absence of cellular inflammatory changes, suggests that injury to the fine vasculature plays a primary role in the development of the characteristic picture of radiation pneumonitis. Fibrin condenses at the alveolar walls to produce the so-called "hyaline membrane." Fibrin membranes were found in 41% of all irradiated lungs, most frequently and most prominently 6 months to 2 yr after doses greater than 2000 R. Increased deposition of fibrillar connective tissue in the alveolar septa was frequent after radiation doses greater than 3000 R and after postirradiation intervals longer than 6 months. The increased accumulation of histiocytes and fibroblasts in the alveoli was seen most frequently at dose levels between 2000 and 5000 R.

Residual radiation damage in lung in the form of sclerosis and reduction of fine vasculature and increased parenchymal fibrosis may not be progressive with time. However, these changes may be additive to similar changes occurring as a result of other insults or normal aging. These combined changes constitute a progressive sclerotic deterioration of the lung parenchyma, with a gradual reduction in the functional reserve capacity, especially in the presence of infections and other stresses. Therefore the clinical problem of pulmonary fibrosis may appear years after irradiation. The dose rate appears to influence the severity of the radiation pneumonitis; at a dose rate of 1000 rads/week, acute reactions are fairly common. When the rate is reduced to 700 rads/week, this reaction becomes of little importance.

III. RADIATION CARCINOGENESIS

Carcinogenesis is likely the single largest concern of radiation exposure. X irradiation induces several types of neoplasms; data from Hiroshima and Nagasaki illustrate the thresholds of gamma-ray doses for leukemia and certain cancers (Table IV). No threshold exists with mixed radiation of gamma rays and neutrons.

A. Risk of Breast Cancer after Irradiation of the Thymus in Infancy

Exposure of the female breast to ionizing radiation in infancy increases the risk of breast cancer later in

TABLE IV
Thresholds for Radiation Carcinogenesis

Dose (rads)	Cancer/leukemia
50	Leukemia, thyroid tumors, breast cancers
128	Lung cancer incidence increased by a factor of 2
1000	Osteosarcoma in children
3000	Osteosarcoma in adults

life. In an average follow-up of 36 years, there were 22 breast cancers among 1201 women X-rayed for enlarged thymus in infancy; there were 12 cancers in 2469 unirradiated controls. This gives an adjusted rate ratio of 3.6. The estimated mean absorbed dose to the breast was 0.069 Gy. There was a dose–response linear relation with a relative risk of 3.48 for 1 Gy of radiation and an additive excess risk of $5.7/10^4$ person-years/Gy.

B. Breast Cancer Risk from Mammography

The risk of breast cancer associated with radiation decreases sharply with increasing age at exposure. Even a small benefit to women of screening mammography outweighs any possible risk of radiation-induced breast cancer. [*See* Radiation Interaction Properties of Body Tissues.]

C. Mechanism of Radiation-Induced Cancer

The mechanism of radiation-induced cancer is not completely understood. Two major hypotheses have been suggested: (1) radiation causes somatic mutation, which is responsible for the malignant transformation of normal cells; and (2) radiation makes the cellular environment compatible for viral replication and viral-induced malignant transformation. The latter hypothesis applies especially to nonhuman animal species; the former one applies more generally. There is no evidence that radiation produces tumors of the ovary, uterus, pituitary gland, adrenal, brain, liver, or alimentary tract in human beings; however, radiation does produce these tumors in animals. Skin cancer is high among persons with chronic radiation dermatitis.

Recent developments in the molecular biology of DNA are leading to new ideas concerning how DNA alterations might be involved in radiation carcinogenesis. Alteration of certain protooncogenes generates oncogenes that initiate the development of malignancy. A chromosome break often occurs at or near the location of a specific oncogene in Burkitt's lymphoma. Such breaks could represent initial lesions in a translocation process that activates the oncogene by inserting it at a new location (e.g., near an active promoter). DNA breakage is one of the principal ways that ionizing radiation affects mammalian cells. It might be involved as an initial event in carcinogenesis.

1. *p53* Suppressor Gene versus Oncogenic Basis of Radiation Resistance

The human gene *p53* is located on chromosome 17p and encodes a 393-aminio-acid nuclear phosphoprotein. Somatic point mutations of this gene are found at high frequency in most human cancers. In mice, knockout of *p53* function permits normal development but confers an enormously elevated risk of developing neoplasia; moreover, *p53* is essential for the normal apoptotic response of thymocytes and intestinal epithelial cells to ionizing radiation. Many of the point mutations in *p53* found in human cancer are associated with accumulation of the protein and alteration of its conformation, making it an attractive target for novel diagnostic and therapeutic approaches. Xenotoxins stabilize the normal *p53* in cells; this may explain the growth arrest induced by such toxins. *p53* is suggested as a "guardian of the genome" in acting to protect cells from genetic damage by inducing a specific cell cycle arrest or apoptosis. [*See* Tumor Suppressor Genes, p 53.]

Rather to the contrary, many oncogenes (over 20 have been described: v-*abl*, c-*fos*, v-*fos*, c-*myc*, v-*myc*, to name a few) are implicated in the development of neoplasia, and these oncogenes play a major role in cellular resistance to ionizing radiation. Normal counterparts of these genes (protooncogenes) are involved in numerous vital cellular functions, and products of these protooncogenes have been shown to interact with one another as components of signal transduction pathways that involve transmission of molecular/biochemical signals from cell membranes to the nuclei to direct the cells to divide or differentiate. The central hypothesis concerning the outcome of τ irradiation of eukaryotic cells suggests that loss of clonogenic capacity and cell death result from damage to the structure and function of genomic DNA.

One cancer family syndrome, a dominantly inherited pattern originally described by F. P. Li and J. F. Fraumeni, Jr., is characterized by a constellation of tumor types, including breast carcinoma, soft tissue sarcoma, brain tumors, osteosarcoma, and leukemias. This specific family involves 18 affected descendants of a single individual through six generations. Neoplasms in 3 members of the family may have been induced by occupational exposure or therapeutic radiation; one member worked in a factory producing heavy water, a second member worked in a foundry, and the third member had radiotherapy for an earlier neurilemmoma and was subsequently diagnosed as having an osteosarcoma in the radiated field. An activated *raf-1* gene was demonstrated in family members.

D. Diagnostic and Therapeutic Radiation Effects

Human exposure to ionizing radiations is ubiquitous. Current estimates of the carcinogenic risks associated with exposure are that radiations are responsible for only a relatively small proportion of all cancers. It is estimated that exposure of 1 million persons to 1 rad of ionizing radiation will induce about 20 leukemias and 100 fatal cancers. On this basis, radiation might well result in about 150 (5%) of the 3000 leukemia deaths a year in England and Wales and 750 (0.6%) of the 120,000 deaths from other malignancies. Less than 3% of cancers in the United States may be attributed to radiation.

1. Aplastic Anemia

Aplastic anemia has been observed after radiation. Deaths caused by aplastic anemia in patients receiving 112–3000 R for the treatment of ankylosing spondylitis were about 30 times higher than expected. The number of observed deaths caused by aplastic anemia was 17 times greater than expected among American radiologists.

2. *In Utero* Exposures

In utero exposure to diagnostic radiation might lead to the development of childhood leukemia and other cancers. A study was made in 1962 in the United States of over 700,000 children born between 1947 and 1954 in 37 maternity hospitals. Deaths from cancer among children in the period 1947 to 1960 were traced, and the frequency of prenatal exposure to radiation among those who died of cancer was contrasted with those of a 1% sample of all children who had been born in the same 37 hospitals. Irradiated children had a mortality rate from both leukemia and other cancers about 45% higher than that of unirradiated children. Cancer risk was directly related to the number of times that X-ray exposure had occurred in the pregnancy. A steady declining risk was found with year of birth.

3. Radiologists

Radiologists have been studied for longer than any other defined population to assess the late effects of exposure to ionizing radiations received as a consequence of their occupation. There is evidence of a substantial excess of cancer; for leukemia, four deaths against 0.65 expected, and for cancers of the skin, six deaths against 0.77 expected.

4. Radiation-Induced Menopause

From the 1930s onward, X irradiation of the ovaries was commonly used to induce artificial menopause among women with benign menopausal bleeding. About 2000 women who had experienced radiation-induced menopause at three Scottish radiotherapy centers between 1940 and 1960 were studied. By 1970, 25% of the women had died, and there had been 7 deaths from leukemia, against 2.7 expected ($P < 0.03$). There was a significant excess of deaths from cancers of the pelvic sites but no excess of other cancers. The excess cancers in the radiated sites became apparent 5–9 years after treatment, and an excess cancer risk persisted beyond 20 years after treatment.

5. Radiation Treatment of Cervical Cancer

Cervical cancer is usually treated by the insertion of radium or by high doses of X rays directed at the cancer or by both methods. No excess of leukemia cases was found in a study of more than 70,000 women treated in this way (16 observed against 14 expected). This finding was confirmed in another study of more than 80,000 women treated with radiotherapy for cervical cancer, in which there were 77 cases of leukemia against about 66 cases expected.

6. X-Ray Treatment of Ankylosing Spondylitis

The study of patients given radiotherapy for ankylosing spondylitis was one of the first to be set up to assess the carcinogenic hazards of radiation exposure. It has been among the most informative for several reasons: (1) a large group of patients (14,000) was followed; (2) there was a long period of follow-up of the irradiated population; and (3) reasonably high radiation doses were used in the treatment of this condition. There was clearly an increased risk of leukemia and aplastic anemia (67 deaths against 6.0 expected) and of cancer (285 deaths against 195 expected). The excess cancers were largely confined to those sites likely to have been directly in the radiation beam.

7. Induction Periods in Radiation Carcinogenesis

The way in which the risk of a radiation-induced cancer varies with time after exposure is easiest to look at in populations in which only one exposure of short duration has occurred. Two groups in this respect are the atomic bomb survivors and patients with ankylosing spondylitis. Neither group, however, is well-suited to examine the risk in the period immediately after exposure because the A-bomb cohort was not defined until 1950. It thus does not include deaths from cancers other than leukemia occurring in the first 5 years after the bombings.

Leukemias appeared in both the spondylitic population and the A-bomb survivors before other cancers. In spondylitics, the leukemia risk was greatest 3–5 years after exposure, subsequently declining, such that by 20 years there was no evidence of an excess risk. Among the A-bomb survivors, the greatest measured risk was 5–9 years after exposure. Subsequently, the excess risk declined, and there was only a small excess mortality in 25 or more years.

8. Dose–Response Relations

Most public health interest centers on the likely effect of low doses of radiation, and it is to these that a substantial proportion of the population may be exposed. The spondylitics were, in general, exposed to large doses of radiation, as were many of the A-bomb survivors, although most of the latter group received doses of less than 10 rads.

The simplest assumption to make is that effects are linearly related to dose and to extrapolate backward on this basis. This has been the general approach adopted for many of the analyses of data from studies of the A-bomb survivors. For leukemia—the cause of death for which the radiation-induced excess is most apparent—this form of dose–response relation fits the data well, although it is a better fit for the Hiroshima data than for the Nagasaki data; also the leukemia risk, for a specified dose, was higher in Hiroshima than in Nagasaki. If the data from the two cities are combined and linear dose–response curves are fitted to leukemia excess mortality, the estimate of the induction rate is 1.9 leukemia deaths/million person-years at risk/rad, and for other cancers the rate is 2.2 deaths/million person-years at risk/rad. Table V provides leukemia incidence from the atomic bombings in Japan, illustrating the higher risk with exposure at older ages and at younger ages. [*See* Leukemia.]

TABLE V
Leukemia Incidence Since the Japanese Atomic Bomb Exposure[a]

Age at exposure (yr)	Years after exposure				
	5	10	20	30	40
15	100[b]	40	3	2	2
15–29	76	40	6	4	2
30–44	36	40	35	20	17
45+	10	40	60	75	78

[a]Data are from A. Castellani (1984). "Epidemiology and Quantitation of Environmental Risk in Humans from Radiation and Other Agents." Plenum, New York, with permission.

[b]Approximate excess risk of leukemia per 100,000 by years after the Hiroshima and Nagasaki atomic bomb exposures.

The greatest risk of leukemia induction is in patients with a mean marrow dose of 100–200 rads. At higher doses the risk appears to be reduced. The data are not well-fitted by a linear dose–response curve, and the assumption that the excess risk is unrelated to dose fits the data better.

It appears that a simple linear dose–response relationship for the induction of leukemia may be incorrect. Little or no excess risk of leukemia among women irradiated for the treatment of cervical cancer occurs. Women given a radiation-induced menopause dose receive a substantially smaller dose to the bone marrow yet show a statistically significant excess leukemia. It is suggested that patients treated for cervical cancer may not be of greatly increased risk of leukemia because the radiation treatment is given in such a way that some of the bone marrow receives a very high dose of radiation (enough to sterilize the marrow) while the dose to the marrow falls off rapidly with distance from the cervix, so that the mean dose to surviving marrow cells may be small. Thus the "effective" dose for leukemia induction in cervical cancer patients under radiation therapy is small.

9. Age at Exposure

Concerning *in utero* radiation exposure, it is suggested that some fetuses may be many times more susceptible than others to radiation carcinogenesis, yet evidence to support this has been unconvincing. There is good evidence, however, that age at exposure is related to radiation risk; the fetus seems to be more sensitive to radiation than the adult. With spondylitics, there is a steep and highly significant increase as the age at exposure increases. The risk for leukemia among children is similar to that among adults up to the age of 50 years, rising slightly after this. It is suggested that radiation may be interacting in multiple ways with other factors that induce leukemia.

For cancers other than leukemia, the situation appears to be similar. Excess risk increases regularly with age at exposure among adults, although the risk of radiation-induced breast cancer is highest among women exposed at 10–19 years of age.

Other authors studying the risks from the atomic bombings in Japan have made a number of points that are important for risk estimation after low-level radiation exposure. First is the issue of linear versus nonlinear models. The purely quadratic model (i.e., risk increasing with the square of the dose) gives a reduced risk when compared with a linear model or a linear-quadratic model. At a dose of 1 rad, the difference in the risk estimate is of one or two orders of magnitude. The second point for which the epidemiological data are lacking is whether there is a reduced risk for protracted or low dose-rate exposures. Information for risk assessment purposes must come from experimental systems to account properly for dose-rate effects. From the available data, there are large differences in the risk estimates depending on how time is treated. Also there is the critical issue of possibly increased effects for individuals exposed at younger ages. Differences of upward of a factor of 20 were observed in the risk estimates for epithelial tumors at the youngest category of 0–9 years. Comparisons were made with BEIR estimates because any differences would be primarily due to differences in modeling approaches. For acute leukemia incidence, the estimates were similar, assuming linearity, but for exposure at the lower ages the ankylosing spondylitis data suggest that the risk may be underestimated. Some mix of linear and quadratic appears to be the model of choice.

For epithelial cancer, the time-linear model appeared more suitable and suggested a somewhat higher risk than given in the BEIR study. If, however, the relative risk model is the correct model, the risks for the youngest age-at-exposure category would have been underestimated. The epithelial cancers were adequately described by a linear dose–response function.

10. A-Bomb Survivors: Further Evidence of Late Effects of Early Deaths

Reanalysis of A-bomb survivor data has shown the following: (1) in the high-dose (>1 Gy) subgroups of the life span study of 5-yr survivors, there is a significant deficit of individuals who were <10 yr or >50 yr at the time of the bomb; and (2) in the cohort of *in utero* exposed children, there is a significant deficit of individuals who were <8 weeks of fetal age when exposed. This reanalysis discusses how this selection bias has affected the perception of three effects of A-bomb radiation: marrow damage, carcinogenesis, and second-generation effects.

11. Synergy of Combined Exposure to Radiation and Paraquat

Combined exposure to radiation and chemical toxicants presents additional possible concerns. To cite one study in this regard, rats given a lengthy oral intake of paraquat at 1/100th of their LD_{50} had depressed antioxidant defenses in their lungs, elevated lipid peroxidation, and elevated lactate dehydrogenase (LDH) activity in lavage fluids. Paraquat treatment combined with a single external whole-body exposure to 4 Gy of radiation produced synergic diminution in antioxidant potential of the lungs and synergically increased lipid peroxidation in them. The lung cells secreting LDH into the bronchoalveolar spaces were more severely damaged than in the case of exposure to paraquat or to radiation alone; moreover, with the combined exposures, there was more marked and more prolonged depression of three lung antioxidant defense properties, including superoxide dismutase, catalase, and nonprotein sulfhydryl groups.

IV. CHEMICAL PROTECTION FROM RADIATION

Thiols, such as glutathione (GSH), limit oxidative stress, thereby protecting cells against radiation. Elevation of GSH to 20 m*M* will protect cells at intermediate oxygen tension.

Vitamin E, an antioxidant, is an efficient inhibitor of free radical-based reactions including lipid peroxidation. Drugs producing hypoxia through reducing blood flow should be effective in reducing radiosensitivity. Calcium antagonists protect mice against lethal doses of ionizing radiation, and recombinant canine granulocyte-colony-stimulating factor protected dogs to otherwise lethal total-body radiation. They had sustained hematopoietic recovery.

Hypothermia reduces metabolic activity, allowing for more complete and efficient repair of radiation damage. Radiation-induced hyperthermia elevates body temperature, by 1–15 Gy in animals, and is mediated via PGE2 and another prostaglandin, possibly PGD2. Body temperature is lowered by 20–200 Gy from a direct action of radiation on the brain. Central administration of WR-2721, *S*-(3-aminopropylamino)-ethylphosphorothioic acid, attenuates radiation-induced hypothermia; this effect is correlated with the inhibition of oxygen uptake.

There is a close correlation in rats between skin pigmentation/coat color and susceptibility to radiation injury.

V. GENETIC EFFECTS OF RADIATION

That radiation induces gene mutation in many organisms dramatizes the hazards of radiation. Evaluation of the genetic risk to human beings of exposure to ionizing radiation rests on a variety of experiences. These include: (1) exposure to diagnostic and therapeutic doses of X ray and radioactive materials; (2) occupationally incurred exposures (e.g., in uranium mining or in the maintenance of nuclear reactors); (3) geographic areas with high natural or man-made background radiation levels; and (4) the atomic bombings of Hiroshima and Nagasaki. Various approaches have been employed to determine the occurrence of mutations; these include a search for changes in the frequency of (1) certain population characteristics, for example, the occurrence of major congenital defects and premature death; (2) sentinel phenotypes; (3) chromosomal abnormalities; and (4) altered proteins.

Genetic surveillance of children born in Hiroshima and Nagasaki began in 1948. The indicators studied include sex ratio, weight, viability, presence of gross malformation, death during the first month of life, and physical development at age 8–10 months. Analysis of data through 1953 revealed marginal findings on the sex ratio and survival of liveborn infants; this prompted a continued collection of data on these two variables, using cohorts matched in age, city, and sex. One study involved all infants liveborn in the two cities between May 1946 and December 1958 one or both of whose parents were within 2 km of the hypocenter at the time of the bombing. The second study used a cohort for which one parent was exposed at 2500 m or beyond and the other similarly exposed or not exposed at all. The third study used a cohort in which neither parent was present in the cities at the time of the atomic bombing.

A. Clinical Findings

From 1948 through 1953, 76,617 pregnancy terminations were studied. The findings identified those pregnancies that terminated with a child with a major congenital defect, who was stillborn, or who died during the first week of life. The increase in frequency of an untoward outcome per sievert of gonadal exposure, as measured by a regression coefficient, was 0.001824 (the standard error was 0.003).

The observed frequencies among all parents of untoward pregnancies were used to define a series of "maternal age–paternal age parity"-specific probabilities of an untoward outcome. These values were used to calculate the expected number of such pregnancies in five dose categories (i.e., less than 1, 1–9, 10–49, 50–99, and $\geq$100 cGy) based on the combined parental exposures. The differences between the observed numbers and the expected numbers in each category, the "excess cases," were then regressed on average dose. The data gave a regression coefficient of 63 untoward pregnancy outcomes per million pregnancies per cGy.

To compare this value with the absolute risk of leukemia or cancers other than leukemia, the genetic risk must be related to time. Because the cancer risks are on an annual basis and because the genetic ones represent the time from the bombing to the average year of birth—6 years, as a first approximation—the pregnancy risk per year per cGy is about 10. This differs less than a factor of 2 from the excess cancer incidence cases per million person-years per rad.

Among the 63,817 persons whose parents' exposures are known, there have been 3786 deaths. Distance from the epicenter is known, but the shielding is not known. Through December 1971, 3231 deaths occurred; 321 more deaths occurred from 1972 to 1980. The impact of parental exposure to one sievert is only one-fifth to one-eighth that associated with a year's difference in the time of birth during these years. Seventy-two of 3552 deaths were ascribed to malignancy, the most common being 35 leukemia cases. The frequency of deaths per 1000 at risk varies from 0.36, both parents exposed, to 1.18, only the father exposed. Cancer of the stomach is common in Japan; there was no trend of its occurrence with atomic bomb exposure.

B. Sentinel Phenotypes

Sentinel phenotypes are those that have a high probability of being caused by dominant mutation. Among these phenotypes are aniridia, chondrodystrophy, epiloia, neurofibromatosis, and retinoblastoma. No relation between maternal radiation and Down's syndrome was demonstrated.

The frequency of sex chromosome aneuploids in the children of parents exposed beyond 2499 m was 13/5058 or 0.00257. The frequency in the children of parents exposed within 2000 m, with an average combined exposure of 0.87 Sv, was 16/5762 or 0.00278; the difference is not significant.

C. Biochemical Variants

Children born to parents exposed in Hiroshima and Nagasaki were examined in 1976 for rare electrophoretic variants of 28 proteins of the blood and erythrocytes. Since 1979 a subset of the children has been further studied for deficiency variants of 11 of the erythrocyte enzymes. Another report had information on 642,004 locus tests in 13,052 children whose parents had an average combined gonadal dose of approximately 0.59 Sv. Three probable mutations were described: (1) a slow-migrating variant of glutamate pyruvate transaminase, (2) a slow-migrating variant of phosphoglucomutase-2, and (3) one involving nucleophosphorylase. Three probable mutations were also seen in 478,803 locus tests on 10,609 children whose parents, one or both, were exposed beyond 2400 m (i.e., received less than 1 cSv). These variants involve the haptoglobin 6-phosphogluconate dehydrogenase and the adenosine deaminase systems. The two estimates of the rate of mutation were 0.47 and 0.63×10^{-5}, respectively. Patently, neither these data nor those on structured variants are sufficiently extensive to warrant strong inferences.

D. Estimates of Risk

Two parameters were used to characterize the risk of radiation-related mutation. One is the probability of a mutation at a specific locus per unit of exposure and the other is the so-called doubling dose, the dose at which the ratio of the spontaneous mutation rate to the slope of the dose–response curve is precisely 2. Estimates of the doubling dose on untoward pregnancy outcome, F1 mortality, and sex chromosome aneuploids were made; as rough approximations to the three standard deviations of the doubling dose estimates, 0.93, 3.88, and 24.26 were obtained, respectively. The weighted average of the estimates is 1.4 Sv, with a standard deviation of 1.6 Sv. An analysis that yields an estimate of the risk of an untoward pregnancy outcome with exposure was 63 per million pregnancies per rad. Of a million pregnancies terminating under the circumstances that prevailed in these years, exposure to 1 cGy would have increased the number of 47,500 "naturally ended" pregnancies by some 63 cases. The relative risk would have increased by 1.001. This pregnancy risk is about twice that of cancer among the survivors. If half the pregnancy risk is attributable to radiation-related mutations, presumably a conservative estimate, the genetic and cancer risks would be approximately equal.

VI. IMMUNOLOGICAL EFFECTS OF RADIATION

Ionization radiation was shown to be immunosuppressive in the early 1900s. Ionizing radiation exposure increases the risk of infectious diseases as a result of tissue invasion by microorganisms, through radiation damage to the hemopoietic and intestinal tissues leading to leukopenia and increased permeability of the intestinal mucosa. Alterations of nonspecific defense mechanisms play important roles in decreasing the resistance to infections.

The radiation-induced impairment of specific immune responses is also a relevant component of the pathogenesis of infectious disease. Studies begun in the early 1950s pointed out the immunological effects of radiation. This area of research continues to define further the direct effects of radiation on immunity as well as to expand the basic understanding of immunity itself.

Radiation strongly influences antibody response, reflecting radiation damage mainly of the inductive rather than the productive phase. Antibody response is depressed when antigen is injected shortly before or immediately after radiation in doses of 200–700 rads. Recovery of the response after sublethal doses starts after about a week and may be complete in 2 months. When antigen is given a week or a month after 25–100 rads, prolonged production and transiently higher antibody peak titers may be observed in irradiated as compared with unirradiated animals.

The antibody molecule itself is indeed radioresistant, in that whole-body radiation with 600 rads failed to change the rate of degradation of passively transferred antibodies, and kilorads of X rays failed to affect the antigen-binding capacity of the antibody molecules. Enhancement of antibody titer occurs either as a true stimulation if antigen is injected before relatively low doses of radiation, 25–300 rads, or by the formation of increased amounts of antibody produced at a lower rate if antigen is injected before relatively large doses of radiation, 500–700 rads. Cellular mechanisms may account for this, for example, disproportionate repopulation of the depleted lymphoid tissues by rapidly dividing antigen-activated cells, the adjuvant effect of bacteria or endotoxins entering the bloodstream from the radiation-damaged intestinal mucosa, or preferential inactivation of suppressor T cells. The secondary response, which takes place after a second exposure to the same antigen, is more radioresistant than is the primary antibody response.

Affinity of antibody for antigen is substantially higher in mice irradiated with 450 rads than in unirradiated controls, although the titer of antibody is lower than in the unirradiated controls. Antibody affinity, however, was up to 20-fold higher in irradiated than in control mice.

Changes in T- and B-lymphocyte populations were studied by determining the *in vitro* mitogenic responses of spleen cells from nonimmunized mice exposed to 450 rads at various times before culture. B cells are relatively more sensitive to radiation than T cells. The recovery of T- and B-cell mitogenic responsiveness was found to be incomplete 8 weeks after irradiation, thus providing an explanation for the lower antibody concentration in radiated than in control mice. The high sensitivity of antibody affinity to radiation-induced alterations of the immune system persisted with even smaller doses of radiation, such as 25 rads 5 days after immunization.

The *in vitro* mitotic responses of spleen cells from these immunized and irradiated mice were measured and again indicated a shift in the recovering lymphocyte population in favor of B cells and a relative lack of suppressor T cells. Enhancement of anti-DNP antibody affinity has also been observed in radiation chimeras. Enhanced affinity may play a role in the protection from radiation-induced immunodeficiencies; it may overcome quantifiable defects of the antibody response.

A. Immunological Effects of Radiation at the Cellular Level

Dysfunctions of the antibody response in radiated animals reflect cellular alterations of the immune system.

1. Auxiliary Cells

Auxiliary cells or antigen-presenting cells are polymorphonuclear leukocytes, reticular cells, and macrophages, all involved in antigen processing after immunization.

Macrophage migration and phagocytosis are not affected by radiation even in the kilorad range. Intracellular catabolism of ingested antigens is variably reduced depending on the antigen used, but only from high radiation, in the kilorad range. However, macrophage intracellular levels of lysosomal enzymes are variably increased after irradiation. Thus radiation-induced dysfunctions of the antibody response are likely to depend mainly on lymphocyte radiosensitivity. [*See* Macrophages.]

2. Lymphocytes

Radiosensitivity of lymphocytes is marked; the peripheral lymphocyte count is significantly depressed by 25 rads and reduced to 25% of control by 100 rads. A small number of thoracic duct lymphocytes can still be collected from mice 4 days after 800 rads. Most of these lymphocytes surviving interphase death are T cells, which, however, are unable to survive after entering mitosis on *in vivo* or *in vitro* antigenic stimulation. Immunofluorescence of T cells has shown the existence of two cell subpopulations, one (8%) of which is extremely radioresistant. B cells are more radiosensitive than T cells. Functional heterogeneity of B cells seems to be associated to some extent with different radiosensitivities. On antigen or mitogen stimulation, both B and T cells acquire some degree of radioresistance, presumably as a result of activation of repair mechanisms preventing interphase death. Antigenically stimulated B cells synthesizing antibodies appear to be much more radioresistant than nonstimulated B cells. Exposure of spleen cells in a diffusion chamber to 10,000 rads during the secondary response to an antigen induces a relative increase in the proportion of plasma cells, the cells that release antibodies into the circulation, but no change in the rate at which they synthesize antibody and no changes in cytoplasmic structures. Estimates of the radiosensitivity of mouse antibody-forming cells provided a survival curve D_0 value of 6000 rads for IgM–PFC (immunoglobulin M–plaque-forming cells) and a D_0 value of 1500 rads for IgG–PFC. A threshold of inactivation was also evident.

Recovery of the lymphocyte populations after sublethal irradiation (5, 50, or 500 rads) to mice starts during the first week and seems to be faster for T than for B cells during the subsequent 3 weeks. However, from 4 to 8 weeks after 450 rads,

B cells recover faster and to a greater extent than the T cells. At 8–9 weeks after radiation, both cell types reach the same degree of recovery, but neither one attains the control values of unirradiated mice. [*See* Lymphocytes.]

Gamma radiation of B cells damages the plasma membrane via highly reactive free radicals. There is a rapid increase in plasma membrane permeability and swelling of the cells, which may play a major role in causing interphase death. Five hundred to 1000 rads caused resting B cells to enlarge slightly; 3000 rads resulted in cell doubling in size within 3–4 hr. Sensitivity to the membrane-damaging effects of gamma radiation was on the order of resting B cells > resting T cells > a long-term L3T4+ T-cell clone > a B-cell lymphoma. The radiation effects could be ameliorated by excluding oxygen at the time of irradiation or by adding the free radical scavenger agent cysteamine.

Mice given 200 rads have normal splenic PFC responses to thymus-independent type I antigen but reduced responses to thymus-independent type II antigen. Single-parameter FACS analyses demonstrated a diminution in both B-cell number and the heterogeneity of membrane antigen expression within the surviving B-cell pool. Multiple-parameter FACS analyses indicate that B cells with the sIgM $\gg$ sIgD ("s" indicating secretory) phenotype are more radiosensitive than B cells of the sIgM $\ll$ sIgD phenotype. Enhanced radiosensitivity of marginal zone B cells is also observed.

The major effects of radiation to a resting B-cell antigen-presenting cell (APC) are a reduction in the effective display of antigen plus class II molecules and a loss in the ability to provide APC-derived costimulatory signals.

3. Selective Effects of Radiation on T Cells

The immune system is a complex network of interacting cellular and soluble elements under genetic control. The antibody response is induced in B cells by macrophage-processed antigens and is modulated by signals passed among different types of immunoregulatory T cells. Cellular members of this network express characteristic membrane structures associated with genetically programmed functions. Some cell types also bear membrane receptors specific for antigen and/or for self-structures encoded by major histocompatibility complex (MHC), Igh-V, and Igh-C genes. Assessment of the radiation effects on helper (T_h) and suppressor (T_s) cell activities represents a major issue in understanding radiation-induced immunological dysfunctions.

a. Helper T Cells

Radiation has selective effects on the T_h-cell populations, leading to inactivation of some but not other cell subpopulations. This selectivity gives rise to abnormal T–B cell cooperation, which apparently selects B cells producing antibodies of higher affinity for the antigen.

b. Suppressor T Cells

Suppressor T cells are one of the major cellular components of the immune network regulating the antibody response. T_s cells are induced during antigenic stimulation and exert their functions by reducing the amplification signals provided by T_h cells or by acting directly on effector B cells.

The radiosensitivity of T_s cells has not been thoroughly investigated, but it is a general consensus that they are more radiosensitive than T_h cells. T_s-cell activity was found to be radiosensitive at early but not at later times after induction. A decrease in number of T_s cells may account for the increase in antibody affinity observed in mice irradiated after immunization. The relative radiosensitivity of antigen-unstimulated precursors of T_s cells compared with precursors of T_h cells may provide an explanation for the increase in affinity observed in mice immunized after irradiation.

4. Recent Radiation Effect Findings on Immunity

a. T-Cell Receptor Genes

Radiation is being used to define T-cell receptor genes. Clonal selection of T-cell receptor V beta repertoire, irrespective of positive or negative selection, appears to occur at the early stage of T-cell differentiation (i.e., on the blast-like $CD4^+$ and $CD8^+$ thymocytes).

The sequential differentiation patterns of thymocytes were observed with cell-surface phenotypes and the expression of T-cell antigen receptor in mice irradiated with 800 rads. Intrathymic radioresistant stem cells for T thymocytes seem to proliferate and differentiate after irradiation with the same pattern as was seen in fetal thymus development.

b. Cytokines and Their Antiradiation Effects

i. IL-1-α, TNF-α, G-CSF, and GM-CSF Interleukin-1-α (IL-1-α), tumor necrosis factor-α (TNF-α), granulocyte colony-stimulating factor (G-CSF), and granulocyte-macrophage colony-stimulating factor (GM-CSF) are molecularly distinct cytokines acting on separate receptors. Administered alone, human recombinant (hr) IL-1-α and hrTNF-α protect lethally irradiated mice from death, whereas murine recombinant GM-CSF and hrG-CSF do not. On a dose basis, IL-1-α was a more efficient radioprotector than TNF-α. The relative effectiveness of TNF-α and IL-1-α probably depends on the genetic makeup of the host. The two cytokines together result in additive radioprotection, suggesting that they act through different radioprotective pathways. Suboptimal, nonradioprotective doses of IL-1-α also synergize with GM-CSF or G-CSF to confer optimal radioprotection. [*See* Cytokines and The Immune Response.]

ii. *TNF on Hemopoietic Reconstitution after Sublethal Irradiation* Pretreatment of mice with murine recombinant TNF-α enhanced hematopoietic reconstitution after sublethal irradiation of mice with 7.5 Gy whole body, suggesting a possible therapeutic potential for this compound in the treatment of radiation-induced myelosuppression.

iii. *IL-1-Enhanced Survival from Radiation* IL-1 enhanced the survival of lethally irradiated mice, treated with isogenic or allogenic bone marrow cells. It may prove clinically useful in patients undergoing bone marrow transplantation.

iv. *Immune Modulators* Immune modulators [glucan-F (GF), glucan-P (GP), krestin (K), lentinan (L), pecibanil (P), azimexone (A), ciamexone (C), MVE-2 (M), picolinic acid (PA), and poly IC:LC (PI)] were studied in female C3H/HeN mice when they were given 20 hr before irradiation with 6.5 Gy. Hematopoietic stem cells were measured by endogenous spleen colony assay. Except for A, C, PA, and PI, all immune modulators increased spleen colony assays in a dose-dependent manner (P > M > GP > K > GF > L). Only P, GP, and GF significantly enhanced survival. These studies suggest that immunomodulated-mediated postirradiation hemopoietic stimulation and survival enhancement can be dissociated.

c. T Cells

Inactivation of naturally occurring radiosensitive T_s cells is associated with increased production of IL-2. The increased levels of IL-2 could account for the radiation-induced enhancement of the host's immune system with a consequent effect on tumor regression. In fact, tumor-activated T cells, but not resting T cells, survive irradiation. They are free to expand in number to cause tumor regression in the absence of radiosensitive T_s cells. T_h cells cooperate with radioresistant T effector cells to produce autoimmune thyroiditis in chickens.

Mice given sublethal irradiation to both the thymus and the peripheral lymphoid tissues have major transient and some persistent disruptions in thymic architecture and in thymic stromal components. This radiation regimen induced both transient and long-term effects, for at least 4 months after total lymphoid irradiation.

Irradiation of the thymic microenvironment during marrow ablative preparative regimens may be partly responsible for some of the immune alterations observed in marrow transplant recipients. It may provide a valuable tool for studying the roles of the thymus on T-cell maturation. Induction of autoimmunity in mice by high-dose lymphoid irradiation can be prevented by injection with normal T cells.

5. Immunotherapeutic Applications of Radiation

Whole-body irradiation has been used with increasing success for nearly two decades to ablate neoplastic cells in the bone marrow. It is followed by allogeneic bone marrow transplantation or by a purged autologous graft.

Total lymphoid irradiation (TLI), with sparing of the bone marrow, is used as a treatment of Hodgkin's disease and autoimmune rheumatoid arthritis, and recently in multiple sclerosis patients. Patients receive 2000 rads, in doses of 150–200 rads per fraction for 2–3 weeks, to their cervical, axillary, mediastinal, and hilar lymph nodes and thymus. Patients are immunosuppressed and show long-lasting remissions in their diseases. Several months after the TLI, patients show a marked decrease in the percentage of T cells and Leu-3 positive cells, decreased mixed lymphocyte reaction (MLR), and decreased lymphocyte proliferation to T-cell mitogens; this is correlated with an increased ratio of suppressor/cytotoxic (CD8) to helper/inducer (CD4) T cells. Even with marked immunosuppression, these patients do not experience any increase in incidence of bacterial infections.

B. Miscellaneous Immunological Effects of Radiation

Delayed-type hypersensitivity (DTH) was not inhibited by irradiation at 200 R but was slightly diminished by 900 R. These results imply that immunosuppressive agents, which inhibit cell division, do not inhibit the initiation of DTH but can affect the amplification phase contributed by the host.

Using DNA content as a quantifiable measure of cellular infiltration, it was shown that selective irradiation does not prevent the development of an inflammatory response. This response does not require T lymphocytes.

Host resistance to alloengraftment develops between 8 and 15 days after irradiation and T-cell-depleted syngeneic bone marrow administration. The mechanism of this resistance is unknown but could involve the level of host engraftment and/or the return of host alloreactivity by the time of allogeneic bone marrow transplantation.

Administration of ATP-$MgCl_2$ (60 μM/kg intravenous) during radiation therapy of 4500 cGy of external pelvic exposure over 4 weeks in random-bred pigs decreased colorectal seromuscular ischemia, skin injury in radiation therapy portals, and intestinal inflammation. ATP-$MgCl_2$ thus apears to offer a significant cytoprotection during preoperative radiation therapy.

Copper (II) (3,5-diisopropylsalicylate) (Cu-DIPS) was shown to have radioprotective activity in C57BL/6 mice injected subcutaneously with 80 mg/kg Cu-DIPS and exposed to 8.0 Gy 3 hr later. Cu-DIPS increased the survival rate at 30 days from 40% to 86% without overt toxicity.

The number of white blood cells in unirradiated mice is $10 \times 10^3/mm^3$ blood, and the reserve leukocytes mobilized by dextran sulfate increase this to $30 \times 10^3/mm^3$. The number of reserve white blood cells is thus $20 \times 10^3/mm^3$. One day after 1 or 3 Gy of radiation the number of reserve cells available falls to 11 or $3 \times 10^3/mm^3$, respectively. Thus leukocyte reserves can serve as a good biological indicator for net impairment of hematopoiesis (i.e., total body burden effect of radiation).

VII. MISCELLANEOUS RECENT FINDINGS ON RADIATION EFFECTS

A. Endothelial Cells and Radiation Injury

Radiation injury in cultured endothelial cells is accompanied by delayed progression through the S phase, blocked progression at G2/M phase, and blocked progression 4 hr before the end of G1 phase. Radiation-induced cell loss was not evident, but increased solute flux as tested by albumin and sucrose levels was observed. Both albumin and sucrose flux was dose dependent over the radiation range of 1, 5, and 10 Gy, being 69% more than controls at 4 hr after 5 Gy and 115% more than controls after 10 Gy. The increase occurs without the presence of inflammatory cells or addition of mediators.

Alpha-difluoromethyl ornithine (DMFO) or low doses of putrescine (P) reversed radiation-induced permeability changes, but addition of 25 mol P did not protect the cells and addition of 25 mol or more P to unexposed cells resulted in morphological evidence of injury. The data suggest that polyamines may be important in oxidant-induced pulmonary endothelial cell injury.

B. Cyclooxygenase Inhibitors and Radiation

Decreased aortic responsiveness to U 46619, a thromboxane A2 mimic, after irradiation of the whole body with 20 Gy may be related to a radiation-induced increase in cyclooxygenase product release. It is concluded that cyclooxygenase inhibitors may be important in attenuating vascular injury seen with radiotherapy.

WR 2721, a radioprotectant, was tested on the radiation-induced decrease in vascular activity to U 46619. The former attenuates such a decrease in rat aorta exposed to 20 Gy whole-body irradiation.

C. Vascular Endothelium and Smooth Muscle

Vascular endothelium (EC) and smooth muscle cells (SMC) respond differently to ionization radiation. SMC cultures show no change in cell number or size after large radiation doses (i.e., 3000 rads). In contrast, after radiation with doses comparable to those of clinical radiotherapy (i.e., 200–400 rads), EC cultures show dose-dependent cell loss and subsequent hypertrophy of the remaining cells. A two- to fourfold increase in protein content 48 hr after radiation is proportional to losses in both cell number and cell size. FACS analysis showed no change in the proportion of cells in G0/1, S, and G2.

D. Cardiac Effects of Radiation

Rats irradiated with 5 Gy showed changes in the relative amount of cardiac myosin enzymes v_1, v_2, and v_3, but not any significant increases in serum levels of the enzymes creatine kinase and lactic dehydrogenase, which typically reveal cardiac damage.

E. Gastric Effects of Radiation

Sublethal doses of ionizing radiation, 300 rads of total body in 100 rads/min, inhibited gastric acid secretion in intact animals by a direct action on the stomach, since abdominal shielding reduced the radiation injury. Radiation damage is not likely to be caused by direct effect on the parietal cell receptors for histamine or gastrin because radiation did not affect the *in vitro* response of the gastric glands.

F. Radiation-Induced Lung Tumors

Six lung tumors at necropsy from Beagle dogs that inhaled particles of either $^{239}PuO_2$ or ^{90}Sr as young adults were examined for the aberrant expression of 22 of the known oncogenes. Sequences similar to the N-*ras* and N-*myc* oncogenes appeared to be constitutively expressed in both the tumor and controlling tissues. In one tumor, obtained from a $^{239}PuO_2$-exposed dog (1145T), the H-*ras*, v-*abl*, *erb*-B, and -*met* hybridizations may have been due to cross-hybridization between homologous tyrosine kinase domains, suggesting that the 1145T tumor was expressing a protein tyrosine kinase not observed in the five other tumors examined or in the control tissues.

G. Residual Hematopoietic Damage after Fractionated Gamma Irradiation

Residual damage in hemopoietic progenitor cells, spleen, and granulocyte-macrophage colony-forming cells (CFU-S and GM-CFC) was detected in mice after 15 daily fractions as low as 0.1 Gy. The injury was dose dependent, and after higher total fractionated doses of 7.5–10 Gy, the CFU-S cells recovered to about 50% of control in 2–12 months. Residual damage was also detected in the stroma in the form of reduced numbers of fibroblastoid colony-forming cells and of CFU-S cells in ossicles under the kidney capsule. The response to a second course of 15 fractions 3 weeks after the end of the first course was similar and additive to the response to the first course in the short term; however, in the long run, recovery levels were similar after either one or two courses.

H. Radiation-Induced Lung Injury: Dose–Time–Fractionation Considerations

An isoeffect formula has been specifically developed for radiation-induced lung injury based on a linear-quadratic model that includes a factor for overall treatment time, allowing for the simultaneous derivation of an α/β ratio and a τ/β time factor. From published animal data, the derived α/β and τ/β ratios for acute lung injury are 5.0 ± 1.0 and 2.7 ± 1.4 Gy/day, respectively, whereas for late damage the suggested values are 2.0 and 0.0 Gy^2/day. Data from two clinical studies, one prospective and the other retrospective, were also analyzed and the ratios were determined. For the prospective clinical study, the resultant α/β and τ/β ratios were 0.9 ± 2.6 and 2.6 ± 2.5 Gy^2/day. Combining the retrospective and prospective data yielded α/β and τ/β ratios of 3.3 ± 1.5 and 2.4 ± 1.5 Gy^2/day, respectively. This isoeffect formula might be applied to both acute and late lung injury.

VIII. SUMMARY

This article discussed the acute, subacute, and chronic effects of ionizing radiation on human beings and experimental animals at the whole-body, organ, cellular, biochemical, and immunological level. Cancer and leukemia risk estimates were presented from data from the atomic bomb exposures in Japan and from other specific human radiation exposures such as radiologists, women at menopause, and patients with ankylosing spondylitis. Concerns about mental retardation effects and genetic effects of ionizing radiation also arose from the atom bomb exposures; studies have been ongoing now for many years to assess these concerns. Radon carcinogenesis is widely discussed among radiation experts, but there is not yet a final consensus on it. Anticarcinogenic states and radioprotection were considered. The rapidly expanding science of the biological response modifiers has provided new approaches for the treatment of radiation sickness. Especially important in human biology is radiation injury to the immune system.

BIBLIOGRAPHY

BEIR Committee (1980). "The Effects on Populations of Exposure to Low Levels of Ionizing Radiations." National Academy of Sciences/National Research Council, Washington, D.C.

Burns, F. J., Upton, A. C., and Silini, G. (1984). "Radiation Carcinogenesis and DNA Alterations." Plenum, New York.

Castellani, A. (1984). "Epidemiology and Quantitation of Environmental Risk in Humans from Radiation and Other Agents." Plenum, New York.

Cerutti, P. A., Najjard, O. F., and Simic, M. G. (1987). "Anticarcinogenesis and Radiation Protection." Plenum, New York.

Diethelm, L., Heuck, F., Olsson, O., Strand, F., Vieten, H., and

Zeippinger, A. (1985). Radiation exposure and radiation protection. *In* "Handbuck der Medizenischen Radiologie (Encyclopedia of Medical Radiology)" (F. Heuck and E. Scherere, eds.). Springer-Verlag, New York.

Doria, G., Agarossi, G., and Adorini, L. (1982). Selective effects of ionizing radiations on immunoregulatory cells. *Immunol. Rev.* **65**, 24.

FASEB Journal (1987–1994). "Abstracts on Ionizing Radiation Effects on Immunology," Vols. 1–8.

Journal of Immunology (1987–1994). "Papers on Ionizing Radiation Effects on Immunology," Vols. 138–153.

Kasid, U., Pirollo, K., Dritschilo, A., and Chang, E. (1993). Oncogenesis of radiation resistance. *Adv. Cancer Res.* **61**, 195.

Livesey, J. C., Reed, K. J., and Adamson, L. F. (1985). "Protective Drugs and Their Reaction Mechanisms." Noyes Publications, Park Ridge, New Jersey.

Maillie, H. D., Simon, W., Watts, R. J., and Quinn, B. R. (1993). Determining person-years of life lost using the BEIR V method. *Health Phys.* **64**, 461.

Neta, R. (1992). Radiation effects on immune system. *In* "Encyclopedia of Immunology" (J. M. Roitt and P. I. Delves, eds.), Vol. 3. Academic Press, New York.

Prasad, K. N. (1974). "Human Radiation Biology." Harper & Row, New York.

Salovsky, P., and Shopova, V. (1993). Synergic lung changes in rats receiving combined exposure to paraquat and ionizing radiation. *Environ. Res.* **60**, 44.

Schultheiss, T. E., and Stephens, L. C. (1992). Invited review: Permanent radiation myelopathy. *Br. J. Radiol.* **65**, 737.

Stewart, A. M., and Kneale, G. W. (1993). A-bomb survivors: Further evidence of late effects of early deaths. *Health Phys.* **64**, 467.

Radiation in Space[1]

R. J. M. FRY
Oak Ridge National Laboratory

GLOSSARY

Absorbed dose Energy imparted to matter by ionizing radiation per unit mass of irradiated material at the point of interest; unit of absorbed dose has been the rad and now, in System International (SI) units, is the gray (Gy) (100 rad = 1 Gy)

Alpha particles Nuclei of helium atoms consisting of two protons and two neutrons in close association. They have a net positive charge of +2 and can therefore be accelerated in large electrical devices similar to those used for protons, and they are also emitted during the decay of some radioactive isotopes

Bremsstrahlung Secondary photon radiation produced by deceleration of charged particles

Equivalent dose (H_T) Absorbed dose averaged over an organ or tissue and weighted for the radiation quality of interest. For general radiation protection purposes, radiation weighting factors, w_R's, for particular types of radiation and based on their relative biological effectiveness have replaced the quality factors (Q's). The unit of equivalent dose is joule per kilogram with the special name sievert (Sv). The relationship of quality factor to linear energy transfer is still used to determine the effectiveness of complex mixtures of radiations as occur in space

Extravehicular activity Any activity undertaken by the crew outside a space vehicle

Fluence Number of particles divided by the cross-sectional area of a sphere that the particles enter

Flux Number of particles or photons per unit time passing through a surface

Gray International system unit (SI unit) of absorbed dose of radiation (1 Gy = 100 rad)

Heavy ions Nuclei of elements such as nitrogen, carbon, oxygen, neon, argon, or iron that are positively charged due to some or all of the planetary electrons having been stripped from them

Inclination of orbit Acute angle that the orbit's trajectory makes with the Earth's equator

Linear energy transfer (LET) Average amount of energy lost per unit of particle track length and expressed in keV μm^{-1}

Protons Positively charged nucleus of the hydrogen atom

Relative biological effectiveness (RBE) Factor used to compare the biological effectiveness of absorbed radiation doses from 250-kVp X rays relative to other different types of ionizing radiation; more specifically, the experimentally determined ratio of an absorbed dose of a radiation in question to the absorbed dose of a reference radiation (e.g., 250-kVp X rays) required to produce an identical biological effect in a particular experimental organism or tissue; if 10 mGy of fast neutrons equaled in effect 20 mGy of 250-kVp X ray, the RBE of the fast neutrons would be 2

Sievert (Sv) SI quantity of radiation equivalent dose; equal to dose in gray times, a quality factor that depends on LET

Z Atomic number; the number of protons in an atomic nucleus or the positive charges in the nucleus

RADIATION IN SPACE IS CONVENTIONALLY DIvided into trapped particle radiation, galactic cosmic radiation, and solar particle radiation. The exposures

[1]Research sponsored by the Office of Health and Environmental Research, U.S. Department of Energy, under Contract DE-AC05-840R21400 with the Martin Marietta Energy Systems, Inc. This manuscript has been authored by a contractor of the U.S. Government under contract No. DE-AC05-840R21400. The U.S. Government's right to retain a nonexclusive, royalty-free license in and to the copyright covering this paper, for governmental purposes, is acknowledged.

ENCYCLOPEDIA OF HUMAN BIOLOGY, Second Edition, VOLUME 7. Copyright © 1997 by Academic Press. All rights of reproduction in any form reserved.

incurred during space missions are influenced by altitude, inclination of the orbit, shielding, and the duration of the mission.

Missions within the Earth's geomagnetic field at low altitudes and inclinations are protected largely from galactic cosmic radiation and solar particle event radiation.

In geosynchronous orbits the dose from radiation is primarily from electrons of the outer radiation belt, which exhibit marked temporal variations in intensity. Galactic cosmic radiation and solar particle events also contribute to dose.

Outside the magnetosphere, galactic cosmic rays (GCR) are the major source of radiation. Although protons are the predominant radiation of galactic cosmic rays, there are helium ions and heavier ions called high-Z and high-energy (HZE) particles. These particles are of particular interest radiobiologically and are of concern for risk estimates of late effects of radiation. The dense ionization along a particle track of the heavy ions that traverses many cells is in contrast to gamma or X rays.

In deep space, solar particle events occur unpredictably. When these events are large, the proton dose rate may rise rapidly to levels that could cause acute radiation damage. The design of shielding to prevent these effects is an important feature in the planning of deep space missions.

The risks from exposure to protons and electrons can be estimated with some confidence; this is not the case for HZE particles because of the density of the ionization and the length of the particle track. Although the HZE component of cosmic rays is small, its contribution to the biological effects of radiation on missions of long duration in deep space is of concern.

I. SOURCES OF SPACE RADIATION

There are three sources of radiation in space: (1) trapped particle radiation, (2) galactic cosmic radiation, and (3) solar particle radiation.

A. Radiation Belts

There is a complex system of magnetic fields within the magnetosphere above the densest part of the Earth's atmosphere that are populated by trapped electrons, protons, and some low-energy heavy ions. These particles are reflected back and forth between regions of magnetic field strength or mirror points in the Northern and Southern hemispheres. The movement of the particles is also complex and, because of the different charges, the electrons drift eastward and the protons westward. In 1960, cosmic-ray detectors were put on Explorer 1 and Explorer 3. Initially, about 600 miles above the Earth's surface, the cosmic-ray flux appeared to decrease dramatically. However, Van Allen and colleagues (1960) deduced and determined that their detectors were saturated by a huge and unexpected flux of trapped electrons and protons now known as the Van Allen radiation belts (Fig. 1). There are two regions of trapped electrons, designated inner and outer, although they are not completely distinct. Considerable dynamic changes occur in the radiation belts, and sometimes there appears to be a third belt. The maximum intensities and energies of the electrons in the outer zone are much higher than those in the inner zone. The inner zone extends out to about 2.8 Earth radii (R_e) (about 18,000 km) and the outer zone from about 2.8 R_e to 12 R_e. In the outer zone, the trapped electrons are affected by changes in the geomagnetic field caused by solar activity. At the altitude of geosynchronous orbits, about 36,000 km, the intensity of electrons shows a diurnal cycle caused by the interaction between the solar wind and the magnetosphere. These interactions produce changes in the magnetic field locally and therefore in the trapping strength; as a result the intensities of the electrons may vary in amplitude by a factor of 10. Smaller variations occur with a periodicity of the solar cycle, and magnetic storms cause sporadic fluctuations. The

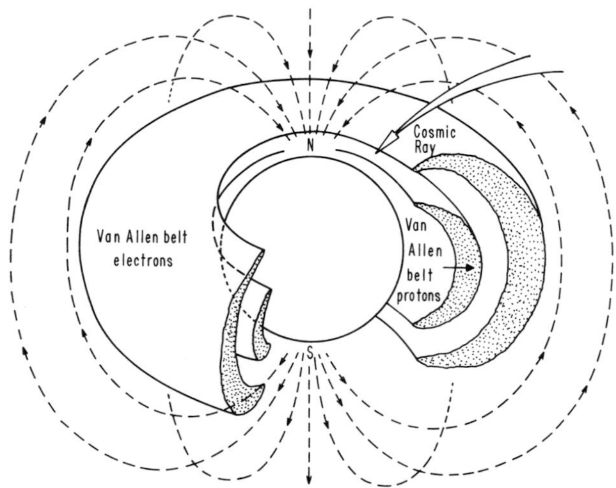

FIGURE 1 The magnetosphere and radiation belts. The inner and outer Van Allen belts are shown as distinct and separate for illustrative purposes, but the separation is not complete.

flux of electrons with energies greater than 45 keV may increase rapidly, in a matter of minutes, by an order of magnitude. So-called substorms occur frequently and result in the injection of elements with energies between 50 and 150 keV from the magnetosphere tail.

The flux of the higher-energy electrons, over 200 keV, remains relatively constant. This degree of detail in the description of the radiation environment is important because short-term missions to altitudes of geosynchronous orbit that may involve extravehiclar activity are likely to be required in the future. Good planning of such activities based on the knowledge of the wide swings in radiation intensity and their periodicity could limit the exposure of the astronauts to radiation markedly. The flux of electrons in the inner zone is less than the peak flux in the outer zone by about a factor of 10 and contributes little to the radiation exposure of space vehicles and crews in low-Earth orbits.

B. South Atlantic Anomaly

The trapped protons, which predominate in the inner radiation belt, extend out to about 3.8 R_e. The higher-energy protons are trapped in a much smaller volume than the electrons but in the range of altitudes that are of importance for low-Earth orbits. In the region between Africa and South America, the spiraling protons reach down closer to the Earth than in other regions. This anomalous distribution of the trapped protons is due to a displacement of the magnetic dipole from the Earth's center that alters the geomagnetic field.

It is the exposure to radiation in the South Atlantic Anomaly of the inner belt that is important for space missions in low-Earth orbits. The altitude and the orbital inclination of the spacecraft determine the rate of exposure to radiation in the South Atlantic Anomaly. Figure 2 shows that the dose rate in the radiation environments of low-Earth orbits is influenced by the altitude of the orbit. Data shown in Fig. 3 all came from missions of the shuttle vehicles, all of which had an orbital inclination of 28.5°. The choice of orbit is influenced by a number of factors, including the site of the launch. Most of the Soviet missions have been at higher orbital inclinations such as 56° and also usually at lower average altitudes. Both of these features of the missions reduce the exposure to the radiation encountered in the South Atlantic Anomaly. The dose rate increases with altitude as the orbit involves increasing radiation exposure in the South Atlantic Anomaly. The solar activity also influences the trapped protons, particularly with lower altitudes of

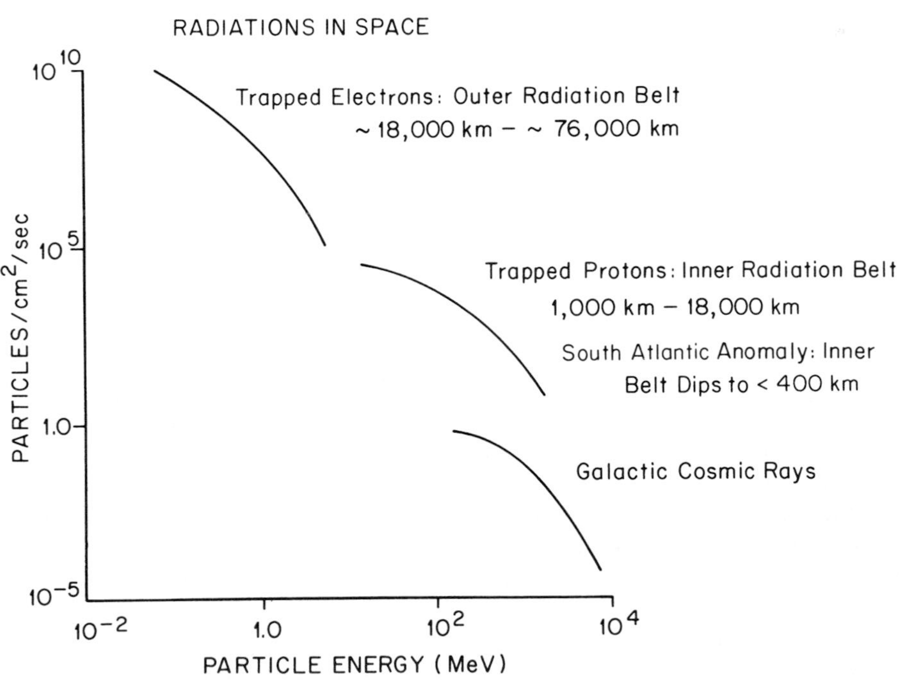

FIGURE 2 The ranges in flux and energy of the particles.

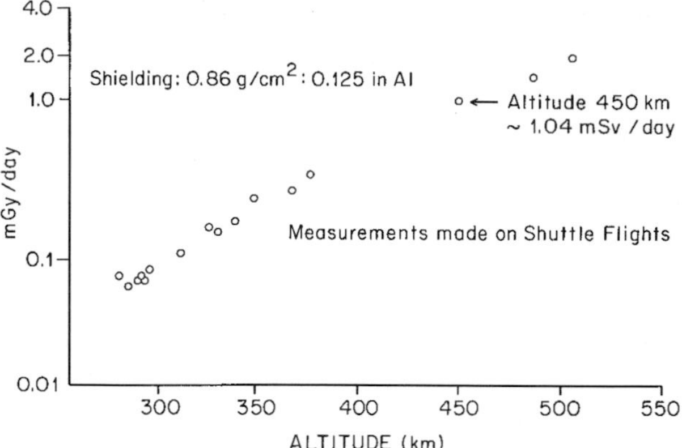

FIGURE 3 Dose rate of radiation as a function of altitude of the space shuttle in an orbit of 28.5°. Note that dose rate is plotted on a log scale indicating the marked influence of altitude. The dose equivalent at 450 km is given in old (rem) and new (mSv) SI units, which take into account the different effects of the radiations involved. [The figure is based on data reported in 1988, *Health Physics* 55, 159–164, with permission.]

the radiation belt. In the active part of the solar cycle, the proton intensities are reduced in contrast to the electrons in the outer region of the radiation belt. There is also a small flux of low-energy trapped heavy ions. The energy of these heavy particles appears to be too low to affect either humans or electronic components of satellites.

On missions to deep space, the space vehicle must pass through the radiation belts. Although the flux of both protons in the inner region and the electrons in the outer region is high, the time to traverse the radiation belts is short, which limits the total dose. The equivalent dose incurred in the low-Earth orbits and traversal through the Van Allen radiation belts to reach deep space has been estimated to be 20 mSv. This equivalent dose is about the same as or less than the dose incurred with a whole-body computerized axial tomography (CAT) scan used for medical diagnoses.

C. Galactic Cosmic Rays

In deep or interstellar space, the radiation environment consists of galactic cosmic rays, solar cosmic rays, anomalous cosmic rays, and solar-flare products. Galactic cosmic rays are the most important component of the deep-space radiations, which range from protons with a Z of 1 to uranium with a Z of 92 (for relative fluxes, see Simpson, 1983; Mewaldt, 1988). GCR arise from a source outside our solar system, but what precisely is the source is still a matter of conjecture; perhaps they arise from explosions of supernova.

In the early part of this century, instruments were sent up in balloons to measure the Earth's natural background radiation at high altitudes. The results were contrary to expectation. It had been assumed that as the instruments got further and further from Earth the background radiation from Earth would decrease correspondingly. The surprise was that the radiation levels increased with altitude. The measured levels rose because of radiation streaming in from the cosmos. The cosmic rays that were the source of the radiation detected by the instruments in the balloons were secondary cosmic rays. The primary cosmic rays are altered in energy and composition by nuclear reactions with atoms in the atmosphere and by the magnetosphere. The cosmic rays reaching the Earth contribute about the same amount of radiation to our background radiation on Earth as the terrestrial radiation that comes from rocks and soil (excluding radon).

In deep space the radiation consists of galactic cosmic rays and solar particle radiation. It is thought that cosmic rays are uniformly distributed outside the solar system. This uniformity results in an equal flux of charged particles that approach the solar system from all directions.

About 98% of the primary galactic cosmic rays are protons and heavier ions. The flux rate is greatest for particles in the energy range of 100 MeV to 10 GeV per nucleus. The radiation consists of 87% protons, 12% helium ions (alpha particles), and 1% heavier ions. Ions heavier than helium, which have been described as HZE particles, are of particular interest to the radiobiologist. The abundance of the various ions varies. Because the capacity for ionization is proportional to Z^2, it is the high-Z particles that are of concern. The fluences of the ions above atomic number of iron (26) are very low. Thus iron with a high Z and a significant fluence is of particular importance.

As noted earlier, the cosmic rays detected on Earth are largely secondary cosmic rays because the magnetosphere and the Earth's atmosphere act as a giant shield. The only gaps in that shield are the polar regions (Fig. 1), where cosmic rays can enter the Earth's atmosphere with less restriction. Thus in low-Earth orbits at 90° orbital inclination, so-called polar orbits, the flux of GCR is greater than at lower inclinations. The contribution from galactic cosmic rays to the total dose in a polar orbit is about three times that in a comparable low-Earth orbit.

Energies of the galactic cosmic rays may reach as high as 10^{20} eV. The Sun has a major influence on galactic cosmic rays. The solar wind formed from highly ionized gas emitted by the Sun carries magnetic fields that point radially away from the Sun. These magnetic fields reduce the intensity of the lower-energy particles, especially during the most active part of the solar cycle, which is called solar maximum. The fluence rate, which is highest during the less active part of the solar cycle, called solar minimum, is about $4/cm^2$, about twice that during solar maximum.

The difference in the radiation level between solar minimum and maximum is sufficient to be of importance to the planning of missions to realms beyond the magnetosphere. However, the potential gain by going during the solar maximum segment of the solar cycle, when the ambient radiation will be at a lower level, comes at the cost of a higher probability of a solar particle event at a magnitude that would contribute significantly to the dose of radiation incurred by crew members.

D. Solar Radiation

The radiation from the sun consists of photonic and particulate emissions and includes electromagnetic radiation spanning radio- to gamma-ray frequencies and particle radiation ranging broadly in flux, fluence, and energy of protons, as well as energetic ions and a small contribution from electron events.

The Sun, a very complex dynamic star, is the constant source of energy that makes life on Earth possible. The solar irradiance received by the Earth varies only about 3%, but the activity of the Sun and variations in solar particle and photonic emissions change over a number of cycles, the best known of which is the approximately 11-year solar cycle. As noted earlier, the solar cycle is divided into solar maximum, the period of greater activity, also referred to as active sun, and solar minimum, a quieter period. For centuries the cyclic behavior of sunspots has intrigued diligent recorders. As understanding of the energy production has increased, so have the concepts about the processes of its release.

The cyclic nature of solar activity is detected most easily by observations of the sunspot and solar flare activity. For well over a thousand years sunspots have intrigued observers, and records of the number of sunspots have been kept. Records over the last two to three centuries made it possible to distinguish and enumerate the individual cycles. We have recently entered what is designated the 23rd cycle. Over this period, the average length of the cycles has been about 11 years, with a period of increasing activity of about 4.8 years and a period of decrease in activity of about 6.2 years. These sunspots result from locally strong magnetic fields.

Solar flare is a term that has undergone changes in meaning over recent years and now includes all electromagnetic and partial emissions associated with an explosive release of energy concentrated in a small volume of the solar atmosphere. During the active period of the solar cycle there are events involving plasma emission and high fluxes of energetic ions. Both types of emissions are important to activities in space as well as on Earth. In days long past, the sighting of the aurora borealis at much lower latitudes than was usual was the visible manifestation of very large electromagnetic storms. As technology has become more widespread and sophisticated, so has the impact of these solar events. The effect on navigation was recognized more than a century ago, and now interference with radio communications and the performance of satellites and even the disruption of the supply of electric power have been experienced and correctly attributed to changes in the emissions of the sun. The energy imparted to the Earth's atmosphere causes heating and expansion. These changes alter the drag on space vehicles and satellites. This effect was seen dramatically in 1972 when Skylab was dragged into an orbit at a lower altitude.

Several classifications of solar particle events have been suggested, one of which is shown in Table I. Any such classification obscures the fact that the distribution of fluences among solar proton events can probably be described by a log normal function. The very large events occur infrequently, about one to three per solar cycle. In the very large events, such as the series of events in 1972, the energy released was estimated to be equivalent to 10^{19} kilowatt hours, many thousand times the current annual energy consumption in the United States.

TABLE I
Characteristics of Solar Particle Events

	Type of event		
	Small	Intermediate	Very large
Frequency (solar cycle dependent)	5–20/year	3–6/year	1–3 cycle
Peak flux (p cm^{-2} sec^{-1}) (>10 MeV)	$10–10^2$	$10^3–10^4$	10^5
Fluence (p cm^{-2}) (>10 MeV)	$10^5–10^7$	$10^8–10^9$	$>10^9$

The other type of solar emission is the release of high fluxes of energetic ions. Although this particle radiation can affect communications in polar regions, it is the potential effects of the showers of energetic protons on the occupants of space vehicles on interplanetary missions that are of concern.

The speculation, if not the understanding, of the source of these particle events has been changing. It had been assumed that solar energetic particles arose from material in or above regions of solar activity. Recent determination of the composition and other evidence suggest an additional source of solar particles. Large-scale coronal mass ejections from the Sun generate shock waves that can accelerate some of the ions in the solar wind, and the accelerated particles move away along the interplanetary magnetic field lines. The current conclusion is that some of the events are associated with solar flares, especially those rich in Fe and ^{3}He. Another class of events, some of which are associated with coronal mass ejections or with X-ray-emitting solar flares, have a higher flux and fluence.

The major concern about very large particle events is related to missions to the Moon or Mars. These events result in a shower of protons varying in energy. The dose rate rises rapidly and, within hours, reaches a rate sufficient to cause acute radiation effects to crews on missions in deep space. Because of the preponderance of low-energy protons in these very large particle events, the dose in the radiosensitive deep organs, such as the bone marrow and gut, fortunately is likely to be very much less than to the more radioresistant skin. There has been a concerted effort to develop methods of predicting the occurrence of these events. The difficulty is in deciding which active region on the Sun's surface will develop into solar flares and the fact that very few of these become large particle events. Predicting the eventual size of the solar flare by monitoring the development of the flare and how the proton dose rate builds up has been more successful. The time taken by energetic protons from a solar flare to reach Earth varies, depending on the position of the origin on the Sun and other factors, between less than 1 hr to several hours. Geosynchronous orbit Earth satellites provide continuous proton monitoring and transmission of data. It is hoped that an X-ray imager can be put in space to detect flares that arise behind the limbs of the Sun. With this knowledge, even in deep space, there would be sufficient warning to allow astronauts to seek a shielded area before the dose rate reached levels that could cause acute radiation effects. It is possible that when a return to the Moon or a visit to Mars is planned, it will be planned to take place in the less active period of the solar cycle, called solar minimum. If not, an adequately shielded storm shelter must be provided.

II. RADIATION AND SPACE MISSIONS

A. Low-Earth Orbits

In low-Earth orbits the radiation exposure rate is considerably greater than on Earth. Increases in radiation exposure are of concern because of the assumption that there is no threshold for the induction of cancer or genetic effects, the so-called stochastic effects of radiation. In the case of acute radiation effects, such as damage to skin or the blood-forming tissues, clinically significant effects do not occur until the threshold doses for the specific effects are exceeded. Threshold doses for noncancer or deterministic effects will not be exceeded on any space mission within the magnetosphere. A most important feature of radiation in space, as well as on Earth, is a low dose rate. All the effects of radiation are reduced by decreasing the dose rate. In the case of nonstochastic effects, the low dose rate allows time for repair of the damage to DNA and, in tissues that undergo cell renewal, time to replace damaged cells. The effect of dose rate on cancer induction in humans is not known adequately, but experimental studies in animals indicate that lowering the dose rate reduces the carcinogenic effect of low linear energy transfer (LET) radiations significantly. [*See* Radiation, Biological Effects.]

Doses incurred by the crews of spacecraft have been low. The highest doses received by U.S. astronauts occurred in the Skylab missions, which had the longest duration of any of the U.S. space missions. The average total bone marrow dose equivalent was about 77 mGy for Skylab 4, which orbited for 84 days at an altitude of 435 km and an orbital inclination of 50°.

In space missions that are within the magnetosphere, the radiation environments can be predicted with considerable confidence. Exposures to crews can be controlled by limiting the duration of the mission, choosing a more benign altitude, and providing appropriate shielding.

B. Deep Space Missions

In deep space, low dose-rate proton radiation predominates, and the risks are relatively well understood. However, there is also the small heavy ion component in galactic cosmic rays, and the risks that these heavy charged particles pose are not well understood. Tobias of the Lawrence Berkeley Laboratory predicted that astronauts would see infrequent flashes of light in deep space because the heavy ions would be capable of stimulating the retina. The astronauts did see flashes. The concern is whether the high-LET radiations with long particle tracks of dense ionizations and accompanying electron tracks that traverse a number of cells can cause damage that is more severe than that induced by low-LET radiations such as X rays. In other words, is the radiobiological effectiveness (RBE) of the heavy ions much greater than X rays for both cancer and noncancer late effects of radiation? The RBEs for a number of effects, such as cell killing, mutation, cancer induction, and cataractogenesis, increase with increasing LET up to about 200 keV/μm. Information about the risks of the late effects of radiation from HZE particles is under study. For example, if the energy deposited by an iron nuclei is 676 times (26^2) greater than that deposited by a cosmic-ray proton, how much greater are the biological effects? Without such information, the risks of missions in deep space, such as to Mars, will be difficult to estimate. [*See* Radiobiology.]

The doses received by the crews on the Apollo missions were low because of the short duration of the missions; the longest mission, Apollo 15, was about 12.5 days. There was a certain amount of luck that the doses were so low in the case of Apollo 15 and 16, because they were undertaken within months of the largest recorded solar particle event. If either mission had coincided with the solar particle event, the doses incurred by the astronauts could have been sufficient to cause serious radiation effects. The highest dose received by any of the crews of the Apollo missions was about 11 mGy on Apollo 14. This dose is of the same order as a dose a patient would receive from a whole-body CAT scan and poses no significant risk for late effects from the radiation exposure.

As discussed earlier, large solar particle events that occur during solar maximum are a hazard for deep space missions, and appropriate shielding, which can be effective against proton radiation, will have to be incorporated.

Although the radiation environments in space are much less benign than on Earth, there is good reason to believe that humans can travel and work in space without unacceptable levels of risk of radiation effects. The design of space vehicles and careful mission planning can provide adequate conditions of radiation safety for missions that involve other considerable hazards.

BIBLIOGRAPHY

Benton, E. V., Adams, J. H., Sr., and Panasyuk, M. I. (eds.) (1996). Space radiation environment: Empirical and physical models. *In* "Radiation Measurements," Vol. 26, pp. 303–539. Pergamon Press, Exeter, London.

Horneck, G., Bücker, H., Cox, A., Todd, P., Yang, T. C., Worgul, B. V., Donlon, M., Atwell, W., Shea, M. A., Smart, D. F., Fry, R. J. M., Townsend, L. W., Curtis, S. B., and Swenberg, C. E. (eds.) (1994). "Advances in Space Research: Life Sciences and Space Research XXV (2)." Pergamon Press, Oxford.

Lee, B. L., Gibson, A. M., Wilson, R. S., and Thomas, S. (1995). Long-term total solar irradiance variability during sunspot cycle-22. *J. Geophys. Res.* **100**, 1667–1675.

Mewaldt, R. A. (1988). Elemental composition and energy spectra of galactic cosmic rays. *In* "Proceedings of Interplanetary Environment," pp. 121–132. J. Feynman and S. Gabriel, eds., Jet Propulsion Laboratory Publication 88-28.

NCRP (1989). Guidance on Radiation Received in Space Activities." NCRP Report 98, National Council on Radiation Protection and Measurements, Bethesda, MD.

Rust D. M. (1982). Solar flares, proton showers and the space shuttle. *Science* **216**, 939–946.

Simpson, J. A. (1983). Elemental and isotopic composition of the galactic cosmic rays. *Annu. Rev. Nucl. Sci.* **33**, 323–381.

Smart, D. F., and Shea, M. A. (1996). Solar radiation. *In* "Encyclopedia of Applied Physics," Vol. 18, pp. 393–429. VCH, Weinheim/New York.

Van Allen, J. A. (1960). "The First Public Lecture on the Discovery of the Geomagnetically Trapped Radiation." Report 60-13, State University of Iowa, Ames, IA.

Radiation Interaction Properties of Body Tissues

DAVID R. WHITE
The Royal Hospitals NHS Trust, London

GLOSSARY

Absorbed dose (D) Mean energy ($d\bar{\varepsilon}$) imparted by ionizing radiation to a given mass (dm) of material ($D = d\bar{\varepsilon}/dm$). [Units: Gray (Gy); 1 Gy = 1 J kg^{-1}]

Compton scattering Interaction of a photon (X ray or γ ray) with an electron of a target atom in which the photon is scattered (deviated from its path) with a reduction in its energy and the electron recoils with additional energy. Compton (recoil) electrons contribute to the absorbed dose in irradiated body tissue.

Electron stopping power (S/ρ) Energy lost (dE) by an electron in traversing a given thickness (dx) of material of mass density ρ [$S/\rho = (1/\rho)(dE/dx)$]. (Units: MeV m^2 kg^{-1})

Exponential attenuation Narrow, monoenergetic beam of photons passing through a material of thickness x is reduced in number according to the exponential relationship $N = N_0 e^{-\mu x}$, where N is the number of transmitted photons, N_0 the number of incident photons, and μ the linear attenuation coefficient

Mass attenuation coefficient (μ/ρ) The fraction of photons (dN/N) that experience interactions in traversing a given thickness (dx) of material of mass density ρ [$\mu/\rho = (1/\rho N)(dN/dx)$]. (Units: m^2 kg^{-1})

Pair production Interaction of a high-energy photon ($\geq$1.02 MeV) as it passes near the nucleus of a target atom. The photon disappears and an electron–positron pair of particles is produced (a positron is a "positive" electron)

Photoelectric absorption Interaction of a photon with an electron of a target material in which all the energy of the photon is transferred to the electron. The electron is ejected from the target atom. These photoelectrons contribute to the absorbed dose in irradiated body tissue

Radiation energy (electron-volt, eV) Kinetic energy gained by an electron when accelerated through a potential difference of 1 V (1 eV = 1.6×10^{-19} J; 10^3 eV = 1 keV; 10^6 eV = 1 MeV)

Tissue substitute Material used to simulate a particular body tissue with respect to a given set of physical characteristics. The physical characteristics may be the radiation interactions in the body tissue or the absorbed dose at a point of interest in the tissue. A volume of a tissue substitute is called a phantom

RADIATION INTERACTIONS THAT OCCUR IN IRRADIATED body tissues are dependent on the type and energy of the radiations being used and the elemental composition and mass density of the tissues. Photons (X rays and γ rays) interact with tissue by an energy-dependent combination of photoelectric absorption, Compton scattering, and pair production. Electrons passing through tissue lose energy by electronic collisions and, to a lesser extent, by the production of bremsstrahlung photons. The magnitude of the radiation interactions is influenced by the compositions of the irradiated tissues. These compositions, which fall into two categories, the soft tissues (adipose tissue, muscle, etc.) and the hard skeletal tissues (cortical bone), vary with the age, nutrition, state of health, and physical activity of the individual.

ENCYCLOPEDIA OF HUMAN BIOLOGY, Second Edition, VOLUME 7. Copyright © 1997 by Academic Press. All rights of reproduction in any form reserved.

I. THE COMPOSITION OF BODY TISSUES

A. Introduction

The composition of body tissues depends on the age, nutrition, state of health, and physical activity of the individual. Changes may be very large as in severe malnutrition or in fatty infiltration of specific organs such as the heart or liver. On the other hand, cortical bone remains constant with respect to elemental composition throughout adult life, although its total mass in the body decreases with the aging process.

To determine elemental composition, the basic components of a tissue specimen are first evaluated by chemical and physical separations. Once the water, lipid, protein, carbohydrate, and mineral contents are established, the overall elemental composition of the tissue can be calculated from the accepted compositions of the components. Mass density can be derived from mass and volume estimations on the fresh specimen. As these measurements are sparse in the literature, mass densities must frequently be estimated by calculation, knowing the densities of the components and their mass proportions in the specimen.

The discussion here is devoted to *human* body tissue. The composition and radiation interaction data given will be predominantly for healthy tissues, with only an outline of the probable variations in composition due to over- and undernutrition, disease, and physical activity.

B. Age Dependence

The chemical composition of the body's organs varies before and after birth. Changes in composition continue until the organs have matured, which may be soon after birth for some (e.g., liver and muscle) or much later for others (e.g., cortical bone).

These changes with age are illustrated in Table I. Adipose tissue, often incorrectly referred to as "fat," is composed of a protein matrix supporting cells (adipocytes) highly specialized for the storage of lipid, a mixture of triglycerides of various long-chain fatty acids. It is the most variable tissue in the body, in particular its contribution to body mass, typically making up approximately 18 and 29%, by mass, of the healthy adult male and female body, respectively. The lipid in the adipose tissue of a full-term newborn baby (34.7% by mass) will more than double by adulthood (74.1% by mass). [*See* Adipose Cell.]

TABLE I
The Components of Adipose Tissue, Muscle (Skeletal), and Cortical Bone (Fetus to Adult)

Body tissue and age group	Percentage by mass				
	Water	Lipid	Protein	Carbohydrate	Minerals[a]
Adipose tissue					
Newborn	59.7	34.7	5.4	—	0.2
Infant (2 days–10 months)	47.6	47.2	5.0	—	0.2
Child (1–18 years)	41.1	55.0	3.7	—	0.2
Adult	21.2	74.1	4.4	—	0.3
Muscle (skeletal)					
Fetus (15 weeks)	90.4	0.9	7.9	—	0.8
Newborn	80.6	2.0	14.0	2.8	0.6
Infant (3 months)	79.2	2.0	17.2	1.0	0.6
Infant/child (6 months–18 years)	(As adult, but lipid increasing 2.1 to 4.2%)				
Adult	74.1	4.2	19.8	1.0	0.9
Cortical bone					
Fetus (20 weeks)	41.2	—	20.4	(5)[b]	33.4
Newborn	21.1	—	24.3	(5)[b]	49.6
Infant (3 months)	23.2	—	24.8	(5)[b]	47.0
Child (1 year)	20.7	—	26.1	(5)[b]	48.2
Child (10 years)	16.9	—	25.8	(5)[b]	52.3
Adult	12.2	—	24.6	5.2	58.0

[a]Includes remaining elements with atomic numbers above 8.
[b]Estimated, not measured, values.

Muscle (skeletal) comprises the connective tissue, blood vessels, blood lymph, and so on usually associated with striated muscle and accounts for some 40 and 30%, by mass, of the adult male and female body, respectively. Fetal muscle has a very high water content (90.4% by mass). During development the percentage of water decreases and the concentration of protein increases. Fatty infiltration also increases with age. [*See* Connective Tissue; Skeletal Muscle.]

Cortical bone is made up of a protein matrix supporting minerals containing phosphorus and calcium, the major compound being hydroxyapatite, $Ca_{10}(PO_4)_6(OH)_2$. The bone mineral steadily increases with age at the expense of the water content. A rapid increase in bone mineral occurs from the twentieth week of gestation (33.4% by mass) to full term (49.6% by mass). This is followed by a slower increase, reaching 58.0% by mass, by 18 years of age and staying at this level during most of adult life. Whole bones will have average compositions that are significantly different from those of cortical bone due to the "diluting" effect of other skeletal components present, such as red and yellow marrow.

The resulting elemental composition of adipose tissue, muscle (skeletal), and cortical bone for the same age groups considered in Table I are listed in Table II. The importance of the hydrogen, carbon, nitrogen, and oxygen elemental groups in the soft tissues and, in addition, phosphorus and calcium in cortical bone is evident from the tabulation.

The trends indicated for muscle (skeletal) in Tables I and II are repeated in the other soft tissues, with time scales being dictated by the rate of maturation of the specific tissues. Once they have reached their mature state, soft tissues have similar compositions, with water contents typically in the range of 60 to 80% by mass (Fig. 1). Higher water contents are found in body fluids such as cerebrospinal fluid and urine, whereas reduced amounts occur in adipose tissue and yellow marrow.

C. Other Influencing Factors

Marked changes in body tissue composition can be caused by both over- and undernutrition. In overnutrition, adipose tissue may be changed considerably, with both the amount of adipose tissue in the body *and* the lipid concentration within the tissue increasing sharply. An obese individual may have adipose tissue with a lipid content of over 85% by mass. Conversely, a lean person may have less than 50% lipid by mass. Severe malnutrition has a large effect on soft tissues such as muscle. Protein is lost and the water content increases, producing elevated concen-

TABLE II

The Elemental Composition and Mass Density of Adipose Tissue, Muscle (Skeletal), and Cortical Bone

Body tissue and age group	Percentage by mass											Mass density (kg m^{-3})
	H	C	N	O	Na	Mg	P	S	Cl	K	Ca	
Adipose tissue												
Newborn	11.1	29.7	0.9	58.0	0.1	—	—	0.1	0.1	—	—	990
Infant (2 days–10 months)	11.2	39.2	0.9	48.4	0.1	—	—	0.1	0.1	—	—	970
Child (1–18 years)	11.3	44.5	0.6	43.3	0.1	—	—	0.1	0.1	—	—	960
Adult	11.4	59.8	0.7	27.8	0.1	—	—	0.1	0.1	—	—	950
Muscle (skeletal)												
Fetus (15 weeks)	10.8	4.9	1.3	82.1	0.2	—	0.1	0.1	0.3	0.2	—	1030
Newborn	10.4	10.3	2.4	76.2	0.1	—	0.1	0.1	0.2	0.2	—	1050
Infant (3 months)	10.3	11.2	2.9	74.8	0.1	—	0.1	0.2	0.2	0.2	—	1050
Adult	10.2	14.3	3.4	71.0	0.1	—	0.2	0.3	0.1	0.4	—	1050
Cortical bone												
Fetus (20 weeks)	6.4	12.8	3.5	57.1	0.1	0.1	6.1	0.2	0.2	—	13.5	1430
Newborn	4.4	15.3	4.1	47.7	0.1	0.2	8.5	0.2	0.1	—	19.4	1720
Infant (3 months)	4.7	15.4	4.2	48.5	0.1	0.2	8.2	0.2	0.1	—	18.4	1680
Child (1 year)	4.5	15.9	4.4	46.7	0.1	0.2	8.7	0.3	0.1	—	19.1	1710
Child (10 years)	4.0	15.9	4.4	45.0	0.1	0.2	9.6	0.3	0.1	—	20.4	1790
Adult	3.4	15.5	4.2	43.5	0.1	0.2	10.3	0.3	—	—	22.5	1920

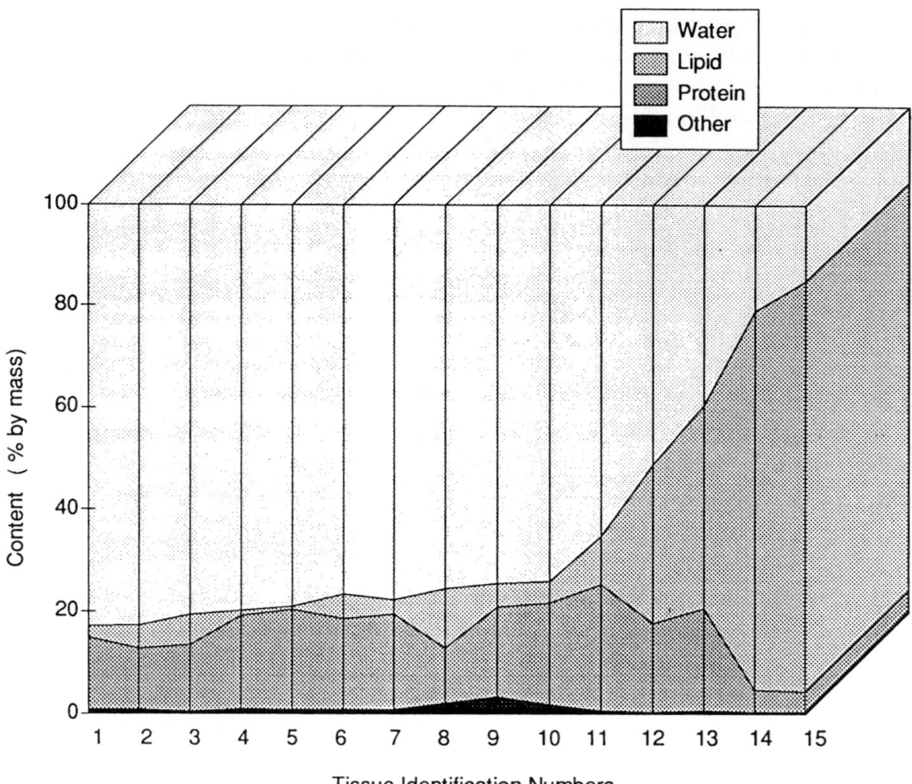

FIGURE 1 The water, lipid, and protein components of adult soft tissues, plotted in order of decreasing water content. Tissue identification: (1) Ovary, (2) testis, (3) GI tract (intestine), (4) lung, (5) blood (whole), (6) kidney, (7) heart, (8) brain (whole), (9) liver, (10) muscle (skeletal), (11) skin, (12) mammary gland, (13) red marrow, (14) adipose tissue, (15) yellow marrow.

trations of sodium and chlorine and suppressed concentrations of potassium and phosphorus. In children, undernutrition may restrict growth and the expected changes in composition with development are curtailed or slowed. [*See* Protein–Energy Malnutrition; Obesity.]

Significant increases in lipid content at the expense of water, and to a somewhat lesser extent protein, can occur in certain disorders. Cirrhosis of the liver due to chronic alcoholism may cause lipid concentrations in the liver to increase from approximately 5 to 19% by mass. Fatty infiltration can occur in and around organs such as the heart. A 12-fold increase in the lipid content of the muscle (skeletal) in a young male suffering from muscular dystrophy has been reported. [*See* Lipids.]

Vigorous exercise over long periods of time may also alter body tissue composition. Total body adipose tissue masses of one-third normal values have been recorded for athletes compared to sedentary people.

II. PHOTON AND ELECTRON INTERACTION PROPERTIES

A. Basic Radiation Interactions

The amount of radiation energy absorbed by an irradiated mass of body tissue (the absorbed dose) depends on the type and magnitude of the radiation interactions that occur. For both photons and electrons, the types of interactions depend on the energy of the incident radiation. The magnitude of the interactions is influenced by the elemental composition and mass density of the tissue.

In the energy interval 10 keV to 10 MeV, photons may undergo a combination of photoelectric absorption, Compton scattering, pair production, and other minor effects. At 10-keV photoelectric absorption, which is strongly dependent on the elemental atomic numbers (Z) present in the tissue, accounts for 93% of the total interactions occurring in muscle (skeletal) at this energy. In cortical bone with its higher Z elements, this proportion rises to 98%. Photon scattering becomes more important as the energy increases. For 1-MeV photons, over 99% of the total interactions in both muscle (skeletal) and cortical bone are due to the Z-independent Compton scattering process. At higher energies the importance of Compton scattering decreases and pair production processes increase. For 10-MeV photons, 22.7 and 31.5% of the total interactions in muscle (skeletal) and cortical bone, respectively, are due to the weakly Z-dependent pair production process, with Compton scattering contributing the remainder.

In the energy interval 10 keV to 10 MeV, electrons lose energy when traversing body tissues by electronic collisions and the production of bremsstrahlung photons or "braking radiation." Up to 500 keV, electronic collisions are the predominant effect. At 10 MeV, bremsstrahlung production becomes more important, contributing 8.4 and 12.4% of the total mass stopping power for muscle and cortical bone, respectively.

B. Photon and Electron Interaction Data

Photon and electron interaction data for a selection of adult body tissues are given in Table III. Data are listed for 10-keV, 100-keV, 1-MeV, and 10-MeV photons and electrons. Total mass attenuation coefficients (μ/ρ), relating to the fractional reduction in the numbers of transmitted photons, are tabulated. Total electron mass stopping powers (S/ρ), relating to the energy lost as an electron traverses a tissue, are also given.

Table III also shows that mass densities range from 950 kg m^{-3} for the fatty tissues such as adipose tissue to 1050 kg m^{-3} for a soft tissue such as muscle (skeletal). Cortical bone has a much higher density of 1920 kg m^{-3}. Larger differences in attenuation (i.e., absorption and scattering) occur in body tissues at low photon energies (e.g., 10 keV), where photoelectric absorption predominates, compared to high photon

TABLE III
Radiation Interaction and Mass Density Data for Adult Body Tissues

	Mass attenuation coefficient, μ/ρ ($m^2\ kg^{-1}$)				Electron mass stopping power, S/ρ (MeV $m^2\ kg^{-1}$)				
	Photon energy				Electron energy				
Body tissue	10 keV ($\times 10^{-1}$)	100 keV ($\times 10^{-2}$)	1 MeV ($\times 10^{-3}$)	10 MeV ($\times 10^{-3}$)	10 keV	100 keV ($\times 10^{-1}$)	1 MeV ($\times 10^{-1}$)	10 MeV ($\times 10^{-1}$)	Mass density (kg m^{-3})
Adipose tissue	3.27	1.69	7.08	2.14	2.33	4.21	1.88	2.14	950
Brain[a]	5.41	1.70	7.04	2.20	2.25	4.11	1.85	2.14	1040
Cortical bone	2.85	1.86	6.57	2.31	1.93	3.61	1.65	2.02	1920
Heart[b]	5.48	1.70	7.01	2.20	2.24	4.09	1.84	2.13	1060
Liver	5.40	1.69	7.01	2.19	2.24	4.08	1.84	2.13	1060
Lung	5.46	1.70	7.01	2.20	2.24	4.08	1.87	2.22	260 (Inflated)
Mammary gland	4.30	1.69	7.03	2.17	2.27	4.13	1.86	2.13	1020
Muscle (skeletal)	5.36	1.69	7.01	2.19	2.24	4.09	1.84	2.13	1050
Red marrow	4.35	1.69	7.02	2.16	2.28	4.14	1.85	2.13	1030
Skin	4.95	1.69	6.99	2.18	2.24	4.08	1.84	2.12	1090
Yellow marrow	3.10	1.69	7.08	2.14	2.33	4.22	1.88	2.14	980
Water[c]	5.33	1.71	7.07	2.22	2.26	4.12	1.86	2.15	1000

[a]Contains 50% gray matter and 50% white matter, by mass.
[b]Blood filled.
[c]Water is included for reference purposes.

energies. At 100 keV and above, equal masses of all soft tissues absorb photons similarly, which is also the case for electrons. Water, included for reference purposes, has similar photon and electron interaction properties to soft tissues.

III. SOME PRACTICAL IMPLICATIONS

A. Medical X-Ray Imaging

Conventional X-ray imaging relies on the differential attenuation that occurs when a beam of X rays passes through a body section. If adjacent tissues within the section attenuate the X rays differently, the transmitted X-ray distribution reflects the tissue pattern, which may be recorded on a photographic film (static image) or a fluorescent screen (dynamic image). By carefully selecting the incident X-ray energies, different parts of the body may be examined in this way.

A particularly challenging application of X-ray imaging is mammography, the radiography of the breast. For a given breast thickness the requirements are for adequate transmission through the breast to the film and maximum differentiation of the soft tissues present, while keeping the absorbed dose to the breast as low as reasonably practicable. X rays generated around 30 kV with suitable filtration maximize the photoelectric interaction processes in the soft tissues to yield the necessary image contrast. Similar methods are adopted to image bone detail in thin body sections.

As the thickness of the body section increases, higher-energy X rays are required to penetrate the tissues. A subsequent decrease in photoelectric processes results and the ability to differentiate soft tissues of similar composition and density is diminished. Computerized techniques are now employed to accentuate these small differences in photon attenuation. Computed tomography (CT) is widely used to provide detailed, enhanced images of thin body slices. [*See* Diagnostic Radiology.]

B. Radiotherapy

Radiotherapy uses X-ray techniques to sterilize the reproductive ability of tumor cells while sparing adjacent normal cells. Megavoltage radiotherapy uses photon energies of 1 MeV or above. The importance of Compton scattering at these energies leads to a more uniform absorbed dose distribution even in thick body sections. In addition, the fact that the maximum absorbed dose does not occur at the skin surface but at an energy-dependent depth in the tissues produces an important "skin-sparing" effect.

Because the major portion of an electron beam's energy is deposited in a limited depth of tissue, high absorbed doses may be given to superficial tumors, while sparing underlying tissues. Changing the electron energy affects the depth of tissue treated.

C. Tissue Substitutes

To investigate experimentally the radiation interactions in body tissues, substitute materials are necessary. Such materials should ideally have absorption and scattering properties that are the same as those of the body tissues being considered. In practice this requirement is approximated, although tissue substitutes are now available that give absorbed doses that differ by no more than 1% from those in the body tissues being simulated. Typical materials that have been used are:

1. Adipose tissue: polyethylene, ethoxyethanol;
2. Muscle (skeletal): polystyrene, water;
3. Cortical bone: aluminum, saturated solution of dipotassium hydrogen orthophosphate.

A number of resin and polymer-based materials containing corrective powdered fillers, simulating a range of body tissues, are also available.

Volumes of tissue substitutes, or phantoms, are used for the measurement of absorbed dose, the assessment of image quality, and the calibration of radiation detectors. At one extreme they may be simple blocks, stacked sheets, or cuboid tanks of water. Alternatively they may be elaborate representations of body sections or whole bodies with embedded skeletons (natural and artificial) and replicated organs.

BIBLIOGRAPHY

Forbes, G. B. (1987). "Human Body Composition. Growth, Aging, Nutrition and Activity." Springer-Verlag, New York.

Hubbell, J. H. (1982). Photon mass attenuation and energy-absorption coefficients from 1 keV to 20 MeV. *Int. J. Appl. Radiat. Isotopes* 33, 1269.

ICRP (International Commission on Radiological Protection) (1975). "Report of the Task Group on Reference Man," ICRP Publication 23. Pergamon, Oxford, England.

ICRU (International Commission on Radiation Units and Measurements) (1984). "Stopping Powers for Electrons and Positrons," ICRU Report 37. ICRU, Bethesda, Maryland.

ICRU (1989). "Tissue Substitutes in Radiation Dosimetry and Measurement," ICRU Report 44. ICRU, Bethesda, Maryland.

ICRU (1992). "Photon, Electron, Proton and Neutron Interaction Data for Body Tissues," ICRU Report 46. ICRU, Bethesda, Maryland.

ICRU (1992). "Phantoms and Computational Models in Therapy, Diagnosis and Protection," ICRU Report 48. ICRU, Bethesda, Maryland.

White, D. R., and Woodard, H. Q. (1988). The effects of adult human tissue composition on the dosimetry of photons and electrons. *Health Phys.* **55,** 653.

White, D. R., Widdowson, E. M., Woodard, H. Q., and Dickerson, J. W. T. (1991). The composition of body tissues. II. Fetus to young adult. *Br. J. Radiol.* **64,** 149.

Woodard, H. Q., and White, D. R. (1986). The composition of body tissues. *Br. J. Radiol.* **59,** 1209.

Radiobiology

SARA ROCKWELL
Yale University School of Medicine

GLOSSARY

Gray (Gy) ISU unit for the absorbed dose of ionizing radiation; 1 Gy = 1 joule of absorbed energy per kilogram of irradiated material

Ionization Removal of one or more orbital electrons from an atom or molecule. "Ionizing" radiations produce ionization in the material that absorbs them.

Linear energy transfer (LET) Factor describing the average rate at which a charged particle transfers its energy to the medium; generally given in kilo electron volts per micron of unit density material

Relative biological effectiveness (RBE) Quantity used to compare the biological effects of different radiations; RBE = D_x/D_r, where D_r is the dose of the experimental radiation needed to produce a specified biological effect and D_x is the dose of the reference radiation (usually 250-kV X rays) needed to produce the same effect

Sievert (Sv) ISU unit of dose equivalent, calculated by multiplying the dose (in Gy) by a quality factor, which adjusts for differences in the relative biological effectiveness of different radiations; 1 mSv = 10^{-3} Sv

RADIOBIOLOGY IS THE STUDY OF THE EFFECTS OF radiation on biological systems. In its broadest sense, radiobiology encompasses many radiations, ranging from radiowaves through visible and ultraviolet (UV) light to ionizing electromagnetic and particulate radiations; these different radiations have widely differing biological effects. The biological systems considered in radiobiology range from biological molecules and subcellular structures to simple organisms (e.g., viruses and bacteria), higher plants and animals, and humans. This article focuses on the effects of ionizing radiations on humans.

I. SOURCES AND KINDS OF RADIATION

Radiation is and always has been an inescapable part of the human environment. Natural background radiation comes from several sources (Table I). Naturally occurring radioactive elements are found in the soil and in rocks. The dose from these radiations varies considerably at different locations; regional averages vary from 0.25 to 0.9 mSv/yr in the United States. In addition, people are exposed to radiation from natural radioactive materials that have been taken into the body in food or inhaled with the air.

One such element, radon, has recently received considerable attention. This radioactive gas is produced by the decay of radium, a radioactive nuclide that is a natural component of the rocks in many areas of the country. Levels of radon in the air inside buildings vary greatly, depending on factors such as the amount of radium in the strata below the building, the characteristics of the water supply, the construction of the building, and the amount and the nature of ventilation in the building. It now appears that radon could be the major source of background radiation in many areas of the country, contributing approximately 2 mSv to the average annual radiation dose equivalent. Because radon and its radioactive daughters emit α particles, which have short ranges, only the cells lining the airways of the lungs are irradiated.

Cosmic rays also contribute to our radiation dose. Cosmic rays are energetic positively charged particles (i.e., atomic nuclei) emitted from the sun and other

ENCYCLOPEDIA OF HUMAN BIOLOGY, Second Edition, VOLUME 7. Copyright © 1997 by Academic Press. All rights of reproduction in any form reserved.

TABLE I
Estimated Annual Effective Dose Equivalents to the U.S. Population in 1980–1982[a]

Source	Average annual effective dose equivalent (mSv)
Natural sources	
Radon	2.00
Cosmic rays	0.27
Radionuclides in body	0.39
Terrestrial and atmospheric sources	0.29
Rounded total	3.00
Medical exposures	
Diagnostic X rays	0.39
Nuclear medicine	0.14
Total	0.53
Occupational exposures	0.009
Nuclear energy	0.0005
Miscellaneous environmental sources	0.0006
Consumer products	
Tobacco	—[b]
Other	0.05–0.13
Rounded total (nonsmokers)[b]	3.6

[a]Data are summarized and condensed from the 1987 report of the National Council on Radiation Protection and Measurement (see Bibliography). The doses are averaged over the entire population of exposed and unexposed persons.

[b]The calculation of the effective dose equivalent from tobacco products is exceedingly problematic, as described in the National Council on Radiation Protection and Measurement report. If the estimated value of ~13 mSv for an average smoker is correct, tobacco represents the largest source of radiation exposure for smokers and raises their total annual effective dose equivalent from 3.6 to ~16 mSv.

celestial sources. Although some of these particles are deflected by the earth's magnetic field and others are absorbed by the atmosphere, some reach the surface of the earth; the intensity of cosmic radiation increases near the poles and at high altitudes (doubling with every 2000 m, from about 0.26 mSv/yr at sea level).

The dose of background radiation that people receive from natural sources varies considerably, as it depends on factors such as the altitude, the local composition of the earth's crust, and the design and ventilation of one's home. People's activities also influence their exposure to natural radiations; for example, a single high-altitude transatlantic flight results in an extra 0.05 mSv of exposure from cosmic radiation.

Medical radiations dominate our exposure from man-made radiation, with diagnostic and dental X rays and diagnostic nuclear medicine procedures contributing most of this dose. Occupational exposure for those working with radioactive materials contributes slightly to the average dose for the total population. Of approximately 1.3 million people in the United States engaged in radiation work in 1980, only one-half actually received measurable exposures. For those exposed, the average dose equivalent received was 2.3 mSv. The nuclear fuel cycle and miscellaneous environmental sources (including fallout from past nuclear weapons testing, emissions from certain non-power reactors, and the release of naturally occurring radionuclides from aluminum, copper, zinc, and lead processing facilities) contribute small doses to the average population exposure.

The dose received from consumer products is varible and difficult to estimate. The doses received from manufactured products that deliberately use radioactive isotopes (e.g., smoke detectors and luminous watches) or produce ionizing radiation (e.g., airport inspection systems and television receivers) are small, estimated at 0.0015–0.012 mSv. Greater radiation exposures result from the release of naturally occurring radionuclides during the combustion of coal or natural gas (0.003–0.006 mSv), and from exposure to radionuclides present naturally in building materials (0.036 mSv), domestic water supplies (0.01–0.06 mSv), and mining and agricultural products (less than 0.01 mSv).

Tobacco products probably contribute the highest radiation dose of all consumer products, although it is difficult to estimate. For an average smoker the decay of ^{210}Po and other radionuclides that are concentrated naturally by tobacco leaves has been calculated to produce average doses to a small region of the bronchial epithelium of ~160 mSv, and a calculated effective (weighted) dose equivalent of 13 mSv. This radiation dose is thought to contribute somewhat to the excess risk of lung cancer in smokers; however, the larger portion of the risk to smokers probably reflects the chemical and physical effects of tobacco products. [*See* Tobacco Smoking and Nicotine, Impact on Health.]

Electromagnetic and particulate radiations are both important in radiobiology. The electromagnetic radiations with enough energy to produce ionization, X rays, and γ rays (Table II) are identical in their characteristics, but are produced in different ways. γ rays are emitted when the nuclei of certain radioactive elements decay. In contrast, X rays are emitted when orbital electrons make transitions between high- and low-energy states, and they carry the energy released in these transitions. X rays are often produced by

TABLE II
Electromagnetic Radiations and Their Effects[a]

Radiation	ν	E	Effects
Radiowaves, radar waves, microwaves	$<3 \times 10^{11}$	<0.00124 eV	Molecular vibrations; heating
Infrared	3×10^{11}–4.3×10^{14}	0.00124–1.77 eV	Molecular vibrations; heating
Visible light	4.3–7.5×10^{14}	1.77–3.1 eV	Absorption by specific molecules, causing excitation, possibly leading to chemical reactions
Ultraviolet light	7.5×10^{14}–3×10^{16}	3.1–124 eV	Absorption by specific molecules, causing excitation, leading to chemical reactions
X rays, γ rays	$>3 \times 10^{16}$	>124 eV	Ionization, nonspecific absorption of energy; nonspecific damage

[a] ν is the frequency of the radiation (in cycles per second). The energy (E) contained in a single photon of radiation is related to the frequency by the relationship $E = h\nu$, where h is Planck's constant.

accelerating electrons, using strong electric fields, so that they carry large amounts of kinetic energy, then stopping these electrons suddenly by bombarding a metal target; the resulting reactions convert the kinetic energy of the electrons into X rays. Particles accelerated to fast velocities (i.e., high energies) also carry enough kinetic energy to produce ionization along their paths. Electrons, protons, neutrons, and a variety of heavier particles can produce ionization when accelerated (Table III).

II. NATURE OF RADIATION DAMAGE

The biological effects produced by electromagnetic radiations vary dramatically with the photon energy and, therefore, with the frequency of the radiation (see Table II). Radiowaves and microwaves, for example, carry only very small amounts of energy in each photon; when a molecule absorbs such a photon, it acquires a quantum of energy that is insufficient to alter its electronic configuration; this energy is dissipated as vibrational energy, or heat. Visible and UV light are absorbed selectively by specific biological molecules having electron orbital structures such that absorption of a photon of the correct energy causes the transition of an electron from a lower-energy state to a higher-energy state. The excited molecule then loses its excess energy through relatively specific and predictable chemical reactions. For example, when DNA is irradiated with UV light ~260 nm in wavelength, the light is absorbed primarily by the rings of the pyrimidine nucleotides, thymidine and cytidine. The major damage to UV-irradiated DNA results from reactions be-

TABLE III
Some Particles of Interest in Radiobiology

Particle	Symbol	Mass (Atomic Mass Units)	Charge
Proton	p	1	+1
Neutron	n	1	0
Electron	e, e^-, β^-	0.00055	−1
α Particle	α	4	+2
Heavy charged particles, cosmic rays	None	Variable	Variable

tween the excited pyrimidines and other nearby pyrimidines, producing intrastrand pyrimidine–pyrimidine dimers; these distort the DNA helix and interfere with normal DNA synthesis, and could be mutagenic or cytotoxic.

The biological effects of these "nonionizing" radiations are different from those of "ionizing" radiations; this reflects differences in the localization of the energy absorbed and in the nature of the resulting damage. The energy carried by a single photon or particle of ionizing radiation exceeds the binding energy of the inner-shell electrons surrounding atomic nuclei. This energy is imparted in random reactions, producing highly localized areas of ionization and intensive damage.

The mechanisms by which electromagnetic radiations interact with matter vary with photon energy and with the composition of the absorbing material. All three absorption processes—photoelectric effect, Compton scattering, and pair production—result in the release of lower-energy photons and energetic "fast" electrons. As these fast electrons interact with atoms along their path, they strip electrons from the shells of the atoms, leaving a track of ion pairs, each of which consists of a negatively charged relatively low-energy electron, plus a positively charged atom. The chemical reactions of these exceedingly reactive ions with nearby molecules lead to the production of large numbers of ions, free radicals, excited molecules, and other chemically reactive species. The chemical reactions between these moieties and critical biological molecules eventually lead to biologically measurable damage.

Fast charged particles (see Table III) produce their biological damage through analogous mechanisms. The initial event is an interaction between the moving charged particle and the electron shell of a nearby atom, which produces an ion pair. A fast charged particle therefore leaves a track of ion pairs along its path that react further, as described earlier.

The physics of the absorption of neutrons is more complex. These uncharged particles interact with atomic nuclei in the absorbing material, through scattering reactions which produce recoil protons, α particles, and heavier nuclear fragments. The interactions of these moving charged particles with nearby atoms then lead to the production of radiation damage.

Two things should be noted about the damage produced by ionizing radiation. First, the interactions between the radiation and the irradiated material are random; unlike visible or UV light, ionizing radiations are not absorbed by specific molecules. Second, many different kinds of biological lesions are produced by ionizing radiation, because most of the damage reflects nonspecific chemical reactions occurring after the initial ionization event. A mammalian cell is approximately 70% water; much of the damage in an irradiated cell therefore reflects the chemical reactions of radiation-damaged water molecules. Molecules of many kinds are damaged during irradiation; most can be replaced with no lasting effects on the cell. However, a few lesions, or even a single, critical lesion, in DNA can alter the function of the cell and can even be lethal. As a result, damage to DNA is the critical factor producing the cytotoxic effects of ionizing radiation.

Irradiation produces many different lesions in DNA and in the chromosomes that provide the structure organizing DNA in human cells (Table IV). These lesions lead to a variety of genetic changes, including simple point mutations, deletions of small or large amounts of genetic information, and DNA or chromosomal rearrangements that alter the expression of genetic information. The genetic changes seen in irradiated cells are nonspecific and are qualitatively similar to those associated with aging or with exposure to nonspecific oxidative and chemical damage. As a result, the spectrum of mutations observed in irradiated cells or in the offspring of irradiated individuals resembles that of spontaneous mutations observed in a natural population. No novel mutations have been observed in large populations of irradiated *Drosophila*, mice, or people.

III. CELLULAR RADIOBIOLOGY

A. Cellular Lethality—Delayed Cell Death

The nature of the damage in irradiated cells underlies both the unique therapeutic uses and the unique haz-

TABLE IV
Some Lesions Produced in the DNA and Chromosomes of Irradiated Cells

Damage to DNA	Damage to chromosomes
Base changes	Deletions
Deletions	Translocations
Single-strand breaks	Inversions
Double-strand breaks	Acentric fragments
Cross-links	Dicentric chromosomes
Adducts	Ring chromosomes

ards of ionizing radiation. Because lethally damaged cells die as a result of DNA damage, they do not die immediately, or even rapidly. Rather, lethally irradiated cells function relatively normally and even divide once or a few times, producing "abortive clones" of daughter cells, all of which eventually die, as the genetic injury becomes manifest. Cells that never attempt to divide might never express their lethal injury; such "radiation-sterilized" cells might function normally for years. This fact provides one of the biological bases for the treatment of cancer with radiation: radiation sterilizes the malignant cells, preventing them from indefinite growth, while the quiescent cells of irradiated normal tissues continue to function relatively normally for long periods. The hazards of high doses of radiation result primarily from the death of dividing cells. As described in detail in Section V,B,1, large doses of radiation are generally required to deplete normal tissues sufficiently to produce overt injury.

In contrast, the hazards of low doses of radiation generally result from the production of DNA damage that is not lethal to the cell but that could become manifest under appropriate circumstances months or even years later. For example, damage to the DNA of a resting hematological stem cell could lie dormant for years before being expressed in leukemic transformation. Similarly, damage to the DNA of a spermatogonium might not kill the developing sperm and might not interfere with fertilization, but could be deleterious to the embryo resulting from conception by a sperm carrying the radiation-induced genetic charges.

The generality that radiation-induced cell death is associated with mitosis and occurs one or more cell cycles after irradiation is not without exception. It has been known for many years that lymphocytes can undergo "interphase death," without dividing, a few hours after irradiation. Interphase death may be partially responsible for the unusual radiosensitivity of lymphocytes and for the rapid immunosuppression seen after irradiation. Recently, it has been shown that high doses of radiation can also induce interphase death among certain other types of cells, apparently by producing damage that triggers the pathway of "programmed cell death" or "apoptosis." The importance and implications of radiation-induced apoptosis are being actively investigated in many laboratories.

B. Cell Survival

The effects of radiation on mammalian cells have been studied extensively using primary cell cultures initiated from normal and malignant tissues or established cell lines. A typical survival curve for cells irradiated by X rays or γ rays is shown in Fig. 1. Because radiation-sterilized cells do not die immediately, cell survival is measured by testing the ability of individual cells to grow into colonies composed of dozens or hundreds of cells. The survival of the cells decreases continuously as the radiation dose increases. Different human cell lines have slightly different survival curves, but the differences are generally relatively subtle.

1. Repair of Radiation Damage

A number of factors modulate the response of cells to radiation. The first of these is the ability of the cell to repair radiation-induced lesions in DNA. Different cells have different intrinsic repair capacities. For example, cells of hematological origin, both normal and malignant, generally are less able to repair DNA damage than are cells from other tissues. Cells from persons with certain genetic diseases associated with defects in DNA repair (e.g., ataxia–telangiectasia) are unusually sensitive to ionizing radiation; people with these diseases might be correspondingly hypersensitive to radiation injury, just as people with xeroderma pigmentosum, a genetic disease that results from a defect in the repair of UV-induced DNA damage, are unusually sensitive to the toxic and carcinogenic effects of the UV radiation present in sunlight. Many factors modulate the ability of the cell to repair radiation damage, including whether the cell is proliferative or quiescent, the position of the cell in the cell cycle, and the environment of the cell (e.g., the extracellular pH and the adequacy of the supplies of nutrients, energy sources, and oxygen). [*See* DNA Repair.]

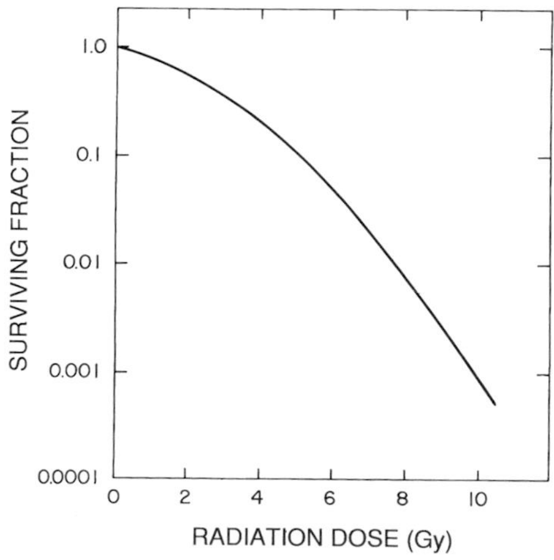

FIGURE 1 Typical survival curve for mammalian cells *in vitro*. Surviving fraction represents the fraction of cells in the population that remain capable of prolonged proliferation.

2. Radiation Sensitizers and Radiation Protectors

Because ionizing electromagnetic radiations produce DNA damage indirectly, through a variety of nonspecific chemical reactions, the production of DNA damage can be altered by chemicals that interact with the reactive intermediates. Radiosensitizers increase the amount of damage produced by a given dose of radiation. One of the most important radiosensitizers for radiobiology is molecular oxygen (O_2). When O_2 is present, this extremely reactive electron-affinic molecule participates in, multiplies, and modifies the reactions of irradiated water and biological molecules. As a result, cells that are aerobic at the time of irradiation are approximately three times more sensitive to the effects of radiation than are severely hypoxic cells. The O_2 concentrations in most human tissues are sufficient to produce maximal radiosensitization by O_2; the chemistry of the radiation reactions in most human tissues, except those with compromised vasculatures, is therefore characteristic of well-oxygenated cells. Because malignant tumors frequently contain hypoxic areas, the radiobiological effects of oxygen are an important consideration in the use of radiation in cancer therapy. Other radiosensitizers of interest in medically oriented radiobiology are the thymidine analogs bromodeoxyuridine and iododeoxyuridine. When incorporated into the DNA of proliferating cells, these halogenated pyrimidines sensitize cells to radiation.

Radioprotectors act by scavenging the relatively long-lived free radicals, thereby reducing the amount of damage produced in biological macromolecules. A variety of radical scavengers, including alcohols and sulfhydryl-containing compounds, are effective radioprotectors. Naturally occurring endogenous intracellular nonprotein sulfhydryls, including glutathione, are thought to play an important role in minimizing the production of DNA damage by radiation and other radical-producing processes, including normal oxidative metabolism. A number of chemical radioprotectors have been tested as possible agents for protecting individuals who are about to be exposed to dangerous levels of radiation; however, the high intracellular concentrations required for radioprotection and the toxicities of the compounds limit their potential usefulness.

3. Effect of Radiation Dose Rate

The response of mammalian cells to radiation is affected by the rate at which the radiation is delivered. The foregoing discussion has assumed that radiation is delivered at the relatively high dose rates characteristic of single therapeutic and diagnostic radiology treatments (i.e., several Grays per minute). However, when a large dose of radiation is fractionated into many smaller treatments, or when the cells are irradiated continuously over a long period at a low dose rate, the radiation is less effective in producing biological damage. As the duration of irradiation increases from a few minutes to a few hours for a given total dose, repair of the damage occurs during irradiation; because of the curvilinear shape of the acute radiation dose–response curve (see Fig. 1), this repair decreases the overall efficacy of the radiation. Moreover, with longer irradiation times the proliferation of the surviving cells increases the size of the cell population during irradiation, thereby decreasing the damage received by any particular cell. In the limit of very low dose rate irradiations, the growth, viability, and characteristics of an irradiated cell population will be indistinguishable from those of an untreated population.

4. Effect of Radiation Quality

The preceding discussion focuses on the radiobiology of electromagnetic radiations. These sparsely ionizing radiations deposit energy randomly and relatively uniformly throughout the cell. The effects of other sparsely ionizing radiations (e.g., high-energy protons and electrons) are similar. Some radiations, however, deposit energy very densely along their paths; these radiations are said to have a high linear energy transfer (LET). High-LET radiations have somewhat different biological effects than sparsely ionizing low-LET radiations, because the ionizations they produce are closely clustered in time and space along the tracks of the particles (Figs. 2 and 3). As a result, some submicroscopic areas are intensely damaged by multiple ionization events, while other areas are essentially unirradiated. High-LET radiations, therefore, produce intense local DNA damage, which is difficult to repair and has a high likelihood of being lethal, whereas low-LET radiations often result in single, simple, reparable lesions. Thus as the LET of the radiation increases, its relative biological effectiveness increases.

Moreover, the indirect damage (i.e., the damage produced through chemical reactions) becomes less important, because the intense local ionization produces sufficient direct damage to the critical structures to render the indirectly produced lesions largely superfluous. As a result, the modulating effects of sensitizers, protectors, cell age, repair processes, fraction-

COLOR PLATE 1 Photograph of a model of a globular protein, α-lactalbumin from the baboon, determined crystallographically. The atoms of the polypeptide backbone are white and those of the side chains are various colors. The spheres corresponding to the individual atoms have only one-third of their van der Waals radii. (The model was constructed by Jonathan Ewbank.) [*See* Proteins.]

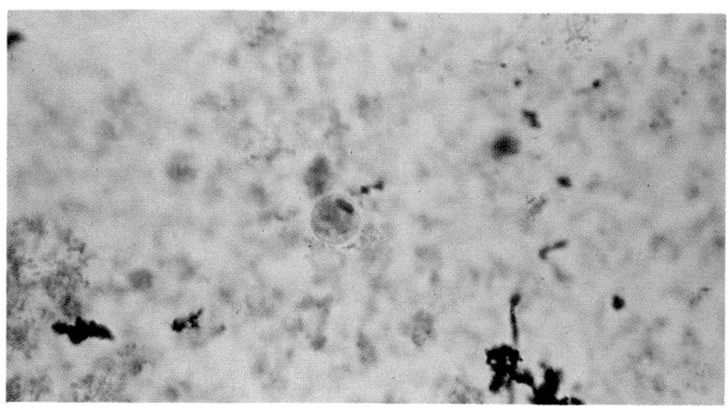

COLOR PLATE 2 *Entamoeba histolytica* cyst in human stool. [*See* Protozoal Infections.]

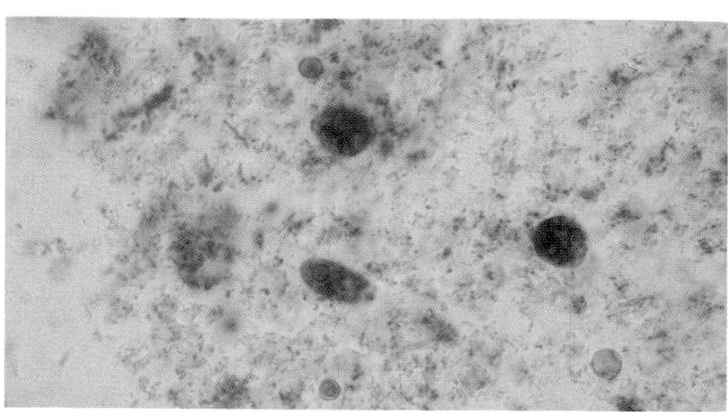

COLOR PLATE 3 *Entamoeba histolytica* trophozoite grown in culture medium at 37°C. [*See* Protozoal Infections.]

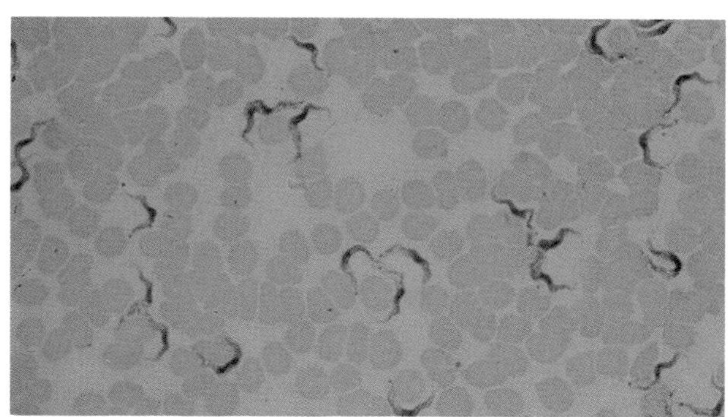

COLOR PLATE 4 *Trypanosoma brucei* trypomastigotes in rat's blood. [*See* Protozoal Infections.]

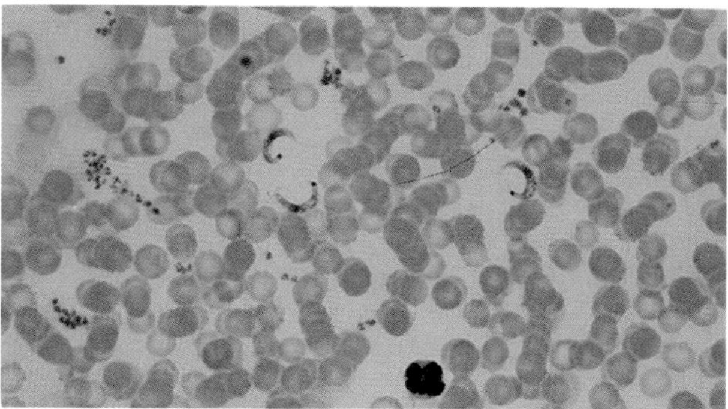

COLOR PLATE 5 *Trypanosoma cruzi* trypomastigotes in human blood. [*See* Protozoal Infections.]

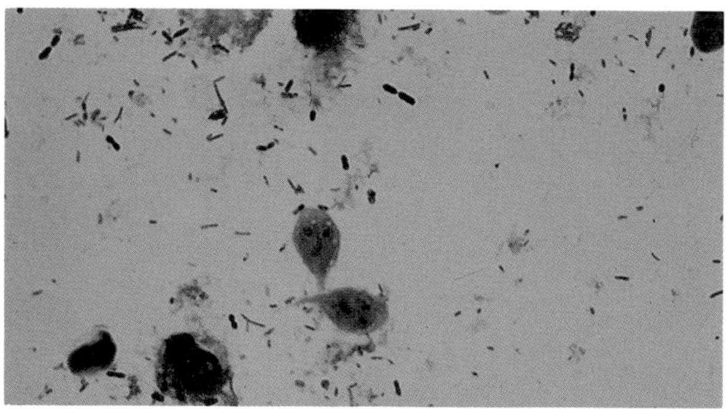

COLOR PLATE 6 *Giardia lamblia* trophozoite in stool. [*See* Protozoal Infections.]

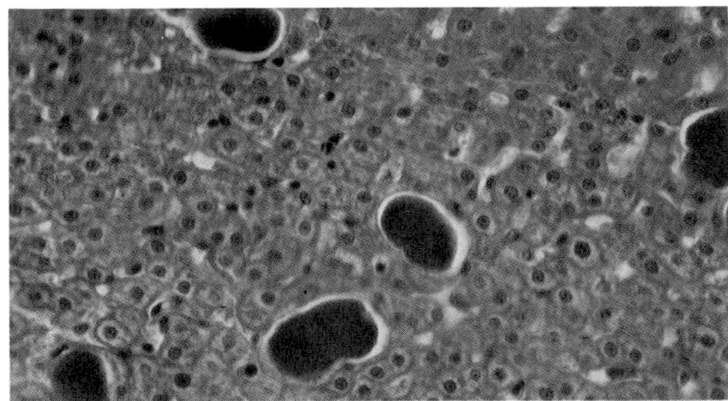

COLOR PLATE 7 *Plasmodium* spp. pre-erythrocytic schizonts in hepatocytes. [*See* Protozoal Infections.]

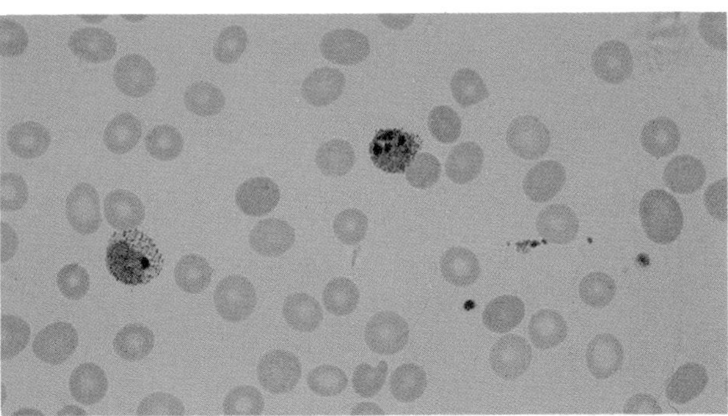

COLOR PLATE 8 *Plasmodium vivax* trophozoite in human blood. [*See* Protozoal Infections.]

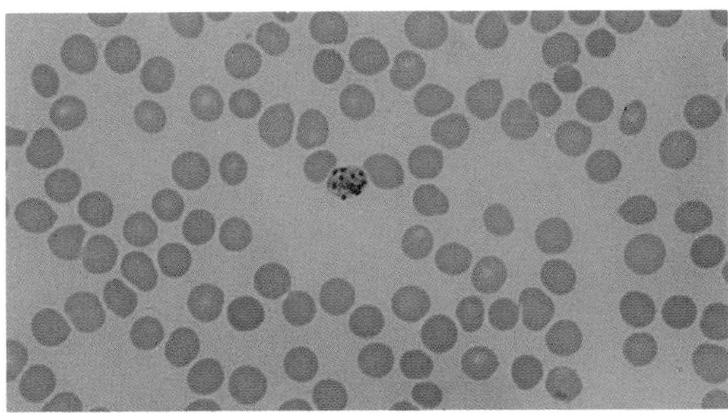

COLOR PLATE 9 *Plasmodium malariae* schizont in human blood. [*See* Protozoal Infections.]

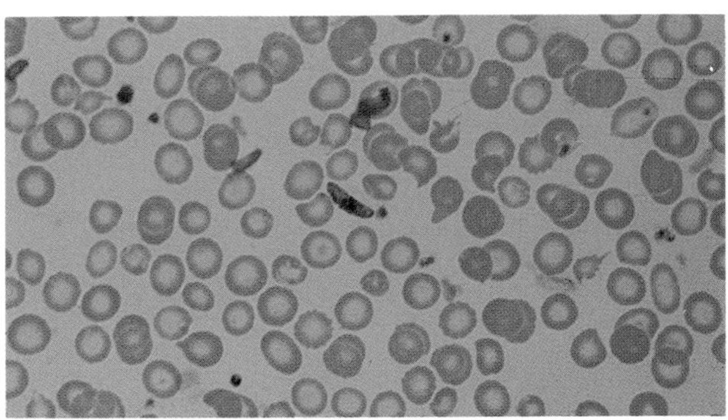

COLOR PLATE 10 *Plasmodium faciparum* gametocyte in human blood. [*See* Protozoal Infections.]

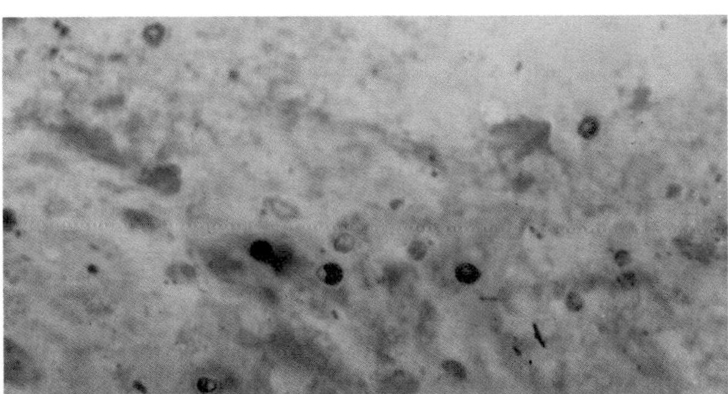

COLOR PLATE 11 *Cryptosporidium* oocysts in stool. [*See* Protozoal Infections.]

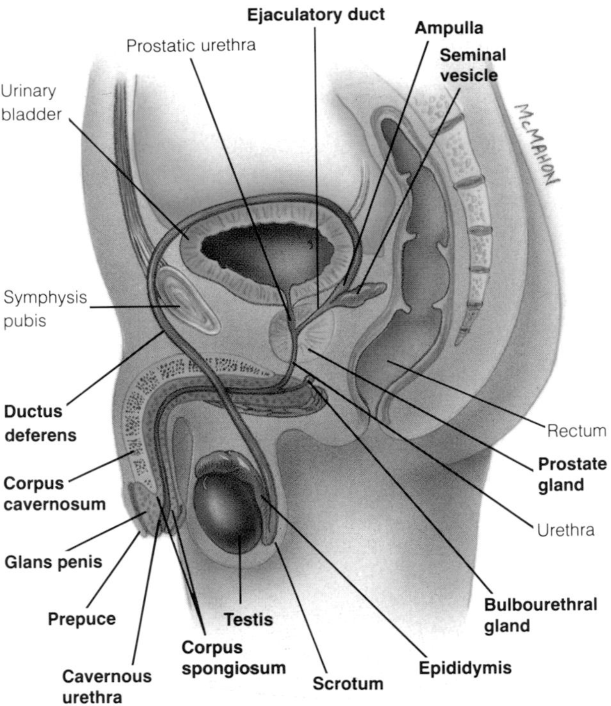

COLOR PLATE 12 Reproductive system of an adult male. [Source: Gaudin, A. J., and Jones, K. C. (1989). "Human Anatomy and Physiology." Harcourt Brace Jovanovich, San Diego, p. 699. Reproduced with permission.] [*See* Reproductive System, Anatomy.]

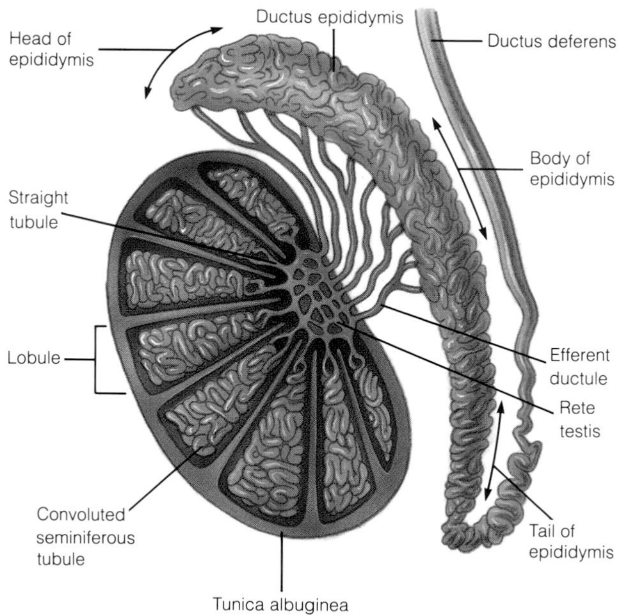

COLOR PLATE 13 The testis and associated ducts. [Source: Gaudin, A. J., and Jones, K. C. (1989). "Human Anatomy and Physiology." Harcourt Brace Jovanovich, San Diego, p. 702. Reproduced with permission.] [*See* Reproductive System, Anatomy.]

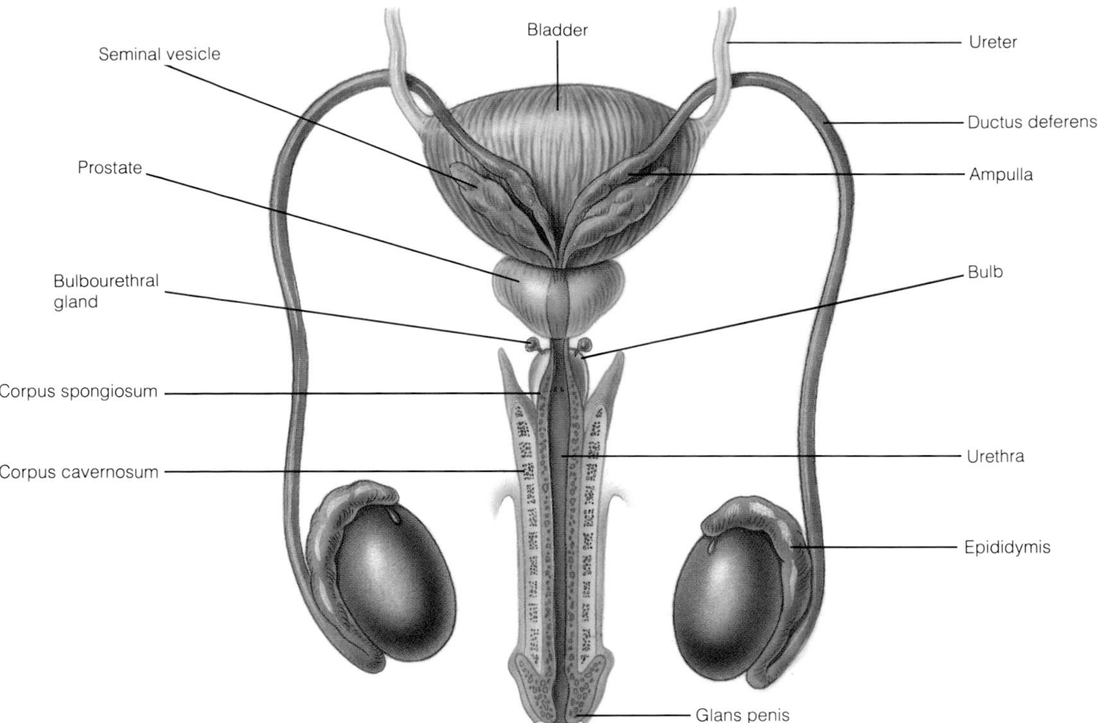

COLOR PLATE 14 Posterior view of the male reproductive system. [Source: Gaudin, A. J., and Jones, K. C. (1989). "Human Anatomy and Physiology." Harcourt Brace Jovanovich, San Diego, p. 703. Reproduced with permission.] [*See* Reproductive System, Anatomy.]

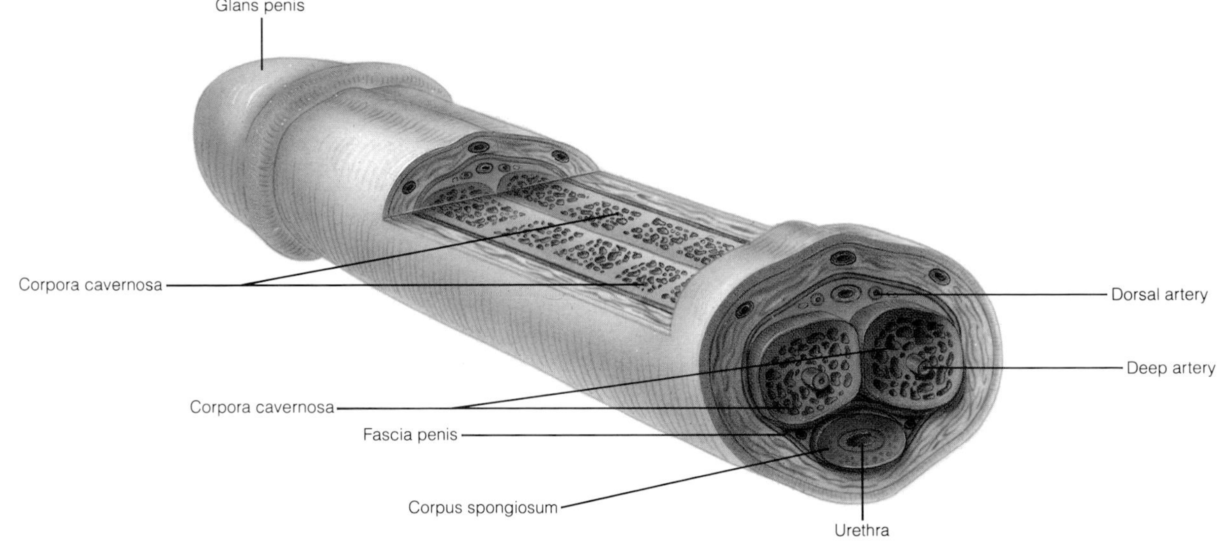

COLOR PLATE 15 Anatomy of a penis. [Source: Gaudin, A. J., and Jones, K. C. (1989). "Human Anatomy and Physiology." Harcourt Brace Jovanovich, San Diego, p. 707. Reproduced with permission.] [*See* Reproductive System, Anatomy.]

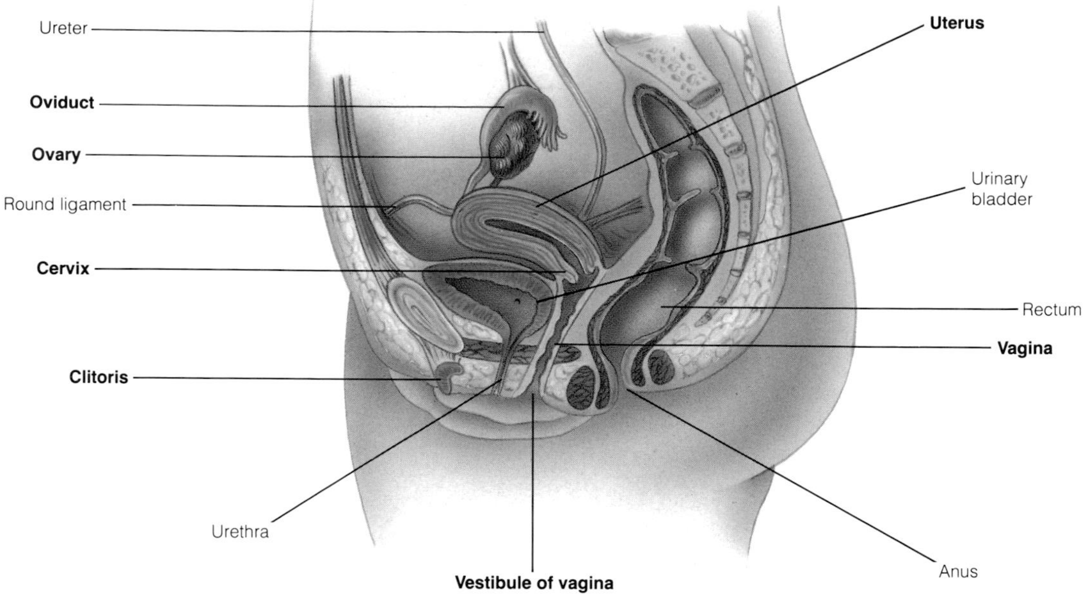

COLOR PLATE 16 Anatomy of the female reproductive tract. [Source: Gaudin, A. J., and Jones, K. C. (1989). "Human Anatomy and Physiology." Harcourt Brace Jovanovich, San Diego, p. 709. Reproduced with permission.] [*See* Reproductive System, Anatomy.]

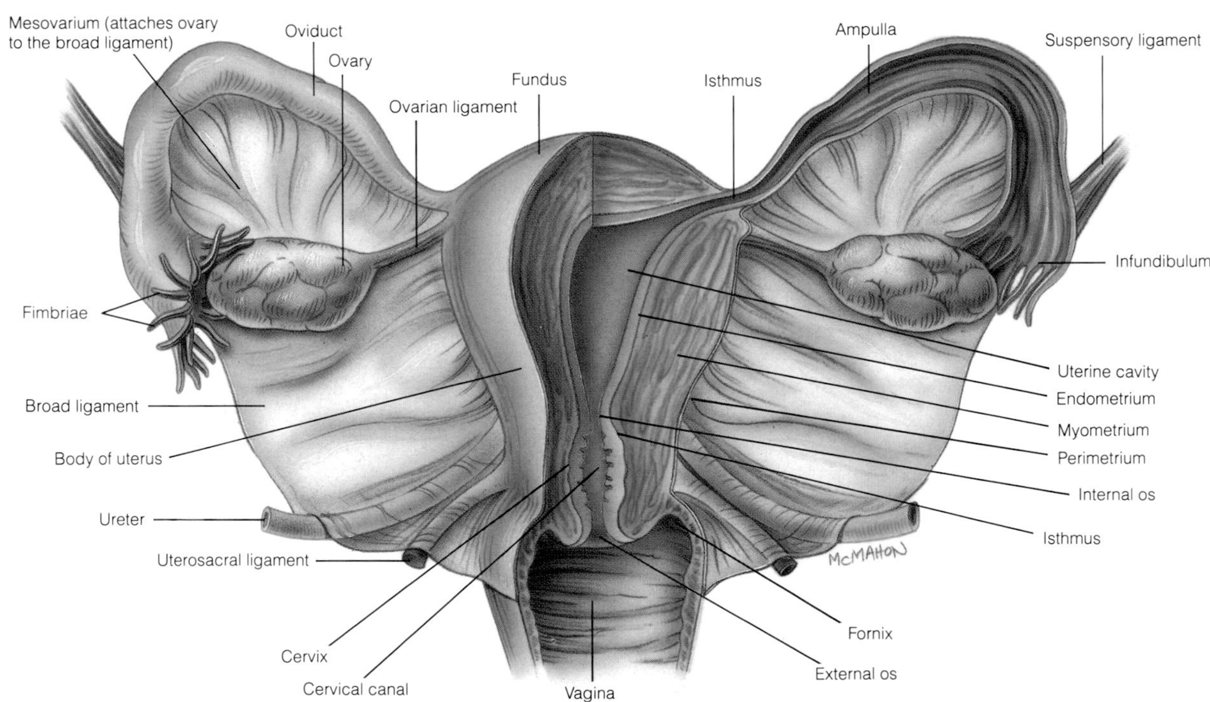

COLOR PLATE 17 Anterior view of the female reproductive tract. [Source: Gaudin, A. J., and Jones, K. C. (1989). "Human Anatomy and Physiology." Harcourt Brace Jovanovich, San Diego, p. 712. Reproduced with permission.] [*See* Reproductive System, Anatomy.]

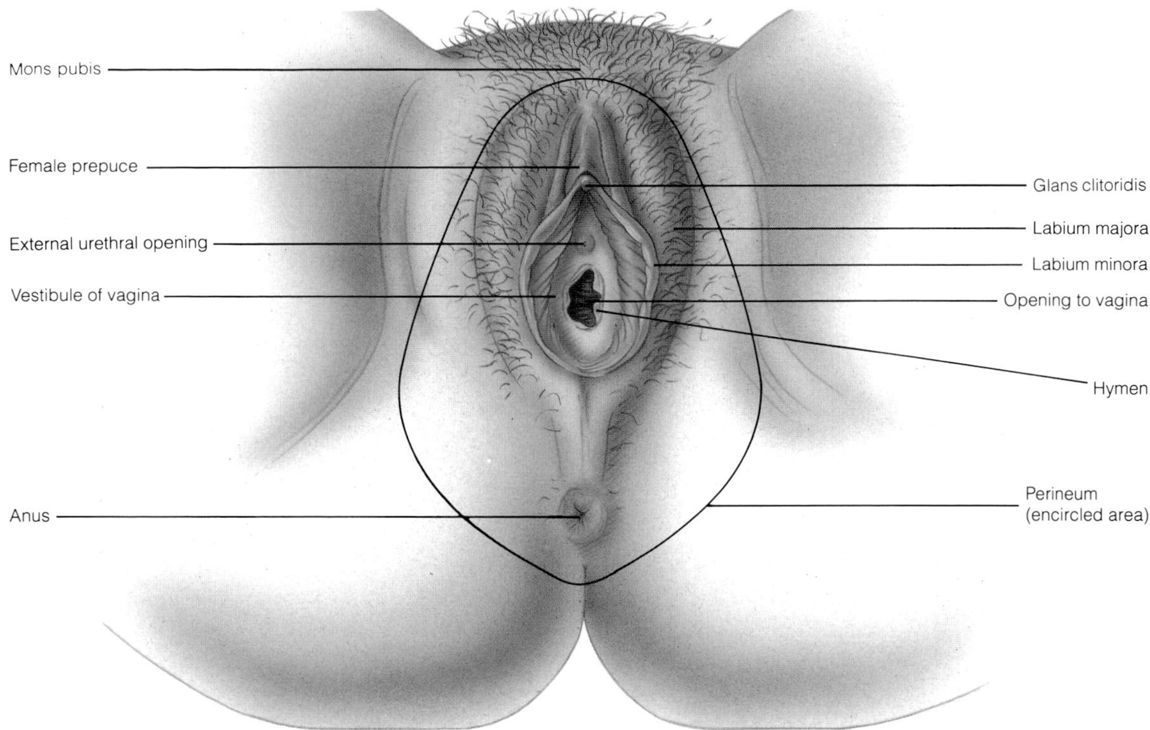

COLOR PLATE 18 External female genitalia (vulva). [Source: Gaudin, A. J., and Jones, K. C. (1989). "Human Anatomy and Physiology." Harcourt Brace Jovanovich, San Diego, p. 715. Reproduced with permission.] [*See* Reproductive System, Anatomy.]

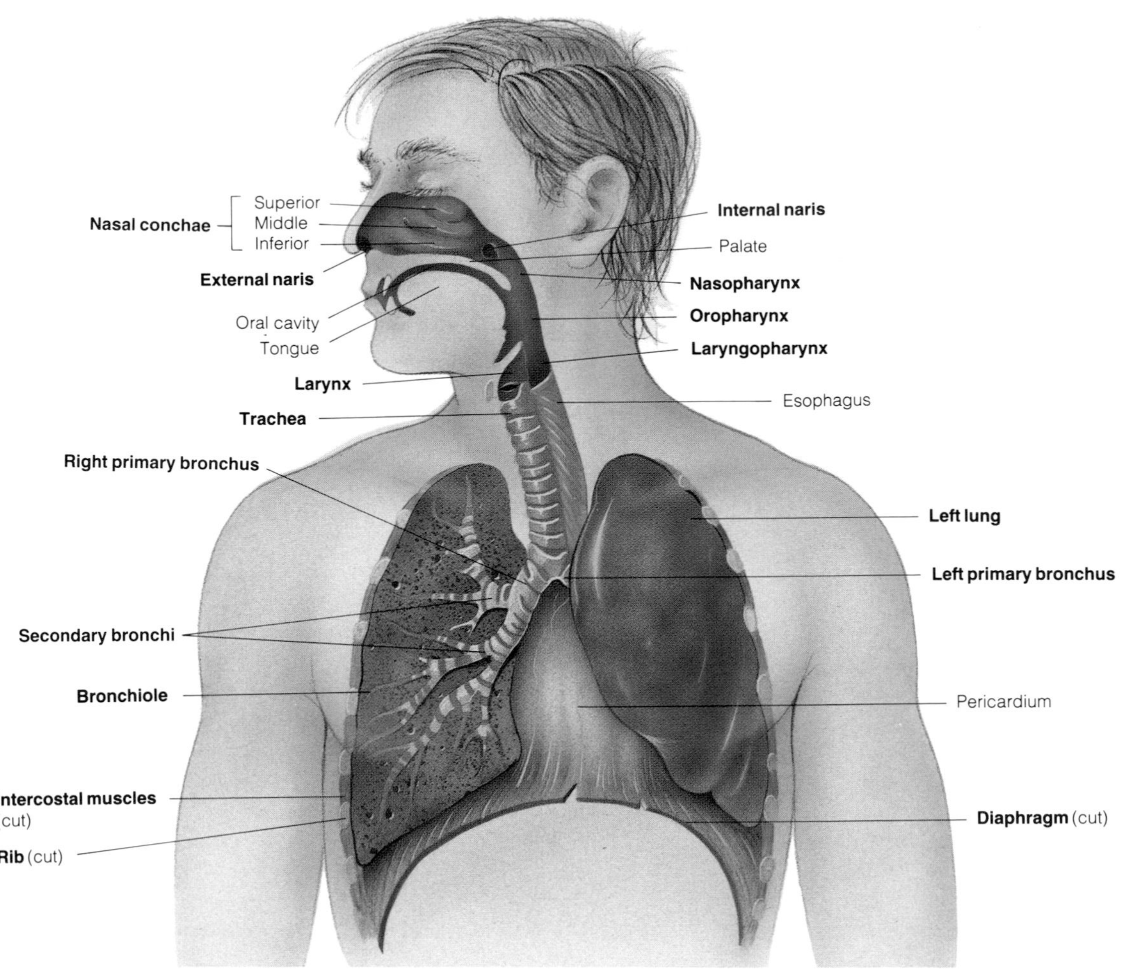

COLOR PLATE 19 The respiratory tract. [Source: Gaudin, A. J., and Jones, K. C. (1989). "Human Anatomy and Physiology." Harcourt Brace Jovanovich, San Diego, p. 505. Reproduced with permission.] [*See* Respiratory System, Anatomy.]

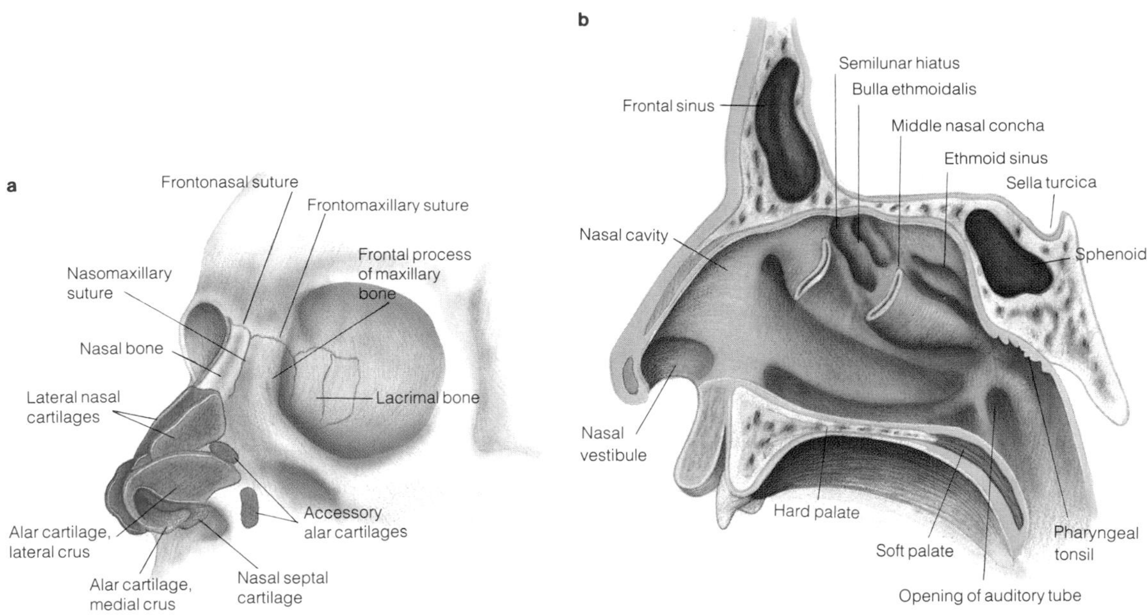

COLOR PLATE 20 The nose. (a) The external nose. (b) The nasal cavity. [Source: Gaudin, A. J., and Jones, K. C. (1989). "Human Anatomy and Physiology." Harcourt Brace Jovanovich, San Diego, p. 507. Reproduced with permission.] [*See* Respiratory System, Anatomy.]

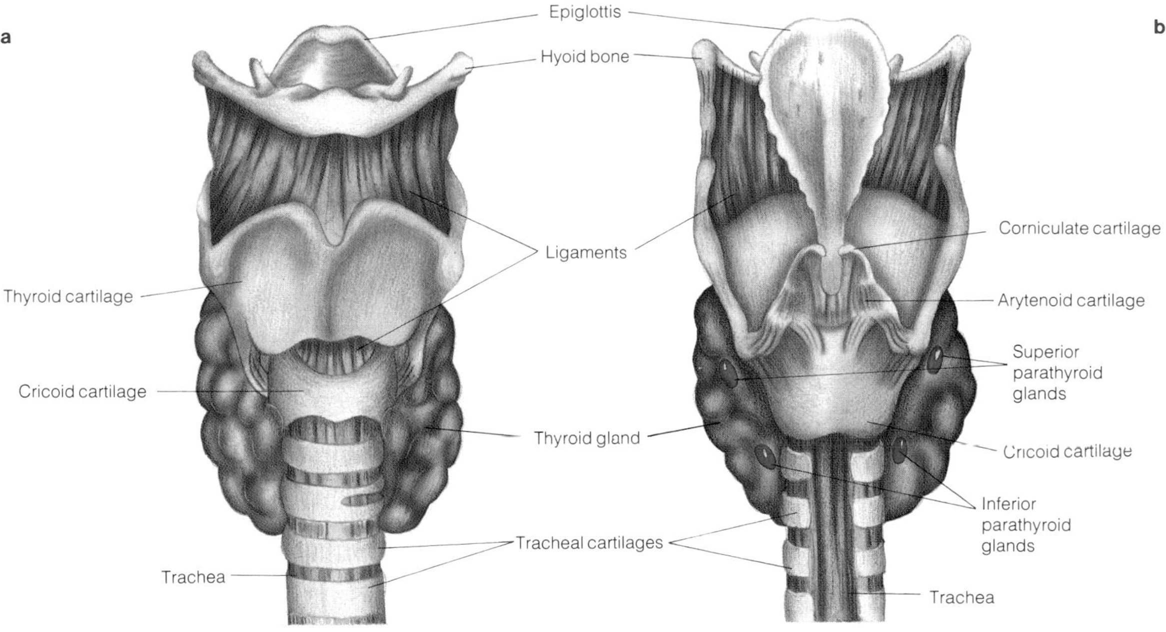

COLOR PLATE 21 The larynx. (a) External anterior view. The isthmus of the thyroid gland has been omitted to show the cricoid cartilage. (b) External posterior view. [Source: Gaudin, A. J., and Jones, K. C. (1989). "Human Anatomy and Physiology." Harcourt Brace Jovanovich, San Diego, p. 510. Reproduced with permission.] [*See* Respiratory System, Anatomy.]

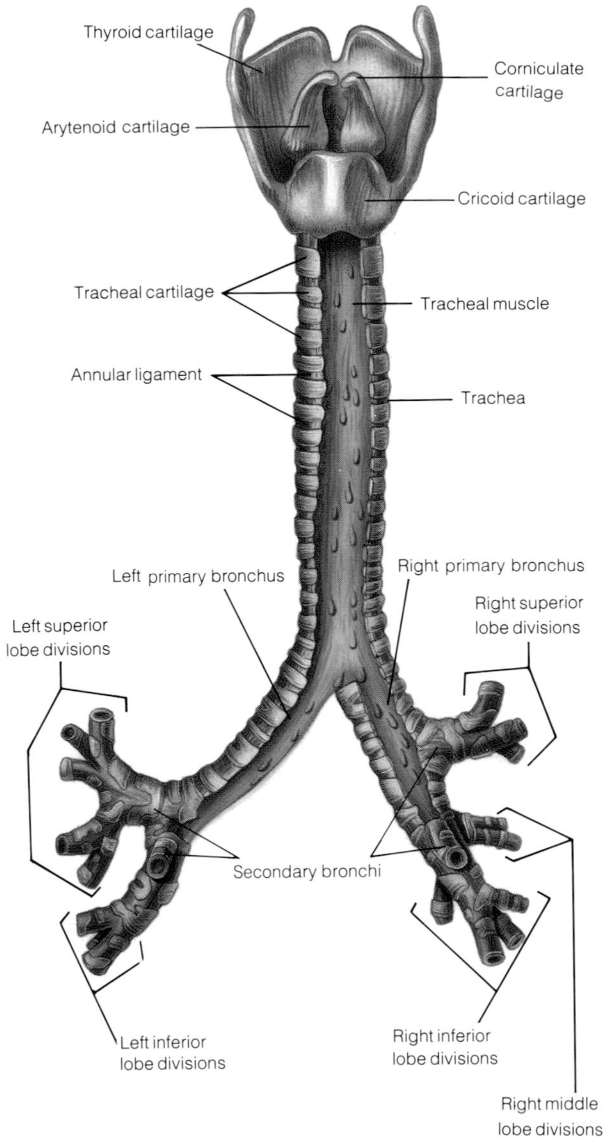

COLOR PLATE 22 The trachea and the bronchi in a posterior view. [Source: Gaudin, A. J., and Jones, K. C. (1989). "Human Anatomy and Physiology." Harcourt Brace Jovanovich, San Diego, p. 511. Reproduced with permission.] [*See* Respiratory System, Anatomy.]

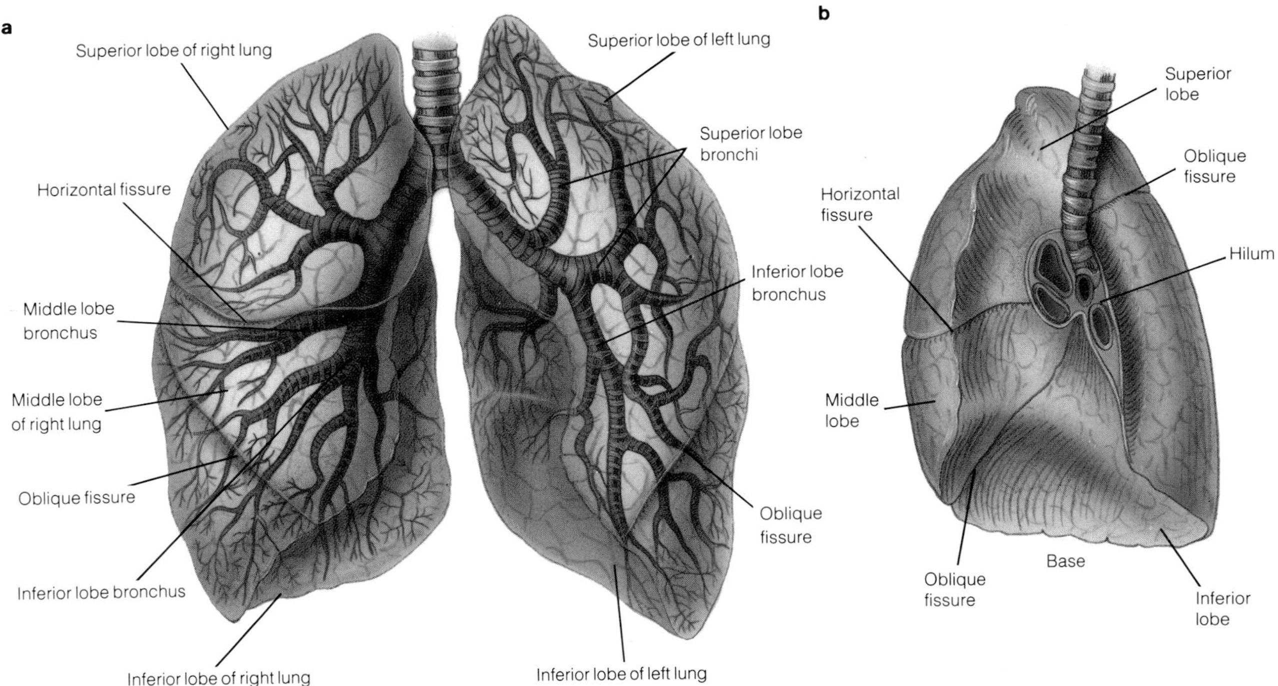

COLOR PLATE 23 (a) Anterior view of both lungs. (b) Medial view of the right lung, showing the hilum. [Source: Gaudin, A. J., and Jones, K. C. (1989). "Human Anatomy and Physiology." Harcourt Brace Jovanovich, San Diego, p. 512. Reproduced with permission.] [*See* Respiratory System, Anatomy.]

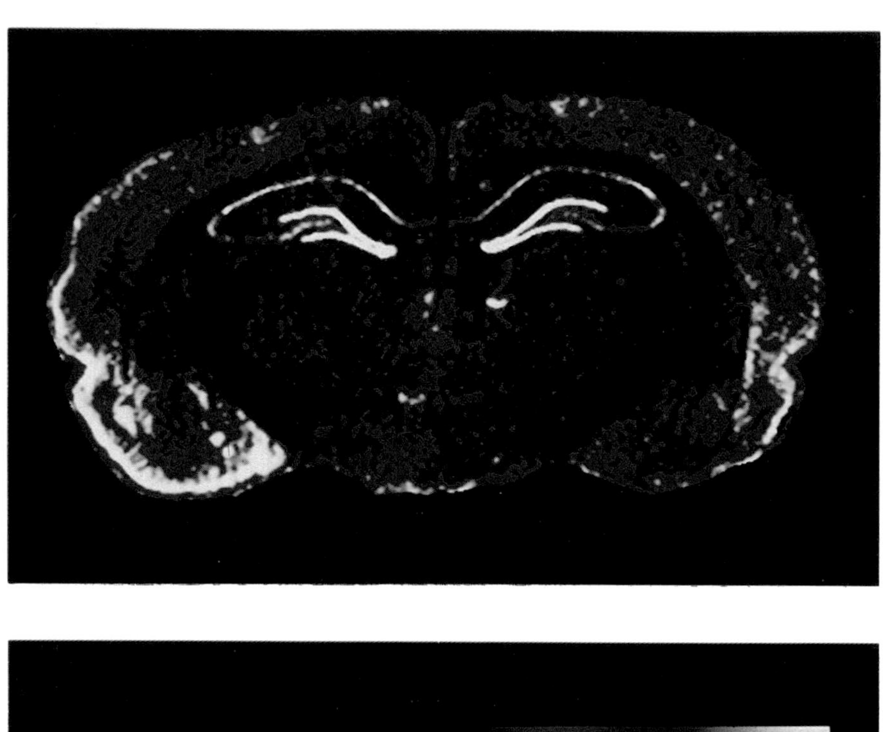

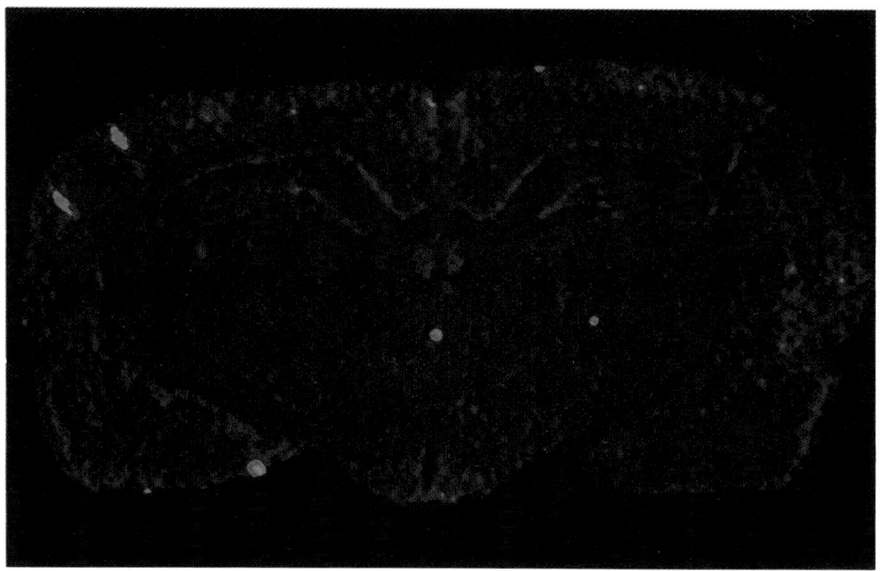

COLOR PLATE 24 Effect of seizures (evoked from area tempestas) on c-*fos* gene expression in various brain regions. Thin coronal sections of brain were exposed to a radioactively labeled probe that selectively recognizes c-*fos* messenger RNA. Brain regions with the highest density of binding to the probe are represented by the color white in the photograph; this reflects the highest c-*fos* expression. The color strip in the center shows the range of colors from lowest (blue) to highest (white) c-*fos* expression. Brain section from a rat that had a convulsive seizure triggered by placing a GABA antagonist in the left area tempestas (top). Similar brain section from a rat that did not have a seizure (bottom). [*See* Seizure Generation, Subcortical Mechanisms.]

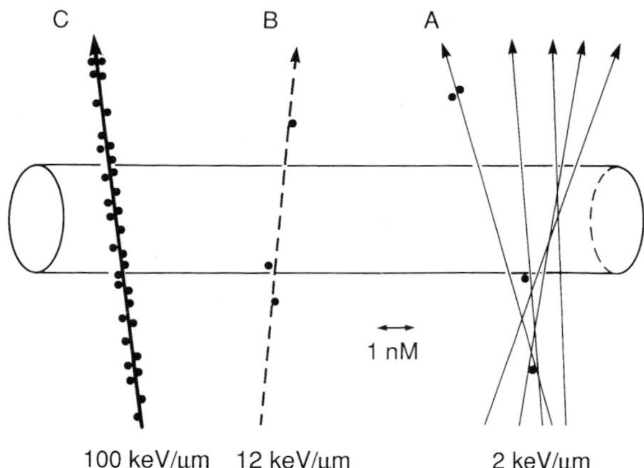

FIGURE 2 Ionization patterns resulting from (A) a low-LET radiation (electrons from 250-kV X rays, LET ~2 keV/μm); (B) a medium-LET radiation (a 15-MeV neutron, track average LET ~12 keV/μm); and (C) a high-LET radiation (2.5-MeV α particle, track average LET ~100 keV/μm). The cylinder represents a double helix of DNA. Each line represents the path of a single photon or particle of radiation. Each dot represents one ion pair.

ation, dose rate, and so on, decrease as the LET of the radiation increases, until, at track average LETs of approximately 200 keV/μm, they become negligible.

IV. MEDICAL USES OF RADIATION

Radiation is widely used in the diagnosis and treatment of human disease. Diagnostic radiology uses low-energy X rays to map differences in tissue density. The basis of their use is the fact that the absorption of low-energy X rays occurs primarily through photoelectric absorption, which is dependent on the atomic number (Z), increasing approximately as Z^3. Small differences in the atomic composition of tissues (e.g., the different calcium levels in bone and muscle) therefore produce large differences in the absorption of the radiation. The uses and the techniques of diagnostic radiology are discussed elsewhere in this encyclopedia. [*See* Diagnostic Radiology.]

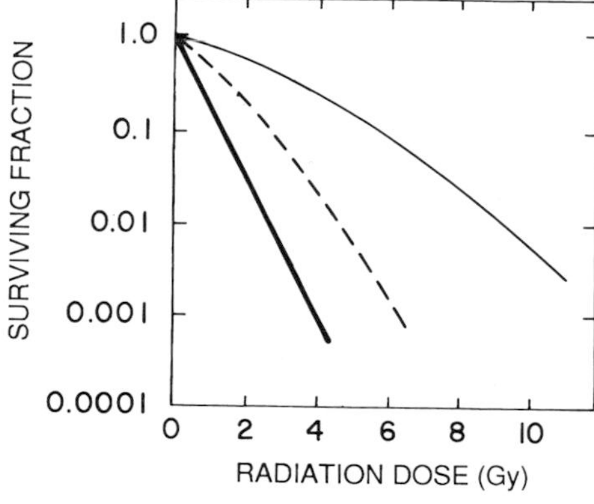

FIGURE 3 Survival curves for aerobic cells irradiated with the low (thin solid line)-, medium(dashed line)-, and high (thick solid line)- LET radiations in Fig. 2. As the LET of the radiation increases, the survival curve becomes straighter and steeper. [These changes are more fully discussed by E. J. Hall (1994). "Radiobiology for the Radiologist," 4th Ed. Lippincott, Philadelphia.]

Nuclear medicine uses injected radionuclides and radiolabeled molecules to diagnose and treat diseases. The tracers used in diagnostic nuclear medicine have been developed and chosen because they localize in specific areas or participate in specific biological reactions. The use of radioactive iodine to study thyroid function and of radioactive gallium or radiolabeled antibodies to locate tumors are examples of routine diagnostic nuclear medicine techniques. Therapeutic applications of nuclear medicine (e.g., the treatment of thyroid cancer or hyperthyroidism with radioactive iodine or the still experimental attempts to treat certain cancers with radiolabeled antibodies) are likewise based on selective localization of the radiation; in these therapeutic applications the large doses of high-specific-activity radionuclides yield local radiation doses large enough to produce cytotoxic effects.

Radiation therapy delivers high doses of localized radiation to treat cancer and certain other life-threatening or debilitating hyperproliferative diseases. The biological basis of radiation therapy is the fact that radiation sterilizes the malignant cells (as described earlier) and thereby destroys their capacity to cause the tumor to grow, invade nearby tissues, and initiate metastases. Radiation, of course, can also injure normal tissues. Therefore, radiotherapy must be carefully planned and delivered, so as to maximize antineoplastic effects and to minimize normal tissue injury.

The choice of radiation source is important. Deep-seated tumors are generally treated using high-energy linear accelerators delivering 4- to 28-MeV X rays. Because these radiations are penetrating and their absorption is relatively unaffected by tissue density, uniform radiation doses can be delivered to tumors deep within the body. At the other extreme, superficial cancers can be treated with low-energy electrons, which penetrate only a few centimeters and therefore spare normal tissues beneath the tumor. Tumors are also treated from several different angles, so that the tumor is always irradiated, but different areas of normal tissue are exposed during each treatment; each area is therefore held to low doses, producing tolerable damage.

The delivery of radiation therapy is carefully planned for each individual patient, based on detailed considerations of the location and pathology of the tumor and of the relative radiosensitivities of the tissues surrounding and involved by the tumor. In addition, the radiation treatment is delivered in many small fractions, generally given each day over ~6 weeks; this allows the normal tissues to recover from radiation injury between each treatment. The differential effect of radiation on tumors and normal tissues is also aided by the differential effect of radiation on proliferating and quiescent cells (described in Section III,A).

Radiation can also be delivered as brachytherapy, the application to or implantation into the tumor of sealed sources of intensely radioactive isotopes, which irradiate the tumor with highly localized radiation continuously over a period of hours, days, or weeks. Brachytherapy is another approach to achieving the localization of radiation dose and the protraction of irradiation that is used in external beam radiotherapy.

Radiation therapy is a valuable modality in the treatment of neoplastic disease. Radiation is used either alone or in combination with surgery and/or chemotherapy in the curvative treatment of many early localized cancers. In some cancers these treatments are extremely effective, allowing over 95% of the patients to be cured and to lead a full normal life. In incurable disease, radiation therapy can also play a valuable role in the palliation of local symptoms and the amelioration of pain. Currently, over 60% of all cancer patients receive radiotherapy at some point during the treatment of their disease. Major research efforts are directed toward the improvement of radiation therapy and the development of improved regimens for combining radiation with other modalities of cancer treatment.

V. HAZARDS OF RADIATION

A. Acute Radiation Injury

1. Injury to Specific Tissues

Acute radiation injuries occur only after large doses of radiation. In general, these reactions reflect the toxicity of radiation to the rapidly proliferating progenitor cells in the tissues. The nature of these acute radiation reactions is well illustrated by the processes occurring in the irradiated intestine. Radiation sterilizes the proliferating cells in the gut crypts, but the quiescent cells of the villi are unaffected and continue to perform their differentiated functions. However, these differentiated cells have a relatively short life span, normally being shed into the lumen of the intestine after 5–8 days. As the differentiated cells are shed, the death of the precursor cells becomes manifest as a deficit in differentiating cells. The result after low doses of radiation is a shortening of the villi and the crypts and a resulting impairment in gut function. Large doses of radiation result in more serious deficits in gut cellularity and more serious impairments in function. Very large doses of radiation can lead to an actual breakdown of the intestinal lining, resulting in serious or life-threatening injury. The biological basis of early reactions in other rapidly renewing tissues (e.g., skin, bone marrow, and hair follicles) is analogous. These reactions occur only after doses thousands of times higher than those used in diagnostic radiology. During radiotherapy their severity is minimized by fractionating the radiation dose into many small fractions, allowing these rapidly renewing tissues to recover between treatments. [*See* Radiosensitivity of the Integumentary System; Radiosensitivity of the Small and Large Intestines.]

2. Acute Effects of Whole-Body Irradiation

The acute effects of whole-body irradiation result primarily from the cytotoxic effects of radiation on rapidly renewing tissues. Our understanding of the mechanism of these effects is derived primarily from animal studies and from extrapolations based on observations of locally irradiated tissues in patients receiving radiotherapy. The effects of total-body exposures on people are known primarily from the medical records of cancer patients and of people receiving dangerous doses of radiation in nuclear accidents (e.g., Chernobyl) and from the fragmentary medical records available in Hiroshima and Nagasaki.

The symptoms of radiation damage can be divided into two phases: a "prodromal syndrome," which occurs minutes or hours after exposure, and the delayed symptoms of the potentially lethal injuries, which develop days to weeks later. The prodromal syndrome is mediated through the response of the autonomic nervous system and includes a variety of gastrointestinal and neuromuscular symptoms. Higher radiation doses produce a more rapid onset, a longer duration, and more severe symptoms. Prodromal symptoms seldom appear at doses below 1 Gy. At doses of 1–3 Gy, symptoms include anorexia, nausea, vomiting, diarrhea, and fatigue. At higher doses these symptoms increase in severity, and other symptoms (e.g., apathy and headache) also occur. At supralethal doses, hypotension, fever, and shock can also occur rapidly. The prodromal symptoms then subside, and the patient enters a relatively symptom-free latent period before the onset of the delayed symptoms, which are manifestations of cytotoxic injury.

The hematopoietic system is the most sensitive tissue to radiation. Doses in excess of 3 Gy result in severe depletion of the hematological progenitor cells. Immunosuppression occurs rapidly. After a latent period of ~3 weeks, decreases in the levels of circulating granulocytes and platelets occur, leading to hemorrhage and secondary anemia. These symptoms resolve as the progenitor cells that survived irradiation proliferate and repopulate the bone marrow and the hematopoietic system recovers. At doses of 4–5 Gy, treatment with whole-blood or platelet transfusions, antibiotics, and supportive care during the hematological crisis could be critical. At doses of 5–8 Gy, survival becomes increasingly problematic, even with intensive medical support, because the severe persistent depletion of the hematological progenitor cells renders recovery of the bone marrow uncertain. The role of bone marrow transplants remains controversial, especially in view of the limited success of this procedure in aiding victims of the Chernobyl accident. [*See* Hemopoietic System.]

At radiation doses of ~10 Gy, radiation damage to the gut becomes manifest ~3–8 days after radiation, reflecting damage to the progenitor cells (described in Section V,A,1). Death from mucosal damage, dehydration, electrolyte loss, and infection may result ~1 week after exposure. Victims surviving gastrointestinal syndrome inevitably develop the hematopoietic syndrome.

At radiation doses of at least 100 Gy, cerebrovascular syndrome results in death within 2 days of exposure. This probably results from acute vascular permeability changes, histamine release, and chemical damage to critical cellular molecules, resulting in overwhelming central nervous system (CNS) injury. The symptoms of the prodromal syndrome merge with those of lethal CNS injury.

B. Late Radiation Hazards

Late radiation injuries can be divided into two kinds: degenerative effects, which result from the killing of large numbers of cells, and stochastic effects, which result from heritable sublethal damage in one or a few cells. Degenerative effects occur only after large doses of radiation, which produce significant tissue damage, such as the doses received by cancer patients treated with intensive radiation therapy or by people involved in very serious radiation accidents. The severity of these injuries increases with the radiation dose. Stochastic effects, in contrast, can result from a single sublethal lesion in a single cell; such effects therefore occur at a certain probability among individuals in a population of people exposed to small doses of radiation. Stochastic effects are probabilistic: They occur with a probability that depends on the radiation dose, but the biological effects per se are the same whenever they occur, regardless of the radiation dose. Stochastic effects of ionizing radiation include carcinogenesis and the induction of mutations.

1. Late Degenerative Injuries

Late radiation damage can develop months to years after irradiation in tissues receiving large radiation doses (e.g., in tissues irradiated with ~40–60 Gy during the treatment of cancer). Injury can result either from depletion of the parenchymal cells (i.e., the cells that perform the unique tissue-specific functions) or from damage to the stromal elements (i.e., the blood vessels and the connective tissue that nourish and support the parenchyma of the tissue). The nature of the injury in a given tissue and the time at which injury develops depend on the relative sensitivity of the parenchymal and stromal cell populations and on the proliferation patterns of these cells.

At one extreme are tissues such as mature muscle and brain, in which the parenchymal cells are totally quiescent and extremely radioresistant; in such tissues late radiation injury may reflect the death of stromal cells, leading to vascular damage, fibrosis, and/or necrosis. In other slowly proliferating tissues, late functional damage appears to reflect the death of relatively quiescent parenchymal cells. For example, it is thought that changes in kidney function that occur months to years after irradiation primarily reflect sterilization of the slowly proliferating, but clonogenic, cells of the proximal tubule, with a resultant gradual loss of tubules, and therefore a decline in tissue function.

In the ovary and testes, delayed radiation injury clearly reflects the depletion of specific radiosensitive parenchymal cell populations. The very different radiation responses of the male and female gonads reflect intrinsic differences in the patterns of cell proliferation in the ovaries and the testes. In males, mature sperm are produced continually by the proliferation, differentiation, and maturation of spermatogonial stem cells. Because mature sperm are relatively radioresistant, the irradiated individual remains fertile as long as these mature sperm remain available. Afterward, a period of decreased fertility or sterility ensues, which reflects the depletion of the more radiosensitive spermatogonial stem cell populations. The severity and duration of this infertility are dose dependent, reflecting the degree of depletion of the stem cell populations and the time required to repopulate these cell populations. High radiation doses result in permanent sterility. Sterility in the male is not accompanied by changes in hormonal profile, libido, or physical capability. In the female, oocyte production is essentially complete a few days after birth. The immature and mature follicles in the ovaries of adult females therefore contain resting oocytes, with no stem cells and no renewal capacity. Irradiation results in a dose-dependent depletion of these follicles. Low doses of radiation therefore produce impaired fertility and high doses result in permanent sterility, which is accompanied by hormonal changes comparable with those occurring during menopause.

Radiation-induced cataracts result from the death of parenchymal cells within the lens of the eye. The lens has an extremely unusual pattern of cell renewal. New cells are produced by mitotic activity in the germinal zone of the epithelium. These cells differentiate into lens fibers and migrate toward the equator. There is, however, no mechanism for the removal of dead cells; they simply accumulate at the posterior pole. If many cells are killed by radiation or by other agents, the accumulation of dead fibers becomes large enough to produce an opacity visible upon ophthalmic examination; greater cytotoxicity produces larger opacities, which could interfere with vision.

Because degenerative late reactions result from the death of large numbers of cells, they occur only in sites receiving large doses of radiation. There is a threshold dose, below which the reactions never occur. In general, the severity of late degenerative reactions increases with the radiation dose and the time after irradiation.

2. Genetic Effects of Radiation

Science fiction books and horror movies have popularized the impression that radiation induces bizarre monstrous mutations. This is not the case. Rather, as described in Section II, radiation produces DNA damage that is similar to that occurring naturally; the effect of irradiation is therefore to increase the frequency of the same mutations that occur spontaneously. The fact that it is impossible to distinguish between radiation-induced and spontaneous natural mutations complicates studies of the genetic effects of radiation. The effect of radiation is seen as an increase in the rate of occurrence of genetic diseases and the rate of changes in specific genetic markers; at low doses of radiation, the effect of irradiation becomes impossible to detect, because it is lost in the background of the spontaneous mutations. [*See* Mutation Rates.]

Studies with human populations exposed to radiation have proved especially difficult, because of the genetic heterogeneity of the populations and also because social, environmental, and physiological factors (e.g., parental age, inbreeding, and exposure to other genotoxic agents) can influence the rates of genetic disease and congenital malformations. None of the epidemiological studies performed on humans has found a statistically significant increase in genetic disease in the offspring of irradiated persons. The most extensive and careful of these studies is that performed on the offspring of the survivors of the atomic bombs at Hiroshima and Nagasaki, where four genetic indicators are being monitored: (1) untoward pregnancy outcome (e.g., stillbirths, major congenital defects, and perinatal deaths); (2) death before age 17; (3) sex chromosome abnormalities; and (4) mutations in blood proteins (i.e., electrophoretic variants in 28 plasma and erythrocyte proteins; activity variants in 8 erythrocyte enzymes). The frequency of these four indicators has not increased significantly in either the low (0.01–0.09 Gy)- or high (over 1 Gy)-dose populations.

Our knowledge of the genetic effects of radiation, therefore, comes from the extensive studies performed on plants (especially *Tradescantia*), insects (especially the fruit fly, *Drosophila melanogaster*), and mice; the mouse appears to be the best model for predicting genetic effects in humans. These studies show that most mutations, whether spontaneous or induced, are deleterious. Studies with mice suggest that there is repair of radiation damage by the gonadal precursor cells and show that delaying conception for ~6 months after radiation exposure decreases the risk of genetic anomalies in the offspring. Because of the differences in the radiation responses of the ovaries and the testes, almost all of the heritable radiation damage is carried by the male rather than the female. Laboratory studies support the concept that there is no threshold for genetic effects: any radiation dose, however small, poses a finite statistical risk of producing genetic damage. In fact, background radiation probably produces ~1–6% of the spontaneous mutations occurring in people. The number of mutations increases with dose. The dose required to double the spontaneous mutation frequency in a population of exposed people has been estimated to be approximately 1 Sv.

3. Carcinogenesis

Radiation induces cancer in people and animals. The mechanism by which this occurs is unclear, and it is possible that several mechanisms contribute to radiation carcinogenesis. First, genetic changes (i.e., mutations or chromosomal rearrangements) induced in somatic cells could start a normal cell down the path of evolution toward malignancy. Second, radiation could induce oncogenes or inactivate oncogene suppressors, thereby causing malignant transformation. In one strain of laboratory mice, induction of an endogenous oncogenic virus by radiation is responsible for radiation-induced leukemia, raising the possibility of a similar mechanism in humans. At very high radiation doses, local chronic tissue damage could contribute to the carcinogenic process; skin cancers in early radiation workers developed primarily on the edges of unhealing radiation ulcers.

At low radiation doses, radiation carcinogenesis appears to have the characteristics expected of a stochastic process: there is no threshold; the radiation-induced cancers cannot be distinguished from spontaneous cancers; and the frequency of cancer increases with dose, but the severity of the induced cancers is independent of dose and indistinguishable from that of spontaneous cancers. As discussed earlier for genetic effects, the fact that radiation-induced cancers cannot be distinguished from the cancers occurring naturally in the population complicates attempts to estimate the risks of radiation exposure.

Carcinogenesis has been studied extensively in people exposed to radiation. These epidemiological studies are subject to all of the problems inherent in defining changes in the rates of relatively common diseases that are also influenced by many other biological, genetic, and life-style factors. Nevertheless, these data are extensive enough that human data provide the framework for predicting the carcinogenic risk of radiation. Populations that have been or are being studied include (1) early radiation scientists exposed to large doses of X rays or natural radioactive materials; (2) patients receiving large doses of localized radiation for the treatment of cancer or benign diseases (including ankylosing spondylosis, ringworm, and postpartum mastitis); (3) survivors of the bombings of Hiroshima and Nagasaki; (4) populations exposed to radiation during atomic bomb tests (e.g., Marshall Islanders); (5) persons exposed to radiation during accidents; (6) women who ingested radium while painting luminous clock dials; (7) patients injected with large doses of radioactive material for the diagnosis or treatment of disease; and (8) patients receiving multiple high-dose fluoroscopies to monitor the efficacy of treatment for tuberculosis. Many of these populations were exposed to very large doses of radiation in the early 1900s, before the delayed hazards of radiation were appreciated and before radiation protection practices were implemented in the workplace or in medical practice.

The conclusions from these epidemiological studies can be summarized as follows: there is a long latent period between exposure to radiation and the development of cancer. Excess cases of leukemias begin to appear ~2 years after irradiation, peak at ~10 years, and essentially disappear by 25 years. The latent period for solid tumors can be much longer and is known to be at least 20–40 years. Because most large cohorts of irradiated people being monitored are still living, it is not yet possible to say whether the excess risk of solid tumors decreases, remains constant, or increases at longer times after exposure.

Many kinds of benign and malignant tumors have been shown to be induced by radiation. Acute and chronic myeloid leukemias account for the excess leukemia incidence observed in irradiated adults, whereas acute lymphocytic leukemia is induced in children. Thyroid carcinoma, breast cancer, lung cancer, stomach cancer, bone cancer, skin cancer, and other malignancies have been documented convincingly in various exposed populations. In general, persons exposed to highly localized radiation (from external sources or from internally deposited radionuclides) seem to be at excess risk only for tumors arising in the irradiated tissues, but not for tumors arising in unirradiated sites.

There is considerable controversy concerning the magnitude of the risk of carcinogenesis, especially at low radiation doses. This results from the nature of the data. Although many of the studies have followed relatively large numbers of people, the actual numbers of observed excess cancers are not large. When the data are subdivided by radiation dose to assess the dose–effect curve for carcinogenesis, the number of cases in each class is small and the error limits on the points are large. As a result the slopes (and even the shapes) of the individual dose–response curves are uncertain and vary from one study to another. In addition, it is difficult to estimate accurately the radiation doses received by the exposed individuals. The identification and follow-up of the appropriate exposed populations also raise serious logistical problems, because the studies must span several decades. Moreover, the natural variation in the spontaneous cancer incidence makes it difficult to establish the appropriate control incidence, while the long latent periods before the cancers appear mean that the total cancer incidence is still evolving in most populations being monitored.

Radiation appears to increase the relative risk of specific cancers; the risk of a radiation-induced neoplasm therefore varies with the incidence of that cancer among the unexposed members of the same population. For example, radiogenic stomach cancer is frequent among irradiated populations in Japan, reflecting the high background incidence of stomach cancer in these populations, but is infrequent in irradiated populations in Europe, where stomach cancer is a rare disease. Evidence that relative risks for the induction of some cancers vary with age, sex, and other factors further complicates the development of risk estimates.

One of the best current evaluations of the risk of radiation-induced cancer is found in the 1990 report of the National Research Council Committee on the Biological Effects of Ionizing Radiation (see Bibliography.) A simplified summary of the committee's major risk estimates is shown in Table V. The uncertainty in these estimates is large because of the complicating factors described here. Moreover, risk estimates for continuously irradiated populations also rely on some exceedingly tenuous assumptions to extrapolate from the effects observed after a large dose of radiation received in a short period of time and predict the effects of small radiation doses accumulating continuously over many decades.

C. Effects of Prenatal Irradiation

The effects of prenatal irradiation are of special concern, because the developing embryo is extremely sensitive to injury from many toxic agents. As there are few data from humans on this critical subject, our knowledge of the effects of prenatal irradiation comes primarily from studies with rodents deliberately irradiated at known times after conception.

The effects of irradiation and the sensitivity of the embryo vary with the stage of development. In the mouse and the rat, preimplantation embryos are exquisitely sensitive to the lethal effects of radiation. A dose as small as 0.05 Gy can kill fertilized mouse eggs. However, the embryos that survive develop normally both before and after birth. The effect of irradiation is therefore seen as a decrease in litter size and an increase in embryonic resorption. Extrapolation to humans suggests that the major effect of radiation during this period (i.e., the first 10 days postconception) would be embryonic death, probably before the mother knew she was pregnant. The limited available data from humans support this concept.

Irradiation of rodents with ~2 Gy during the period of rapid organogenesis results in the production of a variety of structural abnormalities. Specific organs are most sensitive during the beginning of their differentiation, when they contain small numbers of rapidly proliferating cells. Later, during the fetal period, rodents become more resistant to radiation, and the primary effects appear to be generalized growth retardation and developmental abnormalities in late-developing tissues (e.g., the gonads and the CNS). CNS

TABLE V
Estimates of Excess Cancer Mortality in Irradiated Populations
(Excess Mortality per 100,000 Exposed Persons)[a]

	Males		Females	
Irradiation conditions	Leukemia	Nonleukemia	Leukemia	Nonleukemia
Single acute exposure to 0.1 Sv	110	660	80	730
Continuous lifetime exposure 1 mSv/yr	70	450	60	540
Continuous exposure ages 18–65 10 mSv/yr	400	2480	310	2760

[a]Values are summarized from the 1990 Report of the National Research Council Committee on the Biological Effects of Ionizing Radiations (BEIR V) (see Bibliography).

and behavioral damage has been noted in rodents at doses as low as 0.1 Gy.

Extrapolation of these data to humans suggests that irradiation during Weeks 4–11 postconception might produce severe developmental abnormalities, whereas irradiation between Weeks 11 and 16 might lead to a few structural abnormalities in late-developing organs and possibly to stunted growth, microcephaly, and retardation. Because the CNS and the skeleton continue to develop throughout the fetal period and into infancy, irradiation during these periods could also cause growth and mental retardation. The limited data available on humans support these general extrapolations. There are reported instances of structural abnormalities and mental retardation in children whose mothers received pelvic radiotherapy during the first trimester of pregnancy. Children exposed *in utero* to high doses of radiation from the atomic bombs were shorter and weighed less than children receiving little or no radiation. These children also exhibited an increased incidence of microcephaly and mental retardation (judged by intelligence tests and school performance). The incidence of severe retardation increased with dose; significant effects were observed at doses as low as 0.1–0.2 Gy. The effect is highly dependent on gestational age; maximal retardation was seen after irradiation at 8–15 weeks of gestation, smaller effects were seen for 16–26 weeks of gestation, and no significant effects were observed in persons irradiated either earlier or later in gestation. Growth retardation, mental retardation, and learning disabilities have been observed in some children receiving radiotherapy as infants.

These data illustrate the need for caution in the treatment of pregnant and potentially pregnant women with diagnostic or therapeutic radiation. An especially troubling aspect of radiation teratogenesis is the fact that the embryo is most sensitive during the first few weeks, before the pregnancy is obvious or even definitively diagnosed. The doses required for fetal injury are much larger than those received by most occupationally exposed women, but are small enough to mandate special precautions to protect unborn children of radiation workers. Such precautions have been implemented.

BIBLIOGRAPHY

Committee on the Biological Effects of Ionizing Radiations (BEIR V) (1990). "Health Effects of Exposure to Low Levels of Ionizing Radiation." National Academy Press, Washington, D.C.

Hall, E. J. (1994). "Radiobiology for the Radiologist," 4th Ed. Lippincott, Philadelphia.

Johns, H. E., and Cunningham, J. R. L. (1983). "The Physics of Radiology," 4th Ed. Thomas, Springfield, Illinois.

National Council on Radiation Protection and Measurements (NCRP) (1987). "Ionizing Radiation Exposure of the Population of the United States," NCRP Report 93. NCRP, Bethesda, Maryland.

United Nations Scientific Committee on the Effects of Atomic Radiation (UNSCEAR) (1988). "Sources, Effects, and Risks of Ionizing Radiation." United Nations, New York.

Radioimmunoassays

ROSALYN S. YALOW
Veterans Administration Medical Center and The Mount Sinai School of Medicine

GLOSSARY

Double-antibody method Precipitation of antigen–antibody complex with a second antibody directed against the antibody in the complex

Hapten Substance which when coupled to a carrier protein elicits immune response

Immunogen Substance that stimulates antibody response

Iodination Addition of iodine to a molecule

Prohormone Precursor for biologically active hormone

THE MEASUREMENT OF INSULIN IN HUMAN plasma using radioimmunoassay (RIA) methodology was first reported in 1959. Immediately, this methodology was appreciated as having the potential for widespread applicability in endocrinology because, by that time, advances in peptide chemistry had made available highly purified preparations of animal and human peptide hormones. The potential sensitivity and specificity of RIA made possible physiologic studies of dynamic changes in concentrations of circulating peptide hormones that would have been impossible with the bioassays available at that time. By the 1970s, RIA methodology had spread beyond endocrinology into many areas of biologic interest including, among others, pharmacology and toxicology, infectious diseases, oncology, hematology, and neurosciences.

RIA is now employed in thousands of laboratories around the world, even in the less-developed nations, to measure hundreds of substances of biologic interest. At present, commercial kits are available for many substances routinely measured in clinical or nuclear medicine laboratories; however, "in-house" RIAs remain a powerful tool in the research laboratory. Therefore, the value of the underlying RIA principle as well as its problems and pitfalls is appreciable.

I. RADIOIMMUNOASSAY PRINCIPLE

RIA is based on a simple principle (Fig. 1). The concentration of an unknown substance is determined by comparing its inhibition of binding of labeled antigen to specific antibody with the inhibitions observed with a set of known standards. A typical RIA is performed by the simultaneous preparation of a series of tubes containing standards and unknown samples to which fixed amounts of antibody and labeled antigen are added. The fraction of the labeled antigen, which binds to the fixed amount of antibody, is inversely related to the concentration of the unlabeled antigen in standards or in unknown samples. Hence, after an appropriate time, usually ranging from hours to days, the labeled antigen that is bound to antibody (B) is separated from that which is free (F), and the radioactivity associated with each fraction is measured. The ratio of B/F for each standard tube is then plotted as a function of the concentration in that tube. The unknown concentration is determined by comparing its observed B/F ratio with the standard curve (Fig. 2).

A variety of other modes of plotting the data have been used for the standard curve. These include, among others, the percentage of the tracer bound in the absence of unlabeled hormone or even simply the counting rate of the bound fraction as a function of hormone concentration. Linearity of the standard curve is sometimes approached using logarithmic or semilogarithmic plotting.

ENCYCLOPEDIA OF HUMAN BIOLOGY, Second Edition, VOLUME 7. Copyright © 1991 by Academic Press. All rights of reproduction in any form reserved.

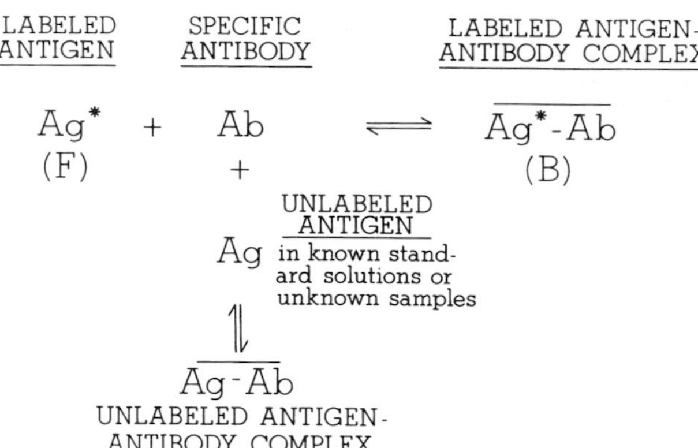

FIGURE 1 Competing reactions that form the basis of RIA. [Reproduced from Yalow (1978).]

The RIA principle is not limited to immune systems but can be extended to systems where, in place of the specific antibody, there is a specific reactor (or binding substance). This reactor could be a specific binding plasma protein (e.g., thyroxine-binding globulin or cortisol-binding globulin) or a tissue-receptor site. In place of the radioisotopic labels, other markers such as enzymes or fluorescent dyes have been used. The general term ligand assay has been applied to these assay systems.

RIA differs from a traditional bioassay in that it is an immunochemical method in which the measurement depends on only the interaction of chemical reagents in accordance with the law of mass action. Immunochemical activity may be identical with or even reflect biologic activity. The validation of an immunoassay procedure simply requires that the concentration in the unknown sample be independent of the dilution at which it is assayed. However, the relationship between RIA and bioassay quantification often requires additional information because in both systems specific and nonspecific factors often play a role.

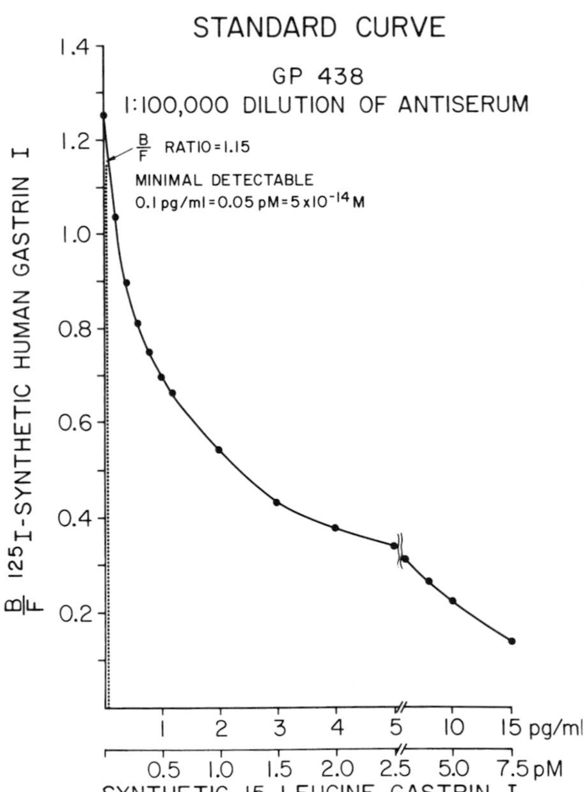

FIGURE 2 Standard curve for the detection of gastrin, a digestive hormone, by RIA. Note that as little as 0.1 pg of gastrin per milliliter of incubation mixture (<0.05 picomolar) is readily detectable. [Reproduced from Yalow (1978).]

II. PREPARATION OF REAGENTS

A. Sensitivity of RIA

The sensitivity in a classic RIA system is dependent on the equilibrium constant of the reaction between antigen and the specific antibody. Furthermore, as evident from Fig. 2, the amount of labeled antigen employed in the assay should generally be no larger than the minimal detectable antigen concentration because the initial portion of the standard curve is the region of greatest sensitivity. Thus, the crucial factors limiting the sensitivity of a RIA procedure are the availability of a suitable antiserum and the specific activity of the labeled antigen.

B. Preparation of Antisera

The availability of a suitable antibody is the *sine qua non* of RIA. Generally, polypeptides of molecular weight >2000 are sufficiently immunogenic when administered as an emulsion in Freund's adjuvant. Lower molecular weight substances (e.g., thyroidal hormones, steroids, prostaglandins, drugs) may be rendered antigenic by coupling to a protein or a large peptide by a variety of methods. Commonly used methods include coupling through carbodiimide to bovine serum albumin or through glutaraldehyde to thyroglobulin. Although a variety of other coupling agents or proteins have been employed, there is no uniform agreement as to which protein or which coupling agent might be optimal for any hapten.

For the preparation of antisera for peptide hormones, the guinea pig has proven to be most satisfactory for a number of reasons. The antigenicity of the immunogen is related to its "foreignness," and many of the guinea pig peptides differ uniquely from those of other mammalian species. Randomly bred guinea pigs have an advantage over inbred animals in that a genetic diversity is more likely, which increases the probability that one of the immunized animals will be a hyperresponder to the immunogen. Obviously, the probability of obtaining a satisfactory antiserum increases with the number of animals immunized, and several guinea pigs can be housed in the same space as a single rabbit. Space considerations are, of course, important for a small laboratory. If the immunogen is a large protein or a hapten coupled to a large protein, a rabbit may be the animal of choice because it is simpler to obtain larger volumes of blood from rabbits than from guinea pigs.

Because the probability of obtaining a satisfactory antiserum increases with the number of animals immunized and the presence of other immunological reactions does not interfere with the reaction of labeled antigen with specific antibody, each animal is immunized simultaneously with several unrelated antigens. This permits reduction of the number of animals to be maintained and bled by a factor equal to the number of immunogens employed.

Animals are injected subcutaneously three times with the antigen(s) homogenized in complete Freund's adjuvant at about 2-week intervals and are tested about 2 months after the first dose. Animals without useful antibody titers at this time are generally abandoned. Additional immunizations at monthly intervals may result in further increases in titer. When the titer appears to have stabilized, immunization is discontinued and the animal is bled regularly. When the titer falls to about one-half, the animal is then reimmunized.

It has been our experience that undiluted antisera can be stored at $-20°C$ for at least three decades without loss of potency. Working stock solutions can be prepared at 1:100–1:2000 dilution in normal saline containing a bacteriostatic agent and can be fortified with 1–2% nonimmune plasma from the same species. These solutions are generally stable for several years when stored at 4°C, but they cannot be frozen.

Monoclonal antibodies, which provide a virtually unlimited amount of homogeneous antibodies against a specific antigenic site, have proven useful in RIA procedures. However, for research laboratories requiring only limited amounts of antisera, the simplicity of antibody production in animals compared with the increased effort required for production of monoclonal antibodies must be considered. [*See* Monoclonal Antibody Technology.]

C. Labeled Antigen

RIA was first applied to the measurement of peptide hormones, most of which contain a tyrosyl or histidyl residue that can readily be radioiodinated. The chloramine-T technique for the oxidation of the radioiodine is commonly employed. This method has the advantage over other methods in that the solutions employed are maintained alkaline, and the radioiodine does not volatilize, an important feature from the perspective of radiation protection.

For more than a quarter of a century, ^{125}I ($T_{1/2}$ = 60 days) has been the isotope of choice for labeling. If there is no more than one ^{125}I atom per molecule, the preparation is likely to be as stable as the uniodinated preparation. Appreciably, iodination at an average of one radioiodine atom per molecule does not mean that all or even the major fraction of the radioactivity is incorporated into molecules containing one radioactive atom. For instance, iodination of insulin in aqueous solution results in the same distribution of iodine atoms among the tyrosyl residues, which is independent of experimental methods and depends only on the average iodine number. Assuming that there is an average of less than one iodine atom per molecule and that iodination occurs only at the two A-chain tyrosyl residues, with an equal probability of iodinating each residue, a Monte Carlo simulation was used to calculate the theoretical distribution of iodine atoms in labeled insulin preparations. The theoretical analysis indicated that even at an average of only 0.8 radioiodine atom per molecule, approximately half of the radioactivity would be in other than monoiodoinsulin. Thus, purification methods that separate the monoiodinated substance from the uniodinated and multiply iodinated forms are required to assure maximally stable products with high specific activity.

For the assay of steroids, drugs, and other substances, which are generally present at quite high concentrations in plasma, labeled tracers of high specific activity are not necessary. Therefore, ^{3}H-labeled tracers were initially employed for many of these assays. In principle, because ^{3}H has a 12-year half-life, the labeled antigen has excellent long-time stability. Currently, however, more commercial kits employ ^{125}I-

coupled tracers to avoid the need for liquid scintillation counting and the problems associated with the disposal of organic scintillation fluids.

D. Separation Methods

The classic immunologic method for separation of antibody bound from free antigen was based on the spontaneous precipitation of antigen–antibody complexes. However, RIA is generally used because of its high potential sensitivity. This requires that the molar concentration of the reagents be so low that spontaneous precipitation does not occur and the antigen–antibody complexes remain soluble. Therefore, a wide variety of other methods have been used to affect separation of antibody-bound and free antigen.

The methods in most common use include (1) precipitation of antigen–antibody complexes with a second antibody directed against the antibody complex (double antibody), (2) the use of organic solvents or salting out to precipitate complexes, (3) adsorption or complexing of antibody to solid-phase material, and (4) adsorption of free antigen to solid-phase material such as cellulose, charcoal, silicates, or ion-exchange resins.

The double-antibody method is generally the method of choice in developing new RIA procedures. However, the cost of the second antibody may make this prohibitively expensive when thousands of samples are to be analyzed. The use of aqueous polyethylene glycol for precipitation of antigen–antibody complexes is often employed after an assay has been validated with double-antibody methodology. The adsorption or complexing of an antibody to solid-phase material has the advantage of being a generally applicable method and is frequently employed in commercial kits; however, it has the disadvantage that, because of chemical alterations in the antibody molecule introduced by the coupling procedure or because of steric hindrance, the assays using solid-phase techniques may be somewhat less sensitive than assays employing the same antiserum when it is not complexed.

For routine procedures, a method dependent on adsorption of free antigen to solid-phase material has been widely used because it is quite inexpensive. Certain common principles apply to all antigen-adsorbent techniques. A given mass of adsorbent is generally more effective if its total surface area is increased, i.e., if the adsorbing particles are made smaller. However, trace amounts of antibody as well as of free antigen may be adsorbed to materials such as charcoal, cellulose, and silicate, unless the concentrations of plasma or other proteins in the incubation mixture are sufficiently high to saturate the binding sites for γ-globulin. The antigen-adsorption methods have proven to be most satisfactory for small antigens, those with molecular weights of 25,000 or less. The greater affinity of the absorbent for the low molecular weight substances in the presence of plasma proteins permits their near-total adsorption even in the presence of virtually undiluted plasma or high concentrations of other proteins.

The fact that a large number of different methods of separation of antibody-bound from free-labeled hormone have been employed is a consequence of the variety of chemical properties of the now hundreds of substances for which RIA has been employed as well as of the experimental predilections of the many independent laboratories that have developed such procedures.

III. PROBLEMS AND PITFALLS

A. Nonspecific Interference in the Immune Reaction

In general, antigen–antibody reactions are optimal in a pH range from 6.5 to 8.5. However, for some antisera to basic peptides such as secretin, the maximum binding of the tracer occurs at a pH of 5, and its binding at pH 4 may be quite comparable to that at pH 7. Furthermore, although it has been shown that the reaction of labeled insulin with many antisera is independent of buffer and pH in the 6.5–8.5 range, the reaction with other antisera may be strongly dependent on both buffer and pH. In general, but not always, an increase in the ionic strength of the buffer decreases the binding of antigen to antibody. Apparently, one cannot predict *a priori* in RIA the optimal pH or buffer to be employed, and these conditions must be determined for each substance and each antiserum.

The binding of antigen to antibody can be inhibited by a variety of substances, including proteins, anticoagulants such as heparin, bacteriostatic agents such as merthiolate (ethylmercurithiosalicylate), and enzyme inhibitors such as Trasylol. Not every substance interferes in all immune reactions to the same extent; therefore, the possible effect of each substance must be tested or, alternatively, it must be assured that the milieu of the unknown sample is identical with that of the known standards. Furthermore, blood and other biologic fluids may contain proteolytic enzymes that can damage labeled antigen or antibody. These nonspecific effects decrease the binding of antigen to antibody and result in falsely high antigen levels.

The factors that interfere with the chemical reaction in a nonspecific fashion must be appreciated but can and should be avoided.

B. Heterogeneity of Antigen

It is now commonly appreciated that many, if not most, peptide hormones are found in more than one form in plasma and in glandular and other tissue extracts. These forms may represent a precursor(s) or a metabolic product(s) of the well-known, well-characterized, biologically active hormone.

Fortunately, the first RIA was described for insulin. The 6000-Da peptide with full biologic activity is the predominant form in the circulation of virtually all subjects in the stimulated state. Only in patients with an insulin-secreting tumor or in those with a rare genetic abnormality that prevents cleavage of the C-peptide does the prohormone appear to predominate. However, there are assays of other hormones in which the usual biologically active form is not predominant or in which more than one biologically active form exists. Some examples of these problems and how they may or may not be dealt with are considered below.

Immunochemical heterogeneity was first demonstrated for the parathyroid hormone (PTH) when it was observed that a factor, which could be used to superimpose a plasma dilution curve on a curve of standards obtained from dilution of an extract of a normal parathyroid gland for two antisera, failed to affect superposability when another antiserum was employed and that the rate of disappearance of immunoreactive PTH after parathyroidectomy depended on the antiserum employed. Subsequently, it was shown that this was a consequence of the presence in plasma of a metabolic COOH-terminal fragment of PTH, which cross-reacted with some, but not all, antisera, and which disappeared more slowly than intact hormone. This fragment is removed from the circulation primarily by the kidney, and its turnover time can be markedly reduced in the presence of impaired renal function. Thus, although the PTH RIA could be clinically useful despite this problem, the results in different laboratories were not quantitatively identical because different antisera with different immunochemical specificities were generally employed. The problem appears to have been solved with the development of a two-site immunoradiometric assay, which appears to be specific for intact PTH (1–84). The assay depends on the preparation of heterogenous antisera to PTH and the subsequent purification of two different antisera on affinity columns: one antiserum is directed to PTH(1–34) and the other to PTH(39–84). To perform the assay, the anti-PTH(39–84) is immobilized on plastic beads. The unknowns or the PTH(1–84) standards in hormone-free plasma are then added. This is followed by the addition of ^{125}I–anti-PTH(1–34). The assay thus depends on the binding of intact hormone to the COOH-terminal immobilized antiserum and its capturing the ^{125}I-antibody directed against NH_2-terminal PTH. This assay appears to have sufficient sensitivity to distinguish among plasmas from subjects with primary hyperparathyroidism, normal function, and the hypercalcemia of malignancy and seems to have solved the problem of the heterogeneity of PTH in plasma. [*See* Parathyroid Gland and Hormone.]

The RIA for gastrin has been complicated by the presence of two biologically active forms of gastrin in plasma. It has been shown that the predominant form of gastrin in plasma collected in the fasted state from normal subjects or from hypersecretors such as patients with Zollinger–Ellison syndrome or pernicious anemia (PA) is generally, but not always, a 34-amino acid precursor peptide (G34), not the 17-amino acid peptide (G17) initially extracted and purified from the antrum by Gregory and colleagues. Both hormonal forms are stimulated by feeding in normal subjects and PA patients. The infusion of equimolar amounts of G34 and G17 results in about the same acid response in the dog. Thus, on the basis of administered dose and resultant biologic effect, G34 and G17 can be claimed equally biopotent. However, the plasma concentrations of G34 during such infusions are about four times higher than those of G17 because of a fourfold slower turnover time. Similar differences in turnover times of G17 and G34 are also found in humans. Thus, a plasma concentration of 50 pg G17/ml is equivalent to an *in vivo* biologic potency equal to a plasma concentration of 200 pg G34/ml. Obviously, however, an occasional patient with marked hyperacidity might present with a tumor that secretes primarily G17 rather than G34, but their plasma gastrin levels are well within what is believed to be the normal range. Therefore, plasma concentrations per se may not be sufficient for diagnostic differentiation between normal subjects and those with a gastrin-secreting tumor. Fractionation of plasma immunoreactive gastrin on Sephadex columns has been used to differentiate between the molecular forms of gastrin. However, the determination of the clinical reason for hypergastrinemia may require additional studies using appropriate tests for the stimulation of gastrin release.

Problems in RIA also arise when different peptides share common amino acid sequences that might result in immunologic and/or biologic cross-reactivity. Such systems include, among others, gastrin and cholecystokinin, which share the same COOH-terminal pentapeptide, adrenocorticotropin and melanocyte-stimulating hormone (MSH), which have similar NH_2-terminal sequences but different biologic activities, and lipotropin, which contains within it the complete structure of β-MSH.

The application of RIA in pharmacology also has problems relating to the specificity of antisera. Structurally related compounds or metabolites may have significant immunoreactivity with some antisera but not with others and may constitute a problem, depending on the purpose of the assay. For instance, if the clinical problem relates to the toxicity of a particular drug, then the question as to whether or not the assay measures only the biologically active form is relevant. If the question relates simply to whether or not a drug had been taken surreptitiously, then the reactivity of metabolites or variation of the immunoreactivity with the exact form of the drug may be irrelevant.

Perhaps the most widely used RIA in therapeutic drug monitoring is that for digoxin. However, a series of reports have suggested the presence of endogenous digoxin-like factors in various clinical conditions as well as the presence in the circulation of metabolites of high immunologic activity but low cardioactivity. These problems with the digoxin assay have not as yet been adequately resolved. Similar problems are likely to occur in assays for other drugs, but such problems have not as yet been fully identified.

IV. CONCLUSION

Generally, the purpose of RIA is to determine quantitatively the presence of biologically active substances. However, evidently the presence of heterogeneous forms of a variety of substances, both hormonal and nonhormonal, may introduce problems of quantification. Similar problems are present in bioassay as well.

For example, stimulation of amylase release from dispersed acini from guinea pig pancreas can be affected by both cholecystokinin and vasoactive intestinal peptide. If both of these peptides were present in a biologic fluid at equimolar concentrations than the apparent biologic activity (stimulated amylase release) to immunologic activity of either peptide would be greater than one. It must be appreciated that biologic activity *in vivo* can be enhanced by the presence of substances other than the one presumably being assayed. For instance, notably several peptides enhance and others suppress the pancreatic secretion of water and bicarbonate. Thus, in any bioassay system in which multiple peptides interact, it is essential that the possible presence of the other substances be considered. Whenever the ratio of biologic and immunologic activity is less than or greater than one, the existence of factors other than the specific molecular form of the peptide to be assayed must be evaluated.

RIA is now used in thousands of laboratories around the world to measure multiple classes of substances of biologic interest, both peptidal and nonpeptidal in nature. Evidently, if there is a need to measure an organic substance of biologic interest and there is no simple method to do so, some imaginative investigator will develop an appropriate RIA procedure. Although new assays are continuously being described and their usefulness documented, it is important to appreciate the problems and pitfalls that limit the precise quantification of substances measured by RIA.

BIBLIOGRAPHY

Chard, T. (1995). "An Introduction to Radioimmunoassay and Related Techniques (Laboratory Techniques in Biochemistry and Molecular Biology), Vol. 6, Part 2, Elsevier Science, Ltd.

Staff, I. (1992). "Developments in Radioimmunoassay and Related Procedures." Unipub.

Yallow, R. S. (1978). Radioimmunoassay: A probe for the fine structure of biologic systems. *In* "Les Prix Nobel en 1977," pp. 243–264. Nobel Foundation, Stockholm.

Yalow, R. S., ed. (1983). "Radioimmunoassay." Hutchinson Ross Publishing Co., Stroudsberg, PA.

Yalow, R. S. (1984). Radioimmunoassay in oncology. *Cancer* **53**, 1426.

Yalow, R. S. (1985). Radioactivity in the service of humanity. *Interdiscip. Sci. Rev.* **10**, 56.

Yalow, R. S. (1985). Radioimmunoassay of hormones. *In* "Williams Textbook of Endocrinology" (J. D. Wilson and D. W. Foster, eds.). Saunders, Philadelphia.

Yalow, R. S. (1987). Radioimmunoassay: A historical perspective. *J. Clin. Immunoassay* **10**, 13.

Radiosensitivity of the Integumentary System

FREDERICK D. MALKINSON
Rush University

GLOSSARY

Collagen Main protein substance (matrix) in connective tissue of skin and certain other structures

D_0 Radiation dose required to reduce the fraction of surviving cells to 0.37 of their previous number

Fibrosis Increased collagen (scar) formation

Gray (Gy) One Gy = 100 rads

Keratinocytes Epidermal cells

Keratoses Superficial scaling, red, premalignant lesions

Orthovoltage In X-ray therapy, voltage in the range of 140–400 kV

Telangiectasia Permanent dilatation of superficial blood vessels

BIOLOGICAL, PHYSICAL, AND CHEMICAL EVENTS IN irradiated skin are best interpreted from the general responses of all cells and tissues to ionizing radiation. This article is a summary of the major radiobiological effects occurring in skin and hair.

I. RADIOBIOLOGY OF THE SKIN: SOME GENERAL CONSIDERATIONS

Our knowledge of the responses of skin to ionizing radiation is derived from two sources: the clinical observations of radiotherapists and experimental animal data. The latter source is of critical importance, since certain data for human subjects can be provided only by incidents of accidental exposure to ionizing radiation. In its response to single and fractionated radiation doses, pig skin reacts similarly to human skin both qualitatively and quantitatively. Mice and rats have also been widely used to study radiation effects in skin.

After a single threshold dose of radiation to human skin is exceeded, a diphasic or triphasic erythema is elicited. After a *single* exposure to 8.0 Gy, an early erythema develops in 1–3 days, followed by increased erythema in the second to fourth week. Hyperpigmentation appears after the third to fourth week and may persist indefinitely. The threshold erythema dose following a single radiation exposure is about 3 Gy for 90 kV of radiation and 6–8 Gy for 200 kV of radiation. If, instead of a single dose, daily *fractionated* doses are given, the wave-like appearance of cutaneous erythema is partially obliterated. The total dose required to produce a threshold erythema increases with increasing numbers of fractions and with increasing overall treatment times. This time–dose effect is related principally to the cellular repair of sublethal damage. Exposure of human skin to high doses of

ENCYCLOPEDIA OF HUMAN BIOLOGY, Second Edition, VOLUME 7. Copyright © 1997 by Academic Press. All rights of reproduction in any form reserved.

fractionated ionizing radiation (2.0 Gy daily orthovoltage radiation over a period of 4 weeks, for a total dose of 40.0 Gy) induces an acute reaction characterized by erythema, epilation, moist desquamation, and erosions, ultimately followed by healing. Somewhat lower doses produce less severe changes of scaling and hyperpigmentation. Over the next few months or years, atrophy, telangiectasia, hypopigmentation, fibrosis, keratoses, ulcerations, and carcinoma may develop. In general, following single-dose or fractionated radiation, acute epidermal changes result primarily from damage to germinative cells. Later chronic changes largely reflect injury to structures deeper in the dermis, particularly collagen and blood vessels.

In vitro experiments revealed D_0 values of 0.92 Gy for human keratinocytes nd 1.39 Gy for mouse keratinocytes. These values correlated well with those obtained in previous *in vivo* studies. [*See* Keratinocyte Transformation.]

The character and magnitude of cutaneous responses to radiation depend on many factors, including total dose, dose fractionation, dose rate, radiation quality, area or volume of tissue irradiated, anatomic site, vascular supply, and age. For single exposures to ionizing radiation, the larger the dose, the greater the skin injury. For multiple exposures over time, however, higher dosages are required to produce the same degree of injury when they are divided into two or more fractions. For example, to obtain a certain severity of acute reaction in pig skin following X irradiation (8 MeV) requires a total dose of 20.2 Gy given as one dose in 1 day, 32.0 Gy given as three equal fractions in 3 days, and 38.0 Gy given as five equal fractions in 5 days. Increased tolerance with fractionation results from the repair of sublethal damage and cell repopulation. In humans, epidermal (keratinocyte) repopulation may be aided by increased expression of epidermal growth factor receptors in these cells. Repair of sublethal damage occurs in less than 24 hr or in one cell cycle after irradiation; repopulation takes days and depends on the tissue's regenerative capacity. [*See* Skin; Skin, Effects of Ultraviolet Radiation.]

The rate at which radiation is administered also affects cutaneous response. Lower dose rates produce less injury, particularly for late radiation effects. Radiation quality also influences the degree of tissue damage. More energetic radiation is more penetrating, increasing the injury to deeper tissues.

In regard to skin surface area, acute reactions to fixed single doses in pig skin were reduced when progressively smaller fields were irradiated. But for irradiated fields larger than 2 cm in diameter, such area effects were lost. These differences appeared to be related to migration rates of epidermal cells, from surrounding normal skin sites, in the healing process.

Animal studies suggest that age factors may also affect acute radiation changes in skin. Postradiation repair rates for radiation-induced DNA damage in proliferating rat epidermal cells diminish with increasing age: for the oldest animals (400 days), DNA repair times were five times longer than in youngest animals (28 days), 107 versus 20 min. There is some evidence in rats that radiation-induced moist desquamation reactions are more extensive in older, compared to younger, animals.

II. EPIDERMAL CELL SURVIVAL: EFFECTS OF SINGLE AND MULTIPLE RADIATION EXPOSURES

Studies with an *in vivo* epidermal cell-cloning technique in mice revealed a linear relationship between X-ray dose and epidermal cell survival. The D_0 was 1.34 Gy in air, 1.12 Gy in hyperbaric oxygen (4 atm), and 3.5 Gy under anoxic conditions. Other studies have shown that D_0 values may vary with anatomic site and can be correlated with gross skin reactions and subsequent healing. With the *in vivo* epidermal cell-cloning technique in mice, two-dose fraction irradiation studies revealed that fractionation intervals beyond 24 hr up to 6 days yielded progressive increases in cell survival. In two human subjects receiving fractionated radiation therapy, D_0 of 4.9 Gy was estimated for epidermal cells.

The higher rate of cell survival, or the diminished intensity of gross epidermal reaction, that occurs when a single dose of radiation is divided into two fractions yielding the same total dose in Gy reflects the repair of sublethal radiation damage. The extent of this repair is related to the magnitude of the shoulder of the cell survival curve. For short time intervals (hours), however, the effects of sublethal damage repair may be modified by induction of a partially synchronous cell population from the first radiation dose, thereby inducing a cell-cycle phase-dependent radiosensitizing or, conversely, radioprotective effect.

III. RADIOSENSITIZERS

The skin is a readily available site for testing augmented radiation responses to radiosensitizers. Studies of skin injury in patients breathing air or exposed to hyperbaric oxygen revealed oxygen enhancement ratios of 1.3 or greater, increasing with higher radiation dosages.

Electron affinic agents (nitroimidazoles) are effective radiosensitizers of hypoxic mammalian cells *in vitro* and of rodent and human skin cells *in vivo,* yielding enhancement ratios of 1.25–2.22, compared with an oxygen enhancement ratio of about 2.6. No radiosensitization occurs in well-oxygenated cells. The older nitroimidazoles are cytotoxic, especially neurotoxic, however, and to date they have not produced significant benefits in tumor therapy. Neurotoxicity results from the high lipid solubility of these compounds, which facilitates their penetration into the myelin sheaths of nerves. New generations of more efficient and less toxic compounds are currently being evaluated, especially since these drugs have also been demonstrated to potentiate cell destruction by several chemotherapeutic compounds, particularly alkylating agents.

Severe cutaneous reactions to ionizing radiation (erythema, moist desquamation) have been reported in patients receiving certain chemotherapeutic agents during courses of radiotherapy for a variety of malignancies. The enhanced frequency of acute radiation reactions, for example, has been seen in breast cancer patients receiving radiation therapy and either cytoxan or the combination of methotrexate and fluorouracil. The latter drug combination also increased the frequency and severity of subcutaneous fibrosis. Chemotherapeutic drugs may also induce moist desquamation well beyond the time that radiation treatment courses have been completed (the "recall phenomenon"). Such drugs include adriamycin, actinomycin D, methotrexate, daunomycin, vinblastine sulfate, and hydroxyurea. Cutaneous necrosis may occur in sites irradiated up to 10 years earlier, and some of these reactions have occurred after relatively low drug dosage administration. Enhanced radiation effects also occur in other tissues or organs. Reduced cutaneous reactions are seen following reduced radiation dosages or increased time intervals between radiation courses and drug administration. With some of these agents (actinomycin D, vinblastine sulfate), true radiosensitization may occur, as shown in drug–radiation studies of hair. For other cytotoxic compounds, however, enhanced radiation effects may only reflect additive cell toxicity of concomitant or sequential administration. Where it does occur, the effects of various chemotherapeutic agents in producing radiosensitization (or, occasionally, radioprotection) can be attributed in part to the ability of these compounds to position proliferating cells in cycle phases conferring relative sensitization (G_2, M) or protection (S) at the time of subsequent irradiation.

IV. RADIOPROTECTORS

The current exploratory therapeutic usages of charged particles (protons, heavy ions, etc.) may involve skin doses that are almost as large as the tumor doses being used. This has stimulated the investigation of, and search for, newer radioprotective agents, especially compounds that are active after topical administration. Thiol and disulfide compounds protect against sparsely ionizing radiation largely by scavenging free radicals. These compounds must be present in tissue at the time of irradiation. They also may promote the repair of radiation-damaged molecules before these molecules can react with oxygen to "fix" the radiation injury. Following the early successful experimental use of cysteine, later studies demonstrated the effectiveness of topically applied cysteamine as a radioprotective agent in mouse skin. Dose reduction factors of up to 1.5 were demonstrable for acute skin reactions. More recent investigations have shown that WR-2721, a cysteamine derivative, is a highly effective radioprotector in laboratory animals, yielding dose reduction factors of up to 2–2.5 for cutaneous ulceration and desquamation. For single-dose or fractionated radiation in mice, optimal protection required drug injection 30–60 min before irradiation. Since this hydrophilic compound penetrates tumors poorly, differential protection of normal tissues can be achieved. To date, however, toxicity, especially hypotension, has largely prevented the use of this agent in humans. These experimental and clinical observations have stimulated the search for more effective, less toxic topical radioprotective agents, as well as studies of certain vehicles, such as dimethyl sulfoxide, that might enhance the transepidermal absorption of radioprotective compounds.

In recent years, prostaglandins and leukotrienes have been shown to be potent radioprotective agents, although their mechanism of action remains obscure. After earlier studies demonstrated the cytoprotective effects of certain prostaglandins for tissue protection against strong acids, alkalis, and other agents, these compounds were shown to be radioprotective for bone marrow and intestinal tract stem cells as well. Subsequent studies revealed that systemically administered or topically applied 16-16 dimethyl prostaglandin E_2 also protected against single-dose and fractionated-dose radiation-induced alopecia in mice. Prostaglandin analogs of naturally occurring biological compounds provide superior protection, up to sixfold, compared to the natural compounds. They protect mice against radiation-induced alopecia, injury to a variety of other tissues, and even lethality from certain chemotherapeutic agents, such as doxorubicin and cytoxan. Additive radioprotective effects occur with addition of other radioprotectors, such as cysteine, and synergistic effects are seen with the addition of WR-2721. Peak protection occurs when prostaglandins are given 1–3 hr before irradiation. No protection is seen with postradiation administration. Studies currently in progress reveal essentially no protection by prostaglandins against chemotherapeutic agents- or radiation-induced effects on a variety of malignant tumors transplanted into mouse skin. All of these investigations suggest that prostaglandins, especially when topically applied, may be useful clinically in providing some protection from radiation injury for normal tissues such as skin, hair, and mucous membranes lying in the path of radiation beams administered for tumor therapy.

Most radioprotective agents are active only when present at or before the time of irradiation. A recent study in rats employing captopril, in clinical use for the treatment of hypertension and congestive heart failure, evaluated effects on skin of the continuous postradiation administration of this drug in the daily diet. Captopril reduced the severity of moist desquamation by about one-third and significantly reduced the number of postradiation tumors. Interestingly, however, the severity of postradiation epilation was unaffected. Captopril has also been found to be a postradiation protective agent for the kidney and lung. The mechanism of action of this agent is unclear.

V. HYPERTHERMIA

Heating skin before, during, or after irradiation enhances the cutaneous response. For skin, thermal enhancement ratios range from 1.1 at 40°C to 2.0 at 43°C.

When heat was applied to mouse skin for intervals of 7.5 min to 4 hr, combined with orthovoltage radiation, an increase in temperature of 0.5°C was equivalent to increasing the dose of radiation by 10–15%. Each rise of 1°C reduced by one-third the time required to enhance the radiation response. For human skin, thermal enhancement ratios of 1.7–2.0 were found in temperature ranges of 43–46°C.

Since the thermal enhancement of radiation damage tends to spare normal tissues but preferentially injures hypoxic cells, which are radioresistant, hyperthermia may play an important role in cancer therapy. For example, laboratory studies of human and mouse melanoma cells have revealed radiosensitization by concurrent or postradiation hyperthermia.

Thermal enhancement of radiosensitivity in radioresistant melanoma is apparently due to hyperthermia-induced reductions in the repair of cellular sublethal damage.

As might be expected, cooling of the skin induces some radioprotection, probably resulting from lowered oxygen tension in vasoconstricted sites.

VI. EARLY VASCULAR EFFECTS

Postradiation erythema results from changes occurring within hours to days after irradiation. These evolve over months and years and, in laboratory animals, can be roughly quantitated by dye leakage, which is most intense 2.5 weeks after irradiation. A dose of several Gy induces dye leakage, and the intensity of extravasation and its time of appearance are dose dependent. With the use of radioisotope tracers, a biphasic pattern of vascular permeability changes is demonstrable. In 24–48 hr postradiation, a decreased, followed by an increased, blood flow was found. Further vasodilatation and vascular permeability changes were noted with the onset of the acute skin reaction, beginning 7 days after irradiation. These changes are affected by the size of the radiation dose and vary depending on the postradiation time interval studied.

Experiments utilizing the Sandison–Clark rabbit ear chamber after single-dose irradiation revealed serial intimal swellings and proliferation appearing along blood vessel walls 3–4 days postradiation. These changes followed endothelial nuclear swelling, enlargement, and pyknosis. Dose-dependent reductions in capillary density presumably impeded oxygen delivery to the skin. Such early events may have significance for the development of telangiectasia and other long-term vascular changes after irradiation.

In vitro studies of endothelial cells in radiation-induced vascular reactions have revealed D_0 values of 1.68–2.40 Gy.

VII. LATE EFFECTS

The clinical use of high-energy, skin-sparing radiation has focused attention on the importance of late radiation damage to connective tissue as an important factor in administering effective cancer therapy. The magnitude of the early acute skin reaction, however, is not a good indicator of the severity of late changes such as pigmentary alterations, atrophy, vascular effects, and fibrosis. β irradiation of pig skin, for example, in doses producing minimal erythema without desquamation, may result in reduced dermal thickness (atrophy) of up to 50% within 2 years. Late radiation changes may be more dose limiting than changes seen in the first few weeks after exposure, perhaps reflecting much slower cell turnover rates and recovery processes in the dermis than in the epidermis. Extensive clinical data suggest that late effects of radiation are usually independent of overall treatment times, unlike early reactions, and that they are more dependent on the size of the daily fraction.

Radiation injury to the vasculature may be the most important factor in late radiation damage in the skin. Endothelial cell damage and complement-mediated inflammation, present for several months postradiation, reduce circulation. These changes may also contribute to the development of progressive fibrosis.

Studies of single-dose exposures in mouse skin demonstrated that increased postradiation collagen synthesis is an early (1 week) change and not only a late radiation effect. By 12 weeks postradiation, collagen synthesis levels were enhanced, in dose-dependent fashion, up to 50% and this effect was sustained for at least 1 year. Increased collagen content of skin and microscopic fibrosis in the lower dermis were associated findings. Subcultures of fibroblasts isolated from irradiated skin 1–48 weeks postradiation were threefold overproducers of collagen. The tissue fibrosis in irradiated skin sites presumably resulted at least partly from cytokines released by radiation-induced inflammatory cell infiltrates. The possible additional roles, for increased collagen synthesis, of direct genetic or other radiation-induced changes in fibroblasts remain conjectural. [*See* Collagen, Structure and Function.]

VIII. RADIOBIOLOGY OF HAIR

Throughout the mammalian life span, hair matrix cells undergo alternate periods of exceedingly active cell proliferation (anagen) and total reproductive inactivity (telogen). The most useful of numerous indices used to evaluate radiation effects on matrix cells are quantitative assessments of hair loss, regrowth and growth rates, the incidence of microscopic morphologic changes, and kinetics studies in irradiated hair matrix cell populations with [^{3}H]thymidine techniques. Hair studies have served as a useful biological indicator system for *in vivo* studies of the effects of ionizing radiation, a wide variety of chemotherapeutic drugs, or different combinations of radiation and pharmacologic agents on rapidly dividing or nonproliferating cell populations. Anagen hairs are 2.5 times more radiosensitive than telogen hairs. Maximum anagen hair loss in rats exposed to 20 Gy was found within 1 week of radiation, whereas the telogen hair loss peak occurred 3 weeks postradiation. Doses producing complete epilation were 12 Gy for anagen versus 24 Gy for telogen hairs. Increasing doses of radiation produce permanent hair loss with follicular atrophy. The threshold dose for this effect in the human scalp is about 5–7 Gy.

Although anagen hairs are approximately 2.5 times more radiosensitive than telogen hairs, anagen hairs repair radiation damage more quickly. Telogen matrix cells sustain and "store" radiation damage, which is then expressed in succeeding cell generations following a mitotic stimulus (plucking). Differences in cell proliferative activity and marked differences in vascular supply between anagen and telogen hair matrices, contributing an oxygen effect, may largely explain their different radiosensitivities. Quantitative postradiation hair loss studies in mice, carried out with photographic techniques, have yielded D_0 values of 492 cGy for telogen hairs and 180 cGy for anagen follicles.

In mice, large radiation doses (20.0–25.0 Gy) permanently alter anagen–telogen cycle times and hair growth rates, and permanently reduce hair matrix cell mitotic indices by 50%. The resultant partial, persistent alopecia presumably reflects radiation destruction of some entire hair matrix cell populations.

Radioprotective drug–radiation effects were first noted when systemic cysteine prevented radiation-induced epilation in dose-dependent fashion in guinea pigs receiving doses of 7.0–9.5 Gy. No protection occurred at doses above 12.0 Gy. Glutathione, cysteamine, or WR-2721 yields dose-reduction factors ranging up to 1.7 for radiation-induced anagen alopecia. Up to threefold greater radioprotection is seen with certain systemically administered prostaglandins, and these compounds are active topically as well. Conversely, radiosensitizing drug–radiation effects inducing increased alopecia have been noted with actinomycin D, colchicine, hydroxyurea, bleomycin, and other compounds. At least some of these effects appear to be synergistic and not simply additive. The time of administration of these agents prior to irradiation is critical, as they act largely by influencing progression of hair matrix cells through the cell proliferation cycle. Their "positioning" of matrix cells at the time of subsequent radiation results in increased or, occasionally, decreased hair loss depending on whether the matrix cells are irradiated in relatively radiosensitive or radioresistant phases of the cell cycle.

The use of hair as a biological indicator system for combined drug–radiation effects has obvious implications for studies in other cell systems and for investigations of therapeutic approaches to experimental and clinical malignancies. Considerations of the various factors involved in combined treatment regimens are complex, however, and several problems have been encountered in the applications of investigative findings to tumor therapy. [*See* Hair.]

IX. RADIOBIOLOGY OF PIGMENTATION

Moderate-dose postradiation hyperpigmentation occurs in skin partly due to enhanced tyrosinase activity in melanocytes and resultant increased melanin synthesis and, later on, increases in the numbers of melanocytes. Hyperpigmentation is directly related to radiation intensity and to total accumulated dose and is characterized by increased melanin deposition throughout the epidermis. However, low single dose or low dose-fractionated radiation also produces increased melanin synthesis, increased transfer of melanin to surrounding epidermal cells, and gross hyperpigmentation. It has been postulated that ionizing radiation may increase melanosomal membrane permeability, thereby increasing the amount of tyrosine available for melanin synthesis. High doses of radiation destroy melanocytes and result in epidermal depigmentation or in graying of hair. Hair matrix melanocytes are destroyed by radiation much more readily than are epidermal melanocytes.

In the mouse, postradiation hair graying is dose dependent and varies with the growth cycle of the hair. In contrast to radiation effects on hair matrix epithelial cells, melanocytes in telogen hairs are far more susceptible to radiation destruction than those in anagen matrices. Melanocytes in resting follicles have a D_0 value of 1.50–2.0 Gy for single-dose radiation. For fractionated radiation, D_0 values range from 2.15 Gy (two fractions) to 4.15 Gy (eight fractions).

Both melanocyte survival studied in hair follicle squashes and depigmentation of hair in laboratory animals have been used to assess quantitatively the combined chemotherapeutic drug and radiation effects, as well as the effects of radioprotective agents.

X. RADIOBIOLOGICAL EFFECTS ON LANGERHANS CELLS

Confirming earlier studies of Langerhans cell destruction by ionizing radiation, more recent electron microscopic investigations revealed that single doses of 20.0 Gy in mice reduced Langerhans cell (LC) numbers to 18% of controls within 10 days. Cell repopulation was rapid by Day 16 and cell numbers were essentially normal by Day 30. LC loss postradiation is dose dependent. When whole-body irradiation was given to mice receiving local skin irradiation in addition, repopulation of the epidermis by LCs was delayed by another 3 weeks. The latter findings largely reflect the bone marrow origin of LC precursors, since mitoses in LCs postradiation were not observed in the epidermis. Additional studies have shown that LCs are relatively radioresistant. The D_0 estimate is 12.5 ± 2.9 Gy.

In regard to late radiation effects, full recovery in LC numbers in 4-month-old mice postradiation was then followed 15 months later by a drop to 70% of LCs found in 19-month age-matched controls, and a further reduction to 60% of normal was noted at 24 months of age. Late changes of radiation-induced fibrosis and impaired circulatory function in skin may impede replacement of LCs from bone marrow precursors. Postradiation LC loss may have implications for the subsequent development of skin tumors and reduced susceptibility to induced contact hypersensitivity.

BIBLIOGRAPHY

Bentzen, S. M., Overgaard, M., Thomas, H. D., Christensen, J. J., and Overgaard, J. (1989). Early and late normal-tissue injury after postmastectomy radiotherapy alone or combined with chemotherapy. *Int. J. Radiat. Biol.* **56**, 711.

Cole, S., Humm, S. A., James, D. R., and Townsend, K. M. S. (1986). Langerhans cells: Quantitative indications of X-ray damage in mouse skin? *Br. J. Cancer* **53** (Suppl. VII), 75.

Geng, L., Hanson, W. R., and Malkinson, F. D. (1992). Topical or systemic 16,16 dm prostaglandin E_2 or WR-2721 (WR-1065) protects mice from alopecia after fractionated irradiation. *Int. J. Radiat. Biol.* **61,** 533.

Hamlet, R., Heryet, J. C., Hopewell, J. W., Wells, J., and Charles, M. W. (1986). Late changes in pig skin after irradiation from beta-emitting sources of different energy. *Br. J. Radiol.,* Suppl. 19, 51.

Hanson, W. R., and Thomas, C. (1983). 16,16-Dimethyl prostaglandin E_2 increases survival of murine intestinal stem cells when given before photon radiation. *Radiat. Res.* **96,** 393.

Maisin, J. R., Albert, C., and Henry, A. (1993). Reduction of short-term radiation lethality by biological response modifiers given alone or in association with other chemical protectors. *Radiat. Res.* **135,** 332.

Malkinson, F. D., and Keane, J. T. (1978). Hair matrix cell kinetics: A selective review. *Int. J. Dermatol.* **17,** 536.

Malkinson, F. D., and Hanson, W. R. (1991). Radiobiology of the skin. *In* "Physiology, Biochemistry, and Molecular Biology of the Skin" (L. Goldsmith, ed.), 2nd Ed. Oxford Univ. Press, New York.

Malkinson, F. D., Geng, L., and Hanson, W. R. (1993). Prostaglandins protect against murine hair injury produced by ionizing radiation or doxorubicin. *J. Invest. Dermatol.* **101** (Suppl.), 135S.

Panizzon, R. G., Hanson, W. R., Schwartz, D. E., and Malkinson, F. D. (1988). Ionizing radiation induces early, sustained increases in collagen biosynthesis: A 48-week study in mouse skin and skin fibroblast cultures. *Radiat. Res.* **116,** 145.

Peel, D. M., Hopewell, J. W., Wells, J., and Charles, M. W. (1984). Non-stochastic effects of different energy beta emitters on pig skin. *Radiat. Res.* **99,** 372.

Peter, R. U., Beetz, A., Ried, C., Michel, G., Van Benningen, D., and Ruzicka, T. (1993). Increased expression of the epidermal growth factor receptor in human epidermal keratinocytes after exposure to ionizing radiation. *Radiat. Res.* **136,** 65.

Sodicoff, M., Lamberti, A., and Ziskin, M. C. (1990). Transepidermal absorption of radioprotectors using permeation-enhancing vehicles. *Radiat. Res.* **121,** 212.

Vegesna, V., Withers, H. R., and Taylor, J. M. G. (1987). The effect on depigmentation after multifractionated irradiation of mouse resting hair follicles. *Radiat. Res.* **111,** 464.

Wambersie, A., and Dutreix, J. (1986). Cell survival curves derived from early and late skin reactions in patients. *Br. J. Radiol.,* Suppl. 19, 31.

Ward, W. F., Molteni, A., Ts'ao, C., and Hinz, J. M. (1990). The effect of Captopril on benign and malignant reactions in irradiated rat skin. *Br. J. Radiol.* **63,** 349.

Withers, H. R. (1967). Dose survival relationship for irradiation of epithelial cells of mouse skin. *Br. J. Radiol.* **40,** 187.

Radiosensitivity of the Small and Large Intestines

ALDO BECCIOLINI
University of Florence, Italy

GLOSSARY

Adventitia or tunica adventitia Part of the connective tissue covering the blood vessels

Cell cycle time (T_c) Time it takes a cell to move through the different cycle phases from one mitosis to another

Circadian pattern Biological parameter that has a rhythmic variability occurring about every 24 hr

Crypt (of Lieberkühn) Localized in the lower part of the villus, it corresponds to the proliferative compartment of the intestinal epithelium

Duodeum First part of the small intestine, about 25 cm long

Glycocalyx Structure made up of mucopolysaccharides that appears as a filamentous excrescence on the microvillous tip of differentiated epithelial cells

Gray (Gy) SI unit of radiation absorbed dose; 1 Gy = 1 J/kg = 100 rad

Hypovolemic shock Shock caused by reduced blood volume that may be due to loss of blood or plasma

Ileum Distal portion of the small intestine localized in the lower abdomen

Jejunum Proximal part of the mesenteric small intestine, occupying the upper portion of the abdominal cavity

Linear energy transfer (LET) Rate at which energy is deposited when ionizing radiation travels through matter. It is expressed in kiloelectron volts per micrometer

Microvilli or striated border or brush border Closely packed, parallel, cylindrical processes visible with the electron microscope

Mucosa Lumenal surface of the small intestine, greatly increased by grossly circular folds (plicae) and by microscopic finger-like structures (the intestinal villi)

Muscolaris Constituted by an external and an internal layer of the muscolaris coat that covers the intestines

Pinocytosis Mechanism of engulfing particles or dissolved materials by a process of vesciculation. Lipid droplets are adsorbed in this way by the epithelial cells of the small intestine

Valvulae conniventes (or valve of Kerckring) Highly visible folds that increase the lumenal surface of the small intestine

Villus Finger-like processes that increase the absorptive surface of the small intestine

GASTROINTESTINAL EPITHELIUM PRESENTS A HIGH radiosensitivity because of its high proliferative activity. The irradiation of the small and large intestines with an acute exposure produces an early appearance of injury due to the fast cell turnover.

When a high enough dose is used, specific clinical conditions defined as gastrointestinal radiation syndrome are observed. The syndrome is characterized by nausea, vomiting, and diarrhea that could cause death within about 3–10 days according to the dose.

When sublethal doses are used, this syndrome is less severe and lasting, in fact, the high proliferative activity of the epithelium alloys, a rapid replacement of dead cells within a few days.

The biology of the small intestine is very complex

ENCYCLOPEDIA OF HUMAN BIOLOGY, Second Edition, VOLUME 7. Copyright © 1997 by Academic Press. All rights of reproduction in any form reserved.

from a morphological and biochemical point of view. For a long time the intestine was considered only as a canal where digestion occurs by means of pancreatic enzymes that continue the salivary and gastric hydrolysis of the food, followed by the absorption of monomeric molecules. In the 1960s it was demonstrated that the intestinal epithelium plays a characteristic role in actively participating in the digestion by means of a membrane digestion process.

The particular structure of the gastrointestinal epithelium, a monolayer of cells, allows the analysis of the localization of proliferative cells, goblet cells, and differentiated cells in different parts of the crypt–villus system when the epithelium is cut according to the longitudinal axis. This particular investigation allows the analysis of the effects of physical and chemical agents on proliferation and differentiation in the initial phase of acute damage and during recovery phase when repopulation occurs.

The experiments on laboratory animals showed that the intestine is a good model to study radiation effects on the mechanisms of proliferation, repair, and cell differentiation.

In humans the gastrointestinal tract is included in the irradiated volume of patients affected by neoplasias localized in the abdominopelvic region and treated by radiotherapy. The total dose administered is quite high, but being divided into daily fractions (2 Gy/day, 5 days/week), it causes damage that can be recovered from within a few weeks after treatment. However, the induced modifications are so severe as to produce temporaneous morphological alterations and a malabsorption of many molecules in the diet.

The irradiation of the alimentary canal can induce later, even severe, damage such as sclerosis, intestinal occlusion, and carcinogenesis.

I. INTRODUCTION

The energy given off by ionizing radiation produces chemical bond breakage and, therefore, modifications in atom organization that can be seen at a molecular level. This series of events takes place in from 10^{-18} sec to a few seconds. It leads to the alteration of macromolecular structures that is later manifested by morphological and functional cell modifications and their consequent biological effects. If these effects are greater than the capacity of the repair mechanisms, clinical damage can be observed both in early and in late phases depending on dose, linear energy transfer (LET), irradiation schedule, and, especially, the proliferative activity of the cell lines that form the tissue. DNA and the enzymes necessary for its replication and repair are thought to be the principal targets for radiation damage. This implies a difference of effect between proliferative cells and differentiated ones, which appear much less sensitive.

Therefore a cell system with a high proliferative activity will be profoundly altered by radiation doses that will bring about only modest lesions in cell systems where the mitoses are rare or absent. The time necessary for the lesion to become clinically evident depends on the turnover of the irradiated cell system as well as on the time necessary for its return to normality.

In a tissue, composed of several types of cells, acute injury is concentrated mainly in that cell component having the highest proliferative activity. Damage to the component with the least mitoitic activity or the slowest cell turnover appears only much later. The same occurs in cell lines such as intestinal epithelium when the dose is high enough to sterilize the proliferative compartment, after which the epithelium disappears and ulceration is produced. Depending on the extent of ulceration and on connective tissue damage, cicatrization may or may not occur with time. When the dose is sublethal the mitotic activity is blocked at first, but later, depending on cell turnover and damage, which is dose dependent, the epithelium returns to normal conditions. Proliferative activity of the epithelium in the gastrointestinal tract is high and therefore the tissue should be considered highly radiosensitive. The small intestine is more radiosensitive than the stomach and large intestine, the duodenum more than the jejunum and ileum, and the colon more than the rectum.

Irradiation in humans can occur accidentally or purposely as in the treatment of neoplasia: here a part of the gastrointestinal tract is included in the irradiated area and therefore receives elevated doses. In other conditions, for example, following radiodiagnostic examinations of the digestive tract or genitourinary tract or for nuclear medicine examinations, the radiation doses absorbed by the intestine are much lower, and thus unable to produce acute injury.

II. ALIMENTARY TRACT

The alimentary tract includes the esophagus, stomach, small and large intestines, and rectum. The mucosa consists of (1) a single layer of epithelium resting on a lamina of connective tissue, (2) a layer of smooth muscle (submucosa), and (3) two layers of muscle tissue, one placed transversally and the other longitudinally.

The structure of the mucosa in the different parts of the alimentary tract is similar, being made up of glands whose proliferative compartment is located in the intermediate region in the stomach and at the base of the crypts in the small and large intestines and in the rectum (see Fig. 1). The size of the proliferative compartment is proportionally greater in the colon and rectum, whereas it is limited to the lower two-thirds of the crypts in the small intestine. In this part of the intestine the proliferation rate is higher.

The proliferative cells go through the different phases of the cycle during their migration toward the villus. Once they reach a certain level of the proliferative compartment, they lose their capacity to divide, change their morphology, and begin to synthesize the molecules that characterize mature cells.

The renewal process of the gastrointestinal epithelium involves proliferation, migration, differentiation, and loss or death of the cells. Studies of these cells is performed by using [^{3}H]thymidine. Information obtained by directly injecting the labeled nucleoside into patients with preterminal illnesses showed that the cell cycle time in the human intestine lasts 1–2 days, depending on the part of the alimentary tract; the S phase (of DNA synthesis) is about 60% of the cell cycle time (T_c). Turnover time (the average life of a cell) in the small intestine epithelium is longer in the proximal tract (about 5 days), where the epithelium is formed by a greater number of cells, than in the distal ileum. In the human, large intestine and rectum turnover time is about 6 days. From a physiological point of view, the muscosa cells actively participate in the digestive process by synthesizing enzymes and by absorbing the hydrolyzed molecules coming from the diet. Furthermore, the epithelium layer acts as a protective barrier against acidity, proteolytic enzymes, and the bacterial flora and their toxins. These agents, together with the continuous microtraumas provoked by the passage of food, necessitate continuous substitution of the epithelial cells. Goblet cells continuously synthesize mucus which coats the luminal surface of the epithelium. For this reason, elevated proliferative activity and rapid turnover exist along the entire gastrointestinal tract and these factors explain why the alimentary canal is highly sensitive to ionizing radiations.

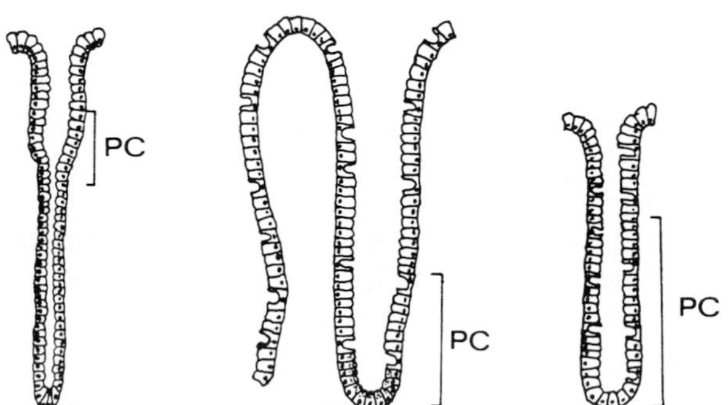

FIGURE 1 Schematic representation of different types of epithelium lining the gastrointestinal tract: (a) stomach, (b) small intestine, (c) colon and rectum. PC, proliferative compartment.

III. ACUTE GASTROINTESTINAL RADIATION SYNDROME

When a mammal is exposed to sufficiently high doses of ionizing radations over its entire body, it manifests a series of symptoms referred to as radiation syndrome. Bone marrow syndrome, gastrointestinal syndrome, and central nervous system (CNS) syndrome appear as the dose increases and therefore within progressively shorter time periods. [*See* Radiation, Biological Effects.]

Whereas the CNS syndrome in humans is poorly defined because of the high dose necessary for its induction, the other two syndromes have been widely studied both in accidently exposed individuals and in patients with tumors who have undergone radiotherapy. More than 2 Gy is needed to produce, in a few hours, the initial symptoms of the gastrointestinal (GI) syndrome: these are similar to air- or seasickness and are influenced by psychological factors and individual susceptibility. Nausea and vomiting follow a state of loss of appetite and apathy, and these symptoms progressively worsen on the second and third day. Then, if a higher dose has been absorbed, all the symptoms characterizing the GI syndrome appear: anorexia, nausea, vomiting, diarrhea, and high fever. A week after exposure, severe dehydration, reduction of plasma volume, and circulatory collapse, as well as profound alterations of the intestinal mucosa, can lead to death. About 5 Gy is sufficient to induce a quite serious GI syndrome and, with whole-body irradiation, there is a marked bone marrow syndrome characterized by lymphopenia, neutropenia, and platelet reduction. Doses above 10 Gy in humans and laboratory animals induce death in 3–8 days, depending on the dose.

Numerous studies of the GI syndrome have been performed but they have not shown whether death is due to a single factor or to a series of factors. The profound alterations of the epithelial cells throughout the entire gastrointestinal tract cause several functional alterations. These include water and electrolyte loss, which can lead to hypovolemic shock, and loss of the mucus layer, the latter permitting greater mucosa sensitivity to microtraumas and autodigestion by gastric and pancreatic enzymes, as well as possible entrance of the intestinal microflora or toxins into the organism. Studies with germ-free mice demonstrate the role of infection: if exposed to a lethal dose, their survival time is about two times greater than that of normal mice because the turnover time in the former is nearly double that found in the latter.

IV. SMALL INTESTINE

The small intestine is devoted to the absorption process. For this reason it is organized in valvulae conniventes, crypt–villus system, and microvilli in the cells or the villus. This complex of structures increases the absorption surface by more than 600 times. The proliferative compartment is localized in the crypt, which is in the lower third of the mucosa (see Fig. 1).

The epithelium is a single layer containing four types of cells: columnar cells (more than 90%), mucus cells (6%), very few enteroendocrine cells, and Paneth cells, located at the base of the crypt. Some recent theories favor the hypothesis of a common origin for all epithelial cells, the totipotent stem cells being located in the first positions at the base of the crypt. These cells are normally included in the nonproliferative compartment in the G_0 phase of the cell cycle. However, when necessary, they become active and produce partially differentiated proliferative cells that subsequently divide and allow the repopulation of the different types of cells that make up the intestinal epithelium: this occurs when the mucosa is damaged by physical or chemical agents.

The proliferative columnar cells undergo more than one division during their migration along the crypt. In laboratory animals and humans, cell cycle time is approximately 14 hr, with differences depending on the position of the proliferative cells in the crypt. Turnover time in humans is about 5 days, whereas in mice and rats turnover is about 3 days. The use of [^{3}H]thymidine, a labeled nucleoside that is taken up by cells in the DNA synthesis phase, has made it possible to demonstrate the irregular distribution of these cells along the crypt. One hour after the injection of the labeling agent the frequency of labeled cells is very low at the base of the crypt but rapidly increases, showing that more than 50% of the proliferative cells are in the DNA replication phase in the lower third portion. In the superior positions, the frequency of S-phase cells decreases rapidly so that in the upper third there are no labeled cells in normal conditions. Since the cells go through the S phase and then the G_2 and M phases, mitotic figures can be observed in positions above the labeled cells.

In the upper third of the crypt, the columnar cells lose their proliferative capacity and synthesize enzyme molecules that characterize the function of the mature cells. Other molecules, such as thymidine kinase and other enzymes, which are involved in DNA synthesis, are, on the contrary, no longer produced.

The main function of the small intestine is the digestion of food and the absorption of the various nutrients that are a part of the diet. A partial hydrolysis of starches and proteins takes place in the mouth and stomach, respectively. Then the pancreatic secretion present in the small intestine continues the digestive process and biliary secretion helps for fat absorption. The simplest molecules are absorbed through active or passive transport mechanisms or through pinocytosis as occurs for fats. Partially broken down molecules must be further hydrolyzed in order to be absorbed. This process takes place in the microvilli of the columnar cells by enzymes, such as tri- and dipeptidases and tri- and disaccharases, located in the brush border and is referred to as *membrane digestion*. The monomers (amino acids, monosaccharides) produced by this hydrolysis are absorbed by the intestinal mucosa. This process of terminal digestion takes place within the glycocalyx under sterile conditions. Other molecules, that is, carriers for active transport, are located in the outer layer of plasmatic membrane near enzyme molecules; in such a way the absorption of monomeric forms takes place easier. The brush border enzymes are synthesized for the first time during the differentiation process in the upper part of the crypt and continues during the migration of the columnar cells along the villus. Brush border enzymes are absent in the stomach, gradually increase from the duodenum to the jejunum, then decrease to reach very low levels in the terminal part of the ileum. They are absent in the colon and in the rectum. These enzymes have a fundamental importance in the digestive process because their congenital or acquired deficiency leads to intolerance and malabsorption, which are particularly severe in childhood and in some intestinal pathologies.

In mammals, brush border enzyme activity is strictly correlated with food uptake and shows marked modifications during the daily light–dark period. In nocturnal animals (e.g., rat, mouse) the activity during the night is twice that during the early afternoon, whereas in humans, brush border enzyme activity is higher during the light period.

Intracellular enzymes, which have no specific dietary function as do the lysosomal enzymes, do not show variations during the light–dark cycle.

A. Experimental Studies

More experimental studies on the effects of ionizing radiations have been performed on the small intestine than on the stomach and on the large intestine. There are many morphological and functional similarities between the small intestine of experimental animals and that of humans. For this reason the small intestine has been widely used as a model in radiobiological studies to obtain information that can then be applied to humans.

Total-body irradiation or irradiation only of the abdomen brings about alterations of the alimentary canal, which have already been described in the discussion of gastrointestinal syndrome. The evolution of acute radiation injury can be analyzed by considering the various levels of organization, as follows:

- Cell level: the proliferative cells die.
- Tissue level: the height of the crypt–villus system is reduced because of the lack of new cells and the physiological loss of cells at the villi apex.
- Organ level: loss of water and electrolytes, malabsorption, disepithelialization.
- Organism level: the preceding conditions lead to a state of physical debilitation, which, for doses greater than 10 Gy, is aggravated by ulcerations, entrance of germs into the body, and a complex of causes that lead to death.

The evolution of intestinal damage in humans is similar, if we consider a certain delay in the sequence of effects caused by the slower turnover.

From a morphological point of view, cell damage becomes evident only after a latent period of about 1–2 hr, in which the sequence of the alterations at the physical, chemical, and biochemical levels takes place. The evolution of radiation injury during the time following irradiation is represented in Figs. 2–4, where the effects on the jejunum of rats exposed to

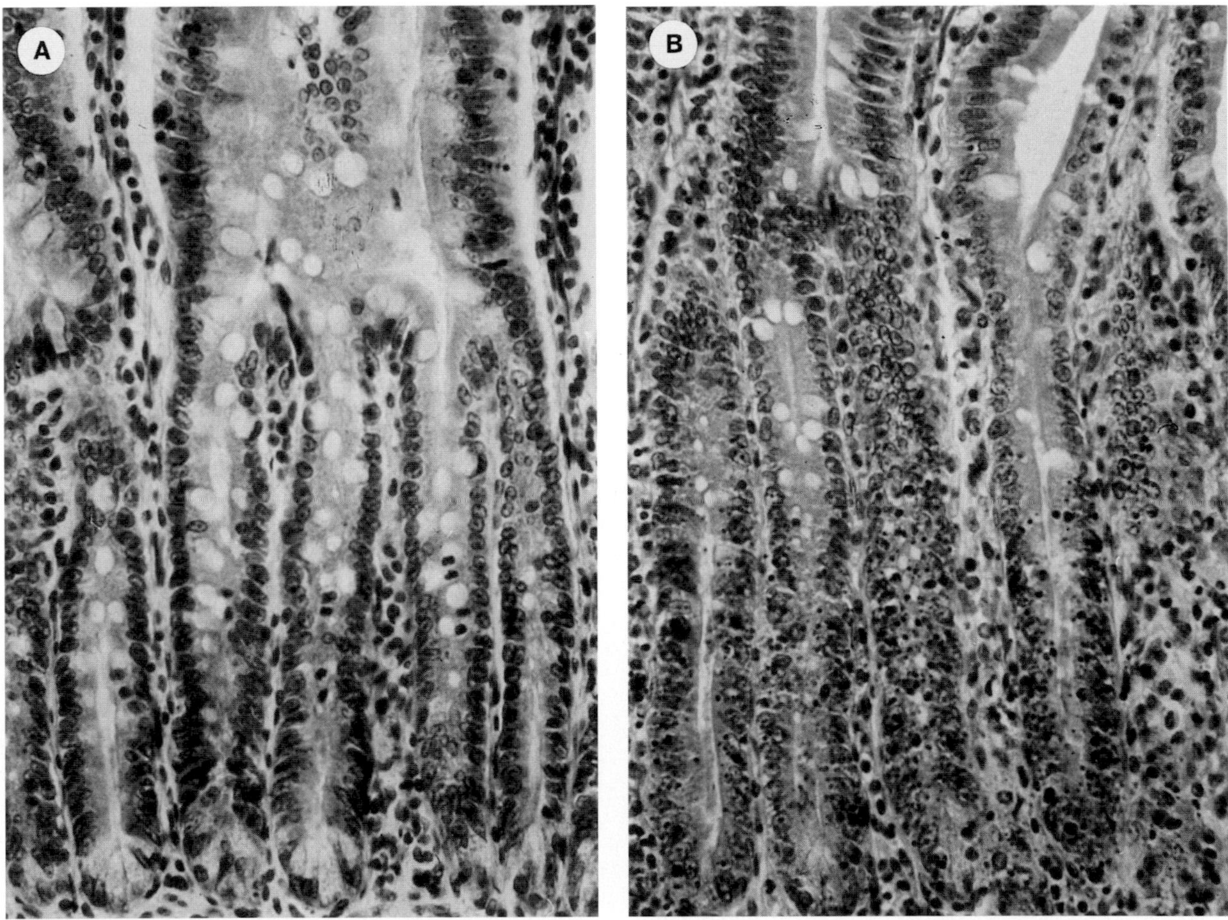

FIGURE 2 (A) Animals sacrificed 0.5 hr after irradiation. The time is too short to observe modifications at the cellular level: cell morphology and the organization of the entire epithelium are absolutely normal and numerous mitoses are present. (B) Four hr after irradiation: in the crypts, the apoptosis phenomenon is well evident, represented by many bodies of nuclear material coming from the death of epithelial cells. The upper part of the crypts and the base of the villi do not show any alteration.

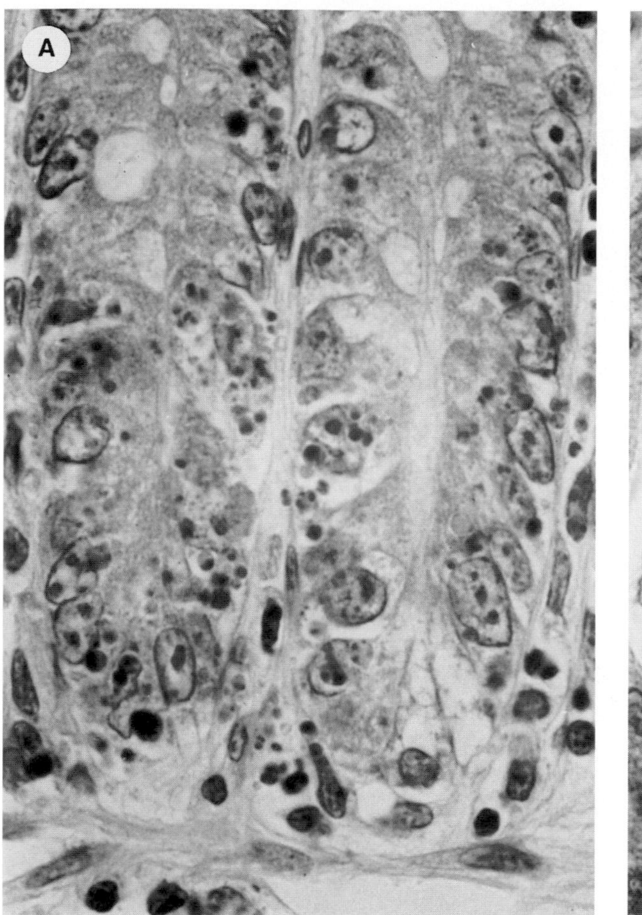
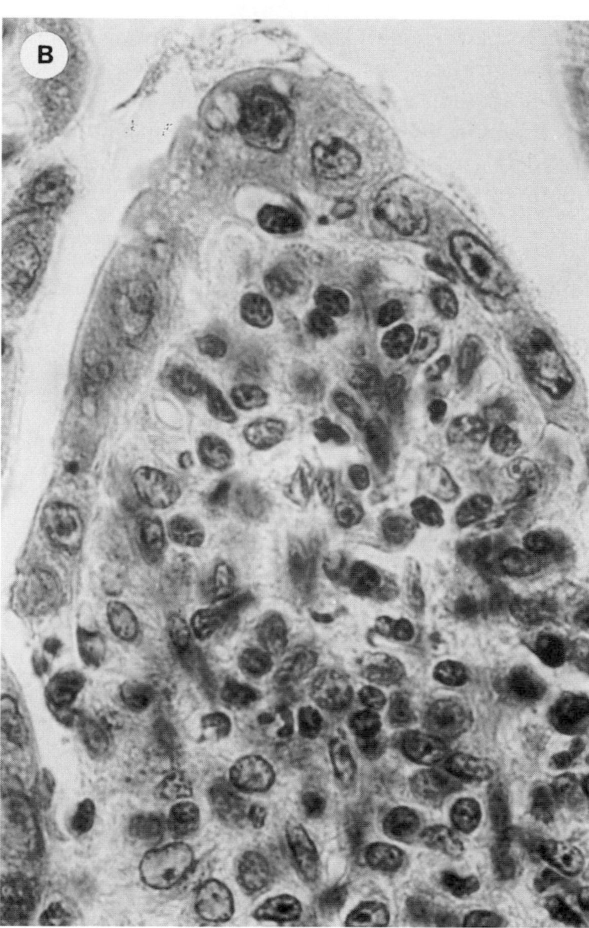

FIGURE 3 (A) Four hours after irradiation, particularly of the crypts: apoptotic bodies and fagocytosis phenomena are evident. Severe morphological alterations are present as cell death and reduction of the number of epithelial cells, which appear enlarged by edema. Mitoses are absent. (B) Seventy-two hr after irradiation, particularly of the villus apex: severe alterations are present in the epithelial cells, which appear few in number and enlarged. In some areas of the left side of the villus, only cytoplasmatic material appears to line the intestine. Abundant flogistic infiltrate is evident in the stroma as the last defense of the organism against the intestinal microflora and microtraumas produced by food.

a sublethal dose of 6 Gy (radiation source, was γ rays from ^{60}Co) are reported. At first the modifications appear in the proliferative cells of the lower part of the crypt, where a block of mitosis and apoptotic cells are observed in the histological sections.

With time, degenerative phenomena (nuclear and cytoplasmic alterations) and cell death involve progressively an increasing part of the crypt and after 1–2 days also the villus. Moreover, the lack of proliferation produces a reduction in the number of cells, which is more severe as the dose increases and reaches a maximum in 48–72 hr with sublethal doses. But already less than 1 day after exposure, endocellular repair phenomena restore the DNA of sublethally damaged cells and the percentage of cells that synthesize DNA and those that are in mitosis rapidly exceed the preirradiation values. The size and the length of the reduction in cell number, as well as the appearance of the successive phase of repopulation, are directly dose dependent, as we can observe from the samples after a 12-Gy dose (Fig. 5).

After approximately 20 hr, an inflammatory infiltration is observed in the stroma, which increases progressively during the entire acute phase of damage. Later the infiltrating cells progressively decrease and the epithelium returns to normal conditions.

On an ultrastructural level, the cell modifications concord with those observed using the optic microscope. Ultrastructural modifications were present in the cells of the lower half of the crypt as lysosome-

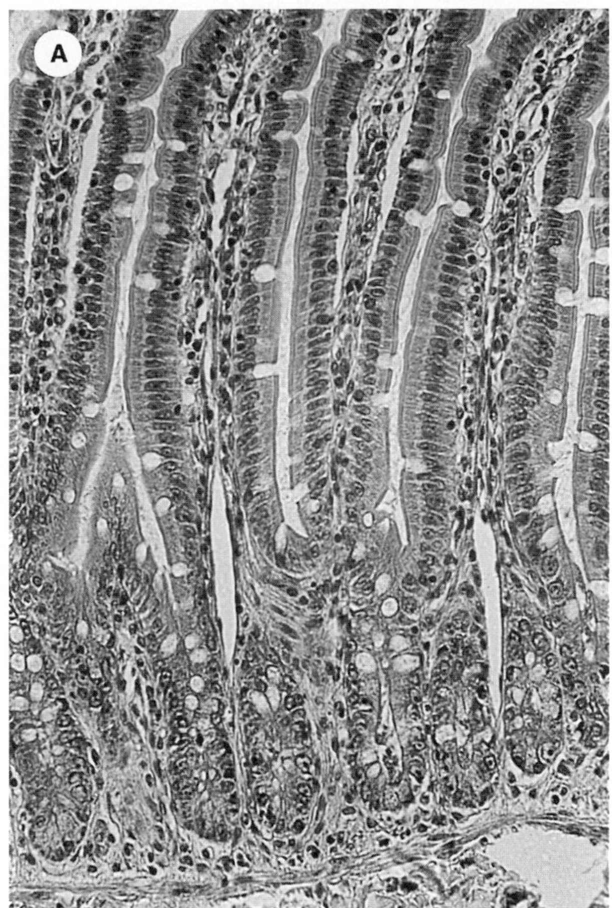

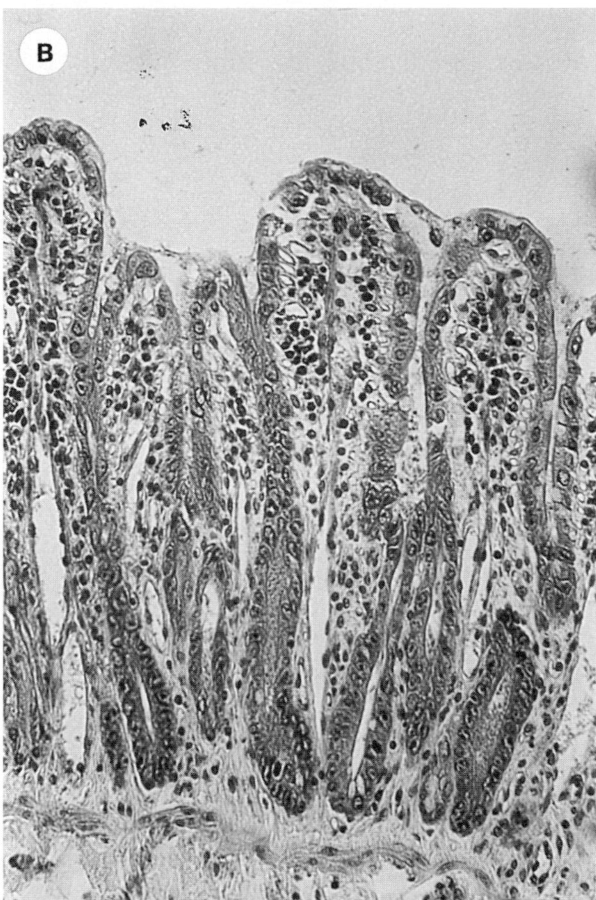

FIGURE 4 (A) Twenty-four hr after irradiation: the apoptosis phenomenon is no longer present. Dead cells are destroyed, whereas repopulation is not evident at this time. For this reason, the crypts appear shortened and lined by a few enlarged cells. Morphological alterations are absent in the villus, which appears to be formed by aligned normal enterocytes with a well evident brush border. The stroma appears to be normally populated. (B) Seventy-two hr after irradiation: morphological and structural modifications are well evident in the whole crypt–villus structure, which appears to be very shortened and covered by misaligned, enlarged cells. In the crypts, the repopulation takes place, cell density is higher than in the villus, and numerous mitoses are present. In the stroma, numerous cells of the flogosis are present and the edema is evidenced by dilated vessels.

like dense bodies and amorphous dense material in the smooth membranes. Progressively the chromatin concentrates as a dense clump close to the nuclear membrane, while the other part of the nucleus becomes transparent. The changes are consistent with the apoptotic process and a number of dense bodies appear in the cytoplasm of the cells. Some dead cells are extruded directly into the lumen. A few hours after irradiation with a sublethal dose, a large number of free or aggregated ribosomes appear in the rough endoplasmic reticulum of the cells in the upper part of the crypt. Phagocytic cells, expansion of the Golgi complex, and enlarged mitochondria are observed in the lower part of the crypt. Later marked alterations are present in the entire crypt–villus system. After a few days, the epithelium recovers and the morphology of the columnar cells seems to return to normal.

Some cell kinetics and functional activities show significant variations with the time of day. The studies on the circadian rhythms are of great interest because the administration of cytotoxic agents at a particular time of the light–dark cycle can produce different effects. Proliferation and particularly brush border enzyme activities show important modifications that in laboratory animals, being nocturnal, reach a maximum during the night and very low levels after midday.

When animals were irradiated at different times of

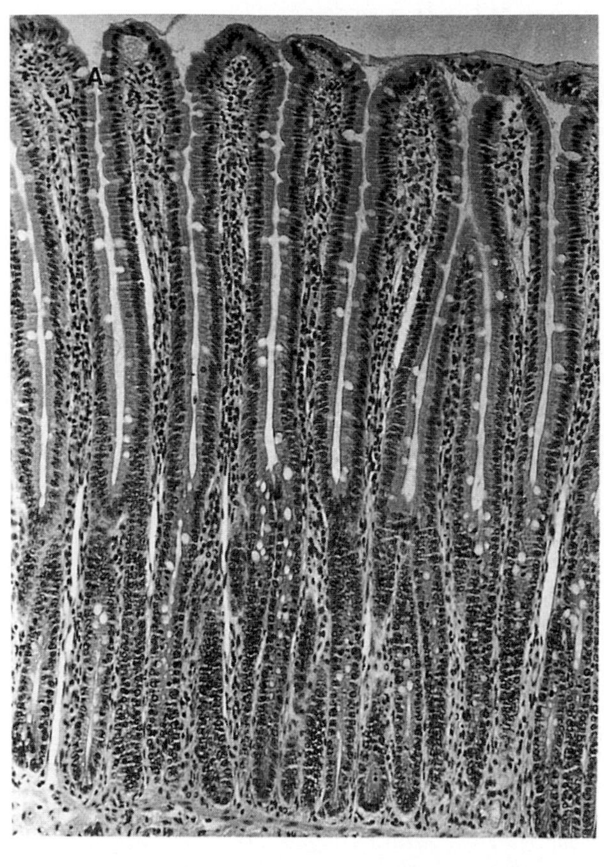
A

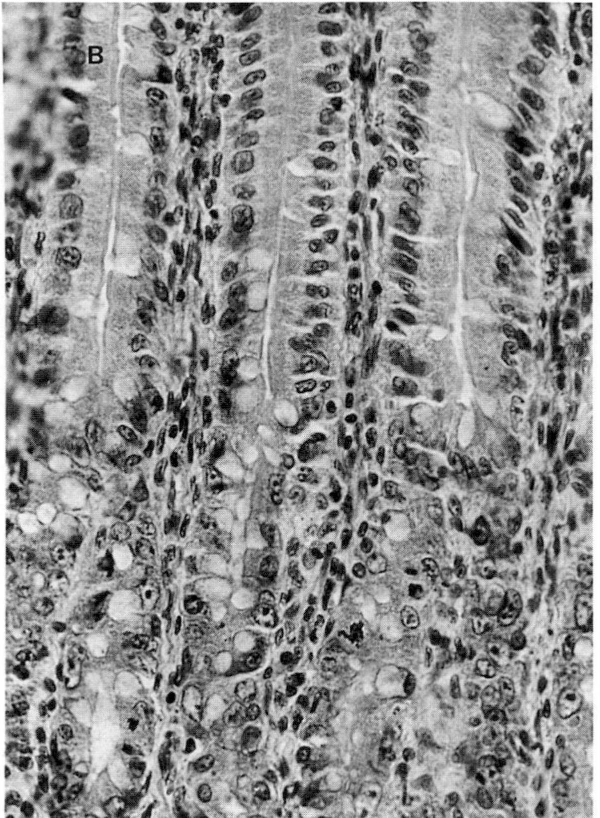
B

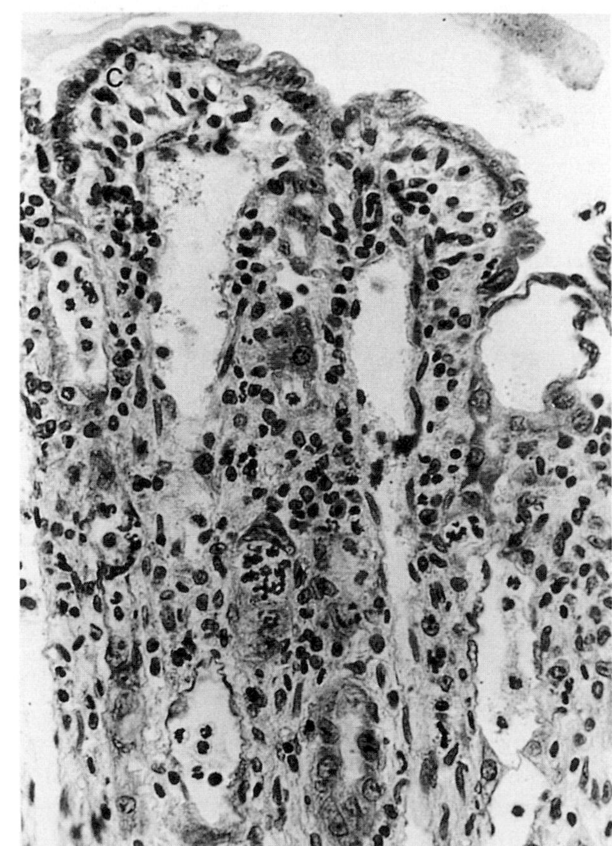
C

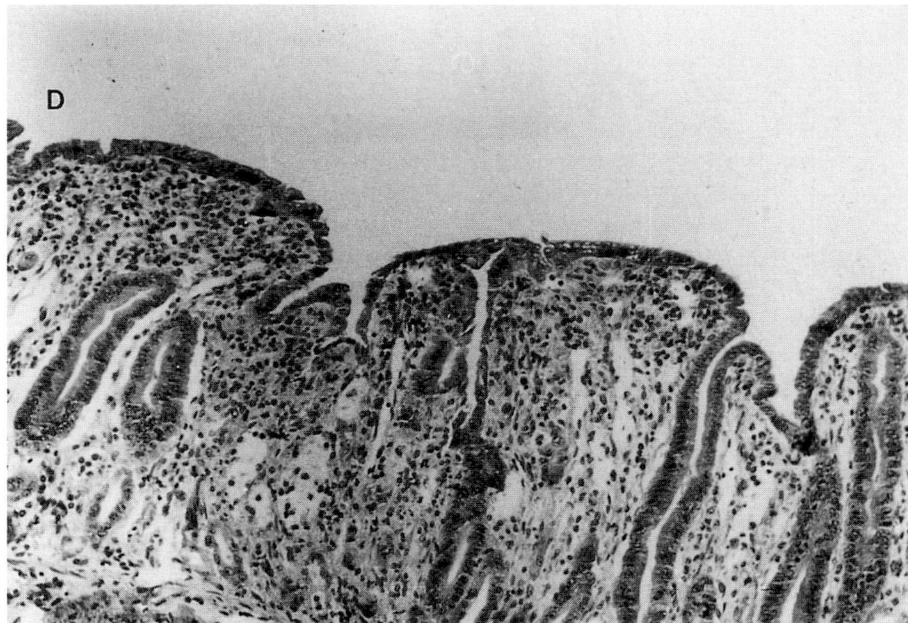

FIGURE 5 The progression of radiation injury in the rat jejunum at different times after 12 Gy abdominal irradiation from a telecobalt unit. The dose is high enough to produce 35% death for intestinal radiation syndrome. (A) Control mucosa: morphology of normal small intestine epithelium. (B) Twenty-four hours after irradiation: epithelial cells lining the whole crypt and the bottom of the villus are very irregular and misaligned, have enlarged nuclei, and are reduced in number. The epithelial cells of the upper part of the villus are still regular in appearance. (C) Seventy-two hours after irradiation: severe morphological alterations are present in the entire epithelium; the villi have lost their individuality and the number of cells is markedly reduced. The vessels are dilated and the lamina propria is hemorrhagic. (D) One hundred and twenty hours after irradiation: the epithelium is flattened and severe morphological alterations are still evident. Marked proliferative activity is observed in some crypts.

the day, a different behavior of proliferation and of brush border enzyme activities is observed. The behavior of the dipeptidase, 1-leucineaminopeptidase (LAP), in the jejunum of rats when irradiation with 3 and 8 Gy was given at midday and midnight is reported in Fig. 6. The results are expressed as a percentage of the values of controls sacrificed at the same time of the day. The absolute values in the control (as mean $\pm$ s.e.m.) were, respectively, 96 $\pm$ 6 at midnight and 68 $\pm$ 5 unit per g of protein at midday. As we can observe, the initial increase of enzyme activity is present in both irradiation groups, and it is more evident with higher dose but less lasting. During the acute phase of intestinal damage, the statistically significant decrease of enzyme activities is greater with higher dose. The return to normal levels is time delayed with respect to the return to a normal morphology of the intestinal epithelium. In the group irradiated at midnight with 8 Gy, control levels are reached many days before the group exposed at midday, which even 30 days after irradiation shows significantly lower activity. Disaccharase activity shows a similar behavior with more pronounced modifications in all phases.

The study of irradiated intestinal epithelium has permitted the formulation of interesting hypotheses with possible clinical applications. As observed with plant tissue and with some tumoral cell lines, the results demonstrate that the ionizing radiations are able to induce an early differentiation, which is very evident 24–36 hr after exposure. This happens when the proliferative cells are damaged sublethally and lose their capacity to divide. They begin to synthesize molecules, for example, the brush border enzymes, which characterize the differentiated state, before the time of their normal differentiation. Altered differentiation is confirmed by a significant increase in the number of goblet cells.

Another interesting aspect is related to the phase of repopulation: at first it appears to involve almost exclu-

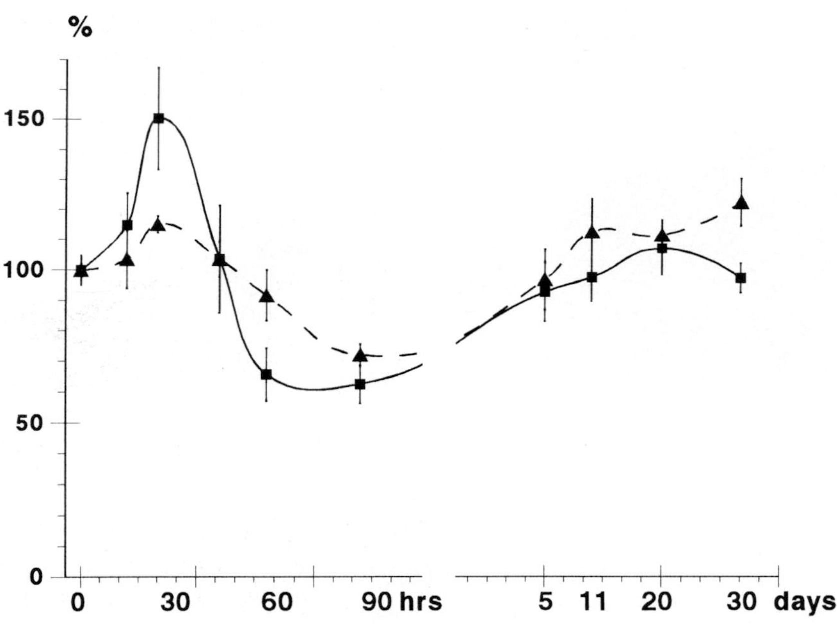

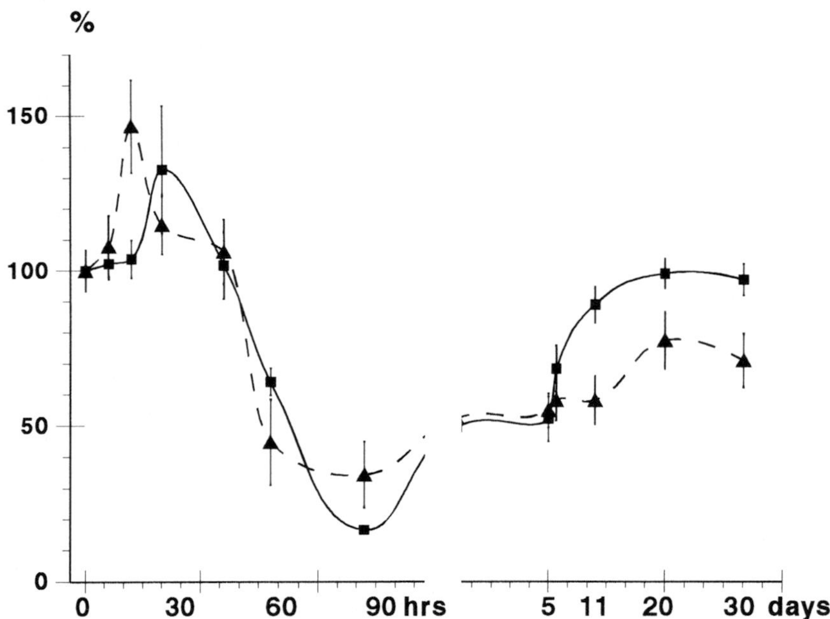

FIGURE 6 Variations in leucine aminopeptidase activity in the second of five segments in the small intestine of rats. Irradiation was given at two different times (midday, dashed line, and midnight, solid line) and sacrificed at different times after exposure to 3 Gy (top) and 8 Gy (bottom). The activity is expressed as a percentage of control values of animals sacrificed at the same time of the day.

sively columnar cells; the goblet cells reappear only later. The late return to a mucosa with normally functioning capacities is confirmed by the activity of the brush border enzymes (Fig. 7). The analysis of some functional aspects, which take place during the time after irradiation, appears to be of particular interest.

The specific digestive–absorption function of the small intestine is modified according to the morpho-

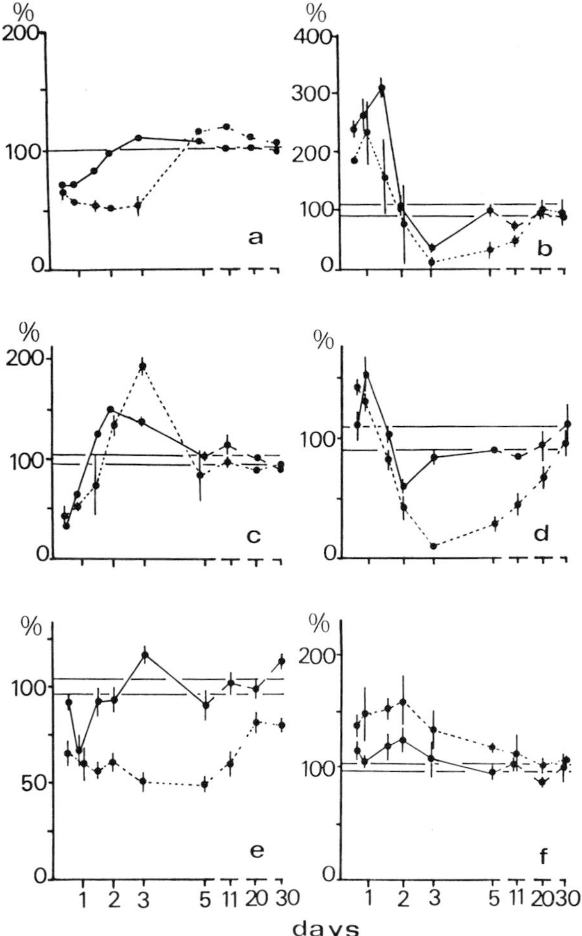

FIGURE 7 Effects of two different sublethal doses on some morphological and biochemical parameters of the small intestine. Rats were exposed to a 3-Gy (solid line) or 8-Gy (dotted line) dose with a γ-ray source (^{60}Co) and sacrificed at different time intervals (on the abscissa) after irradiation. The animals were caged under a constant light/darkness cycle (6:30–18:30). Each point represents the mean value ± SEM of every parameter obtained from five or six rats and the values are expressed as a percentage of controls sacrificed at the same time of the day. The first three parameters [(a) number of epithelial cells, (b) goblet cell index, and (c) labeling index], are related to at least 50 side well-aligned jujunal crypts; the other three [(d) brush border maltase activity, (e) DNA content, and (f) lysosomal β-glucuronidase activity] are referred to the whole small intestine. Three phases are well evident: a few hours after irradiation a reduction in cell number and proliferative activity is observed. During this period, goblet cell index and maltase activity increase, supporting the hypothesis of an early differentiation process. The second phase, acute morphological damage, is characterized by high proliferative activity in the crypts, whereas brush border enzyme activity and goblet cell index are significantly lower than control value. During the last part of this phase, repopulation in the whole epithelium occurs but cells of regular morphology appear not completely differentiated. In fact, brush border enzyme activities are significantly lower than controls and goblet cells are very few. The third phase is characterized by a return to a normal functional ability. Lysosomal enzyme activities increase during acute injury. The higher dose produces higher effects and a delayed return to normal conditions.

logical qualitative and quantitative modifications. A few hours after irradiation and within 36 hr brush border enzyme activities show a statistically significant increase. During the phase of acute radiation injury, brush border enzyme activity is heavily reduced, explaining the malabsorption for disaccharides and dipeptides. Only later do the membrane digestion enzymes return to normal levels.

The determination of lysosomal enzyme activity takes into account the degenerative phenomena that occur in the irradiated mucosa during the time following irradiation. An increase of activity is observed during the acute phase of radiation damage (36–72 hr) when few and altered cells are present in the epithelium and numerous inflammatory cells (granulocytes, plasma cells, macrophages, etc.) are distributed in the stroma. During the repopulation phase, the activity of lysosomal enzymes is progressively reduced and a return to normal levels is observed when the intestinal epithelium reaches a normal morphology. The higher the dose, the higher and more lasting is the increase of lysosomal enzyme activity.

A particular approach to study the organization of the proliferative compartment is by analysis of the distribution in the crypt of the cells that are synthesizing DNA. For this kind of study, a perfect alignment of the sections cut longitudinally to the axis of the crypt–villus structure is necessary. The localization of labeled cells along the crypt is very low at the base of the crypt, where quiescent (G_0 cells) and highly differentiated cells (Paneth cells) are present. The frequency increases in the middle of the crypt, reaching over 50%, then it decreases to values near zero in the higher 20% of total crypt length. Mitosis frequency reaches higher positions but always with very low levels. In this part of the crypt the cells that have lost proliferation capability synthesize brush border enzymes, molecules that characterize differentiation process in the small intestine.

A few hours after exposure, the block of proliferative activity is evidenced by a reduction of S-phase cells that appear to be localized only in the lower half of the crypt. At the next intervals, together with a progressive reduction of epithelial cells and the repopulation phenomenon, the proliferative cells occupy progressively higher positions in the crypt. At 72 hr after an 8-Gy dose, S-phase cells cover the whole crypt. During this period, the severe morphological modifications of the epithelium and the reduction of the differentiation compartment justify the lack of brush border enzyme activity and thus the malabsorption syndrome.

B. Modifications after Fractionated Irradiation

Radiobiological studies have demonstrated that when a dose of radiation is divided into fractions, the resulting effect is inferior to that induced by a single dose of the same size. This phenomenon demonstrates the capability of cells to repair sublethal injury. In fact, in the interval between fractions, endocellular repair occurs that, if the time period is sufficient, will completely eliminate the acute damage. Repopulation of repaired cells restores the normal condition of the epithelium.

The use of dose fractionation in tumor radiotherapy is based on time differences and efficacy differences in sublethal damage repair and recovery between healthy and neoplastic tissues exposed to low-LET radiations. The probability of the destruction of a tumor increases with a higher dose. The problem is related to the tolerance of healthy tissues, which with a high dose can be destroyed. In order to give a dose that is able to sterilize the tumor but that is insufficient to produce the death of healthy tissues, the total dose is divided into fractions separated by idoneous time intervals that produce sublethal damage and restoration by means of endocellular repair mechanisms and repopulation.

Conditioning factors are the dose per fraction and the time elapsed between fractions. The first parameter influences the amount of lethal damage. In fact, the characteristic aspect of the dose–effect curve shows that, after an initial shoulder, a direct relation is observed and the higher the doses the lower the cell survival fraction.

The time elapsed between fractions influences the efficiency of the endocellular repair. When this time is long enough, for example, 24 hr, a complete repair can occur, whereas when the interval between fractions is only a few hours, two different mechanisms take place. For an interval of several hours, the effect of the second dose occurs when the enzyme mechanisms of the DNA repair are activated and thus the damage is more easily restored. When the interval is longer, about 8 hr, the cells that have repaired the first injury are synchronized in the most radiosensitive phase of the cell cycle. For this reason, the second dose produces a more severe injury compared to a single fraction. These time intervals between the fractions differ according to the cell type and the tissues. With the aim of evaluating these phenomena, many experiments have been conducted in the small intestine of rats by using different dose per fraction and time between fractions, but with the same total dose of 6 and 12 Gy.

During fractionation, the injury results from (1) the cell death produced by the first dose, (2) the endocellular mechanisms for the repair of sublethal damage, and (3) the repopulation that occurs in the interval between the fractions. The residual damage and the time elapsed from the different fractions condition morphological and functional modifications in irradiated tissues. The small intestine has been used to evaluate tissue tolerance with this type of irradiation and as a model to study the effects of repeated doses on mechanisms regulating proliferative activity and cell differentiation. The effects of dose fractionation on epithelial cell number in rat jejunal crypt are represented in Fig. 8. The behavior after 2 fractions of 3 Gy and 4 fractions of 3 Gy, both with a 12-hr interval between fractions, is compared with 3- and 8-Gy single doses.

The other morphological and biochemical parameters show the same kind of difference. In general, when low-LET radiations are used, the time between each fraction is particularly important in the manifestation of the effect.

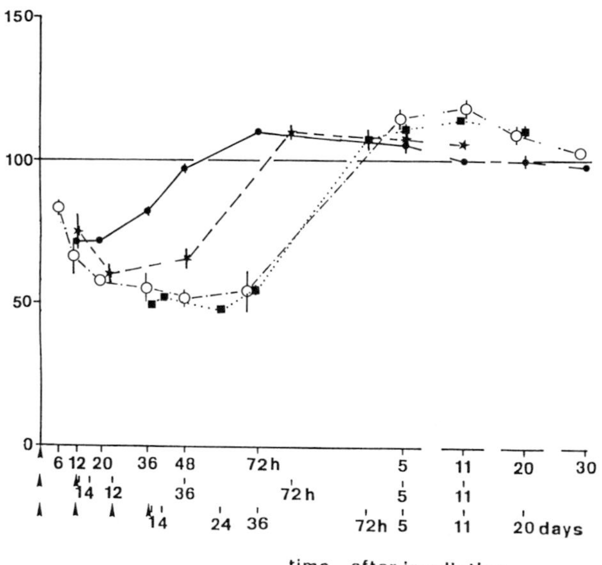

FIGURE 8 Effect of dose fractionation on the number of epithelial cells in the jejunal crypt. The values are represented as percentage ± SEM of control animals. The time of exposure at the 3-Gy fractions is indicated by the arrowheads. The solid line represents the effect of a 3-Gy single dose; the dashed line represents those of two fractions of 3 Gy every 12 hr and the dotted line those of four fractions of 3 Gy every 12 hr. The dot–dash line represents the effect of an 8-Gy single dose. The behavior is similar to that observed in Fig. 7.

When the interval is short (a few hours), the sequence described here is induced after a single dose. If the second fraction is administered during the phase of acute injury, this phase lasts longer because of lesion overlapping. If the fractions are separated by a few days, the initial damage is repaired and the second appears as a function of the size of the dose of the second fraction. The ultrastructural modifications, the determination of brush border enzyme activity, and the lysosomal enzyme activity confirm the observations made after a single dose.

The morphological and biochemical–functional alterations produced by irradiations in a single or fractionated dose lead to a reduction in the weight of the intestine, a highly radiosensitive tissue, which reaches a minimum in laboratory animals after 2–3 days. Body weight also decreases but to a much lesser degree. During the acute phase of injury there is a notable decrease in DNA content, while protein synthesis and protein content in the small intestine decrease to a lesser degree and more slowly. It should be observed that the longer the interval between fractions, or the less the dose per fraction, the less the effect due to the greater efficacy of the repair and repopulation mechanisms.

C. Clinical Studies

Doses received by accidental exposure are only rarely high enough to produce the gastrointestinal syndrome. However, more or less extensive intestinal irradiation is quite frequent in treatment of gynecological tumors, testicular seminomas, and others, and especially in total-body irradiation prior to bone marrow transplant.

Studies conducted between the end of the 1960s and the beginning of the 1970s pointed out that several substances were malabsorbed during the acute phase of damage. However, few studies were adequately performed because the pretreatment levels of the morphological and functional parameters were often abnormal and the tests were only sometimes repeated in the same patient after increasing doses.

The conventional treatment for gynecological or testicular tumors uses fractions of 2 Gy/day, 5 days/week for up to a maximum total dose of 45–55 Gy. The irradiated volume can include the abdominopelvic region, the lower hemiabdomen, or the epigastric region. In these patients, nausea and vomiting generally appear after the first sessions. These symptoms vary greatly in severity and last about 1 week. They are more frequent and more intense when the upper half of the abdomen or the epigastric region have been irradiated. Diarrhea generally appears during the third week (halfway through treatment, 25–30 Gy), especially in women receiving lower abdomen irradiation. Drug administration limits symptom severity.

To evaluate intestinal conditions during treatment, a series of tests were performed. Morphological damage, evaluated by biopsy, shows the same modifications as observed in laboratory animals. When the biopsies are repeated during treatment, the mitotic index progressively decreases until about 40 Gy. Then the values remain constant until the end of treatment. During this phase, the cells in the crypt–villus structure show the previously described alterations and decrease in number. The mucosa returns to normal within the first months after the end of the treatment. Morphological modifications are also observed in the intestinal loops outside the irradiated field.

Radiological investigations with barium sulfate show motility modifications that are directly dependent on the exposed volume. When the region of the lower hemiabdomen is irradiated, acceleration of the transit is very frequently observed, whereas in other cases the transit time changes are modest.

Patients with diarrhea present the most striking alterations. An altered motility modifies the contact time between molecules present in the food and the mucosa, and this can therefore influence absorption.

Oral administration of molecules mixed with a test meal and their assay in blood, urine, and feces permits evaluation of changes in absorption during abdomen irradiation. A state of malabsorption is present in most patients, but this is highly dependent on individual susceptibility. Malabsorption is generally very noticeable halfway through treatment and can worsen when therapy has been completed. It is more serious and frequent in patients irradiated in the abdominopelvic region. Results may differ depending on the substances used in the absorption tests.

Carbohydrate absorption, measured with glucose, which uses active transport as its absorption mechanism, is significantly reduced halfway through and at the end of treatment. Saccharose, instead, must be previously hydrolyzed by the brush border enzymes and is more markedly reduced, whereas xylose, a portion of which is thought to be absorbed by passive transport, shows an evident reduction. The reduction at the end of treatment appears to be 32% for glucose, 40% for saccharose, and 33% for xylose.

Reduction in absorption is much less and of a similar size when proteins and amino acids are administered orally (less than 10%). The reduction of fat absorption is evident but differs for oleic acid and triolein (12 versus 28%, respectively, at the end of treatment). Other substances show a marked malabsorption. At the end of the treatment the remaining absorption rate is improved and the reduction is only of 30%. At the middle and end of treatment, vitamin B_{12} absorption is reduced by 35 and 30%, respectively. There is loss of water and electrolytes, but it does not produce important modifications since diet and suitable drugs can easily compensate. The modest variations observed in the plasma and blood volumes and in the total body water confirm this result.

Changes in biliary and pancreatic secretions are modest and generally return to the normal range of physiological variations. With epigastric irradiation, a reduction of gastric acidity is observed. Bile salts play an important role in inducing diarrhea. These molecules are reabsorbed in the terminal ileum; irradiation of this segment, by provoking a reduced absorption, can be one cause of diarrhea.

An estimate of the intestinal damage by irradiation can be obtained by evaluating the daily quantity of feces. The initial quantity increases twofold during the second week of treatment and remains high until the end of therapy.

Studies of the intestinal microflora demonstrate that in irradiated patients there is a variation in the number of the preexisting microorganisms, rather than in the appearance of new species. Malabsorption increases molecule permanence in the intestinal lumen and therefore produces an osmotic effect. The molecules can be utilized by the microflora, with the possible production of toxic compounds.

In accidental exposure, the administration of water, electrolytes, and antibiotics increases the survival probability. Some studies have demonstrated that administration of an elemental liquid diet immediately before and during radiotherapy of the abdomen and pelvis leads to better alimentation and therefore facilitates absorption of elemental nutrients even during acute intestinal radiation syndrome. The diet must be milk free, gluten free, low fat, and low residue to limit the possibility of diarrhea.

The term radiation enteritis includes many structural and functional alterations in the intestine. The dose of radiation for treatment of abdominal tumors ranges from 30 Gy for non-Hodgkin's lymphoma or seminoma to 80 Gy in the case of endometrium.

About 15–20% of patients undergoing radiotherapy will present enteropathy with higher or lower probability according to the dose and individual characteristics. When external radiotherapy is associated to intracavitary irradiation, the incidence of complications is higher. Submucosa, blood vessels, muscolaris propria, and serosa can show subacute and chronic modifications that appear to be of importance in the development of late injury. Abnormal fibroblasts, submucosa telangiectasia, hyalinization of the vessels, and alterations in endothelial cells with swelling and obliteration appear to be the principal vascular changes. Secondary alterations involve the mucosal component and abnormalities of endothelia, whereas at the microscopical level ulceration, probably caused by progressive vascular sclerosis, hemorrhagic perforation, segmental infarction, and stricture appear.

V. LARGE INTESTINE AND RECTUM

The main features of the large intestine and rectum are (1) the epithelium in the large intestine lacks villi; (2) the crypts decrease in length, going from the colon to the rectum; and (3) there are more goblet cells than in the small intestine. The cell cycle time is about 24 hr and the turnover is about 6 days; the lesser proliferative activity explains the greater tolerance of this segment. The main function of the large intestine is linked to the reabsorption of water and electrolytes and to the elimination of residues.

Radiation injury has the same features as in the small intestine, that is, cell death and inhibition of mitosis a few hours after irradiation. After several days, these alterations involve the entire epithelium, and edema and vascular changes in the submucosal and serosal layers are observed. Hypermotility and tenesmus can be observed and the rectal mucosa appears hyperemic and edematous. A few weeks are necessary for repair. With elevated doses, ulcerations throughout the intestine can be observed.

Between 2 and 12 months following the end of treatment, subacute damage can be observed: the mucosa regenerates but alterations of the endothelial cells remain, the small arteries and capillary vessels are still swollen, and fibrin plugs can cause thrombosis. Lack of circulation is progressive and contemporarily there is the beginning of a chronic collagen degeneration.

The submucosa becomes thicker and fibrotic. These mechanisms can lead to the appearance of atrophy in the glandular mucosa with consequent malabsorption and the formation of ulcers. The progressive degeneration of the connective tissue can lead to obstruction of the canal. For elevated doses, holes through the wall (fistulas) can form. Late intestinal injury can appear even more than 20 years after radiotherapy, but 5 years is considered to be the period of risk.

Radiosensitivity can be expressed on the basis of the *tolerance dose* (TD), that is, the dose administered according to a conventional fractionation schedule (2 Gy/day, 5 days/week) that produces the injury in a certain percentage of cases after a fixed time. The TD 5/5 is the dose that causes severe injury in 1–5% of patients 5 years after radiotherapy. The TD 50/5 is the dose that produces severe injury in 50% of patients in 5 years. The TD 5/5 and the TD 50/5 have been calculated in the alimentary canal by assessing the incidence of ulcers and strictures and are as follows: for esophagus, 60 and 75 Gy; for stomach, 45 and 50 Gy; for small intestine and colon, 45 and 65 Gy; and for rectum, 55 and 80 Gy. The small intestine is the most radiosensitive, but significant injuries are usually avoided because of its mobility.

Evident alterations are observed during intracavitary or external irradiation for carcinoma of the uterus, and they are particularly severe when adhesions are present. Late reactions or complications include inflammation of the rectum, perforation, and vesicovaginal or rectovaginal fistulas. Damage is much more severe when large doses are used for each session. With external radiotherapy of the pelvis for tumors of the cervix, uterus, or bladder, about 70% of the patients have strictures and more than 50% show inflammation of the rectum and colon.

Rectal bleeding and modifications in bowel habits are observed when there are large intestine complications. Sometimes these symptoms are initially attributed to a recurrence because diagnosis is difficult owing to ulcers and necrotic areas. Surgical intervention is difficult in these patients because of the presence of serious fibrosis in the irradiated tissue, which can induce necrosis and fistulas even several days after surgery. The presence of hardened vessels in the area near the irradiated field complicates the possibility of reestablishing normal conditions following surgery.

The rectosigmoid tract of the bowel is less mobile than the rest of the alimentary canal. Therefore, though it is more radioresistant than the small intestine, early and late alterations are similar to those in the other segments, that is, fibrosis, strictures, ulcerations, fistulas, and perforations. In tumors of the uterus and of the prostate gland, the rectum is also irradiated, and injury occurs in this tissue as well as in the urethra and the bladder. Here, too, damage is more probable after laparotomies and previous surgical procedures, due to immobilization of the intestinal loops. The results obtained demonstrate that treatment with ionizing radiations on areas including the intestines must be very carefully conducted. All data confirm the low tolerance of the bowel and that early damage cannot be considered a prognostic factor for late injury.

VI. PREDISPOSING FACTORS IN THE PRODUCTION OF RADIATION INJURY

Generally the radiosensitivity shows little range of variability in normal tissues among different patients. In some cases the doses usually tolerated can produce severe damage.

A particular condition is represented by patients affected by ataxia telangiectasia, Fanconi's anemia, and Down's syndrome, in whom autosomal genetic defects in the mechanism of DNA repair are evidenced.

Besides total dose, time, irradiation modality, and irradiated volume, other risk factors are gender (risk is higher in women than in men), age, hypertension, vascular disease, pelvic chronic inflammation, leanness, and diabetes mellitus. Patients who have undergone previous surgery have adhesions; for this reason the intestine is less mobile, and therefore during the treatment the possibility of repeated irradiation of the same intestinal loops is greater.

Abdominal irradiation of children and infants with tumors produces more severe alterations than in adults, even if lower doses (15–40 Gy) are used. Note that generally the irradiated volume is greater in children than in adults. In 70% of the cases, immediate tolerance is poor and vomiting and diarrhea are present. Surgery and concomitant chemotherapy are also unfavorable factors in these cases. Few observations exist concerning chronic and late effects due to a short rate of survival.

VII. HIGH-LINEAR ENERGY TRANSFER RADIATION

High-LET radiations (α particles, neutrons, protons) are available for radiotherapy in only a few centers. The relative biological effectiveness of neutrons is much greater than that of X rays, γ rays, and accelerated electrons, because they release large quantities of energy per unit track (μm) of the material they traverse. For equal physical doses absorbed, the higher density of interactions leads to much greater cell damage and to less possibility of repair. The effect of high-LET radiations is scarcely influenced by the presence of oxygen or by cell parameters, properties that could make the utilization of this type of radiation valid in the therapy of tumors.

The quality of acute and late lesions induced in the intestinal epithelium is similar to those caused by low-LET irradiations, but the injury is greater. In a high percentage of cases, doses greater than 20 Gy in patients with advanced pelvic tumors lead to serious complications, including fibrosis, obstructions, necrosis, and bowel perforations. However, in about 50% of the cases, thought to be uncurable, complete tumor regression is obtained.

VIII. RADIATION CARCINOGENESIS

The capacity of ionizing radiations to induce tumors is well documented and it is probable that a second tumor will eventually appear in patients undergoing radiotherapy as late injury. However, it is impossible to distinguish between a second spontaneous tumor and a radioinduced tumor. A second tumor in the irradiated volume can appear from 5 to 30 years after exposure, and this makes accurate controls even more necessary in the posttreatment period.

Usually the induced tumors are adenocarcinomas, and their incidence depends on the dose administered. Some studies have demonstrated that irradiation of the abdomen and of the pelvis for primary cervical or ovarian tumors can induce leukemia. The risk of acute leukemia is increased by a factor of 2.1 and that of acute nonlymphocytic leukemia by 4.0 or more. A lower incidence has been observed in women affected by tumors of the cervix or uterus compared to cases irradiated for benign gynecological disorders: in the first case the dose to the bone marrow was higher and resulted in killing bone marrow cells rather than inducing their transformation.

IX. CONCLUSIONS

The modifications produced by ionizing radiations on the alimentary canal have been the object of many studies, but increased understanding of molecular and cellular biology makes continuous research updating necessary.

The intestine is a tissue that has always been widely used in radiobiological studies, and it is considered a valid model. In fact, its morphological and functional characteristics permit easy recognition of the proliferative compartment and, especially in the small intestine, of the differentiated compartment, which is characterized by the synthesis of specific, easily assayed molecules. It is possible, therefore, to obtain accurate analysis of the effects of chemical or physical agents on different cellular activities.

Rapid cell turnover permits manifestation of damage within a brief time period, and the elevated proliferative activity explains the high tolerance of this tissue, which even after high doses of radiations has a rapid repopulation. The induction of an early differentation process and the presence of cells with a regular morphology during recovery, but with incomplete acquisition of their functional capacities, will be an interesting base of study to understand the mechanisms that lead to cell damage and those that intervene in the passage from proliferative to differentiated cells.

Different modalities of dose administration and association with other physical and chemical agents will contribute further to a better understanding of the modifications induced by antineoplastic treatments. The intestinal tract can also be useful for analyzing alterations that take place in the connective tissue and in the vasal endothelium, which are the principal causes of severe late injury.

The knowledge of the cellular mechanisms that lead to the radiation injury and to its recovery is expected to provide information for better effectiveness of therapy for human tumors, and/or to reduce the damage to healthy tissues.

BIBLIOGRAPHY

Becciolini, A. (1987). Relative radiosensitivity of the small and large intestine. *Adv. Radiat. Biol.* **12**, 83–128.

Casarett, G. W. (1980). "Radiation Histopathology." CRC Press, Boca Raton, Florida.

Fajardo, L. F. (1982). "Pathology of Radiation Injury." Masson, New York.

Pottem, C. S., and Hendry, J. H. (1995). "Radiation and Gut." Elsevier, New York.

Quastler, H. (1956). The nature of intestinal death. *Radiat. Res.* **4**, 303–320.

Reasoning and Natural Selection

LEDA COSMIDES
JOHN TOOBY
University of California, Santa Barbara

GLOSSARY

Adaptation Aspect of an organism that was created by the process of natural selection because it served an adaptive function

Adaptive Contributing to the eventual reproduction of an organism or its relatives

Bayes's theorem Specifies the probability that a hypothesis is true, given new data; $P(H|D) = P(H)P(D|H)/P(D)$, where H is the hypothesis and D is the new data

Cognitive psychology Study of how humans and other animals process information

Natural selection Evolutionary process responsible for constructing, over successive generations, the complex functional organization found in organisms, through the recurring cycle of mutation and subsequent increased reproduction of the better design

Normative theory Theory specifying a standard for how something *ought* to be done (as opposed to how it actually *is* done)

Valid argument Argument that is logically derived from premises; a conclusion may be valid, yet false, if it is logically derived from false premises

THE STUDY OF REASONING IS AN IMPORTANT component of the study of the biology of behavior. To survive and reproduce, animals must use data to make decisions, and these decisions are controlled, in part, by processes that psychologists label "inference" or "reasoning." To avoid predators, for example, a monkey must infer from a rustle in the grass and a glimpse of fur that a leopard is nearby and use information about its proximity to decide whether to take evasive action or continue eating. Because almost all action requires inferences to regulate it, the mechanisms controlling reasoning participate in almost every kind of behavior that humans, or other animals, engage in. Human reasoning has traditionally been studied without asking what kind of reasoning procedures our ancestors would have needed to survive and reproduce in the environment in which they evolved. In recent years, however, an increasing number of researchers have been using an evolutionary framework. [*See* Human Evolution.]

I. WHAT IS REASONING AND HOW IS IT STUDIED?

When psychologists study how humans reason, they are trying to discover what rules people use to make inferences about the world. They investigate whether there are general principles that can describe what people will conclude from a set of data.

One way of studying reasoning is to ask "If one were trying to write a computer program that could simulate human reasoning, what kind of program would have to be written? What kind of information-processing procedures (rules or algorithms) would the programmer have to give this program, and what kind of data structures (representations) would those procedures operate on?"

Of course, the human brain was not designed by an engineer with foresight and purposes; it was "de-

ENCYCLOPEDIA OF HUMAN BIOLOGY, Second Edition, VOLUME 7. Copyright © 1991 by Academic Press. All rights of reproduction in any form reserved.

signed" by the process of natural selection. Natural selection is the only natural process known that is capable of creating complex and organized biological structures, such as the human brain. Contrary to widespread belief, natural selection is not "chance"; it is a powerful positive feedback process fueled by differential reproduction. If a change in an organism's design allows it to outreproduce other members of its species, that design change will become more common in the population—it will be *selected for*. Over many generations that design change will spread through the population until all members of the species have it. Design changes that enhance reproduction can be selected for; those that hinder reproduction are selected against.

When evolutionary biologists study how humans reason, they are asking, "What kind of cognitive programs was natural selection likely to have designed, and is there any evidence that humans have such programs?"

A. Mind versus Brain

At present, researchers find it useful to study the brain on different descriptive and explanatory levels. Neuroscientists describe the brain on a physiological level—as the interaction of neurons, hormones, neurotransmitters, and other organic aspects. Cognitive psychologists, on the other hand, study the brain as an information-processing system—that is, as a collection of programs that process information—without worrying about exactly how neurophysiological processes perform these tasks. The study of cognition is the study of how humans and other animals process information.

For example, ethologists have traditionally studied very simple cognitive programs: a newborn herring gull, for instance, has a cognitive program that defines a red dot on the end of a beak as salient information from the environment, and that causes the chick to peck at the red dot upon perceiving it. Its mother has a cognitive program that defines pecking at her red dot as salient information from her environment, and that causes her to regurgitate food into the newborn's mouth when she perceives its pecks. This simple program adaptively regulates how the herring gull feeds its offspring. (If there is a flaw anywhere in the program—if the mother or chick fails to recognize the signal or to respond appropriately—the chick starves. If the flaw has a genetic basis, it will not be passed on to future generations. Thus natural selection controls the design of cognitive programs.)

These descriptions of the herring gull's cognitive programs are entirely in terms of the functional relationships among different pieces of information; they describe two simple information-processing systems. Of course, these programs are embodied in the herring gull's neurological "hardware." Knowledge of this hardware, however, would add little to our understanding of these programs as information-processing systems. Presumably, one could build a silicon-based robot, using hardware completely different from what is present in the gull's brain, that would produce the same behavioral output (pecking at red dot) in response to the same informational input (seeing red dot). The robot's cognitive programs would maintain the same functional relationships among pieces of information and would therefore be functionally identical to the cognitive programs of the herring gull. But the robot's neural hardware would be totally different.

The specification of a cognitive program constitutes a complete description of an important level of causation, independent of any knowledge of the physiological hardware the program runs on. Cognitive psychologists call this position "functionalism," and they use it because it provides a precise language for describing complex information-processing architectures, without being limited to studying those few processes that neurophysiologists presently understand. (Eventually, of course, one wants to understand the neurophysiological processes that give rise to a cognitive program as well.) Cognitive scientists use the term "mind" solely to refer to an information-processing description of the functioning of the brain, and not in any colloquial sense.

II. THE MIND AS SCIENTIST: GENERAL-PURPOSE THEORIES OF HUMAN REASONING

Traditionally, cognitive psychologists have acknowledged that the mind (i.e., the information-processing structure of the brain) is the product of evolution, but their research framework was more strongly shaped by a different premise: that the mind was a general-purpose computer. They thought the function of this computer was self-evident: to discover the truth about whatever situation or problem it encountered. In other words, they started from the reasonable assumption that the procedures that governed human reasoning were there because they functioned to produce valid knowledge in nearly any context a person was likely to encounter.

They reasoned that if the function of the human mind is to discover truth, then the reasoning procedures of the human mind should reflect the methods by which truth can be discovered. Because science is the attempt to discover valid knowledge about the world, psychologists turned to the philosophy of science for *normative theories*—that is, for theories specifying how one *ought* to reason if one is to produce valid knowledge. Their approach was to use the normative theories of what constitutes good scientific reasoning as a standard against which to compare actual human reasoning performance. The premise was that humans should be reasoning like idealized scientists about whatever situation they encountered, and the research question became: To what extent is the typical person's reasoning like an ideal scientist's?

The normative theories of how scientists—and hence the human mind—should be reasoning fall broadly into two categories: inductive reasoning and deductive reasoning. Inductive reasoning is reasoning from specific observations to general principles; deductive reasoning is reasoning from general principles to specific conclusions.

Ever since the work of David Hume in the eighteenth century, induction has carried a heavy load in psychology while taking a sound philosophical beating. In psychology, it has been the learning theory of choice since the British Empiricists argued that the experience of spatially and temporally contiguous events is what allows us to jump from the particular to the general, from sensations to objects, from objects to concepts. Many strands of psychology, including Pavlovian reflexology, Watsonian and Skinnerian behaviorism, and the sensory-motor parts of Piagetian structuralism, have been elaborations on the inductive psychology of the British Empiricists. Yet when Hume, a proponent of inductive inference as a psychological learning theory, donned his philosopher's hat, he demonstrated that induction could never justify a universal statement. To use a familiar example, no matter how many white swans you might see, you could never be justified in concluding "All swans are white," because it is always possible that the next swan you see will be black. Thus Hume argued that the inductive process whereby people were presumed to learn about the world could not ensure that the generalizations it produced would be valid.

Only recently, with the publication in 1935 of Karl Popper's "The Logic of Scientific Discovery," has a logical foundation for psychology's favorite learning theory been provided. Popper argued that although a universal statement of science can never be proved true, it deductively implies particular assertions about the world, called hypotheses, and particular assertions can be proved false. Although no number of observed white swans can prove that "All swans are white" is true, just one black swan can prove it false. Generalizations cannot be confirmed, but they can be falsified, so inductions tested via deductions coupled with observations are on firmer philosophical ground than knowledge produced through induction alone.

This view had broad consequences for psychologists interested in learning. Psychologists who assumed that the purpose of human learning is to produce valid generalizations about the world reasoned that learning must be some form of Popperian hypothesis testing. Inductive reasoning must be used to generate hypotheses, and deductive reasoning coupled with observation must be used to try to falsify them. Furthermore, these reasoning procedures should be general purpose: they should be able to yield valid inferences about any subject that one is interested in.

A broad array of cognitive psychologists, such as Jean Piaget, Jerome Bruner, and Peter Wason, adopted a version of hypothesis testing—often an explicitly Popperian version—as their model of human learning. They used it to set the agenda of cognitive psychology in the 1950s and 1960s, and this view remains popular today. Some psychologists investigated inductive reasoning by seeing whether people reason in accordance with the normative theories of inferential statistics; others investigated deductive reasoning by seeing whether people reason in accordance with the rules of inference of the propositional calculus (formal propositional logic).

A. Deductive Reasoning

Psychologists became interested in whether the human mind included a "deductive component": mental rules that are the same as the rules of inference of the propositional calculus. They performed a wide variety of experiments to see whether people were able (1) to *recognize* the difference between a valid deductive inference and an invalid one, or (2) to *generate* valid conclusions from a set of premises. If people have a "deductive component," then they should be good at tasks like these. For example, in reasoning about conditional statements, one can make two valid inferences and two in valid inferences (Fig. 1).

One of the most systematic bodies of work exploring the idea that people have reasoning procedures

Valid inferences		Invalid inferences	
Modus ponens	Modus tollens	Affirming the Consequent	Denying the Antecedent
If P then Q P	If P then Q not-Q	If P then Q Q	If P then Q not-P
Therefore Q	Therefore not-P	Therefore P	Therefore not-Q

FIGURE 1 "P" and "Q" can stand for any proposition; for example, if "P" stands for "it rained" and "Q" stands for "the grass is wet," then the *modus ponens* inference would read "If it rained, then the grass is wet; it rained, therefore the grass is wet." *Affirming the consequent* and *denying the antecedent* are invalid because the conditional "If P then Q" does not claim that P is the only possible antecedent of Q. If it did *not* rain, the grass could still be wet—the lawn could have been watered with a sprinkler, for example.

that embody the rules of inference of the propositional calculus was produced by Peter Wason and P. N. Johnson-Laird, together with their students and colleagues. Their research provides strong evidence that people do not reason according to the canons of formal propositional logic. For example:

(1) *Recognition of an argument as valid.* To see whether people are good at recognizing an argument as valid, psychologists gave them arguments like the ones in Fig. 1. For example, a subject might be asked to judge the validity of the following argument: "If the object is rectangular, then it is blue; the object is rectangular; therefore the object is blue." In some of the experiments, unfamiliar conditionals were used; in others, familiar ones were used. These experiments indicated that people are good at recognizing the validity of a *modus ponens* inference, but they frequently think *modus tollens* is an invalid inference and that the two invalid inferences in Fig. 1 are valid. Furthermore, they frequently view logically distinct conditionals as implying each other, and they have a pronounced tendency to judge an inference valid when they agree with the conclusion and invalid when they do not agree with the conclusion, regardless of its true validity.

(2) *Generating valid conclusions from a set of premises.* In other experiments, psychologists gave people sets of premises and asked them to draw conclusions from them. Many of the problems requiring use of *modus ponens* were done incorrectly, and most of those requiring the use of *modus tollens* were done incorrectly. It may at first seem puzzling that people who are good at recognizing a *modus ponens* argument as valid would have trouble using *modus ponens* to generate a conclusion from premises. However, analogous experiences are common in everyday life: sometimes one cannot recall a person's name but can recognize it on a list. If humans all had rules of reasoning that mapped on to *modus ponens,* then they would be able both to generate a valid *modus ponens* inference and to recognize one. The fact that people cannot do both indicates that they lack this rule of reasoning. They may simply be able to recognize a contradiction when they see one, even though they cannot reliably generate valid inferences.

Perhaps the most intriguing and widely used experimental paradigm for exploring deductive reasoning has been the Wason selection task (Fig. 2a). Peter Wason was interested in Popper's view that the structure of science was hypothetico-deductive. He wondered if learning were really hypothesis testing; that is, the search for evidence that contradicts a hypothesis. Wason devised his selection task because he wanted to see whether people really do test a hypothesis by looking for evidence that could potentially falsify it. In the Wason selection task, a subject is asked to see whether a conditional hypothesis of the form "If P then Q" has been violated by any one of four instances, represented by cards.

A hypothesis of the form "If P then Q" is violated only when "P" is true but "Q" is false—the rule in Fig. 2a, for example, can be violated only by a card that has a D on one side and a number other than 3 on the other side. Thus, one would have to turn over the "P" card (to see if it has a "not-Q" on the back) and the "not-Q" card (to see if it has a "P" on the back)—that is, D and 7 for the rule in Figure 1a. The logically correct response, then, is always "P and not-Q."

Wason expected that people would be good at this. Nevertheless, he and many other psychologists have found that few people actually give this logically correct answer (<25% for rules expressing unfamiliar relations). Most people choose either the "P" card alone or "P and Q." Few people choose the "not-Q" card, even though a "P" on the other side of it would falsify the rule.

A wide variety of conditional rules that describe some aspect of the world ("descriptive" rules) have been tested; some of these have expressed relatively familiar relations, such as "If a person goes to Boston, then he takes the subway" or "If a person eats hot chili peppers, then he will drink a cold beer." Others have expressed unfamiliar relations, such as "If you eat duiker meat, then you have found an ostrich eggshell" or "If there is an 'A' on one side of a card, then there is a '3' on the other side." In many experiments, performance on familiar descriptive rules is just as low as it is on unfamiliar ones; some familiar rules, however, do elicit a higher percentage of logically

a. Abstract Problem (AP)

Part of your new clerical job at the local high school is to make sure that student documents have been processed correctly. Your job is to make sure the documents conform to the following alphanumeric rule:

"If a person has a 'D' rating, then his documents must be marked code '3'."
(If P then Q)*

You suspect the secretary you replaced did not categorize the students' documents correctly. The cards below have information about the documents of four people who are enrolled at this high school. Each card represents one person. One side of a card tells a person's letter rating and the other side of the card tells that person's number code.

Indicate only those card(s) you definitely need to turn over to see if the documents of any of these people violate this rule.

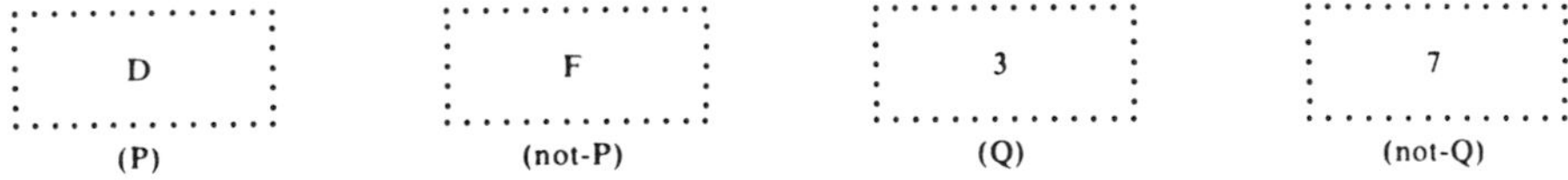

(P) (not-P) (Q) (not-Q)

b. Drinking Age Problem (DAP; adapted from Griggs & Cox, 1982)

In its crackdown against drunk drivers, Massachusetts law enforcement officials are revoking liquor licenses left and right. You are a bouncer in a Boston bar, and you'll lose your job unless you enforce the following law:

"If a person is drinking beer, then he must be over 20 years old."
(If P then Q)

The cards below have information about four people sitting at a table in your bar. Each card represents one person. One side of a card tells what a person is drinking and the other side of the card tells that person's age.

Indicate only those card(s) you definitely need to turn over to see if any of these people are breaking this law.

(P) (not-P) (Q) (not-Q)

c. Structure of Social Contract (SC) Problems

It is your job to enforce the following law:

Rule 1 — Standard Social Contract (STD-SC): "If you take the benefit, then you pay the cost."
(If P then Q)

Rule 2 — Switched Social Contract (SWC-SC): "If you pay the cost, then you take the benefit."
(If P then Q)

The cards below have information about four people. Each card represents one person. One side of a card tells whether a person accepted the benefit and the other side of the card tells whether that person paid the cost.

Indicate only those card(s) you definitely need to turn over to see if any of these people are breaking this law.

	Benefit Accepted	Benefit NOT Accepted	Cost Paid	Cost NOT Paid
Rule 1 — STD-SC:	(P)	(not-P)	(Q)	(not-Q)
Rule 2 — SWC-SC:	(Q)	(not-Q)	(P)	(not-P)

FIGURE 2 Content effects on the Wason selection task. The logical structures of these three Wason selection tasks are identical; they differ only in propositional content. Regardless of content, the logical solution to all three problems is the same: To see if the rule has been violated, choose the "P" card (to see if it has a "not-Q" on the back) and choose the "not-Q" card (to see if it has a "P" on the back). Fewer than 25% of college students choose "P & not-Q" for the Abstract Problem (a), whereas about 75% choose both these cards for the Drinking Age Problem (b)—a familiar social contract. Part (c) shows the abstract structure of a social contract problem. A "look for cheaters" procedure would cause one to choose the "benefit accepted" card and the "cost NOT paid" card, regardless of which logical categories they represent. For Rule 1, these cards represent the values "P & not-Q," but for Rule 2 they represent the values "Q & not-P." Consequently, a person who was looking for cheaters would appear to be reasoning logically in response to Rule 1 but illogically in reponse to Rule 2.

*The logical categories (P and Q) marked on the rules and cards here are only for the reader's benefit; they never appear on problems given to subjects in experiments.

correct responses than unfamiliar ones. Even so, familiar descriptive rules typically elicit the logically correct response in fewer than half of the people tested. Recently, rules expressing causal relations have been tested; the pattern of results is essentially the same as for descriptive rules.

It is particularly significant that performance on the Wason selection task is so poor when the descriptive or causal rule tested is unfamiliar. If the function of our reasoning procedures is to allow us to discover new things about the world, then they must be able to function in novel, that is, unfamiliar, situations. If they cannot be used in unfamiliar situations, then they cannot be used to learn anything new. Thus, the view that the purpose of human reasoning is to learn about the world is particularly undermined by the finding that people are not good at looking for violations of descriptive and causal rules, especially when they are unfamiliar.

B. Inductive Reasoning

The hypotheses that scientists test do not appear from thin air. Some of them are derived from theories; others come from observations of the world. For example, although no number of observations of white swans can prove that all swans are white, a person who has seen hundreds of white swans and no black ones may be more likely to think this hypothesis is worthy of investigation than a person who has seen only one white swan. The process of inferring hypotheses from observations is called inductive inference.

Using probability theory, mathematicians have developed a number of different normative theories of inductive inference, such as Bayes's theorem, null hypothesis testing, and Neyman–Pearsonian decision theory. These theories specify how scientists should make inferences from data to hypotheses. They are collectively known as inferential statistics.

A number of psychologists have studied the extent to which people's inductive reasoning conforms to the normative theories of inferential statistics and probability theory. One of the most extensive research efforts of this kind was spearheaded by Amos Tversky and Daniel Kahneman, along with their students and colleagues. They tested people's inductive reasoning by giving them problems in which they were asked to judge the probability of uncertain events. For example, a subject might be asked to reason about a diagnostic medical test: "If a test to detect a disease whose prevalence is 1/1000 has a false positive rate of 5%, what is the chance that a person found to have a positive result actually has the disease, assuming you know nothing about the person's symptoms or signs?" If the subject's answer is different from what a theory of statistical inference says it should be, then the experimenters conclude that our inductive-reasoning procedures do not embody the rules of that normative statistical theory.

The consensus among many psychologists is that this body of research demonstrates that (1) the human mind does not calculate the probability of events in accordance with normative probability theories, and (2) the human mind does not include information-processing procedures that embody the normative theories of inferential statistics. In other words, they conclude that the human mind is not innately equipped to do college-level statistics. Instead, these psychologists believe that people make inductive inferences using heuristics—cognitive shortcuts or rules of thumb. These heuristics frequently lead to the correct answer but can also lead to error precisely because they do not embody the formulas and calculational procedures of the appropriate normative theory. These psychologists also believe that humans suffer from systematic biases in their reasoning, which consistently lead to errors in inference.

Recently, however, a powerful critique by Gerd Gigerenzer, David Murray, and their colleagues has called this consensus into serious doubt. Their critique is both theoretical and empirical. Gigerenzer and Murray point out that the Tversky and Kahneman research program is based on the assumption that a statistical problem has only one correct answer; when the subject's response deviates from that answer, the experimenter infers that the subject is not reasoning in accordance with a normative statistical theory. However, Gigerenzer and Murray show that the problems subjects are typically asked to solve do not have only one correct answer. There are several reasons why this is true.

1. Statistics Does Not Speak with One Voice

There are a number of different statistical theories, and not all of them give the same answer to a problem. For example, although subjects' answers to certain problems have been claimed to be incorrect from the point of view of Bayes's theorem (but see the following section), their answers can be shown to be correct from the point of view of Neyman–Pearsonian decision theory. These subjects may be very good "intuitive statisticians," but simply applying a different normative theory than the experimenter is.

2. Concepts Must Match Exactly

For a particular statistical theory to be applicable, the concepts of the theory must match up precisely with the concepts in the problem. Suppose, for example, that you have some notion of how likely it is that a green cab or a blue cab would be involved in a hit-and-run accident at night. You are then told that there was a hit-and-run accident last night, and that a witness who is correct 80% of the time reported that it was a green cab. Bayes's theorem allows you to revise your prior probability estimate when you receive new information, in this case, the witness's testimony.

But what should your prior probability estimate (i.e., the estimate that you would make if you did not have the witness's testimony) be based on? It could, for example, be based on (1) the relative number of green and blue cabs in the city, (2) the relative number of reckless driving arrests for green versus blue cab drivers, (3) the relative number of drivers who have alcohol problems, or (4) the relative number of hit-and-run accidents they get into at night.

There is no normative theory for deciding which of these four kinds of information is the most relevant. Yet Bayes's theorem will generate different answers, depending on which you use. If subjects and experimenter differ in which kind of information they believe is most relevant, they will give different answers, even if each is correctly applying Bayes's theorm. Indeed, experimental data suggest that this happens. If one assumes that the subjects in these experiments were making certain very reasonable assumptions, then they *were* answering these questions correctly.

3. Structural Assumptions of the Theory Must Hold for the Problem

Assume that nature had selected for statistical rules; then it also should have selected for an assumption-checking program. For a particular statistical theory to be applicable, the assumptions of the theory must hold for the problem. For instance, a frequent assumption for applying Bayes's theorem is that a sample was randomly drawn. But in the real world there are many situations in which events are not randomly sampled. Diagnostic medical tests, for example, are rarely given to a random sample of people—instead, they are given only to those who already have symptoms of the disease. By their content, certain problems tested invited the inference that the random sampling assumption was violated; given this assumption, the "incorrect" answers subjects were giving were, in fact, correct. Indeed, in an elegant series of experiments, Gigerenzer and his colleagues showed that if one makes the random sampling assumption explicit to subjects, they do appear to reason in accordance with Bayes's theorem.

These experiments and theoretical critiques cast serious doubt on the conclusion that people are not good "intuitive statisticians." Evidence suggests that people are very good at statistical reasoning if the problem is about a real-world situation in which the structural assumptions of the theory hold. Their apparent errors may be because they are making assumptions about the problem that are different from the experimenters', or because they are consistently applying one set of statistical principles in one context and other sets in different contexts. What is clear from the research on inductive reasoning, however, is that the content of the problem matters, a theme we will return to in Section III.

C. Did We Evolve to Be Good Intuitive Scientists?

Good design is the hallmark of adaptation: to demonstrate that human reasoning evolved to fulfill a particular function, one must show that our reasoning procedures are well designed to fulfill that function. If the human mind was designed by natural selection to generate logically valid, scientifically justifiable knowledge about the world, then we ought to be good at drawing correct inductive and deductive inferences. Moreover, this ability ought to be context-independent, to allow us to learn about new, unfamiliar domains. After all, everything is initially unfamiliar.

But the data on deductive reasoning indicate that our minds do not include rules of inference that conform to the canons of deductive logic. The data on inductive reasoning indicate that we do not have inductive-reasoning procedures that operate independently of content and context. We may have inductive-reasoning procedures that conform to normative theories of statistical inference, but if we do, their application in any particular instance is extremely context-dependent, as the issues of conceptual and structural matching show. The evidence therefore suggests that we do not have formal, content-independent reasoning procedures. This indicates that the hypothesis that the adaptive function of human reasoning is to generate logically valid knowledge about the world is false.

III. THE MIND AS A COLLECTION OF ADAPTATIONS: EVOLUTIONARY APPROACHES TO HUMAN REASONING

Differential reproduction is the engine that drives natural selection: if having a particular mental structure, such as a rule of inference, allows an animal to outreproduce other members of its species, then that mental structure will be selected for. Over many generations it will spread through the population until it becomes a universal, species-typical trait.

Consequently, alternative phenotypic traits are selected for not because they allow the organism to more perfectly apprehend universal truths, but because they allow the organism to outreproduce others of its species. Truth-seeking can be selected for only to the extent that it promotes reproduction. Although it might seem paradoxical to think that reasoning procedures that sometimes produce logically incorrect inferences might be more adaptive than reasoning that always leads to the truth, this will frequently be the case. Among other reasons, organisms usually must act before they have enough information to make valid inferences. In evolutionary terms, the design of an organism is like a system of betting: what matters is not each individual outcome, but the statistical average of outcomes over many generations. A reasoning procedure that sometimes leads to error, but that usually allows one to come to an adaptive conclusion (even when there is not enough information to justify it logically), may perform better than one that waits until it has sufficient information to derive a valid truth without error. Therefore, factors such as the cost of acquiring new information, asymmetries in the payoffs of alternative decisions (believing that a predator is in the shadow when it is not versus believing a predator is not in the shadow when it is), and trade-offs in the allocation of limited attention may lead to the evolution of reasoning procedures whose design is sharply at variance with scientific and logical methods for discovering truth.

Although organisms do not need to discover universal truths or scientifically valid generalizations to reproduce successfully, they do need to be very good at reasoning about important adaptive problems and at acquiring the kinds of information that will allow them to make adaptive choices in their natural environment. Natural selection favors mental rules that will enhance an animal's reproduction, whether they lead to truth or not. For example, rules of inference that posit features of the world that are usually (but not always) true may provide an adequate basis for adaptive decision making. Some of these rules may be general purpose: for example, the heuristics and biases proposed by Tversky and Kahneman are rules of thumb that will get the job done under the most commonly encountered circumstances. Their availability heuristic, for instance, is general purpose insofar as it is thought to operate across domains: one uses it whether one is judging the frequency of murders in one's town or of words in the English language beginning with the letter "k." However, there are powerful reasons for thinking that many of these evolved rules will be special purpose.

Traditionally, cognitive psychologists have assumed that the human mind includes only general-purpose rules of reasoning and that these rules are few in number. But natural selection is also likely to produce many mental rules that are specialized for reasoning about various evolutionarily important domains, such as cooperation, aggressive threat, parenting, disease avoidance, predator avoidance, and the colors, shapes, and trajectories of objects. This is because different adaptive problems frequently have different optimal solutions. For example, vervet monkeys have three major predators: leopards, eagles, and snakes. Each of these predators requires different evasive action: climbing a tree (leopard), looking up in the air or diving straight into the bushes (eagle), or standing on hind legs and looking into the grass (snake). Accordingly, vervets have a different alarm call for each of these three predators. A single, general-purpose alarm call would be less efficient because the monkeys would not know which of the three different evasive actions to take.

When two adaptive problems have different optimal solutions, a single general solution will be inferior to two specialized solutions. In such cases, a jack of all trades is necessarily a master of none, because generality can be achieved only by sacrificing efficiency.

The same principle applies to adaptive problems that require reasoning: There are cases where the rules for reasoning adaptively about one domain will lead one into serious error if applied to a different domain. Such problems cannot, in principle, be solved by a single, general-purpose reasoning procedure. They are best solved by different, special-purpose reasoning procedures. We will consider some examples of this in the following sections.

A. Internalized Knowledge and Implicit Theories

Certain facts about the world have been true for all of our species' evolutionary history and are critical to our ability to function in the world: the sun rises every 24 hours; space is locally three dimensional; rigid objects thrown through space obey certain laws of kinematic geometry. Roger Shephard has argued that a human who had to learn these facts through the slow process of "trial and possibly fatal error" would be at a severe selective disadvantage compared to a human whose perceptual and cognitive system was designed in such a way that it already assumed that such facts were true. In an elegant series of experiments, Shepard showed that our perceptual–cognitive system has indeed internalized laws of kinematic geometry, which specify the ways in which objects move in three-dimensional Euclidean space. Our perceptual system seems to expect objects to move in the curvilinear paths of kinematic geometry so strongly that we see these paths even when they do not exist, as in the phenomenon of visual apparent motion. This powerful form of inference is specific to the motion of objects; it would not, for example, help you to infer whether a friend is likely to help you when you are in trouble.

Learning a relation via an inductive process that is truly general purpose is not only slow, it is impossible in principle. There are an infinite number of dimensions along which one can categorize the world, and therefore an infinite number of possible hypotheses to test ("If my elbow itches, then the sun will rise tomorrow", "If a blade of grass grows in the flower pot, then a man will walk in the door"; i.e., "If P then Q," "If R then Q," "If S then Q" *ad infinitum*). The best a truly unconstrained inductive machine could do would be to randomly generate each of an infinite number of inductive hypotheses and deductively test each in turn.

Those who have considered the issue recognize that an organism could learn nothing this way. If any learning is to occur, then one cannot entertain all possible hypotheses. There must be constraints on which hypotheses one entertains, so that one entertains only those that are most likely to be true. This insight led Susan Carey and a number of other developmental psychologists to suggest that children are innately endowed with mental models of various evolutionarily important domains. Carey and her colleagues call these mental models *implicit theories* to reflect their belief that all children start out with the same set of theories about the world, embodied in their thought processes.

These implicit theories specify how the world works in a given domain; they lead the child to test hypotheses that are consistent with the implicit theory, and therefore likely to be true (or at least useful). Implicit theories constrain the hypothesis space so that it is no longer infinite, while still allowing the child to acquire new information about a domain. Implicit theories are thought to be domain-specific because what is true of one domain is not necessarily true of another. For example, an implicit theory that allows one to predict a person's behavior if one knows that person's beliefs and desires will not allow one to predict the behavior of falling rocks, which have no beliefs and desires. The implicit-theory researchers have begun to study children's implicit theories about the properties of organisms, the properties of physical objects and motion, the use of tools, and the minds of others.

B. Reasoning about Prescriptive Social Conduct

The reasoning procedures discussed so far function to help people figure out what the world is like and how it works. They allow one to acquire knowledge that specifies what kind of situation one is facing from one moment to the next. For example, a rule such as "If it rained last night, then the grass will be wet this morning" purports to describe the way the world is. Accordingly, it has a truth value: a descriptive rule can be either true or false. In contrast, a rule such as "If a person is drinking beer, then that person must be over 21 years old" does not describe the way things are. It does not even describe the way existing people behave. It *prescribes:* it communicates the way some people want other people to behave. One cannot assign a truth value to it.

From an evolutionary perspective, knowledge about the world is just a means to an end, and that end is behaving adaptively. Once an organism knows what situation it is in, it has to know how to act, so reasoning about the facts of the world should be paired with reasoning about appropriate conduct. For this reason, the mind should have evolved rules of reasoning that specify what one ought to do in various situations—rules that prescribe behavior. Because different kinds of situations call for different kinds of behavior, these rules should be situation-specific. For example, the rules for reasoning about cooperation should differ from those for reasoning about aggressive threat, and both should differ from the rules for reasoning about the physical world. Recent research by Cosmides and Tooby, Manktelow and Over, and others has explored such rules.

Social exchange, for example, is cooperation between two or more people for mutual benefit, such as the exchange of favors between friends. Humans in all cultures engage in social exchange, and the paleoanthropological record indicates that such cooperation has probably been a part of human evolutionary history for almost 2 million years. Game-theoretic analyses by researchers such as Robert Trivers, Robert Axelrod, and W. D. Hamilton have shown that cooperation cannot evolve unless people are good at detecting "cheaters" (people who accept favors or benefits without reciprocating). Given a social contract of the form "If you take the benefit, then you pay the cost," a cheater is someone who took the benefit but did not pay the required cost (see Fig. 2c). Detecting cheaters is an important adaptive problem: a person who was consistently cheated would be incurring reproductive costs, but receiving no compensating benefits. Such individuals would dwindle in number, and eventually be selected out of the population.

Rules for reasoning about descriptive relations would lead one into serious error if applied to social contract relations. In the previous discussion of the Wason selection task, we saw that the logically correct answer to a descriptive rule is "P and not-Q," no matter what "P" and "Q" stand for (i.e., no matter what the rule is about). But this definition of violation differs from the definition of cheating on a social contract. A social contract rule has been violated whenever a person has taken the benefit without paying the required cost, *no matter what logical category these actions correspond to*. For the social contract expressed in Rule 1 of Fig. 2c, a person who was looking for cheaters would, by coincidence, produce the logically correct answer. This is because the "benefit accepted" card and the "cost NOT paid" card correspond to the logical values "P" and "not-Q," respectively, for Rule 1. But for the social contract expressed in Rule 2, these two cards correspond to the logical values "Q" and "not-P"—a logically incorrect answer. The logically correct answert to Rule 2 is to choose the "cost paid" card, "P," and the "benefit NOT accepted" card, "not-Q." Yet a person who has paid the cost cannot possibly have cheated, nor can a person who has not accepted the benefit.

Thus, for the social contract in Rule 2, the adaptively correct answer is logically incorrect, and the logically correct answer is adaptively incorrect. If the only reasoning procedures that our minds contained were the general-purpose rules of inference of the propositional calculus, then we could not, in principle, reliably detect cheating on social contracts. This adaptive problem can be solved only by inferential procedures that are specialized for reasoning about social exchange.

The Wason selection task research discussed previously showed that we have no general-purpose ability to detect violations of conditional rules—unfamiliar descriptive and causal rules elicit the logically correct response from <25% of subjects. But when a conditional rule expresses a social contract, people are very good at detecting cheaters. Approximately, 75% of subjects choose the "benefit accepted" card and the "cost NOT paid" card, regardless of which logical category they correspond to and regardless of how unfamiliar the social contract rule is. This research indicates that the human mind contains reasoning procedures that are specialized for detecting cheaters on social contracts. Recently, the same experimental procedures have been used to investigate reasoning about aggressive threat. Although there is only one way to violate the terms of a social contract, there are two ways of violating the terms of a threat: either the person making the threat can be bluffing (i.e., he does not carry out the threat, even though the victim refuses to comply), or he can be planning to double-cross the person he is threatening (i.e., the victim complies with his demand, but the threatener punishes him anyway). The evidence indicates that people are good at detecting both bluffing and double-crossing. Similarly, two British researchers, Kenenth Manktelow and David Over, have found that people are very good at detecting violations of "precaution rules." Precaution rules specify what precautions should be taken to avoid danger in hazardous situations.

Situations involving social contracts, threats, and precaution rules have recurred throughout human evolutionary history, and coping with them successfully constituted powerful selection pressures. An individual who cannot cooperate, cannot avoid danger, or cannot understand a threat is at a powerful selective disadvantage in comparison to those who can. More important, what counts as a violation differs for a social contract rule, a threat, and a precaution rule. Because of this difference, the same reasoning procedure cannot be successfully applied to all three situations. As a result, there cannot be a general-purpose reasoning procedure that works for all of them. If these problems are to be solved at all, they must be solved by specialized reasoning procedures. Significantly, humans do reason successfully about these problems, suggesting that natural selection has equipped the human mind with a battery of functionally specialized reasoning procedures, designed to solve specific, recurrent adaptive problems.

IV. SUMMARY

Reasoning procedures are an important part of how organisms adapt. Adaptive behavior depends on adaptive inferences to regulate decisions. Although initial approaches within psychology to the study of human reasoning uncovered many interesting phenomena, the search for a few, general rules of reasoning that would account for human-reasoning performance and explain how humans cope with the world was largely unsuccessful. The recent emergence of an evolutionary perspective within cognitive psychology has led to a different view of how inference in the human mind is organized. Instead of viewing the mind as a general-purpose computer, employing a few general principles that are applied uniformly in all contexts, an evolutionary perspective suggests that the mind consists of a larger collection of functionally specialized mechanisms, each consisting of a set of reasoning procedures designed to efficiently solve particular families of important adaptive problems. In the last decade, a growing body of research results has validated this approach, indicating that humans have specialized procedures for reasoning about such things as the motion of objects, the properties of living things, cooperation, threat, and avoiding danger.

BIBLIOGRAPHY

Carey, S. (1985). "Conceptual Change in Childhood." MIT Press, Cambridge, Massachusetts.

Cosmides, L. (1989). The logic of social exchange: Has natural selection shaped how humans reason? Studies with the Wason selection task. *Cognition* **31**, 187–276.

Cosmides, L., and Tooby, J. (1989). Evolutionary psychology and the generation of culture, part II. Case study: A computational theory of social exchange. *Ethol. Sociobiol.* **10**, 51–97.

Cosmides, L., and Tooby, J. (1992). Cognitive adaptations for social exchange. *In* "The Adapted Mind: Evolutionary Psychology and the Generation of Culture" (J. Barkow, L. Cosmides, and J. Tooby, eds.). Oxford Univ. Press, New York.

Dawkins, R. (1976). "The Selfish Gene." Oxford Univ. Press, Oxford, England.

Dawkins, R. (1986). "The Blind Watchmaker." Norton, New York.

Evans, J. St. B. T. (ed.) "Thinking and Reasoning." Routledge/Kegan Paul, London.

Gigerenzer, G., and Murray, D. (1987). "Cognition as Intuitive Statistics." Lawrence Erlbaum Associates, Hillsdale, New Jersey.

Gigerenzer, G., Hell, W., and Blank, H. (1988). Presentation and content: The use of base rates as a continuous variable. *J. Exp. Psych.: Hum. Percep. Perform.* **14**, 513–525.

Johnson-Laird, P. N. (1982). Thinking as a skill. *Quant. J. Exp. Psych.* **34A**, 1–29.

Kahneman, D., Slovic, P., and Tversky, A. (1982). "Judgment Under Uncertainty: Heuristics and Biases." Cambridge Univ. Press, Cambridge, England.

Keil, F. (1989). "Concepts, Kinds, and Cognitive Development." MIT Press, Cambridge, Massachusetts.

Manktelow, K. I., and Over, D. E. (1987). Reasoning and rationality. *Mind Lang.* **2**, 199–219.

Manktelow, K. I., and Over, D. E. (1990). Deontic thought and the selection task. *In* "Lines of Thinking" (K. J. Gilhooly, M. T. G. Keane, R. H. Logie, and G. Erdos, eds.), Vol. 1. John Wiley & Sons, New York.

Shepard, R. N. (1984). Ecological constraints on internal representation: Resonant kinematics of perceiving, imagining, thinking, and dreaming. *Psychol. Rev.* **91**, 417–447.

Wason, P., and Johnson-Laird, P. N. (1972). "Psychology of Reasoning: Structure and Content." Harvard Univ. Press, Cambridge, Massachusetts.

Williams, G. C. (1966). "Adaptation and Natural Selection." Princeton Univ. Press, Princeton, New Jersey.

Receptor Molecules, Mammalian Fertilization

PAUL M. WASSARMAN
Mount Sinai School of Medicine

GLOSSARY

Acrosome Membrane-bound, lysosome-like vesicle located in the anterior region of the head of mammalian sperm. It contains enzymes that enable motile sperm to penetrate the egg zona pellucida

Acrosome reaction Fusion of sperm outer acrosomal membrane with sperm plasma membrane at numerous sites that results in exposure of the egg extracellular coat to acrosomal enzymes and other acrosomal components associated with the sperm inner acrosomal membrane (a form of cellular exocytosis)

Cortical granules Small, membrane-bound, lysosome-like vesicles located in large numbers in the cortical cytoplasm of mammalian eggs. They contain enzymes and other components that modify the egg extracellular coat shortly after fusion of sperm and egg

Cortical reaction Fusion of egg cortical granule membrane with egg plasma membrane that results in exposure of the egg extracellular coat to cortical granule enzymes and other cortical granule components shortly after fusion of sperm and egg (a form of cellular exocytosis)

Egg-binding protein Molecule located in the sperm head plasma membrane that, together with the complementary sperm receptor on the egg, is responsible for species-specific binding of free-swimming sperm to ovulated eggs

Fertilization Fusion of haploid male and female gametes, sperm and egg, respectively, to form a single diploid cell, the zygote, which is capable of giving rise to a new individual that exhibits characteristics of the species

Polyspermy Fertilization of an egg by more than one sperm; commonly a lethal condition

Sperm receptor Molecule located in the egg extracellular coat (zona pellucida) that, together with the complementary egg-binding protein on the sperm, is responsible for species-specific binding of free-swimming sperm to unfertilized eggs

Spermatozoan (sperm) Male germ cell at the final stage of spermatogenesis that is capable of fertilizing an egg

Unfertilized egg Female germ cell at the final stage of oogenesis that is capable of being fertilized and developing into a new individual that exhibits the characteristics of the species

Zona pellucida Thick extracellular coat, consisting primarily of glycoproteins, that encompasses the plasma membrane of all mammalian eggs and harbors species-specific sperm receptors and acrosome reaction inducer

Zona reaction Remodeling of the mammalian egg zona pellucida by cortical granule components shortly after fusion of sperm and egg that establishes a barrier to polyspermic fertilization

FERTILIZATION OF EGGS BY SPERM, THE MEANS BY which sexual reproduction takes place in nearly all multicellular organisms, is fundamental to the maintenance of life. Fertilization involves the fusion of two highly differentiated germ cells, sperm and egg, into a single cell, the zygote. To maintain speciation, fertilization in both plants and animals exhibits a relatively high degree of species specificity. Such species specificity is attributable in large part to molecules located

ENCYCLOPEDIA OF HUMAN BIOLOGY, Second Edition, VOLUME 7. Copyright © 1997 by Academic Press. All rights of reproduction in any form reserved.

on the surface of male and female gametes. For example, among mammals, species-specific sperm receptors are located in the extracellular coat that encompasses unfertilized eggs. The extracellular coat restricts access of free-swimming sperm to egg plasma membrane. Only sperm that bear compatible egg-binding proteins, complementary to sperm receptors on eggs, can successfully bind to the extracellular coat and, eventually, fertilize the egg by fusing with its plasma membrane. Following fertilization of the egg by a single sperm, sperm receptors are inactivated to prevent fertilization by additional sperm, which would jeapordize normal embryonic development. In this manner, sperm receptors regulate those interactions between male and female gametes that can lead to formation of a zygote and, ultimately, to formation of a new individual that exhibits characteristics of the species.

I. MAMMALIAN FERTILIZATION AND SPERM RECEPTORS

Fertilization in animals involves the fusion of male (sperm) and female (egg) gametes to produce a single cell, the zygote, which is capable of giving rise to a new organism that expresses and maintains characteristics of the species. Various mechanisms have been devised in nature to ensure that, in general, fertilization of eggs by sperm takes place only when gametes are derived from the same (homologous) species. When fertilization of eggs from one species by sperm from a different (heterologous) species does occur (heterospecific fertilization), in most instances, either no progeny are produced or the progeny are infertile (e.g., mules, the offspring of a male donkey and a female horse). [*See* Fertilization.]

It is well established that among animals a major barrier to heterospecific fertilization exists at the level of gamete recognition. Complementary molecules located at the surface of eggs and sperm from the same species recognize one another in much the same way that antibodies recognize specific antigens. As a result, gametes from homologous species interact with one another and fertilization ensues. On the other hand, molecules located at the surface of eggs and sperm from different species do not complement one another and, in most cases, the gametes fail to interact properly and fertilization does not occur. The interaction between eggs and sperm can be likened to that between viruses and their host cells, which exhibits a relatively high degree of species specificity that is attributable to cell surface molecules.

Historically, egg surface molecules that are specifically recognized by complementary sperm surface molecules have been called sperm receptors. The term "sperm receptor" is used in much the same way that the term "virus receptor" (i.e., a molecule located on the surface of host cells) is used in virology. Complementary molecules on the surface of the sperm head have been called egg-binding proteins, as well as other names. Although other factors play a role, the interaction between sperm receptors and egg-binding proteins is largely responsible for species-specific fertilization in both animals and plants.

In general, sperm receptors are associated with envelopes that surround animal eggs and restrict access to the egg plasma membrane. For example, all mammalian eggs are surrounded by a single, relatively thick coat, the zona pellucida (ZP) (Fig. 1). The ZP contains the sperm receptors that restrict heterospecific fertilization. Removal of the ZP from mammalian eggs, thereby exposing egg plasma membrane directly to sperm, largely eliminates the barrier to heterospecific fertilization *in vitro*. For example, human sperm are unable to bind to ovulated hamster eggs *in vitro*, but can actually fertilize hamster eggs when their ZP is removed experimentally [the so-called "hamster test" in *in vitro* fertilization (IVF) clinics]. Thus, sperm receptors are essential for species-specific fertilization. Following fusion of sperm and egg, sperm receptors are inactivated upon completion of the zona reaction (known as the slow block to polyspermy) (Fig. 2). Such inactivation of sperm receptors assists in the prevention of polyspermy (i.e., fertilization by more than one sperm). [*See* Sperm.]

II. PATHWAY FOR MAMMALIAN FERTILIZATION

Of the 50 million or so mouse sperm that begin the journey through the female reproductive tract, only 100 to 200 sperm actually reach the site of ovulated eggs in the oviduct (a number similar to that in human beings). Based primarily on studies with mouse gametes, it can be concluded that the fertilization pathway for many mammalian organisms consists of several steps that occur in a compulsory order. Briefly, these steps include the following:

- Species-specific binding of acrosome-intact sperm to the ZP (Fig. 3). Binding is supported by interactions between sperm receptors and egg-binding proteins.

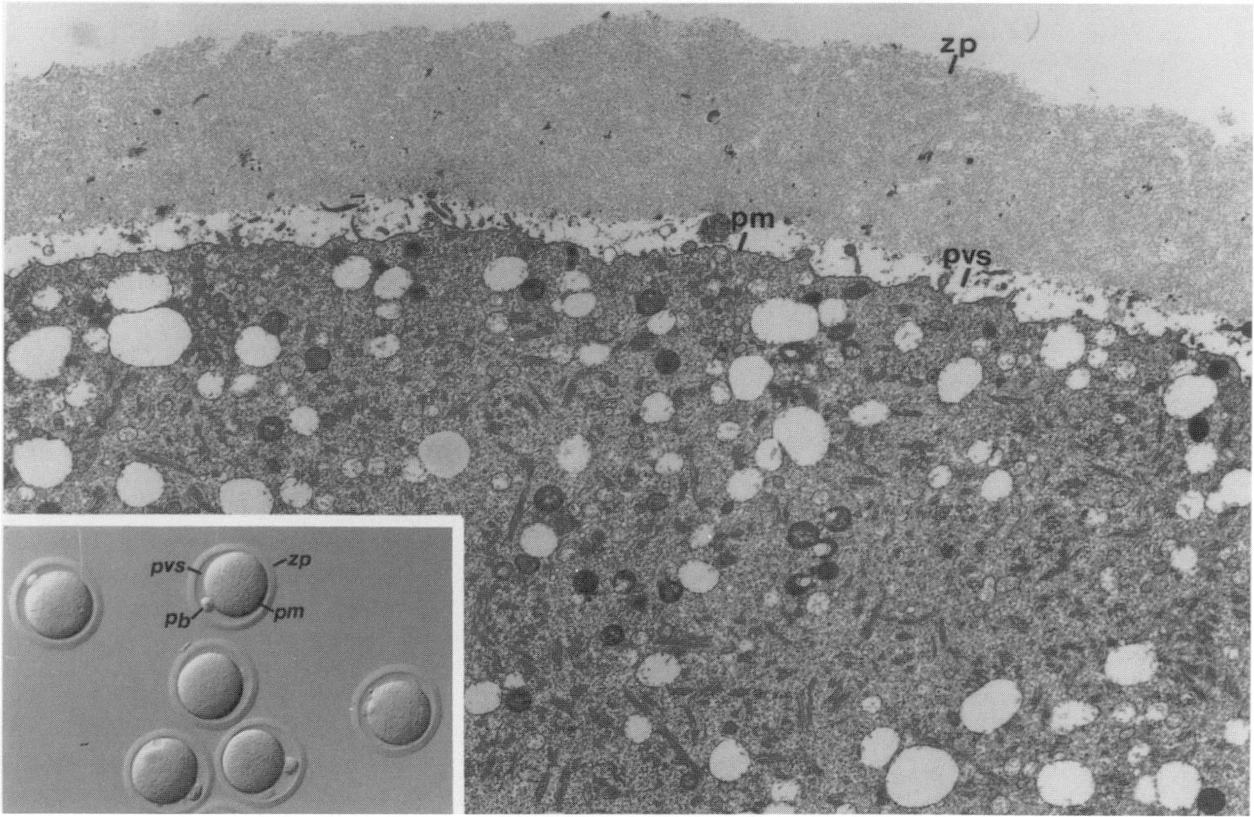

FIGURE 1 Light and electron micrographs of unfertilized (ovulated) mouse eggs. Shown are a transmission electron micrograph of a thin section taken through an unfertilized mouse egg and a light micrograph (inset) of six intact unfertilized mouse eggs. The zona pellucida is about 7 μm in thickness and eggs are about 70 μm in diameter. The light micrograph was taken using Nomarski differential interference contrast (DIC) microscopy. zp, zona pellucida; pm, plasma membrane; pvs, perivitelline space (between the zp and pm); pb, polar body.

- Completion of the acrosome reaction (AR) by sperm bound to the ZP. The mammalian AR consists of fusion of outer acrosomal and plasma membranes at the anterior region of the sperm head, with formation of hybrid membrane vesicles and exposure of inner acrosomal membrane.
- Penetration of the ZP by acrosome-reacted sperm. This may involve participation of one or more sperm proteases, as well as sperm motility.
- Fusion of a single sperm with the egg to form a zygote.
- Establishment of a fast block to polyspermy, at the egg plasma membrane, in response to sperm–egg fusion. This may involve an electrical change (i.e., depolarization/hyperpolarization) at the plasma membrane.
- Completion of the cortical reaction triggered by sperm–egg fusion.
- Completion of the zona reaction (slow block to polyspermy) triggered by cortical granule components. The fertilized egg ZP is more insoluble (harder) than the ovulated egg ZP and, in addition, is refractory to binding of free-swimming sperm (see Fig. 2).

III. TIMETABLE FOR MAMMALIAN FERTILIZATION

Fertilization of mouse eggs *in vitro* follows an approximate timetable. Binding of some sperm to eggs occurs as early as 1 to 2 minutes after combining gametes, with maximum binding observed within 10 to 20 minutes. Once bound to the ZP, some sperm undergo the AR within another 10 to 20 minutes. It takes at least 15 to 20 minutes for a bound, acrosome-reacted sperm to penetrate the ZP and reach the egg plasma

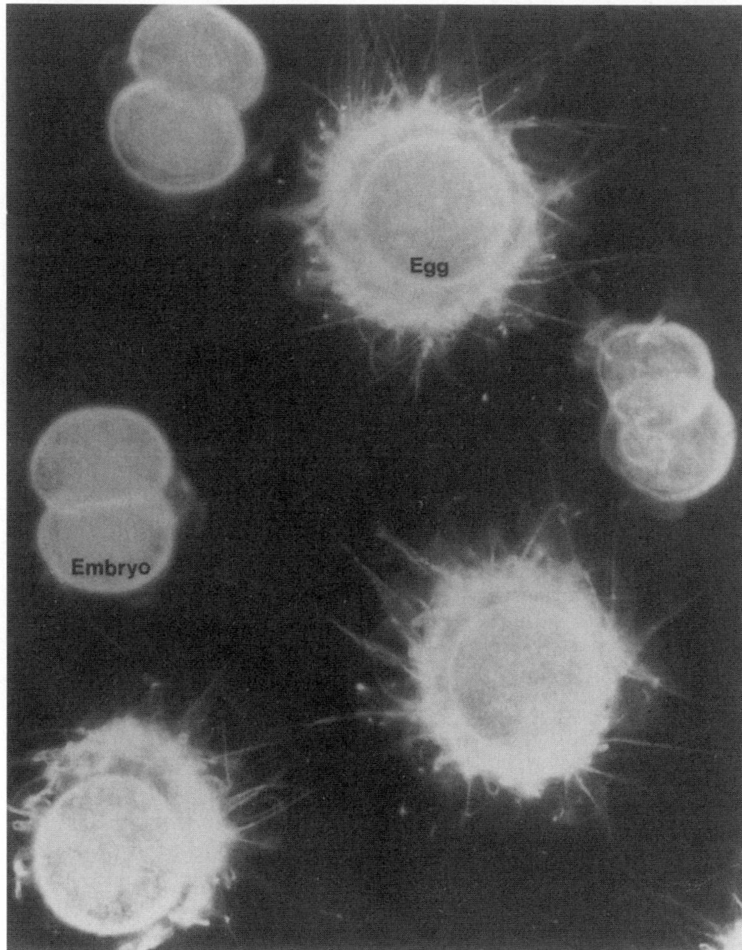

FIGURE 2 Light micrograph of mouse sperm bound to unfertilized (ovulated) mouse eggs in the presence of two-cell mouse embryos *in vitro*. Sperm bind to the ZP of unfertilized mouse eggs, but not to the ZP of two-cell mouse embryos because of inactivation of sperm receptors following fertilization. Such inactivation is part of the zona reaction (slow block to polyspermy). The micrograph was taken using dark-field optics such that the presence of a ZP is not apparent. Preimplantation development of mammals takes place within the ZP. The embryo (expanded blastocyst) hatches from the ZP at the time of implantation in the uterus.

membrane. The sperm head is incorporated into egg cytoplasm (sperm–egg fusion, or fertilization) within another 10 minutes or so. Thus, sperm enter egg cytoplasm as early as 1 to 2 hours after combining mouse gametes *in vitro*. Mouse eggs fertilized *in vitro* give rise to viable offspring following transfer of fertilized eggs to oviducts of foster mothers.

IV. MAMMALIAN SPERM RECEPTOR

The mouse sperm receptor is a glycoprotein, called ZP3, that is found in more than a billion copies in the ZP. Mouse ZP3 consists of a single polypeptide chain (402 amino acids; ~44,000 molecular weight) to which oligosaccharides (a branched or unbranched chain of covalently linked monosaccharides) are covalently linked at certain asparagine, serine, and/or threonine residues. ZP3 contains three or four asparagine-linked (N-linked) oligosaccharides. On gel electrophoresis, ZP3 migrates as a relatively broad band having an apparent molecular weight of about 83,000 (i.e., the presence of oligosaccharides increases the apparent molecular weight of ZP3 by ~40,000). The breadth of the ZP3 band observed on gels is due to the considerable heterogeneity of oligosaccharides

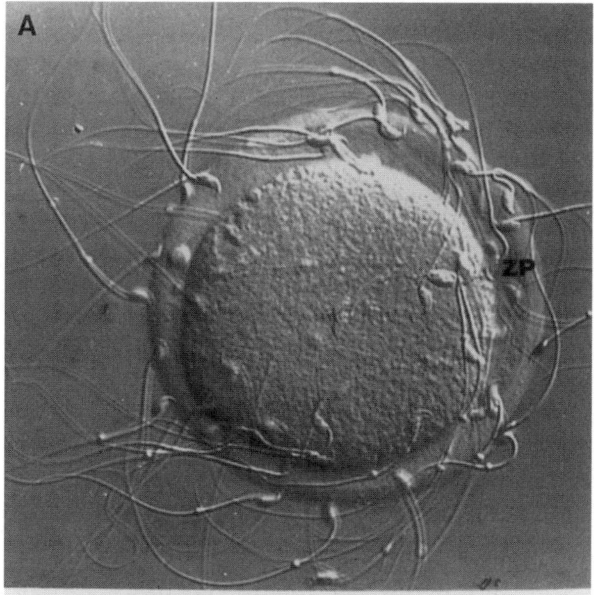

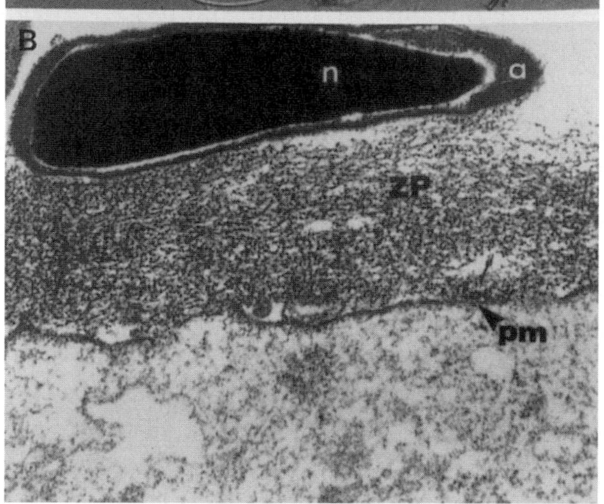

FIGURE 3 Light and electron micrographs of mouse sperm bound to an unfertilized (ovulated) mouse egg *in vitro*. (A) Light micrograph of multiple acrosome-intact sperm bound by their heads to the ZP of an unfertilized egg. Taken using Nomarski differential interference contrast (DIC) microscopy. (B) Transmission electron micrograph of a thin section taken through an acrosome-intact sperm bound by its head to the ZP of an unfertilized egg. Binding occurs between sperm receptors present in the ZP and egg-binding protein (ZP3-binding protein) present in plasma membrane overlying the head of an acrosome-intact sperm. zp, zona pellucida; pm, plasma membrane; n, nucleus; a, acrosome.

attached to the polypeptide chain. The acidic nature of ZP3 is also due to its oligosaccharides, not to its polypeptide chain.

ZP3 has a unique polypeptide chain. The mouse ZP3 polypeptide chain is encoded by a single-copy gene located on chromosome 5 (the human ZP3 gene is on chromosome 7). The transcription unit (~8.5 kilobases) consists of 8 exons and 7 introns. The 8 exons give rise to a messenger RNA with 1272 nucleotides of protein-coding sequence and 45 nucleotides of noncoding sequence. Fully processed ZP3 messenger RNA (~1.5 kilobases) is polyadenylated (~250 residues) at its 3′ terminus. The messenger RNA encodes a 424-amino-acid polypeptide chain, which includes a 22-amino-acid signal sequence that is cleaved from the amino terminus during intracellular processing of the ZP3 polypeptide chain. The ZP3 gene is expressed and ZP3 is synthesized and secreted by growing mouse oocytes, not by any other cell type; that is, ZP3 is an example of sex-specific gene expression.

Molecular probes constructed on the basis of the mouse ZP3 gene sequence, recognize, and hybridize to the analogous ZP3 gene in a variety of other mammals, including human beings (i.e., ZP3 genes have undergone conservative changes during evolution). The sperm receptor polypeptide chain is about the same size (~44,000 molecular weight) for a variety of mammals, from mice to human beings. However, the extent of glycosylation of the ZP3 polypeptide chain varies considerably among different mammals; consequently, the apparent molecular weight of ZP3 from different mammals varies considerably. The ZP3 polypeptide chain is rich in serine, threonine, and proline residues and the positions of its 13 cysteine residues are identical from mouse to human ZP3.

ZP3 has two known functions during fertilization: first, as a sperm receptor during species-specific binding of acrosome-intact sperm to eggs and, second, as an AR inducer following the binding of sperm to eggs. Each acrosome-intact sperm is capable of binding by its head to tens of thousands of ZP3 molecules located at the surface of the ZP. ZP3 is located every 15 nm or so along the filaments that make up the ZP. Interactions between egg-binding proteins, located in plasma membrane overlying the sperm head, and ZP3 in the ZP are sufficient to support gamete adhesion and to trigger the AR (see Fig. 3). Induction of the AR is dependent on the presence of external calcium and apparently requires multivalent interactions between sperm receptors and egg-binding proteins. Binding of acrosome-intact sperm to ZP3 activates sperm signal transduction processes that are analogous to those found in somatic cells.

The sperm receptor (recognition and binding) function of mouse ZP3 is attributable to the glycoprotein's oligosaccharides and, in particular, to a specific class of serine/threonine-linked (O-linked) oligosaccharides. Free-swimming sperm recognize and bind to these ZP3 oligosaccharides. In fact, synthetic oligosaccharides whose structures are related to those on ZP3 can bind to sperm and prevent the sperm from binding to eggs. Therefore, fertilization in mammals is carbohydrate-mediated. The role of carbohydrates in the initial steps of the mammalian fertilization pathway suggests that species specificity of gamete adhesion may be due to variations in the structure of oligosaccharides (e.g., composition, sequence, and conformation) present on sperm receptors of different mammals. It also suggests that inactivation of sperm receptors following fertilization (see Fig. 2) may be due to modification of the oligosaccharides, perhaps by one or more cortical granule glycosidases.

Acrosome-reacted sperm remain bound to the egg by interacting with another ZP glycoprotein, called ZP2, that is present in ZP2–ZP3 dimers along ZP filaments. Thus, ZP3 and ZP2 serve as primary and secondary sperm receptors, respectively. Only bound, acrosome-reacted sperm can penetrate the ZP, possibly by using a protease to digest a path through the extracellular coat. Following fertilization of eggs, both ZP3 and ZP2 are modified by cortical granule enzymes as a result of the cortical reaction and zona reaction. The modified ZP glycoproteins are rendered unable to perform their normal functions and free-swimming sperm are unable to bind to fertilized eggs (block to polyspermy).

V. MAMMALIAN EGG-BINDING PROTEIN

The identity of the egg-binding protein (i.e., ZP3-binding protein) located on mammalian sperm remains problematical. Several candidates, some of which interact specifically with carbohydrates, have been suggested. Currently, two of the primary candidates are the enzyme β-1, 4-galactosyltransferase (GalTase) and the sperm protein sp56.

GalTase, usually considered a Golgi enzyme, is found on the dorsal surface of the mouse sperm head plasma membrane. It is proposed that GalTase supports gamete adhesion by binding specifically to terminal *N*-acetylglucosamine residues of mouse ZP3 oligosaccharides in a lectin-like fashion. Binding, in turn, is thought to cause aggregation of GalTase in the sperm plasma membrane, which leads to activation of signal transduction processes and the AR.

sp56 is a peripheral membrane protein found on sperm head plasma membrane overlying the acrosome. It is proposed that sp56 supports gamete adhesion by binding specifically to terminal galactose residues of mouse ZP3 oligosaccharides in a lectin-like fashion. However, the amino acid sequence of the sp56 polypeptide chain does not contain regions homologous with carbohydrate recognition domains of characterized lectins. Rather, sp56 is thought to be a member of a superfamily of protein receptors that contain multiple consensus repeats, each about 60 amino acids in length, termed Sushi domains.

VI. FUSION OF MAMMALIAN SPERM AND EGGS

Once through the ZP, sperm can make contact with, adhere to, and fuse with ovulated eggs. In mice, plasma membrane over the equatorial segment of the acrosome-reacted sperm head fuses with egg plasma membrane. Apparently, this region of sperm plasma membrane is conditioned as a result of the AR, such that it is only then rendered capable of fusing with egg plasma membrane. Fusion between gametes nearly always involves egg microvillar membrane, probably because microvilli have a low radius of curvature that permits maximum apposition of sperm and egg. In general, egg plasma membrane does not appear to confer species specificity on gamete fusion and there does not appear to be a specific site on the egg surface where gamete fusion must occur. For most mammals, fusion of the sperm head with the egg is followed by incorporation of the sperm tail. Entry of the fused sperm into the egg cytoplasm may be an active process (i.e., the sperm is actively drawn into the egg) carried out by the egg contractile apparatus and microvilli.

A sperm glycoprotein, called fertilin, has been strongly implicated in mammalian gamete fusion. Fertilin is an integral membrane glycoprotein that consists of two highly homologous subunits. Certain structural features of fertilin are consistent with its proposed roles in binding of sperm to and fusion of sperm with egg plasma membrane. For example, in the former context, one fertilin subunit possesses an integrin-binding domain at its amino terminus, typical of a family of extracellular ligands of integrins, called disintegrins. The integrins constitute a family of cell adhesion receptors that participate in cell–cell interactions in a wide variety of biological systems. Since integrins are present on the surface of mammalian eggs, it is possible that the disintegrin domain of sperm fertilin binds to an egg integrin as a prelude to gamete fusion. Interestingly, the other fertilin subunit possesses a short region of polypeptide that resembles a so-called viral fusion peptide. This has led to the suggestion that binding of sperm fertilin to an egg plasma membrane integrin, through its disintegrin domain, could lead to exposure of the fertilin fusion peptide and, consequently, to gamete membrane fusion.

VII. CONCLUDING REMARKS

The mammalian sperm receptor is a specific glycoprotein (protein and carbohydrate) called ZP3 that is present in the mammalian egg extracellular coat, the ZP. ZP3 supports species-specific binding of sperm to eggs and restricts heterospecific fertilization. The protein and carbohydrate (oligosaccharide) constituents of ZP3 play roles in sperm receptor function, with carbohydrate playing a primary role in binding of free-swimming sperm to the receptor. Carbohydrates are currently thought to be intimately involved in several instances of cellular adhesion, for example, in neuronal development, lymphocyte homing, and pathogenic infection. The potential for enormous variation in oligosaccharide structure is a very appealing feature of carbohydrate-mediated cellular adhesion in general since it permits considerable fine-tuning of recognition determinants. The precise nature of the complementary mammalian egg-binding protein (ZP3-binding protein) is not completely clear. However, some of the candidates for this function bind specifically to carbohydrates and are associated with plasma membrane overlying the sperm head (e.g., GalTase and sp56). In most if not all mammals, binding to ZP3 induces sperm to undergo the AR (exocytosis), which is required for both penetration of the extracellular coat and gamete fusion. Binding to ZP3 activates signal transduction processes in sperm that lead to induction of the AR (fusion of plasma membrane and outer acrosomal membrane). Sperm receptors are inactivated following fertilization and the ZP undergoes structural rearrangements, thus assisting in the prevention of polyspermic fertilization. Fusion of mammalian gametes apparently involves the sperm glycoprotein fertilin and egg surface integrins.

BIBLIOGRAPHY

Dunbar, B. S., and O'Rand, M. G. (eds.) (1991). "A Comparative Overview of Mammalian Fertilization." Plenum, New York.

Gwatkin, R. B. L. (1977). "Fertilization Mechanisms in Man and Mammals." Plenum, New York.

Wassarman, P. M. (1988). Fertilization in mammals. *Sci. Am.* **256,** 78.

Wassarman, P. M. (ed.) (1991). "Elements of Mammalian Fertilization," Vols. 1 and 2. CRC Press, Boca Raton, Florida.

Wassarman, P. M. (1995). Towards molecular mechanisms for gamete adhesion and fusion during mammalian fertilization. *Curr. Opin. Cell Biol.* **7,** 654.

Wassarman, P. M., and Litscher, E. S. (1995). Sperm–egg recognition mechanisms in mammals. *Curr. Topics Dev. Biol.* **30,** 1.

Yanagimachi, R. (1994). Mammalian fertilization. *In* "The Physiology of Reproduction" (E. Knobil and J. D. Neill, eds.), pp. 189–317. Raven, New York.

Receptors, Biochemistry

JOHN J. MARCHALONIS
University of Arizona

GLOSSARY

Affinity Strength of binding between a ligand and the receptor combining site; it is usually represented by the association constant for the monovalent binding of one ligand to one site

Agonist Drug that mimics the stimulatory properties of the endogenous component that activates a receptor

Antagonist Drug that mimics the inhibitory capacity of endogenous inhibitors of receptor function and blocks the action of an agonist

Antibody Induced, circulating proteins that bind specifically to foreign antigens and bring about their removal and destruction

Antigen Molecule that is recognized as nonself or foreign by specific lymphocytes of the immune system

Cytokines Intercellular signaling molecules by which infrequently stimulated cells can recruit and activate others; they are involved in the control of a number of physiological processes. Cytokines produced by and affecting white blood cells are termed interleukins; those affecting lymphocytes are termed lymphokines. Receptors for these molecules are designated by the abbreviation for the cytokine followed by R for receptor

Dissociation constant (K_d) Constant that describes the equilibrium state of the reversible reaction between two molecular species A and B such that $A + B \rightleftharpoons AB$. The dissociation constant is

$$\frac{(A)(B)}{(AB)},$$

where (A) and (B) are the concentrations at equilibrium of the free reactants and (AB) is that of the complex. The units of K_d are mol/liter. It is the reciprocal of the association constant

Effector Second attribute of a receptor; following specific binding of ligand, the receptor carries out an effector function such as protein kinase activity or ion transport, which alters cell function

Immunoglobulins Family of proteins to which antibodies belong; the antigen-specific receptors of bone marrow-derived lymphocytes and thymus-derived lymphocytes are also special types of immunoglobulins

Kinase Enzyme that catalyzes the phosphorylation of proteins

Lectins Proteins that are generally of unknown function but are characterized by their capacity to bind certain sugars

Ligand Molecule that is bound specifically by a receptor; this can be a hormone, antigen, or any number of molecules that show specific binding

Receptor Cell-associated molecule that binds ligands specifically; this binding initiates effector mechanisms leading to cell stimulation or regulation

Receptor tyrosine kinase A family of cell surface receptors in which an intrinsic tyrosine kinase is part of the cytoplasic domain

A PROCESS INVOLVING A CELLULAR RECEPTOR IS one in which specific binding of a ligand to this molecule initiates a physiological response. The ligand that binds the receptor can be of either endogenous (e.g., hormones, neurotransmitters) or exogenous (e.g., antigen) origin and may consist of low molecular weight organic molecules (e.g., acetylcholine, adrenaline) or larger molecules such as the polypeptides insulin or growth hormone. Although there are examples of specific cell activation by binding of lectins or microbial

ENCYCLOPEDIA OF HUMAN BIOLOGY, Second Edition, VOLUME 7. Copyright © 1997 by Academic Press. All rights of reproduction in any form reserved.

toxins to plasma membrane glycolipids, the usual situation is one in which the receptor is a protein that shows a high degree of specificity for the particular ligand and binds it with a relatively high-binding affinity. The high-binding affinity corresponds to the fact that relatively low concentrations (nanomolar or less) of ligand such as hormones are required for cell activation. The first attribute of a receptor is its specific binding capacity; the second key attribute is the effector function carried out by the receptor after it has bound the ligand. The binding is generally reversible and can be inhibited, in the case of trace amounts of labeled ligand, by excess native ligand. Extensive literature exists on the pharmacology of drug receptors, which was developed based on demonstration of the ability of particular receptors to be stimulated by a variety of defined agonists and inhibited by a variety of defined antagonists. This article reviews pertinent data on the types, cellular distribution, and biochemical activation of major classes of receptors for neurotransmitters, hormones, and foreign antigens. Although the receptors considered here are often described in the context of their relationship to a particular cell or physiological function, they are found on many types of cells throughout the body. However, differences usually are profound in the number of binding affinity in corresponding receptors expressed on various cell types.

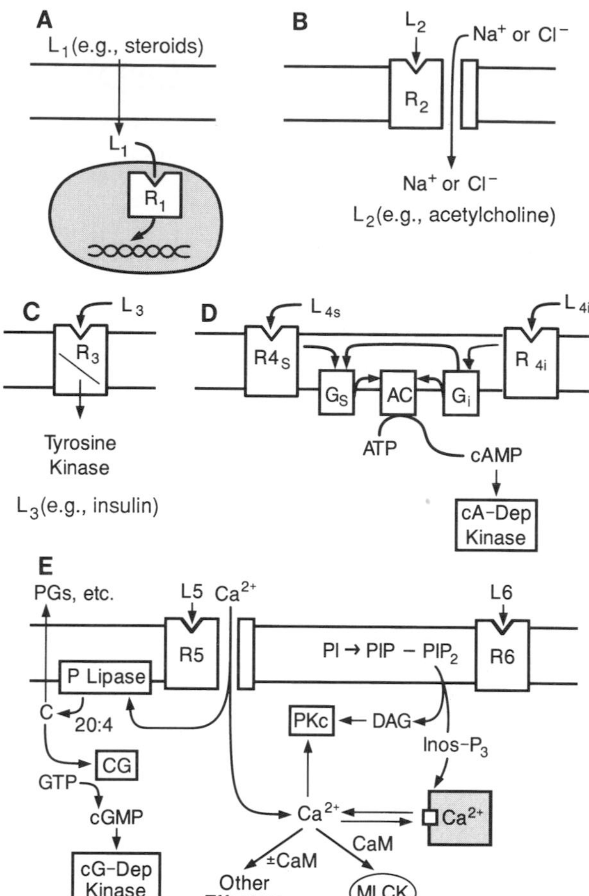

FIGURE 1 Schematic diagram of major types of receptor-mediated regulatory mechanisms. The types of receptors are designated R_1–R_6 and their corresponding ligands L_1–L_6. All of these classes of receptors can coexist on certain individual cells. The abbreviations and identification of receptors are given in the text.

I. TYPES OF RECEPTORS

Figure 1 illustrates the cellular location and models for activation of six characterized types of receptors. Intracellular receptors for steroids (e.g., thyroid hormone, estrogen, and some hormonal lipids such as vitamins A and D) are located within the cell nucleus (Fig. 1A). The hormone binds specifically to the receptor (R_1), and the complex then induces the synthesis of specific proteins by binding to nuclear chromatin and enhancing the transcription of the appropriate genes. The other receptors considered here occur on the plasma membrane of cells. Receptors for several neurotransmitters, such as acetylcholine γ-aminobutyric acid and the amino acid glycine, are either ion channels themselves or are associated with ion channels in the cell membrane. These ion channels open in response to the physiological neurotransmitters (and to its agonists), with the receptors then controlling cell membrane potential and the internal ionic composition. Receptors for several polypeptide growth factors, including insulin (Fig. 1C), are themselves ligand-stimulated, plasma membrane-bound protein kinases. Nine separate subfamilies of vertebrate cell surface receptors containing intrinsic tyrosine kinases (RTPs) in their cytoplasmic domains have been identified. The external portions of the molecules can incorporate immunoglobulin-like domains [e.g., the receptor for platelet-derived growth factor (PDGFR)], cysteine-rich domains [the receptor for epidermal growth factor (EGFR)], the fibronectin type III domain (the Eph-like receptor tyrosine kinases), and covalently linked mixtures of these domains (e.g., the Axl receptor). To date, the ligands for the EphR and AxlR receptors have not been identified. These molecules can all be considered receptors for growth factors that initiate cell division, differentiation, or maintenance of cell viability. The physiological substrates for these kinases are just being identified, but their initial phosphorylaton is localized to tyrosyl instead of to seryl or threonyl residues. The latter is the usual case with subsequent phosphorylation dependent on cyclic adenosine monophosphate (cAMP)-dependent kinases. Numerous receptors (Fig. 1D) such as β- and α-adrenergic receptors act to either stimulate or inhibit adenylate cyclase via a mechanism involving distinct guanosine triphosphate (GTP)-binding regulatory proteins (G_s or G_i, respectively). The cumulative effect of the function of these receptors is either an increase or a decrease in the intracellular concentration of cAMP. Following binding of a stimulatory ligand to $R4_S$, G_s is activated to a molecular form that can stimulate the production of adenylate cyclase by binding GTP. This binding is accelerated by the appropriate agonist–receptor ($R4_S$) complex. Deactivation occurs by the hydrolysis of GTP to guanosine diphosphate. In an analogous fashion, G_i can be activated to a form that inhibits the activation of G_s and most probably inhibits the function or activation of adenylate cyclase. The cAMP nucleotide acts within the cell to stimulate cAMP-dependent protein kinases that catalyze the phosphorylation of various enzymes and other proteins on the seryl residues.

The calcium ion (Ca^{2+}) localized to the cell cytoplasm regulates some functions directly and initiates others only when it is bound to the intracellularly Ca^{2+}-dependent regulatory protein calmodulin (CaM). The calmodulin–calcium complex directly controls the functions of some protein and effects others by activating a distinct group of protein kinases, of which myosin light-chain kinase (MLCK) is an important example. Calcium and cAMP are major intracellular second messengers, but rceptors for many hormones and neurotransmitters (E) apparently can cause the accumulation of multiple second messengers. For example, a fundamental event in some receptor-mediated systems (e.g., lymphocyte receptors for antigen) appears to be the specifically stimulated formation of an inositol-1,4,5-triphosphate ($InosP_3$) and diacylglycerol (DAG) as the consequence of the hydrolysis of a membrane phospholipid, phosphatidylinositol-4,5 by phosphate (PIP_2). The formation of INOS-P_3 is closely linked to the release of Ca^{2+} from intracellular depots. In addition to the key functions of calcium just described, this ion in the presence of DAG also activates a distinct protein kinase (kinase C). Other secondary events that occur in these activations and regulatory cascade invovle the hydrolysis of arachidonic acid ($C_{20:4}$) from membrane phospholipids by calcium-activated phospholipases with the subsequent generation of prostaglandins, prostacycline, leukotrines, and eicosanoids. The formation of these modified active lipids are oxidative events that lead further to the activation of guanylate cyclase, which results in the elevation of the intracellular concentration of guanosine 3′,5′ monophosphate [cyclic CMP (cGMP)]. This cyclic nucleotide is the activator of yet another class of protein kinases, the cG-dependent (cG-DEP) protein kinases.

A number of these receptors can occur on a single cell. Antigen-specific lymphocytes, for example, can express uniquely specific receptors for an individual antigen, receptors for insulin, adrenergic receptors, and receptors for interleukins (polypeptide molecules regulating cell activation and differentiation) simultaneously (Table I). Complex cascades of biochemical reactions can be involved in the activation of single receptors or receptor systems. Moreover, in the cases where a cell has several receptors that utilize a single effector mechanism, multiple extracellular signals may be integrated to yield a cumulative intracellular signal. For example, the submaximal stimulaton of two individual receptors that activate adenylate cyclase and of one receptor that inhibits this enzyme will be finally expressed as a unique rate of synthesis of cAMP. Moreover, receptors that act by different primary mechanisms may be coordinated at other levels (i.e., release of intracellular calcium and the activation of adenylate cyclase can lead to phosphorylation and activation of the same metabolic enzymes or of distinct enzymes with opposing or synergistic functions).

Furthermore, receptors are not the only determinants of acute regulation of physiological and biochemical function, but are themselves subject to regulatory and homeostatic control. The continued stimulation of cells with agonists generally results in a state of desensitization, also known as refractoriness or downregulation, such that subsequent exposure of the cell to the same concentration of ligand results in a markedly diminished response. A documented example is the observation that repeated use of β-adrenergic broncodilators such as isoproterenol (see Section III, A, for the treatment of asthma requires increasing dosages of this agonist). Conversely, hyperreactivity or supersensitivity to receptor agonists is also frequently observed to follow reduction in the chronic level of stimulation of receptors by a particular ligand. Analogous situations of either increased or diminished sensitivity occur following the multiple stimulation of lymphocytes via their antigen-specific

TABLE I
Example of Many Types of Receptors Found on a Single Cell Type (Lymphocytes)

Type of receptor
Receptors for antigen
B cells
Antigen-binding receptor (membrane Ig)
T cells
Antigen-binding T-cell receptor
MHC-restricted T-cell receptors (α/β heterodimers); T3/Ti complex
Triggering
B cells
B-cell growth factor
B-cell differentiation factor
Cytokines (lymphokines)
T cells
IL-2
Mitogens
Cytokines (lymphokines)
Supportive
Hormones
Insulin
Growth hormone
Steroids
Cytokines
Carrier proteins
Transferrin
Low-density lipoproteins
α_2-HS glycoprotein
Transcobalamin
Dihydroxycholecalciferol [1,25$(OH)_2D_3$]
G_c (vitamin D-binding protein)
Miscellaneous
Fc portion of IgG, IgM, IgA, and IgE
Complement component C3
Interferon
β-adrenergic
Histamine
(?) Acetylcholine

receptors. Depending on the manner of presentation of antigen and factor such as dosage, it is possible to observe either an enhanced "secondary stimulation" or a diminished reactivity termed tolerance or immune paralysis.

II. INTRACELLULAR RECEPTORS

A variety of steroid hormones exist. Estrogens, progesterones, and androgens are produced by the male and female sex organs and affect function and development of the sex organs. The adrenocorticotropic hormone stimulates the adrenal cortex, which itself generates a number of steroids that regulate either electrolyte balance (mineralocorticoids) or carbohydrate metabolism (glucocorticoids). All of the steroid hormones are thought to act by controlling the rate of synthesis of proteins by a process that is initiated by the binding of the steroids to receptor proteins in the cytoplasm of sensitive cells, where they form a steroid–receptor complex. Steroids can have profound effects on many cell types; for example, not only are cells of the gonads, placenta, and adrenal tissues sensitive to steroid hormones, but peripheral tissues such as liver, fat, skeletal muscles, and hair follicles can form steroids and are affected by them. Because some lymphocytes are exquisitely sensitive to steroids, this is a commonly used immunosuppressive regime. [*See* Steroids.]

Following binding of the steroid to its intracellular receptor, the complex undergoes a structural modification and moves into the cell nucleus where it binds to chromatin. This binding results in the transcription of mRNA corresponding to specific sets of enzymes and other proteins.

III. CELL MEMBRANE RECEPTORS FOR ENDOGENOUS LIGANDS

A. Adrenergic and Cholinergic Receptors

1. Adrenergic Receptors

Adrenergic receptors function in the physiologic control of circulation and in the response to drugs that act via the sympathetic nervous system and on the vasculature. Two general types of adrenergic receptors, α and β, each of which contains subtypes (α_1, α_2, β_1, β_2), are activated by epinephrine (adrenalin). The α- and β-receptors have opposite effects (inhibitory or stimulatory) on the adenylate cyclase system. Analysis of gene and derived protein sequence has established that the two receptors are homologous to one another, to rhodopsin, and to other receptors that are coupled to the guaninine nucleotide regulatory proteins (G proteins). The molecules contain seven hydrophobic domains, which most probably represent transmembrane-spanning segments. A schematic diagram of the adrenoceptor system is shown in Fig. 1D. Figure 2 depicts the likely extracellular, transmembrane helical and intracellular portions of α_2- and β_2-adrenergic receptors. The specificity for coupling

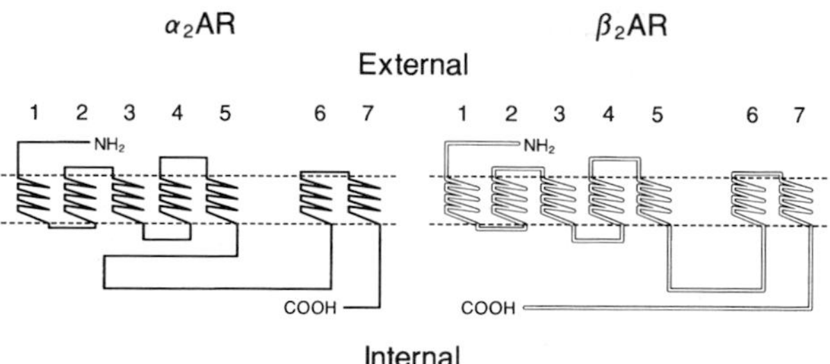

FIGURE 2 Diagram of the α_2-adrenergic receptor (α_2AR) and the β_2-receptor (β_2AR). The hydrophobic domains are depicted as helices that span the plasma membrane. The adrenergic receptors, the muscarinic, cholinergic receptors, and the opsins (e.g., rhodopsins) have a similar overall organization and most probably represent homologous molecules.

to the stimulatory G protein lies within the region extending from the amino terminus of the fifth hydrophobic domain to the carboxyl terminus of the sixth. The major determinants of the α_2- and β_2-adrenergic receptor agonist and antagonist ligand binding are restricted to the seventh membrane-spanning domain. [*See* Adrenergic and Related G Protein-Coupled Receptors.]

Adrenergic receptors are found on neurons and on muscle tissue, including skeletal and heart muscle cells, and on smooth muscle. Activation of β-adrenergic receptors brings about an increase in adenylate cyclase. The β_1- and β_2-receptors can be generally activated by agonists such as isoproterenol or by selective agonists such as norepinephrine, which stimulates β_1-adrenoceptors, and procaterol, which selectively activates the β_2-receptor. Propranolol serves as an antagonist to β-adrenoceptors. In addition to the association of these receptors with the central and autonomic nervous system, they have been found on other types of cells, including circulating lymphocytes.

2. Cholinergic Receptors

Cholinergic receptors were first detected on the basis of their sensitivity to the neurotransmitter acetylcholine. These receptors are prominent in the central nervous system, including motor neurons, hippocampus, and cerebral cortex, and in the spinal cord motor neurons and autonomic ganglia. Such receptors have also been found on cells that are neither nerve nor muscle, such as lymphocytes. These receptors are associated with ion transport channels in the plasma membrane that open in response to agonists and thereby control cellular membrane potential and ionic composition. Two general types of cholinergic receptors were classified on the basis of their reactivity with known agonists and antagonists. There is a set of acetylcholine receptors where the effect of acetylcholine is mimicked by the alkaloid muscarine and is selectively antagonized by atropine. This subset is termed muscarinic receptors. The second broad subset of acetylcholine receptors is that in which the stimulatory effects of acetylcholine are mimicked by the agonist nicotine but are not antagonized by atropine. However, the stimulatory effect of nicotine is selectively blocked by other agents such as tubocurarine or α-bungarotoxin. This subset of cholinergic receptors is termed the nicotinic receptors. Adrenergic, muscarinic, and cholinergic receptors are known to be derived from different genes, but a number of similarities exist among these proteins in terms of overall structure and function. Muscarinic cholinergic and adrenergic receptors as well as others involved in modulating cellular function via guanine nucleotide proteins (G proteins) are apparently members of a multigene family of membrane proteins that has shown considerable conservation in evolution. For example, muscarinic receptors from insects to humans show a high degree of conservation based on immunological and biochemical comparisons. The receptor monomer units consist of approximately 500 amino acids but have larger apparent masses (approximately 80 kDa) by acrylamide gel electrophoresis because of glycosylation and conformational properties. As of this time, five different genes for the muscarinic receptors have been described: m, m_2, m_3, m_4, and m_5.

B. Receptors for Insulin and Insulin-like Growth Factors

Insulin is a polypeptide with a mass of approximately 6000 Da that is made of two chains of amino acids joined by disulfide bonds. This hormone is synthesized in the pancreas. It functions as the primary regulator of blood glucose levels. Insulin acts on cells to stimulate glucose, protein, and lipid metabolism as well as to regulate RNA and DNA synthesis by modifying the activity of a variety of enzymes in transport processes. The understanding of the molecular pathways of insulin action is essential to unraveling the pathogenesis of insulin-dependent diabetes (type I) and non-insulin-dependent (type II) diabetes mellitus. This knowledge is also important in understanding other insulin-resistant states including obesity, uremia, and glucocorticoid and growth hormone excess as well as a variety of rare genetic disorders such as leprechaunism and lipotropic diabetes. Insulin-dependent cell activation and differentiation are initiated by the specific binding of insulin to its plasma membrane receptor. The insulin receptor is present on virtually all mammalian cells with the density varying from as few as 40 receptors per cell to circulating red blood cells to more than 200,000 receptors per cell on fat cells and liver cells. [*See* Insulin and Glucagon.]

The insulin receptor is a heterodimeric glycoprotein comprised of two α-subunits of approximate mass 135 kDa and two β-subunits of approximate mass 95 kDa linked by disulfide bonds to give a β-α-α-β structure. The α-subunit is completely extracellular and contains the insulin-binding site. Approximately one-third of the β-subunit is external to the cell. The β-subunit contains a membrane-spanning helical region of approximately 24 amino acid residues and an intracellular domain that is a tyrosine kinase. Binding of insulin to the external α-subunit initiates an allosteric change that stimulates tyrosine phosphorylation of the β-subunit of the insulin receptor. The insulin receptor is a member of a family of protein kinases that phosphorylate their substrates on tyrosine residues. Two groups of such kinases are related to insulin. The first group includes receptors for growth factors such as epidermal growth factor (EGF), platelet-derived growth factor (PDGF), insulin-like growth factor I, and colony-stimulating factor I. The second group of kinases homologous to the insulin kinase subunit consists of viral oncogene products and their related cellular homologues. These include V-ERB B, V-FMS, V-ABL, V-FES/FPS, V-SRC, V-ROS, and V-YES. The sequence homologies and functional similarities between the two groups of tyrosine kinases suggest that there is a structural similarity between the catalytic domains of growth factor receptors and oncogene products. Moreover, there has been considerable conservation of insulin receptor structure in evolution because the complementary DNA for insulin receptors of humans and the fruit fly *Drosophila* are extremely similar to one another. Figure 3 illustrates the cascade of biochemical mechanisms initiated by the specific binding of insulin to the insulin receptor.

Nine subfamilies of receptor tyrosine kinases resembling insulin receptors and having intrinsic tyrosine kinases in their cytoplasmic domains have been identified. These are often receptors for growth factors including epidermal growth factors (EGF; receptor designated EGFR), platelet-derived growth factors (PDGFr) which include the "steel" subgroup of cytokine receptor, and nerve growth factor (NGFr or trk) receptors.

C. Interleukins

Interleukins or cytokines are polypeptide hormones that regulate the activation and growth of cells involved in the immune system such as macrophages and lymphocytes but can also interact with a variety of nonlymphoid cells. As of this time, more than 10 cytokines have been cloned and characterized molecularly, but this article focuses on one, interleukin 2 (IL-2), because the molecule has been structurally and functionally characterized as has its specific receptor. IL-2 is a small single-chain protein of mass 15.5 kDa that stimulates thymus-derived lymphocytes (T cells) to proliferate following antigenic stimulation. In the case of lymphocytes, the binding of the ligand antigen generally does not by itself stimulate cell division, and interleukins are required to initiate this process. The density of receptors for IL-2 on resting or nonactivated lymphocytes are generally too low to be detectable. However, following activation of T cells by mitogens, for example, IL-2 receptors are demonstrable within 4–8 hr and reach a peak in the range of 30,000–60,000 receptors per cell 48–96 hr following activation. IL-2 receptors can be detected using a monoclonal antibody (anti-TAC) that recognizes the 55-kDa receptor subunit. The majority of resting T cells, bone marrow-derived lymphocytes (B cells), or monocytes do not bind the monoclonal antibody. However, following activation, the receptor becomes detectable. [*See* Cytokines in the Immune Response; Interleukin-2 and the IL-2 Receptor.]

High-affinity (K_d of approximately 10 pM) and low-affinity (K_d of approximately 10 nM, forms of the human IL-2 receptor have been identified. The receptor consists of two IL-2-binding peptides: a 55-kDa peptide that reacts with the anti-TAC monoclonal antibody and a 75-kDa IL-2-binding peptide that does not react with the anti-TAC antibody. The functional receptor probably exists as a dimer composed of one 55-kDa subunit and one 75-kDa molecule, which interact noncovalently in a cooperative manner. Both subunits have an external portion involved in the specific binding of the polypeptide hormone, a transmembrane helical portion and an intracellular domain. There is currently considerable interest investigating the pharmacology of IL-2 receptors because they are essential for the proliferation of antigen-specific activated T cells in immunity and also in malfunctions

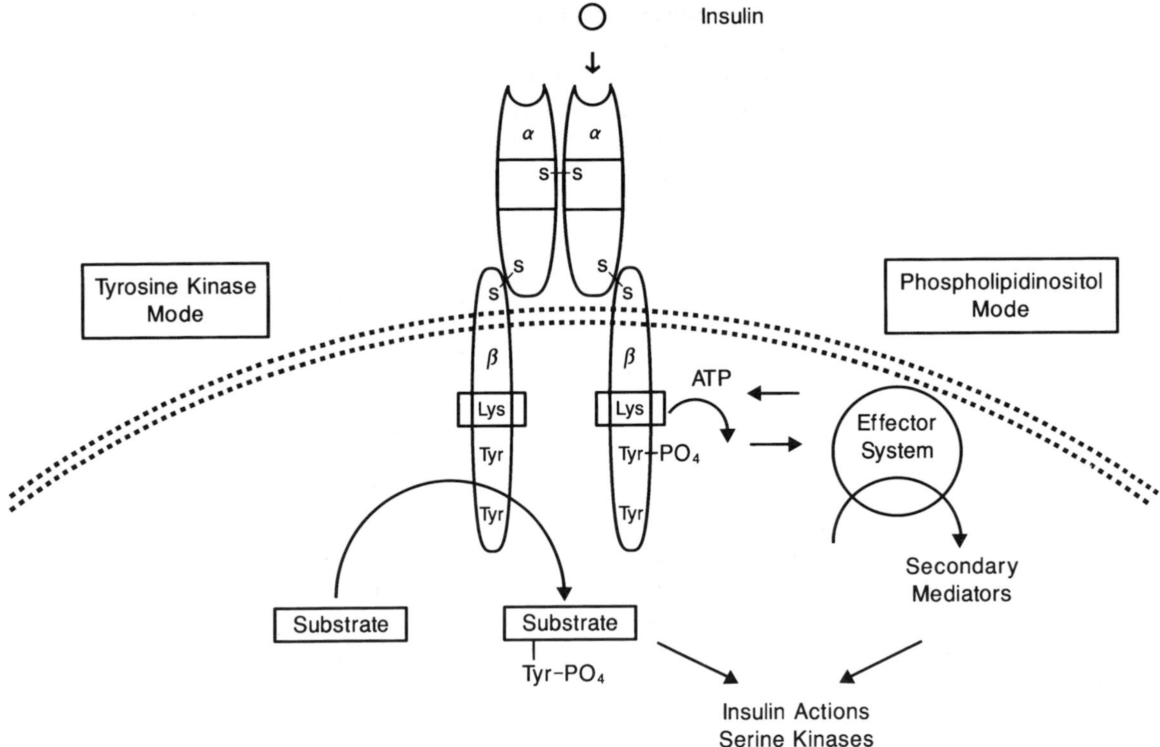

FIGURE 3 Schematic representation of the insulin receptor, its association with the cell membrane, and two general modes of its biochemical function. The binding of insulin is carried out by the α-subunits. The β-subunits are linked covalently to the α-subunits via disulfide bonds. The major portion of the β-subunit is intracellular and is tyrosine (Tyr) kinase. The first mode of activation shown is that involving the tyrosine kinase mechanism. The second mode implicates a phospholipid inositol-glycan head group, which is released in response to insulin that acts via an allosteric mechanism to activate one or more serine kinases.

of the immune system such as T-cell leukemia, where continued stimulation of some leukemic cells is maintained by an autocrine mechanism in which the cells produce both IL-2 molecules and IL-2 receptors in large quantities. As an example of potential therapeutic use, two effective immunosuppressive pharmaceuticals, glucocorticoids and cyclosporin, both mediate their effects by inhibiting the production of IL-2 antigen-activated T cells.

Although cytokines generally induce phosphorylation of protein tryrosines, the large set of cytokine receptors consists predominately of molecules that do not contain intrinsic tyrosine kinase domains. Other examples, shown later, are the cell surface immunoglobulins and T-cell receptors that likewise are not covalently associated with tyrosine kinase. The family of receptors for cytokines contains members of the hematopoietic growth factor superfamily [e.g., receptors for IL-4, IL-5, and IL-6 and granulocyte/macrophage colony-stimulating factor (GM-CSF)], G-protein-coupled receptors (e.g., the receptor for interleukin 8), and immunoglobulin subfamily domains containing RTKs (e.g., the receptor for interleukin 1).

D. Surface Carbohydrates as Receptors for Lectins

Both proteins and lipids associated with the outerface of cell plasma membranes are often glycoslated. The various glycosyl moieties can serve as targets or "receptors" for lectins, which are carbohydrate specific binding proteins found in both the serum or hemolymph and on the cells of many vertebrates and invertebrates and are also common constituent plant seeds. The binding of the lectins to cell-surface glycoproteins or glycolipids is usually of relatively low affinity (e.g., 1–6 μM for binding of concanavalin A to murine lymphocytes), but the binding of certain lectins often initiates cellular activation and differentiation mimicking that shown by interaction of specific ligand with its receptor. Insulin receptors, for example, are glycoproteins that can be activated by lectins reacting with the glycosyl moieties on the chains. A number of characterized surface proteins of human cells are glycosylated and have been shown to bind lectins, particularly concanavalin A, lentil lectin, or phytohemagglutinin to cite some of the more frequently used lectins. The concanavalin A and lentil lectins are of plant origin and have specificities for a α-D-mannose and α-D-glucose. The carbohydrate specificity of phytohemagglutinin has not been established, but this lectin will activate T cells and initiate the cascade of events normally associated with binding of IL-2 by its receptor. Biologically important membrane surface components known to bind lectins include membrane receptor immunoglobulins (IgM_m, IgD_m) of resting B cells; major histocompatibility complex (MHC) products, class I and class II; the α/β heterodimeric antigen receptor of T cells; and the insulin receptor. In addition to the glycoproteins, some glycolipids serve as targets for the binding of lectins or of specific toxins such as cholera toxin. The capacity of glycoproteins or glycolipids to bind lectins is not a specific receptor property, as described earlier, but merely reflects the fact that glycosylated membrane molecules may have the right sugar composition and sequence to combine with the lectins. The lectin molecules are generally multivalent so crosslinking of the membrane molecules occurs with a variety of biochemical consequences, comparable with that shown for specific binding of insulin to its receptor or antigen to its receptor on T cells. Approximately 10% of plasma membrane proteins of lymphocytes are glycosylated and capable of reacting specifically with a major lectin such as concanavalin A. The capacity of cells of various types to bind lectins has provided useful schemes for the fractionation of distinct types of cells and also has facilitated studies of the biochemistry of activation.

IV. LYMPHOCYTE MEMBRANE RECEPTORS FOR ANTIGEN

The receptors considered earlier are found on a number of cell types, and, with the exception of the binding of lectins, function normally in response to endogenous molecular signals. In contrast, the antigen-specific receptors of lymphocytes are restricted to immunologically committed lymphocytes and are adapted to respond to foreign molecular configurations termed antigens. This unique association only occurs with committed lymphocytes because at least two gene rearrangements involving variable (V), joining (J), and constant (C) gene segments are required for expression. [*See* Lymphocytes.]

There are two major classes of lymphocytes that express antigen-specific surface receptors. One class, termed B cells, expresses membrane-associated immunoglobulin as its surface receptor for antigen. Following stimulation and differentiation, these cells and their progeny (plasma cells) eventually produce large quantities of immunoglobulin (antibody) that is secreted into the serum. [*See* B-Cell Activation, Immunology.]

The other major class of lymphocytes, termed thymus-derived lymphocytes or T cells, does not itself produce antibodies, but recognizes antigen and serves a central regulatory role in the generation of antibodies by either helping B cells produce antibodies (helper or amplifier T cells) or suppressing the formation of antibodies (suppressor T cells). T cells usually interact with accessory cells such as macrophages in their regulatory role. This interaction requires compatibility between the cells in the MHC. T cells also carry out antigen-specific cellular immune reactions such as delayed type hypersensitivity, rejection of allografts, destruction of tumors, and elimination of virally infected cells. The capacity of T cells to recognize antigen is considered to be "MHC-restricted or -dependent," and it is suggested that T cells can recognize antigen only in the context of particular determinants of the MHC. MHC products possess a surface groove that allows them to present processes peptide antigens to the low-affinity α/β-receptors on T cells.

An antigen is a molecular structure that is normally foreign ("non/self") and is recognized by antigen-specific receptors of B cells (immunoglobulins) or T cells (α/β or γ/δ heterodimers). Although large structures, including cells and viruses, are often considered to be antigens, only a relatively small portion of their molecular architecture comprises the actual determinants recognized. Antigens can be protein, carbohydrate, or lipid, and the size of the bound determinant (epitope) corresponds to approximately six amino acids or six monosaccharide units. This molecular size corresponds to the size of the combining site of the antibody or T-cell receptor for antigen. In addition, the antigen receptors of T and B cells react to small organic molecules such as 2,4-dinitrophenol, a small molecule termed a hapten. Although the affinity of antibody following boosting or secondary stimulation for antigens such as haptens can be high (nanomolar range), the antigen receptors on T cells and on unstimulated B cells have low affinities for antigens (micromolar range). The low affinity of the T-cell antigen receptor is most probably compensated for by the presentation of certain antigens, such as peptides to helper T cells by the MHC. The number of membrane immunoglobulin antigen receptors on B (IgM_m or IgD_{2m}) or T cells (α/β or γ/δ heterodimers) is in the range of 10,000–100,000 molecules/cell. [*See* T-Cell Receptors.]

A simplified diagram illustrating the properties of antigen receptors of B and T lymphocytes is given in Fig. 4. As depicted in the figure, the surface receptor of B cells is essentially an immunoglobulin that has a combining site and light polypeptide chains identical to those of the secreted antibody. In this particular case, the B-cell receptor is membrane IgM immunoglobulin and the secreted antibody is IgG. IgG molecules appear on the surface of memory B cells (those that are the progeny of cells previously stimulated by antigen and expressing the same specificity) and are the major class of the circulating antibody. Membrane IgM is one of the two major antigen receptor classes on B cells that have not been previously stimulated by antigen. The short transmembrane piece found on the membrane form of immunoglobulin is lacking on the secreted molecules. The antigen receptor on T lymphocytes is a member of the immunoglobulin family and contains a combining site for antigen that is formed by the interaction of variable regions of the α- and β- or γ- and δ-receptors. The α/β or γ/δ T-cell receptors are disulfide bond linked heterodimers that occur in association with a complex of molecules termed the T3 complex, which is itself not antigen specific and appears to be necessary for transmission of a signal to the cell interior. The affinity for binding a foreign antigen by the IgM, IgD, or TCR receptors is generally quite low (micromolecular range), but the affinity of IgG type receptors on memory B cells and circulating antibodies can be substantially higher. In the case of T-cell recognition of antigen, this is usually considered to be restrictd by the MHC because MHC molecules can bind peptide determinants and present these determinants to the low-affinity TCR α/β-receptor.

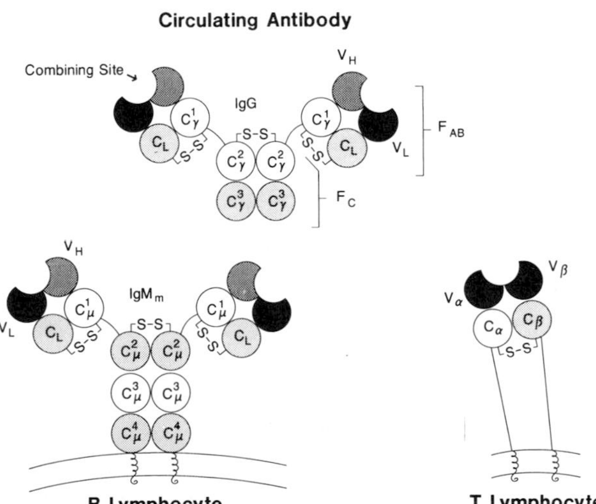

FIGURE 4 Diagram comparing the antigen-specific receptors of B and T lymphocytes with the major class of circulating antibody in humans. The combining site for antigen in all cases is formed via the interaction of the variable (V) domains of heavy and light chains in the case of antibody and B-lymphocyte receptors and V_α and V_β in the case of the α/β T-lymphocyte heterodimeric receptor. The α/β heterodimer depicted here does not include the accessory T3 complex with which is it invariably associated. F_{AB}, antigen binding fragments; F_C, common or crystalizable fragments.

Binding of antigen by the surface immunoglobulin receptors of B cells is in itself not sufficient to activate the B cells. Generally, helper T-cell activity is required in the stimulation and differentiation of B cells. At least two types of mediators known as B-cell stimulating factor and B-cell differentiation factor are required for activation and differentiation to IgG secreting plasma cells (see Table I). These factors are interleukins resembling the IL-2 molecule described earlier. Likewise, for T lymphocytes an antigen-specific response is initiated by the binding of antigen by the T-cell receptor. This process can also be initiated by the binding of molecules that mimic the bonafide ligand such as antibodies to T-cell receptor determinants or mitogenic lectins. Furthermore, monoclonal antibodies directed against the T3 complex can activate T cells in a manner comparable with that observed for antigen or anti-idiotypic antibodies that react directly with the combining site region of the receptor.

Upon binding of ligand to the T-cell receptor, second messengers (Fig. 5) are generated by the phosphatidylinositol (PI) pathway that is initiated by the phosphodiesteratic cleavage by phospholipase C of three phosphoinositides associated with the plasma membrane: PI, diphosphoinositide (PIP), and triphosphoinositide (PIP_2). A cleavage product of each of these phosphoinositides, diacygycerol (DAG), activates a protein kinase (C kinase) that is distinct from the cAMP- or cGMP-activated protein kinases. In addition to its requirement for DAG, protein kinase C

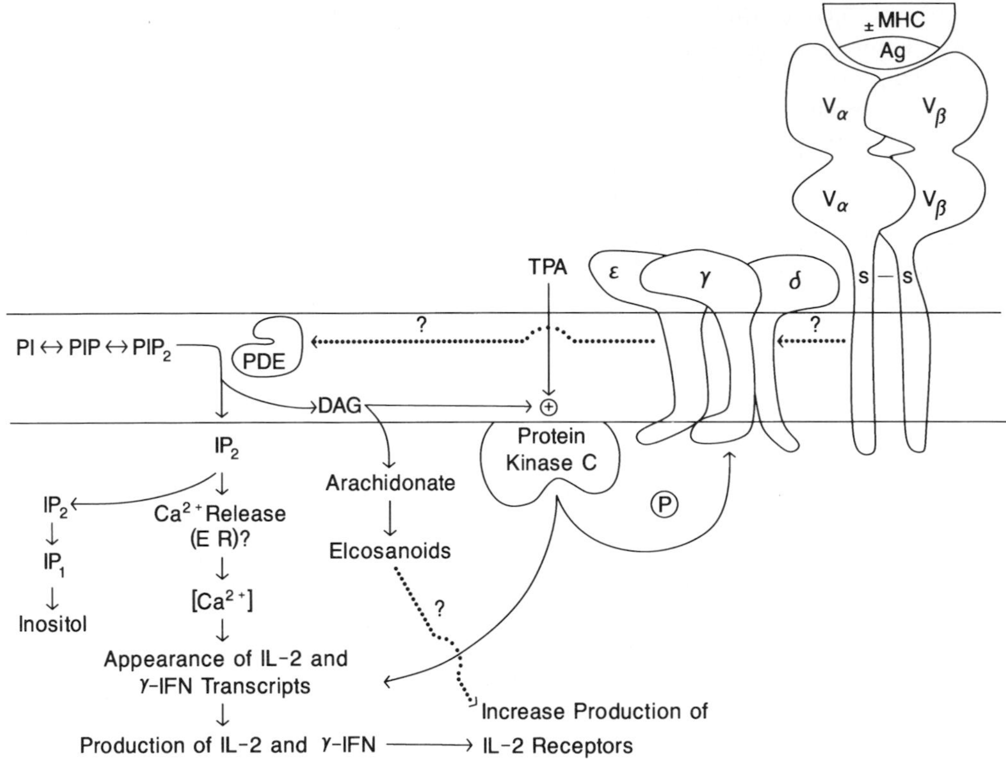

FIGURE 5 Membrane expression of the disulfide-bonded α/β heterodimer existing in association with the T3 complex (δ, γ, ε). This scheme illustrates the membrane association and the processes of transmembrane signaling initiated by binding of either antigen plus MHC to the combining site of the α/β heterodimer (V_α/V_β) or by antibodies against the T3 complex. A detailed description is given in the text. Although the biochemical mechanisms elicited by binding of antigen to the T-cell receptor complex are similar to those associated with the insulin receptor or with the scheme given for receptor 6 in Fig. 1, the rearranging α/β- or γ/δ-receptors are found only in T cells.

requires phospholipids and Ca^{2+} for maximum activity; an increase in cytosolic Ca^{2+} is the result of mobilization of the ion from endoplasmic reticulum as mediated by inositol-1,4,5-triphosphate (IP_3), which is an immediate breakdown produce of PIP_2. The DAG branch and IP branch of the phosphoinositide signal cascade appeared to act synergistically on kinase C to phosphorylate proteins that control the activation process. The protein kinase C phosphorylates the γ-subunit of the T3 complex, but no functional attribute of this phosphorylation has been determined other than it serving as a prerequisite for activation.

The physiological consequences of T-cell activation are blastogenesis, occurring within 48 hr, in which the small resting T lymphocyte is transformed into an immature blast form, the secretion of lymphokines including IL-2 and interferon-γ, and increased production of receptors for IL-2. The overexpression of IL-1 receptors coupled with IL-2 production by the same cell enables it to grow in an "autocrine" manner and allows cell division to take place, resulting in a clonal expansion of the original cell type.

BIBLIOGRAPHY

Ciardelli, T., and Smith, K. A. (1989). Interleukin-2: Prototype for a new generation of immunoactive pharmaceuticals. *Trends Pharmacol. Sci.* **10**, 239–243.

Debets, R., and Savelkoul, H. F. J. (1994). Cytokine antagonists and their potential therapeutic use. *Immunol. Today* **15**(10), 455–458.

Fantal, W. J., Johnson, D. E., and Williams, L. T. (1993). Signaling by receptor tyrosine kinases. *Annu. Rev. Biochem.* **62,** 453–481.

Frielle, T., Kobilka, A. B., Lefkowitz, R. J., and Caron, M. G. (1988). Human β-1 and β-2 adrenergic receptors: Structurally and functionally related receptors derived from distinct genes. *Trends Neurosci.* **11,** 321–324.

Kahn, C. R., and White, M. F. (1988). The insulin receptor and the molecular mechanism of insulin action. *J. Clin. Invest.* **82,** 115–116.

Kerlavage, A. R., Fraser, C. M., Chung, F. Z., and Venter, J. C. (1986). Molecular structure and evolution of adrenergic and cholinergic receptors. *Proteins: Struc. Func. Genet.* **1,** 287–301.

Kobilka, B. K., Kobilka, T. S., Daniel, K., Regan, J. W., Caron, M. G., and Lefkowitz, R. J. (1988). Chimeric α-2, β-2 adrenergic receptors: Delineation of domains involved in effector coupling and ligand binding specificity. *Science* **240,** 1310–1316.

Marchalonis, J. J. (1988). "The Lymphocyte: Structure and Function," 2nd Ed. Dekker, New York.

Marchalonis, J. J., and Schluter, S. F. (1989). Evolution of variable and constant domains and joining segments of rearranging immunoglobulins. *FASEB J.* **3,** 2469–2479.

Raffioni, S., and Bradshaw, R. (1993). The receptors for nerve growth factor and other neurotrophins. *Annu. Rev. Biochem.* **62,** 823–850.

Ross, E. M., and Gilman, A. G. (1985). Pharmacodynamics: Mechanisms of drug action and the relationship between drug concentration and effect. *In* "The Pharmacological Basis of Therapeutics" (Goodman, Gilman *et al.*, eds.), pp. 35–48. Macmillan, New York.

Strader, C. D., Fong, T. M., Tota, M. R., and Underwood, D. (1994). Structure and function of G protein-coupled receptors. *Annu. Rev. Biochem.* **63,** 301–332.

Waldmann, T. A. (1989). Multichain interleukin-2 receptor: A target for immunotherapy in lymphoma. *J. Natl. Cancer Inst.* **81,** 914–923.

Zick, Y. (1989). The insulin receptor: Structure and function. *Crit. Rev. Biochem. Mol. Biol.* **24,** 217–269.

Red Cell Calcium

ROBERT M. BOOKCHIN
Albert Einstein College of Medicine, Yeshiva University

VIRGILIO L. LEW
Physiological Laboratory, University of Cambridge

GLOSSARY

$[Ca^{2+}]_i$ The intracellular (cytoplasmic) concentration of Ca^{2+}, the free ionized (unbound) form of calcium

MAMMALIAN RED CELLS HAVE THE LOWEST TOTAL calcium content of any cell in the body. After loosely bound calcium is washed from their outer membrane surface, human red cells contain less than 5 μmol of calcium/liter cells. This very low total calcium content reflects the absence of any significant calcium-accumulating compartments and calcium-binding structures found within other cells, whose total calcium content is about 100 to 500 times that of mature mammalian red cells.

I. STATES OF CALCIUM

Calcium is found in a variety of physical states within organisms. It can be found precipitated with phosphate-rich compounds in all sorts of hard structures, such as bones, teeth, and shells; it may be tightly or loosely bound to a variety of organic molecules, which are either freely soluble or part of a cell or intercellular structures, or it may exist in free ionized form (Ca^{2+}) within intracellular or extracellular compartments. A dynamic equilibrium exists among these various forms of calcium, with all transfers between the different physical states occurring through the water-soluble, ionized form. The chemical versatility of calcium, whereby it readily forms insoluble salts or reversible high-affinity bonds with organic molecules, allows it to mediate such diverse biological functions as forming skeletons or transmitting transient signals to intracellular targets. It seems a bit ironic that red cells, with no known Ca^{2+}-mediated signaling, exhibit many of the functional features crucial to the signaling process, so that studies of these features in the readily available red cells have provided clues to help understand the messenger functions.

The strategy by which Ca^{2+} transmits messages to intracellular targets involves three distinct processes: (1) the maintenance (and eventual restoration) of large Ca^{2+} gradients across a membrane (this, figuratively, puts the explosive powder in the gun); (2) the opening and closing of Ca^{2+} channels by primary chemical or electrical signals, which causes transient increases in $[Ca^{2+}]_i$ (equivalent to pulling the trigger and converting the potential chemical energy of the gun powder into kinetic energy of the bullet); and (3) the reaction of Ca^{2+} with a target molecule (the bullet hits the target). It is on the first of these processes that work with red cells provided fundamental information.

II. PASSIVE AND ACTIVE Ca^{2+} TRANSPORT IN RED CELLS

A. Calcium Distribution and Permeability

The concentration of Ca^{2+} in the cytoplasm of normal human red cells, $[Ca^{2+}]_i$, was found to be about 100

ENCYCLOPEDIA OF HUMAN BIOLOGY, Second Edition, VOLUME 7. Copyright © 1997 by Academic Press. All rights of reproduction in any form reserved.

n*M*. It is precisely this minute cytoplasmic Ca^{2+} concentration (a feature shared by all cells, with minor variations) that is crucial for the role of Ca^{2+} as a signaling agent. The cells are surrounded by fluids containing at least a hundred thousand fold higher Ca^{2+} concentrations ($\approx$1–1.2 m*M* in mammals). The true inward Ca^{2+} gradients are even steeper than suggested by the concentration differences alone because all cells maintain large electrical potential differences across their membranes, with the inside of the cell negative in relation to the outer surface. The combined electrical and chemical inward Ca^{2+} gradient therefore represents a formidable force. If the cells were permeable to Ca^{2+}, this force would induce a large Ca^{2+} inflow, rapidly dissipating the gradient.

Experiments with calcium tracers (^{45}Ca), and additional experiments demonstrating a requirement of external Ca^{2+} for contractile and secretory cell functions, showed that all cells have a finite permeability to calcium. The red cell is among the least permeable, with an inward passive calcium flow of about 50 μmol/liter cells per hour under physiological conditions. Because human red cells have a normal life span of about 120 days in circulation, such a calcium influx, if unopposed, would raise $[Ca^{2+}]_i$ to over several thousand times the normal value in less than a day, but measurements show little variation in $[Ca^{2+}]_i$ throughout the life of a cell.

B. The Ca^{2+} Pump

To sustain such a huge Ca^{2+} gradient throughout the red cell life span, when the normal "leaks" would tend to dissipate the gradient in a fraction of that time, human red cells utilize a powerful Ca^{2+} pump which derives energy from the hydrolysis of adenosine triphosphate (ATP) to extrude Ca^{2+}. This pump, which is present in the plasma membrane of most animal cells, is part of a family of ATP-fueled pumps that transport sodium, potassium, protons, and calcium ions across biological membranes against electrochemical gradients.

The plasma membrane Ca^{2+} pump is an integral membrane protein with a molecular mass of about 140 kDa. It transports one calcium ion out per molecule of ATP hydrolyzed. An important feature of this pump is that its activity may be regulated by a variety of factors, and we are still learning how this regulation operates in physiological conditions. The main regulatory interactions known to alter the chemical and functional configuration of the pump involve calmodulin, a ubiquitous calcium-binding cytoplasmic polypeptide; calpain, a Ca^{2+}-dependent protease capable of removing enzyme fragments from its C-terminal cytoplasmic domain; a variety of enzymes that attach or remove phosphate groups to specific reactive groups; and possibly the process of self-association into dimers.

In human red cells, the calcium pump is the sole active transporter of calcium ions against their gradient. It is interesting to note, however, that one other such calcium transporter is a sodium : calcium exchanger that uses the energy stored in the normally large inward sodium gradient to transport Ca^{2+} out of the cell. This transporter is found only in the plasma membrane of excitable animal cells, with one exception thus far: in dog red cells, a sodium : calcium exchanger is present that operates as a sodium pump, as the inward Ca^{2+} gradient maintained by the ATP-fueled Ca^{2+} pump is steeper than the sodium gradient. All cells have a tendency to swell, which results from excess impermeant solutes relative to their environment. Unlike plant cells, whose cell walls can sustain hydrostatic pressures, soft-walled animal cells require a mechanism to pump out sodium—a job generally performed by an ATP-fueled sodium pump. In dog red cells, which lack the sodium pump entirely, the combination of a Ca^{2+} pump and a sodium : calcium exchanger thus represents an interesting alternative sodium extrusion strategy compared to that of other animal cells.

Experimentally, divalent cation-selective ionophores can be used to increase the Ca^{2+} permeability of intact red cells in a controlled manner, to load red cells with Ca^{2+}, and to study the properties of the Ca^{2+} pump and the Ca^{2+}-binding components of the cytoplasm. Of the total calcium thus loaded into the cells, 15–30% remains in free ionized form, whereas the remainder is bound to the cytoplasmic calcium buffers. Ca^{2+} activation of the pump is sigmoidal with a very high Ca^{2+} affinity. When saturated, the pump can extrude Ca^{2+} at rates that vary widely between 5 and 25 mmole/liter cells per hour.

III. EFFECTS OF ELEVATED RED CELL $[Ca^{2+}]_i$

A. General Effects

In view of the very low $[Ca^{2+}]_i$ level in normal human red cells, which appears to be rather constant, questions arise about the consequences of any increases in this level, either in special physiological circumstances or in pathological states. Most *in vitro* studies have examined the effects of raising red cell $[Ca^{2+}]_i$ by permeabilizing the cells with the divalent cation ionophore A23187 at various levels of $[Ca^{2+}]_o$. These effects include (i) reduction or depletion of ATP due to increased activation of the $[Ca^{2+}+Mg^{2+}]$ATPase, with accumulation of inosine monophosphate (IMP) in calcium-equilibrated cells, associated with irreversible ATP depletion; (ii) K^+ permeabilization due to activation of the Ca^{2+}-dependent potassium channel; in plasma-like low potassium media, this is accompanied by net KCl and water efflux, and red cell dehydration; (iii) inhibition of the sodium pump by more than 10 μM $[Ca^{2+}]_i$; and (iv) activation of proteases (calpain) or of a red cell transglutaminase capable of cross-linking membrane proteins by $[Ca^{2+}]_i$ levels over 50 μM.

B. Increased Red Cell Calcium in Sickle Cell Anemia

An increased total calcium content has been described in several types of red cell disorders associated with abnormal hemoglobins. Among these, the most extensively studied is sickle cell anemia (SS), which occurs in persons homozygous for the mutant gene directing synthesis of the abnormal hemoglobin S. This example, which illustrates some of the complexities of the distribution and effects of intracellular calcium, even in such a structurally simple-appearing, enucleate cell as the mature human red cell, will be discussed here in some detail.

The fundamental abnormality in SS red cells is the predominance of hemoglobin S. When the SS cells are deoxygenated, the hemoglobin S polymerizes into bundles of long rod-like chains, which produce rigid projections in the cells and often endow it with a curved "sickle" shape. As a result, the cells lose their normal deformability, become mechanically fragile, and tend to be destroyed rapidly in the circulation and to occlude small blood vessels.

Upon reoxygenation in the circulation, hemoglobin S depolymerizes and resumes normal solubility, and the cells "unsickle." A portion of the SS red cells tend to become dehydrated, however, which increases their rigidity and their propensity to sickle. In many of these dense, dehydrated SS cells the cell membrane becomes fixed in a relatively rigid, elongated shape, even when the cells are oxygenated and the hemoglobin is not polymerized; these are the "irreversibly sickled cells" (ISC). [*See* Sickle Cell Hemoglobin.]

Sickle cell anemia cells were found to have an increased calcium content averaging about 50 μmol/liter of red cells, with as much as 200 μmol/liter of red cells or more in the dense, ISC-rich fraction, and to show a temporary increase in calcium permeability (of about four- or fivefold) during deoxygenation-induced sickling. These findings initially prompted the view that all of the effects described earlier in red cells loaded with micromolar levels of Ca^{2+} *in vitro,* using the ionophore, might occur in SS cells, thereby accounting directly for the features of ISCs. Detailed investigations of SS cells, however, revealed none of the expected effects of a continuously high $[Ca^{2+}]_i$. Using methods developed to measure free cytoplasmic Ca^{2+} accurately at low nanomolar levels, it was shown that $[Ca^{2+}]_i$ in oxygenated SS red cells was normal or only minimally increased. Other studies demonstrated that virtually all of the excess (micromolar) calcium in SS cells is compartmentalized in intracellular membrane vesicles with inwardly directed calcium pumps ("inside-out vesicles"), which are largest and most numerous in the ISCs. These vesicles would tend to accumulate calcium during periods of increased calcium influx associated with deoxygenation and sickling, but that accumulated calcium is not accessible to the pump or to the cytoplasmic side of the cell membrane.

Among the effects of experimentally increased $[Ca^{2+}]_i$ described earlier, one proved crucial in explaining the mechanisms by which sickle cells dehydrate. The K^+ channel is activated by submicromolar levels of Ca^{2+}. During sickling-induced calcium permeabilization, small transient increases in $[Ca^{2+}]_i$ are sufficient to activate the Ca^{2+}-dependent K^+ channel in some SS red cells, which causes a net loss of K^+, Cl^-, and water from the deoxygenated cells, accompanied by slight cell acidification (as the increased transmembrane Cl^- ratio is matched by an opposite H^+ ratio). This in turn causes secondary activation of another potassium transporter, the acid-sensitive K : Cl cotransport, which is highly expressed in many reticulocytes and young red cells. Acidification stimulates the K : Cl cotransport, inducing further KCl loss and acidification in a positive feedback cycle. Thus, young cells with a high expression of this cotransport would dehydrate by a fast-track irreversible path, generating a substantial fraction of the ISCs. These findings explained an old puzzle about the young average age of ISCs, compared with normal or dense discocytes, which express little or no K : Cl cotransport activity.

C. Increased Red Cell Calcium in Nonsickling Disorders

An increased total calcium content has been reported in certain other congenitally abnormal red cells, including homozygous hemoglobin C disease (another structurally abnormal hemoglobin which tends to crystallize in the red cells) and β-thalassemia (in which there is deficient synthesis of one of the two polypeptide chains making up the hemoglobin tetramers, the β chains, leaving an excess of α hemoglobin chains, which are unstable when not combined with the β chains). In these two disorders, red cell $[Ca^{2+}]_i$ was measured and, as in sickle cells, was found to be normal or minimally increased. The origin of the increased total calcium and any possible role it may play in the red cell pathology of these disorders have not yet been established.

BIBLIOGRAPHY

Bookchin, R. M., and Lew, V. L. (1983). Red cell membrane abnormalities in sickle cell anemia. *Prog. Hematol.* **13**, 1–23.

Bookchin, R. M., Ortiz, O. E., and Lew, V. L. (1986). Red cell calcium transport and mechanisms of dehydration in sickle cell anemia. *In* "Approaches to the Therapy of Sickle Cell Anaemia" (Y. Beuzard, S. Charache, and F. Galacteros, eds.), Vol. 141. Colloque INSERM, Paris.

Bookchin, R. M., Ortiz, O. E., and Lew, V. L. (1991). Evidence for a direct reticulocyte origin of dense red cells in sickle cell anemia. *J. Clin. Invest.* **87**, 113–124.

Etzion, Z., Tiffert, T., Bookchin, R. M., and Lew, V. L. (1993). Effects of deoxygenation on active and passive Ca^{2+} transport and on the cytoplasmic Ca^{2+} buffering of sickle cell anemia red cells. *J. Clin. Invest.* **92**, 2489–2498.

Lew, V. L., and Garcia-Sancho, J. (1987). Measurement and control of intracellular calcium in intact red cells. *In* "Methods in Enzymology (S. Fleischer and B. Fleischer, eds.), Vol. 125. Academic Press, New York.

Lew, V. L., Hockaday, A., Sepulveda, M. I., Somlyo, A. P., Somlyo, A. V., Ortiz, O. E., and Bookchin, R. M. (1985). Compartmentalization of sickle cell calcium in endocytic inside-out vesicles. *Nature* **315**, 586–589.

Lew, V. L., Tsien, R. Y., Miner, C., and Bookchin, R. M. (1982). The physiological $[Ca^{2+}]$ level and pump-leak turnover in intact red cells measured with the use of an incorporated Ca chelator. *Nature* **298**, 478–481.

Rega, A. F., and Garrahan, P. J. (1986). "The Ca^{2+} Pump of Plasma Membranes." CRC Press, Boca Raton, FL.

Schatzmann, H. J. (1982). The plasma-membrane calcium pump of erythrocytes and other animal cells. *In* "Membrane Calcium Transport" (E. Carafoli, ed.). Academic Press, London.

Red Cell Membrane

STEPHEN B. SHOHET
University of California, San Francisco

GLOSSARY

Band 2.1 (ankyrin or syndein) Globular protein that serves to tie the spectrin network to the bilayer through a high-affinity binding reaction with the intrinsic membrane protein, band 3

Band 3 Important intrinsic membrane protein that spans the bilayer. It contains at least one of the major ion channels (that for anions) and probably has permeability functions in the membrane. Its cytoplasmic segment is connected to the membrane skeleton via band 2.1 and is one of the two major sites for the skeleton

Band 4.1 Globular protein that serves to connect spectrin and red cell actin in the membrane skeleton, and that by a second binding reaction with the intrinsic membrane protein, glycophorin, which is a second site, serves to tie the skeleton to bilayer

Cholesterol Neutral lipid, based on the polycyclic benzanthracene nucleus, that is a major structural component of the membrane bilayer. At the molecular level it fits closely with phosphatide fatty acids in the bilayer and probably provides much of the membrane aqueous barrier function

Intrinsic membrane proteins Proteins that are difficult to extract with detergents. They interact comparatively strongly with the phosphatides and cholesterol of the lipid bilayer. They all have hydrophobic regions and hence, at least partially, are found in the hydrophobic core of the membrane bilayer

Membrane bilayer Double array of structurally bifunctional lipid molecules typically found in the plasma membranes of cells. Comparatively water-soluble portions of the lipids distribute to the aqueous interfaces on the outside and inside of the cell, whereas the water-insoluble fatty acid tails and polycyclic hydrocarbons distribute to a central hydrophobic core in this structure. The bilayer lipids primarily provide a barrier function for the cell. Hydrophobic proteins, which are often also inserted in the bilayer, provide other specialized transport, enzymatic, and recognition functions for the membrane

Membrane skeleton Horizontally extended anastomotic network of extrinsic membrane proteins that lies directly underneath the membrane bilayer and is coupled to its intrinsic proteins. It is essential for bilayer support and membrane integrity, and probably has important roles in membrane deformability and the regulation of cell shape

Peripheral membrane proteins Proteins that are comparatively easily extracted from the membrane and solubilized with nonionic detergents. They are predominantly found in the membrane skeleton

Phosphatide Lipid made up of esterified fatty acids and a phosphorus-linked base on a glycerol backbone. Together with cholesterol, phosphatides are the major lipid components of the membrane

Spectrin Predominant protein component of membrane skeleton. This elongated, high-molecular-weight, fibrillar protein serves as the major structural element in the membrane skeleton network. Its imaginative name reflects its origin from the red cell ghost or specter

RED CELL MEMBRANE IS THE STRUCTURE COMposed of lipids and proteins that separates the interior of the cell from the plasma in which it is suspended. Inherent in its basic role of confining the interior contents of the cell, the membrane is responsible for the structural integrity and many of the physical properties of the cell as well as its shape. In addition, the membrane has important ancillary functions in transporting molecules in and out of the cell and as the site of immunological interactions with other cells and the plasma.

ENCYCLOPEDIA OF HUMAN BIOLOGY, Second Edition, VOLUME 7. Copyright © 1997 by Academic Press. All rights of reproduction in any form reserved.

I. INTRODUCTION

Ever since Antonie van Leeuwenhoek's original observations in the seventeenth century, it has been known that the hemoglobin circulating in the blood is contained in small discrete particles that he called "globules." Subsequent observations rapidly established that these more or less uniform packages of hemoglobin were surrounded by a "delicate pellicle," or skin, which served to prevent the random diffusion of the hemoglobin, thus providing the essential characteristic of cellularity to these "globules." This pellicle, of course, is what is now called the cell membrane. In addition to its primordial function of packaging the hemoglobin into small, highly concentrated, and easily transported elements, this membrane has several important secondary functions and characteristics.

First, the membrane serves as a semipermeable barrier that, while effectively containing the hemoglobin, still allows the passage of smaller electrolytes, carbohydrates, and peptides to provide nutritional and excretory pathways for the cell. Complementary to this, the membrane also contains various "pumps" and enzyme systems designed to facilitate the accumulation or discharge of certain smaller molecules against concentration gradients.

Second, the membrane provides the cell with a unique biconcave configuration that maximizes the efficiency of oxygen transport in this small circulating bag of concentrated hemoglobin. Perhaps derivative to this function, the membrane also contains enzyme systems to repair and remodel itself, and to modulate the asymmetry of its own components.

Third, and in support of its cardinal role in dividing the hemoglobin mass into stable cells, the membrane is wholly responsible for the cell's extraordinary strength and much of its remarkable deformability. These two characteristics are essential to enable the cell to circulate for 120 days and to traverse over 175 miles of capillaries during its lifetime. In contrast to many other cell types, there is no internal structural apparatus beyond the membrane in the red cell to assist in maintaining these characteristics.

Finally, certain proteins contained within the membrane have portions exposed on its external surface that identify the cell immunologically and hence have a role in making the cell a target for various pathological immunological conditions. These membrane proteins may also have an analogous role in directing the normal senescent turnover of red cells by possible immunological recognition mechanisms.

Again starting with Leeuwenhoek, generations of investigators have recognized that the red cell is the most easily biopsied tissue in humans, and that its isolation from the other elements in the blood is fortuitously straightforward. Further, hypoosmotic treatment of red cells causes them to swell and then burst, releasing their hemoglobin; hence, virtually pure membrane can be isolated by simple washing and centrifugation. The ease of obtaining red cells from living subjects, together with the ease of isolating membranes from those cells, has made this membrane extraordinarily valuable as an archetypal model for mammalian cells in general. Indeed, a great deal of what has been learned about membrane structure and function in the red cell has been promptly and directly applied to the plasma membranes of much more complicated cells such as those from kidney, brain, heart, and fat, among many other tissues.

The remainder of this article will be devoted to a brief discussion of the composition and biochemical "anatomy" of the red cell membrane and some selective comments about some preliminary observations of a few apparently typical abnormalities of the membrane in various disease states. When possible an effort will be made to indicate how the composition and disposition of the membrane components may be related to the various membrane functions that have just been mentioned, and the analogous dysfunctions that occur in disease.

II. MEMBRANE LIPIDS

All the lipids in the mature cell are contained within the membrane and are partially responsible for many of its physical characteristics. For example, both the passive cation permeability and the mechanical flexibility of the red cell can be influenced by modifying the lipid composition of its membrane. [*See* Lipids.]

A. Lipid Composition

Approximately one-half of the mass of the human red cell membrane consists of lipids, largely arranged as a bilayer.

Phospholipids and nonesterified cholesterol account for more than 95% of the total lipids. Small amounts of glycolipids, glycerides, and free fatty acids are also present. On a molar basis, the phospholipids and cholesterol are present in nearly equal amounts, and there is evidence that considerable interaction may occur between these major lipid classes within the membrane (e.g., cholesterol "condenses" and stabilizes bimolecular phosphatide leaflets).

B. Phospholipids

The phospholipids are divided into subclasses distinguished, with the exception of sphingomyelin, by the base group that is in phosphodiester linkage to the third carbon atom of their glycerol backbone (Fig. 1). Usually, there are two esterified fatty acids on the 1- and 2-positions of the glycerol backbone, although vinyl ether linkages occur to a substantial extent on the 2-position in some acidic phosphatides, such as phosphatidylethanolamine. The major phospholipids, which are usually named after their bases, and their approximate concentration in human erythrocytes are as follows: phosphatidylcholine (PC), 30%; phosphatidylethanolamine (PE), 28%; phosphatidylserine (PS), 14%; and sphingomyelin (SM), 25%.

Sphingomyelin is distinct structurally and probably dynamically from the rest of the group and may have a more structural role in the membrane. Small, but perhaps physiologically significant, amounts of phosphatidylinositides, phosphatidic acid, and polyglycerol phosphatides are also present in red cells. There are characteristic patterns of esterified fatty acids within each phospholipid class, and these also serve to distinguish subgroups.

Phospholipids containing only one fatty acid are known as lysophosphatides (see Fig. 1). The absence of one of the acyl (fatty acid) groups profoundly influences the physical characteristics of the phospholipid. Although the phospholipids with two fatty acids are highly lipophilic, the lysophosphatide compounds are nearly balanced in terms of lipophilic and hydrophilic characteristics. They tend to concentrate, therefore, at lipid–aqueous phase interfaces. This change in relative solubility increases both their detergent qualities and their rate of exchange between the cell membrane and the plasma. Because of these properties, lysophosphatides in low concentrations (2×10^{-4} M) can cause the lysis of red cell membranes—hence their trivial name. In even smaller concentrations, they can produce profound, eventually irreversible, shape changes ("echinocytogenesis") in the membrane (Figs. 2a–2d). These morphological distortions may be due to small changes in the relative packing density of the phospholipids in the inner and outer leaflets, which operate as a "bilayer couple" with consequent bending of the membrane, analogous to the bending of the bimetallic strip in a thermostat (Fig. 3). Red cell membranes usually contain only small amounts of lysophosphatides and the cell has reacylation and dismutation enzyme systems to convert lysophosphatides back to phosphatides if their concentration begins to rise excessively. These reactions, together with the slow but steady exchange of several important membrane lipids with those of the plasma, also serve to maintain and renew much of the lipid of the membrane. This may account for the remarkable durability of this anuclear but much traveled cell. These lipid renewal reactions are schematically summarized in Fig. 4.

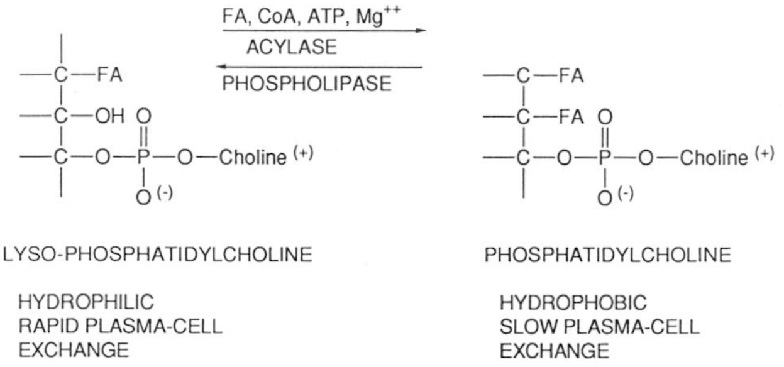

FA, CoA, ATP, Mg^{++} → ACYLASE
← PHOSPHOLIPASE

—C—FA
—C—OH O
—C—O—P—O—Choline (+)
O (-)

—C—FA
—C—FA O
—C—O—P—O—Choline (+)
O (-)

LYSO-PHOSPHATIDYLCHOLINE
HYDROPHILIC
RAPID PLASMA-CELL EXCHANGE

PHOSPHATIDYLCHOLINE
HYDROPHOBIC
SLOW PLASMA-CELL EXCHANGE

FIGURE 1 The acylation reaction basic to red cell membrane lipid renewal. A lysophosphatide is esterified with a free fatty acid to produce a complex phosphatide with profoundly different physical properties. FA, fatty acid; CoA, coenzyme A.

C. Lipid Disposition

Although many structural relationships have been proposed, the precise anatomical localization of all of the lipids within the membranes is still unknown. There is, however, little doubt that a large percentage of the lipid is arrayed in the form of a biomolecular leaflet. In this bilayer disposition, the polar head groups of each lipid layer face away from the center of the membrane into the hydrophilic environments of the cytoplasm and the plasma, while the long acyl tails of the lipids form a central hydrophobic core for the membrane. This hydrophobic core is in a liquid–crystalline state at normal temperatures and may facilitate the physiologically essential flexibility and deformability of the red cell membrane. It has been proposed that many of the protein elements of the membrane are inserted into this lipid matrix in much the same way as icebergs float in the ocean (Fig. 5). In this model, some of the proteins and glycoproteins are confined to one leaflet, whereas others, especially those assumed to have transport and shape-mediating roles, span the entire membrane. In addition, both the inserted proteins and the lipids are relatively free

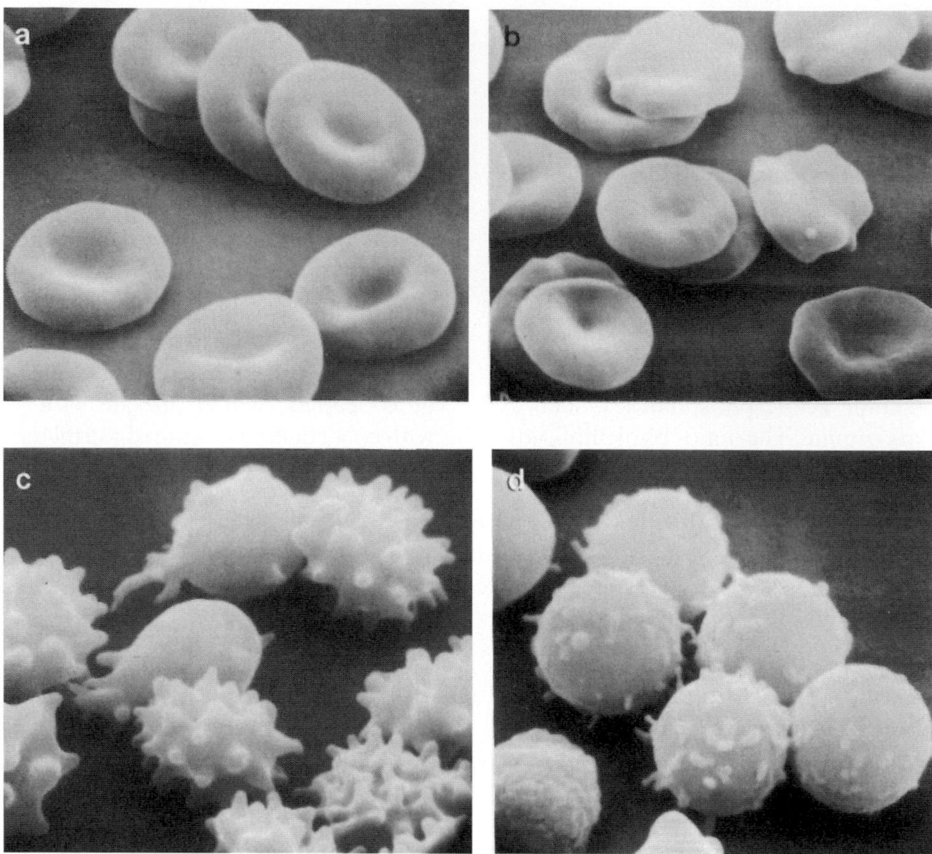

FIGURE 2 Scanning electron micrographs of erythrocytes subjected to increasing concentrations of membrane lysophosphatidylcholine (LPC). (a) Membrane LPC, 0.12 μmol/cm^3 cells; (b) membrane LPC, 0.15 μmol/cm^3 cells; (c) membrane LPC, 0.30 μmol/cm^3 cells; (d) membrane LPC, 0.50 μmol/cm^3 cells. The bulk of the changes in (a) to (c) could be reversed by washing the cells with defatted albumin. The changes seen in the red cells in (d) are irreversible and represent, in part, membrane loss due to microvesiculation induced by the high LPC concentration.

to move laterally within the plane of the membrane at comparatively rapid rates. In contrast, motions across the bilayer from one leaflet to the other are much more restricted. Since it is reasonable to assume that the lipid composition partially determines membrane viscosity, any significant changes in lipid composition that affect the membrane's internal microviscosity (e.g., an increase in cholesterol) might be expected to have some effect on the flexibility of the whole cell. Indeed, in some variants of liver disease, where a distortion of the normal exchange pathway markedly elevates red cell membrane cholesterol, stiff and spiculated "spur" cells are produced that circulate poorly and anemia may occur. However, usually the protein constituents of the membrane are predominant in regulating cell deformability, and calculations of the physical forces involved suggest that lipid–protein interactions must be much more important than pure lipid effects. Further, membrane flexibility is not the only, or even the predominant, requirement for whole-cell deformability. The cell must avoid reductions in surface area and increases in intracellular viscosity to maintain optimal deformability. In the latter case, it is likely that the physical characteristics of the membrane are subordinate to its permeability and water regulation characteristics in maintaining cell deformability.

D. Lipid Asymmetry

The lipids are not symmetrically distributed between the inner and outer leaflets of the membrane. It appears that the majority of phosphatidylethanolamine and phosphatidylserine is contained within the inner or cytoplasmic leaflet of the membrane, whereas the

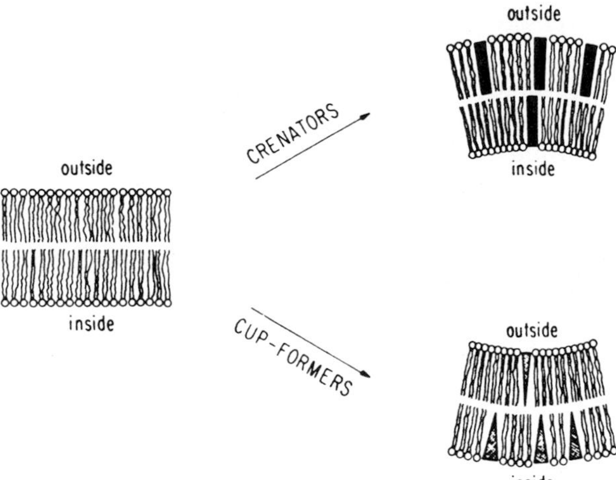

FIGURE 3 The hypothetical treatment of the red cell membrane as a "bilayer couple." Amphipathic compounds, including various drugs and phospholipids, can intercalate preferentially into either the inner or outer leaflet of the membrane because of individual characteristics (e.g., net charge at a given pH). Those that preferentially inserted into the outer leaflet crowd that leaflet, creating a bending moment that tends to induce evagination of the membrane, or crenation; conversely, those that concentrate in the inner leaflet induce invagination of the membrane, or cup formation.

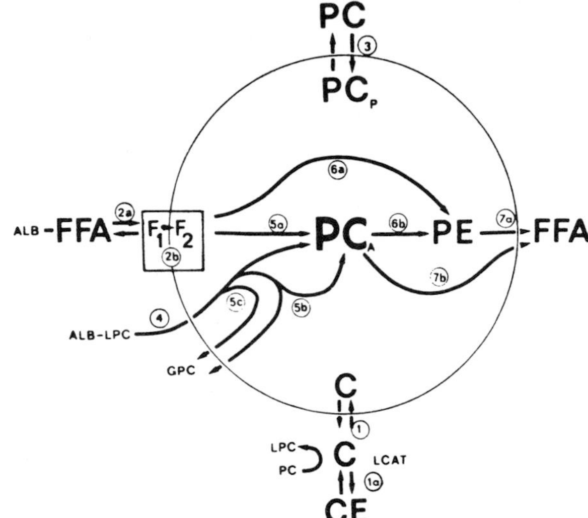

FIGURE 4 Schema of the major exchange and metabolism pathways for lipids in the mature erythrocyte. Alb-FFA, albumin-bound free fatty acid; F_1, membrane surface pool of fatty acid that is freely exchangeable with plasma FFA; F_2, "deeper" membrane pool of free fatty acid used as a source of acyl groups for phosphatides within the membrane; PC_A, phosphatidylcholine actively synthesized within the membrane; PC_P, phosphatidylcholine passively acquired by exchange with the plasma by the membrane; PE, phosphatidylethanolamine; Alb-LPC, albumin-bound lysophosphatidylcholine, which, together with Alb-FFA, serves as the precursor for PC_A; GPC, glycerolphosphorylcholine; C, cholesterol, which exchanges with the membrane; CE, cholesterol ester, which does not exchange with the membrane; LCAT, lecithin cholesterol acyltransferase, which catalyzes the formation of cholesterol ester and hence modulates the rate of the cholesterol exchange.

majority of phosphatidylcholine and the sphingomyelin is contained in the leaflet facing the plasma. The biochemical basis of this asymmetry may be the combined result of some site specificity of the phospholipid exchange and renewal reactions and the sluggish lipid exchange rates between the inner and outer membrane leaflets. In addition, the action of a recently described phospholipid "flipase," which preferentially translocates acid phosphatides from the outer to the inner leaflet of the bilayer, is likely to be involved in maintaining this asymmetry. Although the full extent of the physiological consequences of this asymmetry is not known, at the very least considerable transmembrane charge potential is induced by the excess of positive charges on the inner leaflet. This asymmetry is apparently disturbed in some abnormal states (e.g., sickled and irreversibly sickled cells), and the speculation has been made that such externalization of acidic red cell phospholipids may serve to activate the coagulation system and pathological thromboses. The asymmetric distribution of the typical intramembranous particles seen on freeze-fracture electron microscopy of the cleaved membrane may also be a consequence of lipid asymmetry.

Another potential consequence of lipid asymmetry may involve coupling of the membrane bilayer to an important and novel structure, the membrane skele-

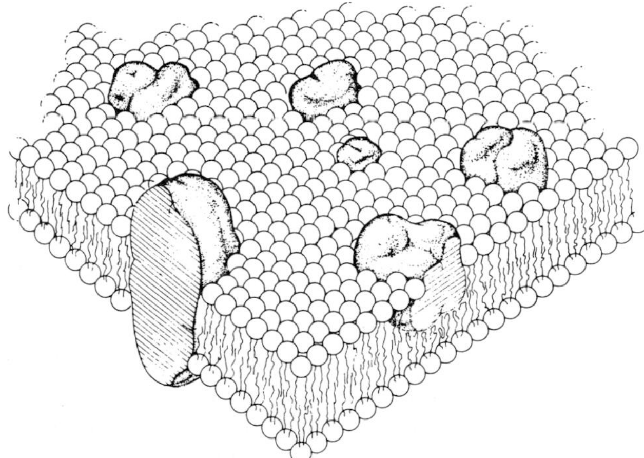

FIGURE 5 A schematic representation of the Singer–Nicholson fluid mosaic model for the structure of cell membranes. Irregular proteins are seen penetrating both into and through the bimolecular leaflet composed of regular arrays of phospholipid molecules. The head groups of the phospholipids face the cytoplasmic and plasma environments, while their acyl tails are enmeshed to form the lipophilic membrane core.

ton. This structure, which will be discussed in detail in Section III, consists primarily of spectrin, actin, and proteins 2.1 (ankyrin) and 4.1. It appears to serve as a scaffolding for the lipid bilayer and to have a dominating role in influencing membrane stability and deformability. Although elegant studies have established specific protein interactions between this membrane "skeleton" and protein components embedded in the lipid bilayer, such as band 3 and glycophorin, spectrin : phosphatidylethanolamine and spectrin : phosphatidylserine interactions have also been observed, as well as band 4.1 : phosphatidylserine interactions. However, it is not clear if the membrane phospholipid asymmetry that places these acidic phospholipids in intimate contact with spectrim at the interface between the inner leaflet of the lipid bilayer and the membrane skeleton is a cause or a consequence of such interactions.

Finally, the predominant location of small quantities of variously phosphorylated phosphatidylinositides in the inner leaflet may be important for the regulation of the affinity of the skeletal protein band 4.1 in red cells, and for their role in the generation of "second messengers" in the membranes of other cells.

III. MEMBRANE PROTEINS AND THE MEMBRANE SKELETON

A. Analysis

Study of red cell membrane proteins was initially difficult because of their insolubility in aqueous media of physiological ionic strength. Indeed, any membrane proteins that did not have such properties would be lost as membranes were washed in the course of preparing hemoglobin-free ghosts. It is possible, however, to dissolve red cell membranes completely in sodium dodecyl sulfate and to accurately analyze protein subunits varying in size from the larger subunit of spectrin (M_r 240,000) to traces of globin monomers (M_r 16,000) with great sensitivity. The study of red cell membrane proteins has been appreciably advanced by this technique and by the uniform numbering of such polypeptides, which most investigators now use (Fig. 6A).

A variety of stratagems have been employed to deduce the relative positions of the various protein subunits in the membrane. For example, it has been possible to determine whether or not a human protein is exposed at the outer or the inner surface of the membrane by labeling membrane preparations on the inside or the outside with radioactive iodine using lactoperoxidase. Also, cross-linking studies have been employed to determine which proteins may be regarded as "neighbors" in the intact membrane. The fact that spectrin and actin (bands 1, 2, and 5) and band 4.1 play a major role in maintaining the shape of the erythrocyte may be deduced from the fact that ghosts of various poikilocytes (abnormally shaped cells) extracted with nonionic detergents such as Triton X-100 maintain their initial poikilocytic shape. These components of the membrane are sometimes designated as the "extrinsic" membrane proteins because they can be released from the membrane by treatment with very low ionic strength, slightly alkaline solutions without disrupting the lipid bilayer. In contrast, the "intrinsic" proteins of the membrane, those that are embedded in the lipid bilayer, are removed only by detergent treatment.

B. Protein Disposition and the Membrane Skeleton

Through these types of studies, the concept of a "membrane skeleton" of proteins that may have a role in modulating cell shape and deformability has emerged. This structure is composed of the extrinsic proteins that are coupled to the bilayer via the intrinsic proteins. In electron micrographs, this skeleton appears to be a moderately dense meshwork of intimately interconnected proteins immediately subjacent to the lipid bilayer. Recent elegant preparations of osmotically stretched or partially extracted membranes show an underlying pattern to this structure, which probably depicts the actual spectrin–actin–band 4.1 connections in a hexagonally arrayed network (Fig. 7). Of course, it should be recognized that these preparations are not native cells, and the treatments that were required for the clarity of these remarkable micrographs may themselves have modified the arrangement of this structure. Nevertheless, for the present, they offer an effective insight into the nature of the skeleton. Whatever the exact biochemical anatomy of this structure, it seems to be clear that it is immediately beneath the bilayer, that it is anchored to the bilayer by at least one and probably two integral membrane proteins (band 3 and glycophorin C), and that it supports the bilayer and acts as an essential scaffolding for that otherwise weak and ephemeral structure.

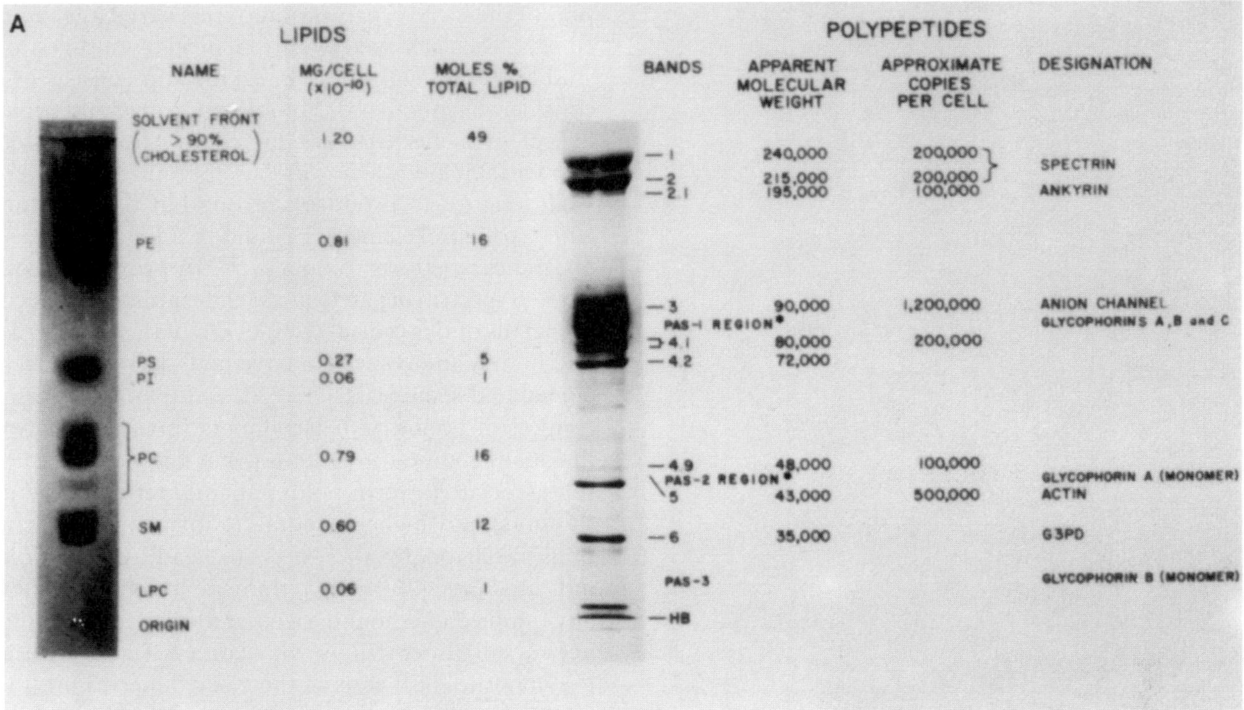

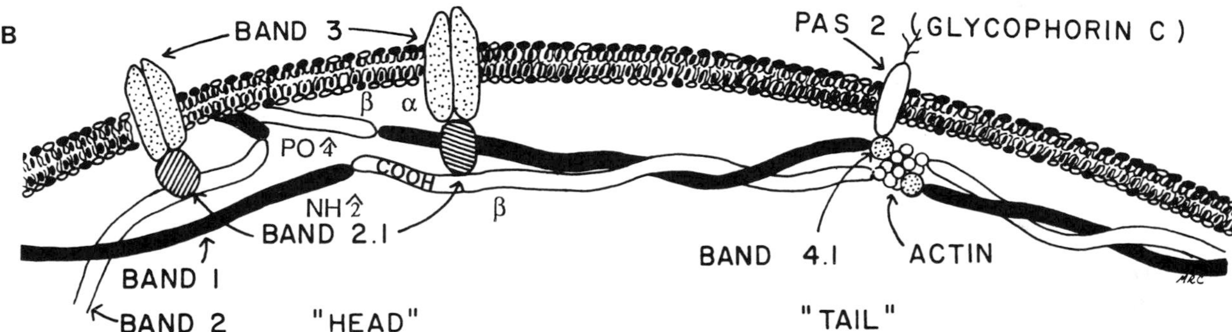

FIGURE 6 Current concepts of red cell membrane composition and organization. (A) Thin-layer chromatogram of red cell membrane lipids (left) and a Coomassie Blue-stained SDS polyacrylamide gel electrophoretogram of red cell membrane proteins (right). (B) Diagrammatic cross section of membrane bilayer and supporting "skeleton." The predominant protein of the membrane, spectrin, occurs as a heterodimer (bands 1 and 2) linked together into a fibrous network. The linkage between the "tail" ends of the dimers appears to be mediated by actin (band 5) and band 4.1. Linkage between the "head" ends of the dimers occurs by direct contact between complementary strands of the heterodimer (the carboxy terminus of the β chain and the amino terminus of the α chain). Attachment of the skeleton to the membrane is produced by a specific association between band 2 of spectrin and band 3 in the lipid bilayer via the spectrin-binding protein, band 2.1 (ankyrin), near the head end of the spectrin dimer. An additional association of the skeletal complex with the lipid bilayer may be provided by a connection between spectrin and another bilayer protein, glycophorin C or A, via band 4.1. Bands 2, 2.1, 3, and 4.1 can be phosphorylated, and some of that phosphorylation is cyclic AMP dependent. The phosphorylation sites of spectrin band 2, which are not shown in this diagram, appear to be close to the carboxy end of the dimer. The outer leaflet of the bilayer is composed predominantly of choline-containing phospholipids (indicated by black head groups), and the inner leaflet is predominantly composed of acidic phospholipids, such as phosphatidylethanolamine and phosphatidylserine (indicated by white head groups). Cholesterol (indicated by black ovals) is shown embedded symmetrically in each leaflet among the fatty acid side groups of the phospholipid, although this has not yet been experimentally verified. The PAS1 and PAS2 bands have been better defined by additional biochemical studies, including gel electrophoresis techniques, with improved resolution. It is now known that the PAS1 region contains the dimer of the sialoglycopeptide glycophorin A and that the PAS2 region resolves into three bands: the dimer of glycophorin B (M_r 47,000), the monomer of glycophorin A (M_r 38,000), and glycophorin C, also called glycoconnectin (M_r 35,000). Unfortunately, there is not, as of yet, a universally accepted nomenclature for these PAS-staining sialoglycopeptides. PE, phosphatidylethanolamine; PS, phosphatidylserine; PI, phosphatidylinositol; PC, phosphatidylcholine; SM, sphingomyelin; LPC, lysophosphatidylcholine; G3PD, glucose-3-phosphate dehydrogenase.

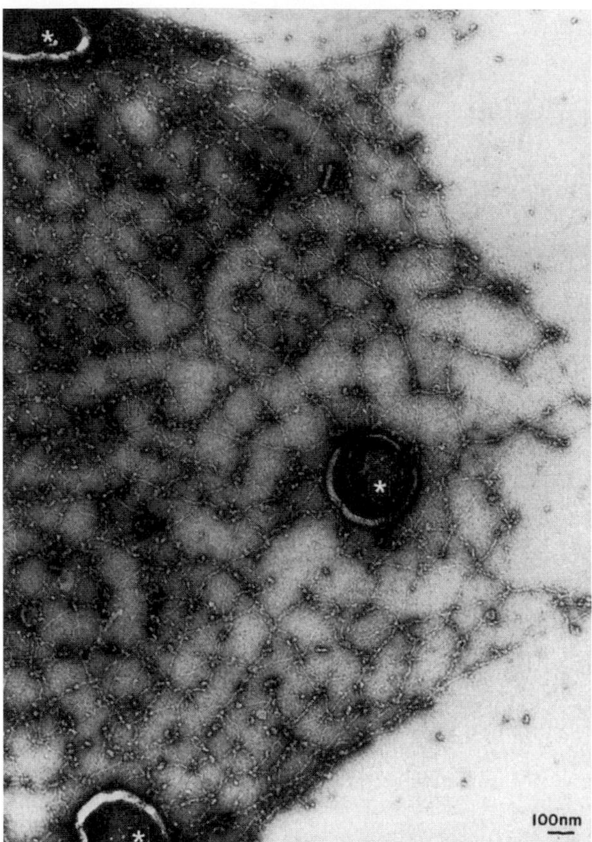

FIGURE 7 Spread membrane skeletal preparation prepared by Triton extraction of ghosts that were expanded by hypotonic treatment. This negative-stained electron micrograph shows the distinctive hexagonal lattice typical of the membrane skeleton in these preparations. The long filaments consist predominantly of spectrin tetramers, whereas the central junctional complexes are most likely actin and band 4.1. The small globular areas decorating the middle of many of the spectrin strands probably represent ankyrin bound near the head groups of the spectrin dimers.

C. Spectrin and Its Interactions

The structural backbone of this skeleton appears to be composed primarily of the protein spectrin, a long fiber-like molecule ideally suited to be the major structural element in a network. Substantial characterization of the basic organization of this very large molecule has been achieved by careful proteolytic digestion studies. As shown in Fig. 8, each chain consists of a considerable number of homologous repeating triple-helical segments that are linked by short, flexible, nonhelical regions. This catenary organization is very well designed to produce a strong but flexible elongated structural element. Fastidious amino acid analyses and early efforts to obtain cDNA probes for both α and β spectrin began to unravel the intimate molecular details of this crucial component of the membrane, and both α and β spectrin have now been completely cloned and sequenced.

Spectrin occurs as heterodimers (made up of two unequal subunits) in solution, but the heterodimers self-associate to form predominantly tetramers (four subunits) and higher oligomeric forms in the membrane skeleton by the "head-to-head" associations indicated in Fig. 6B. The tetramers, in turn, are laterally ramified and bound together into an anastomotic network by interactions with band 4.1 and actin at the opposite (tail) ends of the basic dimeric unit. Importantly, this entire network is then connected to the membrane bilayer by at least one additional linking protein, band 2.1. This protein, also called ankyrin or syndein, has high-affinity binding sites for both spectrin and the integral bilayer membrane protein, band 3. Hence, it serves to "anchor" the membrane skeleton to the bilayer, forming the complete membrane unit. A second anchorage site, which connects band 4.1 to another integral membrane protein, glycophorin C, is also likely. The net effect of this highly orchestrated series of spectrin : spectrin interactions and specific bilayer binding interactions is to produce a strong, flexible, skeletal network that is closely applied to, and coupled with, the undersurface of the bilayer, and that is capable of supporting it and providing it with remarkable physical integrity.

D. Defects of the Membrane Skeleton

As might be expected, a large number of red cell morphological abnormalities and hemolytic anemias

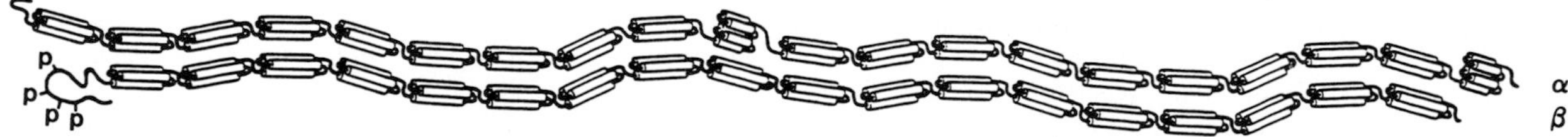

FIGURE 8 The spectrin dimer model: α and β subunits composed of multiple triple helical segments are connected by short nonhelical regions. Each segment contains 106 amino acids and, with the exception of the tenth and perhaps the twentieth subunit in the α chain, they are all homologous in basic structure, though not identical in amino acid composition. The two chains are antiparallel with the NH_2 terminus of the α subunit to the left. The carboxy terminus of the β chain is multiply phosphorylated.

with premature destruction of the circulating cells are associated with abnormalities both in the constituents of this newly defined structure and in their interactions. These abnormalities include quantitative deficiencies of particular proteins, such as spectrin in both recessive and dominant hereditary spherocytosis (hemolytic anemias with reduced surface area red cells) and perhaps in pyropoikilocytosis (severe hemolytic anemia with heat-sensitive fragile red cells), as well as in murine models of spherocytosis. Deficiency of band 4.1 has also been found in cases of severe, possibly homozygous, hereditary elliptocytosis. Also, deficiency of band 2.1 (ankyrin) has been found in a rare heat-sensitive fragmentation hemolytic anemia and in some severe variants of hereditary spherocytosis.

Qualitative abnormalities in the elements of this skeletal structure and in their interactions include defects in spectrin : 4.1 binding in other cases of hereditary spherocytosis and an extensive series of discrete abnormalities in spectrin : spectrin interactions in hereditary pyropoikilocytosis. In addition, qualitative defects in interactions between membrane skeletal elements may be acquired through damage that occurs to red cells in prolonged blood bank storage or during various oxidative stresses. Finally, both short and long variants of band 4.1, and a short variant of spectrin, have been reported in patients with poikilocytic hemolytic anemias, and it is reasonable to propose that some consequent qualitative membrane skeletal malfunction is responsible for the hemolysis in these conditions.

It seems probable that many more abnormalities in the components and interactions of the membrane skeleton will eventually be found when hemolytic disorders of the membrane are exhaustively examined. Indeed, such studies are likely to show that the limited morphological variations that we see in the clinical blood film in hemolytic anemias represent a final common pathway of a much more diverse variety of biochemical abnormalities in this unusual and important membrane structure.

BIBLIOGRAPHY

Bennett, V. (1985). The membrane skeleton of human erythrocytes and its implication for more complex cells. *Annu. Rev. Biochem.* **54**, 273.

Conboy, J., Kan, Y. W., Shohet, S. B., and Mohandas, N. (1986). Molecular cloning of protein 4.1, a major structural element of the human erythrocyte membrane skeleton. *Proc. Natl. Acad. Sci. USA* **83**, 9512.

Fairbanks, G., Steck, T. L., and Wallach, D. F. H. (1971). Electrophoretic analysis of the major polypeptides of the human erythrocyte membrane. *Biochemistry* **10**, 2606.

Knowles, W., Marchesi, S. L., and Marchesi, V. T. (1983). Spectrin: Structure, function and abnormalities. *Semin. Hematol.* **20**, 159.

Kopito, R., and Lodish, H. F. (1985). Primary structure and transmembrane orientation of the murine anion exchange protein. *Nature (London)* **319**, 234.

Lux, S. E. (1979). Dissecting the red cell membrane skeleton. *Nature (London)* **281**, 426.

Lux, S. E., and Palek, J. (1995). *In* "Blood: Principles and Practice of Hematology" (R. I. Handin, S. E. Lux, and T. P. Stossel, eds.), pp. 1701–1818. Lippincott, Philadelphia.

Mohandas, N., Clark, M. R., Jacobs, M. S., and Shohet, S. B. (1980). Analysis of factors regulating erythrocyte deformability. *J. Clin. Invest.* **66**, 563.

Palek, J., and Lambert, S. (1990). Genetics of the red cell membrane's skeleton. *Semin. Hematol.*

Sheetz, M. P., and Singer, S. J. (1974). Biological membrane as bilayer couples. A molecular mechanism of drug–erythrocyte interactions. *Proc. Natl. Acad. Sci. USA* **71**, 4457.

Shen, B. W., Josephs, R., and Steck, T. L. (1984). Ultrastructure of unit fragments of the skeleton of the human erythrocyte membrane. *J. Cell Biol.* **99**, 810.

Shohet, S. B. (1972). Hemolysis and changes in erythrocyte membrane lipids. *N. Engl. J. Med.* **286**, 577, 638.

Singer, S. J., and Nicolson, G. L. (1972). The fluid mosaic model of the structure of cell membranes. *Science* **175**, 720.

Speicher, D. W., and Marchesi, V. T. (1984). Erythrocyte spectrin is comprised of many homologous triple helical segments. *Nature (London)* **311**, 177.

Tchernia, G., Mohandas, N., and Shohet, S. B. (1981). Deficiency of skeletal protein band 4.1 in homozygous elliptocytosis: Implications for membrane stability. *J. Clin. Invest.* **68**, 454.

Reproduction

J. RICARDO LORET DE MOLA
University of Pennsylvania

GLOSSARY

Acrosome reaction The breakdown and merging of the plasma membrane and outer acrosomal membrane of the sperm, allowing the cap-like structure that covers the spermatozoa (acrosome) to release hydrolytic enzymes facilitating the penetration into the oocyte

Capacitation Changes that the sperm undergoes prior to fertilization, including acquisition of hypermotility, binding to the zona pellucida, and the ability to undergo the acrosome reaction

Follicle-stimulating hormone A glycosylated protein consisting of an α and β subunit, and one of two anterior pituitary hormones that control gonadal function

Implantation Process by which an embryo attaches to the uterine wall, penetrating the endometrium first and then invading the circulatory system of the mother, later forming the placenta

Zona pellucida Acellular layer covering the oocyte that is maintained until the embryo reaches the endometrium. It contains receptors for sperm

Zona reaction Changes that the zona undergoes immediately after fertilization to avoid polyploidy (fertilization by more than one spermatozoon)

THE MOST ESSENTIAL PART OF REPRODUCTION IN any species is the egg and sperm interactions resulting in a zygote. It was not until recently that these detailed aspects began to be studied in the human, thanks to the advent of *in vitro* fertilization (IVF) and other assisted reproductive technologies (ART). This article examines the mechanisms involved in gamete transport (sperm and ovum), fertilization, and early implantation.

I. NORMAL REPRODUCTION

The well-being of our society depends directly on the ability to propagate our species, or to reproduce. Fertility denotes the ability of a man and a woman to reproduce. Fecundability is the probability of achieving a pregnancy within one menstrual cycle. Infertility, or the lack of fertility, is defined as a year of unprotected intercourse without conception. Reproduction in the human is not a very efficient event. During each ovulatory cycle, normal healthy couples have only a 25% chance of becoming pregnant; 57% will do so after 3 months of unprotected intercourse and 85% of couples conceive after a year. Spontaneous miscarriage is also a common occurrence in the human. About 40% of spontaneous abortions of up to 28 weeks gestation are chromosomally abnormal, the majority which anomalies incompatible with life. Chromosomal abnormalities occur in about 50% of abortions where development stopped before the formation of a fetus, as compared with 5% among those that reach the fetal stage. Recent changes in demographics have uncovered some interesting trends in reproduction during the past few decades. In particular, a reduction in the birth rate to 15.5 per 1000 population has been seen, compared to the 55 per 1000 reported 200 years ago. These changes in demographics are partially due to deferment of marriage, postponement of pregnancy, and the availability of effective methods of contraception. How these changes and the new treatments for infertility will

ENCYCLOPEDIA OF HUMAN BIOLOGY, Second Edition, VOLUME 7. Copyright © 1997 by Academic Press. All rights of reproduction in any form reserved.

affect the future demographics of our society remains to be seen. [*See* Chromosome Anomalies.]

II. GAMETE PRODUCTION, MATURATION, AND TRANSPORT

A. Spermatogenesis

Spermatogenesis is the process of sperm differentiation in the testis, which begins with the mitotic proliferation of spermatogonia (spermatocytogenesis) and the division and development of spermatocytes (meiosis). The process culminates in the morphologic changes of a round spermatid, with a haploid genetic complement, into an elongated and polarized sperm cell (spermiogenesis) (Fig. 1). All of these processes occur within the seminiferous tubules of the.testis. In addition to the germinal cells present in the seminiferous tubule, somatic cells, known as Sertoli cells, provide the germinal cells with support during the process. [*See* Sperm.]

Primordial germ cells are found in man by the end of the third week of gestation. These cells are first present in the yolk sac and migrate to the genital ridge by ameboid locomotion. Once in the genital ridge they are termed gonocytes. These cells differentiate by mitotic division into fetal spermatogonia by the third month of intrauterine life. Spermatogonia continue to undergo rounds of mitosis, renewing the stem cell population needed to maintain continuous sperm production for a lifetime. The spermatogonia will continue to differentiate into what is known as type B spermatogonia. At this stage, the meiotic process is initiated and the cells become primary spermatocytes. Meiosis involves a single replication event, followed by two successive nuclear divisions resulting in the production of haploidy. [*See* Mitosis, Meiosis.]

Completion of these events yields spermatids, which are destined to undergo substantial morphologic and biochemical transformations as they develop into testicular spermatozoa by a process known as spermiogenesis (Fig. 2). From this point on, the process of spermiogenesis is accomplished without cell division. Considering the morphologic changes seen in the developing spermatids at both the light and electron microscopic levels, the process of spermiogenesis is subdivided into four phases: Golgi, cap, acrosome, and maturation. During the Golgi stage, the spermatids show an extensive Golgi apparatus in their cytoplasm. During the cap phase, there is further development of the acrosome, which attaches to the anterior pole of the spermatid nucleus and flattens in apposition to the nuclear membrane, as well as the flagellar fibrous sheath. In the acrosome phase the cell nucleus begins to elongate and condense. Finally, during the last phase, or maturation, there is a complex transformation process with removal and phagocytosis of most of the residual spermatid cytoplasm by the neighboring Sertolic cells.

The result of this process is the production of the morphologically mature spermatozoa. In the human, this cell is 60 μm long and is subdivided structurally into head and tail. The head consists of the nucleus with its associated acrosome. The entire spermatozoon is covered by a continuous plasma membrane known to be highly polarized in its polypeptide and lipid composition. [*See* Testicular Function.]

B. Sperm Transport

The process of spermatogenesis takes approximately 74 $\pm$ 4 days. This period does not include the added time necessary for sperm maturation in and transport through the epididymis. Epididymal transport time in humans varies, but has usually been estimated to be between 20 and 30 days. Therefore, it requires approximately 80 to 100 days for an individual spermatogonium to give rise to an ejaculated spermatozoon capable of fertilization.

Following the maturation in the testis, the sperm reach the epididymis where they gradually acquire motility and the ability to fertilize an oocyte. Preservation of optimal sperm function during this period of storage requires adequate testosterone levels in the circulation, as well as maintenance of normal scrotal temperatures. The importance of temperature is emphasized by the correlation of reduced numbers of sperm associated with episodes of fever.

The ejaculate contains a combination of sperm and prostate gland secretions. The semen forms a gel almost immediately following ejaculation but then is liquefied in 20–30 min by enzymes derived from the prostate gland. The alkaline pH of semen provides protection for the sperm while in the acidic environment of the vagina. This protection is transient, and most sperm are immobilized within 2 hr; therefore entry into the genital canal must be rapid, and usually occurs within 90 sec of ejaculation. As the ejaculate is placed in the vagina, the spermatozoa quickly gain entrance to the endocervix, and the seminal plasma is left behind in the vagina. The endocervix serves as a reservoir to maintain a constant supply of sperm, sending "waves" of motile sperm to the uterus for up

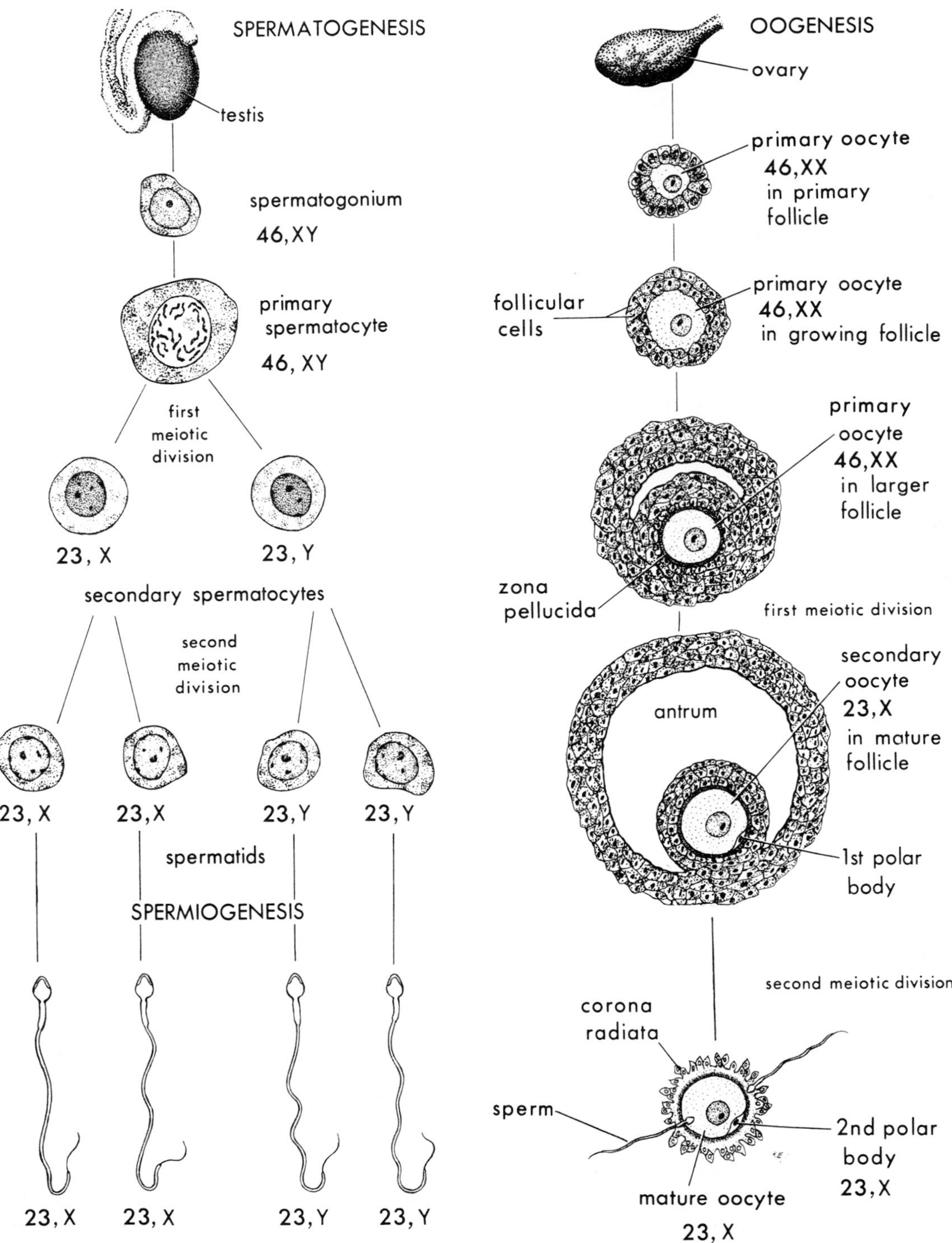

FIGURE 1 Spermatogenesis and oogenesis. (Left) The four sperm are formed from one primary spermatocyte and have a haploid complement at the end of spermiogenesis. (Right) In oogenesis, only one mature oocyte with two inactive polar bodies is formed. From Moore (1988).

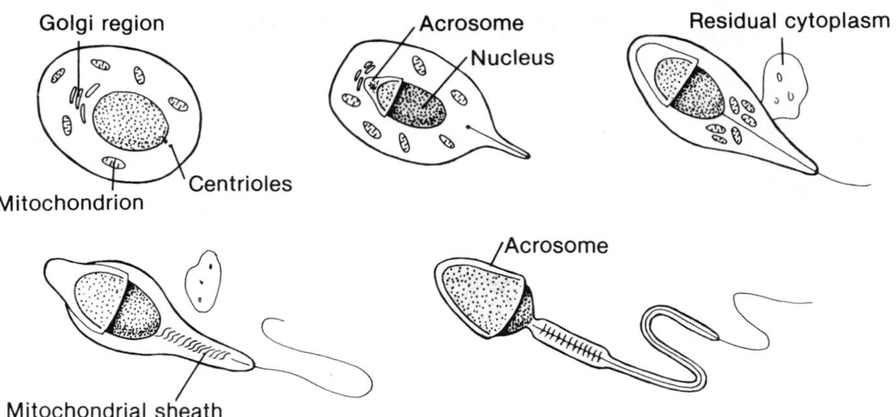

FIGURE 2 During the spermiogenesis process the rounded spermatids are transformed into elongated sperms. Note the loss of cytoplasm, the development of a tail, and formation of the acrosome. The mitochondria arrange themselves end to end in the form of a tight helix, forming a collar-like mitochondrial sheath. Note that the excess cytoplasm is shed during spermiogenesis. From Moore (1988).

to 72 hr after coitus. The exact mechanisms for sperm transport in the genital tract are unknown, but it is believed that contractions of the female reproductive tract occurring during coitus, as well as sperm flagellar motility, participate in this function. Once in the endometrial cavity, the uterus contracts further, propelling the sperm into the fallopian tubes as quickly as 5 min after insemination. In this location, the individual spermatozoan can last for up to 80 hr, maintaining its ability to fertilize. Upon the arrival of the sperm into the fallopian tube, they display a new pattern of movement that has been termed "hyperactivated motility." This motility is thought to be influenced by an interaction with the tubal epithelium, resulting in greater speed. Of an average of 200–300 million sperm present in the ejaculate, only 200 reach the fimbria and remain in proximity to the egg.

C. Capacitation and Acrosome Reaction

The sperm has to undergo certain changes prior to acquiring the capacity to fertilize a normal egg. This poorly understood process is known as capacitation. Capacitation is characterized by the ability of sperm to undergo the following three steps: the acquisition of hypermotility, binding to the zona pellucida, and acrosome reaction. Capacitation changes the surface characteristics of sperm and is associated with decreased stability of the plasma membrane. The membranes undergo more striking modifications when capacitated sperm are close to an ovum or when they are incubated in follicular fluid. Only the sperm that have completed the acrosome reaction can penetrate the zona pellucida surrounding the mature egg. There is a breakdown and merging of the plasma membrane and the outer acrosomal membrane, which is known as the acrosome reaction. This process allows release of the enzyme contents of the acrosome. These enzymes, which include hyaluronidase, proteases, glycosidases, and lipases, are all thought to play roles in the digestion and sperm penetration of the zona pellucida. These changes are also associated with the fusion of the sperm head membranes to the egg membrane. The inner acrosomal membrane fuses with the plasma membrane of the oocyte. By means of capacitation, the sperm acquires hypermotility, which provides changes in movement and velocity to the sperm that may be most critical in the actual penetration of the zona.

The process of capacitation normally occurs inside the female genital tract, but can be induced in simple physiological salt solutions at body temperature. The time required for *in vitro* capacitation of sperm is around 2 hr. Sperm washing procedures probably remove factors that coat the surface of the sperm, one of the initial steps in capacitation. The removal of cholesterol from the sperm membrane is believed to prepare the sperm membrane for the acrosome reaction.

D. Oogenesis

Thanks to several studies, a better understanding of the normal progression and the dynamics of human oocyte and follicular maturation and growth are now available. Oocyte production begins early during fetal development, 3 weeks following conception of a female embryo. At that time, the primordial germ cells, which are located in the epithelium of the yolk sac, migrate by ameboid movement to the region of the developing kidneys (mesonephros) and into the adjacent genital ridge. This area will ultimately become the gonad. It is also the site of primordial germ cell (oogonia) replication. Over the next 5 or 6 months, as a result of mitosis, the number of oogonia will increase to 6–7 million genetically identical diploid daughter cells. This increment in oogonia parallels the rise of fetal hypothalamic gonadotropin-releasing hormone and the fetal release, synthesis, and secretion of luteinizing hormone (LH) and follicle-stimulating hormone (FSH). By the time the female fetus is delivered, the nearly 7 million germinal cells decline to approximately 4 million oogonia, which are now called primary oocytes. All oogonia enter meiosis at the end of the prenatal period. In order to provide the appropriate complement of genetic material at the time of fertilization, the centromeres of homologous chromosomes segregate at MI. Before ovulation, during most of their life span in the ovary, human eggs are immature and arrested in the prophase of the first meiotic division. A number of factors have been implicated in causing the release of such meiotic arrest: gonadotropins, cyclic nucleotides, activators of adenylate cyclase, activators of protein kinase C, releasing hormones, and growth factors. Hypoxanthine, which is present in follicular fluid, has been shown to maintain meiotic arrest, and nonmetabolized hypoxanthine may be responsible for that effect through inhibition of cAMP degradation. The eggs remain arrested at this stage until they participate in the well-orchestrated events that constitute the ovulatory cycle (Fig. 3). [*See* Oogenesis.]

Folliculogenesis normally culminates with the production of a single dominant follicle during each menstrual cycle. These events take place as a result of three successive steps: recruitment, basal or tonic follicular growth, and the selection and maturation of the preovulatory follicle. There are three types of nongrowing follicles, which follow a predetermined transition: (a) primordial follicles, (b) intermediary follicles, and

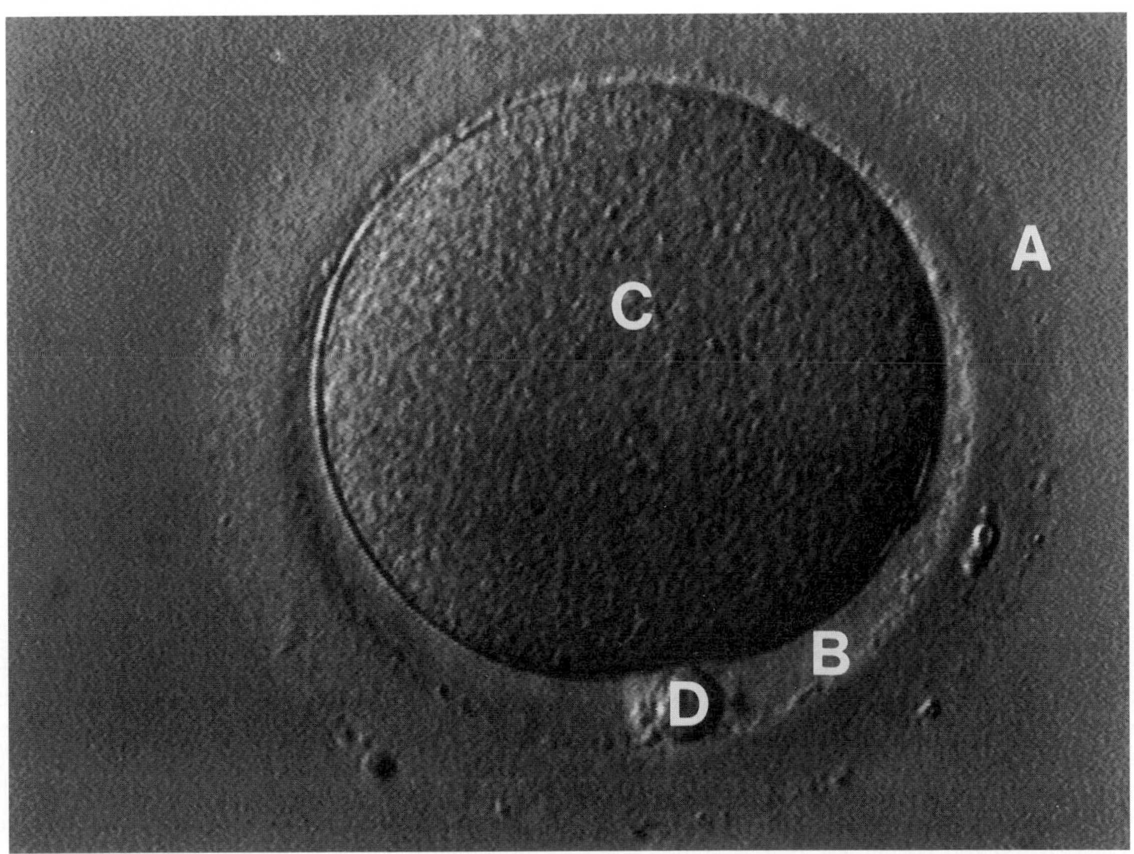

FIGURE 3 Normal mature human oocyte. (A) Zona pellucida, (B) perivitelline space, (C) ooplasm, and (D) polar body.

(c) primary follicles. The sizes of the oocyte and of the germinal vesicle do not change during this phase, which implies that only a maturation process takes place during this time. A large endowment of non-growing follicles is present at birth (at a primordial stage), which decreases significantly with aging. There is an acceleration in the loss of such an endowment from ages 34 to the disappearance of the last non-growing follicle at age 65. The depletion of this initial load of primordial follicles is due to two factors: atresia and initiation of follicular growth. In humans, follicular growth starts when the germinal vesicle of the oocyte reaches 19 μm in diameter and the granulosa cells surrounding the oocyte reach 15 cells in thickness. [*See* Follicle Growth and Luteinization.]

During adult life, gonadotropins appear to control the advanced stages of follicular growth, although conditions of low pituitary function do not completely suppress its initiation. The information available is confusing and sometimes contradictory. There is no evidence of growing follicles in anencephalic human fetuses or hypophysectomized monkey fetuses. In either one, initiation of follicular growth is not blocked. Gonadotropins seem to mediate the change from flattened to cuboidal granulosa cells in mouse and rats (from primordial to primary follicle). Gonadotropin-releasing hormone agonist (GnRH-a) blocks the transition from intermediary to primary follicles in primates.

When the follicle becomes multilayered, it is called a "secondary" or "early growing" follicle. At this stage, the zona pellucida begins to grow and FSH becomes the most important stimulus for growth. Follicle-stimulating hormone receptors are present in the granulosa cells, but lack LH receptors at this stage. As the follicle enlarges, the theca layer stratifies and a vascular wreath develops. At 120 μm in diameter, the theca cells become polyhedral (expressing LH receptors). From this point on, the follicle is defined as preantral, i.e., the first category of growing follicles, with a response to FSH/LH in both the theca and the granulosa. When follicles reach a size of 2–5 mm in diameter, they take an extra 15 days to attain preovulatory size; their granulosa cells are undifferentiated and proliferate, but with little estradiol (E2) output. Follicles can reach this stage without the output of gonadotropins. *In vivo* evidence for this phenomenon is derived from states with low gonadotropin levels, such as infancy, Kallmann syndrome, pregnancy, and oral contraceptive use. In these examples, follicles can reach a size of 2–5 mm and are recruitable (basal or tonic growth), but remain halted at that stage. The largest follicle at the beginning of the follicular phase will be the one selected during that cycle (5.5–8.2 mm). At this stage the incidence of atresia is inversely correlated with gonadotropin levels (highest in luteal phase) as the highest rates of atresia are in the 2–10 mm size follicles. There is greater expression of the aromatase system (which provides the androgen precursors for the secretion of estrogen) and an increment in FSH receptors. Although there is a poor understanding of how these factors act on the process of selection, it could be related to the high FSH receptors and action of nonaromatizable androgens, both of which would amplify the aromatase activity and activate other paracrine systems in a timely manner. During this final growth phase, the follicle undergoes other significant changes. There is increased vascularization, a greater capacity to aromatize androstenedione, and high intrafollicular E2 (440–3850 ng/ml) levels. The high aromatase levels and the increase in LH receptors prepare the follicle for the LH surge, which is the signal to trigger ovulation.

At the time of ovulation, the oocyte is surrounded by granulosa cells (the cumulus oophorus) that attach the oocyte to the wall of the follicle. The zona pellucida, a noncellular porous layer of glycoproteins, separates the oocyte from the granulosa cells (Fig. 1). The granulosa cells maintain metabolic communication with the oocyte by means of gap junctions that expand between the plasma membrane of the oocyte and the cumulus granulosa cells. In response to the midcycle LH surge, maturation of the oocyte proceeds with the resumption of meiosis as the oocyte enters into the second meiotic division and arrests in the second metaphase. Just before ovulation, the cumulus cells retract their cellular contacts from the oocyte. The disruption of the gap junctions induces the oocyte maturation and migration of the cortical granules to the outer cortex of the oocyte.

E. Oocyte Transport

Oocyte transport encompasses the period of time from ovulation to the entry of the egg into the uterus. The oocyte and cumulus mass are very "sticky" and easily adhere to the fimbria during the sweeping of the fallopian tube over the ovary. Within 2 to 3 min after ovulation, the egg is picked up by the fimbria, entering the tubal ampulla where fertilization takes place. In the ampulla of the tube, the cilia beat in the direction of the uterus. Entry into the tube is also facilitated by muscular movements and contractions and ciliated movements of the tubal epithelium, assisting in the migration toward the endometrium. A unidirectional beat is also found in the isthmus of the tube, although most physiologists believe that tubal muscle contractions are mostly responsible for the movement of the egg or early embryo in this portion of the tube. These contractions have a to-and-fro movement rather than a continuous forward progression. In women, oocyte transport through the ampulla requires approximately 30 hr and transit through the entire tube takes around 3 days. Tubal motility and muscle contractions are controlled by the extensive adrenergic innervation of the tube. However, surgical denervation of the tube does not disrupt ovum transport. Several substances also appear to play a regulatory role in tubal function. For example, prostaglandin E relaxes the tubal muscle, whereas prostaglandin F stimulates tubal muscular activity. Why is tubal motility so slow? In most species, residence of the zygote within the tube appears to be a prerequisite for full embryonal development. Indirect evidence in the human for an inadequate uterine environment in the early part of the cycle is provided by studies on patients treated with oocyte donation or transfer of cryopreserved embryos. Those studies have shown the importance of synchrony between development of the lining of the uterus (i.e., endometrium) and the embryo for a successful pregnancy to occur. If the endometrium is in a more advanced stage of development than the embryo, fertility is compromised and vice versa. Thus, the fallopian tube should not be seen as a mechanism for oocyte or embryo transport, but as a structure that provides nutrients and an important holding action for the zygote and early embryo. An interesting difference between human and animal reproduction is the fact that pregnancies can develop outside the *uterus* in the human; the most common site for such an occurrence is the fallopian tube. Humans are predisposed to ectopic pregnancies when there is previous damage to the tube due to infections or surgery. This type of pregnancy is particularly dangerous because the tube cannot hold such a gestation beyond the first trimester, at which time tubal rupture ensues, precipitating severe intraabdominal bleeding.

III. FERTILIZATION

The fertilizable life span of the human oocyte is 12 to 24 hr. In contrast, the human sperm can maintain its ability to fertilize for 48–72 hr, although motility can be maintained after the sperm have lost the ability to fertilize. The zona pellucida plays a dual role in the fertilization process, first by undergoing changes following fertilization leading to a block to polyspermy. It also protects the early zygote on its journey to the uterus prior to implantation. The initial contact between the sperm and the oocyte has the characteristics of a receptor-mediated process. The ligands in the zona pellucida for sperm recognition and induction of the acrosome reaction are the glycoproteins ZP1, ZP2, and ZP3, with ZP3 being the most abundant. Structural alteration of these glycoproteins leads to the loss of receptor ligand; inactivation after fertilization is probably accomplished by one or more of the enzymes present in the cortical granules. Spermatozoa enter the perivitelline space (the space between the zona pellucida and the vitelline membrane), contacting the egg membrane via the sperm equatorial segment. At first, the vitelline membrane engulfs the sperm head, with a subsequent fusion of egg and sperm membranes initiating the cortical reaction. This releases the materials found in the cortical granules (lysosome-like organelles located under the oocyte surface). Changes in the properties of the zona glycoproteins brought about by these enzymes lead to the zona reaction, resulting in the hardening of the extracellular layer, perhaps by crosslinking of proteins, and/or inactivation of sperm receptors. Thus, the zona block to polyspermy is accomplished. [*See* Fertilization.]

These events trigger the activation of the oocyte and completion of meiosis. The second polar body is expelled into the perivitelline space at the time of fertilization, leaving the egg with a haploid complement of chromosomes. The chromatin material of the sperm head decondenses and the male pronucleus is formed. The male and female pronuclei migrate to the center of the egg, where they come into contact, and the chromosomes become rearranged when the first mitotic spindle is formed. The addition of chromosomes from the sperm restores the diploid chromosome complement. Cell division is initiated at this stage (Fig. 4). The oocyte retains it ability to reinitiate cell division due to the stored maternal messenger RNAs. The newly formed embryo will take over this activity at around the four- to eight-cell stage in the human.

IV. IMPLANTATION

A. Preparation of the Endometrium

The endometrium undergoes a series of changes to allow the receptive conditions required for implanta-

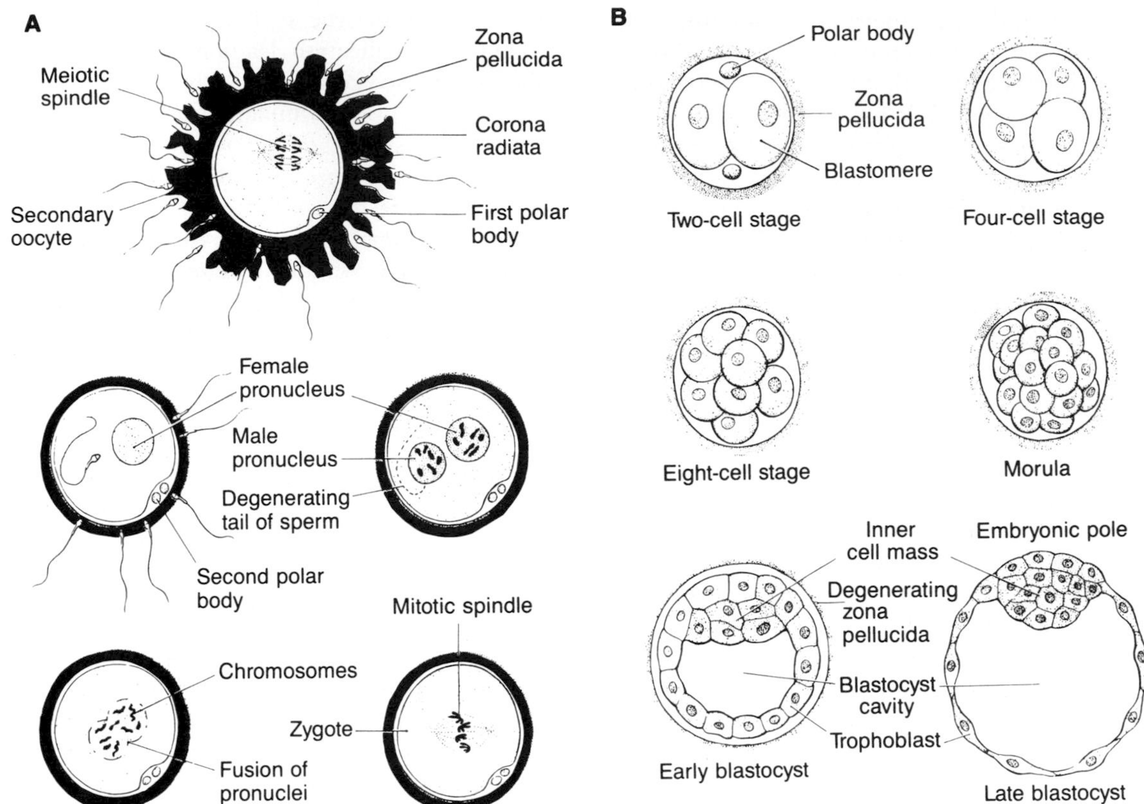

FIGURE 4 The normal events of fertilization beginning when the sperm contacts the secondary oocyte (A) and ending with the formation of a blastocyst (B). From Moore (1988).

tion. The endometrium is 10–14 mm thick in the midluteal phase. Several molecular and biochemical events take place in the midluteal phase. The cells encompassing the endometrial glands are rich in glycogen and lipids, and the secretory activity of these glands has reached its maximum. It is believed that the optimal time for implantation is very narrow, requiring perfect synchrony between the delivery of the embryo and endometrial receptivity. This period of time is otherwise known as the "window of endometrial receptivity," which is restricted to days 16–19 of the 28-day menstrual cycle. Several proteins and embryo products are secreted by the early embryo at the time of implantation. The most widely known is human chorionic gonadotropin (hCG), which is the hormone that is monitored in "pregnancy tests." The human embryo begins to produce hCG before implantation and can be detected in the mother about 6–7 days after ovulation. Because mRNAs for hCG are found in the six- to eight-cell human embryos, it is believed that the early human embryo is capable of signaling the endometrium, ovary, and corpus luteum by means of hCG. This hormone plays an important role in maintaining the corpus luteum beyond its natural life span of 14 days. The corpus luteum is responsible for secreting estrogen and progesterone to maintain the integrity of the endometrium. This is a critical event, as the endometrium is totally dependent on the corpus luteum to maintain its stability, until the placenta takes over steroidogenesis, somewhere between the 7th and the 10th week of gestation. Until that time, the well-being of the pregnancy is dependent on normal luteal function, otherwise the pregnancy may end in a miscarriage. Other hormones and signaling factors, such as early pregnancy factor (EPF), can also be detected in the maternal circulation prior to implantation. EPF plays important immunosuppressive roles and is associated with cell proliferation and growth. Many other proteins have been identified, but their functions remain obscure. [*See* Implantation Embryology.]

Prostaglandins are also important in the earliest stages of implantation and are found at the implantation site. The blastocysts of animals can also produce prostaglandins, and prostaglandin release from human blastocysts and embryos has been demonstrated. The endometrial cells are also capable of prostaglandin synthesis, but the interaction between decidua and the early placenta is poorly understood. Other autocrine and paracrine factors secreted by the endometrium, such as cytokines and a multitude of other growth factors, have been shown to exist at the implantation site. The epidermal growth factor, for example, is highly concentrated in the implantation site of the mouse. Whether it plays a role in minimizing rejection or modulating cellular interactions is not known.

B. Early Placentation

Implantation is defined as the process by which an embryo attaches to the uterine wall and penetrates first the epithelium and then the circulatory system of the mother to form the placenta. The early human embryo implants 2–3 days after entering the uterus or days 18–19 of the cycle (5–7 days after fertilization). The human blastocyst remains in the uterine secretions after its delivery from the fallopian tube for approximately 72 hr prior to implantation. The zona is then lysed, by components of the uterine fluid, by the pressure of the growing embryonal cells, or by a yet unidentified substance produced by the embryo, and then "hatches" out of the zona (Fig. 5). Following these events, the process of blastocyst attachment to the endometrial epithelium then takes place. The site most preferentially chosen for implantation is the upper, posterior wall of the uterus, in the midsagittal plane. The invasion of the early trophoblast is limited by the formation of a decidual cell layer in the uterus. The presence of the zona or lack of developmental maturation in part of the embryo prevents the attachment and implantation in the uterus. Many components of the inflammatory response appear to play important roles in the implantation process. The lymphocyte infiltrate present in the endometrium releases cytokines, activating the cellular lysis of the trophoblast, which is perhaps an important process in limiting cell invasion.

As the embryo comes into close contact with the lining of the uterus, the microvilli on its surface flatten and interdigitate with those on the luminal surface of the epithelial cells. The embryo can no longer be dislodged from the surface of the epithelial cells by flushing the uterus with physiologic solutions. Three types of subsequent interactions between the implanting trophoblast and the uterine epithelium have been described. In the first, trophoblast cells intrude between uterine epithelial cells on their path to the basement membrane. The epithelial cells lift off the basement membrane, allowing the trophoblast to insulate itself with fusion with individual uterine epithelial cells. The embryo secretes collagenases, urokinases, proteases, and plasminogen activators for the digestion of the intercellular matrix that holds the epithelial cells of the endometrium together. Despite its invasive nature, the destruction of maternal cells

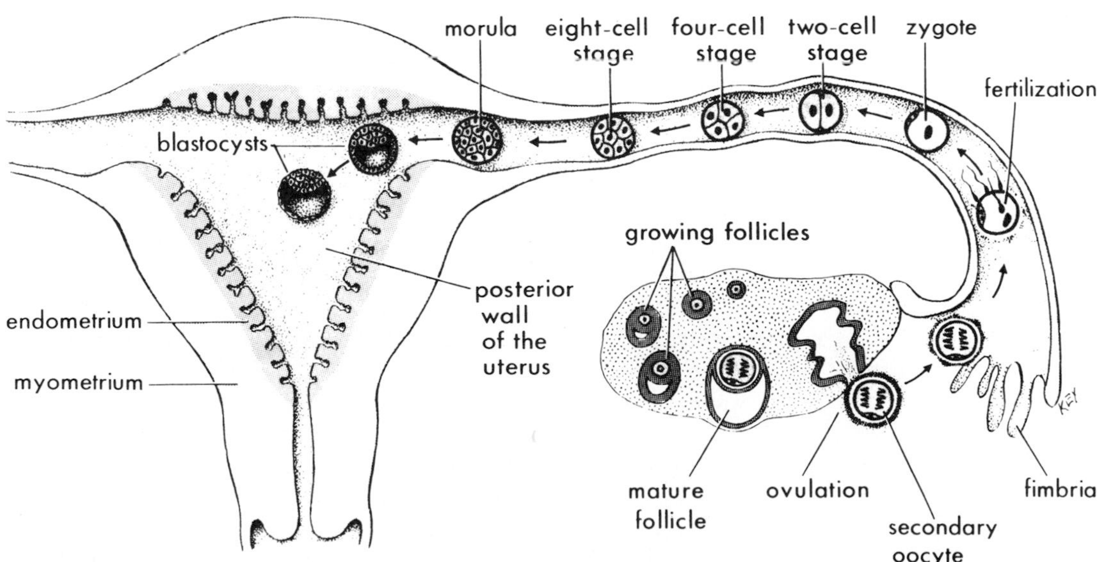

FIGURE 5 Summary of the ovarian cycle, fertilization, and early human development. From Moore (1988).

does not take place nor does it play a role in the process of implantation. The interactions between the invading trophoblast and the endometrium are poorly understood, and their investigation would aid in the understanding of the mechanisms of tissue rejection and acceptance.

V. ASSISTED REPRODUCTION

Since the birth of Louise Brown in 1978, tremendous progress has been accomplished in understanding the normal and abnormal processes of fertilization and reproduction. Techniques such as *in vitro* fertilization (IVF), gamete intrafallopian transfer, zygote intrafallopian transfer, and intracytoplasmic sperm injection have assisted thousands of couples in overcoming a multitude of barriers to reproduction. The first decade of assisted reproduction was characterized by rapid technological developments and fierce debates over ethical issues relating to this procedure, which still continue today.

There were four major developments in IVF during those initial stages of development: stimulation of multiple ovulation, ultrasound technology, improvement of *in vitro* culture media, and cryopreservation. Ovulation induction received a major boost for use in IVF. Ovarian stimulation protocols using human menopausal gonadotropins (hMG) and FSH alone or in combination with GnRH-a have made a significant impact in pregnancy outcomes. Advances in ultrasound technology, particularly the development of transvaginal ultrasound, have simplified the follow-up during ovulation induction. Oocyte retrieval in the early days of IVF used to require a major surgical procedure; with ultrasound-guided follicular aspiration it has now become a simple outpatient procedure, reducing costs and risks. Improvements in embryo culture through a better understanding of embryonal metabolic and nutritional requirements have translated into more "quality" embryos and higher pregnancy rates. The pregnancy rates reported by the American Society of Reproductive Medicine in 1993 show that IVF pregnancy rates currently are 18.3% in the United States and Canada. Cryopreservation of embryos has also allowed more flexibility in patient management. Current pregnancy rates with this technique are 13.3% per procedure. The new preimplantation genetic techniques, such as preimplantation diagnosis of congenital anomalies, provide the tools to diagnose several lethal and crippling disorders at the embryonic level and prior to the transfer of embryos. [*See* In Vitro Fertilization.]

We still have much to learn regarding reproduction in the human. There is no doubt that our understanding grows daily, as new information is readily available by means of scientific journals and meetings. As these technologies expand, new questions will arise, at both a physiological and an ethical level.

BIBLIOGRAPHY

Bavister, B. D., Cummins, J., Roldan, E. R. S. (eds.) (1990). "Fertilization in Mammals," 1st Ed. Serono Symposia, Norwell, MA.

Huisjes, H. J., and Lind, T. (eds.) (1990). "Early Pregnancy Failure," 1st Ed. Churchill Livingstone, New York.

Knobil, E., and Neill, J. D. (eds.) (1994). "The Physiology of Reproduction," 2nd Ed. Raven Press, New York.

Moore, K. L. (ed.) (1988). "The Developing Human," 4th Ed. Saunders, Philadelphia.

Trounson, A., and Gardner, D. K. (eds.) (1993). "Handbook of *in Vitro* Fertilization," 1st Ed. CRC Press, Boca Raton, FL.

Yen, S. S. C., and Jaffe, R. B. (eds.) (1991). "Reproductive Endocrinology," 3rd Ed. Saunders, Philadelphia.

Reproductive Processes, Neurochemical Control

RICHARD E. BLACKWELL
University of Alabama at Birmingham

GLOSSARY

Arcuate nucleus Hypothalamic center that regulates the secretion of gonadotropin-releasing hormone (GnRH)

Catecholamines Neurotransmitters involved in the regulation of GnRH secretion

Dominant follicle Of the 10 to 15 follicles that are recruited at the beginning of the cycle, the dominant follicle is the one that ultimately releases an ovum

Endorphins Body's own endogenous painkillers that are involved in the regulation of GnRH secretion

17-β-Estradiol Principal estrogen produced by the ovary, 95% of which comes from the dominant follicle

Follicle-stimulating hormone (FSH) Glycoprotein secreted from the anterior pituitary gland that acts on the ovary to stimulate follicle growth

Gonadotropin-releasing hormone Decapeptide produced in the base of the brain that stimulates the release of luteinizing hormone and follicle-stimulating hormone from the anterior pituitary gland

Inhibin Produced by the granulosa cells of the dominant follicle, it inhibits the secretion of FSH

Luteinizing hormone Glycoprotein secreted from the anterior pituitary gland that stimulates ovarian stroma to produce androgens. It ultimately brings about the formation of the corpus luteum

Serotonin Neurotransmitter involved in the regulation of GnRH secretion

I. HISTORY OF THE NEUROENDOCRINOLOGY OF REPRODUCTION

In 1932, Holweg and Junkman suggested that a sex center might exist in the brain that would regulate human reproduction. Subsequently, Harris, in 1937, electrically stimulated the median eminence of the brain and produced ovulation. In addition, Westman and Jacobson, in 1937, had demonstrated that a section of the pituitary stalk blocked ovulation. In 1946, Markee showed that direct stimulation of the pituitary gland failed to duplicate this response. These studies strongly suggested that the brain produced some chemical or chemicals that were secreted into the hypothalamic portal system, traveled to the pituitary gland, and regulated the events of ovulation.

The hypothalamic portal system had been described by Popa and Fielding in 1930. Houssay, in 1935, had shown that blood flowed from the brain to the pituitary and in 1950 Harris sectioned the hypothalamic portal vessels and produced target end organ atrophy. It was not until 1955, however, that Guillemin and Rosenberg incubated fragments of the hypothalamus *in vitro* with pituitary tissue and were able to show an increased secretion of the hormone that stimulates the adrenal gland (adrenocorticotropic hormone, ACTH). It was postulated that ACTH secretion was controlled by a small polypeptide and that each of the classic pituitary hormones was controlled by one small protein, thus giving rise to the one peptide–one hormone hypothesis. This was confirmed for the reproductive hormones in 1971 and 1972 when Schally *et al.* and Guillemin *et al.* isolated and presented the structure of gonadotropin-releasing hormone (GnRH). Concomitant with these studies, it was demonstrated by Bergland that blood flowed not only

ENCYCLOPEDIA OF HUMAN BIOLOGY, Second Edition, VOLUME 7. Copyright © 1997 by Academic Press. All rights of reproduction in any form reserved.

from the hypothalamus to the pituitary gland but in a retrograde manner. These studies opened the way for our current understanding of the neurochemical control of the reproductive cycle. [*See* Neuroendocrinology.]

II. ANATOMY OF THE MEDIAN EMINENCE AND ANTERIOR PITUITARY GLAND

The hypothalamus is phylogenetically old and found in mammals throughout evolution. It weighs approximately 10 g and is located at the base of the brain just above the juncture of the optic nerves (optic chiasm). The arcuate nucleus is one of the medial hypothalamic nuclei. It lies just above the median eminence and adjacent to the third ventricle. The median eminence, which is in close contact with the arcuate nucleus, is the final common pathway for the neurohumoral control of anterior pituitary function. It receives peptidergic neurons which contain releasing and inhibiting hormones. The median eminence delivers these hormones to hypothalamic pituitary portal capillaries and these neurochemicals are subsequently transmitted to the anterior pituitary gland, where they act on the gonadotropes to release both luteinizing hormone (LH) and follicle-stimulating hormone (FSH) (Fig. 1). [*See* Hypothalamus; Pituitary.]

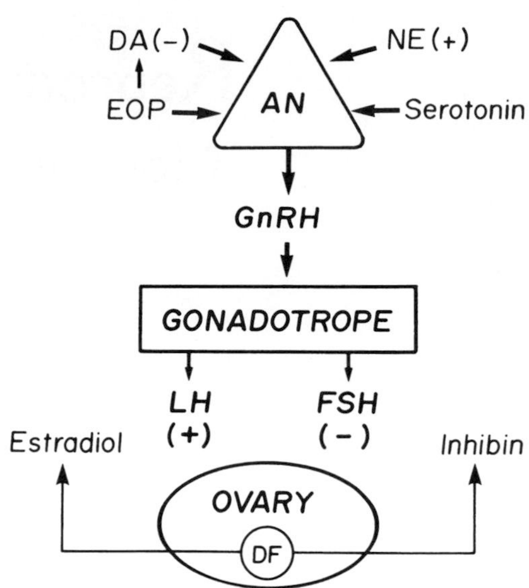

FIGURE 2 Sources and targets of neurochemicals controlling the menstrual cycle. DA, dopamine; NE, norepinephrine; AN, arcuate nucleus; EOP, endorphins; GnRH, gonadotropin releasing hormone; LH, luteinizing hormone; FSH, follicle stimulating hormone; DF, dominant follicle.

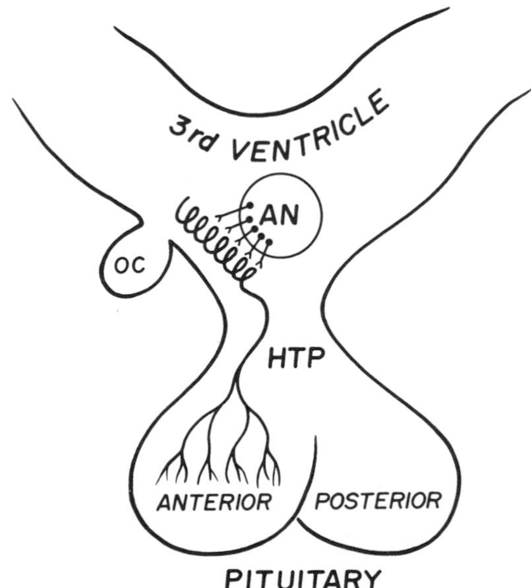

FIGURE 1 The relative positions of pituitary gland, hypothalamus, and brain, and the portal system connecting the median eminence of the hypothalamus with the anterior pituitary. AN, arcuate nucleus; OC, optic chiasm; HTP, hypothalamic pituitary portal capillaries.

III. NEUROCHEMICALS INVOLVED IN REPRODUCTION

At ovulation it was assumed that when increased amounts of LH and FSH were secreted, this was the result of an increased production of GnRH. This notion was dispelled when Knobil *et al.* demonstrated that when GnRH was delivered into the hypothalamic portal system of the rhesus monkey in a fixed amount at hourly intervals, ovulation occurred. It appears that GnRH secretion is controlled by a push/pull mechanism involving the neurotransmitters dopamine and norepinephrine (DA and NE, respectively, in Fig. 2). An infusion of dopamine produces an inhibition of LH secretion whereas norepinephrine has the opposite effect. In addition to catecholamines, the endorphins have a significant effect on gonadotropin secretion. These compounds have a suppressive effect on LH secretion and this is mediated through dopamine neurons. Serotonin, another neurotransmitter, may evoke inhibitory influences on GnRH neuronal activity (Fig. 2).

IV. NEUROCHEMICAL CONTROL OF THE MENSTRUAL CYCLE

As described earlier, the arcuate nucleus is the final integrator of the menstrual cycle. It receives exogenous and endogenous input and releases GnRH into the hypothalamic pituitary circulation which gives rise to hourly pulses of both LH and FSH secretion. These hormones travel to the ovary where LH acts on the stroma to stimulate the production of androgens. The FSH acts on the granulosa cells of 10 to 15 primary oocytes to induce their maturation and growth. The follicle-stimulating hormone increases the absolute number of granulosa cells. And, the size of the follicle induces an enzyme called aromatase which allows the follicle to metabolize androgens to estrogens, the latter being important in follicle growth, and it induces the formation of luteinizing hormone receptors which will ultimately be stimulated to give rise to the corpus luteum. Of the 10 to 15 follicles that are recruited, one becomes the dominant follicle (see Fig. 2). It begins to secrete significant amounts of 17-β-estradiol and inhibin; the latter compound inhibits the secretion of its parent hormone. Subsequently, the follicle begins to produce progesterone. When combined with estrogen, a synergy occurs that triggers the midcycle surge of luteinizing hormone. Following the surge of LH, a myriad of events occurs in the microenvironment of the follicle in preparation for ovulation. Once the follicle has been exposed to the stimulus of the luteinizing hormone for approximately 18–20 hr, the ovum is released. Luteinizing hormone secretion brings about the formation of the corpus luteum with its ever-increasing production of progesterone. Progesterone feeds back on the hypothalamic pituitary central axis and reduces the secretion of GnRH to one pulse every 4 hr. Progesterone and estrogen production increases until approximately the middle of the luteal phase (postovulatory day 8). If conception does not occur at this point, these hormones begin to inhibit secretion from the central axis, leading to the ultimate demise of the corpus luteum and menstruation.

BIBLIOGRAPHY

Blackwell, R. (1993). Neuroendocrinology of reproduction. *In* "Textbook of Reproductive Medicine" (Conn and Blackwell, eds.), pp. 157–169. Appleton Lange, Norwalk, CT.

Blackwell, R. (1994). Neuroendocrinology of the menstrual cycle. *In* "Ovulation Induction (Seibel and Blackwell, eds.), pp. 13–21. Raven Press, New York.

Blackwell, R., and Guillemin, R. (1973). Hypothalamic control of adenohypophyseal secretions. *Annu. Rev. Physiol.* **35**, 357–390.

Fritz, M., and Speroff, L. (1985). The endocrinology of the menstrual cycle: The interaction of folliculogenesis and neuroendocrine mechanisms. *In* "Modern Trends in Infertility and Conception Control" (E. E. Wallach and R. D. Kempers, eds.), Vol. 3, pp. 5–25. Year Book Medical Publishers, Chicago, IL.

Hodgen, G. (1985). The dominant ovarian follicle. *In* "Modern Trends in Infertility and Conception Control" (E. E. Wallach and R. D. Kempers, eds.), Vol. 3, pp. 26–45. Year Book Medical Publishers, Chicago, IL.

Neill, J., Pan, G., Wei, N., and Mulchaney, J. (1990). GNRH receptor regulation. *In* "Neuroendocrine Regulation of Reproduction" (S. S. C. Yen and W. W. Yale, eds.), pp. 249–257. Second Symposia, Norwell, MA.

Yen, S. S. C. (1991). The human menstrual cycle: Neuroendocrine regulation. *In* "Reproductive Endocrinology" (Yen and Jaffe, eds.), 3rd Ed., pp. 273–308. Saunders, Philadelphia, PA.

Reproductive System, Anatomy

ANTHONY J. GAUDIN
KENNETH C. JONES
California State University, Northridge

I. Male Reproductive System
II. Female Reproductive System

GLOSSARY

Atresia Degeneration and disappearance of follicles in the mammalian uterus

Bartholin's glands Two small mucous glands located on each side of the vaginal opening

Bulbourethral glands Two small mucous glands on each side of the prostate gland that secrete part of the seminal fluid

Corpora cavernosa Erectile tissue in both the penis and the clitoris

Corpus spongiosum Erectile tissue surrounding the urethra in the male penis

Cremasteric muscle Thin muscle that suspends and surrounds the testis and the spermatic cord

Endometrium Mucous membrane forming the inner lining of the uterus

Epididymis Convoluted tubule adjacent to the testis that contains mature and maturing sperm

External os The opening of the cervical canal leading into the vagina

Fimbriae Finger-like projections surrounding the abdominal opening into the oviduct

Infundibulum Cavity formed by the fimbriae surrounding the opening into the oviduct

Internal os Opening leading from the cervical canal into the uterus

Isthmus Narrow end of the uterus leading into the cervix

Leydig cells Interstitial cells of the testis that secrete testosterone

Mesovarium A fold of peritoneal tissue that supports the ovary

Myometrium Muscular wall of the uterus

Oocyte Early or primitive ovum prior to development

Seminiferous tubules Numerous coiled tubules in the testis, where sperm are produced

Sertoli cells Cells in seminiferous tubules that support and nourish developing sperm

Tunica albuginea Coat of white fibrous tissue surrounding the testis

Tunica dartos Layer of muscle in the wall of the scrotum that permits shrinking and shriveling of the scrotal skin at cold temperatures

Tunica vaginalis Thin serous membrane surrounding the testis

Vaginal fornix Connection of the vagina to the fornix

UNLIKE OTHER SYSTEMS IN THE BODY, THE REPRODUCTIVE systems differ in two sexes. Although these differences are the primary criteria by which one is identified as male or female, they are not the only ones. The reproductive systems have the related functions of manufacturing sex hormones responsible for male and female characteristics and producing sex cells used to generate offspring. Whereas muscles, nerves, blood, and digestive organs, for example, are all essential to life and the maintenance of healthy internal conditions, reproductive organs are not vital to life, nor are they generally involved in maintaining the internal equilibrium, without which life could not exist. Instead of being essential to the survival of an individual, the reproductive systems are essential to the survival of the species.

I. MALE REPRODUCTIVE SYSTEM

The adult male reproductive system consists of four general regions (see Color Plate 12) The testes are

ENCYCLOPEDIA OF HUMAN BIOLOGY, Second Edition, VOLUME 7. Copyright © 1991 by Academic Press. All rights of reproduction in any form reserved.

paired organs in which sperm (i.e., male reproductive cells) develop. The testes are carried in a small sac, called the scrotum. Sperm leave the testes and are stored in a network of tubules (the second region), through which they pass when they are expelled from the penis. Accessory glands produce most of the fluid that carries the sperm and compose the third major region. The fourth major part of the male reproductive system is the penis, an organ through which sperm are delivered to the female reproductive system during sexual intercourse, of coitus. [*See* Sperm.]

A. Testes

Adult testes, or testicles, are egg-shaped structures. Each testis is about 25 × 50 mm and is divided into 200–300 lobules (see Color Plate 13). Lobular walls are extensions of a thick fibrous covering, called the tunica albuginea, which surrounds each testis. External to the tunica albuginea is a thinner membrane, the tunical vaginalis.

Each lobule in the testis consists of a collection of highly coiled seminiferous tubules. Located between the seminiferous tubules are Leydig cells, also called interstitial endocrinocytes, or interstitial cells of Leydig (Fig. 1). These cells secrete hormones important to sperm production. Microscopic examination of the inner wall of a seminiferous tubule reveals sperm in various stages of development. The periphery of each tubule is marked by a basement membrane. Sperm in the earliest stage lie just inside this membrane, with progressively more mature sperm lying nearer the lumen. Also embedded in the wall of a seminiferous tubule are Sertoli cells, also known as sustentacular, or nurse cells, which provide chemical assistance to developing sperm. [*See* Reproductive System, Anatomy.]

Seminiferous tubules connect and fuse at one side of each testis to form another network of tubules, the rete testis. From the rete testis emerges a smaller collection of larger-diameter tubules, the efferent ductules. These empty into a single tubular structure, the epididymis, located outside of the testis, but within the scrotum.

B. Epididymis

The epididymis is formed by the fusion of efferent testicular ductules into the ductus epididymis, a tightly coiled tube (see Color Plate 13). An epididymis lies alongside each testis, curved in shape to conform to the testis (see Color Plate 14). The upper end is the head, the middle section is the body, and the lower end is the tail. Although each epididymis is only about 4 cm long externally, the ductus epididymis included in it is nearly 6 m long. The duct is only 1 mm in diameter and is tightly coiled. The epididymis is where sperm undergo final maturation and where mature sperm are stored.

C. Scrotum

The scrotum, the sac that holds the testes and the epididymis, is divided into two halves exteriorly by a

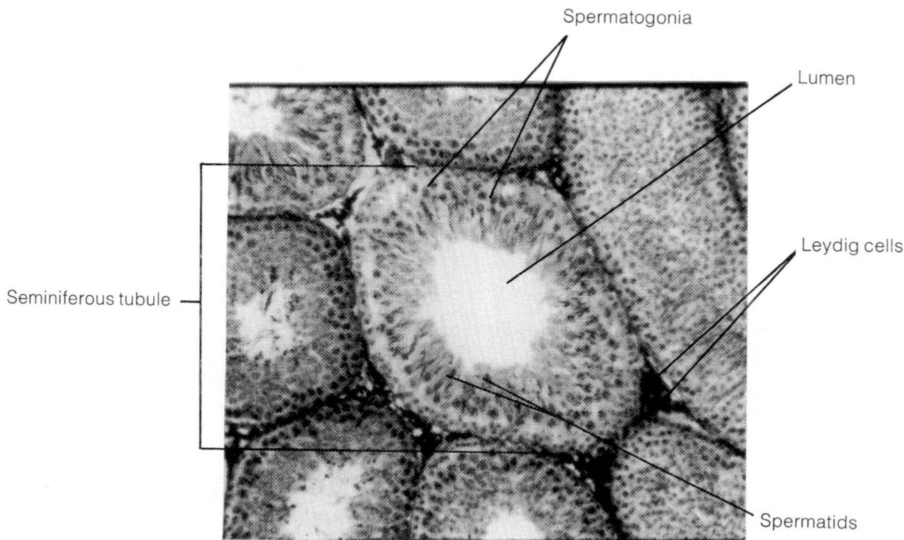

FIGURE 1 A seminiferous tubule in a testis.

median ridge of tissue, the scrotal raphe. Internally, the scrotum is partitioned into two lateral compartments by a septum of connective tissue. Each compartment contains one testis. The scrotal skin covers subcutaneous tissue that contains a layer of smooth muscle, called the tunica dartos. The muscle extends into the septum that divides the scrotum and enables the wall of the scrotum to contract and thicken in response to low temperature. This provides an important mechanism for temperature control in the testes. Sperm development requires temperature lower than that of the body, which is achieved by suspension of the testes outside the body in the scrotum. When the external temperature is too cold for optimal sperm development, the tunica dartos contracts, thickening the scrotal wall and bringing the testes closer to the body, where the temperature is warmer. Another small muscle, the cremasteric muscle, assists in this process. It is located in the inguinal region, above the testes. In warmer temperatures these muscles relax, lowering the testes and allowing the scrotal wall to become thin. Richly endowed with sweat glands, the scrotum also secretes sweat, which cools the testes by evaporation.

D. Ductus Deferens and Ejaculatory Duct

The tail of each epididymis leads to the ductus deferens (or vas deferens), a tube about 45 cm long (see Color Plates 13 and 14). The ductus deferens from each testis passes out of the scrotum through the inguinal canal, a narrow opening in the abdominal wall, and then enters the pelvic cavity. Each duct is accompanied by testicular arteries, veins, nerves, and lymphatic vessels. The entire assemblage, the spermatic cord, is surrounded by the cremasteric muscle and connective tissue sheaths. Once inside the pelvic cavity, the ductus deferens loops over the ureter and passes behind the urinary bladder (see Color Plate 14). The terminal portion is enlarged in diameter, forming the ampulla.

The ductus deferens receives fluids from a seminal vesicle, an accessory gland. The union of the ductus deferens and a duct from the seminal vesicle produces the ejaculatory duct. This short duct, only about 2 cm long, conducts sperm into the urethra.

E. Urethra

The urethra is a single tube extending from the urinary bladder to the tip of the penis (see Color Plates 12 and 14). In males the urethra conducts urine from the bladder during urination and provides a path for sperm during sexual activity. The initial portion of the urethra is the prostatic urethra, a section about 3 cm long that passes through the prostate gland. Prostatic secretions enter the urethra in this section. At the base of the prostate gland, the membranous urethra passes through the urogenital membrane. This portion is less than 2 cm long and leads directly to the cavernous (or penile) urethra, which passes through the corpus spongiosum of the penis. The penile urethra, approximately 16 cm long, terminates at the urethral orifice.

F. Accessory Glands

In addition to fluid produced in the testes, fluid is added to the mixture of sperm by cells of sacs in the ampulla wall of the ductus deferens. Once sperm have passed through the ductus deferens, they are carried in fluid produced by accessory glands, which include the seminal vesicles, prostate gland, and bulbourethral glands.

The paired seminal vesicles are small pouches connected by short ducts to the junction of the ampulla and the ejaculatory duct. Seminal vesicles, each of which is 5–10 cm long, consist of a tightly twisted and convoluted tubule (Fig. 2). Each tubule is lined with secretory cells that release an alkaline fluid rich in fructose, a sugar which the sperm use for energy. The alkaline nature of the secretions helps neutralize

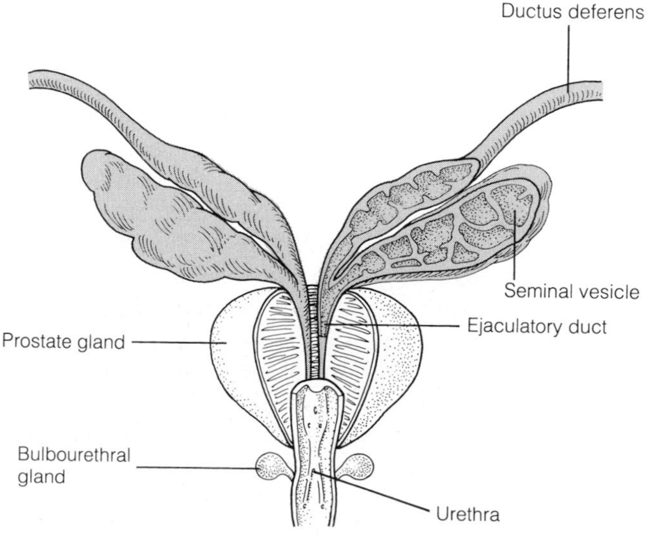

FIGURE 2 The seminal vesicles and the prostate gland.

the acidity of urine remaining in the urethra. The mixture of sperm and secretions is semen, or seminal fluid. Seminal vesicles contribute slightly more than one-half of the seminal fluid volume.

The prostate gland lies below the bladder, where it surrounds the urethra and the two ejaculatory ducts. Shaped like a cored apple, the prostate gland is divided into several branched compartments, each of which opens through a duct directly into the urethra. The compartments are separated from one another by walls of smooth muscle and connective tissue. The entire organ is surrounded by a jacket of connective tissue. The prostate gland produces 1–2 ml of fluid per day, which collects and is released into the urethra just prior to ejaculation.

The paired bulbourethral glands (or Cowper's glands) are located below the prostate gland. Each is about the size of a pea and empties into the cavernous urethra through ducts located close to the base of the penis. Bulbourethral secretions, whose release is triggered by erotic activity, precede ejaculation and provide lubricants for sexual intercourse and the nutrients that sperm require for motility. They also help neutralize the acidity of any urine remaining in the urethra.

Numerous outpocketings lie in the walls of the cavernous urethra, some of which are branched chambers connected to the urethra by short ducts. These structures, called urethral glands, produce mucus that serves as a lubricant to semen.

G. Penis

The penis consists of three cylindrical bodies of spongy tissue, surrounded by sheaths of connective tissue and skin (see Color Plate 15). Two of the spongy bodies compose the corpora cavernosa. These paired structures lie parallel to one another in the superior portion of the penis. The third body, the corpus spongiosum, lies inferior to the corpus cavernosum, extending past the distal ends of the corpus cavernosum into an enlarged tip, the glans penis. The urethra enters the corpus spongiosum at its base and terminates at the glans penis in a slit-like opening, the urethral orifice. The glans penis is covered by a loose extension of the skin of the penis: the prepuce, or foreskin. The prepuce is removed by circumcision.

The two spongy cylinders of the corpora cavernosa are separated from one another for most of their length by a tough inelastic sheath, the tunica albuginea. The corpus spongiosum is also surrounded by a sheath, but one of somewhat more elastic structure. The three bodies, in turn, are surrounded by the fascia penis, a sheath that lies just below the skin. Blood is supplied to the penis by the internal pudendal arteries, which branch from the internal iliac arteries. Each pudendal artery divides further into a network of arteries that provide blood to superficial and deep tissues. [*See* Cardiovascular System, Anatomy.]

II. FEMALE REPRODUCTIVE SYSTEM

The adult female reproductive system consists of five general regions (see Color Plate 16): two ovaries in which eggs, or ova, are produced; two tubular oviducts, through which eggs pass after release; a saclike uterus, in which an embryo develops; a vagina, which leads from the uterus to the exterior; and the vulva, a collective term for the external genitalia. Mammary glands used to provide milk for the newborn infant are also considered part of the female reproductive system.

A. Ovaries

Ovaries are almond shaped paired organs about 2–5 cm long, located on either side of the lower portion of the abdominal cavity, where they are held in place by the mesovarium. This ligament, in turn, connects to and supports other organs of the female reproductive tract. Medially, ovaries are connected to the uterus by the ovarian ligament; laterally, they are attached to the abdominal wall by the suspensory ligament.

Histologically, each ovary consists of an inner medulla (a cortex surrounding the medulla) and a single layer of cells (a modified mesothelium that covers the cortex) (Fig. 3). Nerves, lymph, and blood vessels lie

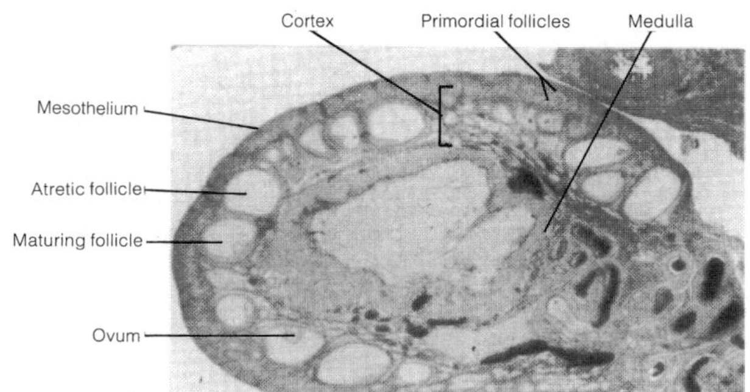

FIGURE 3 A longitudinal section through an ovary.

within the medulla and extend into the cortical region. The cortex and the mesothelium are surrounded by a layer of connective tissue, the tunica albuginea. Embedded within the cortex are 300,000–400,000 undeveloped, but potential, egg cells. In the course of a woman's reproductive years, only about 400–500 of these cells will complete development and be released from the ovary. A mature egg cell, an ovum, is a female gamete capable of fertilization.

Several other structures are also visible in the ovary. Potential egg cells in the cortex of the ovary of a young woman are surrounded by a single layer of cuboidal or columnar cells. The cell, at this stage, is a primary oocyte. The oocyte and the cells surrounding it form a primordial ovarian follicle. Numerous follicles in various stages of development are present, as well as some that failed to develop completely and have degenerated. Follicles that cease developing and subsequently degenerate are said to undergo atresia, and their remains are referred to as atretic follicles.

In addition to producing eggs, the ovaries are important sources of estrogens, steroid hormones responsible for developing the body form and other features characteristic of a female. [*See* Puberty.]

B. Oviducts

The oviducts are about 10–12 cm long and extend from the superior surface of the uterus (see Color Plate 17). The end of the oviduct distal to the uterus and adjacent to the ovary consists of a mass of highly convoluted finger-like projections, called fimbriae. Collectively, fimbriae form a funnel-like opening to the tube, the infundibulum. The infundibulum leads directly into the ampulla, an enlarged portion occupying about one-third to one-half the length of the oviduct. The interior wall of the ampulla is highly convoluted. It is in this region that fertilization normally occurs.

The remainder of the oviduct includes the isthmus, a portion that leads from the ampulla to the uterus, and an interstitial segment that penetrates the uterine wall. There is a progressive decrease in the degree of infolding of the wall and the percentage of ciliated cells from the infundibulum to the interstitial segment. At the end of the uterine tube closest to the uterus, the infoldings are reduced to several longitudinal ridges.

C. Uterus

The uterus is a hollow, pear-shaped, thick-walled sac that rests on the floor of the abdominopelvic cavity, located between the urinary bladder and the rectum (see Color Plates 16 and 17). It is held loosely in place by broad ligaments on either side, uterosacral ligaments that connect the cervix to the sacrum, lateral cervical ligaments that connect the cervix to the pelvic diaphragm, and round ligaments, which are anchored in the tissues beneath the labia majora. Some of the ligaments also carry uterine blood vessels and nerves. The bladder, rectum, and other adjacent organs also help position in the uterus, and because of its flexibility, the uterus can assume a number of different forms and orientations as the bladder and the rectum fill and empty.

The normal uterus in a nonpregnant woman is about 6 cm high, 4.5 cm wide, and 2.5 cm deep. The hollow interior of the uterus, the uterine cavity, connects to the outside through cervical and vaginal canals. Regions within the uterus include the fundus, the dome-like cap on the body (or corpus) of the uterus; the isthmus, a region below the corpus where the uterus narrows; and the cervix, a narrow neck-like extension that protrudes into the vagina. The junction of the uterine cavity and the cervical canal forms a narrow opening, called the internal os (see Color Plate 17). The opposite end of the cervical canal, where it opens into the vagina, is the external os. The uterus is usually bent in anteflexion, a position in which the body of the uterus projects anteriorly over the urinary bladder, and the cervix projects posteriorly, entering the upper end of the vagina at nearly a right angle.

The wall of the body of the uterus consists of an inner endometrium, a complex layer of epithelial cells, glands, and blood vessels; a middle layer, the myometrium, within which lie layers of smooth muscle, connective tissue, and many large blood and lymphatic vessels; and an outermost perimetrium. The perimetrium is continuous with the broad ligaments that support the uterus and the uterine tubes.

The endometrium consists of two functionally and structurally distinct regions: the stratum functionalis and the stratum basalis. Prior to menstruation these two regions can be as thick as 6–7 mm, but during menstruation the cells of the stratum functionalis die and slough off, reducing the endometrium essentially to the stratum basalis, a layer only about 1 mm thick. [*See* Uterus and Uterine Response to Estrogen.]

D. Cervix

Although the cervix is part of the uterus, the structure and function of its walls are quite distinct. The cervical wall is relatively thick, and its inner surface is covered by an elaborate network of ridges and valleys that secrete mucus, which fills the cervical canal, preventing the inward movement of microbes and external substances.

E. Vagina

The vagina, about 10–15 cm long, provides a passage from the uterus to the outside world. The vagina is largely fibromuscular tube and has a wall consisting of three tissue layers. Outermost lies a thick layer of connective tissue, covering a middle layer of muscle. Blood vessels and nerve bundles lie in this wall. The innermost layer of the vagina is a folded layer that secretes mucus.

The connection of the vagina to the cervix forms the vaginal fornix, an enlarged portion of the vagina, in which sperm collect during sexual intercourse. Deposited sperm are thus in close proximity to the external os, enhancing the likelihood of sperm continuing into the cervix.

F. External Genitalia (Vulva)

The vagina leads inferiorly and anteriorly from the cervix and opens into the vulva, or pudendum, which is a collective term for the external genitalia (see Color Plate 18). This opening is surrounded by a membranous ring of tissue, the hymen, which consists of relatively fine and highly vascularized tissue. Normally, an opening exists through the hymen, but sometimes the hymen covers the opening of the vagina completely. In such cases the hymen must be opened surgically once menstruation begins, to permit menstrual flow to escape.

The external genitalia consist largely of modified skin and subcutaneous tissues. The mons pubis is a rather thick layer of fatty tissue that lies over the public symphysis, the junction of the two pubic bones that lie at the base of the abdominopelvic cavity. The mons pubis divides into two thick fatty pads that proceed posteriorly to join again in the space between the openings of the vagina and the anus. These two folds of fatty tissue are the labia majora, which enclose the remainder of the esternal genitalia in the pudendal cleft. Just inside the two labia major are two smaller folds of mucous membrane: the labia minora. These labia join at the anterior end of the pudendal cleft to form a hoodlike structure that covers the clitoris and surrounds the vestibule of the vagina, the cleft between the labia minora. As puberty is reached, the mons pubis and the labia majora become covered with a thick mat of coarse pubic hair. Hair does not develop on the labia minora, which are kept moist by glandular secretions. The entire region enclosed within a roughly diamond-shaped area from the clitoris anteriorly to the coccyx posteriorly and laterally to the region external to the ischial tuberosity of the coccygeal bone is the perineum. (Some authorities, however, restrict their definition of the perineum to only the small area between the anus and the vulva.) The triangular anterior half of the perineum forms the urogenital triangle; the posterior triangle of the perineum is the anal triangle.

From the developmental standpoint the clitoris and the penis are equivalent because they are derived from the same embryonic structure. The clitoris, however, is much smaller and does not contain the urethra. Like the penis, it contains erectile tissue that becomes engorged with blood during periods of sexual stimulation, causing the clitoris to become erect. The spongy tissue responsible for erection of the clitoris is the corporus cavernosum. A corpus spongiosum is absent. The clitoris terminates in a glans clitoridis, which is nearly covered by the folds of the labia minora. The folds covering the glans compose the female prepuce.

The vestibule also contains the opening of the urethra and several mucus-secreting glands that provide lubrication for sexual intercourse. Paraurethral glands open into the area surrounding the urethra; a pair of small round Bartholin's glands lie posterior to the base of each of the labia minora and open through ducts on either side of the vaginal orifice. In addition to paraurethral and Bartholin's glands, the vestibule contains many smaller glands that also contribute mucus.

G. Mammary Glands

Mammary glands are considered part of the reproductive system because of the important role they play in nurturing an infant. Mammary glands begin to secrete milk 2–3 days after a baby is born. The milk is rich in nutrients and immunoglobulins that convey disease resistance to the suckling infant. It is a mixture consisting of about 88% water, 7% lactose, 4% fat, 1% protein, and minute amounts of other valuable nutrients.

A mature lactating mammary gland consists of

15–20 compartments, or lobes, separated by fatty and connective tissue. The lobes consist of subdivisions (i.e., lobules) made up of numerous grape-like clusters of alveoli, which produce the milk. A duct emerges from each cluster of alveoli and fuses with others to form lactiferous (or mammary) ducts, each one carrying the milk produced by the alveoli of a lobe. Each lactiferous duct has an enlarged region, an ampulla, which lies just beneath the nipple. Each ampulla leads into a short continuation of the lactiferous duct that opens at the nipple.

As milk is produced, it accumulates in each breast, collecting in ampullae and in the extensive duct system. This system opens externally in the nipple. The nipple is surrounded by a pigmented ring of tissue called the areola.

BIBLIOGRAPHY

Burger, H. G., and Baker, H. W. G. (1987). The treatment of infertility. *Annu. Rev. Med.* **38**, 29–40.

Cormack, D. H. (1987). "Ham's Histology," 9th Ed. Lippincott, Philadelphia.

Cormack, D. H. (1992). "Essential Histology." Lippincott, Philadelphia.

Fink, G. (1986). The endocrine control of ovulation. *Sci. Prog.* **70**, 403–423.

Gaudin, A. J., and Jones, K. C. (1989). "Human Anatomy and Physiology," Harcourt Brace Jovanovich, San Diego, California.

Longo, F. J. (1987). "Fertilization," Chapman & Hall, New York.

McMinn, R. M. H., and Hutchings, R. T. (1993). "Color Atlas of Human Anatomy," 3rd Ed. New York Med. Pub. Chicago.

Money, J. (1981). The development of sexuality and eroticism in human kind. *Q. Rev. Biol.* **56**, 379–404.

Short, R. V. (1984). Breast feeding. *Sci. Am.* **250**, 35–41.

Respiration, Comparative Physiology

JACOPO P. MORTOLA
McGill University

GLOSSARY

Allometric relationship Any relationship relating a variable to body weight; usually represented in the log-transformed version

Altricial species Species born at an early stage of development, therefore immature at birth

Chemoreceptors Receptors sensitive to chemical variations

Compliance Changes in volume per unitary changes in pressure

Dead space Total volume of the airways that do not participate in gas exchange

Functional residual capacity Amount of air left in the lungs at end expiration

Hematocrit Volume of blood cells expressed in percentage of total blood volume

Hering–Breuer inflation reflex Inhibition of inspiratory activity accompanying lung inflation

Hyperpnea Increase in the absolute level of ventilation

Hyperventilation Increase in ventilation (strictly, alveolar ventilation) relative to metabolic rate (i.e., $\dot{V}_E-\dot{V}_{O_2}$ ratio), irrespective of the absolute value of ventilation

Hypocapnia Decrease in carbon dioxide

Hypoxia Decrease in oxygen

Mechanoreceptors Receptors sensitive to mechanical stimuli, usually pressure or tension

Oxygen consumption Amount of oxygen used per unit time

P_{50} Partial pressure of O_2 (in mm Hg) at which 50% of the hemoglobin is saturated

Perfusion Amount of blood circulating per unit time

Precocial Species born at a late stage of development, therefore mature at birth

Resting volume Volume of the respiratory system when no forces are applied to it

Tidal volume Volume of air inhaled with each breath

Transpulmonary pressure Pressure across the lungs, i.e., the pressure difference between the airways and the pleural space

Ventilation Amount of air exchanged per unit time

Vital capacity Maximal amount of air that can be exhaled after a full inspiration

RESPIRATION IS GAS EXCHANGE, SPECIFICALLY the exchange by a living organism of carbon dioxide (CO_2), a waste product formed during the oxidation of food molecules, for oxygen (O_2), which the organism needs to continue oxidizing its food. At the cellular level, gas exchange occurs by diffusion, according to the partial pressure gradient of the gas. In large cell aggregates and more complex organisms, diffusion alone would not fulfill the minimal cellular metabolic requirements. Therefore, gas convection mechanisms are coupled to diffusion in what constitutes the respiratory apparatus. This article presents briefly the various mechanisms of convection in invertebrates and lower vertebrates. Mammals are then examined in more detail by addressing the question of how a change in body size influences metabolic and ventilatory requirements, and how these requirements are met by modifications in structural design and func-

ENCYCLOPEDIA OF HUMAN BIOLOGY, Second Edition, VOLUME 7. Copyright © 1997 by Academic Press. All rights of reproduction in any form reserved.

tional properties of the respiratory system. Finally, the main modifications dictated by age (newborn) or by life in special conditions (diving, burrowing, and high altitude) are summarized.

I. DEPENDENCE OF OXYGEN: GAS DIFFUSION AND CONVECTION

The respiratory system is designed to provide gas exchange, i.e., to fulfill the cellular necessity for O_2 and to eliminate the by-product of cellular respiration, CO_2. Anaerobic mechanisms to generate energy are available, but they are much less efficient and usually represent emergency or short-term routes of energy production; if anaerobic means are adopted, then aerobically produced energy must be spent to pay back the oxygen debt. Because the pressure of O_2 in the environment is much higher than in the cells (and in the cells the pressure of CO_2 is much higher than in the environment, where it is almost zero), we should expect a continuous diffusion of O_2 into, and of CO_2 out of, the cells. This is indeed the case, and diffusion represents the basic physical process governing respiration. However, the number of molecules that can diffuse in and out per unit time is inversely proportional to the distance between the organelles where cellular respiration takes place (mitochondria) and the environment. Hence, only in very simple organisms can diffusion alone provide the minimum amount of gas exchange per unit time (ml/min; conveniently expressed by O_2 consumption, $\dot{V}_{O_2}$, or CO_2 production, $\dot{V}_{CO_2}$) required for cell survival. In more complex aggregates, gas exchange would take too long to fulfill the metabolic needs. Nature has circumvented this limitation by coupling the diffusion process to convection, whereby the environmental air is effectively brought into contact with all the cells of the organism via finely controlled structures. The way convection operates varies remarkably throughout the animal kingdom, and the design of the convection system is intimately related both to the metabolic needs of the organism and to other nonrespiratory functions of the respiratory structures. In vertebrates, two systems of pipes with their respective pumps convey gases to and from the cells, the respiratory system, which operates with a gaseous or liquid medium, and the cardiovascular system, in which blood represents the convection medium. They are coupled together at the pulmonary (or gill) gas exchange area where gas transfer depends on diffusion, similar to that which occurs at the cellular level (Fig. 1).

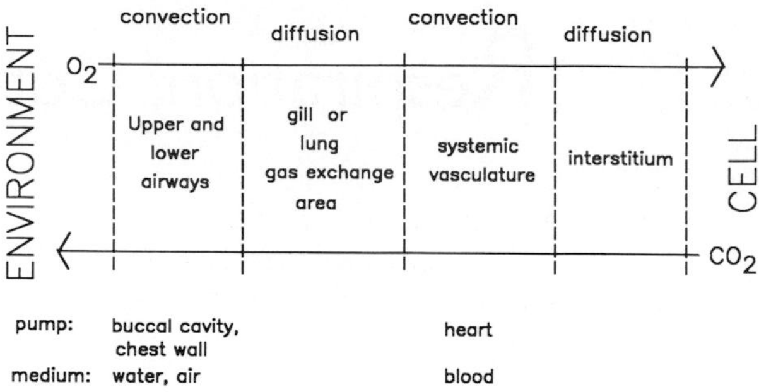

FIGURE 1 Schematic representation of the physical (top) and structural basis (center and bottom) for gas transport between environment and cell in vertebrates.

II. MECHANISMS OF GAS CONVECTION IN VERTEBRATES

A sudden drop in ambient O_2 could be accommodated by changing the environment in the search for better conditions. Indeed, an increase in motility, in some cases with coordinated locomotive responses, represents a common reaction to hypoxia in the simplest living organisms. However, it is apparent that the strategy of "looking for a better place" is of limited value because motion is energetically demanding and because it does not solve problems of tissue hypoxia of internal origin. An alternative solution to an acute drop in O_2 availability would be a decrease in O_2 demand at the cellular level. In fact, decreasing O_2 (O_2 conformity) is a very common phenomenon in hypoxic invertebrates and lower vertebrates; in mammals, it is common only among the smallest species, which have high thermogenic requirements, whereas is not common in larger species, including humans.[1] O_2 conformity is rarely the only response to hypoxia, and some forms of acclimatization, whereby gas diffusion is implemented by convection, can be recognized even in the most primitive living organisms.

In insects, both diffusion and convection are enhanced by the presence of a complex system of tubes, the tracheoles, connecting the body surface to the innermost tissues. In fact, in some cases the tracheoles penetrate the cells, and blood circulation is therefore redundant, at least for the purposes of gas convection. Ventilation ($\dot{V}_E$) is regulated via spiracles, according to control programs which compromise between the

[1]The drop in metabolic rate, and body temperature, during hypoxia is common in newborn mammals, which, in this respect, behave like heterothermic lower vertebrates.

metabolic necessities of the insect and the problem of water loss that would accompany excessive exposure of inner body surfaces.

Enormous increases in gas exchange area can be achieved not only by inward foldings, as in the case of the insect tracheoles, but also by outward protrusions, as in the case of the multiple lamellae of fish gills. In fish, the large diffusion area of the gills is placed along the unidirectional stream of water entering the mouth and exiting via the operculum, behind the gill filaments. The unidirectional flow is usually helped by two aspiration–compression pumps (Fig. 2), in the oral cavity and in the opercular cavity, which act in coordination with the opening and closing of mouth and operculum. This design of unidirectional water flow across the gills from the front to

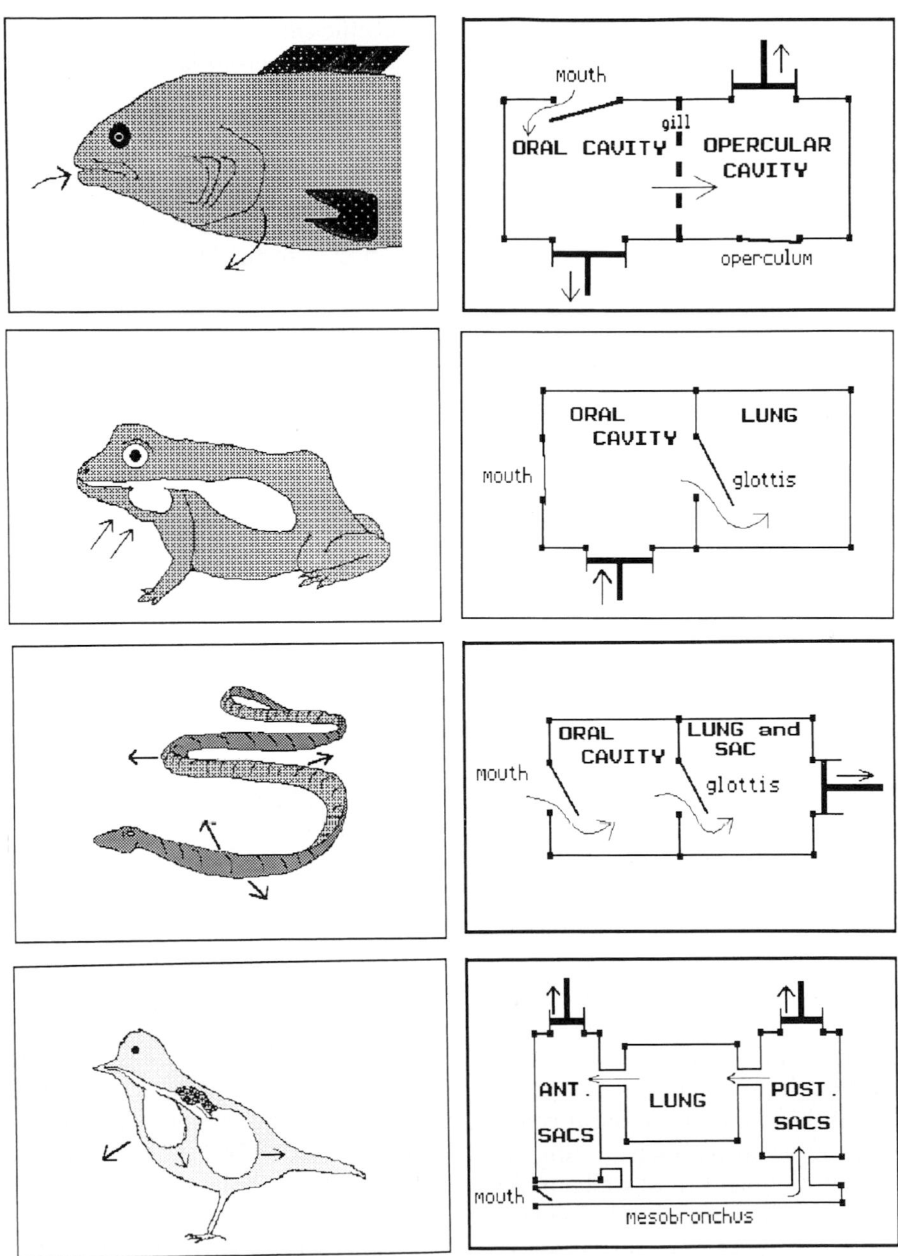

FIGURE 2 Schematic representation of the mechanisms of gas convection in some classes of vertebrates.

the sides has the potential of providing sufficient water convection simply by swimming forward with the mouth open. Indeed, in some fishes, gill ventilation depends more on locomotion than on active pumping, and in these cases (e.g., tuna), continuous forward swimming becomes a fundamental necessity for $\dot{V}_E$. From an energetic viewpoint, one of the best solutions is that of *Remora remora.* By attaching itself to the body of a shark, this fish not only gets a free ride and the leftovers of the shark's meals but, by keeping its mouth open, also a free $\dot{V}_E$. In these cases, the inhibition of pump activity would be mediated by receptors functionally equivalent to the vagal stretch receptors of the mammalian lung. In addition, the unidirectional flow design offers an opportunity for extremely efficient gas exchange. In fact, in the lamellae, the anatomical arrangement is such that the direction of blood flow is opposite to that of the water (countercurrent pattern). The venous blood reaching the gill capillaries is meeting water with progressively higher P_{O_2} (the partial pressure of oxygen, in mm Hg), and the arterialized blood leaving the gill capillaries can have a P_{O_2} higher than that of the water leaving the gas exchange area. This would never be the case with the back-and-forth design of the mammalian $\dot{V}_E$, whereby the arterial P_{O_2} can approach but never exceed alveolar or expired P_{O_2}.

Gas convection in amphibians is a clear example of positive pressure ventilation. The upward lifting of the mouth floor generates a positive pressure which forces air into the lungs (Fig. 2). This mechanism is in many respects simpler than the double-pump arrangement of the fish and is aided only by the valve action of the nares and the glottis. As far as the mechanics of lung inflation, this mechanism of positive pressure ventilation is more efficient than the mammalian suction pump, as chest expansion of the lung by positive pressure eliminates the inefficiency caused by chest distortion. However, the mouth $\dot{V}_E$ of the amphibians has obvious disadvantages. The inflation volume is limited by the size of the mouth, and $\dot{V}_E$ in amphibians must compromise with the nonrespiratory functions of the upper airways much more than it does in reptiles, birds, and mammals.

Turtles were first thought to breathe with a mechanism similar to that of the amphibians, but it is now clear that they operate by negative pressure ventilation as in other reptiles. Contraction of chest wall muscles, laterally located in the turtle, more circularly placed in lizards and snakes, lowers the pressure inside an air sac generating inspiratory flow (Fig. 2). The sac is larger than the functional lung, and in snakes it can be almost as long as the whole body, although the gas exchange area is only in its proximal portion. The reptilian breathing cycle is characterized first by active expiration, then by active inspiration, terminating with closure of the glottis and relaxation, with positive airway pressure in between breaths. Inspired air is therefore sucked through the gas exchange region and retained for a period in the sac. Gas from the sac is forced through the same gas exchange region in the opposite direction during expiration.

The same idea in design, but more sophisticated and efficient, is found in the avian respiratory system. In birds, numerous large avascular sacs, some located anteriorly and others posteriorly, are the compliant structures expanding and compressing with the phases of the breathing cycle, whereas the lungs are small, almost rigid organs connected to the vertebral side of the ribs. Although both anterior and posterior sacs are connected to the lungs, the inhaled air heads almost entirely to the posterior sacs via the mesobronchus (Fig. 2). During inspiration the anterior sacs also expand, but they receive mostly air from the lungs. During the following expiration, the previously inhaled air moves into the lung, while the air stored in the anterior sacs is exhaled. Therefore, the bolus of air inspired takes two full cycles to be exhaled, going from the trachea and the mesobronchus to the posterior sacs during the first inspiration, to the gas exchange area in the first expiration, to the anterior sacs in the second inspiration, and finally out with the expiratory phase of the second cycle. This flow pattern, the direction of which is determined by regional pressure differences rather than anatomic valves, guarantees a continuous unidirectional flow through the lung irrespective of the phase of the breathing cycle. In addition, air and blood flows in the lung are arranged according to a cross-current pattern, which, although not as efficient as the true countercurrent arrangement, is still a better gas exchanger than the $\dot{V}_E$–perfusion matching of the mammalian lung. The result of this sophisticated convection system is that the blood can load more O_2 and does so from an exchange organ with a higher O_2 concentration than that present in the mammalian alveoli. The ultimate test of efficiency may be represented by the ability to survive and perform in hypoxia; the results are unequivocal. When a mammal and a bird of comparable size, metabolic rates, and blood O_2 affinity are exposed to severe hypoxia, the former lays down panting while the latter is still flying. Some birds are known to migrate flying high above the Himalayas, where no mammal would be able to survive.

III. DESIGN OF THE MAMMALIAN RESPIRATORY SYSTEM

Quite differently from that of the avian model, the mammalian solution to the gas exchange problem is a back-and-forth convection of air through the same system of conducting pipes, with the pulmonary gas exchange area located at the terminal end of the conducting airways. The mammalian tidal respiration implies that P_{O_2} and P_{CO_2} in the alveoli will be, respectively, lower and higher than in the environment. Because the alveoli are the terminal structures, the blood leaving the pulmonary capillaries cannot be as well arterialized as with the flow-through system of the avian model. In addition, the common pathway for inspiration and expiration means that part of the inhaled air will not participate in gas exchange,[2] reducing the gas exchange efficiency of $\dot{V}_E$, in comparison with the air and blood flow arrangements of birds. However, that part of the body space which in reptiles and birds is occupied by the sac(s) in mammals is used for the lung itself, resulting in a gas exchange area that is very large relative to body mass and fully protected inside the body.

As in reptiles and birds, the mammalian ventilatory pump operates by a negative pressure suction mechanism. The major difference is the separation of the visceral cavity into a thoracic and an abdominal compartment by a muscular layer, the diaphragm. Contraction of this muscle lowers pleural pressure, which results in lung expansion (caused by the rise in transpulmonary pressure) and the rib cage tending to collapse (caused by the negative pressure difference between inside and outside of the rib cage). To what extent, during diaphragmatic contraction, the inward movement of the rib cage actually occurs depends on a number of factors which include the relative compliance of lung and chest wall and the action of the extradiaphragmatic muscles. Therefore, the respiratory pump in mammals is prone to a great deal of deformation, and the intercostal muscles have the important function of stabilizing the rib cage against the distorting action of the diaphragm.

On the one hand, having the main inspiratory muscle (the diaphragm) facing only one side of the thorax may seem poor mechanical design since a substantial fraction of diaphragmatic force is dissipated not in ventilating the lungs but in distorting the pump itself. On the other hand, the complete separation of the visceral (coelomic) cavity into two compartments decreases the disturbing effects of postural changes, locomotive activities, and variations in abdominal pressure on the distribution of pulmonary $\dot{V}_E$. [*See* Respiratory System, Anatomy; Respiratory System, Physiology and Biochemistry.]

IV. ANIMAL SIZE AND METABOLIC REQUIREMENTS

About 4300 species of mammals are presently known to exist, covering an extremely large range in body size, from the shrew, weighing a few grams, to the African elephant, weighing several tons; marine mammals can have even larger weights,[3] and the blue whale can reach 100–120 tons. Because all mammals are homeotherms (neglecting some special conditions such as hibernation, estivation, and torpor, and the neonatal period) and their body temperature is maintained within a very narrow range (37–39°C), the huge differences in size suggest that thermodispersion, thermoproduction, or both must vary among species. Thermodispersion is mostly determined by body surface. Heat production is determined by the metabolic activities of the cells and hence is proportional to body mass. If, for simplicity, we think of an animal as a sphere, because sphere surface is proportional to the square of length and sphere volume is proportional to the third power of length, it follows that the larger the animal body the smaller its surface-to-volume ratio. Hence, large animals should be much warmer than smaller animals. The fact that this is not so is because the metabolic activity of the cells of large animals is not as pronounced as it is in the smaller species. Indeed, it has long been recognized that, although the $\dot{V}_{O_2}$ of the whole organism increases with the size of the species, after normalization by the animal's weight ($\dot{V}_{O_2}$/kg) the opposite is true: $\dot{V}_{O_2}$/kg progressively decreases with the increase in the weight of the animal (Fig. 3). In other words, if we had 1-g samples of "average" flesh from a 100-g rat, a 20-kg dog, a 70-kg man, and a 5-ton elephant, we would find that the sample $\dot{V}_{O_2}$ progressively decreases from the rat's sample, to the dog's, man's, and elephant's.

[2]The larger the volume of the conductive airways (also called anatomical dead space, V_D), the smaller the fraction of tidal volume (V_T) reaching the gas exchange area (also called alveolar volume, V_A). In many adult mammals during resting breathing V_D approximates one-third of V_T, hence alveolar ventilation V_A equals two-thirds of $\dot{V}_E$.

[3]Mass and weight are used as synonyms throughout the text.

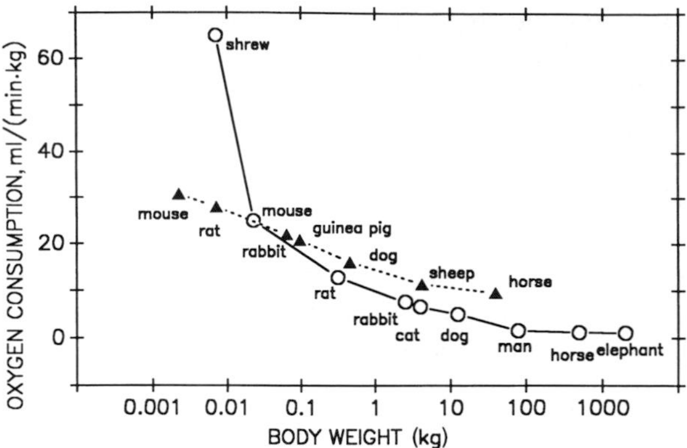

FIGURE 3 Oxygen consumption normalized by body weight in some adult (○) and newborn (▲) mammals. The $\dot{V}_{O_2}$/kg of the newborn is higher than that of the corresponding adult and is usually also higher than that of same-weight adults of different species. Only in the smallest newborn species is $\dot{V}_{O_2}$/kg not as high as expected from the adult curve.

A similar finding would occur if we compared samples from different-sized animals within the same species. In fact, it turns out that $\dot{V}_{O_2}$ *within* a species is proportional to the two-thirds power of body weight ($\propto BW^{0.66}$), which is the exponent to be expected according to the surface–volume argument presented earlier. $\dot{V}_{O_2}$ *among* species, however, scales to $BW^{0.75}$, which is clearly below unity, yet significantly above the expected 0.66. The reason for this discrepancy is not fully understood and may relate to the existence of a systematic variation from the simplistic sphere analogy (possibly related to the influence of gravity in shaping the body of the animal) and in differences among mechanisms of heat dissipation.

V. RESPIRATORY ADAPTATION TO DIFFERENCES IN BODY SIZE AND METABOLIC RATE

Given that $\dot{V}_{O_2}$/kg is not constant among different-sized animals, if the structure and function of the ventilatory apparatus were strictly proportional to size (and therefore not proportional to metabolic rate), major differences among species should be expected in the P_{O_2} and P_{CO_2} of the alveolar gas and blood. This is not the case. It would have been surprising if there were differences as mammalian functions have an optimal operational acid–base range, just as they have an optimal temperature range. Size-related modifications in the mammalian design of the respiratory system occur at both the structural and the functional levels to accommodate the different metabolic needs and will be examined systematically in the following sections.

A. Structure of the Lungs and Respiratory Pump

Comparison of adult mammalian species over a 10^4 range in size have indicated that the mass of the lung is a fixed proportion of the total weight of the animal, approximately 1%. The amount of air present in the lung, at a prefixed transpulmonary pressure, is also directly proportional to body and lung mass, implying that lung compliance (change in volume per unitary change in transpulmonary pressure), whether per unit of lung tissue or per body weight, is an interspecies constant. Indeed, direct measurements of lung compliance show that this is the case; the shape of the pressure–volume deflation curve, and derived mathematical descriptors of the transpulmonary pressures during lung deflation, also vary little among species.

The gross aspects of the internal architecture of the lung are similar among mammals. The volume of the conducting airways, which contribute to the anatomical dead space, is a constant fraction of total lung volume,[4] implying that the volume of the gas exchange region at the lung periphery, or alveolar volume, is a fixed proportion of body and lung size. However, the structural subdivision of this peripheral volume varies among species because the radius of the alveoli is smaller in smaller species. It follows that, per unit of lung volume, small species have a larger number of alveoli and also a greater gas exchange surface area. Indeed, lung surface area is linearly proportional not to body weight or lung volume, but to $\dot{V}_{O_2}$. Therefore, an important adaptation of the respiratory function to the changes in animal size is achieved structurally by varying the internal compartmentalization of the lung. The larger alveoli of the bigger species require slightly longer diffusion times, but this is manageable in the larger species, which breathe at low rates. However, the abundance of surfactant on the alveolar wall maintains low surface tension even in the smallest

[4]This is essentially true even in the giraffe, in which the trachea is very narrow with respect to its length.

alveoli, therefore avoiding the requirement of large inflating pressures.[5]

B. Passive Mechanical Properties

The fact that larger mammals have bigger lungs than smaller mammals yields the obvious expectations that compliance should increase with size, whereas resistance (the pressure difference per unitary change in flow) should decrease. The former is in fact mostly determined by the size of the organ, whereas the latter is largely determined by the length and diameter of the airways (i.e., the longer and narrower the conducting airways, the more resistance they offer).[6] Indeed, both lung compliance and respiratory system compliance are linearly proportional to the body weight of the animal,[7] and the resistance of the respiratory system decreases in bigger animals according to the proportionality $BW^{-0.75}$. The fact that the resistance does not decrease with animal weight exactly in proportion with the increase in compliance implies that the product of these two variables, which is the time constant of the respiratory system, varies among species, gradually increasing in larger animals.[8] This has important functional implications because the time constant of the respiratory system reflects the time of the ventilatory response to a pressure applied: the shorter the time constant, the faster the volume change in response to the pressure generated by the respiratory muscles. Stated differently, for the same muscle pressure generated by a rat or by an elephant, the lungs will inflate more quickly in the rat because, relative to body weight, the respiratory resistance is not as great and the time constant is not as long as in the elephant.

[5]According to the Young–Laplace relationship, the recoil pressure P generated by an air–liquid spherical interface (as it occurs in the alveoli where the wet surface is in contact with air) is proportional to the ratio between surface tension T and the radius of curvature r of the interface. Hence, for a given T, the smaller the r the larger the inflatory pressure required to overcome P.

[6]In laminar flow conditions the driving pressure is directly proportional to flow (Poiseuille's Law), the proportionality constant being $R = (8\beta l)/(\pi r^4)$, where β is viscosity, and l and r represent, respectively, length and radius of the airways.

[7]This neglects a trend for the largest and bulkiest mammalian species to have slightly stiffer chests.

[8]Compliance of the respiratory system (C_{rs}) αBW^1 and resistance of the respiratory system (R_{rs}) $\propto BW^{-0.75}$; hence, the time constant of the respiratory system (τ_{rs}) $= C_{rs}R_{rs} \propto BW^{(1-0.75)} = BW^{0.25}$.

C. Ventilation and Dynamics of Breathing

The amount of air inhaled per unit time is commonly called ventilation.[9] This is the product of the amount of air inhaled with each breath (tidal volume) and the number of breaths per minute (breathing frequency). Because pulmonary $\dot{V}_E$ represents the only means for the gas exchange of the venous blood in mammals, it is not surprising to find a close relationship between $\dot{V}_E$ and $\dot{V}_{O_2}$. If this was not so then animals with a high $\dot{V}_E/\dot{V}_{O_2}$ would have lower alveolar (and arterial) P_{CO_2} than animals with low $\dot{V}_E/\dot{V}_{O_2}$, contrary to the actual observation that alveolar and arterial blood gases are similar among species.

Because $\dot{V}_E \propto \dot{V}_{O_2}$ and $\dot{V}_{O_2} \propto BW^{0.75}$, $\dot{V}_E$ is also $\propto BW^{0.75}$, i.e., relative to animal size, small animals have higher $\dot{V}_E$ than larger animals. For example, a rat, relative to its weight, ventilates 10–15 times more than an elephant. It is interesting that this difference is not due to differences in tidal volume per kilogram, which is approximately 8–10 ml/kg in all mammals, but to differences in breathing frequency, which can be above 100 breaths/min in the rat and less than 10/min in the elephant. Why differences in metabolic requirements, and in $\dot{V}_E$, are met entirely by differences in frequency rather than tidal volume is not immediately obvious, except for the consideration that lung mass is directly proportional to the weight of the animal, and it is probably a good design to have the stroke of a pump proportional to its size. However, it is apparent that the very high values of breathing frequency in the smallest species, possibly above 600 breaths/min in the shrew, require rapid recruitment of the inspiratory motoneurons and immediate muscle responses to quickly lower the pleural pressure and generate very high inspiratory flows per kilogram. It is interesting that these dynamic requirements are accommodated by some structural and functional properties of the respiratory pump. First, as mentioned earlier, resistance · kg is lower in small animals, and the time constant, whether measured in passive conditions or during contraction of the inspiratory muscles, is proportional to $BW^{0.25}$. This is

[9]Inspired $\dot{V}_E$ is usually slightly larger than expired $\dot{V}_E$ because the ratio between CO_2 produced and O_2 used is less than 1, but the difference is very small. Because there is no evidence for a systematic change in $\dot{V}_{CO_2}/\dot{V}_{O_2}$ with animal size, whether inspired or expired $\dot{V}_E$ is considered does not modify the reasoning.

numerically the same exponent, although opposite in sign, of that relating breathing rate to body weight (frequency $\propto BW^{-0.25}$), implying that the product of these two parameters, frequency and time constant, is an interspecies constant. Second, although the mass of the respiratory muscles is directly proportional to the mass of the animal, the diaphragm of smaller animals has a greater proportion of fast twitch fibers, a higher activity of enzymes involved in muscle contraction, and a faster rate of pressure development than larger species.

D. Distribution of Ventilation

The distribution of tidal volume in the lung is not uniform, as it is subjected to the direction of the gravitational vector with respect to the thorax. In a standing subject breathing quietly from functional residual capacity, most of the inspired air, per unit of lung tissue, is directed into the middle–lower lobes and relatively less air reaches the top portions of the lungs. In a supine position, the areas less ventilated would become those located more ventrally, whereas those relatively more ventilated are the gravity-dependent regions of the back.

The factors responsible for these major differences in inspired air distribution are essentially two in number: the curvilinearity of the lung pressure–volume relationship and the gravity-related variation in pleural pressure. In fact, in a standing subject, pleural pressure is more negative (i.e., more subatmospheric) around the upper lobes of the lung than in the dependent regions of the lower lobes, implying that the lung top is more distended than the bottom. Because lung compliance decreases with increased distention due to the shape of the pressure–volume curve, the overdistended top lung regions are less ventilated than the less inflated regions of the lower lobes. The magnitude of the unevenness in $\dot{V}_E$ distribution therefore depends on the curvilinearity of the pressure–volume curve and on the pleural pressure inequalities along the pleural space. The shape of the pressure–volume curve is similar among mammals. However, thoracic dimensions clearly differ, and if the pleural pressure value at the lung surface was strictly determined by the gravitational field, one should expect the presence of huge top-to-bottom pleural pressure differences in the largest mammals and almost no inequalities among regional pleural pressure values in tiny animals like the shrew or the mouse. This is not the case, and large mammals do not have a distribution of the inspired air much worse than small mammals do because the pleural pressure *gradient* (which is the change in pleural pressure per unitary change in lung height) is smaller the bigger the animal. The precise reasons for why this is so are not clear, as it is not clear how gravity determines the regional differences in pleural pressure, but the end result is interesting; small and large mammals have similar *absolute* differences in pleural pressure among lung regions despite the enormous differences in size. Hence, one would expect that tidal volume is unevenly distributed within the lung in just about the same way in all species; measurements available support this prediction.

In the elephant, and possibly in some other very large mammals with very heavy lungs, some direct attachments between the two pleuras (thoracic and visceral) have been described. This limits the lung configurational freedom within the chest, but represents an additional mechanism for an adequate ventilation of all lung regions.

E. Transport of Gases

Although gases dissolve in blood, neither O_2 nor CO_2 are primarily transported in the dissolved form. O_2 is almost entirely carried by the hemoglobin and CO_2 is carried predominantly in the form of bicarbonate ions, HCO_3^-. Because the solubility of CO_2 in the blood far exceeds that of O_2, the transport of O_2 from the lungs to the peripheral tissues is more critical than the elimination of CO_2 from the tissues.

Because red blood cells are very similar among mammals with respect to size and hemoglobin concentration, the amount of O_2 that can be loaded in the blood essentially depends on two parameters: (1) the total mass of circulating blood and (2) the blood concentration of hemoglobin. In addition, because the amount of O_2 that binds to, and is released from, hemoglobin depends on the affinity of this molecule for the gas, it is important to also consider (3) the O_2–hemoglobin dissociation curve. The total mass of circulating blood per unit of animal weight and the hemoglobin concentration are almost constant in all mammals. Blood mass is about 60–70 ml/kg, and the value of hemoglobin concentration oscillates among species around the value of 15 g/100 ml of blood, which, fully saturated, corresponds to an O_2 concentration of about 20 ml/100 ml of blood, implying that the amount of O_2 in the fully arterialized blood is a relatively fixed proportion of the animal's body weight, or about 13 ml O_2/kg. The approximately constant values in hemoglobin concentration and hematocrit probably reflect the optimal compromise between the advantages of a high O_2 concentration in the blood and the energetic disadvantages of an increased work load on the heart. A higher hematocrit would raise blood viscosity, thus the increase in O_2 delivery would be accompanied by an increase in cardiac work. Only in circumstances of chronic low O_2 availability (e.g., at high altitude, in patients with cyanotic heart disease, or in shrews, some bats, and other very small mammals with high metabolic requirements) does the hematocrit increase break the equilibrium between O_2 content and cardiac energetics in favor of the O_2 transport.

Differences in metabolic rates appear to be matched mostly by the way O_2 is unloaded at the tissue level. In almost all healthy mammals, regardless of size, at an alveolar P_{O_2} of 100 Torr (as occurs in resting conditions at sea level) the hemoglobin is almost 100% saturated with O_2. Conversely, the P_{O_2} at which the hemoglobin is 50% saturated (also called P_{50}) increases with decreasing species body weight. Because of their high metabolic requirements, it is important for small species to have a hemoglobin with low O_2 affinity (i.e., a high P_{50}), capable of easily unloading O_2 at the tissue level.

The affinity of hemoglobin for O_2 is influenced by a number of factors, including the blood pH. With high $[H^+]$ (i.e., low pH), the hemoglobin dissociation curve shifts to the right and the hemoglobin affinity for O_2 is decreased (Bohr effect). The $[H^+]$ at the tissue level depends largely on CO_2, according to the reaction

$$CO_2 + H_2O \rightarrow H_2CO_3 \rightarrow H^+ + HCO_3^-.$$

The first of these reactions (the formation of carbonic acid) is catalyzed by carbonic anhydrase, an enzyme present in the red blood cells where most of the HCO_3^- is formed. In the red cells of the smallest mammals, carbonic anhydrase activity is higher than in the red cells of larger mammals; this further decreases the hemoglobin affinity for O_2 and helps unload O_2 in the tissue capillaries.

In summary, the differences in metabolic requirements among mammals are not met by differences in the O_2 content of the arterial blood, which is an almost fixed proportion of body weight, but by differences in the ability to unload O_2 in the tissue capillaries. [*See* Oxygen Transport to Tissues.]

F. Aspects of the Regulation of the Breathing Pattern

Because of the differences in $\dot{V}_{O_2}$/kg and $\dot{V}_E$/kg among species, it is conceivable that the main regulatory feedback loops controlling the breathing pattern vary with body size, either as a result of the differences in $\dot{V}_E$ or the factors contributing to it. Very little information, however, is available for meaningful generalizations.

Because of the parallel mechanical arrangement of the lungs and chest wall in mammals, either structure would be an appropriate anatomical location for receptors designed to sense changes in lung volume. Mechanoreceptors are indeed located both on the chest wall and in the airways, and both sets project to the respiratory center and influence the breathing pattern. The former group, however, is mostly concerned with the reflex control of the respiratory muscles and the integration of their respiratory activity with nonrespiratory functions, including postural control during locomotion. The mechanoreceptors in the airways, specifically the slowly adapting subgroup, send information via the vagus, which is used to regulate the depth and frequency of breathing. As far as is known, all mammals have both sets of receptors with similar general properties. Whether or not their reflex contribution is also similar is difficult to conclude on the basis of the scattered data. The Hering–Breuer inflation reflex is often interpreted as an index of the reflex effect of the stimulation of the vagal slowly adapting pulmonary receptors. This reflex is particularly pronounced in small species, such as the mouse, rat, guinea pig, and rabbit, and is weaker in cats, dogs, and humans. As previously mentioned, because both tidal volume per kilogram and the pleural pressure swing are similar among species, a more powerful Hering–Breuer reflex would cut off inspiratory activity at an earlier time, possibly contributing to the higher breathing rates of the smallest species.

The $\dot{V}_E$ responses to changes in inspired concentrations of O_2 or CO_2 have been studied in many species. Once exposed to hypoxia, $\dot{V}_E$ increases, relative to metabolic rate, in almost all animals tested, and the degree of this hyperventilation, for the same level of hypoxia, is fairly uniform across species. However, in some animals the hyperventilation is entirely, or almost entirely, due to an increase in the absolute value of $\dot{V}_E$ (hyperpnea), whereas in others it is predominantly due to a decrease in metabolic rate. The latter is particularly pronounced in newborn animals or in adults that have high metabolic rates because of high thermogenic requirements. Therefore, in adult mammals, the hypometabolic contribution to hypoxic hyperventilation occurs mostly in the smallest species with a large surface-to-volume ratio (see earlier) or when the metabolic rate is increased because of the response to a cold environment.

The $\dot{V}_E$ response to CO_2 (hypercapnic ventilatory response) is somewhat simpler to compare among species because the metabolic changes are minimal, at least for modest hypercapnic levels. From the studies available, it would seem that the hypercapnic ventilatory response is slightly less in the smallest species; this is true both among adults or among newborns. Whether this represents a difference at the level of the chemoreceptors or a difference in the multiple factors participating in the central integration of their inputs is not known. Nevertheless, the notion that the ventilatory response to CO_2 may differ among species further complicates the interspecies comparison of the ventilatory response to hypoxia.[10]

VI. RESPIRATION IN THE NEWBORN MAMMAL

Because newborns are smaller than adults of the same species, one may expect that differences in the absolute values of structural or functional variables should relate to the previously discussed effects of size on the respiratory function (Section V). In addition, age-related differences occur because of developmental changes not necessarily related to size. One method of separating size-dependent factors from development-dependent factors, either or both of which may cause differences between newborns and adults, is to compare same-sized animals of different ages and species, e.g., comparing a 250-g newborn puppy with a 250-g adult rat. More generally, the allometric relationships[11] of newborns and adults (or any other age) could be compared; were the curves overlapping, the conclusion could be reached that size is the primary factor in the age-related differences of the parameter under consideration; however, differences between the allometric relationships indicate the contribution of maturational aspects not related strictly to body size.

Such an analysis has revealed that in most species $\dot{V}_{O_2}$/kg is higher in newborns than in adults of the same species (Fig. 3); this is not only because of the size difference (e.g., the surface–volume argument presented earlier), but also because of developmental factors related to the large anabolic requirements of the growing organism and differences in thermoregulatory mechanisms. However, in newborns smaller than 80–100 g, $\dot{V}_{O_2}$/kg can be equal to or even less than the value expected in same-sized adults of different species (Fig. 3). A similar pattern occurs for $\dot{V}_E$/kg, since in newborns, as in adults, $\dot{V}_E$ is tightly coupled to the metabolic requirements.

Why the smallest species metabolic and ventilatory rates are not much higher in newborns than in adults, and not in keeping with the newborn–adult differences commonly observed in the larger species, is not immediately clear. It could be noted that if, for example, the newborn mouse was breathing four times faster than the adult mouse, i.e., was maintaining the same newborn–adult ratio as in larger species,[12] it would need to breathe approximately 500 times/min (i.e., four times 120 breaths/min, the approximate breathing rate of the adult mouse). Even assuming a massive recruitment of the inspiratory motoneurons and a very rapid generation of pleural pressure, the time constant of the respiratory system should be extremely short in order to inflate the lungs in the very brief inspiratory time available. In a newborn mouse the time constant is about 60 msec, i.e., much shorter than in newborns of larger species or adults, but this is still too long to permit ventilation at 500 breaths/min. Some species are born at a very immature stage, e.g., the opossum and other marsupials. In these newborns, chest wall compliance is very high and their lungs are voluminous with respect to body size, with large peripheral units. Both aspects do not favor a small value of compliance of the respiratory system and therefore present a structural limit to a further shortening of the time constant. Thus, it is possible that in the smallest newborn species the mechanical

[10] As the hypoxic animal hyperventilates, alveolar and arterial P_{CO_2} decrease. Decreased P_{CO_2} tends to decrease $\dot{V}_E$ and thus tends to counteract the response to hypoxia by a factor that depends on the level of hyperventilation and the sensitivity of the animal to changes in P_{CO_2}.

[11] Allometric analysis is a special case of normalization by body weight (BW) where the effect of animal size (either within a species during growth or among same-age animals of different species) on a given structural or functional parameter Y is examined according to the function $Y = a\text{BW}^b$. The log-transformed version of this function, $\log Y = \log a + b \log \text{BW}$, is particularly useful; when the slope b, which is the exponent of the original exponential function, equals unity, the variable Y increases in direct proportion to the body weight of the animal, with the proportionality factor being a. Slopes higher or smaller than unity indicate that the variable under consideration increases, respectively, disproportionately more or less with the increase in body weight.

[12] For example, during resting conditions a newborn infant breathes with approximately the same tidal volume per kilogram of the adult man (about 8 ml/kg), but its rate, at 50–60 breaths/min, is about four times higher.

characteristics of the respiratory system pose some limit to the ventilatory performance and therefore, possibly, to $\dot{V}_{O_2}$.

The characteristic large chest wall compliance of newborn mammals is an essential structural requirement at birth and favors the mechanics of delivery. After birth, the high ratio of chest wall-to-lung compliance contributes to important differences, with respect to the adult, in the dynamic properties of the respiratory system. The high ratio causes a small resting volume of the respiratory system, which is usually compensated for in the newborn by a dynamic elevation of the end expiratory level. This is achieved via partial closure of the glottis and activity of the inspiratory muscles during expiration, i.e., by prolonging the time of deflation with respect to the neural duration of expiration. The high chest wall-to-lung compliance ratio and the incomplete maturation in the control of the intercostal muscles make the chest wall of the newborn more susceptible to deformation than the adult's. During diaphragmatic contraction and the parallel drop in pleural pressure, a paradoxical inward movement of the thorax during inspiration is a frequent observation in the young of many species, including the human infant.

As may be expected, many of the adult regulatory mechanisms are not completely developed at birth, particularly in the altricial species (the human infant is one of these), resulting in the newborn mammal being more sensitive than the adult to environmental changes and less capable of protecting its own internal environment. The $\dot{V}_E$ response to hypoxia has been extensively studied; newborns, like adults, hyperventilate in hypoxia, but, differently from the adult, they often hyperventilate with little or no hyperpnea, solely by decreasing the metabolic rate. The adaptive hypometabolic response, on the one hand, prolongs their living in acute hypoxic or asphyxic situations; indeed newborns are formidable survivors in poor oxygen conditions. On the other hand, the low metabolic rate reduces and alters body and organ growth, with potentially dramatic short- and long-term consequences.

VII. RESPIRATION IN MAMMALS ADAPTED TO SPECIAL ENVIRONMENTS

A. Diving

For many land mammals, including humans, breath-holding time is of the order of a few minutes. The progressively evolving asphyxia eventually interrupts the $\dot{V}_E$ inhibition, via the stimulation of the chemoreceptors. Some diving mammals, however, can breath-hold for much longer times; dives lasting more than 1 hr have been observed in some seals and whales.

The physiological basis for these achievements is not completely clear, but it is probably the result of many combined mechanisms. The ventilatory response to hypoxia and hypercapnia is usually decreased in adult diving mammals; this decrease is probably acquired with diving experience, since at birth newborn seals have a ventilatory response to hypoxia as brisk as that of other precocial species.

When seals free to swim in a covered tank with only one opening to air are confronted with hypoxia or hypercapnia, they respond by shortening the dive duration and the time intervals between dives rather than by increasing the depth and rate of breathing at the air hole. The suggestion was made that the long breath-holding time may also be favored by a blood-buffering capacity higher than in most land mammals. Other less studied mechanisms that could prolong the resistance to asphyxia include the ability to shut off the perfusion of some body districts, therefore avoiding or delaying systemic acidosis, with an overall reduction in the metabolic rate.

Although some divers, such as otters and seals, can spend a long time at the surface or even out of the water, others, such as dolphins and whales, ventilate only in the short time intervals between dives. This implies the generation of very high inspiratory and expiratory flow rates. The problem of expiratory air flow limitation[13] is reduced by having relatively stiff airways, with solid continuous cartilaginous rings in the larger airways instead of the flexible horseshoe cartilages of most land mammals. During deep dives, the stiff conductive airways also accommodate the gas shifted from the periphery of the lung during chest compression by the hydrostatic pressure. The elimination of air from the gas exchange region into the dead space limits the amount of nitrogen dissolved in the blood and in the tissues, and is therefore an important precaution against potentially fatal embolic problems during the rapid decompression of surfacing. In this

[13]To increase expiratory flow, expiration requires active contraction of the expiratory muscles, which increase intrathoracic (pleural) and alveolar pressures. Although at first the expiratory flow increases, the rise in pleural pressure eventually leads to the narrowing of some intrathoracic airways. The more the expiratory muscles increase pleural pressure, the more the airways narrow, limiting any further increase of expiratory flow (expiratory air flow limitation).

respect it is of interest that, at least in the seal, the chest wall opposes the recoil of the lung only at very small volumes and therefore does not hinder lung compression by the hydrostatic pressure during a dive. Of course, this mechanical arrangement means that, when out of water, the resting lung volume of the seal is very low. However, the potential problems of a low resting volume on land are eliminated by adopting a peculiar breathing pattern, with occlusions of the airways at end inspiration or in the middle of expiration, therefore maintaining the lung volume elevated between breaths. This pattern is rarely seen in adult land mammals, but is very much reminiscent of the strategy adopted by reptiles and by all newborn mammals during the first hours after birth, at a time when the resting lung volume is very low for the mechanical reasons mentioned in the previous section.

B. High Altitude

Many mammalian species manage to survive at altitudes above 3500 m, where barometric and O_2 pressures are below 60% of the sea level values. Some of them are actually permanent inhabitants of these low O_2 regions, e.g., some Camelidae of the Andes (llama, alpaca, vicuña, guanaco), the bovid mountain goat in the Rockies, and yak in the Himalayas. Some mammals are known to visit, at least temporarily, regions above 5000 m. Humans have settled high-altitude regions in the Andes and Tibet between 3500 and 4500 m. However, it is well known that lowlanders exposed to these altitudes often face a number of problems that are life-threatening in some instances and are resolved only by a rapid return to lower levels. The question of what major physiological features permit life at high altitude is of interest. The fact that no clear answer is as yet available probably indicates that there is no unique strategy among species for high-altitude adaptation.

High-altitude populations are characterized by short stature, a relatively large thorax, and big lung volumes. At least some of their pulmonary features are probably not genetic traits but are acquired with the prolonged exposure to hypoxia, although the functional basis for these acquisitions and whether or not they begin during the embryonic and fetal period are not known. Pregnant women at high altitude give birth to small infants with slightly higher values of hematocrit, hemoglobin, and compliance. Highlanders typically have a lower $\dot{V}_E$ and a blunted $\dot{V}_E$ response to hypoxia when they are compared to lowlanders or lowlanders who moved to and lived at high altitude for some time. Very high values of hematocrit and hemoglobin concentration are other characteristics of the native highlanders, who also have a rightward shift of the hemoglobin dissociation curve (with high P_{50}). Whether or not these features can be considered as the physiological prerequisite for high-altitude adaptation is not clear. Many high-altitude animals do not have the same acclimatization properties observed in humans. For example, llamas, yaks, and sheep respond to acute hypoxia with a hyperventilation similar to that of corresponding low-altitude animals. Their values of hematocrit and hemoglobin concentrations are not particularly elevated; in addition, high-altitude Camelidae, rodents, and ruminants have a leftward instead of a rightward shift of the hemoglobin–O_2 affinity curve. All of these characteristics may reflect a better adaptation to extremely hypoxic environments than that observed in humans.

C. Burrowing

The strategy of hiding under ground against predators is common among animals, and many mammals, particularly rodents, marsupials, and insectivores, adopt it either permanently or intermittently. Values of O_2 and CO_2 concentrations in the burrow depend on many factors, including the design of the tunnels and the characteristics of the soil; the concentration of O_2 can be as low as 4–6% in the burrows of the marmot and pocket gopher, and that of CO_2 as high as 6–8% in the burrows of chipmunk, echidna, marmot, or ground squirrel. Hence, as opposed to the hypoxic conditions at high altitude, life in the burrow is characterized by an asphyxic environment, both hypoxic and hypercapnic.

In general, burrowers have a lower $\dot{V}_E$ than nonburrowing mammals, and whether their arterial P_{CO_2} is elevated or not depends on the metabolic rate, which in some burrowers is less than in nonburrowing mammals of a similar body size. The $\dot{V}_E$ response to hypercapnia is often described to be less brisk than in humans or in dogs, and the few data available would also suggest the presence of a blunted $\dot{V}_E$ response to hypoxia. The O_2-carrying capacity, reflected by hematocrit and hemoglobin concentration, is increased in some but not all burrowing species. The finding of a reduced P_{50}, i.e., of a higher affinity of hemoglobin for O_2, seems to be more consistent. The blood-buffering capacity is increased in some burrowing species, whereas in others it is within the range found in nonburrowing mammals. In the first case, blood pH is protected despite hypoventilation, whereas in the latter case the acidosis must be more pronounced, but oxygen delivery to the tissues is improved via the Bohr effect.

BIBLIOGRAPHY

Andersen, H. T. (1966). Physiological adaptations in diving vertebrates. *Physiol. Rev.* **46,** 212–243.

Boggs, D. F. (1992). Comparative control of respiration. *In* "Comparative Biology of the Normal Lung" (R. A. Parent, ed.), pp. 309–350. CRC Press, Boca Raton, FL.

Boggs, D. F., Kilgore, D. L., Jr., and Birchard, G. F. (1984). Respiratory physiology of burrowing mammals and birds. *Comp. Biochem. Physiol.* 77, 1–7.

Dejours, P. (1975). "Principles of Comparative Respiratory Physiology." North-Holland, Amsterdam.

Leith, D. E. (1976). Comparative mammalian respiratory mechanics. *Physiologist* **19,** 485–510.

Mortola, J. P. (1987). Dynamics of breathing in newborn mammals. *Physiol. Rev.* **67,** 187–243.

Mortola, J. P. (1996). Ventilatory responses to hypoxia in mammals. *In* "Tissue Oxygen Deprivation: Developmental, Molecular and Integrated Function" (G. G. Haddad and G. Lister, eds.), pp. 443–477. Dekker, New York.

Schmidt-Nielsen, K. (1986). "Scaling: Why Is Animal Size so Important?" Cambridge University Press, Cambridge.

Tenney, S. M., and Bartlett, D., Jr. (1981). Some comparative aspects of the control of breathing. *Lung Biol. Health Dis.* **17,** 67–101.

Wood, S. C., and Lenfant, C. (1979). Evolution of respiratory processes: A comparative approach. *Lung Biol. Health Dis.* **13.**

Respiratory Burst

ANTHONY J. SBARRA, CURTIS CETRULO, ROBERT KENNISON, MARY D'ALTON, JOSEPH KENNEDY, JR., FARID LOUIS, JOHANNES JONES, CHRISTO SHAKR, and DALE REISNER
St. Margaret's Center for Women and Infants at St. Elizabeth's Medical Center, Tufts University School of Medicine

---

GLOSSARY

Hexose monophosphate pathway Direct oxidative oxidation of glucose

Myeloperoxidase Heme enzyme located in the azurophilic granules of neutrophils

NADPH oxidase Nicotinamide adenine dinucleotide phosphate reduced oxidase, a respiratory coenzyme

Phagocytosis Engulfment of particulate material (viable and/or nonviable) by a cell

Respiratory burst Interaction of material (particulate or nonparticulate) with cells resulting in a burst of oxidative activity

ROS Reactive oxygen species; an electron-reduced species of oxygen

WHEN PHAGOCYTES, MAINLY POLYMORPHOnuclear neutrophilic leukocytes, monocytes, eosinophils, and lung and peritoneal macrophages are exposed to a number of different stimuli, particulate and nonparticulate, they all undergo immediate and dramatic changes in the manner in which they process oxygen. Oxygen uptake undergoes a significant and rapid increase, frequently 50-fold over that of resting cells; direct oxidation of glucose by the hexose monophosphate pathway (HMP) is similarly dramatically increased. These increased oxidative changes noted in cells exposed to particulate material were initially known as the respiratory burst. The major functions of this oxidative burst were earlier thought to generate powerful bactericidal agents by the partial reduction of oxygen. Clearly, however, this respiratory burst has much wider biological application and physiological function.

I. INTRODUCTION

It has been established that the respiratory burst, which all phagocytic cells undergo, is nonmitochondrial during phagocytosis and is due principally to activation of a unique membrane-bound nicotinamide adenine dinucleotide phosphate reduced (NADPH) oxidase. The physiological significance of this oxidative burst was highlighted when it was noted that the absence of the burst in phagocytes from children with chronic granulomatous disease (CGD) were found to lack the ability to kill phagocytized bacteria. This resulted in an increased susceptibility to infections in these children. By studying the leukocytes from children with this disease, it was shown that the primary product of the respiratory burst, the superoxide anion (O_2^-), was significantly reduced. At physiological pH and ionic concentration, the dismutation of superoxide to H_2O_2 and O_2 occurs. The H_2O_2 produced is not in sufficient quantity to kill phagocytized bacteria, but when reacted with myeloperoxidase (MPO) released from the azurophilic granules of the neutrophils, a powerful antimicrobial system is

ENCYCLOPEDIA OF HUMAN BIOLOGY, Second Edition, VOLUME 7. Copyright © 1997 by Academic Press. All rights of reproduction in any form reserved.

formed, which is primarily responsible for the oxygen-dependent antimicrobial activity of the phagocyte. In the absence of oxygen, this antimicrobial system is significantly inhibited. [*See* Phagocytes.]

Although initially the respiratory burst was thought to include only the immediate increase in oxygen consumption, flow of glucose through the HMP, and increased glycolysis during phagocytosis, obviously a more inclusive list of products would be identified and associated with it. The generation of O_2^-, H_2O_2, singlet oxygen, hydroxyl radicals, and some MPO-related products, namely, hypochlorous acid (HOCl) and chloromines, are such products.

II. BIOCHEMICAL AND ENZYMATIC BASIS OF THE RESPIRATORY BURST

A. Oxygen Consumption and the Respiratory Burst Oxidase

The respiratory burst oxidase was at first controversial, but it is now firmly established as NADPH oxidase and not nicotinamide adenine dinucleotide reduced (NADH) oxidase. This is consistent with the activity of the HMP as it is controlled by the generation of nicotinamide adenine dinucleotide phosphate (NADP) from NADPH. Studies with CGD leukocytes have substantiated this finding conclusively. In normal leukocytes, the burst in glucose oxidation by the HMP and in CO_2 production is regulated by NADPH oxidase.

The degree of oxygen stimulation is, to a considerable degree, dependent on the stimulating agent. The generation of superoxide O_2^- by neutrophils during phagocytosis is the result of the one-electron-reduced species of oxygen. The stoichiometry between O_2 consumption and O_2^- generation is controversial owing to the methodology used for assay. Values of 4 : 1 to 2 : 1 between O_2^- generation and O_2 uptake have been reported. The generation of O_2^- is quickly followed by its dismutation to H_2O and O_2. Also, it has been suggested by some, and disputed by others, that the spontaneous dismutation of O_2^- results in generation of singlet oxygen $^1O_2$.

B. Nature and Molecular Structure of NADPH Oxidase

The properties of NADPH oxidase are well defined; its molecular structure, however, is not. Most evidence indicates that the O_2^- generating system of phagocytes is formed by different components forming an electron transport chain:

$$\text{NADPH} \rightarrow \text{flavoprotein} \rightarrow \text{cytochrome } b_{558} \rightarrow O_2$$

The precise structure of components comprising the NADPH oxidase is still uncertain. A major problem in its purification is its extreme lability on detergent extraction and on the usual purification manipulation methods. Despite these and other difficulties, molecular weights of some preparations reportedly are between 150,000 and 1,000,000. Electrophoresis of some of these preparations gave molecular weights of 87,000 and 32,000 for major components. By using different preparation methods, electrophoretic patterns of still different molecular weights were obtained. Apparently, the real nature of NADPH oxidase will not be known until the difficulties of purifying to homogeneity are solved.

C. Intracellular Killing of Microorganisms

MPO is a heme enzyme localized in the azurophilic granules of neutrophils. Release of MPO from these granules to phagolysosomes generates the reactive oxygen species (ROS), which are thought to be responsible for the bactericidal activity of the phagocyte. H_2O_2, at concentrations not normally antibacterial, when acting in concert with MPO and a halide, mainly chloride, forms an effective antimicrobial and cytotoxic system. The major primary product produced by this system is HOCl. The primary product of the respiratory burst is O_2^-, and H_2O_2 is the secondary product generated from the dismutation of O_2^-. The hydroxyl radical (OH·) is also thought to be generated in an iron-catalyzed reaction in which O_2^- and H_2O_2 are reactants (Haber–Weiss reaction). The evidence for OH· involvement in the mediation of the respiratory burst-dependent processes is based on the use of radical scavengers such as mannitol. Oxygen consumed in the respiratory burst is ultimately recovered as H_2O_2. All of the reactive intermediates formed are relatively short-lived; however, the generation of relatively stable and reactive chloramine species is also noted. These are formed from the reaction of HOCl and peptides released from cells during the respiratory burst. The intracellular killing of microorganisms can be separated into phases: (1) microorganisms attach to the cell surface and are then engulfed by the pseudopods of the phagocyte; (2) the intracellular granules fuse with the phagocytic membrane; and (3) the NADPH oxidase is activated. Within the phagosome, O_2^- is dismutated to H_2O_2, which then reacts with a halide and MPO to form HOCl. All of these end products are thought to have antimicrobial and cytotoxic activity. [*See* Neutrophils.]

III. THE RESPIRATORY BURST AND DIFFERENT CELL TYPES

A. Eosinophils

Eosinophils can be stimulated by a number of different agents and respond with an increased oxygen uptake, O_2^- and H_2O_2 production, and stimulation of glucose through the hexosmonophosphate shunt. A major difference between the respiratory burst of neutrophils and eosinophils is that with eosinophils the burst is sensitive to sodium azide. Also, eosinophils reportedly exhibit a higher metabolic burst than neutrophils. Eosinophils have NADPH oxidase activity, and this activity is stimulated in phagocytizing cells. The cells are able to kill a number of different bacteria, including *Staphylococcus aureus* and *Escherichia coli.*

The inhibitory effect of azide on the respiratory burst of eosinophils and not on neutrophils is presently unexplained.

B. Lymphocytes

Generally, lymphocytes are believed to not experience a respiratory burst as do phagocytic cells. However, the respiratory burst and the resulting production of the different ROS by phagocytic cells can affect the lymphocyte. Because these cells are intimately involved in the immunological and inflammatory response, the effects could be highly significant. Presently, it is agreed that the functional capacity of blood T and B lymphocytes is impaired by oxidant damage resulting from the respiratory burst of phagocytizing cells. [*See* Lymphocytes.]

IV. THE RESPIRATORY BURST AND THE METABOLISM OF DRUGS

Theoretically, all ROS generated by the respiratory burst have the ability to metabolize drugs. A number of model systems have been developed and are being used to study the metabolism of drugs. These include the horseradish peroxidase–H_2O_2 system, the catalase–H_2O_2 system, the arachidonic acid metabolism to prostaglandins, and the myeloperoxidase–H_2O_2–halide system. Drugs such as hydroquinones, aromatic amines, hydroxylamines, and hydrazines have been shown to be metabolized by the myeloperoxidase–H_2O_2–halide system of neutrophils.

Recently, it has been shown that products of the respiratory burst, HOCl, and taurine chloramine solutions produced drug-free radical intermediates from chlorpromazine, aminopyrine, and phenylhydrazine. On the basis of these findings, it has been postulated that the oxidation of certain compounds by neutrophils *in vivo* could generate damaging electrophilic free-radical forms. The MPO–H_2O_2–halide system could possibly serve as a unique metabolic pathway for the biotransformation of these drugs *in vivo* and in drug toxicity and chemical carcinogenesis. These observations and postulations provide some evidence that the generation of ROS by the MPO pathway presents a biochemical mechanism *in vivo* for the metabolism of these drugs. Also, the toxic compounds produced could result in tissue damage.

V. THE RESPIRATORY BURST AND ITS EFFECT ON TISSUE

Many studies have been carried out that have clearly demonstrated that the ROS generated by the respiratory burst are implicated in inflammation and tissue injury. The mechanism(s) responsible for these reactions are complex and not precisely defined. Changes in membrane function, brought about by lipid peroxidation, have been reported. Carbohydrates and proteins have also been shown to be modified by ROS. Specifically, O_2^-, H_2O_2, OH·, and HOCl have all been associated with tissue destruction and the inflammatory process. These products are considered to be highly cytotoxic. Cytotoxic activity exhibited by these products is dependent on different factors, including target cell or area of attack and initiating stimulus. An example of tissue injury by ROS is the H_2O_2-mediated lung injury utilizing the glucose–glucose oxidase H_2O_2-generating system. The oxidase is directly added to tissue, and cytotoxicity is subsequently monitored. Interestingly, lung injury can be inhibited by catalase. [*See* Inflammation.]

Overwhelming evidence indicates that ROS produced during the respiratory burst are involved with tissue injury; however, because of the complexity of the system(s) involved, much remains to be learned.

VI. THE RESPIRATORY BURST AND ITS EFFECT ON DIFFERENT PHYSIOLOGICAL STATES

The respiratory burst has been shown to be involved in a number of different diseases and physiological states. Some of these diverse states include diabetes mellitus, atherosclerosis, aging, fertilization, psoriasis, and pregnancy.

Human pregnancy presents a unique example of the involvement of the respiratory burst in another physiological process. It has been suggested that a common denominator exists among the onset of human labor, preterm labor, and premature rupture of the membranes (PROM). That common denominator is phagocytosis and the accompanying respiratory burst. A summary of the events leading to the onset of human labor, preterm labor, and PROM may be seen in Fig. 1. The phospholipase A_2 can be released by microorganisms as well as by lysosomal release as a result of surfactants reacting with the phagocyte. At term, surfactant arising from the fetus is at its maximal concentration. It has been postulated that this surfactant interacting with decidual cells will exhibit a respiratory burst and lysosomal phospholipase A_2 will be released. This enzyme can then initiate the events leading to the onset of normal human labor. Peroxidase can also be released from the cells. This enzyme will form a powerful cytotoxic system with

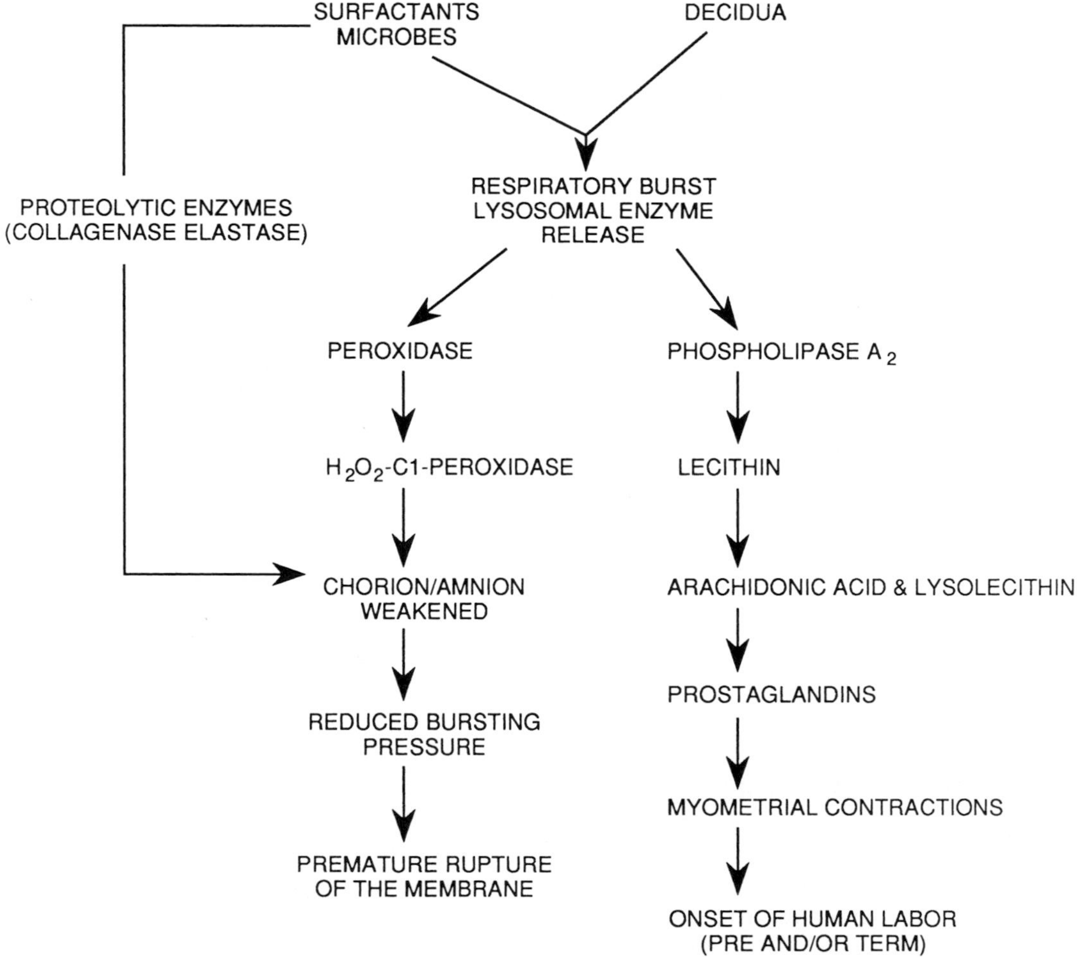

FIGURE 1 Postulated sequence of biochemical events relating the respiratory burst to premature rupture of the membranes and of preterm and term human labor.

metabolic H_2O_2 and a halide. It can attack the chorion amnion and weaken it. This has been observed by the target tissue's significantly reduced bursting pressure when compared with unexposed target tissue. By reducing the bursting pressure of the chorion-amnion, one can expect to see a premature rupturing of the membranes.

VII. SUMMARY

As apparent from the foregoing discussion, the respiratory burst is not limited, as was originally thought, to simply supplying reactive oxygen species (i.e., O_2^-, H_2O_2, and OH·) that are used to kill intracellular microorganisms. It has been shown to have a much wider scope. This is not surprising if phagocytosis and pinocytosis are the mechanism(s) that cells use to eat and drink. Its much wider biological and physiological applications have been presented.

BIBLIOGRAPHY

Babior, B. M. (1992). The respiratory burst oxidase. *Adv. Enzymol.* **65**, 49.

Cohen, M. S. (1994). Molecular events in the activation of human neutrophils for microbial killing. *Clin. Infect. Dis.* **18** (Suppl 2), 5170.

Klebanoff, S. J., and Clark, R. A. (eds.) (1978). "The Neutrophil: Function and Clinical Disorders." Elsevier/North-Holland Biomedical, Amsterdam.

Rossi, F. (1986). The O_2-forming NADPH oxidase of the phagocyte; Nature, mechanisms of activation and function. *Biochim. Biophys. Acta* **853**, 65.

Sbarra, A. J., and Strauss, R. R. (eds.) (1988). "The Respiratory Burst." Plenum, New York.

Segal, A. W., and Abo, A. (1993). The biochemical basis of the NADPH oxidase of phagocytes. *Trends Biochem. Sci.* **18**, 43.

Respiratory System, Anatomy

ANTHONY J. GAUDIN
KENNETH C. JONES
California State University, Northridge

GLOSSARY

Alvelous Microscopic sac that provides a unicellular surface for the movement of oxygen between the blood and the interior of the lung

Epiglottis Cartilaginous flap of tissue that functions to close the glottis during swallowing

Glottis Opening from the pharynx into the larynx

Mediastinum Space between the lungs that contains the heart and several large blood vessels

Trachea Tube that extends from the larynx to the first branch of the primary bronchi

THE ACQUISITION OF OXYGEN AND THE ELIMINATION of carbon dioxide are accomplished by the respiratory tract, a system of passageways that leads from the nose and the mouth to tiny chambers in the lungs. Oxygen diffuses into the blood from the gas in these chambers and is carried to the rest of the body as the blood is pumped through the circulatory system. As oxygen diffuses into the blood, carbon dioxide carried by the blood from active tissues diffuses into the chambers. The respiration accomplished by the lungs consists of pulmonary ventilation, used to move air into and out of the lungs; external respiration, the exchange of gases between the lungs and the blood; and internal respiration, the exchange of gases between the blood and the tissues of the body. Pulmonary ventilation is further divided into either inspiration (or inhalation), the mechanism that draws air into the lungs, or expiration (or exhalation), the mechanism that forces air out of the lungs. A more common term for pulmonary ventilation is "breathing."

I. ORGANIZATION OF THE RESPIRATORY TRACT

The respiratory tract consists of an upper and a lower respiratory tract (see Color Plate 19). The upper respiratory tract includes the external nose, the nasal cavity that lies posterior to the nose, a chamber behind the nasal cavity called the nasopharynx, and the larynx. The mouth can also be involved in inhalation and exhalation, but it is not normally considered part of the respiratory tract. The pharynx is a region common to both the respiratory and digestive tracts and is traversed both by food on its way to the stomach and by air on its way to the lungs.

The lower respiratory tract consists of a network of branching tubes and tubules that lead to a system of sacs, where gas exchange occurs. The first portion of the tubular system consists of the trachea, a tube that descends downward from the larynx. The trachea branches into bronchi, which in turn branch into successively smaller subdivisions called bronchioles. The bronchioles ultimately terminate as blind sacs, called alveoli. Collectively, the network of tubules and alveoli comprises the lungs.

A. Upper Respiratory Tract

The nose provides the opening to the respiratory system. It is formed by cartilage and bone, covered by integument (see Color Plate 20). The two openings

ENCYCLOPEDIA OF HUMAN BIOLOGY, Second Edition, VOLUME 7. Copyright © 1991 by Academic Press. All rights of reproduction in any form reserved.

at its base are the nostrils, or the external nares. Each nostril leads into a vestibule, an enlarged region immediately behind the nostrils. The vestibule leads into the nasal cavity, which extends from the internal portion of the nose into the skull and terminates at the internal nares (or choanae), a pair of openings in the posterior portion of the nasal cavity. The external nares are rimmed by many relatively coarse hairs that filter large particles from inhaled air.

1. Nasal Cavity

The roof of the nasal cavity is formed by the cribriform plate of the ethmoid bone. This plate is perforated by olfactory sensory cells. The floor of the nasal cavity is formed by the superior surface of three structures: the maxilla, the palatine bone of the hard palate, and, posteriorly, the soft palate. The cavity is divided laterally into two nasal fossae by the vertical nasal septum, which extends from the floor of the cavity to the roof. The nasal septum consists of an anterior cartilaginous portion that attaches to the flat perpendicular plate of the ethmoid bone suspended from the roof of the cavity. The lower and posterior portions of the septum consist of another flat perpendicular bone that rests on the floor of the cavity formed by the vomer. Each fossa is also divided into passageways (i.e., meatus) by three bony projections of the lateral walls of the cavity called nasal conchae, or nasal turbinates. [*See* Skeleton.]

The wall of the nasal cavity is a mucous membrane consisting of epithelial tissue richly supplied with blood vessels and mucus-secreting glands. Blood carried in the vessels delivers heat to the inspired air, warming the air as it passes over the tissue. At the same time air is moistened by the evaporation of water from the mucus-covered surface, so air passing through the nasal cavity is warmed (or cooled, depending on the temperature outside), cleaned, and moistened. Mucus is sticky, so it traps much of the material that has passed through the hairs at the nostrils. As the air is moistened, the mucous surface becomes drier and forms semisolid material, in which the fine particulate matter is trapped. This material is carried toward the posterior portion of the nasal cavity by a current of fluid established by ciliated epithelial cells in the tissue lining the cavity. Rhythmic and coordinated motion of these cilia moves the mucus out the posterior exit of the nasal cavity and into the oropharynx. Once in the oropharynx, the material is either swallowed or spat out.

During development, outcroppings from the nasal membrane delineate cavities surrounded by the bony tissue of the developing skull. After development is complete, the cavities remain as the paranasal sinuses. Maxillary sinuses, the largest of the paranasal sinuses, are located in the body of the maxillary bones on either side of the nasal fossae. Frontal sinuses are located in the frontal bone above the eyes, and the ethmoid sinuses are found in the ethmoid bone in the lateral wall of the nasal cavity (Fig. 1). Unlike other sinuses, ethmoid sinuses consist of numerous individual air cells, clustered in groups that collectively compose the sinuses. The sphenoid sinus is a single chamber in the sphenoid bone, just above and behind the junction of the ethmoid bone and the vomer in the posterior portion of the nasal septum (see Color Plate 20).

Each sinus is lined with epithelial tissue similar to that of the nasal cavity itself. The tissues also produce a mucus that drains through ducts that pass from the sinus into the nasal cavity. These ducts are relatively narrow, and when they swell as a result of infection, drainage of the mucus is blocked. When this happens, the mucus and other fluids collect in the sinuses, causing a buildup of pressure that can be very painful.

2. Pharynx

The pharynx is a chamber about 12 cm long that connects the oral and nasal cavities with the esophagus and the larynx. As such, it provides a common pathway for food on its way to the digestive tract and air on its way to the respiratory tract (Fig. 2). Air passing through the nasal fossae exits the nasal cavity through the internal nares and enters the upper part

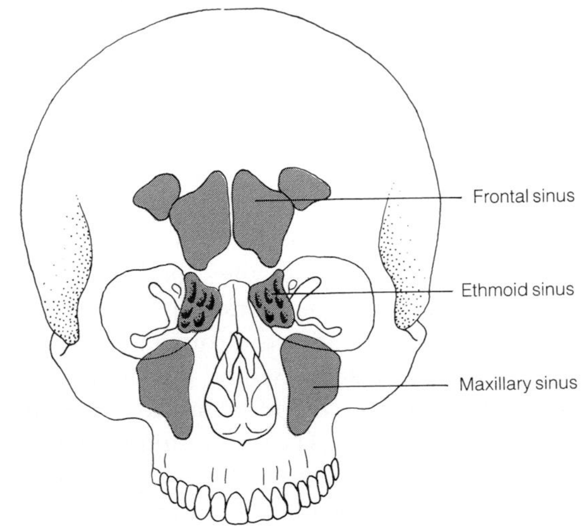

FIGURE 1 The sinuses. [From A. J. Gaudin and K. C. Jones (1989). "Human Anatomy and Physiology," p. 508. Harcourt Brace Jovanovich, San Diego.]

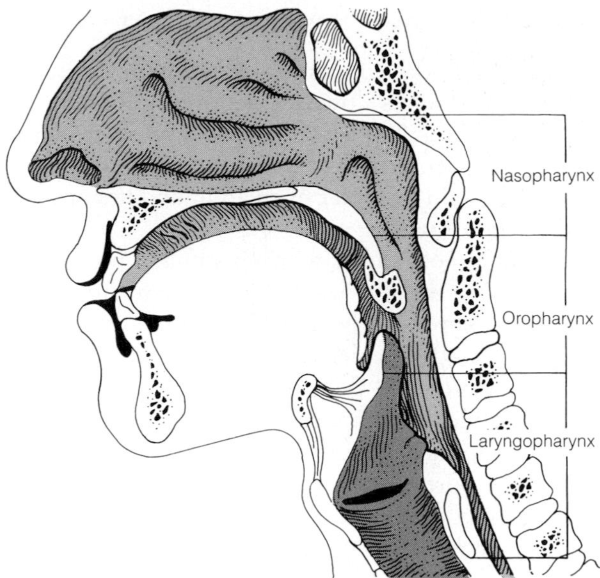

FIGURE 2 The pharynx. [From A. J. Gaudin and K. C. Jones (1989). "Human Anatomy and Physiology," p. 509. Harcourt Brace Jovanovich, San Diego.]

of the pharynx in a region above the soft palate, called the nasopharynx.

The roof of the nasopharynx is formed by the sphenoid bone at the base of the skull, and its floor is formed by the soft palate. The lateral walls of the nasopharynx contain the openings of the auditory tubes, called the pharyngeal apertures. The posterior portion of the nasopharynx contains a broad flat mass of lymphoid tissue, called the pharyngeal tonsils (or adenoids). Infection can cause the adenoids to swell and interfere with the passage of air and mucus through the nasopharynx.

At the posterior edge of the soft palate, the nasopharynx opens into the oropharynx. The oropharynx communicates anteriorly with the oral cavity, making it directly visible through the open mouth. The oral cavity communicates with the oropharynx through the isthmus of the fauces, a narrow passageway formed by the base of the tongue, the soft palate, and two curved folds of tissue that lie anterior and posterior to the palatine tonsils. The anterior folds form the palatoglossal arch and mark the boundary between the oral cavity and the oropharynx. The oropharynx leads downward into the laryngopharynx, a short tube that lies posterior to the larynx.

3. Larynx

After passing from the nasal cavity into the nasopharynx, inhaled air passes through the oropharynx and enters an opening in the anterior wall just below the base of the tongue: the laryngeal aperture. Separating the aperture from the base of the tongue is a leaf-like cartilaginous flap of mucous membrane-covered tissue, called the epiglottis. During swallowing, the larynx is elevated, forcing the epiglottis down over the opening into the larynx and preventing food or liquid from passing into the respiratory tract.

Air passing past the epiglottis and through the laryngeal aperature enters the larynx. The larynx acts as a valve that controls access to the tubular system that lies below it (see Color Plate 21).

The larynx consists of bony and cartilaginous structures held together by ligaments, muscles, and other tissues. The larynx lies below the horseshoe-shaped hyoid bone. Suspended from the hyoid bone by ligaments and membrane is the thyroid cartilage, the largest unit of the larynx. The thyroid cartilage produces a bulge in the front of the neck called the laryngeal prominance, better known as the Adam's apple.

Suspended from the thyroid cartilage by ligaments and muscle is the cricoid cartilage, a ring-like structure that nests at the base of the thyroid cartilage. The cartilages give the larynx its form and support the tissues used by the larynx in acting as a valve and a sound-producing organ.

The interior of the larynx is lined with mucous membrane that forms folds of tissue that extend into the passageway (Fig. 3). Just within the laryngeal aperture is an expanded region called the vestibule, which is limited inferiorly by two such folds, called the vestibular folds, or "false vocal cords." Just below these folds there is another widening of the channel, followed by two more folds of the membrane. These are the vocal folds, or "true vocal cords," and are responsible for producing sound, as air passing over them from the lower respiratory tract causes them to vibrate. The space between the two vocal folds is known as the glottis. Inferior to the vocal folds, the laryngeal channel widens and continues into the trachea, the next organ in the tract.

The larynx contains several muscles that control the diameter of the passageway through it. With the exception of the posterior cricoarytenoid muscles, which lie on the anterior surface of the base of the larynx and dilate the passage, a spasm of the laryngeal muscles can close the folds and prevent airflow into the lungs.

4. Voice and Singing

The vestibular folds are controlled by muscle tissue in the wall that can be contracted to open and

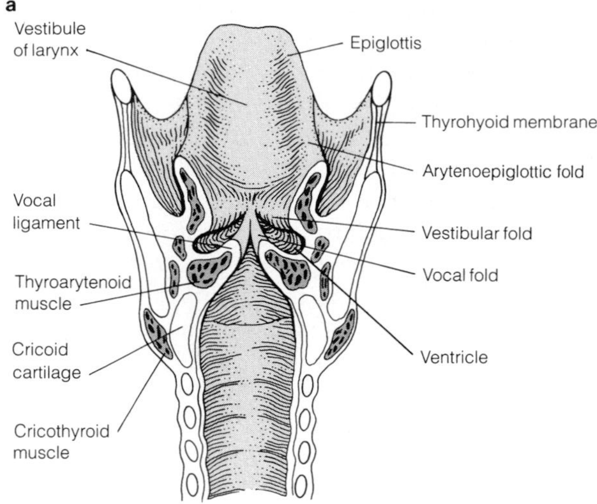

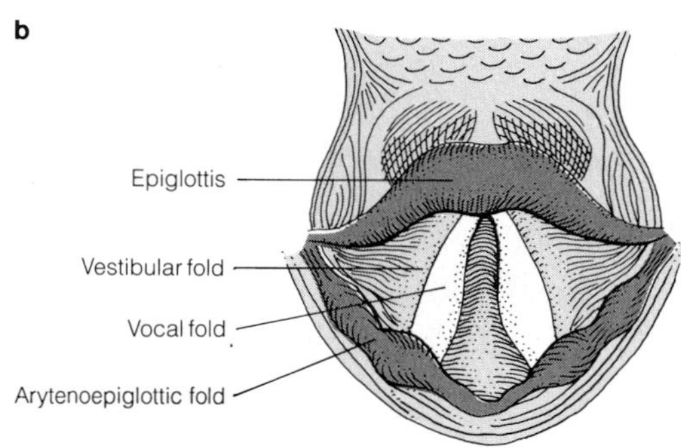

FIGURE 3 Interior organization of the larynx. (a) Posterior view. (b) View from the pharynx, with the glottis open. [From A. J. Gaudin and K. C. Jones (1989). "Human Anatomy and Physiology," p. 510. Harcourt Brace Jovanovich, San Diego.]

close the glottis. These muscles help close larynx to the expulsion of air from the lungs when holding one's breath, lifting a weight, or otherwise straining.

The vocal folds are rimmed by vocal ligaments that border the glottis. Within the folds themselves are muscles that control the tension on the ligaments and control the size of the opening between the folds. Several other laryngeal muscles are also involved in determining the shape and orientation of the vocal folds, by adjusting the orientation of the various cartilages. (The quality of sound produced by the vibrating folds is determined in this way.) The muscles primarily involved in establishing tension in the vocal ligaments are the paired cricothyroid and thyroarytenoid muscles.

B. Lower Respiratory Tract

The trachea, more commonly called the windpipe, is a tube that extends downward from the base of the larynx and branches into two primary bronchi (see Color Plate 22). The trachea is about 12 cm long by about 2.5 cm in diameter in an adult and lies anterior to the esophagus.

Unlike the eosphagus, which is soft and pressed flat when empty, the trachea is kept permanently open. This is accomplished by 16–20 horizontally oriented C-shaped bands of cartilage in the tracheal wall. The open portion of the "C" is spanned by trachealis muscle. The bands are connected to one another by intervening annular ligaments and are oriented so that their open portions are on the posterior side of the trachea, next to the esophagus.

The inner surface of the trachea is lined with ciliated mucous membrane laid on a supporting submucosa. This lining is similar in composition to the mucosal lining of the lower portion of the larynx, the nasal cavity, and the nasopharynx. It also helps to filter particulate matter from air passing through the trachea. Externally, the trachea is surrounded by a sheath of connective tissue.

II. GROSS MORPHOLOGY OF THE LUNGS

The lungs are located in the chamber of the chest formed by the ribs laterally, the vertebral column posteriorly, the sternum anteriorly, and the diaphragm inferiorly. This region, the thoracic cavity, also contains several other structures, including the heart, the esophagus, the thoracic lymphatic duct, major nerves, and major arteries and veins that carry blood to and away from the lungs. These organs are located in the mediastinum, the medial portion of the thoracic cavity. The lungs themselves lie on either side of the mediastinum, enclosed in pleurae, a pair of double-walled sacs of serous membrane (see Color Plate 23). The pleura on each side of the thoracic cavity is a single "bag" that folds back on itself, as if one were to push a fist into a partially inflated balloon until the opposing inner surfaces were in contact.

Each pleura consists of a thin serous membrane that lies tightly pressed to the inner wall of the rib cage, folding back on itself in the region of the mediastinum and continuing back out and around the outer surface of the lung, to which it is also closely pressed. The parietal pleura consists of that portion of the serous bag that is not closely adherent to the lungs. The portion that adheres to the lung surfaces is the pulmonary, or visceral, pleura. The portion of the parietal pleura that adheres to the thoracic wall, the costal pleura, goes on to adhere to a portion of the diaphragm, where it forms the diaphragmatic pleura. The latter folds upward from the diaphragm and outlines the mediastinum. This portion is called the mediastinal pleura.

Returning to the analogy of the partially filled balloon with a fist in it, this fist corresponds to the lungs, and the inner surfaces of the balloon in contact with one another are analogous to the surfaces of the parietal and plumonary pleurae. The portion of the balloon where the fist has entered and that would surround the wrist is equivalent to a region known as the pulmonary hilum. This is a roughly triangular hole, through which the primary bronchus, a pulmonary artery and vein, bronchial arteries and veins, nerves, and lymphatic vessels pass into the lung (see Color Plate 23). Collectively, these structures compose the root of the lung.

The two lungs are not mirror images of one another, the left lung being somewhat smaller than the right. Each lung has deep fissures on its surface, which divide it into lobes. The right primary bronchus branches from the trachea and gives rise to three secondary bronchi, each of which enters a lobe. In contrast, the left primary bronchus gives rise to only two secondary bronchi, each of which enters a lobe. Consequently, the right lung has three lobes and the left lung has two. The pulmonary pleura follows the contours of the lobes, descending to the base of each fissure and out again, adhering closely to the surface of the lungs at all points except at the root (Fig. 4).

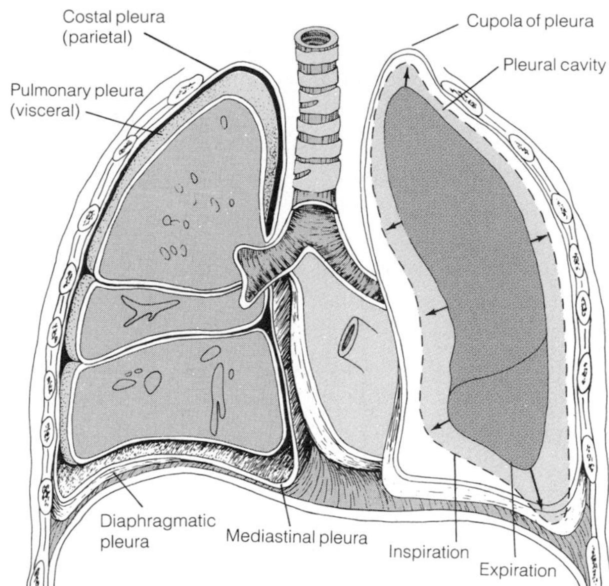

FIGURE 4 Organization of the pleurae. [From A. J. Gaudin and K. C. Jones (1989). "Human Anatomy and Physiology," p. 513. Harcourt Brace Jovanovich, San Diego.]

III. PULMONARY TREE

Inspired air passes into the lungs through the primary bronchi and is distributed throughout the lungs by an elaborate system of tubules that composes the pulmonary tree (Fig. 5). The pulmonary tree consists of two functionally distinct regions, one in which air passes and another in which gas exchange occurs. The first of these is the conducting division and the second is the respiratory division of the pulmonary tree.

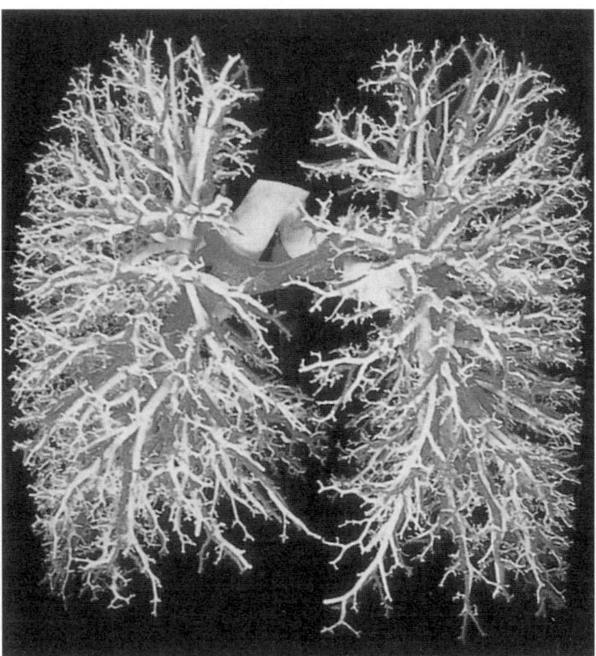

FIGURE 5 The pulmonary tree. [From A. J. Gaudin and K. C. Jones (1989). "Human Anatomy and Physiology," p. 513. Harcourt Brace Jovanovich, San Diego.]

A. Conducting Division

The two primary bronchi descend downward and laterally into the chest cavity for a few centimeters, before passing through the hilum. The right branch extends a somewhat shorter distance than the left one and also has a somewhat larger diameter than the left branch and is oriented more vertically. As a result, if a foreign body succeeds in passing through the larynx, it tends to fall through and lodge in the right bronchus more readily than in the left.

The bronchial wall in this region has a cartilaginous skeleton similar to that of the trachea, although the bands are not so regular. In the region where the bronchus passes into the lung, the bands become even more irregular and are gradually replaced by cartilaginous plates of variable shape and size. Continued branching produces tubes with reduced cartilaginous support, and after a few branchings the cartilage is missing entirely. At that point the tubules are referred to as bronchioles. Bronchioles continue to branch smaller and smaller until they become terminal bronchioles. Normally, 16 branch points occur from the trachea to the terminal bronchioles, producing nearly 66,000 terminal bronchioles. These are distributed approximately equally between the right and left lungs.

Lacking a cartilaginous skeleton, bronchioles are softer and more flexible than bronchi. The bronchioles also contain smooth muscle, the constriction of which reduces their diameter and the amount of air that can pass through them.

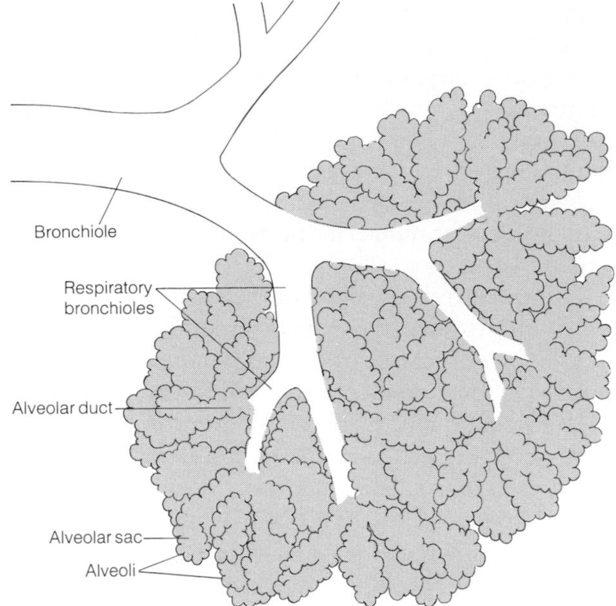

FIGURE 6 A primary lobule. [From A. J. Gaudin and K. C. Jones (1989). "Human Anatomy and Physiology," p. 514. Harcourt Brace Jovanovich, San Diego.]

B. Respiratory Division

Terminal bronchioles branch still further to produce a network of respiratory bronchioles. Typically, three levels of branching of respiratory bronchioles occur, with the final respiratory bronchiole in the sequence branching to produce a pair of alveolar ducts. Each alveolar duct can branch further, terminating in an alveolar sac. The walls of the passageways from the respiratory bronchioles to the alveolar ducts are perforated by openings that lead into small chambers, called alveoli. Alveoli are the functional units of the lung, where gas exchange occurs. A respiratory bronchiole and its subsequent divisions collectively compose a unit of the lung called a primary lobule (Fig. 6).

There are typically 23 levels of branching between the trachea and the alveoli, giving rise to about 150 million alveoli in each lung. The average diameter of an alveolus is about 300 μm, so that the entire functional surface of the lungs through which gas exchange occurs amounts to about 75 m^2, equal to a little over 800 square feet, or a square measuring more than 28 feet per side. [*See* Respiratory System, Physiology and Biochemistry.]

IV. STRUCTURE OF THE ALVEOLAR WALL

The alveoli are supported by a network of fine bundles of smooth muscle fibers and fibrous protein that lie in the walls formed where alveoli come into contact with one another. These walls are called interalveolar septa. The septa also have the important function of carrying the capillaries in which blood flows as it absorbs oxygen from the alveolar air and releases carbon dioxide into it.

Figure 7 shows the organization of one of these extremely delicate septa. The surfaces on either side of the septum that face the interior of an alveolus are covered primarily by an epithelium that is so thin that much of it can only be visualized by the electron microscope. The only parts that are thick enough to see are the regions where the nuclei are located. The surface toward the septum rests on a basement membrane of glycoproteins and mucopolysaccharides secreted by the epithelial cell.

Sandwiched between the epithelial cells on either

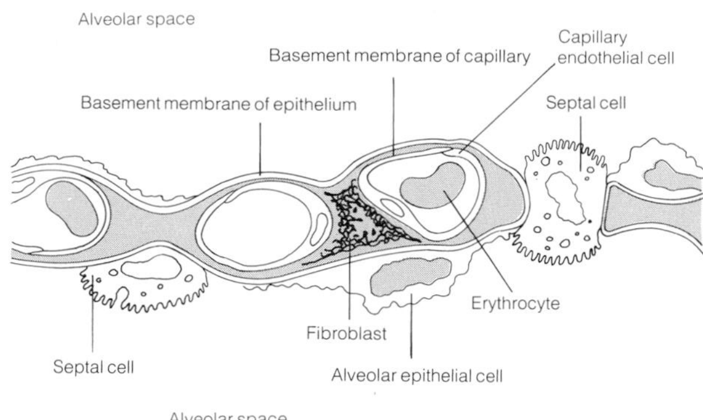

FIGURE 7 An interalveolar septum. [From A. J. Gaudin and K. C. Jones (1989). "Human Anatomy and Physiology," p. 515. Harcourt Brace Jovanovich, San Diego.]

side of the septum are the capillaries. They are fine tubular vessels formed by flat endothelial cells wrapped in cylindrical form; blood passes through these cylinders. Like the epithelial cell of the alveolus, the endothelial cell of a capillary produces a thin basement membrane that, at some points, merges with that of the epithelium.

Also included in the septum are occasional septal cells that secrete a phospholipid material that adheres to the alveolar surface of the epithelial cells. This substance acts as a surfactant, which reduces the surface tension of the moisture on the cell surface. This, in turn, facilitates the diffusion of oxygen into the epithelium from the alveolus. Septal cells are roughly cuboidal and can be present on only one surface of a septum or can extend all the way through, providing surfactant for both sides of the wall.

An alveolus contains macrophages that remove inhaled microorganisms and particulate matter. These macrophages are unusual in that they spend considerable time in the alveolus, foraging over the epithelial surface, where they ingest dust and soot particles, as well as microbes. Once macrophages have ingested this foreign matter, they are carried out through the bronchi to the pharynx and swallowed. Digestive chemicals within the stomach destroy the macrophages and the materials they contain. [*See* Macrophages.]

BIBLIOGRAPHY

Ball, D. (1987). Black lungs and black walls. *New Sci.* **113,** 32.

Cormack, D. H. (1987). "Ham's Histology," 9th Ed. Lippincott, Philadelphia.

Cormack, D. H. (1992). "Essential Histology." Lippincott, Philadelphia.

Ganong, W. F. (1993). "Review of Medical Physiology," 16th Ed. Lange, Los Altos, California.

Gaudin, A. J., and Jones, K. C. (1989). "Human Anatomy and Physiology." Harcourt Brace Jovanovich, San Diego.

Kelly, D. E., Wood, R. L., and Enders, A. C. (1984). "Bailey's Textbook of Microscopic Anatomy," 18th Ed. Williams & Wilkins, Baltimore.

Mines, A. H. (1993). "Respiratory Physiology," 3rd Ed. Raven, New York.

Nadel, J. A., and Barnes, P. J. (1984). Autonomic regulation of the airways. *Annu. Rev. Med.* **35,** 451–468.

VanBolde, L. M. G., Batenburg, J. J., and Robertson, B. (1988). The pulmonary surfactant system: Biochemical aspects and functional significance. *Physiol. Rev.* **68,** 374–455.

Weisfeldt, M. L., and Chandra, N. (1981). Physiology of cardiopulmonary resuscitation. *Annu. Rev. Med.* **32,** 435–442.

Respiratory System, Physiology and Biochemistry

JOHN B. WEST
University of California, San Diego

GLOSSARY

Bohr effect Increase in the oxygen affinity of hemoglobin as a result of increasing the pH of blood

Chemoreceptor Sensor that responds to a change in the chemical composition of the blood or other fluid surrounding it

Dead space Volume of the lungs not participating in gas exchange

Fick principle Method for measuring cardiac output based on the fact that the amount of oxygen entering the lungs via the mouth is the same as the amount of oxygen leaving the lungs via the blood

Fick's law of diffusion Principles determining the rate of diffusion of a gas through a thin tissue sheet

Haldane effect Increase in carbon dioxide concentration of the blood for a given partial pressure of carbon dioxide, caused by a reduction in the hemoglobin oxygen saturation

Hypoxemia Condition in which the partial pressure of oxygen in the arterial blood is abnormally low

Partial pressure Proportion of the total gas pressure occupied by one gas species

Plethysmograph Airtight box like a telephone booth, in which a subject sits for measurements of lung volume and other respiratory variables

Shunt Admixture of poorly oxygenated and well-oxygenated blood, which results in hypoxemia

THE FUNCTION OF THE RESPIRATORY SYSTEM IS TO move oxygen from the air of the environment to the mitochondria of the body cells, where it is utilized, and to move carbon dioxide in the opposite direction. This is called gas exchange. The various links in the overall process include (1) pulmonary ventilation, or the moving of oxygen from the air into the alveoli in the depth of the lungs (and carbon dioxide in the opposite direction); (2) pulmonary blood flow, which moves oxygen out of the lungs after it has been taken up by the blood; (3) pulmonary gas exchange, or the movement of oxygen and carbon dioxide across the blood–gas barrier in the lungs; (4) blood gas transport, or the carriage of oxygen and carbon dioxide in the blood; (5) the mechanics of breathing, or the forces involved in supporting and moving the lungs and the chest wall; (6) the control of ventilation, or the mechanisms that regulate the gas exchange function of the lungs; and (7) peripheral gas exchange, or oxygen delivery to cells and intracellular respiration.

I. PULMONARY VENTILATION

Pulmonary ventilation is the process of getting gas to and from the alveoli of the lungs. From a physiological standpoint the airways of the lungs can be divided into a conducting portion and a respiratory region

ENCYCLOPEDIA OF HUMAN BIOLOGY, Second Edition, VOLUME 7. Copyright © 1997 by Academic Press. All rights of reproduction in any form reserved.

(Fig. 1). The function of the conducting airways is to deliver inspired air to the alveoli of the respiratory region. The conducting airways do not contain alveoli, where gas exchange can occur, and therefore constitute the "anatomic dead space." The volume of dead space in human lungs is about 150 ml. Beyond the conducting airways, increasing numbers of alveoli line the walls, and gas exchange can therefore occur across the thin blood–gas barrier around the pulmonary capillaries (Fig. 2). The volume of the respiratory zone in the human lung is about 2.5–3.0 liters. [*See* Respiratory System, Anatomy.]

A normal inspired volume of air (i.e., tidal volume) is about 500 ml, although there is considerable variation. Since 150 ml of this volume remains behind in the anatomic dead space, only 350 ml penetrates to the alveoli, where gas exchange can occur. The volume of fresh gas entering the alveoli (350 ml in this example) multiplied by the respiratory frequency is known as the alveolar ventilation. This is lower than the total ventilation, which is the tidal volume multiplied by the respiratory frequency.

Lung volumes can be measured with a spirometer, a light bell-shaped container immersed in a water tank

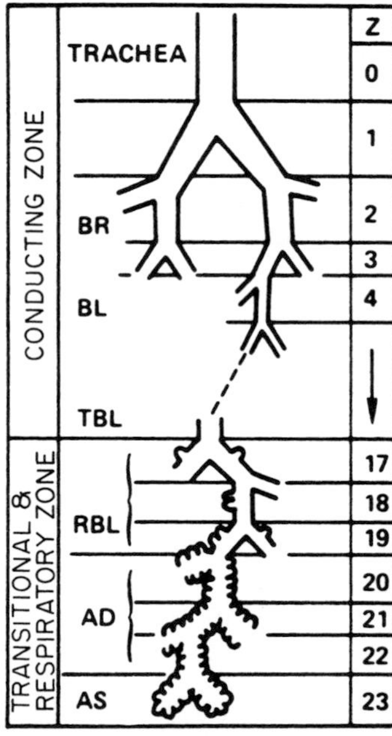

FIGURE 1 Idealization of the airways of the human lung. The first 16 generations (*z*) make up the conducting airways, and the last 7 compose the respiratory, or transitional and respiratory, zone. BR, bronchus; BL, bronchiole; TBL, terminal bronchiole; RBL, respiratory bronchiole; AD, alveolar duct; AS, alveolar sac. [From E. R. Weibel (1984). "The Pathway for Oxygen." Harvard Univ. Press, Cambridge, Massachusetts.]

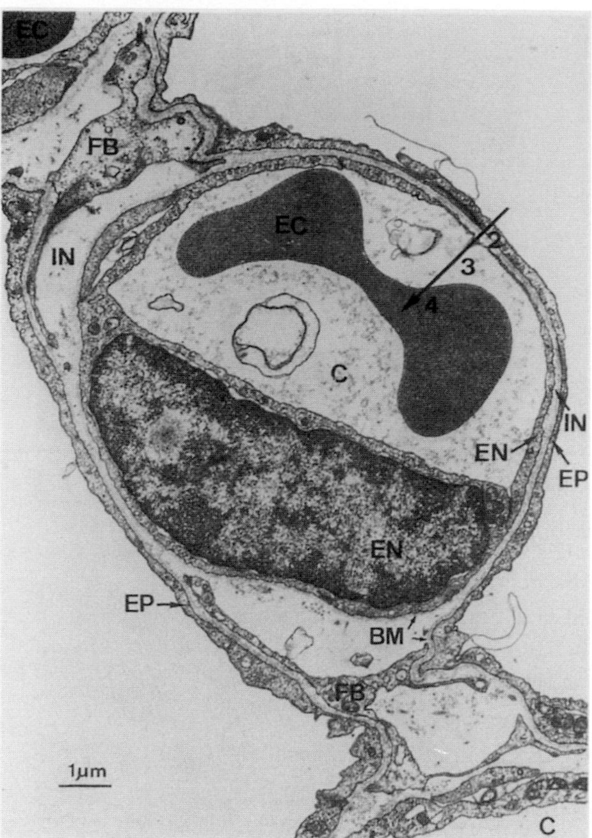

FIGURE 2 Electron micrograph showing a pulmonary capillary (C) in the alveolar wall. Note the extremely thin blood–gas barrier (less than 0.5 μm). The large arrow indicates the diffusion pathway from alveolar gas to the interior of the erythrocyte (EC) and includes the layer of surfactant (not shown), alveolar epithelium (EP), interstitium (IN), capillary endothelium (EN), and plasma. Also seen are parts of structural cells called fibroblasts (FB), basement membrane (BM), and a nucleus of an endothelial cell. [From E. R. Weibel (1970). *Respir. Physiol.* **11**, 54–75.]

(Fig. 3). As the subject exhales, the bell moves up and the pen moves down, marking the chart. In Fig. 3, normal breathing, giving the tidal volume, is followed by an inspiration to total lung capacity, which is followed by a maximal exhalation to residual volume. Not all of the air can be forcibly exhaled from the lungs. The maximal volume of air that can be exhaled from total lung capacity is called the vital capacity. The functional residual capacity is the volume of gas in the lungs at the end of a normal expiration.

Neither the functional residual capacity nor the residual volume can be measured with a simple spirometer. One way of obtaining these values is to connect the subject to a spirometer of known volume, containing a known concentration of the very insoluble

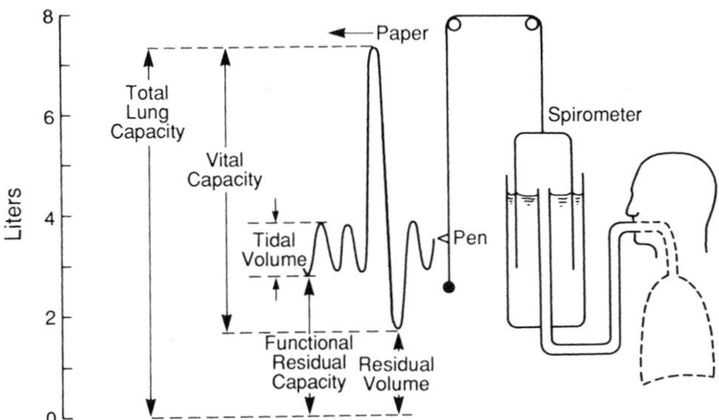

FIGURE 3 Lung volumes, as measured with a spirometer. Note that the functional residual capacity and residual volume cannot be measured without additional techniques. [From J. B. West (1995). "Respiratory Physiology—The Essentials," 5th Ed. Williams & Wilkins, Baltimore.]

gas helium. The subject then breathes in and out of the spirometer until the helium concentrations in the spirometer and the lungs are the same, and the volume of the lungs can then be derived. Another technique is to place the subject in a large airtight box, known as a plethysmograph, and to use Boyle's law (i.e., pressure multiplied by volume is constant at constant temperature) to derive the lung volume.

II. PULMONARY BLOOD FLOW

Just as pulmonary ventilation brings oxygen to the blood–gas barrier, where gas exchange occurs, so the pulmonary circulation picks up the oxygen in the alveoli and delivers it to the left side of the heart, from which it is distributed to the rest of the body.

The pulmonary circulation begins at the main pulmonary artery, which receives blood pumped by the right ventricle of the heart. This artery then branches successively, following the system of airways (see Fig. 1) ultimately feeding the capillaries. These form a dense network in the alveolar walls, giving an efficient arrangement for gas exchange. The oxygenated blood is then collected from the capillaries by small pulmonary veins, which eventually unite to form large veins, which drain into the left atrium of the heart.

The pulmonary circulation is characterized by the low pressures within it. For example, the mean pressure in the pulmonary artery in humans is only about 15 mm Hg, whereas the mean pressure in the aorta fed by the left heart is on the order of 100 mm Hg. A corollary is that the vascular resistance of the pulmonary circulation, defined as (pulmonary artery pressure—pulmonary venous pressure)/blood flow, is also very low.

The pulmonary circulation has a remarkable facility for accepting increases in cardiac output, with only small rises in pulmonary artery pressure. This is because some capillaries are normally closed or contain nonflowing blood, and these can open when capillary pressure rises, a phenomenon known as recruitment. In addition, the capillaries can distend when the pressure inside them rises. This is called distension.

Pulmonary blood flow, and therefore cardiac output, can be measured by the Fick principle. This states that the total amount of oxygen taken up by the lungs is equal to the cardiac output multiplied by the oxygen concentration difference between the mixed venous blood entering the pulmonary artery and the blood that leaves the lungs. This principle is simply a statement of the conservation of mass. In practice, systemic arterial blood is sampled with a needle to measure oxygen in the blood leaving the lungs, and a thin tube (i.e., a catheter) is passed into the pulmonary artery via a peripheral vein to sample mixed venous blood.

Because the pressures in the pulmonary circulation are so low, the hydrostatic differences between the top and the bottom of the lungs are significant and result in a larger blood flow through the bottom of the lungs than through the top. For example, in normal upright humans, the apical (i.e., uppermost) region of the lungs is only just perfused with blood. Since these regional differences of flow are determined by gravity, this difference disappears when a person lies flat. But then a difference in blood flow is seen between the uppermost and lowermost parts of the supine lungs. Upon exercise in the upright position, the topographical differences in blood flow become less obvious, because the pulmonary artery pressure increases.

An important active response of the pulmonary circulation is hypoxic pulmonary vasoconstriction. This refers to the fact that, when the alveolar gas is made hypoxic (e.g., when breathing a low-oxygen mixture or at high altitude), the small pulmonary arteries (arterioles) of the hypoxic region constrict. If the ventilation of a small region of the lungs is reduced, the vasoconstriction diverts blood flow from it. The mechanism of hypoxic vasoconstriction is not understood, but clearly does not depend on central nervous system connections, because it occurs in the lungs when they are isolated from the rest of the body.

Small amounts of fluid continually move across the walls of the pulmonary capillaries into the interstitial spaces of the lungs. If the pulmonary capillary pressure is increased, the rate of fluid movement rises, causing interstitial pulmonary edema. If the rate of fluid loss from the capillaries is greatly increased, fluid can spill over into the alveoli, causing alveolar edema. This is a potentially life-threatening situation and can occur, for example, as a result of a heart attack (i.e., myocardial infarction), when the left heart fails to pump blood as it should and the pressures in the pulmonary circulation consequently rise. Normally, the small amount of fluid draining from the pulmonary capillaries is carried away in the pulmonary lymphatics.

The pulmonary circulation has important metabolic functions, as well as its primary role in gas exchange. For example, the relatively inactive polypeptide angiotensin I is converted to the potent vasoconstrictor angiotensin II during the passage of blood through the lungs. This conversion is catalyzed by an enzyme (i.e., angiotensin-converting enzyme) that is located on the walls of the capillary endothelial cells. A number of vasoactive substances are completely or partially inactivated during passage through the lungs, including bradykinin, serotonin, and some prostaglandins.

III. PULMONARY GAS EXCHANGE

The exchange of oxygen and carbon dioxide between the alveolar gas and the pulmonary capillary blood occurs across the thin blood–gas barrier shown in Fig. 2. Both gases pass across the tissue sheet by simple passive diffusion, and they obey Fick's law:

$$\dot{V}_{\text{gas}} = \frac{A}{T} D(P_1 - P_2)$$

where $\dot{V}_{\text{gas}}$ is the volume of gas per unit of time moving across the sheet, A is the area of the sheet, T is its thickness, D is a diffusion constant, and P_1 and P_2 are the partial pressures on either side of the sheet. Clearly, the properties of a sheet that would enhance diffusion are a large surface area and small thickness. The area of the blood–gas barrier in the human lungs is between 50 and 100 m^2, and the thickness is less than 0.5 μm in many places. Therefore, its geometry is ideally suited to rapid diffusion.

The diffusion constant (D) is proportional to the solubility of the gas and inversely proportional to the square root of its molecular weight. Carbon dioxide diffuses about 20 times faster than oxygen across a tissue sheet, owing to its much higher solubility and the small difference in molecular weights.

When mixed venous blood enters the pulmonary capillaries, its partial oxygen pressure (P_{O_2}) is only about 40 mm Hg. The P_{O_2} of the alveolar gas is approximately 100, so there is a large pressure difference to promote diffusion. In fact, the diffusion of oxygen across the blood–gas barrier occurs so rapidly that the P_{O_2} of the capillary blood has almost reached that of the alveolar gas within about 0.25 sec. Since the blood spends about 0.75 sec in the capillary under resting conditions, there is ample time for diffusive equilibration. Upon exercise the time spent by the blood in the capillary is reduced, because of the greatly increased cardiac output. Even so, equilibration between the P_{O_2} of alveolar gas and end-capillary blood is believed to occur in the human lungs under all but the most exceptional conditions. The rate of combination of oxygen with hemoglobin (see Section IV) is also believed to delay the loading of oxygen onto the blood to some extent.

The diffusion characteristics of the human lungs can be measured by having humans inhale a low concentration of carbon monoxide and measuring the rate at which it is removed by the blood. It can be shown that the uptake of this gas is limited by the diffusion properties of the blood–gas barrier (i.e., thickness and area). Diseases in which the thickness of the blood–gas barrier is increased typically reduce the diffusing capacity of the lungs for carbon monoxide.

Various factors reduce the efficiency of the lungs for gas exchange. If the arterial blood leaving the chest has an abnormally low P_{O_2}, the condition is known as hypoxemia. One cause is a reduced alveolar ventilation, because this results in a low alveolar P_{O_2} and a consequent depression of the arterial value. Causes include paralysis of the respiratory muscles and the depression of ventilation by the effects of drugs on the central nervous system. Hypoventilation, as this is called, also results in an increased arterial P_{CO_2}, because carbon dioxide elimination is also affected.

Another cause of hypoxemia is shunt (i.e., the presence of channels that allow blood to bypass ventilated regions of the lungs). An example is a communication between the right and left sides of the heart, as in patients with cyanotic congenital heart disease ("blue" babies). This allows the admixture of poorly oxygenated venous blood with well-oxygenated arterial blood on the left side of the heart.

A common cause of inefficient pulmonary gas exchange is the mismatching of ventilation and blood flow in different regions of the lungs. It can be shown that the gas exchange that occurs in any lung unit depends on the ratio of its ventilation to blood flow. In the normal upright human lungs, the differences in ventilation and blood flow of various levels result in a higher Po_2 at the apex of the lungs (e.g., compared with the base). In diseases such as chronic bronchitis and emphysema, the architecture of the lung is so disrupted that the normal matching of ventilation and blood flow is disturbed. This is known as ventilation–perfusion inequality. Arterial hypoxemia is then inevitable, and an increased arterial Pco_2 is also often seen.

If pulmonary gas exchange becomes inefficient, respiratory failure is said to occur. This is usually characterized by a low arterial Po_2 and an increased arterial Pco_2. Patients with respiratory failure are often treated by raising the inspired oxygen concentration, and sometimes by connecting them to a mechanical ventilator.

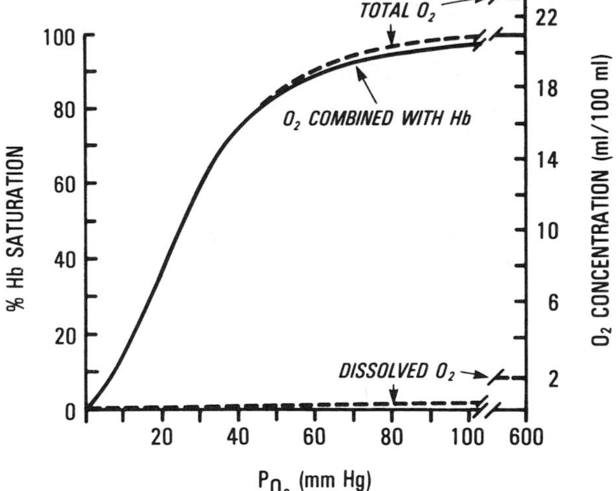

FIGURE 4 Typical oxygen dissociation curve for blood, showing dissolved oxygen and oxygen combined with hemoglobin (Hb). The total oxygen concentration is also shown for a hemoglobin concentration of 15 g/100 ml of blood. [From J. B. West (1995). "Respiratory Physiology—The Essentials," 5th Ed. Williams & Wilkins, Baltimore.]

IV. BLOOD GAS TRANSPORT

Both oxygen and carbon dioxide must be transported in the blood between the lungs and the peripheral tissues. Most oxygen is carried in the blood in combination with hemoglobin; only about 1% is carried in solution, owing to its low solubility (0.003 ml of O_2/100 ml of blood/mm Hg of Po_2).

Hemoglobin consists of an iron–porphyrin compound, heme, joined to the protein globin, consisting of four polypeptide chains. The chains are of two types, α and β, and differences in their amino acid sequences give rise to various types of human hemoglobin. Normal adult hemoglobin is known as A. Hemoglobin F (fetal) has a high affinity for oxygen and is especially suited for transporting oxygen in the relatively hypoxic intrauterine environment. Hemoglobin S (sickle) is poorly soluble in its deoxygenated form, and patients with sickle-cell disease are prone to obstruction in their blood vessels.

Oxygen forms an easily reversible combination with hemoglobin to give oxyhemoglobin:

$$Hb + O_2 \leftrightharpoons HbO_2$$

Figure 4 shows the relationship between the oxygen concentration in the blood and the Po_2. The hemoglobin saturation is the proportion of the available binding sites of hemoglobin that are combined with oxygen. At a normal arterial Po_2 of 100 mm Hg, approximately 97% of the sites are bound to oxygen.

The curved shape of the oxygen dissociation curve (see Fig. 4) has several physiological advantages. The flat upper portion means that even if the Po_2 of alveolar gas falls somewhat, loading of oxygen will be little affected. In addition, as the red blood cell takes up oxygen along the pulmonary capillary, a large partial pressure difference between the alveolar gas and blood continues to exist, even when most of the oxygen has been transferred. An advantage of the steep lower part of the dissociation curve is that peripheral tissues can withdraw large amounts of oxygen for only a small decrease in capillary Po_2. This maintenance of capillary Po_2 assists the diffusion of oxygen into the tissue cells.

Several factors alter the affinity of hemoglobin for oxygen, that is, they shift the dissociation curve leftward or rightward. Increases in temperature, Pco_2, hydrogen ion concentration, and 2,3-diphosphoglycerate (2,3-DPG) all shift the curve to the right, that is, they reduce the oxygen affinity of the hemoglobin. The effect of hydrogen ion on the dissociation curve is known as the Bohr effect. The first three factors exert their effects in exercising muscle, which is relatively hot and has a high Pco_2 and an increased hydrogen ion concentration. This reduced oxygen affinity assists the unloading of oxygen from peripheral capillaries.

2,3-DPG is an end product of red blood cell metabolism. An increase in its concentration occurs in several conditions, characterized by chronic hypoxia. Blood stored for transfusion shows a slow decrease in 2,3-DPG, with the result that after the transfusion of large quantities, the release of oxygen to tissues may be impaired. This change in 2,3-DPG in stored blood can be retarded by additives.

Carbon monoxide competes with oxygen for the same binding sites on hemoglobin. Furthermore, it has about 250 times the affinity of oxygen for hemoglobin and can therefore combine with the same amount of hemoglobin when the P_{CO} is 250 times lower than the P_{O_2}. For this reason, small amounts of carbon monoxide can tie up large amounts of hemoglobin in the blood, making it unavailable for oxygen carriage. A heavy smoker can have 10% of his hemoglobin combined with carbon monoxide.

Carbon dioxide is carried by the blood in three forms: dissolved, as bicarbonate, and in combination with proteins. Since carbon dioxide is much more soluble than oxygen in blood, dissolved carbon dioxide plays a significant role in its carriage. However, most of the carbon dioxide in blood is carried as bicarbonate, which is formed according to the reaction

$$CO_2 + H_2O \overset{CA}{\leftrightarrow} H_2CO_3 \leftrightarrow H^+ + HCO_3^-$$

The first reaction is very slow in plasma, but fast within the red blood cell, which contains the enzyme carbonic anhydrase (CA). Carbonic acid can also combine with blood proteins, especially globin, forming carbamino compounds. Some 30% of the carbon dioxide released as the blood passes through the lung capillaries comes from carbamino compounds.

The relationship between carbon dioxide concentration and P_{CO_2} in blood is known as the carbon dioxide dissociation curve, which is considerably steeper than that for oxygen. One consequence of this is that the P_{CO_2} difference between mixed venous and arterial blood is only about 5–7 mm Hg, whereas it is about 60 mm Hg for oxygen. As blood loses oxygen in peripheral capillaries, it is better able to load carbon dioxide. This phenomenon is known as the Haldane effect.

The carriage of carbon dioxide by the blood has an important effect on the acid–base status of the body, because dissolved carbon dioxide forms carbonic acid. For example, a patient whose lungs are diseased, and therefore unable to properly excrete carbon dioxide, develops an increased hydrogen ion concentration in the blood, a condition known as respiratory acidosis. By contrast, if a normal subject hyperventilates and blows off carbon dioxide, he develops respiratory alkalosis. (This occurs in a newcomer to high altitude.) The level of arterial P_{CO_2} can be regulated rapidly by changes in ventilation, and therefore the lungs play an important role in maintaining the correct acid–base status of the blood.

V. MECHANICS OF BREATHING

In normal breathing at rest, inspiration is active, but expiration is passive. The most important muscle of inspiration is the diaphragm, a thin dome-shaped sheet of muscle supplied by two phrenic nerves originating from the spinal cord in the neck. When the diaphragm contracts, the abdominal contents are forced downward, and the vertical dimension of the chest cavity is increased. The rib cage moves out at the same time. The action of the diaphragm is assisted by external intercostal muscles, which connect adjacent ribs and slope downward and forward. When these muscles contract, the ribs are pulled upward, thus increasing both the lateral and anteroposterior diameters of the thorax. However, paralysis of the intercostal muscles alone does not seriously affect breathing, because the diaphragm is so effective. Accessory muscles of inspiration include neck muscles, which assist inspiration during vigorous exercise.

The most important muscles of expiration are those of the abdominal wall. When these contract, intraabdominal pressure is raised and the diaphragm is pushed upward. This action is apparently assisted by the internal intercostal muscles, the action of which is opposite that of the external intercostal muscles.

The lungs are elastic and are normally expanded by a reduction of pressure in the intrapleural space between the lungs and the chest wall. This space normally contains only a few milliliters of fluid to lubricate the surfaces of the two pleural membranes, but can enlarge if air, for example, enters it (i.e., a pneumothorax). The relationship between the intrapleural pressure around the lungs and its volume is shown in Fig. 5. Here, a lobe of the lungs is inflated by reducing the pressure around it, and its volume is measured with a spirometer. The pressure–volume curve is flatter at high states of lung inflation, and the lungs follow a different pathway during inflation compared with deflation. This is known as hysteresis. In this example the pressure inside the lungs remains atmospheric.

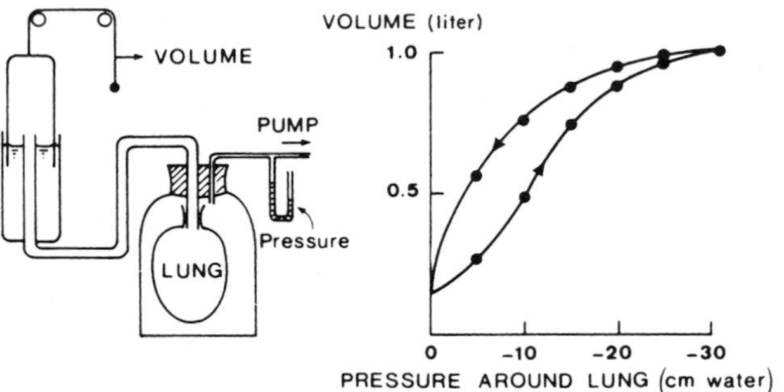

FIGURE 5 Measurement of the pressure–volume curve of a lung lobe. The lung is held at each pressure for a few seconds, while its volume is measured. Note the nonlinearity of the curve and that the inflation and deflation curves are not the same. [From J. B. West (1995). "Respiratory Physiology—The Essentials," 5th Ed. Williams & Wilkins, Baltimore.]

The difference between alveolar and intrapleural pressures is sometimes called transpulmonary pressure.

The slope of the pressure–volume curve is known as the compliance, often measured over the normal working range of the lungs; for the total human lungs the value is about 200 ml per centimeter of water. In other words, a normal tidal volume of about 500 ml requires a decrease in the intrapleural pressure of only about 2.5 cm of water. The compliance of the lungs is reduced in some diseases, for example, if fibrous tissue is deposited in the alveolar walls (i.e., pulmonary fibrosis). The compliance is increased by age and also by emphysema. In both instances an alteration in the elastic tissue of the lungs is probably responsible. The elastic behavior of the lungs can be partially attributed to elastic fibers in the lungs, including collagen and elastin. However, the extraordinary distensibility of the lungs has probably less to do with the simple elongation of these fibers than with changes in their geometrical arrangement. An analogy is a nylon stocking, which is very distensible because of its knitted makeup, although the individual nylon fibers are very difficult to stretch.

Another important factor in the pressure–volume behavior of the lungs is the surface tension of the liquid film lining the alveoli. Surface forces generate pressures in curved surfaces (e.g., bubbles), and these pressures are particularly large when the bubbles are small. This could be a serious problem for the lungs, because the alveoli have a diameter of only about 0.3 mm. Fortunately, some cells lining the alveoli secrete material that profoundly lowers the surface tension of the alveolar lining fluid. This surfactant includes the phospholipid dipalmitoyl phosphatidylcholine, and it is secreted by the type II alveolar cells.

The effects of this material on surface tension can be measured outside the lungs in a surface balance, and such studies show that the material can reduce the surface tension to extremely low values. This has the advantage of stabilizing the small alveoli, reducing the work required to expand the lungs and also reducing the tendency to pulmonary edema. Some premature babies are born with an immature surfactant system, because this develops relatively late in fetal life. This condition, known as the respiratory distress syndrome, is characterized by unstable stiff lungs. Recent work indicates that these babies can be treated by instilling artificial surfactant into their lungs after birth.

The lungs are contained within the chest wall, which is made up of the rib cage and the diaphragm. Like the lungs, the chest wall is elastic. At the end of a normal expiration (i.e., functional residual capacity), the tendency of the lungs to collapse because of their elastic recoil is balanced by the tendency of the chest wall to spring out. Indeed, the balance of these two forces is what determines the resting volume of the lungs and the chest wall. If one lung collapses, for example, as a result of a pneumothorax (i.e., air in the pleural space), the chest wall on that side springs out to some extent.

In order for inspiration to occur, the inspiratory muscles contract, intrapleural and alveolar pressures decrease, and air is drawn into the alveoli along the system of airways (see Fig. 1). During quiet breathing, expiration is passive. When the inspiratory muscles cease to contract, the lungs and the chest wall return to their resting positions. During expiration, intrapleural pressure becomes less negative (see Fig. 5), alveolar pressure rises slightly, and air moves from the alveoli to the mouth. Airway resistance can be calculated from the difference between alveolar and mouth pressures divided by flow.

At one time it was thought that the major site of airway resistance was in the very small airways. This was natural, because flow in the peripheral airways is laminar and Poiseuille's equation states that, under such conditions, resistance is inversely proportional to the fourth power of the radius. However, it is now known that because of the prodigious number of peripheral airways arranged in parallel, they actually contribute little to overall airway resistance, the main site of resistance being in the medium-sized bronchi. The fact that the peripheral airways contribute so little resistance is important in the detection of early airway disease. It is likely that the first abnormalities in chronic bronchitis, for example, occur in the small airways, but these changes are difficult to detect.

Airway resistance is sometimes increased by the contraction of muscle in the airway walls, as in asthma. The degree of contraction of the muscle is under the control of the autonomic nervous system. Drugs that stimulate the sympathetic nervous system reduce airway constriction. Airway resistance increases at low lung volumes and when gases of high density are breathed, as in scuba diving.

During a maximal forced expiration, some airways within the lungs are compressed by the high pressures developed by the expiratory muscles, and the expiratory flow rate is therefore limited. Under these conditions the expiratory flow rate cannot be raised by increasing the strength of contraction of the expiratory muscles, that is, flow is independent of effort. Though this situation occurs only in healthy humans during forced expiration, it may limit the ventilation of patients with lung disease during even moderate exercise.

Work is required to move the lungs and the chest wall, but the oxygen cost of ventilation at rest is small. During heavy exercise the work of breathing might increase so much that the oxygen cost becomes a significant proportion of the total oxygen requirements of the body. [*See* Exercise.]

VI. CONTROL OF VENTILATION

The level of ventilation is remarkably closely controlled, with the result that during normal activity, both at rest and during exercise, the arterial P_{CO_2} changes by only 2 or 3 mm Hg. This is in spite of the fact that, during heavy exercise, carbon dioxide production and oxygen uptake can increase 10-fold or more.

The three basic elements of the respiratory control system are the *sensors* that gather information, feeding it to the *central controller* in the brain. The controller coordinates the information and sends impulses to the *effectors* (i.e., respiratory muscles), which cause ventilation.

The central controller is located in the pons and the medulla of the brain. Part of the controller is the medullary respiratory center, which contains cells that generate the normal respiratory rhythm. This normal automatic process of breathing can be overridden voluntarily; for example, we can elect to hold our breath if we wish.

The sensors include the central and peripheral chemoreceptors. This term denotes a sensor that responds to a change in the chemical composition of the blood or other fluid around it. The central chemoreceptor is located near the ventral surface of the medulla and responds to changes in hydrogen ion concentration of the extracellular and cerebrospinal fluids in its vicinity. Carbon dioxide diffuses from cerebral blood vessels into the cerebrospinal fluid, releasing hydrogen ions, which stimulate the chemoreceptor.

There are also peripheral chemoreceptors located near the common carotid arteries and the arch of the aorta (Fig. 6). These receptors respond to decreases in arterial P_{O_2} and pH and increases in arterial P_{CO_2}. For example, they are responsible for the increase in ventilation that occurs at high altitude, when these receptors are stimulated by the low arterial P_{O_2}.

Other receptors have their nerve endings in the lungs and the upper respiratory tract. For example, the pulmonary stretch receptors are stimulated by inflation of the lungs and then tend to inhibit further inspiratory muscle activity. Irritant receptors in the airway walls cause airway narrowing when stimulated by noxious gases or cigarette smoke. Juxtacapillary receptors are located in the alveolar walls and can be stimulated when fluid leaks out of the capillaries.

The integrated responses of these receptors result in increased ventilation in response to increased levels of carbon dioxide, low levels of inspired oxygen, or acidification of the blood (as in uncontrolled diabetes). However, the large increase in ventilation during

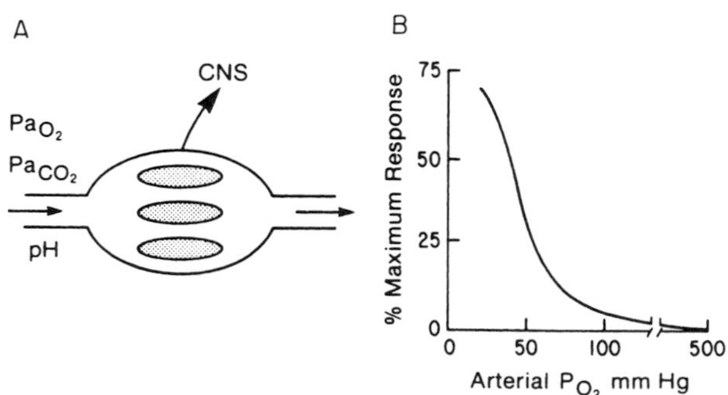

FIGURE 6 (A) A peripheral chemoreceptor (e.g., carotid body), which responds to changes of P_{O_2}, P_{CO_2}, and pH in arterial blood. Impulses travel to the central nervous system (CNS) through a small nerve. (B) The nerve impulse response of the chemoreceptor to arterial P_{O_2}. Note that the maximum response occurs below a P_{O_2} of 50 mm Hg. [From J. B. West (1995). "Respiratory Physiology—The Essentials," 5th Ed. Williams & Wilkins, Baltimore.]

exercise is not easily explained by these receptors and remains something of a mystery.

Sometimes breathing becomes unstable and follows a periodic pattern, with waxing and waning of respiration over several breaths, followed by a few seconds of breathholding. Such periodic breathing frequently occurs at high altitudes and is also seen in some types of lung and heart disease.

VII. PERIPHERAL GAS EXCHANGE

When oxygen reaches the peripheral capillaries of the body via the circulation, it diffuses to the mitochondria, where it is utilized. There is good evidence that the Po_2 at the site of utilization is low, probably less than 1 mm Hg. Therefore, the much larger Po_2 in the capillaries ensures an adequate diffusion head of pressure. Facilitated diffusion (which requires a chemical combination of oxygen with hemoglobin or myoglobin) can occur in skeletal muscle cells, which contain myoglobin; this helps to move the large amounts of oxygen consumed during heavy exercise.

Diffusion distances in tissues are typically much longer than in the lungs (see Fig. 2). Capillaries can be on the order of 50 μm apart, resulting in very low Po_2 values in the center of the core of tissue supplied by adjacent capillaries, particularly near the venous end of the capillaries. This has been referred to as the "lethal corner."

The process of oxygen utilization within the mitochondria is known as oxidative phosphorylation. This series of reactions produces large amounts of energy, with 6 mol of ATP for every mole of oxygen consumed. [*See* ATP Synthesis in Mitochondria.]

If oxygen is in short supply, some energy can be obtained by anaerobic glycolysis with the formation of lactate. However, only relatively small amounts of energy can be obtained from this reaction, and the lactic acid released perturbs the acid–base balance. This mechanism is chiefly reserved for short periods of heavy work, but it can also occur in abnormal states in which oxygen delivery is compromised (e.g., by a defective peripheral circulation).

Oxygen utilization by peripheral tissues can be hampered by a low Po_2 in the arterial blood (e.g., caused by lung disease), by a reduction in the amount of hemoglobin in the blood (i.e., anemia), by reduced local blood flow to the tissues (as in shock), and by tissue poisons (e.g., cyanide) that prevent the uptake of oxygen in the respiratory chain.

BIBLIOGRAPHY

Crystal, R. G., West, J. B., Weibel, E. R., and Barnes, P. J. (eds.) (1997). "The Lung: Scientific Foundations," 2nd ed., 2 vols. Raven, New York.

Weibel, E. R. (1970). Morphometric estimation of pulmonary diffusion capacity. *Respir. Physiol.* **11**, 54–75.

Weibel, E. R. (1984). "The Pathway for Oxygen." Harvard Univ. Press, Cambridge, Massachusetts.

West, J. B. (1995). "Respiratory Physiology—The Essentials," 5th Ed. Williams & Wilkins, Baltimore.

West, J. B. (1992). "Pulmonary Pathophysiology—The Essentials," 4th Ed. Williams & Wilkins, Baltimore.

Reticuloendothelial System

MARIO R. ESCOBAR
Virginia Commonwealth University

HERMAN FRIEDMAN
University of South Florida College of Medicine

GLOSSARY

Dendritic cells Cells of macrophage lineage present in lymph nodes (follicular and interdigitating), spleen, other lymphoid organs, and skin (Langerhans cells) that bear class II major histocompatibility markers, have Fc receptors, and can process antigens for an immune response

Histiocytes Phagocytic cells of the macrophage series that are fixed in tissues

Interleukin-1 Macrophage-derived substance (MW 15,000) that has multiple biologic properties, including the ability to promote short-term growth of T cells (previously called leukocyte-activating factor)

Marginated monocytes Monocytes that adhere to endothelial cells when a chemoattractant is produced in intravascular sites

Opsonization Process of enhancing phagocytosis (e.g., by aggregated antibody or activated complement)

Phagolysosome Membrane-limited cytoplasmic vesicle formed by fusion of a phagosome and a lysosome

Phagosome Membrane-limited vesicle containing phagocytosed material

Professional phagocytes Phagocytic cells that possess specialized membrane receptors for IgG or C3

Respiratory burst Metabolic function of phagocytic cells that provides the oxygen radicals employed by these cells for the destruction of susceptible bacteria

ELIE METCHNIKOFF ASCERTAINED IN 1892 THAT the physiologic role of phagocytic cells was to promote the resistance of the host. The term macrophage (so coined because these cells have the ability to ingest large particles) was introduced by him. Although since then many names (e.g., reticulohistiocytic system, lymphoreticular system, monocyte–macrophage system) have been considered for this group of highly phagocytic cells, the two most widely accepted are reticuloendothelial system (RES) and mononuclear phagocyte system (MPS). The term RES was first introduced by Aschoff in 1924 and was defined, according to functional and morphologic studies, as the multiorgan collection of wandering and sessile mononuclear phagocytes, in addition to a variety of lymphatic and sinusoidal cells. Fibroblasts and endothelial cells, which take up colloidal gold by endocytosis, were later added to the list of reticuloendothelial cells by other investigators. In the early 1970s, a number of scientists suggested that RES should be substituted by a more precise term. Accordingly, in 1975 R. van Furth, Langevoort, and Schaberg proposed the more restrictive term MPS, which consists only of bone marrow promonocytes, circulating blood monocytes, and both mobile and tissue macrophages. MPS as a conceptual framework for these cells excludes the vascular endothelium, reticulum cells, and dendritic cells of lymphoid germinal centers. One of the macrophage characteristics that some of these cells lacks was, for example, the presence of prominent phagoly-

ENCYCLOPEDIA OF HUMAN BIOLOGY, Second Edition, VOLUME 7. Copyright © 1997 by Academic Press. All rights of reproduction in any form reserved.

sosomes (e.g., reticulum cells). Recent studies, however, have revealed that many of the phenotypic differences among macrophage types and between macrophages and related cells are environmentally induced or due to variations between normal and experimental conditions. Hence, this article will maintain a broader perspective and focus on the RES concept.

I. ONTOGENY, ORGANIZATION, AND BIOLOGIC ROLE

A. Ontogeny

Mononuclear phagocytes arise in the bone marrow from a pluripotential stem cell common to all hematopoietic cells, including erythrocytes, megakaryocytes, granulocytes, and mononuclear phagocytes. As the stem cell becomes more committed through successive divisions, the mononuclear phagocytes and the cells of the granulocytic series continue to share a common committed stem cell. In culture, these bone marrow cells generate mixed colonies of granulocytes and macrophages under some conditions and monocytic colonies under others. The first progenitor cell, which can be recognized as part of the MPS, is the monoblast. This cell measures 10–12 μm in diameter and has a small rim of basophilic cytoplasm containing a few granules. Monoblasts are phagocytic and adhere to glass. They display Fc receptors as well as the esterase cytochemistry typical of the more mature progeny and are distinct from myeloblasts, which are the precursors in the granulocytic series. Each monoblast divides once, giving rise to the promonocyte with a cell-cycle time of approximately 12 hr. [*See* Hemopoietic System; Phagocytes.]

The promonocytes measure about 15 μm in diameter and have an indented nucleus that occupies more than half of the cell. They share with the monoblast the typical features of mononuclear phagocytes, including prominent storage granules that strain azurophilic in smears, some of which are also positive for myeloperoxidase. The azurophil storage granules are synthesized only through this stage of maturation.

The promonocytes mature into monocytes, which have decreased numbers of peroxidase-positive granules and an increased ratio of cytoplasm to nucleus. In contrast to the neutrophils, the bone marrow reserve of preformed monocytes is small. In humans, they are released into the blood within 60 hr of their production, where they circulate with a half-life of about 8.5 hr, leaving randomly (i.e., unrelated to age) from the circulation to the extravascular pool. Daily monocyte turnover is approximatley 7×10^6 cells/hr/kg body weight. The ratio of circulating to marginated monocytes in humans is approximately 1 : 3. During inflammation, the proliferation of monocytes is increased by the expansion of the promonocyte pool; their cell-cycle time is decreased, and they are released more rapidly into the circulation. [*See* Neutrophils.]

Tissue macrophages arise by maturation of monocytes that have emigrated from the blood and by replication of immature macrophages in the resident macrophage population. This process is under the control of specific growth factors, termed colony-stimulating factors (CSF), which are produced by fibroblasts and lymphocytes. Of these, the best characterized is CSF-1, or M-CSF, which is lineage-specific for mononuclear phagocytes. Mononuclear phagocytes also proliferate and differentiate in response to CSFs that affect other hematopoietic lineages. These include GM-CSF and G-CSF, which control both myeloid and mononuclear phagocyte lineages, and interleukin-3, which also affects myeloid, erythroid, and lymphoid lineages. [*See* Macrophages.]

Under normal conditions, >50% of the circulating monocytes probably settle in the liver as Kupffer cells, with another 15% settling in the pulmonary alveoli. The life span of the mature macrophage is estimated to be several months. During the inflammatory response, both the influx of blood monocytes and the local proliferation of tissue macrophages increase sharply, and in some granulomas the macrophage turnover may also be increased. During the process of inflammation, free-tissue macrophages may become activated, leading to structural and functional changes in response to mediators, such as gamma, interferon released by antigen-stimulated lymphocytes and complement components. Multinucleated giant cells arise by either fusion of macrophages or failure of cytokinesis during mitosis. The epithelioid cell—another form of mature inflammatory mononuclear phagocyte—has decreased phagocytic and digestive capacities and increased endoplasmic reticulum, which may indicate that it has secretory roles. [*See* Inflammation.]

Macrophages are seen early in the development of the lymphoid system and are known to play a role in tissue resorption associated with embryogenesis. Parallel with the maturation of the lymphoid system, the mononuclear phagocytes show increasing development during fetal and neonatal life, and a number of their functions are relatively immature at birth.

B. Organization

The RES or MPS is part of the so-called lymphoreticular system, which includes additionally granulocytes, platelets, and lymphocytes. The progenitors of these cells are pluripotential hematopoietic stem cells located within the bone marrow, fetal liver, and yolk sac of the fetus. Like the mononuclear phagocytic cells, lymphocytes are also mononuclear and interact closely with the cells of the RES but are nonphagocytic. Functionally, phagocytosis in humans is carried out as part of the nonspecific immune response primarily by cells of the RES as well as by neutrophiles and, to a lesser extent, by eosinophils. These types of cells have been referred to as professional phagocytes because their membranes possess specialized receptors for the Fc portion of IgG molecules (IgG1 and IgG3 subclasses) and for the activated component of complement C3. These receptors increase the efficiency of phagocytosis by assisting in the ingestion of microorganisms with IgG or activated C3 on their surfaces. On the other hand, nonprofessional, or facultative, phagocytes include endothelial cells, epithelial cells, fibroblasts, and other cells that will ingest microorganisms under specified conditions but do not possess specialized membrane receptors for IgG or C3. Mononuclear phagocytes, in contrast to the neutrophiles or polymorphonuclear leukocytes, show much greater diversity in function and response. This diversity of structure and function is a result of the progressive maturation of these cells from their bone marrow precursors, their experiences with endocytosis, and their interaction with T lymphocytes. [*See* Lymphocytes.]

C. Biologic Role

The primary role of the phagocytic cells in the body economy is the localization and removal of foreign substances, such as microorganisms. Several integrated functions may be required to achieve these goals. First, the phagocytic cells must reach the site of foreign configuration by a process called chemotaxis, which is a mechanism of unidirectional locomotion of the phagocytes toward an increasing gradient of a chemotactic stimulus or chemoattractant. The phagocytes must then ingest the foreign substance by phagocytosis. The process of phagocytosis is part of the nonspecific immune response and represents the host's initial encounter with non-self. Endocytosis is a more general term and includes both phagocytosis or the ingestion of particles and pinocytosis or the uptake of nonparticulates (e.g., fluid droplets). Both processes involve the engulfment and uptake of particles or fluid from the extracellular milieu. Finally, following one of a number of alternative mechanisms—each of which involves a series of biochemical events—the phagocytes must destroy the foreign substance(s) or inhibit the replication of the challenging microorganisms (i.e., microbial killing). The microbicidal activity of macrophages may vary depending on their source as well as on the type of parasite and its virulence. In general, bacteria of low virulence are not susceptible to killing. For example, in the facultative intracellular parasites, *Listeria monocytogenes* appears the most susceptible and mycobacteria the most resistant. The most microbicidal macrophages are the Kupffer cells of the liver, followed by the fixed macrophages of the spleen. Resident peritoneal macrophages and alveolar macrophages have much less activity, but even these cells can kill a substantial proportion of *L. monocytogenes* or opsonized *Salmonella typhimurium*. There is, however, no evidence that normal macrophages can kill pathogenic bacteria, even of such attenuated strains as *Mycobacterium bovis* bacillus Calmette-Guérin.

II. ORGAN-SPECIFIC CELLULAR COMPONENTS AND THEIR INDIVIDUAL BIOLOGIC CHARACTERISTICS

A. Organ-Specific Cellular Components

Two general classes of mononuclear phagocytes of the RES are recognized: wandering and fixed. The wandering cells include the circulating monocytes in the peripheral blood and the free macrophages in the sinusoids of lymphoid organs or in the stroma of many other organs. In their normal state, these macrophages may be localized as temporary residents in connective tissue (histiocytes), lung (alveolar macrophages), and serous cavities (pleural and peritoneal macrophages). In contrast, the fixed cells are permanent residents in various tissue locations and constitute the majority of mononuclear phagocytes, which are distributed strategically throughout the body. They are found lining the microcirculation of the liver (Kupffer cells or sinusoid lining cells), spleen (sinusoid lining cells, reticular cells, dendritic macrophages), and lymph nodes, bone marrow, adrenals, and thymus (fixed tissue macrophages). In addition, they are present at other sites such as the intraglomerular mesangium of the kidney (mesangial macrophage), joints (synovial type A cells), bone (osteoclasts), and brain (microglia), or they are exposed to the external environment in the respiratory, gastrointestinal, and genitourinary tracts.

B. Individual Biologic Characteristics

1. Pulmonary Macrophages

Like the other cells of the MPS, the pulmonary or alveolar macrophages originate from the bone marrow. The differences between the majority of the tissue macrophages and the alveolar macrophages appear to be related to the distinctive functional milieu of the lung. In contrast to the phagocytes of the liver sinusoids and the peritoneum, the alveolar macrophages operate in an aerobic atmosphere. In addition, faced with a constant threat of an unexpectedly high load of inhaled particulates, the system must respond quickly with an enormous increase in cellular output. Apparently, these functions are supported by a labile population of interstitial macrophages, a cellular compartment between the circulating blood monocytes and the free alveolar cells. In this compartment, cells may undergo biochemical adaptation prior to emigration into the air sacs and, in response to demand, adaptive proliferation of interstitial cells may supplement the normal population of alveolar macrophages.

2. Peritoneal Macrophages

Present knowledge indicates that under normal steady-state conditions, peritoneal macrophages form a self-replicating population. The unstimulated peritoneal cavity may contain a number of monocytes, which can be considered to form a transient (traveling) population of cells different from that of the nontraveling and self-sustaining pool of resident peritoneal macrophages. When an inflammation is induced in the peritoneal cavity, there is an increased influx of bone marrow-derived monocytes from the peripheral blood into the peritoneal cavity, these monocytes differentiating to mature (exudate) macrophages at the site of the inflammation. Because the monocyte-derived macrophages differ in a number of respects (e.g., functionally) from the resident peritoneal macrophages, macrophages from a peritoneal exudate form a heterogeneous population of cells. Moreover, the way in which the peritoneal exudate is induced and the nature of the inducing agent itself may change the biologic properties of the macrophage population. For these reasons, doubt concerning the use of peritoneal exudates as a source of macrophages is unavoidable.

3. Liver Macrophages

Liver macrophages, specifically known as Kupffer cells, are easily recognized on the basis of their fine structural characteristics and their distribution preferentially in the periphery of the liver lobule, where they maintain a close contact with endothelial cells. Although Kupffer cells can also be seen in direct contact with other sinusoidal elements, such as fat-storing cells, pit cells, and reticulin fibers, contact between two Kupffer cells is seldomly observed. The three other types of liver sinusoidal cells differ from Kupffer cells with regard to morphology, reaction to experimental conditions, and endocytic capacity. Kupffer cells are endowed with specific functions, such as their strong reaction to foreign particles, uptake of endotoxin, and endocytosis of circulating tumor cells, bacteria, cellular debris, antigens, immune complexes, and fibrin degradation products.

4. Spleen Macrophages

In the spleen of an adult human, 3 million erythrocytes are destroyed per second, in addition to large numbers of neutrophiles, eosinophils, and platelets. This reveals the extraordinary activity of the macrophages in the splenic red pulp, which alone are responsible for all of this cytolysis. Contrary to earlier belief, most cells in the spleen (previously called reticulum cells) correspond to stages in the transformation from monocytes to macrophages. It would be too simplistic to attribute a bone marrow origin to all the red pulp macrophages as has been suggested for Kupffer cells or pulmonary macrophages. The spleen is a very active hematopoietic organ during the fetal period and, therefore, it has been proposed that some of the macrophages seen in the cord at birth originate from monocytes produced locally. On the other hand, the macrophages seen in the germinal centers, which develop in the white pulp as a result of a humoral immune reaction, are different from those in the red pulp (Billroth cords). These white pulp macrophages have the unique property of phagocytizing lymphocytes. Lymphophagocytosis has been observed only in the germinal centers of the lymph nodes, spleen, thymus, tonsils, and appendix. Other cells, whose categorization as macrophages is still debatable, can be found at the junction of the red and white pulp, and in the white pulp, but are not associated with germinal centers. These are elongated and possibly stellate cells that take up and eventually completely degrade the antigenic material transported to the spleen by lymphocytes.

5. Lymph Node Macrophages

The concept of the RES does not distinguish clearly between reticulum cells and macrophages. In this classification, bone marrow-derived fixed macrophages are included in the cellular reticulum and, therefore, belong to the stationary elements or framework of the node in which lymphocytes reside for a certain time. However, the reticulum cells, which manufacture and sustain the intercellular skeleton, are essentially different from macrophages, which have the morphological characteristics of mononuclear phagocytes. According to the concept of the MPS, the cellular reticulum together with the intercellular skeleton forms the framework in which lymphoid cells and macrophages home and interact. The reticulum cells and macrophages in the various compartments are more or less specialized. The reticulum cells in the paracortex most resemble fibroblasts, forming a delicate reticulum throughout the area, except in the region of the germinal centers. In these structures, the antigen-retaining dendritic reticulum cells are highly specialized with a hypertrophic cell membrane, forming an intricate web in which antigen is localized. The macrophages in the outer cortex, or B-cell compartment, are strikingly different from those in the inner cortex, or T-cell compartment. In the outer cortex, the macrophages are large, actively phagocytic cells in various stages of development. They appear to migrate from the marginal sinus through the marginal zone and the follicles toward the base of the germinal centers in the region of the capillary venules. In the T-cell compartment, the characteristic type of macrophage is the interdigitating cell with a corresponding auxiliary role for T-cell function.

6. Bone Marrow Macrophages

In terms of structure, the macrophages or reticular cells of the bone marrow have many features in common with macrophages elsewhere. A distinctive feature is the intimate contact made by these cells with every other element in the bone marrow. Their processes partially invest the walls of sinusoids and, in places, penetrate the lumen; they have a central position in islands of erythroblasts and relate to developing leukocytes. The significance of these relationships lies in the functions of marrow macrophages. Phagocytosis and digestive functions are employed in the removal of effete red cells and particulate matter from the bloodstream, in the disposal of extruded erythroid nuclei, and in the culling of defective cells during development. In association with their phagocytic and metabolic activities, they can act as storage cells, for example, in relation to inert particles, lipid, and iron; in certain circumstances, crystalloid material may be found in them. Their extensive relationships may enable the marrow macrophages to play an important role both in mediating hemopoiesis and in controlling the delivery of cells into the circulation. The phagocytic, digestive, and storage functions of normal marrow macrophages are highlighted by the changes in some pathological disorders involving the MPS.

III. Molecular Basis of Biologic Activities

The most prominent functional property of the macrophage is endocytosis. During phagocytosis, particles are bound to specific or nonspecific membrane receptors, then surrounded by the cell membrane, forming phagocytic vesicles. Receptors that bind the Fc portion of immunoglobulins and the C3 component of complement endow the macrophage with the ability to recognize opsonized particles.

There are at least four classes of Fc receptors. First, a proteinase-sensitive receptor can bind monomers or complexes of IgG of certain subclasses (i.e., IgG1 and IgG3 in humans). Antibodies binding to this receptor are cytophilic; they bind to the macrophage before interacting with antigen. Second, there is a proteinase-resistant Fc receptor that binds to and mediates endocytosis of antigen–antibody (immune) complexes or aggregates of IgG subclasses (IgG2 and IgG4 in humans). A third type of proteinse-resistance Fc receptor specific for IgG3 has been demonstrated in the mouse. Fourth, macrophages also have a receptor that binds IgE. During phagocytosis mediated by these Fc receptors, the receptors are cleared from the membrane, gradually returning over a subsequent period of 6–24 hr. The number of Fc receptors may vary depending on the status of the host. For instance, it may increase four times during inflammation and in certain diseases.

The receptors for complement are independent of the Fc receptors. In unstimulated monocytes or macrophages, the C3 receptors are much more efficient at mediating binding than at mediating ingestion. It is likely that, *in vivo*, the C3 and Fc receptors function synergistically. During macrophage activation, the complement receptors acquire the ability to mediate ingestion on their own. There at least three complement receptors on human mononuclear phagocytes. Complement receptor 1 is specific for C3b and complement receptor 2 for C3d. C5a is chemotactic for mononuclear phagocytes, and it is likely that this third complement receptor is expressed on these cells as well as on granulocytes.

Macrophages also have receptors for lymphokines, which are involved in macrophage activation, and for CSF, which regulate macrophage proliferation. Receptors for insulin have also been demonstrated on macrophages. In addition, macrophages have other receptors that recognize complex carbohydrates and fucosyl- and mannosyl-terminal glycoproteins. These receptors may be important in the clearance of glycoproteins and in the recognition of senescent cells, heterologous erythrocytes, yeasts and other fungi, bacteria, and parasites. Macrophages recognize α_2-macroglobulin–proteinase complexes, which may be important in *in vivo* clearance of enzymes such as thrombin, plasmin, kallikrein, and activated complement components. Receptors for proteins containing iron may play a role in the secretion of iron by macrophages. Receptors for fibrin–fibrinogen complexes may play an important role in the clearance of fibrin from the circulation or inflammatory sites. The receptor for fibronectin may aid in the adhesion of monocytes to areas containing breaches in the integrity of the endothelial lining of the vessels, and fibronectin may also act as an opsonin for certain particles. Finally, macrophages may play an important role in the regulation of triglyceride and cholesterol metabolism through their receptors for normal and altered lipoproteins.

With regard to chemotaxis, macrophages contain on their surfaces and secrete proteolytic enzymes active at tissue pH that may be relevant to their capacity to migrate *in vivo*. Besides the chemotactic component of complement C5a or anaphylatoxin, there are other chemotactic substances such as bacterial products (e.g., *N*-formyl-methionyl peptides) and products from stimulated B and T lymphocytes that attract mononuclear phagocytes to sites of inflammation and delayed hypersensitivity reactions.

Factors produced by fibroblasts, fragments of collagen, elastin, and denatured proteins may help attract macrophages to sites of tissue injury. Incidentally, at least two classes of substances inhibit the random migration of macrophages and, thus, prevent migration away from sites of inflammation. Lymphokines (e.g., macrophage migration inhibitory factor, macrophage activation factor) and proteolytic enzymes produced during activation of complement (e.g., factor Bb) and of the fibrinolytic system (e.g., plasmin) are among these factors.

Although phagocytosis was recognized over a century ago, only in the last decade has the importance of macrophages as secretory cells been determined. Over 50 secretion products of macrophages have been identified. This secretory function of macrophages is under complex control and varies with their physiologic state. Some of these include lysozyme, complement components (C1–C5), arachidonic acid metabolites, acid hydrolases, neutral proteinases, plasminogen activator, elastases, collagenases, gelatinases, arginase, and lipoprotein lipase. Macrophages also secrete a variety of plasma proteins, many of which have previously been identified as secretion products of hepatocytes. These include α_2-microglobulin, α_1-proteinase inhibitor, tissue inhibitor of metalloproteinases, fibronectin, transcobalamin II, apolipoprotein E, tissue thromboplastin, and coagulation factors V, VII, IX, and X. Finally, among many other products produced by macrophages are those that serve to regulate the functions of other cells such as interleukin-1, angiogenesis factor, and interferon.

Although the biochemistry and metabolism of phagocytosis are beyond the scope of this article, it is worth noting that, even though the respiratory burst is intimately connected with phagocytosis, it is not essential. Recent evidence suggests that free-tissue macrophages and newly recruited monocytes—but not fixed-tissue macrophages—can response to lymphokines and phagocytic stimuli by mounting a respiratory burst. The failure of fixed-tissue macrophages, such as Kupffer cells, to produce active metabolites of oxygen may be important in protecting tissues from damage during the scavenger functions of the macrophage. Many soluble agents, including antigen–antibody complexes, C5a, ionophores, and tissue promoters, can trigger the respiratory burst without phagocytosis. The respiratory burst can also be triggered by opsonized particles or surfaces when phagocytosis is frustrated by the use of a drug such as cytochalasin B. Phagocytosis can also proceed without the repiratory burst. In particular, phagocytosis mediated by complement C3 or C3bi receptors does not trigger release of hydrogen peroxide or arachidonic acid metabolites. [*See* Respiratory Burst.]

IV. PHYSIOLOGIC INTERRELATIONSHIPS WITH OTHER SYSTEMS OF HOST DEFENSE, TISSUE INJURY, AND HOMEOSTASIS

Although macrophages were originally characterized by their phagocytic property, they are not unique in this regard because many other cell types are also phagocytic at one time or another during their development. For example, most endothelial cells are not phagocytic; however, many types of endothelial cells can manifest phagocytic behavior under special conditions, such as being stressed by small particulate challenges. Furthermore, in addition to this biologic activity, macrophages also play many other important roles: They contribute to host defense, tissue hygiene, wound healing, and general homeostatic mechanisms by interacting with many other cells and tissue fluids within the body. The unique versatility and easy adaptability of the macrophages to their tissue environments endow these cells with the ability to influence many processes negatively, positively, or in complex feedback loops. The invasion of a host by pathogenic microorganisms results in a pattern of cellular and humoral defense reactions. As will be explained later in this section, a pattern of systemic metabolic responses accompanies the mobilization of the immunological host-defense mechanisms.

As expected for a cell that (in evolutionary terms) preceded specific immunity, macrophages play a protean role in systemic host defense. As such, they are involved in resistance to infection, radiation injury, trauma, shock, and the metastatic spread of tumors. While the predominant role of neutrophils is the destruction of microoganisms, particularly extracellular pathogens that rely on the evasion of phagocytosis for survival, macrophages are concerned mainly with the control of those microorganism that are able to survive intracellular residence and against which neutrophils are ineffective. In the host defense against infection with intracellular pathogens, the principal effector cells are monocytes and macrophages. Monocytes may serve as a backup system to neutrophils in acute infections, but they phagocytize less efficiently and lack many of the potent bactericidal systems of the neutrophil. Macrophages are much more important in chronic infections with intracellular pathogens, such as certain bacteria (e.g., *Mycobacterium leprae, M. tuberculosis, Legionella pneumophila, L. monocytogenes, Salmonella typhi, Brucella abortus*), some fungi (e.g., *Candida albicans, Cryptococcus neoformans, Histoplasma capsulatum*), many protozoa (e.g., *Toxoplasma gondii,* malarial *Plasmodia, Leishmania donovani, Trypanosoma cruzi*), and most viruses. The growth of these microorganisms is not stopped by phagocytosis, and further steps are required to inhibit their multiplication and spread. These steps involve the accumulation of macrophages within infective foci and their activation. Once activation is achieved, the growth of the pathogen can be stopped. Sensitized lymphocytes play the central role in this process, thus enhancing the bactericidal activities of macrophages through direct cell-to-cell contact or by the intervention of soluble mediators, such as lymphokines, interleukins, and gamma interferon. Cellular immune reactions not only occur against the microbes listed here but may also be generated against microbial products or against soluble proteins. The cellular landmark in some of these immune reactions involving cellular mechanisms is granuloma formation. The macrophages forming the granuloma are called epithelioid cells because they adhere closely to each other, adopting an epithelioid-like characteristic. Some of the macrophages, may fuse with each other to form multinucleated giant cells.

Apart from their historical and classic role in pathologic processes, such as inflammation and host resistance to microbial agents, macrophages represent one of the major—and, indeed, perhaps primary—effector mechanisms of the host in the surveillance and destruction of tumor cells. They have been identified as a cellular component of experimental as well as human tumors, constituting an appreciable portion of the total tumor mass (8–54%). Enhancement in the content of tumor macrophages has been interpreted as being associated with tumor regression and decreased incidence of metastases. Similarly, because macrophages possess the ability to exert an antiproliferative effect, the mitotic activity of tumor cells may also be influenced by the degree of infiltration of a tumor by macrophages. The validity of these findings with regard to human tumors can be supported by histologic observations where regression was characterized by a significant host-cell infiltration.

One of the most important contributions of macrophages to host defense is their central role in the initiation and regulation of the immune response, both *in vivo* and in culture. Macrophages may achieve this role in a number of ways with varying degrees of specificity. By a rather nonspecific mechanism, they can either improve the viability of lymphocytes or suppress their proliferation through thymidine, arginase, complement cleavage products, prostaglandin E, and interferon. Macrophages may also alter the function of lymphocytes more specifically by the interleukin-1 (formerly called leukocyte-activating factor) pathway. Yet another more specific function of macrophages is their role in both humoral and cell-mediated immune responses. As stated previously, they are involved both in the initiation of responses as antigen-presenting cells and in the effector phase as inflammatory, tumoricidal, and microbicidal cells, in addition to their regulatory function. The uptake of antigens by macrophages is the first step in the processing of antigen leading to the production of circulating antibody. In such cases, antigen is not completely degraded by the macrophage but becomes bound to macrophage RNA or membrane. The macrophage is not the cell that recognizes antigen as foreign, but the macrophage nonspecifically processes the antigen so that it may be recognized by specific antigen-reactive lymphocytes. Processed antigens are expressed on the surface of antigen-presenting macrophages in conjunction with self-surface markers (class II major histocompatibility complex or MHC markers) that are recognized by T-cell receptors for antigen and for self-class II MHC. This function requires that T cells and macrophages display the same MHC-fencoded class II determinants (HLA-DR antigens in humans). However, not all monocytes and macrophages express such determinants. The precise mechanism underlying this process at the molecular level is still not well understood.

Only in the last two or three decades have relevant methodological advances been useful to the quantitative assessment of the biochemistry and physiology of mononuclear phagocytes. These advances have included sophisticated isotopic and nonisotopic tracer techniques, improved methods for the isolation and purification of macrophages, refinement of cell- and tissue-culture procedures, and the advent of powerful immunologic approaches. The production of monoclonal antibodies, the development of molecular biology, and the applications of scanning electron microscopy and flow cytometry, as well as the use of sophisticated approaches in immunochemistry and immunohistology, have expanded considerably over still relatively limited capabilities to perform quantitative and well-controlled studies on isolated cell populations.

Much of this progress in biotechnology and its application to studies on the RES have confirmed Aschoff's suggestion (made in his classic Janeway lecture in 1924) that the RES plays an important role in the metabolic functions of the host. In addition to the broad spectrum of activities displayed by macrophages, as described earlier, they also participate in a number of systemic responses involving carbohydate, lipid, and protein metabolism. These systemic metabolic responses to infection and endotoxicosis are accompanied by either a generalized catabolic state with wasting of body tissue to supply amino acids and fatty acids as energy sources during the febrile state or, conversely, a selective anabolic state with protein synthesis of acute stress proteins and immunoglobulins.

V. DISORDERS OF THE RES

A detailed discussion of the pathobiology of the RES is beyond the scope of this article; the reader should refer to the bibliography section for further information. In brief, there are a number of disorders associated with either quantitative and/or functional abnormalities of monocytes and macrophages. For example, whereas an enlargement of the spleen and lymph nodes, which are rich in these cells, may be a result of the monocytosis or reactive hyperplasia produced in response to infection with *M. tuberculosis* in normal individuals, these cells may proliferate abnormally to extremely high levels in individuals with monocytic leukemia or other malignant histiocytic proliferative disorders.

One group of disorders of the RES develops as a result of the ingestion of a nondigestible substance, the overloading of iron as in hemosiderosis, or inborn errors of metabolism in which a specific genetic defect of macrophage enzyme function has occurred (e.g., Gaucher's disease, Hurler's disease). These disorders fall under the category of storage diseases.

Another group of macrophage disorders includes certain genetic abnormalities, such as chronic granulomatous disease, in which both macrophages and polymorphonuclear leukocytes lack as enzyme important in the respiratory burst associated with phagocytosis. In this disease, the patient's phagocytes are unable to kill those pathogens that are susceptible to reactive metabolites of oxygen. Defects in synthesis and secretion of complement components can cause macrophage dysfunction because macrophages constitute a major source of certain complement components, and the products of complement activation are important for macrophage functions, such as phagocytosis and chemotaxis. In osteoporosis, significantly decreased numbers of osteoclasts can be detected in bone, so that bone resorption is abnormal.

A number of iatrogenic or environmentally induced disorders also exist, such as the high concentration of glucocorticosteroids and ionizing radiation, that may interfere with the macrophage defense system, including macrophage migration into tissues as well as macrophage proliferation. These acquired defects may lead to the onset of frequent opportunistic infections.

Although genetically determined differences in macrophage responses to lipopolysaccharides have been reported in animals, a similar heterogeneity in lipopolysaccharide responsiveness may likely occur in humans as well.

BIBLIOGRAPHY

Abbas, A. K., Lichtman, A. H., and Prober, J. S. (1994). "Cellular and Molecular Immunology," 2nd Ed. Saunders, Philadelphia.

Bellanti, J. A. (1985). "Immunology III." Saunders, Philadelphia.

Friedman, H., Escobar, M. R., and Reichard, S. M. (gen. eds.) (1980–1988). "The Reticuloendothelial System—A Comprehensive Treatise," Vols. 1–10. Plenum, New York.

Hiemstra, P. S., Eisenhaer, P. B., Harwig, S. S. L., van den Barselaus, M. T., Van Furth, P., and Lehner, R. P. (1993). Antimicrobial proteins of murine macrophages. *Infect. Immunity* **61,** 3038–3046.

Janeway, C. A., Jr., and Travers, P. (1994). "Immunology." Garland, New York.

Jawetz, E., Melnick, J. L., and Adelberg, E. A. (eds.) (1993). "Review of Medical Microbiology," 19th Ed. Appleton & Lange, Los Altos, California.

Klein, J. (1990). "Immunology." Blackwell Scientific, Cambridge, England.

Roit, I. M., Brostoff, J., and Male, D. K. (1989). "Immunology," 2nd Ed. Mosby, St. Louis.

Sell, S. (1987). "Immunology, Immunopathology and Immunity," 4th Ed. Elsevier, New York.

Stites, D. P., Stobo, J. D., and Wells, J. V. (1987). "Basic and Clinical Immunology," 6th Ed. Appleton & Lange, Los Altos, California.

Szentivanyi, A., and Friedman, H. (eds.) (1986). "Viruses, Immunity, and Immunodeficiency." Plenum New York.

Retina

JOHN E. DOWLING
Harvard University

GLOSSARY

Action potentials Transient, all-or-none potentials that usually serve to transmit information along nerve cell axons; also called impulses or spikes

Amacrine cells Axonless neurons whose processes are confined to the inner plexiform layer

Bipolar cells Output neurons of the outer plexiform layer that carry information to the inner plexiform layer

Fovea Specialized retinal region of highest visual resolution, containing only cones

Ganglion cells Third-order neurons in retina whose axons form the optic nerve and carry visual information from the eye to rest of brain

Graded potentials Local, sustained potentials whose amplitude is graded according to stimulus strength, generated mainly in nerve cell dendrites and sensory receptors

Horizontal cells Neurons whose cell bodies sit along the distal margin of the outer nuclear layer and that extend processes mainly in the outer plexiform layer

Interplexiform cells Neurons whose cell bodies reside in the proximal part of the inner nuclear layer and that extend processes in both plexiform layers

Plexiform layers Regions consisting principally of neuronal processes, where synaptic interactions take place

Receptive field Area of retina that when illuminated influences the activity of a cell

Synapses Sites at which neurons make functional contact

Visual pigments Light-sensitive molecules in photoreceptors consisting of 11-*cis*-retinal (vitamin A aldehyde) and protein (opsin)

THE RETINA IS A THIN LAYER OF NEURAL TISSUE that lines the back of the eye. It is a true part of the brain (central nervous system) displaced into the eye during development. In addition to the light-sensitive photoreceptor cells, the retina contains five basic classes of neurons and one principal type of glial cell, the Müller cell. The neurons are organized into three cellular (nuclear) layers, which are separated by two synaptic (plexiform) layers. Virtually all the junctions (synapses) between the retinal neurons are made in the two synaptic layers, and all visual information passes across at least two synapses, one in the outer plexiform layer and another in the inner plexiform layer, before it leaves the eye.

Processing of visual information occurs in both plexiform layers. The outer plexiform layer separates visual information into on- and off-channels and carries out a *spatial*-type analysis on the visual input. The output neurons of this layer, the on- and off-bipolar cells, demonstrate a center–surround antagonistic receptor field organization. The inner plexiform layer is concerned more with the *temporal* aspects of light stimuli. Many cells receiving input in this layer respond with transient responses and respond better to moving stimuli than to static spots of light. The output neurons of this layer, the ganglion cells, reflect the processing of information in either the outer plexiform layer (i.e., the cells respond in a sustained fashion to appropriately positioned stimuli) or the inner plexiform layer (i.e., the cells respond better to moving stimuli than to static ones).

Figure 1 is a light micrograph of a piece of human retina. The photoreceptors are located farthest from the front of the eye, at the top of the micrograph. Light, entering the eye, passes through the transparent retina and is captured by the pigment-containing outer segments of the photoreceptors. (Overlying the photoreceptors is the pigment epithelium, which serves to absorb stray light and to prevent backscatter of light

ENCYCLOPEDIA OF HUMAN BIOLOGY, Second Edition, VOLUME 7. Copyright © 1997 by Academic Press. All rights of reproduction in any form reserved.

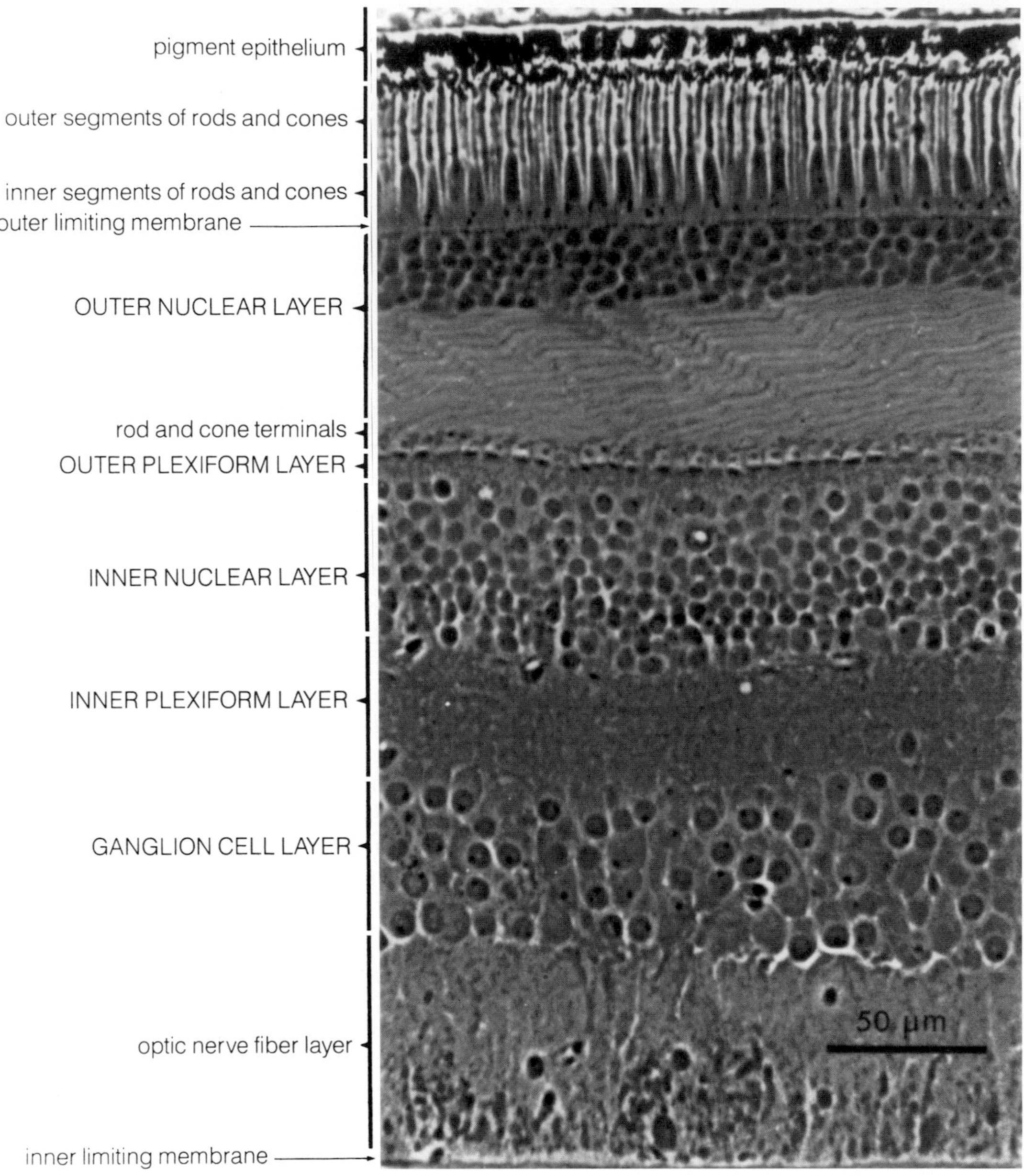

FIGURE 1 Vertical section through human retina. Micrograph shows an area about 1.25 mm from center of the fovea. In the foveal region of the retina, inner layers of the retina are pushed aside so that light can impinge more directly on receptors. Thus, around the fovea and for some distance away (as shown here), receptor terminals are displaced laterally from the rest of the photoreceptor cell.

into the retina.) The cell bodies of the photoreceptors are located in the outer nuclear layer, whereas the cell bodies of four of the basic classes of retinal neurons—horizontal, bipolar, amacrine, and interplexiform cells—are in the inner nuclear layer. The cell bodies of the ganglion cells make up the most proximal cellular layer. The outer and inner plexiform layers are interspersed, respectively, between the outer and inner nuclear layers and the inner nuclear and ganglion cell layers.

In many primates, including humans, a small region of the retina is specialized for high-acuity vision. It is called the fovea and is centrally located (i.e., on the visual axis of the eye). The layers of the retina below the photoreceptor inner and outer segments are displaced aside from the fovea so that light can impinge directly on the photoreceptors. Only cones are present in this area, and the foveal cones are the thinnest and longest photoreceptors in the retina. The rod-free area is about 0.3 mm in diameter and contains approximately 35,000 cones. No blood vessels are found in the fovea and, furthermore, there are few blue-absorbing cones in the center of the fovea. These specializations serve to improve the visual resolution of the fovea. [*See* Eye, Anatomy.]

I. PHOTORECEPTORS

Vertebrate retinas typically contain two types of photoreceptors, rods and cones, differentiated on the basis of their outer segment shape (see Fig. 1). This criterion is not always reliable; in primates, for example, the outer segments of the cones found in the fovea show no significant taper (i.e., they are rod-shaped). Rods mediate dim-light vision, whereas cones function in bright light and are responsible for color vision.

A. Visual Pigments

Light sensitivity of the photoreceptors results from the presence of visual pigment molecules contained within their outer segments. One rod pigment, called *rhodopsin,* and three cone pigments are in the primate retina. Rhodopsin absorbs light maximally in the blue-green region of the spectrum (500 nm), whereas the primate cone visual pigments absorb maximally in the blue (420 nm), green (530 nm), and red-yellow (560 nm) regions of the spectrum. The cone pigments are segregated into separate classes of cones; thus blue, green, and red-yellow sensitive cones are in the primate retina. Color-blind individuals are missing or produce an altered visual pigment. Red-blind individuals (protanopes) are missing or possess an altered red-yellow pigment; green-blind individuals (deuteranopes) are missing or have altered the green pigment; blue-blind patients (tritanopes) are missing or have altered the blue visual pigment. In the primate eye, there are more red and green cones than blue cones, and as noted earlier, blue cones are extremely rare in the high-acuity foveal region of the retina. [*See* Color Vision.]

The genes encoding for rhodopsin and the cone pigments have been identified and isolated in humans and a number of other species. The red- and green-sensitive pigment genes in humans are on the X chromosome (i.e., they are sex-linked), whereas the genes for the blue-sensitive pigment and rhodopsin are on autosomes. The red- and green-sensitive visual pigments are highly homologous (~95%), whereas there is about 40% homology between the red and green pigments and the blue pigment, and about the same homology between all the cone pigments and rhodopsin. Studies of the color vision pigment genes in red- and green-blind individuals have shown that color blindness is caused by a loss or alteration of one or another of the genes. Red-blind individuals have an altered gene for the red-sensitive pigment, whereas green-blind individuals have an altered green-sensitive pigment gene or are lacking the gene altogether. When the gene for the red- or green-pigment is altered, usually no red- or green-sensitive cone cells develop in the retina. Sometimes, red- or green-sensitive cones form, but the pigment within the cell, and therefore color vision, is abnormal (a condition termed anomalous trichromacy).

The visual pigment molecules are concentrated to a high degree in the outer segments of the photoreceptors. The outer segments contain numerous transverse membranous discs (Fig. 2), and virtually all the visual pigment molecules are contained within the disc membranes. A typical outer segment may have as many as 2000 transverse discs and may contain 10^9 visual pigment molecules.

1. Visual Pigment Chemistry

All visual pigments have a similar chemistry. They consist of two components, retinal (vitamin A aldehyde), termed a *chromophore,* bound to a protein called opsin. Different visual pigments have different opsins, and this accounts for the variations in their color sensitivity. The light sensitivity of the visual pigments is due to the retinal chromophore. When a visual pigment molecule absorbs a quantum of light, several molecular transformations occur, first in the chromophore and then in the protein (opsin) part of the molecule. These transformations lead to the excitation of the photoreceptor cell and also to the separation of the retinal chromophore from opsin. This latter process is called *bleaching* because it results in the loss of color of the visual pigment molecules and their ability to absorb visible light.

Retinal can exist in different shapes (i.e., several cis–trans isomers of the molecule are possible). All visual pigments require one particular isomer, the 11-cis, for their synthesis, and this form of the chromophore combines spontaneously with opsin to form visual pigment. When a visual pigment molecule ab-

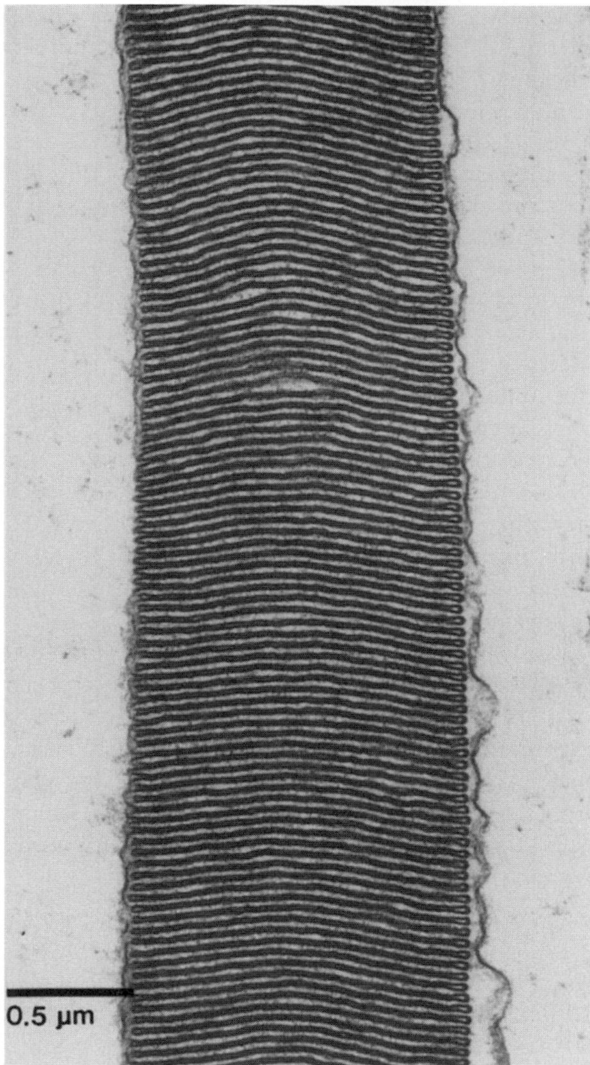

FIGURE 2 Electron micrograph of a portion of a cone outer segment. Contained within the structure are numerous transverse membranous discs.

sorbs a quantum of light, the first transformation is the isomerization of the chromophore from the 11-cis to the all-trans form. Indeed, this is the only action of light in the visual process, to change the shape of the chromophore. The chromophore-shape change initiates a series of conformational changes in the opsin, and this leads to excitation of the photoreceptor cell and release of the chromophore from opsin.

A series of intermediates have been identified between the absorption of light by visual pigment molecules and the release of the chromophore from opsin. One of the intermediates, metarhodopsin II, appears responsible for excitation of the photoreceptor cell (i.e., it is the photoactive intermediate). Figure 3 shows a simplified scheme of the visual cycle for the rod visual pigment rhodopsin.

The retinal chromophore of rhodopsin derives from vitamin A, and it represents a slightly oxidized form of the vitamin. During the operation of the visual cycle, some retinal is lost and must be replaced from body stores of the vitamin. In vitamin A deficiency, this replenishment fails, a full complement of visual pigment can no longer be synthesized in the photoreceptors, and light sensitivity of the rods and cones is decreased. The loss of light sensitivity is more obvious in the dark, and hence the condition is known as night blindness. Refeeding of vitamin A to a vitamin A–deficient animal or patient usually restores visual sensitivity. [*See* Vitamin A.]

B. Photoreceptor Responses

Excitation of visual pigment molecules leads ultimately to a change of potential across the membrane

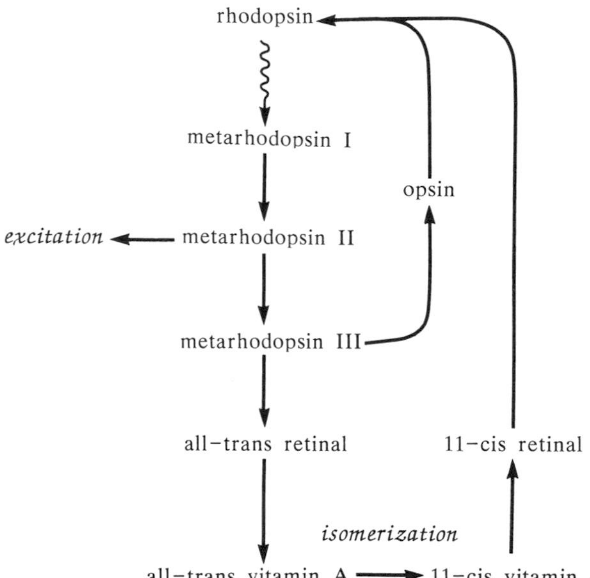

FIGURE 3 Scheme of the sequence of events that occurs following the absorption of a quantum of light by the rod visual pigment, rhodopsin. Light initiates the conversion of rhodopsin to retinal and opsin through a series of metarhodopsin intermediates. Metarhodopsin II is the active intermediate leading to excitation of the photoreceptor cell. Eventually, the chromophore of rhodopsin, retinal, separates from the protein opsin and is reduced to vitamin A (retinol). For the resynthesis of rhodopsin, the shape of vitamin A must be changed (isomerized), from the all-trans to the 11-cis form, and this isomerization takes place in the pigment epithelium overlying the receptors (see Fig. 1). Vitamin A is replenished in the eye from the blood.

surrounding the photoreceptor cell. Membrane potential controls the release of neurotransmitter molecules from synapses, and in this way light influences the exchange of information between photoreceptors and second-order neurons in the retina.

All vertebrate photoreceptors hyperpolarize in response to light (i.e., the membrane potential becomes more negative). Why photoreceptors hyperpolarize in the light is as follows: In darkness, the membrane of the outer segment is leaky to Na^+. Because Na^+ levels are higher outside the cell than inside, positive Na^+ ions enter the cell in darkness, causing the cell to be partially depolarized (i.e., the membrane potential is more positive than is typically the case for neurons at rest). Light decreases the conductance (leakiness) of the outer segment membrane to Na^+, thereby decreasing the flow of positive ions into the cell and causing the cell to become more negative (i.e., to hyperpolarize).

The conductance of the outer segment membrane to Na^+ is controlled by a second-messenger molecule, cyclic GMP, which maintains channels in the membrane in an open state. In the light, levels of cyclic GMP fall because an enzyme (called *phosphodiesterase*), which breaks down cyclic GMP, is activated, and this causes the channels in the outer segment membrane to close. Phosphodiesterase in turn is activated by a protein called *transducin* (a so-called G protein), and transducin is activated by metarhodopsin II, the photoactive visual pigment intermediate. The cascade between the visual pigment molecule and the Na^+ channel in the outer segment membrane is shown in Fig. 4.

One reason for the cascade between the light activation of visual pigment molecules and the closing of Na^+ channels in the outer segment membrane is for amplification of the signal. That is, one photoactive visual pigment molecule can interact with many transducin molecules (as many as 500), and one phosphodiesterase molecule can break down about 2000 cyclic GMP molecules per second. The cascade of reactions between photon absorption and cyclic GMP inactivation can result in an amplification of about 10^6.

The channels in the outer segment membrane controlled by cyclic GMP also allow some Ca^{2+} to enter the photoreceptor cell, and Ca^{2+} appears to play an important regulatory role in the phototransduction process. Ca^{2+} strongly inhibits an enzyme, guanylate cyclase, that promotes the synthesis of cyclic GMP. In the dark, the Ca^{2+} entering the cell inhibits this enzyme, and cyclic GMP synthesis is low. In the light, when the channels in the outer segment membrane

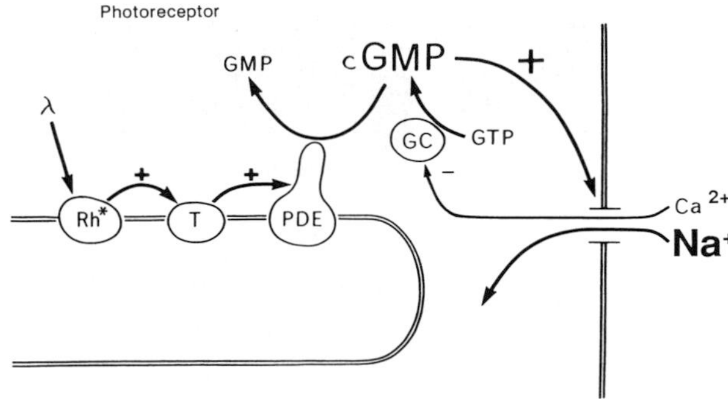

FIGURE 4 Summary diagram of interactions occurring in the rod outer segment during phototransduction. Light-activated rhodopsin (Rh*) activates transducin (T), which in turn activates the enzyme phosphodiesterase (PDE). These interactions occur in the disc membrane. Activation of PDE leads to breakdown of cyclic GMP (cGMP) to an inactive product (GMP). Cyclic GMP maintains channels in the outer segment membrane in an open configuration, thereby allowing both Na^+ and Ca^{2+} to enter the cell in the dark. With a fall in cyclic GMP levels in the light, channels in the outer segment membrane close. The resulting fall of Na^+ levels causes the cell to hyperpolarize. Decrease of Ca^{2+} levels enhances guanylate cyclase (GC) activity, an action that counters the effects of light and increases cyclic GMP levels in the outer segment.

are closed, Ca^{2+} entry into the cell decreases and intracellular Ca^{2+} levels fall. This results in an increased synthesis of cyclic GMP, which serves to counter the effect of light of lowering cyclic GMP levels. Thus, in continuous light, the photoreceptor response recovers partially to dark levels, enabling the photoreceptor to continue to respond even in bright light. This process is termed *adaptation,* and photoreceptor light and dark adaptation plays an important role in the ability of the visual system to respond over a wide range of ambient illumination.

II. CELLULAR AND SYNAPTIC ORGANIZATION

A. Cellular Organization

Most of what is known about the classes of retinal cells has come from light microscopic studies of retinas processed by the silver-staining method of Golgi. This technique enables investigators to see the extent and distribution of the processes of the cells within the retina, to classify the cells, and to construct schemes of the cellular organization of the retina such as that shown in Fig. 5 for the primate retina. Although there are just five major classes of neurons in the retina, there are many morphological types and subtypes in

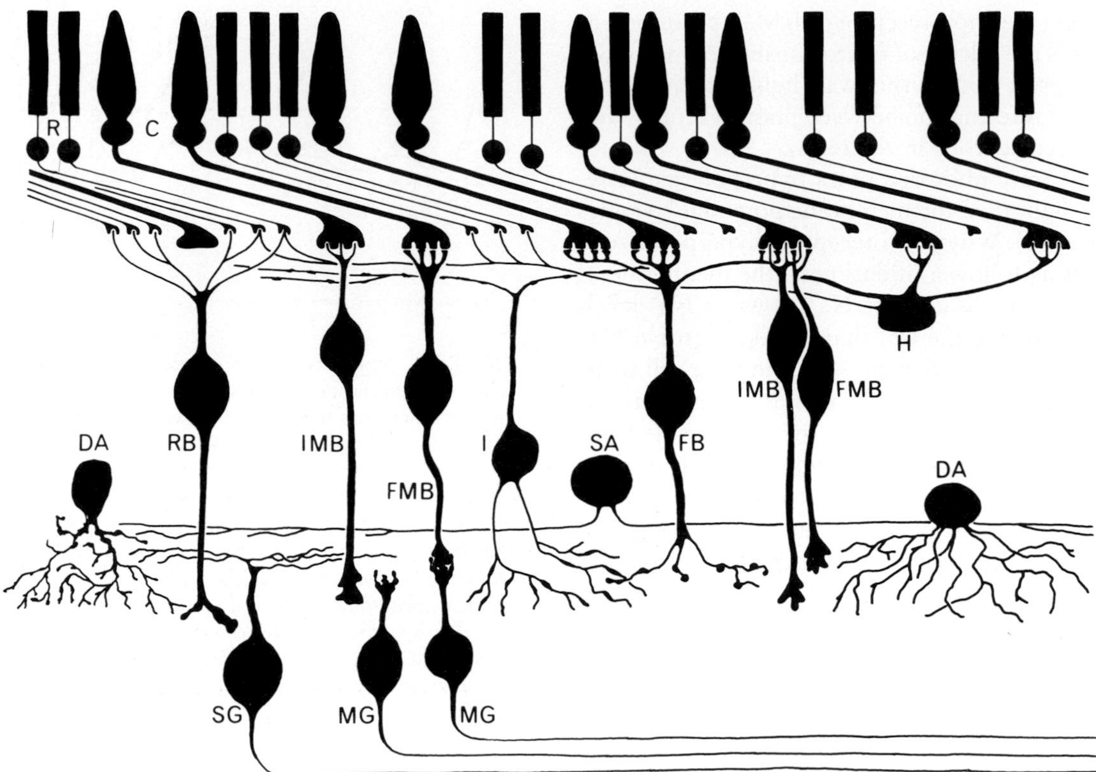

FIGURE 5 Major cell types found in primate retina as viewed in vertical sections of Golgi-stained retinas. See text for description of cells: R, rods; C, cones; RB, rod bipolar cell; IMB; invaginating midget bipolar cell; FMB, flat midget bipolar cell; FB, flat bipolar cell; H, horizontal cell; I, interplexiform cell; DA, diffuse amacrine cell; SA, stratified amacrine cell; SG, stratified ganglion cell; MG, midget ganglion cell.

each major cell class, and even today the total number of morphological types is not known. It is clear that some types are much more common than others, and the focus here will be on the major types of cells in the mammalian and primate retinas.

In the outer plexiform layer, two cell classes—horizontal and bipolar cells—receive input from the photoreceptors. The bipolar cells are the output neurons for the outer plexiform layer; all information passes from outer to inner plexiform layers via these neurons. Horizontal cells, however, extend processes widely in the outer plexiform layer, but their processes are confined to this layer. Their role is to mediate lateral interactions within the first synaptic zone.

Bipolar cell terminals provide the input to the inner plexiform layer, and three classes of cells—amacrine, interplexiform, and ganglion—are ultimately activated. Amacrine cells, like horizontal cells in the outer plexiform layer, spread processes widely in the inner plexiform layer, but their processes are confined to this layer. Interplexiform cells, however, extend processes in both plexiform layers. Ganglion cells are the output neurons for the retina; their axons run along the margin of the retina, collect at the optic disc to form the optic nerve, and carry all the visual information to higher visual centers.

1. Outer Plexiform Layer Neurons—Horizontal and Bipolar Cells

Most retinas contain two basic types of horizontal cells: a cell with a short axon that runs 400 μm or so before ending in a prominent terminal expansion and an axonless cell. The axonless cell has not been seen in the primate retina.

Horizontal cells of the cat have been studied in particular detail, and Fig. 6 shows examples observed by looking down on a flat mount of the whole retina. Electron microscopy has shown that the processes of the axonless cell (Fig. 6b) and the proximal dendritic processes of the short axon cell (Fig. 6a) connect exclusively with cones, whereas the axon terminal processes of the short axon cell end exclusively in the rod

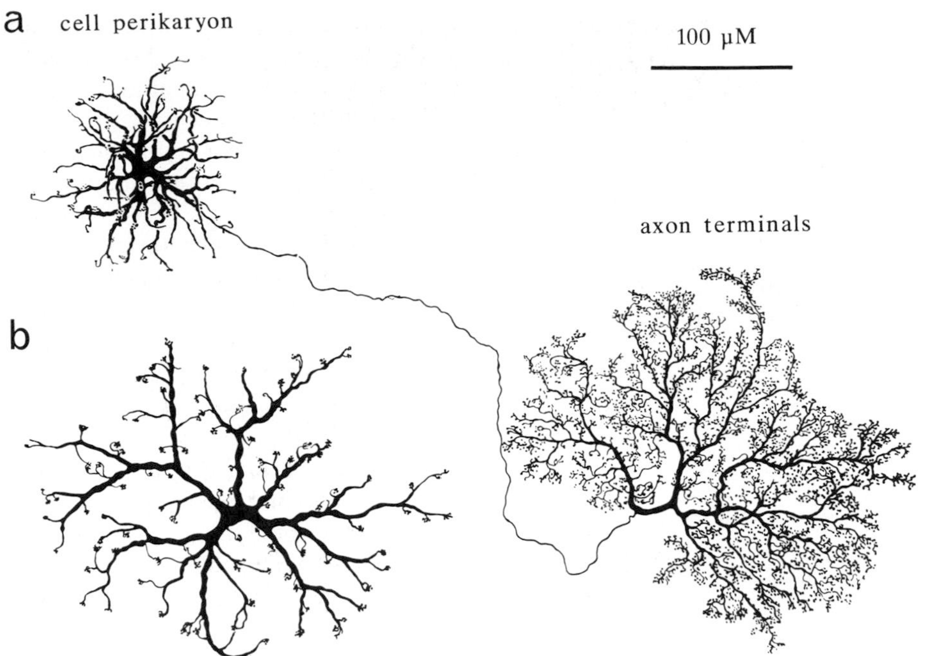

FIGURE 6 Drawings of Golgi-stained horizontal cells from the cat, viewed by looking down on a flat mount of retina. There are two types of horizontal cells in cat retina: an axonless cell (b) and a cell with a short axon (a).

terminals. Why the input to the short axon cells is organized this way is not clear; an appealing suggestion is that this allows for segregation of rod and cone responses in different regions of the cells. In fishes, for example, there are separate rod and cone horizontal cells.

In primates, five principal types of bipolar cells are distinguished: one type is exclusively connected to rods and four types are exclusive to cones. The rod-related bipolars extend their dendrites into the rod synaptic terminals, and their axon terminals end deep in the inner plexiform layer. The rod bipolars contact as many as 30–50 rod terminals.

Two of the cone-related bipolar cell types contact only a single cone terminal. These cells, called *midget bipolar cells,* make different kinds of synaptic contacts with the cone terminals—invaginating and flat contacts (see Section II,B). Their axon terminals end at different levels within the inner plexiform layer, and they appear to be related to the generation of either on- or off-responses to light in the retina. Every cone terminal in the primate retina probably makes connections with both kinds of midget bipolar cells (see Fig. 5). In addition to the cone midget bipolar cells, there are also bipolar cells that contact several cones, probably as many as six or seven. They are called *flat bipolars* or *diffuse invaginating bipolars* based on the type of connection they make with the photoreceptors (see Section II,B).

2. Inner Plexiform Layer Neurons—Amacrine, Interplexiform, and Ganglion Cells

Amacrine cells have no axonal processes; all their processes usually look similar. Amacrine cells are diverse in terms of the extent and distribution of their processes, and it is possible to describe a large number of amacrine cell types in most species. Investigators typically classify amacrine cells into two major types: diffuse and stratified amacrine cells. Diffuse amacrine cells extend their processes throughout the thickness of the inner plexiform layer, whereas the stratified cells extend their processes on one or a few levels in the layer. This simple classification scheme can be expanded to include narrow- and wide-field diffuse or stratified amacrine cells, depending on how far their processes extend, and mono-, bi-, or multistratified cells, depending on whether their processes are confined to one, two, or several levels in the inner plexiform layers.

Interplexiform cells have been recognized as a separate class of retinal neuron only recently. Their peri-

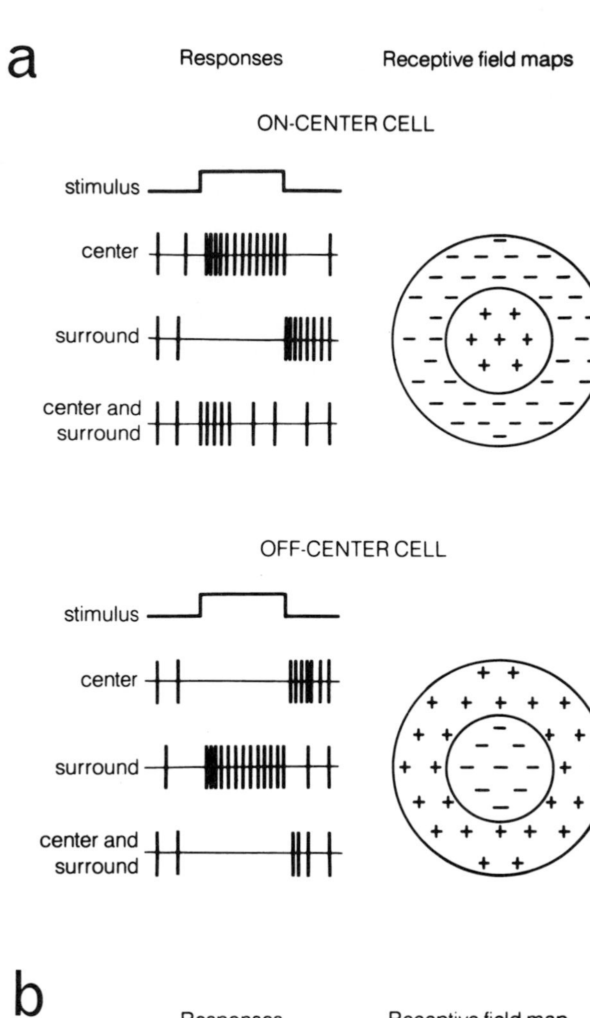

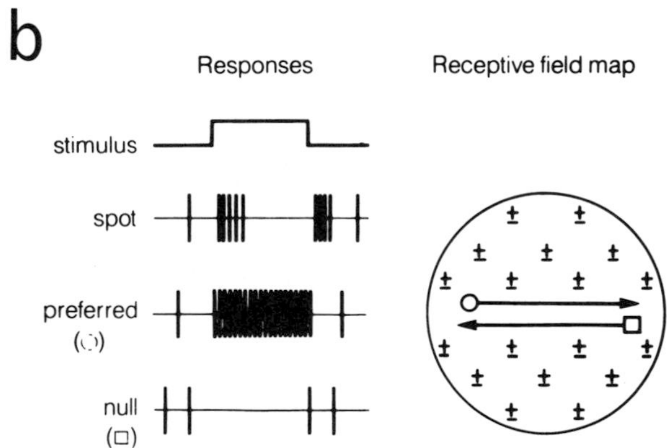

FIGURE 7 (a) Idealized responses and receptive field maps for on-center (top) and off-center (bottom) contrast-sensitive ganglion cells. Drawings on the left represent hypothetical responses to a spot of light presented in the center of the receptive field, in the surround of the receptor field, or in both center and surround regions of the receptive field. A + symbol on the receptive field map indicates an increase in firing rate of the cell (i.e., excitation); a − symbol indicates a decrease in firing rate (i.e., inhibition). (b) Idealized responses and a receptive field map for a direction-sensitive ganglion cell. Such cells respond with a burst of impulses at both onset and termination of a spot of light presented anywhere in the cell's receptive field. This response is indicated by ± symbols all over the map. Movement of a spot of light through the receptive field in the preferred direction (open circles) elicits firing from the cell that lasts for as long as the spot is within the field. Movement of a spot of light in the opposite (null) direction (open squares) causes inhibition of the cell's maintained activity for as long as the spot is within the receptive field.

karya (cell bodies) sit among the amacrine cells, but they send processes to both plexiform layers. Input to these cells is in the inner plexiform layer, whereas most of their output is in the outer plexiform layer. Hence they appear to be mainly a centrifugal type of neuron, carrying information from inner to outer plexiform layers.

Ganglion cells, like amacrine cells, are also diverse in their morphology. They too are classified into diffuse and stratified cells. In primates, particularly in the central region of the retina, ganglion cells are observed with limited dendritic fields. These cells, termed *midget ganglion cells,* receive input from one midget bipolar cell. The dendrites of these midget ganglion cells ramify either in the upper or lower parts of the inner plexiform layer, so that they receive input from one or the other of the two midget bipolar cells described earlier. This implies that one cone in the central (foveal) part of primate retina can send two separate messages to the rest of the brain via two midget ganglion cells. One cell is believed to signal increases in illumination of the cone (i.e., it is an on-cell); the other signals decreases in cone illumination (i.e., it is an off-cell).

3. Retinal Processing of Visual Information—Ganglion Cell Responses and Receptive Field Organization

As noted earlier, ganglion cells signal brightness and darkness information to the rest of the brain, but they also communicate much more than that. Indeed, two basic kinds of processing appear to be occurring in the retina: one carried out mainly in the outer plexiform layer and the other in the inner plexiform layer. Ganglion cells convey information to the rest of the brain that reflects these two stages of processing.

Ganglion cells typically respond to illumination of a restricted but relatively large region of the retina. This region is called the *receptive field* of the cell and is typically about 1 mm in diameter. The most common ganglion cell in the mammalian retina shows evidence of spatial processing of visual information in its responses. Ganglion cells of this type are often called *contrast-sensitive cells,* and they are subdivided into two mirror image classes: on-center, off-surround cells and off-center, on-surround cells. Each has a receptive field that is organized into two concentric zones that are antagonistic to each other (Fig. 7a).

On-center cells respond to an increase in the illumination of the receptive field center with a sustained burst of nerve impulses, whereas illumination of the surround inhibits the firing of nerve impulses by the cell for as long as the light is on. Off-center cells respond in the opposite way; illumination of the receptive field center inhibits the cell, whereas surround illumination provides a sustained excitation of the cell. In both cases the center and surround zones are antagonistic; if center and surround areas are simultaneously illuminated, the cell responds only with a weak response that usually reflects the central response. These ganglion cells mainly reflect the processing that occurs in the outer plexiform layer of the retina.

Other types of ganglion cells respond with more transient responses to retinal illumination, regardless of the position of the illuminating spot in the receptive field, and they reflect more the responses of inner plexiform layer neurons. These cells typically respond with more vigorous responses to moving stimuli than to static spots of light. The receptive fields of some of these cells are organized into antagonistic center and surround regions like the sustained contrast–sensitive cells, but others give short on–off bursts of impulses to spots of light positioned anywhere in the receptive field. Some of the latter cells (particularly in nonmammalian species) show direction-selective properties (Fig. 7b). Movement of a light spot in one direction vigorously excites the cell, whereas movement in the opposite direction inhibits the cell.

As noted earlier, midget ganglion cells receive input from a single midget bipolar cell, which in turn receives input from a single cone. The center of the receptive field corresponds to this direct pathway and is therefore cone-specific. The antagonistic surround response, however, probably reflects mainly horizontal cell activity, and the horizontal cells receive input from many cones. The center of the receptive field, therefore, is small, and it has spectral sensitivity different from the surround mechanism. Such ganglion cells are termed *color-opponent cells.*

B. Synaptic Organization

Two types of chemical synaptic contacts are observed in both plexiform layers. One of these is similar in morphology to known chemical synapses seen throughout the brain, and it is termed a *conventional synapse.* It is characterized by an aggregation of synaptic vesicles in the presynaptic terminal clustered close to the membrane. In the retina, conventional synapses are made by horizontal, amacrine, and interplexiform cells. The other type of synapse is characterized by an electron-dense ribbon or bar in the presynaptic process and is called a *ribbon synapse.* Photoreceptor and bipolar cells make ribbon synapses in the retina. Electrical (gap) junctions are also observed in both plexiform layers of the retina, and they are believed to mediate direct electrical interactions between certain retinal neurons.

Figure 8a shows a drawing of a bipolar ribbon synapse and an amacrine cell conventional synapse in the inner plexiform layer. Typically, there are two postsynaptic processes at the ribbon synapses of bipolar cells, whereas at conventional synapses, only one postsynaptic process is found. In this drawing, the amacrine cell synapse is made back onto the bipolar terminal that is making a synapse on it. Thus, a reciprocal or feedback synaptic arrangement is suggested, which is commonly seen between bipolar cell terminals and amacrine cell processes. Amacrine cell synapses are also made on the processes and cell bodies of ganglion, interplexiform, and other amacrine cells.

Photoreceptor cell synapses are particularly complex, and a summary drawing of a part of a cone synaptic terminal is shown in Fig. 8b. The synaptic ribbons are found above invaginations of the basal surface of the terminal. Processes from horizontal cells and invaginating midget and invaginating diffuse bipolar cells penetrate into the terminal invaginations. Horizontal cell processes lie lateral to the synaptic ribbons, whereas the dendrites of the invaginating bipolar cells are centrally positioned. In addition to the ribbon synapse, the cone photoreceptor terminals make a second, unusual synaptic contact with the flat bipolar cells, which is called a *flat or basal junction.* No ribbon or vesicle cluster is associated with this junction, but some specializations of the membranes on both sides of the synapse are seen. The flat midget bipolar processes are typically found immediately adjacent to the invaginating midget bipolar processes, whereas the processes of the other flat bipolar cells are positioned away from the invaginations. Rod photoreceptor terminals do not make basal junctions; all rod bipolar dendrites penetrate into invaginations of the rod terminal.

Figure 9 is a simplified summary diagram of the synaptic organization of the primate retina. Each cone

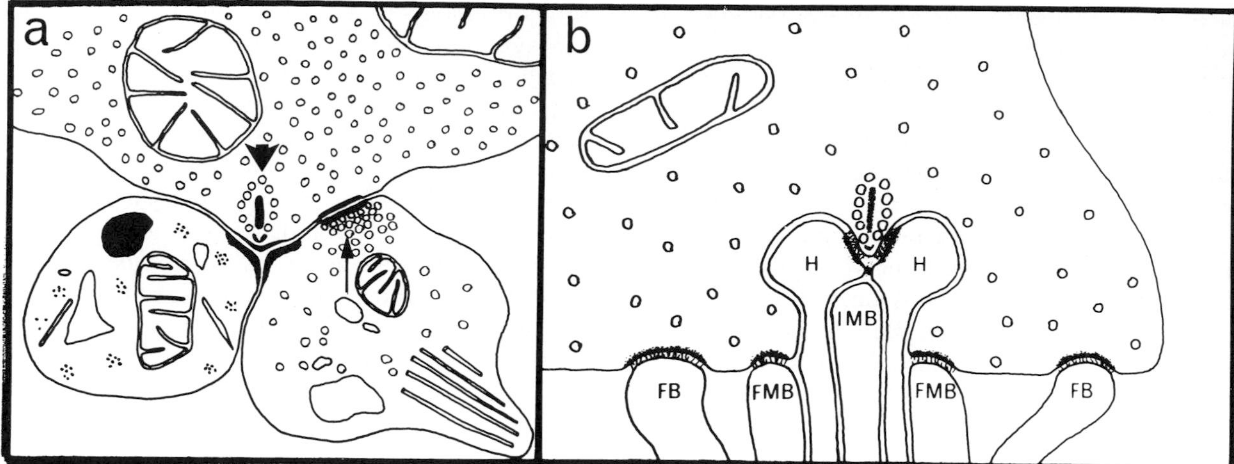

FIGURE 8 (a) Schematic drawing of a bipolar cell ribbon synapse (arrowhead) and a conventional amacrine cell synapse (arrow) back onto the bipolar cell terminal. One postsynaptic process at the ribbon synapse is an amacrine cell process (right); the other is a ganglion cell dendrite (left), a typical arrangement at cone bipolar cell terminals in primate. (b) Schematic drawing of synapses made by cone terminals in primates. See text for details. H, horizontal cell process; FB, flat bipolar cell dendrite; FMB, flat midget bipolar cell dendrite; IMB, invaginating bipolar cell dendrite.

in the primate makes connections with two midget bipolar cells, one that makes invaginating-type junctions and a second one that makes flat junctions (left side of figure). In the central region of the primate retina, these two bipolar cell types synapse on separate midget ganglion cells. In the periphery of the primate retina, midget bipolar cells synapse on ganglion cells that receive input from a few to many midget bipolar terminals.

All cone terminals in primates also make synapses with diffuse cone bipolars and the proximal (dendritic) processes of horizontal cells. The axonal processes of the horizontal cells extend to the rod terminals. Synapses made by the horizontal cells have been observed on bipolar cells in many species, but very rarely back onto the photoreceptor terminals. (There is, however, good physiological evidence for synapses from horizontal cells into photoreceptors in many species.) Another interesting feature of the cone photoreceptor terminals is that they make junctions with each other and with adjacent rod terminals. Evidence in many species, including primate, suggests that these junctions are small electrical synapses.

All bipolar terminals make synapses onto amacrine cell processes, and these processes may make a reciprocal synapse back onto the terminal, as in Fig. 8a. Amacrine cell processes also make junctions on ganglion cells, interplexiform cells, and other amacrine cells.

Rod bipolar terminals in the mammalian retina do not contact ganglion cells directly (see right side of Fig. 9). Rather amacrine cell processes are always postsynaptic at the ribbon synapses of the rod bipolar terminals. One of these makes a feedback synapse onto the terminal, whereas the other belongs to a special amacrine cell that makes both gap junctional (electrical) and conventional (chemical) synapses with cone bipolar terminals. These cone bipolar terminals then contact the ganglion cells. Thus all rod information in the mammalian retina passes through an amacrine cell before it is transmitted to the ganglion cells. Why this is so is not clear; it has been suggested that the amacrine cell serves to amplify the rod signal (see discussion of amacrine cell responses in Section III,D).

Finally, interplexiform cells receive their input from amacrine cells, and they make some synapses in the inner plexiform layer on amacrine and ganglion cells processes. Most of their synapses, however, are made in the outer plexiform layer on bipolar and horizontal cells.

It should be noted that Fig. 9 is highly simplified. It is unlikely that any one amacrine cell makes the variety of contacts shown for either of the amacrine cells drawn in the figure. For example, amacrine cells involved in the rod pathway make gap junctions only with diffuse invaginating bipolar terminals.

III. NEURONAL RESPONSES

In many nonmammalian species, intracellular recordings can be made routinely from most of the

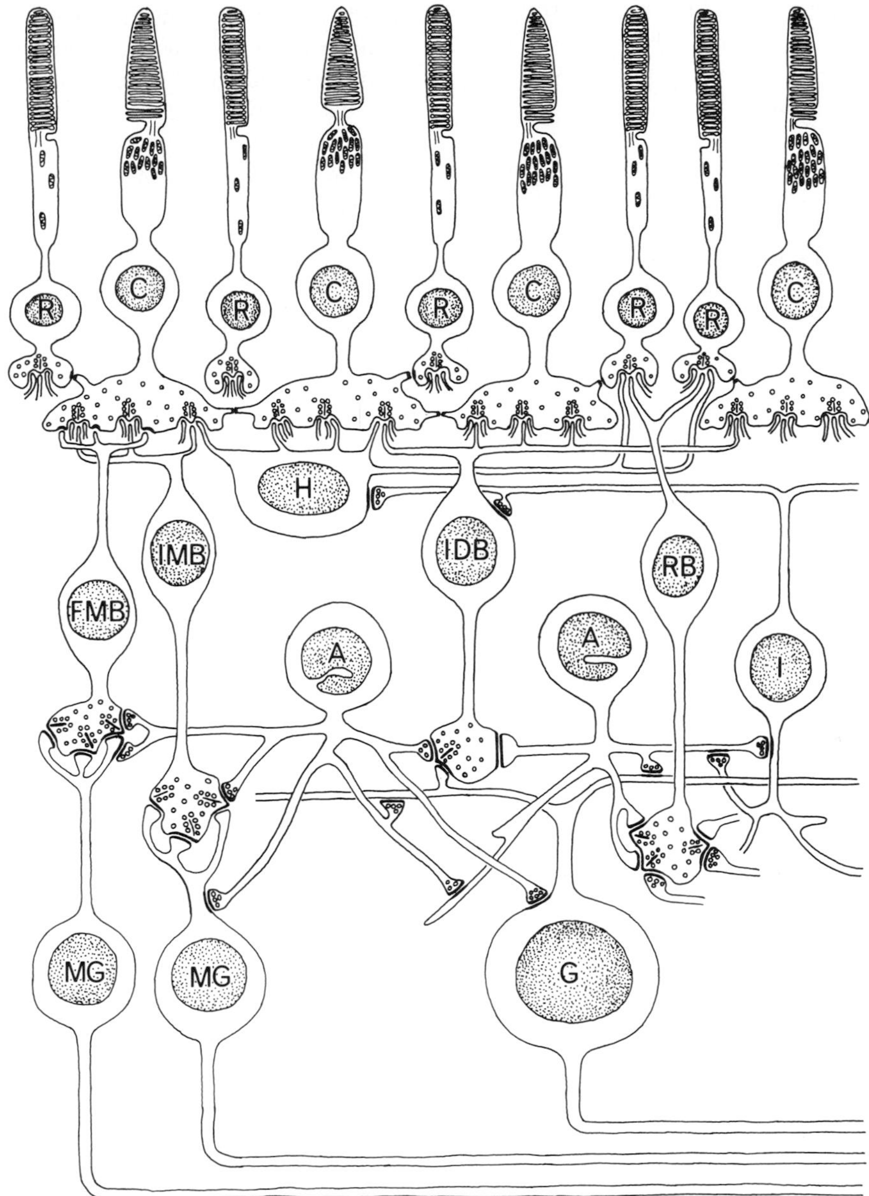

FIGURE 9 Summary diagram of the synaptic organization of the primate retina. See text for details. R, rod; C, cones; FMB, flat midget bipolar cell; IMB, invaginating midget bipolar cell; H, horizontal cell; IDB, invaginating diffuse bipolar cell; RB, rod bipolar cell; I, interplexiform cell; A, amacrine cell; G, ganglion cell; MG, midget ganglion cell.

retinal cells. In mammalian retinas, intracellular recordings are much more difficult to make, and relatively few have been reported from the primate retina. The following discussion is based mainly on recordings from nonmammalian species, but the recordings made so far from mammalian retinal neurons, especially from rabbit cells, are similar.

The distal retinal neurons–receptors, horizontal cells, and bipolar cells—respond to light with sustained, graded membrane potential changes (Fig. 10). Unlike most neurons found elsewhere in the brain, they do not generate action potentials. This may be the case because these neurons have relatively short processes, and they do not need to transmit information over long distances; in other words, passive spread of potential along the cell membrane is sufficient to transmit information from one end of the cell to the other. A second reason that the distal retinal

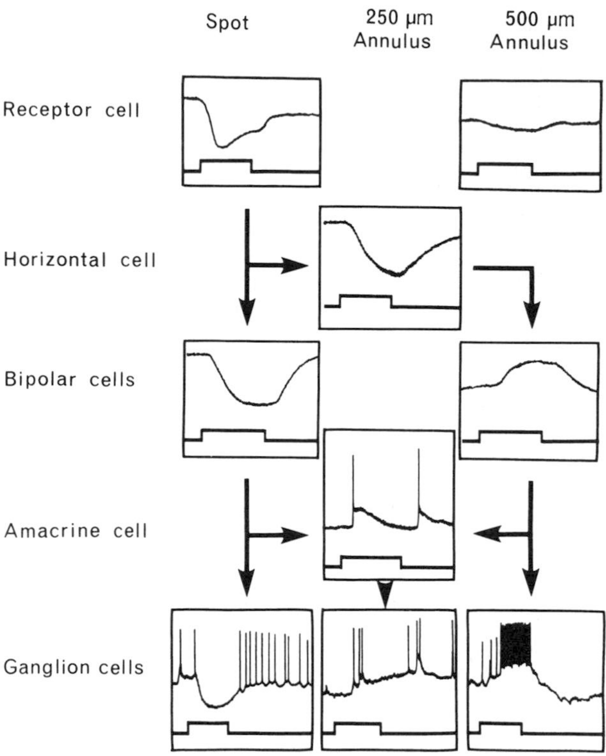

FIGURE 10 Intracellular responses from receptor, horizontal, bipolar, amacrine, and ganglion cells of mudpuppy retina. Distal retinal neurons (receptor, horizontal, and bipolar cells) respond to illumination with sustained graded potentials; proximal retinal neurons show both sustained and transient potentials and action potentials. Receptor, bipolar, and ganglion cells respond differently to center (spot) and surround (annular) illumination. Horizontal and amacrine cells usually respond similarly to spot and annular illumination; here responses to a small annulus (250 μm) are shown that stimulate both the center and surround of the receptive field. The bipolar cell illustrated is a center-hyperpolarizing cell, the amacrine cell shown is a transient amacrine cell, and the ganglion cell is an off-center cell. Arrows indicate in a general way how the responses are synaptically generated. That is, receptor cells directly drive horizontal and and bipolar cells. Bipolar cell responses evoked with a large annulus (right side) are generated by horizontal cell activity. Bipolar cells provide the major input for amacrine cell responses and responses of the off- and on-center ganglion cells. The transient amacrine cells provide the major input for the on–off ganglion cells (center record).

cells may function with graded potentials is that such potentials are capable of discriminating a wider range of signals than can all-or-none events (i.e., action potentials).

Another unusual feature of the distal retinal neurons is that most of them respond to light with hyperpolarizing potentials; on illumination, the cell's membrane potential becomes more negative. Elsewhere in the nervous system, hyperpolarizing potentials are usually associated with inhibition because they prevent neurons from generating action potentials. In the distal retina, the neurons do not generate action potentials; thus hyperpolarizing potentials can reflect excitation, and the photoreceptors are a good example of this.

A. Photoreceptors

Both rod and cone photoreceptors only hyperpolarize in response to illumination, but rods are more sensitive to light than are cones by about 100 times in the primate and other species. Why photoreceptors hyperpolarize in response to light and why in darkness the membrane potential of the cell is partially depolarized was discussed earlier. A typical resting value for the membrane potential of photoreceptors in the dark is −30 mV, and in response to bright illumination, the membrane potential goes to −60 mV, a typical resting potential for most neurons at rest. Thus, in the distal retina, photoreceptors and other neurons behave as though darkness is the stimulus and light turns them off (i.e., it hyperpolarizes them). Why the system works this way is not known.

As noted earlier, there are electrical synapses between photoreceptors, but their role is not well understood. They may relate to the functioning of the photoreceptors rather than as a pathway for information flow. It has been proposed that electrical coupling between photoreceptors can reduce membrane noise, hence improve signal detection, and also that it may increase the amplification of signals transmitted from the photoreceptors at their synapses. Whatever the function of the electrical coupling between photoreceptors, it does increase the receptive field size of the cell somewhat. That is, the receptors respond to illumination over a wider area of the retina than that of a single receptor. Nevertheless, photoreceptors typically have the smallest receptive fields of any of the retinal neurons.

B. Horizontal Cells

Horizontal cells, like photoreceptors, have relatively low resting membrane potentials in the dark (−30 mV), and most often they only hyperpolarize in response to light (see Fig. 10). Some horizontal cells respond with small depolarizing responses to certain wavelengths of light and with hyperpolarizing responses to other wavelengths. These cells appear to be involved in color processing and are termed *chromaticity* (*C*) *cells*. Horizontal cells that only hyperpolarize to illumination are called *luminosity* (*L*) *cells*.

The receptive fields of horizontal cells are characteristically large, often several millimeters in diameter. Thus, the receptive field size usually exceeds the dendritic spread of the cells. Horizontal cells typically make extensive electrical junctions with each other, and the large receptive fields of these cells can be explained on the basis of the extensive electrical coupling between the cells.

C. Bipolar Cells

Bipolar cells, like receptors and horizontal cells, respond to light with sustained graded potentials (see Fig. 10). However, two types of bipolar cells are found in all retinas: those that depolarize in response to central spot illumination and those that hyperpolarize to such stimuli. Furthermore, the bipolar cell receptive field is organized into antagonistic zones such that illumination of the surround antagonizes the response to spot illumination. Thus, bipolar cells show a center-surround receptive field organization, and there are separate on- (center depolarizing) and off- (center hyperpolarizing) bipolar cells in all species so far examined.

D. Amacrine Cells

In most retina two basic types of amacrine cell responses—transient and sustained—are observed. Transient amacrine cells usually give on- and off-depolarizing responses to illumination presented anywhere in their receptive field (see Fig. 10), but there are also transient amacrine cells that respond only at the on or off of illumination. Transient amacrine cells typically generate action potentials, and these potentials are observed on the transient on- and off-depolarizations. Usually only one or two action potentials are observed on the transient depolarization, and thus it has been proposed that the action potentials may serve as a local amplifying mechanism for the amacrine cell potentials, rather than as the signal transmitted along the cell, as is the case for many neurons.

Sustained amacrine cell responses resemble horizontal cells responses. They may be either hyperpolarizing or depolarizing in polarity, and the amplitudes of the two types of responses are comparable. Usually sustained amacrine cells, like transient amacrine cells, give similar responses to spot illumination anywhere in their receptive field.

E. Interplexiform Cells

Only a few recordings from interplexiform cells have been reported, and they have all been from the retinas of nonmammalian species. The potentials are sustained with, in some cases, transient components. In other words, they are like amacrine cell responses.

F. Ganglion Cells

The receptive field properties of ganglion cells were discussed earlier, and it was pointed out that two basic types of ganglion cell responses are recorded in most retinas: sustained and transient responses. Intracellular recordings reveal the underlying potentials that give rise to these two response types. For example, Fig. 10 illustrates a sustained off-center, on-surround ganglion cell. With central spot illumination (left side of the figure), a sustained hyperpolarizing potential was evoked in the cell, and the cell was inhibited from discharging action potentials for the duration of the stimulus. With annular illumination (right side of the figure), a sustained depolarizing potential was produced, which elicited a steady discharge of action potentials from the cell for the duration of the stimulus.

Figure 10 also illustrates an on–off transient ganglion cell response (center response). With small annular illumination, the cell responded with transient depolarizing potentials at the onset and offset of the light. Each depolarization evoked a short burst of action potentials, but the durations of both the depolarizing potentials and action potential discharge were always shorter than the light stimulus. Such cells give similar responses to illumination anywhere in the receptive field.

G. Functional Organization of the Retina

Figure 10 shows in a simplified way how some of the potentials and certain of the receptive fields of the retinal neurons may be produced by the synaptic interactions occurring within the retina. The figure correlates the basic connections of the retinal cells with intracellular responses recorded from an amphibian (mudpuppy) with a spot of light, small annulus, or large annulus.

As noted earlier, receptors have small receptive fields. They respond well to spots of light centered over the receptor but poorly to surrounding annuli (i.e., to light that does not directly strike the cell). The anatomy of the retina indicates that bipolar and horizontal cells are both activated by the receptors and Fig. 10 shows that both cell types respond with sustained graded potentials that resemble in waveform the photoreceptor response. The figure further shows that the horizontal cells interact with bipolar cells and that this interaction is opposite in terms of sign from the receptor–bipolar interaction. In Fig. 10 the receptor causes the bipolar cell to hyperpolarize, whereas the horizontal cell causes the bipolar cell to depolarize. Horizontal cells have a much larger lateral extent than do bipolar cells; thus, a center-surround receptive field organization is observed in the bipolar cell response. The center response of the bipolar cell (left side of figure) is mediated by direct receptor–bipolar cell interaction, whereas the surround response (right side) reflects input to the bipolar cell from the horizontal cell.

As noted earlier, two types of synapses made by the cone terminals have been observed: invaginating and flat junctions. It appears that in many species the flat synapses result in the center-hyperpolarizing responses in bipolar cells, whereas the invaginating junctions result in the center-depolarizing bipolar cell responses. (Mammalian rod terminals make only invaginating contacts, thus all rod bipolars are center depolarizing.) The surround responses in all bipolar cells are provided by the horizontal cells, and these interactions may be mediated by direct horizontal-to-bipolar cell synapses or by inhibitory feedback synapses onto the photoreceptor terminals. [*See* Cell Junctions.]

The bipolar cell terminals carry the visual signal from the outer to inner plexiform layers. Depolarizing or on-bipolar cell terminals are found in the inner half of the plexiform layer, whereas hyperpolarizing bipolar cell terminals are in the outer half of the layer. Thus there is a division of the inner plexiform layer into on- and off-strata. On-responses of amacrine and ganglion cells are generated in the lower part of the layer, whereas off-responses are generated in the upper part. Thus, on-center ganglion cells have their processes in the lower or on-strata of the inner plexiform layer, off-center cells in the upper or off-layer, and on–off ganglion cells spread processes in both halves of the layer.

The responses of the two basic types of ganglion cells found in the mudpuppy (and other retinas) appear to be closely related to the responses of the input neurons to the ganglion cells (see Fig. 10). The sustained on- or off-center ganglion cells appear to receive much of their synaptic input directly from the bipolar cells; their responses resemble those of the on- or off-bipolar cells, and their receptive fields primarily reflect the processing that occurs in the outer plexiform layer. The on–off transient ganglion cells, however, resemble in their properties the transient amacrine cells, and they appear to receive most of their input from amacrine cells. These cells reflect more the processing that occurs in the inner plexiform layer. As noted earlier, the transient ganglion cells respond better to moving than to static stimuli, and some of these cells show complex receptive field properties such as direction selectivity. Substantial anatomical evidence shows that complex ganglion cell receptive field properties, such as motion and direction sensitivity, are mediated in the inner plexiform layer, predominately as a result of interactions between amacrine and ganglion cells. That is, in those species that have many motion- and direction-selective cells, there are more amacrine cell synapses per unit area of the inner plexiform layer than there are in species that have relatively few motion- and direction-selective ganglion cells.

In addition to the two basic types of ganglion cells described here, many ganglion cells have a mix of transient amacrine and bipolar cell characteristics. The Y-type ganglion cell of the cat is an example. The Y cells show a center-surround receptive organization like the bipolar cells, but their responses to both center and surround stimulation are quite transient and they are sensitive to moving stimuli. These cells appear to receive more of a mix of bipolar and amacrine cell synaptic input than the more sustained on-center or off-center ganglion cells.

It should be noted, finally, that all ganglion cells receive some amacrine cell input. The sustained ganglion cells, however, appear to receive less amacrine cell input than do the transient ganglion cells, and it seems likely that much of their amacrine cell input is from sustained amacrine cells. This discussion neglects the interplexiform cells and the role they may play in retinal function, which will be discussed in the next section.

IV. PHARMACOLOGY

The retina, like other regions of the brain, uses a large number of neuroactive substances. At the present time at least 15 substances are believed to be released from retinal neurons during retinal activity (Table I). These substances may be classified into two general catego-

TABLE I
Neuroactive Substances Found in the Retina

Amino acids
L-Aspartate
γ-Aminobutyric acid (GABA)
L-Glutamate
Glycine
Amines
Acetylcholine
Dopamine
Serotonin
Peptides
Cholecystokinin
Enkephalin
Glucagon
Neurotensin
Neuropeptide Y
Somatostatin
Substance P
Vasoactive intestinal peptide

ries: neurotransmitters and neuromodulators. Neurotransmitters act directly on retinal neurons, by altering membrane permeability to one or several ions. The ions move across the cell membrane causing a change of potential in either the depolarizing or hyperpolarizing direction. These changes in potential are rapid, and thus neurotransmitters are responsible for the fast excitatory and inhibitory pathways in the retina or brain. Neuromodulators, however, do not directly affect membrane permeability. Rather, they usually activate enzyme systems, and they modify neuronal activity biochemically. Neuromodulators do not usually initiate neural activity; rather they modify activity initiated by the neurotransmitters.

Relatively few substances appear to serve as neurotransmitters in the retina. L-Glutamate, acetylcholine, and perhaps L-aspartate mediate fast excitatory pathways in the retina, whereas GABA and glycine mediate fast inhibitory pathways. The bulk of the neuroactive substances released from retinal neurons (see Table I) appear to be neuromodulatory in nature, although little is known about the action or role of most of these substances.

A. Amino Acids

Both photoreceptors and bipolar cells appear to employ L-glutamate as their transmitter. L-Glutamate depolarizes both horizontal cells and off-center (hyperpolarizing) bipolar cells, whereas it hyperpolarizes the on-center (depolarizing) bipolar cells. Neuroactive substances are released from neurons when they are depolarized, and because photoreceptors are maintained in a depolarized state in the dark, they release L-glutamate in the dark. When illuminated, photoreceptors hyperpolarize, and transmitter release is decreased. Thus, the light responses of the horizontal and bipolar cells reflect the withdrawal of transmitter from the cell. Horizontal cells are depolarized in the dark because of the dark release of L-glutamate from the photoreceptors; their light response is a hyperpolarization, reflecting the decrease in transmitter release from the photoreceptors in light. The same explanation holds for the off-center bipolar cells. These cells are depolarized in the dark; they hyperpolarize in light as transmitter release from the photoreceptor decreases. The on-center cell, however, is hyperpolarized in the dark, and it depolarizes in the light in response to the decreased release of L-glutamate from the photoreceptor.

Although the same substance, L-glutamate, mediates the photoreceptor input to horizontal cells and both types of bipolar cells, the receptor proteins with which the glutamate interacts differ somewhat in the three types of cells. Thus, it is possible to block the responses of one or another of the three cells with pharmacological agents while leaving the responses of the other cells intact. One substance, 2-amino-4-phosphonobutyric acid (APB), blocks specifically the on-center bipolar cells, and this results in the loss of all on-activity throughout the visual system. Monkeys treated with APB are unable to distinguish increases of illumination, although they can discriminate decreases of retinal illumination.

The amino acids GABA and glycine serve as inhibitory neurotransmitter agents in both the inner and outer plexiform layers. Many horizontal cells contain GABA, and approximately 80% of the amacrine cells in many retinas contain either glycine or GABA. Both GABA and glycine powerfully inhibit ganglion cells by opening channels in the cell membrane that cause hyperpolarization of the cell and inhibition of action potential generation.

GABA and glycine act on many retinal neurons, and they mediate specific inhibitory effects. For example, by blocking the effects of GABA in a retina with pharmacological agents, direction-sensitive ganglion cells lose their directional selectivity. Such cells now respond to spots of light moving in any direction across the retina. It is believed, therefore, that direction sensitivity is mediated by GABAergic amacrine cells in the inner plexiform layer.

B. Amines and Peptides

Most amines and neuropeptides appear to function as neuromodulatory agents in the retina, although little is known about the function of most of these agents, particularly the neuropeptides. The exception is acetylcholine, which functions in the retina as an excitatory neurotransmitter in the inner plexiform layer. Acetylcholine is found in amacrine cells, and some of these cells spread processes in the on-region of the inner plexiform layer and mediate transient excitatory responses in ganglion cells at the onset of illumination. Others extend processes in the off-region of the inner plexiform layer and mediate transient depolarizing responses in ganglion cells at the offset of illumination. The cell bodies of the on-acetylcholine-containing amacrine cells are found among the ganglion cells, whereas the cell bodies of the off-acetylcholine-containing amacrines are in the inner nuclear layer. The acetylcholine-releasing amacrine cells are believed to provide excitatory input to the transient on–off ganglion cells, especially those that are direction sensitive.

The other amines, dopamines and serotonin, and the neuropeptides have also been principally localized to amacrine cells. These cells are, for the most part relatively scarce cells, accounting for no more than a few percent of the total number of amacrine cells. These cells usually spread their processes widely in the inner plexiform layer, and so they are capable of exerting wide effects. It is also the case that coexistence of neuroactive substances occurs in many amacrine cells. It has been reported that two peptides can coexist in the same amacrine cell, that a peptide and a monoamine coexist, or finally that a peptide or monoamine and an inhibitory amino acid are in the same cell. Some evidence has been provided that three or even more agents may be colocalized in the same neuron. As yet, the significance of the colocalization of two or more neuroactive agents in a single neuron is not understood.

C. Interplexiform Cells and Dopamine

Much of what we know about the action of neuromodulators in the retina has come from the study of dopamine in the teleost retina. In the teleost, dopamine is present in interplexiform cells and so these studies have also shed light on the role of these cells in retinal function. A brief summary of the findings are presented here as a model for the action of neuromodulators and interplexiform cells in the retina.

In teleosts, the synaptic output of the interplexiform cells is mainly on the cone-related horizontal cells. Two effects of dopamine on these cells have been observed; a loss of light responsiveness (i.e., light responses are reduced in amplitude after dopamine application to the retina), and a decrease in electrical coupling between horizontal cells. Dopamine does not exert these effects by acting directly on horizontal cell membrane channels; rather it interacts with a membrane receptor protein linked to the enzyme adenylate cyclase. This catalyzes the formation of the second-messenger molecule cyclic AMP (Fig. 11). Cyclic AMP, in turn, activates an other set of enzymes called *kinases* that add phosphate groups to specific proteins. This process, called *phosphorylation,* serves to activate or inactivate cellular processes.

The kinases activated by cyclic AMP in horizontal cells appear to phosphorylate both the glutamate channels (i.e., the channels activated by the photoreceptor transmitter) and the gap junctional channels. The phosphorylation of these channels serves to alter their properties. In the case of the gap junctional channels, phosphorylation decreases the time that the channels remain open, thereby decreasing the flow of current that passes across the junction. However, phosphorylation of the glutamate channels appears to modify the frequency of opening of the channels, which again modulates ion flow across the membrane.

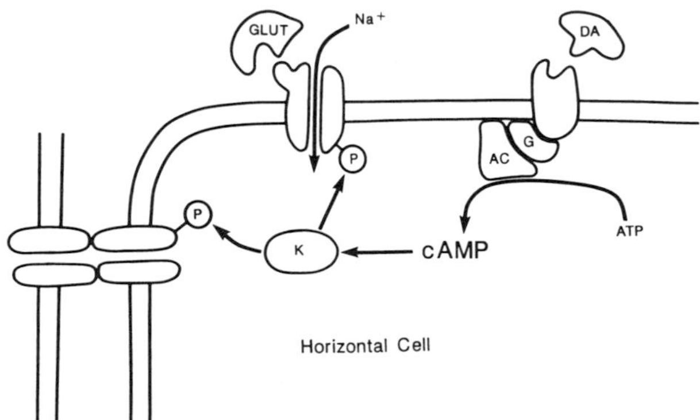

FIGURE 11 Summary scheme showing how dopamine (DA), acting via cyclic AMP, may influence responsiveness of horizontal cells to L-glutamate (the photoreceptor transmitter) and electrical coupling between horizontal cells. DA interacts with receptors that are linked to the enzyme adenylate cyclase (AC) via a G protein. Activation of adenylate cyclase results in conversion of ATP to cyclic AMP. Cyclic AMP interacts with kinase (K) that phosphorylate (P) glutamate (Glut) channels or the gap junction channels.

The action of dopamine on horizontal cells is not to initiate activity, but to modify the cell's response to the photoreceptor transmitter and the interactions between the cells. Furthermore, these effects are slow; they take many seconds to develop and they last for minutes.

Overall the effect of dopamine is to decrease the effectiveness of the horizontal cells in mediating lateral inhibitory effects in the outer plexiform layer. Decreasing light responsiveness of the cell and shrinking its receptive field size are effective ways of lessening its influence. As noted earlier, horizontal cells form the antagonistic surround response of bipolar cells, and thus a decrease in bipolar cell surround responses is observed after dopamine application to the retina.

What is the significance of the modulation of lateral inhibition and surround antagonism by dopamine and the interplexiform cells in the retina? It has long been known that after prolonged periods of time in the dark the antagonistic surround responses of ganglion cells are reduced in strength or even eliminated. An obvious speculation is that interplexiform cells and dopamine play such a role and regulate the strength of lateral inhibition and center-surround antagonism in the retina as a function of adaptive state. In fish, evidence in favor of this has been provided. After periods of prolonged darkness, horizontal cell receptive field size and light responsiveness are substantially decreased.

BIBLIOGRAPHY

Baylor, D. A. (1987). Photoreceptor signals and vision. *Invest. Ophthalmol. Vis. Sci.* **28**, 34–49.

Boycott, B. B., and Dowling, J. E. (1969). Organization of the primate retina: Light microscopy. *Philos. Trans. Roy. Soc. London, Ser. B* **255**, 109–184.

Dowling, J. E. (1987). "The Retina: An Approachable Part of the Brain." Harvard Univ. Press, Cambridge, Massachusetts.

Ehinger, B., and Dowling, J. E. (1987). Retinal neurocircuitry and transmission. *In* "Handbook of Chemical Neuroanatomy, Vol. 5. Integrated System of the CNS, Part I" (A. Bjorklund, T. Hokfelt, and L. W. Swanson, eds.), pp. 389–446. Elsevier, Amsterdam.

Gouras, P. (ed.) (1991). "The Perception of Colour: Vision and Visual Dysfunction." Macmillan, London.

Kolb, H. (1994). The architecture of functional neural circuits in the vertebrate retina. *Invest. Ophthalmol. Vis. Sci.* **35**, 2385–2404.

Polyak, S. L. (1941). "The Retina." Chicago Univ. Press, Chicago.

Stryer, L. (1986). Cyclic GMP cascade of vision. *Annu. Rev. Neurosci.* **9**, 87–119.

Wässle, H., and Boycott, B. B. (1991). Functional architecture of the mammalian retina. *Physiol. Rev.* **71**, 447–480.

Werblin, F. S., and Dowling, J. E. (1969). Organization of the retina of the mudpuppy, *Necturus maculosus*. II. Intracellular recording. *J. Neurophysiol.* **32**, 339–355.

Retroviral Vaccines

DANI P. BOLOGNESI
Duke University Medical Center

GLOSSARY

Adjuvants Substances that increase the potency of responses to immunogens by making them more easily recognizable to cells of the immune system

Antibodies Molecules produced by immune B lymphocytes that circulate in the blood and bind to foreign antigens with exquisite specificity

Antibody-dependent cell cytotoxicity (ADCC) Instance where antibodies associate with Fc receptors on cells that themselves have cytolytic properties. The antibodies provide the specificity of the reaction by binding to specific cell-surface antigens and thereby direct the effector cells (usually macrophages or natural killer cells) to the target

Antigen Any substance that can evoke a humoral and/or cellular immune response

Cell-mediated immunity Cellular arm of the immune response, represented by a number of elements, among which are helper T cells, cytotoxic T cells, macrophages, and natural killer cells

Cytotoxic T cells (CTL) T lymphocytes that become activated by helper T lymphocytes and develop the capacity to destroy cells that exhibit foreign antigens at their surface

Enhancement Process by which immune responses, rather than blocking infection or disease, actually exacerbate it

Epitope Precise molecular configuration that defines a particular antigen

Fc receptor Structure found on the surface of cells to which antibodies are able to attach by their Fc portion

Fusogenic event When membranes of different organelles coalesce and allow the contents of the respective structures to become unified. Many viruses use such a process to introduce their contents into cells in order to establish infection

Glycoprotein Protein that, in addition to a chain of amino acids, also consists of side chains of sugar (glyco) molecules

Helper T cells Cells of the immune system that, in response to appropriately presented antigens, send out chemical signals called lymphokines, which help activate other lymphocytes and amplify the response to the antigen

Humoral immunity Humoral arm of the immune response, mediated by antibody molecules produced by B cells

Immunogen Substance that is able to provoke an immune response

Immunological memory Constitutes the generation of long-lived "memory" cells (T cells, B cells) of the immune system that are capable of recognizing an immunogen and stimulating a generalized response to it long after vaccination

Inflammatory reaction Refers to the use of adjuvants that provoke a strong response at the site of immunization, which in turn attracts many other components of the immune system to focus attention on the immunogen

Integration Insertion of a foreign DNA (e.g., a viral genome) into the DNA of the host cell

Interference Process by which cells infected by a given virus become resistant to second infection. This is usually but not solely related to competition for viral receptors on the surface of the susceptible cell

Neutralization Process by which antibodies directed against viral components, usually on the virion surface, are able to block infection of susceptible cells

Recombinant subunits Subcomponents of the virion expressed in bacterial or mammalian cells using recombinant DNA technology

Recombinant vectors Same as recombinant subunits except that the respective genes are incorporated into live attenuated viruses or bacteria for delivery and expression

ENCYCLOPEDIA OF HUMAN BIOLOGY, Second Edition, VOLUME 7. Copyright © 1997 by Academic Press. All rights of reproduction in any form reserved.

Retrovirus Belonging to a family of viruses whose genetic material is ribonucleic acid (RNA) but that replicate through a DNA intermediate. The process is catalyzed by a virus-encoded enzyme termed reverse transcriptase

Secretory antibodies Unique antibody molecules that are produced by cells of the secretory immune system, which is associated with mucous membranes and is the first line of defense against invading organisms

Syncytium End result when a number of cells fuse with one another

Virulence Feature of infectious organisms that results in disease and is usually linked to rapid replication and dissemination of the virus in the host

VACCINATION HAS HISTORICALLY PROVEN TO BE the most simple, cost-effective, safest, and efficacious means to prevent and thereby control the spread of disease in both animals and humans. Among the most successful vaccines have been those prepared against viruses. The resounding triumphs of the smallpox and polio vaccines are of landmark proportions, but major advances have also been achieved in the control of diseases such as yellow fever, measles, mumps, and rubella through vaccination. More recently, impressive strides have been made in development of vaccines against more complex viruses such as hepatitis. On the horizon, one can envision eventual successes against certain members of the human herpes and papilloma virus families. This article will focus on vaccines against retroviruses, beginning with the experiences with animal retroviruses and ending with what is perhaps the most formidable and pressing challenge of all: a vaccine against the human immunodeficiency virus (HIV).

I. GENERAL PRINCIPLES OF VACCINE DEVELOPMENT

The roots of vaccination lie in folk practices, dating back to before the seventeenth century in the Far and Middle East, which indicated that limited "inoculation" of the disease-causing organism itself by an "unnatural route" or exposure to it at a propitious age could lessen and even prevent disease. When Edward Jenner realized that exposure to the cowpox virus, which was similar to smallpox immunologically but was unable to cause disease in humans, resulted in a comparable outcome, the platform for the science of vaccination and the principles of immunization were established. Indeed one can now name successful vaccines that represent derivations of a pathogenic organism, ranging from the organism itself, treated physically or chemically such that its ability to cause disease is severely limited, to actual subunits of the agent that are completely noninfectious and can even be produced by genetic engineering, as is the case for hepatitis B (Table I).

In order to be effective, these derivations must be capable of inducing protective immunity. It follows that those components of the disease organism that represent its critical targets for immune attack must be presented to the vaccinee in recognizable form.

TABLE I
Current and Future Virus Vaccines[a]

Contemporary		Developmental and future	
Kind	Example	Kind	Virus targets
Live	Poliovirus	Subunits, peptides, recombinant, infectious vector, other (?)	Herpes (varicella zoster, simplex)
	Measles		Cytomegalovirus
	Mumps		Rotavirus
	Rubella		Dengue
	Yellow fever		Human papilloma viruses
	Vaccinia		Human T-cell leukemia viruses
	Adenovirus		HTLV-, HTLV-II
Killed (whole virus)	Poliovirus		HIV
	Influenza		
	Rabies		
Subunit (natural and recombinant)	Hepatitis B		

[a]Courtesy of Maurice Hilleman, Merck, Sharpe and Dohme.

Otherwise, the disease organism will be invisible to the immune system. However, not all components of a pathogen represent beneficial immunological targets. In actuality, some parts of the organism may be undesirable elements because they elicit inappropriate immune responses that can sometimes mask those that are protective. Modern vaccine approaches thus take great pains to define what is effective and eliminate what may be deleterious.

Though antibodies can directly recognize regions on native molecules, the cellular arm of the immune system requires prior processing of a given immunogen by what are known as antigen-presenting cells. These cells must internalize the antigen and modify it in a way such that it can associate with major histocompatibility (MHC) molecules and be recognized by antigen receptors on helper T lymphocytes. Antigen processing is also important for recognition at the surface of target cells by killer T lymphocytes, again in association with MHC.

Various means are used to enhance the recognition of an antigen by the immune system. In general, the more particulate the antigen, the more effective it is as an immunogen. By contrast, smaller subunits of a pathogen in the form of proteins or even peptides require formulation with "adjuvants" in order to achieve sufficient recognition. Complexing with adjuvants is largely empirical, but formation of complex aggregates of antigen and adjuvants seems to be a guiding principle. The role of the adjuvant is that of a powerful immune stimulant in its own right; it establishes an inflammatory reaction at the site of inoculation, thereby attracting the attention of the immune system to the antigen. Today, various approaches are being devised to circumvent the "inflammatory response approach," which has been used most successfully in animals and for which a good counterpart does not exist in humans because of serious adverse effects that have been experienced. This is being done by developing vehicles that will carry the antigen directly to antigen-presenting cells, including nonpathogenic infectious vectors (usually innocuous bacteria or viruses) that would transmit the actual genes coding for a given antigen and allow their synthesis from within the antigen-presenting cells. As these methods are perfected, they will have great bearing on development of human immunodeficiency virus (HIV).

Experience with both existing and certain experimental vaccines dictates that complete blockade against infection of the host by the pathogen is *not* a prerequisite for successful vaccination. Indeed, most vaccines protect against disease rather than infection per se. One can interpret this to signify that while viruses may replicate in a multitude of different cells in the body, the targets for disease are usually rather specific. For instance, in the case of polio, replication of the virus in epithelial cells of the gut is tolerable, whereas penetration of the virus within the central nervous system (CNS) leads to disease. One can then envision at least two interdependent barriers against poliovirus infection. One line of defense would be mounted by the secretory immune system meeting the virus as it crosses mucous membranes. Although secretory antibodies do not prevent infection by poliovirus, they play the important role of limiting it to a level manageable by the systemic immune response. The latter is represented by both antibodies and immune lymphocytes found in the circulation and it is their role to prevent the virus from establishing a significant infection in the CNS. This is achieved by neutralization of virus infectivity and destruction of virus-infected cells, thereby clearing the infection before perceptible damage has occurred. A persistent infection in a sanctuary like the CNS, which is shielded from the immune system, generally results in serious consequences.

Finally, a successful vaccinated individual possesses an immune system that is primed to respond rapidly upon encounter with the organism. A primed immune system consists of lymphocytes that retain the property of immunological memory for the lifetime of the individual. When such lymphocytes see the immunological target, they initiate a cascade of events that result in the generation of a full-blown effector phase of the immune response characterized by killer lymphocytes, neutralizing antibodies, and other protective elements that can effectively attack and clear the invading pathogen. [*See* Lymphocytes.]

Turning to the issue of retroviruses, a question often asked is whether it is even possible to vaccinate against these agents. These viruses exhibit features that pose a number of new challenges for vaccinologists. First and foremost, retroviruses can integrate their genome into the cells of the host and thereby establish a permanent infection unless all of the infected cells are removed. A hallmark of these viruses is that they induce various forms of cancers in animals and a cell harboring an integrated retroviral genome can, in certain instances, become a malignant cell even in the absence of virus production. This is because certain regulatory elements in retroviruses can exert their effects not only on the viral genes but also on cellular genes. Thereby this added genetic information can disrupt normal cell growth and differentiation patterns, which can lead to malignant transformation. This issue is of extreme importance for vaccines based on use of live attenuated or inactivated virus, particularly if they are destined for humans. Human retroviruses not only possess the traditional gene regulatory elements but are replete with "accessory genes" that are able to impact even more extensively on the expression of normal cellular genes. Therefore, one must consider the use of vaccines containing retrovirus genetic material as potentially dangerous, unless this is done in individuals already infected as postexposure prophylaxis.

A successful vaccine against any retrovirus would thus represent an important milestone toward the development of a counterpart against disease causing human retroviruses. What has the experience been thus far? In short, there has been little progress, but a growing number of animal retrovirus models are receiving attention in this regard and such information is contributing to the evolving strategies toward vaccines for human use.

II. VACCINES AGAINST LEUKEMIA RETROVIRUSES OF MAMMALS

Until recently there has been relatively little motivation to develop vaccines against animal retroviruses principally because the diseases they caused in natural circumstances have not been of sufficient impact to mount large-scale efforts. In one particular case, feline leukemia, vaccines have been approved for veterinary use but these have been largely ineffective. In the case of bovine leukemia, a disease of growing economic importance, an effective vaccine would prove of considerable value but other measures, such as eradicating infected and diseased animals, are proving successful for disease control.

Nevertheless, the efforts to develop experimental vaccines against such viruses are mounting rapidly, particularly because of their value as models for developing vaccines against human retroviruses.

A. Murine Retroviruses

The experimental studies with animal retrovirus vaccines date back to the early 1970s, when the murine models received the greatest attention because of the extensive knowledge about the virion structure, its antigenic composition, and the practicality of a small and plentiful laboratory animal model.

The viral component responsible for the salient immunobiological features of a retrovirus is its major external glycoprotein (gp). First the glycoprotein is required for infection and mediates the attachment of the virus to the host cell. It is also the viral component that is directly involved in the phenomenon of interference, which reflects competition for viral receptors at the cell surface. Finally, the glycoprotein specifies the pattern of neutralization by antiviral antibodies. Consistent with these properties is its strategic location on the outer envelope of the virion (Fig. 1).

Another component found on the surface of the virion is a hydrophobic transmembrane protein (tmp), which noncovalently anchors the glycoprotein to the virion. The transmembrane protein can either contain or be devoid of carbohydrates, but in every case the degree of glycosylation is considerably less than in the exterior glycoprotein. The transmembrane protein can also be a target for neutralization by antibodies.

Animals infected with retroviruses usually respond with easily detectable antibodies against the exterior glycoprotein. For the most part this immunity is specific for the infecting agent. Others represent determinants which are common to all viruses of a given species or may even extend to those found in widely different species. It is only in rare cases that animals respond to these latter domains under natural conditions. On the other hand, one can immunize animals with virus or purified glycoprotein and obtain antibody responses that are much broader in their reactivity than natural antibodies. Studies have also been done toward generating cytotoxic T cells (CTL) through use of recombinant vectors such as vaccinia virus, into which genes were incorporated coding for retroviral structural antigens. As with other viral systems that have been studied, CTL could be generated against both external (envelope) and internal (core) components and these mirrored CTL arising during natural infection.

As noted earlier, immunization with glycoprotein elicits strong neutralizing antibodies. Mice immunized with purified glycoprotein can, indeed, resist substantial challenges of infectious leukemogenic virus. Though monomeric glycoprotein is capable of inducing protective immunity, relatively large quantities of purified antigen were required to accomplish this reproducibly. On the other hand, the use of glycoprotein linked to the transmembrane protein so as to form multimeric aggregates resulted in a much more effective immunogen (see Fig. 1).

Although such experiments demonstrate that antibodies may be sufficient for protection, it is important

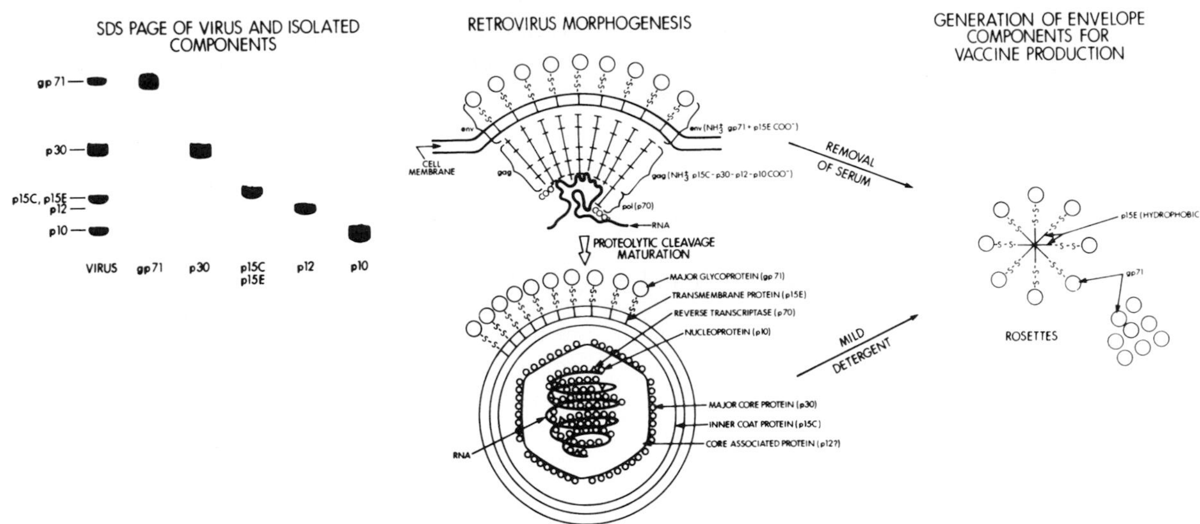

FIGURE 1 Morphogenesis, structure, and composition of a typical murine leukemia retrovirus. The surface components gp71 (gp) and p15E transmembrane protein (tmp) derived from the *env* gene of the virus can be recovered in homogeneous form as either multimers (rosettes) or monomers as a result of shedding from the cell surface or treatment with mild nonionic detergents.

to note that many of these results were obtained under experimental conditions that would not be applicable for widespread vaccination. Moreover, experimental infection may be quite different than what occurs during natural transmission. To this end, it is important to determine what other immune mechanisms might constitute a protective response. In this regard, there are concrete examples that cellular immunity constituting both helper and cytotoxic lymphocytes can exert powerful protective immunity against infection and disease.

B. Feline Retroviruses

Many of the principles emanating from studies with murine retroviruses were rapidly applied to the feline system with the intention of developing a practical vaccine for use against naturally occurring infections that were severely debilitating to household as well as free cat populations. The first licensed vaccine employed a mixture of viral antigens collected from infected cell cultures. Although effective in experimental tests, it failed to generate significant protection when applied in veterinary clinics.

The second generation of commercial feline leukemia virus (FeLV) vaccines was based on inactivated whole-virus preparations in various proprietary adjuvants. Although the respective manufacturers reported better than 80–90% efficacy for these vaccines, independent evaluations of these commercial inactivated whole-virus vaccines failed to demonstrate any protective efficacy in carefully controlled vaccine trials according to USDA guidelines. An experimental FeLV vaccine based on purified gp70 resulted in an apparent enhancement of FeLV infection upon exposure to virus, and a vaccine based on recombinant vaccinia virus expressing FeLV gp70 failed to elicit detectable gp70-specific antibodies, presumably due to poor replication of vaccinia virus in cat cells. Taken together, these results accumulated over nearly a decade of research indicated that the practical development of an effective FeLV vaccine was not the easy task predicted initially.

During recent years, however, several new FeLV vaccines have been described and reported to provide at least 90% protective efficacy in experimental vaccine trials. These vaccines include at least two inactivated whole-virus vaccines, one of which is produced from a molecular infectious clone of FeLV. Also included are subunit vaccines containing recombinant FeLV gp70 produced in *Escherichia coli* and mixed with QS-21 adjuvant or composed of FeLV-immunostimulating complexes (ISCOMS), a vaccine based on canarypox virus recombinant expressing the FeLV Env and Gag proteins, and vaccines using feline herpesvirus and baculovirus as vectors for the *internal* and *envelope* genes of FeLV. The protective efficacy of the commercial whole-virus and subunit vaccines has been confirmed in independent vaccine trials. These results demonstrate the protective efficacy of the various FeLV vaccines against experimental challenge given fairly soon after the last immunization. However, none of the studies appears to address the important question of the duration of protective immune responses in vaccinated cats.

Finally, comparison of the various vaccines provides some insights into potential immune correlates of protection. In some FeLV vaccine studies, a correlation has also been reported between protection and the levels of virus-neutralizing antibodies. However, recent vaccine studies examining whole-virus, subunit, and live recombinant vaccines not only fail to demonstrate such a correlation, but indicate that protection from experimental FeLV challenge can be achieved in the absence of detectable FeLV neutralizing antibody responses. It should be emphasized here that these results do not mean that neutralizing antibodies cannot contribute to protection, but that other immune responses (humoral and cellular) can provide efficient protection even in the absence of detectable virus neutralizing antibody responses. In this regard, it will be important to characterize in more detail the cellular immune responses to FeLV envelope and internal proteins in cats that control virus replication after experimental infection and in cats immunized with effective FeLV vaccines.

C. Bovine Leukemia

The studies that have been conducted with the antigenic structure of the bovine leukemia virus (BLV) focus attention on its envelope components as the principal targets for vaccine design, much like the experience with murine and feline viruses. However, in contrast to the studies with subunit FeLV vaccines, which were deliberately presented in denatured form, the experience with BLV points to critical epitopes for neutralizing antibodies that depend on the native configuration of the envelope. This concept also derives from studies demonstrating that natural antibodies to the envelope were protective against BLV transmission from mother to the offspring. Careful mapping of the various epitopes responsible for elicitation of neutralizing antibodies has been carried out. The outcome indicates that the regions responsible for induction of the strongest protective immunity may be involved in the process of viral entry through a fusogenic event with the cell membrane. BLV infection also results in cell/cell syncytium formation, which is one of the hallmarks of HIV infection. Finally, BLV is an important model for the human T-cell leukemia virus (HTLV-I) because of the striking resemblance in both genomic organization and the positioning of neutralization of T-cell epitopes.

III. VACCINES AGAINST ANIMAL LENTIVIRUSES

As with the animal retroviruses associated with malignancies, vaccine efforts against various prominent members of the lentivirus family have been lagging, again despite their involvement in naturally transmitted chronic diseases that affect domestic animals such as sheep (Visna/Maedi), goats (caprine arthritis, encephalitis virus), and horses (equine infectious anemia virus). More recently they have also been implicated in immunodeficiency syndromes of subhuman primates. Indeed, a major impetus for further studying these models is that HIV is a member of this group of retroviruses and shares, to different degrees, certain of the unique properties of these agents. Among them are genomic organization, pathogenic processes, and, most poignantly, their high variability and propensity to escape from immune defenses.

A. Equine Infectious Anemia Virus

The natural mechanism of transmission of equine infectious anemia virus (EIAV) is via insect vectors. The disease is typified by progressive cycles of febrile illness with severe consequences to the affected animals, including death. The initial infection is met by a vigorous neutralizing antibody response, but variants emerge that are refractory to its effect. As the variant replicates, a second round of antibody is produced that is specific for the new virus; but again new variants emerge. By repeating this cycle of neutralization, mutation, and escape, the virus is able to complete its pathogenic mission. Antigenic variation and escape from neutralizing antibody also occur with Visna virus, but only very low and hardly measurable levels of neutralizing antibody are ever found with caprine arthritis encephalitic virus.

Several experimental EIAV vaccines have been evaluated, including attenuated live virus, inactivated whole virus, and subunit vaccines composed of purified viral envelope glycoproteins or a baculovirus expressed recombinant SU protein. The results of these vaccine trials demonstrate a remarkable range of efficacy from complete protection to severe enhancement of virus replication and disease.

A donkey leukocyte-attenuated EIAV vaccine has

been widely used in the field in China and Cuba to control EIAV. The use of this attenuated vaccine is reported to provide better than 90% protection, but this claim remains to be confirmed by independent vaccine studies. Concerns about the possible reversion to virulence with this highly mutable virus and the lack of a marker to differentiate vaccinated from naturally infected horses have prevented wider application of attenuated EIAV vaccines.

These potential problems have led to efforts to develop a practical EIAV vaccine focusing on inactivated whole-virus and subunit vaccine strategies that have been evaluated in the EIAV/Shetland pony system. The results of initial experimental vaccine trials demonstrated that an inactivated whole-virus vaccine was able to elicit nearly 100% protection against a rigorous homologous virus challenge. The whole-virus vaccine, however, failed to protect against infection by "heterologous" challenge that was derived as an antigenic and pathogenic variant, although it did protect the ponies from the development of clinical symptoms, evidently by greatly suppressing the levels of virus replication.

A second EIAV vaccine composed of lectin affinity purified viral envelope glycoproteins was also shown to provide 100% protection against the standard homologous virus challenge, but failed to protect against infection by the heterologous virus challenge. In contrast to the whole-virus vaccine, however, immunization with the viral gp subunit vaccine not only failed to prevent the development of disease but appeared to enhance the clinical symptoms in about one-half of the immunized ponies.

The final EIAV experimental vaccine evaluated to date consisted of a baculovirus-expressed glycoprotein (rgp90). Vaccination of ponies with the EIAV rgp90 failed to elicit protection from infection by either the homologous or heterologous EIAV challenges. In addition, it was demonstrated that the rgp90 vaccine resulted in a marked enhancement of virus replication and exacerbation of disease in immunized ponies exposed to the heterologous virus challenge.

These observations indicate that immune responses to EIAV are a double-edged sword that can either provide protection against virus infection and strictly control virus replication or markedly enhance both virus replication and disease. The range of efficacy observed with the panel of EIAV vaccines provides a novel system for examining correlates of protective and enhancing immune responses to the same virus. Initial studies of potential immune correlates of protection and enhancement have involved mainly the characterization of humoral immune responses and only limited analyses of cellular immune responses. On the other hand, all attempts to vaccinate against Visna virus or caprine arthritis, encephalitis virus (CAEV) have led to enhancement irrespective of the type of vaccine used.

B. Simian Immune Deficiency Virus

Simian immune deficiency virus (SIV) represents the closest relative thus far known to the human immunodeficiency virus. Their ability to produce an acquired immunodeficiency syndrome (AIDS)-like illness in small subhuman primates (such as macaques, sooty monkeys, and mandrills) within months after inoculation provides a valuable animal model in which to exercise vaccine strategies.

Initial vaccine studies in this model focused on the use of whole inactivated virus preparations that demonstrated varying degrees of protection against infection and disease. However, the inability to establish a clear correlate of protection based on antiviral responses, particularly neutralizing antibodies, led investigators to search for other explanations. The surprising results that emerged indicated that protection was mediated by cellular antigens derived from the host cell in which the virus was grown.

In parallel with these studies, numerous attempts to obtain protection with viral subunits, recombinant vectors bearing viral genes, virus-like particles, and various combinations thereof met with nearly uniform failure. Furthermore, no protection was observed despite the ability of some of these vaccines to induce high levels of neutralizing antibodies and CTL. There was, however, some success in SIV models where the virus was less pathogenic and this correlated with neutralizing antibody responses.

Much more success was achieved with molecularly derived live attenuated vaccines. Specific viral mutants were generated lacking specific genes but nonetheless able to establish a low-grade infection in macaques that did not progress to disease. When animals were infected with such genetically modified viruses, they were able to resist large doses of highly pathogenic SIV. This study design excludes the role of cellular antigens and establishes a model from which correlates of protection based on viral antigens might possibly be derived.

In summary, the vaccine studies carried out to date in the SIV model have uncovered two vaccine approaches with acceptable success: whole inactivated viruses bearing xenogeneic transplantation antigens or live attenuated viruses. Neither represents an easily applicable approach toward vaccination of HIV in humans and to date there is no identifiable correlate of protection for successful vaccination. Yet, because of practicality and various possibilities for disease development, this primate model continues to provide the best experimental system for evaluating vaccine candidates against an immunodeficiency virus that are relevant toward development of an HIV vaccine for humans.

IV. VACCINES AGAINST HUMAN RETROVIRUSES

Although animal studies are of extreme importance for providing concepts and strategies for vaccine development, none fully represents the host–virus relationships that exist between human retroviruses and humans. In fact, it is the discovery of human retroviruses that has accelerated the studies of their animal counterparts.

A. Human T-Cell Leukemia Viruses

Although the HTLV viruses are important human pathogens, it seemed, for some time, that they could be controlled by public health and education measures. Given the low attack rate of these viruses and the low incidence of disease following infection (1 in 100), the risk/benefit ratio of a vaccine would be relatively low. However, it is now clear that the distribution of these viruses is much wider and extends to individuals (i.e., drug users) where education methods and public health measures are ineffective. Moreover, these viruses are responsible for other diseases as well, such as tropical spastic paraparesis (TSP) and possibly other central nervous system disorders. Thus, it can be anticipated that a large number of HTLV carriers exists in the population who can further transmit the virus through various means, and that vaccination programs are therefore eminently justifiable to eradicate the infection.

Because of the morbidity and mortality risks associated with virus infection, approaches that would use whole inactivated virus have not been pursued. Instead, most of the attention has been focused on the envelope glycoproteins of the virus and the role of neutralizing antibodies and cytotoxic lymphocytes in the protective response. Various sites that represent important biological and immunological epitopes have been mapped. A vaccine consisting of an envelope subunit of HTLV produced in *E. coli* generated protective immunity in cynomolgus monkeys against primary infection by HTLV-I. Of interest is that this study employed HTLV-I-infected cells as the challenge vehicle, which is a step closer to natural transmission than virus itself. Protection correlated with the presence of neutralizing antibodies, indicating that humoral immunity can be an effective barrier against infection.

More intensive studies are currently being carried out in a rabbit model of HTLV-I infection. This approach will enable the optimization of candidate HTLV vaccines in terms of immunogenicity and efficacy. Experimental vaccines able to induce cellular responses (CTL) are also under study.

B. HIV Vaccines

The barriers standing before the development of a vaccine against HIV are formidable and for some time have clouded the thinking and dampened the enthusiasm as to how such a task might be approached. However, as a result of progress along several fronts, there is cautious but growing optimism, which stems mainly from the demonstrations of efficacy of vaccines for both SIV and HIV in animal models. Complementing such efforts are significant advances in understanding the variability of the virus and in how to design immunogens that can induce T- and B-cell memory to critical target epitopes of the virus. Finally, not only animals but humans are able to respond favorably to certain candidate immunogens.

1. Special Considerations for a Vaccine against HIV

A few key properties of HIV can be singled out in considering the development of a vaccine (Table II). First and foremost, like other members of the retrovirus family, HIV is able to integrate its genetic information in the genome of its target cells. The survival of the virus is thus directly linked to the survival of the cell. Although HIV-1 can kill certain cells it infects, most notably T cells, other cells (e.g., monocyte/macrophages) can tolerate virus replication for much longer periods without succumbing. Thus, permanent reservoirs of infection can be established to serve as factories for production of virus, which can then carry out its destructive effects on the host. [*See* Acquired Immunodeficiency Syndrome, Virology.]

The virus can also establish latent infections, that

TABLE II
Major Obstacles toward a Vaccine against HIV

1. Natural transmission includes free virus and infected cells
2. Virus can be transferred covertly and efficiently from cell to cell
3. Virus can establish latent infections in T cells and macrophages
4. Virus replicates as a swarm, generating numerous antigenic variants
5. Virus both impairs and destroys the immune system
6. Virus resides in the central nervous system
7. No HIV animal model exists for both infection and disease
8. No epidemiological evidence that antiviral immune response during natural infection is protective

is, the integration of its genome in that of the target cell without synthesis of virus or viral gene products. Latency is a well-known property of viruses of the herpes family, against which vaccines are still being sought. Latently infected cells are invisible to the immune system, unless signals are applied that are able to activate the expression of viral genes. For HIV these signals can be other viruses but also host factors, including elements that regulate immune function. From this alone, one can glean how intimately virus and target cell functions are intertwined, providing considerable selective advantage to the survival of the virus.

Another feature of HIV-1 that bears some attention is its mode of transmission, which can occur through either free virus or virus-infected cells. Virus can also be transmitted covertly from cell to cell through the process of fusion and be invisible to immune defenses. HIV-infected monocyte/macrophages contain large concentrations of virus in intracellular vesicles. The virus is released within such structures rather than at the cell surface. Macrophages could disseminate the virus to other cells by a "trojan horse" fashion, as it occurs with other lentiviruses. Moreover, if macrophages carrying sacks filled with virus are ruptured, large quantities of HIV-1 would be liberated and could further propagate. Thus, immune attack on such cells may bear negative consequences.

HIV-1 is also notorious for its ability to escape immune attack through mutation within its envelope gene. In fact, large numbers of variants of this virus exist in the population. As is the case with EIAV, variants may be selected that are resistant to immune attack. This property is reminiscent of those of influenza viruses, but is much more extreme.

The very fact that macrophages can be infected by HIV-1 raises the question of the role of a special class of antibodies, the enhancing antibodies. Such antibodies would bind the virus and bring it to the macrophage surface by attachment to Fc receptors present on such cells, thereby enhancing the infectivity of the virus.

Thus, HIV is a formidable adversary. It attacks the immune system upon which a vaccine depends. It is able to hide from immune defenses by establishing latent infections and by developing covert mechanisms of transmission. It can infect sites forbidden to the immune system such as the central nervous system. The extensive variation of the virus allows it to escape immune defenses. These features have led to the speculation that a successful vaccine must be able to totally prevent infection. If this is indeed the case, it would place demands far above what has ever been required of vaccine for other pathogens, where complete blockade of infection is not a requirement for protection against disease.

2. Progress in HIV Vaccine Research

The foregoing issues and uncertainties notwithstanding, vaccine developers initially focused on vaccine approaches based on the virus envelope. Several independent studies have demonstrated that recombinant envelope products were effective in preventing HIV infection in chimpanzees and that antibodies were the best correlate of protection. The hypothesis that threshold levels of neutralizing antibodies might protect people against HIV infection became plausible, and a number of clinical trials were initiated to evaluate the safety and immunogenicity of envelope-based candidate vaccines. The best performance was achieved by two recombinant gp120 vaccines prepared in mammalian cells. In terms of magnitude, breadth, and duration of the neutralizing antibody responses to several laboratory strains, the results in humans surpassed even those achieved in the chimpanzee model. In both low- and high-risk volunteers, these vaccines also proved to be very well tolerated.

It was at this point that the question arose of proceeding to larger trials in high-risk volunteers to evaluate efficacy. Before this could be done, it was necessary to demonstrate how well they matched the target viruses in the population virologically, immunologically, and genetically. The initial approach used to determine this was to evaluate their ability to induce antibodies capable of neutralizing fresh patient isolates by use of peripheral blood mononuclear cells (PBMC) as targets (Fig. 2). Although quite effective

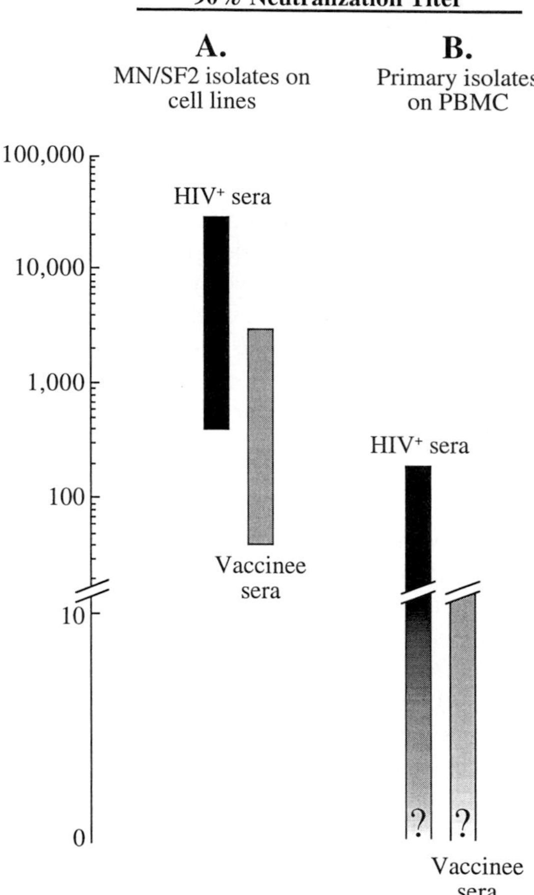

FIGURE 2 Neutralization of laboratory strains versus primary isolates. The bars indicate the range of neutralization titers of sera from HIV-1-infected individuals and uninfected volunteers who have been vaccinated with MN and SF-2 envelope glycoproteins on (A) laboratory isolates grown on T-cell lines (MN, SF-2) and (B) primary isolates (panel of 10) representative of clade B (origin of MN and SF-2) passaged only on peripheral blood mononuclear cells (PBMC). Note that sensitivity of primary isolates to neutralization is much lower with both HIV+ sera and sera from vaccines but, although a fraction of sera HIV-1-infected individuals can still neutralize some of the isolates, none of the vaccine sera tested register positive readings. The minimum positive reading in our assay systems is a titer of 1 : 10.

in their ability to induce neutralizing antibodies to HIV isolates that were adapted to T-cell lines, which actually overlapped with those found in HIV-infected individuals, the immune responses elicited by these vaccines have failed to neutralize fresh patient isolates on PBMCs. A flurry of studies ensued to determine whether this was an assay problem or whether it reflected a fundamental difference between primary and laboratory isolates. The outcome of these and continuing efforts points away from this being an assay problem and points to significant differences as to how these viruses exhibit their neutralization targets. These surprising findings contributed heavily to a decision to delay efficacy trials in the United States with these vaccines.

We have thus witnessed unexpected developments and perhaps learned valuable lessons as the first wave of vaccines approached the all-important milestone of efficacy trial testing. These vaccines were based on the hypothesis that neutralizing antibodies could represent a correlate of protection. For the present, the definitive clinical studies to answer this question have been put on hold until more information becomes available. There are other hypotheses, such as cellular immunity or mucosal immunity, that could either stand alone or, as is much more appealing to many, be combined through imaginative vaccine approaches in order to elicit a more comprehensive immunity. These are currently in development and could be ready for large-scale trials over the next several years (Table III). They include complex recombinant pox vectors that, in addition to the envelope, include other structural as well as regulatory gene products of the virus that have proven capable of inducing both neutralizing antibodies and CTL. The added benefit of

TABLE III

HIV Vaccine Candidates in Development[a]

Second Generation (being considered for clinical trials)
- Complex pox vectors (vaccinia, avipox) consisting of multiple HIV gene products (*gag*, *pol*, *env*, *nef*)
- VLP[b]
- Pseudovirions[c]
- Peptide cocktails (including oral formulations)

Third Generation (in research)
- Live attenuated viruses
- Whole inactivated viruses
- Naked DNA
- Viral vectors (adenovirus, poliovirus, mengovirus, rhinovirus)
- Bacterial vectors (BCG, *Salmonella*, *Shigella*)
- Chimeric proteins[d]

[a]For review see Walker and Fast (1993).
[b]Nucleic acid-free, noninfectious virus-like particles (VLP) that self-assemble in yeast, insect, and mammalian expression systems involving *env*, *gag*, and *pol* gene products of HIV.
[c]Nonreplicating multiply mutated HIV.
[d]Immunogenic protein sequences from other organisms coupled to HIV proteins or peptides.

immunity at mucosal sites can be achieved with other viral and bacterial vectors such as adenovirus, poliovirus, mengovirus, and salmonella. Other approaches include vaccination with DNA and the live attenuated virus vaccines. The latter is thus far the most successful vaccine strategy and now a major research topic to determine if it can be made safe and what mechanism is responsible for its superior efficacy.

3. Remaining Challenges in HIV Vaccine Development

Thus, despite substantial efforts to develop a vaccine against HIV, it is evident that many important questions still remain unanswered. Continued efforts are needed to understand basic features of HIV infection and pathogenesis, to improve animal models, and to identify correlates of protection. One must also consider the significance of a growing number of instances in which protection against HIV infection occurs without recognizable correlates of immunity. For instance, there are the examples in chimpanzees, where protection occurs in the absence of neutralizing antibodies. Similarly, studies with several different vaccine approaches against HIV-2 reveal little or no neutralizing antibodies or CTL at the time of challenge, and yet protection against infection is achieved. Finally, from studies in cohorts HIV exposed (repeatedly) to nonseroconverters and in monkeys exposed to low-dose rectal challenges, one again finds indications of possible protection in the absence of classic immune correlates.

What are these situations telling us about host control of virus replication and pathogenesis? Are there host defense mechanisms that are not easily detectable as classic effector mechanisms (e.g., neutralizing antibodies and CTL) that when supplemented with vaccination can effectively block infection? Investigation of these and related examples of protection in the absence of identifiable correlates would seem worthwhile, and such studies may eventually show that even the measurable correlates that are familiar to us may be only markers but not the actual mechanism of protection.

It is now evident that a great deal is expected of an HIV vaccine or it approaches the milestone of an efficacy trial. Badly needed are acceptable guidelines designed to best forecast the likelihood that vaccine candidates will prove efficacious. In the absence of immune correlates, more emphasis might be placed on vaccine efficacy in several animal models with varying degrees of virus virulence and disease. Similarly, appropriately designed Phase II studies in individuals at risk for HIV infection may help to guide entry into large definitive efficacy trials.

V. SUMMARY

From the studies reviewed here it is evident that experimental vaccines are possible against a number of retroviruses. Therefore, the features of viral integration and latency appear not to be a barrier to vaccine efficacy in and of themselves. There are, however, determinants of virulence that make the development of vaccines difficult. In general, these would be factors that allow the virus to overwhelm host defenses, including those induced by vaccination. For instance, if one added the properties of variability and immunopathogenicity to vaccines for susceptible viruses such as murine leukemia viruses, thereby adding mechanisms for immune escape, one could anticipate agents that would be much more resistant to vaccination. If the virus is not easily suppressed by vaccine-induced immune defenses either because of antigenic variants or because virus populations are difficult to neutralize (e.g., CAEV), sufficiently rapid replication could permit the virus to expand before other host defenses could mount immune responses sufficient to clear the infection. Moreover, rapid replication can generate more antigenic variants that would increase in advance of the immune response, as with lentiviruses such as Visna and EIAV, but probably also with viruses like SIV and HIV. Since the latter have the additional property of immunopathogenicity, they can further impair the ability of the host to control the virus. Thus, one can highlight the features of rapid replication, variability, and immunopathogenicity as the dominant determinants of virulence that represent the major barriers to vaccination for immunodeficiency viruses and properties such as variation and immune enhancement for the other lentiretroviruses. Although not recognized as a principal feature of the pathogenesis of the immunodeficiency viruses, immune enhancement similar to that induced by subunit envelope vaccines in the EIAV model could also play an important role with the corresponding HIV vaccines.

Insofar as correlates of protection are concerned, conclusions cannot be drawn that are generally applicable to all models and in several cases even in a single model. Although in many instances, neutralizing antibodies appeared to be the best correlate of protection and actually were shown to be sufficient for protection in some cases, there are clearly important contributions by the cellular arm of the immune system. However, the effector arm of such responses remains undefined, as well as whether they are directly vaccine induced or represent a host response to the challenge virus. As with other viruses, retroviral vaccines probably serve to establish initial barriers of protection along with long-term immunological memory that permit the host to recognize the pathogen and mount a response that repels the infection.

The most difficult challenge is to develop a safe and effective vaccine against HIV. The absence of an animal model for HIV-1 infection and disease continues to pose a major obstacle for development of a vaccine against this virus. In a sense, it forces one to turn to humans as the experimental model, which means clinical trials of vaccines in large numbers of volunteers at risk for HIV infection. Such trials are expensive, difficult to carry out, fraught with concerns for safety, and surrounded by thorny ethical and societal issues. The overarching problem at the present time is that with over a dozen vaccine candidates in clinical trials and an infrastructure that will soon be ready to accept vaccines for large-scale efficacy trials, the HIV vaccine establishment is in the awkward position of having to decide how to proceed with its most promising candidates in the absence of correlates of immunity. Without a directional beacon coupled with a generally low confidence level in current vaccine candidates, the already difficult practical issues associated with conducting such trials magnify considerably. And yet, it stands to reason that one can learn a great deal from such studies even if usable vaccines do not emerge from the first attempts; they could lay the groundwork for a vaccine development effort that ultimately will be successful. Examples would be better clues about correlates of protection or the possibility that vaccines can impact on the course of disease without necessarily blocking infection and thereby render some benefit. It will probably require some balance between these two contrasting alternatives to ultimately shape the decision of when and with what vaccine one should proceed toward large-scale efficacy trial testing.

BIBLIOGRAPHY

Bolognesi, D. P. (1993). Human immunodeficiency virus vaccines. *Adv. Virus Res.* **42,** 103–148.

Desrosiers, R. C., Wyand, M. S., Kodama, T., Ringle, D. J., Arthur, L. O., Sehgal, P. K., Letvin, N. L., King, N. W., and Daniel, M. D. (1989). Vaccine protection against simian immunodeficiency virus infection. *Proc. Natl. Acad. Sci. USA* **86,** 6353–6357.

Haase, A. T. (1986). Pathogenesis of lentivirus infection. *Nature (London)* **322,** 130–136.

Matthews, T. J. (1994). Dilemma of neutralization resistance of HIV-1 field isolates and vaccine development. *AIDS Res. Hum. Retroviruses* **10,** 631–632.

Montelaro, R. C., Ball, J. M., and Rushlow, K. E. (1993). Equine retroviruses. *In* "The Retroviridae" (J. Levy, ed.), Vol. 2, pp. 257–360. Plenum, New York.

Nakamura, H., Hayami, M., Ohta, Y., Ishikawa, K., Tsujimoto, H., Kiyokawa, T., Yoshida, M., Sasagawa, A., and Honjo, S. (1987). Protection of cynomolgus monkeys against infection by human T-cell leukemia virus type-1 by immunization with viral env gene products produced in *Escherichia coli. Int. J. Cancer* **40,** 403–407.

Pedersen, N. C., and Johnson, L. (1991). Comparative efficacy of three commercial feline leukemia virus vaccines against methylprednisolone acetate-augmented oronasal challenge exposure with virulent virus. *JAVMA* **199,** 1453–1455.

Portetelle, D., Limbach, K., Burny, A., Mammerickx, M., Desmettre, P., Riviere, M., Zavada, J., and Paoletti, E. (1991). Recombinant vaccinia virus expression of the bovine leukemia virus envelope gene and protection of immunized sheep against infection. *Vaccine* **9,** 194.

Schafer, W., and Bolognesi, D. P. (1977). Mammalia C-type oncornaviruses: Relationships between viral structural and cell-surface antigens and their possible significance in immunological defense mechanisms. *Contemp. Top. Immunobiol.* **6,** 127–167.

Schultz, A. M., and Hu, S. L. (1993). Primate models for HIV vaccines. *AIDS (Phila.)* **7,** S161–S170.

Varmus, H. (1989). Retroviruses. *Science* **240,** 1427–1435.

Walker, M. C., and Fast, P. E. (1993). Human trials of experimental AIDS vaccines. *AIDS (Phila.)* **7,** S147–S159.

Wong-Staal, F., and Gallo, R. C. (1985). Human T-lymphotropic retroviruses. *Nature (London)* **317,** 395–403.

Retroviruses as Vectors for Gene Transfer

RALPH DORNBURG
Thomas Jefferson University

GLOSSARY

Antigen Structure (e.g., a protein or polysaccharide) that can be recognized by an antibody

Anti-sense RNA RNA transcript of a gene in the opposite orientation to the gene (complementary to the normal mRNA)

cDNA Complementary DNA copy transcribed from a messenger RNA template rather than a DNA template

Enhancer DNA sequence that increases the efficiency of transcription from a promoter

Long terminal repeats Sequences located at the 5′ and 3′ ends of retroviral DNA

Oncogene Gene whose expression results in the malignant transformation of a cell

Promoter DNA sequence that is recognized by RNA polymerases and drives gene expression

Protooncogene Cellular gene (most probably involved in differentiation and/or cell divisions) from which an oncogene evolved

Receptor Cell-surface protein that binds a specific ligand (e.g., a protein or a virus)

Single-chain antibody Fragment of an antibody comprising the variable domains of both the light and heavy chains connected by a peptide bridge

Transfection Introduction of genetic material into a cell by experimental procedures

Vector Genetically engineered DNA construct to introduce and to express a gene in a target cell

RETROVIRUSES ARE RNA VIRUSES THAT REPLICATE through a DNA intermediate that is integrated into the host cell genome. Retroviruses are widespread in nature and can be associated with different forms of malignant tumors. Thus, they are also termed RNA tumor viruses. Many tumorigenic retroviruses contain a gene (oncogene) in their genome, in addition to or substituting for their replication genes. The expression of the oncogene results in the malignant transformation of the infected cells. Oncogenes are of cellular origin and were picked up by retroviruses in the course of earlier infections. Thus, retroviruses act as vehicles to transfer cellular genes from cell to cell and from organism to organism. This ability and the high efficiency and accuracy of retrovirus replication makes these viruses useful for the genetic engineering of transfer vectors for studying a large variety of biological procesess. Moreover, because of their properties, retroviral vectors are also being used in the first human gene therapy trials.

I. RETROVIRUS GENOME AND LIFE CYCLE

In general the retroviral life cycle resembles that of many other animal viruses, in that it can be divided into virus attachment and entrance into cells, a synthesis period, and a period of virus assembly and release. However, the molecular mechanism of replication is unique for this family of viruses (and a few related viruses) and follows a rather complicated but highly efficient pathway.

A. Virions, Genome, and Taxonomy

Retrovirus virions are medium-sized particles with a diameter of about 100 nm, consisting of a core structure surrounded by a lipid bilayer membrane (Fig. 1). In electron micrographs, the shape of the core varies among different retroviruses, ranging from spherical

ENCYCLOPEDIA OF HUMAN BIOLOGY, Second Edition, VOLUME 7. Copyright © 1997 by Academic Press. All rights of reproduction in any form reserved.

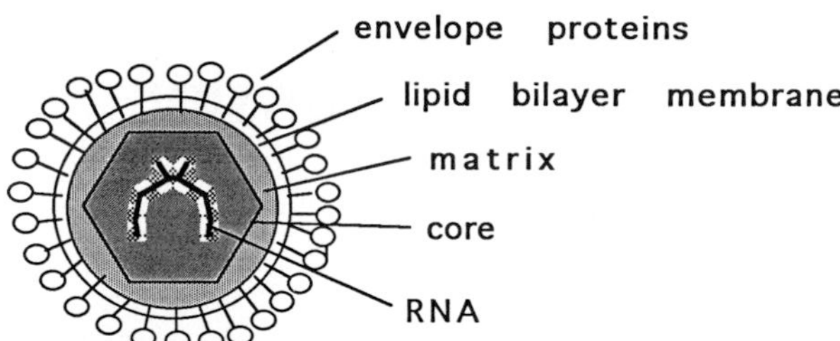

FIGURE 1 A retroviral particle. Two genomic RNA molecules are encapsidated in a core structure, which is surrounded by a lipid bilayer membrane. Envelope proteins are embedded in the bilayer membrane. For a more detailed explanation see the text.

to rod-like forms. The core consists of viral proteins called gag (group-specific antigen), the viral DNA polymerase and integration protein, two identical copies of the viral genomic RNA, and associated tRNA. The lipid bilayer of the virion is derived from the host cell membrane and contains viral envelope proteins that look like spikes in electron microscope pictures. These proteins recognize and interact with host cell-surface receptors and enable the virus to penetrate the cell.

The RNA genome of retroviruses resembles an eukaryotic mRNA, has a 5′ cap, and contains a poly(A) tail. It has the same polarity as the viral mRNAs, and therefore it is called plus-strand RNA. In simple retroviruses it contains three genes coding for proteins (Fig. 2). The *gag* gene usually codes for four core proteins: the matrix protein (MA), which is mainly found in the matrix and probably forms the junction between the core and the lipid bilayer; the capsid protein (CA), which is the major component of the core; the nucleocapsid protein (NC), which is tightly bound to the RNA; and the protease protein (PR). They are also designated with the letter p and a number that reflects the molecular weights in thousands (e.g., p20). In all retroviruses, the core proteins are derived from a single polypeptide precursor by proteolytic cleavage by the viral protease. The *pol* gene codes for the viral polymerase—an RNA-dependent DNA polymerase (i.e., reverse transcriptase)—and the viral integration protein.

DNA polymerase and integration protein are first translated with the core proteins into a single polypeptide and then are cleaved into the two active enzymes by the viral protease. The products of the *env* gene are two envelope proteins, surface protein (SU) and transmembrane protein (TM), which are found on or in the lipid bilayer membrane, respectively. They are also derived from a single precursor protein, cleaved by a cellular enzyme, and glycosylated as are cellular membrane proteins. Hence, they are also termed glycoproteins (e.g., gp70); the number reflects the molecular weight in thousands. Some retroviruses such as the human T-cell leukemia viruses (HTLV) and the human immunodeficiency virus (HIV) carry additional genes other than *gag, pol,* and *env.* Some of these genes have essential functions in regulatory viral gene expression.

In addition to the protein-coding genes, the retroviral RNA genome contains several regulatory sequences (also called cis-acting sequences) required for efficient viral replication. These sequences are located primarily at the ends of the genome.

In modern molecular biology, the nucleotide sequence of the retroviral genome serves as the yardstick for classification. Therefore, the most commonly studied retroviruses are organized in six groups: the avian leukosis-sarcoma viruses, the avian reticuloendotheliosis viruses and mammalian leukemia-sarcoma viruses, the mouse mammary tumor viruses, the primate Type D viruses, the human T-cell leukemia related viruses, and the lentiviruses. Sequence comparisons on the nucleotide and amino acid level have led to the establishment of evolutionary trees.

B. Life Cycle

To enter a cell, the virus first has to attach to the cell membrane of the target cell (Fig. 3). This attachment

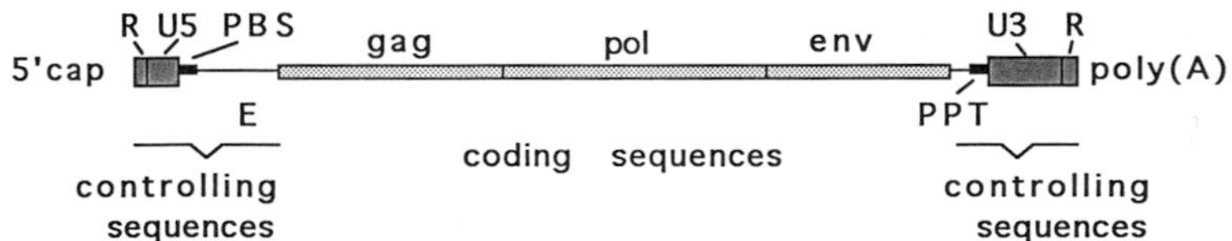

FIGURE 2 Organization of a C-type retrovirus RNA genome. The genome contains three protein-coding genes. Regulatory sequences are located mainly at the ends. For a more detailed explanation see the text. U3, U5, unique in all retroviral RNAs at the 3′ and 5′ ends, respectively; R, repeated region; PPT, purine-rich region; E, encapsidation sequence.

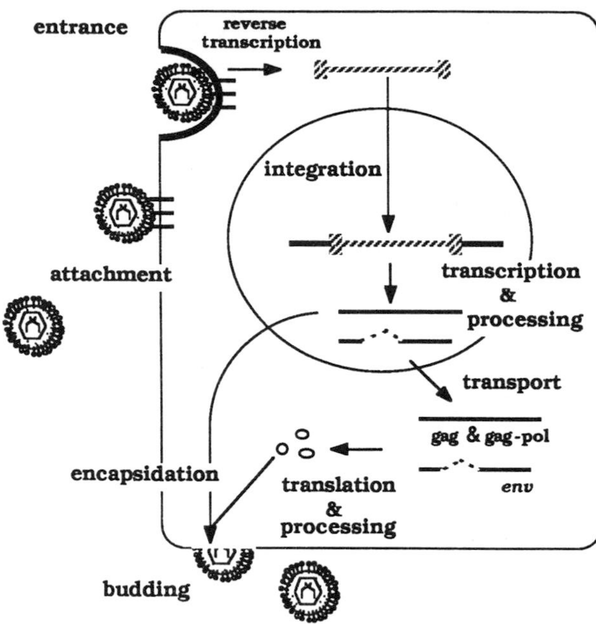

FIGURE 3 Simple oncoretrovirus life cycle. After entrance of the virion into the cell, the retroviral RNA is copied into a double-standard DNA, which is integrated into the host genome. DNA transcription and RNA processing result in genomic RNA, as well as retroviral mRNAs. Virus assembly takes place at the cell membrane and results in the budding of retroviral virions. For a more detailed explanation see the text.

is mediated by the SU protein, which recognizes a specific receptor (usually a membrane protein) of the target cell. The interaction of the envelope with the cellular receptor is very specific and determines the host range of the virus. For example, the human immunodeficiency virus (HIV-1) binds only to the human CD4 molecule expressed on T cells. Other retroviruses utilize household receptors (e.g., an amino acid transport protein) expressed on many different tissues. After attachment, retroviruses enter cells by two different mechanisms: in the case of most retroviruses, the viral and cellular membranes fuse at the cell surface and the retroviral core particle is released into the cytoplasm of the target cell (e.g., HIV). In the case of other retroviruses, the virus particle is absorbed by endocytosis and membrane fusion is triggered by a pH shift in the endosome. Once released into the cytoplasm, the virus has access to the cellular nucleotide triphosphates and starts to copy its RNA genome into a double-stranded DNA. In this complicated process, all enzymatic reactions are carried out by the viral reverse transcriptase (Fig. 4).

As a result of this replication process, the protein-coding regions of the retroviral genome are flanked by long terminal repeats (LTRs). The LTRs carry specific sequences (short inverted repeats) at their ends that are essential for the efficient integration of the provirus into the host genome: sequences at the 5′ end of the left LTR and sequences at the 3′ end of the right LTR form the attachment site (att) recognized by the viral integration protein, which carries out this process. Thus, integration is sequence specific in regard to the viral DNA. However, no sequence specificity is known regarding the nucleotide sequence of the integration site in the host genome, although "hot spots" of integration have been reported. As a result of the integration, four to six nucleotides of the chromosomal DNA are duplicated, depending on the viral integration protein. The integrated DNA is called the provirus.

After integration of the provirus, RNA transcription produces genomic retroviral RNAs. RNA transcription is performed by cellular RNA polymerase II and is driven by a promoter and enhancer present in the U3 region of the LTRs. 3′-end RNA processing is regulated by viral sequences present in the R (repeat) or U3 (depending on the particular virus) and U5 regions. In any case, the RNA transcript starts at the first nucleotide of R in the left LTR and ends with the last nucleotide of R in the right LTR. As a result, RNA transcripts are identical to the original genomic viral RNA. Gag proteins and reverse transcriptase are translated from genomic RNA. Expression of the envelope proteins is from spliced genomic RNAs and follows the pathway of cellular membrane proteins. [*See* DNA and Gene Transcription.]

To complete the viral life cycle, genomic viral RNAs are encapsidated by gag proteins to form core structures. The selective encapsidation of genomic viral RNAs is mediated by specific encapsidation sequences (called E in avian retroviruses or ψ in murine retroviruses) located 3′ of the primer binding site (PBS) on the viral RNA. Core structures interact with viral envelope proteins that are embedded the cell membrane. As a result of this interaction, virus particles bud from infected cells to give progeny virus. Finally, proteolytic cleavage of precursor proteins takes place immediately after the budding process, resulting in virions that are ready to infect fresh target cells.

As a result of this replication process, retroviruses often do not kill or lyse the infected cells. Instead, the cell machinery is continuously used for virus production. Moreover, after the provirus becomes a part of the cell genome, both daughter cells carry the provirus and produce retrovirus particles. Retroviral replication can be efficient: many thousands of virus particles

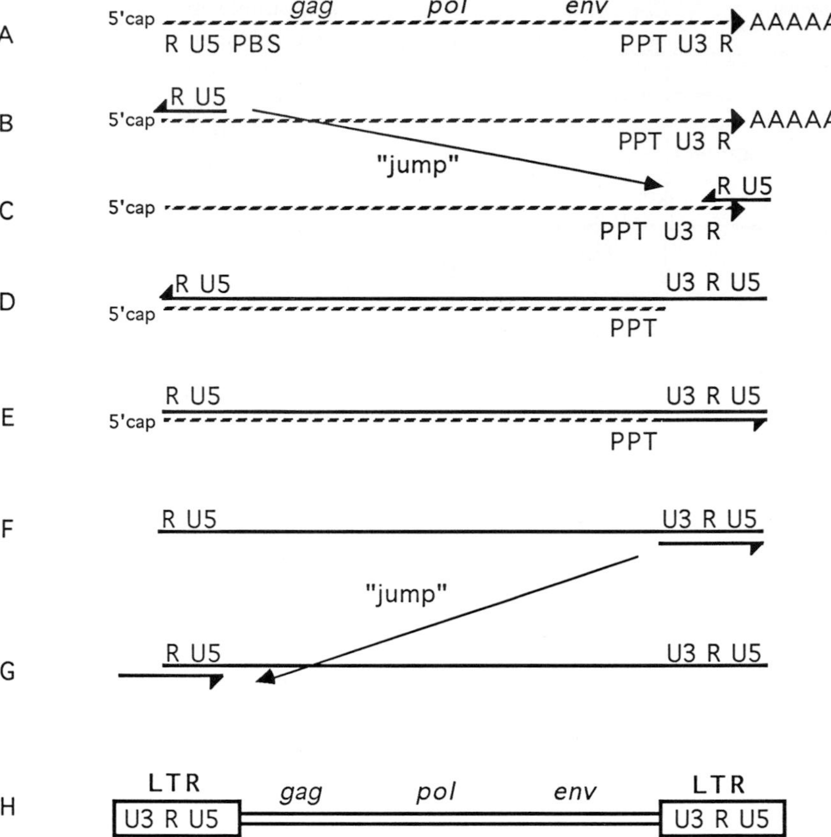

FIGURE 4 Reverse transcription of the retroviral RNA. (A) Retroviral genomic RNA. (B) A short cDNA copy (solid line) of the viral RNA is synthesized using a cellular tRNA as primer, which hybridizes to the primer binding site (PBS) located near the 5′ end of the viral RNA. (C) Next, the RNA moiety of the resulting double-stranded DNA–RNA hybrid is removed, allowing the short single-stranded cDNA copy to "jump" to a repeated region (R) at the 3′ end of the second RNA molecule and to hybridize to it. (D) This cDNA serves as a primer for the cDNA synthesis of the complete viral genome. As a result of this "jump," a region designated U5 (because it is unique in all retroviral RNAs at the 5′ end) is now attached to the 3′ end of the single-stranded cDNA copy. Next, the RNA moiety of the DNA–RNA hybrid is removed up to a purine-rich region (PPT). (E) This polypurine track serves as a primer for the second strand of DNA synthesis. (F) Then, all of the remaining RNA is degraded. (G) The double strand is melted, enabling the short second strand to "jump" to the other end of the first cDNA strand. (H) The single-stranded regions are filled in to complete the synthesis of a double-stranded DNA molecule. In the course of this DNA synthesis, a region called U3 (unique at the 3′ end of retroviruses) is duplicated. As a result of this replication mechanism, the protein-coding regions are flanked by long terminal repeats (LTR). All enzymatic reactions are carried out by the viral reverse transcriptase.

can be produced per day from a single cell infected with one provirus.

C. Endogenous Retroviruses

Retroviruses are also able to infect germline cells or their precursors in preimplantation embryos. As a result, the progeny carry retroviral proviruses in all body cells. Such retroviruses are called endogenous to distinguish them from those resulting from exogenous infections. Endogenous retroviruses are found in all vertebrates (including humans) that have been investigated thoroughly.

There are also other cellular DNA sequences with

partial nucleotide sequence homology to retroviruses. These and other repeated cell sequences most probably arose by the reverse transcription of RNAs. They are estimated to make up 10% of the mammalian genome.

II. NATURAL RETROVIRAL VECTORS

Many retroviruses contain another gene (i.e., an oncogene) in their genome. Expression of the oncogene results in the malignant transformation of the infected cells. Such retroviruses are called highly oncogenic retroviruses, and they arose by recombination of the viral genome with cellular protooncogenes. As a result of this process, in most oncogenic retroviruses the viral oncogene substitutes for parts of the protein-coding regions of the viral genome (Fig. 5). Thus, the viral oncogene product is often a fusion protein of virus and cell sequences. Oncogenes are always expressed from transcripts originating and terminating in the viral LTRs. Most highly oncogenic retroviruses cannot synthesize all of the proteins necessary for retroviral replication, thus they are replication defective. [*See* Oncogene Amplification in Human Cancer.]

Highly oncogenic retroviruses are natural gene transfer vehicles (vectors) that carry a nonviral gene. Defective highly oncogenic viruses can infect only fresh target cells if their host cell is also infected with a wild-type virus called a "helper" virus. The helper virus provides those viral proteins that the defective retrovirus cannot synthesize as a result of the deletion and substitution in its genome. However, highly oncogenic retroviruses still contain all cis-acting sequences necessary for retroviral replication. Thus, their genomic retroviral RNA is encapsidated into virions (provided by the helper virus) to form infectious retroviral particles.

Retroviruses without an oncogene can also induce tumors. It has been documented in several cases that insertion of a provirus near a protooncogene can result in the uncontrolled and/or increased gene expression of that oncogene, leading ultimately to the transformation of the infected cell.

III. RETROVIRAL VECTORS

A. Vectors, Helper Cells, and Experimental Design

Oncogenic retroviruses are used as a model in the construction of retroviral gene transfer systems, which usually consist of two components: the retroviral vector that contains the gene of interest replacing retroviral protein-coding sequences and a helper cell that supplies the retroviral proteins for the encapsidation of the vector genome.

Retroviral vectors have been derived mainly from chicken and murine retroviruses and are constructed in several molecular cloning steps. First, a retrovirus provirus is cloned in a bacterial plasmid in which it can be amplified to obtain large quantities of DNA for genetic engineering purposes. Next the viral protein-coding genes are removed and replaced by the gene(s) of interest. Figure 6 shows some examples of retroviral vectors. Genes can be inserted in the same or reverse orientation to the vector. They can be expressed by the LTR promoter, additional internal (inducible) promoters, and/or spliced RNAs. Genes that are inserted in the reverse orientation must be expressed from an internal promoter. The inserted genes can contain introns. Introns of genes that are in the same orientation as the vector are lost as a consequence of retroviral replication. In many vectors, selectable marker genes (usually resistance genes to antibiotics)

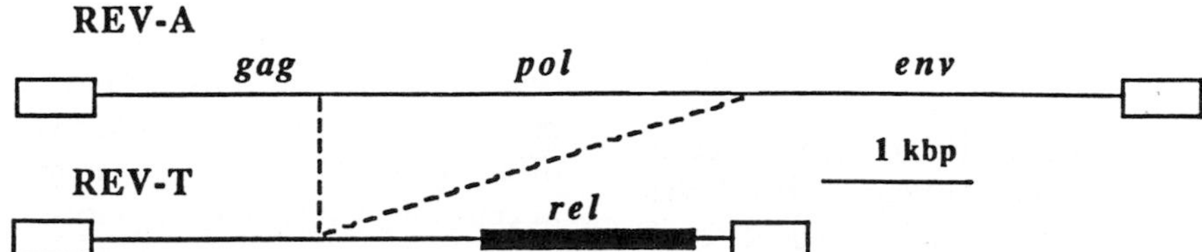

FIGURE 5 A highly oncogenic retrovirus. A provirus of a replication competent avian retrovirus, reticuloendotheliosis virus A (REV-A), is shown at the top. Recombination with a cellular protooncogene (called c-*rel*) resulted in the highly oncogenic retrovirus REV-T. In most highly oncogenic retroviruses the oncogene substitutes for viral protein-coding sequences. Retroviruses that carry an oncogene are naturally occurring gene transfer vectors.

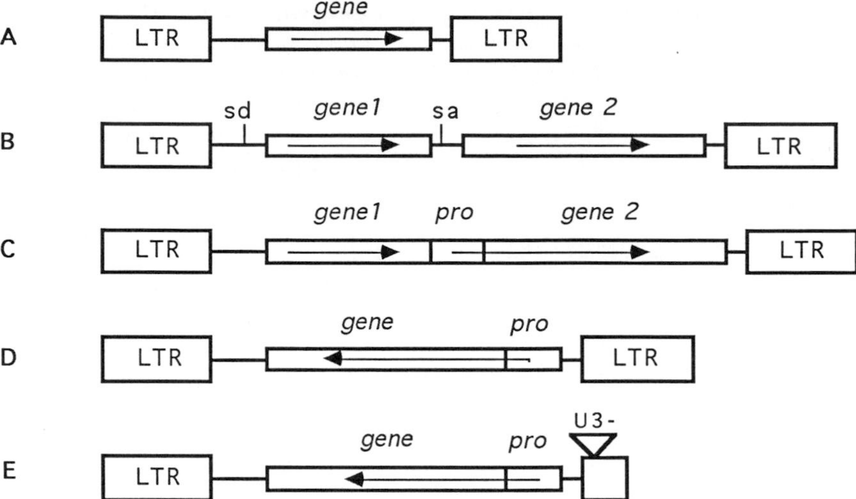

FIGURE 6 Retroviral vectors. All kinds of genes can be inserted into a retroviral vector in the same orientation (A–C) or in the reverse orientation to the vector (D and E). Genes can be expressed from the LTR promoter (A–C), from spliced mRNAs (B), or from internal promoters (C–E). Retroviral vector constructs with deleted U3 region in the right LTR result in a provirus without an LTR promoter after one round of replication. Gene expression is only performed from the internal promoter (E). sd, splice donor site; sa, splice acceptor site; pro, internal promoter. The vector constructs are in bacterial plasmids. The plasmid sequences that abut the LTRs are not shown.

have been inserted to obtain high virus titers (see Section III,B).

Helper cells carry a defective retrovirus provirus that is constitutively expressing retroviral proteins, but cannot encapsidate its own genomic RNAs as a result of deletion of the encapsidation sequence in the viral nucleic acid. Helper cells have also been made by the transfection of plasmid constructs into established cell lines to express retroviral proteins from nonretroviral promoters. Such helper cells avoid the risk of recombinations between the retroviral vector and the helper sequences.

Formation of a helper-free retroviral vector for gene transfer is outlined in Fig. 7. Helper cells are transfected with retroviral vector DNA constructs that carry appropriate encapsidation sequences. Thus, RNA transcripts of such constructs are packaged into virions provided by the helper cell. The virus produced from the helper cell is called "helper-free" since it contains no replication-competent helper virus. If the vector has a selectable gene, helper cells can be selected for the expression of the marker gene. As a result, each selected helper cell produces virions containing the RNA transcript of the transfected vector. Supernatant tissue culture medium is used to infect fresh target cells. Titers of up to 10^7 infectious virus particles per milliliter of tissue culture medium have been obtained. The efficiency of introducing a gene by other methods is several orders of magnitude lower than that by retroviral infection.

B. Tissue Culture Experiments with Retroviral Vectors

A large variety of different eukaryotic, bacterial, and viral genes have been inserted and expressed in retroviral vectors. So far, there is no report that any gene could not be expressed.

Retroviral vectors have been used mainly in tissue culture experiments to investigate various aspects of the retroviral life cycle and to study gene expression and the effects of gene products of nonretroviral genes (e.g., oncogenes or protooncogenes) in particular cells. They were also used to introduce genes for the production of antisense RNAs to investigate the effect of antisense RNAs on gene expression, particularly in cells infected with other viruses.

C. Experiments with Retroviral Vectors in Animals

Several experiments with retroviral vectors have been performed in early embryos (mainly in mice) to tag

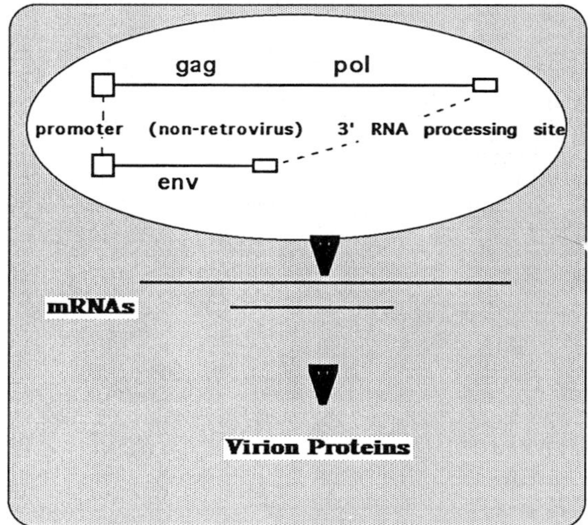

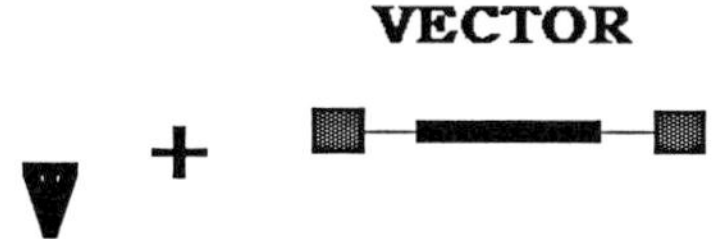

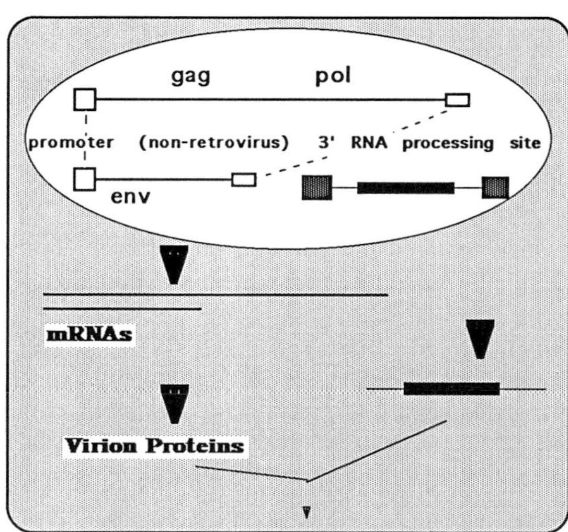

FIGURE 7 Formation of a helper-free retroviral vector gene transfer. (A) In helper cells, retroviral proteins are expressed from different plasmid DNAs. These RNA transcripts do not contain encapsidation sequences, thus they are not encapsidated into retroviral particles. (B) Such helper cells are transfected with a retroviral vector plasmid construct. The RNA transcript of the retroviral vector contains an encapsidation sequence and, therefore, is encapsidated into virions supplied by the helper cell. Supernatant tissue culture medium is used to infect fresh target cells.

chromosomal locations of developmental genes. The purpose of these experiments is to destroy genes involved in cellular differentiation by the insertion of a vector provirus. Animals in which the insertion has taken place show a characteristic developmental defect. Molecular cloning of the chromosomal sequences surrounding the integrated provirus has led to the discovery of developmental genes. Infection of murine embryos with retroviral vectors was also used for cell lineage studies, because all cells derived from a single infected progenitor cell carry the vector provirus at the same location.

In other experiments, bone marrow cells (i.e., stem cells) infected or marked with retroviral vectors have been injected in lethally irradiated mice to study hematopoiesis and the development of the immune system. The marked stem cells undergo differentiation in the injected animal. Comparison of the location of integrated vector proviruses in different fully differentiated cells from such animals led to insights into how cells of the immune system develop.

Cell transplantation experiments were also performed to study the possible application of retroviral vectors in gene therapy. For example, the gene coding for human adenosine deaminase (ADA) has been introduced into mouse lymphocytes by retroviral vectors followed by reinjection of the cells into the animals. It was found that the human gene was expressed in the animal. In other experiments, the gene coding for the human blood clotting factor VIII was introduced by a retroviral vector into rat fibroblasts in tissue culture. Cells expressing factor VIII were then transplanted in rats, and subsequently human factor VIII was detected in the blood of rats. In addition, infection of preimplantation embryos with retroviral vectors is being considered for use in farm animals as a substitute for gene transfer by microinjections.

Retroviral vectors have also been used extensively in mice to study the effect of various therapeutic genes for the treatment of various cancers. For example, tumor cells were infected with retroviral vectors carrying interleukin genes followed by reinjection into the animal. It was found that this treatment caused a stimulation of the immune system, which led to tumor regression in not all but a significant percentage of treated animals.

D. Retroviral Vectors in Human Gene Therapy

Today, approximately 3000 human diseases are known that result from a single defect gene. Some diseases are very rare; others occur with relatively high frequencies (e.g., blood clotting factor VIII deficiency and sickle-cell anemia). It is estimated that 1 to 2% of newborns are affected with single-gene disorders. The present therapies are mainly based on special diets, injection of proteins, and/or blood transfusions (e.g., for the supplement of clotting factors). However, in most cases such therapies are ineffectual and unsatisfactory. Thus, the only possibility for curing such genetic disorders is the introduction of a functional gene into the body cells of such patients. The functional gene could be expressed even while the inherited defective gene persists.

As outlined earlier, gene therapy experiments with retroviral vectors have been performed in animals (mostly mice and rats) with genes for adenosine deaminase, globin, and blood clotting factors. The results of these experiments have been promising enough to justify the initiation of clinical trials in humans. At the end of 1991, the first human patients affected with Severe Combined Immunodeficiency Syndrome (SCID), an autosomal recessive genetic disorder caused by a deficiency of the enzyme adenosine deaminase, were treated with retroviral vector transduced cells, which contained a functional copy of the gene coding for this enzyme. The success of these first clinical trials has spurred a series of other clinical trials to introduce genes into humans to cure not only genetic diseases but also cancer and AIDS. Many clinical trials are now in progress and results are eagerly awaited.

In spite of the initial success of human gene therapy trials, many questions and concerns are being raised. Besides technical problems (see Section III,E), the questions concern safety, efficiency, and social effects. For instance, can the insertion of a retroviral vector into the genome lead in some cells to protooncogene gene expression and to the development of tumors (see Section II)? Is there a chance that infectious viruses will be produced by recombination with retroviral sequences in the helper cell that can infect other people and/or germline cells? What are the costs in relation to the efficiency of the treatment? A committee of the National Institutes of Health has been discussing these and related questions since 1984.

E. Problems, Limitations, and Possible Solutions

As a consequence of the properties of retroviruses, there are several limitations in the use of retroviral vectors. There appears to be a maximum limit on the size of the RNA that can be encapsidated into a retrovirus virion. Many eukaryotic genes contain large introns and/or are regulated by sequences far upstream or downstream of the coding sequences. Thus, several genes may be too large to be inserted in a retroviral vector. Moreover, gene regulation from internal promoters can interfere with virus production or the LTR promoter can suppress or stimulate the internal promoter. Attempts have been made to overcome this shortcoming with the construction of "suicide vectors" (see Fig. 6E). In addition, retroviral vectors often behave in unpredictable ways. Many sequences inserted into a retroviral vector appear to be unstable, and there is sometimes strong selection of spontaneous mutations against particular constructs.

All retroviral vectors currently used in human gene therapy trials have been derived from amphotropic (ampho) murine leukemia virus (MLV). As outlined earlier, ampho-MLV has a very broad host range and can infect cells of various tissues of many species. This also applies to ampho-MLV-derived retroviral vectors. If injected directly into the bloodstream, the chances that the vector particles would infect their actual target cells are extremely low. Thus, the gene delivery has to be performed *ex vivo:* the target cells of interest are removed from the patient and cultivated in tissue culture flasks. After infection with the retroviral vector, the cells are reinjected into the patient. However, *ex vivo* gene therapy has several limitations and cannot be applied on a broad clinical basis; besides extremely high costs (about $50,000 to $100,000 per patient!), it is still not clear whether the cultivation of cells in a nonphysiological environment alters their natural long-term behavior (termed the "homing problem").

The solution to these problems is a cell-type-specific gene delivery system that would allow the injection of the gene delivery vehicle directly into the patient's bloodstream or tissue of interest. Thus, experiments have been initiated in several laboratories to develop cell-type-specific retroviral vector particles. In particular, experiments have been performed to substitute the receptor binding domain of the envelope glycoprotein with another ligand to bind the vector virus particle only to a receptor specific for ligand displayed on the vector particle.

The most versatile ligand for this approach is the antigen binding site of an antibody, because antibodies can be raised against virtually any cell receptor. [*See* Antibody–Antigen Complexes: Biological Consequences.] However, complete antibodies are very bulky and are not suitable for this approach. The problem has been solved using single-chain antibody

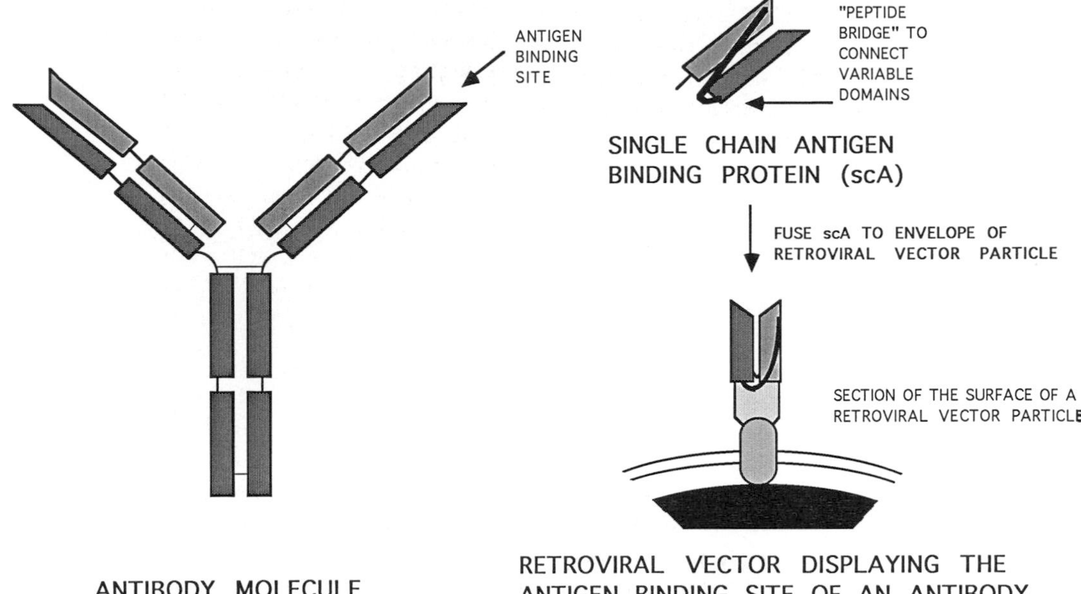

FIGURE 8 Construction of retroviral vector particles displaying the antigen binding domain of an antibody. A complete IgG antibody molecule contains four peptide chains that form one molecule with two identical antigen binding domains. The antigen binding site is formed by the amino-terminal variable domains of the heavy and light chains. Using recombinant polymerase chain reaction technology, single-chain antibody (scA) genes have been constructed that express the complete antigen binding domain only. The variable domains are connected by a peptide bridge encoded in the scA gene. Such single-chain antibodies have been fused to the envelope of retroviral vectors to make such gene transfer vehicles specific for distinct target cells: the retroviral vector will bind to and infect only cells that express an antigen recognized by the antibody moiety of the envelope.

technology (Fig. 8). Single-chain antibodies (scA) comprise the variable domains of both the heavy and light chains of an antibody molecule connected by a peptide bridge. This peptide bridge is encoded in a spacer region inserted between the coding regions of the two variable domains (see Fig. 8). Most recently, it has been shown that retroviral vectors that display an scA on the viral surface are competent for infection. However, many questions regarding specificity and efficiency still remain to be answered and much research is still necessary to bring this technology into the clinic.

BIBLIOGRAPHY

Chu, T. T.-H., and Dornburg, R. (1995). Retroviral vector particles displaying the antigen binding site of an antibody enable cell type-specific gene transfer. *J. Virol.*

Cournoyer, D., and Caskey, C. T. (1993). Gene therapy of the immune system. *Annu. Rev. Immunol.* **11**, 297.

Dornburg, R. (1995). Reticuloendotheliosis viruses and derived vectors. *Gene Ther.* **2**, 301–310.

Eglitis, M. A., and Anderson, W. F. (1988). Retroviral vectors for introduction of genes into mammalian cells. *BioTechniques* **6**, 608.

Hunter, E., and Swanstrom, R. (1990). Retrovirus envelope glycoproteins. *Curr. Top. Microbiol. Immunol.* **157**, 187–253.

Katz, R. A., and Skalka, A. M. (1994). The retroviral enzymes. *Annu. Rev. Biochem.* **63**, 133–173.

Miller, A. D. (1990). Retrovirus packaging cells. *Hum. Gene Ther.* **1**, 5–14.

Morgan, R. A., and Anderson, W. F. (1993). Human gene therapy. *Annu. Rev. Biochem.* **62**, 191–217.

Temin, H. M. (1986). Retrovirus vectors for gene transfer: Efficient integration into and expression of exogenous DNA in vertebrate cell genomes. *In* "Gene Transfer" (R. Kucherlapati, ed.). Plenum, New York.

Varmus, H., and Brown, P. (1988). Retroviruses. *In* "Mobile DNA" (M. Howe and D. Berg, eds.). American Society of Microbiology, Washington, D.C.

Weiss, R., Teich, N., Varmus, H., and Coffin, J. (eds.) (1985). "RNA Tumor Viruses: Molecular Biology of Tumor Viruses," 2nd Ed. Cold Spring Harbor Laboratory Press, Cold Spring Harbor, New York.

Ribonucleases, Metabolic Involvement

DUANE C. EICHLER
University of South Florida

GLOSSARY

Ankyrin Membrane cytoskeleton protein

Ataxia Loss of muscular coordination

Eosinophil White blood cell, a type of leukocyte

Interferon Substance capable of inducing a state of resistance in response to infection by a virus

Intrathecal With the spinal cord

Metastatic Shifting of a disease from one part of the body to another

Processive To process along the same RNA chain without dissociating

Ribonucleoprotein particle RNA with bound protein

Retrovirus Virus that contains genomic RNA and whose life cycle has a stage in which its RNA is copied to DNA

RIBONUCLEASES (RNases) ARE ENZYMES THAT hydrolyze phosphodiester linkages in RNA. Because RNA can be perceived as the more transient of the nucleic acids molecules when compared to DNA, enzymes that degrade or affect RNA structure can play a key role in the metabolism of RNA and gene expression. Nevertheless, far more knowledge and understanding has been gained about the enzymology of cellular RNases than is known about their actual *in vivo* function. In general, there are three areas of metabolism in which ribonucleases can be considered to participate: (1) removal of RNA species by degradation, (2) maturation of an RNA species by affecting the formation of a functionally active RNA from an inactive precursor, and (3) synthesis of RNA and DNA.

I. BACKGROUND

Like other polymer-degrading enzymes, ribonucleases may be distinguished and therefore classified by their mode of attack, mechanism of cleavage, or their substrate specificity. For example, the term *exoribonuclease* defines an enzyme that requires recognition of an RNA terminus in order to initiate attack, and includes not only enzymes that release mononucleotides as degradation proceeds from a terminus but also enzymes that release oligonucleotides as well. Because the RNA polymer has direction, known as polarity, the direction of attack may be used to further distinguish an exoribonuclease as either a $5' \rightarrow 3'$ or $3' \rightarrow 5'$ exonuclease. In contrast, *endoribonuclease* refers to a ribonuclease that does not require recognition of a free terminus to initiate attack; instead, cleavage is initiated at internal phosphodiester linkages.

Ribonucleases may also discriminate between structural regions of an RNA molecule. One ribonuclease may preferentially cleave only in the single-stranded regions of an RNA molecule, whereas another ribonuclease may prefer the based-paired or helical regions of secondary structure in an RNA molecule.

Finally, ribonucleases can be distinguished by their mode of attack at the phosphodiester linkage. Ribonucleases that require the 2′-OH group of the ribose moiety to initiate attack at the adjacent phosphodiester bond are termed *phosphotransferase-type RNases* (or cyclizing RNases) (Fig. 1). By necessity, this group of RNases shows absolute specificity for RNA and the products of hydrolysis contain either a 3′-phos-

ENCYCLOPEDIA OF HUMAN BIOLOGY, Second Edition, VOLUME 7. Copyright © 1997 by Academic Press. All rights of reproduction in any form reserved.

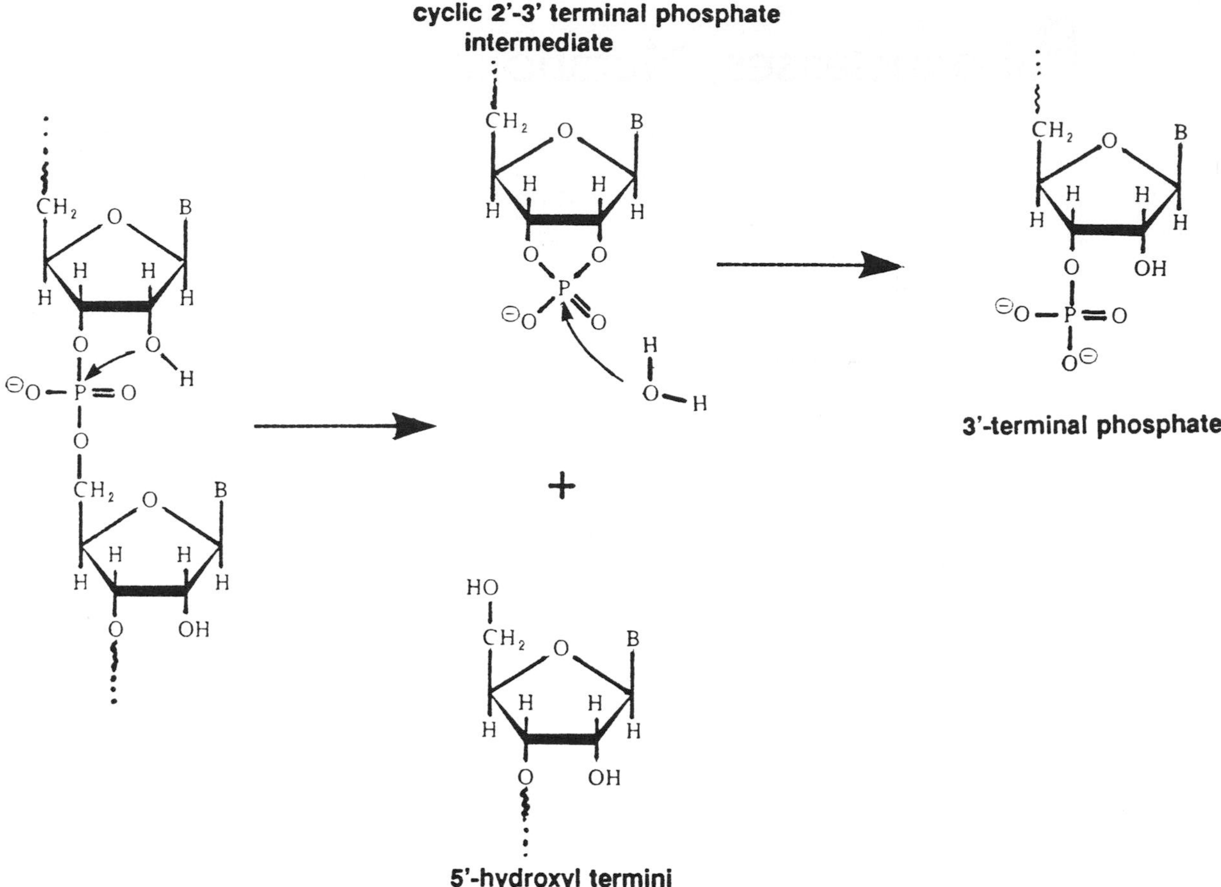

FIGURE 1 Phosphotransferase-type RNAses (cyclizing RNases).

phate or 2′,3′-cyclic phosphate terminus. This group often includes RNases that demonstrate base specificity as well. For example, pancreatic RNase is a phosphotransferase-type RNase that is specific for pyrimidine bases.

Ribonucleases that utilize water as the nucleophilic agent to attack the phosphodiester linkage are termed *phosphodiesterase-type RNases* (Fig. 2). Cleavage of the phosphodiester bond by a phosphodiesterase-type RNase results in products with either a 3′- or 5′-terminal phosphate depending on which side of the phosphodiester bond cleavage was initiated. Phosphodiesterase-type RNases also include enzymes that can cleave both RNA and DNA, for example, spleen phosphodiesterase.

A third mode of attack distinguishes the *phosphorylase-type RNases* (Fig. 3). This type of RNase utilizes inorganic phosphate as the nucleophilic agent to attack the phosphodiester linkage. Phosphorylase-type RNases, although interesting, have not been very well characterized in mammalian tissue.

Categorization of ribonucleases by their specificity and/or mechanism of action may tell us something about the enzymatic properties of the ribonuclease, but it does not address the metabolic role of the ribonuclease. Therefore, rather than listing groupings of RNases based on their catalytic properties, this article will focus on the metabolic roles of ribonucleases. In this regard, cleavage of a phosphodiester bond by a ribonuclease can be considered as playing one of two possible roles metabolically. Ribonuclease cleavage may result simply in the degradation of an RNA species where the resulting products of degradation are used to replenish ribonucleotide pools, or cleavage may instead affect the synthesis or maturation of an RNA species resulting in formation of a biologically functional molecule.

The purpose of this article is to give the nonspecialist an overall view of the status of information relating to ribonuclease function in mammalian tissue. Descriptions of specific ribonucleases will be limited; therefore, references in the Bibliography are intended

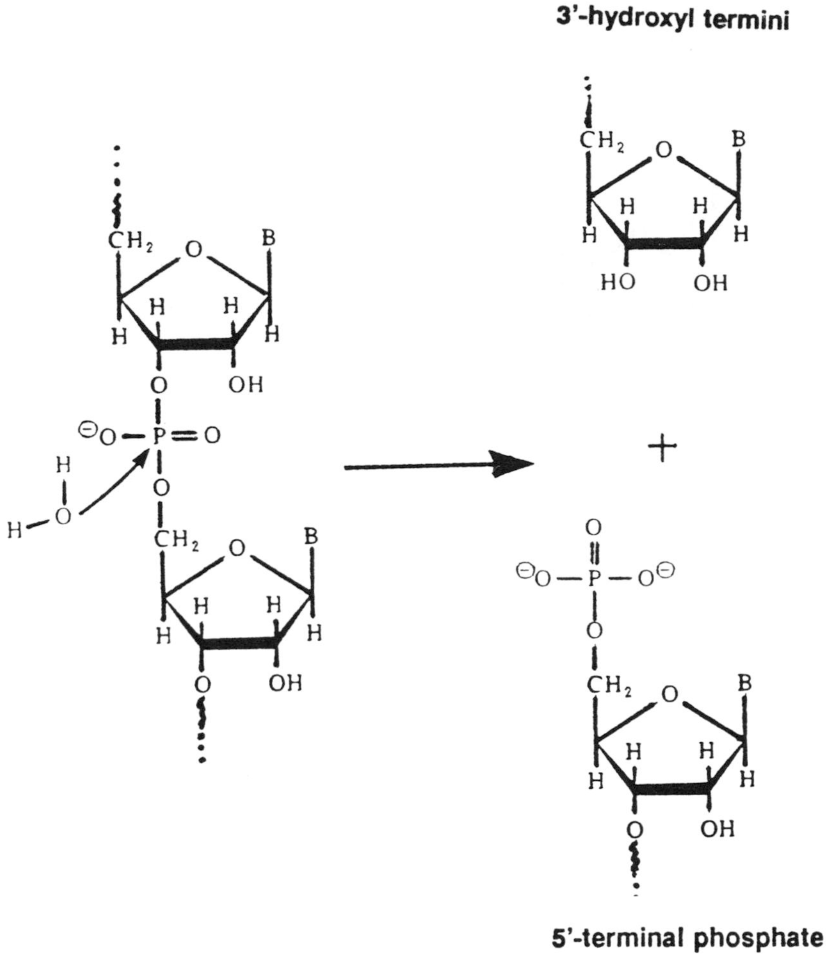

3'-hydroxyl termini

5'-terminal phosphate

FIGURE 2 Phosphodiesterase-type RNases.

to provide a resource for more detailed information relating to known RNases. For convenience, specific examples of RNases that are discussed in this article will be referred to by their most common names, and these examples were chosen not to be all-inclusive but rather to provide a general understanding of the metabolic role of ribonucleases in mammalian tissue and physiology.

II. PYRIMIDINE-SPECIFIC RIBONUCLEASES

A. Pancreatic-like Ribonucleases

In mammalian tissue, two major groups of *pyrimidine-specific ribonucleases* have been distinguished as *secretory* and *nonsecretory* RNases. The secretory RNases have been extensively studied, resulting in a large volume of literature concerned with their characterization. This group of designated secretory RNases can be distinguished by their preferential hydrolysis of the homoribopolymers poly(C) and poly(U), and a pH 7.5 optimum for the hydrolysis of RNA. Secretory RNases, exemplified by pancreatic ribonuclease, are found mainly in secretory organs such as pancreas and submaxillary glands.

In contrast, nonsecretory RNases are found in substantial amounts in tissue such as liver, kidney, and spleen and are generally less well characterized. Nonsecretory RNases hydrolyze RNA most effectively at pH 6.5 and show lower activity toward poly(C) than do secretory RNases. Both groups, however, still share extensive amino acid sequence homology and common properties such as the endonucleolytic cleavage of RNA and 2′,3′-cyclic pyrimidine products.

Although extensive work has been dedicated to the characterization of pyrimidine-specific RNases, their

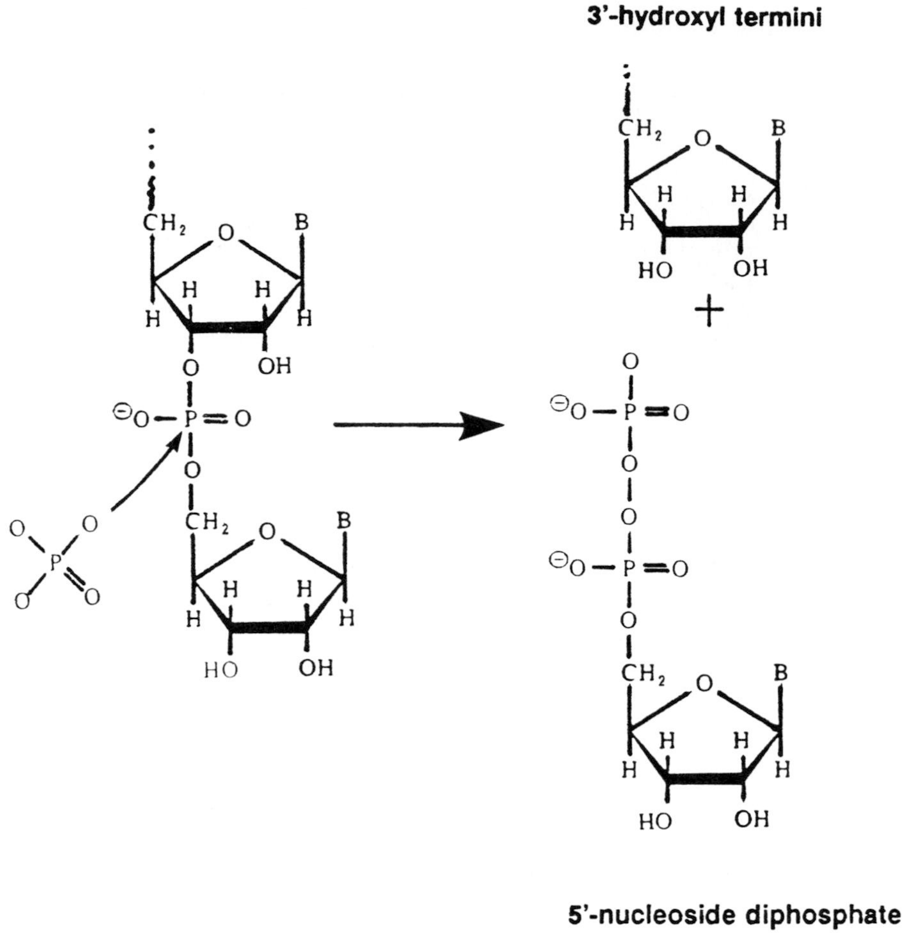

3'-hydroxyl termini

5'-nucleoside diphosphate

FIGURE 3 Phosphorylase-type RNases.

metabolic role in mammalian tissue has not been well established. [*See* Purine and Pyrimidine Metabolism.]

B. Inhibitor of Pancreatic-like RNases

The regulation of pancreatic-like RNases in mammalian tissue may be correlated with a 50-kDa protein inhibitor that binds to the RNase, rendering it inactive. The most extensively studied inhibitor comes from human placenta and is often referred to as *placental RNase inhibitor* (PRI). This protein inhibitor forms a very tight complex with *pancreatic-like RNases* but can be rapidly inactivated by sulfhydryl modification reagents such as *N*-ethylmaleimide to restore RNase activity. *In vivo,* inactivation of the specific RNase inhibitor has been suggested as a way of controlling the latent form (active form) of nonsecretory RNase activity.

Nevertheless, the intracellular inhibitor may have distinctly different functions depending on the tissue and tissue growth in regulating RNA degradation. For example, a number of observations suggest that the inhibitor may serve to protect a cell from noncytoplasmic RNases. Extracellular pancreatic-like RNases may enter the cytoplasmic compartment of the cell and arrest protein synthesis by virtue of their cleavage of ribosomal RNA or tRNAs. The cytotoxic effect of these extracellular RNases upon entry to the cytoplasm could therefore be prevented by the presence of the intracellular inhibitor.

C. Angiogenin

The primary structure of tumor-derived *angiogenin,* a protein that stimulates formation of blood vessels, was shown to be very similar to that of the pancreatic ribonucleases. In keeping with this similarity, the angiogenic protein was also shown to possess ribonuclease activity. The ribonuclease activity of angiogenin, however, could be distinguished from that of human pancreatic RNase, particularly in its limited ability to cleave either 18*S* or 28*S* rRNA, and the lack of activity under conditions of a standard pancreatic RNase assay. Interestingly, the cytotoxicity of angiogenin appears to relate to its selective degradation of tRNAs once taken up by the cell.

Both the RNase activity and angiogenin activity of this protein are inhibited by human placental RNase inhibitor. Therefore it has been proposed that similar tissue inhibitors may regulate the activity of angiogenin *in vivo*. Inhibition of angiogenin RNase activity by placental RNase inhibitor also blocks the angiogenic activity, suggesting that the two activities are interrelated.

D. Eosinophil-Derived Neurotoxin and Eosinophil Cationic Protein

When injected intrathecally into rabbits, both the *eosinophil-derived neurotoxin* and the *eosinophil cationic protein* produce the Gordon phenomenon, a neurological syndrome characterized in the rabbit by stiffness, ataxia, muscle weakness, and muscle wasting. Interestingly, both the eosinophil-derived neurotoxin and eosinophil cationic protein were first associated by their structural similarity with human pancreatic RNases; however, further analysis of sequence suggested that they were better related to the nonsecretory ribonucleases. In addition, the observed amino acid similarities also suggested that eosinophil-derived neurotoxin and eosinophil cationic protein are encoded by genes that evolved by the process of gene duplication giving rise to the variety of proteins in the *pyrimidine-specific ribonuclease supergene family*.

E. Cytotoxic Effects of Members of the Pancreatic-like RNases

Members of the pancreatic-like ribonuclease family appear to have diverse activities, some of which may be related to host defense and others to physiological cell death. Recently, the *cytotoxic* properties of some of these pancreatic-like RNases have been investigated with regard to the therapeutic treatment of cancer. Interestingly, a member of this pancreatic-like RNase superfamily, isolated from frog (*Rana pipiens*) oocytes and now called *onconase,* is in stage II of testing as an antitumor agent. Similar to angiogenin, onconase apparently inhibits protein synthesis by depleting the cell of tRNAs. In addition, bovine seminal RNase, the only dimeric protein of the superfamily, has been investigated for its selective cytotoxic action toward malignant cells. The essential elements for bovine seminal RNase that have been demonstrated for antitumor action are a functional catalytic center and dimerization. Bovine seminal RNase has been shown to inhibit both *in vivo* and *in vitro* the growth of epithelial tumor cells with no appreciable effects on normal cells.

The cytotoxic potential of these RNases has also been explored indirectly by coupling the RNase to a cell binding ligand for the purpose of engineering them into cell-type-specific cytotoxins. For example, the cytotoxic potential of bovine pancreatic RNase A was explored by coupling it to transferrin or to antibodies directed to the transferrin receptor. These RNase hybrids were shown to be a thousand times more cytotoxic to human erythroleukemia cells *in vitro* than the uncoupled RNase.

III. RIBONUCLEASE INVOLVEMENT IN RNA STABILITY

Although transcriptional regulation may often be the major determinant of gene expression, the stability of an RNA molecule can ultimately determine how long it can function. For this reason the rate at which an RNA is degraded can serve as a major control point in the regulation of gene expression in mammalian tissue. In fact, the wide diversity in decay rates of RNA seen in mammalian tissue appears to be determined to a large extent by interactions between factors, including endonucleases and internal mRNA structures.

The existence of RNA stability factors supports the idea that RNA decay rates may be determined by multiple interactions. In fact, it has been suggested that the regulation of *RNA stability* may be affected by two types of interactions. First, sequence within the RNA would determine the "intrinsic" or "unregulated half-life," independent of regulatory factors. These sequences might influence the susceptibility of the RNA to ribonucleases. In contrast, some sequences within the RNA might interact with regulatory factors that modify the intrinsic half-life. In fact, in the cell it may well be a combination of both intrinsic and regulated processes that ultimately determines the overall decay rate of an RNA molecule.

A. mRNA Decay

The half-lives of eukaryotic mRNAs vary greatly. Whereas globin mRNA has a half-life in excess of 50 hr, the half-life of the *c-myc* oncogene mRNA is less than 20 min. Many transiently expressed cellular growth regulators are also encoded by mRNAs with half-lives as short as 1 hr or less. Such short but regulatable half-lives, in conjunction with changes in transcription rate, allow mRNAs to be produced in a transient burst or to reach a new steady-state level very rapidly.

Recent studies indicate that particular 5′- or 3′-untranslated sequences, specific mRNA degradative activities, and cellular ribonuclease inhibitors regulate mRNA decay. Although the specific RNases responsible for mRNA turnover have not been identified, there is some recent evidence to suggest that eukaryotic mRNAs may be degraded in a $3' \rightarrow 5'$ manner by processive exoribonucleases analogous to the prokaryotic enzymes *RNase II* and *polynucleotide phosphorylase*. Several studies also suggest that the rate-limiting step in mRNA degradation results from an endonucleolytic cleavage near the 3′ end of the mRNA that leaves the mRNA susceptible to $3' \rightarrow 5'$ exonucleolytic attack. An endonuclease that attacks the 3′ stem-loop structure of histone mRNAs may be such an enzyme. There are also examples for which the decay rate of a specific mRNA was found to be modulated by hormones and other physiological effectors related to cellular differentiation, DNA replication, and viral infection.

Some features of the mRNA decay process in mammalian cells suggest the involvement of a ribosome-bound ribonuclease that responds to very specific signals. One clear example of the relationship between translation and mRNA decay is the observation that nonsense mutations in a gene can reduce the abundance of the mRNA transcribed from that gene in a process that has been termed *nonsense-mediated mRNA decay*. Nonsense mutations cause premature termination of translation, and factors associated with nonsense-mediated mRNA decay appear to scan the mRNA from the 3′ side of the nonsense condon (the premature stop signal). This interaction and scanning promote an altered RNP structure that renders the mRNA susceptible to cleavage very near the 5′ terminus of the mRNA, leading to decapping. Subsequently, the uncapped nonsense-containing mRNA is degraded by a $5' \rightarrow 3'$ exoribonuclease.

B. RNase Inhibitor

Many mammalian tissues have been found to contain an inhibitor of pancreatic-like RNases (see Section II,A). Thus, it has been suggested that the association and dissociation between RNases and their inhibitors play a role in regulating the rate of mRNA decay. HeLa cells possess a factor that resembles human placental RNase inhibitor in that the mRNA-protecting activity is effective against pancreatic-like RNase activities and that treatment of a HeLa extract with *N*-ethylmaleimide completely destroys the protective activity of the inhibitor. Interestingly, though, purified human placental RNase inhibitor cannot inhibit the ribonuclease activity responsible for mRNA decay found in the cytoplasmic extract of HeLa cells.

C. The (2′,5′)-Oligoadenylate Synthetase–Nuclease System

When exposed to viruses or certain chemical inducers, mammalian cells try to protect themselves from further infection or damage by producing interferons. Interferons, through their interaction with specific cellular receptors, mediate a wide variety of biochemical events that result in the modification of cellular functions as part of the protective response. Two interferon-induced enzymes, the *2′,5′-oligoadenylate synthetase* and the p1/eIF-2α protein kinase, have been identified that play important roles in inhibiting viral and cellular protein synthesis. These interferon-induced enzymes must be activated, and both synthetic and natural double-stranded RNAs (dsRNA) can fulfill the activation requirement *in vitro*.

Three enzymes have been shown to play important roles in the 2′,5′-oligoadenylate synthetase system ($2',5'$-A_n system): a synthetase that catalyzes the formation of the novel oligonucleotides possessing 2′,5′-phosphodiester bonds; an endoribonuclease (*RNase L*) that is activated by $2',5'$-A_n; and a phosphodiesterase that catalyzes the hydrolysis of oligonucleotides possessing 2′,5′-phosphodiester bonds. The only biological function known for $2',5'$-A_n is its role in the activation of RNase L.

Activated RNase L catalyses the cleavage of both viral and cellular RNA on the 3′ side of -UpXp- sequences (predominantly UA, UG, and UU) to yield products with -UpXp- at the 3′ termini. Many different types of single-stranded RNA substrates are cleaved by activated RNase L, including synthetic polyuridylic acid and various natural RNAs; however, ribosomal RNA is the best-characterized type of cellular RNA observed to be cleaved both *in vitro* and *in vivo*.

RNase L has an estimated molecular weight of 75,000 to 85,000 and binds 2′,5′-A_n. This enzyme has been identified in cytoplasmic extracts prepared from a variety of mammalian sources. The level of latent 2′,5′-A_n-dependent RNase L present in most types of cells in culture differs less than twofold between untreated and interferon-treated cells. However, culture conditions have been described in which the functional activity of the active form of RNase L is regulated by parameters other than the availability of 2′,5′-A_n. For example, in NIH 3T3 cells, the 2′,5′-A_n-dependent RNase L can be independently induced by interferon or by growth arrest. RNase L has also been observed to be regulated during cell differentiation.

Recently, the mechanism by which 2′,5′-A_n activates RNase L has been elucidated. The results suggest that the 2′,5′-A_n-dependent RNase L molecules form homodimers upon binding 2′,5′-A_n. In the absence of 2′,5′-A_n, a region of RNase L that contains nine ankyrin-like repeats interacts with the catalytic domain to block ribonuclease activity. Thus, the principal function of 2′,5′-A_n is to alter the structure of RNase L in order to allow formation of the active, dimeric form of the enzyme.

IV. RNA PROCESSING RIBONUCLEASES

In eukaryotic cells, most nuclear transcripts are synthesized as larger precursors that must undergo a series of cleavage and/or modification reactions in order to form smaller functional RNA species. In some cases, the shortening of the primary transcripts results from direct cleavage of phosphodiester bonds; in other cases, the shortening results from a series of intramolecular rearrangements that collectively excise the nonexpressed sequences (*introns*) and ligate the flanking expressed regions (*exons*). Some processing steps appear to be completely dependent on the occurrence of earlier steps and may be governed by the order in which the substrate recognition signals become available. Recognition sites may also be composed of unique combinations of secondary and/or tertiary structure, and may result from packaging in the form of a *ribonucleoprotein particle* (RNP). It is also important to consider that in many instances a large proportion of the primary transcript is discarded during posttranscriptional processing. Therefore, a ribonuclease may play either a degradative or processing role, and perhaps in some instances both. The involvement of ribonucleases in the maturation of eukaryotic transcripts is of primary importance, particularly since these cleavage events may serve as critical signals for the overall maintenance of RNA expression in mammalian tissue.

Difficulties in specifically defining processing ribonucleases have been overcome by our ability to develop adequate *in vitro* systems that mimic the *in vivo* situation. In addition, *antibodies* that specifically cross-react with components of nuclear processing systems, obtained from patients suffering from systemic lupus erythematosus (SLE), have provided extremely powerful tools to elucidate the structure of processing complexes. Our knowledge of RNA processing has therefore expanded rapidly in the last few years, and a step-by-step analysis of the components and mechanisms involved in RNA processing is being pursued.

A. Ribosomal RNA Processing

Synthesis, processing, and assembly of ribosomal RNA occurs in nucleoli. Transcription of mammalian ribosomal DNA by *RNA polymerase I* yields a large primary transcript of approximately 14 kilobases, which is subsequently cleaved to yield the mature 18*S*, 5.8*S*, and 28*S* *ribosomal RNA* (rRNA) species. The temporal sequence of processing events that lead to the formation of functional rRNA species has been extensively studied, and the processing events that release the 5′ and 3′ termini of the mature 18*S*, 5.8*S*, and 28*S* rRNA species often occur in a polar fashion, from the 5′ to the 3′ end of the transcript. However, the precise order of the cleavages and intermediates generated may vary with tissue growth rates and between species.

Processing of the ribosomal RNA precursor is a nonconservative process since nonutilizable sequences released from the rRNA precursor during maturation are degraded. Moreover, when protein synthesis is inhibited, synthesis of the precursor RNA still continues, but little or no mature rRNA emerges from the nucleolus. The precursor or even partially processed precursor rRNA is degraded when it cannot be assembled into mature ribosomes due to the lack of ribosomal proteins. In contrast, the synthesis of ribosomal RNA in resting lymphocytes appears to exceed the capacity to be utilized in ribosome assembly. Thus, significant turnover or "wastage" of *precursor rRNA* occurs in the nucleolus until the lymphocytes are stimulated to divide, which increases the demand for functional ribosomes. Taken together, these types of observations support the proposal that ribonucleases play an important role not only in the processing the precursor rRNA, but also in the degradation of nonutilizable pre-rRNA transcripts.

Although the sites of rRNA processing are reasonably well defined, the mechanism and proteins involved have proven elusive. Development of an *in vitro* system for characterizing these ribosomal RNA maturation processes, similar to those used to analyze the processing of mRNA precursors, has lagged behind. In fact, there are only a few reports where limited processing of precursor rRNA has been achieved *in vitro*.

Nevertheless, from these limited studies several activities have been implicated in mammalian ribosomal RNA processing. A single-strand specific endoribonuclease isolated from mouse nucleoli has been shown to specifically recognize and cleave at an early processing site (+650 site) in mouse ribosomal RNA (comparable to the +414 site in human ribosomal RNA). In addition, two distinct activities have been characterized that together carry out the processing step that produces the mature 5′ end of human 18*S* rRNA. The first activity, performed by an endonuclease, cleaves one or two nucleotides upstream of the mature 5′ end of the 18*S* rRNA sequence in the precursor. The second activity, performed by a 5′ → 3′ exonuclease, trims the remaining nucleotide(s) to produce the mature 5′ end.

Most recently, a nuclear-encoded ribonucleoprotein endoribonuclease, known as RNase MRP, has been shown to function in both the nucleolus and the mitochondria. In mitochondria, RNase MRP catalyzes the endonucleolytic cleavage of the RNA primer during mitochondrial DNA replication. In the nucleus, where most of the RNase MRP activity resides, fractionation studies demonstrated that the MRP RNase is localized to the nucleolus and is associated with pre-rRNA processing complexes. Its precise role in ribosomal RNA processing has not been elucidated, but early studies suggest that it may be involved in cleavages that release the 5.8*S*–IVS2–28*S* rRNA intermediate from the 18*S* rRNA sequence.

B. Maturation of mRNA

Because one of the first characteristics used to distinguish them from other RNAs in the nucleus was the heterogeneity of their sizes, *RNA polymerase II* transcripts in the nucleus were known as *heterogeneous nuclear RNA* (hnRNA) molecules. Many of these transcripts leave the nucleus as messenger RNA (mRNA) molecules, and before leaving are covalently modified at both their 5′ and 3′ ends in ways that clearly distinguish them from transcripts made by the other cellular RNA polymerases. In addition, most primary transcripts contain regions that are not found in the mature functional mRNAs. These regions are called introns and are removed during the maturation of the transcript to mRNA by a process termed *splicing*. Unlike typical phosphodiester cleavage reactions, the splicing reaction involves an intramolecular rearrangement of the phosphodiester bond. These reactions are catalyzed by a structure called a *spliceosome*, which assembles the boundaries of the intron sequence and coding sequences (exons) into a complex that permits the specific rearrangements to occur. The spliceosome is composed of small *ribonucleoprotein particles* called snRNPs and other protein factors. The order and assembly of the snRNPs and protein factors into the spliceosome and their participation in splicing have, in part, been worked out; however, the actual catalytic participants have not. Noncoding sequences (introns) that are removed during splicing are very rapidly degraded, presumably due to the uncapped nature of their 5′ ends, suggesting participation of a 5′ → 3′ exoribonuclease that recognizes free uncapped 5′ ends to initiate degradation.

Capping of the 5′ ends of transcripts occurs either concurrently or immediately after transcription and involves the condensation of GTP with the diphosphorylated 5′ end of the initial transcript. The guanosine cap is then methylated at the 7 position to complete this modification. The function of the 7-methylguanosine cap is not completely known, but seems to serve an essential role in the efficient assembly of a translation initiation complex and in protecting a transcript from 5′ → 3′ exonucleolytic degradation.

The 3′ ends of most RNA polymerase II transcripts are defined not by termination of transcription, but by a second modification in which the growing transcript is cleaved at a specific site and a poly(A) tail is added by a separate polymerase to the cut 3′ end. For mammalian mRNAs, this process occurs posttranscriptionally in a three-step reaction. First, the primary transcript is cleaved by an endonuclease at the polyadenylation site, and then the upstream cleavage product slowly receives an oligo(A) tail of 10 to 12 nucleotides. In a third step, the oligo(A) tail is elongated in a fast and processive reaction, resulting in approximately 200 adenosine residues being added to the newly made 3′ end.

Sequence elements are required for the polyadenylation reaction. The highly conserved hexamer sequence (AAUAAA) located 10 to 30 nucleotides upstream of the poly(A) addition site designates and positions the polyadenylation complex. Less well defined GU-rich and U-rich sequences located downstream of the poly(A) site have also been shown to influence 3′ end formation; however, it has not been demonstrated that these downstream sequences play a role specific for each pre-mRNA species.

Neither the *poly(A) polymerase* activity nor the endoribonuclease activity responsible for providing the available 3′ end separates on extensive purification, suggesting that the component that contains the poly(A) polymerase activity is also required for the cleavage reaction. Similar to the splicing reactions, a variety of experiments have produced evidence indicating that snRNPs are involved in the cleavage–*polyadenylation* reaction. In particular, RNA fragments containing the AAUAAA sequence have been immunoprecipitated by anti-snRNP antibodies and by antibodies directed against the trimethyl cap structure of U-type small nuclear RNAs.

The function of the poly(A) tail is not known for certain, but there is evidence to support a role in the export of mature mRNA from the nucleus, in mRNA stability (see Section III,A), and in the initiation of translation.

C. RNA Polymerase III Products

Like RNA polymerase II, *RNA polymerase III* acts on a number of different cellular genes. It catalyzes the synthesis of ribosomal *5S RNA*, all *tRNAs*, and certain other small RNAs. Similar to other nuclear RNAs, the tRNAs are synthesized in precursor forms that are processed posttranscriptionally to yield mature tRNA molecules. Precursor tRNAs are longer than the final products and the extra polyribonucleotides are removed from both the 3′ and 5′ ends. The maturation of tRNA also entails changes in the purine and pyrimidine bases. Similarly, the 5*S* rRNA is synthesized as a larger precursor and cleaved to yield an approximate 150-nucleotide product.

In mammalian cells, most tRNA genes are transcribed as monomers and only a few extra nucleotides on either side of the mature sequence are removed in processing. *RNase P* removes, with great accuracy, the extra nucleotides from the transcripts of tRNA genes to yield the correct 5′ terminus of the mature tRNAs. Although this activity was first defined in bacterial cells, recent evidence has supported the presence of this activity in human cells as well.

RNase P is a particularly interesting enzyme because the catalytic element of the enzyme is RNA, not protein, and the reaction catalyzed is not a *transesterification* (rearrangement reactions) like the splicing reactions involved in mRNA processing but rather a hydrolytic reaction. No covalent linkage is formed between the enzyme and the substrate, nor are there any intramolecular bonds formed transiently in the substrate during the reaction.

Studies of RNase P from mammalian cells have proceeded much more slowly than similar studies carried out in bacteria because the enzymatic activity is more labile and less abundant. Anti-RNase P antibodies, derived from patients with SLE, were used to initially identify the RNA species that copurifies with RNase P derived from human cells.

V. RIBONUCLEASES INVOLVED IN RNA OR DNA SYNTHESIS

One of the unique specificities observed for ribonucleases is the recognition of the RNA strand of an RNA/DNA hybrid structure. Enzymatic activities capable of specifically degrading the RNA complement of the hybrid RNA/DNA duplex have been designated as *RNase H*-type activities, where the letter "H" refers to hybrid. RNA/DNA hybrid structures are typically transient structures and are observed during transcription and during *DNA replication* at both the origin of replication and the replication fork. Also, the unique pathway of replication of retroviruses necessitates the formation of an RNA/DNA hybrid structure in order for the virus to convert its genetic information from RNA to DNA.

A. Cellular Replication

During replication, *DNA polymerase* synthesizes continuously along the parental template strand in the 5′ → 3′ direction; however, the lagging strand DNA polymerase requires a primer to synthesize each short, newly replicated DNA fragment. A process is therefore required not only to produce the RNA primer but also to remove the base-paired primer in order to permit repair and ligation of the newly synthesized DNA fragments. Synthesis of *RNA primers* involves an enzyme called *primase*. These primers are about

10 nucleotides long in eukaryotes. They are made at intervals on the template lagging strand and are elongated by the DNA polymerase to initiate each newly replicated DNA fragment (Okazaki fragment). The synthesis of each *Okazaki fragment* ends when the DNA polymerase runs into the RNA primer attached to the 5′ end of the previous adjacent fragment. The role for RNase H, then, is in the removal of the RNA primer, hydrolyzing the RNA component of *the primed DNA fragments*.

The mechanism for removal appears to be a two-step process. RNase H makes a single cut in the primer RNA, one nucleotide 5′ of the RNA–DNA junction. Subsequently, a double-strand specific 5′ → 3′ exonuclease removes the remaining monoribonucleotide. In contrast to the viral RNase H activity, the mammalian RNase H activity is an endonuclease releasing di- to oligonucleotides containing 5′-terminal phosphate groups. When the RNase H activity removes the RNA primer, it is replaced with DNA, and *DNA ligase* then joins the 3′ ends of the adjacent DNA fragments to produce a continuous DNA chain from the many newly synthesized DNA fragments made on the lagging strand. Consistent with its proposed role in DNA replication, RNase H activity has been reported to rise and fall in concert with replication in whole cells. [*See* Nucleotide and Nucleic Acid Synthesis, Cellular Organizations.]

B. Retrovirus Replication

Ribonuclease H also turns out to be a ubquitous activity of retrovirus and is an integral component of the virion RNA-dependent DNA polymerase known as reverse transcriptase. The infecting RNA of retroviruses serves as a template for the synthesis of a single-stranded DNA complement. The activity responsible for the synthesis of DNA from an RNA template during retroviral infection is the viral-specific enzyme, reverse transcriptase. *Reverse transcriptase* has three associated enzymatic activities: it copies an RNA molecule to yield double-stranded DNA/RNA; it copies a single strand of DNA to form double-stranded DNA; and it degrades RNA in the DNA/RNA hybrid. The latter activity is the associated RNase H activity. Recent evidence suggests that the DNA polymerase and RNase H activities reside within separate domains of a single polypeptide chain. Viral RNase H is an exonuclease, which catalyzes the processive attack of the RNA strand in either a 5′ → 3′ or 3′ → 5′ direction to liberate oligonucleotides or 2 to 8 residues with 5′-terminal phosphate groups. RNase H is also associated with other classes of tumor viruses, but in some instances its viral origin has not been firmly established.

C. Transcription

Despite extensive investigation, the cellular functions of the various RNase H-type activities in human tissue have not been clearly established. One particular role for an RNase H-type activity, not initially anticipated, was the involvement in transcription. Recent studies have shown that there are a large number of genetic elements and protein factors needed in conjunction with purified RNA polymerase II for accurate and efficient transcription of different genes. Although the majority of known factors are implicated in correct initiation of transcription, additional factors appear to be important in the elongation and termination phases of transcription as well.

One such protein factor that influences transcriptional elongation by RNA polymerase II is factor SII. This factor is a 35- to 38-kDa polypeptide that has been purified from human tissue and shown to facilitate RNA polymerase II readthrough at transcription arrest signals. Although the exact mechanism by which SII affects readthrough is not well understood, the properties of SII interaction with RNA polymerase II elongation complex have suggested the following mechanism. SII binds to RNA polymerase II and stimulates a ribonuclease activity that is part of the RNA polymerase II elongation process. This ribonuclease shortens the nascent transcript from the 3′ end by a few nucleotides. Subsequently, the RNA polymerase II elongation complex reextends the nascent RNA chain in a process referred to as *cleavage–resynthesis*. Apparently, the alterations in the RNA polymerase II elongation complex that accompany the cleavage of the transcript at the 3′ end in some manner permit renewed chain elongation. This suggests that the overall synthesis of newly synthesized transcripts does not necessarily proceed in a direct manner, and that transcriptional elongation requires ribonuclease involvement in order to promote continued transcription. [*See* DNA and Gene Transcription.]

BIBLIOGRAPHY

Belasco, J., and Brawerman, G. (1993). "Control of Messenger RNA Stability." Academic Press, San Diego.

Deutscher, M. P. (1988). The metabolic role of RNases. *Trends Biochem. Sci.* **13**, 136–139.

Eichler, D. C., and Craig, N. (1994). Processing of eukaryotic ribosomal RNA. *Prog. Nucl. Acid Res. Mol. Biol.* **49**, 197–239.

Katz, R. A., and Shalka, A. M. (1994). The retroviral enzymes. *Annu. Rev. Biochem.* **63,** 133–173.

Pestka, S., Langer, J. A., Zoon, K. C., and Samuel, C. E. (1987). Interferons and their actions. *Annu. Rev. Biochem.* **56,** 727–777.

Reines, D., Ghanouni, P., Gu, W., Mote, J., Jr., and Powell, W. (1993). Transcription elongation by RNA polymerase II: Mechanism of SII activation. *Cell. Mol. Biol. Res.* **39,** 331–338.

Roberts, R. J., Linn, S. M., and Lloyd, S. (eds.) (1993). "Nucleases," end Ed. Cold Spring Harbor Laboratory Press, Cold Spring Harbor, New York.

Sachs, A. B. (1993). Messenger RNA degradation in eukaryotes. *Cell* **74,** 413–421.

Sierakowska, H., and Shugar, D. (1977). Mammalian nucleolytic enzymes. *Prog. Nucl. Acid Res. Mol. Biol.* **20,** 59–130.

Wahle, E., and Keller, W. (1992). The biochemistry of 3′-end cleavage and polyadenylation of messenger RNA precursors. *Ann. Rev. Biochem.* **61,** 419–440.

Weickmann, J. L., and Glitz, D. G. (1982). Human ribonucleases: Quantitation of pancreatic-like enzymes in serum, urine, and organ preparations. *J. Biol. Chem.* **257,** 8705–8710.

Ribosomes

IRA G. WOOL
University of Chicago

GLOSSARY

Genome Complete set of genes in the organism

Homology Relation implying that the two proteins under consideration were derived from a common ancestral gene. Related proteins have a certain percentage of similar or identical amino acid residues at the same positions but are not necessarily homologous

Ribosomes Ribonucleoprotein particles that bind mRNA, aminoacyl-tRNA, and initiation, elongation, and termination factors and catalyze peptide bond formation; they synthesize the protein encoded in the mRNA, which is a transcript of a gene

Transcription Process in which the enzyme RNA polymerase copies the DNA of genes into the complementary RNAs that then serve either as stable transcripts (transfer and ribosomal RNAs) or as messenger RNAs, which encode proteins

Translation Conversion of the amino acid sequence encoded in mRNA into a protein; the process occurs on ribosomes

RIBOSOMES ARE RIBONUCLEOPROTEIN CELLULAR organelles that link the genotype to the phenotype by translating a sequence of nucleotides in a transcript of a gene into a sequence of amino acids in a protein. They catalyze protein synthesis in all of the organisms in our biosphere; indeed, they are at one and the same time universal, essential, and complicated. The ribosomal proteins and the ribosomal nucleic acids provide the specific binding sites (for messenger RNA, transfer RNA, and initiation and elongation factors) and the catalytic activities required for the synthesis of a protein. The grand imperative is to know the structure of ribosomes so as to be able to account for their function. This has as a prerequisite knowledge of the chemistry of the constituents. Human ribosomes are composed of two subunits, which are designated by their sedimentation coefficients: the smaller is 40*S* and the larger 60*S*. The subunits associate—they are held together by noncovalent bonds, perhaps by magnesium salt bridges—and form the functional 80*S* ribosome. The 40*S* subunit has a single molecule of RNA, designated 18*S* rRNA, and 33 proteins; the 60*S* subunit has 3 molecules of RNA—5*S*, 5.8*S*, and 28*S*—and 47 proteins. A great deal is known about prokaryotic (principally *Escherichia coli*) ribosomes: the primary structures of the nucleic acids and of the proteins; the secondary structures, from comparative sequence analysis, and preliminary proposals for the tertiary folding of the rRNAs; the binding sites for the ribosomal proteins on the rRNAs; the topography of the proteins from neutron scattering and from immune electron microscopy; and there are even preliminary, low-resolution X-ray crystallographic data for the structure of the ribosomal subunits. Less is known of eukaryotic ribosomes because their structure is more complicated, because of the difficulty of applying genetic analysis, and because of the lack of a means for reconstituting ribosomal subunits. Nonetheless, progress is being made with mammalian ribosomes including those from humans. What we know supports the contention that human ribosomes closely resemble in both structure and function the particles

ENCYCLOPEDIA OF HUMAN BIOLOGY, Second Edition, VOLUME 7. Copyright © 1997 by Academic Press. All rights of reproduction in any form reserved.

from other mammals. For this reason, this article draws on studies of both rat and human ribosomes.

I. STRUCTURE OF RIBOSOMES

It has become very important to determine the structure of ribosomes because such knowledge is believed to be essential for a rational, molecular account of the function of the organelle in protein synthesis. First, however, we must establish the sequences of nucleotides and of amino acids in the constituent nucleic acids and proteins.

A. The Sequences of Nucleotides and the Secondary Structure of Ribosomal RNAs

The sequences of nucleotides in the four species of RNA—5*S*, 5.8*S*, 18*S*, and 28*S*—have been determined for human ribosomes and for a large number of other eukaryotic species. Indeed, the sequences of nucleotides for some 2660 16*S*/18*S*-like and 258 23*S*/28*S*-like rRNAs have now been determined. This library of sequences provides the data for the determination of the secondary structure using comparative (phylogenetic) sequence analysis. In this procedure, one begins with the assumption that two rRNAs, say *E. coli* 16*S* and human 18*S*, have the same secondary structure. The primary structures are aligned using regions of sequence identity as a guide and putative helices are constructed. Compensated base changes in the nucleotide sequences in the helices are taken as evidence for the structure, whereas uncompensated changes leading to mismatches are evidence against it. Generally, at least two independent examples of compensated base changes are required to establish the existence of a helix. In principle, the phylogenetic method amounts to having the data from a preexisting genetic experiment in which all the organisms in the biosphere are considered pseudorevertants of the various mutations in rRNA genes that have occurred during evolution. Most of the structure derived by phylogenetic comparison has now been confirmed by direct experimental test, principally by modification of the rRNA in intact ribosomes with chemical reagents followed by identification of the altered nucleotides by the extension of primers complementary to nucleotides in specified regions of the RNA with reverse transcriptase. Modified bases, generally in single-stranded sequences, interrupt transcription by the enzyme. The conclusion is that the rRNAs from all species, including humans, can be folded into the same secondary structure.

B. The Sequences of Amino Acids in Ribosomal Proteins

An effort is being made to determine the sequences of amino acids in all of the proteins of mammalian ribosomes. The conviction that underlies this large undertaking is that the data are an absolute requirement for the resolution of the structure, and furthermore that all analytical and structural chemistry is laden with theory. That this information is already available for *E. coli* ribosomes in no way diminishes the importance of the enterprise. Indeed, each of the sets of amino acid sequences, mammalian and bacterial, is likely to enhance the value of the other just as subsequent sequences of rRNAs were increasingly valuable because they made possible phylogenetic comparisons and, hence, led to the secondary structures of the nucleic acids. One is not so naive as to believe that the determination of the primary sequences will lead inevitably to the structure of the particle or that the structure will tell us inevitably how the organelle functions in protein synthesis. However, it is a near certainty that without the basic chemistry it will be impossible to fully comprehend the structure and to fully account for function. There are other uses of the sequences of amino acids in ribosomal proteins: they may help in understanding the evolution of ribosomes, in unraveling the function of the ribosomal proteins, in fathoming the reasons for the discrepancy between the number of ribosomal proteins in *E. coli* (54) compared to that in eukaryotes (80), in defining the rules that govern the interaction of the ribosomal proteins and the rRNAs, and in uncovering the amino acid sequences that direct the ribosomal proteins to the nucleolus for assembly on nascent rRNA. Indeed, it is difficult to predict all of the uses for the data, and their greatest value may be for some purpose we cannot envision now.

The sequences of amino acids in 77 human ribosomal proteins have been determined (75 are complete and 2 are partial). Most of the human ribosomal protein amino acid sequences are a by-product of the human genome project and most were recognized by their near identity with rat sequences. It is likely that there are at least 3 additional proteins, L29, L34, and L36, since the rat homologs are known. [Ribosomal proteins are designated by their coordinates on two-dimensional gels; those of the small (40*S*) subunit are preceded by an S and numbered from left to right in successive tiers on the gel; large (60*S*) subunit proteins are preceded by an L.] If the bookkeeping is correct and no new ones are uncovered, human ribosomes have 80 proteins.

In humans there are two transcribed genes for the ribosomal protein S4, one on the X and a second on the Y chromosome. What is extraordinary is that the amino acid sequences encoded in the S4X and S4Y alleles differ at 19 of 263 positions; extraordinary since there are seldom more than a few amino acid differences in homologous mammalian ribosomal proteins and this great a deviation is more than one would expect to find in a comparison with *Xenopus laevis* proteins. The S4Y gene is located in the sex-determining region of the Y chromosome and S4X escapes X inactivation. Both genes are transcribed in human cells; mRNAs specific for S4X and S4Y can be demonstrated by Northern hybridization and ribosomes from male humans have 90% S4X and 10% S4Y. Thus there are male- and female-type ribosomes. [*See* X Chromosome Inactivation.]

It is possible to describe an average human ribosomal protein: it has a molecular weight of 18,500 (the range is from 47,280 for L4 to 3454 for L41) and contains 164 amino acids (the range is from 421 to 25). The protein is very basic; it has a pI of 11.05 (the range is from 4.07 for P1 to 13.46 for L41) and contains 22 mol% arginine and lysine and only 9 mol% aspartic and glutamic acids. The protein is likely to contain a number of clusters of basic residues and several short amino acid repeats.

The amino acid sequences of related rat and human ribosomal proteins are near identical: the average for 72 comparisons of complete sequences is 99%; for 32 it is 100%. There are two isoforms of S24 derived from alternate splicing of the mRNA; the first is identical to rat S24 and the second lacks the carboxyl-terminal three amino acids.

C. Internal Duplications in Ribosomal Proteins

A number of human ribosomal proteins have duplications of amino acid sequences. For example, protein L7 has four repeats of a segment of 12 amino acids arranged in tandem near the NH_2 terminus. The repeats are very basic; four to six of the residues are lysyl or arginyl and there are no acidic amino acids. The occurrence of multiple, related, generally basic repeats in ribosomal proteins insinuates that they have functional significance, but there is no certainty as yet what this might be. Possibilities are that they play a role in the interaction with ribosomal RNA or that they are involved in directing the proteins to the nucleolus for assembly of ribosomes. There is evidence that information specifying the localization of proteins in the nucleus is encoded in short sequences of amino acids, although it is not yet possible to derive a consensus sequence nor to formulate general rules for the structure of the peptide. In the best-characterized examples, entry into the nucleus is contingent on a consecutive sequence of several basic amino acids often preceded by a prolyl residue. For example, the first 21 amino acids of yeast ribosomal protein L3, which is needed for entry into the nucleus, contain a sequence of this type. Several of the repeats in human ribosomal proteins also have these characteristics. Although there is no definitive experimental evidence available to evaluate the proposal, the repeats have sufficient similarity to known nuclear localization sequences as to require consideration of the possibility that they serve the same function.

D. The Characteristics of Ribosomal Protein mRNAs

Human ribosomal protein mRNAs are relatively short, on the average about 700 nucleotides in length, since they encode small proteins. A typical mRNA has in its 5′ UTR (untranslated region) an initial sequence of 4 to 20 pyrimidines followed by a GC-rich stretch of approximately 40 nucleotides and ending in an AUG codon for the initiation of translation. The 3′ UTR begins with a termination codon, has an AU-rich region of approximately 50 nucleotides, and ends with a long poly(A) tail.

Pyrimidine sequences are found at the immediate 5′ end of most, if not all, human ribosomal protein mRNAs. Many have the sequence CTTTCC or a variant ($CT_{2\text{ or }3}C_{1\text{ or }2}$). This motif may be a promoter that binds a trans-acting factor that accounts for the regulation (and perhaps the coordination) of the translation of ribosomal protein mRNAs. Only a relatively small number of nonribosomal protein mRNAs have a 5′ polypyrimidine sequence. Downstream of the polypyrimidine stretch, the 5′ UTR is GC-rich (60 to 80% G and C). Thus, the region is likely to be structured and, of course, GC helices can be formidable barriers to translation. How these are overcome is not known, but encumbered leader sequences strongly imply that the synthesis of ribosomal proteins is regulated at the initiation of translation. The initiation codon occurs most frequently in the context (A/G)

(A/C)CAUGG, a close approximation to the consensus for all vertebrates and to the experimentally derived optimum (A/G)CCAUGG. Codon usage in human ribosomal protein mRNAs is not unusual.

The frequency of the usage of the three termination condons in rat ribosomal protein mRNAs is: UAA, 56%; UGA, 28%; and UAG, 16%. This is distinctive since in vertebrates in general, and in other human proteins in particular, UGA is the most common termination codon. The more frequent use of UAA for termination may reflect the general character of the 3′ UTR of human ribosomal protein mRNAs, that is, they are more AU-rich than most vertebrate nonribosomal protein mRNAs. The generalization for human ribosomal protein mRNAs is that the 5′ UTR is GC-rich and the 3′ UTR is AU-rich. The 3′ UTR of most human ribosomal protein mRNAs has the hexamer AAUAAA, which directs posttranscriptional cleavage–polyadenylation of the 3′ end of the precursor mRNA. The poly(A) tract usually begins about 14 nucleotides from the signal hexamer.

There are multiple copies of mammalian ribosomal protein genes, the average being 12. However, in no instance (with the exception of the separate alleles encoding S4X and S4Y) has it been shown that more than one of the genes is functional; the presumption is that the others are retroposon pseudogenes. The intron-containing functional human ribosomal protein genes are being mapped; the chromosomal locations of 50 have been determined.

II. FUNCTION OF RIBOSOMES

The synthesis of proteins on ribosomes involves a series of reiterative cycles—the overall process is divided for convenience of description into initiation, elongation, and termination. It is easiest to begin with elongation, since it is prototypical for the others.

The growing nascent polypeptide is attached by an ester linkage to a transfer RNA (tRNA) just as amino acids are attached to tRNA in aminoacyl-tRNA (aa-tRNA). The latter is the substrate for protein synthesis. The peptidyl-tRNA (pep-tRNA) alternately occupies one of two adjacent sites on the ribosome termed P for peptidyl and A for aminoacyl. The A and P sites are the crucial postulates of a paradigm proposed by James Watson in 1964 to explain the biochemistry of protein synthesis; a coherent exposition would be difficult without invocation of the A and P sites, however, the regions of rRNA that comprise their structure have only recently been delineated.

The initial step in elongation is recognition: it is the process wherein a trinucleotide codon in the mRNA residing in the recognition region of the A site, that is, the decoding domain, specifies the binding of a cognate aa-tRNA by base-pairing with the complementary anticodon. The aa-tRNAs are synthesized in a coupled reaction. First, the amino acids are activated by formation of an enzyme-bound aminoacyladenylate with the liberation of inorganic pyrophosphate (PP_i); second, the aminoacyl group is attached to a cognate transfer RNA by a high-energy ester linkage and enzyme and AMP are released. These reactions are catalyzed by a set of 20 enzymes (aminoacyl-tRNA synthetases), one for each amino acid, and require ATP. This reaction is the most critical for the fidelity of protein synthesis—it is essential that the correct amino acid is attached to a cognate tRNA.

The recognition step requires an accessory nonribosomal protein termed elongation factor 1 (EF-1). EF-1 forms a ternary complex with aa-tRNA and GTP and this complex binds noncovalently to the ribosome in a codon-specific reaction. The aa-tRNA is positioned in the A site, GTP is hydrolyzed, and the EF-1 · GDP complex and P_i are released.

During peptidyl transfer, which is the second step in elongation, the nascent polypeptide in the P site replaces an ester bond to tRNA by a peptide bond with the α-amino group of the aa-tRNA in the A site; the peptide is increased in length by one amino acid and is transferred from the P to the A site. This reaction is catalyzed by peptidyl transferase, which is located in 60*S* ribosomal subunits but whose molecular identity is not known. The energy for the synthesis of this peptide bond is derived from the high group-transfer potential of the ester linkage in the aa-tRNA. Peptidyl transfer follows immediately on the binding of aa-tRNA to the A site.

The final step in elongation is translocation. The pep-tRNA is transferred from the A to the P site and the mRNA is dragged along with it. The deacylated tRNA left by the peptidyl transfer reaction is displaced from the ribosome and the next codon is brought into apposition with the A site. The ribosome is now ready to repeat the process in a recursive manner until the synthesis of the protein is completed. Translocation is catalyzed by a binary complex of GTP with a second elongation factor, EF-2. A possible mechanism for translocation is considered in some detail later.

The initiation of protein synthesis has the same biochemical scenario as elongation. The prime purpose served by the initiation reactions is the proper framing of the mRNA on the ribosome so that the initial AUG codon is translated in phase. The initiation codon specifies a methionyl residue and that amino acid is esterified to a tRNA that is used only to begin the synthesis of a protein. The initiator methionyl-tRNA forms a ternary complex with GTP and a particular initiation factor, termed eIF-2, rather than with EF-1 as occurs during elongation; the ternary complex binds to the 40*S* ribosomal subunit and then mRNA and 60*S* subunits join the complex. The initiation met-tRNA is base-paired to the AUG codon nearest the 5′ end of the mRNA and positioned in the P site. These latter reactions are promoted by a number of other initiation factors. The 80*S* ribosomal initiation complex is now competent to begin the elongation cycle.

Termination is the process by which the completed protein is releasd from the tRNA that bore the carboxyl-terminal amino acid. Chain termination requires a specific codon (UAA, UAG, or UGA) at the 3′ end of the reading frame in the mRNA and a specific factor termed release factor (RF). The hydrolysis of the ester linkage between the tRNA and the completed polypeptide is catalyzed by peptidyl transferase, the same enzyme that is responsible for peptide bond formation. The enzyme can form a peptide bond with an amino group as the nucleophilic agent attacking the ester bond of aa-tRNA or the enzyme can catalyze hydrolysis with water as the nucleophilic agent. It appears that the interaction of RF with the termination codon in the A site not only activates peptidyl transferase but favors water as the nucleophilic agent; the result is the release of nascent peptide.

III. EVOLUTION OF RIBOSOMES

It has been apparent for some time that individual ribosomal proteins from different eukaryotic species are derived from common ancestral genes. The sequences of amino acids are closest for related mammalian ribosomal proteins; for example, rat, mouse, and human proteins L32 have exactly the same sequence. The homologies are also obvious when the sequences of amino acids in human and yeast ribosomal proteins are compared. Despite the two species being evolutionarily distant eukaryotes, 65% can be correlated. The percentage of amino acid identities in the alignments ranges from 40 to 80%. The data are sufficient to provide confidence that most if not all of the eukaryotic ribosomal proteins are homologous and that it will be possible to establish a protein-to-protein correlation for all species.

Until relatively recently, common wisdom held that with only a few exceptions there are no close sequence similarities between eukaryotic and prokaryotic ribosomal proteins, and it was suggested that a relationship might only be found in their three-dimensional structure. The thinking with regard to this issue was strongly influenced by the results of the comparison of the structure of ribosomal RNAs. The theme for these molecules is conservation of secondary structure rather than primary sequence. This is not to say there are no conserved nucleotide sequences; most assuredly there are. The conserved sequences tend to be (but are not exclusively) in nonhelical regions, and there is evidence that they are important for function.

Although some had argued from the start for common ancestors for eukaryotic and eubacterial ribosomal proteins, the question was whether one was going to be able to trace the chemical spoor. It has been possible to accrue convincing evidence for the relationship only in the past few years. What has made the difference is the accumulation of near-complete sets of amino acid sequences of mammalian (rat and human) and eubacterial (*E. coli*) ribosomal proteins. Twenty-four human and *E. coli* ribosomal proteins can be correlated directly; for 8 additional pairs, an initial statistically weak correlation is reinforced by pairwise comparisons using the amino acid sequences of either the yeast or the archaebacterial counterpart or both.

What does homology of a significant number of human and eubacterial ribosomal proteins say about the evolution of the particle? It supports the argument that the ancestor of contemporary ribosomes was a ribonucleoprotein particle substantially similar to the organelle we know today—that most of the ribosomal proteins were added to the rRNAs before the divergence into eubacteria, archaebacteria, and eukaryotes. Any other scenario requires an unacceptable number of a priori assumptions.

Since eubacterial ribosomes have 54 proteins and eukaryotes approximately 80, it follows that some of the latter must be unique. It is now possible to make a preliminary identification of those proteins. Of the 80 human ribosomal proteins, 32 have a recognizable homolog in the eubacterial and archaebacterial kingdoms, another 17 have homologs in the archaebacterial but not the eubacterial kingdoms, and 31 ribosomal proteins appear to be unique to eukaryotes. This raises the following question: What was the reason for the addition of the extra proteins? Eukaryotic ribosomes are larger; they contain a greater number of proteins and an additional molecule of RNA. Albeit the latter is a difference more apparent than real; prokaryotes lack 5.8*S* rRNA but have a related sequence at the 3′ end of 23*S* rRNA. Thus prokaryotes have not gone to the trouble of processing the 5.8*S*-like RNA out of the primary rRNA transcript, whereas eukaryotes have. The individual protein and RNA molecules in eukaryotic ribosomes are, on the average, larger than those in prokaryotes. There is no explanation for these differences that is entirely convincing. Moreover, it is a paradox in the sense that eukaryotic and prokaryotic ribosomes perform precisely the same function (the catalysis of protein synthesis) and, more importantly, that they do so by appreciably the same biochemical means, even though the initiation of protein synthesis is more complicated in animal cells. The critical question is what was the evolutionary pressure for the accretion of the extra proteins and for the increase in the size of the proteins and of the RNAs?

One can, of course, look at the problem the other way around and ask why prokaryotic ribosomes are smaller. One answer is that they had to be streamlined. During log-phase growth, as much as 35% of the protein in a bacterium is ribosomal; the bacterium might have difficulty supporting a particle with as much protein as eukaryotic ribosomes have. Human ribosomes are not only larger than those of bacteria, but 50% of the mass, rather than 35%, is protein. The assumption implicit in this suggestion is that prokaryotic ribosomes at one time had a greater number of proteins, a number comparable to that in eukaryotes. Prokaryotic ribosomes then responded to selective pressure by discarding proteins and reducing the size of those that were retained—the advantage being a greater number of smaller ribosomes capable of performing the same function as their larger progenitors. This would have had to have happened after the divergence of primitive eukaryotes. The argument does not change the fundamental nature of the problem, however, it merely alters the way it is put. What ribosomal proteins could be dispensed with by prokaryotes without loss of function, and if the extra proteins could be discarded with impunity, why have they been retained by eukaryotes?

One reconciliation suggests that the earliest cells, progenitors of prokaryotes and eukaryotes, had nuclei and eukaryotic-like ribosomes and that a number of the proteins were needed, not for protein synthesis, but to manage the complicated traffic between nucleus and cytoplasm. The assembly of eukaryotic ribosomes is more complicated than that of prokaryotic ribosomes, at least in part because of the extensive intracellular traffic it requires. Ribosomal proteins are synthesized in the cytoplasm and then transported to the nucleus, where they are assembled on nascent rRNA transcripts; after the processing of the rRNA, which most likely occurs during assembly, the ribosomal subunits must pass through nuclear pores to get to the cytoplasm. Thus, eukaryotic ribosomal proteins are likely to contain amino acid sequences that serve as zip codes for nuclear and nucleolar localization; moreover, some ribosomal proteins may be involved only in the assembly process and some may be needed only for the processing of the 45*S* rRNA precursor to mature 18*S*, 28*S*, and 5.8*S* rRNAs; in addition, 5*S* rRNA, perhaps in a complex with ribosomal protein L5, must be recruited to the nucleolus from another site in the nucleus, where it is transcribed. Obviously, bacteria no longer require either the nuclear or nucleolar localization amino acid sequences, and may not require assembly, processing, or transport proteins.

Another rationalization of the paradox derives from the momentous discoveries concerning RNA by Thomas Cech and others. The initial finding was that ribosomal RNA has the capacity to mediate self-splicing and self-ligation, that is to say, the removal of a sequence of nucleotides (an intron) from a rRNA and the religation of the ends so as to reestablish the integrity of the molecule. Before this observation, nucleic acids were considered to be relatively inert chemically—capable of serving in information transfer by providing a template for polymerization reactions (as with messenger and transfer RNAs) and capable of providing the scaffolding for proteins in ribonucleoprotein particles and organelles (as in ribosomes) but not of catalysis. The latter was assumed to be the sole province of proteins.

Cech's experiments, which confounded common wisdom and even intuition, have changed forever our ideas about nucleic acids and, like all momentous discoveries, has had far-reaching consequences. Perhaps the single most important of these is the bearing his discovery has had on theories of molecular evolution. It is apparent now that RNA can be the repository of information that can be transmitted and that it can have enzymatic activity. This makes it possible, even likely, that the precellular biological world was dominated by RNA, that is, that RNA preceded DNA and protein. [*See* Human Genome and Its Evolutionary Origin.] It is of relevance that Cech's work lends support to the idea that ur-ribosomes had only RNA as had been proposed early on by Francis Crick and Leslie Orgel. The concept is that the basic biochemistry of protein synthesis (the binding of aminoacyl-tRNA, peptide bond formation, and translocation) is an intrinsic property of ribosomal RNA; that the ribosomal proteins, a later evolutionary embellishment, facilitate the folding and the maintenance of an optimal configuration of the ribosomal RNA and in this way confer on protein synthesis both speed and accuracy. Perhaps then eukaryotic ribosomes, which have far more RNA than prokaryotes (approximately 1300 more nucleotides), need more proteins to tune their nucleic acids. Yet there is still another scenario: If the ur-ribosome had only RNA and divergence of eukaryotes and prokaryotes occurred before the emergence of a complete set of ribosomal proteins, then one would not expect in the amino acid sequences the close resemblance that is apparent in the structure of the rRNAs; rather one would expect homology only of those ribosomal proteins that had evolved before divergence.

IV. BIOGENESIS OF RIBOSOMES: REGULATION OF THE SYNTHESIS OF THE MOLECULAR COMPONENTS AND THE ASSEMBLY OF THE PARTICLES

The assembly of ribosomes is an extraordinarily complex process that must test the cell's capacity for regulation and coordination. Biogenesis of human ribosomes requires the synthesis of equimolar amounts of the four rRNAs and of at least 80 proteins. (Human ribosomes are presumed, in the absence of definitive evidence, to have molar amounts of most proteins; the exceptions are likely to be the large subunit acidic phosphoproteins P1 and P2.) In addition, the formation of the rRNAs and of the ribosomal proteins has to be precisely balanced. What is more, human cells have as many as a million ribosomes. The coordinated synthesis in the nucleolus of 5.8*S*, 18*S*, and 28*S* rRNAs is easily accomplished since their genes are in a single large transcription unit. However, the synthesis of 5*S* rRNA occurs outside of the nucleolus, so its transcription must be coordinated with that of the other rRNAs and, in addition, it must be delivered to the nucleolus. The coordination of the synthesis of the ribosomal proteins is more complicated than that of the rRNAs: the transcription of 80 unlinked genes must be balanced. In exponentially growing cells, the balanced synthesis of the various ribosomal proteins is due primarily to their mRNAs being present in similar amounts and to their being translated with similar efficiencies. Thus, coordination of synthesis would seem to be determined in the first instance, and most importantly, by regulation of transcription of the individual ribosomal protein mRNAs, although there is also evidence for regulation of pre-mRNA processing and of translation. This implies that the source of the coordination resides in cis-acting elements of the promoters of ribosomal protein genes and in trans-acting factors.

In *E. coli,* the 54 ribosomal protein genes are organized into 20 operons; the largest has 11 genes and the smallest a single gene. Almost half of these genes, all of which are present in single copies, are at one locus (*Str*) on the chromosome; the remainder are scattered over the genome. The ribosomal protein operons are complex transcriptional units, with some containing nonribosomal protein genes and some having two promoters. In bacteria, the main control of ribosomal protein formation is not of transcription of mRNAs but is mediated by autogenous regulation of their translation. This regulation exploits the arrangement of the genes in operons and specific interactions of the ribosomal proteins with RNAs. To wit: certain ribosomal proteins when synthesized in excess function as repressors of the expression of their own operons by binding to a control region on the polycistronic mRNA rather than to rRNA. This follows from similarities in the structure of the control region of the polycistronic mRNA and the repressor ribosomal protein binding site on rRNA. Of course the affinity for the latter has to be greater than that for the former. As to the synthesis of rRNA, there probably is regulation (repression) of transcription by nontranslating ribosomes and perhaps by guanosine tetraphosphate (ppGpp).

Information on the organization of ribosomal protein genes in eukaryotes is just beginning to accumulate. What is certain is that both their structure and the regulation of their expression are different than in prokaryotes. The number of copies of the genes for individual ribosomal proteins varies among eukaryotes. In yeast there are either one or two copies, most often two, of the genes that have been analyzed (more than 30 of them); when there are two, they are both functional. In *Xenopus* there are two to six genes; no nontranscribed ribosomal protein genes have been identified. In mammals there are 7 to 20 copies of each, but only one is transcribed; moreover, the genes for different ribosomal proteins are not clustered.

A distinctive feature of the structure of ribosomal protein genes in mammals (most of the analyses have been of mouse genes, but the few human genes studied are very similar) is the lack of a canonical TATA box. In the region where one would be expected there is a 6- to 7-base-pair element that contains 5 or 6 AT pairs. Presumably, some aspect of this sequence pattern or the novel organization of the cap region, or both, assumes the function usually served by the TATA box, that is, to position RNA polymerase II for the initiation of transcription. Although precise initiation of transcription in the absence of a canonical TATA box is rare, it does occur; it has been described for a few viral and cellular genes. The latter are, like the ribosomal protein genes, members of the "housekeeping" class. Conceivably, a novel promoter structure may confer special regulatory properties on this class of genes. [*See* Genes.]

Another striking characteristic of mammalian ribosomal protein genes is the strucutre of the cap site, that is, the region where transcription is initiated. It is embedded in a $\geq$12-nucleotide stretch of pyrimidines flanked by blocks of greater than 80% GC content. The cap site pyrimidine tract has the motif 5′-CTTCCYTYYTC-3′; initiation of transcription is at the C at position 4 or 5. The promoter region of mammalian ribosomal protein genes contains, in addition, a number of cis-acting control elements, indeed, maximal expression requires 200 base pairs in the 5′ flanking region. Within this region there are at least five discrete elements that affect expression and that bind transcription factors. Although the general pattern of the structure of the promoters in individual mammalian ribosomal protein genes is similar, they do have distinct characteristics.

As has already been indicated, the regulation of transcription is likely to be embedded in the structure of the promoters of mammalian ribosomal protein genes; more importantly this is also likely to be the site of coordination of the transcription of the 70 to 80 ribosomal protein genes. The assumption, until proven otherwise, has to be that coordination is mediated by a set of common or overlapping trans-acting factors affecting in some complex way a pattern of cis-acting sequences in the promoters. The analysis to date has provided information on the architecture of these promoters in a small number of ribosomal protein genes, but we still lack knowledge of the exact pattern of the control elements, of whether there is a common architecture for all mammalian ribosomal protein gene promoters, and of the identity of the factors that bind to cis-acting sequences.

V. FUNCTION OF RIBOSOME DOMAINS

Little is known of the function of individual ribosomal components or even of ribosomal domains. None of the ribosomal proteins or nucleic acids has activity when separated from the particle. To circumvent this impediment, advantage has been taken of the activity of toxins and antibiotics. The value that derives from an analysis of their mechanism of action is in directing our attention to regions of the ribosome where our efforts to comprehend functional correlates of structure are likely to be rewarded.

α-Sarcin is a small, basic, cytotoxic protein produced by the mold *Aspergillus giganteus* that inhibits protein synthesis by inactivating ribosomes. The inhibition is the result of the hydrolysis of a phosphodiester bond on the 3′ side of residue G-4325, which is near the 3′ end of 28*S* rRNA. The cleavage site is embedded in a purine-rich, single-stranded segment of 12 nucleotides that is near universal. This is the longest and most strongly conserved sequence of nucleotides in rRNA and, indeed, the ribosomes of all the organisms that have been tested, including the producing fungus, are sensitive to the toxin. α-Sarcin catalyzes the hydrolysis of only the one phosphodiester bond and this single break accounts entirely for its cytotoxicity. This remarkable specificity is peculiar to α-sarcin; treatment of ribosomes with other ribonucleases causes extensive digestion of rRNA.

The finding that cleavage of a single phosphodiester bond inactivates the ribosome implies that this sequence in the α-sarcin domain is crucial for function since ribosomes ordinarily survive mild treatment with nucleases despite many nicks in their RNA; indeed, some organisms physiologically divide their 28*S* rRNA into domains. Thus, intact rRNA per se is not essential for protein synthesis. The presumption that the α-sarcin region of 28*S* rRNA is critical for ribosome function has gained considerable reinforcement from the elucidation of the mechanism of the action of ricin. Ricin, which is among the most toxic substances known (a single molecule is sufficient to kill a cell), is an RNA *N*-glycosidase and the single base in 28*S* rRNA that is depurinated is A-4324, that is, the nucleotide adjacent to the α-sarcin cut site.

There are good reasons to suspect that the α-sarcin/ricin domain is involved in EF-1-dependent binding of aminoacyl-tRNA to ribosomes and EF-2-catalyzed GTP hydrolysis and translocation. This supposition follows from the findings that these are the partial reactions most adversely affected by α-sarcin and by ricin, respectively, and that cleavage at the α-sarcin site interferes only with the binding of elongation factors. The most convincing evidence for this interpretation comes from the demonstration that elongation factors footprint in the α-sarcin/ricin domain. EF-Tu protects only four of the nucleotides in prokaryotic 23*S* rRNA against chemical modification and these correspond in eukaryotic 28*S* rRNA to A-4324 (ricin), G-4325 (α-sarcin), and G-4319 and A-4329, all of which are in the universal sequence. EF-G also protects only four nucleotides and three are the same as the ones protected by EF-Tu, namely, the bases that correspond to G-4319, A-4324, and G-4325.

An attempt is being made to determine how α-sarcin recognizes a single phosphodiester bond in a particular domain in rRNA; this is part of an effort, unfortunately but necessarily oblique, to understand how ribosomal proteins, which like α-sarcin are small and basic, recognize specific sites in rRNA. At the same time information is being sought on the contribution that the α-sarcin site RNA makes to ribosome function.

The secondary structure of the α-sarcin domain RNA encompasses a helical stem of 7 base pairs with a single bulged nucleotide and a single-stranded loop of 17 nucleotides. Information on the requirements for toxin recognition has come from analysis of their effects on a synthetic RNA that has the sequence and the secondary structure of the α-sarcin/ricin domain stem and loop and of a set of mutant oligoribonucleotides. Recognition by the toxins requires the stem (but not the bulged nucleotide) and the single-stranded loop in which the sequence of at least 12 nucleotides (the universal sequence) affect binding and enzymatic activity; the stem needs only 3 base pairs and the identity of the pairs does not have an influence. The identity elements for ricin devolve into a GAGA tetraloop (corresponding to positions 4323 through 4326 in 28*S* rRNA) and, it would seem, little more. Recognition by α-sarcin, on the other hand, is critically dependent on a single guanosine in the loop (it corresponds to the base at position 4319).

It had been assumed that the structure of the toxin domain in 28*S* rRNA is more complex than is depicted in the usual two-dimensional cartoons; this assumption has now been substantiated. The conformation of the α-sarcin/ricin domain RNA was determined by nuclear magnetic resonance spectroscopy. The RNA has a compact structure that contains several purine · purine base pairs, a GAGA tetraloop, and a bulged guanosine (the nucleotide critical for α-sarcin recognition) adjacent to a reverse Hoogsteen A · U pair. It is stabilized by an unusual set of cross-strand base-stacking interactions and imino proton to phosphate oxygen hydrogen bonds.

The secular grail in this research is an understanding of how the α-sarcin/ricin district of 28*S* rRNA functions in protein biosynthesis. The relevant observation here is that the two toxins have different, perhaps even incompatible, recognition elements; nonetheless, they affect adjacent nucleotides in rRNA. This seeming paradox is best reconciled by assuming that the conformation of the α-sarcin/ricin domain RNA changes during each turn of the elongation cycle and that one of the alternate conformers is recognized by ricin and the other by α-sarcin. The putative switch in conformation of the RNA might then contribute to, or even account for, translocation. What is envisaged is an allosteric transition in RNA conformation with the effectors being either the elongation factors or GTP or both. The reasoning is by analogy with allosteric proteins: small changes in conformation of each of the subunits of an allosteric protein due to the binding of a ligand can produce large changes in their relationship. Similarly, a change of a few angstroms in the structure of 18*S* and of 28*S* rRNAs might affect subunit contacts and produce a relatively large translational movement between them. This movement could drive the vectorial displacement of the mRNA one codon across the surface of the ribosome and could account for the translocation of peptidyl-tRNA from the A to the P site.

This proposal provides a bonus, namely, a plausible answer to a question that has long vexed ribocentrics: Why do ribosomes have subunits? The answer may be that translocation requires moving parts, as Alexander Spirin first suggested, and that the motor that drives the displacement of the subunits is the reversible changes in the conformation of rRNA, especially and particularly that of the α-sarcin/ricin domain in 28*S* rRNA, but also perhaps of a complementary change in the conformation of the small subunit rRNA. A candidate structure is the 625 loop in 18*S* (or the 530 loop in 16*S*) rRNA. Harry Noller and colleagues have demonstrated that tRNA bound to the A site of *E. coli* ribosomes protects nucleotides clustered in the 1400/1500 region and in the 530 loop of 16*S* rRNA. However, in the models of the three-dimensional structure, these two regions are too distant from each other for both sets of protected nucleotides to make direct contacts with the anticodon of tRNA. Since the 1400/1500 region is where the codon–anticodon interaction occurs, the changes in the 530 loop are more likely to reflect a conformational transition induced in the latter site by binding to the former. The 530 loop is situated in 70*S* ribosomal couples near where EF-Tu and EF-G bind to the 50*S* subunit. Thus one or the other or both of the factors may simultaneously (or sequentially) change the conformation of the rRNAs in the small and large subunits. If this interpretation is a reflection of the actual biochemistry, it is easier to understand the catastrophic effects of the toxins on ribosome function. Cleavage of the phosphoribose backbone by α-sarcin or depurination by ricin in the α-sarcin/ricin domain of 28*S* rRNA is seen as abolishing the capacity to bind the elongation factors and hence to drive translocation by reversibly switching conformations.

VI. CODA

With respect to the structure and function of human ribosomes, all that remains are the difficult tasks: the tertiary folding patterns of the rRNAs, the three-dimensional structure of the proteins, and ultimately the structure of the subunits at atomic resolution. These are formidable problems, but progress in studies of macromolecular structure has been so astonishing in the last several years as to encourgae optimism. Regarding the function of the organelle, we lack information *only* on how tRNAs align on the ribosome, on how a peptide bond is made, and on how movement of mRNA and of peptidyl-tRNA is catalyzed.

BIBLIOGRAPHY

Hardesty, B., and Kramer, G. (eds.) (1985). "Structure, Function, and Genetics of Ribosomes." Springer-Verlag, New York.

Hill, W. E. (ed.) (1990). "Structure, Function and Evolution of Ribosomes." American Society of Microbiology, Washington, D.C.

Mager, W. H. (1988). Control of ribosomal protein gene expression. *Biochim. Biophys. Acta* **949,** 1–15.

Nierhaus, K. H., *et al.* (eds.) (1993). "The Translational Apparatus. Structure, Function, Regulation, Evolution." Plenum, New York.

Noller, H. F. (1984). Structure of ribosomal RNA. *Annu. Rev. Biochem.* **53,** 119–162.

Perez-Bercoff, R. (ed.) (1982). "Protein Biosynthesis in Eukaryotes," Nato Advance Study Institutes Series. Plenum, New York.

Raué, H. A., Klootwijk, J., and Musters, W. (1988). Evolutionary conservation of structure and function of ribosomal RNA. *Prog. Biophys. Molec. Biol.* **51,** 77–129.

Trachsel, H. (ed.) (1991). "Translation in Eukakryotes." CRC Press, Boca Raton, Florida.

Wittmann, H. F. (1982). Components of bacterial ribosomes. *Annu. Rev. Biochem.* **51,** 155–183.

Wittmann, H. F. (1983). Architecture of prokaryotic ribosomes. *Annu. Rev. Biochem.* **52,** 35–65.

Wool, I. G. (1979). The structure and function of eukaryotic ribosomes. *Annu. Rev. Biochem.* **48,** 719–754.

RNA Replication

CLAUDE A. VILLEE
Harvard University

GLOSSARY

Codons and anticodons The specific sequence of three nucleotides in messenger RNA (codon) and the complementary sequence of three nucleotides in transfer RNA (anticodon) in which genetic information is stored and transferred, determines the base pairing between specific codons and specific anticodons, and, in turn, determines the order in which amino acids are synthesized into peptide chains

Exons and introns The initial RNA product of the transcription of DNA is a large molecule termed heterogeneous nuclear RNA. This is cut and spliced by specific enzymes to yield a much shorter messenger RNA composed of only a portion of the original heterogeneous nuclear RNA. The pieces of RNA that are included in the final messenger RNA are termed exons. The other, unused, portions are termed introns, or intervening sequences. It is not at all clear what function, if any, these intervening sequences may have. For the heterogeneous nuclear RNA to be converted into a functional messenger RNA, the introns must be deleted from the molecule and the exons must be precisely spliced together to form a continuous message that codes for a specific protein

Helix-destabilizing proteins Two strands that constitute DNA must be physically separated in order for replication to proceed; the unwinding of the double strand is catalyzed by DNA helicases, and the separated strands are then bound by helix-destabilizing proteins, which bind to single-strand DNA and prevent the reestablishment of the double helix until each strand has been copied

Okazaki fragments Relatively short DNA chains formed on the lagging strand; each is initiated by a separate primer and then is extended toward the 5′ end of the previously synthesized fragment by DNA polymerase

Retroviruses A genomic RNA that is a single plus strand, which may serve as a template to direct the formation of a DNA molecule, which, in turn, may act as a template for making messenger RNA. The virion RNA is copied into a single strand of DNA, which then forms a complementary second-DNA strand; this double-strand DNA is integrated into the chromosomal DNA of the infected cell and the integrated DNA is then transcribed by the cell's own machinery into RNA that either acts as a viral messenger RNA or becomes enclosed in a virus. The so-called central dogma of biology that information is transferred only in the direction of DNA → RNA → protein had to be amended after the discovery of retroviruses in which information is transferred from RNA to DNA

Ribozymes A new class of biological catalysts that, as nucleic acids, can act as enzymes; the RNA that composes the introns may have the ability to splice itself without the assistance of protein catalysts

Sense strands and antisense strands Of the two strands that comprise the double helix of a DNA molecule, only the sense strand contains a sequence of nucleotides that can be read out to form a protein. The complementary strand, termed the antisense strand, has a sequence of nucleotides that, if read out, would give either a garbled or a totally lacking messenger RNA

THE CENTRAL DOGMA OF MOLECULAR BIOLOGY, unchallenged for several decades, was that genetic information always flows from DNA to RNA to protein. A seminal exception to this rule was discovered in 1964 by Howard Temin. Temin's experiments showed that infection of cells by certain cancer-causing viruses is blocked by inhibitors of DNA synthesis and by inhibitors of DNA transcription. These experi-

ENCYCLOPEDIA OF HUMAN BIOLOGY, Second Edition, VOLUME 7. Copyright © 1997 by Academic Press. All rights of reproduction in any form reserved.

ments indicated that synthesis and transcription of DNA are required for the multiplication of RNA tumor viruses and that in these systems information flows in the opposite direction, from RNA to DNA. Temin suggested that a DNA provirus was an essential step in the replication of RNA tumor viruses. In these organisms an enzyme was required that would synthesize DNA using RNA as a template. In 1970 Temin and Baltimore identified just such a DNA polymerase, one that uses RNA as a template in the synthesis of DNA. This RNA-dependent DNA polymerase, now termed "reverse transcriptase," has been shown to be present in all RNA tumor viruses. Other RNA viruses that do not form tumors replicate themselves directly, without using DNA as an intermediate.

The replication of RNA is a process basically similar to that of the replication of DNA and the transcription of DNA to form RNA. In all of these processes, one polynucleotide strand serves as a template, and specific polymerases use specific nucleotide triphosphates to add nucleotides to the strand in a specific order. The specific base added in sequence is determined by the complementary base pairing of the bases in the initial template strand and the newly forming strand. The mechanism by which double-strand DNA undergoes replication to form two double helices has been understood for a long time. That RNA can serve as a template to form complementary DNA strands was recognized somewhat later, and, more recently, it was realized that RNA itself can undergo replication. It can serve as a template for the formation of a complementary strand of RNA. The complementary pairing of purine and pyrimidine bases is basic to the transfer of genetic information and the synthesis of specific proteins composed of sequences of amino acids. Specific base pairing is determined by the nature of the hydrogen bonds joining the bases. Two bonds join adenine and thymine, and three bonds join guanine and cytosine. The pairs are always A–T and G–C.

Our understanding of the process of RNA replication has developed relatively recently whereas our understanding of DNA replication developed much earlier. In fact, the knowledge of DNA replication made possible our appreciation of the mechanisms involved in RNA replication.

I. NUCLEIC ACID STRUCTURE

Nucleic acids are of two types: deoxyribonucleic acid (DNA) and ribonucleic acid (RNA). Each consists of large molecules composed of four kinds of nucleotides: adenylate (A), thymidylate (T), guanylate (G), or cytidylate (C). In RNA, the T is replaced by U (uridylate). Each nucleotide consists of a nitrogenous base (adenine, thymine, guanine, and cytosine) plus a sugar (deoxyribose in DNA and ribose in RNA) plus orthophosphate. The nucleic acids are of fundamental importance in biology because of their role in transmitting biological information from one generation to the next and in transcribing that genetic information in the synthesis of proteins. RNA serves as the genetic material by which information is transferred from one generation to the next in certain small viruses. All other viruses and all bacteria, plants, and animals use DNA as the genetic material.

A. DNA Structure

The DNA molecule consists of a very long double helix. The two chains that make up the helix extend in opposite directions and are paired by the so-called Chargaff rules, i.e., in nitrogenous base in one chain is paired to a base in the other chain such that an A (adenine) in one chain always pairs to a thymine (T) in the other chain and a G (guanine) pairs with C (cytidine). The specific base pairing is determined by the nature of the hydrogen bonds joining the bases: two bonds join A and T and three bonds join G and C. To fit into the double helix, one large base, A or G, must pair with a smaller base, T or C. An A–G pair would be too large and a C–T pair would be too small to fit in the available space. In DNA replication (as shown subsequently), the two chains separate momentarily and, with the aid of a number of proteins (some of which are enzymes), a new chain is formed by base pairing to each of the original chains. The product is two DNA molecules, each composed of a double helix [*See* DNA and Gene Transcription.]

B. RNA Structure

RNA functions primarily in translating the genetic information in DNA into the sequence of amino acids that will comprise the protein coded by that sequence of nucleotides in the gene. There are three distinct types of RNA, each with a particular function in bringing about the synthesis of specific proteins. Ribosomal RNA (rRNA) serves as an integral part of the ribosomes (structures inside the cell on which protein synthesis occurs). Transfer RNA (tRNA) consists of relatively small nucleic acid molecules, each with a specific binding area of three nucleotides (the anticodon) and with a specific amino acid attached to the opposite end of the tRNA molecule. The third type of RNA, messenger RNA (mRNA), is produced by the transcription of the sequence of nucleotides in DNA to form a comparable sequence of ribonucleotides in RNA. Each set of three nucleotides in a molecule of mRNA constitutes a condon, which undergoes specific base pairing with the three nucleotides in the anticodon of tRNA. This ensures that the amino acids will be lined up in the proper sequence, yielding the correct protein molecule.

Because similar codes are used in both DNA and RNA, in certain retroviruses, such as the acquired immunodeficiency disease virus, information is transferred from RNA, which is the genetic material of the retroviruses, to DNA, which becomes an intermediate in the synthesis of the next generation of RNA. Information is transferred directly from RNA to RNA in certain RNA viruses such as the influenza virus or the poliomyelitis virus.

II. DNA IS THE GENETIC MATERIAL

Experiments have shown that the transforming principle of pneumococci is DNA and not protein. Analyses of a variety of cells then showed that eggs and sperm, which are haploid (containing one set of genes), also contain only half as much DNA as somatic diploid cells (two sets of genes). In 1952, radiolabeled precursors demonstrated that viruses multiplying within bacteria (bacteriophages) inject their DNA, but not their protein, into the bacterial cell. The injected DNA undergoes replication, resulting in the presence of many DNA molecules within the bacterial cell. Despite these findings, most biologists did not accept the ideal that the genetic material was DNA. However, in 1953, Watson and Crick published their famous paper in *Nature* describing their model of DNA as a double helix with the nitrogenous bases on the inside of the helix. This immediately suggested a mechanism by which DNA could undergo replication, and very quickly biologists accepted the theory that DNA was the genetic material and that it was a double helix of two nucleotide chains.

The Watson–Crick model explained many previous findings such as Chargaff's analyses demonstrating that in a great variety of cells the amounts of A and T were equal, and also the amount of G and the amount of C were equal. It explained the X-ray crystallographic pictures of DNA molecules taken by Wilkins and Franklin.

A. Information Transfer: DNA to DNA

The Watson–Crick model also suggested a mechanism by which the information in DNA could be copied precisely. Because nucleotides pair with each other in a complementary fashion, i.e., A to T and G to C, each of the nucleotide strands in the DNA molecule could serve as a template for the synthesis of the opposite strand. When the hydrogen bonds joining the two strands are broken, the two chains can separate. Each chain can then pair with complementary nucleotides to form the corresponding strand. This results in two DNA double helices, both of which are identical to the original one. Each consists of one original strand from the parent molecule and one newly synthesized complementary strand.

The Watson–Crick model also suggested a mechanism by which DNA could undergo a mutation. For a long time it had been known that genes can undergo mutations (sudden, heritable changes) that are inherited in subsequent generations. The Watson–Crick double-helix model suggested that a mutation could simply involve a change in the sequence of bases in the DNA. If the DNA is copied by a mechanism using complementary base pairing, then any change in the sequence in the bases on one strand would result in a new sequence of complementary bases being paired during the next cycle of replication. The new sequence of bases would then be transmitted to the daughter molecules by the copying mechanism that had been used to copy the original genetic material.

1. DNA Replication is Semiconservative

The replication of DNA is termed semiconservative because each of the original strands is conserved in one of the daughter strands and constitutes one-half of the daughter helix. Direct evidence that the replication process is semiconservative was provided by experiments using the bacterium *Escherichia coli*. For several generations, bacteria were grown in a medium containing heavy nitrogen (^{15}N), labeling the parent strands of DNA in the bacterium. The presence of the ^{15}N atoms increased the density of the DNA molecules so that they can be distinguished by appropriate techniques from strands containing regular nitrogen (^{14}N). Some of the bacteria containing the ^{15}N-labeled DNA were transferred to a medium containing the usual nitrogen isotope, ^{14}N, and permitted to undergo additional cell divisions. The newly synthesized DNA strands were less dense because they incorporated ^{14}N bases. After one generation, the DNA molecules of the cells had a density intermediate between bacteria containing the ^{15}N bases and normal bacteria containing only ^{14}N bases. After a further cycle of cell division, the DNA in the density gradient sedimented at levels, indicating that about half consisted of hybrid DNA helices containing equal amounts of ^{15}N and ^{14}N; the remaining half contained only ^{14}N DNA. From these results, each strand of the parental double helix was shown to be conserved in a different daughter molecule just as predicted by the semiconservative replication model.

Because the two strands of DNA in the double helix are physically intertwined, they must be separated for replication to proceed. The separation of the strands has proved to be a complex process. Replication can occur in these very long DNA molecules only if the strain of the unwinding strands is relieved. The unwinding is catalyzed by DNA helicases, enzymes that move along the helix, unwinding the strands as they move. The separated strands are bound by helix-destabilizing proteins. These bind to single-strand DNA, preventing the reestablishment of the double helix until each strand has been copied.

The DNA polymerases that link together the nucleotide subunits in the replication of DNA add nucleotides to the 3′ end of a polynucleotide strand that is paired to the strand being copied. The substrates for the DNA polymerases are deoxyribonucleoside triphosphates. As the nucleotides are joined, two of the phosphates are removed, and this provides the energy to drive the synthetic reaction. A new polynucleotide chain is elongated by the addition of the 5′ phosphate group of the next nucleotide subunit to the 3′ hydroxyl sugar at the end of the growing strand; thus, DNA synthesis proceeds in a 5′ → 3′ direction. DNA polymerases can catalyze the addition of nucleotides only at the 3′ end of an existing DNA strand. This leads to the question of how the synthesis of DNA can be initiated when the two strands are separated. This is accomplished by utilizing a short piece of an RNA primer that is synthesized by an aggregate of proteins called a primosome. The RNA primer pairs with a single strand DNA template at the point of initiation of replication. DNA polymerase can then synthesize the new chain by adding nucleotides to the 3′ end of the RNA primer. When DNA synthesis has proceeded to an appropriate extent, the RNA primer is filled in by DNA polymerase.

2. DNA Replication: Leading and Lagging Strands

One of the initial puzzles regarding the replication mechanism stemmed from the complementary DNA strands extending in opposite directions, but DNA synthesis can proceed only in the 5′ → 3′ direction. The strand being copied is being read in a 3′ → 5′ direction; hence, it would follow that only one strand can be copied at a time. Experiments demonstrate clearly that DNA replication begins at specific sites on the DNA molecule, called origins of replication, and that both strands are replicated at the same time. One, termed the leading strand, is formed continuously after the process has been initiated; its complementary lagging strand is synthesized in short pieces, which are subsequently joined to make a complete DNA chain. DNA is synthesized beginning at a Y-shaped structure called a replication fork. The lagging strand is synthesized in relatively short fragments (Okazaki fragments), each of which is initiated by a separate primer and is then extended toward the 5′ end of the previously synthesized fragment by a DNA polymerase. DNA polymerases are complex enzymes that serve several functions. As the growing fragment approaches the fragment synthesized previously, one part of DNA polymerase degrades the previous RNA primer, allowing other polymerases to fill in the gap between the two fragments. These are then linked by DNA ligase, which joins the 3′ end of one end of the fragment to the 5′ end of another by a phosphodiester bond.

When double-strand DNA is separated, two fork-like structures are created. The molecule is replicated in both directions from the origin of replication. In eukaryotic chromosomes, each of which is composed of a linear DNA molecule, there usually are multiple origins of replication. Each replication fork proceeds until it meets one coming from the opposite direction. This results in the formation of a chromosome containing two DNA double helices. [*See* Chromosomes; DNA Replication.]

B. Information Transfer: DNA to RNA

The transcription of DNA to form RNA and the replication of RNA are processes that are basically similar to that of DNA replication. In all of these, one polynucleotide strand serves as a template and specific polymerases use specific nucleoside triphosphates to add specific nucleotides. The specific base added in sequence is determined by the complementary base pairing of the bases in the initial strand and the forming strand.

The synthesis of proteins using the genetic information present in DNA involves two stages. In the first stage, the information present in the specific sequence of nucleotides in one of the DNA strands (called the minus strand) is copied and a complementary plus strand of mRNA is produced. This transcription process is very similar to the process by which DNA is replicated: the plus mRNA strand is formed by complementary base pairing with the minus strand of DNA. The genetic information in DNA is transcribed to yield mRNA.

C. Information Transfer: RNA to Protein

In the second stage of protein synthesis, information that has been transcribed into mRNA is used to determine the amino acid sequence in the protein. Clearly, this process involves the conversion of the nucleic acid code of the mRNA into an amino acid code of the protein; hence, it is termed a process of translation. The information present in the mRNA is in the form of a genetic code composed of three bases forming a codon. Each group of three bases in mRNA determines the presence of a specific amino acid. Thus, each codon in the mRNA specifies one of the amino acids in the protein. The codon that specifies the amino acid tryptophan is UGG. Any of six different codons (CGU, CGC, CGA, CGG, AGA, and AGG) may specify the amino acid arginine.

III. PROTEIN SYNTHESIS: mRNA AND tRNA

The information contained in the condons of mRNA is translated by a very complex process that takes place in eukaryotic cells within the ribosomes. The recognition and decoding of the codons in the mRNA are accomplished by tRNAs. The anticodon of the tRNA molecules recognizes a codon in mRNA by complementary base pairing. The amino acid specified by the anticodon is attached to the other end of the tRNA molecule.

The synthesis of a specific protein guided by the genetic information located in the DNA requires the joining of amino acids in the correct order by chemical bonds. This occurs on the ribosomes. The complex process of protein synthesis involves first the attaching of the ribosomes to the 5′ end of the mRNA. The ribosome then travels along the mRNA, joining the codon of the mRNA with the anticodon of the tRNA so that the amino acids are lined up in the proper sequence.

A. Sense and Antisense Strands

DNA is a double helix, composed of two complementary antiparallel chains, and only one of the two chains, termed the minus or sense strand, normally undergoes transcription to form mRNA. The other DNA strand, the antisense strand, is usually not transcribed but may undergo transcription by DNA-dependent RNA polymerases in certain cells to yield a plus or antisense RNA. The RNA polymerase that catalyzes the transcription process recognizes a specific promotor sequence of bases at the 5′ end. The mRNA is synthesized by the addition of nucleotides one at a time, to the 3′ end of the growing molecule.

The antisense RNA has been shown to regulate the expression of certain genes in both prokaryotes and eukaryotes. The antisense RNA can pair with mRNA, thus inhibiting the translation of the mRNA. It seems possible that antisense RNAs could be prepared and introduced into human cells where they could inactivate specific genes. Thus, antisense RNAs could be used in the treatment of cancer and viral diseases such as AIDS. The antisense strand of DNA, the one that is normally not transcribed, is complementary in sequence to the sense or template strand. Thus it is identical in base sequence to the product, RNA, but it is composed of T's instead of U's.

The promoter sequences of different genes may be different, and this could determine which genes will be transcribed at any given time. A promotor sequence of bacteria is typically about 40 bases in length and is located about 8 bases upstream (i.e., toward the 5′ end), at the point that RNA transcription begins. Three of the specific codons in mRNA serve as stop signals for the RNA polymerase and bring about the termination of transcription.

B. Modification and Processing of mRNA: Introns and Exons

The mRNAs of bacteria can be used directly after they have been transcribed without any further processing. In contrast, the mRNA molecules of eukaryotic organisms undergo posttranscriptional modifications and processing. A molecule of 7-methyl guanylate, an unusual nucleotide, is added as a cap at the 5′ end of the mRNA chain. This may be the basis of the greater stability of eukaryotic mRNAs, which have half-lives ranging as long as 24 hr, whereas the half-life of prokaryotic mRNAs is about 15 min. A long tail of polyadenylic acid, composed of 100–200 adenine nucleotides, is joined to the 3′ end of the mRNA.

A third step in the modification of eukaryotic mRNA molecules involves cutting and splicing the mRNA at specific sites. Interrupted coding sequences are present in eukaryotic DNA and may be quite long; they do not code for the amino acids present in the final protein product. These noncoding regions within the gene are called intervening sequences, or introns, as opposed to exons (which are expressed sequences and parts of the protein-coding sequence). The number of introns present in a gene can be quite variable. The β-globin gene, which produces one of the components of hemoglobin, contains two introns. The ovalbumin gene, which determines one of the proteins in egg white, contains 7 introns and the gene for another egg white protein, conalbumin, contains 16 introns. The combined lengths of the introns may be considerably longer than the combined protein-coding exon sequences. The ovalbumin gene contains about 7700 bp, whereas the sum of its coding sequences is only 1859 bp.

The transcription of a gene containing both introns and exons yields a large RNA transcript termed heterogeneous nuclear RNA (hnRNA). For the hnRNA to be converted into a functional mRNA, the introns must be deleted from the molecule and the exons must be spliced together to form a continuous message that codes for a specific protein. The splicing reactions are mediated by special base sequences within and to either side of the introns. The splicing may involve the association of small nuclear ribonucleoprotein complexes, which bind to the introns and catalyze the cleavage and splicing reactions. In some species, the RNA within the intron may have the ability to splice itself without the assistance of protein catalysts. These nucleic acids can act as enzymes. This new class of biological catalysts has been termed ribozymes. The final product, the functional mRNA, has had its introns removed and the exons spliced together; it has a 7-methyl guanylic cap at the 5′ end and a poly(A)tail at the 3′ end. It is then ready to pass out of the nucleus into the cytoplasm and, when attached to a ribosome, serve as the blueprint for the synthesis of a protein with its specific sequence of amino acids.

IV. VIRAL DNA AND RNA

Much of what is known about genetic machinery in general and the replication and transcription processes in particular has been derived from studies of viruses. These small (most of them are too small to be seen in a light microscope) and relatively simple organisms have lent themselves to a great variety of studies of replication. The viruses that infect animal cells exhibit a tremendous variety of shapes, sizes, and genetic strategies.

The generalization gradually emerged that all viruses contain nucleic acids. This further suggested that viruses and genetic material had similar functions. These speculations received confirmation by studies of bacterial viruses or bacteriophages. As shown by Hershey and Chase, only the bacteriophage DNA, not the bacteriophage protein, enters the bacterial host cell and initiates replication in the host cell. This leads to the production of several hundred progeny viruses. Viruses thus are genetic elements enclosed by a protective coat and are able to move from one cell to another. They have been termed mobile genes by some investigators. As studies proceeded, it became clear that some viruses have RNA, but not DNA. Thus, in these organisms it was concluded that RNA must be the viral genetic component. This theory was confirmed by the finding that purified RNA preparations from tobacco mosaic virus were infectious even in the absence of any tmv protein. Subsequently, many other viruses, such as polio, influenza, and measles, were found to contain RNA, but not DNA. The potential of RNA for carrying genetic information and undergoing replication is clear. Viral DNA or RNA codes not only for the coding proteins of the virus, but also for the enzymes that are required to replicate the viral nucleic acid. The smaller DNA viruses, such as the monkey SV-40 virus, and the very small bacteriophage ϕX174 contain much less genetic information and must rely to a much greater extent on enzymes from the host cells to carry out the synthesis of proteins and DNA. These viruses do contain coding for enzymes that initiate their own DNA synthesis selectively. For a virus to be successful when it invades the cell, it must override the cellular control signals that would otherwise prevent the viral DNA from doubling more than once in each cell cycle.

Viruses originally were classified by the names of the diseases they caused or the animals or plants that they infected; however, many different kinds of viruses can produce the same symptoms and appear to be the same disease states. A dozen or more different viruses can produce red eyes, runny noses, and sneezing. Viruses are now classified according to the sequence of reactions by which the mRNA is produced.

In this classification system, a viral mRNA is termed a plus strand, whereas its complementary sequence, which cannot function as mRNA, is termed a minus strand. Furthermore, a strand of DNA complementary to a viral mRNA is termed a minus strand. Production of a plus strand of mRNA requires that a minus strand of RNA or of DNA be used as the template. This permits the identification of six classes of animal viruses. In each of these classes, the nucleic acid of the virion (the infective particle) ultimately becomes the mRNA of the virus (Fig. 1).

A. Classes of Animal Viruses

Class I viruses have double-stranded DNA. The adenoviruses and the SV-40 virus are class I viruses. The DNA of these viruses typically enters the nucleus of the host cell where the enzymes that normally are responsible for producing cellular mRNA are diverted to producing viral mRNA. The pox viruses, another group of class I viruses, are large viruses that have their own enzymes for making mRNA; they undergo replication in the cell cytoplasm.

The paroviruses, members of class II, are simple viruses that contain a single strand of DNA. Some parvoviruses enclose both plus and minus strands of DNA within their capsules, but they are present in separate virions. Others enclose only a minus strand within the capsid; this is copied within the cell into double-strand DNA, which is then copied to yield mRNA.

B. Information Transfer: RNA to RNA, RNA Replication

The other four classes of animal viruses contain RNA genomes. A wide range of animals, including human beings, are infected by viruses in each of these four classes. Viruses of class III contain a double strand of RNA. The minus RNA strand acts as a template for the synthesis of a plus strand of mRNA. The virions of class III viruses have segmented genomes containing 8–12 double-strand RNA segments, each of which codes for a specific polypeptide. These viruses contain a complete set of enzymes that can produce mRNA. The viruses of class IV contain a single plus strand of

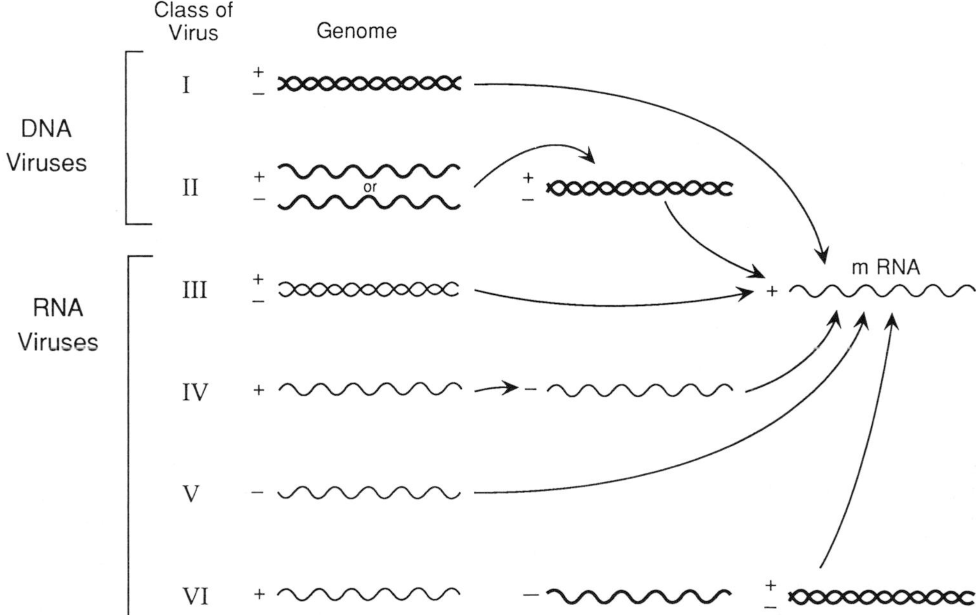

FIGURE 1 Classification of animal viruses. Heavy lines indicate DNA; light lines, RNA. Class I: The minus strand of DNA is transcribed to yield mRNA. Class II: The single strand of DNA is replicated to yield double-strand DNA, whereas its minus strand is translated to yield mRNA. Class III: The minus strand of the genome is copied to yield a minus strand of RNA, which is then copied to form the plus strand of RNA. Class V: The minus strand of the single-strand RNA in the genome is copied directly to yield a plus strand of mRNA. Class VI: The single-strand RNA genome is copied to form a minus strand of DNA. This is copied to yield double-strand DNA. The minus strand of this double-strand DNA is copied to form a plus strand of mRNA.

RNA. Because the viral genomic RNA is identical to the mRNA, the virion (genomic) RNA by itself can initiate the process of infection when introduced into a cell. The mRNA is copied into a minus strand, which then produces more plus strands. Two types of class IV viruses are recognized. The RNA molecule in the virion of poliomyelitis virus serves as the mRNA to encode all of the viral proteins. The individual proteins are first synthesized as a single, very long polypeptide strand, which is subsequently cleaved to yield the various functional proteins. The mRNA of these viruses is the same length as the genome RNA. Viruses of class IVb, called togaviruses because the virions are surrounded by a lipid envelope, synthesize at least two forms of mRNA in the host cell. One of these mRNAs is the same length as the virion RNA, whereas the other corresponds to the third of the virion RNA at the 3′ end. Class IVb includes many rare insect-borne viruses that cause encephalitis in human beings.

Class V viruses contain single minus strands of RNA. The RNA in the virion has a base sequence complementary to that of the mRNA. Thus, the virion contains a template for making mRNA but that template does not itself encode proteins.

Two subdivisions of class V can be distinguished. Class Va viruses have a genome that is a single molecule of RNA. A virus-specific polymerase contained in the virion synthesizes several different mRNAs from different parts of this single RNA template strand. Each of the class Va viral mRNAs encodes one protein. Class Vb viruses, exemplified by influenza virus, have segmented genomes. Each segment is a template for the synthesis of a single mRNA. As with class Va viruses, the virion contains a virus-specific RNA polymerase required to produce the mRNA. The minus strands of class V nucleic acids alone, i.e., in the absence of the virus-specific polymerase, are not infectious. The influenza virus RNA polymerase initiates the transcription of each mRNA by a unique mechanism. The polymerase begins mRNA synthesis by borrowing 12–15 nucleotides from the 5′ end of the cellular mRNA or mRNA precursor in the nucleus. This oligonucleotide serves as a primer for the replication of RNA catalyzed by the viral RNA polymerase. Individual mRNAs made by class Vb viruses generally encode single proteins, but some of the mRNAs can produce two distinct proteins (by reading different sequences of triplets within the same mRNA).

The multiplication of RNA viruses involves the formation of complementary strands. Many RNA viruses studied, such as polio virus, have a single-strand RNA polynucleotide chain. Other viruses have double helical RNA viral chromosomes; this is seen in reoviruses that infect a great variety of organisms. RNA replication is mediated by specific RNA-dependent RNA polymerases (called replicases), which are coded for in the viral RNA chromosome.

C. Retroviruses: Information Transfer from RNA to DNA

The viruses of class VI, called retroviruses, have a genomic RNA that is a single plus strand. This serves as a template that directs the formation of a DNA molecule. This, in turn, acts as the template for making the mRNA. First, an enzyme present in the virion, reverse transcriptase, transcribes the virion RNA into a single strand of DNA, which serves as a template for the synthesis of a complementary second strand. The double-strand DNA is integrated into the chromosomal DNA of the infected cell as a provirus. Finally, the provirus is transcribed by the cell's own machinery into RNA that either acts as a viral mRNA or becomes enclosed in a virus. This completes the retrovirus cycle (see Fig. 2). If the retrovirus contains cancer genes, the cell that it infects may be transformed into a tumor cell. [*See* Retroviruses as Vectors for Gene Transfer.]

V. EVOLUTION OF COMPLEMENTARY BASE PAIRING

The evolution of the process of complementary base pairing was a key step in the evolution of life, for it permitted the development of mechanisms for the transfer and storage of genetic information. Most scientists now believe that RNA is the more ancient means of transferring genetic information and that RNA replication preceded DNA replication in the course of evolution. Thus, RNA replication may have been the evolutionary precursor of DNA replication.

The replication of RNA and DNA, the transcription of DNA to RNA, and the translation of RNA to form proteins all involve this important basic principle of complementary base pairing determined by specific hydrogen bonds between base pairs A to T and G to C. In protein synthesis the lining up of the amino acids is determined by the complementary base pairing of codon and anticodon.

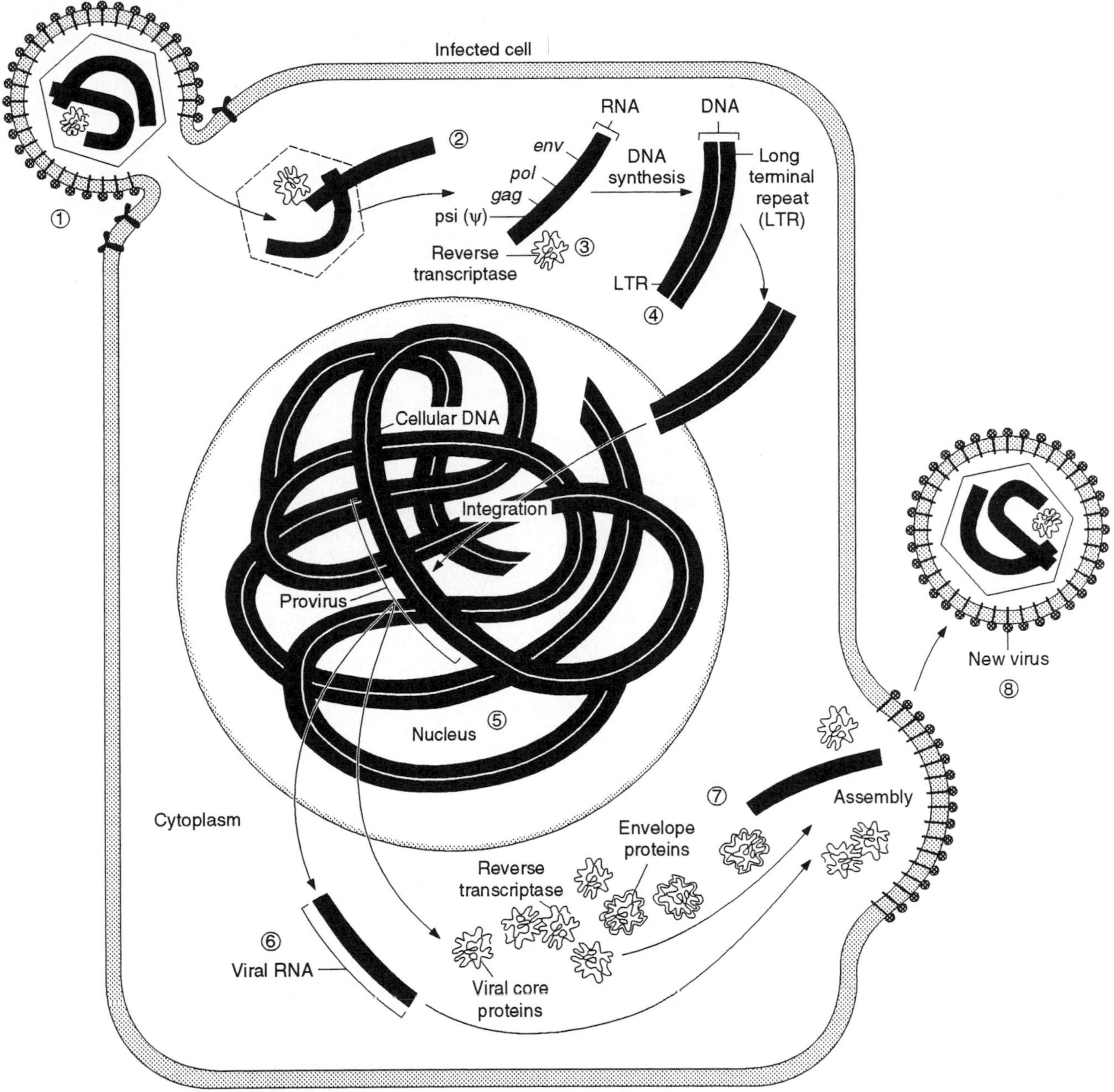

FIGURE 2 Life cycle of an RNA tumor virus, a retrovirus. (1) The virus particle containing RNA attaches to the cell membrane and releases the RNA into the host cell. (2) The reverse transcriptase (3) makes a DNA copy of the RNA, after which the DNA strand is replicated to produce a double-stranded molecule (4). The DNA provirus is inserted into the DNA of the host cell's chromosome (5). The viral DNA uses the cell's transcription apparatus to make RNA copies (6). These are released into the cytoplasm and translated to make essential proteins for the virus particles (7). The RNA and proteins then bud from the cell, forming new virus particles that (8) can infect other cells, completing the life cycle.

The course of evolution of nucleic acids may have been:

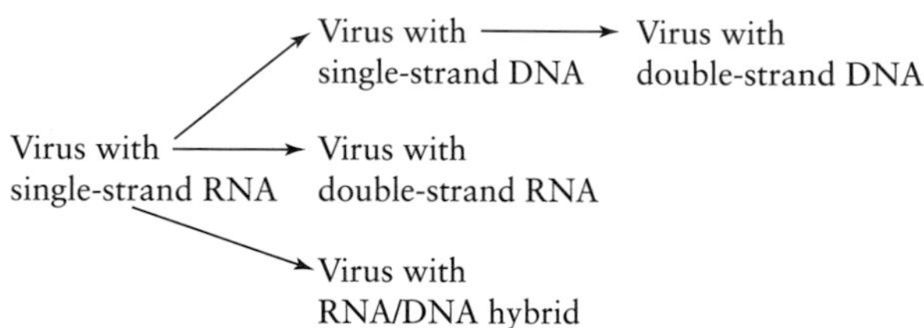

VI. EVOLUTION OF NUCLEIC ACIDS

The idea that life on planet Earth began with the emergence and evolution of RNA molecules able to replicate themselves has been gaining acceptance. Hypotheses postulate the development of RNA molecules with the ability to catalyze the enzymatic production of more RNA molecules (RNA replication). Such molecules could undergo evolution by natural selection to form larger RNA molecules with specific heritable properties. Although initially small and having relatively few enzymatic properties, the forces of natural selection on the ribozymes would lead to the emergence of complex RNA molecules with a variety of enzymatic properties, including the ability to synthesize complex molecules of DNA and RNA with the catalytic property of producing nucleotides and nucleic acids. Just where along this sequence of events "life" appeared remains an unanswered question.

BIBLIOGRAPHY

Alberts, B., Bray, D., Lewis, J., Raff, M., Roberts, K., and Watson, J. D. (1995). "Molecular Biology of the Cell," 3rd Ed. Garland Publishing, New York.

Darnell, J. E., Jr. (1985). RNA. *Sci. Am.* October, **253**, 75.

Felsenfeld, G. (1985). DNA. *Sci. Am.* October, **253**, 90.

Gallo, R. C. (1986). The first human retrovirus. *Sci. Am.* December, **255**, 88.

Hogle, J. M., Chow, M., and Filman, D. J. (1987). The structure of polio virus. *Sci. Am.* March, **256**, 101.

Lewin, B. (1994). "Genes V," Oxford University Press, Oxford.

Lodish, H., Baltimore, D., Berk, A., and Darnell, J. (1995). "Molecular Cell Biology," 3rd Ed. Freeman, New York.

Watson, J. D., Hopkins, N. H., Roberts, J. W., Steitz, J. A., and Weiner, V. M. (1992). "Molecular Biology of the Gene," 5th Ed. Benjamin Cummings, Menlo Park, CA.

Salivary Glands and Saliva

STEVEN D. BRADWAY
MICHAEL J. LEVINE
State University of New York at Buffalo

I. Salivary Gland Anatomy
II. Salivary Gland Physiology
III. Composition of Saliva
IV. Function of Saliva
V. Salivary Dysfunction and Therapy

GLOSSARY

Acylation In the context of this article, the covalent incorporation of fatty acids into protein

Cariogenic Substance that produces carious lesions in teeth

Hypotonic Solution that has a lower osmotic pressure (contains less solutes) than serum

Disulfide cross-linked proteins Proteins that contain either intermolecular or intramolecular covalent bonds between the sulfur groups of their cysteine amino acids

Electrolyte In the context of this article, the inorganic ions found in serum and saliva: sodium, chloride, potassium, calcium, bicarbonate, phosphate, etc.

Genetic polymorphism Differential expression of the same gene or gene family in different individuals

Glycosylation Covalent coupling of carbohydrate to protein

Parenchyma Functional cells of an exocrine gland; in the context of this article, the acinar and ductal cells of salivary glands

Phosphorylation Addition of phosphate groups onto proteins

Structural modulation Alteration of a salivary component by chemicals or enzymes after it has been secreted into the oral cavity

SALIVA IS A COMPLEX EXOCRINE SECRETION, which coats all oral surfaces and plays a major role in maintaining the homeostasis of the oral cavity. A constant supply of saliva (1–2 liters/day at a flow rate of 1–3 ml/min) is produced by four major groups of glands, which generate characteristic secretions in response to a diverse group of stimuli from both the oral and the extraoral environment. The proteins and glycoprotein components of saliva are synthesized by conventional biosynthetic mechanisms in acinar cells to form primary saliva. As the primary saliva passes through the secretory ducts, some electrolytes are readsorbed by cells lining the ducts, and the secretion is discharged into the oral cavity as a hypotonic solution. Many salivary components, by virtue of their physical characteristics, are well adapted to form protective films called pellicles on oral surfaces. These pellicles function to lubricate and moisten oral tissues, mediate selective microbial attachment, and act as barriers against noxious substances. Other salivary components and electrolytes maintain oral pH, aid in digestion, exert antimicrobial activity, and mediate mineralization processes. The significance of these functions can be seen in salivary dysfunction, where decreased or absent salivary flow is associated with rampant tooth decay, yeast infections, and inflammation of the oral mucous membranes. Over 300 therapeutic pharmacologic agents as well as local and systemic diseases are known to adversely affect salivary function. Indeed, salivary dysfunction is increasingly recognized as a significant clinical problem in the therapy of many patient groups. Extensive research to characterize the functional characteristics of saliva has been performed since the mid-1970s, and the information obtained is now being applied to create functional salivary replacements for individuals with salivary dysfunction.

ENCYCLOPEDIA OF HUMAN BIOLOGY, Second Edition, VOLUME 7. Copyright © 1991 by Academic Press. All rights of reproduction in any form reserved.

I. SALIVARY GLAND ANATOMY

A. Gross Anatomy

There are three sets of anatomically distinct major salivary glands and a number of minor salivary glands (Fig. 1). The largest of these glands, the parotid gland, has a superficial and deep lobe, which grossly forms an inverted three-dimensional triangle positioned at the posterior border of the mandible. Secretions of the parotid gland are carried to the mouth through Stensen's duct, which has an orifice in the buccal mucosa adjacent to the maxillary second molar. The submandibular gland is round in shape (≈2–3 cm in diameter) and is found just below and inside the lower angle of the mandible. The submandibular duct (Wharton's duct) empties through orifices under the most anterior frenum of the tongue. The sublingual gland is composed of a major and multiple minor lobes, which lie just under the mucosal lining of the floor of the mouth. The major lobe drains through a common duct (Bartholin's duct) below the tongue, and the minor lobes drain through individual ducts in the lingual fold close to the base of the tongue. All of the major glands are innervated by parasympathetic and sympathetic fibers of the autonomic nerve system.

The fourth group of salivary organs are collectively referred to as the minor salivary glands. These reside in the submucosa of the oral mucous membranes and are given regional names such as the labial, buccal, and palatine glands to designate their location in the lips, cheek, and palate, respectively. The glandular parenchyma of the minor salivary glands is small, unilobular, and generally empties into the oral cavity through a single common duct. The ducts of minor salivary glands are often associated with cells of the immune system called the gland-associated lymphoid tissue.

B. Microanatomy

Salivary glands, like other exocrine organs, have a glandular parenchyma encapsulated by fibrous connective tissue, which partially divides the gland into lobes. The functional structure of the parenchyma is composed of a highly branched system of progressively smaller epithelial ducts that end in a semicircular cluster of acinar cells. The organelles of the acinar cells reflect their synthetic function and consist of abundant rough endoplasmic reticulum, Golgi apparatus, and secretory granules. The acinar cells can be functionally and histologically segregated into serous and mucous types. Serous cells have small, dense secretory granules confined to the apical portion of the cell, whereas the secretory granules of mucous cells are large, lucent structures that occupy most of the cellular cytoplasm. The parotid gland is composed primarily of serous cells, whereas the submandibular and sublingual glands are composed of mixed populations of serous and mucous cells. In contrast, the minor salivary glands contain purely mucous acinar cells. In the mixed acinar populations, the mucous cells form a slightly elongated acinus, which ends in

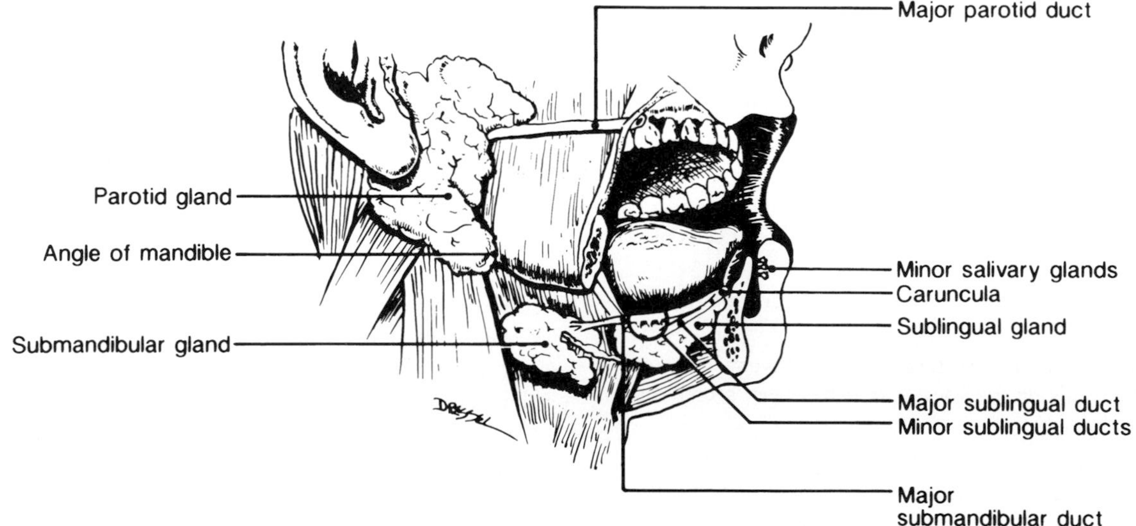

FIGURE 1 Gross anatomy of the salivary glands.

a semicircular cap of serous cells (serous demilune). Adjacent acinar cells are connected at discrete areas by an intercellular seal called the junctional complex, which is composed of tight junctions, adhering junctions, and desmosomes. An additional cellular connection, the gap junction, provides a porous junction, which allows communication between the acinar cells. The intercellular space not occupied by the junctional complex is termed the intercellular canaliculi. The basal membranes of the acinar cells are highly folded as an adaptation for ion and fluid transport. The acinar cell complex is capped with contractile cells called myoepithelial cells, and the entire complex of cells is surrounded by capillaries.

The salivary gland ducts are classified in order of ascending size as intercalated, striated, and main excretory ducts. The ductal cells possess highly folded basal membranes, which increase the basal surface of these cells. This is thought to be a functional adaptation for the transport of electrolytes from primary saliva. These cells also perform a minor secretory function and contain small numbers of secretory organelles. The ductal cells are connected by junctional complexes and gap junctions but are also separated by intercellular canaliculi. In combination, the acinar and ductal cells form a semipermeable membrane, termed the saliva–blood barrier, that separates acinar and ductal lumen from the glandular stroma.

II. SALIVARY GLAND PHYSIOLOGY

Saliva, like other excretory secretions, is derived from two separate functional events: (1) the synthesis, storage, and secretion of salivary macromolecules and (2) the fluid and electrolyte transport from the salivary capillary beds. Both synthesis and fluid transport are mediated by acinar cell mechanisms, and the products of these events are combined in the acinar lumen to produce an isotonic secretion called primary saliva. Electrolytes are then readsorbed and/or secreted into the primary saliva in the salivary duct system to finally produce a hypotonic secretion. However, the concentrations of proteins, water, and electrolytes are also dependent on the time of day as well as the duration and type of the secretory stimulus. Integration of this system is primarily controlled by the autonomic nerve system, which is stimulated by taste and mechanical receptors in the mouth. This control can be modulated by neural influences (anxiety, fear, etc.) from the central nerve system as well as by hormones and drugs, which interact with receptors on the parenchymal cells of the salivary gland.

A. Protein Synthesis and Secretion

The synthesis, packaging, and secretion of proteins by salivary parenchymal cells proceed by the same mechanisms of transcription and translation found in other exocrine cell types. In this process, a "message" in the form of mRNA is transcribed from genes in the acinar cell nucleus and exported to the cell cytoplasm. There, the message directs the synthesis of a protein (i.e., is "translated") into the lumen of the rough endoplasmic reticulum, a hollow cytoplasmic organelle. During and after synthesis, the protein can be further modified in a process termed posttranslational modification by the covalent addition of carbohydrate (glycosylation), sulfate (sulfation), phosphate (phosphorylation), and/or lipid (acylation). Concomitant with protein processing, the rough endoplasmic reticulum undergoes a continuous transformation, first into the smooth endoplasmic reticulum and then into an organelle termed the Golgi apparatus. The Golgi apparatus finally gives rise to spherical structures, called secretory granules, that contain a mixture of the processed proteins, which have been highly concentrated in preparation for secretion as salivary products. Following stimulation of the salivary glands, the secretory granules migrate to the luminal surface of the acinar cell, fuse with the plasma membrane, and empty their contents into the acinar lumen.

B. Fluid and Electrolyte Transport

The fluid phase of saliva is derived from a two-step process. Initially, water and electrolytes are transported from extraglandular interstitial fluid into the acinar lumen to produce primary saliva. The electrolyte concentration of the primary saliva is then modified in the ductal system to produce a hypotonic secretion. Salivary fluid and electrolytes are derived from capillary-derived interstitial fluid. Fluid transport into the salivary gland is thought to function in agreement with the solute–solvent coupling hypothesis, which can be conceptualized as three chambers defined by two permeable barriers (Fig. 2). The interstitial tissue surrounding the acini represents the first chamber, which contains serum transudates, whereas the acinar lumen and salivary duct system comprise the second and third chambers, respectively. The acinar cells act as a complex barrier that serves two functions: (1) a semipermeable membrane allowing free passage of water but not salt and (2) a pump for the active transport of electrolytes into the acinar lumen. During secretion, electrolytes, pumped from the interstitial

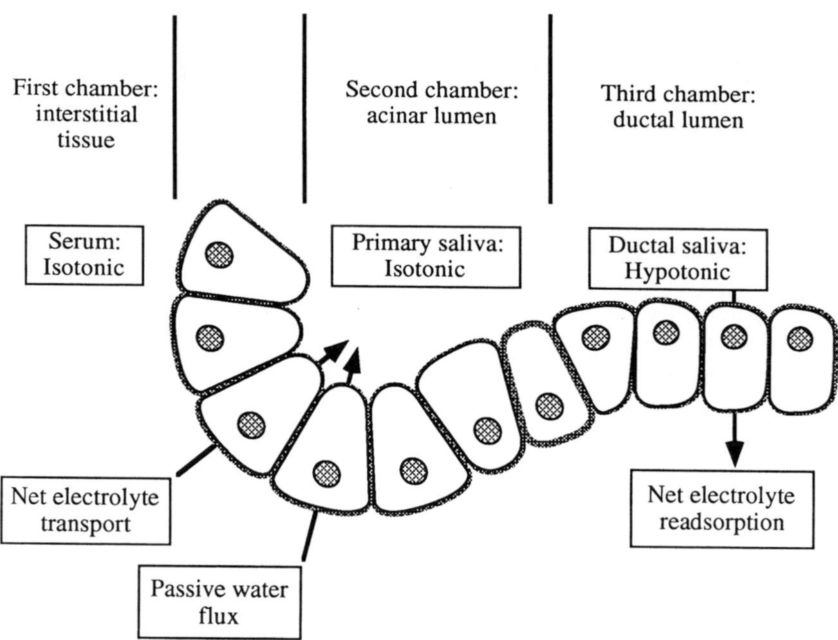

FIGURE 2 Fluid and electrolyte transport.

fluid, raise the osmotic pressure in the acinar lumen and attract a passive influx of water. This combined action creates a force that drives the primary saliva against the hydrostatic pressure (the second "barrier") created by saliva already in the ductal system.

As unstimulated or resting saliva passes through the salivary duct system, ion concentrations relative to serum are modified such that sodium, chloride, and bicarbonate are decreased and potassium is increased. Sodium is readsorbed across the luminal membrane of the duct cells by an energy-requiring mechanism in a direct or indirect exchange for potassium. As the sodium is pumped into the extraductal stroma, it attracts chloride across the ductal barrier by a transcellular route through the junctional complex. However, with glandular stimulation, sodium and chloride concentrations increase to near serum levels, the bicarbonate ion concentration increases to levels far above those of serum, and the potassium ion concentration decreases slightly. As salivary flow increases, the ductal transport mechanisms become saturated and allow sodium and chloride ions to be discharged with the salivary secretion. Collectively, these processes result in a hypotonic secretion (Table I).

C. Neural Control of Salivary Secretion

Protein secretion and fluid transport are triggered by the synergistic influence of both sympathetic and parasympathetic neurons of the autonomic nerve system. Neural impulses, however, cannot cross the plasma membrane and must be converted to intercellular chemical messages by the process of signal transduction. Salivary glands possess two functionally distinct signal transduction mechanisms: the adenylate cyclase and the phosphatidylinositol systems. These can be segregated into three components: (1) an extracellular receptor for neural transmitters; (2) an enzyme system, which synthesizes intracellular chemical messengers; and (3) coupling proteins, which regulate the receptor–enzyme interaction.

Separate functions of acinar and ductal cells appear to be activated by different types of autonomic neurons, which activate the synthesis of specific intracellular messengers. Stimulation of β-adrenergic (sympathetic) neurons, for instance, initiates protein synthesis by activating the adenylate cyclase system to produce the messenger cyclic adenosine monophosphate (cAMP). In contrast, stimulation of α_1-adrenergic (sympathetic) and parasympathetic neurons results in fluid and electrolyte transport following the synthesis of the intercellular messenger inositol triphosphate (IP_3) by the phosphatidylinositol system. Similarly, ductal readsorption and secretion of salivary electrolytes are controlled by the effect of second messengers on the ductal epithelium following the selective stimulation of sympathetic as well as parasympathetic neurons. [*See* Autonomic Nervous System.]

TABLE I
Composition of Major Human Salivary Secretions[a]

	Constituent	Parotid[b]	Submandibular[b]	Sublingual[c]	Plasmma
Electrolytes (mEq/liter)	Potassium	21.0/24.0	17.0/14.4	13.2	4.0
	Sodium	36.0/1.3	45.0/3.3	32.7	140.0
	Chloride	28.0/22.0	25.0/12.0	26.2	105.0
	Bicarbonate	30.0/1.1	18.0/4.0	10.9	27.0
	Calcium	1.6/1.1	2.4/1.56	2.1	5.0
	Magnesium	0.12/0.16	0.04/0.07	?	2.0
	Phosphate	3.7/9.0	5.5/5.6	4.1	2.0
			Submandibular/sublingual		
Organics[c] (mg/100 ml)	Protein	221.0	132.0		7000
	Lipids	8.0	8.0		600
	Carbohydrate[d]	31.0	15.0		100–140

[a]From Ferguson (1989), Levine *et al.* (1978), and Slomiany *et al.* (1982).
[b]Mean values for stimulated/unstimulated saliva.
[c]Mean values for stimulated salivas.
[d]Carbohydrate content of glycoproteins.

III. COMPOSITION OF SALIVA

The total protein, carbohydrate, and lipid concentrations of the major human salivary secretions are considerably less than that in serum (Table I). Salivary gland secretions contain at least 40 proteins and glycoproteins, many of which can be grouped into at least seven families whose members are genetically and structurally related (Table II). In addition, singular components may have relatives in other exocrine secretions. The secretions of all salivary gland groups, combined with oral microflora, gingival secretions, and sloughed or desquamated oral epithelium, result in a mixture termed whole saliva. Only the glandular products will be discussed in the following sections.

A. Families of Salivary Proteins

Multiple members of families of salivary proteins within the same individual have been shown to result from a small group of gene transcripts by posttranscriptional and/or posttranslational modifications. The family of proline-rich proteins (PRPs) has been well studied and serves to illustrate this phenomenon. To date, more than 20 PRPs have been grouped into three major categories: acidic, basic, and glycosylated species. All are related by their high content of proline, glycine, and glutamine, but each group is distinguished by repeated amino acid sequences or posttranslational modifications that endow them with a distinct isoelectric point (acidic or basic) and/or carbohydrate content. The PRPs are now thought to be derived from the transcripts of six gene loci through differential gene splicing. These proteins may then undergo additional posttranslational modifications in which carbohydrate as well as phosphate groups are added. In addition, some of the acidic PRPs are thought to undergo proteolytic processing in the salivary duct by the enzyme kallikrein.

Other salivary families are composed of groups of proteins that share similar structural and functional characteristics. For instance, human salivary mucins exist as two distinct, highly glycosylated molecules designated MG1 and MG2. MG1 is composed of multiple, disulfide-linked subunits comprising a molecule of >1000 kDa. The MG1 oligosaccharides may be sialylated (i.e., terminate with a negatively charged sialic acid residue) or sulfated and are located in densely glycosylated regions, which are separated from nonglycosylated or "naked" peptide regions. MG1 also contains a small amount of covalently linked fatty acids (i.e., acylated) and possesses hydrophobic domains in the naked core regions. In contrast, MG2 is a smaller molecule of 120–130 kDa, which contains a single polypeptide chain uniformly glycosylated with small oligosaccharide chains to give this molecule a "bottle brush" configuration. Another family of glycoproteins, the amylases, is synthesized as two subfamilies composed of glycosylated and nonglycosylated isoenzymes. Each of these subfamilies contains isoenzymes, which may have only slight charge variations produced by posttranslational deamidation of glutamine residues in the parent molecule. Two additional families of enzymes, the salivary

TABLE II
Salivary Families

Family name	Function	Biochemical composition
Mucins	1. Selective clearance and adherence of microflora 2. Tissue-coating and formation of intraoral pellicles 3. Lubrication at hard and soft tissue interfaces 4. Microbial nutrient source 5. Digestion and taste 6. Complexing with lysozyme, cystatin, and sIgA	Glycoproteins
Acidic and basic proline-rich proteins	1. Selective clearance and adherence of microflora 2. Tissue-coating and formation of intraoral pellicles 3. Lubrication at hard and soft tissue interfaces 4. Microbial nutrient source 5. Modulation of mineralization processes on tooth surfaces 6. Complexing with albumin	Phosphoproteins and glycoproteins
Cystatins	1. Antimicrobial activity 2. Complexing with other salivary molecules to coat oral surfaces 3. Modulation of mineralization processes on tooth surfaces 4. Thiol protease inhibition	Proteins and phosphoproteins
Histatins and statherin	1. Antimicrobial activity 2. Tissue-coating and formation of intraoral pellicles 3. Modulation of mineralization processes 4. Buffering of salivary pH	Proteins and phosphoproteins
Amylase	1. Digestion of complex carbohydrates 2. Tissue-coating and formation of intraoral pellicles 3. Selective clearance and adherence of microflora 4. Antimicrobial activity 5. Digestion and taste 6. Complexing with other salivary molecules to coat oral surfaces	Proteins and glycoproteins
Carbonic anhydrases	1. Reduce salivary pH by catalyzing the formation of bicarbonate from carbon dioxide	Glycoproteins
Salivary peroxidases	1. Catalysis of the formation of products that are toxic to some oral bacteria	Glycoproteins

peroxidases and the carbonic anhydrases, each contain members that differ slightly in physicochemical characteristics but perform the same catalytic functions. Other families of salivary proteins, such as the cystatins and histatins, are categorized by common compositional characteristics. Cystatins, for instance, contain a characteristic content of cysteine amino acids, whereas histatins are rich in histidine residues.

B. Singular Salivary Components

The singular salivary components (Table III) include kallikrein, lactoferrin, lysozyme, secretory IgA (sIgA), and fibronectin. With the exception of sIgA, these components are synthesized primarily by ductal cells; however, some are synthesized in lesser quantities by acinar cells. Secretory IgA is synthesized as monomeric IgA by plasma cells surrounding the acinar cells. The salivary acinar cells then import monomeric IgA, process it, and secrete the modified immunoglobulin, secretory sIgA, into saliva. Kallikrein is a serine protease that is secreted by the striated ducts of the major salivary glands. Both lactoferrin and lysozyme are antimicrobial agents that share identity with similar agents in tears, gastric mucosal secretions, and specific granules of neutrophils. Fibronectin, a substance that mediates intercellular adherence interactions in most human tissues, also appears to be present in saliva.

IV. FUNCTION OF SALIVA

In general, salivary constituents play a protective role either as individuals or in molecular complexes. Each of these constituents also appears to possess unique physicochemical characteristics that allow them to

TABLE III
Individual Salivary Molecules

Molecule	Function	Biochemical composition
Fibronectin	1. Tissue-coating and formation of intraoral pellicles 2. Mediates microbial adherence	Glycoprotein
Kallikrein	1. Posttranslational processing of proline-rich proteins and cystatins	Glycoprotein
Lactoferrin	1. Antimicrobial activity 2. Complexing with other salivary molecules to coat oral surfaces	Glycoprotein
Lysozyme	1. Antimicrobial activity 2. Complexing with other salivary molecules to coat oral surfaces	Protein
Secretory IgA	1. Complexing with other salivary molecules to coat oral surfaces 2. Mediates clearance and adherence of microflora 3. Antimicrobial activity	Glycoprotein

express their biologic function either free in solution or when adsorbed to oral surfaces. The functional characteristic(s) of individual salivary components is dependent on their structural characteristics; thus, alteration of their structure by host and/or microbial enzymes alters their functional characteristics. The collective result of all of these functions serves to (1) maintain the microbial ecology of the oral cavity, and (2) prepare food for swallowing and digestion, and/or (3) preserve the integrity of oral tissues.

A. Tissue Coating

Salivary components adsorb to tooth enamel, dental materials, microbial surfaces, and epithelial cells with a selectivity that depends on the physicochemical characteristics of the individual adsorbent surfaces as well as that of the salivary component. Adsorbed components, in turn, can complex with other salivary constituents to form protective films called pellicles, which function as lubricants, permeability barriers against acids, moisture retainers, and modulators of microbial adherence. Complexing among salivary components may act to concentrate and possibly enhance the functional characteristics of an individual salivary component. For instance, salivary mucins coat surfaces and also form complexes with antimicrobial factors such as sIgA, lysozyme, and cystatins. This may serve to localize and concentrate these substances on oral surfaces and increase antimicrobial activity. Additionally, proline-rich glycoprotein, a salivary lubricant, provides enhanced lubrication when complexed with human serum albumin. Other functional characteristics of salivary pellicles depend on the carbohydrate moieties of their salivary constituents. Moisture retention, for instance, is primarily mediated by the carbohydrate moieties. Highly glycosylated components such as mucin and proline-rich glycoprotein also provide better lubrication than that of less glycosylated salivary components such as lactoferrin, amylase, and secretory IgA. Enzymatic alteration of salivary proteins prior to or after they are adsorbed to oral surfaces (i.e., structural modulation) may also affect their functional characteristics. For instance, a relatively nonpathogenic bacteria, *Streptococcus sanguis,* utilizes a sialic acid-binding adhesion on its surface to bind to oral surfaces coated with salivary glycoproteins containing terminal sialic acid residues. Other oral bacteria produce an enzyme, neuraminidase, which may cleave these sialic acid residues to expose underlying galactose residues. Subsequently, putative oral pathogens such as *Actinomyces viscosis* or *Streptococcus mutans,* which possess galactose-binding proteins, are then able to adhere to these surfaces and initiate a disease process.

B. Antimicrobial Activity

Salivary peroxidase, lysozyme, lactoferrin, histatins, and sIgA all exert antimicrobial activity. The salivary peroxidase system utilizes dietary thiocyanate ions (SCN^-) and bacterial hydrogen peroxide to synthesize hypothiocyanate, a substance that reversibly inhibits bacterial growth and metabolism. Lysozyme cleaves carbohydrate linkages in the cell wall of gram-positive bacteria, making them susceptible to changes in osmotic pressure and causing them to burst in the hypotonic environment of the oral cavity. Lactoferrin is a noncatalytic iron-binding molecule that exerts bacteriostatic effects by binding iron, an essential bacterial nutrient. Additional studies suggest that lactoferrin may possess a direct, iron-independent, bacterocidal effect on various strains of streptococci. The histatins can inhibit the viability of the oral pathogen *Candida albicans* and can also inhibit the growth of *S. mutans*. Secretory IgA (sIgA) is the primary component of the oral mucosal immune system. sIgA is thought to produce antimicrobial action by specifically binding oral microbes to prevent them from adhering to and colonizing oral surfaces.

C. Posttranslational Processing

In vitro studies have shown that salivary kallikrein cleaves the long forms of the acidic proline-rich proteins to produce a shorter form of these proteins and a small C-terminal peptide. Similarly, the N-terminal portion of salivary cystatins may also be processed. The immunohistochemical localization of kallikrein suggests that this cleavage takes place in the ductal system. The biological significance of these events remains to be determined.

D. Digestion

Digestion is primarily an enzymatically mediated function of the lower alimentary canal. Saliva contains two enzymes that contribute to this process. The first is α-amylase, one of the most abundant salivary components, which hydrolyzes dietary starch into smaller fragments. Lipase, the second enzyme, is secreted by the minor salivary glands of Von Ebner at the base of the tongue and is thought to play a role in the initial digestion of lipids. In addition, saliva assists in the hydration and dispersion of the food particles during the mastication or chewing process. This aids in the formation and lubrication of the food bolus in preparation for passage through the esophagus and provides a fluid medium that is beneficial to the processes involved in taste.

E. Buffering Capacity

Food residue left on teeth after meals is rapidly converted by oral bacteria to organic acids which can cause decalcification of tooth structure and subsequent tooth decay (dental caries). Saliva provides several buffering mechanisms to counteract this process. Salivary bicarbonate, the primary buffering agent in saliva, is produced in the salivary ductal cells. Bicarbonate may also be formed directly in the oral cavity by the action of carbonic anhydrases. Once in the oral cavity, bicarbonate ions may be complexed to salivary mucins, which are adsorbed to oral surfaces. This may enhance the protective nature of mucins by producing a buffered barrier against acid penetration to oral mucosa and tooth enamel. Other salivary components may act as buffers by virtue of their amino acid content. Histidine-rich peptides (i.e., histatins) contain a high content of the basic amino acid histidine, which may act to neutralize acidic by-products of bacterial metabolism.

F. Mineralization Processes

Tooth enamel is composed of a relatively insoluble calcium-phosphate mineral termed hydroxyapatite. Under normal conditions of ionic strength and pH, this mineral will slowly dissolve in a saliva that is devoid of its protein constituents. However, with normal salivary flow and composition and minimal exposure to bacterial acids, individuals do not lose their teeth to dissolution over long periods of time. In fact, decalcified enamel associated with an early carious lesion will remineralize if the tooth surface is regularly cleaned and allowed to come in contact with saliva. Evidence suggests that the acidic residues, including phosphate, of salivary phosphoproteins may bind calcium in saliva to produce a much higher concentration of calcium than would otherwise be possible. This supersaturation constitutes a thermodynamic driving force that favors the formation of calcium phosphate salts (i.e., remineralization) on the surface of the teeth. [*See* Dental and Oral Biology, Anatomy.]

V. SALIVARY DYSFUNCTION AND THERAPY

A. Etiologies of Salivary Dysfunction

Some variations in salivary flow rates and composition are part of normal salivary physiology. In contrast, sustained or permanent alterations in these parameters are considered pathologic and are referred to as salivary dysfunction. Salivary glands may be affected by local or systemic factors, which result in a transient or permanent glandular dysfunction, possibly glandular destruction. Factors that cause salivary gland dysfunction include (1) physical obstruction of salivary duct, (2) destruction of glandular paren-

TABLE IV
Factors Affecting Salivary Gland Function[a]

Category	Condition	Effect on salivary gland
Factors that directly affect the gland	Ductal obstruction	Pressure atrophy of the acinar cells
	Acute and chronic inflammation	Immune-mediated destruction of glandular parenchyma
	Head and neck radiation therapy	Destruction of the glandular parenchyma by radiation
	Trauma	Neural or parenchymal destruction
	Benign and malignant neoplasm	Neoplastic infiltration of the glandular parenchyma
Systemic diseases	Sjögrens's syndrome	Autoimmune destruction of the glandular parenchyma
	Graft versus host disease	Inflammatory destruction of the glandular parenchyma
	Viral infection (such as mumps)	Inflammatory destruction of the glandular parenchyma
	Cystic fibrosis	Increase salivary viscosity, altered electrolyte readsorption
	Dehydration in diabetic acidosis, uremia, etc.	Decreased salivary flow

[a]Adopted from Mandel (1980) and Baum *et al.* (1985).

chyma, and/or (3) pharmacologic alteration of secretory mechanisms (Tables IV and V).

Ductal obstruction often completely blocks the salivary duct but does not stop acinar secretion. This results in an increased intraductal pressure, which can lead to atrophy of the acinar cells and loss of glandular function. Primary factors that lead to obstruction of the salivary duct are the formation of mineralized stones and/or trauma. These conditions are often treated by removal of the obstruction or repair of the salivary duct. Obstructive dysfunction may also be a manifestation of systemic disease such as cystic fibrosis or tumor involvement of the salivary duct. Cystic fibrosis is a genetically inherited disease that is associated, in part, with a highly viscous secretion in mucin-producing glands, which results in obstruction of the salivary duct by "mucin plugs." Salivary gland or nonsalivary head and neck tumors can also compress or involve the salivary gland ducts. In both of these cases, treatment of the primary disease usually dictates the resolution of the salivary dysfunction. [*See* Cystic Fibrosis, Molecular Genetics.]

Destruction of the salivary gland parenchyma may take place as the result of autoimmune disease or radiation therapy, or as a consequence of infection. Autoimmune involvement of salivary glands is most commonly associated with the primary and secondary forms of sicca syndrome. The primary form of sicca syndrome is characterized by involvement of the lacrimal and salivary glands, which is accompanied by symptoms of both dry eyes and dry mouth. The secondary form (Sjögrens's syndrome) involves at least one additional connective tissue disease, which may include systemic lupus erythematosus, scleroderma, chronic hepatobiliary disease, and Raynaud's phenomenon. Both forms of sicca syndrome are characterized by a localized inflammatory infiltrate that results in swelling of the salivary gland and progressive destruction of the glandular parenchyma. While sicca

TABLE V
Effects of Pharmacologic Agents on Salivary Glands[a]

Effect on salivary gland	Drug category	Representative drug
Increased salivary flow	Centrally acting drugs	Strychnine Reserpine
	Parasympathomimetic drugs	Muscarine Neostigmine Physostigmine
	Sympathomimetic drugs	Norepinepherine Epinepherine Ephederine
Decreased salivary flow	Centrally acting drugs	General anesthetics Barbituates
	Parasympatholytic	Atropine Scopolamine Antihistamines
	Sympatholytic	Phentolamine Ergotamine Clorporomazine
	Ganglionic blockers	Psychoactive drugs Hexamethonium
Altered salivary composition	Cardiac glycosides	Digitalis
	Cancer chemotherapy drugs	Methotrexate Vincristine Cytoxan

[a]Adopted from Mandel (1980), Baum *et al.* (1985) and Sreebny and Broich (1987).

syndrome patients may retain partial salivary gland function, this disease often progresses to glandular destruction and almost complete loss of function. An immune-related glandular dysfunction similar in histological appearance to sicca syndrome has also been reported in patients receiving bone marrow transplants for diseases such as leukemia. In these patients, all immune component cells are killed by whole-body radiation after which a patient receives new immune cells in the form of a normal marrow transplant. In some patients, the transplanted immune cells recognize the new host as foreign and mount an immune response against the salivary glands among other tissues. This response has been associated with symptoms of dry mouth and decreased secretion of sIgA in minor salivary glands.

All of the salivary glands can also be damaged by the doses of radiation used in the treatment of head and neck tumors with each gland varying in its sensitivity to radiation. Immediately following radiation therapy, <5% of parotid gland function remains. Due to a higher resistance to radiation, the submandibular-sublingual glands initially retain approximately 25% of their function. However, their residual function decreases to <5% after 3 years.

Infection of salivary glands may be the result of a primary systemic infection (e.g., mumps) or localized retrograde infections, which are secondary to other glandular dysfunction. Mumps is a viral infection transmitted through direct or indirect contact with saliva containing the mumps virus. The primary target of this virus is the parotid glands, but it may also infect the gonads, pancreas, central nervous system, and heart. Mumps is characterized by parotid gland swelling and focal necrosis of the glandular parenchyma. This is a transient self-limiting disease that runs its clinical course in 3–7 days and does not result in permanent glandular dysfunction. Localized retrograde infections are generally secondary to reduced salivary flow. The flow of saliva through the salivary duct provides a natural countercurrent barrier against the migration of oral bacteria into the salivary gland. During decreased salivary flow, bacteria can migrate up the salivary ducts and produce acute or chronic suppurative infections of the glandular parenchyma. This type of infection can occur with any condition that produces reduced salivary flow and commonly occurs during reduced salivary flow associated with dehydration in infant, elderly, or debilitated patients.

Many pharmacologic agents are known to affect the autonomic nerve system by altering central nervous function by affecting either nervous impulses at autonomic ganglia and/or postganglionic synapses. At the postganglionic level, sympathomimetic and parasympathomimetic drugs generally increase salivary flow by stimulating fluid transport and acinar secretion, whereas sympatholytic and parasympatholytic drugs decrease salivary flow (Table V). Many drugs used to treat serious conditions such as Parkinson's disease, various psychoses, and hypertension produce decreased salivary flow or possibly altered salivary composition. However, in a therapeutic context, stimulatory drugs such as pilocarpine have been tested in clinical trials to treat patients with decreased salivary flow.

B. Therapy of Salivary Dysfunction

In many cases, patients with significant salivary flow report a perception of dry mouth, whereas others with little salivary flow have no sensation of dryness. Thus, the treatment for salivary dysfunction should only be initiated after determining the extent of salivary flow. In many instances, a differential diagnosis should also include radiographic and biopsy procedures as well as salivary flow measurements. The information obtained will enable the clinician to determine if patients with deficient salivary function can be stimulated to produce additional saliva (i.e., intrinsic therapy) or if the patient will require topical replacement of lost salivary flow (i.e., extrinsic therapy). Intrinsic therapy is based on a pharmacologic approach in which agents such as dilute citric acid, pilocarpine, and bromhexine are used to stimulate any residual function in the salivary glands. Therapy with drugs such as pilocarpine produces inherent side effects that have prevented this method of therapy from progressing beyond the experimental stage. The extrinsic approach is the most widely practiced method of therapy and is predicted on periodic oral rinsing with salivary replacements. The most commonly used preparations are formulated from carboxymethylcellulose and the sugar alcohols, xylitol and sorbitol. However, these preparations exhibit low oral retention as well as poor taste and mucosa-irritating characteristics. A more therapeutic preparation should have sustained effects that would include lubrication, tissue-coating and moistening properties, and selected antimicrobial function. Research directed at the chemical and physical characterization of salivary proteins and glycoproteins now suggests that discrete portions of these molecules endow them with their functional characteristics. This knowledge is now being used to design and synthesize composite molecules that will contain multiple protective functions. This approach should result in saliva replacements that have enhanced functional characteristics compared to authentic saliva.

BIBLIOGRAPHY

Abdel-Latif, A. A. (1986). Calcium-mobilizing receptors, polyphosphoinositides, and the generation of second messengers. *Pharm. Rev.* **38,** 227–272.

Arnold, R. R., Brewer, M., and Gauthier, J. J. (1980). Bacterial activity of human lactoferrin: Sensitivity of a variety of microorganisms. *Infect. Immun.* **28,** 893–898.

Baum, B. J. (1987). "Regulation of salivary secretion." *in* "The Salivary System" (L. M. Sreebny, ed.), Chapter 6, CRC Press, Boca Raton, FL.

Baum, B. J., Bodner, L., Fox, P. C., Izutsu, K. T., Pizzo, P. A., and Wright, W. E. (1985). Therapy-induced dysfunction of salivary glands: Implications for oral health. *Spec. Care Dent.* **5,** 274–277.

Bobek, L. A., and Levine, M. J. (1992). Cystatins, the cysteine proteinase inhibitors. *CRC Rev. Oral Biol.* **3,** 307–332.

Bobek, L. A., Tsai, H., Biesbrock, A. R., and Levine, M. J. (1993). Molecular cloning, sequence and specificity of expression of the gene encoding the low molecular weight human salivary mucin (MUC7). *J. Biol Chem.* **268,** 20563–20569.

Cohen, R. E., and Levine, M. J. (1989). Salivary Glycoproteins. *In* "Human Saliva: Clinical Chemistry and Microbiology" (J. O. Tenovuo, ed.), Vol. 1, Chapter 4. CRC Press, Boca Raton, FL.

Douglas, W. H., Reeh, E. S., Ramasubbu, N., Bhandary, K. K., Raj, P. A., and Levine, M. J. (1991). Statherin: A major boundary lubricant of human saliva. *Biochem. Biophys. Res. Commun.* **180,** 91–97.

Ferguson, D. B. (1989). Salivary electrolytes. *In* "Human Saliva: Clinical Chemistry and Microbiology" (J. O. Tenovuo, ed.), Vol. 1, Chapter 3. CRC Press, Boca Raton, FL.

Gardner, E. A., Gray, D. J., and O'Rahilly, R. (1986). "Anatomy: A Regional Study of Human Structure," 5th Ed., Chapters 58 and 59. Saunders, Philadelphia.

Goedert, M., Naby, J. I., and Emson, P. C. (1982). The origin of substance-P in the rat submandibular gland and its major ducts. *Brain Res.* **252,** 327–333.

Iacono, V. J., MacKay, B. J., Direnzo, S., and Pollock, J. J. (1980). Selective antibacterial properties of lysozyme for oral microorganisms. *Infect. Immun.* **29,** 623–632.

Izutsu, K. T. (1987). Salivary fluid production in health and disease. *In* "The Salivary System" (L. M. Sreebny, ed.), Chapter 5. CRC Press, Boca Raton, FL.

Knauf, H., Lubcke, R., Kreutz, W., and Sachs, G. (1982). Interrelationship of ion transport in rat submaxillary duct epithelium. *Am. J. Physiol.* **242,** F132–F139.

Levine, M. J. (1993). Development of artificial salivas. *Crit. Rev. Oral Biol. Med.* **4,** 279–286.

Levine, M. J. (1993). Salivary Macromolecules: A structure/function synopsis. *Ann. N.Y. Acad. Sci.* **694,** 11–16.

Levine, M. J., Aguirre, A., Hatton, M. N., and Tabak, L. A. (1987). Artificial salivas: Present and future. *J. Dent. Res.* (Special Issue) **66,** 693–698.

Levine, M. J., Herzberg, M. C., Ellison, S. A., Shomers, J. P., and Sadowski, G. A. (1978). Biochemical and immunological comparison of monkey (*Macaca arctoides*) and human salivary secretions. *Comp. Biochem. Physiol. B* **60,** 423–431.

Mandel, I. D. (1980). Sialochemistry in disease and clinical situations affecting salivary glands. *CRC Crit. Rev. Clin. Lab. Sci.* **12,** 321–366.

Minaguchi, K., and Bennick, A. (1989). Genetics of human salivary proteins. *J. Dent. Res.* **68,** 2–15.

Oppenheim, F. G., Xu, T., McMillian, F. M., Levitz, S. M., Diamond, R. D., Offner, G. D., and Troxler, R. F. (1988). Histatins, a novel family of histidine-rich proteins in human parotid secretion: Isolation, characterization, primary structure and fungistatic effects on *Candida albicans*. *J. Biol. Chem.* **16,** 472–477.

Raj, P. A., Soni, S. D., and Levine, M. J. (1994). Solvent induced helical conformation of an active candidacidal fragment of salivary histatins. *J. Biol. Chem.* **269,** 9610–9619.

Scannapieco, F. A., Torres, G., and Levine, M. J. (1993). Salivary α-amylase: Role in dental plaque and caries formation. *Crit. Rev. Oral Biol. Med.* **4,** 301–307.

Schenkels, L. C. P. M., Gururaja, T. L., and Levine, M. J. (1996). Salivary mucins: Their role in oral mucosal barrier function and drug delivery. *In* "Oral Mucosal Drug Delivery" (M. J. Rathbone, ed.), Chapter 9, pp. 191–220. Dekker, New York.

Sicher, H., and DuBrul, L. E. (1970). "Oral Anatomy," 5th Ed., pp. 191–196. Mosby, St. Louis.

Slomiany, B. L., Murty, V. L. N., Aono, M., Slomiany, A., and Mandel, I. D. (1982). Lipid composition of human parotid and submandibular saliva from caries-resistant and caries-susceptible adults. *Arch. Oral Biol.* **27,** 803–808.

Sreebny, L. M., and Broich, G. (1987). "Xerostomia (dry mouth)." *In* "The Salivary System" (L. M. Sreebny, ed.), Chapter 9. CRC Press, Boca Raton, FL.

Sreebny, L. M., and Valdini, A. (1987). Xerostomia: A negleced symptom. *Arch. Intern. Med.* **147,** 1333–1337.

Tenovuo, J., and Pruitt, K. M. (1984). Relationship of the human salivary peroxidase system to oral health. *J. Oral Pathol.* **13,** 573–584.

Salmonella

M. A. ROUF
University of Wisconsin, Oshkosh

GLOSSARY

Carrier Apparently healthy person harboring a pathogen

Facultative anaerobe Organism that is able to grow in the presence or absence of oxygen

Phagovar Subdivision of a serovar based on the sensitivity to a series of bacteriophages at appropriate dilutions

Plasmid Small piece of circular DNA besides bacterial chromosomal DNA

Salmonellosis Infection of the gastrointestinal tract by *Salmonella*

Serovar (serotype) Subdivision of a species based on its antigenic analysis

THE SALMONELLAE ARE PATHOGENIC FOR humans, causing enteric fever, gastroenteritis, and septicemia. They also infect many animal species besides humans. Animals are the main reservoirs of *Salmonella* species. About 2200 serovars of *Salmonella* have been identified based on "O" and "H" and "Vi" antigens. The serovars and species, although equated, are not the same. The O antigen, also known as somatic antigen, is the outer portion of the endotoxin complex. The H antigen (flagellar antigen) is used in conjunction with O antigen to identify serovars. The Vi antigens occur primarily in *Salmonella typhi*. Each serovar of *Salmonella* is characterized by its unique structure and arrangement of simple sugars in the O-specific polysaccharide chain.

Salmonella cells enter the animal body by ingestion of contaminated food or drink or via person-to-person contact. The infection begins in the intestinal tract. Disease manifestations by *Salmonella* depend on the virulence and the type of the serovars, general health of the animal, and the number of viable cells ingested. Typhoid fever is caused by *S. typhi*. It is a systemic infectious disease resulting in prolonged fever. Paratyphoid fever is relatively mild and of shorter duration compared with typhoid fever. Although paratyphoid fever may be caused by any serovars of *Salmonella*, the serovars most often responsible are *S. paratyphi* A, *S. schottmulleri, S. hirschfeldii,* and *S. senadi*. The most effective treatment for enteric fevers is administration of antibiotics. Typhoid vaccines are available and offer important protection. Any of the 2200 serovars that are ubiquitous in nature can cause gastroenteritis. Most commonly *S. typhimurium* and *S. enteritidis* are responsible for foodborne infection. Poultry, meat, and eggs, and recently milk and cheese, are the main sources of transmission of nontyphoidal salmonellosis. Septicemia is most often caused by *S. choleraesuis,* followed by *S. typhimurium*.

I. CHARACTERISTICS AND CLASSIFICATION

Salmonella is a genus of bacterium that belongs to the family Enterobacteriaceae. The organisms are facultatively anaerobic, gram-negative, non-spore-forming, straight rods, and 0.7–1.5 millimicron in width and 2–5 millimicron in length. Most are motile by means of peritrichous flagella (flagella arising out of

ENCYCLOPEDIA OF HUMAN BIOLOGY, Second Edition, VOLUME 7. Copyright © 1997 by Academic Press. All rights of reproduction in any form reserved.

the entire cell surface). Colonies in typical culture media are 2–4 mm in diameter.

The typhoid bacillus (*S. typhi*) was the first organism in this genus to be observed in tissue of dead patients by Eberth in 1880. The organism at that time was known as *Eberthella.* The genus is now named after the American bacteriologist D. E. Salmon.

In 1884, Gaffky was able to culture typhoid bacteria from mesenteric lymph nodes, and Pfeiffer cultured them from a fecal specimen. In 1885, Salmon and Smith isolated *Salmonella* from cases of swine fever and named the organism *Bacillus cholera suis.* Swine fever, however, is caused by a virus; the authors mistakenly thought it was a bacterial disease. *Salmonella enteritidis* was isolated by Gaertner in 1888 from patients who ate contaminated meat and developed food poisoning.

As most etiological agents of various infectious diseases were isolated and identified during the period leading up to the 1930s, the number of organisms in this genus increased rapidly in number, creating a confusion in naming of the species, which exists even today. Presently, the bacteria in the genus *Salmonella* are classified according to antigenic relationship. About 2200 serovars have been described, and new ones are being added from time to time. Although antigenic relationship is useful in classifying *Salmonella,* it is taxonomically unsound.

A. Taxonomy

Based on biochemical characteristics, the genus *Salmonella* is presently subdivided into five subgenera, namely, subgenera I–V (Table I). The subgenus V includes organisms that grow in the presence of potassium cyanide (KCN) (like those of subgenus IV) but have other biochemical differences from all other subgenera. They are further divided into about 2200 serovars (serotypes) based on the Kauffman-White schema. Each is given a specific epithet. For example, *S. typhi* is the name of a serovar, not the name of a species, although it has a specific epithet. The use of a species name to describe a serovar (serotype) has led to much confusion and misconception of equating serovars to species, which is a definite taxonomic term.

TABLE I
Differential Characteristics of the "Subgenera" of the Genus *Salmonella*[a,b]

	Subgenus				
	I	II	III	IV	V[c]
β-Galactosidase (ONPG test)	−	− or x	+	−	+
Acid production from:					
Lactose	−	−	+ or x	−	−
Dulcitol	+	+	−	−	+
Mucate	+	+	d	−	+
Galacturonate[d]	−	+	d	+	+
Utilization of:					
Malonate	−	+	+	−	−
d-Tartrate	+	− or x	− or x	− or x	−
Gelatin hydrolysis (film method)	−	+	+	+	−
Growth in presence of KCN	−	−	−	+	+
Habitat of the majority of strains:					
Warm-blooded animals	+	−	−	−	−
Cold-blooded animals and environment	−	+	+	+	+

[a]Symbols: +, positive for 90% or more of strains in 1–2 days; d, positive for 11–89% of strains in 1–2 days; −, positive for 0–10% of strains in 1–2 days; x, late and irregularly positive (3–7 days). The temperature for all reactions is 37°C.

[b][Reprinted, with permission, from L. Le Minor (1984). Genus III *Salmonella. In* "Bergey's Manual of Systematic Bacteriology" (N. R Krieg and J. G. Holt, eds.), Vol. 1, p. 427. Williams & Wilkins, Baltimore.

[c]From L. Le Minor, M. Veron, and M. Popoff (1982). *Ann. Microbiol.* (*Inst. Pasteur*) **133B,** 223–243.

[d]From Le Minor *et al.* (1979). Monophasic serovars of "subgenus" III are galacturonate-negative; Diphasic serovars are positive.

The naming of *Salmonella* is controlled by an international agreement. According to this system, a serovar is named after the place where it was first isolated (e.g., *S. bangkok, S. dakar, S. karachi, S. london, S. miami*).

The classification of *Salmonella* remains unsatisfactory. In "Bergey's Manual of Systematic Bacteriology" (1984), it is stated that employing specific names for *Salmonella* serovars is extremely useful, although serovars and species should not be regarded as equivalent. It is further stated that DNA relatedness data of organisms distinctly belonging to the five subgroups (subgenera) show that they belong to a single genetic *Salmonella* species. Therefore, it is proposed as "a single diverse species with five subspecies." The suggestion is made to designate "*S. enterica*" as the only species. It was also suggested that in medical bacteriology Latin binomials (generic and specific names) could be used for serovars in subgenus I and for the named serovars of the subgenera II and IV. Serovars in subgenera II, III, and IV, which have not been named, could be listed as subspecies followed by antigenic formulae. On the basis of DNA relatedness, the editorial board of "Bergey's Manual of Systematic Bacteriology," without any specific recommendation, stated that the genus *Salmonella* should consist of a single species, *S. choleraesuis,* with six subspecies as suggested by LeMinor, Veron, and Popoff. The following organisms, respectively, were proposed as the type species for the six subspecies: *S. choleraesuis, S. salamae, S. arizonae, S. diarizonae, S. houtenae,* and *S. bongori.*

Because of the complexity and confusion in the taxonomy of *Salmonella* in terms of species, some microbiologists recognize only three species, *S. choleraesuis, S. typhi,* and *S. enteriditis* (containing approximately 2000 of about 2200 of *Salmonella* serovars).

The sensitivity of *Salmonella* isolates to various bacteriophages is also used to classify them into phagovars. A serovar can be divided into biovars based on different sugar fermentation patterns. Biovars may serve as markers to identify pathogens in epidemic outbreaks.

B. Physiological and Biochemical Characteristics

The salmonellae have simple nutritional requirements. They grow on most routine bacteriological growth media without enrichment or special supplement. They are facultatively anaerobic (i.e., they do not require oxygen for growth but grow well in both presence or absence of oxygen). Their optimum growth temperature is 37°C, but they grow reasonably well at room temperature.

Salmonella does not ferment lactose except for some organisms in the subgenus *Salmonella* III (*arizonae*). Also, they usually do not ferment sucrose, salicin, inositol, and amygdalin. Glucose is fermented usually with the production of gas, except in *S. typhi,* which does not produce gas from either glucose or other carbohydrates. Typically, H_2S is produced and citrate utilized as the sole source of carbon. They are indole- and urease-negative and unable to liquefy gelatin. Lysine and ornithine decarboxylase reactions are usually positive. The differential characteristics of the various subgenera are given in Table I.

C. Antigenic Differentiation

All *Salmonella* are identified antigenically by a simple agglutination test using intact cells and monospecific antisera. The salmonellae possess three major antigens called the O, H, and Vi antigens. The detection of Vi antigen, which occurs primarily in *S. typhi* and a few other serovars, can be used as a screening test for typhoid carriers.

The O antigen, also known as somatic antigen, is the outer portion of the endotoxin complex. The endotoxin complex of *Salmonella* and other gram-negative organisms is a lipopolysaccharide (Fig. 1) that joins the outer membrane structure to the peptidoglycan cell wall. Although the endotoxin component of the cell wall is of low toxicity, it may play an important role in the pathogenesis of gram-negative bacteria. Endotoxins can evoke fever (pyrogenic), cause disturbance in capillary permeability and associated changes affecting circulation and blood pressure, and activate serum complement, kinin, and the clotting systems, among a multitude of effects. This endotoxin is made up of three covalently linked parts: (1) the outer O-specific chain, (2) the middle oligosaccharide R-core that links to the O-specific chain, and (3) the inner lipid A layer bound to the cell wall.

The O-specific chain (O antigen) is heat-stable and composed of oligosaccharides (groups of sugars), which extend like whiskers from the membrane surface to the outer environment. The O chains are made up of repeating units of identical oligosaccharides. Each serovar of *Salmonella* is characterized by its unique structure and arrangement of simple sugars in the O-specific chain. Often, but not always, the terminal sugar residue of the O-specific chain is immunodominant. The smooth virulent "strains" of *Salmo-*

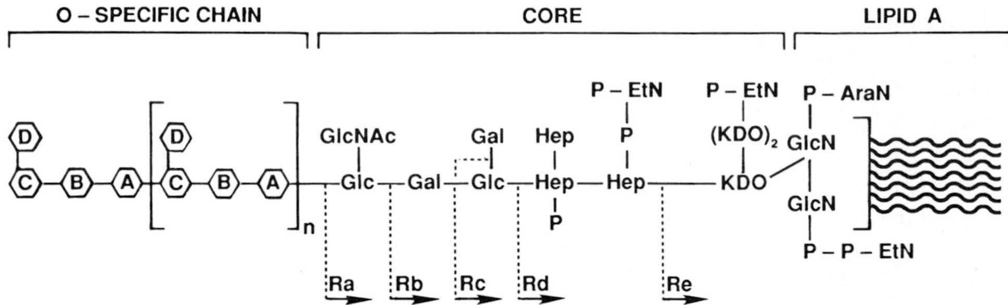

FIGURE 1 General structure of a *Salmonella* lipopolysaccharide (LPS). Key: A–D, sugar residues; Gal, D-glucose; GlcN, D-glucosamine; GlcNAC, *N*-acetyl-D-glucosamine; Hep, L-glycero-D-*manno*-heptose; KDO, 2-keto-3-deoxy-D-*manno*-octonate; AraN, 4-amino-L-arabinose; P, phosphate; EtN, ethanolamine; ~, hydroxy and nonhydroxy fatty acids; Ra–Re, incomplete R-form LPS. [Reprinted, with permission, from O. Luderitz, C. Galanos, H. Mayer, and E. Th. Rietschel (1986). *In* "Proceedings of the 10th International Convocation on Immunology" (H. Kohler and P. T. Lo Verde, eds.), p. 78. John Wiley & Sons, New York.]

nella possess a full complement of O sugar repeating units, whereas the "rough" avirulent or less virulent *Salmonella* lack a complete sequence of O sugar repeating units. The O antigens are designated by numbers 1–67. "Smooth" and "rough" derive from the appearance of colonies on agar.

Besides the importance of the O-specific chain (O antigen) of the endotoxin in determining the specificity of serovars, R-core is also an important entity in that antibodies directed against R-core may protect against infection by a wide variety of gram-negative bacteria. This is probably due to the fact that the R-core structure has less diversity than O-chains and is common to other Enterobacteriaceae. In addition to the *Salmonella* R-core, other core types have also been identified in Enterobacteriaceae.

Because *Salmonella* are motile by peritrichous flagella (Fig. 2), flagellar antigen, also known as H antigen, has been used in conjunction with O antigen to identify serovars. There are two classes of H antigen designated as phase 1 and phase 2. The phase 1 antigen is specific, occurs in only a few serovars, and is designated by a small letter, a–z. Recently identified phase 1 antigens are designated as $Z_1, Z_2, Z_3, \ldots, Z_{60}$, for lack of additional letters available for designations. There are fewer kinds of phase 2 antigens but they are much more widely distributed than phase 1 antigens. Many *Salmonella* have common phase 2 antigen; therefore, they exhibit antigenic relationships. The phase 2 antigens are designated by arabic numerals.

The Vi (capsular) antigen is a surface antigen most commonly found in *S. typhi* and *S. hirschfeldii* (*S. paratyphi* C). The Vi antigen can be removed from cells by hot saline or trichloracetic acid treatment. Although several types of Vi antigens are recognized, they are immunologically related. Presently, the specific name given to a *Salmonella* is in reality a designation for a particular serovar. Though most serovars are named binomially, some are designated only by their antigenic formulas. A serovar with antigenic formulae 6, 7 : r : 1, 7 represents O antigen 6 and 7 : phase 1 H antigen r : phase 2 H antigens 1 and 7 and is named *S. colindale*. In the Kauffman-White schema, serovars with common O antigens are arranged into groups. Table II lists the names, O-antigen groups, and antigenic formulas of some common salmonellae infecting humans. For a detailed listing of all serovars, see "Bergey's Manual of Systematic Bacteriology."

D. Bacteriophage Typing

A number of *Salmonella* serovars, which are indistinguishable by serological or biochemical test, can be divided into phage type based on the sensitivity of cultures to a series of bacteriophages. Phages are bacterial viruses that infect bacteria and eventually cause lysis of the bacteria. *Salmonella typhi, S. hirschfeldii, S. paratyphi,* A, *S. schottmulleri, S. typhimurium,* and, to a much lesser extent, other serovars have been typed using various bacteriophages.

The preponderance of one or another of the phage types of *S. typhi* may characterize a geographical origin and could help determine the origin of an epidemic.

E. Habitat

Although *Salmonella* serovars are widely distributed in nature, they are obligate parasites (i.e., live inside animal hosts) and are not found elsewhere. They may be adapted to single host or ubiquitous human and

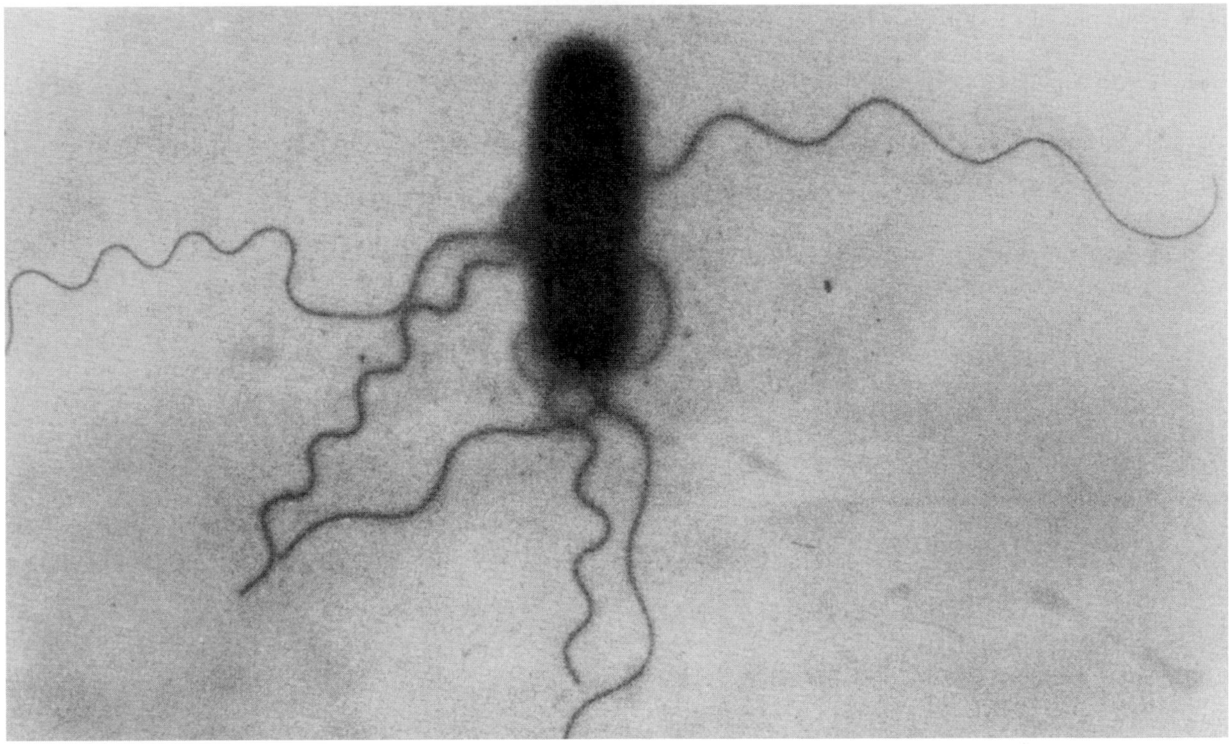

FIGURE 2 A negatively stained cell of *Salmonella* sp. showing peritrichous flagellation. [Courtesy of W. L. Dentler. Reprinted, with permission, from V. T. Schuhardt and T. W. Huber (1978). *In* "Pathogenic Microbiology," p. 262. Lippincott, Philadelphia.]

animal pathogens, and some may be of still unknown pathogenicity. They may be primary etiological agents or secondary invaders. They are found in many ecosystems that have been polluted by humans and animals. A large number of *Salmonella* are parasites of lower animals such as rodents, birds, reptiles, and even insects.

Some serovars are strictly adapted to one particular host. For example, *S. typhi, S. sendai,* and *S. paratyphi* A are adapted to humans, *S. abortusovis* to sheep,

TABLE II
Antigenic Grouping and Formulas of Some Common *Salmonella*

Serovar	O group	Somatic antigen	Flagellar (H) antigen		Disease
			Phase 1	Phase 2	
S. paratyphi	A	1, 2, 12	a	1, 5	Enteric fever (paratyphoid), gastroenteritis
S. schottmulleri	B	1, 4, 5, 12	b	1, 2	Enteric fever (paratyphoid), gastroenteritis
S. typhimurium	B	1, 4, 5, 12	i	1, 2	Most frequent agent of gastroenteritis, septicemia
S. hirschfeldii	C	6, 7 (Vi)	c	1, 5	Enteric fever (paratyphoid, endocarditis)
S. choleraesuis	C	6, 7	c	1, 5	Enteric fever (paratyphoid), gastroenteritis, septicemia
S. typhi	D	9, 12 (Vi)	d	—	Enteric fever (typhoid)
S. enteritidis	D	1, 9, 12	g m	1, 7	Gastroenteritis

and *S. gallinarium* (*S. pullorum*) to poultry. Many *Salmonella* are also ubiquitous and can be found in a variety of animals at any given time. A majority of the serovars show no host specificity, and most common in this group is *S. typhimurium*.

II. PATHOGENICITY FOR HUMANS

Salmonellosis is an important infectious disease of humans. Countries with poor sanitation and hygiene have the highest incidence of salmonellosis. Salmonellosis can be caused by any of the 2200 serovars; however, 10 serovars account for about 73% of the infections. *Salmonella* infection may range from mild gastroenteritis to enteric fever to severe fatal septicemia. Millions of people worldwide are affected annually by *Salmonella* infection, and worldwide financial loss from salmonellosis is in the billions of dollars.

A. Mechanism of Pathogenicity

Various events leading to infection are summarized in Fig. 3. *Salmonella* cells enter the animal body by ingestion of contaminated food or drink or via person-to-person contact. The infection begins in the intestinal tract. Disease manifestations by *Salmonella* depend on the virulence and the type of the serovars, general health of the animal, and the number of viable cells ingested. Approximately 10^5–10^9 organisms are needed to produce disease in >50% of individuals. However, some serovars under certain circumstances may produce disease with a much lower number of organisms.

Following ingestion of *Salmonella*, organisms pass through the stomach, where they are subjected to gastric acidity. Conditions within the stomach regulate the number of bacteria that enter the small intestine. In the acid-deficient stomach, a larger number of viable bacteria are discharged into the small intestine, thereby increasing the possibility of disease production. From the stomach, bacterial cells are transported to the ileum and colon, where the cells adhere to the target cells by various mechanisms that are not completely understood. The adhesion process is influenced by bacterial motility and chemoreactants produced by intestinal cells (which bring *Salmonella* cells into closer contact with mucosal receptors). Some salmonellae such as *S. typhi* produce adhesion, free of O, H, and Vi antigen, that is made of 85% cell-surface protein and Rd_1P^+ lipopolysaccharide, which specifically binds to mannose-like receptors on epithelial cells. Several outer membrane proteins are known to be virulence factors in mice infected with *S. typhimurium*. Production of adherence factors may also be enhanced by specific plasmids. For example, in *S. typhimurium*, a 60-MDa plasmid has been found to be responsible for adhesive and invasive properties.

Studies of invasion of guinea pig ileum by *S. typhimurium* indicate that bacterial adherence to the brush border of the cells leads to the damage or degeneration of microvilli at the site of attachment (Fig. 4). This degeneration enables bacteria to be engulfed into the cells by a process termed receptor-mediated endocytosis (RME). The epithelial cells that are involved in this process are known as M cells, and they overlay Peyer's patches (lymphoid organs) and other mucosal lymphoid follicles. Only viable bacteria are transported by this process, which is known as translocation, an important initial step in *Salmonella* pathogenicity. It is interesting to note that other intestinal bacteria are not taken up along with *Salmonella*. After colonization is established, the organisms invade the intestinal epithelium and proliferate within the epithelial cells and lymphoid tissue.

After RME, the organisms are found in vacuoles within the epithelial cells. Those salmonellae causing typhoid-like disease seem to be transported through the cells in the vacuoles and then enter the lamina propria, which is under the epithelium, via the basal cell membrane. Once in that area, the inflammatory response consists primarily of infiltration by monocytic cells. Finally, organisms colonize the cells of the reticuloendothelial system, and presumably, when a certain population is reached, organisms break out into the bloodstream (secondary bacteremia), producing enteric fever symptoms. *Salmonella* serovars that cause gastroenteritis remain localized in the mucosa. Further spread of nontyphoidal salmonellosis is prevented by polymorphonuclear cell (PMN) response.

Most *Salmonella* elicit an inflammatory response and ulceration in the lamina propria, with fluid secretion and diarrhea. Generally, there are two types of *Salmonella* diarrhea: (1) toxigenic diarrhea caused by enterotoxin (cholera-like toxin) and (2) acute *Salmonella* diarrhea, which occurs during the invasion process. Additionally, some salmonellae produce Shiga-like cytotoxin (yet to be isolated), which leads to the ulceration of intestinal epithelial tissues. *Salmonella* enterotoxin activates adenylate cyclase (AC), causing elevation of cyclic adenosine monophosphate (cAMP). That in turn leads to ion flux disruption of the intestinal epithelial cells and cascade events leading to secretion of chloride and bicarbonate ions and

FIGURE 3 Mechanisms of pathogenicity of *Salmonella*.

inhibition of sodium absorption, resulting in diarrhea. Salmonellae that do not produce enterotoxin cause diarrhea by invading intestinal epithelia and stimulating an inflammatory reaction, which evokes fluid secretion and diarrhea. However, some invasive strains do not evoke fluid secretion and others secrete fluid before inflammation. Thus, it seems that more than one mechanism is involved in this type of diarrhea.

Endotoxin is a potent inflammatory-inducing factor and a strong chemotactic factor. The attracted PMNs phagocytize bacterial cell, and endotoxin is released from the dead bacteria. The released endotoxin causes damage to surrounding cells and induces release of prostaglandins, mediators of inflammation, and platelet-activating factors (PAFs). The prostaglandins from inflamed tissue also activate adenylate cyclase, leading to elevation of cAMP, which leads to fluid secretion, as explained earlier.

PAF is a mediator of allergic and inflammatory reactions produced by various blood cells. It acts as a vasodilator, elevates vascular permeability, and stimulates the release of lysosomal enzymes. It is also a potent ulcerogenic necrotizing enterocolitis factor. Following neutrophil aggregation, other vasoconstrictors such as thromboxane A_2 and noradrenaline, free radicals, and lysosomal enzymes are released in associ-

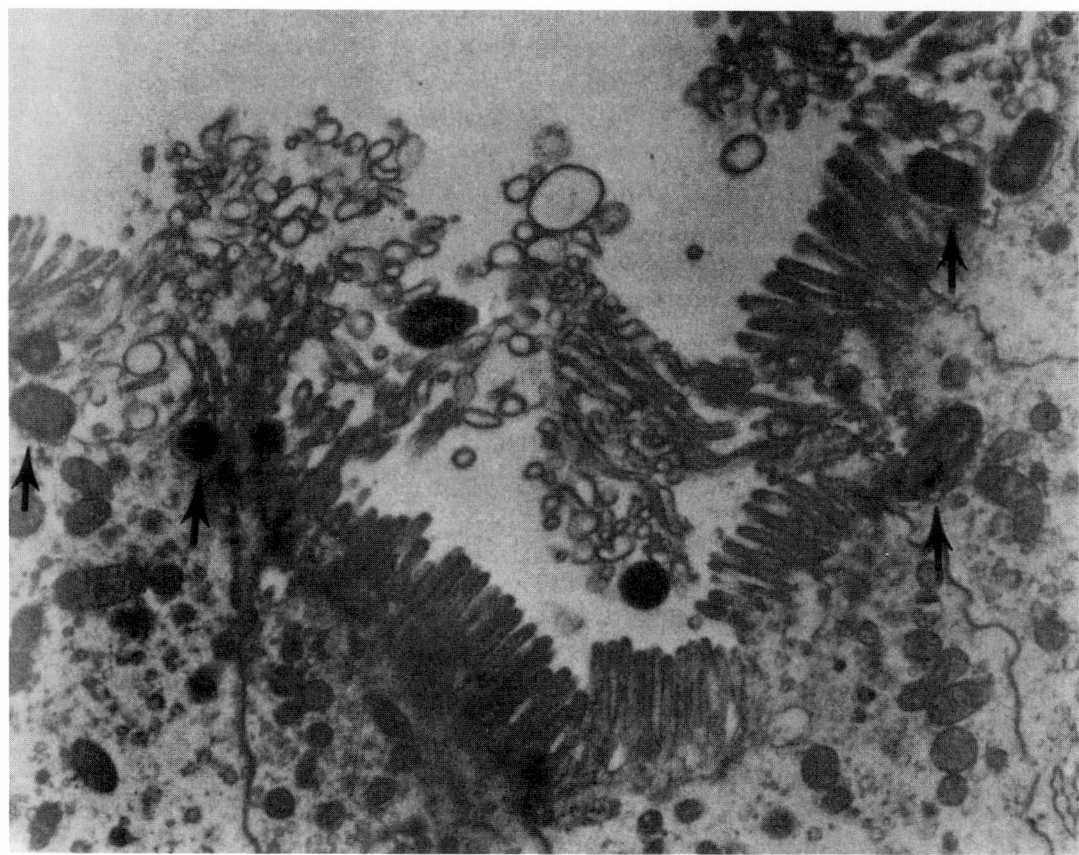

FIGURE 4 Electron photomicrograph showing the invasion by *S. typhimurium* of the epithelial lining of small intestine in guinea pig. Arrows point to invading *Salmonella* organisms. [Reprinted, with permission, from A. Takeuchi (1975). *In* "Microbiology," p. 176. American Society for Microbiology, Washington, D.C.]

ation with PAF, causing local tissue ulceration. The result of these actions leads to severe colitis with diarrhea that may be associated with bloody mucous stool. Blood leaks out of capillaries as a result of alteration of intestinal permeability, which causes marked inhibition of fluid absorption and active colonic secretion.

B. Dominant Serovars in Various Diseases

Salmonella serovars may be adapted to a particular host or can be ubiquitous. Often, a particular serovar is involved in producing the specific symptoms of a disease. However, a serovar can occasionally produce symptoms of any of the various disease caused by *Salmonella*. For example, *S. typhi* causes typhoid fever and *S. enteritidis* (causes gastroenteritis; *S. paratyphi, S. schottmuller, S. hirschfeldii,* and *S. choleraesuis* all can cause paratyphoid fever and gastroenteritis; *S. typhimurium* causes gastroenteritis and septicemia; *S. choleraesius* can also cause septicemia or focal infections.

The salmonellae can be divided into three groups on the basis of their host preference. The first group includes those highly adapted to humans (*S. typhi, S. paratyphi,* and *S. senadi*) and commonly causing enteric fever. The second group includes those adapted primarily to a particular animal host. Of the organisms in this group, *S. dublin* and *S. choleraesuis* are also pathogenic to humans. Infection by the latter can be quite severe in children. The third group includes unadapted serovars. This group includes over 2000 serovars that are ubiquitous in nature and that seemingly attack humans and other animals with equal facility. Many serovars in this group cause gastroenteritis and account for 85% of all *Salmonella* infections in the United States. In developing countries, *Salmonella* is the major cause of bacterial diarrhea.

C. Enteric Fever—Typhoid

Typhoid fever is caused by *S. typhi*. It is a systemic infectious disease resulting in a prolonged fever. Humans are the only reservoir for *S. typhi*, which is pathogenic only for humans. The organism is transmitted by food or drink contaminated with human excreta. It is prevalent in developing countries with improper sewage disposal, contaminated water, and poor hygienic habits. In developed countries, foods contaminated by carriers are primarily responsible for transmissions of the organisms. Flies and other insects also spread the organism from feces to food. Occasional transmission results from homosexual activity (anal–oral route), by direct contact in children, and by improper use of toilet paper and contamination of hands with feces. The disease is most severe in children and older adults. In the United States, over 400 cases of typhoid per year have been reported to the Centers for Disease Control (CDC) in Atlanta, under the surveillance program of the last 10 years. Throughout much of the world, typhoid is endemic, and at times it becomes epidemic, causing suffering to millions and death to many.

1. Symptoms and Diagnosis

The incubation period for typhoid fever depends on virulence and the number of organisms ingested. It is usually 7–14 days but may be as short as 3 or as long as 60 days. The clinical symptoms are gradual onset of continuous fever, headache, aches and pains, enlargement of the spleen, inflammation of the intestine, tender abdomen, rose-spot rash, leukopenia, arthralgias, diarrhea, pharyngitis, constipation, anoxemia, abdominal pain, and tenderness. Normally, diarrhea is not a typical symptom of typhoid fever. Recent findings in Asian countries indicate diarrhea in patients with positive blood culture for *S. typhi* or *S. paratyphi* A, but without any other pathogen capable of producing diarrhea. This diarrhea is watery and contains large quantities of leukocytes and protein. Although not a typical symptom, diarrhea in typhoid may also occur as a result of double infection from a diarrheogenic pathogen. If untreated, the fever rises, reaching a plateau in 2–3 days and remains high (39.4–40°C) for another 1–2 weeks, falling to normal in 4–5 weeks. The bacteria pass from the intestine to the mesenteric lymph nodes, where they multiply and eventually reach the bloodstream. They may be found in urine, the gallbladder, bile ducts, and bone marrow. Complications in untreated patients may result in intestinal perforation and acute cholecystis, hepatitis, pneumonia, or abscess formations and neuropsychiatric disturbances. Bacterial localization in different organs (metastatic infection) can appear during acute or convalescent periods of infection, or months to years after the infection. The most common sites of infections are the bones, bone marrow, and joints, but any part of the body may be involved. Osteomyelitis (bone infection) may develop as long as 6–7 years after typhoid fever. Relapses occur in about 10–20% of patients. The disease symptoms and the treatment are the same as the first infection. Occasionally, a second relapse occurs.

Initial diagnosis may be based on continuous fever and other symptoms, but ultimate laboratory diagnosis is based on the isolation of the *Salmonella* serovar from the blood. Within the first 10 days, the majority of the cases show positive blood culture, which drops off during the later stages of infection. Stool cultures are positive during the third to fifth weeks. Urine cultures are also positive at the same time as stool cultures, however at a much lower percentage. The organism can also be isolated from bone marrow and rose spots. A persistent positive stool culture in a patient beyond 3–6 months indicates a 90% possibility of the patient becoming a carrier.

2. Treatment and Control

The most effective treatment for typhoid fever is administration of antibiotics. Ampicillin, chloramphenicol, and trimethoprim-sulfamethoxazole (TMP-SMZ) are primarily used. Chloramphenicol is the drug of choice for treating typhoid fever. In recent times, isolates of *Salmonella* have increasingly shown drug resistance to antibiotics owing to the R plasmids, which carry genes for resistance and can transfer their resistance to sensitive strains. They are responsible for the rapid rise of multiple-drug resistance. *Salmonella typhi* resistant to ampicillin, chloramphenicol, and TMP-SMZ have been observed in many parts of the world. Cefotaxime, a third-generation cephalosporin, has recently been used with success to treat typhoid and nontyphoid salmonelloses and may be a prime candidate for the treatment of multiresistant systemic salmonelloses. Proper treatment and disposal of sewage and appropriate purification of water are the two major means of controlling typhoid and *Salmonella* outbreaks in general. Also, pasteurization of milk and other drinks, proper hygienic habits, and control of carriers from working in the food industry where they come in contact with food would help reduce the transmission of *Salmonella*. Surveillance programs would also be helpful in controlling epidemics. In appropriately treated situations, the mortality rate is negligible. [*See* Antibiotics.]

3. Typhoid Vaccine

Presently, three typhoid vaccines are available for general public use in the United States: a parenteral heat-phenol-inactivated vaccine that has been in use for a long time; a newly approved parenteral vaccine available as of now made from Vi capsular polysaccharide (ViCPS); and an oral live attenuated vaccine (Ty21a). A fourth vaccine made from acetone-inactivated whole cells used parenterally is currently available only to the armed forces. Typhoid vaccines have been used for immunization since the turn of the century. Vaccination alters the character of the epidemics and may lessen the severity of the illness but are not 100% effective. Vaccine protection can also be overwhelmed by large inocula of *S. typhi*. Even with these drawbacks, the vaccines offer the best available protection.

Routine typhoid vaccination is not recommended for U.S. residents. Only travelers to the developing countries with risk of exposure should be immunized. Presently, no vaccine is available for nontyphoidal salmonellosis.

The parenterally administered whole-cell inactivated vaccine causes substantially more adverse reactions but is no more effective than the Ty21a and ViCPS vaccines. The Ty21a vaccine should not be used for immunocompromised individuals or individuals on antibiotics at the time of vaccination. The three vaccines each have different lower age limits among children. Also, for continued protection against repeated exposure to *S. typhi*, booster doses of the particular vaccine are required to maintain immunity.

D. Enteric Fever—Paratyphoid

Paratyphoid fever is relatively mild and of shorter duration compared with typhoid fever. It has the same symptoms as typhoid fever, and both are known as enteric fever. Although paratyphoidal enteric fever may be caused by any serovars of *Salmonella*, the serovars most often responsible are *S. paratyphi* A, *S. schottmulleri* (*S. paratyphi* B), *S. hirschfeldii* (*S. paratyphi* C), and *S. senadi*. Paratyphoid is often marked by the sudden onset of chills, but otherwise the pathogenesis and the manifestations are very similar to those of typhoid. The incidence of paratyphoid, compared with typhoid based on hospital records, is about 1:10. Proper clinical diagnosis involves isolation, identification, and serological typing of the organisms. Treatment is similar to that for typhoid fever.

E. Gastroenteritis

Any of the over 2000 serovars that are ubiquitous in nature can cause gastroenteritis. Most commonly, *S. typhimurium* and *S. enteritidis* are responsible for gastroenteritis. The disease symptoms appear within 12–48 hr and often overnight after the ingestion of food contaminated with *Salmonella*. A relatively large inoculum is usually required (10^8–10^9 cells) to produce clinical symptoms, but infection may occur at smaller inoculum. The disease symptoms are sudden onset of nausea, acute vomiting, abdominal cramp and pain, and watery diarrhea occasionally containing mucus, blood, or both. A slight fever of 38.3–38.9°C is seen in over half the cases. The peak incidence occurs in children under the age of 6 years and in persons over 50 years of age. In newborn and young children, symptoms are diverse and range from grave typhoid-like illness and septicemia to asymptomatic infection. The mortality rate in very young children may be as high as 20% from *Salmonella* sepsis. The disease is self-limiting, usually mild, lasting 1–4 days, and in a majority of cases treatment is not required. However, at times it can be debilitating and protracted. Fatalities are rare except in nursing homes (7–8%) and among infants (6–7%). In a pediatric ward, the infection is transmitted by the hands of personnel.

Salmonellosis may also occur in patients with an established diagnosis of acquired immunodeficiency syndrome, or it may occur during the first manifestation of this disorder. Whereas the incidence of enteric fever has gone down dramatically, the salmonellosis of the gastroenteritis type has gone up astronomically. The Committee on *Salmonella* of the National Research Council in 1969 conservatively estimated that 2 million cases of salmonellosis occur in the United States per year. The present estimate is 4–5 million cases per year. The true magnitude of the problem is impossible to determine. The rapid rise of foodborne gastroenteritis is blamed, among other things, on the use of antibiotics in farm animals, which has given rise to antibiotic-resistant *Salmonella* serovars.

1. Various Food and Feed Sources Implicated

The major source of the *Salmonella* problem is food of animal origin. Poultry, meat, and eggs are the main sources of nontyphoidal salmonellosis, accounting for about 36% of the outbreaks. A miscellaneous variety of processed foods are also known to contain *Salmonella*. Many cases of salmonellosis in which the source of infection was not established may have been foodborne. [*See* Food Microbiology and Hygiene.]

The main source of *Salmonella* problems is poultry, which accounts for 17% of salmonellosis. The U.S. Department of Agriculture reports that 37% of chicken carcasses carry *Salmonella*. This figure has remained constant for more than a decade. Chickens eat each other's droppings, and in this way a large percentage of animals become inoculated. Four percent of beef and 12% of pork is also contaminated with *Salmonella*.

Salmonellosis associated with egg and egg products led to the passage of the Egg Products Inspection Act in 1970 by the United States. Since this legislation, the CDC has not received any report of major outbreaks associated with bulk egg products.

However, recent findings indicate that *S. enteritidis* contamination of eggs can lead to widespread outbreaks and serious public health hazard manifesting in various symptoms. Bacteria can be found in the shell, inside the egg, and in muscle tissue of layer hens. Also, in enteritidis-infected eggs, bacteria appear to reside but not grow in egg white until the temperature fluctuates, during holding, transportation, and refrigeration of eggs. In that case the barrier surrounding the egg yolk breaks down and the organisms proliferate in yolk. Additionally, rapidly grown broiler chickens may become immunocompromised and can carry the *S. enteritidis* phage Type 4 organisms in their muscle tissue, causing potential human outbreaks.

The contamination of Grade A eggs with *S. enteritidis* through the hen's ovary remains a major source of infection. Presently, eggs and egg products, and to a lesser extent chicken, still constitute a significant source of salmonellosis.

Raw or improperly pasteurized milk, milk products, and powdered milk are other sources of *Salmonella* infections. In 1985, a major outbreak of *Salmonella*, the biggest in U.S. history, which caused food poisoning to about 200,000 people in the United States, occurred from improperly pasteurized milk. Various cheeses made from unpasteurized or improperly pasteurized milk have also been involved in *Salmonella* infections. Miscellaneous products, such as dried yeast, smoked fish, bakery products, cream-filled desserts, ice cream, sauces and salad dressings, bread mix, coconut, frog legs, spices, gelatin, sandwiches in vending machins, and vegetables, have been incriminated at times as the vehicle of *Salmonella* transmission and gastroenteritis.

Another major source of salmonellosis is improper kitchen hygiene, such as improperly cleaned cutting boards, and the contamination of fruits, salads, vegetables, beef, ham, and so on from chicken drippings happening either at the grocery store or at home.

By-products of animal origin, milk and milk products, egg products, and protein concentrates of vegetable origin are some of the numerous sources of *Salmonella* in animal feed. Foodborne salmonellosis can be controlled only by eradication of *Salmonella* from feed and farm environments, pasteurization of raw foods of animal origin, and consumer education to the danger of salmonellosis. The U.S. Department of Agriculture and the FDA believe that eradication of *Salmonella* from feed and feed ingredients is too costly for the gained benefit. Therefore, feed will continue to be a source of *Salmonella* infection to animals.

Most gastroenteritis symptoms subside promptly in healthy individuals and require no treatment. If needed, the illness is treated symptomatically except in high-risk patients who are treated with antibiotics.

2. Detection of *Salmonella* in Food

Detection of *Salmonella* in food is complicated by a lower population of *Salmonella* and other bacteria in food, injury of surviving cells from processing, and the inherent variability encountered in the analysis of various types of food. Presently, detection of *Salmonella* involves preenrichment in nutritive nonselective media, selective enrichment into a special broth, and finally selective plating onto suitable solid media. Suspicious colonies are cultivated further and examined for typical reactions.

F. Septicemia

Although any *Salmonella* serovar can cause septicemia or localized disease or focal infections (liver, gallbladder, etc.), *S. choleraesuis* infection is most frequent, followed by *S. typhimurium*. The clinical symptoms vary with the serovar involved. In general, from the intestinal tract, the organisms reach the bloodstream in a manner similar to that in *S. typhi* and multiply in the bloodstream, producing recurrent high fever, chill, and loss of appetite. From the bloodstream, organisms may reach various parts of the body and produce pneumonia, myocardial abscesses, osteomyelitis, meningitis, and arthritis. They may also invade the endocardium and the endothelium of large arteries, thereby causing intravascular lesions. Septicemia by *Salmonella* can occur with or without bacteremia. In bacteremia, the organism can be isolated from blood samples but not from stools.

Salmonella choleraesius is most prone to invade the bloodstream, even when it causes other clinical forms of salmonellosis. Isolation of this organism from the blood usually indicates the presence of an abscess. The fatality rate in nontyphoidal bacteremia is high, from 15 to 20%. The illness is particularly severe in children. Ampicillin or chloramphenicol are effective in treating bacteremia.

III. *SALMONELLA* INFECTION OF ANIMALS (OTHER THAN HUMANS)

Salmonella infections of cattle and poultry are of common occurrence. About 10% of the 2200 serovars are routinely isolated from animals. Although some serovars show specificity to certain animals producing characteristic clinical symptoms, the majority do not. Infection symptoms may vary from acute septicemia to diarrhea to abortion. Many infections are asymptomatic.

The most common cattle isolates are *S. typhimurium, S. dublin,* and *S. newport.* In calves, *S. typhimurium* is associated with serious clinical symptoms of acute salmonellosis. *Salmonella* infection in cows may also cause abortion either during or after the acute phase of illness, or even without frank clinical symptoms of illness. *Salmonella abortusequi, S. abortusovis,* and *S. choleraesuis* may cause severe illness and abortion in horses, sheep, and pigs, respectively. Furbearing animals such as dogs, cats, rodents, ferrets, bats, and marsupials are also infected with *Salmonella,* producing a variety of salmonellosis symptoms. Poultry constitutes the greatest host reservoir for *Salmonella* and an important indirect source of salmonellosis in humans. *Salmonella pullorum* once caused pullorum disease in chicks 1–14 days old, resulting in death to almost 100% of the hatch. This serovar, which can be transported by eggs, was eradicated by slaughter of infected adult birds, whose infection had been detected by blood tests. Fowl typhoid caused by *S. gallinarium* also used to cause severe financial loss in poultry operation, and it has also been eliminated by blood-testing programs. The occurrence of salmonellae among cold-blooded vertebrates such as snakes, turtles, and lizards has been known for many years. Pet turtles have been the cause of concern in the last decade. In 1975, the FDA prohibited the distribution of red-eared turtles in the United States because of documentation of thousands of cases of salmonellosis per year from this source.

IV. DETECTION OF *SALMONELLA* IN CLINICAL SPECIMEN

Diagnosis of salmonellosis ultimately depends on the isolation and identification of *Salmonella* from clinical specimens. The organisms may be isolated from circulating blood in earlier phases of enteric fever and septicemia. In later stages organisms are found in stool, urine, and other fluids.

Stool specimens are normally inoculated into an enrichment medium such as gram-negative broth or selenite F broth and incubated for 12–18 hr at 35°C. Several drops of this enrichment broth are then used to inoculate moderately selective agars such as Hektoen enteric or XLD agar, or highly selective media such as bismuth sulfite and brilliant green agar.

Most other clinical specimens are inoculated into a supportive medium such as blood, chocolate, or MacConkey agar. After incubation, the media should be checked for growth by gram stain and streaked on suitable media as just described for isolation of *Salmonella.* Suspected colonies from various media are then identified using routine biochemical tests for *Enterobacteriaceae* and *Salmonella.* It is important to do differential tests for the identification of related organisms. Next, tentatively identified organisms are further identified by serological typing using O antisera for serogroups A to E, H antisera for flagellar antigens, and Vi antiserum as warranted. Further identification of *Salmonella,* including the complete serological identification, is normally carried out by a specialized laboratory.

V. CARRIER STATUS

Pathogenic microbes can be transmitted to healthy people by a person who has the disease, one who has recovered and carries the organism, or one who has no clinical symptoms and was asymptomatically infected. Such a person is called a carrier. In case of enteric fever, *Salmonella* is present in the patient's stool in the later stages, and a small percentage of these patients can discharge the organisms for a period of 3–10 weeks after the onset of the illness. A certain small portion of these people in turn continue to excrete *Salmonella* for from several years to throughout their entire life. About 3% of untreated or recovered typhoid patients are believed to be chronic carriers. Carriers harbor a high population of *S. typhi,* up to 10^{10} organisms per gram of feces. A persistent fecal carrier usually has infection in the gallbladder. Urinary carriers are also found in certain parts of the world. In many of the asymptomatic carriers, bacteria persist for a relatively short period and the excretion of *Salmonella* is intermittent. Food handlers constitute a high percentage of carriers. Infants frequently become long-term carriers and excrete *Salmonella* longer than adults. It is estimated that there are about 2000 typhoid carriers in the United States, most of them elderly females who have chronic biliary disease. The proportion of typhoid carriers must be high in developing countries, but it varies from different geographic locations and for different time periods.

Duration of carrier state is variable in different age groups and in both humans and animals. Most serovars causing gastroenteritis do not usually establish a chronic carrier state. There are no general estimates of the number of carriers of unadapted serovars; it is estimated to be very high.

All types of carriers can be treated with antibiotics to remove the foci of infection; however, until the underlying anatomical problems are corrected by surgical intervention, antibiotic treatment is of little value.

VI. EPIDEMIOLOGY OF *SALMONELLA*

Salmonellosis is one of the most important epidemiological problems in both developed and developing countries, the former from nontyphoidal salmonellosis and the latter from both typhoidal and nontyphoidal salmonellosis. Although all the serovars of *Salmonella* are potential pathogens for humans, only a handful of serovars are most frequently involved in salmonellosis. The 10 most frequently reported serovars from human and nonhuman sources reported to the CDC in 1987 represent 73.1 and 57.5% of the total isolates (Table III) from human and nonhuman sources, respectively.

Salmonella typhimurium is the most frequently isolated serovar from both human and nonhuman sources. Salmonellosis, excluding typhoid, is most prevalent in infants <1 year old and children from 1 to 4 years of age. This group constitutes about 32% of the reported cases in the United States. Older adults, ≥60 years old, represent about 11% of the cases. The incidence of nontyphoidal salmonellosis as

TABLE III

Ten Most Frequently Reported *Salmonella* Serotypes from Human Sources Reported to the CDC in 1987 and from Nonhuman Sources Reported to the CDC and USDA in 1987[a]

	Human 1987			Nonhuman 1987		
Rank	Serotype	Number	Percent	Serotype	Number	Percent
1	typhimurium[b]	10,462	23.5	typhimurium[b]	1,246	13.5
2	enteritidis	6,950	15.6	heidelberg	1,124	12.2
3	heidelberg	5,714	12.8	choleraesuis[c]	667	7.2
4	newport	2,858	6.4	reading	438	4.8
5	hadar	2,170	4.9	hadar	352	3.8
6	infantis	1,136	2.5	senftenberg	331	3.6
7	agona	1,080	2.4	newport	302	3.3
8	montevideo	1,037	2.3	montevideo	301	3.3
9	thompson	635	1.4	enteritidis	270	2.9
10	braenderup	548	1.2	anatum	266	2.9
	Subtotal	32,590	73.1		5,297	57.5
	Total	44,609			9,208	

[a]Courtesy of Centers for Disease Control, Atlanta.
[b]Typhimurium includes var. copenhagen.
[c]Choleraesuis includes var. kunzendorf.

reported to the CDC has about doubled in the last 10 years from 13 to 21 per 100,000 population (Fig. 5). However, the real estimate is impossible to make and there are conservatively estimated to be 4–5 million cases per year in the United States alone.

The major source of this problem in humans is food of animal origin, predominantly poultry, beef, and pork. Inadequate cooking or processing of contaminated products, cross-contamination of working surfaces in the kitchen environment, and a lack of human hygiene (and, more specifically, lack of consumer education) are most contributory to the salmonellosis problem.

Salmonella typhi, the causative organism for typhoid, is highly adapted to humans, and humans are the only source of this organism. Human carriers excrete the typhoid bacillus in stool and occasionally in urine. The bacillus can then be transmitted to healthy individuals via contamination of food and drink. In the rural areas of developing countries, typhoid epidemics usually result from fecal contamination of drinking water. Typhoid fever in general is on the decline throughout the world owing to better drinking water, sanitary disposal of wastes, and the processing of milk. In the United States, the incidence of typhoid fever has gone down from 1/100,000 in 1955 to 0.2/100,000 population in 1987 (Fig. 6). The sources of infection in the United States are already infected people, carriers, or contraction of infection while overseas. Healthy carriers in food-handling situations are the most important link in the chain of transmission of this infection. Milk can also be a very important source of *Salmonella* infection. For example, a recent massive outbreak of salmonellosis by antimicrobial-resistant *Salmonella* was traced to pasteurized milk that affected about 200,000 people in Illinois. Flies and pests—rats, mice, cockroaches, ants—can also spread the organism from feces to food.

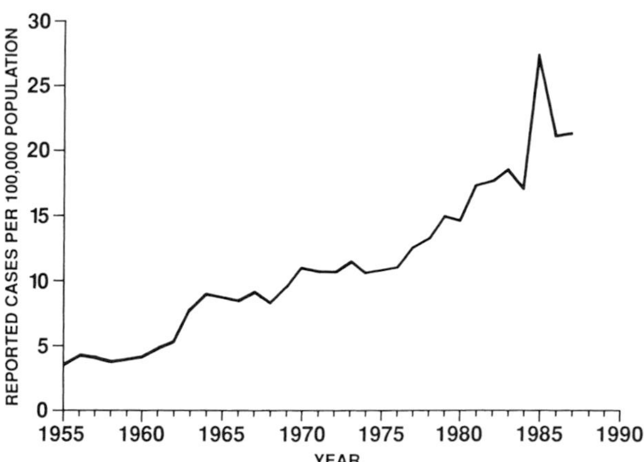

FIGURE 5 Salmonellosis, by year, in the United States, 1955–1987. (Courtesy of Centers for Disease Control, Atlanta.)

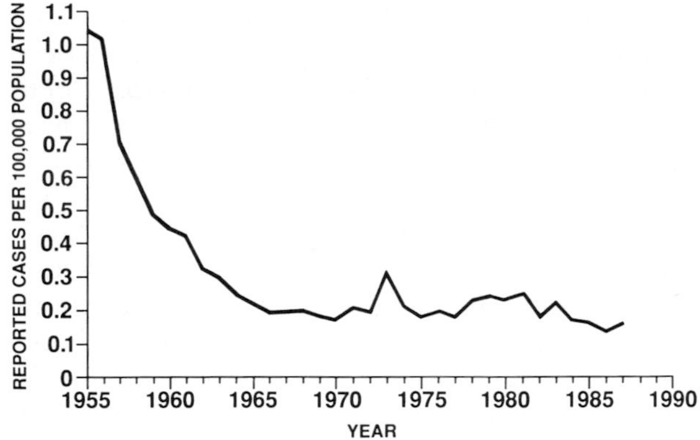

FIGURE 6 Typhoid fever, by year, in the United States, 1955–1987. (Courtesy of Centers for Disease Control, Atlanta.)

BIBLIOGRAPHY

Centers for Disease Control (1994). *Morbidity and Mortality Weekly Report* **43**, No. RR-14. U.S. Govt. Printing Office, Washington, D.C.

Committee on Salmonella (1969). "An Evaluation of the *Salmonella* Problem," publication No. 1683. National Academy of Sciences, Washington, D.C.

LeMinor, L. (1984). Genus III *Salmonella. In* "Bergey's Manual of Systematic Bacteriology" (N. R. Krieg and J. G. Holt, eds.), Vol. I, pp. 427–458. Williams & Wilkins, Baltimore.

Lund, B. M., Sussman, M., Jones, D., and Stringer, M. F. (eds.) (1988). "Enterobacteriaceae in the Environment and as Pathogens," The Society for Applied Bacteriology Symposium Series No. 17, Nottingham, England, U.K., July 7–9, 1987. Blackwell Scientific Publications, Oxford, England.

Rubin, R. H., and Weinstein, L. (1977). "Salmonellosis: Microbiologic, Pathologic and Clinical Features." Stratton International Medical Book Corporation, New York.

Silliker, J. H. (1980). Status of *Salmonella*—Ten years later. *J. Food Prot.* **43**, 307–313.

Soe, G. B., and Overturf, G. D. (1987). Treatment of typhoid fever and other systemic salmonelloses with cefotaxime, ceftriaxone, cefoperazone, and other newer cephalosporins. *Rev. Infect. Dis.* **9**, 719–736.

Salt Preference in Humans

GARY K. BEAUCHAMP
Monell Chemical Senses Center

GLOSSARY

Addison's disease Primary failure or insufficiency of the adrenal cortex to secrete hormones

Bartters syndrome Rare condition characterized by renal potassium wasting and elevated renin and aldosterone levels, with normal or low blood pressure

Diuretic Agent that increases the volume and flow of urine, often with associated electrolytes

SALT (NaCl) IS UBIQUITOUS IN DIETS IN DEVELOPED countries. A major reason for this is that people prefer salted foods to the same foods without added salt. Although little evidence exists for genetic determination of individual differences in consumption and preferred level of salt, more research in this area is necessary. Considerable data support the view that the optimal level of salt in the diet is determined in part by the level an individual is currently consuming; increasing or decreasing customary salt intake, as long as the salt is tasted, increases or decreases, respectively, the preferred level of salt in food. Although these data are consistent with a hypothesis that optimal salt preferences are learned, other data, from both animal models and human developmental studies, suggest that salt preference has an innate component. Furthermore, early experience with low- or high-salt diets may have a long-term impact on preferred salt levels. A preference for salt, like a preference for sweets, has an innate basis that can be modified by individual experience.

I. INTRODUCTION

Because sodium is an essential nutrient, it is not surprising that regulatory systems have evolved to ensure discovery, recognition, and consumption of sufficient sodium to meet that requirement. A nexus in these systems is the sense of taste. Salty (NaCl) taste sensation is thought to be one of a small number of primary taste qualities, others being sweet, bitter, sour, and perhaps those involved in detecting amino acids. Most humans express preferences for familiar salty foods over those same foods without salt. Consequently, many of the foods available for purchase contain relatively high levels of salt added by the manufacturers to increase acceptability.

The major medical issue concerning salt intake is the apparent relationship between high salt consumption and hypertension. Though some animal model studies have linked salt preference with the genetic tendency to develop elevated blood pressure or hypertension, few reliable data indicate a positive relationship between individual differences in human salt sensitivity or preference and the probability of developing hypertension. Although a few studies are positive, the bulk of the data indicates that hypertensive individuals are neither more nor less sensitive to the taste of salt and their level of preference for salt is not different from those individuals with normal blood pressure. However, because many individuals consume salt at levels in the range thought by many investigators studying hypertension to be excessive, high salt preference may be medically counterproductive for those susceptible to hypertension. For this reason alone, it is important to understand the factors controlling and

ENCYCLOPEDIA OF HUMAN BIOLOGY, Second Edition, VOLUME 7. Copyright © 1997 by Academic Press. All rights of reproduction in any form reserved.

modifying salt preferences and salt intake. [*See* Hypertension.]

II. FUNCTIONS OF SALT IN FOOD

For humans, salt appears to serve two flavor purposes when it is added to, or occurs naturally in, food. First, it adds its characteristic salty taste. Second, it modifies other flavors and thereby enhances the palatability of foods. For example, bread contains a relatively high level of salt and unsalted bread is quite unpalatable. Yet, few perceive bread as salty or even notice the salt when their attention is called to it. Recent research suggests that one mechanism involved in the flavor-enhancing activity of salt is the ability of the Na^+ ion to differentially suppress off-flavors. Specifically, Na^+ is a potent inhibitor of many bitter compounds that are often offensive to humans yet are found in many foods, particularly vegetables.

III. DETERMINANTS OF HUMAN SALT PREFERENCE

That humans have a preference or liking for salt, particularly in food, is obvious. The factors that influence and control this preference and the ways it can be modified are, however, complex and not fully understood. The determinants of salt preference are divided here into physiological, genetic, psychological, and developmental factors or influences. In reality, these categories of explanation interact and are not mutually exclusive.

A. Physiological Factors

Most adults in the United States consume 6–12 g NaCl/day (100–200 meq), an order of magnitude or more above physiologically determined requirements. Although it seems unlikely that this level of consumption would reflect some sort of physiological need, that has been suggested.

If people on very low sodium diets were actually in a chronically depleted state, they might be expected to crave salt. The history of salt traders, the levying of taxes on salt, and the frequency with which wars have been fought over salt all attest to the intense desire for this substance. Events in India represent only one example of this history. During the British occupation, salt was so heavily taxed that the poor in some states had to part with a substantial portion of their income just to purchase what they felt to be sufficient salt. Huge fortifications were built to stifle salt smuggling and Gandhi used the issue of salt monopolies to strike at the heart of the British Empire. All of these historical observations, not only from India but from Africa, South America, and elsewhere, indicate that when access to salt it restricted, people will go to great lengths and pay large sums to get it. This suggests that the desire for salt is a powerful motivating force and that a craving for salt ought to exist when low-sodium diets are consumed, either chronically or acutely.

Arguing against this suggestion is the observation that some people live on as little as 1 g/day of NaCl or less without apparent harm and perhaps without craving. At the extreme, the Yanomamo Indians are reported to consume approximately 0.06 g Na/day, as determined by urinary excretion (extra loss in sweat is possible). Other unacculturated peoples also have been reported to chronically consume very low sodium diets. In most cases, when these peoples are faced with the opportunity to consume diets high in salt, their salt consumption increases, but evidence that this is due to a particular craving for salt is equivocal.

Indeed, exceedingly little direct human evidence indicates that even an extremely low sodium diet and sodium depletion are followed by a greatly increased desire for salt. Evidence from clinical studies (Table I) indicates that excessive salt preference, even under conditions of presumably extreme sodium loss (e.g., Addison's disease, Bartters syndrome), is rare in humans and, when reported, is almost always of childhood onset.

Acute experimental studies of salt depletion have also found little evidence for elevated salt preference or salt appetite. When human volunteers have been placed on very low sodium diets and depleted of sodium in other ways (e.g., heavy sweating, treatment with diuretics), they have not expressed a strong hunger for salt. However, when asked in one study to rate how much they would like to eat foods varying in salt content, sodium-depleted subjects did express a greater desire for salty foods, and in taste tests they tended to prefer higher levels of salt in food when depleted. These data do not conclusively demonstrate a physiologically determined heightened salt appetite following sodium depletion in adult humans because a psychological explanation is also possible: extremely low sodium diets are bland and unpalatable, when contrasted with the previously experienced food, and

TABLE I
Clinical Reports of Salt Appetite[a]

Subject(s), description	Age at onset of salt appetite	Amount of NaCl consumed	Medical diagnosis	Comments
A. 15-yr-old diabetic boy (three other diabetic children referred to—one seemed also to have a salt appetite)	Unknown, <15 yr	60–90 g/day	Diabetes	Consumed salt to satisfy "an abnormal craving for salt." Excess salt consumption resulted in elevated blood pressure.
B. 20-yr-old woman	"Present since early childhood"	130–195 g/day (self-report), 138 g/day (clinically determined)	Primary pulmonary arteriosclerosis	First noticed appetite for salt when large enough to climb into chair to reach it. Carried rock salt with her at all times. Refused to go on low-salt diet for >4 days. Threshold for salt taste low relative to control values.
C. 3.5-yr-old boy	First noticed at 1 yr old	Unknown, probably substantially >20 g/day	Corticoadrenal insufficiency	First began eating pure salt at about 18 months of age. "Salt" among first words learned. Did not like sweets.
D. 10 of 64 patients with Addison's disease	Unknown	Unknown	Addison's disease	"Increased desire for salt and salty foods." According to Richter, one patient covered food with salt and eats very salty foods, salting such things as oranges and lemons; he disliked sweets.
E. 19% of patients with Addison's disease	Unknown	Unknown	Addison's disease	Salt appetite *not* listed among nine diagnostic signs and symptoms of disease. Not one of the 12 adult cases described had elevated salt appetite listed as symptom.
F. 36-yr-old male	Apparently recent	0.25 kg salty olives each day	Addison's disease	Also exhibited an appetite for licorice sweets, known to contain a palliative for Addison's disease. In another case, an adult craved licorice but denied craving salt.
G. The Southwood-Gannon family (n = 20 total)	Unknown, disease onset in childhood	Unknown	Genetically determined periodic paralysis with normalkalemia	No direct evidence of salt appetite. Many family members consumed a large quantity of salt daily, perhaps as a therapeutic agent as attacks are moderated.
H. 5-yr-old boy	"The patient had always been a poor eater and craved salt"	Unknown	Bartters syndrome	Another child was not noted to have salt appetite. Most often clinical descriptions of Bartters syndrome do not mention salt appetite; most involve observations on adults.

(*continues*)

TABLE I (*Continued*)

Subject(s), description	Age at onset of salt appetite	Amount of NaCl consumed	Medical diagnosis	Comments
I. 8-yr-old boy	Unknown, <8 yr	Unknown	Pleoconial myopathy	Desired salt; covered food with it and "frequently will eat a teaspoon of salt directly." A 12-yr-old sibling also exhibited salt hunger. No relationship among sodium loading, sodium deprivation, aldosterone antagonism, and attacks.
J. 13-yr-old male	"Present all his life"	Unknown	Mitochondrial myopathy	Ate sandwiches of bread and salt and salted bananas. Mother had to forbid him from pouring salt from store containers directly onto food.
K. 16 of 43 children with sickle-cell disease	Unknown, "as early as 2.5 yr"	Unknown	Sickle-cell hemoglobinopathies	Six had "abnormal" and 10 had "exceedingly increased salt appetite" relative to rest of family. Latter category included children who salted apples, oranges, and peaches, as well as one 2.5-yr-old who salted rock candy, licked off salt, and discarded candy.
L. 33-yr-old female	Adult, recent	"By the shakerful"	Iron deficiency	Iron replacement therapy resulted in a cessation of salt eating.

[a]Adapted from Beauchamp *et al.* (1990).

it may not be the salt taste these subjects desire so much as the more flavorful food associated with saltiness. However, because preference for sweet foods and sucrose in food declined during depletion, a simple desire for more orosensory stimulation cannot explain these results. This issue needs further investigation.

Related work has shown that normal subjects administered a hydrochlorothiozide diuretic that causes excretion of sodium exhibited a compensatory increase in sodium intake. Apparently, subjects expressed no obvious desire for increased salt and, in fact, it was not possible to determine the source of the increased intake. Interestingly, this effect was not observed with the diuretic amiloride, so it would seem that simple sodium loss could not account for the observed results.

It would appear that sodium depletion should be the adequate and necessary condition stimulating a heightened salt preference. However, depletion of other nutrients may also stimulate avid salt consumption. Some research suggests that calcium and protein depletion stimulates salt appetite in experimental animals. It is generally believed that individuals in cultures in which a primarily vegetarian diet is consumed need and crave more salt than do primarily meat-eating peoples. The traditional explanation for this has been that meat-eaters naturally obtain substantial amounts of sodium in their diet. Although it is certainly true that meat diets generally contain more sodium than most vegetable diets, little direct evidence indicates that the low-sodium diets that might exist in vegetarian cultures lead to an elevated desire for salt. Vegetarian diets in some populations are not only low in salt but may also be low in protein; perhaps that, as well as or instead of sodium content of the diet, underlies an increased desire for salt.

B. Genetic Factors

Genetic factors could account for differences between groups and between individuals in their salt requirements, intakes, and preferences. A number of animal model studies implicate genetic control over salt intake and preference. For example, both inbred strains of rats and inbred strains of mice differ in their salt preference. However, for humans, the evidence favoring a genetic influence is not yet strong.

It has been speculated that Blacks may have a different mechanism than Whites for handling sodium. In particular, they may be less efficient in dealing with excess salt. This could contribute to the high prevalence of hypertension among Blacks when salt is readily available. Although one report indicates that Black adolescents and adults prefer higher levels of salt than do Whites, little evidence indicates a race difference in salt intake or avidity. Additionally, it is problematic whether or not genetic differences would necessarily underlie a race difference in preference.

More direct evidence for a role for genes could come from studies comparing identical and fraternal twins. Twin studies so far have failed to find a heritable component to salt taste preference. The studies conducted to date cannot be considered definitive, however, owing to methodological deficiencies. Consequently, a genetic explanation for individual differences in salt taste preference remains a possibility. More work is needed both in human populations and with animal models.

C. Psychological Factors

Experimental studies have examined the effects of alterations in consumption of salt on salt taste preference. In general, the amount of salt in food required to make the food taste best (optimal level) has been measured. A series of studies has demonstrated that if salt intake is reduced by about 50% over a period of time, the optimal level of salt in foods declines. The amount of salt required to optimize food flavor is decreased to about 65–75% of the prediet amount. Thus, if the optimal level of salt in soup for an individual was 1% while he or she was on a diet containing 150 meq Na/day, decreasing sodium consumption to about 75 meq/day would result, on average, in a reduction in optimal salt levels in soup to 0.75%. Importantly, this change is gradual, taking probably 1–2 months to reach asymptote.

For several reasons, this change in preference probably is due to the altered (decreased) experience of tasting salty foods rather than to a physiological response to the change in amount of sodium the body must handle. First, the gradual nature of the effect, as noted earlier, would suggest that experience plays the major role, because physiological responses to changes in sodium consumption (e.g., changes in the hormones of sodium balance such as renin and aldosterone) are quite rapid, occurring within hours or days. Second, one study reported that if salt consumption was increased by requiring subjects to add approximately 10 g of salt to their food, preferred levels of salt in food increased. However, if the same amount of additional salt was consumed as tablets and, thus, not tasted, there were no changes in taste preferences. Apparently, a sensory adaptation to different levels of salt accompanies a change in salt intake. To some extent, people like the level of salt they taste rather than, or in addition to, choosing the intensity of the taste of salt they like.

The third reason for believing that this change in optimal salt level following dietary change is a psychological phenomenon comes from another study in which dietary salt was reduced by 50%, but continued *ad libitum* use of table salt was permitted. Under these conditions, use of added salt increased almost fourfold but did not approach compensation for the amount removed. In fact, they only replaced 20% of the decrement so that there was an overall decrease of approximately 40% in sodium intake. In this instance, salt was presumably added "to taste," implying that although the salt content of their regular food was in a range that the subjects found palatable, it was unnecessarily high, probably because it was dispersed throughout the food, rather than all being on its surface and easily accessible to the taste receptors. This study suggested a novel technique to reduce salt intake in that individuals were able to use as much salt as they wished while still reducing total salt intake. For this to be practical, however, there should be wide availability of all categories of lowered sodium foods. In terms of control of salt preference, the importance of this research is that there were no changes in taste preference, even though salt intake was decreased 40%. By adding salt to the outside of their food, individuals apparently obtained a sufficient salty taste experience to prevent any preference changes.

In sum, a substantial number of studies support the conclusion that changes in salt taste preferences following changing of salt intake are mediated by hedonic–cognitive expectations based on current dietary experience. Physiological changes associated with alterations in the amount of NaCl available to the body do not appear to play a role.

D. Developmental Factors

It is in infancy and childhood that the strongest claims for an experiential influence on human salt taste preference have been made. Interestingly, both high salt intake and extremely low salt intake in infancy have been postulated to increase later salt preference. However, actual data concerning the early development of salt preference and how experiences may serve to modify or perhaps permanently establish heightened preferences are conflicting and confusing.

Newborn infants are either indifferent to or avoid moderate to high concentrations of saline solution relative to water. By the time children are 2–3 or more years of age, preferences for salty foods over those same foods without salt are common. These two observations have led some to conclude that salt preference is learned, although that conclusion does not necessarily follow. In fact, several studies indicate that a preference for salt solutions over plain water, although not evident in infants <4 months of age, is evident in infants 4–23 months of age. A developmental change in response to salt in human infants may represent, in part at least, postnatal maturation of the ability to taste salt. This interpretation is consistent with a substantial body of evidence from animal models that demonstrates that the neurophysiological response to salt exhibits a postnatal developmental change. In both rats and sheep, responses to salt in newborn animals are less robust than responses in adults.

Even if there is a postnatal maturation of salt taste perception and preference, experience with salt could modulate this preference in infants just as it does with adults. There are reports of a correlation between exposure to salted food and relative preference for saltiness in that food; however, these data are not conclusive as they are based on data from very few infants. Further longitudinal studies on the development and early modification of salt taste preference are needed. In particular, the issue of the relationship between level of exposure to salt and salt preference should be explored at several ages to understand the origin of the very high salt preferences. In this regard, a number of studies now suggest that young children and adolescents actually prefer higher levels of salt in food than do adults. It is unclear why this is the case, although it is possibly due to altered nutritional requirements during growth.

Contrasting with the hypothesis that high-salt intake in infancy and childhood may elevate later preferences, some animal experimental data and human clinical reports suggest that sodium depletion early in life may have profound effects on later salt preference. A body of evidence, primarily derived from studies with rats, now strongly indicates that the salt taste system is plastic during its development. Salt taste perception, as measured electrophysiologically, pharmacologically, and behaviorally, matures postnatally. Also, early restriction of dietary sodium alters subsequent response to NaCl. Finally, an episode of sodium depletion induced over a period of 1 or several days by a very high dose of diuretic combined with salt-free diets produces permanent elevation in salt intake weeks or months after recovery; this effect appears especially robust if the deletion occurs during early development.

The mechanisms responsible for these effects likely have both peripheral and central components. Changes in peripheral activity of salt-sensitive neurons may be involved. The persistent elevation of salt appetite following extreme sodium depletion, however, seems more likely to be of central origin, possibly a consequence of the elevation of renin and angiotensin that follows sodium depletion.

The analysis of available human clinical studies suggests that early sodium depletion may have profound effects on subsequent salt appetite in humans as well (see Table I). Evidence that sodium depletion (occurring in clinical entities such as Bartters syndrome and Addison's disease) produces a craving for salt that might be homologous with salt appetite induced in some experimental animals following severe depletion is not extensive. However, when human salt appetite had been reported, the onset was almost invariably in childhood. The most dramatic case was of a 3.5-year-old boy with adrenocortical insufficiency causing an obsessive salt appetite. This child died from sodium depletion following his placement on a controlled hospital diet. Prior to this he had consumed salt in very large amounts sufficient to replace the massive sodium loss that occurred in the absence of adrenal hormones that ensure sodium conservation. Several other similar cases of early childhood onset of salt appetite have been reported (see Table I). The absence of persuasive evidence for an adult onset of salt appetite suggests that in humans, too, early depletion may have especially potent effects on later salt taste perception and preference. This area is ripe for further research.

IV. CONCLUSIONS

Humans prefer to consume substantially more salt than is necessary for the maintenance of normal physiology. In salt-sensitive hypertensives, this may lead to illness, and it may also contribute to the worsening of other illness (e.g., congestive heart failure, hepatic cirrhosis with fluid retention, hypertensives on antihypertensive medication). Some sensory and behavioral factors responsible for this high consumption have been identified, but a full explanation of the mechanisms is still lacking. To the extent that one understands the determinants of salt intake, one may be better able to provide guidance for successful reduction of dietary salt intake if this is medically indicated.

ACKNOWLEDGMENTS

Preparation of this essay was supported by the National Institutes of Health, Grant Number DC 00882, and Chemosensory Clinical Research Center, Grant Number 00214.

BIBLIOGRAPHY

Beauchamp, G. K. (1987). The human preference for excess salt. *Am. Sci.* **75**, 27–34.

Beauchamp, G. K., Bertino, M., and Engelman, K. (1990). Human salt preference. *In* "Chemical Senses: Appetite and Nutrition" (M. I. Friedman, M. Tordoff, and M. R. Kare, eds.). Marcel Dekker, New York.

Breslin, P. A. S., and Beauchamp, G. K. (1996). Suppression of bitterness by sodium: Variation among bitter stimuli. *Chem. Senses* **20**, 609–623.

Denton, D. (1982). "The Hunger for Salt." Springer-Verlag, Berlin.

Mattes, R. D. (1984). Salt taste and hypertension: A critical review of the literature. *J. Chronic Dis.* **37**, 195–208.

Shephard, R. (1988). Sensory influence on salt, sugar and fat intake. *In* "Nutrition Research Reviews" (J. W. T. Dickerson, A. G. Low, D. J. Milward, and R. H. Smith, eds.), Vol. 1. Cambridge Univ. Press, Cambridge, England.

Scanning Electron Microscopy

DAVID C. JOY
University of Tennessee

GLOSSARY

Convergence angle Cone angle formed by electron beam as focused onto the specimen

Electron gun Device to produce high-energy electrons for imaging

Pixel Smallest element of image on the display or record screen

Raster Rectangular pattern of scan lines generated during imaging

Scintillator Material that emits light when struck by electrons

Working distance Free space between the sample and the microscope lens

THE SCANNING ELECTRON MICROSCOPE (SEM) IS the most widely used form of electron microscope in the health and biological sciences. The SEM is popular because it combines some of the simplicity of the optical microscope with much of the performance of the more expensive and complex transmission electron microscope. Most important of all, the SEM can look at real, solid specimens such as whole cells, pieces of tissue, and bones and even complete organisms such as viruses and bacteria. Its ability to produce high-resolution, high depth-of-field images of three-dimensional surfaces makes it an instrument of unique value.

I. INTRODUCTION

The SEM was originally developed in Germany in the 1930s by A. Knoll and M. Von Ardenne. Later, important improvements were made by Zworykin, Hillier, and Snyder at the RCA Research Laboratories in the United States in the 1940s. The design and performance of their instrument anticipated much that was found in later microscopes, but their success was ultimately limited by the poor vacuum conditions under which they had to work. The current form of the instrument is the result of the work of C. Oatley and his students at Cambridge University between 1948 and 1965. The first commercial SEM, the Cambridge "Stereoscan," was produced by Cambridge Instruments in the United Kingdom in 1965. Today nearly a dozen companies manufacture scanning microscopes for the international market, with prices varying from US$40,000 to in excess of US$500,000.

II. PRINCIPLES OF THE SEM

Figure 1 shows schematically the basic principle of the SEM. Two electron beams are used simultaneously: the incident beam strikes the specimen to be examined; the second electron beam strikes a cathode ray tube (CRT) viewed by the operator. As a result of the impact of the incident beam on the specimen, a variety of electron and photon emissions are produced. The chosen signal is collected, detected, amplified, and used to modulate the brightness of the second

ENCYCLOPEDIA OF HUMAN BIOLOGY, Second Edition, VOLUME 7. Copyright © 1997 by Academic Press. All rights of reproduction in any form reserved.

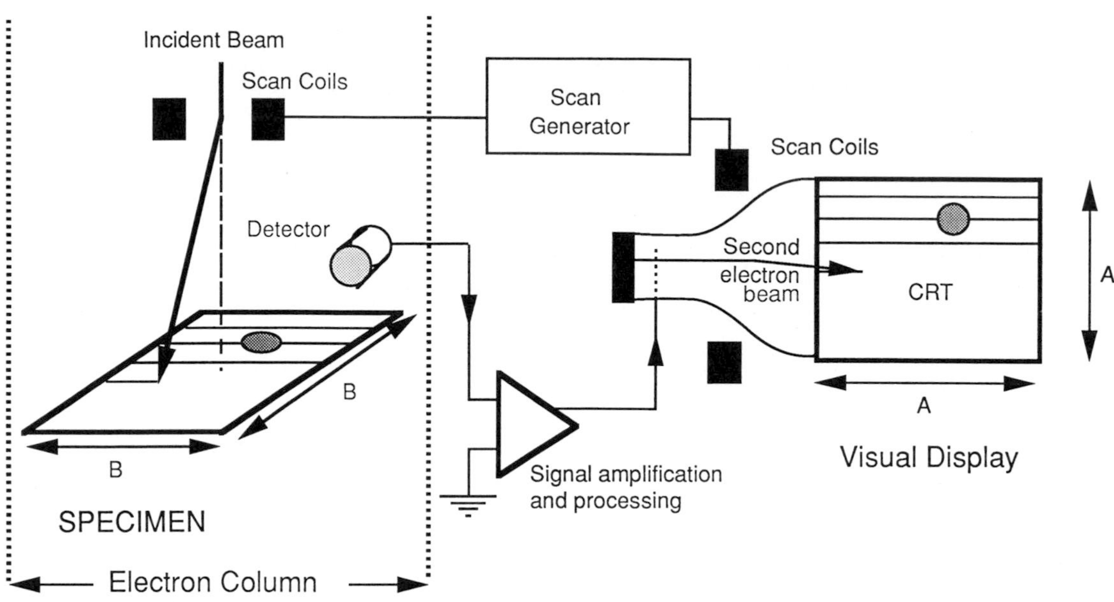

FIGURE 1 Schematic illustration of the principle of operation of the SEM.

electron beam, so that a big collected signal produces a bright spot on the CRT while a small signal produces a dimmer spot. The two beams are scanned synchronously so that for every point scanned on the specimen there is a corresponding point on the CRT. Typically, the beams scan square patterns on both the specimen and the CRT. They start at the top left-hand corner of the area, scan a "line" of points parallel to the top edge, and then, when they reach the end of the line, they fly back to the starting edge and scan a second line, and so on until the whole square area has been "rastered." Each complete image is conventionally called a frame. If the display area of the CRT tube is $A \times A$ in size and the area scanned on the specimen is $B \times B$ in size, then variations in the signal from the specimen will be mapped onto the CRT as variations in brightness with a linear magnification of A/B. Thus, a magnified map or image of the specimen is produced without the need for any imaging lenses.

This method of imaging offers several important advantages:

1. Magnification is achieved in a purely geometric manner and can be varied by simply changing the dimensions of the area scanned on the specimen.

2. Any emission that can be stimulated from the specimen under the impact of the incident electron beam—for example, electrons, X rays, visible photons, heat, sound—can be collected, detected, and used to form an image. The SEM is therefore not restricted to imaging with radiations that can be focused by lenses.

3. Several different types of image can be produced and displayed simultaneously from the same area of the sample, enabling different types of information to be correlated; only a suitable detector, amplifier, and display screen for each signal of interest are necessary. Furthermore, these signals can be mixed with each other to generate new types of imaging information.

4. Because the picture on the screen is formed from an electrical signal, which varies with the position of the beam and hence with time, the image can be electronically processed to control or enhance contrast.

A consequence of this arrangement is that a fundamental limit to the imaging performance is set by the display CRT screen. The smallest feature that can be discerned on the CRT is equal to the size of the electron spot on the display screen. Conventionally it is assumed that 1000 scan lines, each containing 1000 picture elements, or pixels, make up each image frame scanned. Therefore, each picture is formed from a pattern of 1000×1000 (i.e., 1 million) pixels. When the SEM is operating at a magnification of M, then the resolution in the image (i.e., the smallest detail on the specimen that can be observed) is equal to the pixel size divided by the magnification. Because the size of the spot on the CRT is typically 100–200 μm (0.2 mm), magnifications of a few hundred times the resolution are limited to 1 μm or so. Only at high magnifications is the resolution limited by more fundamental electron-optical considerations.

III. COMPONENTS OF THE SEM

The main components of an SEM are contained in two units: the electron column, which contains the electron beam scanning the specimen, and the display console, which contains the second electron beam, which impinges on the CRT. The high-energy electron beam incident on the specimen is generated by an electron gun, two basic types of which are in current use. The first (Fig. 2A) is the thermionic gun in which electrons are obtained by heating a tungsten or lanthanum hexaboride cathode or filament to between 1500 and 3000K. The cathode is held negative at the required accelerating voltage E_0 with respect to the grounded anode of the gun so that the negatively charged electrons are accelerated from the cathode and leave the anode with an energy of E_0 kiloelectron volts (keV). Thermionic guns are in wide use because they may safely be run in vacuums of 10^{-5} Pa (i.e., 10^{-7} Torr) or worse. The alternative source (Fig. 2B) is the field emission gun in which a sharply pointed wire of tungsten is held close to an extraction anode to which is applied a potential of several thousand volts. Electrons tunnel out of the tungsten wire, which can be at room temperature, into the vacuum and are then accelerated as in the thermionic gun toward the anode. Field emission guns require an atomically clean emitter surface, thus they must be operated under ultrahigh vacuum conditions, typically in a vacuum of 10^{-7} Pa (i.e., 10^{-9} Torr) or better. For either emitter the entire length of the electron column traveled by the electron beam from the gun to the specimen chamber must also be pumped to an adequate vacuum using oil-diffusion, turbo-molecular, or ion pumps individually or in combination.

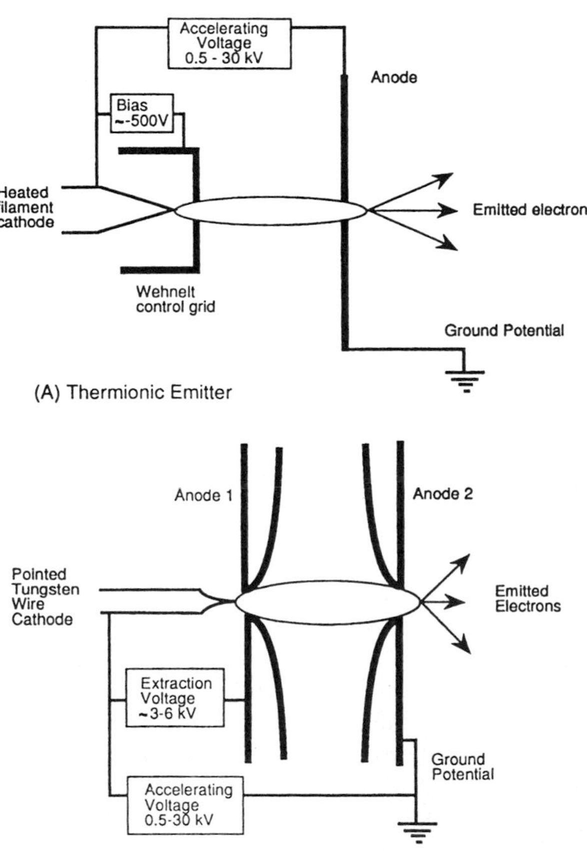

FIGURE 2 Electron sources used in the SEM. (A) Thermionic emitter using a tungsten or LaB_6 cathode. (B) Field emitter.

IV. PERFORMANCE LIMITS OF THE SEM

The performance of the SEM depends on a number of related factors, perhaps the most important of which is the output of the electron source. The source is quantified by its brightness, β, which is the current density (amps/m^2) it delivers into unit solid angle (steradian). The brightness increases linearly with the accelerating voltage of the microscope but also varies greatly from one type of source to another. At a given energy, a field emission gun is between 10 and 100 times as bright as an LaB_6 thermionic emitter, which is, in turn, between 3 and 10 times brighter than a tungsten thermionic emitter. Typically at 20 keV, a field emission gun has a brightness in excess of 10^{12} amps/m^2/sterad.

The diameter of the electron beam from the gun is reduced or demagnified by passing it through two or more lenses before it reaches the sample surface (Fig. 3). An electron lens, which consists of a coil of wire carrying a current, focuses the electron beam in exactly the same way as a glass lens focuses light, but it has the convenient property that the focal length can be varied by changing the magnitude of the current flowing through the solenoid. By varying the excitations of the lenses, the beam diameter at the specimen

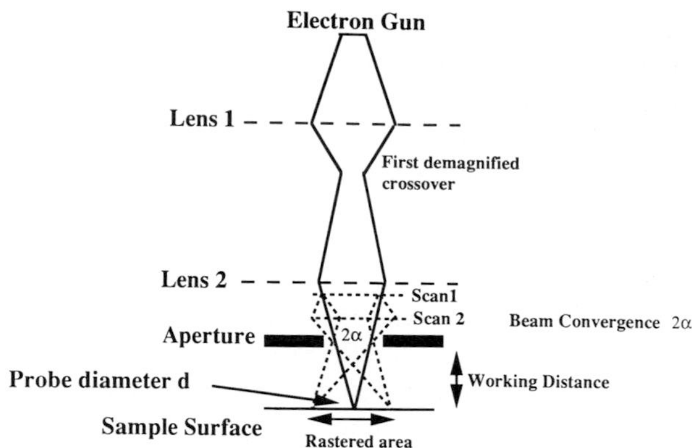

FIGURE 3 Diagram illustrating the ray paths taken by electrons traveling in the column of the SEM.

can be set to any desired value from the source size downward.

The electron beam is usually scanned by first deflecting it across the optical axis in one direction, and then immediately deflecting it in the opposite sense through twice the angle. This arrangement ensures that all of the scanned rays pass through a single point. An aperture can then be placed at this position to define the beam convergence angle α, where α will be equal to the diameter of aperture divided by twice the working distance (see Fig. 3).

The gun brightness is constant throughout the electron-optical system, and hence the value β measured for the focused probe of electrons impinging on the specimen is the same as the value that would be measured at the source. If the probe diameter is d, if the incident beam current is I_B, and if the convergence angle of the beam is α, then by definition the brightness β at the sample, and hence at the gun, is

$$\beta = \frac{4 I_B}{\pi^2 d^2 \alpha^2} \tag{1}$$

For typical operation, α is fixed by the choice of aperture size, and β is constant, thus

$$I_B = \left(\frac{\pi^2}{4}\right)(\beta \cdot \alpha^2) d^2 \tag{2}$$

which shows that as the diameter (d) of the probe is made smaller, the current contained in the probe falls as d^2. The spatial resolution of the SEM (i.e., the smallest detail on the sample that can be observed) cannot be significantly less than the probe diameter. The so-called brightness equation, Eq. (2), therefore indicates that the available source brightness will set a limit to the resolution of the microscope because, as discussed in the following, a certain minimum incident current is required to form an image.

The depth of field, D_f, of the image, defined as the vertical focusing range outside of which the image resolution is visibly degraded, is given as

$$D_f \approx \frac{\text{pixel size}}{\alpha} \tag{3}$$

Because α is between 10^{-3} and 10^{-2} radians, the depth of field is typically several hundred times the pixel size. At low magnifications, therefore, the depth of field can be on the order of several millimeters, giving the SEM an unrivaled ability to image complex surface topography and to produce images with a pronounced three-dimensional quality to them. Notably, however, this fortunate effect results from the fact that electron-optical lenses must be "stopped down" to very small apertures (i.e., small values of α) in order to work, and thus very few of the electrons leaving the source actually reach the specimen. An exact optical analogy would be the "pinhole" camera, which also has a large depth of field but requires long exposure times.

The image in the SEM is built up from an electrical signal, which varies with time. If the average signal level is S, and if as the beam scans across some feature the signal changes by some amount, ∂S, then the feature is said to have a "contrast" level C given by

$$C = \frac{\partial S}{S} \tag{4}$$

Changes in the signal can also occur because of statistical fluctuations in the incident beam current and in the efficiency with which the various emission processes take place within the specimen. Thus, repeated measurements across the same feature on a specimen will give signal intensities that will vary randomly around some mean value. These inherent statistical variations constitute a "noise" contribution to the image, which therefore has a finite "signal-to-noise" ratio. For image information to be visible, the magnitude of the signal change ∂S occurring at the specimen must exceed the magnitude of the random fluctuations by a factor of five times or so. This leads to the concept of threshold current I_{TH}, which is the minimum incident beam current required to observe a feature of contrast level, C. I_{TH} is given by the relationship

$$I_{TH} = \frac{4.10^{-12}}{C^2 \tau} \text{ amps} \tag{5}$$

where τ is the time in seconds required to record one frame of the image (assumed to contain 10^6 pixels). The observation of low-contrast features therefore needs high-beam currents or long exposure times. Typically, images are produced in 1 sec or less, but images to be photographed are recorded for 30–100 sec to improve the signal-to-noise ratio. The threshold current requirement sets a fundamental limit to the performance of the SEM in all modes of operation. A comparison of Eq. (5) with Eq. (2) shows that because the beam current varies rapidly with the beam diameter, the resolution limit of the SEM will depend on the contrast of the feature being observed.

V. ELECTRON–SOLID INTERACTIONS

The interaction of an electron beam with a solid specimen produces a wide variety of emissions, all of which are potentially useful for imaging. Each of these signals is produced with a different efficiency, comes from a different volume of the sample, and carries different information about the specimen. Figure 4 shows schematically the energy distribution of the electrons produced by an incident beam of energy E_0. The distribution displays two peaks: one peak at an energy close to that of the incident beam and a second peak at a much lower energy. The high-energy peak is made up of electrons that have been "backscattered," or reflected, by the sample. The fractional yield of backscattered electrons, that is, the number of backscattered electrons per incident electrons, is called the backscatter yield and usually is identified as η. The average energy of these electrons is about 0.5–0.6 of the incident energy E_0, and the backscatter yield η is typically 0.2–0.4.

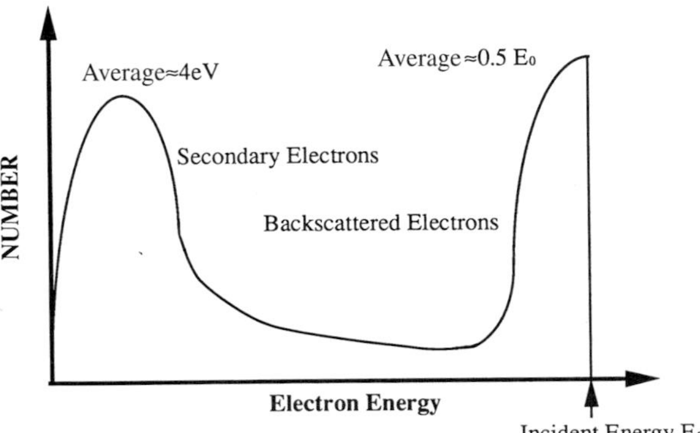

FIGURE 4 The energy distribution of electrons emitted from a solid specimen irradiated by an electron beam of energy E_0.

The lower-energy peak, which lies within the energy range 0–50 eV, is made up of what is usually called the secondary electron signal. As before, we can define a relative secondary yield, that is, the number of secondary electrons per number of incident electrons, identified as δ. The average energy of the secondary electrons is about 4 eV, independent of the incident beam energy, but the secondary yield δ varies rapidly with accelerating energy, being on the order of 0.1 or less for most materials at 30 keV, but on the order of unity for energies of 1 or 2 keV. Because the secondaries are low in energy, they cannot travel more than a few nanometers through the sample to reach the surface and escape, so the secondary signal images the surface region of the specimen.

The total current flowing into and out of the specimen must balance to zero, thus

$$I_B = \eta I_B + \delta I_B + I_{SE} \tag{6}$$

where I_B is the incident beam current and I_{SE} is the current flowing in the ground connection to the sample. This "specimen current" contains information about both the secondary and backscatter signals and can form the basis of an important imaging mode. Equation (6) shows that if $\delta + \eta$ is unity, then no current flows to earth. At this condition, the incident electron beam is neither injecting charge into the specimen nor extracting charge from it. When the sample being examined is an electrical conductor, this situation is not of much significance, but when the specimen is not a good electrical conductor then no current can flow to earth, so any excess (or deficit) charge is retained in the specimen. At high-incident electron energies (greater than a few kiloelectron volts), the total yield $\delta + \eta$ is less than unity, so charge is injected into the specimen, which therefore charges negatively. This reduces the effective incident energy of the beam so $(\delta + \eta)$ increases, but charging will continue until the effective incident energy reaches a value E2 at which $\delta + \eta$ becomes unity. At this energy, each electron in produces, on average, one electron out, so no further charge is deposited and the specimen potential stabilizes. Therefore, at this E2 energy it is possible to form an image from even an insulating material without the need to make the surface an electrical conductor by coating it with metal. Typically, E2 is on the order of 1–3 keV for materials such as polymers or ceramics, but low-density materials such as dried, unstained, biological tissue can have E2 values as low as a few hundred electron volts. Because operation at the E2 energy avoids the necessity of coating the sample, there is increasing interest in low-voltage scanning microscopy in the biological field.

VI. SECONDARY ELECTRON IMAGING

Secondary electrons are the most popular choice of interaction with which to form an image. There are two main reasons for this:

1. Because they are low in energy, it is possible to collect most of the secondaries produced by the sample by biasing the detector to a modest positive potential so that it attracts electrons to itself. Efficient collection is possible even when the detector is not in the line of sight of the sample.
2. The main contrast mechanism associated with secondary electrons produces images that are readily interpretable by analogy with reflected light images in the macroscopic world.

Figure 5 shows a schematic drawing of the detector that is now standard for secondary electron detection. The detector is based on a disc of scintillator, which emits light under the impact of electrons. The light travels along a light pipe, through a vacuum window, and into a photomultiplier, where it is converted back into an electrical current. Because the amount of light produced by the scintillator directly depends on the energy of the electron that strikes it, secondary electrons, which have only a few electron volts of energy, would produce only a very little signal. Therefore, to increase the efficiency a bias of +10 kV is applied to the front face of the scintillator to accelerate all electrons to at least this energy. This high potential may, however, distort or deflect a low-energy incident beam so the scintillator is shielded by a Faraday cage, made of open mesh wire and carrying a potential of only +200 V or so. This is still sufficient to attract and collect 50% or more of the available secondary electrons.

The dominant imaging mechanism for secondary electrons is topographic contrast. It has been estimated that >90% of all scanning micrographs rely on this mode. The effect arises because an increase in θ, the angle of incidence between the beam and the surface normal, will lead to an increase in the yield of secondary electrons as shown in Figure 6. This can be understood by noting that as θ is increased, the fraction of secondary electrons produced within the escape region of the surface also increases. If an electron beam moves over a rough surface, then the local angle of incidence between the beam and the surface normal will change and produce a corresponding change in the secondary signal.

Figure 7 shows examples of this type of imaging, from cultured rat basophilic leukemia cells that have been fixed but not metal-coated, at magnifications ranging from a value (600×) that is comparable with that usually associated with an optical microscope to an upper value (100,000×) that is equivalent to that obtained on a transmission electron microscope. Figure 8 shows another pair of examples of this type of image, this time from human prostate cancer cells that have been coated with a 3-nm film of gold-palladium.

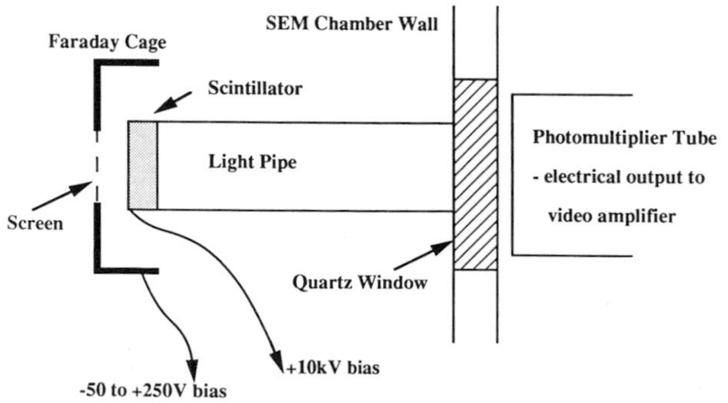

FIGURE 5 The Everhart-Thornley secondary electron detector.

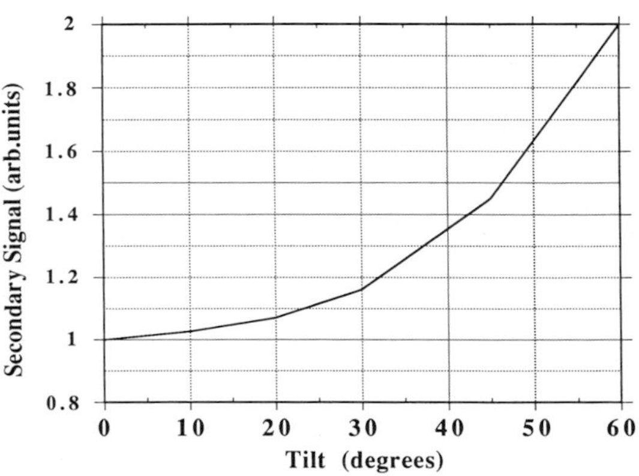

FIGURE 6 Variation in secondary electron yield with θ, the angle of incidence of the beam.

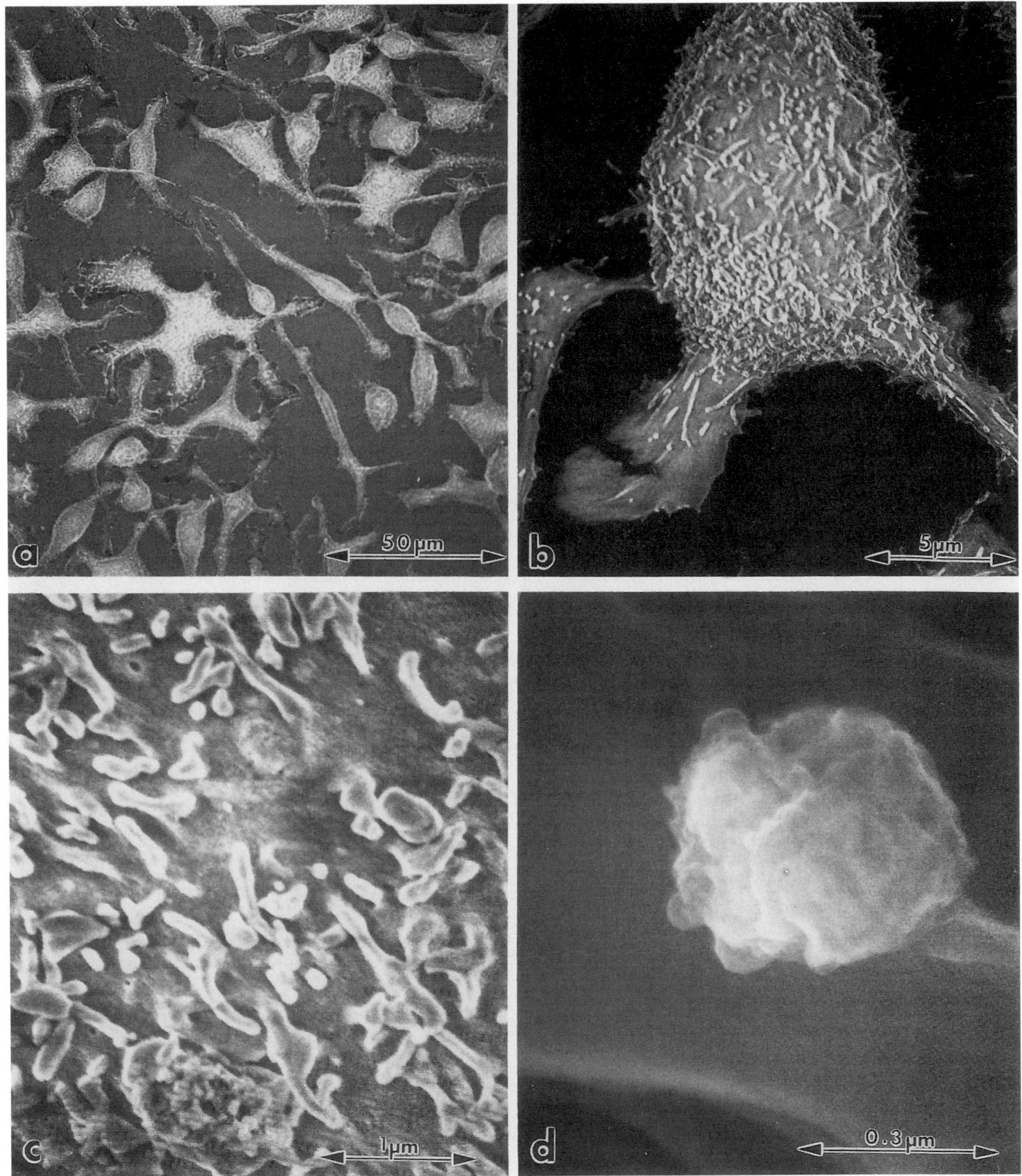

FIGURE 7 Images of rat basophilic leukemia cells at original magnifications of (a) 600×, (b) 6000×, (c) 25,000×, and (d) 100,000×. Micrographs were recorded at a beam energy of 5 keV on a Hitachi S-900 field emission SEM.

Again the benefits of a wide magnification range, three-dimensional view, and high depth of field are evident. Perhaps surprising is the ease with which this type of image can be understood and interpreted. In fact, the view of a surface obtained by the secondary electron image in the SEM is analogous to that which would be obtained if the observer were to look down the column at the specimen illuminated by a light

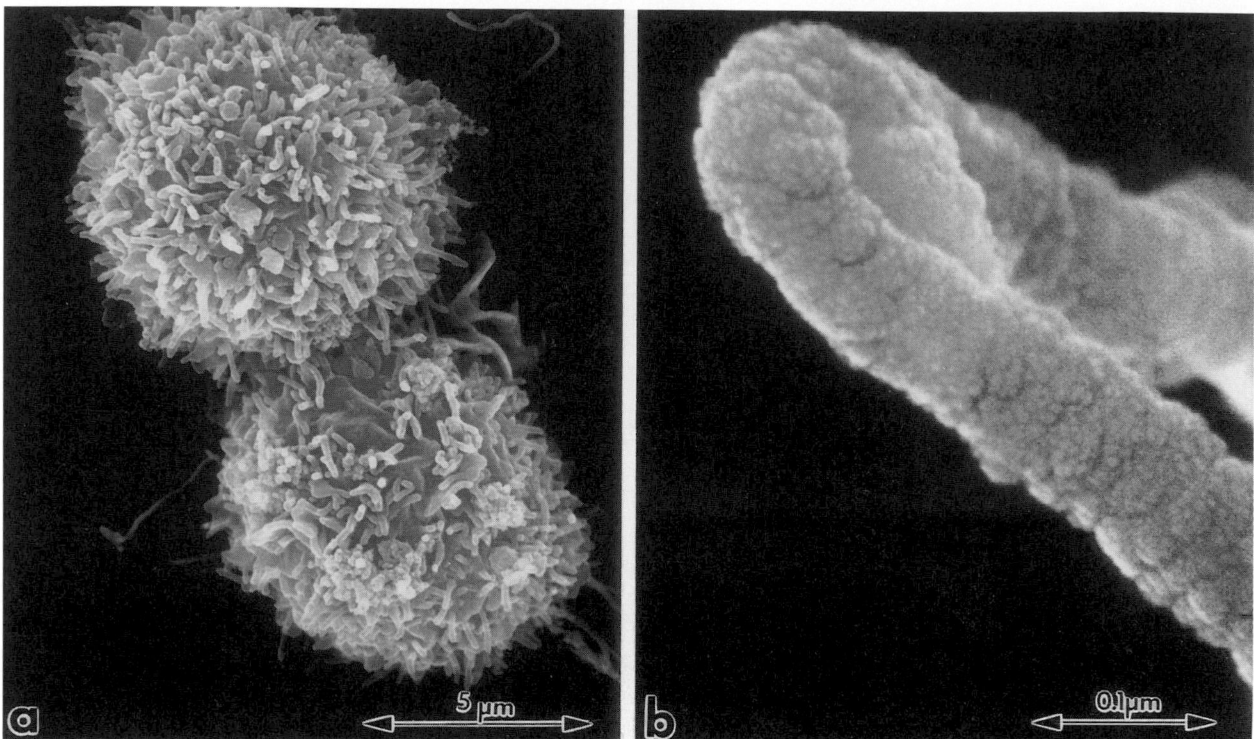

FIGURE 8 Secondary electron images of human prostate cancer cells at original magnifications of (a) 6000× and (b) 180,000×. The specimen was coated with a 3-nm-thick film of gold-palladium. The fine structure visible on image (b) is the grain of the coating. Images were recorded at 20 keV on a Hitachi S-900 field emission SEM. [Micrographs courtesy of Dr. Carolyn Joy, University of Tennessee.]

source placed at the detector. Faces at a high angle of inclination to the beam and facing the detector will be bright, whereas surfaces normal to the beam will be darker. Most microscopists also rotate the micrograph so as to place the illumination at the top of the picture; then all bright surfaces are tilted up and facing toward the detector, while darker surfaces are horizontal or facing away from the detector. Our brain, based on a lifetime of experience in using such clues, can then reconstruct the surface topography with a high degree of confidence. This, coupled with the three-dimensional quality that comes from the high depth of field, explains much of the popularity of the SEM and secondary electron imaging.

The spatial resolution of the secondary electron image under optimum conditions can approach 1 nm (10 Å) with modern, high-performance field emission SEMs. In general, however, the spatial resolution of biological images is limited by the preservation of detail in the tissue after it has been stabilized, dried, and fixed. The typical biological material is mostly water, so it cannot simply be placed in the SEM and examined because the rapid drying that would occur in the vacuum would lead to gross shrinkage and distortion. Instead, the water must first be exchanged carefully with alcohol, and then the alcohol removed by freeze-drying, critical-point drying, or vacuum-aided chemical drying. Because dry tissue is very fragile, it is also usually necessary to fix the sample—either chemically or by freezing—before it is observed. Each of these essential steps can produce artifacts in the specimen and obscure or destroy fine detail. Thus, high-resolution secondary imaging involves as a prerequisite painstaking care with the protocols of specimen preparation.

VII. BACKSCATTERED IMAGING MODES

Backscattered electrons are those incident electrons that have been scattered through an angle >90° within the sample and, thus, can leave it again. They typically have an energy that is on the order of 0.5 of their original incident energy. Consequently, the backscattered electrons can emerge from considerable depths within the specimen, an estimate for this escape depth, R_{BS}, being

$$R_{BS} = \frac{70\,E^{1.67}}{\rho}\,\text{nm} \qquad (7)$$

where ρ is the density and E is the beam energy in kiloelectron volts. For a beam energy of 20 keV, backscattered electrons can carry information about regions 5–10 μm below the surface of dry tissue compared with just a few nanometers for the secondary electrons. However, because the diameter over which the backscattered electrons emerge from the surface is of the same order as R_{BS}, the spatial resolution of the image will be worse than that of the secondary signal.

Backscattered electron imaging modes are complementary to secondary modes and produce unique information of their own; however, they have so far attracted somewhat less attention because of the problem of signal collection. Whereas the low-energy secondary electrons are readily collected by the application of a small bias field, the high-energy backscattered electrons travel in straight lines from the specimen and must therefore be collected by placing a suitable detector in the path of the electrons. A typical arrangement (Fig. 9) places a ring of scintillator concentrically around the beam and directly above the specimen. No bias is applied to the scintillator so only high-energy backscattered electrons produce an output. For beam energies over a few kiloelectron volts, this type of arrangement is highly efficient (>50%).

The yield η of backscattered electrons varies with both the atomic number of the target and the energy of the incident electrons. For energies over about 5 keV, the backscattering yield is almost independent of the accelerating voltage, and η can be approximated by the function

$$\eta = -0.0254 + 0.016Z - 0.000186Z^2 \qquad (8)$$

where Z is the atomic number of the target. η varies between about 0.05 for carbon ($Z = 6$) and about 0.5 for gold ($Z = 79$). The yield is also somewhat dependent on the topography of the surface, but for a large backscattered detector placed symmetrically above the specimen surface this effect is small enough that such backscattered images contain little topographic information.

Because the yield is simply related to the atomic number Z, the backscattered image displays contrast that is directly related to the atomic number of the area sampled by the electron probe. This technique has found several important applications in biology. For example, regions of tissue or cells can be identified by heavy metal (e.g., osmium, silver, tungsten, gold) stains, which bind preferentially to specific features of interest. Though the metal might be invisible in the secondary image because of the strength of the topographic contrast, the large difference in effective atomic number between the stain and the predominantly carbon matrix provides a high-contrast signal in the backscattered image mode. A comparison of the secondary and backscattered images then permits the stain to be localized. An extension of this technique—cell-surface labeling—uses small (2–20 nm diameter) particles of colloidal gold bound either to a chemical entity, which has a specific affinity for some site or sites on a cell surface, or to an antibody. Even though the gold particles are small, they are readily visible in the backscattered image and, hence, by again comparing secondary and backscattered images, the binding sites for the labeled compound can

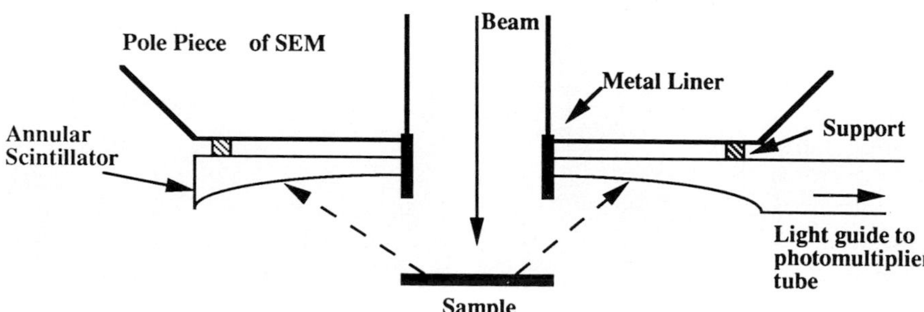

FIGURE 9 Typical detector for backscattered electrons using a large annular scintillator placed above the specimen and concentric with the beam axis.

be identified. Figure 10 shows an example of this type of imaging in which the receptor sites for prolactin on porcine oocytes are identified by the 12-nm gold labels visible as bright dots against the dark carbon background.

VIII. UNWANTED BEAM INTERACTIONS

Although interactions between the sample and the beam of incident electrons provide the information carrying contrast of the image, other interactions are equally potent but less beneficial. In particular, the electron beam is a powerful source for ionizing radiation. To give an idea of how significant this effect is, the specimen in the SEM is subjected to as much radiation as an individual standing 3 m from a 10-megaton H-bomb. This is very harmful to biological materials because ionization can lead to the breaking and cross-linkage of bonds. This radiation damage is a severe limitation to the ability of the SEM to efficiently image such materials. A deposited dose of only 1000 e^-/nm^2 (160 coulombs/m^2) is sufficient to destroy the crystal structure of a soft protein, yet this is less than one-tenth of the dose normally required to take a single high-resolution SEM image. An X-ray spectrum requires a greater dose by a further factor of 100 times. At the moment, there is no way of protecting against such damage although, for some samples, the rate of damage is known to be reduced by a factor of two to five times if the specimen can be held at very low (below 10K) temperatures. Therefore, it is necessary to plan microscopy so as to minimize the electron dose to which any part of the sample is exposed—for example, by working at lower magnifications, shooting blind on previously unexposed areas, and accepting noisier images and spectra—and to be aware of the kinds of artifacts that radiation damage can cause.

IX. CONCLUSION

Although the SEM is a simple instrument, its mode of operation gives it some unique advantages. It combines the low-magnification capability and simplicity of the optical microscope with the high resolution of

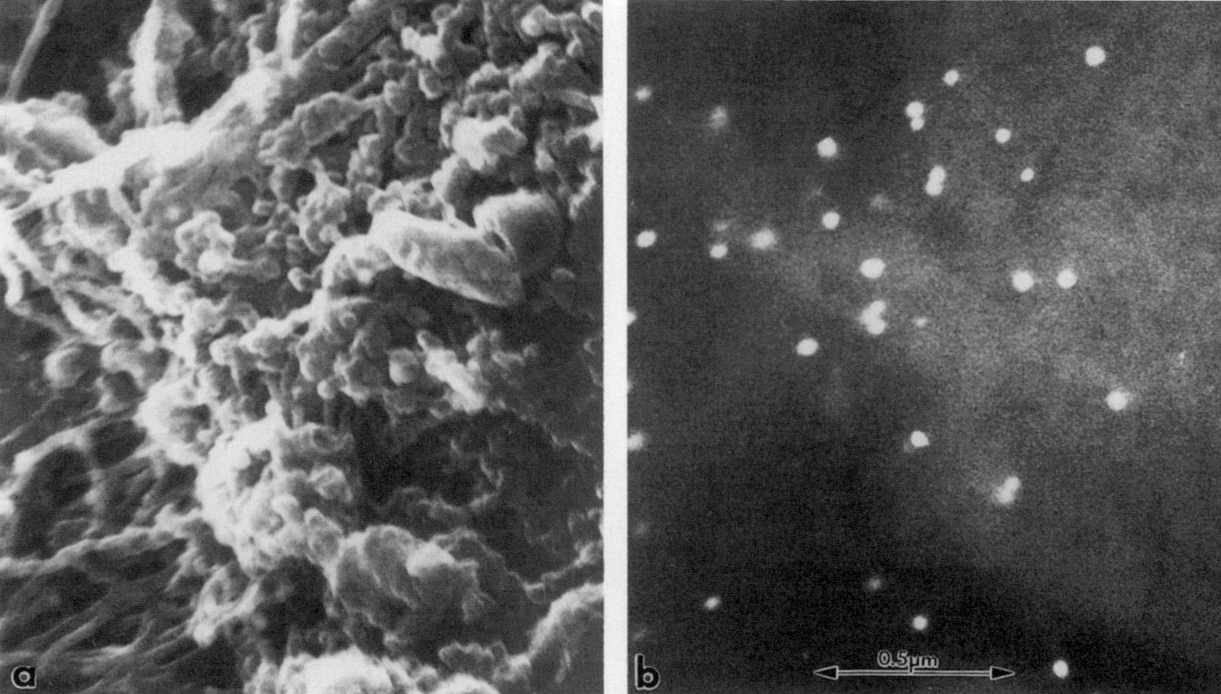

FIGURE 10 (a) Secondary and (b) corresponding backscattered electron images of 12-nm colloidal gold-labeled prolactin on the zona pellucida of a porcine oocyte. Images were recorded at 20 keV energy and a magnification of 60,000× using a Hitachi S-900 field emission SEM. [Micrographs courtesy of Dr. D. Wininger, University of Tennessee.]

the transmission electron microscope and produces images of solid specimens, which are both readily interpretable and aesthetically pleasing. Thus, the SEM is a tool of growing importance in the biomedical sciences.

BIBLIOGRAPHY

Echlin, P. (1992). "Low Temperature Microscopy and Analysis." Plenum, New York.

Goldstein, J. I., Newbury, D. E., Echlin, P., Joy, D. C., Romig, A. D., Lyman, C. E., Fiori, C., and Lifshen, E. (1992). "Scanning Electron Microscopy and X-ray Microanalysis." Plenum, New York.

Joy, D. C., Romig, A. D., and Goldstein, J. I. (1986). "Principles of Analytical Electron Microscopy." Plenum, New York.

Newbury, D. E., Joy, D. C., Echlin, P., Fiori, C. E., and Goldstein, J. I. (1986). "Advanced Scanning Electron Microscopy and X-ray Microanalysis." Plenum, New York.

Oatley, C. (1982). The early development of the SEM. *J. Appl. Phys.* **53**, R1–R13.

Watt, I. M. (1995). "The Principles and Practice of Electron Microscopy." Cambridge Univ. Press, London.

Scanning Optical Microscopy

C. J. R. SHEPPARD
University of Sydney

GLOSSARY

Autofocus technique Method of producing projections in confocal microscopy by selecting the peak signal in an appropriate direction

Beam scanning Type of scanning optical microscope in which the beam is scanned by, for example, galvomirrors. The beam thus transverses the optical system off-axis during scanning

Confocal microscopy Technique of scanning optical microscopy, often performed in a fluorescence mode, but also in bright-field reflection or transmission. Confocal microscopy results in optical sectioning of thick objects, and improved resolution and contrast

Differential phase contrast Technique of scanning optical microscopy, achieved by using a detector split into two halves, giving an image of phase gradients in the object. It can be performed in either reflection or transmission

Extended focus technique Method of producing projections in confocal microscopy by averaging the signal in an appropriate direction

On-axis scanning Type of scanning optical microscope in which the specimen (or objective lens) is scanned mechanically so that the beam always travels along the axis of the instrument, thus avoiding off-axis aberrations and shading.

SCANNING OPTICAL MICROSCOPY IS A RECENTLY introduced method which exhibits a range of advantages compared with conventional optical microscopy, while retaining the noninvasive nature of optical microscopy. Thus specimens can be observed with the minimum of preparation and need not be exposed to a vacuum environment as in electron microscopy. Living tissue can be observed in its natural watery condition. Specimens can be observed without staining, taking advantage of electronic contrast enhancement or methods of phase imaging. Alternatively, specific components can be stained using immunogold or immunofluorescence techniques. Perhaps the most well known at present of the various scanning methods available is confocal fluorescence microscopy, which is considered in detail in Section III,C.

I. INTRODUCTION

A. Advantages of Scanning

Broadly, the advantages of scanning optical microscopy stem from two main properties. First is the fact that the image is measured in the form of an electronic signal, which allows a whole range of electronic image processing techniques, both analog and digital, to be employed. These include image enhancement techniques such as frame averaging, contrast enhancement, edge enhancement, and image subtraction to show changes or movement, image restoration techniques for resolution enhancement, and noise reduction and image analysis techniques such as feature recognition and cell sizing and counting.

Second is the property that imaging in a scanning microscope is achieved by illuminating the object with a finely focused light spot. This allows a number of novel optical imaging modes to be employed such as confocal imaging or differential phase contrast, but also introduces the possibility of imaging modes in which the incident light spot produces some related effect in the specimen that can be monitored to produce an image. In addition, by restricting the size of the photosensitive detector, the noise level is reduced

ENCYCLOPEDIA OF HUMAN BIOLOGY, Second Edition, VOLUME 7. Copyright © 1991 by Academic Press. All rights of reproduction in any form reserved.

and the measurement accuracy can be greatly improved.

B. Methods of Scanning

In a scanning microscope the object is illuminated with a focused light spot which is scanned relative to the object (Fig. 1). This can be achieved either by scanning the light spot or by scanning the object itself. Most commercial instruments at present scan the light spot. The two most widely used methods employ galvomirror scanners for both the x and the y scan, or a Nipkow disk as is used in the tandem scanning microscope. Galvomirrors are either of the resonant variety, which can oscillate at high speed but are fixed frequency devices, or more usually of the feedback-stabilized type, which can scan at a line frequency of about 1 kHz. Nipkow disk scanners, in which a disk with an array of holes is rotated, allow the use of a white light source because they have multiple apertures. Their main disadvantages are that signal level can be low and that a television camera must be used in order to produce an electronic image signal. Alternative beam-scanning systems include polygon mirror scanners, which achieve high scanning speeds but are difficult to synchronize, and acoustooptic scanners, which allow TV scanning rates but suffer from chromatic variations which rule them out for fluorescence applications without use of special imaging geometries.

The main advantage of object scanning is that the optical system remains completely unchanged during scanning so that the imaging properties are unvarying across the field. Beam-scanning systems, however, can experience brightness variations across the field, a fall-off in resolution at the edges of the field, and also noticeable curvature of field. Object scanning is thus preferable for quantitative work or in image reconstruction. The disadvantages of object scanning are that imaging can be rather slow, taking perhaps a few seconds to record a single frame, and that use of electrical probes is more difficult. The speed consideration is perhaps not so important when it is realized that often long scan times are necessary in order to collect sufficient light to produce a low-noise image. We can perhaps anticipate a revival of object-scanning methods in the future.

The final method of scanning to be considered is mechanical scanning of the objective lens. This again has the advantage that the light travels on-axis through the optical system, thereby giving good quantitative imaging. However, speed is again rather low.

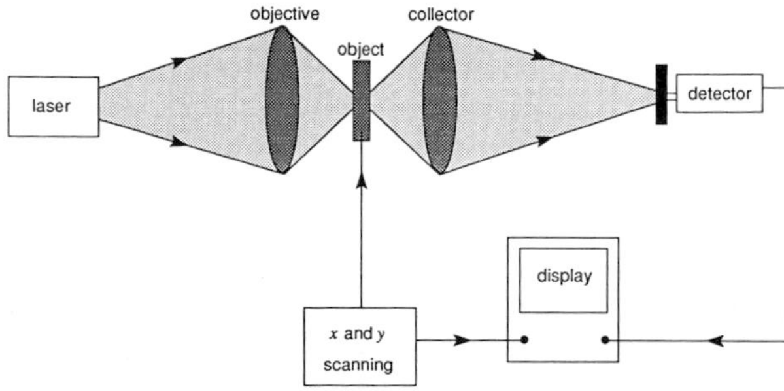

FIGURE 1 A schematic diagram of a scanning optical microscope.

C. Design of Scanning Systems

A number of alternative arrangements are used for beam-scanning microscopes. Usually a microscope eyepiece, or lens of similar design, is used to produce a focus that subsequently illuminates the objective lens. In order that the beam fills the objective lens aperture, the axis of rotation of the scanning mirrors must be situated close to the plane of the entrance pupil of the eyepiece. A single mirror placed here and scanned in both the x and y directions is the simplest design. Alternatively, two separated galvomirrors can be used, coupled by a telecentric system consisting of two lenses or mirrors. Finally, two close-coupled galvomirrors can be used if the axis of rotation of one is offset so that it both translates and rotates.

In an object-scanning system, the optical system is simplified as the system is operated in an on-axis condition. It is found that it is possible to optimize performance by small corrections to the effective tube length of the objective. Mechanical object scanning is usually achieved using electromechanical devices. One feature that must receive some attention is the necessity of ensuring that the plane of scanning is accurately located. This can be achieved by mounting the specimen stage on leaf springs or stretched wires.

D. Alignment

An optical system gives its best performance only if it is accurately aligned. In particular, a confocal microscope must be aligned correctly or it will exhibit artifacts, especially with a small pinhole size. First of all the beam-expanding system must be adjusted and the beam collimated so that the objective is used at its correct tube length. The best way of aligning the system is to examine a planar object with a pinhole of large diameter. Without refocusing the object, the pinhole size is reduced and its axial and transverse position is adjusted for maximum signal.

E. Photodetectors

Many types of detectors may be used for scanning optical microscopy. If the signal level is high, as for example in transmission or in reflection from a surface such as bone or tooth, a simple photodiode may be employed. For confocal fluorescence applications, the signal level is low, and is reduced further as the thickness of the optical section is reduced. Then highly sensitive detectors such as photomultiplier tubes, perhaps cooled, and with photon counting are advantageous.

The relevant properties of a detector are quantum efficiency, sensitivity, dynamic range, and linearity. Conventional photocathodes have a quantum efficiency of only about 20%, but this figure can be raised to above 30% by using negative electron affinity photocathodes. Silicon detectors of the photodiode, avalanche photodiode, or charge coupled device (CCD) variety can have quantum efficiencies greater than 80%. Avalanche photodiodes in principle combine high quantum efficiency with the ability to photon count with pulse height analysis for rejection of dark current. However, at present there are difficult practical problems in their use. Charge-coupled devices combine high quantum efficiency with the facility for integration within the device and the possibility of cooling to improve noise performance. CCDs are usually of the area or linear array variety, and in some cases additional information can be extracted from this spatial information.

F. Choice of Objectives

At present there are no objective lenses specially designed for scanning microscopy, and so objectives must be chosen from commercially available types. For laser microscopy the objective must be corrected for only one wavelength (or two, including a fluorescent wavelength), which should allow lenses of increased numerical aperture or working distance to be developed. Similarly, in microscopes with on-axis scanning, off-axis aberrations are unimportant, giving further flexibility in the design. For example, an immersion lens, designed for one wavelength and with a limited field of view for on-axis scanning, with a numerical aperture of 1.4 and a working distance of 1 mm, seems feasible.

Achromat lenses have some advantages over apochromat lenses for some applications as they have lower loss, as well as being considerably cheaper. We have found fluorite lenses to be a good compromise for on-axis scanning. However, for beam scanning the off-axis aberrations and field curvature are too large. Again with beam-scanning confocal microscopes, there is a need for high-aperture lenses (to give good collection efficiency) of low magnification. These are not necessary with on-axis microscopes as one objective can be used to cover the whole range of magnification, simply by altering the amplitude of scan. Confocal microscopes are more sensitive to aberrations than conventional instruments, particularly for imaging in the depth direction. So incorporation of a correction collar is useful. Finally, for many biological studies, water immersion objectives are preferable, because with uncovered specimens this removes both the reflection and aberrations produced by the cover glass–specimen interface.

G. Image Processing for Scanning Microscopy

First of all it is worth pointing out that many of the early scanning laser microscopes did not use a computer for image manipulation, but rather used analog electronics together with a long-persistence display. The main impetus for employing digital techniques was to provide image storage to improve the real-time observation of images, but in fact very high-quality images with a resolution of several thousands of lines can be produced by photographing directly the cathode ray tube display. Similarly, many forms of image processing, such as contrast enhancement, filtering, image addition, and subtraction, can be achieved using analog methods. The use of digital methods of course greatly extends the flexibility of the system, but analog-processing facilities are worth retaining in order to improve the dynamic range of the recorded data.

A single 512×512 image contains a quarter of a megabyte of information. This means that a three-dimensional (3D) image consisting of many sections requires large memory and also processing time for 3D manipulation. For this reason, images that consist mainly of "empty space" can be stored and processed alternatively using a vector scan method in which only the nonzero contributions to the image are stored with their coordinates. Actually, good stereoscopic effects can be obtained from few sections, and it is also possible to store projections directly, rather than the sections themselves, which can also greatly reduce the quantity of data.

Most scanning microscopes generate a single image in a few seconds, which means that with a commercial imaging system a slow-scan input is necessary. Alternatively, personal computers nowadays have large enough memories to permit direct storage of the images in their random access memory (RAM). Personal computers can provide most of the functions necessary for scanning microscopes, but are not fast enough for real-time manipulation of 3D images.

A range of standard image enhancement methods can be used with advantage in scanning optical microscopy. These include contrast enhancement by linear stretching or histogram equalization, edge enhancement by filtering in either the spatial or the Fourier domain, low-pass or median filtering to reduce noise in images, image averaging, and so on. Most of these methods can be employed with 3D as well as 2D data sets.

Once a 3D image has been stored, projections in arbitrary directions can be produced by image rotation. This is a computationally intensive process, and an alternative is to stack sections with an appropriate pixel offset between adjacent sections. The sections may be stacked by summation, corresponding to the extended focus method described in Section III,B, or by selecting the peak signal, corresponding to the autofocus method.

II. IMAGING MODES OF SCANNING MICROSCOPY

A. Introduction

In a scanning microscope the object is illuminated with a focused spot of light. Such an arrangement is extremely versatile as an image can be generated from a wide range of different effects of this illuminating spot. Examples include photoelectron imaging, photoacoustic imaging, photothermal imaging, and photodesorption studies. Another method of great application in the semiconductor industry involves detecting the current (or voltage) generated by the incident radiation. This technique could also have applications in the biological area. Here we discuss in more detail two particular imaging modes. The first involves the detection of light at a different wavelength from that incident. The second uses detector arrays to give differential phase contrast.

B. Spectroscopic Methods

An advantage of using scanning methods for spectroscopic imaging is that imaging, performed with the incident radiation, is separated from wavelength selection and analysis of the emitted radiation, thus simplifying system design and resulting in superior performance. The detection system, because it does not have to image, may also have greater sensitivity. This is of great advantage in fluorescence microscopy, which also results in the further advantage that the resolution is determined by the shorter incident wavelength rather than the longer fluorescence wavelength.

Fluorescence, or luminescence, microscopy can give information concerning spatial variations in excitation states, binding energies, band structure, molecular configuration, structural defects, and the concentration of different atomic and molecular species.

Use of a pulsed laser allows investigation of transient effects such as the lifetime of excited states and capture and emission cross-sections. Other examples of spectroscopy that may be performed using scanning techniques include absorption spectroscopy, Raman spectroscopy, resonance Raman spectroscopy, coherent anti-Stokes Raman spectroscopy (CARS), two-photon fluorescence, photoelectron spectroscopy, and photoacoustic spectroscopy. [*See* Photoelectric Effect and Photoelectron Microscopy.]

C. Differential Phase Contrast

Image formation in a scanning microscope (nonconfocal) is in principle identical to that in a conventional microscope. However, one difference is that the detector sensitivity distribution of a scanning microscope can be made of negative strength, which is not true for the source intensity distribution of a conventional microscope. An example is the arrangement shown in Fig. 2, in which the detector is split into two halves.

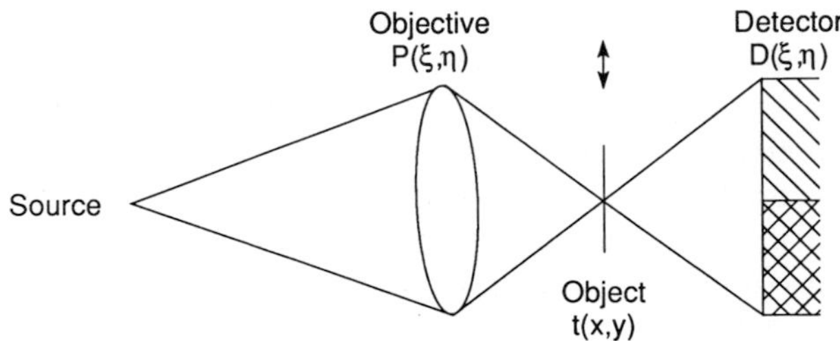

FIGURE 2 The split detector method for achieving differential phase contrast.

In the absence of a specimen, each half receives an equal signal so that if they are subtracted one from the other there is no net signal. If an object consisting of a phase wedge (i.e., a prism) is inserted, the beam on the detector is deflected, and a net signal results that is proportional to the phase gradient. Adding the two signals gives a conventional bright-field image. The split detector technique allows the observation of weak phase structure. It is extremely sensitive: for example, it can show up the edge of a monomolecular film. It has various advantages over the alternative Nomarski differential interference contrast (DIC) method, most notably that the phase information can be more easily extracted. By electronic integration of the phase gradient information, an absolute phase image can also be produced. The method can also be used in reflection to show up surface topography. By altering the configuration of the detector, the system can be optimized for different properties, e.g., weak phase gradients or fine detail.

III. CONFOCAL MICROSCOPY

A. Confocal Imaging

In a conventional microscope the object is illuminated using a large-area incoherent source via the condenser, and each point of the object is imaged by the objective lens. The objective is responsible for determining the resolution of the system. If the image is now measured point by point by a detector of small aperture, the image is unchanged, providing the detector is small enough. In a scanning microscope, however, there is a point source and a large-area detector instead of a large-area source and a point detector. Reciprocity argues that image formation is identical in a scanning microscope and a conventional microscope. However, now the first lens (also termed the objective, but sometimes the projector) determines the resolution.

In the confocal microscope, both a point source and a point detector are used, which is achieved by placing a pinhole in front of the detector so that in this case both lenses take part equally in the image formation process and the resolution is improved by a factor of about 1.4. Other advantages of confocal microscopy are (1) that out-of-focus information is rejected by the pinhole so that an optical section is imaged and (2) that unwanted scattered light is also rejected by the pinhole. Although the system shown in Fig. 1 is a confocal transmission system, commercial systems usually operate in the reflection mode. This makes the operation of the system much easier as the point detector can be arranged to coincide with the image of the point source in the beam splitter, resulting in coincidence of the illuminating and detection spots.

B. Optical Sectioning and Three-Dimensional Microscopy

Light emanating from regions of the specimen separated from the focal plane are defocused at the pinhole plane and are hence rejected, thus resulting in an optical sectioning effect. By scanning in the depth direction, sequential sections can be studied and a complete three-dimensional image built up. Three-dimensional data can, for example, be stored in a computer for subsequent processing and display.

Of course, 3D images are difficult to display directly so instead we can extract sections oriented in some arbitrary direction. For example xz images are sections parallel to the system axis. We can also produce projections in an arbitrary direction, in which the depth information is suppressed. Projections can be projected either by the summation of sections, resulting in an extended focus image, or by detecting the peak signal in depth, giving an autofocus image. Either approach can be achieved using digital or analog methods, and an extended focus image can also be generated directly by photographic integration on a cathode ray tube. Two projections produced at slightly differing angles by either the extended focus or the autofocus method give a stereoscopic image pair.

As well as recording the peak signal in depth, as in the autofocus method, the depth position can be recorded in order to locate its position in depth. This depth information can be displayed as a surface profile image in gray levels or color, or can be combined with the peak information to produce color-coded images or reconstructed views. The sensitivity of the depth measurement is about 50 nm, or can be better than 0.1 nm using confocal interference techniques.

Projections in different directions can be computed and stored for animated display, but the information content of a full 3D image at present is too great to permit real-time rotation.

By using a flat reflecting surface such as a mirror as the specimen, the imaging performance in the depth direction can be investigated by axial scanning and observation of the resulting defocus signal. This can alternatively be achieved by (1) tilting the mirror slightly, scanning in a transverse plane, and examining the line-scan signal or (2) producing an xz image and again extracting a line scan. The defocus signal gives much information about the performance of the optical system. Ideally it should be a smooth, narrow response with weak sidelobes. In practice it is made broader as a result of the finite size of the pinhole and is often degraded by the lack of alignment or the presence of aberrations. In particular the presence of strong sidelobes or auxiliary peaks is very detrimental to three-dimensional imaging performance. Confocal microscopy is very sensitive to the presence of small amounts of aberration such as spherical aberration or astigmatism, which can be introduced by the use of incorrect cover glass thickness, or the effects of the mounting medium or focusing deep into the specimen. In these cases, observation of the defocus signal allows optimization of the imaging performance by altering the tube length at which the objective is used or by inserting correction lenses.

C. Confocal Fluorescence Microscopy

At present, most commercial confocal microscopes are designed for fluorescence operation, so that to many people confocal and fluorescence are almost synonymous. However, it should be stressed that many useful nonfluorescent applications can be made using confocal bright-field techniques. The main advantages of the fluorescence mode of confocal microscopy is that specific stains can be introduced, but also high-quality 3D images can be produced without problems arising from coherent optical noise (speckle).

As described earlier, resolution of a scanning fluorescence microscope is superior to that in a conventional fluorescence microscope. There is a further improvement for confocal fluorescence microscopy, so that, for example, if the fluorescence wavelength is 1.5 times the primary wavelength the cutoff in the spatial frequency response is 1.5 times as great in a scanning compared with a conventional fluorescence microscope, but 2.5 times as great in a confocal fluorescence system. The price one pays for this dramatic improvement in resolution is the decreased signal strength of the confocal arrangement.

Most commercial confocal fluorescence systems use an argon ion laser to provide a main line at 488 nm, which is close to the absorption peak of fluorescein isothiocyanate. The other strong line of the argon ion laser at 514 nm can be used to excite rhodamine or Texas Red. An alternative laser for fluorescence work is the helium cadmium laser, which gives lines at 442 and 325 nm. The former can be used with acridine orange, fluorescein, and Feulgen-Schiff staining, whereas the latter can be used with Hoechst or DAPI. Either helium–neon (633 nm) or krypton ion (647 nm) lasers can be used to excite chlorophyll b. The krypton ion laser can give a large number of lines throughout the visible and ultraviolet. A further laser of interest is the frequency-doubled neodymium YAG laser at 532 nm.

D. Effects of Pinhole or Slit Size

As the pinhole of a scanning microscope is reduced in size the imaging performance more nearly approximates true confocal imaging, but the signal strength also decreases so that in practice some compromise setting is necessary. The important parameter is the size of the pinhole compared with the diameter of the Airy disk in the pinhole plane so that the absolute size of the pinhole needed to obtain a specific effect depends on the geometry of the system. We introduce a normalized pinhole radius u_d given by

$$v_d = \frac{2\pi r_d \, \mathrm{NA}}{\lambda M},$$

where r_d is the true radius of the pinhole, NA is the numerical aperture of the objective, λ is the wavelength, and M is the magnification between the pinhole and object planes. Use of a pinhole with v_d equal to about 3 gives a good compromise in performance.

As the pinhole size is increased, the resolution in both transverse and axial directions is degraded. For the axial direction, resolution for a planar object decreases monotonically with increasing pinhole size, and for large pinholes the width of the response increases linearly. However, for a point-like object, axial resolution does not become significantly worse for v_d greater than about 10. This explains how 3D restoration techniques can be used for some forms of object with conventional fluorescence microscopes.

Some designs of microscope use a slit aperture rather than a cricular pinhole. Although this can give an increased signal relative to a pinhole of diameter equal to the slit width, the resolution is also degraded and no benefit is gained overall in this respect. Nevertheless, slit apertures can be used to advantage in various ways to obtain real-time image formation.

E. Confocal Bright-Field Reflection Microscopy

Confocal microscopy in the bright-field (nonfluorescence) reflection mode has many applications in the materials science and industrial areas, and can also be exploited in biological studies. The tandem scanning microscope is often used in a bright-field mode as signal levels are frequently low in the fluorescence mode. Reflected bright field has the advantage of giving extremely sharp depth imaging, sharper than in confocal fluorescence, but in some cases there can be difficulties in interpretation as a result of coherent noise (speckle). This is not usually a problem when the specimen consists of reflecting surfaces, but can be when it is necessary to investigate refractive index variations within a semitransparent object. The coherent noise is reduced by using the autofocus or, more effectively, by the extended focus technique, and in some cases the visibility of the image features can be improved by filtering.

A further important technique of confocal reflection microscopy is the use of immunogold probes. In this case the gold particles are spatially well separated so that coherent interactions are not important, and the gold particles can be located in three-dimensional space. Scanning microscopes can be very sensitive for detecting scattering from these gold particles, and preparations with particles as small as 5 nm in diameter without silver enhancement have been imaged. In many cases, however, the lower limit to the size of the particles is set by the requirement that the scattering, which varies with the fourth power of the diameter, is strong compared with scattering by the refractive index variations in the specimen itself.

F. Confocal Transmission Microscopy

Although much of the early work on confocal microscopy was performed in the transmission mode, this is now not nearly as frequently used as the confocal fluorescence technique. The reason for this is that there is extreme difficulty in aligning and maintaining alignment during scanning, caused by refractive effects in the specimen. This can be to some degree alleviated by using a double-pass method, where the transmitted light is reflected back through the specimen and detected using the usual reflected light detector. This method can also be used to increase the signal in confocal fluorescence microscopy.

Confocal transmission retains the resolution improvement of confocal imaging and also results in an improvement in depth imaging, but not to such a marked degree as in confocal reflection. Confocal transmission microscopy can also be performed in a differential phase-contrast mode and, by mixing with a coherent reflection signal, can result in a further improvement in depth imaging.

G. Confocal Interference Methods

Scanning techniques are well suited to interference methods, and in particular confocal interference microscopy is a powerful technique. This is possible because confocal imaging is a coherent process and is further simplified experimentally by the long coherence length of lasers. Confocal methods have the major advantage over interference microscopy in that the shape of the reference beam wavefront is immaterial, only its phase and amplitude at the detector pinhole are important, removing the requirement for matched optics and making alignment much less critical. Furthermore, it is possible to combine the system with multiple detectors and real-time processing to extract image information. Confocal interference microscopy can be used for investigating the refractive index or surface height variations or for obtaining the signal phase for the restoration of images or measuring system aberrations.

Confocal interference microscopy can be performed in transmission, using a Mach-Zehnder arrangement, or in reflection, using a Michelson geometry. High sensitivity can be achieved using either phase-shifting or heterodyne techniques. The heterodyne technique also exhibits an imaging property, which allows nonconfocal images to be formed without an objective lens being necessary. This may prove useful in focusing deep into a specimen, which would be impossible with a real lens. Alternatively, combining illumination of the object by a focused spot and heterodyne detection can result in confocal imaging without the use of a physical pinhole.

As described earlier, the axial imaging performance of a confocal microscope can be investigated by observation of the variation with defocus in signal using a reflecting plane as the specimen (the defocus signal). Confocal imaging is a coherent technique, so that if only the intensity of the defocus signal is measured, the phase information is lost. However, if interference methods are used to extract the phase and amplitude, the aberrations of the imaging system can be determined by a simple Fourier transformation of the defocus signal. This is useful for optimization of the imaging system or to provide information for image reconstruction.

BIBLIOGRAPHY

Inoué, S. (1986). "Video Microscopy." Plenum, New York.

Kino, G. S., and Corle, T. R. (eds.) (1996). "Confocal Scanning Optical Microscopy and Related Imaging Systems." Academic Press, San Diego.

Pluta, M. (1988). "Advanced Light Microscopy. Specialized Methods," Vol. 2. Elsevier, Amsterdam/New York.

Sheppard, C. J. R. (1987). Scanning optical microscopy. *Adv. Opt. Electr. Microsc.* **10**, 1–98.

Wilson, T. (ed.) (1990). "Confocal Microscopy." Academic Press, New York.

Wilson, T., and Sheppard, C. J. R. (1984). "Theory and Practice of Scanning Optical Microscopy." Academic Press, New York.

Schistosomiasis

PAUL J. BRINDLEY
Queensland Institute of Medical Research and Australian Centre for International & Tropical Health & Nutrition

GLOSSARY

Cercaria Tailed, aquatic larva of the schistosome. It escapes from infected snails into fresh water. The cercaria (plural cercariae) is the infectious stage of the schistosome parasite for humans.

Granuloma Lesion formed by host inflammatory cells, primarily lymphocytes, macrophages, and eosinophils, around schistosome eggs entrapped in organs of the body of the person infected with schistosomes

Miracidium Ciliated larva of the schistosome that is released from schistosome eggs after the eggs are excreted from the human body into freshwater streams and ponds. The miracidium (plural miracidia) is a free-living larva that penetrates and infects the intermediate host snail in order to perpetuate the life cycle of the parasite

Organomegaly Enlargement of the liver and spleen, often characteristic of chronic schistosomiasis

Praziquantel Drug used in the chemotherapy of human schistosomiasis that is active against all forms of human schistosomiasis. Praziquantel is a safe drug that is taken as a single dose by mouth and has minimal side effects

Schistosomes Trematode worms belonging to the zoological phylum Platyhelminthes. Adult schistosomes are obligate parasites of humans and other mammals. Larval schistosomes are parasites of snails

Schistosomulum The immature form of the schistosome parasite that develops in the bloodstream from a cercaria that has managed to successfully invade human skin

Swimmer's itch Dermatitis caused by larvae of species of nonhuman schistosomes such as species of *Trichobilharzia* that are parasites of aquatic and migratory birds

SCHISTOSOMIASIS (ALSO KNOWN AS "BILHARZIA") is an infectious, parasitic disease of tropical and subtropical regions. The disease is caused by infection with blood flukes of the genus *Schistosoma,* which live in the blood vessels of the intestines or bladder. Infection is acquired in contaminated water when people come into contact with the aquatic larva of the schistosome parasite. Chemotherapy with drugs including praziquantel is used to treat human schistosomiasis and to control the spread of the disease. Chemotherapy with praziquantel is safe and effective. Other control measures center on public health education, improved sanitation, and molluscicing to remove the intermediate host snails which transmit the infection.

I. INTRODUCTION

Schistosomiasis is the most important of the human helminthiases, i.e., infections with parasitic worms. Various forms of schistosomiasis afflict more than 200 million people in tropical and subtropical Asia, Africa, South America, and the Caribbean, and of these probably more than 20 million suffer clinical morbidity or disability. In addition, schistosomiasis is responsible for the death of tens of thousands of people each year. A further 600 million people are at risk of infection in endemic regions and, moreover, schistosomiasis is spreading in some regions. In highly endemic areas, 75% of the children may be infected, and in terms of consequent disability-adjusted life

ENCYCLOPEDIA OF HUMAN BIOLOGY, Second Edition, VOLUME 7. Copyright © 1997 by Academic Press. All rights of reproduction in any form reserved.

years lost among the parasitic diseases of humans, schistosomiasis ranks second only to malaria.

The disease is transmitted by contact with contaminated water and is caused by various species of blood flukes of the genus *Schistosoma* that live and lay eggs in the veins of the intestines (*Schistosoma mansoni, S. japonicum, S. mekongi,* and *S. intercalatum*) or bladder (*S. haematobium*). Mass and targeted population chemotherapy provides the most effective method for controlling schistosomiasis as no vaccine is available and because sewage disposal, snail control (molluscicides, drainage), and attempts to change human water contact behavior have often failed, usually through high cost and negligence. Ironically, water resource developments and engineering in the form of dammed rivers, reservoirs, and irrigation systems have favored the expansion of snail populations, thereby contributing to the spread of schistosomiasis. Moreover, recent movements of refugees and the movements of rural peoples in urban slums in Africa, the Eastern Mediterranean, Asia, and Brazil are causing the introduction, spread, and aggravation of schistosomiasis. In well-characterized exceptions, multifacted approaches of all available control strategies during the last several decades have successfully eradicated schistosomiasis from Japan and from the Caribbean island of St. Lucia.

II. LIFE CYCLE OF HUMAN SCHISTOSOMES

There are around 20 species of schistosomes, of which three are major pathogens of humans (Fig. 1). These three are *S. mansoni, S. japonicum,* and *S. haematobium*. In addition, *S. mekongi* and *S. intercalatum* also cause morbid disease in humans. The life cycle of the human schistosomes includes both an invertebrate snail host and the definitive human host (or any of an array of other mammalian species in the case of *S. japonicum*). Schistosomiasis only occurs therefore in regions where suitable intermediate host snails are endemic. Eggs are released from the reproductive tract of the female schistosome into the venules of the blood vessels supplying the intestines (*S. japonicum, S. mansoni*) or bladder (*S. haematobium*) from where they migrate into the lumen of the bowel or bladder and are excreted and from where they carry on the transmission of the infection. In addition, a large percentage (probably as many as 50% of them) are retained

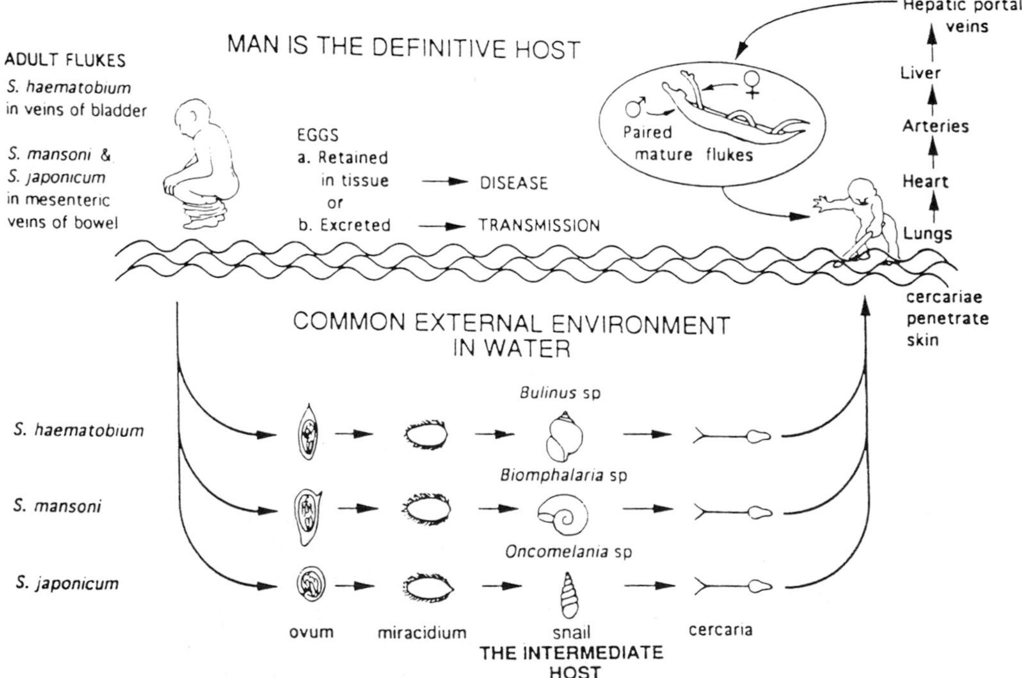

FIGURE 1 Schematic representation of the basic life cycle of the three principal species of schistosomes that infect humans. [After Jordan *et al.* (1993), with permission.]

in organs and tissues where they provoke an intense inflammatory response; indeed the inflammatory and fibrotic responses to these tissue-embedded schistosome eggs are the cause of most of the morbidity associated with schistosomiasis. The egg matures after it is released from the female worm, and by the time it leaves the infected person (or reservoir host animal), it contains a fully developed larva known as a miracidium. The miracidium is induced to hatch from the egg by the osmotic change experienced in freshwater compared to the physiological conditions of the human host. The miracidium is a self-reliant, free-living organism that is motile (it is ciliated) and seeks out and penetrates the flesh of an appropriate intermediate host snail. Snail-specific fatty acid molecules shed by the snails are believed to be among the chemical signals that the miracidia use to locate the snail. After penetrating the flesh of the snail, probably by employing both mechanical and proteolytic enzyme mechanisms, the miracidium transforms into the next larval stage, which is termed the primary sporocyst. Schistosome parasites undergo asexual multiplication in their snail hosts so that after a couple of weeks secondary sporocysts have formed within the primary sporocysts. The secondary sporocysts migrate to the digestive gland of the snail where they develop further and where further asexual multiplication of the schistosome takes place. The next larval stage in the life cycle, called the cercaria, is produced in large numbers within the secondary sporocyst.

The cercaria is the larval stage that is infective to humans. When mature, the cercariae escape from the tissues of the snail into the water where the snails are living and cercariae seek out the definitive host of the life cycle: the human and other mammals. The cercaria is mobile, it bears a tail, but only survives for about a day after its release from the mollusc. If the cercaria locates a human in the water (Fig. 2), it penetrates the skin directly, often through a hair follicle. It employs a proteolytic enzyme that is produced in its acetabular glands to affect penetration of the skin of the human host (Fig. 3). During or soon after penetrating the skin and dermis, the cercaria sheds its tail and undergoes other physiological changes that enable it to take up its parasitic existence in the human. At this stage, the larva is known as a schistosomulum. The schistosomulum enters a small blood vessel (or rarely the lymphatics) and is thereby transported to the lungs, where it remains for a day to a week or so depending on the species. The gut of the schistosomulum has developed by this time to the point where it begins to feed on red blood cells. Hemoglobin obtained from ingested red blood cells provides the amino acids necessary for the growth, development, and reproduction of the schistosome. Subsequently, the schistosomulum migrates to the liver, developing into young male or female worms within the portal veins. After 4 to 6 weeks, worm pairs mate and the paired parasites migrate to their predilection sites, which for *S. mansoni, S. japonicum, S. mekongi,* and *S. intercalatum* are the mesenteric veins of the intestines and for *S. haematobium* are the veins of the vesicle plexus. In these locations, they begin to release their hundreds to thousands of eggs per day. Adult schistosomes survive for long periods, perhaps up to 35 years, although often not more than 5 years.

III. GEOGRAPHICAL RANGE OF HUMAN SCHISTOSOMES

The geographic range of schistosomiasis is dependent on an array of factors but central among these is the presence of species of appropriate intermediate host snails. In turn, the presence of these snails is dependent on appropriate biotic conditions, including suitable rainfall. *S. haematobium* is transmitted by snails of the genus *Bulinus* which are pulmonate, aquatic snails. It occurs throughout much of Africa, both north of and south of the Sahara, and on Madagascar and other Indian Ocean islands. Further, its range extends north and east from Africa into Syria, Iraq, Iran, Saudi Arabia, Oman, Turkey, and Portugal. The related schistosome *S. intercalatum* occurs in several central African countries, including Nigeria, Zaire, Gabon, and Sao Tome and Principe, and is also transmitted by species of *Bulinus* snails.

Like *S. haematobium, S. mansoni* is also transmitted by aquatic pulmonate snails, but these snails belong to the genus *Biomphalaria*. Infection with *S. mansoni* is widespread in Egypt, the Sudan, Libya, and the Middle East as well as in sub-Saharan Africa where it occurs in the equatorial countries as well as those of southern Africa. *S. mansoni* also occurs in Central and South America, primarily in Brazil but also in Venezuela and Surinam and in a number of Caribbean islands, including Puerto Rico. *S. mansoni* probably was transported to the New World in African slaves transported there during the 17th and later centuries. In Africa, the snail intermediate hosts include *Biomphalaria pfeifferi, B. sudanica, B. alexandina,* and *B. choanomphala.* In South America, the important intermediate host snail is *B. glabrata*, al-

FIGURE 2 Infection is acquired by contact with freshwater infested with schistosome cercariae. Occupational contacts such as laundering the family clothes in a stream involve a high level of exposure. This woman is bathing and washing clothes in a creek in central Leyte, the Philippines (Brindley original, 1993).

though *B. straminea* and *B. tenagophila* can transmit the infection.

In contrast to the aquatic pulmonate snails that transmit S. *mansoni* and *S. haematobium*, amphibious (land and water dwelling) prosobranch snails of the genus *Oncomelania* transmit *S. japonicum* in China and the Phillipines whereas the prosobranch of the genus *Tricula* transmits *S. mekongi* in Loas, Cambodia, and Thailand. The distribution of *S. japonicum* is rather patchy within both the Philippines and China. In the Philippines, the geographical range of *S. japonicum* includes southern Luzon and the islands of Leyte, Samar, Bohol, Mindoro, and Mindinao. In China, the infection is found in southern provinces around and south of the Yangtze River, including Anhui, Jiangsu, Hubei, and Yunnan provinces. A wide variety of mammalian hosts act as reservoirs of *S. japonicum* infection, including the pig, dog, goat, rat, water buffalo, and cattle.

IV. PATHOLOGY

The progression of schistosomiasis can be divided into three phases. The initial phase, lasting from 4 to 12 weeks after exposure to the cercariae, is characterized by fever and allergic reaction. This phase is also known as Katayama fever. Subsequently, during the intermediate phase occurring several months to several years after infection, pathological changes in the intestinal or urinary tracts occur as the eggs of the schistosomes are passed from the body of the infected person to the environment. In the final phase, complications of the gastrointestinal, renal, and/or other

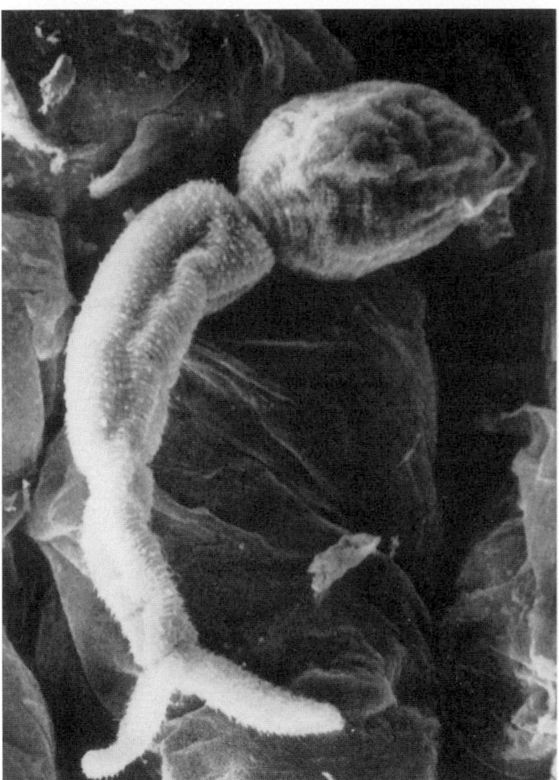

FIGURE 3 Scanning electron micrographs of cercariae of *Schistosoma mansoni* penetrating human skin. (Left) Cercaria on the surface of the skin. (Right) A cercaria has penetrated the skin so that only the tail is now visible. The tail is shed from the body of the cercaria at the time of skin penetration or soon after. [After McKerrow *et al.* (1989), with permission.]

organs take place. During this time, the eggs may not be shed, or shed only in small numbers, and they are often lodged in organs other than the liver, intestines, or bladder.

The significant pathological changes in schistosomiasis are caused by the eggs and, in particular, by host inflammatory responses to the eggs lodged in the liver, bladder, and ureters. During infection with *S. mansoni* and *S. japonicum,* the eggs move through the intestinal wall and are liberated into the lumen of the bowel. Eggs are also carried in the bloodstream and are deposited in various organs, mainly in the liver and intestines. Antigens released from the schistosome eggs in all these sites generate host inflammatory responses, which result in the formation of fibrosis and granulomata around the trapped eggs.

In *S. haematobium* infections, granulomatous lesions result in damage to the urinary tract, revealed by blood in the urine. Urinary function is impaired, urination becomes painful, and the uterus may be affected. Obstruction in urinary function can lead in turn to the impairment and failure of kidney function. A further complication in advanced cases of urinary schistosomiasis can be the development of cancer of the bladder. Intestinal schistosomiasis usually develops more slowly, with the damage to the intestines and liver leading to splenomegaly and hepatomegaly and hypertension of the abdominal blood vessels. Subsequently this can cause the enlargement of other vessels, from which rupture and bleeding can be fatal. The extent of pathology depends primarily on the number of eggs and sites at which they are lodged. The egg production per day per pair of schistosomes varies among the species. *S. mansoni* produce around 300 eggs per day whereas *S. japonicum* produce about 3000 eggs per day. Moreover, eggs of *S. japonicum* reach the brain more often than eggs of the other schistosomes, and neurologic complications, including paralysis and coma, occur in about 9% of cases. Because of the larger output of eggs and these kinds of complications, infection with *S. japonicum* may often be more severe than that caused by other schistosomes. In contrast, its geographical range is not as widespread as *S. mansoni* and hence smaller numbers of people are infected, or at risk of infection, with *S. japonicum*.

Cercarial dermatitis or "swimmer's itch" is a patchy red pinpoint skin rash associated with itching on parts of the body that have been in contact with water, either fresh or marine water. The major cause of swimmer's itch is schistosome-like parasites of aquatic and migrating birds. These parasites cannot develop in the human body, unlike the human schistosome species. Swimmer's itch occurs in both freshwater and marine coastal environments. The reaction is principally a sensitization resulting from previous exposure. Treatment is usually not necessary but mild antihistaminic or mild corticosteroid creams can be beneficial. [*See* Immunology of Parasitism.]

V. DIAGNOSIS AND TREATMENT OF SCHISTOSOMIASIS

Schistosomiasis can be directly and quickly diagnosed by parasitologic methods, which simply involve finding the eggs of the schistosome in the urine or feces. In particular, *S. haematobium* eggs, which have a terminal spine, are identified in the urine. A simple concentration test for diagnosis is sedimentation in a conical urinalysis glass of a day's output of urine. *S. mansoni*/*S. japonicum* infections are confirmed by finding eggs in the stool. The commonly used tool for this diagnosis is termed a Kato/Katz slide. For this technique, a small amount of feces is placed on a modified microscope slide, the fecal sample is sealed with tape, and the slide is examined under a microscope. Although it is a sensitive yet simple technique, several Kato/Katz slides may need to be prepared for each patient to diagnose light infections. The eggs of *S. mansoni* have a large lateral spine. *S. japonicum* eggs do not have an obvious spine, unlike the terminal spined eggs of *S. haematobium* and the lateral spined eggs of *S. mansoni*. In addition to locating eggs in the urine or stool, infections with all human species of schistosomes can be diagnosed by microscopic examination of minute biopsies of the rectal mucosa. Ultrasonography has become a useful tool for the diagnosis of schistosomiasis, as the various forms of schistosomiasis cause pathological lesions that can be visualized by this procedure (Fig. 4). Diagnosis of schistosomiasis by serological methods is possible with methods that detect circulating antigens.

Chemotherapy is employed to treat schistosomiasis and to control its spread. Chemotherapy of schistosomiasis began at the time of the first World War when tartar emetic, potassium antimony tartrate, was administered to patients with schistosomiasis haematobia. The subsequent history of schistosomiasis chemotherapy has involved the employment of successively better drugs, including lucanthone, hycanthone, and niridazole. At present, three antischistosomal drugs—metrifonate, oxamniquine, and praziquantel—appear on the World Health Organization's list of essential drugs. Metrifonate is employed for the treatment and control of infection with *S. haematobium*, oxamniquine is used for *S. mansoni*, and praziquantel is active against all schistosome species parasitizing humans, including *S. japonicum*. Treatment failures occur in schistosomiasis and, in some instances, can be ascribed to drug-resistant schistosomes. Indeed, resistance to oxamniquine is now present in natural populations of *S. mansoni* in Brazil. Perhaps more worrying are recent reports of the failure of praziquantel chemotherapy in Africa, which suggests that schistosomes may be developing tolerance to this most important drug. A major concern with the widespread use of schistosomicides is the development of drug-resistant parasites. Given the enormous catastrophe that drug resistance to chloroquine and other compounds has become in the treatment and control of malaria, it is also likely that widespread drug resistance to schistosomicides would be disastrous with respect to schistosomiasis. However, chemotherapy may fail in infected individuals for other reasons besides infection with drug-resistant parasites, including noncompliance, drug malabsorption, or inadequate immunological status. [*See* Chemotherapy, Antiparasitic Agents.]

VI. METHODS OF CONTROL

Chemotherapy is important in controlling schistosomiasis at the community level, particularly in reducing morbidity and disease in infected persons. Control strategies can involve mass chemotherapy of whole populations of endemic areas. Selective population chemotherapy is often undertaken where the targeted population is usually school-age children. In some instances, chemotherapy has been targeted to those individuals in the endemic area who are most heavily infected.

Transmission of schistosomiasis requires contact with water contaminated with cercariae. In turn, this means that transmission involves the presence in water of appropriate species of intermediate host snails. Continued transmission can usually only be main-

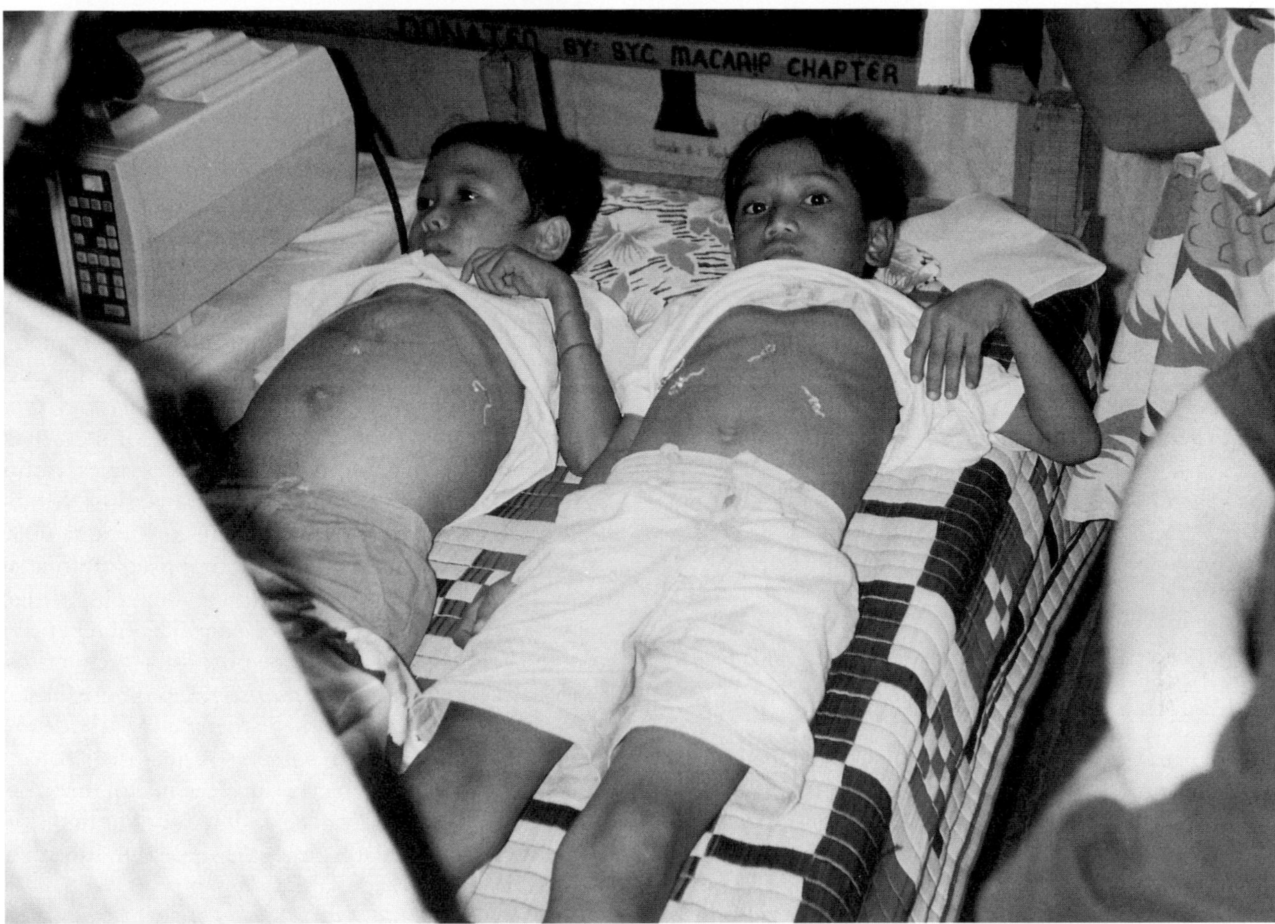

FIGURE 4 Boys of about 10 years of age, from the village of Macanip on the island of Leyte, the Philippines, undergoing ultrasonographic examination. Note the organomegaly characteristic of schistosomiasis japonica on the child on the left, next to the ultrasonography instrument (Brindley original, 1993).

tained where lack of hygienic sanitation allows water contamination with feces or urine. Accordingly, reducing the contact of people with contaminated water and/or reducing the contamination of water with human feces or urine, or excreta from reservoir host species, can block the transmission of schistosomiasis. Integrated public health measures, including health education, environmental modification, mollusciding to kill the intermediate host snails, construction of hygienic latrines, and the use of footbridges are all used to block the transmission and the life cycle of the schistosomes.

Modifications of the environment, such as redigging or resurfacing irrigation canals used for crops, may make it difficult for the snails to survive and to therefore act as competent transmitters of the disease. However, the different species of snail intermediate hosts all present different problems in respect to control and disease transmission. A major advantage of engineering-based control methods is the simultaneous reduction in other infectious diseases that are associated with poor hygiene and sanitation. Mollusciding, usually with niclosamide, is not only expensive but can also have detrimental effects on other species and on the ecology at large.

Vaccination offers the potential for controlling schistosomiasis, although no vaccine is available yet for use in humans. The development of vaccines against schistosomiasis is considered essential for the long-term control of the disease because reinfection frequently occurs following successful treatment with antischistosomal drugs. The feasibility of vaccination is supported by field studies that have demonstrated that humans acquire a natural resistance to reinfection, at least in the cases of *S. mansoni* and *S. haematobium*. A major goal of the vaccines currently under development is to lower the morbidity rate from granulomata formation around the eggs deposited in the liver, bladder, and other organs. This can be addressed by attempting to reduce the number of adult schistosomes surviving in the blood vessels and laying eggs, by reducing the number of eggs that the pairs of adult schistosomes succeed in producing, and/or by immunomodulatory strategies directed at reducing the host inflammatory responses to the eggs in the tissues and organs. It is now thought that the level of protection afforded by an antischistosome vaccine will not need to be 100% because the parasite does not replicate in the human host. Rather it is thought that a somewhat lower efficacy, maybe only 70%, could reduce the morbidity associated with schistosome infection to an insignificant level.

VII. SCHISTOSOME MORPHOLOGY, GENES, AND GENOME

Schistosomes are classified zoologically as belonging to the genus *Schistosoma*, family Schistosomatidae, order Strigeata, subclass Digenea, class Trematoda, and phylum Platyhelminthes. Species of the genus *Schistosoma* have seven pairs of autosomal chromosomes and one pair of sex chromosomes (ZZ for a male worm and ZW for a female), comprising a haploid genome size of approximately 2.7×10^8 bp in the case of *S. mansoni*. Schistosomes are diploid organisms. Adult schistosomes are dioecious and indeed they exhibit dramatic sexual dimorphism, with the male being much larger than the female parasite. Adult male schistosomes are 10–20 mm in length and 0.3–1.1 mm in width, depending on the species. Adult female schistosomes are 10–30 mm in length and 0.15–0.3 mm in width. The sexual dimorphism is only expressed in the mature, adult parasites. Both males and females have oral and ventral suckers, and a blind gut. The female schistosome spends her adult life lying in a longitudinal groove of the male schistosome. This groove is known as the gynecophoric canal, and the grooved body of the male schistosome provides us with the name (Greek) of these organisms: "schistosome" meaning "split body." Schistosome eggs are nonoperculate, unlike those of other species of trematodes. Eggs of *S. mansoni* are ovoid in shape, about 60×140 μm in size, and have a prominent lateral spine. *S. japonicum* (and *S. mekongi*) eggs are round, 60×100 μm in size, and have a very reduced lateral spine. Eggs of *S. haematobium* and *S. intercalatum* are ovoid, about 60×160 μm in size, and have a prominent terminal spine. Cercariae of all the schistosomes are similar in appearance and size, about 1 mm. They are just visible to the naked eye. [*See* Helminth Infections.]

The genome size of the schistosome is about one-tenth the size of the human genome and is about 10 times larger than the genome of the malaria parasite *Plasmodium falciparum*. It is anticipated that each of the life cycle stages will express about 20,000 protein-encoding genes, many of which will be stage specific. Schistosome genes usually contain introns and promoter sequences, in the normal fashion of metazoan eukaryotes. Like nematodes and trypanosomes, trans-spicing occurs in some schistosome transcripts. Modified bases are rare if present at all, and about 60% of the schistosome genome is composed of single-copy gene sequences. The genome includes a high percentage, maybe as much as 30 or 40%, of repetitive sequences including retrotransposon-like sequences. The GC content of the *S. mansoni* genome is about 45%. The genomes of the schistosomes *S. japonicum* and *S. mansoni* are the subject of genome projects currently funded by the World Health Organization and, accordingly, many gene and deduced amino acid sequences from schistosomes are available in public databases such as GenBank.

BIBLIOGRAPHY

Amiri, P., Locksley, R. M., Parslow, T., Sadick, M., Rector, E., Ritter, D., and McKerrow, J. (1993). Tumour necrosis factor-a restores granumolas and induces egg-laying in schistosome-infected SCID mice. *Nature* **356**, 604–607.

Bergquist, N. R. (1995). Controlling schistosomiasis by vaccination: A realistic option? *Parasitol. Today* **11**, 191–194.

Brindley, P. J. (1994). Drug resistance to schistosomicides and other anthelmintics of medical significance. *Acta Trop.* **56**, 213–231.

Brindley, P. J. (1994). Relationships between chemotherapy and immunity in schistosomiasis. *Adv. Parasitol.* **34**, 134–161.

Jordan, P. (1985). "Schistosomiasis: The St. Lucia Project," pp. 1–422. Cambridge University Press, Cambridge.

Jordan, P., Webbe, G., and Sturrock, R. F. (eds.) (1993). "Human Schistosomiasis," pp. 1–465. CAB International, Wallingford, UK.

Mahmoud, A. A. F., and Wahib, M. F. A. (1990). Schistosomiasis. *In* "Tropical and Geographical Medicine" (K. S. Warren and A. A. F. Mahmoud, eds.), 2nd Ed., pp. 456–472. McGraw-Hill, New York.

McKerrow, J. H., *et al.* (1989). *In* "Models in Dermatology" (H. Maibach and N. Lowe, eds.), Vol. 4, pp. 276–284. Karger, Basel.

Tanaka, M., Hirai, H., LoVerde, P. T., Nagafuchi, S., Franco, G. R., Simpson, A. J. G., and Pena, S. D. J. (1995). Yeast artificial chromosome (YAC)-based genome mapping of *Schistosoma mansoni*. *Mol. Biochem. Parasitol.* **69**, 41–51.

World Health Organization (1995). Tropical Disease Research. Progress 1975–1994. Highlights 1993–1994. Twelfth Programme Report of the UNDP/World Bank/WHO Special Programme for Research, pp. 77–86. Special Programme for Research and Training in Tropical Diseases (TDR), Geneva.

Wynn, T. A., Cheever, A. W., Jankovic, D., Poindexter, R. W., Caspar, P., Lewis, F. A., and Sher, A. (1995). An IL-12-based vaccination method for preventing fibrosis induced by schistosome infection. *Nature* **376**, 594–596.

Schizophrenia, Psychosocial Treatment

KIM T. MUESER
Dartmouth Medical School

PATRICK W. CORRIGAN
University of Chicago

DAVID L. PENN
Louisiana State University

GLOSSARY

Assertive community treatment Model of case management in which interdisciplinary teams of mental health professionals with low staff-to-patient ratios and shared (rather than individual) caseloads work actively with patients and families in their natural environment to ensure that their needs are met, to minimize time spent in the hospital, and to improve overall quality of life

Behavioral or psychoeducational family therapy Intervention provided to families to improve their ability to manage schizophrenia more effectively and to minimize the negative impact of the illness on family members

Cognitive rehabilitation Intervention provided to either remediate or compensate for cognitive deficits present in patients with schizophrenia, such as poor memory, vigilance, or conceptual skills

Expressed emotion Stressful communication of negative emotions from relatives to the family member with schizophrenia, which is hypothesized to increase the risk of those patients to symptom relapses

Milieu therapy Remediation of social deficits in psychiatric patients through modification of the immediate environment to create a "therapeutic community" conducive to patients' acceptance of social responsibility

Relapse Return of schizophrenic symptoms to a patient whose symptoms were previously in remission, or the exacerbation of symptoms in a chronically symptomatic patient

Social learning theory Employed in behavioral approaches to psychosocial interventions, a set of principles assumed to govern the acquisition of socially appropriate behavior in the natural environment; behaviors are learned through a combination of observation of others (modeling), positive social reinforcement for certain behaviors (reward), and negative consequences for certain behaviors (punishment)

Social skills training Intervention developed to improve the social competence of persons through behavioral rehearsal (practice) of skills and programmed generalization of skills into the natural environment

Stress–vulnerability–coping skills model Model postulating that the outcome of schizophrenia is determined by a dynamic balance between biological vulnerability, environmental stress, and the ability of patients to cope effectively with the effects of stress

Supported employment Model of vocational rehabiliation for psychiatric patients that emphasizes rapid placement in competitive, integrated work settings accompanied by provision of ongoing supports, rather than extended vocational skills training or work in sheltered workshops

Token economy Environmentally based treatment approach whereby desirable social behaviors are rewarded by the provision of tokens or credits, which are exchangeable for material goods and privileges, and undesirable behaviors are suppressed by fining patients tokens or credits

ENCYCLOPEDIA OF HUMAN BIOLOGY, Second Edition, VOLUME 7. Copyright © 1997 by Academic Press. All rights of reproduction in any form reserved.

PSYCHOSOCIAL TREATMENTS FOR SCHIZOPHRENIA, when combined with appropriate neuroleptic medication, improve the course and outcome of the illness by helping to remediate the social impairments and decreased quality of life that result from the disorder. In the process, psychosocial treatments diminish the impact of life stressors that exacerbate the course of the illness. Various techniques, applied to the patient, his or her family, and their living environment, have been developed to meet these goals. In addition, some psychosocial treatments help to reduce the stress and burden experienced by family members. In this way, family members are able to join the treatment team as partners in the care plan.

I. INTRODUCTION

A. The Course of the Illness

Schizophrenia is a psychotic illness that typically begins during adolescence or young adulthood. The disorder is marked by characteristic symptoms such as hallucinations, delusions, and incoherence, as well as significant social impairments. Symptoms and social impairments tend to wax and wane during the course of the illness. Emil Kraepelin, at the beginning of the twentieth century, believed that schizophrenia invariably resulted in long-term unremitting symptoms and chronic deterioration of interpersonal and self-care skills. As a result, the prognosis was thought to be bleak and treatment was mostly custodial. More recent research, however, suggests that schizophrenia is best described by several different course types. Eight of these different courses are summarized in Fig. 1, which range from the relatively benign to the most severe. Note that far more than two-thirds of persons with schizophrenia exceed Kraepelin's prognostications. Psychosocial interventions embody this more optimistic outlook by helping persons with schizophrenia attain more independent living and a better quality life.

The mutable course of the disorder in the *individual* may be understood via a stress–vulnerability–coping skills model. According to this model, symptoms and their associated social impairments occur because biologically vulnerable individuals are overwhelmed by normal life stressors. The biologically inherited vulnerability may be exacerbated by abusing alcohol or psychotomimetic drugs. The noxious effects of life stressors are modulated by a person's social competence and by the network of social support they avail.

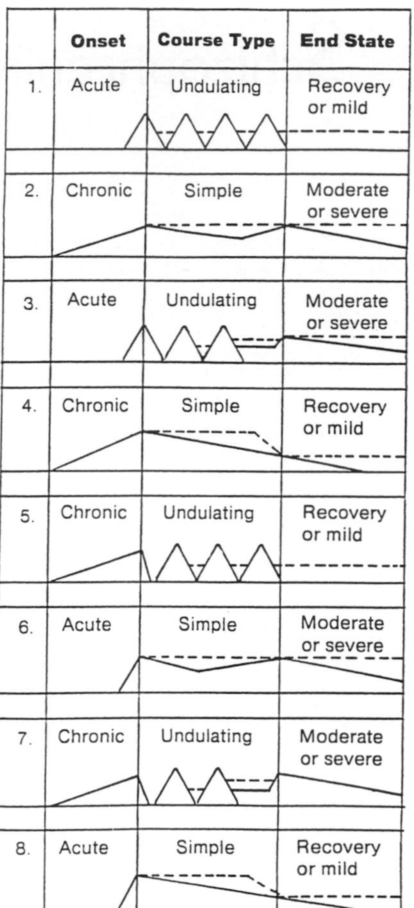

FIGURE 1 Different longitudinal outcomes for the long-term course of schizophrenia are depicted. Rather than the disorder invariably following a chronic, deteriorating course, many different outcomes are possible, including partial or complete recovery.

Either too much environmental change, stress, or ambient tension, or not enough coping skills and social support, may lead to a symptom exacerbation and loss of social and occupational functioning. [*See* Schizophrenic Disorders.]

B. The Role of Psychosocial Treatment

The significance of the stress–vulnerability–coping skills model of symptom formation lies in the emphasis given to the active role of the patient's coping skills and support system, both of which suggest targeted objectives and modalities for therapeutic intervention. The clinician can prescribe neuroleptic drugs to buffer the underlying biological vulnerability. When necessary, environmental modification may be employed to ameliorate the negative effects of stressors on a vulnerable person. In such instances, hospitalization may be beneficial because the patient is temporarily removed from stressors in family and community settings. Alternatively, treatment may emphasize strengthening the patient's social support network, as is the case in family and group therapies and self-help clubs. Finally, increasing the patient's resilience through training in social and problem-solving skills may reduce the likelihood of symptom relapse.

Psychosocial treatments are an essential component of rehabilitation for schizophrenics. Although treatment with neuroleptic medications reduces the risk of relapse, approximately 35–40% of schizophrenic patients who are compliant with neuroleptic medications still relapse within a year. Furthermore, the side effects of neuroleptics reduce adherence, even in patients responsive to the beneficial effects of the drugs. Most importantly, drugs cannot teach life and coping skills, nor can they improve the quality of a person's life, except indirectly through suppression of symptoms. Most patients with schizophrenia need to learn or relearn social and personal skills for surviving in the community and reducing the risk of symptom relapses. Many patients also require psychosocial interventions to reduce their abuse of alcohol or drugs, which may worsen the illness.

A final role for psychosocial intervention is to remediate basic social impairments due to the illness. Social inadequacy correlates with poor symptomatic and behavior outcomes, as well as rehospitalization. Interpersonal problem-solving appears to be markedly deficient in patients with schizophrenia, especially in limitations to alternative ways of responding to situational challenges. Deficits in social skills may include misperception of relevant social cues and poor cognitive processing of these cues, leading to inadequate generation of response alternatives and inappropriate behavioral responses to others in the situation. Poor social problem solving may partly reflect core attentional and psychophysiological impairments in schizophrenia, important targets for psychosocial treatment programs.

II. THE SPECTRUM OF PSYCHOSOCIAL INTERVENTIONS

Psychosocial interventions encompass a broad array of different psychotherapeutic methods that are applied in a variety of settings, ranging from the hospital or clinic to out in the community, at home, or in a social club. The focus of intervention can be on the individual as in one-to-one therapy, group therapy, family therapy, or a total milieu. Methods of intervention can derive from one or more explicit or implicit theoretical orientations, such as cognitive-behavioral, psychodynamic, family system, client centered, or supportive therapy. It is recognized that the treatment methods may overlap considerably, and that much of the therapeutic impact of any psychosocial treatment derives from nonspecific effects that are inherent in therapy that is offered in a credible, hopeful, and positive manner. Patients with schizophrenia typically participate in multiple psychosocial interventions over time, as the specific needs of the patient and available resources change.

Psychosocial interventions target a wide range of goals and objectives. Some treatments aim primarily at maintaining a person at a marginal level of functioning by minimizing stress and risk of relapse. Other treatments provide crisis counseling and assertive outreach to meet immediate survival needs in the community. Still others attempt to build social and independent living skills. Thus, psychosocial interventions comprise a multifaceted set of therapeutic techniques, applied to a range of different modalities, with a multiplicity of goals and objectives. Specific psychosocial treatments for schizophrenia will be reviewed according to the focus of each intervention: individual treatment, milieu therapy, group therapy, and family therapy. Recent advances in vocational rehabilitation will then be summarized, and we will conclude with considering the integration of psychosocial and pharmacological treatments.

III. INDIVIDUAL TREATMENT

A. Traditional Psychotherapy

A range of different psychotherapies have been tried for patients with schizophrenia, but many approaches have met with limited success in improving the outcome of the illness. The clinical efficacy of psychodynamic treatment for schizophrenia has been debated since soon after the development of psychoanalysis in the early twentieth century. Over the past several decades, a number of controlled research studies have examined the effect of psychoanalytic or exploratory psychodynamic treatment for outpatients with schizophrenia on objectively defined domains such as symptomatology, occupational and social functioning, and recidivism. The results of these studies have consistently demonstrated that psychodynamic therapies confer little benefit to patients and may actually worsen the course of the illness and adaptive functioning of some persons. The results of long-term psychodynamic interventions conducted in carefully crafted residential milieus for schizophrenia have not been more encouraging than in outpatient treatment studies. Thus, insight-oriented psychodynamic therapies do not currently have an empirically supported role in the residential, inpatient, or outpatient treatment of schizophrenia. Similarly, client-centered or purely supportive approaches to psychotherapy, while less extensively studied, have also not been found to improve functioning or symptoms in persons with schizophrenia.

Cognitive therapies, which can be delivered in either an individual or group setting, will be described in the next section. Social skills training, which can be provided individually but is most often given in groups, will be described under group therapy.

B. Cognitive Treatment

There are two general approaches to addressing cognitive dysfunction in schizophrenia: *process* and *content* approaches. The goal of process interventions is remediation of basic information processing skills, such as memory, vigilance, and conceptual skills. This approach is commonly referred to as *cognitive rehabilitation.* The rationale underlying cognitive rehabilitation is that relapses may be prevented by addressing those cognitive deficits that either serve as vulnerability markers for future psychotic episodes or that militate against adequate coping skills being utilized. Furthermore, as cognitive impairments are strongly associated with the ability to meet basic needs, enjoy meaningful interpersonal relationships, and vocational success, cognitive remediation may also enhance functioning in a range of other domains.

A number of different treatment strategies fall under the rubric of cognitive rehabilitation. The first strategy is to help the individual learn or relearn a deficient skill through practice. This strategy often utilizes computer technology analogous to video games, with tasks that are graduated in difficulty. The second strategy is to employ intact functions to replace the skills that have been damaged, such as encouraging individuals with memory impairment to elaborate on to-be-remembered material. The third strategy is to use prosthetic strategies to compensate for cognitive impairment. An example of this approach is a thought-disordered patient who is encouraged to carry around his schedule to prevent being late for appointments. The final strategy involves adapting the social and physical environment to optimally address the person's strengths and weaknesses. For example, group therapy with highly distractible patients is often conducted in rooms free of distraction with posters that summarize educational material.

There is a growing consensus that conceptual skills can be improved with a combination of monetary reinforcement and instructional cues. Findings also indicate that cognitive processes such as attention, vigilance, and memory are amenable to repeated practice techniques, although the durability of these gains remains unclear. The future challenge is to determine the stability of cognitive remediation, the extent to which it generalizes to more complex levels of functioning, such as social behavior, and the most effective way of integrating cognitive remediation with the range of other psychosocial interventions provided to persons with schizophrenia, such as social skills training, family intervention, or vocational rehabilitation.

Content approaches focus on changing the nature of, or one's response to, the content of dysfunctional thoughts, such as delusions. Examples of this approach include modifying thoughts or beliefs associated with delusions, such as the belief that one's thoughts can be heard by others, or teaching ways to cope with auditory hallucinations, such as humming to music. Content interventions, which typically focus on residual psychotic symptoms, are "coping-oriented." Therefore, content approaches tend to focus on stress management by reducing the unpleasant effects of persistent psychotic symptoms.

Content approaches borrow techniques from traditional cognitive-behavioral therapy in ameliorating residual symptoms. For example, fixed delusions may be addressed with the verbal challenge, a nonconfrontational approach in which the internal consistency of the patient's belief system is questioned and alternative explanations for the belief are offered. Reality testing, another cognitive-behavioral strategy, requires formulation and implementation of a behavioral experiment in which an activity is performed that could challenge or invalidate the delusion. For example, a patient whose threatening auditory hallucinations lead to the belief that someone wants to kill her may be encouraged to wear heavy industrial earmuffs; if she can still hear the voice, then it must have been internally generated.

Preliminary research findings suggest that content approaches to persistent psychotic symptoms reduce the conviction in, and preoccupation with, delusional beliefs. Furthermore, the frequency and intensity of hallucinations are reduced by distraction and attributional training, such as teaching the patient to reattribute voices to an internal rather than external source. These encouraging findings need to be replicated in more controlled treatment trials in order to better understand the impact of content approaches to cognitive treatment, its limitations, and which patients are most capable of benefiting from intervention.

C. Case Management

The comprehensive treatment of schizophrenia must include a range of medical and psychiatric services to remediate primary symptoms, to rehabilitate social and self-care skills, to provide basic medical, housing, and social services, and to maintain continuity of care through various stages of the illness. Case management is the process of linking these needs for treatment and community support with resources available in the hospital or community. Assisting persons with schizophrenia to maximize their use of existing resources can enable them to increase their independence and the quality of their lives. Case management also helps to ensure accountability, accessibility to resources, efficiency, and continuity of care. Effective case management requires the following component services: client identification and outreach; individual assessment; service planning; linkage with required services; monitoring service delivery; and patient advocacy.

An important facet of case management for schizophrenia is the development of social supports that can act as a buffer against the aversive effects of environmental stress. Persons with schizophrenia are particularly vulnerable to disruptions in their social networks. Patients recovering from an acute episode of their illness often fail to reestablish their former network because of the social stigma of the illness and deficits in social skills. Because a fundamental aspect of schizophrenia is that patients have difficulty or are unable to identify and articulate many of their most basic needs, the case manager assumes this vital role as coordinator of psychosocial and somatic treatments. In some programs, the process of case management is extended beyond the traditional inpatient or outpatient treatment setting to locales in the community where patients reside. Such assertive outreach programs are described next.

D. Assertive Outreach Programs

To address the difficulty that patients with schizophrenia have in advocating for themselves, assertive outreach programs (often referred to as *assertive community treatment* or ACT programs) have been developed to aid their functioning in the community. The hallmark of these approaches is the locus of treatment, which is almost always in the community, at locales such as homes, streets, day programs, and community rehabilitation facilities and residences where the focus is on helping patients meet their basic living needs. In addition, intervention is provided on an as-needed basis, by an interdisciplinary team, seven days a week, with intensive case management, crisis intervention, and patient advocacy as the core ingredients of treatment.

The most widely disseminated and emulated ACT program has been the Training in Community Living (TCL) program in Madison, Wisconsin. Since the development and first controlled investigation of the TCL program in the 1970s, numerous similar programs for persons with schizophrenia and other severe psychiatric disorders have been developed, and over twenty controlled studies of ACT have been conducted. Research on assertive outreach programs has pointed to several consistent trends across most studies. The most dramatic effect of ACT programs is to reduce the number of psychiatric hospitalizations and time spent in the hospital, which, due to the high cost of inpatient treatment, frequently translates into net savings in cost. In addition to reduced hospital utilization, ACT programs have been found to successfully stabilize the housing of homeless patients, to have a moderate effect on reducing symptoms, and to modestly increase the capacity to work.

ACT programs have been found to produce a range of clinical benefits, but these gains are usually not maintained if the program is terminated. This finding underscores the need to provide psychosocial support indefinitely for most patients with schizophrenia, just as neuroleptic medications must be taken for this illness on a long-term basis. It remains to be seen whether social learning principles applied within the context of ACT programs can improve the acquisition, durability, and generalization of living skills.

IV. MILIEU THERAPY

A. Elements of Effective Milieus

The locus for milieu therapy is a living, learning, or working environment. Examples include the inpatient psychiatric unit, day hospitals, psychosocial rehabilitation clubs, board-and-care homes, community-based residential alternatives to hospitals, and sheltered workshops. The defining characteristics of treatment and rehabilitation milieus are the use of a team to provide treatment and the large amount of time spent by the patient in the environment. Recent adaptations of milieu therapy have included 24-hour programs that are situated in the community locales frequented by patients and that provide support, case management, and training in living skills.

Milieu therapy, or the therapeutic community, may be based on any one of a number of modalities ranging from structured behavior therapy to spontaneous, humanistically oriented approaches. Most programs encompass the following attributes: (1) emphasis on group and social interaction, (2) rules and expectations that are mediated by peer pressure for normalization of adaptation, (3) blurring of the patient role by viewing patients as responsible human beings, (4) emphasis on patients' rights for involvement in setting goals, for freedom of movement, and for informality of relationships with staff, and (5) emphasis on interdisciplinary participation and goal-oriented, clear communications.

Data have emerged that describe the elements of effective therapeutic milieus for patients with schizophrenia. These elements can be divided into those related to milieu structure and those related to treatment procedures. Structural elements associated with favorable outcomes include (1) small size of milieu and patient census, (2) high staff-to-patient ratios and staff stability, (3) heterogeneity of patient population, with an optimal mix of two-thirds higher functioning patients to one-third acutely ill patients, and (4) clarity and consistency in status and roles among staff. Other programmatic elements that appear to be related to successful treatment are (1) active participation by patients and nursing staff, (2) administrative commitment to short stays (i.e., 3 months or less for the average hospital episode), and (3) outplacement of patients requiring long-term, custodial care.

Treatment process variables correlated with good outcome include high levels of staff–patient interaction, with the focus of interactions on adaptive practical aspects of everyday behavior, rather than on symptomatic and psychodynamic issues. Treatment units that set clearly defined and time-limited goals with patients are more effective, as are units that organize and schedule prosocial activities for most waking hours. A variety of psychosocial treatment models have been developed and tested and are considered more effective than standard or comparison approaches. These models include the social learning–token economy program and therapeutic communities such as the Fountain House social club (described in Section IV, C).

B. Social Learning–Token Economy Programs

For chronic, treatment-refractory, long-stay patients with schizophrenia who have resisted all efforts at deinstitutionalization, the psychosocial strategy of choice is social learning–token economy. Utilizing behavioral assessment and therapy, highly trained paraprofessionals and nursing staff have been successful in remediating the bizarre symptoms and social and self-care deficits of most severely ill patients. In a rigorous study conducted on the token economy, a social learning–token economy program was compared with an equally intensive therapeutic milieu and a customary care group.

Over 100 severely debilitated and chronically institutionalized patients were treated by a paraprofessional staff who rotated between the units to control for nonspecific personalities of the therapists. Staff-to-patients ratios were similar to those used in custodial institutions. The social learning program employed a highly specific token economy with many hours of structured educational activities throughout the day and evening. Patients were rewarded with tokens for engaging in socially adaptive behaviors, which could then be exchanged for material goods or privileges. The therapeutic milieu followed principles of peer pressure and democratic decision making. Both programs were offered in 28-bed units in a regional psychiatric hospital. Most of the patients received inpatient care for at least 1 year and then obtained aftercare for 6 months after their discharge into the community.

A multimodal assessment battery revealed impressive and clear-cut results favoring the social learning–token economy approach. Improved functioning, enabling long-term community tenure, occurred in 97% of the social learning patients. The therapeutic milieu program was less effective, but its 71% release and community maintenance rate was still a favorable outcome when compared with the 45% rate of patients released from custodial care and living in the community for 18 months or longer. The success in sustaining patients in the community was mirrored by the significant clinical and behavioral improvements, which even produced a minority of patients who, by direct behavioral observational ratings, could not be distinguished from a normal population. After only 14 weeks of treatment, every resident in the social learning program showed dramatic improvements in overall functioning, regardless of usual prognostic indicators, such as duration of hospitalization.

By the end of the second year of programming, fewer than 25% of residents in either experimental program were on maintenance psychotropic drugs. Two clinically significant conclusions reached by the study were that (1) most older chronic mental patients, when provided active and structured psychosocial therapies, did not require maintenance neuroleptic drugs, and (2) clinical improvements resulted not from *how much* attention and enthusiasm was offered to patients by staff but *how* that attention was given. Patients in the therapeutic milieu program received more overall attention but improved less than their counterparts in the social learning program.

Social learning principles have been extended into the community where treatment is provided at day hospitals and community support programs. One such program, associated with a veterans administration hospital, offered structured and scheduled classes in which social, vocational, and survival skills were taught. Rehospitalization rates were only 10% for the patients participating in the behavior-oriented educational program, whereas the rates were 53% in a comparison group receiving traditional aftercare services. In a similar program located in a community mental health center, specific goal-setting and active behavioral training of social and community living skills led to increasing numbers of clinical goals attained during a 24-month follow-up period, whereas matched patients from a more traditional day hospital showed decrements in goal attainment over the 24-month period. Treatment procedures based on this program have been widely adopted by community mental health centers throughout the United States.

C. Therapeutic Communities

Some efforts to develop a milieu conducive to improving the social and independent living skills in patients with schizophrenia have resulted in therapeutic communities in which patients are involved in the administration and ongoing operation of the program. An early effort to demonstrate that a comprehensive learning-based therapeutic community could be effective with chronic mental patients used an approach in which the hospital was a base for initial training of patients in interpersonal skills, effective decision making, and group governance and cohesion. Following this training, the patients were assisted in gradually assuming full responsibility for its maintenance and operation. They bought and prepared food, kept financial records of income and expenditures, and consulted community-based physicians and agencies for their medical, psychiatric, and support needs. Live-in staff faded themselves out of the picture until the lodge residents were functioning autonomously. In addition, the patients earned money by starting a local business that provided janitorial services, yard work, general hauling, and painting. A 40-month follow-up indicated that the patients trained to live and work in a community lodge sustained significantly greater time outside the hospital and in gainful employment.

Another approach to the therapeutic community, termed psychosocial rehabilitation, emerged during the late 1940s when ex-patients began to meet together in a social club in New York City to satisfy their needs for acceptance and emotional support. Emphasizing self-help, mutual interdependence, and reliance on assets, the movement led to the establishment of Fountain House, which has spawned hundreds of similar programs. The main assumption of this approach is that patients have a fundamental right to work and that employment facilitates community adjustment and reduces symptoms. Employment opportunities are provided both in the clubhouse (e.g., food preparation, switchboard) and by transitional jobs available in the community, with no limitation on the length of participation in the program. These transitional jobs are opportunities for club members to work temporarily en route to full-time employment elsewhere or to work on a longer-term basis in the entry-level position.

An 18-month follow-up evaluation of club members working in transitional jobs revealed that 16% were employed independently on a full-time basis and an additional 45% continued part-time work in the transitional program or were attending school or other training programs. Only 2% were in a psychiatric hospital at the time of the 18-month follow-up. Another evaluation of the psychosocial club model found that 38% of the members were rehospitalized during a 2-year follow-up period, in comparison to a 60% rehospitalization rate for a contrast group. Members of the club also had significantly lower rehospitalization rates 5 years later; those who were hospitalized from the club spent 40% fewer days in the hospital than did the rehospitalized control subjects. Despite these positive results, methodological problems with self-selection, lack of diagnostic clarity, and non-randomly assigned control groups limit the conclusions that can be drawn from these studies.

Other approaches to vocational rehabilitation are described in Section VII.

V. GROUP THERAPY

A. Modes of Therapy

There are as many schools of group psychotherapy as there are of individual therapy. Group therapy can be characterized by its theoretical and operational qualities and procedures. Some group therapies are highly structured and employ a behavioral orientation, other groups are unstructured with a psychodynamic and insight-oriented focus, and still others are primarily supportive in nature. Most therapy groups are provided in the hospital, clinic, or private office, although multiple family groups have met in homes and storefronts, and groups emphasizing peer support and normalization often meet in community centers, schools, and churches.

The goals of therapy groups overlap considerably, with varying degrees of emphasis placed on insight, behavior change, skill development, social support and maintenance, and participation in recreational activities. What cuts across most modalities is the group leader's use of the naturally developing interactional dynamics in groups, such as cohesion, to strengthen the group process and to improve the outcomes. A large body of evidence collected from groups serving a spectrum of patient populations suggests that cohesion has a generally favorable impact on group therapy, one that is similar to the therapeutic alliance between patient and therapist in individual therapy.

With the trend toward brief inpatient hospitalizations for psychiatric patients, followed by continuing care in the community, almost all group psychotherapy takes place in the aftercare period. Exploratory and psychodynamic group therapy during the inpatient period may worsen the clinical state of patients who are still floridly psychotic and, thus, vulnerable to overstimulation and hyperarousal from their treatment environment. Group therapy is likely to be more beneficial when offered after symptoms, such as delusions and hallucinations, have been controlled, and if it focuses on practical, everyday problems of living experienced by patients trying to adjust to the community. Most outpatient groups aim at supporting a patient's stabilization and community tenure, assisting the patient in coping with stressful life events, and facilitating efforts at longer-term rehabilitation.

Although controlled research studies have shown that insight-oriented group therapy is not efficacious for patients with schizophrenia, the beneficial effects of more socially interactive groups on outcome criteria such as symptomatology, rehospitalization, or vocational or social adjustment remains to be established. The consensus of clinicians, buttressed by controlled studies, is that employing group therapy during the aftercare, outpatient phase of treatment is more effective than during inpatient treatment. Group therapy formats have been adapted broadly for use in providing services such as social skills training, medication evaluations, occupational and recreational therapy, patient and family education, patient government, self-help, and mutual support.

B. Social Skills Training

1. Training Methods

The most highly structured form of group therapy for patients with schizophrenia is social skills training. The goals are explicit, the session agendas are usually planned in advance, the procedures follow written guidelines often derived from a manual, and *in vivo* practice (i.e., in the "real world") and homework assignments are emphasized. Social skills can be defined as those interpersonal behaviors required (1) to attain instrumental goals necessary for community survival and independence and (2) to establish, maintain, and deepen supportive and socially rewarding relationships. Schizophrenia disrupts one or more of the affective, cognitive, verbal, and behavioral domains of functioning and thereby impairs a person's potential for enjoying and sustaining interpersonal relationships, which are the essence of the social quality of life. Recurring schizophrenic disorders pose enduring social disruptions for affected persons. These disorders involve symptoms that adversely affect patients' social quality of life and also evoke impairments that hamper learning or relearning adaptive social behaviors. In applying behavior analysis principles to identify and remediate deficits in social behaviors, clinicians have developed treatment packages, termed *social skills training* (SST), that have proved effective in patients with schizophrenia.

In virtually all published reports of SST, role playing (a simulated social encounter) is the vehicle used both to assess patients' pretreatment social competence and to train targeted behavioral excesses or deficits during treatment. Training scenes are selected either on the basis of the individual's past difficulties or from problem situations that have been found to apply to most patients. Training sessions vary in length from 15 to 120 minutes, depending on the number of patients participating and on their level of functioning. Although the group format provides vicarious learning opportunities through observation of other patients' behavior, as well as from amplified reinforcement from peers, the group experience is sometimes supplemented by individual training; such training allows more intensive focus on a single patient's behavior and provides an opportunity for more practice within sessions.

Participants in the role playing include the target patient(s), a respondent, and the therapist. A combination of focused instructions, modeling (demonstration), feedback, and social reinforcement are applied as a "package" to remediate deficits in social behavior. Modeling and feedback are provided by group participants or through videotape playback. Target behaviors selected for change usually include both nonverbal and paralinguistic behaviors (e.g., eye contact, voice loudness or intonation, response latency, smiles) and content behaviors (e.g., requests for change, highlighting the importance of a need, empathic responses, compliance, hostile comments, irrelevant remarks). The efficacy of SST lies in the specific, functional, and goal-oriented nature of the behaviors that are targeted for change. The sequence of steps in a SST session is outlined in Table I.

2. Research Findings

Extensive research has examined the effects of SST for patients with schizophrenia. These studies have typically compared SST with control treatments such as discussion groups or exercise, and a number of conclusions are evident. First, a wide range of social skills can be taught to patients, ranging from simple behaviors such as eye contact and voice volume to more complex behaviors such as assertiveness and conflict resolution. Second, SST is associated with a reduction in social anxiety in social situations. Patients with schizophrenia report feeling less anxious after SST. Third, generalization of social skills to novel situations does not spontaneously occur; it must be built into the natural environment by prompting and reinforcing target behaviors. This typically involves recruiting family members and significant others to participate in treatment and follow-up. Fourth, the durability of acquired social skills is a function of the duration of treatment and the patient's degree of symptomatology; overlearning associated with repeated practice of skills promotes retention. Skill maintenance is unlikely to occur for brief training durations of less than 2–3 months of twice-weekly sessions or if the patient has persistent psychotic symptoms. Fifth, acquisition of social skills is predicted by patients' cognitive functioning; better memory and attention facilitate the learning of new skills.

SST has been found to improve a variety of indices of social functioning. These include perceived physical attractiveness, social competence during interactions with others, and social adjustment, such as role functioning and quality of interpersonal relationships. Furthermore, there is some evidence indicating that SST is associated with a reduction in relapse rates. Less compelling evidence, however, has been reported for the role of SST in reducing symptomatology. Finally, controlled research has yet to examine whether SST impacts quality of life in the community or vocational adjustment.

There are a number of unanswered questions regarding SST. Little is known about the key treatment "ingredients" that are necessary to evince clinical change. These treatment ingredients include the optimal patient-to-therapist ratio for group formats, appropriate length of session duration, and the frequency and intensity of follow-up sessions. Another unresolved issue concerns the most effective strategy for managing cognitive deficits that can interfere with the acquisition of social skills. Cognitive remediation or teaching compensatory strategies for cognitive deficits may augment the effects of SST, but the combination of these approaches has not yet been evaluated. Finally, little is known regarding at what stage(s) of the illness SST should be initiated. Most research has examined the effects of SST with older, more chronically ill, patients. It is plausible that the provision of SST at an earlier stage of schizophrenia might arrest some of the clinical deterioration that occurs over the first several years after the onset of the illness.

VI. FAMILY THERAPY

A. Rationale for Family Intervention

Early theories of the etiology of schizophrenia postulated that the family played an important role in the development of the illness. As evidence mounted that schizophrenia was biological in nature and was transmitted genetically rather than interpersonally, the focus of research and treatment shifted to examining family factors that influence the course of the illness. In a series of carefully designed, cross-culturally replicated studies, critical, intrusive, or emotionally over-

TABLE I
Procedures for Structured Social Skills Training in Groups

1. Specify the interpersonal problem in each group member in turn by asking the following:
   a. What emotion, need, or communication is lacking or not being appropriately expressed?
   b. With whom does the patient want and need to improve social contact?
   c. What are the patient's short- and long-term goals?
   d. What are the patient's rights and responsibilities?
   e. Where and when does the problem occur?
2. For each patient in the group, formulate a scene that simulates or recapitulates the features of the problem situation. The scene should include the following characteristics:
   a. Constructed as a positive goal
   b. Functional for the patient
   c. Frequently occurring in the patient's life
   d. Specific
   e. Consistent with the patient's rights and responsibilities
   f. Attainable
3. Observe while the patient and surrogate role players (other group members) rehearse the scene. During this "dry run," therapists should position themselves close to the action so that they can make an assessment.
4. Identify the assets, deficits, and excesses in the patient's performance during the dry run. Praise assets and efforts and solicit positive feedback from other group members.
5. Assess and train "receiving" and "processing" skills by asking the patient:
   a. What did the other person say?
   b. What was the other person feeling?
   c. What were the patient's short-term goals?
   d. What were the patient's long-term goals?
   e. Did the patient obtain these goals?
   f. What other alternatives could the patient use in this situation?
   g. Would one of these alternatives help the patient reach his/her goals?
6. Employ modeling, using therapist or other group members to demonstrate potentially effective alternatives using expressive and adaptive "sending" skills.
7. Highlight the desired behaviors being modeled and review "receiving," "processing," and "sending" skills, and ask other group members and targeted patients to repeat this procedure.
8. Rerun the scene with the patient, giving positive feedback to reinforce progress and effort and soliciting further positive feedback from other group members.
9. Use coaching and nonverbal prompts to shape behavioral changes and improvements in small increments, starting at the patient's present level.
10. Focus on all dimensions of social competence in training "sending" skills:
    a. Topical content and choice of words and phrases
    b. Nonverbal behaviors
    c. Timing, reciprocity, and listening skills
    d. Effective alternatives
11. Generalize the improvements in competence by the following procedures:
    a. Repeating practice and overlearning
    b. Selecting specific, attainable, and functional goals and scenes
    c. Providing positive feedback for successful transfer of skills to real-life situations
    d. Prompting the patient to use self-evaluation and self-reinforcement
    e. Fading the structure and frequency of the training
    f. "Programming" for generalization in the natural environment

involved attitudes and feelings of relatives—termed high *expressed emotion*—have consistently been found to be a powerful predictor of relapse in schizophrenia. This communication of negative affect from family members to the patient, which increases the patient's risk of relapse, is a consequence, in part, of the burden that relatives experience while caring for a chronically ill patient. The caregiving burden on relatives has increased dramatically since the discovery of neuroleptic medications in the 1950s, which allowed the majority of patients with schizophrenia to be treated in the community.

Several modes of family therapy have been designed and empirically tested for their ability to change the emotional climate of the family, reduce the burden of the illness on relatives, and improve the outcome of the illness. These new, behavior-oriented, psychoeducational approaches to family therapy have a distinctively different rationale than the earlier clinical studies that implicated family structure, communication, and relationships in the etiology of schizophrenia. In fact, as part of the initial contacts with the family group, the newer family therapies emphasize that no etiological link exists between family relations and the development of schizophrenia. Instead, the stress–vulnerability–coping skills model of schizophrenia is described to explain how the stress of an already established major mental illness can place burdens on patient and relatives alike, thereby raising tension levels in the family. Given the fragile coping capacity of the index patient, the increased stress reverberating throughout the ambient family emotional climate can lead to a relapse of symptoms. The dynamic interplay among family stress, patient vulnerability and symptomatology, and burden of the illness on the family is illustrated in Fig. 2.

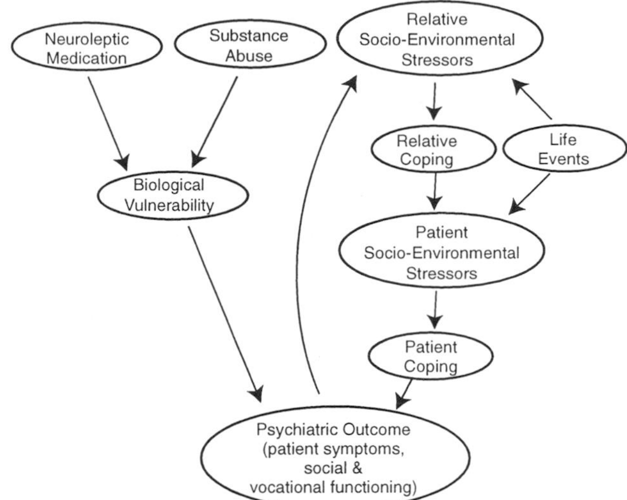

FIGURE 2 The stress–vulnerability–coping skills model of schizophrenia. The outcome of the illness (symptoms, social and vocational functioning) is determined by patients' biological vulnerability and their ability to cope with environmental stressors impinging on them. Biological vulnerability may be increased by the use of illicit substances (e.g., amphetamines, cocaine) and decreased by neuroleptic medications. The ability of patients to cope with socioenvironmental stressors, such as life events or a hostile emotional climate in the family, impacts on the illness. Just as stress has a negative impact on the patient, severe stress on the family, such as that caused by a floridly ill patient at home, can overwhelm the coping efforts of the family, leading to greater stress on the patient and an increased risk of symptom relapse. Social skills training focuses on improving patients' ability to cope effectively with stress. Behavioral family therapy focuses on decreasing tension in the home by improving family members' ability to cope with stressors and to manage the illness of schizophrenia.

B. Treatment Methods

A common feature of effective approaches to family therapy is an emphasis on educating the patient and family members about the nature of schizophrenia and its available treatment. Time is spent on demystifying the varied symptoms, signs, and prognoses associated with the disorder and on translating the neurobiological underpinnings into lay terminology. The role of neuroleptic drugs in the treatment and prophylaxis of schizophrenia is highlighted, and an effort is made to improve adherence to the pharmocotherapeutic regimen. Some therapists prefer to meet with the relatives alone for initial sessions and to invite the patient to join in later when acute symptomatology has been controlled and the span of attention increased. Therapists provide educational and other interventions with individual families in multiple family groups, and during day-long or evening "survival skills workshops."

The methods of family therapy run along a continuum in terms of the systematic use of behavioral learning principles and the use of family systems theory; however, they have in common educating relatives and patients about the illness, teaching family members skills for communicating effectively and solving problems together, and using the principles of social skills training (Table II). In contrast to purely educational family treatments, the behavioral method does not assume that stressful negative affective communication by relatives can be lowered by the provision of information alone and aims instead to improve the quality of family interactions. As in social skills training with patient groups, behavioral family therapy sessions are highly structured with preplanned agendas and homework assignments to facilitate the acquisition of skills. Assessments are routinely conducted on all family members and the family as a unit, with the principal goal of improving each member's ability to achieve their own personal goals.

TABLE II
Component Interventions Comprising Behavioral Family Therapy[a]

Behavioral analysis of all members of the family
- Defining individual and familywide problems
- Setting goals for each individual and family system
- Identifying reinforcers
- Pinpointing assets and resources within individuals, family, and community

Education of all family members about schizophrenia and currently available treatment and rehabilitation modalities

Training in communication skills
- Expressing positive feelings to others and acknowledging when others do or say something positive to you
- Active, reflective listening
- Making positive requests and asking for what you want
- Expressing negative feelings in constructive ways

Training in problem-solving skills
- Be specific and objective in describing the problem
- Express how you feel directly and subjectively about the problem
- Listen to each other actively and reflectively as the problem is described and feelings are expressed
- Help each other generate alternatives and options in dealing with the problem
- Weigh the potential consequences or outcomes (risks and benefits, pros and cons) of each alternative
- Choose a reasonable alternative
- Decide how to implement the alternative

Behavioral interventions for specific problems
- Contingency management for negative symptoms
- Job-finding skills training
- Friendship skills training
- Independent living skills training

[a]Adapted from K. T. Mueser and S. M. Glynn (1995). "Behavioral Family Therapy for Psychiatric Disorders." Allyn & Bacon. Needham Heights, Massachusetts.

C. Research on Family Therapy

Since 1980, a growing number of controlled research studies have been conducted on long-term (at least 9 months) family intervention for schizophrenia. Two types of research questions have been addressed: (1) Is family therapy for single families more effective than not providing family therapy? (2) Does the efficacy of family therapy provided in a multiple-family group format differ from that in a single-family format? The results of studies addressing these questions are summarized in Table III.

Inspection of the table indicates that family therapy, conducted in either single- or multiple-family group format, is associated with a lower rate of relapse or rehospitalization over 1.5–2 years. Furthermore, single- and multiple-family group formats do not differ in their impact on reducing relapses and rehospitalizations. Although less extensively examined, family therapy studies that have evaluated the impact of intervention on other areas of functioning have generally found beneficial effects of therapy on social adjustment, symptoms, and lower family burden. Evidence is also accumulating that indicates that family treatment for schizophrenia is cost-effective because of the decreased utilization of expensive inpatient hospital services.

Much is still not known about family therapy for schizophrenia. For example, the durability of family interventions has not been established, it is unclear whether patients and their families can be "matched" to specific models of family intervention to achieve better outcomes, and the optimal timing for treatment remains unexplored. In addition, the mechanisms responsible for the improved outcomes associated with family therapy are not known. Despite the many questions that remain about family intervention for schizophrenia, it is one of the most potent psychosocial treatments currently available for the illness.

VII. VOCATIONAL REHABILITATION

One of the most common social consequences of schizophrenia is unemployment. Disturbing symptoms, socially inappropriate behavior, cognitive impairment, lack of drive and initiative, and the stigma of mental illness all conspire to produce extremely low employment rates for persons with schizophrenia and other chronic psychiatric disorders. Unemployment among these patients is associated with a lower quality of life and self-esteem, fewer economic resources, and greater symptomatology. It is not surprising, therefore, that the majority of patients with schizophrenia express an interest in employment, usually competitive employment in integrated work settings.

A variety of different vocational rehabilitation models have been developed to address the needs of persons with severe psychiatric disorders. Four different approaches to improving vocational outcomes are most commonly practiced. The *vocational skills training* approach involves systematically teaching patients a wide array of different skills thought to be critical to successfully finding and maintaining competitive employment, such as job interviewing and assertiveness skills. The *sheltered workshop* approach assumes that improvements in vocational capacity are

TABLE III
One-and-a-Half- to Two-Year Cumulative Relapse or Rehospitalization Rates (in Percentages) of Schizophrenia Patients Receiving Long-Term Single-Family Therapy, Multiple-Family Group Therapy, or Routine Treatment

Principal investigator	Theoretical orientation	Sample size	Single-family therapy	Multiple-family group therapy	Routine treatment
Falloon (1985)	Behavioral	32	17	—	83
Leff (1985)	Supportive	24	14	—	78
Tarrier (1989)	Behavioral	42	33	—	59
Leff (1990)	Supportive	23	33	36	—
Hogarty (1991)	Family systems	57	32	—	67
McFarlane (1993)	Behavioral	172	42	28	—
Randolph (1993)	Behavioral	41	10	—	40
Schooler (1994)	Behavioral/supportive	313	29	35	—
Xiong (1994)	Behavioral	63	44	—	64
Zhang (1994)	Supportive	78	15	—	54
Total/mean		845	27	33	64

best developed by providing work opportunities to patients in sheltered settings that are populated by psychiatric patients and staffed by mental health professionals. These settings establish low expectations for work productivity and a high tolerance for the mental health needs of the patients, and tend to transition patients to competitive employment only very gradually and sometimes not at all. The *clubhouse model*, described previously in Section IV,C, seeks to improve vocational functioning by engaging patients in a community in which they first assume responsibilities for helping to run the community (such as cooking and cleaning), followed by temporary competitive transitional employment experiences, followed then by regular competitive employment. The *supported work* model helps patients obtain employment by focusing on rapid placement in competitive work, patient advocacy with employers, and follow-along support to enable patients to remain employed while continuing to get their mental health needs met. In contrast to other models of vocational rehabilitation in which vocational services are frequently brokered out to other agencies, the supported employment model involves the provision of employment services by the same professionals who are treating the psychiatric disorder, facilitating the integration of different service components.

Research on the effectiveness of the first three vocational rehabilitation models has been equivocal. However, in recent years several controlled studies have demonstrated that the supported employment approach results in superior employment outcomes for patients with schizophrenia or other chronic psychiatric illnesses. An example of a recently completed study of one supported employment model developed at the New Hampshire–Dartmouth Psychiatric Research Center, the Individual Placement and Support (IPS) model, follows. The principles of the IPS model are summarized in Table IV.

Patients with severe psychiatric disorders were randomly assigned to either an IPS program or a private rehabilitation program, called Group Skills Training (GST), which provided skills training and help at job placement. GST differed from IPS in terms of both service organization (i.e., separation of mental health and vocational rehabilitation agencies in GST) and approach to prevocational activities. Whereas IPS staff tried to match patients with jobs immediately, GST staff prepared patients for work by instruction in relevant skills for choosing, getting, and keeping jobs. Patients in this study were 143 unemployed adults with severe mental illnesses living in two New Hampshire cities. Vocational outcomes consistently favored IPS over GST throughout the 18-month follow-up period by a ratio of approximately 2 : 1. IPS patients were engaged in competitive employment at approximately a 40% rate from the third month of the study through the eighteenth month, whereas GST

TABLE IV
Principles of Individual Placement and Support Model of Vocational Rehabilitation for Severely Mentally Ill Patients[a]

1. Rehabilitation is an integral component of mental health treatment rather than a separate service.
2. The goal is to help patients attain competitive employment in integrated work settings.
3. People with severe mental illness can obtain jobs rapidly without prevocational training.
4. Vocational assessment is a continuous process used to facilitate job matching rather than to screen out patients.
5. Follow-along supports are often necessary to sustain employment.
6. Vocational services are based on patients' preferences and choices.
7. Services are usually provided in the community, rather than the clinic.
8. A team approach to vocational rehabilitation promotes integrated services.

[a]From D. R. Becker and R. E. Drake (1993). "A Working Life: The Individual Placement and Support (IPS) Program." New Hampshire–Dartmouth Psychiatric Research Center, Concord.

patients took longer to attain a stable rate of approximately 20%. IPS patients were also more likely to attain jobs that involved at least 20 hours of work per week (47% versus 22%). By the end of 18 months, they worked more hours and earned more wages than patients in GST.

VIII. INTEGRATING PSYCHOSOCIAL AND PHARMACOLOGICAL TREATMENTS

Although pharmacological and psychosocial treatments have a long history in ameliorating the various dysfunctions and disabilities of schizophrenia, theoretical and practical aspects of the disorder that have emerged from each camp have been strangely divergent. Psychosocial treatments are grounded in social learning views of severe mental illness. Pharmacological treatments reflect a more biological perspective. Research suggests that drug and psychosocial treatments account for separate and synergistic effects on therapeutic outcome. Figure 3 illustrates some of the effects that medication has on psychosocial treatments and that psychosocial interventions have on pharmacology.

As suggested from Fig. 3, psychosocial interventions improve medication use. Patients with schizophrenic symptoms who learn interpersonal coping skills are better able to collaborate with their mental health care team. As a result, they are more likely to administer medication according to prescription. Moreover, the psychiatrist is able to select and titrate medications more effectively when patients are providing information about their symptoms and side effects in an assertive manner.

Neuroleptic medication directly affects neurophysiological processes that, in turn, influence several psychological processes, including cognition, arousal, and emotion. Changes in psychological processes may enhance the individual's skill learning abilities so that his or her repertoire of interpersonal skills is subsequently improved. Researchers have been especially interested in the mediating effects of cognition. Hence, neuroleptic medication that improves the patient's ability to attend to and remember information is likely to enhance participation in psychosocial interventions. Unfortunately, high doses of neuroleptic medication, and antiparkinsonian medication given for side effects, have actually been shown to diminish the attention and memory of patients. Hence, health care professionals have to carefully titrate medication dose to yield the best effects on psychosocial treatments. [*See* Neuropharmacology.]

Other pharmacological agents may also undermine psychosocial treatments and exacerbate schizophrenia. Research has suggested that patients with severe mental illness abuse alcohol or illegal drugs at a higher rate than do comparison groups. Several reasons for this have been posed: patients are self-medicating symptoms of the disorder, are attempting to counteract the sedating effects of neuroleptic medication, or are using substances to socialize with others. Whatever the cause, psychosocial treatment providers must regularly assess and treat substance use disorders in schizophrenic patients. Recent efforts have attempted to extrapolate interventions developed for a substance abuse population for patients who also suffer severe mental illness by integrating substance abuse and mental health treatments. Promising interventions include an effort to match various psychosocial interventions with the patient's stage of substance abuse. For example, patients who deny substance abuse are provided empirical evidence of the dangers of their use habits. Patients who acknowledge a problem are taught coping strategies to deal with stress that leads to substance abuse. They are also taught interpersonal skills that help them deal with peers who encourage substance abuse. Patients who have learned to cope with their abusive behaviors are taught relapse prevention skills to diminish the likelihood of backsliding.

Finally, psychosocial treatment should focus on stressors in the environment and deficits in personal characteristics that seem to play specific roles in relapse and community maladjustment. Relapse in schizophrenia is common even when drug compliance is firmly established. Nor is there any evidence that a patient's level of psychopathology at hospitalization or discharge predicts subsequent relapse. The best explanation, based on converging lines of evidence from empirical studies, is that the patient's personal assets and deficits, social environment, and the type of psychosocial therapy are the most powerful influences on relapse, even in the face of reliably administered maintenance medication.

BIBLIOGRAPHY

Becker, D. R., and Drake, R. E. (1993). "A Working Life: The Individual Placement and Support (IPS) Program." New Hampshire–Dartmouth Psychiatric Research Center, Concord.

Bellack, A. S., and Mueser, K. T. (1993). Psychosocial treatment for schizophrenia. *Schizophrenia Bull.* **19,** 317–336.

Bellack, A. S., Mueser, K. T., Gingerich, S., and Agresta, J. (1997). "Social Skills Training for Schizophrenia." The Guilford Press, New York.

Birchwood, M., and Tarrier, N. (eds.) (1992). "Innovations in the Psychological Management of Schizophrenia: Assessment, Treatment and Services." John Wiley & Sons, Chichester, England.

Burns, B. J., and Santos, A. B. (1995). Assertive community treatment: An update of randomized trials. *Psychiatr. Services* **46,** 669–675.

Corrigan, P. W., and Liberman, R. P. (eds.) (1994). "Behavior Therapy in Psychiatric Hospitals." Springer, New York.

Corrigan, P. W., and Penn, D. L. (1995). The effects of antipsychotic and antiparkinsonian medication on psychosocial skill learning. *Clin. Psychol. Sci. Practice* **2,** 251–262.

Drake, R. E., Bartels, S. B., Teague, G. B., Noordsy, D. L., and Clark, R. E. (1993). Treatment of substance abuse in severely mentally ill patients. *J. Nervous Mental Dis.* **181,** 606–611.

Mueser, K. T., and Berenbaum, H. (1990). Psychodynamic treatment of schizophrenia: Is there a future? *Psychol. Med.* **20,** 253–262.

Mueser, K. T., and Gingerich, S. (1994). "Coping with Schizophrenia: A Guide for Families." New Harbinger Publications, Oakland, California.

Mueser, K. T., and Glynn, S. M. (1995). "Behavioral Family Therapy for Psychiatric Disorders." Alllyn & Bacon, Needham Heights, Massachusetts.

Paul, G. L., and Lentz, R. J. (1977). "The Psychosocial Treatment of the Chronic Mental Patient." Harvard Univ. Press, Cambridge, Massachusetts.

Schizophrenic Disorders

ERMIAS SELESHI
JERRY G. OLSEN
ALEXANDER L. MILLER
JAMES W. MAAS[1]
University of Texas Health Science Center at San Antonio

GLOSSARY

Anticholinergic Blocking effect of nervous system function mediated by the neurotransmitter acetylcholine

Extrapyramidal system Part of the central nervous system controlling certain aspects of voluntary motor behavior

Neuroleptic Drug possessing antipsychotic effect. The term means to "grasp the neuron" from its action on the nervous system

Paranoia Specific fear or suspicion not founded on reality, often associated with other disturbances of thought

Pathognomonic Characteristic that is a unique or specific identifying feature of a disease

Psychosis Disturbances of thought characterized by false beliefs (delusions), perceptions (hallucinations), and loss of order in thought processes

Psychotomimetic Capacity or potential of inducing psychotic symptoms

Tardive dyskinesia Abnormal involuntary movements of voluntary muscles such as the tongue, face, neck, diaphragm, and limbs, most commonly associated with prolonged use of neuroleptic drugs

[1]Deceased.

SCHIZOPHRENIA IS A DISORDER OF BEHAVIOR characterized by disturbance of perception, thought processes, and reality testing. The term, as originally coined, meant split mindedness, to describe the dissociation between thought, emotion, and behavior. This led to the popular misconception of split personality, equating schizophrenia to the condition currently known as multiple personality disorder. An estimated 1% of the population of the United States are victims of schizophrenia, which costs approximately 2% of the gross national product, mostly because of recurrent hospitalizations and secondary loss of the productive capacity of patients. Only a tiny fraction of research funds are allocated to the study of schizophrenia when compared with other less common and less costly diseases. Social stigma and gradual socioeconomic decline of affected individuals and their families give the distorted impression that the disease is more prevalent among the poor and minority ethnic groups. Available data, however, indicate that schizophrenia has no socioeconomic or ethno-cultural boundaries. Aggregate data obtained from empirical observations, biological studies, and the wide variability in the severity, course, and treatment response indicate that schizophrenia is the manifestation of a group of heterogeneous disorders. Genetic and other environmental factors may predispose an individual to develop the disorder or modify its clinical manifestation.

I. CLINICAL FEATURES

A. Early Description and Terminology

The term schizophrenia was introduced in 1911 by Eugene Bleuler, one of the two European psychiatrists who are credited for the early description of the clini-

ENCYCLOPEDIA OF HUMAN BIOLOGY, Second Edition, VOLUME 7. Copyright © 1997 by Academic Press. All rights of reproduction in any form reserved.

cal picture and phenomenology of the disorder. However, many of the other frequently used terminologies associated with schizophrenia (e.g., paranoia, hebephrenia, and dementia praecox) did appear in the latter half of the nineteenth century. The Swiss psychiatrist Emil Kraepelin applied the term dementia praecox to describe various psychotic states with an early age of onset in the teens or twenties and a progressive decline of mental function, resulting in a clinical state reminiscent of dementia. He distinguished schizophrenia from manic-depressive illness, which is characterized by episodic psychotic disturbance with symptom-free intervals and a lack of progressive mental deterioration. In addition to providing a detailed description of the symptoms of schizophrenia, Kraepelin wrote of a small percentage of patients who may experience complete recovery or run a benign favorable course.

Bleuler's notions differed from those of Kraepelin in several ways. He applied psychoanalytic interpretations to the behaviors manifested by schizophrenic patients and focused on the severity and duration of the illness rather than its clinical course. He proposed four cardinal diagnostic features of psychotic disturbance (i.e., looseness of associations of thought content, autism, disturbance of affect, and ambivalence), often referred to as the "four A's" of Bleuler. He categorized hallucinations, delusions, and catatonia as nonspecific accessory symptoms. Thus Bleuler's criteria were broader and more inclusive. Both pioneering workers believed in an underlying organic cause of schizophrenia.

B. Clinical Diagnosis

None of the symptoms of schizophrenia is specific or pathognomonic for the illness. Thus diagnosis is based on the presence and duration of a cluster of symptoms, on the clinical course over time, and, when possible, by documentation of certain personality traits or prodromal changes in behavior antedating the onset of symptoms. The disorder usually begins in late teens or early twenties. Unlike in early editions, in the current "Diagnostic and Statistical Manual of Mental Disorders, Fourth Edition," the uncommon onset after the fifth decade of life is accepted. The age of onset and clinical course have both therapeutic and prognostic implications.

The cardinal features of psychotic symptoms in schizophrenic disorders include disturbances of thought, perception, language, psychomotor function, mood, and affect. Additionally, eating disorders and so-called soft neurologic signs could be present. Soft neurologic signs cannot be attributed to a specific area of brain abnormality and are therefore referred to as nonlocalizing. They include asymmetric coordination difficulties with fine motor tasks (e.g., clumsiness in counting the fingers rapidly on one side of the body). The core category of symptoms is referred to as negative symptoms and will be discussed later. These contrast with the positive symptoms of thought and perceptual disturbances. Impairment of judgment and insight is often associated with both acute psychotic symptoms and the residual phase of the illness.

The disturbance of thought can be broken down into form and content. Form refers to the overall structure of the thought process. The schizophrenic patient may manifest looseness of associations, whereby the verbalization of the patient's thoughts lack logical order of cause and effect. Topics are abruptly changed and bear no clear association to the issue being discussed. The patient's trends of thought may abruptly be interrupted for a few seconds, and the patient may be unable to continue from the point of interruption. Language function can also be affected. The patient may use complex terms inappropriately (stilted speech), create new words (neologism), repeat words or statements several times (echolalia), or be verbose without conveying much information (poverty of content). The ability to understand and interpret abstract concepts becomes impaired, and ideas are oversimplified (concrete thinking). Although memory and orientation are preserved, the acutely psychotic patient may manifest impairment of attention.

The disturbance in the content of thought is broadly referred to as delusion, in which the patient displays firm beliefs that have no basis in reality. These may take the form of paranoia, whereby the patient believes he or she is being persecuted by familiar persons, institutions, government agencies (especially law enforcement), or alien beings, such as extraterrestrial creatures, or the devil. The delusion can be extremely elaborate and often bizarre, taking on a science fiction quality in which patients may describe how they are being spied on through various equipment (i.e., TV, radio, ovens, etc.) in their surroundings. Not uncommonly, real life events, such as being arrested or being fired from a job because of abnormal behavior, get incorporated into the delusional system and reinforce the belief. Another form of delusion is ideas of reference, in which patients infer a special significance about events around them. These may include the ways others communicate with them, the arrangement of furniture in a room, contents of news broadcasts, or prevailing weather conditions. Patients may believe that their ideas and thoughts are removed from their minds (thought withdrawal) or broadcast on news media (thought broadcasting). In addition, they may believe that alien thoughts or feelings are being forced on them (thought insertion) or that they possess telepathic powers to understand and influence the behavior of others. At times the content of delusional material takes on a grandiose quality. In such instances, patients believe they possess tremendous wealth and physical or political power, have a special religious or political mission, or have the ability to influence events around them (magical thinking). Somatic delusions refer to the patient's belief in suffering from terminal illness, deformity of the body, or destruction of internal organs. In summary, delusional thought contents can take on many forms, from a well-organized and elaborated system with some internal logic to bizarre and fragmented beliefs much more diverse than what has been highlighted here.

Perceptual disturbances consist of perversion of sensory modalities in the absence of actual environmental stimuli. These symptoms are referred to as hallucinations and may involve any of the five senses (i.e., hearing, touch, vision, smell, or taste). Auditory hallucinations are the most common form, and the last three are more frequently associated with an organic brain disorder. A single patient may have more than one form of hallucination simultaneously or at different points in time. The hallucinations may be elaborated and be part of the patient's delusional system. As an example, a patient may believe that he or she receives verbal instructions from controlling forces, such as the President or God. Frequently patients report one or more continuous, critical, praising, or threatening voices in reference to the patient (i.e., a running commentary). Voices instructing patients to carry out certain tasks are called command hallucinations. Not uncommonly, patients may only hear single words, such as their names, muffled voices, foreign languages, and animal or machinery noises. Visual hallucinations consist of either fully formed animated appearances of humans, animals, or alien beings or fragments of such (e.g., faces, skeletons, shadows, flying objects). Patients do not usually readily admit the presence of delusions or hallucinations, thus inferences are made based on their outward behavior. They may carry on a conversation or appear to be responding to an unidentifiable stimulus.

Various types of deviant psychomotor behavior are encountered in schizophrenia. Catatonia is a type of motor behavior in which patients assume a still position for long periods of time (catatonic stupor) or are extremely agitated and in constant vigorous movement (catatonic excitement). Catatonic patients may resist attempts to move their limbs (negativism) or maintain any position to which a limb is moved (waxy flexibility). Catatonic features are less frequent now than they used to be, probably because of the advent of antipsychotic drugs and early treatment of patients. Other abnormal motor behaviors that are encountered include stereotyped, repetitive gesticulations and echopraxia, in which the patient will imitate movements and positions assumed by the examiner. Although signs and symptoms of depression or anxiety may be associated with schizophrenia, in general the mood and affect displayed by the patients may not be reflective of the thought content, and this is referred to as inappropriate affect. Nonspecific behavioral changes are frequently seen and may represent grossly disorganized behavior such as a decline in the patient's personal appearance, hygiene, and social grace. These features are often associated with the so-called negative symptoms of schizophrenia. Three characteristic negative symptoms are included in the definition of schizophrenia, namely: affective flattening, alogia, and avolition. Patients with affective flattening may show little or no emotion while talking about issues that would be expected to generate strong feelings (e.g., fear, sadness, happiness) in the average person. Alogia describes a decrease in speech fluency and productivity and does not simply reflect a patient's desire not to speak. Avolitional behaviors represent difficulties initiating and sustaining new goal-directed activities. Schizophrenics with moderate to severe negative symptoms may also exhibit social withdrawal, anhedonia, and cognitive impairment that are not a response to an exacerbation of positive symptoms. These groups of symptoms are responsible for a major part of the disabling effect of the illness, as they impair patients in their educational, occupational, and social functioning. Impairment in executing sound judgment, as well as insight about the illness, are a common feature in both the acutely psychotic and chronic schizophrenic patient.

C. Course and Long-Term Outcome

Schizophrenia runs a chronic relapsing and remitting course. In general, unlike mood disorders, there is no return to premorbid level of function during periods of remission. It is not uncommon to observe symptoms of depression when the patient is recovering from an acute psychotic episode. Ten percent of schizophrenics die from suicide, and four times as many attempt it during their lifetime. Although treatment noncompliance is probably one of the most common causes of relapse, other factors may be contributory. Environmental stress and concurrent use of alcohol or other illicit drugs play some role. [*See* Depression: Suicide.]

There appears to be a cumulative effect of deterioration during the first few years of recurrent relapse, which stabilizes as time goes on. Positive symptoms become less marked, although up to one-third of patients will continue to experience them in a substantial way. The progressive deterioration, associated with emergence of the functionally and socially disabling negative symptoms, is sometimes referred to as deficit state. Socioeconomic decline of the schizophrenic patient has given rise to the downward drift theory. This phenomenon may be explained by the affliction of young patients, before they achieve their educational and vocational potentials, and subsequent failure of the majority to become self-sufficient because of recurrent psychotic relapses. Many gravitate downward into poverty, and a sizable portion of the homeless in urban centers in the United States are schizophrenics.

The disorder, however, is not invariably bleak in outcome. Various stidies have indicated that up to a quarter of patients achieve significant recovery and lead relatively normal lives, and an equal number function with mild to moderate symptoms. Nonetheless, the remaining 50% are significantly impaired for life. Recovery and overall improvement have been reported among chronically ill, deinstitutionalized patients over a long period of prospective follow-up. In general, factors such as older age and acute manner of onset associated with apparent precipitant factors when symptoms initially appear, favorable premorbid function, significant positive, depressive, and paranoid symptoms, as well as being married with a good social support system, confer favorable outcome. Family history of schizophrenia, history of birth injury, and younger age of onset with a chronic, relatively unremitting course are associated with poor prognosis.

II. CLASSIFICATION

Over the years many terms have been applied to describe clinical subtypes of schizophrenia based primarily on clinical presentation. Some of these have prognostic significance in their treatment response or long-term course. Paranoid schizophrenia is characterized by prominent positive symptoms of paranoid delusions. Catatonic schizophrenia or catatonia is the subtype with the peculiar psychomotor symptoms of stupor or excitement. Both paranoid and catatonic categories have a favorable outcome in their response to treatment. The patient with hebephrenic schizophrenia manifests gross disorganization of behavior and thought processes with a disinhibited, inappropriate, and silly affect. Simple schizophrenia, however, has prominent features of negative symptoms with minimal and occasional, if any, hallucinations and delusions. The hebephrenic and simple subtypes carry poor prognosis. The term latent or borderline schizophrenia has been applied to patients with enduring odd personality traits and occasional thought disorders. Pseudoneurotic schizophrenics have prominent anxiety symptoms with some underlying thought disorders, phobias, and sexual identity conflicts.

The DSM-IV introduced in 1994 is the current guideline for the diagnosis of psychiatric disorders in the United States. However, there are other closely related guidelines, such as the Research Diagnostic Criteria and the Schedule for Affective Disorders and Schizophrenia, primarily used in clinical research settings. In addition to the presence of a constellation of symptoms, the DSM-IV requires temporal criteria to be met to subtype diagnostic groups, as well as to classify the course of illness. Prodromal and residual symptoms of change in behavior, as well as impairment of function, are also included in the scheme. To diagnose schizophrenia, a combination of positive symptoms and prodromal or residual features should be documented for at least 6 months. The clinical picture over the long course is further subclassified depending on periods of improvement or worsening of symptoms. Course specifiers may be given after 1 year from the onset of active symptoms. These include: Single episode in partial or full remission; episodic with or without interepisode residual symptoms; continuous; or other or unspecified pattern. "With prominent negative symptoms" may be added if they are notably present during the course of illness. Five subtypes of the disorder are noted. The paranoid and catatonic types closely resemble the subtypes in older criteria. The disorganized type is equivalent to hebephrenia. Patients with the undifferentiated type manifest prominent positive symptoms or grossly disorganized behavior and do not fulfill criteria for other subtypes. Unlike the old subtype of simple schizophrenia, positive symptoms are prominent in the undifferentiated group. The fifth subtype, residual type, is used for patients who show little or no positive symptoms but display an apparent impairment from negative symptoms. The criteria emphasize that organic causes, including drug abuse and other phychiatric diagnoses such as manic-depressive illness, be excluded before the diagnosis of schizophrenia is made.

III. GENETICS

There is compelling evidence that genetic predisposition confers vulnerability to the development of schizophrenia. The pattern of inheritance is not clearly understood but is believed to be a Mendelian pattern with incomplete penetrance. The search for a chromosomal marker has been elusive to date. Investigators are searching for genes related to schizophrenia by attempting to correlate neurobiological systems as well as closely spaced genomic markers with the presence or absence of schizophrenia. Utilizing restriction fragment length polymorphism studies of families with multigeneration expression of the disease, several promising regions have been identified for further study. Use of the polymerase chain reaction has allowed researchers to focus upon the short arms of chromosomes 6 and 8, the long arm of chromosome 22 as well as the human leukocyte antigen system. Many family and twin studies have been carried out in both the United States and Europe, supporting the contribution of genetic factors. The findings indicate that even in adopted apart siblings and twins raised in separate environments, the risk of developing the disease remains the same as those raised by their biological parents. Increasing genetic load confers higher likelihood of becoming schizophrenic. The risk of a nontwin sibling of a schizophrenic patient is 8%, compared with the 1% prevalence rate in the general population. This rate rises to 12% if a parent or a dizygotic twin is affected. The child of two schizophrenic parents has a 40% chance of becoming schizophrenic, while the concordance rate in monozygotic twins is 50%.

IV. EPIDEMIOLOGY

The wide diversity of diagnostic criteria applied has resulted in a wide range of incidence and prevalence rates of schizophrenia worldwide. The estimated incidence of schizophrenia in the United States is one per 1000 population, and the lifetime prevalence is approximately 1%. However, only a small fraction of the schizophrenic patient population receives any form of treatment, and the majority of the patients receiving treatment require hospitalized care. Although there is no difference in prevalence between the sexes, the peak age of onset in men, 15–25 years old, is 10 years younger than in women. The disease rarely manifests itself during the first or after the fifth decade of life. Some reported differences in incidence rates among races and socioeconomic classes are attributed to probable biased application of diagnostic criteria and caused by the downward drift concept. Significant correlation with population density in large but not in medium or small urban centers has been applied to support the role of environmental stress as a cause for schizophrenia. Yet demographic factors that may promote the aggregation of patients in large cities have not been adequately addressed. Higher incidence of schizophrenia among new immigrants in urban centers has raised a question about the possible contribution of cultural factors in the etiology of the disease. Similarly, the higher incidence of the disorder in industrialized nations and the growing incidence in developing nations undergoing industrial changes have been explored.

The birth of a disproportionately high percentage of schizophrenic patients during the winter months has led to the development of environmental etiologic factor theories such as viral infections, the effect of the maternal immune response to infection on fetal nervous system development, and nutritional deficiencies. Schizophrenics as a group have a higher incidence of birth complications than the general population and have an unexplained higher rate of mortality from various medical disorders.

V. METAPSYCHOLOGICAL THEORIES

The psychoanalytic school pioneered by Sigmund Freud has its central focus on the disturbance of ego function. According to Freud, ego function disintegrates and returns to a more primitive stage of development. The term *primitive* refers to earlier stages of life of the individual. The structural theory, formulated by Freud, divides psychic apparatus organization into the id, ego, superego. The ego, driven by narcissistic needs, develops in response to demands placed on it by the environment, represented in part by the superego. In return, it has the executive function of monitoring the id, which is governed by innate drives of sex and aggression. Sustained, overwhelming conflict between id impulses and superego values are believed to be the source of psychological symptoms.

The ego's failure to master and resolve these conflicts in an appropriate way results in its disintegration and return to an infantile level in which reality testing is impaired. Freud also proposed that specific symptoms were reflective of underlying conflicts (e.g., paranoia representing ambiguity of sexual identity). More recent theories emanating from this school suggest that various symptoms have symbolic significance, representing the fears and wishes of the individual, and the regressive phenomenon may be an adaptive event for reorganization of the psyche in creating new reality. [*See* Psychoanalytic Theory.]

The interpersonal theorists focus on the influence of parental figures. The schizophrenic patient fails to master the separation–individuation phase of development when the infant starts identifying itself as a separate entity from the mother and incorporates object constancy. This development enables the infant to feel secure in the mother's availability whenever she is not within visual range. The failure of development in this area is believed to be generated by ambiguous and conflicting messages that the mother may convey to her child. Harry Sullivan postulated that the behavior of overanxioius mothers may generate the same anxious feelings in their babies who later develop schizophrenia.

Various family theories have been advanced as a cause of emotional disturbance in the growing child that later becomes schizophrenic. Learning theorists believe that schizophrenic patients develop irrational perceptions and reactions by imitating emotionally disturbed parents. The double-bind concept refers to a setting in which emotionally disturbed families place the child in a constant situation of making decisions between ambiguous choices, producing confusion. Another theory about the balance of relationships among members of a family relates to the abnormal over-closeness of the child with the parent of the opposite sex or the parent playing the dominant role in the family. The pattern and style of expression of various emotions among members of a family have also been postulated to influence the expression of schizophrenic symptoms.

Despite the compelling evidence supporting an underlying, genetically determined organic cause of schizophrenia, many psychodynamic and psychosocial models have been proposed to explain disease etiology as well as means of therapeutic intervention. Although few traditional centers focus on a psychotherapeutic approach as a primary modality of treatment, most currently use a combined approach that includes the use of antipsychotic drugs. Successful comprehensive treatment programs incorporate some aspects of the preceding theoretical models in their attempt to rehabilitate patients to various levels of independent functioning within the community. The effort centers on modifying maladaptive interpersonal relationships by fostering and rewarding appropriate behaviors, while discouraging dysfunctional ones. [*See* Schizophrenia, Psychosocial Treatment.]

VI. CURRENT BIOLOGICAL TREATMENT

First of all, it must be appreciated that there is no definite cure for schizophrenia, and the syndrome can only be controlled, not cured. It should also be recognized that in our present state of knowledge we are not aware of the etiology of schizophrenia or psychosis. In fact, there is some debate regarding the relation between schizophrenia and psychosis. Some psychiatrists, for example, feel that the neuroleptic drugs are useful in the treatment of both schizophrenia and psychosis, whereas others are of the opinion that they are helpful only in managing psychosis. Thus, if the drugs selectively treat the positive symptoms, leaving the core negative symptoms untouched, any attempt to construct an etiologic theory of schizophrenia based on response to drugs rests on shaky ground. This is because the dopamine (DA) receptor blocking effect of typical neuroleptic drugs has been central to the so-called dopamine hypothesis of schizophrenia.

The care of the psychotic patient, nonetheless, was revolutionized as of 35 years ago when the neuroleptics were first introduced to the North American continent. These drugs adequately control agitation, delusions, hallucinations, and other miscellaneous thought disorders of patients, whether they are in a schizophrenic, manic, organic, or drug-induced clinical setting.

The drugs are now most often referred to as neuroleptics or antipsychotics, but in the early history of their use they were also called major tranquilizers. They are divided into several chemical classes. The phenothiazine class of drugs is divided into three subgroups (i.e., the aliphatic, piperidine, and piperazine derivatives). Chloropromazine (Thorazine) is the most prominent member of the aliphatic group and one of the first drugs used. Thioridazine (Mellaril) is a commonly used piperidine derivative. Examples of piperazine phenothiazines include trifluoperazine (Stelazine) and fluphenazine (Prolixin). Haloperidol (Haldol), a butyrophenone class of neuroleptic, is probably the most commonly prescribed neuroleptic today. Thiothixene (Navane) belongs to the thioxanthene class. Molindone (Moban) is a dihydroindolone and loxapine (Loxitane) is a dibenzoxazepine. In addition to these conventional classes, a growing number of atypical antipsychotics (e.g., clozapine, risperidone, olanzapine, sertindole) are becoming available for use in the United States. These drugs appear to have decreased risks for actue and long-term neurological side effects at standard doses. They have shown promise against negative symptoms apart from their relief of positive symptoms. Clozapine (Clozaril) has been approved for use in schizophrenia unresponsive to other neuroleptics. Because of a 1.3% first-year cumulative incidence of agranulocytosis (i.e., a marked reduction in the number of white blood cells), strict guidelines for frequent blood testing and patient monitoring must be followed. Despite the relatively high costs of drug and blood monitoring, clozapine has been shown to be cost effective in the treatment of refractory schizophrenia. Risperidone (Risperdal) exhibits a favorable response profile on measures of positive and negative symptoms. It is approved for both first-episode as well as chronic schizophrenia. Olanzapine (Zyprexa) and Sertindole (Serlect) offer the benefit of once-daily dosing and very favorable side effect profiles. Compliance with taking medication is improved with these low side effect drugs. In addition, olanzapine shows little drug-drug interaction via hepatic metabolic enzyme systems.

The antipsychotic drugs are also classified according to their clinical potency. This is based on comparison of potency with 100 mg of chlorpromazine equivalents and appears to parallel the degree of DA receptor blocking activity of the drugs. The drugs are usually given orally in pill or liquid form and take about 100 minutes to achieve peak plasma level. Parenteral preparations for intramuscular (IM) injection are available for many of the drugs. Administration by injection results in peak plasma levels within 30 minutes and is used in emergent situations when rapid tranquilization of a severely agitated or violent patient is desired. Haloperidol and fluphenazine are available in oil-based depot injection forms. These preparations are slowly absorbed from the tissue and can be given at 2- to 4-week intervals. They are useful in ensuring compliance, instead of oral preparations that have to be taken on a daily basis. The choice of drug is usually based on side-effect profile and desired clinical response. Clozapine is currently the drug of choice for treatment resistant patients. The high-potency drugs such as haloperidol or fluphenazine are less sedating but are associated with a high incidence of extrapyramidal side effects (EPS). However, the low-potency drugs such as chlopromazine or thioridazine produce significant sedation and postural hypotension (i.e., a significant drop of blood pressure in the upright position) and possess potent anticholinergic action. This property may actually have a protective effect from EPS, as anticholinergic drugs are frequently prescribed with the high-potency drugs to treat EPS. Because their reduced side effect profiles may enhance patient compliance, it is likely the new atypical antipsychotics will become increasingly popular.

Treatment of the schizophrenic patient can be divided into the acute and chronic phases. Objectives of treatment during the acute phase include controlling agitation and acute psychotic symptoms. The type of drug, dose, and route of administration depend on the prevailing clinical state. If marked agitation or violent behavior is encountered, a more sedating agent is given. The first few doses may be given by IM injection if needed. The autonomic side effects, especially postural hypotension, may limit the use of these drugs at adequate doses in the elderly or in those with underlying cardiovascular problems. Although the high-potency drugs are less sedating, EPS can be severe and intolerable especially at higher doses. It can, however, be effectively controlled with the use of anticholinergic drugs such as benztropine (Cogentin), trihexyphenidyl (Artane), or diphenhydramine (Benadryl). These agents are usually given by mouth but can be administered by injection for rapid effect. As emergence of side effects may interfere in the patient's acceptance of the treatment, they should be promptly and effectively treated to ensure long-term compliance. The antipsychotic effects of the neuroleptics may take 6 weeks or longer to be maximally effective. Patients on clozapine may show additional improvement after the first six weeks with up to 50% of the response occurring between six weeks and six months. Therefore, an aggressive increase of neuroleptic dose, more than is recommended, is probably not appropriate during the first few weeks of treatment. Such dose escalation is associated with an increasing incidence of the unpleasant and troublesome side effects without significantly benefitting the target symptoms. Not surprisingly, rapid neuroleptization (i.e., a treatment strategy of giving high doses of haloperidol or other high-potency agents by IM injection at one to two hourly intervals to achieve rapid tranquilization of the acutely psychotic patient) has lost favor in recent years.

The objectives of the long-term or maintenance phase of treatment are control of psychotic symptoms with lower maintenance dose of neuroleptics. In addition to minimizing the acute dose-related EPS, the cumulative exposure to neuroleptics, presumed to be responsible for tardive dyskinesia (TD), will be controlled. If psychotic symptoms have been completely absent for at least 6 months on maintenance dose after the treatment of an acute episode, attempts are made to gradually decrease and discontinue the neuroleptic, as long as the patient remains symptom-free. Although a higher incidence of relapse is likely with low-dose therapy, recurrences of psychotic symptoms are usually rapidly controlled with dose adjustments. Studies of blood neuroleptic level measurement for dose determination and clinical response monitoring have produced conflicting results and have generally been disappointing. This is partly due to the multitude of metabolites of undetermined significance that these drugs have, as well as problems with the various laboratory techniques used to assay drug concentrations. In general, the findings of the various studies support the presence of minimum threshold blood drug concentration that should be present to achieve clinical response. Higher drug concentrations are associated with increasing incidence of side effects.

Several neuroleptic treatment-related side effects should be noted. The acute ones take several forms. EPS involves the extrapyramidal part of the central nervous system, where a high concentration of DA neurons is localized. It usually has features of Parkinsonism, including muscle rigidity, bradykinesia (i.e., slowness of movements), and gait disturbances. Akathisia is a peculiar sense of motor restlessness in which the patient is unable to remain still or relax for any length of time. It is often confused with agitation and mismanaged by giving more of the offending agent. Beta blocker drugs, such as propranolol, are useful in controlling this symptom. A unique form of EPS is acute dystonia, which is a painful sustained spasm of neck, limb, extraocular, and sometimes oropharyngeal and laryngeal muscles. It is an emergent condition that responds to parenteral anticholinergic drug administration. In long-term typical neuroleptic-treated schizophrenics, TD will develop in approximately 5% of patients for each year of drug use. This is usually a cosmetically undesirable, but sometimes a physically disabling, movement disorder. It manifests as an uncontrollable stereotyped movement of the tongue and mouth (chewing, lip pursing), face (grimacing), neck and spine (contorting movements), or the hands and feet (writhing). The condition affects women more frequently than men. Although most patients with TD have an average of 3–5 years of chronic neuroleptic exposure, TD has been reported in some cases after only a few months of neuroleptic use. It is believed to be related to increased sensitivity of DA receptors deprived of normal DA stimulation. Its onset is usually insidious and may first be observed during neuroleptic dose reduction, oftentimes prompting dose escalation, which temporarily masks the manifestation. Although there is potential for full recovery, the condition is irreversible in many patients. The neuroleptic malignant syndrome is a life-threatening medical emergency of undetermined cause, associated with both acute and maintenance treatment of high-potency neuroleptics. It is characterized by marked elevation of body temperature, severe muscle rigidity, profuse sweating, elevated muscle enzymes and white blood cell counts, and alteration of consciousness. Elevated blood myoglobin level may cause renal failure. The treatment is mainly symptomatic, including lowering body temperature. The symptoms can persist from 10 days to several weeks, depending on the duration of action of the offending neuroleptic.

Adjunct treatments to neuroleptics in schizophrenics include lithium carbonate, antidepressants, benzodiazepines, and the anticonvulsant drugs carbamazepine and valproic acid. Although these agents can be of some benefit in selected patients with concurrent specific symptoms, such as mood disturbance or anxiety, they are of equivocal value on schizophrenic symptoms. Other forms of therapy for schizophrenia include electroconvulsive therapy (ECT), psychosurgery, and insulin coma. The last one is obsolete and is only of historical interest. Psychosurgery is used rarely to control violent or compulsive behavior, when all other means fail. The surgery involves destruction of the frontal lobe or its connections. ECT was once widely used in catatonic patients, and although it appears to control psychotic symptoms, the benefits are short-lived. There is some recent resurgence of interest in ECT use to treat schizophrenia.

VII. CURRENT BIOLOGICAL THEORIES

The etiology of schizophrenia basically remains unknown. In a situation in which a chronic and destructive disease such as schizophrenia exists, it is only natural that there would be a great number of theories or hypotheses about its origin. Some of these are interesting but have no support whatsoever. Others fall into the "where there's smoke, there's fire" idea that there is supportive evidence of the possibility of what may be wrong but no firm evidence. Unfortunately, there is no hypothesis for which there is absolutely firm and substantial data, both from clinical and basic science observations, to indicate the type of underlying abnormality in schizophrenia. An example of the first of these types of theories is the vitamin C deficiency theory, which was proposed by Dr. Linus Pauling. However, tests of this hypothesis have yielded only equivocal results.

In the second category, there is suggestive evidence, but no firm data, to clearly support theories that deal with neurotransmitters, with constitutional factors, and with structural problems. Much of the work on the biology of schizophrenia has examined the effects of antipsychotic drugs on neurotransmission. During the 1970's and 1980's the focus was primarily on the dopamine system. The discovery that clozapine, a drug superior in efficacy to typical neuroleptics, has major effects on multiple receptor systems, has spawned a host of new studies on the roles of other neurotransmitters in schizophrenia.

The neurotransmitter for which there is considerable convincing evidence at the basic science level is DA. This is a neurotransmitter that is a precursor of norepinephrine (NE). In certain regions of the brain there is no enzyme necessary to convert DA to NE, and DA itself becomes the primary neurotransmitter. Probably the best-known example of the effects of a deficiency of DA would be Parkinson's disease. Work with schizophrenia was given a particular impetus by the fact that amphetamine can produce paranoid-like psychosis. It causes the release of DA and NE. The balance tipped in favor of DA, rather than NE, when it was found that almost all the typical neuroleptics used to treat schizophrenia bind to dopamine receptors with a potency that is correlated with doses used clinically. In contrast, the potency with which the drugs bind to NE or serotonin receptors is different from their clinical potency. Furthermore, certain experiments with animals can produce behaviors that are analogous to stereotypy in the schizophrenic when the DA neuronal systems of these animals are manipulated. Interest in the DA system was occasioned by a report some 30 years ago from a Swedish group led by Carlsson. They gave neuroleptics to mice and then measured the metabolic products of DA in the brain. They found that when neuroleptics were administered to the mice, there was an increase in the metabolites of DA, and they postulated that this was because of a blockade of DA receptors in the brain, with concurrent feedback that caused more DA to be released. This basic *in vivo* study has been replicated many times, and there is little question of its validity, in that the administration of neuroleptics is associated with increased DA metabolites. Whether this effect is mediated by receptors on the cell body or by feedback loops is perhaps more controversial and is not really necessary to the theory's credibility. This observation evolved into the so-called DA hypothesis of schizophrenia. Recent work indicates that psychoses of many etiologies appear to involve overactivity of DA systems and that antipsychotic drugs block a family of related DA receptors (D2, D3, D4) in proportion to their clinical potency.

There are distinct neuronal DA systems within the brain (i.e., tubero-infundibular, mesocortical, mesolimbic, and nigrostriatal). The binding of neuroleptics to receptors in the brain can be determined in the laboratory or can be determined *in vivo* by positron emission tomography (PET), a novel neuro-imaging technique. Because a substantial portion of the DA-like ligands or their antagonists need to bind to receptors to be visualized by PET, the question has been left open as to whether there may be deficit in the mesolimbic and mesocortical systems that cannot be visualized because of the low density of DA receptors in these areas. There are two opposing results from PET studies, one from Johns Hopkins University and the other from the Karolinska Institute in Stockholm. The group at Johns Hopkins has shown an increased density of D2 receptors in brains of schizophrenics, even in those who were not on, or have never been on, medication. Whereas at the Karolinska Institute, it was demonstrated that the number of receptors is the same in both healthy controls and patients. Although both groups used PET, different receptor ligands and mathematical models were used to label receptors and to calculate their density. Hence, the controversy is not resolved, and we will have to wait for newer techniques or different models of measuring binding to answer this important question. Neuroanatomically, it has been postulated that the deficit in the schizophrenic patient may be in the mesocortical or mesolimbic system. This was originally thought to be the case because the neostriatum, which is visualized by PET, seems to be concerned mostly with motor functions or the control of movement. However, more recent work has suggested that the caudate, a part of the neostriatum, may be implicated in cortical functioning; hence, it may be that lesions in the schizophrenic brain will be found in the striatum.

Complex regulation of dopaminergic and noradrenergic function in the cortex and midbrain has been the subject of intensive investigation over the last twenty years. Early investigations of the serotonin system, utilizing brain lesion, immunohistochemical, and serotonin (5HT2) blocking studies helped form the bases for hypotheses regarding serotonin-dopamine interactions. Current PET evidence in baboons supports the concept of serotoninergic inhibition of dopamine function in the frontal and prefrontal cortex. These effects are thought to occur directly or by indirect influences on γ-aminobutyric acid (GABA) and cholinergic activity. Phencyclidine (PCP), an excitatory *n*-methyl-D-aspartate or NMDA-type glutamate receptor blocker can produce psychotic symptoms (both positive and negative) that closely resemble schizophrenia. In theory this may reflect a functional decline in GABA modulation of dopaminergic function in the cortical and limbic systems, leading to cortical hypofrontality as well as limbic system dysregulation and an overall increased likelihood of developing positive symptoms. By taking advantage of these putative neurotransmitter system interrelationships, researchers are developing drugs to improve frontal cortical and striatal dopamine function, decrease negative symptoms, and reduce the likelihood of EPS.

There is stronger clinical evidence supporting a role for NE in psychotic states than for DA. This neurotransmitter is absent in the striatum but is found widely distributed throughout the entire cortex. However, the results from animal studies are less convincing for NE than for DA. For example, the potency with which receptors in the central nervous system bind NE-like ligands or neuroleptics has no relation to their clinical potency as antipsychotic drugs. Three separate studies have reported elevated CSF NE levels in psychotic patients. A study of levels of NE in neuroanatomic structures indicates that there are elevated NE levels in limbic structures of paranoid patients. Lastly, there is evidence that 3-methoxy-4-hydroxyphenylglycol (MHPG) levels in urine and plasma are elevated in schizophrenic patients and the level of MHPG correlates directly with the severity of psychotic symptoms. MHPG is a major metabolite of NE in brain. It is possible that both DA and NE brain systems are involved in the genesis of psychosis and that NE and DA are functionally related. However, the linkage of the two systems is at present poorly understood.

Various structural abnormalities have been reported in postmortem studies of the brain of schizophrenic patients, as well as by the modern imaging techniques of computerized axial tomography (CAT) and magnetic resonance imaging (MRI). The postmortem findings are varied and inconsistent. They include atrophy of various parts of the brain, including the neostriatum, the cerebellum, and temporal lobes, as well as enlargement of the cerebral ventricles. It is unclear if any of the changes in brain tissue are a reflection of the disorder, changes associated with neuroleptic exposure, or postmortem artifacts. Similarly, CAT and MRI findings are nonspecific and not universally found. Abnormal findings are more frequently associated with the so-called type II schizophrenia, or deficit state, than with type I, which is characterized by positive symptoms. In conclusion, despite the availability of these powerful tools, some theoretical framework is needed to search systematically and to identify specific findings. [*See* Magnetic Resonance Imaging.]

BIBLIOGRAPHY

Buchsbaum, M. S., and Haier, R. J. (1987). Functional and anatomical brain imaging: Impact on schizophrenia research. *Schizophr. Bull.* **13**, 115–132.

Davis, K. L., Kahn, R. S., Ko, G., and Davidson, M. (1991). Dopamine in schizophrenia: A review and reconceptualization. *Am. J. Psychiatry* **148**, 1474–1486.

Johnson, K. M., Jr. (1987). Neurochemistry and neurophysiology of phencyclidine. *In* "Psychopharmacology: The Third Generation of Progress," (Meltzer, H. Y., ed.). Raven Press, New York.

Kane, J. M. (1987). Treatment of schizophrenia. *Schizophr. Bull.* **13**, 133–156.

Kaplan, H. I., and Sadock, B. J. (eds.) (1989). "Comprehensive Textbook of Psychiatry," 5th Ed. Williams & Wilkins, Baltimore.

Kapur, S., and Remington, G. (1996). Serotonin-dopamine interaction and its relevance to schizophrenia. *Am. J. Psychiatry* **153**, 466–476.

Kleinman, E. J., Casanova, M. F., and Jaskiw, G. E. (1988). The neuropathology of schizophrenia. *Schizophr. Bull.* **14**, 209–222.

Maas, J. W., Contreras, S. A., Miller, A. L., Berman, N., Bowden, C. L., Javors, M. A., Seleshi, E., and Weintraub, S. E. (1993). Studies of catecholamine metabolism in schizophrenia/psychosis—I. *Neuropsychopharmacology* **8**, 97–109.

Maas, J. W., Contreras, S. A., Miller, A. L., Berman, N., Bowden, C. L., Javors, M. A., Seleshi, E., and Weintraub, S. E. (1993). Studies of catecholamine metabolism in schizophrenia/psychosis—II. *Neuropsychopharmacology* **8**, 111–116.

McGlashan, T. H., and Carpenter, W. T., Jr. (eds.) (1988). Long-term followup studies of schizophrenia. *Schizophr. Bull.* **14**, 497–673.

Meltzer, H. Y. (1987). Biological studies in schizophrenia. *Schizophr. Bull.* **13**, 77–111.

Sherrington, R., Brynjolfsson, J., Petursson, H., *et al.* (1988). Localization of a susceptibility locus for schizophrenia on chromosome 5. *Nature* **336**, 164–167.

Seizure Generation, Subcortical Mechanisms

KAREN GALE
Georgetown University Medical Center

GLOSSARY

Amino acid neurotransmitters Amino acids released by specific neurons in order to transmit information across synaptic junctions to the postsynaptic target neuron. Some are excitatory (causing depolarization of the postsynaptic target, thereby increasing the chance of its firing), such as glutamate and aspartate, whereas others are inhibitory (causing hyperpolarization of the postsynaptic target, thereby reducing the chance of its firing), including γ-aminobutyric acid (GABA) and glycine. Each amino acid neurotransmitter exerts its action via highly specific and selective receptors located on the postsynaptic target neuron. The "NMDA receptor" refers to a subtype of excitatory amino acid receptor that is selectively activated by the drug *N*-methyl-D-aspartate

Basal ganglia Includes corpus striatum (caudate nucleus, putamen, and globus pallidus or entopeduncular nucleus) and substantia nigra. The caudate and putamen (neostriatum) receive input from widespread regions of the cortex and, after integrating this input, send projections out via relays in entopeduncular nucleus and substantia nigra. The basal ganglia are important for the initiation of movements, regulation of posture, and sensory–motor integration

Brain stem (also referred to as hindbrain) Part of the brain containing the midbrain, pons, and medulla. The part of the brain is especially important for controlling general arousal and regulating states of consciousness, waking and sleeping, breathing, and heart rate

Clonic convulsion Seizure manifestation consisting of repetitive, rhythmic jerking of the limbs, trunk, head, and/or facial muscles

Electroencephalographic recording Recording of variations in electrical potentials of the brain by means of electrodes placed on the surface of the scalp, skull, dura, or within deep structures of the brain. This activity is amplified and then recorded by the movement of a pen on a paper chart passing beneath it. The resulting record represents the summation of a variety of complex electrical potentials in the cortical or subcortical areas being monitored

Forebrain Part of the brain containing the telencephalon (cerebral cortex, corpus striatum, limbic system) and the diencephalon (thalamus and hypothalamus). In general, the most recently evolved areas of the brain are located in the forebrain

Limbic system Group of anatomically interconnected brain regions that regulates emotional experience and expression. Major forebrain components include hippocampal formation, septum, amygdala, olfactory bulbs, hypothalamus, piriform and entorhinal cortex, and nucleus accumbens. Also known as the "visceral brain," this system subserves functions essential for individual and species survival such as feeding, fight, and flight, mating, and caring of offspring. This system also plays a crucial role in learning and memory and goal-directed behavior

ENCYCLOPEDIA OF HUMAN BIOLOGY, Second Edition, VOLUME 7. Copyright © 1997 by Academic Press. All rights of reproduction in any form reserved.

Maximal electroshock Electroconvulsive shock, when applied at a current strength above that sufficient to cause tonic hindlimb extension in all subjects (usually rats or mice), is referred to as *maximal electroshock*

Subcortical All brain structures other than the cerebral cortex (i.e., neocortex). Although the cerebellar cortex is not properly "subcortical," it will be included in the present discussion

Tonic convulsion One type of motor manifestation of a generalized seizure, it consists of coordinated contractions of the musculature, especially evident in the limbs. Often it involves the contraction of extensor muscles leading to *tonic extension* of the forelimbs and/or hindlimbs

I. INTRODUCTION

The abnormal paroxysmal discharge of neurons in the central nervous system is referred to as a *seizure* and is usually detected by electroencephalographic (EEG) recording. If a seizure is confined to a group of neurons within a circumscribed brain region, it may not necessarily produce any behavioral symptoms. When seizure activity is conducted to distant brain regions, it is said to be propagated and the manifestations of the seizure depend on the particular neural circuits involved. Most forms of seizures involve activation of motor systems and often are referred to as convulsive seizures, as distinct from seizures that cause impairment of consciousness without motor activation. Motor manifestations of seizures range from automatisms and spasms to tonic or clonic jerking, often accompanied by impaired consciousness. The precise motor components and their patterns of expression are closely related to the brain areas engaged in the seizure. Seizures can be categorized by their behavioral symptoms and by the nature and extent of the brain areas involved in their generation.

When seizures occur recurrently and spontaneously, the term *epilepsy* is used to describe the disorder. Epilepsy develops as a secondary consequence of various chronic neurological conditions, especially when damage and scarring of brain tissue are present, as in the case of cerebral palsy, brain tumors, head injury, or cerebrovascular insults. The anatomical site of origin of the seizure is referred to as a focus and, when a defined focus exists, the pathways for seizure conduction are determined by the anatomic connections of that focus. [*See* Epilepsy.]

In the normal brain, acute convulsive seizures can be provoked by any of a number of conditions. Specific drugs are capable of inducing seizures, and such drugs are generally referred to as chemoconvulsants. Severe hypoglycemia (as can be produced by excess insulin), hypoxia, sleep deprivation, nutritional deficiencies (especially in vitamin B_6 or magnesium), fever, poisoning, and electroconvulsive shock all can precipitate seizure activity. Convulsive seizures are also one component of a withdrawal syndrome that occurs when continuous long-term use of alcohol, barbiturates, benzodiazepines, and certain other sedative–hypnotic drugs is abruptly discontinued.

This article focuses on convulsive seizures as they occur in animal models, because it is only in experimental animals that we have had the ability to probe and analyze subcortical brain regions that influence seizure initiation and/or propagation. Although it is not known to what degree seizure mechanisms in animals are equatable with those in humans, it is likely that significant similarities exist. This assertion is supported by the fact that many of the same acute conditions that provoke convulsive seizures in humans do the same in experimental animals, and the drugs that are anticonvulsant in humans are also effective in many of the experimental animal seizure models. Moreover, prolonged seizures have been documented to cause damage to brain tissue in both humans and experimental animals, and the anatomic pattern of this damage exhibits some cross-species similarities.

Based on studies in experimental animal models, a number of different brain regions have been implicated in the initiation and modulation of convulsive seizure activity. The particular brain regions and pathways that are engaged by a seizure vary according to how the seizure is elicited. For any seizure there are brain regions responsible for (1) initiating the seizure, (2) propagating the seizure activity, and (3) controlling or suppressing seizure activity once it has been initiated. At present, we do not have a complete neuroanatomical map of all these components for any particular seizure model. At best, we have a partial list of subcortical brain regions that have been experimentally identified as contributing to seizure initiation, propagation, and/or control in one or more experimental seizure models in animals.

II. FOREBRAIN REGIONS INVOLVED IN SEIZURE INITIATION: THE LIMBIC SYSTEM

The subcortical brain regions that have received the greatest attention in connection with propagated seizures are those associated with the limbic system. In particular, the amygdala and hippocampus are a part of circuits involved with the initiation of certain types of propagated seizures with motor manifestations. These seizures, characterized by automatisms and/or clonic movements of the mouth, face, and upper extremities, have been referred to as "limbic motor seizures" as they are accompanied by afterdischarges and electrographic seizure activity throughout the limbic system. Electrical stimulation of either amygdala or hippocampus (among other forebrain regions) when applied repeatedly over a period of several days can lead to *kindling,* the process by which a subconvulsant stimulus, when applied repeatedly, can come to trigger a convulsion. Once an animal has been kindled (e.g., by electrical stimulation in the amygdala), the animal will reliably exhibit convulsive seizures when stimulated (in the amygdala) and this response will be maintained indefinitely. Moreover, an animal kindled from one limbic region (e.g., amygdala) will require very few stimulations in another limbic region in order for the kindling to be transferred to that second region (e.g., hippocampus). The behavioral characteristics of these seizures are relatively constant, regardless of the limbic structure (septum, olfactory bulb, entorhinal cortex, amygdala, or hippocampus) used for the stimulation, suggesting that a common seizure-generating pathway or circuit can be engaged by stimulation of any of a number of sites in the limbic system. No single limbic region appears necessary for producing limbic motor seizures because attempts to interfere with these seizures by destroying various individual structures have been generally unsuccessful. Lesion studies must be viewed with extreme caution, however, because over hours or days, intact brain tissue adjacent to or connected with the damaged region often can change its activity to compensate for the damage. In some instances, this occurs because inhibitory neurons are damaged, thereby disinhibiting other groups of neurons outside the lesioned area. Therefore, a particular brain region may in fact play a crucial role in seizure initiation and/or propagation, even though seizure activity continues to occur following its removal. One way to minimize this problem is to use focal drug injections into the brain area of interest in order to rapidly and reversibly suppress the activity of the area. Anticonvulsant effects can then be evaluated before many of the adaptive and compensatory changes set in. [*See* Hippocampal Formation.]

A very small region within the deep anterior piriform ("prepiriform") cortex has been identified that triggers limbic motor seizures when stimulated by the application of minute quantities of certain drugs. This functionally defined area, coined the "area tempestas," is regulated by inhibitory and excitatory amino acid neurotransmitters in both the rodent and the monkey. Blockade of inhibitory neurotransmission [mediated by γ-aminobutyric acid (GABA)] or augmentation of excitatory neurotransmission (mediated by glutamate or aspartate) within this site in one hemisphere initiates bilaterally synchronous motor and electrographic seizures that resemble those evoked by kindling of limbic structures. The area tempestas is most likely part of a seizure-generating circuit that connects the piriform cortex, entorhinal cortex, hippocampus, thalamus, and amygdala. This area can also be kindled rapidly, requiring relatively few electrical stimulations (as is true for amygdala) and exhibiting an especially rapid rate of kindling transfer to/from amygdala. Moreover, evidence suggests that drug-induced inhibition of transmission within the area tempestas can significantly elevate the threshold for seizures triggered from amygdala in kindled rats. The area tempestas influences limbic system excitability via projections to the piriform cortex, perirhinal cortex, and entorhinal cortex, regions known to be particularly prone to the generation of epileptic seizure activity.

Whereas the area tempestas may be concerned with the *initiation* of limbic seizure activity, the substantia innominata is a region concerned with the *expression* of the motor components of limbic seizures. Inhibition of activity within this brain region prevents the convulsions evoked by amygdala stimulation in kindled animals but does not prevent the afterdischarge recorded electrically from the amygdala. It has been suggested that the substantia innominata serves as a relay between the amygdala and cortical areas participating in the organization of motor components of the seizure response. Additionally, the substantia innominata sends projections to midbrain regions that can influence motor systems via descending pathways (see Fig. 2).

The mediodorsal thalamus (MD) is a midline thalamic region that is intimately and reciprocally connected with most nuclei of the limbic system as well as several regions of temporal and frontal cortex. It is therefore ideally situated to funnel signals back and forth between several components of the forebrain seizure network, and to regulate these signals. Not surprisingly, inhibition of activity in MD reduces susceptibility to limbic seizures, perhaps by interrupting several neural loops that are needed to distribute the seizure discharge.

Another prominent region of the limbic network, the hippocampus (together with closely related nuclei referred to as the hippocampal system) has received a great deal of attention in the study of seizure generation in animals and humans. The local network of neurons and synapses *within* (i.e., intrinsic to) the hippocampus has been characterized and analyzed in slices of tissue removed from the brain and maintained in a dish (*in vitro*). The local neural network in the hippocampus is especially prone to entering into rhythmic bursting patterns of electrical discharge that resemble the activity occurring in association with a seizure (also see discussion under Section IX). This propensity of the circuits intrinsic to the hippocampus to generate seizure-like discharge is also present in intact humans and experimental animals. Accordingly, during the initial development of a seizure, the hippocampus is one of the first brain regions to exhibit electrical signs of seizure discharge. However, when seizure discharge is limited to the hippocampus itself, it is not associated with any behavioral or motor manifestations or other signs of clinical seizure response. One interpretation of the early hippocampal involvement is that the seizure starts in this region; this interpretation is especially popular in the field of epilepsy research and clinical practice. An alternative view is that the seizure is triggered elsewhere in the brain, and that the hippocampus, by virtue of its intrinsic circuitry is an especially responsive target. In experimental animals, the induction of local seizure discharge within the hippocampal circuitry can, in fact, *prevent* the propagation of seizures through the larger network of the limbic system. This means that activation of hippocampal circuitry may help to *resist,* not promote, seizure development. Thus, by being a highly sensitive seizure detector, the hippocampus can function as a "fuse" in the system, sending out signals that prevent the organization of seizure activity in the larger limbic network. A damaged hippocampus may lose the ability to serve such a protective role, which may explain why hippocampal damage associated with seizure disorders can worsen the outcome.

One type of surgical treatment for a subset of patients with severe seizure disorders involves the removal of brain tissue in the medial temporal lobe, including the hippocampus, surrounding areas (amygdala, perirhinal cortex and/or entorhinal cortex), and connecting fibers. Many of the limbic system connections illustrated in Fig. 2 are severed by this surgery, thus interrupting important routes of seizure propagation. As researchers develop a greater understanding of the specific pathways and connections that are crucial for generating a clinical seizure, it may be possible to target the surgical treatment to more limited and selective components of the limbic circuitry.

III. FOREBRAIN AND MIDBRAIN REGIONS INVOLVED IN SEIZURE PROPAGATION AND CONTROL: THE BASAL GANGLIA AND RELATED NUCLEI

The major components of the basal ganglia, i.e., the caudate-putamen (striatum), globus pallidus (entopeduncular nucleus in particular), and substantia nigra, have all been demonstrated to influence seizure susceptibility in numerous experimental seizure models. The best studied of these structures is the substantia nigra, the only region that has been examined in connection with more than 10 different experimental seizure models. Treatments that either increase inhibitory transmission (mediated by the neurotransmitter GABA) in substantia nigra or block excitatory transmission (mediated by glutamate or certain neuropeptides such as substance P) in this nucleus in both hemispheres prevent or attenuate convulsive seizures induced by several chemoconvulsants (pilocarpine, kainic acid, bicuculline, and flurothyl, among others), by maximal electroshock, by kindling of amygdala, by drug application into the area tempestas, and by acoustic stimulation in rats susceptible to audiogenic seizures due to either ethanol withdrawal or genetic determinants. In addition, inhibition within the substantia nigra decreases susceptibility to nonconvulsive electrographic seizures that occur either spontaneously (in genetically predisposed strains of animals) or as a consequence of treatment with drugs such as pentylenetetrazol in low doses.

The integrity of the substantia nigra is not required for seizure induction, indicating that this structure is not part of a crucial seizure-conducting pathway. Instead, the substantia nigra is part of a seizure-suppressing circuit that becomes engaged by seizure discharge. Viewed in this way, the substantia nigra and associated nuclei of the basal ganglia act to maintain a homeostatic balance of brain excitability by creating a resistance to seizure spread and generalization.

Certain key structures of the forebrain and midbrain may work in concert with the substantia nigra in this seizure-resisting capacity. The striatum is one major source of neural input to the substantia nigra, whereas the superior colliculus (deep layers) is one important target of neural projections coming from the nigra. One of the most prominent pathways connecting the striatum and substantia nigra is inhibitory and utilizes GABA as its transmitter. Likewise, the substantia nigra sends an inhibitory GABA-containing projection to the superior colliculus. Consequently, it might be expected that the stimulation of the neurons in the striatum that give rise to the GABA inputs to the substantia nigra would be anticonvulsant (see Fig. 1) because this would enhance GABA transmission in the substantia nigra. The experimental evidence supports this proposal; electrical or drug-induced excitatory stimulation in striatum tends to exert an anticonvulsant action in the experimental seizure models that have been examined so far.

Within substantia nigra, inhibitory GABAergic transmission acts to suppress the activity of output projections to superior colliculus. This relationship predicts that blockade of GABA transmission in the nigral projection target area of superior colliculus should be anticonvulsant. Again, the experimental evidence is consistent with expectations. Blockade of GABA receptor-mediated transmission in the deep layers of superior colliculus is anticonvulsant against both clonic and tonic forms of convulsive seizure activity, as well as against nonconvulsive spike-and-wave electrographic seizure discharge. This illustrates a fundamental principle of central nervous system organization: disinhibition. Inhibitory transmission within the substantia nigra acts to reduce the activity of the nigral outputs to the colliculus (which are themselves inhibitory), resulting in the withdrawal of inhibition, or *disinhibition,* of neuronal targets in the superior colliculus (see Fig. 1). Presently, it is not understood how neuronal activity within the superior colliculus acts to impede seizure progression. Some of the colliculus neurons that are activated by nigral disinhibition may be those that project to the brain stem reticular formation where they can relay with both ascending and descending neural pathways. Stimulation of certain regions of the reticular formation is capable of interfering with the development of

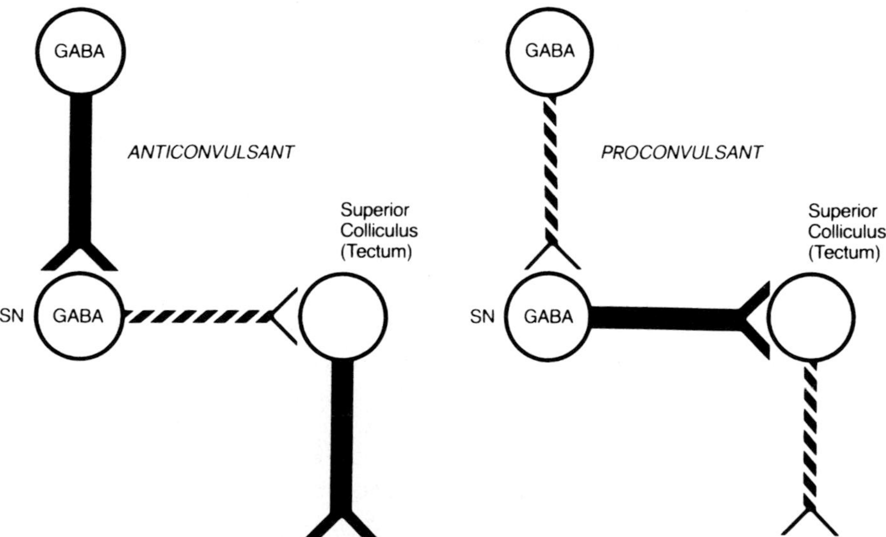

FIGURE 1 Serial inhibitory neuronal links in the basal ganglia. From the striatum to the substantia nigra (SN), a GABA-containing neural pathway projects. From substantia nigra to superior colliculus (tectum), there is also a GABA-containing neural pathway. Within the substantia nigra, the first of these pathways (striato-nigral) inhibits the second (nigro-tectal). When more GABA is released by the first pathway, there is more inhibition in substantia nigra and, consequently, the activity of the nigro-tectal pathway is suppressed (broken lines). The suppression of this second pathway results in the removal of inhibition (i.e., disinhibition) in the superior colliculus. This is associated with anticonvulsant effects (see text). The opposite effect (i.e., the effect of decreasing the activity of the striatonigral GABA pathway) is portrayed on the right side of the figure. Heavy solid lines indicate increased neural firing. Broken lines indicate decreased firing.

a synchronized pattern of neuronal discharge in cortex. By engaging these desynchronizing influences of the reticular formation, the superior colliculus (Fig. 1) could disrupt the generation of seizures, which require synchronized neuronal bursting.

The entopeduncular nucleus, also known as the medial segment of the globus pallidus, is similar to the substantia nigra in terms of function, morphology, and neuroanatomic connections. Together with the substantia nigra, this nucleus relays neuronal outputs from the striatum. It is therefore not surprising that the same treatments that suppress seizure propagation when placed in the substantia nigra exert a similar action in the entopeduncular nucleus. This brain region has not yet been studied in the multiplicity of seizure models that have demonstrated a regulatory role for substantia nigra. Nevertheless, it appears that the influence of the entopeduncular nucleus may be limited to limbic motor seizures. This region does not share the widespread influence exerted by substantia nigra.

The influence of the basal ganglia on seizure generalization is not confined to the motor manifestations of the seizures. In instances in which electrographic signs of seizures have been monitored, anticonvulsant manipulations of basal ganglia structures induce a corresponding suppression of electrographic seizure discharge in cortical and subcortical structures. This is consistent with the role of basal ganglia circuits in the control of cortical excitability and the regulation of synchronization of neuronal discharge in widespread regions of the forebrain.

IV. DIENCEPHALIC REGIONS INVOLVED IN THE PROPAGATION OF PENTYLENETETRAZOL-INDUCED SEIZURES

Seizures induced in animals by the chemoconvulsant agent pentylenetetrazol (PTZ) have been used extensively as one model for screening potential antiepileptic drugs. Metabolic mapping studies have shown that neural projections from the mammillary bodies of the hypothalamus to the anterior thalamus exhibit markedly increased activity during PTZ-induced seizures. Lesions of this mammilothalamic pathway can abolish both the behavioral and the electrographic expression of PTZ seizures, as can focal microinjections of drugs that enhance GABA transmission in the anterior thalamus. The mammilothalamic tract is believed to be part of a subcortical pathway that interconnects the thalamus with the midbrain tegmentum, a route that may be significant for the behavioral expression of the seizures. The cortical projections to and from the thalamus are most likely responsible for the cortical electrographic manifestations of the seizures. It is noteworthy that lesions of the thalamus have been found to suppress cortical seizure discharge in other seizure models in which seizure activity is elicited by focal drug or electrical stimulation of the cortex itself.

V. HINDBRAIN REGIONS IMPLICATED IN SEIZURE INITIATION

Convulsive seizures can be elicited by appropriate stimulation of brain stem regions, but these seizures are readily distinguished from those elicited by the stimulation of forebrain structures. In the rodent, forebrain-evoked seizures usually involve clonic movements of the face, neck, and forelimbs, whereas hindbrain-evoked seizures lack this feature and instead consist of explosive running and bouncing clonus and/or tonic extension of forelimbs and hindlimbs. Electrical stimulation of the reticular formation of the midbrain, pons, or medulla can elicit self-sustained tonic convulsive seizures. These seizures do not necessarily engage forebrain circuitry and, moreover, their expression does not require intact connections with the forebrain. Animals with complete transection of the brain (precollicular transections), disconnecting forebrain from hindbrain circuits, are fully capable of exhibiting tonic seizures as well as explosive running and bouncing clonus in response to brain stem electrical stimulation, systemic injection of chemoconvulsants such as PTZ, or electroconvulsive shock applied via corneal or ear-clip electrodes. However, the seizure responses associated with forebrain limbic circuitry (see above) cannot be elicited in animals with precollicular transections that disconnect the forebrain from the hindbrain.

A brain stem site from which running/bouncing clonic seizures can be triggered is the inferior colliculus. Electrical stimulation of the inferior colliculus in the rat, as well as focal application of drugs that stimulate excitatory amino acid transmission in this site, elicits sudden, explosive bouts of leaping, running, and bouncing on all four limbs. Blockade of GABA transmission in the inferior colliculus will have the same convulsive effect. This behavior resembles that associated with sound-induced (audiogenic) seizures in susceptible rodent strains.

The inferior colliculus, as part of the auditory sensory pathways, is probably a crucial afferent relay station for converting acoustic stimulation into convulsive discharge, and subconvulsant stimulation of this region can render otherwise normal animals susceptible to audiogenic seizures. Direct chemical or electrical stimulation of this brain region does not typically evoke *tonic* convulsions, unless ascending noradrenergic projections have been compromised by lesions or depletion of norepinephrine.

VI. HINDBRAIN REGIONS REQUIRED FOR SEIZURE EXPRESSION

Lesions that produce bilateral damage in the pontine reticular formation block tonic convulsions induced by electroshock and chemoconvulsants. At the same time, these lesions do not interfere with the facial and forelimb clonus associated with limbic seizures. A key structure damaged by the pontine lesions is the nucleus reticularis pontis oralis (RPO). Selective lesions of the RPO will also prevent the tonic convulsive components of chemoconvulsant and electroshock seizures. Together with the stimulation effects described earlier, these observations demonstrate that the brain stem is both necessary and sufficient for the initiation and development of tonic convulsions.

VII. HINDBRAIN PROJECTIONS ASSOCIATED WITH AUTONOMIC ACTIVITY: INFLUENCE ON SEIZURE SUSCEPTIBILITY

Afferent vagal activation can influence cortical excitability, producing electroencephalographic synchronization or desynchronization depending on the stimulation parameters. It has been demonstrated in both animals and humans that electrical stimulation of the afferent vagus can attenuate a variety of seizures, ranging from limbic motor seizures to tonic convulsions. The nucleus tractus solitarius (NTS) is the major relay for vagal afferents, and this nucleus, located in the brain stem, probably mediates the anticonvulsant action of vagal stimulation. The NTS has a widespread influence on subcortical and cortical circuits, most likely exerted via projections to the brain stem reticular formation (especially the parabrachial area), hypothalamus, and amygdala. It remains to be determined which of the projections from NTS are crucial for influencing seizure susceptibility and seizure progression.

VIII. CEREBELLUM

Removal of the cerebellum selectively inhibits tonic components of convulsive seizures, such as tonic hindlimb extension produced by maximal electroshock. Cerebellectomy does not interfere with the expression of clonic seizure components and may actually facilitate focal seizure discharge in cortex and certain types of clonic seizure activity.

Consistent with the results of cerebellar removal, damage to the cerebellar peduncles of both hemispheres also blocks tonic convulsions. The cerebellar peduncles contain the major efferent projections from the cerebellum to the brain stem reticular formation and other motor nuclei.

IX. INSIGHTS FROM METABOLIC AND ELECTROGRAPHIC MAPPING STUDIES

The information discussed earlier derives largely from studies utilizing selective stimulation or lesions of specific brain regions or pathways. Another approach to elucidating brain regions involved in seizure generation involves the use of radiolabeled 2-deoxyglucose (2-DG). This method for estimating brain glucose utilization by contact autoradiographic analysis of brain sections allows for a survey of all brain regions in any given experimental animal. In addition to its use in animal studies, this metabolic mapping procedure has been applied to human seizures by using positron emission tomography (PET) scanning to detect the distribution of radiolabeled 2-DG. In human studies, analyses concentrate on patterns of metabolic alterations in cortical structures because the resolution of most PET scans is not sufficient to allow a detailed evaluation of subcortical regions. The contact autoradiographic analyses used for animal studies, however, are sufficient to resolve even some of the smallest subcortical nuclei.

During seizures evoked by the stimulation of structures within the limbic system, the pattern of 2-DG uptake depends on the severity of the seizure. Partial seizures are associated with increased 2-DG uptake in the closest target regions in which the stimulated brain region directly projects. During full seizures, in which a complete pattern of convulsive seizure activity is manifest bilaterally, 2-DG uptake is increased in hippocampus, amygdala, nucleus accumbens, substantia nigra, entorhinal cortex, substania innominata, and the anterior and periventricular nuclei of the thalamus (see Fig. 2 for some of the interconnections between these regions).

Seizure-evoked activation of 2-DG uptake limited to the hippocampus is associated with staring and behavioral arrest. When bilateral activation of amygdala, substantia nigra, and thalamic nuclei occurs, strong facial and forelimb clonus with rearing is usually evident. A consistent finding from all such studies is that seizures propagate along known neuroanatomical pathways. Moreover, the neuroanatomical pathways that are metabolically activated during a seizure correspond to the circuitry connected with the stimulated region, the type of seizure evoked (e.g., limbic), and the type of convulsive behavior produced.

When seizures are elicited by the stimulation of discrete regions of neocortex, the resulting focal motor seizures are associated with a pattern of 2-DG uptake and behavior that is distinct from the limbic seizures described earlier. In this case, the basal ganglia (caudate-putamen, globus pallidus, substantia nigra) and paraventricular and ventral thalamic nuclei are involved, even with mild and unilateral seizure activity. Larger areas of thalamus and contralateral frontal cortex become involved when seizures are more severe, and the activation of medial thalamic nuclei is associated with bilateral seizure spread. It has been suggested that when the seizure discharge activates a sufficient neuronal population in medial thalamus, transsynaptic target areas in medial-frontal or orbito-frontal cortex can become bilaterally recruited.

The results obtained using metabolic mapping are generally in agreement with electrographic recording from subcortical structures. Although electrographic recording can only sample activity from a limited number of brain regions in any one subject, it has the advantage of providing information concerning the temporal sequence of seizure spread that is not provided by metabolic mapping. Typically, the first limbic system region to exhibit electrographic signs of seizure discharge is the hippocampus, perhaps because this region is especially susceptible to developing synchronous bursting patterns of neuronal discharge. Although the hippocampus can readily generate electrographic seizure activity, seizures evoked within this brain region do not easily spread to engage other brain regions. As long as the seizure discharge is confined to hippocampal circuits, there may be little or no signs of motor seizure activity. Thus, the fact that in various experimental models, as well as in human seizure disorders, the earliest signs of seizure activity are often detected in hippocampus does not necessarily mean that this is the site of seizure initiation. In fact, experimental evidence indicates that the hippocampus usually manifests the first signs of electrographic seizure discharge even when seizures are evoked from elsewhere in the brain (e.g., the area tempestas in the deep anterior piriform cortex). Thus, the hippocam-

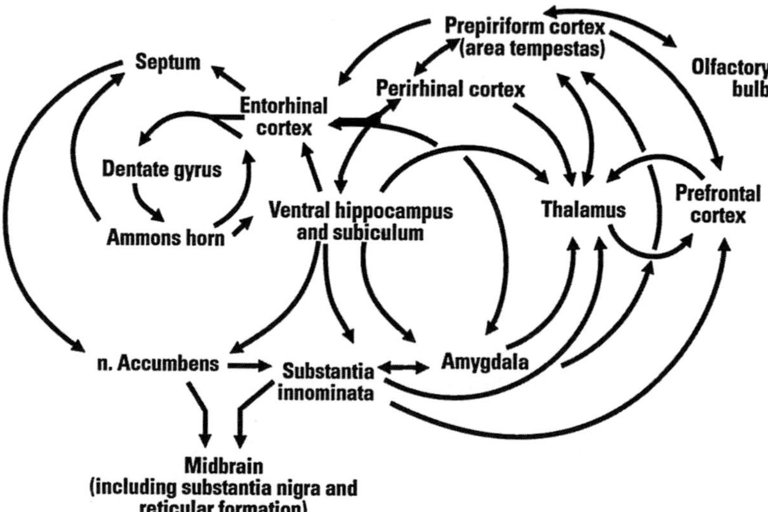

FIGURE 2 Schematic diagram of some limbic system regions and their connections that are involved in seizure generation and propagation.

pus appears to be an easily recruited target of epileptogenic stimuli triggered from sites located elsewhere in the forebrain.

X. EFFECT OF SEIZURES ON IMMEDIATE EARLY GENE EXPRESSION IN BRAIN REGIONS

Activation of neural cells can lead to the induction of new programs of gene expression as part of the adaptive and trophic processes that continually occur in the brain. For one class of genes, increased transcription occurs within minutes of stimulation; this class is referred to as immediate early genes (IEGs). These genes, which include c-*fos*, c-*jun*, jun-B, and zif/268, among others, are thought to control the expression of late response genes whose products influence specific structural and functional aspects of the neurons in which they are expressed. Some of the IEGs encode transcription factors that bind to DNA and can facilitate (or suppress) expression of nearby genes. [*See* DNA and Gene Transcription.]

Convulsive seizures provoke an increase in expression of the c-*fos* gene and other IEGs in neural cells in the brain. It appears that, at least for certain neuronal populations, induction of IEG expression can be used as a marker for increased excitatory activation of neuronal cells during convulsive seizures.

With seizures evoked from area tempestas or by systemic treatment with chemoconvulsants such as kainic acid, pentylenetetrazol, and picrotoxin, the most pronounced increases in mRNA for c-*fos* are seen in the piriform cortex, perirhinal amygdala, entorhinal cortex, olfactory bulbs, and throughout the hippocampus (Color Plate 24). A similar pattern of increase occurs with mRNAs for zif/268, c-*jun*, and jun-B. There is a consistent lack of detectable increase in IEG expression in superior and inferior colliculus, geniculate bodies, substantia nigra, and cerebellar cortex in these circumstances. Increases in IEG expression reach a peak at 30 min after seizure initiation; by 4 hr, mRNA levels for the IEGs return to baseline.

Based on the electrographic and behavioral characterization of convulsive seizures evoked from the inferior colliculus, we would not expect to see marked activation of IEG expression in forebrain areas in association with these seizures. Stimulation of inferior colliculus by focal application of bicuculline evokes explosive running–bouncing convulsions in the absence of forebrain electrographic seizure activity, and this type of convulsion, as discussed earlier, depends on hindbrain, but not forebrain, circuitry for its expression. In fact, little or no elevation of IEG expression is detected in hippocampus, entorhinal cortex, or piriform cortex following convulsions evoked from inferior colliculus. In general, no marked increases in IEG expression are evident in any forebrain area thus far surveyed. Similarly, there are no striking increases in c-*fos* expression in forebrain regions in response to audiogenic seizures; this is another seizure model predominantly associated with hindbrain mechanisms. Interestingly, in the audiogenic seizure model, several subcortical auditory nuclei, including inferior colliculus, exhibited marked increases in c-*fos* expression. As these nuclei do not exhibit increased c-*fos* expression in association with seizures evoked by chemical stimulation of inferior colliculus, it is likely that they are part of circuits involved in processing of the seizure-eliciting sensory (i.e., auditory) input and are not substrates involved in seizure propagation or expression.

It is interesting to compare the results obtained with 2DG accumulation and IEG expression. With seizures associated with the forebrain, both techniques reveal marked activation of the olfactory bulb, piriform and entorhinal cortex, and hippocampus. These areas are interconnected via excitatory synaptic links, which, based on these mapping studies, appear to be highly activated by the seizures. There are also several regions in which marked seizure-evoked increases in 2DG accumulation occur without obvious changes in IEG expression. For example, the substantia nigra and selected thalamic nuclei have dramatic increases in the 2DG signal while appearing unchanged when examined for expression of c-*fos* or other IEGs. In these instances, increased 2DG accumulation is probably associated with the activation of inhibitory inputs into these structures, such as the inhibitory GABA-containing terminals innervating the substantia nigra. Increased inhibitory input to a region would not be expected to evoke an increase in IEG expression in postsynaptic cells, thus accounting for the discrepant patterns with the two techniques.

XI. NEUROTRANSMITTER MECHANISMS

It has become clear that no single neurotransmitter can be related to the genesis of seizures. Rather, neurotransmitters and neuromodulators interact in a complex pattern to determine seizure susceptibility and the characteristics of seizure spread. This interaction is anatomically specific and depends on the configuration and identity of the synapses in a given brain locus. γ-Aminobutyric acid (GABA), the most prevalent inhibitory transmitter in brain, exerts an anticonvulsant influence in certain brain regions whereas in other regions this same neurotransmitter acts in a proconvulsant fashion. Likewise, glutamate, the most prevalent excitatory transmitter in brain, has both seizure-inducing and seizure-suppressing actions in different brain areas.

At the same time, pharmacological studies have demonstrated that the overall net effect of enhancing GABA transmission in the central nervous system (CNS) is to reduce convulsive seizure susceptibility and, conversely, a general reduction in GABA transmission can evoke convulsive seizures. In humans, the drug γ-vinyl-GABA, or vigabatrin, an agent that interferes with the metabolic breakdown of GABA, thereby elevating the GABA content in brain, has clinically effective anticonvulsant activity. Conversely, isoniazid (a drug used in the treatment of tuberculosis) interferes with GABA synthesis and, consequently, can cause convulsive seizures. It is believed that the anticonvulsant action of the benzodiazepines (e.g., valium) in both experimental animals and humans is due to the ability of these drugs to enhance GABA-mediated neurotransmission. However, general (systemic) treatment with drugs that block glutamate transmission is anticonvulsant, whereas the dominant action of drugs that stimulate glutamate receptors is to induce seizure activity.

Another neurotransmitter that has been found to have a general influence on seizure susceptibility is norepinephrine. Depletion of this transmitter, or lesions of a major norepinephrine-containing cell group, the locus coeruleus, usually causes a decrease in seizure threshold. Likewise, drugs that block CNS receptors for norepinephrine (γ receptors in particular) can lower seizure threshold in experimental animals as well as in humans. This may account for the increased seizure susceptibility occasionally observed in patients treated chronically with certain antipsychotic drugs such as phenothiazines.

Opioid drugs have complex influences on seizure susceptibility. Depending on the mechanism of seizure induction, morphine and other compounds that stimulate the μ opiate receptors exert either anticonvulsant or proconvulsant actions: anticonvulsant effects are obtained against electroshock seizures, whereas proconvulsant effects can be seen with most chemoconvulsant seizures. Stimulation of μ opiate receptors can induce electrographic seizure discharge in hippocampus, and when placed directly into the area tempestas or into the ventral hippocampus, convulsive seizures are evoked. When placed into the substantia nigra, morphine and related opioid peptides produce anticonvulsant actions against electroshock seizures.

In experimental animals, high doses of drugs that stimulate muscarinic cholinergic transmission tend to induce convulsive seizures. One such agent, pilocarpine, has been used to create an animal model of limbic motor seizures. These seizures are especially sensitive to blockade by inhibition of activity in the area tempestas (see earlier discussion). Since muscarinic stimulation in the area tempestas can evoke convulsive seizures, it is possible that this region is an important site of origin of seizures evoked by pilocarpine or similar agents. In addition, stimulation of muscarinic receptors by drug treatment directly in amygdala can be used instead of electrical stimulation to produce kindling of this region.

Finally, it should be recognized that convulsive seizures do not necessarily require activation of brain circuitry for their occurrence. It is possible to induce a type of tonic convulsion by a direct action on spinal cord neurotransmission. This is best exemplified by the convulsant drug strychnine, which acts by blocking receptors for glycine, an inhibitory neurotransmitter in the spinal cord. Strychnine induces tonic convulsions without the appearance of cerebral electrographic seizure discharge and without any other convulsive components. [*See* Neurotransmitter and Neuropeptide Receptors in the Brain.]

BIBLIOGRAPHY

Avoli, M., Gloor, P., Kostopoulos, G., and Naquet, R. (eds.) (1990). "Generalized Epilepsy," Neurobiological Approaches." Birkhauser, Boston.

Dichter, M. (ed.) (1988). "Mechanisms of Epileptogenesis: From Membranes to Man." Plenum, New York.

Gale, K. (1989). GABA in epilepsy: The pharmacologic basis. *Epilepsia* **30**(Suppl. 3), S1–S11.

Gale, K. (1985). Mechanisms of seizure control mediated by gamma-aminobutyric acid: Role of the substantia nigra. *Fed. Proc.* **44**, 2414–2424.

Gale, K. (1992). Subcortical pathways involved in seizure generation. *J. Clin. Neurophysiol.* **9**, 264–277.

Schwartzkroin, P. A. (1993). "Epileptic Seizures: Models and Mechanisms." Cambridge University Press, New York.

Wada, J. A. (ed.) (1981). "Kindling," Vol. II. Raven Press, New York.

Wada, J. A. (ed.) (1986). "Kindling," Vol. III. Raven Press, New York.

Wada, J. A. (ed.) (1990). "Kindling," Vol. IV. Plenum Press, New York.

Selenium in Nutrition

GERALD F. COMBS, JR.
Cornell University

GLOSSARY

Lipid peroxidation Process of oxidative degradation of polyunsaturated lipids, involving the generation in the vicinity of their 1,4-pentadiene structures of carbon-centered free radicals followed by the attack on those centers by molecular oxygen and the subsequent formation of chain cleavage products

Keshan disease Juvenile cardiomyopathy associated with severe selenium deficiency and prevalent in areas of endemic selenium deficiency in China

Selenium Group VIA element that is an essential nutrient for animals and humans, in which it functions as the active center of the antioxidant enzyme glutathione peroxidase

Selenocysteine Selenium-containing amino acid analog of cysteine, in which form selenium occurs in all known selenoenzymes

Selenoenzyme Enzyme in which selenium (as selenocysteine) is an essential component; these enzymes include several selenium-dependent glutathione peroxidases and type I iodothyronine 5′-deiodinase

Vitamin E Membrane-resident factor with antioxidant function because of its ability to reduce free radical species without propagating lipid peroxidation

NOT UNTIL THE LATE 1950s WAS THE ELEMENT selenium (Se) thought to play a role in normal metabolism. Until that time, the biomedical significance of this metalloid had been recognized only for its toxic properties. However, in 1957 it was discovered that trace amounts of Se could alleviate necrotic liver disease and capillary leakage in vitamin E-deficient animals, suggesting that Se spared the need for that fat-soluble vitamin. Resarch in the 1960s culminated in the recognition that Se is more than just a vitamin E-sparing factor, that it is an essential nutrient required for the antioxidant enzyme glutathione peroxidase. Since that time, an increasing understanding has emerged of the metabolic functions and health implications of this trace element. At present, Se is generally regarded as having prime importance in the metabolic protection from oxidative stress, with special relevance to disease of the heart muscle and drug metabolism. In addition, the results of animal tumor model studies and several human epidemiological investigations support the hypothesis that Se may have a role in the protection against carcinogenesis. It is the purpose of this review to summarize present understanding of the role of Se in human nutrition and health.

I. SELENIUM IN FOODS AND HUMAN DIETS

A. Selenium in Foods

The Se contents of foods vary widely due mainly to the amount of biologically available Se in the environment (e.g., the soluble Se content of the soil for plant species; the biologically available Se content of the diet for animal species). Because of the intimate relationship between plants and animals in food chains, the Se contents of foods from both plant and animal origins tend to be greatly influenced by the local soil Se environment. Thus, foods of all types tend to show

ENCYCLOPEDIA OF HUMAN BIOLOGY, Second Edition, VOLUME 7. Copyright © 1997 by Academic Press. All rights of reproduction in any form reserved.

geographic patterns of variation in Se content reflecting, in general, local soil Se conditions. Because almost all Se in plant and animal tissues is bound to proteins, the Se contents of foods also tend to be correlated with its protein content.

Variation in the Se contents of foods due to geochemical differences is readily seen by comparing the Se contents of like foods from different countries. For example, whole wheat grain may contain more than 2 ppm Se (air-dry basis) if produced in the Dakotas of the United States, but as little as 0.11 ppm Se if produced in New Zealand, and only 0.005 ppm Se if produced in Shaanxi Province, China. On a global basis, foods with the lowest Se contents are found in the low-Se regions of China, in particular, the provinces of Heilongjiang, northern Shaanxi, and Sichuan. Ironically, foods containing the greatest concentrations of Se have also been found in the same country, though in different locales. Indeed, endemic selenosis of animals and humans has been identified in a few mountainous communities in Hubei and southern Shaanxi provinces.

The Se contents of foods of animal origin depend largely on the Se intakes of livestock. Food animals raised in regions with low-Se feeds deposit relatively low concentrations of the mineral in their edible tissues and products (e.g., milk, eggs), whereas animals raised with relatively high Se nutriture yield food products with much greater Se concentrations. Because livestock needs Se to avoid debilitating deficiency syndromes, Se (usually in the form of Na_2SeO_3) is used as a feed supplement in animal agriculture in many parts of the world. This practice became widespread in North America and Europe only within the last 10–15 years and has reduced what would otherwise be a stronger geographic variation in the Se contents of animal food products.

Within the normal ranges of Se supplementation of livestock diets, muscle meats from most species tend to contain 0.3–0.4 ppm Se (fresh weight basis). Organ meats usually accumulate greater concentrations of Se; the livers of most species generally contain about four times as much Se as skeletal muscle, and the kidneys of steers, lambs, and swine have been found to accumulate 10–16 times the amounts in muscle.

B. Selenium in Human Diets

Because of differences in geography, agronomic practices, and food availability and preferences, most of which are difficult to quantify, evaluations of Se intakes of specific human population groups are often not precise. General comparisons can be made, however, of the Se contents of different food supplies by using the average Se concentrations determined within specific major classes of foods in different locales. Table I presents the typical Se contents of the major classes of foods from several countries. For the most part, these values are based on actual analyses of foods from each country.

The average per capita daily Se intakes of adults in various countries of the world have been estimated (Table II). They vary widely among different regions, the lowest being only 7–11 μg Se per person per day in the areas of Se-responsive human diseases in China. In contrast, the Se intakes of residents of countries with well-recognized endemic Se deficiency disorders in livestock (i.e., Finland and New Zealand) are estimated to be at least threefold those in the Se-deficient regions of China. Residents of the so-called low-Se parts of the United States (e.g., Ohio and the southeastern seaboard) have estimated Se intakes approximately two- to fivefold those of Finns or New Zealanders.

The most important sources of Se in the diets for most people are cereals, meats, and fish. Dairy products and eggs contribute small amounts of Se to the total intakes in most countries, although these can represent large percentages of the total Se intakes in countries where the rest of the diet provides little Se (e.g., Finland, New Zealand). Vegetables and fruits are uniformly low in Se (when expressed on a fresh weight basis) and provide only small amounts (less than 8% of the total intake) of the mineral in most human diets.

Core foods for Se in U.S. diets have been identified in a USDA study that compiled data from the USDA 1977–1978 Nationwide Food Consumption Survey with published values for the Se contents of American foods. That analysis revealed that five foods (beef, white bread, pork, chicken, and eggs) contributed about 50% of the total Se in the "typical" American diet, and that 80% of the total dietary Se was provided by a core of only 22 foods (Table III).

Differences in patterns of food consumption, whether general ones due to cultural influences or specific ones due to personal preferences and food availability, can significantly affect Se intake. In view of this potential variation in Se intake, it is not surprising to find considerable variation among individuals who are able to select their own patterns of food consumption. In a study of the Se intakes of free-living individuals in Maryland, the mean daily intake of Se was found to be 81 μg per person; however,

TABLE I
Typical Se Contents of Major Classes of Foods from Several Countries

Food	U.S.A.	England	W. Germany	Finland	New Zealand	China			Venezuela
						Se deficient[a]	Moderate	High-Se[b]	
Cereal products	0.03 –0.66	0.02 –0.53	0.03 –0.88	0.005–0.12	0.004–0.09	0.005–0.02	0.017–0.11	1.06– 6.9	0.132–0.51
Vegetables	0.001–0.10	0.01 –0.09	0.04 –0.10	0.001–0.02	0.001–0.02	0.002–0.02	0.002–0.09	0.34–45.7	0.002–2.98
Fruits	0.002–0.01	0.005–0.01	0.002–0.04	0.002–0.03	0.001–0.004	0.001–0.003	0.005–0.04	—	0.005–0.06
Red meats	0.05 –0.27	0.05 –0.14	0.13 –0.28	0.01 –0.07	0.01 –0.04	0.01 –0.03	0.05 –0.25	—	0.17 –0.83
Organ meats	0.43 –1.90	0.20 –2.46	0.09 –0.95	0.06 –1.71	0.05 –2.03	0.05 –0.10	0.05 –1.00	—	0.36 –0.83
Poultry	0.04 –0.15	0.05 –0.15	0.05 –0.15	0.05 –0.10	0.05 –0.10	0.02 –0.06	0.05 –0.10	—	0.10 –0.70
Fish	0.19 –1.9	0.10 –0.61	0.24 –0.53	0.18 –0.98	0.03 –0.31	0.03 –0.20	0.10 –0.60	—	0.32 –0.93
Milk products	0.01 –0.24	0.01 –0.08	0.01 –0.10	0.01 –0.09	0.003–0.025	0.002–0.01	0.01 –0.03	—	0.11 –0.43
Eggs	0.06 –0.20	0.05 –0.20	0.05 –0.20	0.10 –0.20	0.24 –0.98	0.02 –0.06	0.05 –0.15	—	0.50 –1.5

[a]Keshan disease-endemic areas.
[b]Areas of endemic selenosis.

17% of diets provided less than 50 μg Se per person per day, whereas 5% provided more than 150 μg Se per person per day.

Differences in Se consumption by women in different parts of the world are reflected as differences in the Se contents of human milk and, therefore, in the Se intakes of breast-fed infants (see Table II). Large differences in the Se nutriture (i.e., estimated intakes from 7.5 to 212 μg per infant per day) are seen. However, it is apparent that the differences in Se intakes are generally much less than those of adults in the same countries. Limited analyses of the Se contents of human milk in the low-Se areas of China indicate that the Se intakes of nursing infants, particularly in areas of endemic Keshan disease, may be as little as 33–40% of those of breast-fed infants in Finland and New Zealand.

C. Bioavailability of Dietary Selenium

The utilization of dietary Se involves several physiological and metabolic processes that convert a portion of ingested Se to certain metabolically critical forms that are necessary for normal physiological function. Ingested Se (normally in the form of Se-containing proteins and inorganic compounds) is subject to several potential losses en route to the metabolic production of its metabolically active form(s). These losses include those associated with the digestion and enteric absorption of ingested Se; thus Se compounds that are insoluble under conditions of the lumenal environment of the small intestine, as well as Se-containing proteins of low digestibility, will pass through the animal to be eliminated in the feces. It is probable that, under normal circumstances, there is only a small enterohepatic circulation of absorbed Se and, therefore, that fecal Se represents only that amount of ingested Se that was not absorbed. In general, the apparent absorption of Se in foods, inorganic compounds, and selenoamino acids is good (ca. 70%); however, it is highly variable both between and within single sources.

Not all absorbed Se is physiologically important. Some is metabolized to methylated forms that are

TABLE II
Estimated Per Capita Daily Intakes of Se (μg) of Adults and Infants in Several Countries

Age group	U.S.A.	England	W. Germany	Finland	New Zealand	China			Venezuela
						Se deficient[a]	Moderate	High-Se[b]	
Adults	60–220	50–120	60–150	30–100	30–80	7–11	60–120	750–4990	200–350
Infants, 3 months	12–18	10–15	18–24	5–10	7–9	2–3	12–18	180–250	—

[a]Keshan disease-endemic areas.
[b]Areas of endemic selenosis.

TABLE III
Core Foods for Se in American Diets

Rank	Food	Percentage contribution to total Se intake	
		Individual	Cumulative
1	Beef	17.2	17.2
2	White bread	14.2	31.4
3	Pork/ham	8.2	39.6
4	Chicken	6.5	46.2
5	Eggs	4.8	51.0
6	White rolls	4.0	55.0
7	Whole wheat bread	3.3	58.3
8	Noodles, pasta, etc.	3.0	61.3
9	Whole milk	2.8	64.1
10	Canned tuna	2.1	66.2
11	Milk, 2% butterfat	1.8	68.0
12	White rice	1.7	69.7
13	Macaroni and cheese	1.3	71.0
14	Luncheon meat	1.2	72.2
15	Spaghetti w/meat sauce	1.1	73.3
16	Mayonnaise	1.1	74.4
17	Meat loaf (beef)	1.0	75.4
18	Hamburger on bun	1.0	76.4
19	Oatmeal	1.0	77.4
20	Cracked wheat bread	1.0	78.4
21	Rye bread	1.0	79.4
22	Turkey	.9	80.3

readily excreted. For example, trimethyl selenonium cation comprises 10–20% of urinary Se, and the volatile dimethyl selenide is readily excreted across the lung in expired air, although the production of the latter species is probably important only under conditions of high intakes of Se. In addition to these forms, Se is normally metabolized to several species, most notably the Se analogs of the S-containing amino acids and, hence, to Se-containing polypeptides and proteins. Of these, the Se-dependent glutathione peroxidase (GPX) is the only physiologically critical species known. However, several other Se proteins with undefined function havee been identified. One or more of these may prove to be physiologically important; however, because Se is metabolized in many cases indiscriminately as an analog of S, it is likely that at least some of these Se proteins are physiologically inert, serving only as reserves of Se to be made available for metabolism to critical forms as they turn over.

In general, the bioavailability of Se compounds can be described as follows: (1) the more reduced (and insoluble) inorganic forms of Se have very low bioavailabilities; (2) the common selenoamino acids (i.e., selenomethionine, selenocysteine) and Se in most plant materials have reasonably good bioavailabilities (i.e., approaching that of sodium selenite); and (3) Se in many animal products has low to moderate bioavailability. In addition, other factors can significantly influence the bioavailability of Se by affecting the utilization of ingested Se in either the digestion/absorption or the metabolism/excretion phases (Table IV). Factors (e.g., vitamin E and other antioxidants) that increase the enteric absorption of Se and/or increase the metabolism of absorbed Se to the physiologically critical forms will positively affect Se bioavailability. Alternatively, factors (e.g., heavy metals) that decrease the enteric absorption of Se and/or increase the metabolism of Se to more readily excreted forms (e.g., the methylated forms) negatively affect Se bioavailability.

D. Dietary Needs for Selenium

The quantitative dietary requirements of humans for Se are subject to some debate for the reason that, unlike most other essential nutrients, no clear-cut deficiency syndrome has been identified in deficient humans. Nevertheless, there is general agreement that most humans require about 0.85 μg Se/kg body weight/day to support maximal tissue activities of GPX. For most adults, this corresponds to intakes of 55–70 μg Se/person/day (Table V).

II. BIOCHEMICAL FUNCTIONS OF SELENIUM

Although the nutritional role of Se has always been associated with that of vitamin E and, hence, with antioxidant activity, the metabolic basis of this nutritional role remained unknown until the early 1970s when Se was discovered to be an essential part of the antioxidant enzyme glutathione peroxidase. In the late 1980s that finding was extended by discoveries that Se is essential for several isoforms of GPX and the extrathyroidal 5′-deiodinase (5′DI) involved in the metabolic activation of thyroxine to triiodothyronine, and that it is present in each of these selenoenzymes as the selenoamino acid selenocysteine (SeCys).

Selenium is incorporated into the SeCys in selenoenzymes by a highly specific mechanism involving the

TABLE IV
Factors Affecting the Bioavailability of Dietary Selenium

Type of factor	Good Se bioavailability	Poor Se bioavailability
Chemical form of Se	Oxidation to selenate or selenite Presence as selenoamino acids (Se-cysteine, Se-methionine)	Reduction to insoluble selenide or elemental Se
Food sources of Se	Se-enriched yeasts Wheat Most plant products	Meat and fish products Soy protein products
Dietary factors	Restricted food intake Vitamins E, C, and A (at high levels) Supplemental methionine Synthetic antioxidants (e.g., ethoxyquin)	Heavy metals (e.g., Cd, Hg) Deficiencies of pyridoxine, vitamin E, riboflavin, methionine Excess sulfur, arsenic

cotranslational insertion of SeCys residues directed by a UGA codon. This mechanism has been well studied, particularly for *Escherichia coli* formate dehydrogenase, in which case four genes (*sel*D, *sel*A, *sel*C, and *sel*B) and their products have been identified in this process. The molecular biology of eukaryotic systems is thought to be similar. This process involves the reaction of a reduced Se metabolite (e.g., selenide) with tRNA-bound phosphoserine to yield bound selenocysteine; the product is inserted into the growing polypeptide chain only at sites specified by a discrete stem loop mRNA configuration near the UGA codon (which otherwise signals termination of peptide synthesis).

The best-characterized selenoenzyme is glutathione peroxidase. It catalyzes the reduction of hydrogen peroxide (H_2O_2) or fatty acid hydroperoxides using reducing equivalents of reduced glutathione (GSH). Human GPX has been purified and been found to be a homologous tetramer, each subunit of which contains one Se atom (in the form of SeCys) located at the active site. The metabolic function of GPX is as part of the cellular defense system against damage induced by free radicals, particularly those of oxygen. The addition of electrons to molecular oxygen can yield a number of radical intermediates, some of which can interact with membrane lipids to alter membrane structure and function. Two such radicals, superoxide anion ($O_2^{\cdot-}$) and peroxide ($HOO^{\cdot}$), can be formed by endogenous enzyme systems and by exogenous agents such as drugs and pesticides. Although superoxide itself is relatively unreactive with the polyunsaturated fatty acids (PUFAs) of membrane phospholipids, in the presence of ferric iron (Fe^{3+}) it can interact with peroxide to form very highly reactive species such as hydroxyl radical (HO^-) and singlet oxygen ($^1O_2$). The latter can initiate lipid peroxidation by abstracting hydrogen atoms from PUFAs, causing reactions that branch and become autocatalytic. Whereas the formation of peroxides and free radicals is essential to certain physiological functions (e.g., the bactericidal activities of polymorphonuclear leukocytes and macrophages; intermediary steps in prostaglandin synthesis), uncontrolled lipid peroxidation is detrimental to the cell.

Cellular defense against damage due to free radical chain reactions involves attacks on these processes at several different stages. The proximal reactive oxygen species, superoxide, is converted by a family of enzymes called superoxide dismutases to H_2O_2. That product is then converted either to water by GPX or

TABLE V
1989 National Research Council Recommended Dietary Allowances for Se

Group	Age (years)	μg Se/day
Infants	0–0.5	10
	0.5–1	15
Children	1–6	20
	7–10	30
Males	11–14	40
	15–18	50
	19+	70
Females	11–14	45
	15–18	50
	19+	55
Pregnant		65
Lactating		75

to water and O_2 by the heme-enzyme catalase. Because GPX is the more widely distributed of these enzymes within the cell (i.e., it is found in the cytosol and mitochondrial matrix space, whereas catalase is limited to the peroxisomes), it is most important in removing H_2O_2 and, thus, diminishing the amount of that species available to react with superoxide to initiate free radical reactions. Free radicals that are formed can be scavenged by vitamin E, which can donate a hydrogen atom to quench the radical. This explains the long-observed relationship of vitamin E and Se, and why either one is sufficient to prevent a variety of deficiency diseases of animals.

Substrates of GPX include hydroperoxides of free fatty acids, as well as hydrogen peroxide, which suggests that GPX may also serve in the antioxidant defense system by interrupting the autocatalytic phase of lipid peroxidation. However, GPX cannot reduce lipid peroxides when they are acylated (i.e., in phospholipids of biological membranes); therefore, peroxidized fatty acids must be rendered free of their glyceryl components by a membrane-associated lipase in order for GPX to function in this manner. Otherwise, it acts only to reduce hydrogen peroxide in preventing the initiation of lipid peroxidation.

Selenium also plays an important role in the metabolism of thyroid hormone. That role involves its function in the form of SeCys as an essential constituent of the type I iodothyronine 5′-deiodinase (T_4-DI_I), which catalyzes the conversion of thyroxine (T_4) to the metabolically active triiodothyronine (T_3). In humans, the combined activity of this selenoenzyme in liver and muscle is estimated to account for some 80% of T_3 production. Thus, it is likely that deprivation of Se may contribute to the etiology of iodine deficiency disorders.

Whereas the clinical signs of Se deficiency in animals appear to relate to losses of GPXs and/or T_4-DI_I, the physiological effects of losses of other SeCys proteins are unclear. These effects include changes in hepatic enzymes and substrates involved in the metabolism and detoxification of drugs and other foreign compounds; they are not necessarily of the same magnitude or direction for all enzymes. For example, in vitamin E-adequate mice, the activities of certain cytosolic enzymes (e.g., some GSH-*S*-transferases, GSSG reductase) were elevated by long-term Se deprivation, while others (e.g., GSH-thioltransferase and -sulfotransferase) were decreased. Similarly, some microsomal enzyme activities were increased (e.g., cytochrome P_{450}-dependent hydroperoxidase, heme oxygenase, UDP-glucuronyl transferase), some were reduced (e.g., NADPH-cytochrome P_{450}-reductase, flavin-monoxygenase), and others were unchanged (e.g., NADH-cytochrome-b_5-reductase, aniline hydroxylase, aminopyrine-*N*-demethylase). Glutathione-*S*-transferase activities have been shown in several species to be elevated in Se deficiency.

Changes in nutritional Se status have been shown to increase the toxicities of some compounds but to decrease the toxicities of others via effects related to the changes in drug metabolic capacity outlined earlier. For example, deprivation of Se can increase the toxicities of compounds such as nitrofurantoin (a urinary tract antibacterial agent) and paraquat (a dipyridinium herbicide), both of which are metabolized to free radicals that generate reactive oxygen species. Dietary Se protects against the acute toxicities of these types of compounds through its function as GPX in removing H_2O_2 and, thus, preventing lipid peroxidation that would otherwise be stimulated by these prooxidants. In contrast, Se deprivation can render other compounds (e.g., iodipamide, acetaminophen, aflatoxin B_1) less toxic. In these cases, protection is thought to be due to the increases in GSH-*S*-transferase activities and GSH concentrations that occur from Se deficiency. At least one of the isoforms of GSH-*S*-transferase also has GPX activity; it has been suggested that its increases in Se-deficient animals may compensate for losses in GPX. In addition, GSH-*S*-transferases serve important detoxification functions by conjugating foreign compounds with GSH, there by promoting their excretion.

Several SeCys-containing proteins in addition to GPX and T_4-DI_I have been identified in animal tissues. Whether any has enzymatic activity or any other physiologic function is not known. The levels of one selenoprotein found in rat plasma and liver, that is, "selenoprotein-P," reflect changes in Se nutriture and has been proposed as a possible transporter of Se; it may prove useful as an index of nutritional Se status. Selenium can also be incorporated nonspecifically into proteins if it is consumed in the form of the Se analog of methionine, selenomethionine (SeMet). Unlike SeCys, which itself cannot charge $tRNA^{Cys}$, SeMet can compete with methionine for charging $tRNA^{Met}$. Therefore, the consumption of Se as SeMet (e.g., from foods derived from plants) results in the nonspecific incorporation of Se in the form of SeMet into several proteins (e.g., albumin, β-galactosidase, several muscle proteins). There is no evidence that SeMet/Met replacement affects the functions of these proteins;

the physiologic relevance of the phenomenon is in providing a reserve of Se with relatively slow turnover.

III. EVALUATING SELENIUM STATUS IN HUMANS

Analyses of the Se concentrations of human tissues demonstrate a strong geographic variation in Se status correlating with the geographic variation in the Se contents of food supplies (Table VI). Populations with the lowest apparent intakes of Se (i.e., the Se-deficient regions of China, New Zealand, and Finland) also have the lowest concentrations of Se in their blood. Blood Se concentrations have revealed substantial geographic variation within the United States: in the 1960s the mean Se concentration of whole blood from blood banks in 19 cities was found to vary from 157 to 256 ppb. That variation appeared to relate to differences in local intakes of Se, as the cities ranking at the extremes of that range were located in areas known to be either high (e.g., northern Great Plains states) or low (northeastern states) with respect to Se. The levels of Se and GPX in whole blood show a good correlation at blood Se concentrations less than 100 ppb; however, at higher levels (such as are generally found in Americans), the correlation is relatively weak, suggesting that Se resides in forms other than active GPX in increasing amounts in individuals with adequate or better Se intakes.

Nutritional Se status can be assessed on the basis of the Se contents of hair or nails, which Chinese studies have shown to be highly correlated with the level of Se intake. This approach has the obvious advantage of ease of sample availability, storage, and preparation; however, in the case of hair sampling, standardization of the scalp location and amount of hair is important to ensure reproducibility. It is also important to screen subjects to ascertain who may have recently used Se-containing antidandruff shampoos, as their hair will bear Se residues that do not reflect nutritional Se status. The renal clearance of Se is important in the homeostasis of the element; therefore, measurement of urinary Se excretion can provide useful information for the assessment of nutritional Se status in humans. The urinary excretion of Se is generally greater than the fecal excretion of the element; urine Se content is a function of the level and form of Se intake, the nature of the diet, and the Se status of the individual. Urinary Se excretion is closely correlated with the Se concentrations of blood or plasma but can, therefore, be influenced by the Se consumed with a recent meal.

Plasma Se concentrations are 10–45% lower in pregnant women than in nonpregnant women. It has been suggested that this difference may be a factor in the etiology of preeclampsia; however, blood Se levels in normal and preeclamptic pregnancies have been found to be similar. It is likely that at least a portion of the decrease in plasma and whole-blood Se is due to the hemodilution of pregnancy and the demands of the growing fetus for Se. Blood Se concentration rapidly returns to nonpregnancy levels after delivery. Changes in plasma Se concentration during pregnancy have been associated with changes of similar magnitude in the plasma activity of GPX.

Premature infants are generally born with Se status comparable to those of full-term infants at the time of delivery. However, because hospitalized premature infants are frequently maintained with low-Se parenteral nutrient solutions or are fed low-Se formula diets for as long as several weeks, they frequently show progressive declines in blood Se concentrations and GPX activities. Because they also have low plasma levels of vitamin E, their low Se status with its attendant reduction in antioxidant protection may have

TABLE VI
Typical Blood Se Concentrations (ng/ml) of Residents of Several Countries

Tissue	U.S.A.	England	W. Germany	Finland	New Zealand	China			Venezuela
						Se deficient[a]	Moderate	High-Se[b]	
Plasma	65–180	80–110	32–88	46–109	43–54	—	—	—	—
Whole blood	75–265	—	—	56–130	—	10–21	33–136	3200–3480	355–813

[a]Keshan disease-endemic areas.
[b]Areas of endemic selenosis.

significant physiological consequences (e.g., increased adverse drug reactions).

IV. CONSEQUENCES OF LOW SELENIUM STATUS

Two diseases of children are recognized as being associated with severe nutritional Se deficiency: a cardiomyopathy named Keshan disease and a chondrodystrophy named Kaschin-Beck disease. Both diseases occur in rural areas in the belt of endemic Se deficiency of China. The endemic distributions of both diseases are similar but not always identical; both correspond to the distribution of endemic Se deficiency in foods.

Keshan disease has been diagnosed in more than a dozen Chinese provinces, almost exclusively among people living in mountainous areas where the soil is low in Se (i.e., <125 ppb Se, of which less than 2.5% is water-soluble). Accordingly, locally produced foods are exceedingly low in Se content (e.g., grains generally contain <40 ppb Se). Selenium/vitamin E-deficiency diseases of livestock (e.g., "white muscle disease" in lambs and "mulberry heart disease" in pigs, a skeletal myopathy and a cardiomyopathy, respectively) are also endemic in these areas. Humans living in those areas typically show the lowest tissue Se levels of any free-living populations anywhere on earth (e.g., blood Se <25 ppb; hair Se <100 ppb), indeed, less than half of the corresponding values for New Zealanders or Finns. Keshan disease has been more prevalent among farming families in rural districts than among residents of urban centers; this difference probably relates to the more monotonous dietary habits of rural residents and to their stronger dependency on food produced in their immediate locales.

Keshan disease is a multifocal myocarditis occurring primarily in children between the ages of 2 and 10 years and, to a lesser extent, among women of childbearing age. Infants are rarely affected; their intakes of Se from breast milk, even in these deficient areas, appears to be about 3 μg per day, which is thought to be sufficient to protect them. The first signs of the disease are normally not observed until after children are weaned to solid foods, at which time their intakes drop to about 1.5 μg Se per child per day. Keshan disease shows marked seasonal and annual variations in incidence, suggesting that other factors must be involved in its etiology.

Keshan disease is manifest as acute or chronic insufficiency of cardiac function, cardiac enlargement, gallop rhythm, cardiac arrhythmias, and electrocardiographic (EKG) and radiographic abnormalities. Subjects may show cardiogenic shock or congestive heart failure; embolic episodes from cardiac thromboses have been reported. Four clinical subtypes of Keshan disease have been identified: an acute type with sudden onset of dizziness, malaise, substernal discomfort, and dyspnea in otherwise apparently healthy children with no history of cardiac disorders; a chronic type, characterized by chronic congestive heart failure with varying degrees of cardiac insufficiency; a subacute type, with signs and symptoms varying with the degree of cardiac insufficiency but with an accelerated course; and a latent type, characterized by normal heart function with mild cardiac dilation and associated EKG changes. The case fatality of Keshan disease in China was greater than 80% in the 1940s, but has been reduced in recent years to about 30%, primarily as the result of better medical care.

Hypotheses for the etiology of Keshan disease have proposed that it may be caused by food and/or waterborne toxins (e.g., mycotoxins, heavy metals, nitrite), certain infectious agents (e.g., cardiophilic viruses), or specific nutrient deficiencies (e.g., Se, Mo, Mg, riboflavin). Available experimental evidence provides support for only the last two of these hypotheses. Intervention studies have demonstrated that Se can be very effective in the prevention of Keshan disease. Improvements in the Se status of children have been achieved by the use of either table salt fortified with Se (e.g., 10–15 ppm as Na_2SeO_3) or oral tablets containing Na_2SeO_3; both vehicles have been used to produce dramatic reductions in the incidence of Keshan disease. Although the etiology of Keshan disease is still not fully elucidated, it is clear that Se deficiency is at least a contributing factor and is responsible for defining the endemic distribution of the disease.

Cardiomyopathies associated with low Se status have been reported in a few cases outside of China. However, the extent to which Se deficiency may have been involved in the pathogenesis of those cases has not been determined. Certainly, low Se status is not a general feature of cardiomyopathy patients in the United States.

Kaschin-Beck disease is an osteoarthropathy with an endemic distribution corresponding to that of Keshan disease. The disease affects primarily the epiphyseal cartilage, articular cartilage, and epiphyseal growth plates of growing bones. The long bones are most frequently affected; however, the cartilage tissue associated with any bone in the skeleton may be involved. Affected cartilage shows atrophy and necrosis with repair and disturbance in endochondral ossification. The most striking histological feature is chondronecrosis with proliferation of surviving chondrocytes in clusters, categorized as a coagulation necrosis. The condition results in enlarged joints (especially of the fingers, toes, and knees), shortened fingers, toes, and extremities, and, in severe cases, dwarfism.

Studies of the effects of Se supplementation in the prevention and therapy of Kaschin-Beck disease are few but have yielded some encouraging results. Two controlled Se-intervention studies showed that the Se tablets reduced the severity of the disease and facilitated improvement of prevalent cases. The etiology of Kaschin-Beck disease is presently unclear. Its coincident distribution with that of severe Se deficiency and the reported effectiveness of supplemental Se in reducing its incidence and/or severity suggest that Se may play some role in its etiology.

It is likely that Se deficiency may be a factor in the etiology of iodine deficiency disorders in at least some areas. It is clear that the chief production of thyroid hormone (T_3) requires a selenoenzyme (T_4-DI_I), the activity of which is reduced by Se deprivation. Studies in central Africa found that the prevalences of goiter and myxedematous cretinism were each greater among populations with relatively low plasma Se levels than those with greater plasma Se levels. Such a relationship suggests that the efficacy of iodine supplementation programs may be limited in Se-deficient populations, in which cases treatment with both Se and I would be indicated.

Altered Se status has been found consistently in association with relatively few human diseases. Reduced blood Se levels (by 21–60%) have been measured in patients with alcoholic liver disease and cirrhosis; however, these effects may be without significant biological consequences inasmuch as they have not been found to be associated with reductions in blood GPX. Patients with neuronal ceroid lipofucsinosis, acrodermatitis enteropathica, kwashiorkor, and chronic renal failure have been found to show reductions in blood Se by 17–45%, whereas patients with muscular dystrophy have been found with increased (by ca. 40%) plasma Se levels. Multiple sclerosis patients have shown reductions of blood GPX by as much as 23%, yet their plasma Se concentrations appear not to be affected. The plasma Se concentrations of patients with Down's syndrome are consistently reported to be abnormally low (by ca. 27%), although these individuals also show increased (by ca. 55%) erythrocyte GPX activities. Many of these apparent differences in Se status between diseased and healthy persons may relate to differences in diet (i.e., Se intake). Such an effect is to be expected in cases such as kwashiorkor, in which inadequate intake of protein, the vehicle for most dietary Se, results in the disease. No evidence indicates abnormalities in Se metabolism in any of these diseases.

Finnish studies have shown low serum Se levels to be associated with increased risk to cardiovascular disease. The apparent protective effect was associated with increased concentrations of high-density lipoprotein-bound cholesterol (by 43%) in serum. Low Se status, as indicated by very low concentrations of the element in blood cells and/or plasma, has been identified in infants with the inborn errors of amino acid metabolism, maple syrup urine disease, or phenylketonuria. However, low Se status is not related directly to any of these metabolic diseases. Because these children are fed purified diets low in protein and, hence, low in Se, they show serum Se levels (e.g., as low as 5 ppb) with erythrocyte GPX activities of only 10–20% of those of healthy children. Parenteral nutrition fluids based on amino acid mixtures contain negligible amounts of Se (e.g., <5 μg Se per 1000 kcal) if they are not supplemented with the element. Therefore, it is not surprising that low Se status has been observed in patients maintained with total parenteral nutrition (TPN) for extended periods of time. To prevent this from happening, it is recommended that TPN solutions be supplemented to provide 25–30 μg Se/day.

V. SELENIUM AND CANCER

Several Se compounds have been demonstrated to inhibit or retard carcinogenesis in a variety of experimental animal models. The available information concerning experimental carcinogenesis as affected by either nutritional or pharmacological levels of Se reveals different general effects that may relate to dose-related differences in Se metabolism. Numerous studies have evaluated the effects of pharmacological levels of Se on experimental carcinogenesis using several systems, including chemical, viral, and transplantable tumor models. Of those, two-thirds have found that high-level Se treatment reduced the development of tumors at least moderately (i.e., by 15–35% from control levels) and in most cases very substantially (i.e., by more than 35%). Only a few studies have found Se treatments of animals to be without effect on tumor outcome. Taken collectively, these results indicate that Se may have less antitumor efficacy when used at levels less than the equivalent of 1 ppm in the diet. Although relatively few studies have evaluated the effect of nutritionally deficient levels of Se on carcinogenesis in experimental animal models, the available results indicate that dietary deficiencies of Se generally do not affect carcinogenesis of the colon or liver in the rat, or of the mammary gland in the mouse, but may actually enhance carcinogenesis of the mammary gland in the rat and of the skin in the mouse.

Selenium status has been evaluated as a putative factor in the etiologies of several human cancers. Ecological studies have tended to show that Se status and cancer mortality were inversely correlated. These have concerned cancers at several sites: lymphomas, peritoneum, lung, breast, colon, rectum, and liver. A recent ecological study of Se status and cancer mortality in the United States indicated that intermediate- and high-Se counties had lower rates than low-Se counties of cancers of all sites combined and, specifically, in both sexes, of the lung, colon, rectum, bladder, esophagus, and pancreas. Inverse associations were also found for cancers of the breast, ovary, and cervix. Cancers of the liver, stomach, Hodgkin's disease, and leukemia showed positive associations with forage Se level in either sex.

A number of case–control studies have been conducted to test the hypothesis that Se status may be related to cancer risk. One conducted in eastern North Carolina found that subjects ranking in the lowest decile of plasma Se concentration (with a mean of 84 ppb) had a relative risk of nonmelanoma skin cancer (e.g., basal cell or squamous cell carcinomas) greater than 3 in comparison to subjects ranked in the highest decile (mean plasma Se of 221 ppb). Further, when cases and controls were stratified on high versus low plasma retinol and total carotenoids, the results indicated that low plasma Se was associated with elevated risk of skin cancer in patients low with respect to either of those variables.

A prospective case–control study was conducted to evaluate the relationship of serum Se level and total cancer mortality in a random population sample of more than 8100 persons in eastern Finland. During a 6-year period of observation, the subjects that developed cancers were found to have had significantly lower serum Se concentrations at the beginning of the study than those of matched controls who did not develop cancer (50.5 ± 1.1 ppb versus 54.3 ± 1.0 ppb). Another prospective case–control study found that the relative risk of subjects in the lowest quintile of serum Se (i.e., <115 ppb) was twice that of subjects in the highest quartile (i.e., >154 ppb). In addition, the relative risk in low-Se subjects appeared to be greatest if they were also relatively low in vitamins A and E (as indicated by low plasma concentrations of retinol and α-tocopherol). A third prospective case–control study showed that the mean serum Se concentrations were lower in terminal cancer patients than in matched controls (i.e., 53.7 ppb versus 60.9 ppb). The relative risk of death to cancer was 5.8 among subjects in the lowest tertile of serum Se concentration when compared with subjects in the highest tertile. Again, an interaction of Se vitamin E was detected; subjects in the lowest tertiles with respect to serum concentrations of both nutrients had a relative risk of fatal cancer of 11.4 compared to subjects in the highest tertiles.

Some epidemiological investigations of Se and cancer have failed to detect such relationships. Nevertheless, the *plausibility* of the hypothesis that low Se status may increase human cancer risk is supported by the results of the ecological correlational studies and cross-sectional case–control studies discussed here. One recent clinical intervention trial found selenium supplementation to reduce cancer risks in older Americans by as much as 65%. It should be noted that this hypothesis is different from that derived from studies with animal tumor models, which would hold that pharmacologic levels of Se may be anticarcinogenic. In fact, experimental animal studies have indicated that there are likely to be multiple mechanisms by which Se can affect carcinogenesis, although these, too, are yet to be elucidated. Therefore, it is likely that the dose–response relationship between Se and tumor development is not linear (i.e., that the nature of the low Se status effect that is suggested by human epidemiological studies may be quite different from the high-dose Se effect indicated by the results of animal studies).

Antitumorigenic effects of Se have been realized (in animal tumor model studies) only at dose levels much greater than those required to support the known selenoenzymes at maximal levels. Therefore, it appears that the cancer-protective effects may be due to Se metabolites produced appreciably at high-Se intakes. Two such metabolites are likely candidates: selenodiglutathione, which is formed by the thiol-dependent reduction of selenite or selenate when Se is provided in such form, and methylselenol and/or its further metabolites (e.g., dimethylselenide, trimethylselenonium), which are general products of Se metabolism. Each of these candidates has been found to have antitumorigenic activity in animal tumor models. Though the mechanisms of their antitumorigenic activities are not clear, several have been suggested: protection of critical macromolecules by the occupation of hydrophobic sites, formation of methylselenylated bases in nucleic acids, and direct cytotoxicity.

VI. SELENOSIS

The potential for Se toxicity was first recognized well before the nutritional role of the element was discovered. In the 1930s, Se was found to be a cause of a dermatologic disorder ("alkali disease") that occurred in cattle and horses in northern Nebraska and South Dakota. Investigations revealed that those animals grazed on pasture plants that accumulated high levels of Se from seleniferous soils. Although Se toxicity has subsequently been studied in several animal species, only a few cases of human exposure to hazardous levels of Se have been reported. Most of these cases have involved occupational exposures (e.g., of workers in copper smelters or Se-rectifier plants) from the inhalation of Se aerosols. Some cases, however, have involved the oral consumption of high levels of Se in various forms: selenious acid (H_2SeO_3) in gun blueing preparations (these can contain as much as 2% H_2SeO_3, i.e., >12,000 ppm Se), sodium selenate (Na_2SeO_4), selenium dioxide (SeO_2), extremely high Se foods in a particular locale of endemic selenosis in China, a particular nut (coco de mono) that appears to accumulate Se, and an over-the-counter supplement that was erroneously formulated with excessive Se. These cases have demonstrated that acute exposure to high levels of Se can produce hypotension (resulting from vasodilation), respiratory distress, and a garlic-like odor of the breath (due to the exhalation of dimethylselenide). They also show that chronic exposures can produce gastrointestinal disturbances (e.g., dyspepsia, diarrhea, anorexia) as well as garlic breath as the major signs; more severe cases show skin eruptions, pathological nails, and hair loss.

The lethality of acute Se intoxication has been established in animals and has been documented in a few human cases. In two cases (a 3-year-old boy who drank gun blueing and a 17-year-old man who consumed an unknown amount of SeO_2), the subjects died within only a few hours of exposure. In one case (a 52-year-old woman who drank gun blueing), the subject was revived, only to die 8 days later of respiratory failure. Reports of nonfatal acute selenosis (including one of a 2-year-old girl who drank Se-containing gun blueing) indicate that the signs and symptoms of the toxicity can be reversible upon cessation of Se exposure, and that complete recovery without sequelae can be expected.

Naturally occurring chronic selenosis was identified in the 1960s among residents of Enshi County, Hubei Province, China. It appears to have resulted from exceedingly high concentrations of Se in the local food supplies and, in fact, throughout that environment. In that particular area, the local soils were found to contain nearly 8 ppm Se, and coal (the ash of which was used to amend the soil) was found to contain as much as 84,000 ppm Se. Consequently, locally produced foods contained the highest concentrations of Se ever reported: corn, 6.33 ppm Se, and rice, 1.48 ppm Se. Even the water, which leached through seleniferous coal seams, contained unusually high concentrations of Se (e.g., 54 ppb Se). In the five most heavily affected villages, morbidity was about 50%. Almost all residents showed signs, the most common of which were losses of hair and nails. Some also showed skin lesions (e.g., erythema, edema, eruptions, intense itching), hepatomegaly, polyneuritis (e.g., peripheral anesthesia, acroparesthesia, pain in the extremities, convulsions, partial paralysis, motor impairment, hemiplegia), and gastrointestinal disturbances. One death was attributed to selenosis (a postmortem evaluation of that case was not made). In a village that had a history of high prevalence of these signs and symptoms, it was estimated that local residents consumed 3200–6690 μg Se per person per day. It should be noted that this level of intake is approximately 100 times the nutritionally significant level for Se. Recent studies in these regions have indicated that risk of selenosis may be significant among individuals with chronic intakes above 750–850 μg Se from that seleniferous food supply.

Signs of chronic selenosis have also been reported among users of certain oral Se supplements. Although the consumption of Na_2SeO_3 at rates as great as 1 mg Se/person/day for at least short periods of time appears to produce no toxic signs or symptoms, clear signs of selenosis have been reported in a man who took an oral supplement that provided that level of Se daily for 2 years. That subject had garlic breath and thickened and fragile nails; these symptoms subsided when he stopped taking the Se supplement.

Thirteen cases of selenosis were identified among consumers of a commercial Se supplement that, through a manufacturing error, contained an average of 27 mg Se per tablet. The cases showed variable signs according to the number of tablets actually consumed.

The individual that consumed the greatest amount of total Se in this group developed total alopecia, severe nausea and vomiting, severe diarrhea, and garlic breath. Some cases showed irritability, fatigue, and paresthesia; peripheral neuropathy was documented in two cases.

Selenosis was reported in a woman who had used a SeS_2-containing antidandruff shampoo two to three times weekly for 8 months. Her symptoms appeared one day within an hour after using the shampoo: eruptive scalp and mild nonrhythmical tremors of the arms and hands progressing to generalized tremors of increasing severity. Within 2 hours she had a metallic taste of the mouth and garlic breath. Two days later she was weak and anorectic; the lethargy persisted for a couple of days, during which time she was nauseous and had porphyrinuria. She recovered gradually thereafter, having ceased using the shampoo; her recovery was complete in 2 weeks.

Present knowledge of the toxicology of Se in humans is incomplete. Although the proximal biochemical lesion(s) involved in Se toxicity are not clear, it is thought that these involve the oxidation and/or binding of critical sulfhydryl groups in proteins and/or nonprotein thiols by Se species present in excessive concentrations. It has been proposed that the signs of Se toxicity are related to changes in intracellular concentrations of reduced glutathione and/or other nonprotein sulfhydryls.

In humans, garlic breath as well as dermatologic and gastrointestinal signs are the earliest indicators of Se intoxication. The lowest Se dosage that has been found to be associated with the manifestation of these signs is about 1000 μg Se per day, which was achieved by the daily use of an oral supplement of 900 μg Se (as 2 mg of Na_2SeO_3) and a daily intake of about 100 μg Se from other dietary sources. That level may very likely be near the threshold dose for toxic responses, as the signs associated with it were mild and were not manifest for nearly 2 years of exposure. Therefore, it has been suggested that, for adults, levels of 550 and 750 μg Se per person per day be used as upper limits for safe exposure to inorganic (e.g., Na_2SeO_3) and organic (e.g., selenoamino acids) Se compounds, respectively. Because foods do not contain such high concentrations of Se except in very unusual circumstances, the risks of exposure to potentially intoxicating amounts of the element are limited to its improper use as a food supplement (i.e., formulation errors) and the improper consumption of seleniferous products not intended for human consumption (e.g., accidental consumption of Se reagents).

BIBLIOGRAPHY

Arthur, J. R., Nicol, F., and Beckett, G. J. (1993). Selenium deficiency, thyroid hormone metabolism, and thyroid hormone deiodinases. *Am. J. Clin. Nutr.* **57**, 236S–239S.

Burk, R. F., and Hill, K. E. (1993). Regulation of selenoproteins. *Annu. Rev. Nutr.* **13**, 65–89.

Clark, L. C., Combs, G. F. J., Turnbull, B. W., Slate, E. H., Chalker, D. K., Chow, J., Davis, L. D., Glover, R. A., Graham, G. F., Gross, E. G., Krongrad, A., Leshev, J. L., Park, H. K., Sanders, B. B., Smith, C. L., and Taylor, R. J. (1996). Effects of selenium supplementation for cancer prevention in patients with carcinoma of the skin. *J. Amer. Med. Assoc.* **276**, 1957–1963.

Combs, G. F., Jr. (1988). Selenium in foods. *In* "Advances in Food Research" (C. O. Chichester and B. S. Schweigert, eds.), Vol. 32, pp. 85–113. Academic Press, New York.

Combs, G. F., Jr. (1989). Selenium. *In* "Nutrition and Cancer Prevention: Investigating the Role of Micronutrients" (T. E. Moon and M. S. Micozzi, eds.), pp. 389–420. Marcel Dekker, New York.

Combs, G. F., Jr. (1994). Essentiality and toxicity of selenium: A critique of the Recommended Dietary Allowance and the Reference Dose. *In* "Risk Assessment of Essential Elements" (W. Mertz, C. O. Abernathy, and S. S. Olin, eds.), pp. 167–183. ILSI Press, Washington, D.C.

Combs, G. F., Jr., and Combs, S. B. (1986). "The Role of Selenium in Nutrition." Academic Press, New York.

Combs, G. F., Jr., Spallholz, J. E., Levander, O. A., and Oldfield, J. E. (eds.) (1987). "Selenium in Biology and Medicine," Vols. A and B. AVI Publishing Co., Westport, Connecticut.

Levander, O. A. (1986). Selenium. *In* "Trace Elements in Human and Animal Nutrition" (W. Mertz, ed.), 5th Ed., Vol. 2, pp. 209–279. Academic Press, New York.

Levander, O. A. (1987). A global view of human selenium nutrition. *Annu. Rev. Nutr.* **7**, 227.

Schubert, A., Holden, J. M., and Wolf, W. R. (1987). Selenium content of a core group of foods based on a critical evaluation of published analytical data. *J. Am. Diet. Assoc.* **87**, 285–299.

Wendel, A. (ed.) (1989). "Selenium in Biology and Medicine." Springer-Verlag, New York.

Yang, G. Q., Wang, S., Zhou, R., and Sun, R. (1983). Endemic selenium intoxication of humans in China. *Am. J. Clin. Nutr.* **37**, 872–881.

Yang, G. Q., Yin, S., Zhou, R., Gu, L., Yan, B., Liu, Y., and Liu, Y. (1989). Studies of safe maximal daily dietary selenium intake in a seleniferous intake in a seleniferous area in China. II. Relations between Se-intake and the manifestations of clinical signs and certain biochemical alterations in blood and urine. *J. Trace Elem. Electrolytes Health Dis.* **3**, 123–130.

Yang, G. Q., Zhou, R., Yin, S., Gu, L., Yan, B., Liu, Y., and Li, X. (1989). Studies of safe maximal daily dietary selenium intake in a seleniferous area in China. I. Selenium intake and tissue selenium levels of the inhabitant. *J. Trace Elem. Electrolytes Health Dis.* **3**, 77–87.

Sensation-Seeking Trait

MARVIN ZUCKERMAN
University of Delaware

GLOSSARY

Augmenting–reducing (of the cortical evoked potential) Tendency of the amplitude of the cortical evoked potential to increase (augment) or decrease (reduce) as a function of the intensity of a stimulus; individuals differ reliably in this function and may be characterized as augmenters or reducers

Cortisol One of the corticosteroids from the adrenal cortex that regulates the release of glycogen from the liver into the bloodstream, fat metabolism, striated muscle strength, blood pressure, and lymphoid tissue (anti-inflammatory effect); it is released by stress acting through the adrenocortico-tropic hormone from the pituitary gland

Estrogens (estradiol, estrone) Hormones produced by the ovaries in the female that control maturation of the female reproductive system and development of secondary sex characteristics; they function in vascularization of the vaginal tissue and the lubrication of the vagina during stimulation, but are not an essential factor in sexual arousability; some estrogen is produced in males, largely through conversion from androgens

Monoamine oxidase Enzyme involved in the catabolic deamination of the monoamines norepinephrine, dopamine, and serotonin

Orienting reflex Involuntary physiological response to the first presentation of a stimulus that tends to habituate or diminish in strength with subsequent presentations of the same stimulus

Testosterone Hormone produced by the testes in the male that controls maturation of the male reproductive system, sperm production, and secondary sex characteristics; also found in females, where the major source is the adrenal cortex; it affects sex drive and sexual arousability in both sexes

SENSATION SEEKING IS A PERSONALITY TRAIT DEfined by the disposition to seek varied, novel, complex, and intense sensations and experiences and to take physical and social risks for the sake of such experiences. Operationally, the sensation seeking scale (SSS) or other test variants related to the construct (e.g., Monotony Avoidance, Venturesomeness, and Arousal Seeking scales) are used to define the personality dimension or groups of subjects characterized as "high" or "low" sensation seekers.

I. THEORY AND TRAIT DESCRIPTION

The theory and initial trait assessment of sensation seeking emerged in the context of experimental research on sensory deprivation in humans. The wide variation in responses to periods of sensory deprivation or invariant sensory environments suggested the possibility that people have optimal levels of stimulation and arousal that influence their behavior and preferences in many areas of life activities. A trait of sensation seeking was defined (see Glossary) and a questionnaire, the SSS, was designed to assess the general trait.

Further development of the SSS revealed four stable factors based on the item content of the scale. These factors have been replicated in many different populations around the world. Three of the factors describe

ENCYCLOPEDIA OF HUMAN BIOLOGY, Second Edition, VOLUME 7. Copyright © 1997 by Academic Press. All rights of reproduction in any form reserved.

different modes or styles of sensation seeking while the fourth factor, Boredom Susceptibility, represents the reaction to a lack of varied stimulation. The factors may be defined as follows:

1. *Thrill and Adventure Seeking* (TAS) consists of items expressing a desire to engage in sports or activities involving an above-average level of physical danger or risk, such as mountain climbing or parachute jumping.
2. *Experience Seeking* (ES) contains items describing the seeking of new experiences by living in a nonconforming life-style with unconventional friends and/or an involvement with travel, art, music, and drugs.
3. *Disinhibition* (Dis) was named for the items describing the disinhibiting of behavior in the social sphere by drinking, partying, and seeking variety in sexual partners.
4. *Boredom Susceptibility* (BS) items describe an aversion for repetitive experience of any kind, routine work, or even dull or predictable people. Other items indicate a restless reaction when things are unchanging.

Form V of the SSS has a total score based on the sum of the four factor-derived scales in place of a general scale used in previous versions.

II. PHENOMENAL CORRELATES

The validity of a personality scale is usually assessed in terms of its predictions or correlations with phenomena outside of itself or other self-report tests. The SSS was first applied to the prediction of responses to experimental sensory deprivation. Male college student subjects were exposed to two 8-hour sessions of confinement in a sound-proof room on two different occasions. One condition involved sensory deprivation (darkness and quiet), whereas the other involved social isolation and confinement but the room was lighted and minimal visual and auditory stimuli were provided. The order of conditions was counterbalanced. Nonpurposive movements were measured using a pressure transducer connected to the air mattress on which they were reclining. Sensation seekers (highs on SSS) were found to show more restlessness than low-sensation seekers in both kinds of confinement conditions as a function of time in the monotonous environment. In another experiment, sensation seekers tended to respond more for the reward of visual stimuli, again as a function of time in the invariant environment.

Highs in a college population reported having had more varied sexual activities with more partners than lows. They also reported using alcohol more heavily and admitted more use of illegal drugs than lows. Studies of drug use in normal and drug-abusing populations show that sensation seeking correlates not only with drug use but also with the variety of drugs used. Some evidence indicates that sensation seekers are somewhat more attracted to amphetamine and hallucinogenic drugs than to suppressant drugs, but users of a wide range of illegal drugs, including opiates, tend to be higher sensation seekers than nonusers. The relationships with alcohol and drug use are obtained even when items referring to a desire to use, or actual use of, these substances are removed from the scales. High-sensation seekers are also more likely than lows to smoke tobacco. Food preferences are related to sensation seeking: senation seekers like foreign, crunchy, and spicy foods, whereas lows tend to prefer bland, soft, and sweet foods. Vegetarians are lower sensation seekers than gourmets.

High-sensation seekers tend to volunteer for unusual types of experiments or activities such as hypnosis, sensory deprivation, or meditation on the expectation that they will have interesting experiences. However, when these activities turn out to be boring (completely predictable, or non-arousing types of experiences) they persist less than lows. Highs like designs that are novel, complex, and asymmetrical, whereas lows show a relatively greater preference for familiar, simple, and symmetrical designs. Highs like paintings that are expressionistic, ambiguous, or surrealistic, whereas lows prefer quiet pastoral scenes. Highs tend to watch less television than lows. When offered a choice of channels to watch, highs tend to switch from channel to channel, whereas lows watch one program. Highs like films with explicit sexual and violent themes, whereas lows tend to avoid sexual and morbid themes in films and reading. Highs like rock or jazz; lows like more bland popular or film sound track types of music.

Highs are attracted to occupations involving interactions with people, whereas lows tend to be attracted to solitary business or clerical occupations. Female highs are attracted to nontraditional occupations, whereas female lows prefer the more traditionally feminine ones. Highs more than lows are dissatisfied in nonstimulating occupations, such as assembly-line jobs. Highs given boring tasks in the laboratory report more dissatisfaction than lows.

Highs tend to engage in risky sports involving speed (auto racing), unusual sensations (parachuting, scuba diving), or skiing. Among novice skiers, those higher on sensation seeking tend to have more accidents, indicating their propensity for risk-taking before they have developed the requisite skills. Highs also tend to engage in risky body contact sports more than lows. Other sports such as bowling, tennis, or gymnastics are equally likely to be practiced by highs and lows. Highs drive their cars faster than lows and the more aggressive highs have more accidents and are more likely to drive when drunk.

III. PSYCHOPATHOLOGY

Sensation seeking is a normal dimension of personality unrelated to traits of anxiety or neuroticism but positively related to traits such as impulsivity, social dominance, surgency, sociability, autonomy, and exhibitionism and negatively related to traits such as deference, nurturance, and orderliness. Most highs or lows on sensation seeking do not have psychiatric disorders.

A problem with using test results from subjects who are currently imprisoned or hospitalized is that their current scores on the tests may reflect the current state of the disorders or the influence of incarceration rather than their personalities in normal conditions and states. Data were obtained from a large-scale survey of 2115 persons in the general community. As part of a demographic survey, they were asked about past treatments and hospitalizations for psychological problems and their diagnosis. The drugs reportedly prescribed were used as an additional check on their self-reported diagnoses. While 17% of the sample reported having received some diagnoses, 11% of the total sample were reliably classifiable within the categories listed in Table I. Each subject in each diagnostic group was matched for sex, age, and education (as closely as possible) with a control subject from the same pool of survey respondents who reported no history of psychological problems. Those who matched the controls were blind as to the diagnoses and SSS scores of the subjects with positive histories and the SSS scores of the controls.

A subgroup of bipolar disorders (manic-depressive) scored significantly higher than matched controls on

TABLE I
Comparisons of Diagnostic Groups and Matched Control Groups[a]

Groups	Ns[b]	Mean sensation-seeking scores					Age	Ed[c]	% male
		TAS	ES	Dis	BS	Total			
Bipolar (manic-depressive)	19	8.2	7.6	6.7	5.4	27.8	30.3	5.0	37
Controls	19	5.6	6.1	5.2	2.0	21.0	30.5	4.7	37
t test		3.53**	2.28*	1.54	2.01	4.79***			
Major depressive	39	6.5	7.1	5.9	4.5	24.0	31.0	5.9	28
Controls	39	7.2	7.1	6.1	4.2	24.6	31.5	5.9	28
t test		1.25	0.00	0.26	0.36	0.64			
Antisocial substance abuse	17	9.0	8.4	7.2	4.7	29.2	30.8	5.0	65
Controls	17	7.7	6.2	5.4	3.7	22.7	29.7	5.1	65
t test		2.44*	3.10**	2.52*	0.91	2.75*			
Neurotic	100	6.2	7.2	6.1	5.0	25.1	35.8	6.3	34
Controls	100	6.8	7.1	5.6	4.5	24.1	35.7	6.2	34
t test		1.58	0.24	1.23	1.42	0.37			
Personality trait disorders	15	7.4	7.7	5.9	5.3	26.3	33.8	5.5	27
Controls	15	7.1	6.6	5.8	4.0	23.5	34.0	5.5	27
t test		0.26	0.99	0.12	1.62	0.95			
Schizophrenic	20	6.7	7.0	5.0	4.8	23.4	28.5	5.2	45
Controls	20	6.6	6.9	5.9	4.0	23.3	27.8	5.2	45

[a]From M. Zuckerman and M. Neeb (1979). Sensation seeking and psychopathology. *Psychiatry Res.* **1**, 255. Copyright Elsevier/North-Holland Biomedical Press. Reproduced with permission.
[b]Ns, number of subjects.
[c]Education is on an 8-point scale: 1 = grade school, 2 = 1–3 years of high school, 3 = high school graduate, 4 = 1–3 years of college, 5 = college graduate, 6 = some graduate shool, 7 = master's degree, 8 = doctorate degree.
$^{*}P < 0.05$, $^{**}P < 0.01$, $^{***}P < 0.001$.

the total SSS, ES, and BS, but unipolar major depressive disorders did not differ from their matched controls. The results are congruent with psychometric results showing consistent correlations between the hypomania score on the Minnesota Multiphasic Personality Inventory in normal and patient populations, as well as a number of common biological correlates of sensation seeking and bipolar disorders. A group loosely labeled antisocial personality, including sociopathic and substance abusers, also scored higher than their controls on the total SSS and TAS, ES, and Dis subscales. These results are consistent with previous findings showing that prisoners diagnosed as psychopathic scored higher than nonpsychopathic prisoners on all of the SSS scales. They are also consistent with many studies showing a relationship between sensation seeking and drug and alcohol abuse. No differences were found among neurotics, personality trait disorders (other than antisocial types), or schizophrenics and their respective control groups. The lack of difference between neurotics and controls is consistent with the absence of correlation between scales of anxiety or neuroticism and sensation seeking in normal and abnormal populations. One type of schizophrenic, a chronic behaviorally retarded type, has been reported to have low SSS scores, but these subjects were hospitalized at the time of testing.

IV. BIOLOGICAL CORRELATES

A large-scale study on twins shows a relatively high degree of heritability for sensation seeking compared with other personality traits. Identical twins separated shortly after birth and raised in different families show nearly as much resemblance in sensation seeking as those reared in the same family. This finding suggests that shared parents and family environment are less important in this trait than shared genes and the specific environment outside of the home (e.g., peers), which may be different for each twin as well as for nontwin siblings. Because we do not inherit traits as such, but only their biological bases, the data suggest that we might expect to find some biological bases for the sensation-seeking trait. Such correlates have been found and many have proven to be replicable.

A. Orienting Reflex

The orienting reflex (OR) in humans is measured as a physiological response to a novel stimulus (see Glossary). The amplitude of electrodermal response (EDR), measured as skin resistance to a current applied across electrodes on the palmar surface of the hand or fingers, provides a sensitive index of changes in interest or emotions elicited by stimuli. High-sensation seekers showed a stronger EDR than lows to the first presentation of a simple but novel stimulus but did not differ from lows on subsequent presentations of the stimulus as response habituated in both groups. Although the phenomena proved difficult to replicate with simple, meaningless stimuli, auditory or visual stimuli with strong content appealing to sensation seekers yielded consistent electrodermal OR differences between highs and lows. Using heart rate (HR), more consistent differences have been obtained between high and low scorers on the Disinhibition (Dis) subscale. When heart rate is measured beat-by-beat following a stimulus, the extent of the initial deceleration provides a measure of OR. With an auditory stimulus of moderate intensity, high-Dis subjects tend to show stronger ORs. Lows showed either weaker ORs or a defensive or startle response, as indicated by an acceleration of HR to the same stimulus. Stimulus novelty and intensity interact in producing the reaction related to Dis. Figure 1 shows the contrasting HR reactions of high and low disinhibitors to a 70-dB tone.

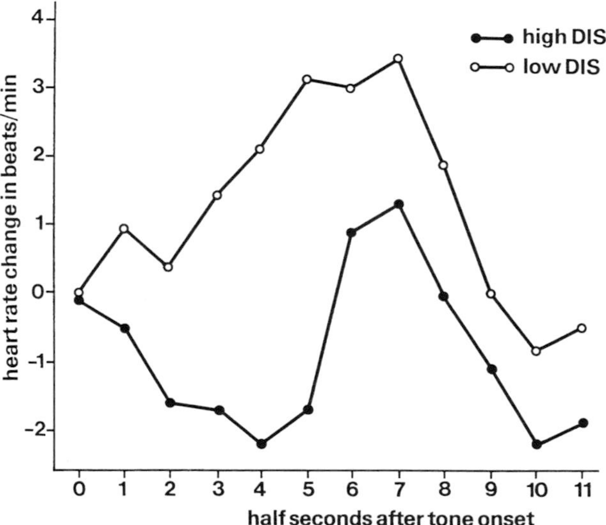

FIGURE 1 Mean heart rate changes during 5.5 sec following stimulus offset for subjects with high and low disinhibition (Dis) scores averaged over the first three trials. Stimulus was a 1000-Hz 80-dB tone. High versus low Dis × trials × time interaction significant, $P < 0.05$. [Reprinted, with permission, from J. F. Orlebeke and J. A. Feji (1979). The orienting reflex as a personality correlate. *In* "The Orienting Reflex in Humans" (H. D. Kimmel, E. H. van Olst, and J. F. Orlebeke, eds.). Copyright 1979 by Lawrence Erlbaum Associates.]

B. Augmenting–Reducing of the Cortical Evoked Potential

To obtain a measure of the cortical response to a stimulus, a brief stimulus, such as a light flash or a tone, is presented to the subject a number of times while the electroencephalogram (EEG) is recorded. The half-second periods of the EEG subsequent to the stimulus are averaged by a computer at selective points in time. The resulting waveform represents the evoked potential (EP) for that individual in response to the specific stimulus. Studies of identical twins have shown a very high similarity of the complex waveforms, indicating a high heritability for the EP. The augmenting–reducing paradigm compares the relationships between the amplitude of an early EP component, at approximately 100–140 msec after stimulus presentation, reflecting the initial impact of the stimulus on the cortex, and the intensities of stimuli (see Glossary).

A significant relationship has been found in many studies between the Disinhibition subscale of the SSS and the augmenting–reducing continuum of visual and auditory EPs. Figure 2 shows the results of a study using the visual EP, and Fig. 3 shows the results of a study using the auditory EP. In both studies, the high disinhibitors tend to show an increasing cortical reaction proportional to increasing intensities of the stimuli. In contrast, the low disinhibitors show little increase in EP amplitude with increases in stimulus intensity, and, in the case of the visual stimuli, they show a marked reduction in amplitude at the brightest intensity of stimulation. Studies in humans and cats (who show a similar individual variability in the phenomena) show that augmenting–reducing reflects brain and not peripheral receptor differences.

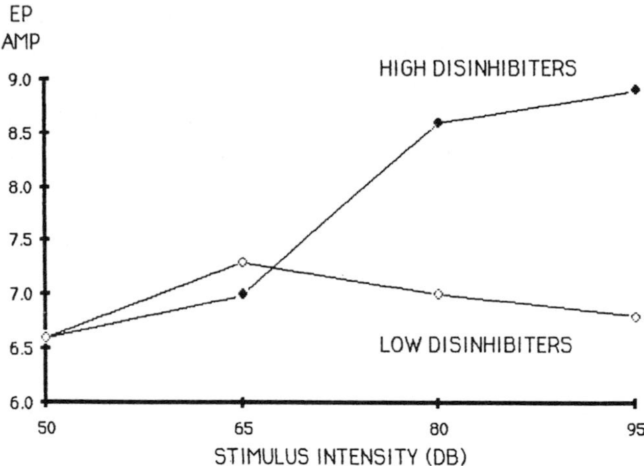

FIGURE 3 Mean auditory EP amplitudes (μV) for groups scoring high and low on the Dis subscale of the SSS, form V, at each of four stimulus intensities (50, 65, 80, 95 dB). High versus low Dis × stimulus intensity interaction significant, F(df 3, 111) = 3.25, $P < 0.05$. [Reprinted, with permission, from M. Zuckerman, R. F. Simons, and P. G. Como (1988). Sensation seeking and stimulus intensity as modulators of cortical, cardiovascular, and electrodermal response: A cross-modality study. *Personal. Individ. Diff.* 9, 368.]

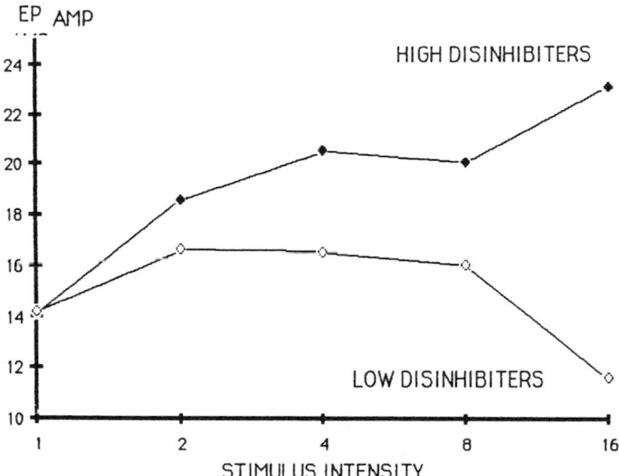

FIGURE 2 Mean visual EP amplitudes (in arbitrary millimeter deflection units; each unit = 0.42 μV) for groups scoring high and low on the Dis subscale of the SSS, form IV, at each of five stimulus intensity settings for the Grass photostimulator. High versus low Dis × stimulus intensity interaction significant, F (df 4/104) = 2.83, $P < 0.05$. [Reprinted, with permission, from M. Zuckerman, T. Murtaugh, and J. Siegel (1974). Sensation seeking and cortical augmenting–reducing. *Psychophysiology* **11**, 539. Copyright 1974 by The Society for Psychophysiological Research.]

Cats who are classified as augmenters by the EP method show analogous types of behavior to human sensation seekers, tending to be active and exploratory, to approach novel stimuli, and to show strong response for reward. Reducer cats show slow adaptation to a novel environment and fear and avoidance of novel stimuli, but they learn better than augmenter cats on a task requiring inhibition of response in order to obtain reward. The correlation of the slope measure of EP augumenting and days to learn a task requiring the animal to maintain a low rate of response for reward was +0.73 ($P < 0.05$); the higher the slope (augmenting) of a cat, the longer it took to learn to delay its responses. The impulsivity shown by augmenter cats and the constraint shown by reducers is also consistent with the finding that augmenting–reducing of the EP is related to the trait of impulsivity as well as sensation seeking in humans.

An inbred strain of rats showing EP augmenting is characterized by exploratory and aggressive behavior, an easily acquired taste for alcohol, and more responsivity to strong stimulation of "reward centers" in the brain. Another inbred strain, characterized by EP reducing, is less exploratory and aggressive and more fearful, and has little taste for alcohol. As with cats, augmenter rats show traits analogous to those of human sensation seekers.

C. Gonadal Hormones

The relationships between gonadal (sex) hormones and sensation seeking have been studied in males. Females have not been extensively studied because of the problems of variation due to the menstrual cycle. Both testosterone and estrogens in males correlated with sensation seeking, particularly with the Dis subscale. Table II shows the mean values for plasma testosterone, estradiol, estrone, and progesterone in males scoring high or low on the Dis subscale of the SSS. High-Dis males had significantly higher levels of testosterone, estradiol, and estrone than low-Dis males. In addition to sensation seeking, testosterone tended to correlate with sociability, impulsivity, and extent of heterosexual experience in males. [*See* Hormonal Influences on Behavior.]

D. Other Hormones

Cortisol in cerebrospinal fluid (CSF) has been found to be negatively correlated with sensation seeking. Cortisol tends to be elevated in severe depressions, and low levels of cortisol in CSF may be related to the lack of inhibition in the disinhibitory type of sensation seeking. Similarly, thyroid function is lower in high- than in low-sensation seekers. Thyroid is related to tension and anxiety and thus may inhibit behavior that raises arousal levels.

TABLE II

Gonadal Hormone[a] Levels of Male Subjects[b] Scoring in High and Low Ranges of the SSS Disinhibition (Dis) Subscale[c]

	High Dis		Low Dis		
Hormone	Mean	SD	Mean	SD	t
Testosterone[d]	943.6	210.3	711.1	278.3	2.98*
17-Estradiol[e]	38.0	11.1	27.2	6.4	3.78**
Estrone[e]	40.1	18.1	24.6	11.4	3.23*
Progesterone[e]	56.4	25.3	52.5	16.7	0.57

[a]Obtained from plasma, venous puncture.

[b]ns = 40 high-Dis and 40-low Dis subjects.

[c]From R. Daitzman and M. Zuckerman (1979). Disinhibitory sensation seeking, personality and gonadal hormones. *Personal. Individ. Diff.* **1**, 103. Reproduced with permission.

[d]ng/100 ml.

[e]pg/ml.

$*P < 0.01$, $**P < 0.001$.

E. Monoamine Oxidase

Monoamine oxidase (MAO) is an enzyme found in neurons of the monoamine systems in brain. It regulates the levels of neurotransmitters in these systems by breaking down the neurotransmitter after it is taken up into the cell from the synaptic cleft. It regulates the levels of neurotransmitters and, therefore, the degree of neural transmission by keeping a balance between production and amount stored in the neuron. In living humans, type B MAO is measured from blood platelets. This type of MAO is found primarily in dopamine-containing neurons in the primate brain. Levels of platelet MAO show high heritability and are related to individual differences in activity in newborn infants. MAO has been found to be negatively correlated with general sensation seeking in a number of studies. Although the typical correlations are not high ($r = -0.25$), they are usually significant and always in a negative direction. High-sensation seekers tend to have low levels of MAO. This is consistent with higher levels of MAO in women than in men and the tendency of both brain and platelet MAO to rise with age. Sensation seeking is higher in men than in women and scores on the SSS fall with age. Studies of high- and low-MAO individuals in the general population show that low-MAO types tend to be more sociable, less law-abiding, and more prone to use drugs (Table III). Drug users, alcoholics, and bipolar disorders have low levels of MAO, consistent with their high levels of sensation seeking. MAO is related to EP augmenting–reducing: augmenters have low levels of MAO relative to reducers. Studies of high- and low-MAO monkeys in natural colonies show analogous relationships to social behavior. Low-MAO monkeys are more playful, aggressive, and sexually active; high-MAO monkeys are passive and relatively unsociable (Table IV).

F. Monoamines: Their Metabolites and Enzymes

The findings relating MAO to sensation seeking suggested that there might be a relationship between at

TABLE III
Significant Differences between Low and High (Upper and Lower 10%) Platelet MAO Subjects[a,b]

MAO	Males		Females	
	Low	High	Low	High
No. hours socializing on averge weekday and weekend	13.2	8.5**	15.2	11.4*
Seen psychologist/psychiatrist for personal problems	37%	12%*	44%	29%*
Convictions for offenses other than traffic violations	37%	6%*	0%	0%

[a]The investigators mention other differences for which significance levels but no incidence figures are given: More low-MAO males than high-MAO males had used illegal drugs ($P = 0.008$), including depressants ($P < 0.05$) and stimulants ($P < 0.05$), and were currently using more stimulants ($P < 0.05$) and hallucinogens ($P < 0.05$). Low-MAO males smoked more cigarettes than highs ($P < 0.01$).

[b]From R. D. Coursey, M. S. Buchsbaum, and D. L. Murphy (1979). Platelet MAO activity and evoked potentials in the identification of subjects biologically at risk for psychiatric disorders. *Br. J. Psychiatry* **134**, 372.

$^*P < 0.05$, $^{**}P < 0.01$.

TABLE IV
Partial Correlations, Removing Age Effects, between Mean Behavior and Platelet MAO Activity in Rhesus Monkeys Living in Natural Conditions[a]

Activities	Males (n = 17)	Females (n = 26)	All subjects (n = 43)
Forage	−0.14	−0.04	−0.08
Move	−0.22	−0.18	−0.20
Lookout	0.13	0.03	0.07
Play	−0.71*	0.26	−0.17
Give grooming	−0.28	0.16	−0.01
Receive grooming	−0.21	−0.41*	−0.34*
Self-grooming	0.69**	0.33	0.49***
Inactive	0.15	0.01	0.06
Rest–sleep	0.32	0.40	0.37*
Dominant–agonistic	−0.55*	−0.14	−0.31*
Submissive	0.38	0.09	0.20
Social contact	−0.54*	−0.18	−0.33*
Alone	0.59*	0.45*	0.51***

[a]From D. E. Redmond, Jr., D. L. Murphy, and J. Baulu (1979). Platelet monoamine oxidase activity correlates with social affiliative and agonistic behaviors in normal rhesus monkeys. *Psychosom. Med.* **41**, 87. Copyright by American Psychosomatic Society, with permission.

$^*P < 0.05$, $^{**}P < 0.01$, $^{***}P < 0.001$.

least one of the monoamine systems and the trait. The primary monoamine systems in brain are identified with one of the neurotransmitters norepinephrine (NE), dopamine, or serotonin. Although it is impossible to assess brain levels of the neurotransmitters in humans, their activity may be gauged by levels in CSF, plasma, and urine. Ordinarily, the metabolites of the monoamines are assessed, although CSF levels of NE may furnish a direct index of activity in one NE brain system that sends efferents down the spinal cord. In one study, CSF NE was negatively correlated with general sensation seeking ($r = -0.51$; partial r controlled for age, height, and weight $= -0.49$; both r's significant; $P < 0.01$). Consistent with this finding were negative correlations between plasma levels of the enzyme dopamine-beta-hydroxylase (DBH) and sensation seeking in this study ($r = -0.44$; partial $r = -0.60$; $P < 0.05$) and others. DBH is the enzyme converting dopamine to NE in the NE neuron. A low level of the enzyme would inhibit the production of NE. As yet, no correlations have been found between the metabolites of serotonin and dopamine in CSF and sensation seeking; however, the serotonin metabolite is low in impulsive persons.

V. A PSYCHOBIOLOGICAL MODEL

The approach to the psychobiology of sensation seeking has thus far been largely correlational. Experimental work on the role of biological factors in determining behavior in humans is difficult. It is unethical to produce irreversible brain alterations in humans in order to study the effects on behavior and personality; however, using data from biological psychiatry, where potent drugs are used to ameliorate disturbed behavior, and from comparative experimental work with animals, where the brain may be selectively lesioned or stimulated, it may be possible to fit the correlational data from normal humans into a biological model.

The strong OR found in high-sensation seekers is a product of novelty because differences between highs and lows disappear when stimuli are repeated. The strong OR trait may reflect the disposition in a nervous system adapted to process novelty and approach novel stimuli. This is a characteristic of sensation seekers at the behavioral level. They constantly seek novel information and tend to avoid repetitious or redundant information. In low-sensation seekers, novelty tends to be threatening, and in the face of novel stimuli they tend to manifest defensive physiological responses as if in preparation for flight or planning escape.

In contrast to the OR, which is a response to novelty at low to moderate levels of intensity, the augmenting or reducing of the EP is a reaction to the intensity of stimulation, because it is defined by the relationship between EP amplitude and intensity of stimulation. Augmenters seem to have a "strong nervous system," to use a Pavlovian theoretical term. The task performance of augmenters is less affected than reducers by stressful distractions. Reducing is essentially a protective reaction, but it may result in cortical inefficiency as a consequence of cortical inhibition in response to a high intensity of stimulation. Augmenting could also represent a vulnerability in many situations because the reflexive protection against high levels of stimulation is lacking. In behavioral terms, the link between augmenting and disinhibition can be seen in the tolerance of high-sensation seekers for loud music and highly arousing stimuli of all kinds. The vulnerability of augmenters can be seen in their disposition to affective disorders of the bipolar type. In manic states, bipolars seek intense stimulation without relent, and the stimulation serves to further excite them. The combination of low MAO and augmenting in persons at risk for this kind of affective disorder suggests that the lack of regulation on one or more of the monoamine systems predisposes them to the failure of cortical inhibition mechanisms and the loss of behavioral control seen in mania.

Gonadal hormones seem to play an important role in activating social, sexual, and sensation-seeking behavior. One of the causal paths for this effect may be their reducing effect on MAO. Low MAO is consistently linked to a variety of sensation-seeking related behaviors, from activity in newborn infants to interest and participation in mountain climbing in adults, but the role of this enzyme is not clear. Low levels of type B MAO could be associated with high activity or underegulation of dopaminergic systems in the brain. Dopaminergic systems in the limbic brain are involved in the initiation of appetitive approach behavior. Whether released by direct stimulation or stimulant drugs, such as amphetamine and cocaine, it produces a state of high energy and euphoria in humans. In rats, dopamine maintains self-stimulation through electrodes placed in the nucleus accumbens. High levels of dopaminergic activity may underlie the hedonistic pursuits of human sensation seekers and their attraction to drugs, which increase dopaminergic activity. Such drugs also increase activity in the other catecholamine systems prominent in the brain, the noradrenergic ones. Ascending norepinephrine pathways innervate the entire cortex and serve to arouse the cortex in response to novel signals or those of biological significance. Sensation seekers may engage in exciting, sometimes risky, activities in order to activate these catecholamine systems to an optimal level. Low-sensation seekers may already be at their optimal level of catecholamine system activity and too intense or novel stimulation might push them over an optimal level into a zone characterized by anxiety and even panic. Both sensation seeking and drug use may represent an attempt to find an optimal level of catecholamine activity, producing a state somewhere between boredom and terror.

Serotonergic systems are related to behavioral control in animals, and low levels in humans are associated with impulsive and aggressive behavior. A weak serotonergic system may underlie the lack of inhibition in the impulsive sensation seeker and the willingness to take risks in order to experience intense or novel stimulation.

Environmental factors obviously must play some role in producing high or low levels of the trait, but the studies on twins suggest that it is not the shared family environment but rather the environment that is specific for each member of a family that influences sensation seeking. The peer environment, for instance, may be more influential than the level of stimulation in the home shared by all family members. This is not to say that *prolonged periods* of low or high stimulation during development may not influence the expression of the inherited component of sensation seeking. Subjects coming out of sensory deprivation experiments of several days or more show reduced cortical activity and a temporary lack of motivation to seek stimulation, as if their optimal level of stimulation had moved downward. However, such extreme stimulus reduction is not very common during development.

Peer or parental modeling or reinforcement may influence the phenotypic forms that sensation seeking will take, more than the broad basic tendency itself. Needs for sensation may be modified by attitudinal and religious training, but it is just as likely that sensation seeking will modify learned attitudes if they are not in accord with the basic temperament. The child and adolescent who conform and those who rebel against parental standards may be following their own biological destinies and seeking from their peers and environment that which fits their own temperaments. There is much to learn about the interactions of the genotype and environment in producing personality traits. We select what we need from the range of available environments and are in turn influenced by what we select as well as that which we cannot control.

BIBLIOGRAPHY

Zuckerman, M. (1979). "Sensation Seeking: Beyond the Optimal Level of Arousal." Lawrence Erlbaum Associates, Hillsdale, New Jersey.

Zuckerman, M. (ed.) (1983). "Biological Bases of Sensation Seeking, Impulsivity and Anxiety." Lawrence Erlbaum Associates, Hillsdale, New Jersey.

Zuckerman, M. (1983). Sensation seeking: A biosocial dimension of personality. *In* "Physiological Correlates of Human Behavior: Vol. 3, Individual Differences" (A. Gale and J. Edwards, eds.). Academic Press, New York.

Zuckerman, M. (1984). Sensation seeking: A comparative approach to a human trait. *Behav. Brain Sci.* 7, 175.

Zuckerman, M. (1985). Biological foundations of the sensation seeking temperament. *In* "The Biological Bases of Personality and Behavior" (J. Strelau, F. H. Farley, and A. Gale, eds.), Vol. I. Hemisphere, Washington, D.C.

Zuckerman, M. (1988). Brain monoamine systems and personality. *In* "Neurobiological Approaches to Human Diseases" (D. Hellhammer, I. Florin, and H. Weiner, eds.). Hans Huber Publishers, Toronto.

Zuckerman, M. (1990). The psychophysiology of sensation seeking. *J. Personality* **58**, 313–345.

Zuckerman, M. (1994). "Behavioral Expressions and Biosocial Bases of Sensation Seeking." Cambridge Univ. Press, New York.

Zuckerman, M., Buchsbaum, M. S., and Murphy, D. L. (1980). Sensation-seeking and its biological correlates. *Psychol. Bull.* **88**, 198.

Zuckerman, M., Ballenger, J. C., and Post, R. M. (1984). The neurobiology of some dimensions of personality. *In* "International Review of Neurobiology" (J. R. Smythies and R. J. Bradley, eds.), Vol. 25. Academic Press, New York.

Sensory–Motor Behavioral Organization and Changes in Infancy

HENRIETTE BLOCH
Laboratoire de Psycho-Biologie de l'Enfant

GLOSSARY

Allometry Study of the relative growth of different parts of the body; their relation is called *allometric*

Altricial Pertaining to a species that is born immature

Automatic walking Forward stepping reflex triggered by handling a newborn under its armpits, upright, with the soles of its feet in contact with a plane surface. This reflex disappears spontaneously around the third month

Behavioral state "Temporary stable conditions of neural and autonomic functions, known as sleep and wakefulness" (H. F. R. Prechtl). With behavioral criteria such as open/closed eyes, breath regular/irregular rhythm, gross movements/no movement, vocalization/no vocalization, and concomitant cues of polygraphic recording, five behavioral states are distinguished: (1) quiet sleep; (2) agitated sleep; (3) quiet alertness; (4) agitated alertness; (5) cries

Biomechanical constraint Any limitation to moving due to mechanical arrangement among body components, such as muscles, joints, skeleton elements, or fate deposition

Cephalocaudal law Universal principle of development in organized species. Development begins from the cephalic segment and then, step by step, reaches the extremities

Circular reaction Organized act by which any positive effect leads to its subsequent reelicitation

Gestational age Fetus or infant age computed from the last menses of the mother

Habituation When a stimulation perdures or is repeated, the initial response it provokes decreases with time. Such a decline differs from saturation or forgetting because the initial rate of response can be recovered with partial changes in the stimulation or a delay

Moro reflex Reflexive extension of the arms, followed by an abduction on the chest. This reaction responds to a sudden loss of support. It seems to have a vestibular origin

Myelination Discontinuous sheath of myeline that encircles progressively many nerve fibers in the vertebrate central nervous system and trains a faster conduction of nervous impulse

Posture Antigravity positioning of the whole body and positioning of its mobile parts with respect to each other

Reversibility Main property of a logical structure in which invariants are formed by reciprocal ties between two actions or intellectual operations oriented in opposite directions

Saccadic eye movement Jump of the gaze. To fixate a target, an adult directs his/her gaze by an initial jump of the eyes, matched to the target distance. The newborn's jump is of standard amplitude. A series of jumps is necessary to point a fixed target or to track and suit a mobile

LIKE OTHER MAMMALS, THE HUMAN INFANT ARrives at birth still immature. Nevertheless, it is equipped with instruments that can ensure relations with the external world. Its sense organs are morphologically developed, and its sensory systems, although they are at different levels of maturation, are all functional. Their functioning implies a motor participation: Every sense organ can change its orientation either by a self, local movement or by a movement of the head, which contains all the sense organs except the tactile ones. If anything prevents a sense organ

ENCYCLOPEDIA OF HUMAN BIOLOGY, Second Edition, VOLUME 7. Copyright © 1997 by Academic Press. All rights of reproduction in any form reserved.

from moving, perception will be altered. It has been shown in kittens that by cutting off the ophthalmic branch of the trigeminal nerve, one eye can be impeded to move in the orbit, and that trains alterations in the functional array of cells in the striate cortex and causes defects in the binocular visual field.

Moreover, the neonate's motility includes more than rapid stereotyped reflex reactions and mass agitation. Slow, aimed movements and general, isolated movements are performed; some of them appear gracious and delicate, as manipulation of fingers, which can be observed in wakefulness and in so-called agitated sleep. Touching a part of the body and reaching to some object in the immediate environment may carry out a gaze shift toward the touched or reached surface. Therefore, we can assume that sensory and motor activity are linked in several ways from the first days of life. Observations on preterm-born infants and on fetuses show that such linkages are prewired. Neurophysiological studies conducted with animals invite us to think that early sensory–motor behavior is controlled by a common structure in the midbrain brain stem, the colliculus. It may be that some other structures in receptive and motor cortices and in the cerebellum take part in the command process. The image of the impulsive newborn living in chaotic confusion has been shattered; such a view was based on our ignorance. However, the human neonate is altricial. Many limitations and constraints bear heavily on its sensory and motor behaviors (i.e., fragile behavioral states, biomechanical constraints due to allometry, and coaction effects caused by a possible unclear differentiation between systems). The organization of early behavior appears different in the newborn and the young infant and from what it will be later in childhood and adulthood.

To understand the initial features and the mechanisms of transition, we first must examine separately the panel of sensory and motor activity and then their possible coordinations.

I. THE NEONATE REPERTOIRE

A. Newborn Sensory Activity

Sensory activity is devoted to gathering and processing information from the external environment. A primary condition to do that is to detect separate stimulations, that is, to specify where is the stimulus and what it is. Because the maternal womb is a narrow and confined milieu, it had been supposed that the fetus experienced only contact sensitivities, such as touch and taste, and that the neonate, in consequence, would be sensible to contact stimulations rather than to distal ones, such as sight and sound. Such a hypothesis has been disconfirmed. The orientation response can be obtained from the first hours after delivery for visual and auditive stimulations as well as for contacts. The orientation response is observed in healthy prematures as early as 30 weeks gestational age. In visual perception, which is the most often studied modality, peripheral receptors of the retina are activated and trigger a head-eye turning toward the source of stimulation. The gaze can stare at the visual target, but holding the fixation seems to be difficult if not impossible, because of the foveal immaturity and a lack of control of the rapid, little eye movements that accompany ocular fixation. These are considered strong indications that human newborn perception could be very poor. However, the neonate can distinguish a shape on the ground, when the shape is a closed, structured, and complex one and when the contrast between the ground and the shape is high enough. Perceptual habituation is also possible for auditory, olfactory, gustative, and visual stimulus. This allows us to assume that external information is brought and processed. However, many parameters of perceptual behavior are yet unknown.

Perceptual discriminations are now studied systematically. Some of them appear to be better in neonates than later in infancy, such as speech sounds and odors, and some of them seem rather less sharp. Perceptual exploration and the organization of activity also seem to be greatly different than at later ages. For instance, visual tracking is jumpy and is called saccadic, and visual pursuit of any mobile object is discontinuous. The visual field is less extended in the neonate than at later ages. So, we can summarize this evidence in concluding that the perceptual world in the beginning of life is very different from that of the adult. However, it does exist. [*See* Perception.]

B. Motor Activity

Motor activity of the newborn involves several categories of movements and positionings. Any sort of movement refers to the postural framework. Posture is defined as the spatial orientation of the whole body or the limbs for holding an attitude or executing a movement. In vertebrates, posture is performed by the action of muscular systems on the skeleton. Every positioning and movement can be described as a part of any postural organization and in relation to it. Postural maturation is rudimentary at birth. Nevertheless, human fetuses and neonates can adopt preferential postures that might have an influence on some aspects of later motor development.

Spontaneous motility can be observed with different frequencies in every behavioral state as shown in Fig. 1. A perceptual stimulus can induce more than one movement, depending on its signification and value. Rhythmical reactions often are given as immediate responses. Adapted responses are slower than later in the life.

Part of the motor repertoire present at birth is going to disappear in the first postnatal months. Many reflexes (e.g., the Moro reflex, the grasping, the automatic walking) are called *archaic reflexes*. Some of them can be considered to have influence on later, more integrated responses, whereas some of them will not. Grasping, for example, was first considered as an archaic, rudimentary form of prehension; but recent studies about early hand-reaching have convinced us that grasping has to be viewed as acting against reaching and smooth prehension. However, automatic walking seems to exert influence on later autonomous locomotion. Training in automatic walking accelerates the accession to autonomous walk. So the importance of early reflexes for motor development is controversial.

Biomechanical constraints are heavy at the beginning of extrauterine life. The weight of the head, which is comparatively greater than the trunk and the limb weight, pushes the body to flexion and provokes head inclinations, even falls. Antigravitational reactions are not formed, so the newborn has only weak control of its own posture. Its postures are arranged by the adult caregiver. Despite this obvious immature aspect, preferential positions adopted by neonates are in continuity with fetal positions. The preferential positions are asymmetric and seem to be triggered by the asymmetric tonic neck reflex. Therefore, the spontaneous motility, which has to be referred to posture, is different in quantity and in quality among the newborns. Nevertheless, it reveals some lateral differences and invites us to think that the two hemispheres are not involved in the same way in any motor act.

The motor repertoire of the newborn is, however, considerable, and some movements that are spontaneous can be used as aimed movements in a coordinated act. This is the case for head movement: In supine position, a baby can turn its head in response to a perceptual stimulation, and head movement is a part of the orienting reaction. However, it is not performed

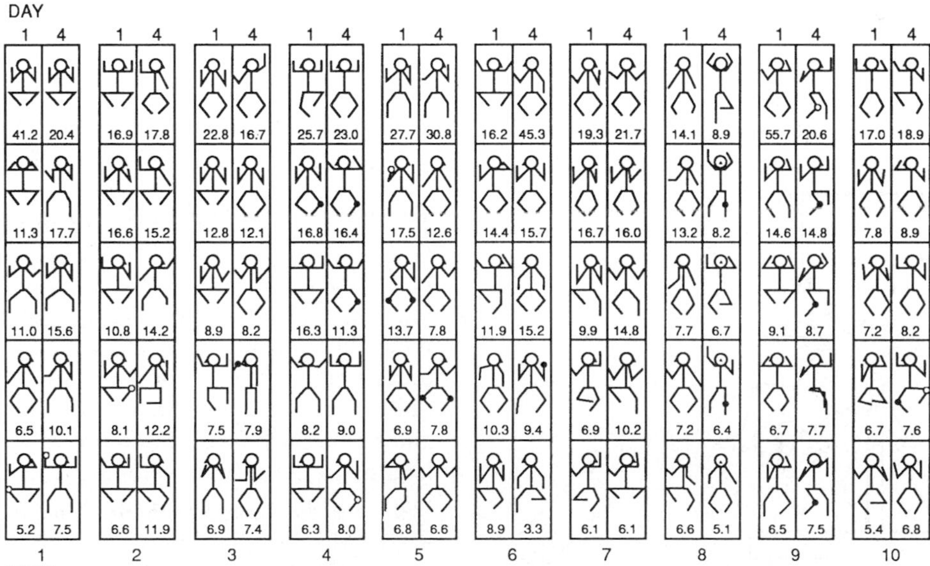

FIGURE 1 The five most frequently occurring postures per baby on Day 1 and Day 4 after birth. [From G. Cioni, F. Ferrari, and H. F. R. Prechtl (1989). Posture and spontaneous motility in fullterm infants. *Early Hum. Dev.* 18, 247–262, with permission.]

simultaneously with eye movement, and it can be impeded by an eccentric eye displacement, as if it is controlled by a musculo-external muscular locking.

C. Newborn Sensory–Motor Coordinated Behaviors

Because of the evidence of prewired sensory and motor abilities, newborn capacity to couple sensory–motor acts is currently being explored in some observational and experimental settings. Such studies, unfortunately, are still not numerous enough to provide useful information about available conditions and suitable criteria of early couplings. Attention is presently paid to eye–head, eye–hand, and hand–mouth coordinations. Observations in every day life environments show that such coordinations are neither obvious nor frequent when the infant is lying in its cradle. However, they can be observed when the newborn is held in a sitting position, with the head upright. Thus posture appears to be a determinant condition for coordinated behavior.

As studied in the experimental design of Fig. 2, with respect to good postural condition, neonate coordinated behavior relies on a global synchrony between perception and movement toward any object in the proximal environment. However, it has specific features that raise questions and controversies:

1. Movement is not always triggered by the perception of the target. In eye–head coordination, head movement can precede a gaze shift; hand movement toward an object can be performed with the eyes closed.

2. Attainment of the target, considered as the goal, is neither frequent nor accurate. The approach movement of the head or of the hand stops before encountering the target. In hand–mouth coordination, it has been shown that the hand can touch some parts of the face before or without going to the mouth, which stays open.

3. Perception and movement are not closely and continuously coupled. For instance, velocities and amplitudes of head movement are not in harmony with eye movement. In hand-reach-like movement, the baby does not visually control its hand, except when the hand approaches the vicinity of the target.

4. Perceptual pointing seems to be better than the associated movement. The neonate is able to point accurately to a target by moving the eyes and to inhibit the saccadic jump when the target is not too eccentric (in regard with his/her own position). The eye movement appears to be well organized. Head and hand movements admit many different patterns, for they have more degrees of freedom that have first to be reduced.

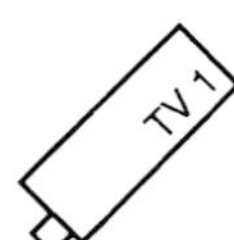

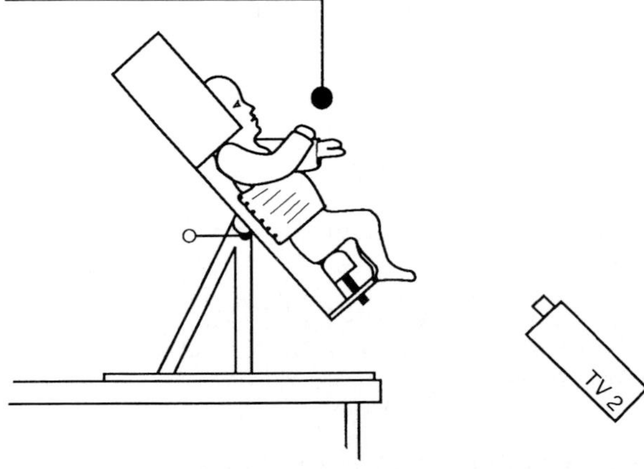

FIGURE 2 Experimental situation for studying newborn eye–hand coordination. [From C. Hofsten (1982). Eye–hand coordination in the newborn. *Dev. Psychol.* **18**, 450–461, with permission.]

II. BASES AND DETERMINANTS OF SENSORY–MOTOR DEVELOPMENT

The 1-year-old infant's behavior appears to be both very different from the newborn's and more similar to child and adult behavior: The infant can stay standing up, it begins to walk, it recognizes persons, it remembers positions of objects, it searches for and hides objects, it can compare and associate several objects and build some new configurations with them, and so on. Such large differences have led us and still lead us to consider development in the first year as made through successive qualitative changes rather than a linear, quantitative progression. At the beginning of the twentieth century, the very commonplace opinion was that this development was mainly sensory–motor and implied some changes in nervous functioning and in the postural framework. The latter were easier to study than the internal changes.

A. Postural Maturation and Its Consequences

Posture develops in humans within the first 2 years after birth, when body positions are generally mastered, following a cephalocaudal law and a proximo–distal principle. Four stages have been distinguished from the neonate prone position:

1. Around the third month, the infant becomes able to hold its head upright. It is able to prevent forward and lateral inclinations or "falls" of the head and can turn the head into a horizontal plane without participation of the torso and the shoulders. The main consequence of this change is that the line of the gaze can remain constant.
2. The sitting position is attained at 4–5 months. When the infant can stay seated without any back support, its arms are "liberated." So, the space of manual prehension extends and may be structured with a stable reference to the body median vertical axis.
3. The infant begins to stand up at about the ninth month. First, it stands up holding a solid stationary object such as a wall or a chair; then it can stay up without any material support. This stage appears to be transition to walking.
4. At last, the infant begins to walk by itself at about 13–15 months, first leaning on a support and then without any support. The autonomous locomotion represents achievement of a general motor coordination and implies a stable organization of movements. It greatly enlarges the space of action.

This postural development includes other local transitions and steps, such as creeping and crawling, which show a continuity. Cultural influences can shorten a stage, but they cannot modify the successive order of the main stages. However, sensory or motor handicaps can be responsible for severe retardations. For instance, blind-born infants attempt the sitting and standing position and walking later than normal infants when they do not receive any compensatory help, such as an ultrasonic guide, but they attempt these stages in the same order as do normal infants. Therefore, the postural changes in human infancy appear similar to the postural changes in all the superior vertebrates and are considered maturational changes. They can be considered organizers of the environmental space: Every sense organ has a specific receiving area, which catches the stimulations from a spatial limited part of the surroundings. Movement makes this part mobile, and a whole coordinated system of movement makes the successive parts homogeneous. [*See* Motor Skills, Acquisition.]

B. Changes in Nervous Functioning

The infant central nervous system differs from that of the child and the adult in its morphology and its functioning. It is the seat of large, rapid, and antagonist transformations, which are not yet totally clear, during the first years of life. All the structures of the brain are formed at birth, but they include populations of cells and synapses that can be more or less numerous than they will be later; their spatial array is also different, and some of them seem not to be implicated in early behaviors (e.g., the associative, prefrontal, and frontal cortices). Two major developmental nervous processes have been predicted to cause changes in sensory–motor organizations and behaviors: myelination, which was viewed for a long time as a single, general criterion of nervous maturation, and corticalization.

1. Myelination

Myelination of several motor and sensory pathways begins before birth in humans but continues after birth. The pyramidal pathways, which play the main role in motor-coordinated acts and in the spatial orientation of the head and the body, are not myelinated at birth. Their myelination begins at about the normal term of birth and accelerates at about the third month. It is not achieved before the end of the second postnatal year and occurs earlier in the corticospinal paths than in the cortico-moto-neural ways that command voluntary movements. Among the sensory systems, the visual system has been the most studied. Although it develops after the others, its rhythm of maturation is rapid after birth. Nevertheless, it is far from a full myelination during the first months postnatally. The subcortical pathway, which goes from the peripheral retina to the superior colliculus, is fully myelinated at 3 months, having started at the sixth month of gestation. In contrast, the central retina-to-cortex tract begins to myelinate only after birth; the myelination increases rapidly up to 4 months and then decelerates, but is not achieved before the end of the second year. Other input connections to the primary visual cortex remain poorly myelinated up to 3–4 months. The myelination of extrastriate areas and intracortical neurons takes a long course, as long as 7 years.

According to the myelination process, voluntary movements, such as those involved in prehension, could not be performed before the fourth postnatal month. Such requirements of prehensive behavior, for example, the continuous control of movement from its initiation and the anticipatory shaping of the hand in relation to the perceived size and shape of the object to be reached, would not be possible before this time.

2. Corticalization

Corticalization is an evolutionary process that characterizes development in primates and is highly important in humans. Regarding sensory–motor capacities, it presents the following features:

- An enlargement of the lateral parts of the cerebellum, which are involved in the motor control of limbs;
- An increasing of the prefrontal cortex, which is involved in the motor control of manuality;
- A functional specification of the parietal associative cortex, which commands the accurate visual guidance of hand movements; and
- A large development of the associative areas of the frontal cortex, which are concerned with the processing of external information and self-movement-produced information. [*See* Cortex.]

The corticalization process is commonly supposed to be important in the first months after birth. According to this view, perceptual analysis of form, performed by the foveal vision, would truly enhance around the second month. The eye–head coordination would also appear and would be controlled by the corticofugal pathways and not by the colliculus. Changes in the oculomotor organization of eye movements, namely, in the saccadic components, have been observed at this time.

Later, the corticalization seems to be the main factor in the onset of autonomous walking, which represents a synthesis of voluntary movement, including a control of alternative leg movement, parallelism of the lower limbs, vertical equilibrium of the body, and visual control of direction and distance. The main characteristic of cortical functioning is a broad distributed activation of different areas. For example, it has been shown that visual perception involves not only the primary visual areas but also the temporal and the parietal cortices. That leads one to presume there is a permeable frontier between sensitivity and motoricity. Moreover, this is considered to be a source of flexibility and of rapidly adjusted behavioral responses.

C. Open Questions

Questions arising from the general bases and determinants of sensory–motor development in human infancy will now be discussed. First, these bases and determinants do not allow us to understand all the changes observed in behavior. Second, the sensory–motor repertoire in the first few months appears more rich and organized than should be presumed from the myelination and the corticalization processes. That does not mean that these processes have no influence on behavior, but that they are not its only determinants and that the relations are not simple and direct. The temporal coincidences between neural and behavioral changes provide some opposite evidence. For instance, if we accept that eye movements are first controlled by the colliculus and cannot include a foveal fixation, we cannot explain the visual habituation phenomenon in newborns, which requires a fixation to hold for a long time, often more than 3–5 seconds; nor can we explain some fine discriminations of form features performed by newborns as distinction between curvilinear and linear contours of visual forms. If we accept that any aimed arm–hand movement can be generated before the pyramidal pathways are myelinated, we cannot understand the neonate reaching behavior.

III. COURSE AND PROCESSES OF SENSORY–MOTOR DEVELOPMENT

Sensory–motor development has been studied, with different approaches, since almost the first century. However, its course and its processes still remain partly unclear. The obscurities and discrepancies seem to derive from three causes. First, hypotheses were different for motor and for perceptual development and induced separate research. Motor changes in infancy and childhood were approached with maturationist hypotheses, and perceptual changes with cognitive and learning ones.

Second, it is difficult to detect and describe successive changes without any model or theory about their directionality or without a reference to a stable state. The reference taken in motor studies was more often biological and phylogenetical. The aim was to arrange the changes in a hierarchical continuity from embryology. The studies of perception referred to adult perceptual behavior and were used to understand how a stable world is built.

Third, for understanding sensory–motor coupled

development, the research of a single principle predominated for a long time. Three general theories have been proposed.

Because some questions remained unanswered within such general, theoretical frames, current research has moved away from them and tried to accumulate more empirical data, without any explicit *a priori* model. During the past 20 years, research has focused on collecting data and inventing efficient methods of investigation. Sometimes theoretical points were discussed or revisited. Presently, it seems that psychologists again feel a need for theoretical overviews. Such a pendulum of historical movement compels us to review general models, to summarize the data, and, at last, to consider current interpretations.

A. Review of General Theories of Sensory–Motor Development

All the general theories that have been proposed consider the sensory–motor development as a part of ontogenesis and as the first level in a constructive process.

1. The Learning Model of Accommodation

J. M. Baldwin was the first psychologist who paid attention to the relation between perception and action from birth. He described the course of sensory–motor development as a series of changes into "circular reactions." What he called circular reactions are sensory–motor coupled responses to external stimulations, which appear during the first year of life. Movement would be first, such as reflexive movement in the newborn. Any spontaneous movement can have a positive or a negative effect and can be felt by the infant as pleasant or annoying. According to Thorndyke's law of effect, any act that has a positive or pleasant effect would be repeated. For instance, sucking is a reflexive reaction for ingesting milk. Then sucking movement would be repeated in a vacuum and would constitute a "primary circular reaction."

Later, a "secondary circular reaction" would appear when two parts of the body can be associated in a single act: When the hand of the baby encounters a visible bauble by accident, the sound its movement produces would train the act again. The infant learns that its own movement has a consequence on an external object, so it can learn to adjust its movement to replicate the sound. Such an adjustment reflects an accommodation. The secondary circular reaction appears around the fourth month postnatally and implies a coordination between vision and hand-reaching.

A third step would be when the infant becomes able to associate more than two systems. A "tertiary circular reaction" would appear at about 8–10 months, with complex or sequential actions and several objects.

Such a model allies sensory–motor development to a learning process. It supposes that the human infant has a "hedonic"-oriented capacity as soon as birth. That has been confirmed by L. P. Lipsitt in his research on neonate sucking behavior.

2. The Constructivist Theory

Jean Piaget started from Baldwin's description of circular reactions. However, he did not accept that they give evidence of accommodation. He considered that sensory–motor coordinations allow the infant to create means–goal relationships, which develop through active assimilation: The infant would perform mental work for linking its own actions and the consequences they have. When it is placed in a new situation, it can distinguish known and unknown components and apply a strategy in acting. The successive behaviors toward a hidden object provide evidence about an assimilation, for instance, when a 9-month-old infant searches for a hidden object at the place where it was found previously and not at the place where the object disappeared.

Therefore, sensory–motor development represents the first stage of cognitive development. This stage would be achieved when the fundamental organization that governs both action and cognition is built. Piaget demonstrated that such a structure is formed when the infant is able to walk, because autonomous walking implies what he called "reversibility." According to this theory, logic would be first in action, then it would pass into reasoning in concrete situations. Active assimilation could be inferred only from changes in a repeated act. Sensory–motor learning would not only be a gain in performance but would involve a transformation of a mental structure that controls the behavioral response.

3. The Ecological Theory

J. J. Gibson and E. J. Gibson assumed that sensory–motor development is not a process of enrichment but rather depends on perceptual differentiation. First, they considered that the perceptual systems can take and treat simultaneously a lot of information specified in any sensorial array, which contains object and space features and has an ecological meaning.

This information is not referred to the modality of input, but to the physical environment. Second, they claimed that there are functional equivalences between exteroceptive and proprioceptive information. So, intermodal relation can exist and perception is intrinsically meaningful. Perception–action couplings are determined by what the Gibsons called "affordances." Affordances are properties of objects or events in the surroundings that respond to the needs of the perceiver. They are together physical and psychological and they are, according to the Gibsons, "ecological." These properties are immediately translated into action terms. The perception of a nipple implies the act of sucking, and by sucking a nipple, the infant learns its elasticity, its texture, its form, and its size. Objects can be distinguished from the possibility of action they offer. A young infant does not know that a projected slide cannot be taken in hand, so it reaches for the picture as for a "real" object. By doing that it learns to discriminate pictures and solid objects; so it learns to reach only for solids.

B. Data on the Course of Sensory–Motor Development

The course of sensory–motor development does not appear as a linear progression. In many cases, it includes stops and apparent regressions.

The example of hand-reaching is illustrative: (1) The neonate reacts to contact of any object placed in the palm of its hand by a grasping response. Its fingers close upon the object by an immediate, sudden, and sharp reflexive movement. This reaction is described as stereotyped and serves as an item in the first postnatal neurological examination.

In the first days after birth, reaching movement has been shown toward distal visible objects. Movement is an arm–hand slow projection, without any grasping. The hand remains open during the movement. More often, it stops before encountering the object, although the infant is looking at it. Such a movement happens only when the infant is in a quiet alert state and is promoted when the infant is seated, with the head upright. Timing of movement is slow (about 1 second for reaching the vicinity of an object put at 20 cm from the body); velocity is not regular but shows accelerations and decelerations. Decelerations are coincident with changes in direction. At about the end of the first month, the frequency of such reaching hand movements decreases. The movement is made with the hand closed; when the hand contacts the object, there is no manual exploration. In this period, movement can be described as ballistic. At about 3 months, another change occurs: The arm–hand movement becomes again slower, visual control of it can be observed, the hand opens before the encounter of the object, and there is an anticipatory shaping according to the size of the object. However, the speed of reaching is often too rapid in the final phase, and the hand pushes the object and cannot easily ensure a prehension. Contact is performed by trial and error. Constant directionality and distance of the hand-aimed movement are observed at about the sixth month.

This development has a U shape and that leads us to consider that the successive stages are attained through reorganizations. The decline of the reaching behavior between the second and the fourth months may be due to an increasing importance of visual search and attention, which would inhibit the movement. Because of different rhythms of maturation in sensory and motor systems, it can be supported that the earliest sensory–motor behaviors cannot be harmoniously coupled. A good synchrony would suppose reciprocal control and take time to be built.

Continuity from grasping to prehension is not obvious. On the contrary, grasping appears as a transitory constraint, which is not a component of the early reaching movement. That can be interpreted as a change in the composition of elements, with a change in the function of behavior, or as an effect of inhibition by higher nervous centers.

(2) Development of prehension is not achieved at 6 months postnatally. Several later changes will occur: The reaching movement becomes less rigid, with smooth final braking, the manual gesture to take an object becomes more and more accurate, and more direct than tentative. The thumb–index finger grip, which appears at about 9–10 months, allows the child to take very small objects. In the same time, visual guidance is alleged, and visual pointing is very precise. However, the little child remains unhandy for several years. Child performance in ordinary sensory–motor tasks, such as catching a drop, lacing shoes, and buttoning up a coat, and in sensory–motor skills, such as sewing, drawing, and writing, increases up to 7 or 9 years with active practice and may also improve by modeling. However, performance is still inferior to the adult performance. Qualitative differences can be shown in the organization of such behaviors between 7-year-old children and adults, namely, in their spatiotemporal unfolding. Nevertheless, the commonplace opinion is that the course of sensory–motor development ends at about 2 years, because basic

local and general sensory and motor structures are stabilized and are homogeneously connected. So, the changes that happen later are considered as gains provided by some kinds of learning. Discussion generally centers on acquisition of motor skills and focus on learning processes in this field.

However, some currently conducted research induces some doubt about that. In studying the running and jumping abilities of a child, it has been shown that some postural and dynamical constraints are responsible for specific features and determine transient compositions of movements in relation to perceptual control, which are not overtopped with only practice. We need more information about them. However, they lead us to consider that the course of sensory–motor development is longer than we thought and could mean that sensory–motor development does not entirely depend on cognitive structures and functioning that emerge around the second year, but would continue in a parallel direction to cognitive development.

C. Concluding Remarks

For summarizing what we know about sensory–motor development, the following statements can be made:

1. Despite sensory and motor systems that do not mature at the same speed and rhythm, perception–action couplings are performed very early in the neonatal period. They appear asymmetric and transitional and are followed by several deep changes.

2. Several kinds of transition are observed in infancy and are presumed in childhood. Some of them involve a change of function: The development from early reaching to prehensive behavior exemplifies such a change. Early reaching seems not to be devoted to manual prehension or to tactile exploration, but rather to help stabilize vision. Some changes appear as structural ones, when connection between perceptual and motor components becomes closer and when new nervous networks can be inferred from behavioral analysis. Some changes can be both functional and structural.

3. The general theories assigned a cognitive aim to sensory–motor development and suggested that sensory–motor organizations and behaviors are more and more controlled by cognitive goals. However, some changes appear to be more adaptive than truly cognitive, and such a single principle of development cannot be currently supported. Even if it is obvious that the well-organized perception–action coupled responses serve to master the external would in relation to body stabilization, many kinds and models of mastering have to be considered.

4. New perspectives are now open in psychobiological studies of development, with the dynamic systems approach and with a renewal of functionalist approaches that use precise control procedures for detecting the aim of behavior and for relating subjective goal to proceedings of attainment.

BIBLIOGRAPHY

Blass, E. M. (ed.) (1986). "Handbook of Behavioural Neurobiology: Developmental Processes in Psychobiology and Neurobiology." Plenum, New York.

Bloch, H. (1994). Intermodal participation in the formation of action in the infant. *In* "The Development of Intersensory Perception: Comparative Perspectives" (D. Lewkowicz and R. Lickliter, eds.), pp. 309–333. Lawrence Erlbaum Associates, Hillsdale, New Jersey.

Bloch, H., and Bertenthal, B. I. (eds.) (1990). "Sensory–Motor Organisations and Development in Infancy and Early Childhood," Noto Series, Vol. 56. Kluwer, Dordrecht.

Gibson, J. J. (1979). "The Ecological Approach to Visual Perception." Houghton Mifflin, Boston.

Piaget, J. (1983). Piaget's theory. *In* "Handbook of Child Psychology" (W. Kessen, ed.), Vol. 1, pp. 103–128. John Wiley & Sons, New York.

Sequence Analysis of DNA, RNA, and Protein

HEIDI SOFIA
STEPHEN SHAW
National Institutes of Health

GLOSSARY

BLAST Computer algorithm (Basic Local Alignment Search Tool) that is the basis for a family of popular sequence analysis programs

Coding sequence Region of a gene that specifies the sequence of amino acids in a protein

Domain Structural unit of a protein composed of a stretch of amino acids that, over evolutionary time, can be assembled into new proteins to create new functions

Entrez Graphical retrieval tool that links databases of references, sequences, genetic maps, and three-dimensional protein structures

GenBank® Comprehensive public database of nucleotide and protein sequences that includes annotations describing aspects of these sequences

Gene Basic unit of heredity found in the chromosomes. A gene can encode a protein or an RNA molecule

Gene family Set of genes with sufficient similarity so that they can be inferred to have evolved from an ancestral gene existing at some time in the past

Genome All the DNA in a single (haploid) set of chromosomes in an organism

Homology Relationship between genes or proteins that are inferred to be derived from a common ancestral gene

Identity Exact match between sequences. Unlike homology, which is all or none, identity can be expressed as a percentage

MEDLINE® Large bibliographic database of medical and biological literature provided by the National Library of Medicine at NIH

Mutation Process by which a gene undergoes a structural change or the change resulting from that process

Ortholog Gene that is directly related to another gene through the evolutionary tree

Paralog Gene that is related to another gene through the process of gene duplication

Residue Amino acid that is found within the polypeptide chain of a protein

Sequence Order of nucleotide bases in DNA or the order of amino acid residues in a protein

Similarity Matches between two or more proteins owing to amino acid residues that are chemically similar, but not necessarily identical. Also refers to partial matches between nucleotide sequences

String Term used in computer science to describe a linked character sequence

GENETIC SEQUENCE ANALYSIS IS THE STUDY OF the information contained in the nucleotide sequences of DNA and RNA and in the amino acid sequences of protein. Its purpose is to extract important biological knowledge and understanding out of the code found in these biomolecules. This discipline has become very exciting as the rate of sequence determination has accelerated and more human disease genes have been discovered. Moreover, as computers have been linked up around the world through the Internet, the tools for sequence analysis have become easily available to all. This article outlines both practical and conceptual issues concerning this topic to inform the general

ENCYCLOPEDIA OF HUMAN BIOLOGY, Second Edition, VOLUME 7. Copyright © 1997 by Academic Press. All rights of reproduction in any form reserved.

reader and to empower anyone with a special interest to begin sequence analysis on his or her own. Readers can find their own starting place in this article. A foundation is laid and theoretical issues are addressed in the first two sections, and so starting at the beginning gives a good overview of genetic sequence analysis before considering specific instances. However, an excellent way to learn this material is by examples, so feel free to start with Section III on biological examples and refer back to the first two sections.

I. THE INFORMATION LANDSCAPE OF BIOLOGY

Biologists are finding it increasingly important to learn computer analysis methods as a standard component of their work. Computer technology has become essential for scientists to keep pace with the deluge of information produced by large-scale sequencing projects. Computer hardware and software have become part of a revolution in biological research, in which discoveries are being made about genes, proteins, and unique living systems at an ever-increasing rate. Figure 1 illustrates this rapid growth in several kinds of biological information, including bibliographic, sequence, map, and structure data. Nucleotide databases are growing the most explosively, fueling in turn the expansion of the others. To organize, understand, and exploit this flow of information, biologists have turned to computer methods. This application of computer science to the life sciences has created the exciting new discipline of computational biology.

Computers have begun to change the very nature of the research that biologists can do. The human brain is not well suited to analyzing long data "strings," but computers excel at this task. Many new types of questions can be answered using computational methods. As a result, unprecedented breakthroughs can be expected in the next several decades in biology, medicine, and technology. The most revolutionary aspect of these developments is the open access to scientific resources that now exists. Important tools and data necessary to make discoveries in biology are no longer confined to any particular laboratory but rather are found in the public domain, allowing a new paradigm to emerge in the learning and discovery process. A novel information landscape is now available for exploration by scientists across the world, or by any motivated individual. This information space is accessible through computer access of public databases via the Internet.

Although the discussion here is not a tutorial, it is designed to enable the reader to explore these

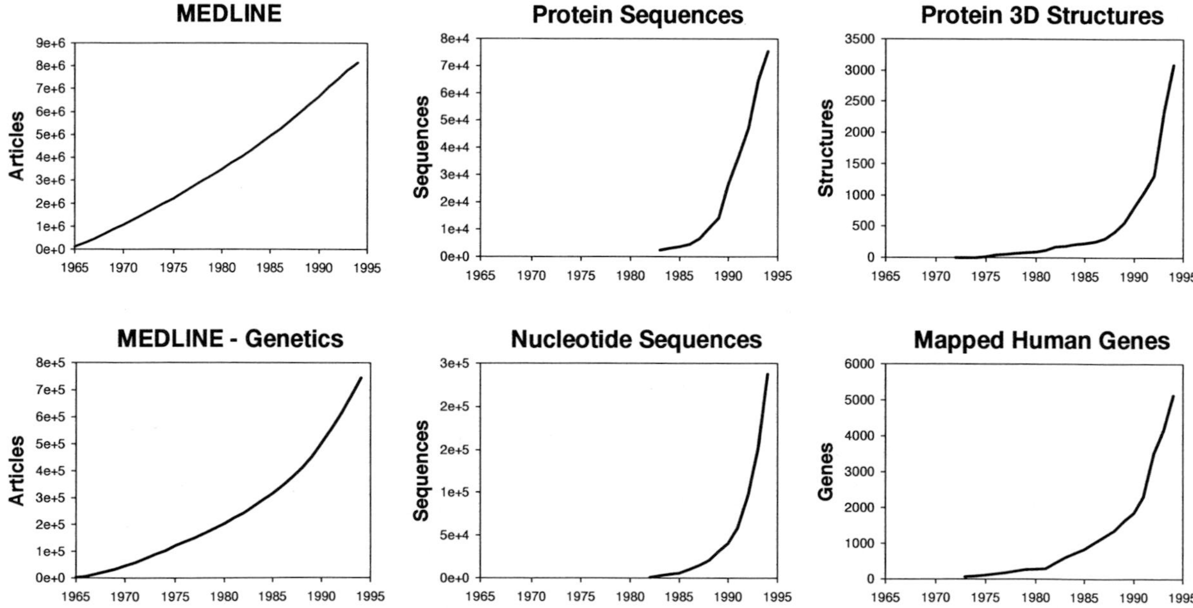

FIGURE 1 Growth in sequence information. Each panel reflects the growth over the last three decades of the indicated kind of biological information.

exciting new resources independently. Basic concepts and illustrations are provided to serve as a compass in the discovery process. Some useful topics for getting started are described, such as where biological information is kept (databases) and how to get at the information (retrieval tools), as well as how to understand the information and make discoveries with it (analysis tools). Particular examples of human genes and proteins are analyzed by several basic computational methods as an illustration of these approaches.

A. The Application of Computers to Genetic Sequence Data

1. Sequence Analysis

By their very nature, genetic sequences are an excellent application for computer analysis. There are vast quantities of genetic data and this data has a simple coding structure. In the human genome alone, there are 3 billion base pairs, approximately 100 million per chromosome, and about 80,000 genes. Scientists are currently sequencing the human genome and a collection of important model organisms in large-scale automated projects. These model organisms span the evolutionary tree and are essential for understanding the human data. They include *Escherichia coli* (bacteria), *Saccharomyces cerevisiae* (yeast), *Arabidopsis thaliana* (plant), *Caenorhabditis elegans* (worm), *Drosophila melanogaster* (fly), and *Mus musculus* (mouse).

A wide range of tasks related to DNA and biology are facilitated by computer methods. With current methods, DNA sequence can be determined approximately 500 base pairs at a time. These small pieces must then be put together into larger "contigs." The assembly of raw sequence into a contiguous string based on regions of partial overlap is an important computational problem in itself. Once assembled, important features in the sequence—difficult or impossible to locate with just the human eye—are easily identified using a computer. For example, genes are often found split up into "exons" that are widely separated in the DNA sequence; these gene pieces are assembled in the messenger RNA during gene expression. The biologist can pull out such a gene segment from a long sequence using a computer analysis with "codon bias statistics" that detects subtle differences in the DNA composition between regions.

Pattern seeking in general is a very important application of computers. Motifs in the DNA sequence mark the sites where gene pieces fit together. A consensus for such a sequence motif can be defined and used in a computer search to find sites that match the pattern. Other consensus motifs in DNA sequences include restriction sites, certain repetitive sequences, and control elements such as promoters and enhancers. Proteins show consensus patterns at ligand-binding sites.

Once a genetic sequence has been assembled, other studies, such as similarity searches, evolutionary analyses, and protein structure calculations, can provide the biologist with a wealth of information. These studies are often performed using distant computers through the Internet.

2. Sequence Similarity Searching

The sequence similarity search is a basic computational tool that is crucial to the work of many biologists, and consequently thousands of searches are performed each day against the public databases. Searching databases for homologous genes and proteins is a powerful way for a scientist to extract biological information out of sequence data, and it is typically the first step taken in characterizing a new sequence. Waiting for the results is an exciting moment for the investigator, who may have spent many long hours in the laboratory cloning and sequencing a gene. In a matter of seconds the DNA sequence can be compared to a massive accumulation of biological knowledge stored in the public databases. Dramatic new clues may be obtained about the identity of the sequence and its evolutionary history, which can suggest many new experiments.

In a similarity search, a query sequence is compared sequentially to a large database of other sequences; alignments between sequences are then scored with statistical parameters. A search can detect relatedness among genes that span over a billion years of molecular evolution. Human genes are often found to resemble sequences from "primitive" organisms. Decades of research on model organisms is proving to be of great value in the new era of genome studies because sequence similarity is a powerful way to make an educated guess about the function of a gene.

3. Human Disease Genes

It can take years to capture a particular gene implicated in a human disease—through painstaking genetic linkage analysis of affected families, and then cloning and sequencing efforts—but the gene function is often rapidly elucidated by a similarity search. A particular example of this phenomenon is the ataxia telangiectasia (AT) gene implicated in a premature-aging syndrome. One person in 40,000 is born with this disease. These individuals have inherited a copy of the defective AT gene from both their mother and their father, and are therefore homozygous for the mutation. The 1% of the population who are heterozygous for the AT gene mutation may have increased tumor incidence. When a gene implicated in the AT syndrome was isolated, it was found by similarity searching to encode a particular enzyme (phosphatidylinositol 3-kinase) that is critical for cell growth and DNA repair. [*See* Ataxia-Telangiectasia and the ATM Gene.]

Inferences about the function of a gene often are an important step toward developing a treatment or a cure for a disease. For example, Alzheimer's disease is poorly understood and few treatments exist; finding the key genes and proteins involved in this malady would allow doctors to design drugs and treatments that are directed at specific biochemical targets. Genetic information may fundamentally change the structure of medicine, as it becomes more apparent that virtually everyone has predisposing genes for some condition or another.

A collection of human disease genes have been identified using genetic, molecular, and computational methods. The cystic fibrosis (CF) gene mutation illustrates the often quirky nature of evolutionary history, as seen from the human perspective. The CF gene normally encodes a transporter protein necessary for ion homeostasis in the cell. A mutated form may have survived in certain populations where cholera was endemic by providing a resistance mechanism to the disease. Thus a single gene can cause one disease and protect from another. This principle is also illustrated by the sickle-cell syndrome and malaria. In both disorders, cystic fibrosis and sickle cell anemia, an individual who is heterozygous for the gene mutation may have an advantage in certain environments, but a homozygous individual has an affliction. [*See* Cystic Fibrosis, Molecular Genetics.]

In cancer research, a gene responsible for one form of hereditary colon cancer has been found to be homologous to a yeast gene encoding an enzyme that repairs mismatching bases in DNA. In human aging studies, genes associated with cellular degeneration have been identified. A gene resulting in amyotrophic lateral sclerosis (ALS), also known as "Lou Gehrig's disease," is a mutated form of an enzyme (superoxide dismutase) normally required to protect against toxic effects of oxygen. [*See* DNA Markers as Diagnostic Tools.]

4. Genome Comparisons

In a short time, sequences of complete genomes will be available for comparisons between closely related and distant species. Genome comparisons are very suggestive for what lies ahead in computational biology. For example, although there are dramatic phenotypic differences between humans and chimpanzees, the difference between these two species at the level of the DNA sequence is very small. A few key changes have resulted in language use, abstract thought, and the ability to make and use complex tools. It is also interesting to compare the human genome to a distant relative, the pufferfish. Although both species have approximately the same number of genes, the pufferfish has only 400 million base pairs in its genome, or about 7.5 times less DNA. Thus, this odd fish has all the instructions necessary to produce a complex animal with a similar arrangement of genes as in the human. Some of the "extra" DNA in the human is composed of simple repetitive elements such as Alu's, which form patterns unique to every individual, often useful in forensic work. Sometimes referred to as "junk DNA," these intergenic sequences may have roles that simply have not yet been characterized. Computational biology is certain to be important in the elucidation of these biological questions. [*See* Human Genome.]

B. What Are Genetic Sequence Data?

1. DNA Contains a Code

DNA is a unique double-stranded, double-helical biopolymer. The DNA structure was first elucidated by James Watson and Francis Crick in 1953. These young scientists were able to win a competitive race against the Nobel laureate Linus Pauling, using the X-ray crystallography data of Rosalind Franklin. The DNA structure changed the thinking of a generation of scientists because it immediately suggested mechanisms for the most fundamental biological events, including replication and gene expression.

The blueprint for life is based on a simple code written in the four bases of deoxyribonucleic acid (DNA). The information from the DNA code generates the living process by specifying what is built with the biological machinery that is inherited from a precursor cell, how much is built, and when. These cellular instructions are packaged as genes, which are the basic unit of heredity. Genes contain both a coding sequence that determines the structure of the gene product and other "control" sequences that regulate the timing and the level of gene expression. An RNA copy is made from the coding sequence of a gene during gene expression. Depending on the gene, this copy may fold into "structural RNA" such as is found in ribosomes and tRNA, or it may be translated into protein.

The simple pattern of the genetic code is based on the sequence of the four DNA bases: adenine (A), thymine (T), cytosine (C), and guanine (G). The exact order of these bases (A, T, C, G) in a gene specifies the gene product, which may be either a protein or a structural RNA molecule. This genetic code is analogous to the most basic language in a computer, which contains only two elements, the numbers 0 and 1.

To decipher the code, the DNA bases in a gene are read three at a time. Each triplet of bases is a short genetic word called a "codon," which contains the information for one of the 20 amino acids. For example, the sequencc "ATG" encodes the amino acid "methionine," which begins every protein. The codon "GAG" encodes glutamate, "CAT" specifies histidine, and "TAC" tyrosine. The sequence "TAA" is called a "stop codon" because it signals the end of the coding region in the sequence. Putting these codons together results in the DNA sequence "ATGGAGCATTACTAA." This sequence string would be read by the cell as ATG-GAG-CAT-TAC-TAA, and translated into the short polypeptide methionine-glutamate-histidine-tyrosine. Our sequence would be written as "MEHY" using the single-letter abbreviations for residues in a protein.

2. The Protein Code

As we have seen, the DNA sequence is translated into a second important biological code—the amino acid sequence in a protein. To convert the DNA code to the protein code, a temporary nucleotide copy called "messenger RNA" (mRNA) is made from the gene. The mRNA copy serves as a physical template for the synthesis of the protein on a ribosome. The genetic code is read one codon at a time from the mRNA as each amino acid residue is added to the growing polypeptide chain.

A newly created protein is a long, linear molecule that must "fold" into its mature form, which is a more compact and stable arrangement. The information for protein folding is found in the "primary structure" of a protein, which is the amino acid sequence. The rules that govern protein folding are not fully understood, but local interactions between amino acids are an important element in the process. An example of this local "secondary structure" is the alpha helix, which is a compact spiral typically formed by a stretch of 10 to 20 amino acids. "Beta sheets" and "turns" are two other elements. Secondary structure can often be predicted from the amino acid sequence using computation methods. The final conformation of a protein is known as the "tertiary structure" in which the sheets, turns, and helices are compactly arranged. Determining the three-dimensional structure of a protein through X-ray crystallography or nuclear magnetic resonance (NMR) spectroscopy is hard work, but it is very useful for the biologist because the location of key residues determines the activity and function of that protein in the cell.

Thus, proteins are synthesized as long chains that contain within them the information necessary to fold into mature active forms. In this way, all the activity of the cell is ultimately captured in the simple linear code found in DNA sequence.

C. Databases for Managing Genetic Sequence Data

1. Background

All types of important biological data have been collected into large public databases. A database is simply an organized collection of information. It may have an elementary format, as on index cards, or it may exist in a more complex electronic form on a computer. Whether simple or sophisticated, a database inevitably filters access to data through a set of rules that are implicit in its structure. The larger a database is, the more important its design features become in making data accessible. Retrieval tools are designed to assist an investigator in using information stored within the database. Analysis tools then give the biologist the power to manipulate the data and to extract new information by making novel connections.

The public databases that exist today for use by biologists contain literature references, sequence data, gene map data, and protein structure information. These are readily available to any interested person and are not difficult to use after a brief introduction. A few of the most useful and important databases are described next.

2. Examples of Databases

The first resource that an investigator might want to use in starting a project is the biological and medical literature. MEDLINE® is an international bibliographic database provided by the National Library of Medicine (NLM) at the National Institutes of Health (NIH). It provides electronic access to nearly 10 million records, each containing title, author, and journal information, and in most cases an abstract of the article. MEDLINE references can be accessed with a program called PubMed that is part of the retrieval software Entrez described in Section II,A.

GenBank® is a nucleotide sequence database provided by the National Center for Biotechnology Information (NCBI) at the NLM, and is part of an international collaboration to collate the world's nucleotide sequence data. GenBank exchanges daily updates with the two other major international sequence databases, the European Molecular Biology Laboratory (EMBL) database and the DNA Data Bank of Japan (DDBJ), and with GSDB, the Genetic Sequence Data Bank. Many journals require authors to submit new sequences to these public databases and obtain accession numbers prior to publication. Although most of the sequences in GenBank are human, over 30,000 species are represented.

The dbEST database, also operated by the NCBI, contains a special type of nucleotide sequences called "ESTs" (Expressed Sequence Tag). These sequences are short, somewhat less accurate sequences derived from random cDNA clones and are being produced at an extraordinary rate by large-scale genome projects. EST sequences provide a survey of the genes expressed by an organism in a given tissue and developmental stage. The dbEST database also contains precomputed information about similarity of the entries to other sequences. The dbSTS is a similarly formatted database for STS sequences, or "Sequence Tagged Sites," which are short sequences that have been uniquely mapped on the chromosome.

SWISS-PROT, PIR, and GenPept are protein sequence databases. Although the complete or partial sequences of a small number of purifiable proteins have been determined directly, especially N-terminal sequences, this experimental work is labor-intensive. Today almost all reported protein sequences are derived indirectly through coding DNA sequences as conceptual translations.

The Brookhaven National Laboratory Protein Data Bank (PDB) is a different type of database than those just described. It contains three-dimensional protein, DNA, and soon RNA structure data derived from X-ray crystallography and NMR studies. PDB also provides atomic-resolution descriptions of protein–ligand interactions. If a three-dimensional structure is available for a protein, then it can be used to make predictions about homologous proteins. Protein structural data allow the biologist to check whether functionally important residues in binding and catalytic sites are conserved.

II. NAVIGATING THE WORLD OF BIOLOGICAL INFORMATION

The databases described here are workhorses that biologists rely on to organize and store the growing knowledge describing living systems. However, isolated facts are not especially useful. It may be convenient to group all nucleotide sequences together and manage them in a single database, yet to gain any meaning from them, the biologist must be able to pull sequences out of the database and apply them to a biological model or an analysis, and connect them with other biological data. The investigator must be able to retrieve and integrate complex information.

An essential part of a biologist's job is the analysis of what is alike and different in a biological system. Thousands of computer searches are performed every day to determine whether nucleotide or protein sequences are homologous to other sequences. Often a biologist doing a homology search can benefit from the hard work of other scientists concerning the sequences found in the search. The results obtained about one system may be extended to others that have been less well characterized.

A. "Entrez" to the Information World

An excellent starting point for any person beginning to explore the world of biological information is a tool called Entrez, created by the National Center for Biotechnology Information at the NLM. It provides an easy way to navigate a number of important biological databases within one environment. This software links almost a billion characters of sequence, structure, taxonomy, and map data, as well as the molecular genetics literature, as illustrated in Fig. 3.

Like many retrieval tools, Entrez supports Boolean searching of its databases with AND, OR, and NOT, and other terms as well. However, Entrez is also able to integrate information across a collection of databases. Once an item—a sequence or reference—is retrieved from a database, related items can be pulled out as well with just a few mouse clicks. For example, each MEDLINE record is linked with nucleotide and

	HYDROPHILIC											HYDROPHOBIC								
	NEGATIVE				ALCOHOL		POSITIVE					LARGE				SMALL		AROMATIC		
	D	N	E	Q	S	T	H	R	K	C	P	M	I	L	V	A	G	F	W	Y
D	6	1	2	0	0	-1	-1	-2	-1	-3	-1	-3	-3	-4	-3	-2	-1	-3	-4	-3
N	1	6	0	0	1	0	1	0	0	-3	-2	-2	-3	-3	-3	-2	0	-3	-4	-2
E	2	0	5	2	0	-1	0	0	1	-4	-1	-2	-3	-3	-2	-1	-2	-3	-3	-2
Q	0	0	2	5	0	-1	0	1	1	-3	-1	0	-3	-2	-2	-1	-2	-3	-2	-1
S	0	1	0	0	4	1	-1	-1	0	-1	-1	-1	-2	-2	-2	1	0	-2	-3	-2
T	-1	0	-1	-1	1	5	-2	-1	-1	-1	-1	-1	-1	-1	0	0	-2	-2	-2	-2
H	-1	1	0	0	-1	-2	8	0	-1	-3	-2	-2	-3	-3	-3	-2	-2	-1	-2	2
R	-2	0	0	1	-1	-1	0	5	2	-3	-2	-1	-3	-2	-3	-1	-2	-3	3	-2
K	-1	0	1	1	0	-1	-1	2	5	-3	-1	-1	-3	-2	-2	-1	-2	-3	-3	-2
C	-3	-3	-4	-3	-1	-1	-3	-3	-3	9	-3	-1	-1	-1	-1	0	-3	-2	-2	-2
P	-1	-2	-1	-1	-1	-1	-2	-2	-1	-3	7	-2	-3	-3	-2	-1	-2	-4	-4	-3
M	-3	-2	-2	0	-1	-1	-3	-1	-1	-1	-2	5	1	2	1	-1	-3	0	-1	-1
I	-3	-3	-3	-3	-2	-1	-3	-3	-3	-1	-3	1	4	2	3	-1	-4	0	-3	-1
L	-4	-3	-3	-2	-2	-1	-3	-2	-2	-1	-3	2	2	4	1	-1	-4	0	-2	-1
V	-3	-3	-2	-2	-2	0	-3	-3	-2	-1	-2	1	3	1	4	0	-3	-1	-3	-1
A	-2	-2	-1	-1	1	0	-2	-1	-1	0	-1	-1	-1	-1	0	4	0	-2	-3	-2
G	-1	0	-2	-2	0	-2	-2	-2	-2	-3	-2	-3	-4	-4	-3	0	6	-3	-2	-3
F	-3	-3	-3	-3	-2	-2	-1	-3	-3	-2	-4	0	0	0	-1	-2	-3	6	1	3
W	-4	-4	-3	-2	-3	-2	-2	3	-3	-2	-4	-1	-3	-2	-3	-3	-2	1	11	2
Y	-3	-2	-2	-1	-2	-2	2	-2	-2	-2	-3	-1	-1	-1	-1	-2	-3	3	2	7

FIGURE 2 Substitution matrix for amino acids. The matrix shows a scoring system for how favorable or unfavorable it is to find a particular pair of amino acids in a comparison of two sequences. Data shown are for the matrix called BLOSUM62. Positive values indicate similarity between the two amino acids, whereas negative values indicate dissimilarity. Boxes have been placed around groups of amino acids considered similar; a simple description of the group appears at the top.

protein sequences that were published in the paper. Nucleotide sequences are connected to the proteins that are derived from them by conceptual translation. Because of "hard links" such as these, Entrez makes it easy to collect all the known facts that a biologist is likely to need about a system.

Entrez makes other more creative connections between data through the process of "neighboring." Additional links between data items in Entrez have been precomputed and are stored within the system until an inquiry is made. Retrieving these connections is like getting the computer to respond to a question it had already considered some time in the past. For example, the biologist can ask the computer to show sequences that are similar to the one on the screen. Protein sequences are shown if they match with a score not expected to appear by chance. Nucleotide sequences are shown if they overlap at one end or if one is completely contained within the other. When a reference is retrieved, the biologist may request to see related papers. Entrez can determine the "relatedness" of a reference using a statistical method that analyzes term frequencies in the abstract and title.

Two other features of Entrez are especially useful to the biologist. Entrez has a genomes division with a collection of complete genomic sequences organized in a graphical format. Entrez also includes access to structure data for proteins with three-dimensional molecular graphics representation of biomolecules.

Entrez can link biological information, find "neighbors" in sequences and references, and show protein structures as well as the map locations of genes within chromosomes. This software acts as a "virtual assistant," assisting the researcher in similarity searches, retrieving potentially relevant papers from the literature, recording the opinions of thousands of experts, drawing maps, and displaying structures.

B. Finding Homologous Sequences

The term "homology" in molecular biology means inferred common ancestry. It refers to sequences that are similar enough to make it likely that they share a common gene ancestor. Borrowed from evolutionary biology, homology originally described a relationship between anatomical structures, for example, the human arm and the whale's fin, which show structural similarity due to descent from a common ancestor.

In molecular biology, the term homology is commonly misused to mean similarity between sequences, and similarity searches are often incorrectly termed homology searches. Since homology has been defined as an all or nothing phenomenon—either a common ancestor exists or it does not—there cannot be a percentage of homology between two sequences.

The goal of a similarity search is to find sequences that are alike enough to justify an inference about the biological properties of the gene or protein. Homology searching is an attempt to make discoveries by studying the "fingerprint" left by the evolutionary history of a genetic sequence. Random mutation and natural selection act in opposition to each other in evolution to shape gene copies. Mutagenic events cause random base substitutions, insertions, and deletions, acting through mechanisms such as copying errors and genotoxic agents. However, sites in a gene that are essential for function are conserved through successive generations by natural selection, and so conserve a pattern that can be detected by similarity searching.

A similarity search may result in no hits or it may produce a long list of strong matches to the query sequence. The sequences found may be well characterized or poorly understood. Even if nothing is known about the members of the hit list, it is still useful to identify the family members of the sequence across species boundaries and in different biological contexts. A researcher working on a particular problem may end up providing an important clue to a seemingly unrelated biological system through a similarity search.

C. Gene Families

Gene families arise in evolution when two exact copies of a sequence diverge from each other through "genetic drift." For example, consider myoglobin, an oxygen transport protein from muscle that is found in both human and rat. A single ancestral form of the myoglobin gene existed in an ancient mammal that preceded both primates and rodents. When new species evolved from this progenitor, copies of the myoglobin gene were genetically isolated in separate species and underwent independent mutation. Different sequence changes accumulated and were passed on through successive generations in separate branches of the evolutionary tree that led ultimately to human or rat. In this way, the process of speciation has created many gene families. A few examples of important gene or protein families include the actins, collagens, hemoglobins, histones, immunoglobulins, and keratins.

Numerous branches to a gene family can be added through "gene duplication." As a rare event, a gene may be reproduced within a germ cell, resulting in two copies. When this occurs, one gene is still available to perform the original function in the organism while the other is released from this requirement. The second copy is free to evolve by mutation and develop a new function or a new pattern of expression.

Gene family members that arise through the formation of new species are called "orthologs." Genes that are similar as a result of a gene duplication event are called "paralogs."

D. How a Sequence Similarity Search Is Done

1. Similarity Search

In any sequence similarity search, a query sequence (nucleotide or amino acid) is compared sequentially with all known sequences in a database using a computational algorithm or method. Often the biologist sends a sequence through the Internet to a distant computer that has been installed with a program for database searching. The report typically has two parts: the "hit" sequences are ranked in a summary list in order of their score and the alignment of the query sequence with each hit is shown. Statistical parameters help the scientist evaluate how good the match is between the query and each "hit" sequence. Many different programs for database searching are available, employing different algorithms, for example, FASTA, BLITZ, BLAZE, FLASH, FASTDB, and the BLAST family.

Search programs are often designed to look for "local alignments" between sequences by breaking them into high-scoring segments, for example, the BLAST algorithm described in the next section. A "global alignment," which extends over the entire length of the sequences, may seem more intuitive at first but may be more difficult to interpret statistically. Local alignments can be produced by an "optimal" local alignment algorithm, but faster "heuristic" methods can speed these searches at the cost of missing an occasional similarity that meets the search criteria. Heuristic algorithms rely on certain shortcuts, for example, forbidding the introduction of gaps. Two heuristic algorithms in wide use are the BLAST (Basic Local Alignment Search Tool) and FASTA (Fast Alignment Search Tool Algorithm) methods.

2. BLAST Searches

The NCBI offers BLAST searching through the Internet as a public service, making it convenient to do similarity searches either by electronic mail or at a World Wide Web (WWW) site. The BLAST algorithm permits rapid searching of the public databases while at the same time producing scores that have a rigorous statistical interpretation. BLAST programs find ungapped alignments between segments of the sequences by pairwise mapping of the query sequence onto the database sequence. These segments, called "HSPs" for "High-Scoring Segment Pairs," are then considered together for statistical significance of the overall match between the two sequences. Multiple segments can still result in a significant score when considered together even if they are not significant in themselves.

A collection of BLAST programs is available and in wide use. The "blastn" program searches the nucleotide databases and "blastp" is the corresponding tool for protein. The "blastx" program translates a nucleotide query in all six potential open reading frames before searching the protein databases. This program is especially useful for finding coding sequences in a nucleotide sequence and for detecting frameshift errors in the sequence that mask these regions. The "tblastn" and the "tblastx" searches are both powerful programs that translate entire databases as part of the search.

E. Sequence Identity and Similarity

Typically when nucleotide sequences are compared, only perfect matches between bases in the sequence alignment contribute to scoring. The overall match between the sequences is described using a percentage of identity. For example, two genes that have exactly the same bases at half the positions are said to be "50% identical." However, in protein sequences, matches between residues may be either "identical" or "similar." The term "similar" in this context refers to amino acid residues that are chemically related. For example, phenylalanine and tyrosine are similar because they both contain aromatic rings. Other examples scored as similar include the acidic amino acids glutamate (E) and aspartate (D); the basic amino acids lysine (K) and arginine (R); and the hydrophobic amino acids isoleucine (I), leucine (L), and valine (V).

The concept of similarity is useful in understanding the pattern of amino acid changes allowed during evolution. A mutation that swaps a "similar" residue for another is considered a "conservative substitution" because protein function is often undisturbed. Similar amino acids tend to show partial interchangeability with each other during evolution. Substitutions of these amino acids with each other in evolution are "allowed" because the change often does not interfere with function. Programs that analyze sequence similarity use such information in the form of a "substitution matrix" that tabulates the likelihood of every possible substitution. The most commonly used matrix is Blosum62, shown in Fig. 2, but others are indicated for special purposes. The matrix was created by the systematic analysis of amino acid replacements allowed by evolution in functionally similar proteins. Each combination of amino acids has a score. A positive score indicates identity or a relatively conservative substitution between structurally similar amino acids. Negative scores represent unfavorable substitutions.

In the following section, similarity searching and other computational tools are applied to specific biological examples. These illustrate in detail some of the concepts concerning genetic sequence analysis that have been described.

III. BIOLOGICAL EXAMPLES

A. Identity

Example 1: The CD9 protein.
Principle: A similarity search can link divergent areas of biology in unexpected and revealing ways.

The cell-surface protein CD9 is an abundant protein found on the outer membrane of many cell

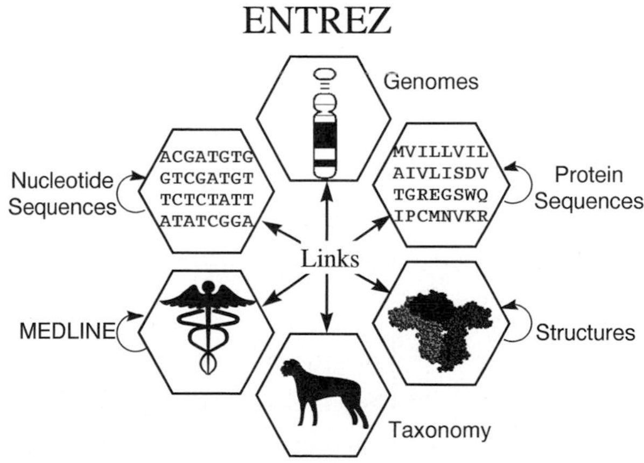

FIGURE 3 Schematic of information in the Entrez databases. Each of the hexagons indicates a kind of data organized within the Entrez world. The arrows indicate links between types of data that facilitate easy exploration from one kind of information to another.

types in humans and other species. It is expressed at particularly high levels on platelets and is important in their ability to act in blood clotting. CD9 turns on adhesion properties of platelets, causing them to stick together in order to plug broken blood vessels and stop bleeding. The CD9 gene has been cloned and the sequence is in GenBank under the accession number M38690.

A Japanese group working on a separate project conducted a biological screen for proteins that regulate the motility and metastasis of tumor cells. They isolated a protein that they named "motility regulating protein." When they cloned and sequenced the gene, a similarity search showed that CD9 and the new protein were identical.

Finding a "new" sequence that turns out to be already in the public databases under a different name is an increasingly common occurrence now that so many genes have been sequenced. More and more, results like these are providing a pivotal and definitive link between multiple areas of research that had previously not been thought to be related.

B. A Typical Search Report

Example 2: The SIL protein.
Principle: The rapid growth of sequence databases often makes it important to repeat similarity searches on a regular basis to keep the results up-to-date.

1. A blastp Search of the SIL Protein

In this example, a biologist searches the protein databases with the human SIL protein sequence from the GenBank record M74558. The SIL gene is of interest because of its association with leukemia. It is frequently found to be disrupted as a result of genomic rearrangement in patients with the T-cell acute lymphoblastic form of this disease.

```
Format of      From: any person
e-mail query   To: blast@ncbi.nlm.nih.gov

instructions   PROGRAM blastp
to program     DATALIB nr
               FILTER SEG+XNU
               BEGIN

Protein ID:    >gi|338088|gp|M74558|HUMSIL_1 SIL gene product [Homo sapiens]
Sequence:      MEPIYPFARPQMNTRFPSSRMVPFHFPPSKCALWNPTPTGDFIY
               LHLSYYRNPKLVVTEKTIRLAYRHANENKKNSSCFLLGSLTADEDEEGVTLTVDRFDP
               GREVPECLEITPTASLPGDFLIPCKVHTQELCSREMIVHSVDDFSSALKALQCHICSK
               DSLDCGKLLSLRVHITSRESLDSVEFDLHWAAVTLANNFKCTPVKPIPIIPTALARNL
               SSNLNISQVQGTYKYGYLTMDETRKLLLLLESDPKVYSLPLVGIWLSGITHIYSPQVW
               ACCLRYIFNSSVQERVFSESGNFIIVLYSMTHKEPEFYECFPCDGKIPDFRFQLLTSK
               ETLHLFKNVEPPDKNPIRCELSAESQNAETEFFSKASKNFSIKRSSQKLSSGKMPIHD
               HDSGVEDEDFSPRPIPSPHPVSQKISKIQPSVPELSLVLDGNFIESNPLPTPLEMVNN
               ENPPLINHLEHLKPLQPQLYDEKHSPEVEAGEPSLRGIPNQLNQDKPALLRHCKVRQP
               PAYKKGNPHTRNSIKPSSHNGPSHDIFEKLQTVSAGNVQNEEYPIRPSTLNSRQSSLA
               PQSQPHDFVFSPHNSGRPMELQIPTPPLPSYCSTNVCRCCQHHSHIQYSPLNSWQGAN
               TVGSIQDVQSEALQKHSLFHPSGCPALYCNAFCSSSSPIALRPQGDMGSCSPHSNIEP
               SPVARPPSHMDLCNPQPCTVCMHTPKTESDNGMMGLSPDAYRFLTEQDRQLRLLQAQI
               QRLLEAQSLMPCSPKTTAVEDTVQAGRQMELVSVEAQSSPGLHMRKGVSIAVSTGASL
               FWNAAGEDQEPDSQMKQDDTKISSEDMNFSVDINNEVTSLPGSASSLKAVDIPSFEES
               NIAVEEEFNQPLSVSNSSLVVRKEPDVPVFFPSGQLAESVSMCLQTGPTGGASNNSET
               SEEPKIEHVMQPLLHQPSDNQKIYQDLLGQVNHLLNSSSKETEQPSTKAVIISHECTR
               TQNVYHTKKKTHHSRLVDKDCVLNATLKQLRSLGVKIDSPTKVKKNAHNVDHASVLAC
               ISPEAVISGLNCMSFANVGMSGLSPNGVDLSMEANAIALKYLNENQLSQLSVTRSNQN
               NCDPFSLLHINTDRSTVGLSLISPNNMSFATKKYMKRYGLLQSSDNSEDEEEPPDNAD
               SKSEYLLNQNLRSIPEQLGGQKEPSKNDHEIINCSNCESVGTNADTPVLRNITNEVLQ
               TKAKQQLTEKPAFLVKNLKPSPAVNLRTGKAEFTQHPEKENEGDITIFPESLQPSETL
               KQMNSMNSVGTFLDVKRLRQLPKLF

"hit" list     Database:  Non-redundant PDB+SwissProt+SPupdate+PIR+GenPept+GPupdate
of proteins               165,296 sequences; 48,834,109 total letters.
from BLAST
search                                                                   Smallest
                                                                            Sum
                                                                 High   Probability
               Sequences producing High-scoring Segment Pairs:   Score  P(N)      N

               pir|A41685|A41685     SIL protein - human >gp|M74558|HUM...  6508  0.0      1
               sp|P32334|MSB2_YEAST  MSB2 PROTEIN (MULTICOPY SUPPRESSIO...    66  0.049    5
               gp|S75826|S75826_1    MFBA [Lentinula edodes]                  64  0.17     3
               pir|S26661|S26661     major histocompatibility complex-b...    51  0.25     5
               sp|P15265|MCS_MOUSE   SPERM MITOCHONDRIAL CAPSULE SELENO...    44  0.30     4
```

FIGURE 4 Format for an e-mail query of proteins similar to the SIL gene and a simplified "hit list" generated by BLAST.

The researcher sends by e-mail a search request containing the SIL protein sequence to the GenBank BLAST server at the following electronic address: blast@ncbi.nlm.nih.gov. The message is organized according to the simple format required by the computer as shown in Fig. 4 (top). The reader can try this search independently after studying this example. (For more detailed information on using the BLAST e-mail server, the reader is directed to the final section of this article; for example, using e-mail help for BLAST, you can look up the use of the filters SEG and XNU used in this query.) The results come back as an e-mail message, typically within minutes.

The search report returned by the BLAST e-mail server contains four basic sections: (a) header information, (b) a histogram, (c) a "hit" list, and (d) alignments. The complete report contains a large amount of information that is useful for the more advanced user but can be confusing to the beginner. For simplicity, only a few key features of the SIL protein search report are described here. Although the reader may not understand every aspect of the BLAST search report he or she will be able to focus on the portions that contain the basic results.

2. How to Read the Report

The most basic summary of the search results is found in the "hit" list, which shows the sequences found by the search in order of best to worst. The "hit" list for the SIL protein search is shown in Fig. 4 (bottom). Several parameters are reported that can be used to evaluate each match. The score (S) and the P value [shown under the column "Smallest Sum Probability $P(N)$"] measure the quality of the alignment between the query sequence (the SIL protein) and each sequence in the list. The actual alignments would be found in the last section of the complete report. Examples of alignments are shown and discussed later in this article.

The P value is often the easiest way to evaluate the significance of a match found in a BLAST search. This statistical parameter is always a number between 0 and 1. The P value is defined as the probability that the similarity between the sequences could have resulted from chance alone given a database of the current size and composition. A small P value (close to zero) for an alignment is associated with a strong hit, whereas a larger P value (close to one) is probably not significant.

To find out more about the matches, the reader must decipher the text in each line. At the beginning of each line in the first field, an abbreviation names the database and the accession number for each "hit" sequence. This information allows the reader to retrieve the sequence record from the database. The second field gives a short description of the sequence called the "definition line," which may be truncated depending on its length. This information allows the investigator to make an educated guess about which sequences are worth further effort.

Sometimes an alignment between sequences is broken into more than one piece. The last column (N) in the list indicates how many segments of local alignment, or "High Scoring Segment Pairs" (HSP), were used in evaluating the match. Multiple segments can "add up" to make the overall match significant, even if the isolated individual segments are not.

3. The Search Results

a. The Best Hit (A41685)

As expected, the strongest match of the query sequence M74558 in the database is with itself. The first "hit" sequence in the list is shown as "pir|A41685|A41685 SIL protein – human >gp|M74558|HUM. . . ." This is merely an abbreviated way to say the following:

> A human sequence has been deposited in the PIR protein database with Accession Number A41685. It is identical to the translated SIL gene from GenBank Accession Number M74558.

Retrieving the GenBank record for M74558 would provide other useful information, for example, a journal reference that might describe the biology of the sequence and the name of a scientist who could supply a clone to a collaborator.

b. The Second Hit (P32334)

The next hit in the list is the yeast protein MSB2 with Accession Number P32334, which shows a P value of 0.049. This score is based on five segments from the alignment as can be seen by the N value. This is a modestly significant match; possibly enough to provide some useful information, or possibly just random noise, but certainly not enough for any definitive answer. The other hits in the list are of even less significance.

After investigation, the biologist finds that the MSB2 protein is a multicopy suppressor to a temperature-sensitive mutation in the CDC24 gene, which is required for polarity establishment and bud formation in yeast. This biological information provides only indirect evidence for the MSB2 protein's function, and it is even more tenuous when used to try to guess at the function of the SIL protein. However, as more puzzle pieces are collected, this information may, or may not, eventually fall into place in light of the overall picture.

This search did not provide any direct clues as to the function of the query sequence. However, considering the rate at which sequence databases are growing, the search should be repeated often. New sequences are constantly being added that could be more informative and allow the identification of the SIL protein from human. You, the reader, could repeat the search and possibly be the first person to determine the function of this sequence.

C. Gene Families

Example 3: The chemokine family.
Principle: Insights on new genes can come from assigning them to a family of related sequences.

1. The Chemokine Family

Sometimes a database search returns a long list of significant matches. This occurs when a gene sequence is very similar to multiple sequences from the same organism or from a number of organisms. Genes can be assigned to a "family" based on sequence similarity, as can the proteins they encode. In the example that follows, we explore a gene family by analyzing a collection of human genes important in the function of the immune system.

One large gene family, recently the focus of much excitement, encodes signaling proteins called "chemokines" that communicate messages to cells in the immune system. Chemokines are small single-domain proteins that recruit white blood cells (leukocytes) to sites of bacterial invasion. These proteins are secreted as CHEMical signals from one cell to cause movement (KINEsis) in other cells, hence the name "chemokine." Overproduction of specific chemokines is thought to contribute to certain chronic human diseases in which particular kinds of white cells migrate into skin, gut, or joints and cause damage. Researchers hope that by understanding the biology of these messenger proteins—answering questions such as what stimulates their secretion and what are their cellular targets—and by using this knowledge to develop inhibitory drugs, these diseases can be treated more effectively.

2. BLAST Search of MIP-1a

In this example, a human macrophage inflammatory protein called "MIP-1a" is the starting point for an analysis of the chemokine family. A BLAST search of the protein databases with MIP-1a returns a long list of very strong hits with closely related proteins from human, rat, mouse, and rabbit (Fig. 5). We have shortened the list by deleting 45 sequences from the middle as shown and truncating the end. The *P* values from this search are expressed in scientific notation because they are so close to zero. Even the weakest similarity shown (human interleukin-8) has less than one chance in 1,000,000 ($P = 0.00000078$) of being due to chance!

Two representative hits have been shown to illustrate sequence similarities among the chemokine family: MIP-1b near the top of the "hit" list and IL-8 at the bottom. To see the exact nature of the similarities, you need to look at the second part of the BLAST search shown in the bottom half of Fig. 5. The major point can be seen at a glance, namely, that MIP-1a is similar to MIP-1b over its entire length and the same is true of IL-8. In the case of IL-8, that alignment is displayed as two separate pieces.

A similarity search of the MIP-1a protein identifies the members of the chemokine family and provides the raw materials for further analysis. To extract more information out of BLAST results like these, three other important computational methods—a multiple sequence alignment, a dendrogram of sequence similarity, and a dot plot analysis—are described in the following sections.

D. Multiple Alignment of Protein Sequences

Example 4: Branches in the chemokine family.
Principle: Multiple alignment can define the branches of a protein family by revealing the patterns that are shared between the members.

A protein similarity search returns alignments that show the details of similarity between any two given sequences. A multiple alignment extends the comparison to a group of sequences as illustrated in Fig. 6. Although the BLAST algorithm picks out the strongest regions of local similarity, a multiple alignment tool such as CLUSTALW, available from the European Bioinformatics Institute's WWW site by ftp (ftp://ftp.ebi.ac.uk/pub/software), seeks the best "global alignment" over the entire sequence. Often to get good alignments, the program has to introduce "gaps" (shown by dashes). Such gaps are reasonable because, during evolution, proteins can either gain or lose one or more amino acids by mutational events.

Two subfamilies of the chemokine family—alpha and beta—are revealed when the sequences from

```
First        BLASTP 1.4.8MP [20-June-1995] [Build 13:58:02 Oct 17 1995]
Section      Query=  gi|127078|sp|P10147|MI1A_HUMAN MACROPHAGE INFLAMM...
                     (92 letters)
             Database:  Non-redundant PDB+SwissProt+SPupdate+PIR+GenPept+GPupdate
                        162,157 sequences; 47,736,760 total letters.
                                                                               Smallest
                                                                                 Sum
                                                                       High  Probability
             Sequences producing High-scoring Segment Pairs:           Score  P(N)      N

             sp|P10147|MI1A_HUMAN MACROPHAGE INFLAMMATORY PROTEIN 1-AL...   476  1.6e-62   1
             gi|961440|gp|D63785|HUMLD78AB_1 LD78 alpha beta [Homo sap...   409  4.7e-53   1
             gi|512875|gp|A17101|A17101_1 LD78 synthetic gene gene pro...   364  1.1e-46   1
             sp|P16619|MI10_HUMAN TONSILLAR LYMPHOCYTE LD78 BETA PROTE...   357  8.3e-46   1
             gi|790633|gp|U22414|RNU22414_1 macrophage inflammatory pr...   297  2.3e-37   1
             sp|P10855|MI1A_MOUSE MACROPHAGE INFLAMMATORY PROTEIN 1-AL...   295  4.4e-37   1
             pir|A30552|A30552 T-cell activation protein alpha precurs...   291  1.6e-36   1
             gi|512890|gp|A17117|A17117_1 MIP-1-alpha gene gene produc...   220  8.8e-36   2
             sp|P13236|MI1B_HUMAN MACROPHAGE INFLAMMATORY PROTEIN 1-BE...   282  3.0e-35   1
             pir|B30574|B30574 T-cell activation protein precursor (cl...   282  3.0e-35   1
             gi|459150|gp|U06435|RNU06435_1 macrophage inflammatory pr...   282  3.0e-35   1
             pir|A40978|A31767 T-cell activation protein 2 precursor -...   278  1.1e-34   1
             gi|512905|gp|A17134|A17134_1 Human ACT-2 synthetic gene g...   278  1.3e-34   1
             gi|515149|pdb|1HUM|A Human Macrophage Inflammatory Protei...   278  1.4e-34   1
             . . . . . . . . . . .(45 sequences deleted). . . . . . . . . .
             gi|995911|gp|U26426|MMU26426_1 eotaxin precursor [Mus mus...    98  2.6e-07   1
             sp|P36925|IL8_SHEEP INTERLEUKIN-8 PRECURSOR (IL-8). >pir|...    76  3.9e-07   2
             sp|P26894|IL8_PIG INTERLEUKIN-8 PRECURSOR (IL-8). >pir|A5...    75  4.8e-07   2
             sp|P22951|AMC1_PIG ALVEOLAR MACROPHAGE CHEMOTACTIC FACTOR...    75  4.8e-07   2
             gi|516197|gp|X61151|SSIL8_1 interleukin-8 [Sus scrofa]          75  5.2e-07   2
             gi|33959|gp|Z11686|HSINTLK8M_1 interleukin 8 [Homo sapiens]     72  7.8e-07   2
Second
section

match to     >sp|P13236|MI1B_HUMAN MACROPHAGE INFLAMMATORY PROTEIN 1-BETA PRECURSOR (MIP-1
MIP-1b                   BETA) (T-CELL ACTIVATION PROTEIN 2) (ACT-2) (PAT 744) (H400)
                         (SIS-GAMMA) (LYMPHOCYTE ACTIVATION GENE-1 PROTEIN) (LAG-1) (HC21)
                         (SMALL INDUCIBLE CYTOKINE A4). >pir|JH0319|JH0319 lymphocyte
                         activation protein 1 precursor - human >pir|A37411|A37411 probable
                         cytokine HC21 precursor - human >gi|32036|gp|X16166|HSHC21_1
                         cytokine 21 [Homo sapiens] >gi|34218|gp|X53683|HSLAG1CDN_1 LAG-1
                         gene product [Homo sapiens] >gi|178018|gp|J04130|HUMACT2A_1 LAG2
                         gene product [Homo sapiens]
                         Length = 92

one region     Score = 282 (131.7 bits), Expect = 3.0e-35, P = 3.0e-35
               Identities = 48/70 (68%), Positives = 61/70 (87%)

             Query:     22 SASLAADTPTACCFSYTSRQIPQNFIADYFETSSQCSKPGVIFLTKRSRQVCADPSEEWV 81
                           SA + +D PTACCFSYT+R++P+NF+ DY+ETSS CS+P V+F TKRS+QVCADPSE WV
             Sbjct:     23 SAPMGSDPPTACCFSYTARKLPRNFVVDYYETSSLCSQPAVVFQTKRSKQVCADPSESWV 82

             Query:     82 QKYVSDLELS 91
                           Q+YV DLEL+
             Sbjct:     83 QEYVYDLELN 92
match to     >gi|33959|gp|Z11686|HSINTLK8M_1 interleukin 8 [Homo sapiens]
IL-8                     Length = 97

first          Score = 72 (33.6 bits), Expect = 7.8e-07, Sum P(2) = 7.8e-07
region         Identities = 12/34 (35%), Positives = 18/34 (52%)

             Query:     52 ETSSQCSKPGVIFLTKRSRQVCADPSEEWVQKYV 85
                           E+    C+    +I        R++C DP E WVQ+ V
             Sbjct:     56 ESGPHCANTEIIVKLSDGRELCLDPKENWVQRVV 89

second         Score = 38 (17.8 bits), Expect = 7.8e-07, Sum P(2) = 7.8e-07
region         Identities = 12/48 (25%), Positives = 19/48 (39%)

             Query:      2 QVSTAALAVLLCTMALCNQFSASLAADTPTACCFSYTSRQIPQNFIAD 49
                           +++ A LA   L + ALC         +A        C      S+       FI +
             Sbjct:      4 KLAVALLAAFLISAALCEGAVLPRSAKELRCQCIKTYSKPFHPKFIKE 51
```

FIGURE 5 Representative BLAST results for MIP-1a show that it belongs to the chemokine family of genes. In each case there is a "header" of multiple lines giving the name of the sequence in full, along with other annotations (such as names of identical sequences in the database). Then information is provided about each region of similarity (one for MIP-1b and two for IL-8). There are two lines tabulating parameters that describe the similarity in each region of similarity. Look at the parameter called "Identities" in each section. MIP-1b has a single long region of similarity that is 68% identical (47 of 70 positions) to the query sequence. In contrast, the more distantly related IL-8 has two shorter regions of 25–35% similarity (12 of 34 identities in one region and 12 of 48 in the other). Finally, the printout shows a matchup or "alignment" for the regions, query on top and subject below. Where they are identical, that amino acid letter is shown between them. Where they are similar, a "+" is shown between them (see preceding discussion of similarity). Sometimes the similarity is too long to show on one line and is broken up into two lines. Simply by inspecting the alignments of the query with MIP-1b and IL-8, you can see that the former has many more "+" markers and shared letters than the latter.

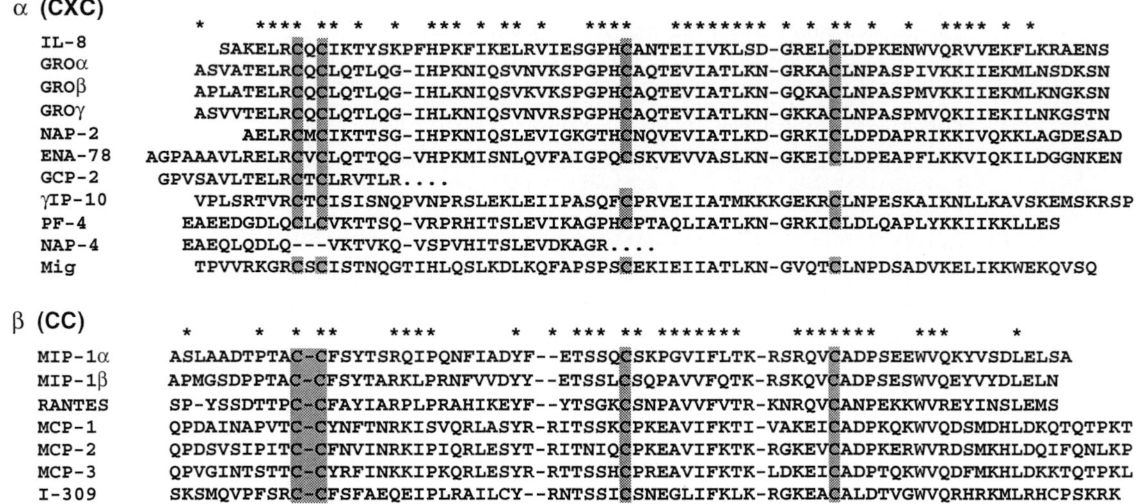

FIGURE 6 Multiple alignment of human MIP-1a with chemokines from both the alpha and beta subfamilies from humans. Shaded areas indicate conserved cysteines. Asterisks indicate positions at which the majority of sequences in each family are identical. Dashes indicate gaps that have been inserted to optimize alignment. [From P. M. Murphy (1994). The molecular biology of leukocyte chemoattractant receptors. *Annu. Rev. Immunol.* **12**, 593–633.]

the MIP-1a similarity search are analyzed by multiple alignment. These subfamilies show different patterns of absolutely conserved cysteine residues as shown by the shaded areas in Fig. 6. The spacing of cysteine residues is important for the shape and stability of extracellular proteins through disulfide bond formation. The MIP-1a protein is a member of the chemokine-beta group, which shows an absolute conservation of four cysteine residues (C) at positions 10, 11, 33, and 49. MIP-1b, which was our previous example of the strongest homology in Fig. 5, is likewise a member of the chemokine-beta subfamily. The chemokine-alpha group differs by the insertion of one residue between the first two cysteines. IL-8, which was at the bottom of our "hit" list in Fig. 5, can be seen to be a member of the chemokine-alpha subfamily.

Biologists have found that the alpha family of chemokines generally signals to one kind of human white blood cell, inducing granulocytes to migrate into sites of bacterial invasion. The beta family of chemokines typically signals to other kinds of human white blood cells, recruiting monocytes and lymphocytes into sites of viral infection. The investigator gains many leads regarding the probable function of the MIP-1a gene by finding that it is a member of the beta chemokine subfamily.

E. Dendrogram

Example 5: Evolutionary history of the human chemokine family.

Principle: A sequence family can contain many branches of orthologous and paralogous genes.

The pattern of sequence similarities in a family of proteins suggests the manner in which they evolved. By studying this pattern, a model can be built of the evolutionary events by which a family evolved from a common ancestral gene. Such a model is usually illustrated using a figure called a dendrogram, which is a drawing in the form of a branching tree. Specialized software such as Paup or Phylip can construct a dendrogram through multiple rounds of pairwise comparison of nearest neighbors. Programs such as these weigh the similarity between every pair of sequences and after a number of iterations assign a position in the tree to each sequence. Many multiple alignment programs also create a simple tree in order to simplify the alignment task. These programs pre-compute the sequence similarity between each pair of proteins and using these similarity scores to create a clustering order, which can be represented as a dendrogram.

Figure 7 illustrates a simple dendrogram for the human chemokine family, which was generated by a multiple alignment program called PILEUP from the

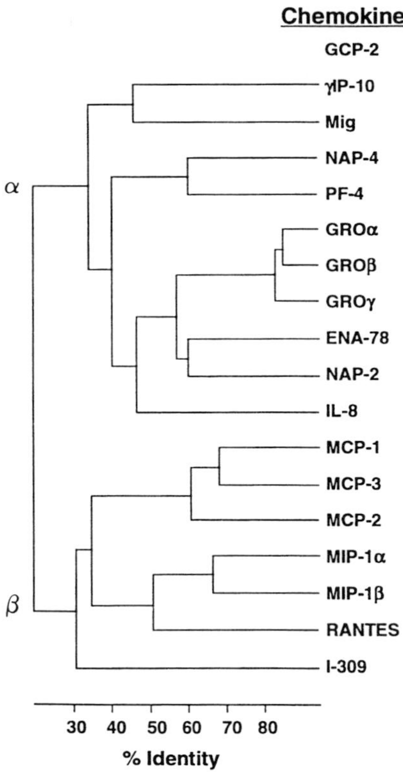

FIGURE 7 Dendrogram for chemokines illustrating hierarchy of similarities among the sequences. See text for detailed explanation. [From P. M. Murphy (1994). The molecular biology of leukocyte chemoattractant receptors. *Annu. Rev. Immunol.* **12**, 593–633.]

GCG suite of programs. Chemokines share a common ancestral protein not shown on the dendrogram. The gene for this primordial protein was duplicated long ago and gave rise to the paralogous α and β chemokine genes, both of which underwent subsequent evolution and reduplication. The β gene diverged into at least 7 separate genes, all of which are clustered on human chromosome 17. The α gene diverged into at least 11, all of which are clustered on human chromosome 4. Travelling on a horizontal line from right to left represents going back in time. In this figure, the closest similarity is between GRO-α and GRO-β because the vertical line that connects them is the farthest to the right. In contrast, the first vertical line that links GROα to any of the β family is all the way to the left.

F. Multidomain Proteins

Example 6: The selectin family.
Principle: Proteins have a modular structure made up of domains.

Unlike the simple chemokines, which have only one domain, many proteins consist of multiple domains strung together. Domains are structural segments in a protein, typically encompassing approximately 100 contiguous amino acid residues. Domains in a protein are semi-independent both functionally and structurally. They serve as "building blocks" in evolution that can be assembled into a variety of configurations to form new proteins with new functions. A computational tool called the "dot plot" is introduced in the next section as a useful way to analyze multidomain proteins.

The L-selectin protein, also called CD62L, plays a pivotal role in the ability of white blood cells to exit the bloodstream at lymph nodes during infection. L-selectin has a mosaic structure of three domain types that are shared by other proteins: two copies of a unit called "complement regulatory protein repeat," one copy of "epidermal growth factor" domain (EGF), and a "lectin" domain which binds sugar molecules, typically attached to other proteins. The family was named after this particularly "SElective LECTIN" domain. Although only the lectin domain has a clearly assigned function, it has proven to be a breakthrough in research on inflammation.

1. Dot Plot Analysis

A dot plot is a simple graphical way to look at similarity results and find protein domains. In this plot, the sequence of the query protein is represented on the *y* axis of a graph and compared to another protein on the *x* axis. A dot is drawn wherever there is a stretch of amino acids that is similar between the two proteins. These dots merge into diagonal lines if there is strong similarity between the two sequences. For example, a dot at the coordinate (x=50, y=60) indicates that the amino acids in the x-axis sequence around position 50 (x axis) are similar to those around position 60 in the query sequence (y axis). A protein sequence that is plotted against itself shows a continuous diagonal line. When plotted against a dissimilar sequence, a blank graph is obtained. A dot plot has the advantage of showing the pattern of similarity between two proteins at a glance without reading numerical coordinates. This is especially important in cases of complex multidomain proteins such as L-selectin, which we now discuss.

2. Domains in L-selectin

A dot matrix program called "Dotter" is used to analyze the domain structure of L-selectin from the similarity search results (Fig. 8). This program is freely

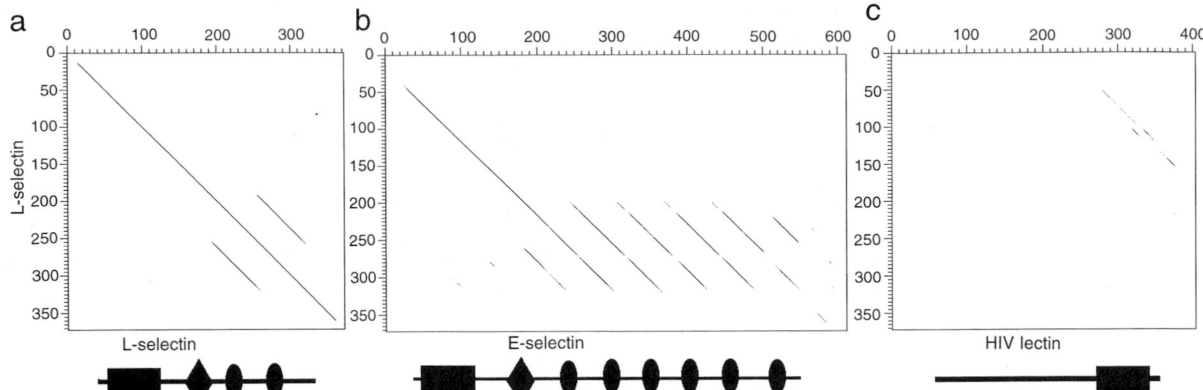

FIGURE 8 Combinatorial use of domains in lectins. The three panels illustrate dot matrix comparisons of a multidomain protein, L-selectin, (a) to itself, (b) to a closely related selectin, E-selectin, and (c) to a C-type lectin from HIV. Below each panel is a block diagram showing the organization of the three domains in each molecule: lectin (rectangle), epidermal growth factor (diamond), and short consensus repeat (oval).

available from the ftp site of the Sanger Centre in England (ftp://ftp.sanger.ac.uk/pub/dotter). As expected, when the sequence of L-selectin is compared with itself, we see a long diagonal descending from left to right that shows identity (Fig. 8a). In addition, we see two short diagonals spanning the regions 194–255 and 256–317. Thus, this protein has two stretches of amino acids that are very similar to each other, called an "internal repeat." Stretches of amino acids similar to these two occur in many other proteins, and this pattern has been called a "complement regulatory protein repeat," depicted as an oval in the schematics below each dot plot.

A similarity search of L-selectin reveals two closely related proteins, E-selectin and P-selectin (also known as CD62E and CD62P). A dot plot of L-selectin against E-selectin shows a long diagonal, which means that these proteins are similar throughout much of their length (Fig. 8b). The multiple shorter diagonals indicate that the "complement regulatory protein repeat" region of L-selectin at 194–255 is repeated six times in E-selectin, represented by six ovals in the E-selectin schematic.

The N terminus of L-selectin is homologous to a large lectin family. A dot plot of L-selectin against a lectin from the HIV virus that causes AIDS is shown in Fig. 8c. The similarity observed is not nearly as strong as that between E-selectin and L-selectin, but the pattern of conserved amino acids fits very well the pattern of similarities previously found between lectins. The lectin domain is shown as a rectangle and is found toward the N terminus of the selectins but toward the C terminus of the HIV lectin.

Cell-binding studies have shown that the L-selectin protein binds to special sugars found only on the walls of blood vessels where white cells are targeted for migration. Thus both functional studies and structural studies indicate that L-selectin is a sugar-binding molecule. Intense efforts are under way to develop sugar-based drugs for autoimmune diseases that might be improved by interfering with the binding of selectins.

IV. ACCESS TO INFORMATION TOOLS

Global computer networks have brought state-of-the-art computational biology tools to anyone with access to a personal computer and a link to the Internet. Many other useful software tools are available in the public domain in addition to the ones described in this article. Connection can be made through electronic mail, the World Wide Web, client-server software, or a locally established search capability.

The most basic access to BLAST similarity searching is through electronic mail. To request instructions on how to perform BLAST homology searches, simply send electronic mail to blast@ncbi.nlm.nih.gov with the word "help" in the body of the message (without the quotes). Within minutes, the user will receive a number of documents, including one explaining the BLAST e-mail service, a detailed manual on BLAST features and parameters, and a third that addresses frequently asked questions. The query sequence can be entered into the e-mail message either by typing or more simply by the computer's "copy and paste"

TABLE I
Resources on the World Wide Web That Are Particularly Useful in Sequence Analysis

Organization	Abbreviation	Location	Home page	Comment
DNA Databank of Japan	DDBJ	Japan	http://www.nig.ac.jp/	Japanese arm of International Collaboration
EMBL/Heidelberg	EMBL	Germany	http://www.embl-heidelberg.de/	Computing tools for structural biologists
European BioInformatics Institute	EBI	England	http://www.ebi.ac.uk/	European arm of International Collaboration
Genome Data Base	GDB	U.S.A.	http://www.gdb.org/	WWW links; provides searching of interrelated datasets
Genome Net		Japan	http://www.genome.ad.jp/	Sequence interpretation tools and links to other sites
National Center for Biotechnology Information	NCBI	U.S.A.	http://www.ncbi.nlm.nih.gov/	U.S. arm of International Collaboration; home of Entrez, BLAST, dBEST
The Sanger Centre		England	http://www.sanger.ac.uk/	Mapping the human and other genomes; unix software
University of Geneva	ExPASy	Switzerland	http://expasy.hcuge.ch/	Home of SwissProt, library of links, text, and search tools

function. The BLAST server is only one example of the many useful e-mail servers that are available on the Internet.

In addition, the World Wide Web is an excellent way to get access to information and sequence analysis tools. The best way to learn about the WWW is to explore it. A list of especially useful locations is shown in Table I. BLAST and Entrez services can be used through the World Wide Web with network software such as Netscape or Mosaic or other browsers. The following URL (http://www.ncbi.nlm.nih.gov/) is the location of the NCBI home page, which contains a large collection of useful information. Many other sites, some listed in Table I, also offer useful tools and links to other databases.

The WWW is also a good place to find more advanced documents about sequence analysis concepts, tools, and tutorials. For example, three excellent samples are included in the Bibliography. Unlike published literature, information on the WWW may "disappear" whenever the responsible person or institution moves that document or no longer offers that service, although some appear to be here to stay. However, the rate of growth of WWW offerings indicates that better breadth and depth of offerings will more than compensate for material removed. Moreover, the Internet search tools available make it surprisingly easy to track down relevant resources. For example, an Internet search from the location http://www2.infoseek.com/ reveals hundreds of documents containing all three terms (sequence, homology, search).

Other important resources include network newsgroups and mailing lists that are useful for making contact with other interested people. Listservers organized around topics are useful because they are a good place to get questions answered by other people with similar interests.

The collection of resources described in this article is enough for anyone with minimal access and reasonable persistence to master the basic tools of computational biology.

BIBLIOGRAPHY

From the World Wide Web

Barton, G. J. "Protein Sequence Alignment and Database Scanning." Available from http://geoff.biop.ox.ac.uk/papers/rev93_1/rev93_1.html/

Baxevanis, A. D., Boguski, M. S., and Ouellette, B. F. Computational analysis of DNA and protein sequences, Chap. 25 of "Genome Analysis" to be published by Cold Spring Harbor Press in 1997. Available from http://www.ncbi.nlm.nih.gov/Baxevani/CSH/

Robison, K. "Sequence Analysis Tutorial." Available from http://twod.med.harvard.edu:80/seqanal/

Traditional Publications

Altschul, S. F., Boguski, M. S., Gish, W., and Wootton, J. C. (1994). Issues in searching Molecular Sequence Databases. *Nature Genet.* **6,** 119–129.

Andreas, D., Baxevanis, A. D., Boguski, M. S., and Ouellette, B. F. (1997). "Genome Analysis: A Laboratory Manual" (B. Birren, E. Green, P. Hieter, S. Klapholz, and R. Myers, eds.). Cold Spring Harbor Laboratory Press, Cold Spring Harbor, NY.

Schuler, G. D., Epstein, J. A., Ohkawa, H., and Kans, J. A. (1996). Entrez. *In* "Methods in Enzymology," Vol. 266, pp. 141–162.

Watson, J. D., and Crick, F. H. C. (1953). Molecular structure of nucleic acids: A structure for deoxyribose nucleic acid. *Nature* **171,** 737–738.

Serotonin in the Central Nervous System

JOAN M. LAKOSKI
Pennsylvania State University College of Medicine

BERNARD HABER
University of Texas Medical Branch at Galveston

GLOSSARY

5-Hydroxyindole acetic acid (5-HIAA) Principal metabolite formed by deamination via the enzyme monoamine oxidase

Receptor subtypes Multiple membrane-associated recognition sites for a neurotransmitter

Serotonin 5-Hydroxytryptamine (5-HT) is an indolealkyl amine that is localized in neurons, where it functions as a neurotransmitter

Serotonin agonist Compound recognized as serotonin-like by the receptor

Serotonin antagonist Compound that blocks the interaction of serotonin with the receptor

Tryptophan Essential amino acid transported into brain tissues, where it is the primary substrate for the synthesis of serotonin

Tryptophan hydroxylase Rate-limiting enzyme in the synthesis of serotonin that hydroxylates the precursor substrate tryptophan

SEROTONIN (5-HT; 5-HYDROXYTRYPTAMINE) IS AN endogenously produced indolealkyl amine located in platelets, mast cells, enterochromaffin cells, and specific neurons of the central nervous system. Within the nervous system, serotonin is located in discrete neuronal cell groups located in the midbrain and pons. These parts of the brain provide both ascending and descending neuronal projections to innervate most all nervous system structures. The neurochemistry, pharmacology, and physiology of serotonin support its role as a major neurotransmitter. Recent data obtained with a variety of radioligands have identified multiple selective receptor subtypes (5-HT_{1A}, 5-HT_{1B}, 5-HT_{1C}, 5-HT_2, 5-HT_3, 5-HT_4, 5-HT_5, 5-HT_6, 5-HT_7). A critical role for serotonin in the nervous system is linked to the regulation of multiple processes, including sleep, pain, anxiety, depression, migraine, and abuse of substances such as hallucinogens and stimulants. Serotonin continues to emerge as a key modulator of central processes regulating cardiovascular function, neuroendocrine regulation of hormone secretion, and the development, plasticity, and aging of the nervous system. The future understanding of the physiological specificity of serotonergic effects will be provided by detailed knowledge of the receptor subtype structures within the cell membrane and their linkage to signal transduction mechanisms.

I. NERVOUS SYSTEM DISTRIBUTION OF SEROTONIN

Serotonin was first isolated in the blood in 1948. Subsequently, in the mid-1960s, serotonin-containing neurons were first discovered and mapped in the central nervous system with the application of the Falck–Hillarp fluorescence immunohistochemical technique. Characteristic of the serotonergic system, these neurons are grouped in clusters of cells lying in or near

ENCYCLOPEDIA OF HUMAN BIOLOGY, Second Edition, VOLUME 7. Copyright © 1997 by Academic Press. All rights of reproduction in any form reserved.

the midline region of the midbrain and pons (upper brain stem). Nine serotonin-containing nuclei (termed B_1–B_9; collectively called "raphe nuclei") have been characterized as the principal groups of 5-HT neurons. These neurons, in turn, send axons to most all regions of the nervous system; the more caudal nuclei (B_1–B_3) send theirs to the spinal cord, whereas the more rostral nuclei (B_4–B_9) ascend to provide widespread innervation of the pituitary, hypothalamus, thalamus, amygdala, hippocampus, and cortical brain regions.

II. SEROTONIN RECEPTOR SUBTYPES

A. Neurochemical Profile

Serotonin is synthesized in the nervous system from the precursor amino acid tryptophan via the enzyme tryptophan hydroxylase to form the intermediate compound 5-hydroxytryptophan; this product is then decarboxylated to form serotonin. This neurotransmitter is metabolized primarily by monoamine oxidase (type A) to form 5-hydroxyindole acetic acid (5-HIAA). Levels of this serotonin metabolite are often assayed in the cerebrospinal fluid or circulating blood to assess the clinical status of serotonergic function in the brain. It is important to note that serotonin cannot enter the brain owing to its inability to cross the blood–brain barrier. Rather, the precursor tryptophan can be actively taken up into nervous system tissues to then serve as the substrate for formation of this neurotransmitter.

As a chemical signaling substance or neurotransmitter in the nervous system, serotonin is synthesized in the nerve terminal, stored in vesicles located in the cytoplasm, and upon stimulation is released into the synaptic cleft. In the extracellular domain located between two adjacent neurons, serotonin binds to a selective recognition site, or receptor, which is located on the plasma membrane of the adjacent neuron to complete the process of chemical neurotransmission across the synapse. Importantly, the action of serotonin released into the synaptic cleft is terminated by being taken up again into the nerve terminal from which it was originally released, or by reuptake into adjacent glial cells. The transport of serotonin back into the nerve terminal, from which it was released, is mediated through the action of specific membrane carrier or uptake pumps located on the presynaptic nerve membrane, the serotonin transporter. Once inside the neuron, serotonin may be stored in vesicles for reuse or metabolically degraded to an inactive form.

B. Pharmacological Profile

Since the early 1950s it has been recognized that 5-HT receptors may exist in multiple forms. With the use of radioligand binding techniques in the 1970s, ligands with high affinity for 5-HT receptors revealed a marked pattern of heterogeneous distribution of binding sites for serotonin in the central nervous system. Initially, two central serotonin binding sites were proposed: a 5-HT_1 site labeled by [^{3}H]5-HT and a 5-HT_2 site labeled by [^{3}H]spiperone, with the hallucinogen lysergic acid diethylamide (LDS) possessing similar affinity for both ligands. Since these observations, evidence has emerged for 5-HT_1, 5-HT_2, 5-HT_3, 5-HT_4, 5-HT_5, 5-HT_6, and 5-HT_7 receptor subtypes.

The 5-HT_1 family of receptors is characterized by a nanomolar affinity for 5-HT and 5-carboxyamidotryptamine and only micromolar affinity for 5-HT_2 and 5-HT_3 antagonists. This receptor class can be further subdivided into receptor subtypes, including 5-HT_{1A}, 5-HT_{1B}, 5-HT_{1D}, 5-HT_{1E}, and 5-HT_{1F}. Each of these receptor subtypes is characterized by a distinct pharmacological profile and pattern of distribution in the brain. The aminotetralin 8-hydroxy-2-(di-*n*-propylamino)tetralin (8-OH-DPAT) has a 1000-fold higher affinity for the 5-HT_{1A} receptor than serotonin. Thus, [^{3}H]8-OH-DPAT is used as a radioligand for 5-HT_{1A} receptors and has revealed the highest density of 5-HT_{1A} receptors to be in the hippocampus and dorsal raphe nucleus. The highest density of 5-HT_{1B} receptors is in the substantia nigra and globus pallidus. The 5-HT_2 receptor has the highest density of binding sites in the choroid plexus of the cerebral ventricles.

With the introduction of the quinazolinediane derivative ketanserin, a selective 5-HT_2 antagonist, characterization of the 5-HT_2 binding site was carried out. The highest density of 5-HT_2 sites is found in the medial prefrontal region of the cortex as identified in a variety of mammalian brain tissues, including humans. In general, this binding site is characterized by a relatively low affinity for serotonin agonists but a high affinity for 5-HT_2 antagonists. Recently several phenylaklyamines, such as DOM, have been identified as potent 5-HT_2 agonists.

The recent introduction of selective 5-HT_3 antagonists, including MDL 72222, ICS 205-903 and GR 38032F, and the selective 5-HT_3 agonist, 2-methyl serotonin, has prompted extensive studies of 5-HT_3 binding sites in the central nervous system. Central 5-HT_3 binding sites have been identified in the brain with high concentrations in cortical regions and the area postrema of the brain stem. Such distribution of binding sites may prove to underlie an apparent therapeutic role of antagonists specific for this receptor as antiemetic drugs.

The 5-HT_4 receptor has been identified on neurons in the central nervous system with high densities in the hippocampus and other limbic areas. A broad range of roles, including participation in cognition, affective disorders, and other mental illnesses, have been suggested to be mediated by this serotonergic receptor subtype. The 5-HT_4 receptor is also located in neurons in the alimentary tract and regulates contractile responses of smooth muscles underlying the peristaltic reflex. The identification of indoles (5-HT, 5-methyoxytryptamine) and benzamides (cisapride, renzapride) as selective agonists as well as the selective antagonist GR 113808 has facilitated the recent pharmacological and physiological characterization of 5-HT_4 receptors in the nervous system.

Classification of 5-HT receptors by cloning techniques has also identified novel receptors, including the 5-HT_5, 5-HT_6, and 5-HT_7 receptors. Functional correlates and pharmacological tools to characterize these receptors are, to date, limited. However, preliminary studies support a central nervous system distribution of the 5-HT_6 and 5-HT_7 receptors.

C. Signal Transduction Mechanisms

Functional correlates of the activation of the multiple 5-HT recognition sites defined via receptor binding techniques also require consideration. The identification of the biochemical transducing systems that are associated with each site has lent further support to the concept of multiple 5-HT receptor subtypes.

One major signaling pathway involving adenylate cyclase activation can be stimulated by 5-HT to induce an increase in the formation of cyclic AMP, a second messenger which, in turn, activates specific protein kinases. Serotonin-induced stimulation of adenylate cyclase activity is the signaling mechanism that is coupled to 5-HT_4-mediated responses. In contrast, actions of serotonin linked to 5-HT_{1A} and 5-HT_{1B} receptors are negatively coupled to adenylate cyclase, such that activation of these receptors will result in a decrease in the formation of cyclic AMP.

A second major signaling pathway which mediates the actions of serotonin on nervous system function is the phospholipase C/phosphoinositide hydrolysis mechanism utilizing inositol lipids located in the plasma membrane for signal transduction. Ultimately, transduction of this receptor-mediated signal will include effects on calcium ions and formation of diacylglycerol, which, in turn, stimulates protein kinase C. This phosphoinositide signaling pathway is also involved in 5-HT-mediated responses with selectivity for the 5-HT_2 receptor. A number of 5-HT_2 antagonists block serotonin-induced activation of the phosphoinositide hydrolysis system.

The 5-HT_1, 5-HT_2, and 5-HT_4 receptors are members of the G protein superfamily. Molecular cloning techniques show that 5-HT receptor genes encode single-subunit proteins whose structure is characterized by seven transmembrane domains and the ability to activate G protein-dependent processes, such as adenylate cyclase or phosphoinositide signaling mechanisms.

The signaling mechanism of the 5-HT_3-mediated response is distinctly different from that of the foregoing. This receptor is a member of the ligand-gated ion channel superfamily, such that 5-HT_3 receptor activation is directly coupled to ion flow through a Na^+ channel.

D. Physiological Functions

Multiple physiological effects of 5-HT were first identified in studies of smooth muscle contraction in the gut. In the central nervous system, the technique of drug microiontophoresis, as applied to different brain regions, established the presence of both inhibitory and excitatory responses to the application of 5-HT. Autoreceptors, which are inhibited by 5-HT, have been identified in midbrain raphe cell groups, with dorsal raphe responses most sensitive to 5-HT_{1A} agonists and median raphe responses most sensitive to 5-HT_{1B} agonists. *In vitro* brain slice preparations coupled with intracellular recording techniques have identified hyperpolarizing effects, depolarizing effects, or direct ion-gated depolarizing, rapidly desensitizing effects on resting membrane potential in association with 5-HT_{1A}-, 5-HT_2-, and 5-HT_3-mediated responses, respectively.

III. ROLE OF SEROTONIN IN HUMAN DISEASE

A. Pain and Nociception

Central serotonin neurons have been implicated in pain and the actions of analgesic drugs. It has been recognized for some time that serotonergic neurons project to and comprise components of ascending and descending neuronal pathways known to mediate nociceptive responsiveness. In addition, reciprocal projections between 5-HT neurons and neurons containing the peptide enkephalin, as well as colocalization of these neurotransmitters, may underlie nociception. In general, direct or indirect enhancement of serotonergic function has been demonstrated to have antinociceptive activity in animal models. For example, 5-HT reuptake inhibitors have been demonstrated to enhance morphine analgesia. Although clinical usefulness of serotonergic related compounds is at present limited with respect to the therapeutic treatment of pain, this neuronal system has a critical role in the central regulation of pain pathways. [*See* Pain.]

B. Sleep

Serotonin has long been implicated in the complex behavior of sleep. In general, enhancement of 5-HT activity has been reported to increase total sleep time, whereas the inhibition of 5-HT synthesis with *p*-chloroamphetamine results in insomnia. Dose-dependent effects of the 5-HT reuptake inhibitor fluoxetine have also been reported to lengthen the latency to the onset of the rapid eye movement (REM) stage of sleep in animals. [*See* Sleep.]

C. Depressive Disorders

A role for serotonin in depressive disorders has emerged from numerous clinical studies of decreases in 5-HT and its metabolite, 5-HIAA, seen in brain tissue from depressed patients who have commited suicide. Similarly, monoamine oxidase inhibitors and 5-HT reuptake inhibitors, which effectively increase central levels of 5-HT, are therapeutically useful antidepressants. Several more selective 5-HT reuptake inhibitors have been recently developed, including fluoxetine, sertraline, citalopram, and paroxetine, which block neuronal reuptake of 5-HT with minimal effects on dopamine or noradrenergic reuptake processes. These compounds have been found to be effective antidepressant drugs with the added advantage of lacking several side effects often associated with use of tricyclic antidepressant drugs. Thus, the effectiveness of selective 5-HT reuptake inhibitors as antidepressant drugs supports the hypothesis that enhancement of synaptically available 5-HT will ameliorate symptoms of depression. [*See* Depression, Neurotransmitters and Receptors.]

D. Anxiety

Serotonin has long been implicated for a role in mediating anxiety. Though early studies hypothesized a direct interaction of central 5-HT with the anxiolytic actions of benzodiazepines (such as valium), a role for several 5-HT receptor subtypes has emerged. Both behavioral and clinical studies have identified the selective 5-HT_2 antagonist ritanserin to have efficacy as an anxiolytic agent. However, most attention has focused on buspirone, a nonbenzodiazepine anxiolytic drug with high affinity for the 5-HT_{1A} receptor. Several 5-HT_{1A} agonists, including 8-OH-DPAT, gepirone, and ipsapirone, have all been demonstrated to have similar anxiolytic properties, but the actual neuronal mechanisms and pathways mediating their effects remain to be established.

E. Obesity and Eating Disorders

Brain serotonergic neurons have been demonstrated to mediate, in part, the hypothalamic regulation of food intake. Drugs that act either directly or indirectly to increase stimulation of serotonergic neurons decrease food intake in rats. Such drugs include serotonin-releasing drugs, such as *p*-chloroamphetamine and (±)fenfluramine, which have been marketed as anti-obesity drugs in the clinical setting. In addition, the selective 5-HT uptake inhibitors fluoxetine and zimelidine have been demonstrated to decrease body weight in nondepressed obese patients. Conversely, several direct-acting 5-HT_{1A} agonists, including 8-OH-DPAT, buspirone, and ipsapirone, have recently been reported to increase food intake in rats. Clearly, serotonin has a role in mediating the responses associated with food intake; however, the neuronal circuitry and receptors mediating the regulation of food intake are not yet established. [*See* Eating Disorders; Obesity.]

F. Psychopharmacology of Hallucinogens

Historically, the original classification of multiple 5-HT responses by Gaddum including both "D" and "M" receptors, with the former being antagonized by LSD. This observation, coupled with the known structural similarity of 5-HT to the psychedelic drug LSD, has prompted extensive investigation into the role of serotonergic neurons in mediating hallucinations. As identified in numerous biochemical and physiological studies, the central effects of LSD include direct effects on 5-HT_{1A} autoreceptor function and on 5-HT_2 receptors. Most recently, the selective 5-HT_2 antagonists, such as ritanserin, have proved successful in blocking the biochemical, physiological, and behavioral effects of LSD in animal studies. Such interactions contribute to our understanding of the complex phenomenon of hallucinations and may prove to be a neuronal basis that is common to the actions of all hallucinogens.

G. Psychopharmacology of Stimulants

Central nervous system stimulants, including amphetamine and cocaine, produce numerous effects that range from euphoria to convulsions. The central effects of stimulants are mediated by effects on dopamine, noradrenergic, and serotonergic neuronal systems. Specifically, amphetamine, which acts by releasing stores of a given neurotransmitter, will enhance the availability of 5-HT in the synapse and ultimately prolong the agonist effect. While acting by a different mechanism, cocaine also enhances the action of available 5-HT, thereby acting as an indirect serotonin agonist. Specifically, in addition to known actions at Na^+ channels, cocaine blocks the reuptake of 5-HT into the presynaptic terminal. Cocaine has also been recently demonstrated to alter the cell firing of spontaneously active 5-HT neurons recorded in the dorsal raphe nucleus as well as modulate the neuroendocrine regulation of several hormones under tonic serotonergic regulation. Although the receptor subtypes that may mediate the actions of cocaine in the central nervous system are as yet unresolved, vasoconstrictive actions of serotonin in peripheral tissues, which are mediated via a 5-HT_3 receptor, are attenuated by cocaine. Indeed, the structure of cocaine has been successfully modified to yield new potent and selective 5-HT_3 antagonists, such as MDL 72222 and ICS 205-930.

H. Central Cardiovascular Function

Serotonin has the ability to constrict blood vessels and amplify the vasoconstrictor responses to other compounds, as well as enhance blood platelet aggregation. These actions, coupled with the known serotonergic innervation of cerebral vasculature, have helped to identify a role for serotonin in normal and pathogenic cardiovascular function.

Several centrally acting 5-HT agonists and peripheral 5-HT antagonists have been suggested to be effective hypertensive agents. Several 5-HT_1 agonists have been reported to decrease blood pressure. However, the antihypertensive properties of ketanserin, a 5-HT_2 receptor antagonist, are due not to actions on a 5-HT receptor but rather an α-adrenergic receptor. Another potential therapeutic use of 5-HT_2 receptor antagonists with respect to cardiovascular dysfunction is emerging in the treatment of thrombotic obstructions of the aorta. Likewise, 5-HT_3 receptor density is most strikingly high in brain stem regions, for example, the area postrema, with critical roles in the central pathways regulating cardiovascular function.

I. Etiology of Migraine

A role for 5-HT in the pathogenesis of migraines has been implicated by the effectiveness of selective 5-HT antagonists in the treatment of migraine. Serotonin receptors are located on sensory nerves, in central pain pathways, and on cerebral blood vessels, which may be the site of their antimigraine action via either 5-HT_{1A}, 5-HT_2, or 5-HT_3 receptors. Effective antimigraine drugs, such as methysergide and ergotamine, have high affinities for a 5-HT_{1A} receptor. However, these compounds may also block vascular 5-HT_2 receptors and/or act as partial agonists that potentiate serotonin-induced contraction of intracranial blood vessels. Recent reports have identified the 5-HT_{1D} agonist sumatriptan to be effective in the treatment of migraine, thereby implicating a role for this receptor subtype in the etiology of migraine. [*See* Headache.]

IV. NEUROENDOCRINE REGULATION

The regulation of the secretion of hormones from the pituitary is a complex process that involves the interaction of the hypothalamic–pituitary–adrenal and/or gonadal axis. In its role as a neurotransmitter in the nervous system, 5-HT-containing neuronal systems are one of several systems identified to have specific modulatory roles in a variety of neuroendocrine functions. Not only has a role for 5-HT been identified in the regulation of neuroendocrine function in depressive disorders, this neurotransmitter is emerging as a key factor in the normal processes that regulate reproductive processes such as ovulation, menstrual cyclicity, sexual behavior, and fertility. [*See* Neuroendocrinology.]

A. Anterior Pituitary Hormones

The secretion of hormones from the anterior pituitary, including luteinizing hormone (LH), prolactin, growth hormone (GH), thyroid-stimulating hormone (TSH), and adrenocorticotropin hormone (ACTH), is closely regulated by specific hypophyseal hormones. These hormones enter into the portal circulation at the median eminence of the hypothalamus to regulate the production of a given hormone by the pituitary. For the listed hormones, these hypophyseal hormones include luteinizing-releasing hormone (Gn-RH), "prolactin-releasing factor" (PRF), somatostatin, thyroid-releasing hormone (TRH), and corticotropin-releasing factor (CRF), respectively. These are, in turn, closely regulated by a variety of neurotransmitters located in the hypothalamus and in extrahypothalamic brain regions. The specific roles 5-HT in regulating neuroendocrine function in the nervous system are mediated via the modulation of these hypophyseal hormones. [*See* Hypothalamus.]

Serotonin has been demonstrated to have an inhibitory role on the secretion of LH. Electrical stimulation of serotonergic neurons in the dorsal raphe nucleus, as well as intraventricular administration of 5-HT, suppresses LH in both intact male and ovariectomized female rats. Administration of a 5-HT synthesis inhibitor or a selective 5-HT neurotoxin also attenuates proestrus increases in LH as well as ovulation in the rat. Although the neuronal site of action of 5-HT in mediating these responses is not yet clear, the presence of dense serotonergic innervation in the median eminence and in the anterior pituitary has been identified. With respect to identification of a role for specific receptor subtypes, evidence for a multiplicity of serotonergic influences on the release of LH in the rat supports both stimulatory and inhibitory roles of both 5-HT_1 and 5-HT_2 receptors in a steroid hormone-dependent manner. Indeed, estrogen has been demonstrated to alter 5-HT receptor binding characteristics with specific decreases in 5-HT_1 and 5-HT_2 binding sites reported in various brain regions following hormone pretreatment.

The regulation of prolactin secretion, though under inhibitory regulation from dopamine neurons located in the arcuate nucleus of the hypothalamus, is under a stimulatory influence by serotonergic mechanisms. The stimulatory effects of 5-HT on prolactin release occur at the level of the hypothalamus. Lesion of the paraventricular nucleus will block the serotonergic stimulation of this hormone. The specific neuronal pathways mediating this response include the midbrain raphe nuclei, as destruction of 5-HT cell bodies in the dorsal raphe nucleus decreases circulating levels of prolactin. With respect to the role of receptor subtypes that may mediate these effects, 5-HT_{1A} agonists (8-OH-DPAT, gepirone, ipsapirone) and 5-HT_2 agonists (TFMPP [1-(3-trifluormethylphenyl-perazine], MK-212) have been reported to stimulate prolactin release. These serotonergic agonists may ultimately prove useful as probes in clinical neuroendocrine challenges studies.

Serotonergic neurons outside of the dorsal raphe nucleus have been demonstrated to stimulate the secretion of ACTH and β-endorphin in the rat. The 5-HT_{1A} agonist 8-OH-DPAT, the 5-HT_{2A} agonist *m*-chlorophenylpiperzine (*m*-CPP), and the 5-HT_{2C} agonist 1-(2,5-dimethoxy-4-iodophenyl)-2-aminopropane (DOI) produce marked, dose-dependent increases in ACTH levels. In addition, recent evidence suggests that 5-HT_2 receptors mediate CRF secretion from the hypothalamus. Thus, stimulation of the hypothalamic–pituitary–adrenal axis occurs by multiple 5-HT receptor subtypes.

B. Posterior Pituitary Hormones

The hormones released from the posterior pituitary, which include vasopressin and oxytocin, are synthesized in the supraoptic and paraventricular hypothalamic nuclei. Serotonergic innervation to the paraventricular nucleus has been demonstrated to provide a facilitatory effect on the secretion of vasopressin. The stimulatory input on vasopressin secretion is provided by the 5-HT neurons, which originate in the dorsal raphe nucleus. To date, the relative contributions of various 5-HT receptor subtypes to vasopressin secretion are unclear.

C. Other Endocrine Glands

Serotonin has also been implicated in the regulation of endocrine glands other than the pituitary. Such endocrine glands are either controlled via circulation of bloodborne substances or regulated via direct neuronal innervation. The juxtaglomerulosa cells of the kidney, which secrete renin in response to changes in sympathetic output, are one such endocrine gland. The secretion of renin is centrally regulated by 5-HT, such that a stimulatory effect on renin secretion is mediated via the serotonergic neurons of the dorsal raphe nucleus. Stimulation of 5-HT_2 receptors is also consistent with a stimulatory effect on renin secretion, whereas 5-HT_{1A} agonists have little apparent effect in regulating this neuroendocrine response.

V. DEVELOPMENT, PLASTICITY, AND AGING

Serotonin is one of the earliest developing neurotransmitter systems in the brain. Serotonergic neurons influence the growth of other neurons with which they come into contact, depending on the type of cell-to-cell interaction during development. Thus, this neuroactive substance plays an important role in the development of neural circuits where 5-HT will be later used as a chemical signal important for brain function.

Serotonin has also been found in early embryos, where it is present outside the nervous system in regions undergoing active morphogenesis (formation of tissues and body structures), such as the heart, intestine, and craniofacial region (nose, face, ears) locations where congenital malformations are most common. Animal studies suggest that certain psychoactive drugs, including some antidepressants, tranquilizers, and antipsychotics, which act by altering the functions of serotonergic neurons can cause structural and/or functional abnormalities in the brain when given to pregnant mothers during critical periods of gestation. Animal studies suggest that manipulations of serotonergic systems during the prenatal period may have permanent effects on the developing nervous system.

Serotonin also has a fundamental role in plasticity of the central nervous system. In addition to normal nervous system development, changes in 5-HT-mediated responses have been identified following injury to the brain or as a consequence of neurochemical-induced lesions of this transmitter system.

The role of serotonin in aging continues to emerge as a strikingly important one with respect to age-related changes in central nervous system function. Receptor binding studies have revealed significant decreases in the number of 5-HT binding sites with age in the brain regions involved with cognitive processes, such as the cortex and hippocampus. Furthermore, significant age-related alterations in the synthesis and turnover of 5-HT have been identified in the brain, as well as changes in the cellular physiological activity of 5-HT-containing neurons. These observations confirm that a key role of 5-HT is in mediating age-related changes in normal and pathological aging.

VI. SUMMARY

The role of 5-HT as a major neurotransmitter substance in the central nervous system is clearly established. Serotonin contributes to the regulation of numerous biological states, such as sleep, eating, and perception of pain, as well as affective disorders, including anxiety, depression, and the mood-altering properties of several widely abused drugs and migraine. Recent progress has been made in identifying multiple recognition sites for 5-HT in the brain and in understanding their functional role in the nervous system. The future understanding of the physiological specificity of serotonergic effects will continue to expand as our knowledge of the molecular structure of these multiple receptor subtypes emerges. Ultimately, knowledge of the role of 5-HT in the nervous system will provide new therapeutic approaches toward the treatment of human disease.

ACKNOWLEDGMENT

Publication No. 71C supported by U.S.P.H.S. Grant PO1 AG10514 awarded by the National Institute of Aging (J.M.L.).

BIBLIOGRAPHY

Aghajanian, G. K., Sprouse, J. S., and Rasmussen, K. (1987). Physiology of the midbrain serotonin system. *In* "Psychopharmacology: The Third Generation of Progress" (H. Y. Meltzer, ed.), p. 141. Raven, New York.

Elgen, R. M., Wong, E. H. F., Dumuis, A., and Bockaert, J. (1995). Central 5-HT_4 receptors. *Trends Pharmacol.* **16**, 391.

Fuller, R. W. (1994). Uptake inhibitors increase extracellular serotonin concentration measured by brain microdialysis. *Life Sci.* **55**, 163.

Glennon, R. A. (1987). Central serotonin receptors as targets for drug research. *J. Med. Chem.* **30**, 1.

Hoyer, D., Clarke, D. E., Fozard, J. R., Hartig, P. R., Martin, G. R., Mylecharane, E. J., Saxena, P. R., and Humphrey, P. P. A. (1994). VII. International union of pharmacology classification of receptors for 5-hydroxytryptamine (serotonin). *Pharmacol. Rev.* **46**, 157.

Lakoski, J. M., Keck, B. J., and Dugar, A. (1996). Neurochemical lesions: Tools for functional assessment of serotonin neuronal systems. *In* "Paradigms of Neural Injury" (J. R. Perez-Polo, ed.), Methods in Neurosciences, Vol. 30, Chap. 8. p. 115. Academic Press, San Diego.

Owens, M. J., and Nemeroff, C. B. (1994). Role of serotonin in the pathophysiology of depression: Focus on the serotonin transporter. *Clin. Chem.* **40**, 288.

Van de Kar, L. D. (1991). Neuroendocrine pharmacology of serotonergic (5-HT) neurons. *Annu. Rev. Pharmacol. Toxicol.* **31**, 225.

Sex Differences, Biocultural

ROBERTA L. HALL
Oregon State University

GLOSSARY

Gathering and hunting economy Economic organization that dominated the evolutionary history of the hominids from 4 million until 12,000 years ago; plant and animal resources were obtained through gathering, scavenging of dead animals, and hunting; patterned tools are associated with cultures of the last 2 million years

Genotype Total complement of an individual's genetic material

Hominids All species, living and extinct, of the genera *Australopithecus* and *Homo;* they share a common ancestor with the chimpanzee and gorilla but are not ancestral to either

Prehistory The period, prior to 5000 years ago, for which no written records exist; it is studied by archaeology, which analyzes material remains of human activity

Sex ratio Number of males to 100 females at particular points in the life cycle; *primary* refers to the ratio at conception, *secondary* to the ratio at birth, and *tertiary* to the ratio at some other specified time

Sexual dimorphism Patterned differences between males and females of one species; it may include color, size, shape, body composition, and anatomical features

ANTHROPOLOGISTS HAVE OBSERVED THAT SOME cultures emphasize distinctions between the sexes whereas others mute them. At various times in history, the biological differences between males and females have been considered either as major determinants of individual and social patterns or as minor phenomena. A biocultural perspective considers the evolution of male and female characteristics within a social context. The values of a society affect behaviors such as nutrition, education, and levels of physical activity; those in turn may have effects on the development of sex-specific characteristics. In addition, a society's values influence the cultural interpretation of differences between the sexes.

I. INTRODUCTION

A biocultural perspective cannot define for all time the core differences between men and women. Instead, this perspective suggests that absolute definition is not an appropriate goal. Certainly, a biocultural perspective will not provide a blueprint of proper behavioral patterns for the two sexes, nor will it provide answers to ethical questions. However, such a perspective examines diverse solutions to the problem of male–female definition that have been developed by various species and societies. By going beyond the boundaries of one culture, one time period, or species, it puts the problem of sex definition within the domain of research so that the subject can be approached by the intellect rather than by the emotions.

II. ANIMAL INSIGHTS

At its most inclusive level, sexual reproduction encompasses all types of reproduction in which an individual

ENCYCLOPEDIA OF HUMAN BIOLOGY, Second Edition, VOLUME 7. Copyright © 1997 by Academic Press. All rights of reproduction in any form reserved.

offspring receives genetic contributions from more than one parent. Whereas microbial species that meet this definition minimally do not have defined male or female roles, organs, or individuals, more complex sexual species can be categorized by whether the reproductive cell provides only genetic material (male) or provides both nourishment and genetic material (female). Some organisms, including many common plants, such as corn, contain both male and female organs and functions; some groups, for example, aphids (insects that prey on plants), have the capacity to reproduce either sexually or asexually; and other species, including all primates, develop individuals as either male or female by structuring their reproductive capacity to one or the other style. Except among mammals, who bear their young alive and provide care for their young during infancy, the female parent is not always a caregiver. Both parents in many species of birds devote much of their energy to providing for their young; among other vertebrates, caregiving by either sex is rare. Examples of paternal care are uncommon and tend to be repeated in the literature (e.g., the male sea horse tends its young). Parental care is simply one among many possible behavioral adaptations that may be advantageous to a species and hence become part of its innate behavioral repertoire. For species lacking a complementary anatomical program such as exists in mammals, parental care is neither obligatory nor inherently superior.

Among mammals, females not only house and nourish the developing embryo but feed the offspring after birth. Whereas involvement of the mother is obligatory, the involvement of the male is optional. However, from an evolutionary standpoint, a male that participates in reproduction will have to follow the accepted behavioral pattern in its species. Most bears, for example, do not have a paternal role. Male and female bears have a courting period in which individuals permit another adult of their species to become intimate for several days, but their normal relationship with other adults is avoidance or conflict. Indeed, adult male bears threaten the security of cubs. Wolves, who are members of the same taxonomic order as bears, have developed the opposite strategy. Male adults, the father and other related males of the same pack, provide food for the pack's mother and for wolf pups. In addition, they act as babysitters and teach skills such as hunting. These examples illustrate the wide range of options open to mammalian species, but this does not mean that the same range is open to individual bears or wolves. Females and males must perform their respective roles, which appear to be genetically programmed to a large extent but may vary with individual social experience. Problems that captive animals have with reproduction illustrate the dependence of sexual behavior on maintenance of a species' socioecological patterns.

We do not know specifically how a male is programmed to act like a father. Male wolves and coyotes are "good" fathers, but male dogs characteristically reserve their loyality for their human masters. Female coyotes and wolves who are sexually stimulated to court with dogs pay a stiff price when their suitor abandons them and they must try to raise puppies alone. How much of the paternal behavior of the canids is controlled by its genes and how much by its own life experience? Some flexibility must exist in canids, because coyote family behavior differs greatly according to whether they are protected from humans. Where hunted, pups are reared for a short time but family groups are neither large nor permanent. Where coyotes are protected, family ties persist. Both males and females adapt their behavior to necessity; even though pack life may be the preferred option within the coyote repertoire, it is not obligatory if the environment demands a solitary existence.

III. PRIMATE PATTERNS

Social carnivores such as the canids provide the best nonprimate analog for human social patterns, but biologists and anthropologists seeking models for the evolution of human sex differences have usually turned to the order Primates, in which *Homo sapiens* belongs. However, great diversity in reproduction, sex roles, and sexual dimorphism is found among the primates, and scholars have long debated which groups make the most appropriate model. Some have focused on chimpanzees and gorillas on the grounds that the groups that share the most recent common ancestor with humans are most appropriate. Interestingly, these two groups differ greatly between themselves in social organization, including sex roles, and in male–female size differences.

Chimpanzee males and females show no greater overall size differences than do human males and females. The two species of chimps—*Pan troglodytes,* the common chimp, and *Pan paniscus,* known as the pygmy chimp but actually not much smaller than the other species—differ in sex roles. (In 1994, geneticists suggested that *Pan troglodytes* might actually consist of two different species, but the argument over taxonomy does not affect the major division between *troglodytes* and *paniscus*.) Common chimp troops live in territories within which females mate with various males who are organized into a shifting hierarchy. Females are more likely than males to move from troop to troop, and they tend to take their young with them; females and their grown and adolescent offspring form one part of the social structure while liaisons among males form another. Pygmy chimps are reported to have a different and more human-like social structure with more female bonding, more sexual and affiliative behavior, even between members of the same sex, and occasional use of the face-to-face copulatory position. It is unfortunate for many reasons that so few chimpanzees—particularly *Pan paniscus*—exist in the wild. If more did, one benefit would be further studies to elucidate the range of social behaviors in free-ranging chimps. Of all primates, these species might be expected to show the greatest number of sex role options. Biochemical studies suggest that the common and pygmy chimps have been separate species for 1.5 million years, and it would be valuable to know the degree to which their sex differences are encoded. Such an understanding might provide models for the role of sex differences in the evolution of prehistoric hominids (see the next section).

Three subspecies of gorilla are conventionally recognized, with a growing tendency to consider these subdivisions as species distinctions. Though they differ in size, all three show a large degree of sex differences. Gorillas live in troops consisting of one elder male, a band of females, and their young. Their social pattern is markedly different from that of chimpanzee species.

A third great ape species, the orangutan of Southeast Asia, has yet a different form of social structure in the wild. Although orangs are apparently quite socially responsive when reared in captivity, in the wild they tend to be solitary as adults, with the exception of the courtship period. The basic social group is the adult female and young. Biochemical studies show that orangs are less closely related to chimps, gorillas, and humans than these three are to each other. Orangs, gorillas, and chimpanzees are similar in placing almost all of the responsibility for rearing infants on the mother. Although male chimps and gorillas may provide defense against attack by outsiders, the burdens of feeding and providing emotional security rest with the female. In chimps and gorillas, older siblings may be of some help to infants, but not predictably or consistently. One theory of human origins holds that the human innovation of a father who was responsible for providing food and protection permitted human ancestors to reproduce faster and more effectively and thus gave them the edge over other apes. Though this theory is too simplistic, it does call attention to problems that the great apes have in maintaining their populations when habitat is lost and when adult females are killed by humans.

Some anthropologists have argued that the great apes are inappropriate social models for humans. This argument is based on the idea that, in splitting from the apes, human ancestors left the forests and chose semiopen or savanna habitat, and accordingly adjusted their behavior. By this reasoning, primates of the savanna offer more useful models of early human sex differences. Baboons in Africa and macaque species in Asia, both highly social, intelligent, and successful monkeys, have thereby received considerable attention. They and other monkeys of Africa and Asia have diversified greatly in the last several millions of years and now occupy a wide range of habitats from the cold and snowy mountains of northern Japan to the lush tropics and arid rocky deserts of Africa. Many monkeys live in multimale groups in which adult males play a role in rearing young. Differences exist between closely related species, but in general females and their young form the core of society while males tend to move, usually every several years, from troop to troop. In some arid regions, the baboon foraging unit consists of a single adult male and a small group of females and their young. The adult male guides their foraging and defends the females against predators and against being taken over by other males. This organization has been quite inaccurately called a harem; a luxury for the well-to-do male it definitely is not! Rather it is an adaptation to a harsh environment.

In richer savannah areas, baboons live in large troops that consist of many males and females of all ages. In these troops, males and females show consistent differences in growth patterns, reproductive styles, and social behavior. Females are smaller than males and mature at a younger age. The longer growth period of males means that they grow larger and can thus defend the periphery of the troop; if they survive to adulthood they start their reproductive careers later. Meanwhile, females start to produce offspring at a young age. Reductions in the nutritional needs of females because of small size are balanced by the demands of reproduction (which occurs every second year) and lactation. Several decades ago, studies of troops of savanna baboons portrayed their societies as male-dominated and rigidly hierarchical. However, long-term studies have established that females are the core of baboon life. These studies show intertroop mobility, mainly of adolescent and adult males, and have identified social mechanisms for changes in troop hierarchies. In addition, long-term studies of monkey troops, like those of great apes, have revealed considerable personality differences among members of the same troop.

Although each troop's environmental and historical situation may require its males and females to have certain behavioral traits, not all individuals comply. The existence of variation permits a population to shift its social norms to meet new environmental or historical constraints. Studies of our primate relatives have suggested the roots of human diversity but have not provided a blueprint of male and female behavior. [*See* Primates.].

IV. PREHISTORIC HOMINIDS

Identification of an individual's sex from its preserved skeleton, which often is quite fragmentary, is never certain. This lack of certainty is the primary impediment to developing models of the history of hominid sexual dimorphism. Teeth are the most frequently preserved fossil remains of hominid skeletons. Fortunately, teeth, particularly the canines, show consistent male–female size differences in nonhuman primates and, to a lesser but statistically significant extent, in modern people as well. Nonetheless, identification of fossil specimens by sex (or by species) is seldom agreed upon by all experts, so disputes about primary data categories persist.

One of the most important early hominid groups, identified at two sites in eastern Africa, is *Australopithecus afarensis,* dating from 3 to 4 million years ago. Donald Johanson has identified two sexes at these sites, the male being considerably larger than the female. A contrary hypothesis holds that the finds represent two species of different size. The Johanson model suggests the possibility of divergent sex roles, and perhaps different growth and maturity patterns for males and females, such as were noted in some contemporary monkeys. It is uncertain whether this implies a female-centered social structure and intratroop promiscuity like that of the monkeys, or a society like that of the gorillas, which is dominated by an elder male. An alternative may be a social structure perhaps closer to that of some modern cultures that live in small bands in which one family may include one male, several wives, and all their children. What role, if any, males played in child-rearing is a frequently asked, but unanswerable, question.

The australopithecine period of 1.5 to 5 million years ago ends several hundred thousand years after the first fossils of our own genus, *Homo,* are known. *Homo erectus* specimens are found in the period from 1.8 to 0.3 million years ago, but their sparse skeletal remains are insufficient to establish sex differences in size. Behavioral models of the species in East Asia, Africa, and Europe are based on the archaeological record, primarily on stone tools and evidence of hunting. Behaviors of males and females in the *Homo erectus* period have been hypothesized primarily from ethnographic patterns described over the past 100 years in the technologically simplest gathering and hunting cultures of *Homo sapiens.* Such speculation (or model building) is acceptable if used as a hypothesis, but there is a danger that such descriptive pictures will take on a false reality and subsequently will be used to make unsubstantiated inferences or to "explain" the modern data from which they have been derived.

The Neandertal population that occupied Western Europe from about 125,000 to 40,000 years ago is one of the best known and most argued about skeletal populations in prehistory. In contrast to *Homo erectus* samples, there are enough data to at least hold conversations about Neandertal body size and shape, and about sex differences. Compared with modern people, Neandertals were short and very muscular. Both males and females appear to have had broad pelvises, but explanations of this finding differ. One hypothesis is that their locomotory adaptations differed from ours, that is, that Neandertals were less efficient in their stride. A second idea is that the broad pelvis is simply an aspect of their robusticity, which evolved as an adaptation both to extreme cold and to facilitate combat at close range with large mammals such as the cave bear and mammoth. A third hypothesis is that females were selected for bearing robust young with large heads; a variant on this hypothesis proposed a gestation period of perhaps 11 months and the birth of precocious young, more physiologically mature than modern infants, presumably as an adaptation to the extreme climatic conditions of Europe during the Wurm glaciation. A nineteenth-century hypothesis that Neandertals were not upright was shelved decades ago. Notably absent from Neandertal theories is the petite female, protected by one or more robust males, for skeletal remains indicate that both male and female Neandertals were robust and muscular. From the Neandertal period on, size sexual dimorphism is considered to vary to the degree in which it varies in modern people. As the next section illustrates, this variation offers some insights into general principles of human growth and development.

V. GATHERING AND HUNTING HUMANS COMPARED WITH EARLY FARMERS

Study of the sexual dimorphism in prehistoric populations of *Homo sapiens* is possible to a degree far exceeding that in premodern hominids due in large part to confidence in the primary data. Within *Homo sapiens,* our own species, it is possible to use recognized criteria to determine the sex of a skeletal individual with more than 90% confidence of accuracy. Although modern *Homo sapiens* appeared in Asia and in Europe at least 40,000 years ago, and in Africa at least 100,000 years ago, there are no large samples of more than 15,000 years antiquity. Large samples are required in the study of sex differences to be sure the data are not skewed by individual anomalies.

Thus, when we discusss the degree of sexual dimorphism in modern *Homo sapiens,* we really are referring only to the most recent populations of *Homo sapiens sapiens,* and then only to some selected samples, luckily preserved and studied. With this caveat, based on studies of North American populations of the last several thousand years and of Eurasian populations of less than 10,000 years antiquity, a few generalizations can be made.

Sexual dimorphism in skeletal populations of modern people appears to depend on a number of factors, including the type of economy (gathering and hunting, village farming, preindustrial states in which farming was the main occupation, etc.); the degree of class structure, or lack of it, and the social class of the sample; and the ecological and nutritional status of the skeletal population. A hunting life-style appears to select for larger, more robust males if all other things, such as the ecological richness of the habitat, are equal. A farming life-style tends to select less for robusticity than for physical perseverance. Farming permits, and in fact encourages, increased population density, but when drought or other climatic changes occur it selects for small body size, because this reduces caloric needs and allows survival when food is scarce. Research indicates that females tend to be more programmed in their growth patterns and that males tend to be more susceptible to increase in size when nutrition is plentiful, and to reduction in growth when it is not. Because adult females in nonindustrialized cultures, and certainly females in prehistoric societies, could expect to spend most of their reproductive years either pregnant or lactating, selection has tended to buttress the female against the nutritional stress of child-bearing and nursing. One way to do this is to restrain total growth; a second way concerns body composition, to be discussed later.

VI. ENVIRONMENTAL AND CULTURAL VARIATION

Over the past century, many unplanned experiments that involve rapid changes in the environment and culture of populations have occurred. Franz Boas was one of the first anthropologists to seize the opportunity presented by migration to North America of peoples from all over the world to compare the physical traits of migrants and those who remained in the original territory. He found that migrants tended to have proportionately broader heads; later studies have found an increase in body size, often occurring to a greater extent in males than in females. Many sophisticated studies in recent decades have compared migrants with two groups of stay-at-homes, those who continue to practice traditional culture and those who adopt a Western life-style. Predictably, the change in life-style is more important than migration itself.

Attitudes toward body size and shape undergo change as access to nutrition changes. Peasant cultures tend to value fat deposition as an indication of stored wealth and as a hedge against hard times, but in a cosmopolitan industrialized culture, fat is superfluous. Activity patterns also differ; in most nonindustrial cultures, women perform hard physical labor and develop muscles, whereas women in industrialized societies have to exercise during leisure time to achieve muscularity. In the latter societies, personal choice concerning activity levels, as well as in selection of food, assumes a major role, along with the genotype, in determining the body's shape.

Many scholars of women studies have pointed out that just as women's average activity levels and body shapes have changed with the advent of industrial society, so have attitudes toward women's roles. The "helpless female" concept did not arise in traditional cultures, where women's work was essential to family survival. This stereotype appears to have been a short-lived cultural phenomenon, an aristocratic status symbol that the Western middle class could afford for only a few decades. Environmental and cultural changes will continue to test the degree of flexibility of the human frame.

VII. BODY COMPOSITION

Body composition of contemporary male and female adults varies most notably in the percentages of muscle and fat. Whereas young adult males average 15% of body weight in fat, females average about 27%; in muscle, the comparison is 52% (males) to 40% (females). Hormonal control of fat patterns in females has been shown to influence the nutrient reserves required for successful completion of pregnancy and lactation.

Although useful in comparing some male–female traits, these blanket average statements fail to take behavior and biological variability into account. Recent research has focused on the trained female athlete who reduces her fat composition and increases relative muscle composition. The institution of a training regimen during adolescence can postpone menarche, and institution of an extensive exercise program after menarche can produce periods of amenorrhea. These observations have contributed to understanding the role of relative fat in triggering the onset of menarche and how the trend toward taller adults and a younger age at the adolescent growth spurt has been produced by overnutrition.

Although sex differences in body composition become acute during the adolescent growth spurt, they originate long before adolescence. Variation occurs among individuals as well as between different body sites, where fat and muscle tissue are deposited. Reduction of fat tissue in females in industrialized societies, such as during periods of famine in war-torn countries, may lead to a reduction in the birth rate. This process is due to prolonged periods of amenorrhea in a substantial fraction of the female population. Such a condition may also be self-induced, as in the case of anorexic women. In populations of hunter-gatherers, of which few remain in the late twentieth century, such conditions may be related to prolonged lactation or to undernutrition. Restoration of a critical percentage of fat appears to be required before menstrual cycling resumes.

Young boys, by contrast, may lose muscle mass under conditions of malnutrition. In modern society, girls are not alone in experiencing eating disorders, such as anorexia and bulimia, in which the affected persons have complicity in their bodies' inability to take in and digest food. Boys involved in wrestling have been known to use artificial means to reduce their weight in order to qualify in a lower weight class. Like bulimics, they willingly vomit food and lose nourishment. Additional research is needed to determine the long-term consequences of these behaviors in adolescence. [*See* Eating Disorders.]

The consequences of taking steroids to achieve the opposite effect, that is, to put on muscle and weight, have been documented, but so long as dire consequences have only a probabilistic basis they fail to discourage all young men from steroid intake.

In cosmopolitan industrial societies, enhancement of muscle tissues is considered fashionable by both men and women. This slight exaggeration of a male pattern of body composition is favored over enhancement of the female pattern of fat deposition. A biocultural perspective relates this fashion to the abundance of food and food storage technology; it is no longer necessary or even healthful to store calories as fat. Success, symbolized in the appearance of strength and apparent youthfulness, counts more than stored food, perhaps in part because the symbol is the scarcer commodity.

VIII. SEX RATIO

In most populations, the number of newborn male children exceeds that of female children by 5 to 8%. The sex ratio at birth is termed the secondary sex ratio; the proportion at conception, the primary sex ratio of males to females, is believed to be considerably higher. R. A. Fisher's theory of the sex ratio proposes that evolution works to produce an equal sex ratio at the age of procreation; thus, given the higher mortality of male embryos, infants, and children, there must be an excess of them at conception. The degree of vulnerability of males in comparison to females varies among societies. It may be either enhanced or compensated for by cultural practices such as male participation in warfare or other occupational sex differences with attendant levels of risk, marital patterns such as polygyny, differential age at marriage for males and females, and the existence of social institutions such as monasteries and nunneries. In recent decades, much attention has been focused on the greater life expectancy of females and its relationship to their lower level or later onset of chronic diseases of old age. Hormonal and genetic factors appear to account for many of these sex differences, but behavioral factors also contribute. Significantly, outside of societies in which many young men perish in warfare, the greater longevity of females appears to be due to their less than maximal fertility, a relatively recent development. By contrast, the very high fertility of some village farming communities puts women at greater health risk than men not only in childbirth but by producing stresses that predispose them to other disorders.

In some traditional farming communities, both males and females married at a young age and immediately started a family in which children made an economic contribution. Newly formed families often moved to the frontier, and in this way vast areas of Eurasian forest and steppe were converted to farmland several thousand years ago. In the Americas, immigrants produced a similar ecological change within a hundred years. In these agrarian societies, the two sexes usually had different roles but both were constrained and their life cycles were relatively symmetric. In societies in which new land was not available, other patterns held. For example, many traditional African societies practiced polygyny, but a male had first to prove himself before marriage, whereas females married young. The different age rules for marriage produced asymmetry due to attrition of the older males, but polygyny supplied a balance.

In contrast to traditional societies, industrialized society supplies no social customs to maintain a balanced adult sex ratio. Age at marriage varies widely from decade to decade, and apparently is affected by political conditions such as war, by erratic economic conditions, and by wide pendulum swings in attitudes toward women. Whereas in traditional societies members of neither sex are permitted to set their life course, males in modern society tend to be granted this right more consistently than are women. Although the trend in the twentieth century has been toward increasing the choices offered women, particularly in the workplace, the pattern is an erratic one involving both advances and setbacks. Modern society is capable of rapid swings in public policy, which can be brought about by well-organized political forces. An example of shifting attitudes toward women and the choices they are allowed to make in their own lives can be found in changes in the legal status of abortions in the United States. Not illegal until the late nineteenth century, abortions were legalized by a Supreme Court decision in 1973, but court actions, following political pressure during the 1980s, produced restrictions on this female province.

IX. MORBIDITY AND MORTALITY

Biological factors, such as sex differences in hormones and immunologic competency, and behavioral factors, such as activity patterns and cultural roles, act—and interact—to produce different patterns of morbidity and mortality in males and females and different demographic profiles in various age groups throughout life. Although greater life expectancy of women in modern industrial society is a well-established fact, greater female longevity is neither inevitable nor universal in the human species. Prior to the reproductive age, females do appear to have a biological advantage, however, and appear to be less affected by diseases related to nutritional deprivation and infectious agents.

Evidence suggests that male fetuses have a higher probability of spontaneous abortions than do female fetuses, though the nature of data on spontaneous abortions makes caveats necessary. Evidence from periods of childhood is more conclusive: male infants face a higher risk of mortality than female infants, particularly if environmental stress is occurring. During their growth period, girls appear to be better buffered against environmental stress, maintaining a more stable growth trajectory than do boys under adverse dietary conditions. This sex difference occurs in spite of the observation that nutritional advantages often are bestowed upon male children in traditional cultures. Biological benefits tend to be offset by opposing risks. Whereas women benefit from a more vigorous antibody response to infectious agents, they also have increased susceptibility to autoimmune disorders such as systemic lupus erythematosus and rheumatoid arthritis.

Cross-cultural and prehistoric studies provide some information that is useful in partitioning biological from behavioral components of morbidity and mortality. The greater reproductive burden in women in societies that do not practice contraception is associated with higher morbidity and mortality. In preindustrial societies that have very high fertility, female longevity tends to be less than that of males, opposite to the current pattern in industrial societies with reduced fertility. Men in some societies have faced higher probability of death by violence and accidents, which have become common causes of injury and death in young men in the contemporary United States.

Interactive effects of biology and culture are particularly evident in periods and cultures undergoing rapid change, such as contemporary postindustrial societies. The category "Western diseases" refers to illnesses common in industrialized and largely sedentary populations in which most people survive to middle and advanced ages, unlike much of prehistory when life expectancy was short. These diseases, such as heart disease, ulcers, and diabetes mellitus, have behavioral and biological etiologies. Heart disease has received considerable scrutiny. Patterns of daily routine that contribute to risk of heart disease include stress, inadequate exercise, and a diet rich in fats. These risk factors have been identified in males for many decades, whereas estrogen and menstrual loss of blood are considered possible biological contributors to women's lowered risk of heart disease in premenopausal years. Clinicians looking only at biological factors tend to see estrogen supplements for women as a way to maintain low risk of heart disease in postmenopausal years. By contrast, a biocultural approach focuses on the interaction of hormonal patterns with changing job status and activity patterns and considers behavioral as well as chemical interventions. Studies of the subtle interactions of etiological factors producing "Western diseases" in men and women should provide additional understanding of sex differences in patterns of illness and mortality over the next few decades.

X. COGNITIVE AND BEHAVIORAL DIFFERENCES

All studies of cognitive differences between the sexes show that great overlap exists between males and females; group averages for either sex offer little predictive power regarding the performance of any one individual. Still, some patterned behavior differences have been identified and deserve mention. A second type of cognitive study focuses on sex differences in the anatomy of the brain; a third type looks at the influence of hormones.

Sensory and motor abilities of boys and girls differ in that girls tend to have greater auditory acuity whereas boys excel in visual keenness, particularly in tracking moving objects. These traits were studied in the development of the verbal skills of speaking and reading, in which girls tend to outperform boys, and of spatial and mathematical skills, in which boys tend to outperform girls. Because reading is considered an essential skill for all citizens, most boys are strongly encouraged to succeed as readers. The same pressure is rarely applied to girls to assure they achieve mathematical competence. The role of socialization in developing or maintaining these divergent patterns was hotly debated in the mid-twentieth century with no final resolution. Although socialization is unlikely to be a total cause of observed sex differences, it probably serves a supportive role. If, as suggested, boys are forced to devote extra effort to gain essential skills in reading, but girls are not forced to gain essential skills in mathematics, socialization pressures contribute to the disadvantaging of females in certain professions. [*See* Sex Differences, Psychological.]

Anatomical comparisons of male and female brain structure are in their infancy. Researchers have focused on differences between the brain's two hemispheres and on links between them. Language processing tends to be localized in the left hemisphere and rapid pattern analysis in the right. However, research has not shown that one sex shows more development in one hemisphere while the other sex favors the opposite. Some studies have indicated greater laterality in males—that is, that a greater proportion of males than females have brains in which one hemisphere dominates the other—but this finding is not universal. An innovative research project undertaken in the late 1980s found the area of fibers that connect the two halves of the brain to be larger in women than in men. Researchers related this finding to the greater verbal fluency of women, but until further studies are done these results are tentative and any inferences are speculative.

Hormonal effects on sex development have been studied experimentally in animal models and in humans with chromosomal abnormalities. Though these studies demonstrate that hormones produce significant effects, they do not provide ready models for subtle behavioral variation in genetically normal individuals. Studies conducted in 1988 showed that variation of hormone levels in healthy women is correlated with performance of tasks involving motor skills and spatial reasoning. During times when the female sex hormones estrogen and progesterone were highest, subjects tended to perform better on fine motor tasks, but their performance of spatial skills improved when these hormones were low. In addition to confirmatory tests, studies of the performance on similar tasks by men in relation to circulating levels of the male hormone testosterone are needed before these preliminary findings can serve as a basis for interpreting male–female variation in motor and spatial skills.

XI. INDIVIDUAL VARIABILITY

The process of sexual reproduction ensures that each individual (excluding monozygotic twins) is genetically unique. Individual variability originates in the genotype but can be fostered, or muted, socially. Traditional societies are known for channeling the abilities of their people; as anthropologist Ruth Benedict termed it, they are "personalities writ large." Benedict studied the processes whereby traditional cultures shape the perspectives and behavior of their members at a time when her own culture (U.S. society between the two world wars) was experiencing uncertainties about its own attitude toward individuality and the changing roles of males and females. In her work she became concerned about the misfit, the aberrant individual whose personality or talents simply do not incline toward the culture into which he or she is born. Benedict's hope was that modern society could accommodate diverse personalities more than traditional societies had done.

Is Benedict's hope feasible? A biocultural perspective on differences between males and females shows that biological differences between the two groups are real, but that the extent of the differences depends both on environmental factors (such as nutrition) and behavioral factors (such as activity level). A historical biocultural review shows that interpretations of these patterns do change within a culture, over time. Historically, many different perspectives have waxed and waned, but few cultures at any time have been truly tolerant of aberrant personalities.

So far in human history it appears that the dominant perspective toward differences between males and females at any one time rests upon the economic, political, and historical traditions of the culture more than upon scientific understanding. Scientific data are used by many politicians in attempts to prove points, rather than in a dispassionate attempt to seek the truth. Indeed, most ethicists would argue that scientific understanding is not an appropriate base upon which to rest values or legislation. Though science can help us to perceive the natural phenomena—in this case the biology and behavior of males and females, and their interactions—it is people who must strive to treat individuals ethically. If a society does evolve that treats males and females fairly, and allows both sexes the opportunity for innovative personal growth, it will be because its members have chosen to value individual productivity more than uniformity.

BIBLIOGRAPHY

Hall, R. L. (ed.) (1982). "Sexual Dimorphism in *Homo sapiens:* A Question of Size." Praeger Scientific, New York.

Hall, R. L., Draper, P., Hamilton, M. E., McGuinness, D., Otten, C. M., and Roth, E. A. (1985). "Male–Female Differences: A Bio-Cultural Perspective." Praeger Scientific, New York.

Kimura, D. (1992). Sex differences in the brain. *Sci. Am.* **267,** 118–125.

Langer, L. M., Warheit, G. J., and Zimmerman, R. S. (1991). Epidemiological study of problem eating behaviors and related attitudes in the general population. *Addictive Behav.* **16,** 167–173.

McGuinness, D. (1985). "When Children Don't Learn." Basic Books, New York.

Pickford, M., and Chiarelli, B. (eds.) (1986). "Sexual Dimorphism in Living and Fossil Primates." Il Sedicesimo, Firenze, Italy.

Stinson, S. (1985). Sex differences in environmental sensitivity during growth and development. *Yearbook Phys. Anthropol.* **28,** 123–147.

Yesalis, C. E. (1992). Epidemiology and patterns of anabolic–androgenic steroid use. *Psychiatr. Ann.* **22,** 7–18.

Sex Differences, Biological

SUSUMU OHNO
Beckman Research Institute of the City of Hope

GLOSSARY

Inhibin Gonadal peptide hormone that acts on the hypothalamus–pituitary axis

Müllerian inhibiting substance Substance secreted by testicular Sertoli cells to destroy Müllerian ducts

Müllerian (paranephric) duct Embryonic duct system that develops into female internal reproductive organs

Y-linked *Sry* gene Putative testis determining gene

Y-linked *Zfy* gene Former candidate to be the testis determining gene

A CLEAR GENETIC DIFFERENCE BETWEEN MAN AND woman that is responsible for sexual dimorphism resides solely in the presence or absence of the male-specific Y chromosome. However, most of the small number of genes present on the Y chromosome have their counterparts on the X.

A great majority (but not all) of mammalian species are sexually dimorphic, adult males being noticeably taller and heavier than adult females. In these species, adult males are more often than not endowed with variously conspicuous signs of masculinity (e.g., antlers of the stag and mane of the lion). Our own species is no exception. According to the "Statistical Abstract of the United States, 1986," the average height and weight of men between 25 and 34 years of age were 176.7 cm and 78.5 kg during the period 1976–1980, whereas those of women were 163.0 cm and 64.4 kg. These differentials are about the same as one sees between stallions and mares. In view of the fact that in certain seal species, adult males weigh nearly 10 times more than adult females, sexual dimorphism of our own species is not outrageously conspicuous as far as height and weight differentials are concerned.

I. IS *Sry* THE Y-LINKED *Tdf* GENE?

If one is to seek genes that are present exclusively in one sex, they can be sought only on the Y chromosome. The X and Y chromosomes in the common ancestor of all mammals were homologous, with the same genetic content. The X chromosome has since been conserved so that whatever gene resides on the human X chromosome automatically resides on the X chromosome of all other placental mammals. In contrast, the Y chromosome underwent extensive genetic degeneration, accumulating nongenic DNA base sequences and becoming reduced in size. It follows then that the number of functioning genes still residing on the Y must necessarily be small, and that the majority of them must still find their counterparts (alleles) on the X.

The X and the Y still pair with each other and exchange genetic materials during meiosis of spermatocytes in the testis. This very short, homologous segment resides at the tip of the short arm of the X as well as the Y (Fig. 1). Whereas the short arm of the X is quite substantial, that of the Y is barely visible under the light microscope. The still functioning Y-linked genes appear to be concentrated in the vicinity of this tiny homologous segment, while their counterparts on the X may or may not be in the vicinity.

ENCYCLOPEDIA OF HUMAN BIOLOGY, Second Edition, VOLUME 7. Copyright © 1997 by Academic Press. All rights of reproduction in any form reserved.

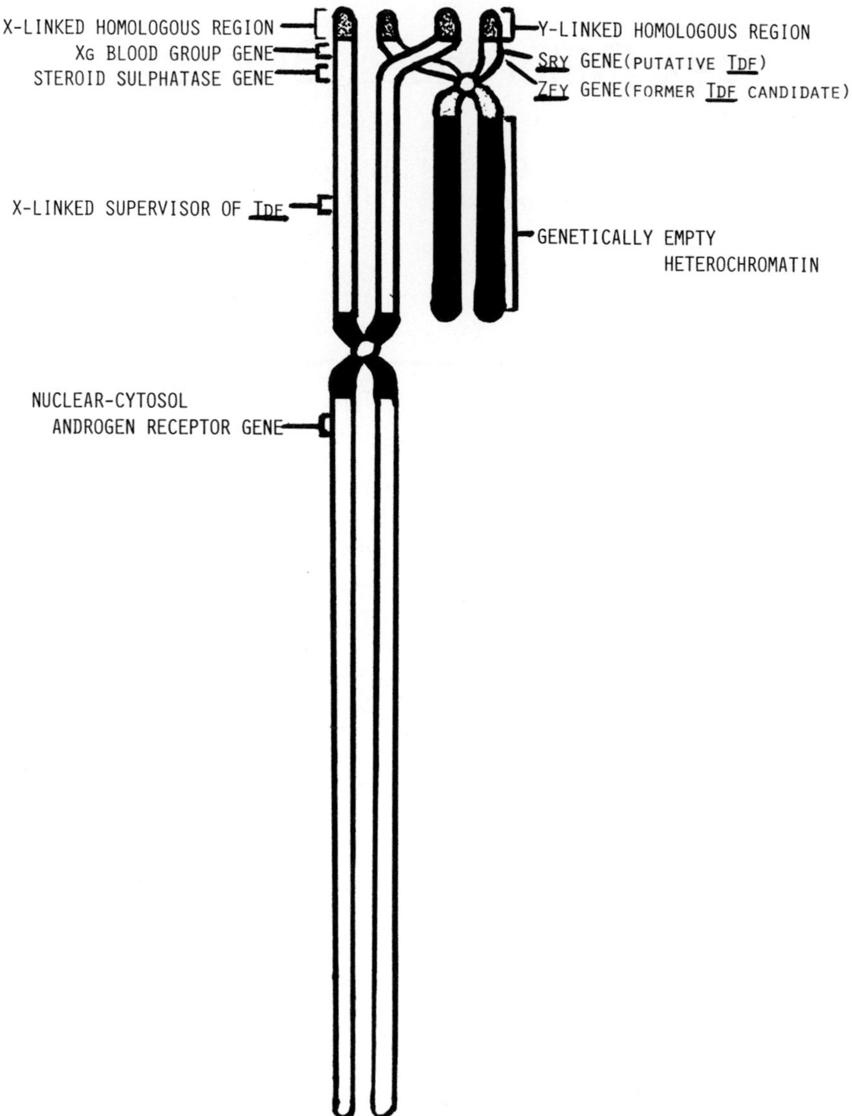

FIGURE 1 Schematic illustration of the human X (left) and Y (right) chromosomes. Each chromosome is divided into its short and long arms by a small circle that denotes a centromere, the attachment site to spindle fibers during cell division. The X chromosome is of a substantial size, containing roughly 1.8×10^8 base pairs of DNA. It carries about as many genes as other chromosomes (autosomes) of comparable size such as chromosomes 11 and 12. The number of genes carried by the X should, therefore, approach 6000. Most of these genes, however, have nothing whatsoever to do with sex determination and sex differentiation. The Y chromosome, in contrast, is a minute element of variable size, usually one-third of the X. A bulk of it is made of heterochromatin, shown as solid black, which contains no genes. The variable amounts of heterochromatin carried by individual Y chromosomes explain their size differences. The shaded area of the tip of the short arm of the X and Y is the homologous segment with which the X and Y pair with each other and regularly exchange genetic materials during meiosis (reduction division) of spermatocytes in the testis of adult males. In the vicinity of the homologous segment of the X are two genes; one for Xga blood group antigen and the other for steroid sulfatase. In somatic cells of females, one of the two X chromosomes is inactivated, but these two genes are not. In the vicinity of the homologous segment of the Y are two genes: *Sry* and *Zfy*. The former is the current favorite to be *Tdf* (testis determining gene), whereas the latter was the previous candidate to be *Tdf*. Interestingly, both of them have their ancient homologues on the X chromosome: *Srx* and *Zfx*. Although their precise locations on the X chromosome are not known, they make it clear that the Y, now largely degenerate, was once a homologue of the X. Located in the middle of the short arm of the X chromosome is the gene supervising the activity of Y-linked *Tdf*. The long arm of the X also carries a gene important for sexual development; it encodes the nuclear-cytosol androgen-receptor protein, which mediates all the divergent responses of various target cells to testosterone and 5α-dihydrotestosterone.

Those in the vicinity of the pairing segment escape the X-inactivation mechanism that inactivates one of the two X's in female somatic cells. Such Y-homologous genes on the X include the one for Xg blood group and the other for steroid sulfatase.

Although these two genes have nothing to do with the mechanism of sex determination, steadily accumulating evidence suggests that the Y-linked *testis determining gene (Tdf),* which decides the fate of embryonic indifferent gonads to be a pair of testes instead of ovaries, also resides in the vicinity of the pairing segment on the Y. Because of this rather inadvisable position of *Tdf* gene on human Y chromosome, in an evolutionary sense, there is an inherent danger that an exchange between the X and the Y occurring slightly deeper than usual might transfer *Tdf* from the Y to the X, thus producing XX males. Indeed, unique to human species, XX males occur with a high incidence of one in every several thousand male births. It was thus reasoned that human XX males offer the ideal opportunity to identify the elusive *Tdf.* Of the variable length of Y-derived DNA that the paternally derived X of these human XX males carry, the shortest Y fragment should still contain *Tdf.*

Following this rationale, a succession of genes thought to be *Tdf* were identified in XX men. When the first edition of this encyclopedia was published in 1991, the strong candidate to be *Tdf* was the Y-linked gene encoding a DNA-binding protein containing zinc-binding finger domains. The fall of this Y-linked *Zfy* gene from the pedestal was very quick, for it was shown that the presence of *Zfy* was neither sufficient nor necessary for testicular development. At the moment, the very strong candidate to be *Tdf* exists in the form of the Y-linked *Sry* gene, which also encodes a DNA-binding protein but of HMG (High Mobility Group) variety. Introduction of this gene alone to mouse XX fertilized eggs transformed a good proportion of these so-called transgenic XX mice to males with testes. At first glance, this appears to constitute an indisputable evidence that *Sry* is indeed the long-sought *Tdf.* However, it should be pointed out that the transgenic treatment of mouse XX fertilized eggs with an autosomal gene encoding MIS (Müllerian Inhibiting Substance) also gave rise to males with testes. MIS is actively secreted by Sertoli cells of the newly organized testis immediately after its transformation from an embryonic indifferent gonad. Although one known function of MIS is to destroy a pair of Müllerian ducts to prevent the development of oviducts and uterus in males, it may also have an autocrine testis-organizing function.

We shall now ask the pertinent question: "Is the presence of *Sry* sufficient as well as necessary for the testicular organization?" In the embryonic development, *Sry* gene is activated at the right time in the right place for testicular organization. The "time" is when the fate of indifferent gonads is to be decided and the "place" is the XY indifferent gonad itself. But for this very reason, *Sry* gene clearly occupies the position of a middle manager in the regulatory hierarchy—the high authority deciding the time and the place of *Sry* gene activation.

It is thus expected that in the mutational abolishment of this higher authority, the presence of Y-linked *Sry*, albeit intact, would not suffice for testicular organization. Indeed, dysfunctioning mutations sustained by *Sry* cannot account for all the known human XY females. However, the presence of this higher X-linked authority, which by mutating can render the normal *Sry* powerless, is spectacularly demonstrated by the Scandinavian wood lemming *Myopus schisticolor.* Fertile XY females are as common as normal XX females in this rodent species. The Y chromosomes, and therefore the *Sry* genes, of these XY females are a priori normal, for each is inevitably derived from their fertile father. The fault lies in a gene or genes residing in the middle of the short arm of the X.

II. TESTICULAR ORGANIZATION WITHOUT *Sry*

If the presence of *Sry* is insufficient for testicular organization, is *Sry* absolutely necessary for this process? The answer is, again, "No." Since the *Sry* gene of nonhuman mammals resides in more sensible positions of the Y chromosome away from its pairing segment, compared to in humans, XX males arc not as prevalent in other mammalian species. Nevertheless, they do occur and the cause cannot be sought in the transfer of *Sry* from the Y to the X during male meiosis. Even with regard to humans about 10% of the XX males show no trace of *Sry* or any other Y-derived segment. Indeed, testicular organization in the absence of *Sry* was proven in *polled* goats. In the goat, the dominant autosomal mutation *polled* renders the male and the female alike hornless and testicular organization ensues when XX embryos become homozygous for *polled.* As expected, there was no trace of *Sry, Zfy,* or any other Y-derived DNA in these XX male goats. It should be recalled that XX mouse embryos transfected with autosomal gene encoding MIS hormone may also organize testes.

III. WHY THERE IS MORE THAN ONE *Tdf*

The *magnum dictum* of evolution in recent years is: "Always the same gene for similar construction." For example, one would not think of the vertebrate eye and the compound eye of the insect as homologous organs. Yet the same gene *Pax 6* that maintained 95% sequence identity between human and the fruit fly controls the development of both. The sex determination is one great exception, for entirely different genes are involved in the sex-determining mechanism of mammals and that of the fruit fly. This exception becomes no exception, however, if we forfeit the notion of testis *versus* ovary, and regard both as mere variations of the same organ: the gonad. Indeed, one gene, *Ftz*-F1, controls development not only of the gonad but also of the adrenal cortex. Inasmuch as somatic elements of these two organs are only cell types of the body engaged in steroid hormone production, it is fitting that this gene also controls activities of many genes encoding steroid hormone-synthesizing enzymes. Unquestionably, the gonad is one organ, a testis and an ovary being its minor variations. Once this is realized, it is no longer surprising that the task of deciding whether to organize an ovary or a testis can be relegated to any of a number of genes in the middle managerial position. Although the Y-linked *Sry* is most likely to be the long-sought *Tdf* of mammals, its presence *sensu stricto* is neither necessary nor sufficient for testicular organization, as we have seen.

IV. THE TESTIS AND THE OVARY AS HOMOLOGOUS ORGANS

The development of the main reproductive organ is intimately associated with a state in the development of the kidneys. Mammalian embryos develop three kinds of kidneys in succession, the last one persisting as the functional kidney. Development of the gonads is associated with the *mesonephros,* and to a certain extent with the pronephros. A pair of gonadal ridges, from which future testes or ovaries will evolve, develop at the inner sides of the mesonephros, very close to the base of the mesentery.

The all-important *primordial germ cells* arise outside the embryo in its yolk sac and migrate toward the newly formed gonadal ridges mainly through the mesentery. As soon as the migration of primordial germ cells to the gonadal ridges is completed, the fate of gonads is determined, for normally male XY embryos of our own species organize the testis between the 43rd and 49th days of gestation. Primordial germ cells become encased within tubular structures (future seminiferous tubules) together with one somatic element—*Sertoli cells*—whereas the other somatic element left outside the tubules becomes *Leydig's cells* (Fig. 2). Soon, these two somatic elements begin hormonal activities, thus determining the fates

FIGURE 2 Schematic illustration of the human embryonic testis (left) and fetal ovary (right) as homologous organs. Large cells with clear cytoplasm are germ cells that have migrated from the yolk sac: primordial germ cells of the male (left) and oogonia and oocytes of the female (right). Cells with black cytoplasm are Sertoli cells of the testis (left) and future granulosa cells of the ovary (right). Cells with shaded cytoplasm are Leydig's cells of the testis (left) and future theca cells of the ovary (right). Other somatic elements are shown without nuclei. Outlined ducts shown at the base of both the testis and the ovary are rete tubules. After puberty, rete tubules of the testis perform the important function of collecting and transporting spermatozoa made in seminiferous tubules toward the epididymis. Rete tubules of the ovary, on the other hand, have no known function. In the male gonad, testicular cords begin to take shape as early as the 49th day of gestation and establish themselves as definitive seminiferous tubules by the 60th day. Their ovarian counterparts, ovigerous cords, on the other hand, are not formed until the 100th day of gestation; they are not convoluted as testicular cords and are open-ended at the top. Nevertheless, ovigerous cords contain germ cells and future granulosa cells in the same manner as testicular cords encase germ cells and Sertoli cells. In the testis, the germ cells remain primordial until the neonatal stage, when they differentiate into definitive spermatogonia, which remain quiescent until puberty (about the 12th year). In the ovary, in contrast, germ cells begin differentiation as soon as ovigerous cords are formed. After a brief period of multiplication as oogonia, they differentiate into oocytes and enter the first meiotic prophase. As they do so, they move more deeply into the ovigerous cords. Meiosis is interrupted at the diplotene stage and each oocyte then surrounds itself with a single layer of follicular (granulosa) cells. This formation of primordial follicles breaks up the ovigerous cord structure from the deepest part.

of two pairs of duct systems related to the kidney: the Wolffian and Müllerian.

Wolffian ducts would regress without a trace unless exposed to testosterone synthesized by the Leydig's cells of newly organized testes. In the presence of testosterone, this duct system would differentiate into epididymis, ductus deferens, and seminal vesicles. Although the concentration of circulating testosterone is very low until the 112th day of gestation, masculinization of Wolffian ducts can proceed because they apparently receive testosterone directly from the testis. In fact, rare individuals manage to develop a testis on only one side, generally the right. In such an individual, Wolffian duct derivatives persist only on the testicular side. Leydig's cells of newly organized testis have yet one more assigned function: to induce the *urogenital sinus* into a penis and scrotum, instead of a vagina. The circulating testosterone level of male embryos of that stage is not high enough to saturate the androgen receptors that mediate its effect in the urogenital sinus. For this reason, cells of the urogenital sinus are endowed with the enzyme *5α-reductase,* which converts testosterone to *5α-dihydrotestosterone.* The aforementioned androgen receptor has a considerably higher binding affinity for the latter compound. If there is a genetic defect of this enzyme in XY persons, their external genitalia look more like a vagina, although they have testes with completely masculinized Wolffian duct derivatives. These individuals are initially raised as girls. With the approach of puberty, their circulating testosterone levels become high enough to masculinize the urogenital sinus without the help of 5α-reductase. Because the "girl" becomes a man at age 12, this trait is known as "penis at 12 syndrome." Interestingly, the presence of this enzyme in the urogenital sinus derivatives appears to render the vaginal area of women susceptible to testosterone as well. The circulating testosterone level of adult women is roughly one-twentieth of that of adult men. Yet, the growth of pubic hair in women is apparently dependent on this minuscule amount of testosterone, which becomes significant when converted *in situ* to 5α-dihydrotestosterone.

Whereas the Wolffian (paranephric probably of pronephros origin) duct system perishes without testosterone, differentiation of the Müllerian duct system (mesonephric) to female reproductive tracts, fallopian tubules, uterus, and upper part of the vagina is an automatic process that requires no hormonal induction. There are no Müllerian duct derivatives in normal men because the Müllerian duct system is destroyed during embryonic development. This destruction is carried out by the Sertoli cells of the newly organized testes. The glycoprotein excreted by Sertoli cells for this purpose is Müllerian inhibiting substance, which is about 550 amino acid residues long. This protein is related to transforming factor "β" as well as to the β chain of inhibin.

In view of the noted specific functions assigned to Leydig's and Sertoli cells of newly formed testes, it may appear that these two cell types are truly male or testis specific, having no counterparts in the female reproductive organ, the ovary. However, this is not the case. When a pair of gonadal ridges begin to differentiate toward the ovary instead of toward the testis, they too form tubular structures. These ovigerous cords are not as conspicuous nor as convoluted as testicular cords, which give rise to seminiferous tubules. Yet, the ovigerous cords, which open toward the surface, contain primordial germ cells and one somatic cell type, the *follicular cells,* which after puberty are called granulosa cells. The other somatic cell type left outside the ovigerous cord becomes *theca cells.*

Whereas male germ cells in the testis remain inactive until the onset of puberty, primordial germ cells in the ovary soon cease their multiplication and enter into meiosis. Each oocyte, upon completing the diplotene stage of the first meiotic division, becomes encased by a single layer of follicular cells, thus becoming the *primordial ovarian follicle.* It is the formation of follicles that breaks up the structure of the ovigerous cords, as shown in Fig. 2. Thus, testicular Sertoli cells have their counterparts in ovarian follicular (or granulosa) cells, and testicular Leydig's cells in ovarian theca cells. Indeed, ovarian *granulosa cells* also produce MIS, although at a later period, after the Müllerian duct system became indestructible by MIS. As already noted, inhibin is related to MIS and the function of this hormone is to prevent, by negative feedback, the pituitary secretion of FHS (follicle-stimulating hormone). This too is produced by testicular Sertoli cells and ovarian granulosa cells alike. Ovarian follicular cells are rich in one microsomal enzyme, *aromatase,* which converts male hormone testosterone to the female hormone *estradiol.* In women of reproductive age, the aromatase function is essential because each menstrual cycle is initiated by a rise in estradiol concentration. However, testicular Sertoli cells are also rich in aromatase. Nevertheless, the testis does not become an estradiol-producing organ because it is sequestered inside seminiferous tubules. Thus testosterone secreted by Leydig's cells is not accessible to them. The separation between Sertoli cells and Leydig's cells breaks down when the former become malignant and begin to propagate profusely; testicular Sertoli cell tumors are rather common in dogs. When they occur, the testis becomes the feminizing estradiol-producing organ.

If ovarian theca cells are the equivalent of testicular Leydig's cells, can the ovary become the testosterone-producing organ? They probably do, as seen in female rabbits immunized with testosterone to obtain the antibodies that are used in clinical radioimmunoassay of testosterone. The ovaries of these rabbits begin to produce as much testosterone as male rabbits, probably because all the testosterone they produce is inactivated by antibodies. Indeed, testicular Leydig's cells and ovarian theca cells are homologous to each other, as are testicular Sertoli cells and ovarian granulous cells.

Thus, the testis and the ovary should be viewed as homologous organs representing two sides of the same coin. In fact, many fish species readily exchange the ovary with the testis or vice versa in their lifetime. The most notable is the perch-like tropical coral fish *Anthias squaminpinnis*. This is a conspicuously, sexually dimorphic species (a rather unusual event in fish). The female is a bright yellow while the mature male is bright reddish with a blue sash across his chest. Each group consists of only one male and a number of females. When the male is killed or removed, the most dominant of the females becomes a male, changing its ovary to a testis and the yellowish coloration to red with a blue sash within a matter of weeks.

V. THE FEMALE AS THE PRIMORDIAL SEX AND THE MALE AS A MODIFIED FEMALE

As far as mammals are concerned, the basic plan of development is to be feminine. It will be recalled that differentiation of Müllerian (mesonephric) ducts to fallopian tubules and uterus of the female reproductive tract is an automatic process that requires no intervention, as is the urogenital sinus's differentiation toward the vagina. In sharp contrast, Wolffian (paranephric) ducts would perish if left alone, their differentiation to male reproductive tracts requiring testosterone; differentiation of the urogenical sinus to penis and scrotum requires not only testosterone but also its *in situ* conversion to 5α-dihydrotestosterone.

Even primordial germ cells are inherently planned to be eggs rather than sperm. It will be recalled that primordial germ cells originate outside the embryo in the yolk sac. During their migration toward the gonadal ridges, which emerge on the surface of the mesonephros, a number of them mismigrate and home into the nearby adrenals. In certain inbred strains of laboratory mice, these mismigrated primordial germ cells are numerous and readily identifiable. In the adrenals, even XY germ cells of male embryos behave as eggs and undergo up to the diplotene stage of the first meiotic prophase; then each surrounds itself with a single layer of adrenal cortical cells, forming a misguided primordial follicle.

For professional biologists, sex starts with differentiation of the gonads, but for others the gonads are of little interest. What matters are external appearances and behaviors. From animal experiments, it seems likely that if given ample amounts of testosterone at three critical stages of development, XX individuals with the ovary would develop as men in both appearance and behavior. The critical stages are: (1) 50th to 80th day of gestation to develop Wolffian ducts into male reproductive tracts and the urogenital sinus to penis and scrotum; (2) days around the parturition to imprint the central nervous system, to ensure the later manifestation of masculine behavior; and (3) 12th year onward to gain the height and weight of a man and to maintain masculine attributes (e.g., beards) and behavior. A number of reports have described sex differences in spatial perception and certain mathematical abilities. If they exist, they are not likely to be the direct consequences of a genetic difference but of the presence or absence of perinatal imprinting on the central nervous system. [*See* Sex Differences, Psychological.]

Conversely, if protected from the effect of one's own testosterone, XY individuals with testes should develop as females. The effect of testosterone as well as 5α-dihydrotestosterone on target cells is mediated by a *nuclear-cytosol androgen-receptor protein,* which is encoded by a gene on the X chromosome (see Fig. 1). When this locus becomes dysfunctional by mutation, the entire body of XY individuals becomes totally nonresponsive to the normal amount of testosterone produced by their testes. This is known as the complete form of *testicular feminization syndrome.* An affected individual has neither male nor female reproductive tracts because the Wolffian ducts, unable to respond to testosterone, perish, while the Müllerian ducts are actively destroyed by the hormone MIS, secreted by testicular Sertoli cells. Not being able to respond to 5α-dihydrotestosterone converted from testosterone *in situ,* the urogenital sinus develops as the vagina. Thus, the external appearances of such XY individuals are female. After puberty, testosterone increases in amount but it is not utilized by the target cells; it is thus eventually converted to the female hormone estradiol by the already mentioned aromatase, which is also present in extragonadal tissue, noticeably in fat tissue. Because these individuals are normally responsive to estradiol, they develop well-shaped breasts as well as hips. In fact, these XY individuals with the complete form of testicular feminization syndrome are invariably more beautifully feminine than their normal XX sisters. Such is the consequence of the female being the primordial sex.

One tell-tale sign of the complete form of testicular feminization is the total absence of pubic and axillary hair. This suggests that even in sexually mature normal women, the hair development is maintained by testosterone, as already noted.

BIBLIOGRAPHY

Behringer, R. R., Cate, R. L., Froelick, G. J. Palmiter, R. D., and Brinster, R. L. (1990). Abnormal sexual development in transgenic mice chronically expressing Müllerian inhibiting substance. *Nature* **345**, 167–170.

Gropp, A., Winking, H., Frank, F. M., Noack, G., and Fredga, K. (1976). Sex-chromosome abberations in wood lemmings (*Myopus schisticolor*). *Cytogenics Cell Genet.* **17**, 343–358.

Just, W., Cabral D'Almeida, J. C., Goldschmidt, B., and Vogel, W. L. (1994). The male pseudohermaphrodite XX poled goat is *Sry* and *Zfy* negative. *Hereditas* **120**, 71–75.

Koopman, P., Gubbay, J., Vivian, N., Goodfellow, P., and Lovell-Badge, R. (1991). Male development of chromosomally female mice transgenic for *Sry. Nature* **351**, 117–121.

Luo, X., Ikeda, Y., and Parker, K. L. (1994). A cell specific nuclear receptor is essential for adrenal and gonadal development and sexual differentiation. *Cell* **77**, 481–490.

Ohno, S. (1987). Conservation *in toto* of the mammalian X-linkage group as a frozen accident. *Chromosomes Today* **9**, 147–53.

Page, D. C., Mosher, R., Simpson, E. M., Fisher, E. M. C., Mardon, G., Pollack, J., McGillvray, B., de las Chapelle, A., and Brown, L. G. (1987). The sex-determining region of the human Y chromosome encodes a finger protein. *Cell* **51**, 1091–1104.

Pereira, E. T., Cabral D'Almeida, J. C., Gunha, A. C. Y. R. G., Patton, M., Taylor, R., and Jeffery, S. (1991). Use of probes for ZFY, SRY and Y-pseudoautosomal boundary in XX males, XX true hermaphrodites and XY females. *J. Med. Genet.* **28**, 591–595.

Quiring, R., Walldorf, U., Kloter, U., and Gehring, W. (1994). Homology of the *eyeless* gene of *Drosophila* to the *small eye* gene in mice and *aniridia* in human. *Science* **265**, 785–787.

Sinclair, A. H., Berta, P., Palmer, M. S., Hawkins, J. R., Griffiths, B. L., Smith, M. J., Foster, J. W., Frischauf, A-M., Lovell-Badge, R., and Goodfellow, P. N. (1990). A gene from the human sex-determining region encodes a protein with homology to a conserved DNA binding motif. *Nature* **346**, 240–244.

Takahashi, M., Hayashi, M., Manganaro, T. F., and Donahoe, P. K. (1986). The ontogeny of müllerian inhibiting substance in granulosa cells of the bovine ovarian follicle. *Biol. Reprod.* **35**, 447–453.

Tilley, W. D., Marcelli, M., Wilson, J. D., and McPhaul, M. J. (1989). Characterization and expression of a cDNA encoding human androgen receptor. *Proc. Natl. Acad. Sci. USA* **86**, 327–331.

Sex Differences, Psychological

LEE WILLERMAN
University of Texas at Austin

GLOSSARY

Dimorphism Differences in form or structure between members of the same species

PSYCHOLOGICAL SEX DIFFERENCES ARE BROADLY construed to refer to behavioral differences between the sexes. Though the unmistakable accent is on difference, considerable overlap exists for most of the behaviors to be described, and personal characteristics other than sex usually provide a firmer foundation for making socially important decisions that affect individuals.

Sexual dimorphism is generated by a testis-determining factor located on the short arm of the Y chromosome that produces maleness. The absence of this factor results in a female. Differences in the brain accompany the more conspicuous sex differences in body structure and size. Women's brains weigh about 100 g less than men's brains, even after statistically correcting for sex differences in body size. Women appear less prone to aphasic disturbances following damage to their left hemisphere, except in the left anterior frontal lobe, where damage is more likely to produce aphasia in women. The sexes also differ in dendritic arborization patterns in the visual cortex and hippocampus, areas not previously thought to be sexually dimorphic. There is sex dimorphism in the size of certain nuclei in the anterior hypothalamus, and only the hypothalamus of females responds to rising plasma titers of estradiol with the cyclic release of luteinizing hormone-releasing hormone. Anterior hypothalamic nuclei are targets of the organizing influence of prenatal testosterone and are involved in regulating sexual behavior in lower animals.

Observations such as this have led to theories that the brain of the two sexes is somewhat differently organized. Experiments in some strains of rats reveal that male fetuses whose mothers were stressed during pregnancy often had a smaller amount of tissue in the preoptic area of the hypothalamus; as adults they were less effective copulators and had lower testosterone levels than nonstressed control animals. The relatively smaller preoptic area in stressed males is believed to arise from a 1-day delay in the onset of a normal testosterone surge during the last trimester of pregnancy. These findings confirm the centrality of the hypothalamus and the importance of testosterone in regulating sexual behavior, although it is unwarranted to generalize from results such as these to all aspects of sexual behavior. For example, these stressed male fetuses often display more feminine behavior than control males, but they do not show a sexual *preference* for other male rats.

If psychological sex differences were a direct result of biological sex differences, controversy surrounding the following findings would be considerably lessened. The shifting magnitude and, occasionally, even the direction of sex differences on many psychological measures over time make it difficult to draw hard and fast conclusions, as does the joint effect of biology and culture. Some experts argue that even posing questions about "built-in versus acquired" sex differences misses the mark by ignoring their interaction. Their analogy is to first language learning, the capac-

ENCYCLOPEDIA OF HUMAN BIOLOGY, Second Edition, VOLUME 7. Copyright © 1997 by Academic Press. All rights of reproduction in any form reserved.

ity for which is innate, but the specific language learned depending entirely on the language spoken in the home. The more traditional approach argues that variations in constitutional factors can be pitted against variations in environmental factors to estimate the relative importance of biology versus social environment in a given population. This relative importance approach is more dominant, although the joint effects of constitution and environment are drawing increasing attention.

Initial biological differences often are correlated with socialization practices that prepare children for their adult sex roles. These treatment differences make it difficult to determine the degree to which constitution alone produces predictable variations in later behavior. With only two sexes, explanatory power is greatly constrained, because any sex-related biological or rearing difference becomes a potential candidate to explain any psychological sex difference. So great is the sheer number of these potential candidates that hypotheses can easily proliferate. For example, toddlers' predilection for rough-and-tumble play could be a rehearsal strategy for adult intermale competition (analogous to the playful pouncing of kittens in preparation for later predation) and/or the result of the higher rates of rough-and-tumble father–child interactions observed with male, as opposed to female, infants.

I. CHILDHOOD DIFFERENCES

Gender identity, the ineffable sense of one's own femaleness or maleness, evolves early in life and, in most normal children, seems to be established by 4 years of age. Exceptions occur when children with gender identity disorders are brought to professional attention having repudiated the manifestations of their sex—for example, asserting that genitalia will appear or disappear, or reporting feelings of being trapped in the body of the wrong sex. Most of these children eventually develop a gender identity congruent with their genetic sex, but perhaps two-thirds of such boys become homosexually oriented despite their development of a masculine gender identity (the rates for extremely masculine girls are less certain). Gender identity, however, should not be thought of as dichotomously male or female. A significant fraction of people, not all of whom are homosexuals or transsexuals, have dreams or fantasies of being the other sex or wonder about having a sex-change operation, suggesting the existence of intermediate gender identities.

Gender role refers to the outward manifestation of gender identity (i.e., the behaviors conventionally associated with being male or female). Distress about gender role adequacy is much more common than gender identity disturbance, predictably so in teenagers if puberty or dating skills are delayed, and in adults if the inability to fulfill stereotypic gender roles is accentuated (e.g., when previously employed men lose their jobs, or when women want to but cannot conceive).

Some sex dimorphisms are inevitable (e.g., men inseminating, women bearing children), others are completely arbitrary (e.g., dresses versus trousers), and still others are intermediate dimorphisms usually more common in one sex or the other but expressible by either sex under appropriate circumstances (e.g., sleeping mothers being more sensitive than fathers to an infant's crying). Among the most psychologically interesting of these latter differences are energy expenditure, physical competitiveness, fighting off predators in defense of territory, and erotic arousal by visual imagery (more commonly observed in males) and retrieving, cuddling, and rehearsal play with dolls or playmates (more frequently observed in females).

By 3 years of age, most toddlers can distinguish the two sexes in pictures, although some may be uncertain about their own sex remaining forever constant or the hegemony of anatomy in sex determination. They also show preferences for interacting with same-sex playmates, the patterns of interaction being quite different within the sexes. In comparison to girls, boys more often prefer rough-and-tumble activities and mock fighting. Boys also accede less often to protests from girls not to take their toys, even at 3 years of age.

Sex-segregated subcultures of the preschool year persist into the early school years, but the source of that persistence is unknown. It is possible to identify masculine and feminine preschoolers of both sexes, that is, children whose behaviors tend to resemble the stereotype of one sex or the other. Nevertheless, most children, regardless of masculinity or femininity, play mainly with same-sex peers, despite occasional efforts by adults to encourage mixed-sex interactions. Sex segregation in the early school years is sustained in part by peers who tease about being in love if another child shows an interest in someone of the opposite sex.

It has been difficult to demonstrate compellingly that differential parental socialization of the sexes or even extrafamilial social influences such as the media cause, rather than correlate with, the gender segregation and play style differences. The absence of decisive evidence on this matter has complicated environmental theories about the origins of psychological sex differences. Biological theories about the play style differences probably are more established because of behavioral homologies in lower animals.

Whatever the causes, sex-related preferences in early childhood appear to affect the matrix of experiences encountered and influence later mate choice. For example, virtually no marriages exist between classmates in Israeli kibbutzim, where like-aged children live in mixed-sex group homes from infancy on. Classmates often marry members of the same kibbutz, but they nearly always come from younger or older classes, the early familiarity apparently discouraging romantic interests within their age cohort. Exceptions occur if one of the children enters the cohort after 6 years of age, suggesting that very early playmate experiences have important functions in later mate selection. If early familiarity discourages later sexual interests, then these results may also help to explain the absence of widespread sibling incest in ordinary families.

II. INTEREST AND PERSONALITY DIFFERENCES

Sex differences in interest patterns are among the largest and most stable psychological sex differences known. These differences are telling because they reflect preferences rather than obligations and, therefore, reveal more about what people really like to do rather than what they must do. Studies of reading preferences in elementary and high school students indicate that boys are 3 times more likely than girls to prefer adventure and mystery stories, whereas girls are 10 times more likely to prefer stories about home and school life. These differences persist into adulthood, when women are much more likely than men to read romance novels. However, sex differences in interests tend to wane slightly in middle age with males moving in the feminine direction.

Despite decades of increasing educational and occupational opportunities for women in industrialized societies, interest differences do not seem to have diminished appreciably. For example, the Strong-Campbell Interest Inventory was constructed by asking adults in many occupations about their likes and dislikes (e.g., decorating a room with flowers, cabinetmaking), to see whether or not interest pattern differences relate to the likelihood of success in various occupations (they do). The most recent restandardization in the 1970s revealed sex differences on 149 of the 325 items, and the differences are still similar to those indicated in the original 1930s standardization. In one item, "liking to decorate a room with flowers," 79% of women lawyers in the original standardization indicated a liking as opposed to 21% of the men lawyers; in the most recent survey, 62% of the women lawyers and 15% of the men lawyers indicated a similar liking, a modest decrease in overall liking, but nevertheless a similar large difference between the sexes.

Most readers could have guessed the direction of the sex difference in liking to decorate with flowers. Early masculinity–femininity scales of the 1930s were empirically constructed from such items to identify psychological differences between the sexes, the assumption being that masculinity–femininity was a single bipolar dimension; high femininity perforce meant low masculinity, and vice versa.

More recently, perhaps motivated by increasing career flexibility for women, experts devised independent scales of masculinity and femininity from the perspective of two separate personality dimensions, thus allowing people to be high or low on both, rather than masculine *or* feminine. These new masculinity–femininity scales focus on socially desirable personality traits such as emotional expressiveness, nurturance, and interpersonal orientation in women, and instrumentality (task as opposed to person orientation) and competitiveness in men. Despite meaningful average differences between the sexes on these personality scales, a tight linkage among gender-congruent items does not exist—masculine men often express tender feelings, and feminine women often enjoy the competition in careers formerly occupied by men. The merit of these scales, however, was to demonstrate the wide variety of possible psychological adaptations to one's gender.

Perhaps the absence of a tight linkage among personality items is not so surprising if one assumes that sexual selection affected physiology or behaviors more directly related to reproductive success—physical attractiveness and nurturance in women, size and aggression in men. Other sex-related adaptations probably arose because of their correlations with the primary targets of sexual selection (e.g., a greater preference for decorating a room with flowers because "home" has more significance for women, or men more often finding the idea of sex without love erotic because of their relatively low necessary investment in the costs of gestation and postnatal nurturance).

Recent cross-cultural research has shown fairly consistent patterns of sex differences in mate preference that seem related to signs of reproductive potential in women and the potential of men to provide resources for mother and child. Questionnaires completed by respondents of both sexes in 37 industrialized and preindustrialized societies revealed, in nearly every case, that women more than men preferred ambitious and financially secure potential spouses, whereas men valued more highly than women a good-looking spouse. In every society, both sexes independently agreed about the desirability of men marrying younger women, confirming the social worth of youth in female reproductive potential and male resource capacity in mate choice.

Differences between the sexes on most self-report personality inventories are quite modest if scales of masculinity–femininity are excluded. The recent massive restandardization of the 462-item California Personality Inventory (CPI), one of the most popular instruments of this genre for the normal adolescent and adult population, found no substantial sex differences for 19 of 20 personality traits. Among the personality traits measured were dominance, empathy, and sociability, and none revealed mean differences between the sexes of more than one-fifth of a standard deviation, suggesting nearly complete overlap of the distributions of scores for the two sexes. The CPI femininity scale, however, revealed that females scored nearly two standard deviations higher than males, a difference of this magnitude indicating that the average female scores at the 95th percentile of the male distribution. High scorers of either sex on this scale are rated as sympathetic, sensitive to criticism, interpreting events from a personal point of view, and reporting feelings of vulnerability; low scorers of either sex are viewed as decisive, action-oriented, taking the initiative, and unsentimental. Descriptions such as these seem to capture stereotypes about the major personality differences between the sexes.

III. COGNITIVE DIFFERENCES

A review of cognitive differences between the sexes must acknowledge that tests are designed by people who first must decide on the types of items to include in the domain of interest. Most tests do not ask about tracking game, surviving in the desert, or making dyes from plants, perfectly legitimate items in preindustrialized societies. Conventional intelligence tests attempt to exclude items that rely on material explicitly taught in school. The tests generally include items that call for problem-solving, reasoning, and abstraction. Verbal intelligence test items might ask a child about the erroneous thinking in "Mary's jeans were so tight that she had to put her pants on over her head" or explain how sewing machines and automobiles are alike. By contrast, pure achievement test items ask about material explicitly taught in school, although some achievement and intelligence test items do resemble each other when they require reasoning based on school-related subject matter. Nonverbal intelligence tests require no specific knowledge of language; completing jigsaw puzzles and constructing block designs after model pictures are exemplars and can even be pantomimed for people who are congenitally deaf or come from preliterate societies.

A. General Intelligence

Evaluating sex differences in general intelligence using standardized tests is usually not meaningful, because most tests are designed to be equally applicable to both sexes, and items initially showing sex differences in percent passing were either excluded from final versions or counterbalanced by items with the opposite pattern of sex difference. Men do perform relatively better than women on items that require elementary geographic or scientific knowledge (e.g., Why does water boil in the mountains at a lower temperature than in the valleys?), but this is most likely to be a result of discrepant interest patterns. Perhaps one tends to remember the gas laws only if they jibe in some way with interests or are reinforced in specific courses. A widely used nonverbal intelligence test (Raven Progressive Matrices) ignored sex differences in its construction and found none afterward, suggesting that sex differences in general intelligence are either negligible or nonexistent.

B. Specific Abilities and Scholastic Achievement

Tests designed to tap specific abilities reveal moderately interpretable differences between the sexes. Females outdo males on tests of verbal fluency (e.g., in one minute say as many words as possible that begin with the letter "R") and perceptual speed (e.g., quickly cross out all the words beginning with the letter "e" in a long string of words). Achievement tests tapping written language, grammar, and *arithmetic computation* also show a female advantage.

None of these differences, however, has attracted quite so much attention as the conspicuous male advantage on *mathematical reasoning tests,* such as the Scholastic Aptitude Test (SAT) mathematics section. The sex difference is most apparent at the highest levels of achievement with males comprising 96% of those with the maximum possible mathematics score of 800. The cause of the *mean* advantage is not clear, but sex differentials in high school dropout rates and low-ability males being less likely to take the test are partial explanations; however, no selection differential can account for the remarkable overrepresentation of males at the very highest level, and many hypotheses have been proposed to explain the result. One hypothesis has focused on the greater nonconformity of boys in school (i.e., they are less well behaved), which paradoxically gives them an advantage in solving math problems that require nonconventional reasoning instead of direct application of explicitly taught and well-practiced strategies. Because males now show a slight advantage on the SAT verbal test as well, where nonconformity would not seem to have been advantageous, that hypothesis seems doubtful.

The male advantage on the SAT mathematics test is accentuated at the top of the score distribution because males are also more variable. Greater male variability needs to be explained, and the current biological focus is on genetic and hormonal differences between the sexes. Nonbiological approaches focus on differential expectancies about the role of mathematics in one's future (e.g., attitudes toward mathematics or taking different courses), but they have not gotten very far in accounting for the difference. The same could be said for sex differences in self-perceptions (e.g., females tending to attribute success to luck and failure to personal inadequacy), and thus a tendency exists for the sake of nothing better, to find biological approaches more appealing.

Males perform better than females on tests of spatial visualization, especially when three-dimensional mental rotation is required. The reason for this advantage also is not clear, but an analogous sex difference in learning mazes that require some form of spatial problem-solving is found in several rodent species.

Provocative recent findings suggest that performance on some tests is affected by menstrual cycle phase. For example, women do relatively better during menstruation than during midcycle on tests of spatial ability and better during midcycle than during menstruation on tests of speed of articulation and manual coordination. The midcycle phase is associated with higher levels of estrogen and progesterone, whereas menstruation is associated with lower levels of these sex hormones. Performance on spatial tests, which normally favor males, is positively affected by low levels of female hormones, and performances on which females normally have the edge, like speed in articulation and manual coordination, are enhanced by higher levels of female sex hormones. These, and related observations, suggest that female sex hormones are not simply activating or depressing performance generally, but have differential effects on different types of activity. Human females with congenital adrenal hyperplasia (CAH), involving overproduction of androgen from the adrenals prior to birth, appear to be better than normal females at three-dimensional rotation. Because the overproduction of androgen is curtailed by cortisone treatment beginning soon after birth, partial prenatal masculinization of the CAH female brain is a viable hypothesis for this advantage. A subsidiary hypothesis is that interest and activity patterns of females with high spatial visualization scores (or males with low spatial visualization scores) may be more like those of the other sex. Thus, it might be the interest–activity patterns that provide test-relevant experiences that influence spatial visualization, rather than something in the hormonal environment directly affecting spatial visualization.

Studies offer modest support for this hypothesis. Some species of voles show differences in ranging behavior between the sexes, although all voles are monomorphic. Male voles of high-ranging species show better spatial maze-learning ability than males in low-ranging species. Young human males are also more likely to range farther from their caretakers, suggesting some experiential basis to the sex difference in spatial ability. Both normal and CAH females with masculine activity patterns tend to do better on spatial visualization tasks, but the effect is too small to eradicate the overall mean sex difference in spatial visualization scores, and the direction of effects has not been determined; does higher spatial visualizing ability lead to more masculine interest–ranging patterns, or vice versa?

Lower spatial visualization scores in men with idiopathic hypogonadotropic hypogonadism (IHH) would seem to reinforce the role of pubertal testosterone in the male spatial visualization advantage, especially because these men show no deficits in verbal ability. IHH is characterized by a failure to enter puberty because of a hypothalamic deficiency in gonadotropin-releasing hormone. Although IHH males appear to have a near-normal prenatal hormonal environment, this is not certain, and none of these males was tested prior to his teenage years. Moreover, studies of mathematically precocious preadolescents reveal a disproportionate number of males scoring at the very highest levels, suggesting hormonal or experiential influences prior to the pubertal testosterone surge. In short, evidence exists for prenatal and pubertal testosterone influences on spatial ability; their relative influence, however, remains uncertain.

Other evidence for an early biological influence on spatial and mathematical ability comes from the study of females with Turner syndrome, a disorder caused by the complete or partial absence of one X chromosome. Even as children, many show a significant deficit in spatial and mathematical ability despite normal verbal intelligence. These females have low levels of estrogen because of defective or absent ovaries and do not enter puberty unless estrogen supplements are provided. Although several somatic problems are associated with this condition, including very short stature, affected females are usually within the normal range of mental health and are heterosexually oriented. No brain defects have yet been identified, and a prenatal hormonal deficiency effect on spatial and numerical ability is quite plausible, although the specific hormonal mechanism has not been clarified.

IV. SEXUAL ORIENTATION

Aside from gender identity, the largest psychological gender difference known is in the sex of the preferred mate. All mammalian species are heterosexual, and there is even controversy about whether homosexual orientation exists in lower mammals. The debate is not about whether or not homosexual behavior naturally occurs—it does, but mostly when males are blocked from access to females. Studies of mate choice in some lower mammals reveal that early castration of males or testosterone administration to females can partially or completely reverse sexual orientation, so that hormonal factors must figure predominantly.

Heterosexual and homosexual orientation refer, respectively, to the tendency to become erotically aroused by the opposite or same sex. Sexual orientation is not solely determined by sex acts with opposite or same-sexed partners; heterosexual teenage boys often engage in mutual masturbation, for example, but it is friction rather than homosexual fantasy that sparks their sexual arousal.

Current research suggests that the matrix out of which homosexuality emerges is evident early in life. Gender-incongruent behavior even in preschool children, especially extreme femininity in boys, appears to augur later homosexuality; tomboyism in girls is not nearly as good a predictor of later lesbianism because it occurs with a high base rate even among future heterosexuals. Although homosexuality runs in families, and identical twins are more often concordant than fraternal twins for homosexual orientation, specific genetic factors have not been firmly identified. One report has suggested that brothers concordant for homosexuality may become so because of a sex-linked gene, but nothing is known about genetic transmission of female homosexuality. Homosexuality is unlikely to have a single cause and there are likely to be multiple genetic and environmental determinants.

Social environmental explanations have been unsatisfactory thus far because distinctive peculiarities in the early lives of homosexuals have not been discovered. Recollections of childhood can be faulty, however, and large, prospective longitudinal studies that begin with monitoring events during pregnancy and observing the development of the children hold more promise. Current hypotheses emphasize prenatal endocrine factors, perhaps arising from stresses to the mother or fetus reducing androgen levels in males, and excess prenatal androgen increasing the risk for eventual lesbianism in females. Unfortunately, trustworthy evidence for these hypotheses is lacking in humans. Whereas some studies have reported that heterosexual and homosexual males differed in their luteinizing hormone levels following an estrogen challenge, with homosexual males tending to respond somewhat like females, other studies have failed to replicate this and no consensus has been achieved about the reliability or meaning of these results.

Other evidence that neuroendocrine factors contribute to sexual orientation comes from people whose prenatal hormonal milieu has been altered by accidents of nature. But describing the psychological characteristics of people exposed to unusual prenatal hormonal environments can be complicated, because phenotypic, and not genetic, sex determines whether children are reared as boys or girls. Testicular feminization syndrome provides an example of appearance as being decisive in the sex of rearing, while not shedding much light on the role of prenatal hormonal factors in shaping sexual orientation. This syndrome arises because an autosomal recessive gene leads to a reduced or complete inability of receptors in the brain and elsewhere to respond to circulating testosterone. In genetic males, the outcome is a feminine phenotype, feminine gender identity, and sexual attraction to males. Because they are reared as girls—subjectively and phenotypically female except for the absence of ovaries and uterus and the presence of undescended testes—it is misleading to regard their sexual attraction to males as homosexual.

A more informative condition is CAH, which produces nearly comparable levels of prenatal androgen in affected female and in normal male fetuses. The external consequence in the affected female is an enlarged clitoris, correctable by surgery; excess secretion of androgen from the adrenals after birth is controlled by cortisone administration. Though reared as females, they are often tomboys, and some become lesbians. This outcome points to a significant role for the prenatal hormonal milieu in influencing sexual orientation. Because only a minority become lesbians, however, uncertainty remains about the possible influence of other factors such as the timing or degree of prenatal androgen excess or postnatal experiences on sexual orientation.

Another interesting syndrome in this context involves a genetic inability to produce 5-alpha-reductase, an enzyme that converts testosterone to dihydrotestosterone (DHT). DHT is necessary for the early masculinization of the external genitalia but has no known function in the brain. Genetic males with 5-alpha-reductase deficiency have ambiguous external genitalia, and more than half are reared unquestioningly as girls. However, a gonadal testosterone surge at puberty causes a deepening of the voice, muscularity, the descent of undescended testes, and phallic enlargement so that erection and intromission can occur. Virtually all of these girls develop erotic interests in females, eventually cohabiting and foresaking their earlier identity for a masculine gender identity and gender role. Because this disorder runs in families, parental knowledge about the risk for the disorder could have biased the girls' rearing, but parents typically report shock and amazement when their child exhibits sexual interests in females. Because nearly all such females developed erotically in ways consistent with their prenatal testosterone levels despite unambiguous rearing as girls, hormonal influences are implied to play a powerful role in sexual orientation. It is premature, however, to argue strongly on behalf of any one hypothesis to explain the origins of homosexuality, and multifactorial causation is a distinct possibility.

V. SEX DIFFERENCES IN PSYCHOPATHOLOGY

There are many sex differences in the rates of psychopathology. In childhood, the rates of attention-deficit hyperactivity disorder, conduct disorder, dyslexia, stuttering, some forms of mental retardation, and early infantile autism are much more prevalent in boys. These disorders run in families, although most cases occur sporadically. Explanatory hypotheses have focused on X-linked genes or sex chromosome anomalies, sex differences in prenatal vulnerability to anoxia or infection, mother–male fetus antibody reactions, and sex differences in child-rearing practices.

The fragile X syndrome, occurring in 1 of 2000 male births, provides an example of a disorder that produces an excess of mental retardation in males and may also account for some of the male excess among those with early infantile autism. Identified by a constriction on the X chromosome in lymphocytes cultured in a medium deficient in folic acid and thymidine, this syndrome produces mental retardation in affected males (a few have normal IQs) and a much more variable outcome ranging from mild retardation to normal intelligence in females. In both sexes, an elongated face and large ears are associated with the fragile X syndrome, and enlarged testicles are common in men. Although most people with fragile X syndrome lack remarkable behavioral features, with the exceptions of mental retardation and perhaps hyperactivity, 12 to 25% of males with early infantile autism have the fragile X anomaly, suggesting behavioral pleiotropy of the fragile site. [*See* Fragile X Syndromes.]

Among adults, men are much more likely than women to have a history of antisocial personality disorder and alcohol abuse or dependence, with a sex ratio ranging from 4 : 1 to 6 : 1 in epidemiologic surveys. The male excess for hyperactivity, alcoholism, and antisocial personality disorder has been attributed to interactions of genetic factors with sex differences in prenatal vulnerability and to social factors. Disorders with by far the greatest male preponderance are paraphilias such as fetishism. Paraphilias are characterized by recurrent intense sexual urges and fantasies that require nonhuman objects, humiliation to self or others, or nonconsenting adults or children for sexual satisfaction. Fetishism specifically involves intense preoccupations with nonliving objects (e.g., gloves, panties) for sexual satisfaction. In comparison to thousands of male fetishists, only three female fetishists have ever been reported, and all three were homosexually or bisexually oriented, implying the possibility of early hormonal factors figuring in the etiology. A sex difference of such magnitude is truly remarkable and an explanation could bear on sex differences in other realms, including the preponderance of males (whether homosexual or heterosexual) as purchasers of pornography.

Women are more likely than men to have a history of major depressive episode (2 : 1 or 3 : 1), agoraphobia (3 : 1), and simple phobia (2 : 1). Other disorders show smaller, less reliable sex differences in prevalence. Included among these are nonalcoholic drug abuse and dependence, with men predominating, and dysthymia (chronic depression), somatization (an unusually wide variety of inexplicable bodily complaints), panic disorder, and obsessive–compulsive disorder, with women predominating. Many hypotheses have been proposed for the female excess, including feelings of powerlessness, less reluctance to admit vulnerability and fear, a greater tendency to internalize stress, and a sex bias in diagnostic criteria. All of these hypotheses have found support to various degrees, but more refined hypotheses are required to handle the sex ratio heterogeneity from disorder to disorder. [*See* Depression; Mental Disorders; Nonnarcotic Drug Use and Abuse.]

Many of the disorders with sex differences are heritable; for example, adoption and/or twin studies indicate that hyperactivity, antisocial personality disorder, major depression, alcoholism, and somatization have heritable components. Except for a putative X-linked dominant form of manic-depressive disorder and X-linked mental retardation, no psychopathological studies indicate X-linkage, suggesting that sex differences in rates for specific disorders could arise from sex differences in vulnerability thresholds (e.g., female stutterers are rare but, if affected, must have required a greater dose of the relevant genes because they are more likely than affected men to produce similarly affected offspring). Another possibility is sex-specific symptom expression of the same diathesis (e.g., antisocial males often have sisters with somatization disorder).

VI. THE FUTURE OF SEX DIFFERENCES

Modern society no longer requires that physical strength be the final arbiter in the division of labor. Very few occupations are necessarily limited to one sex, and as external barriers to equal opportunity diminish, occupational choices will be made more often on the basis of personal interests and talent. Sex differences in occupational interests have remained fairly stable for several decades and, if the past is a guide, we should not expect much change in preferences for various occupations.

Because the onus of reproduction falls largely on women, any correlates of differential parental investment such as different attitudes about sexual behavior might be difficult to eliminate. Perhaps emblematic of these differences are findings in a large British survey that while three-quarters of unmarried college women think that the idea of a sex orgy is disgusting, only 18% of a corresponding group of men do and that 73% of the men versus 35% of the women believe it is all right to seduce people who are old enough to know what they are doing. Another study reported that a third of men said they might rape if there were no chance of getting caught. Differences such as these can be perennial sources of intersexual conflict, at least for a large proportion of people.

It seems likely that sex differences in preference and style during sexual intercourse also will not change much, with conflicts arising that often affect other aspects of the relationship. Women frequently complain that their partners have a narrow concept of affection, moving too speedily toward intromission while short-circuiting whole-body caressing. On the other hand, men often complain that their mates are sexually witholding. Observational studies of sexual behavior in homosexual couples provide insights into how the sexes differ in preferred styles of sexual intercourse because they need not make as many compromises as heterosexual couples. Results from these studies indicate that homosexual males tend to spend relatively less time in foreplay in comparison to lesbians and tend to leave the bed more quickly after orgasm, in contrast to lesbians, who often fall asleep holding one another. Moreover, cohabiting male homosexual couples are also more likely than lesbian couples to be involved in sexual infidelities. Some of these sex differences could also be a source of tension within heterosexual couples. [*See* Sexual Behavior.]

Whatever their origins, differences in sexual behavior appear to be fairly sturdy and not necessarily rational from the perspective of an intelligent automaton. For example, the mammalian sex difference in longevity is tied to testosterone since male neonatal castration lengthens life span and testosterone administration to females shortens life span. This difference in longevity, about 7 years in humans, combined with an average difference of about 3 years in age of spouses, means that the average woman will be widowed for 10 years. If other things were equal, a demographic reversal in the average age at marriage would increase the duration of couple companionship.

Sex dimorphism with respect to body size predicts overall patterns of mating across many species: monomorphic species are monogamous, and dimorphic species are polygynous or promiscuous. In the monomorphic gibbon, pairs mate for life; in the extremely dimorphic gorilla, the dominant male has a harem, and most males are perforce heterosexually celibate. In a moderately dimorphic species such as humans, the relevant evolution is unmistakably closer to a heritage of intermale competition, but our smaller degree of dimorphism suggests some divergence, as does the humanly distinctive pubertal growth of breasts in the absence of pregnancy.

Human males also have moderately sized testes relative to other primates, which means a limited reservoir of sperm. Consequently, human males have not been equipped by evolution for closely spaced bouts of sperm-producing intercourse and the sex drive is probably somewhat less predictably imperious. Interpersonal ambiguities arising from being intermediate on these evolutionary factors probably figure importantly in some tensions between the sexes. These considerations may lead one to ask how relations between the sexes and psychological sex differences might alter, absent of any changes in brain biology, if the sex difference in strength were reversed or if men bore children.

ACKNOWLEDGEMENTS

The author thanks Professors David B. Cohen, Stephen Finn, Judith Langlois, and John C. Loehlin for reviewing earlier versions of this article.

BIBLIOGRAPHY

Benbow, C. (1988). Sex differences in mathematical reasoning ability in intellectually talented preadolescents: Their nature, effects, and possible causes. *Behav. Brain Sci.* **11,** 160.

Buss, D. (1989). Sex differences in human mate preferences: Evolutionary hypotheses tested in 37 cultures. *Behav. Brain Sci.* **12,** 1.

Campbell, D. P., and Hanson, J.-I. C. (1981). "Manual for the Strong-Campbell Interest Inventory," 3rd Ed. Stanford Univ. Press, Palo Alto, California.

Gough, H. G. (1987). "California Psychological Inventory: Administrator's Guide." Consulting Psychologists Press, Palo Alto, California.

Ilai, D., and Willerman, L. (1989). Sex differences in WAIS-R item performance. *Intelligence* **13,** 225.

Kimura, D. (1987). Are men's and women's brains really different? *Can. Psychol.* **28,** 133.

Kimura, D., and Hampson, E. (1993). Neural and hormonal mechanisms mediating sex differences in cognition. *In* "Biological Approaches to the Study of Human Intelligence" (P. A. Vernon, ed.), pp. 375–397. Ablex Publishing, Norwood, New Jersey.

Maccoby, E.E. (1988). Gender as a social category. *Dev. Psychol.* **24,** 755.

Reinisch, J. M., Rosenblum, L. A., and Sanders, S. A. (eds.) (1987). "Masculinity/Femininity: Basic Perspectives." Oxford Univ. Press, New York.

Stewart, J. (ed.) (1988). Sexual differentiation and gender-related behaviors. *Psychobiology* **16,** whole issue.

Wilder, G. Z., and Powell, K. (1989). Sex differences in test performance: A survey of the literature. *College Board Report* No. 89-3.

Willerman, L., and Cohen, D. B. (1990). "Psychopathology." McGraw–Hill, New York.

Sexual Behavior

JOHN D. BALDWIN
JANICE I. BALDWIN
University of California, Santa Barbara

GLOSSARY

Erotic stimulus Any stimulus that can elicit the sexual reflex, be it a conditioned stimulus or the unconditioned stimulus that is biologically associated with the sexual reflex

Myotonia Heightened muscle tension, adequate to cause various subcomponents of the sexual response

Pavlovian conditioning Process by which stimuli that are paired with and predictive of a reflexive response become conditioned stimuli that can elicit certain aspects of that reflexive response

Sexual disorder Deviation from the accepted functioning of the sexual reflex that causes a person clear distress or interpersonal problems

Sexual orientation Degree to which a person is sexually attracted to the other sex, the same sex, both sexes, or neither sex

Vasocongestion Increased blood flow into and congestion of the blood vessels that activate parts of the sexual response

HUMAN SEXUAL BEHAVIOR INCLUDES ALL THE voluntary and reflexive activities that can lead to sexual reproduction, though these activities are not always used in ways that result in pregnancy and childbirth. The reflexive and inborn facets of sexual behavior are not adequate for assuring successful reproduction, and considerable learning is needed for an individual to become reproductively capable. That learning can be influenced significantly by social, cultural, and individual variables, leading to substantial variability in human sexual activities.

I. THE SEXUAL RESPONSE OF HEALTHY ADULT MALES AND FEMALES

The human sexual response has been described in several ways. One of the more widely accepted descriptions divides the continuous series of changes of the sexual response into four general phases: excitement, plateau, orgasm, and resolution. Another categorization describes three phases: desire, excitement, and orgasm, with desire occurring before biological arousal. Since many people have problems of sexual desire (i.e., a lack of interest in sexual fantasies and activity), it is an important topic and will be considered before the other four phases.

A. Desire

There is extensive variation among people in the amount of sexual activity that they desire. People with no sexual desire are sometimes described as "asexual" or "nonsexual." People with low sexual desire may think about sex from time to time, but they have little interest in sexual fantasies or activity and may not become easily excited in sexual situations. People with

ENCYCLOPEDIA OF HUMAN BIOLOGY, Second Edition, VOLUME 7. Copyright © 1997 by Academic Press. All rights of reproduction in any form reserved.

high sexual desire think about sex often and may become sexually aroused in nonsexual situations. Desire for sexual activity is not an essential prerequisite for the remaining phases of the sexual response: people can progress through the biological phases of the sexual response without having sexual desire.

There is no single "normal" level of sexual desire. Two people with high sexual desire may be very happy with their sexual life in having sexual interactions every day; and two people with low sexual desire may have an equally satisfactory sex life, engaging in intercourse only once a month. Levels of sexual desire are only considered to be problematic when they lead to personal or interpersonal problems, for example, when two sexual partners have different levels of sexual desire, that is, "discrepant sexual desire." The greater the discrepancy between two people's level of sexual desire, the greater their sexual problems can be. The partner who wants to have frequent sexual interactions may feel very frustrated by the lack of interest that his or her partner shows; and the person with low sexual desire may feel pressured to engage in sex more than he or she wants. Such problems can create considerable tensions in a relationship.

B. Excitement

Any of a variety of stimuli can lead to sexual excitement. Sometimes people become excited without planning it or desiring it. Other times, a clear sexual desire leads them to seek out the stimuli that elicit sexual excitement. The central nervous system is structured from birth such that touch to the genitals leads to sexual excitement, if it is of the appropriate pressure and movement pattern. Stimuli from any sense modality—along with sexual thoughts and fantasies—can take on the ability to trigger sexual excitement.

In the female, the first sign of sexual excitement is vaginal lubrication; in the male it is penile erection. These early signs of excitement are easily reversible if there is no further sexual stimulation. However, with continued effective stimulation, there can be further developments of sexual excitement. In the female, the inner two-thirds of the vagina expands, the uterus lifts up, the major lips (or labia majora) pull back, and the minor lips (or labia minora) and clitoris increase in size. In the male, the testes enlarge and are elevated somewhat closer to the body.

C. Plateau

With continued effective sexual stimulation, people reach a higher level of sexual arousal, the plateau phase. In the female, the outer third of the vagina swells, narrowing this portion of the vaginal barrel, while the inner two-thirds expands more, and complete elevation of the uterus occurs. The clitoris retracts under the clitoral hood (but is still sensitive to stimulation around it). In the male, the testes continue to swell, becoming 50 to 100% larger than normal. They are pulled closer to the body. Usually at this time, the Cowper's glands secrete a fluid that neutralizes the acidity of the urethra. The fluid may also contain sperm left from the last ejaculation or nocturnal emission.

D. Orgasm

After sufficient effective sexual stimulation, orgasm occurs. In both the female and male, the length and intensity of orgasm can vary, depending on the level of sexual arousal, the type of stimulation obtained during orgasm, health, age, and other variables. Typically, orgasm lasts 3 to 15 sec.

In the female, the primary orgasmic response consists of rhythmical contractions of the muscles surrounding the outer third of the vagina. The first several contractions occur at approximately every 0.8 sec; but subsequent contractions are spaced further apart and are less intense. These contractions are experienced as pleasurable, often being described as waves, surges, or pulses of pleasure. Females have the potential for multiple orgasms: if they continue to desire and receive effective stimulation, they may alternate between plateau and orgasm several times in close succession.

In the male, orgasm is divided into two clearly recognizable stages: ejaculatory inevitability and ejaculation. Ejaculatory inevitability occurs a few seconds prior to ejaculation. At this time, contractions of the prostate, vas deferens, and seminal vesicles cause semen to be released into the urethra. After this pleasurable several-second-long process, the semen is expelled from the male's body by rhythmical contractions of the perineal muscles, which contract at intervals of 0.8 sec for the first several contractions, though later contractions are spaced further apart and gradually become less intense. The contractions are most pleasurable when they are most intense—near the beginning. After ejaculation, the penis may become very sensitive and further stimulation may feel painful.

E. Resolution

After orgasm is over, the various physical changes that develop during excitement, plateau, and orgasm reverse, and the body gradually returns to the unaroused condition. Resolution from the plateau phase take longer if there is no orgasm: the occurrence of orgasm hastens the changes during resolution.

After ejaculation, males enter a period called the refractory period, in which further stimulation does not produce a sexual response. The refractory period can last from a few minutes to many hours, depending largely on a person's age—being longer in older males. After the refractory period is over, sexual stimulation can once again lead to erection, rising sexual arousal, and orgasm.

II. THE MECHANISMS OF THE SEXUAL RESPONSE

A. Reflexes

The sexual response is based on complex reflex mechanisms that are mediated by the sacral, lumbar, and parts of the thoracic vertebrae in the lower spinal cord. The primary neural receptors that trigger the reflex come from the genitals (or nearby erogenous areas). The unconditioned stimuli that activate the reflex are touch to the genitals. The reflex mechanisms also send signals to the cerebral cortex and are responsive to signals from the cerebral cortex. Through cerebral mediation and Pavlovian conditioning, additional stimuli, such as fantasies, can come to elicit the sexual response. Finally, the sexual reflex can be suppressed by any of a variety of aversive stimuli, such as worry, guilt, anxiety; and much of sexual therapy focuses on helping people reduce the aversive components of their sexual lives, while increasing the aspects that are sexually arousing and exciting.

B. Vasocongestion and Myotonia

During the early phase of sexual stimulation, sacral reflexes are activated that cause increased blood flow to the genitals and restrict the veins that allow blood to leave the genital area. The resulting increase in blood in the genitals is called vasocongestion, and it causes an increase in the size of various parts of the genitals. Upon early stimulation, the penis becomes larger, firmer, and more erect. In the female, vasocongestion around the vaginal barrel causes transudation—or lubrication—on the inside walls of the vagina. In addition, vasocongestion causes the labia minora to increase in size while the clitoris becomes firmer and slightly larger.

Among the changes mediated by myotonia (muscle tension) are heightened muscle tone in various parts of the body, the elevation of the testicles close to the body, and, with further stimulation, there are muscular contractions of the auxiliary organs that operate during ejaculatory inevitability to prepare the ejaculate for expulsion from the body. Myotonia is also responsible for the muscle contractions at orgasm that are experienced as waves of pleasure in both sexes and cause the ejaculation of the semen in the male.

C. Cerebral Cortex

The neural mechanisms that connect the reflex centers to the cerebral cortex connect with the sensory centers, allowing people to feel touch and pressure in the genitals. In addition, some of these neural systems innervate the pleasure and pain (i.e., reinforcement and punishment) centers in the brain. This assures that certain forms of sexual stimulation are experienced as pleasurable; and people usually learn to repeat these activities owing to positive reinforcement. Other forms of sexual stimulation—for example, strong pressure to the gonads—are painful and people learn to avoid these forms of stimulation. These reinforcement and punishment mechanisms help assure that most people, if they have appropriate life experiences, learn how to obtain effective stimulation and avoid aversive stimulation. The muscle contractions at the time of orgasm are usually experienced as especially pleasurable.

Nerves that connect the cortex to the sexual reflex mechanisms in the lower spinal cord allow for the "psychogenic" stimulation of the sexual reflex. Namely, when people see or fantasize about erotic stimuli, the cortical activity can activate the sexual reflexes. For some people, psychogenic stimulation produces only a mild sexual response, such as partial penile erection or slight lubrication of the vagina; but in other people, the response can be quite strong, in some cases leading to orgasm. There is a great deal of variation among people in the types of fantasy and erotic stimuli that produce such psychogenic responses, and these differences are due to differences in prior sexual learning experiences, especially those involving Pavlovian conditioning.

D. Pavlovian Conditioning

The sexual reflex is among the reflexes capable of being conditioned via Pavlovian conditioning. Namely, when a neutral stimulus is paired repeatedly with the unconditioned stimulus of appropriate touch to the genitals, the neutral stimulus gradually becomes a conditioned stimulus capable of eliciting the sexual reflex. In common parlance, these conditioned stimuli are called erotic stimuli or sexual "turn-ons." Usually, the more frequently that an erotic stimulus or turn-on has been paired with the unconditioned stimulus of touch to the genitals, the more capable the conditioned stimulus is of eliciting sexual responses.

Almost any stimulus that is frequently paired with and is predictive of the sexual reflex can become an erotic stimulus or sexual turn-on through Pavlovian conditioning. The specific stimuli that people respond to as sexually exciting differ from person to person, depending on each individual's prior sexual experiences. Some people learn to respond to fantasies or pictures of nude bodies as erotic stimuli, if these have been paired with effective sexual stimulation in the past. Others come to find "talking dirty" to be sexually arousing, if such talk has been a common part of their exciting sexual experiences. Although many people would be offended by such language and do not understand how anyone could find it sexually exciting, the mechanisms of Pavlovian conditioning allow us to understand that almost *any* stimulus can become an erotic stimulus or sexual turn-on, if it is frequently paired with and is predictive of the sexual response.

When people learn to imagine their erotic stimuli (even in the absence of any external stimuli), they gain fantasy access to these stimuli. Some people cultivate sexual fantasies because of the pleasure and sexual arousal that they bring. During sexual intercourse, fantasies can add to the total excitatory sexual stimulation and enhance the sexual response. Fantasies also serve to distract people from some of the stimuli that can inhibit the sexual response (e.g., guilt, fears about the adequacy of one's sexual performance, or fears of not being attractive enough).

Any of a variety of stimuli can become associated with sexual anxiety, depending on each individual's prior history of Pavlovian conditioning. For one person, fears of becoming pregnant can interfere with the sexual response; whereas another person has no fears of pregnancy. Given the pervasive social attention to physical attractiveness—especially for females—some people may fear removing their clothes to make love because they fear that their partner will not think that their body is attractive enough. A given individual's sexual fears can be understood only in terms of that individual's prior learning experiences.

E. Social and Cultural Influences

Each individual is strongly influenced by his or her location in the social matrix. People who grow up in homes that accept sexuality as a natural part of life are likely to receive little or no socialization that induces guilt or fear about sexuality; and they may learn many positive thoughts and activities that help them have a gratifying sexual life (with the minimum of problems). On the other hand, people who grow up in homes where sexuality is never discussed other than in ways that clearly associate it with sin, shame, or guilt may learn many negative associations with sexuality, creating a condition called erotophobia. Such people often feel sufficiently guilty about sex that they avoid sex education material and do not learn effective means of birth control. This can leave them very vulnerable to sexual accidents if their phobia about sex is overcome by curiosity about sex or pressure from another person to experiment with sex. While erotophobic people can become pregnant and contract sexually transmitted diseases as easily as other people, their lack of knowledge about sex leaves them more vulnerable to such mishaps since they have not planned ahead for sex and have less knowledge about ways to avoid the problems.

III. SEXUAL ORIENTATION

Although most people are attracted to and sexually aroused by people of the other sex, heterosexuality is not universal in humans or in other species. In almost all societies that have been studied, homosexuality has been found; and there is also evidence of bisexual and nonsexual people in many studies.

The diversity of sexual orientation has been viewed as reflecting the type of erotic stimuli that excite a person's sexual thoughts, feelings, and desires. Figure 1 shows that a person can have varying degrees of sexual responsiveness to other-sex and same-sex stimuli. People who have little or no sexual response to either same- or other-sex stimuli are labeled nonsexual. In a society that places as much emphasis on sexuality as does the United States, there can be negative associations with this word; and people sometimes feel uncomfortable with their lack of sexual interest or their choosing to be celibate at certain parts of life. However, some poeple never become interested in sexual activity; and others may pass through periods of their lives, such as after a divorce or the death

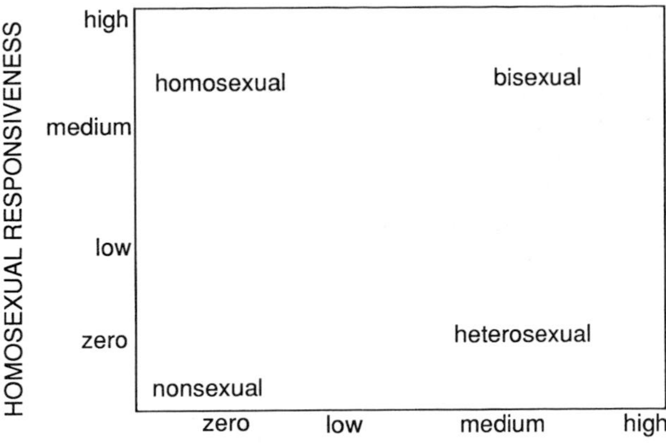

FIGURE 1 Four variations on sexual orientation (heterosexual, homosexual, bisexual, and nonsexual) are related to an individual's levels of hetero- and homosexual responsiveness—that is, desire and preference for hetero- and homosexual partners.

of a spouse, in which they have no sexual interest or desire and want to lead celibate lives.

Individuals whose primary erotic and amorous emotional interest is in members of the other sex, even though that interest may not be overtly expressed, are called heterosexuals. However, it should be clear from the figure that there is actually a range of variation within the heterosexual population: Some people respond to a few heterosexual stimuli as sexually interesting and arousing, whereas other people are attracted to and sexually excited by a broad range of heterosexual stimuli. Thus heterosexual sexual desire can range from low to medium to high, sometimes changing across an individual's lifetime, depending on many variables.

Individuals whose primary erotic and amorous emotional interest is in members of the same sex, even though that interest may not be overtly expressed, are called homosexuals. Again, it should be clear from the figure that there is actually a range of variation within the homosexual population: Some people respond to a few homosexual stimuli as sexually interesting and arousing, whereas other people are attracted to and sexually excited by a broad range of same-sex stimuli. Thus homosexual sexual desire can also range from low to medium to high, sometimes changing across an individual's lifetime, in response to many variables. A person can be a homosexual and never engage in sexual relations with a member of the same sex. However, many homosexuals can respond sexually to the other sex though they find that their emotional and psychological attachment is to the same sex.

People who respond erotically to both male and female stimuli and enjoy engaging in sex with both sexes are labeled bisexuals. Their level of sexual interest can range from low to medium to high—changing across their lifetimes. Not surprisingly, there is a great deal of variation within this category, and too few studies have been done to date to provide an accurate overview of the range of bisexual life-styles and activities.

Although most people believe that individuals are *either* heterosexual *or* homosexual, Fig. 1 reveals a much greater range of patterns of sexual orientation. In spite of the fact that there has been a great deal of research on the causes of various sexual orientations, there are at present no conclusive answers. It is possible that there are biological factors that predispose individuals to be heterosexual, homosexual, bisexual, or nonsexual. But environmental factors, personal life experiences, and various forms of learning, including Pavlovian conditioning, could also be part of the explanation. More research is needed on this topic.

IV. PATTERNS OF SEXUAL BEHAVIOR

The sexual reflex functions from the first days of life: Soon after birth, many boys have penile erections; and some girls experience vaginal lubrication and clitoral tumescence within the first day of birth. Even before birth, ultrasound studies suggest that erections may occur in baby boys. Sexual behavior, in the form of masturbation, often emerges in the first year. There is a considerable range of variation among children in early sexual behavior, depending in part on their parents' response to the behavior. Some parents punish and suppress early sexual behavior, whereas others allow considerable sexual exploration. Because of the hormonal influences of puberty, sexual stimulation becomes more pleasurable, and teenagers typically become increasingly interested in sexuality. During the teen and early adult years, the diversity of people's sexual activities tends to increase as they explore sexuality; but by middle or late adulthood, many people show a declining interest in and diversity of sexual activities.

A. Masturbation

During early infancy, babies explore their environments and their own bodies, touching and looking at many parts of their anatomy with as much interest as they have in the world around them. When they touch the genitals, the sexual response may be elicited, along with the pleasurable feelings associated with it. The pleasurable sensations reinforce touching the genitals, and young children may learn habits of engaging in genital stimulation. Of course, children who are swaddled, kept completely dressed most of the time, or punished for genital exploration may not learn to touch their genitals. Punishment may in fact condition inhibitions and fears about touching themselves. However, there are documented cases of babies learning to masturbate to orgasm before 1 year of age (though boys do not ejaculate any fluid before puberty).

Sometimes children, inquisitive to learn what other bodies look like, engage in mutual sexual exploration; and this can also lead to touching that elicits sexual responses. Such sexual exploration can lead to sexual interactions, including same-sex and/or other-sex exploration. Though parents may fear that same-sex play may lead to homosexuality, this is not usually the case.

In cultures where people sleep together in a common room or hut, children may see their parents making love and later imitate coital activities during play with other children. There is a great deal of variabilitiy among cultures in childhood sexuality, but usually less variation within a cultural group. In most Western industrial societies, childhood sexuality is usually suppressed, since punishment of sexual exploration is commonplace. In societies where parents do not inhibit their children, sexual exploration and play can continue throughout childhood: there is no biologically determined "latency period" for sexuality.

As genital stimulation and sexual arousal become more pleasurable during puberty, adolescents find it more rewarding to think about and/or to do. Boys are much more likely than girls to learn to masturbate to orgasm during the adolescent years, reflecting, in part, the stronger punishment and social inhibitions that are placed on girls.

B. Fantasy

Thoughts about sex are most likely to become erotic fantasies when they are paired with the sexual arousal induced by masturbation or coitus; but sexual thoughts can also become very exciting by being treated as a "forbidden fruit"—something that is exciting, mysterious, but not permitted before marriage. Fantasies of the forbidden fruit are tempting because of their prohibited status. Fantasies that are paired with sexual stimulation—masturbation or coitus—are exciting because they become conditioned erotic stimuli (through Pavlovian conditioning). Teens and adults who have masturbated to images of the other sex become sexually aroused by seeing people who somewhat resemble those images. The more that people masturbate to different images, the more erotic stimuli they learn to respond to; thus there are numerous stimuli that can elicit their sexual arousal. People who develop an extensive and exciting fantasy life respond to many stimuli as sexually interesting, which increases their desire for further sexual experience. In contrast, people with little masturbatory or coital experience often have few sexually exciting thoughts and low sexual desire, since little they see or hear leads to sexual arousal.

C. Erotic Touching and Foreplay

Most modern Western societies allow teenagers to explore intimate contact before marriage, though it has not always been this way. Because physical touch to the skin, lips, hair, and other body parts are all pleasurable, it is not surprising that couples enjoy discovering the sensations of touching, fondling, kissing, and hugging. In such circumstances, pressure and touch to the genital areas are quite likely; and the sexual pleasures of such contact reinforce the learning of activities that may be labeled as erotic touching or foreplay, depending on whether the touching precedes intercourse. The learning of these erotic forms of stimulation may be inhibited in societies that make young people feel guilty about doing such things; or they can be facilitated in societies that inundate their youth with pictures, movies, and videos of people doing sexual activities. Societal messages also influence the next step of sexual exploration. Societies with clear messages that sex must be reserved until marriage may suppress further sexual exploration until marriage (or at least engagement); but such prohibitions are not very strong in most Western, industrial societies.

D. Sexual Positions

As erotic touching becomes more intimate and clothing is shed, it is easy for most people to discover coitus. For young people who have been told (or shown in the media) that there are "certain best sexual positions," they would be very likely to explore those positions. Without cultural guidelines, patterns of sexual exploration are more random and unpredictable.

There are numerous sexual positions, and people may discover them through sexual exploration or by learning about them from books, friends, movies, therapists, or other sources. Erotophobes and people who have been made to feel guilty about sex may feel inhibited about exploring various sexual positions. However, less inhibited people often explore a variety of sexual positions and learn which ones feel good, which are unexciting, and which are painful. Because people commonly cease exploring and settle down into a few favorite positions after they have explored a range of possibilities, older people usually engage in less sexual exploration than do younger people. If married couples report that they are losing interest in sex because it has become too routine, sexual therapists and marriage counselors sometimes give couples books that show multiple sexual positions and encourage them to try novel ones. There are several good books, such as "Sexual Awareness" by McCarthy and McCarthy," "How to Make Love to the Same Person for the Rest of Your Life, and Still Love It" by O'Connor, and "The New Joy of Sex" by Comfort, that contain pictures and information on how to improve sexual enjoyment and pleasure.

V. SEX DIFFERENCES IN MOTIVATION FOR SEXUAL INTERACTION

In most cultures, males tend to have a greater sexual interest and motivation than females. From the perspective of sociobiology, it is possible to argue that these sex differences are innate, since males with high sexual interest and motivation can reproduce dozens if not hundreds of their own kind but females cannot. However, the difference could also arise from social–psychological factors. For example, women learn that careless sexual exploration leads to the loss of virginity and perhaps premarital pregnancy (which can hurt a woman's chances of marriage) and even sterility (induced by sexually transmitted diseases). These undesirable side effects of sexual behavior often condition fears and anxiety about sex that can inhibit female sexuality—including masturbation, fantasy, sexual desire, and sexual experimentation.

VI. SEXUALITY AND THE LIFE CYCLE

A. Premarital Sex

Many cultures have strict rules against premarital sex. Since most people throughout history have not had effective means of birth control, premarital sexual activity often led to pregnancy, and young unmarried mothers usually could not find marriage partners. Males often did not want to marry a pregnant woman or a woman with a child because the males wanted to devote their child-rearing efforts to raising their own children, not someone else's. In such cultures, virginity can be very important, and a man may give a woman back to her family if she is found (or suspected) not to be a virgin. Naturally, there is much variation among cultures, and some societies allow a certain amount of premarital sex, perhaps with a fiancé, just before marriage.

With the development of increasingly effective forms of birth control, the decreasing influence of strict religious codes, and other recent changes, many modern industrialized nations have experienced significant increases in premarital sexuality. In the United States, this trend has led to an increasing amount of teenage pregnancy and teenage abortion, in part because many parents and teachers have been reluctant to provide good birth control information to teenagers. In Western Europe and Canada, where rates of teenage premarital sexual activity are similar to those in the United States, the teen pregnancy and abortion rates are considerably lower. For example, in the United States, the pregnancy rate is 96 per 1000 15- to 19-year-olds, compared to 35 in Sweden and 14 in the Netherlands. In modern Western European countries, teenagers are given better sexual education and better access to birth control services than in the United States. There is also less of a traditional religious influence.

Although the prevalence of premarital sexual intercourse has gradually increased, the age of onset has gradually declined. For example, Alfred Kinsey, a pioneer sex researcher, found that half of the women he interviewed some 45 years ago had intercourse before marriage; but Morton Hunt found that 81% of 18- to 24-year-old married women had done so in the early 1970s. In 1971, only 27% of unmarried 17-year-old women had engaged in intercourse; but this percentage increased to 41% in 1976 and 61% in 1979. In the 1980s, this trend for increasingly early sexuality continued. In 1990, 67% of 17-year-old women had engaged in intercourse. Between 1970 and 1992, the typical age of marriage was postponed by about 4 years, shifting from 20.6 years to 24.4 years in females and 22.5 to 26.5 years in males. Thus, many young people are spending a longer portion of their lives as sexually active but unmarried. Some individuals spend this time with a small number of partners, whereas others have numerous partners.

B. Cohabitation

Although cohabitation is less common in the United States than in France or Sweden, increasing numbers of people are living with a partner without being married. In the United States, approximately 4% of unmarried adults are cohabiting, in contrast to 1% in 1960. Those who cohabit tend to be more liberal, more likely to use drugs, and less religious than people who do not cohabit. However, cohabitation is becoming increasingly routine before marriage, though there is no evidence that it helps people make better choices of marital partners. Most cohabiting unions do not lead to marriage, dissolving relatively quickly. There is preliminary evidence that this living arrangement may be one factor reducing the number of people marrying. People who cohabit before marriage have a higher risk of divorce than noncohabitants. This may result from their having a more skeptical view than noncohabitors have about the institution of marriage.

C. Marriage

Most married people report that sexual relations are most fulfilling in the first year after marriage and that the quality of their sexual experience declines thereafter. In their 20s and 30s, married couples typically engage in coitus two to three times a week. The frequency of intercourse usually declines gradually as people age. People also report a decline in communication, listening, respect, and romantic love after marriage. These changes are not inevitable, but they do indicate that many people do not devote much attention to making their marriages work. They also suggest that many people do not have the time, effort, or skill to solve their marital and sexual problems and/or that the multiple other demands of adult life—jobs, careers, children, socializing—are so compelling or rewarding that the marriage becomes neglected.

Couples commonly engage in noncoital sex play before penile–vaginal intercourse. Foreplay typically lasts some 5 to 15 min, though the time varies from couple to couple and from time to time for any given couple. Various stimulative techniques are used in foreplay, including touching, caressing, and kissing of the genitals, breasts, lips, and other body parts. Couples use a variety of positions for engaging in intercourse. While the male-on-top, face-to-face position is very common, couples are increasingly using the female-on-top position. Side-by-side positions and combinations of female superior and side-by-side positions are also used. Rear-entry positions are less commonly explored. Coitus usually lasts approximately 10 min.

Many couples engage in oral–genital stimulation. Ninety percent of people married less than 25 years have engaged in oral sex during the preceding year. Anal intercourse is not a common practice for most couples: for married people between 18 and 35, approximately 50% of men and 25% of women have engaged in anal intercourse on at least one occasion.

D. Extramarital Sex

Although marriage vows require faithfulness, studies reveal that a considerable number of people violate the traditional marital contract. A 1992 survey of a national sample revealed that 25% of married men and 15% of married women had had at least one extramarital relationship. However, the rates for younger women were higher than those of younger men. Recent magazine surveys of their readers indicate higher rates of extramarital sex, in one 60% of the women and 75% of men had extramarital sexual experience. There is historical and cross-cultural evidence that extramarital sex has been common at other times and in other societies. Although most individuals attempt to keep their extramarital affairs secret, they do not always succeed; and the discovery that there has been extramarital sex can be very damaging to the quality of a marriage. Such a major breach of trust may make it difficult for the cheated partner to completely trust the other again.

Some people recognize that secret affairs are common yet potentially deleterious to marriage and advocate "open marriage" as an alternative. Built on the assumption that most people want both a committed relationship and freedom, open marriage is designed to allow a couple to see other people and decide if this should include sexual relations or not. The incidence of open marriage is unknown. One study found that in 15% of marriages, both partners agreed that extramarital sex was acceptable under certain circumstances.

VII. SEXUAL DISORDERS

It has been estimated that approximately 50% of couples will have a major sexual disorder, such as erectile or orgasmic disorder, at some time during their lives.

Many people do not seek professional help or even read self-help books, and a substantial number of these people do not overcome their sexual problems. This is unfortunate, since modern sexual therapies and medical interventions can help people even after they have had a sexual problem for a decade or more. Though perhaps 25 to 35% of problems have an organic basis, most problems result from people having too little effective stimulation to elicit the sexual reflex and/or experiencing guilt, anxiety, feelings of inadequacy, or other negative emotions that suppress the sexual reflex. Therapy consists of diagnosis and treatment of any physical condition and/or providing correct sexual information, teaching effective sexual techniques, and resolving the negative emotions that suppress the sexual reflex.

A. Female Problems and Therapies

The most common female sexual problem is orgasmic disorder in which females regularly experience delays in reaching orgasm—or do not reach orgasm. This can be lifelong (if a woman has always had this disorder) or acquired (if a woman once did not have this problem). Situational orgasmic disorder indicates that a woman can have orgasms in certain situations but not others (e.g., during masturbation but not during coitus). Global orgasmic disorder means the problem occurs in all situations. Other common and closely related problems are low sexual desire, difficulty becoming excited, difficulty staying excited, and not having orgasms as often or as easily as desired. Although orgasmic disorder is usually not caused by organic factors, a variety of physical problems, such as neurological or pelvic disorders, can result in female orgasm disorder. The therapies for psychologically induced orgasmic problems focus on increasing communication, giving women and men useful knowledge about female sexuality, and teaching the sexual skills needed to assure that the female receives effective sexual stimulation. Therapies typically proceed in gradual steps from exercises in nonsexual body touching (called sensate focus) to gradually adding more sexual stimulation. Counseling also focuses on the specific types of myths, fears, anxieties, or guilts that a woman might have that are suppressing her sexual responsiveness and enjoyment.

Many women experience dyspareunia—or painful intercourse—from time to time; some experience it often. Dyspareunia can result from any of dozens of causes, including inadequate lubrication, overly vigorous and overly lengthy penile stimulation, vaginal infections (such as monilia, gardnerella, or trichomoniasis), pelvic infections, irritations or infections of the vulva, and thinning of the vaginal walls. Psychological factors can also result in painful intercourse, usually relationship problems. Only after a careful analysis of the specific causes that are relevant to a given individual's problem is a therapy arranged to solve that problem.

A less common female problem is vaginismus, in which the muscles surrounding the outer third of the vagina contract involuntarily and either completely prevent penile intromission or make it painful. This strong, involuntary muscular response can be due to any of a variety of aversive experiences related to sex, including a history of painful intercourse, past physical or sexual assault, a religious socialization stressing sexual guilt, and fear or dislike of her partner. The woman receives counseling to help her cope with the emotional causes of the problem, such as dealing with memories of her childhood sexual assault. To help her overcome the involuntary muscular contractions, therapists give the woman a small dilator and have her insert it in her vagina—in privacy at home—after she does a series of muscle relaxation exercises. Once she can successfully insert a small dilator, she is given slightly larger dilators with which to practice. Over multiple therapy sessions, as she becomes used to the exercises, she is given larger dilators until she can accept one the size of a penis. Once she has learned to keep her vaginal muscles relaxed with a large dilator, she is gradually reintroduced to sexual intercourse in ways that help her remain relaxed during early intromission.

B. Male Problems and Therapies

When a man cannot attain or sustain an erection firm enough for sexual activities and this causes him clear distress or interpersonal problems, he is said to have an erectile disorder. This problem is sometimes called impotence, but therapists prefer the less threatening and more accurate label of erectile disorder. This can be lifelong or acquired, global or situational. Common organic causes for erectile problems are diabetes and alcoholism. Other causes include spinal cord injuries, multiple sclerosis, and hormone deficiencies. Psychological causes can usually be traced to the presence of anxiety, guilt, or other negative emotions. The therapy for psychological causes parallels the therapy for female orgasmic problems: Therapists give the man and his partner information about male sexuality in order to overcome myths and anxieties people have about the male sexual response. The male and his partner are taught more effective means of providing sexual stimulation for him, and they receive counseling that can reduce whatever negative emotions may be suppressing the sexual reflex. For example, it is common for males to have "performance anxiety" during intercourse, since they know that the penis must be firm to attain intromission; therefore, therapists have developed various exercises to reduce performance anxiety. Sensate focus exercises are assigned as homework during the beginning of therapy to take the pressure off the male, since there is no need to perform sexually; and if he does have an erection, he is not allowed to use it. As therapy progresses, the clients are introduced to "teasing," in which the man's partner alternates between stimulating the penis to erection, then allowing the erection to disappear by ceasing stimulation. After having his erection appear then disappear on numerous occasions, the man begins to lose his performance anxiety, since he learns that the sexual reflex will occur quite automatically with adequate stimulation. As his performance anxiety declines, he has fewer problems with attaining erection.

Men who have delayed or absent orgasm that causes them clear personal distress or interpersonal problems are described as having male orgasmic disorder. This problem can be lifelong or acquired, global or situational. This disorder can result from organic causes, with drugs and neurological problems being the most common. However, most cases result from psychological factors, such as a very negative sexual socialization or traumatic sexual experiences. In addition to sensate focus exercises, the therapy for psychological problems uses stages of successive approximation that start with the man masturbating by himself to orgasm. Next, his partner joins him to watch and learn from him. After the man can successfully ejaculate in front of his partner and is comfortable with this, his partner learns to masturbate the man to ejaculation. Finally, the partner stimulates the man to the point of ejaculation, then inserts the penis into her vagina before ejaculation. After several repetitions of this, many men can overcome the problem. Counseling can help reduce any fears, guilt, or anxiety a man might have about ejaculating inside the partner's vagina.

The most common sexual problems for young men is early ejaculation, in which ejaculation regularly occurs after little sexual stimulation—be it before, at the moment of, or soon after intromission and before the man wants to ejaculate, causing clear personal distress or interpersonal problems. The disorder can be lifelong or acquired, global or situational. Early ejaculation can be characterized as a lack of voluntary control over the timing of ejaculation. This disorder is rarely caused by organic problems. It often emerges when a male learns sexual habits of attaining sexual stimulation very rapidly, causing the sexual reflex to progress to orgasm in a very brief period of time. There are several forms of therapy, which focus on slowing down the sexual reflex. The "stop–start" method teaches the man to alternate between starting sexual stimulation and stopping. During the stop phases, his sexual arousal level will decline a little; and by inserting enough stop periods, he will be able to extend his sexual response to a much longer duration. The "tease–squeeze" method consists of slowing a man's sexual response by applying a rather firm squeeze to the base of the glans or the base of the shaft of the penis. The squeeze reduces a man's sexual arousal and erection temporarily, then he can resume stimulation. These therapies are usually performed with a cooperative partner.

Dyspareunia—or painful intercourse—in men is not common; but infections or inflammations of the penis, urethra, prostate, or testes can be among the causes leading to painful intercourse. Psychological influences may also be involved. After careful diagnosis to determine the cause, specific therapies are designed.

BIBLIOGRAPHY

Baldwin, J. D., and Baldwin, J. I. (1997). Gender differences in sexual interest. *Arch. Sex. Behav.* **26**, 181–210.

Heiman, J., and LoPiccolo, J. (1988). "Becoming Orgasmic: A Sexual Growth Program for Women." Simon & Schuster, New York.

Hyde, J. S., and DeLamater, J. (1996). "Understanding Human Sexuality," 6th Ed. McGraw–Hill, New York.

Laumann, E. O., Gagnon, J. H., Michael, R. T., and Michaels, S. (1994). "The Social Organization of Sexuality." Univ. of Chicago Press, Chicago.

Leiblum, S. R., and Rosen, R. C. (eds.) (1988). "Sexual Desire Disorders." Guilford, New York.

Masters, W. H., Johnson, V. E., and Kolodny, R. C. (1995). "Human Sexuality," 5th Ed. HarperCollins, New York.

Masters, W. H., Johnson, V. E., and Kolodny, R. C. (1994). "Heterosexuality." HarperCollins, New York.

McCarthy, B., and McCarthy, E. (1989). "Female Sexual Awareness: Achieving Sexual Fulfillment." Caroll & Graf, New York.

McWhirter, D. P., Sanders, S. A., and Reinisch, J. M. (eds.) (1990). "Homosexuality/Heterosexuality. Concepts of Sexual Orientation." Oxford Univ. Press, New York.

Reiss, I. L., with Reiss, H. M. (1990). "An End to Shame: Shaping Our Next Sexual Revolution." Prometheus, Buffalo, New York.

Zilbergeld, B. (1992). "The New Male Sexuality." Bantam, New York.

Sexuality, Evolutionary Perspectives

MARY S. McDONALD PAVELKA
University of Calgary

GLOSSARY

Bipedalism Locomotion on two legs (as opposed to quadrepedalism, which is locomotion on four legs)

Climacteric Term used to refer to the "change of life" for women of approximately 50 years of age, as they make the transition from the reproductive to the postreproductive stage of the life course; also known as the menopause

Isosexual Sexual interactions between individuals of the same sex; sometimes used as an alternative to the term homosexual when referring to nonhuman animals

Natural selection Differential reproductive success of individuals, which leads to changes in the frequency of biological and behavioral traits in future generations

Primates Order of mammals that includes prosimians, monkeys, apes, and humans

Proximate mechanism Immediate trigger or motivation for an individual's behavior

Sexual selection Theory proposed by Charles Darwin to explain features of sexual dimorphism, such as the peacock's tail, which gives individuals increased reproductive success through increased access to mates

Ultimate or evolutionary explanation Explanation of the selective forces that may have been operating on a behavior in the evolutionary past

UNDERLYING THE GREAT DIVERSITY OF HUMAN beliefs and behaviors in the area of sex and sexuality are evolved biological and behavioral potentials, constraints, and hormonal regulators. *Homo sapiens* are members of the order Primates, and with our primate relatives we share certain basic features of primate sexuality. Several elements of human sexuality are actually species-specific manifestations of primatewide biological and behavioral traits. Others appear to be unique to our species. We need to understand the variation both between and within species in order to gain a fuller appreciation for the expressions of sexuality in ourselves. Some aspects of human sexuality that directly lend themselves to the insights of an evolutionary and cross-species perspective are menopause, estrus and the hormonal regulation of sexual behavior, homosexual behavior, female orgasm, and the role of aggression in sexuality.

I. MENOPAUSE

The first feature of human sexuality that needs to be understood in evolutionary terms is the uniquely human form of reproductive senescence characteristic of human females: the menopause. Though often regarded as a medical disorder, the menopause is actually a salient feature in human female life history, one that is a fundamental part of women's sexuality. The term menopause refers to the permanent cessation of menstruation and menstrual cycles that occurs in women who are usually around the age of 50.

Both oocyte depletion and failure of the ovaries to respond to the pituitary hormones are germane to the cessation of reproductive function in women. The number of potential egg cells available is fixed at birth and this number steadily declines thereafter. The end of reproductive ability is a result of the depletion of potential ova for fertilization. The menstrual cycle, including monthly periods of menstrual bleeding, ceases secondarily as a result of the reduction in estro-

ENCYCLOPEDIA OF HUMAN BIOLOGY, Second Edition, VOLUME 7. Copyright © 1997 by Academic Press. All rights of reproduction in any form reserved.

gen and progesterone. The situation for men is quite different, since they produce new sperm throughout life, and many are able to reproduce into extreme old age. [*See* Menopause.]

Human female menopause must be regarded not only from the physiological and hormonal point of view, but also from a life history perspective that brings to light some rather striking aspects of this condition. As a life history characteristic of human females, menopause is universal, it occurs halfway through the maximum life span of the species, and it consistently occurs at approximately age 50 in different populations. Menopause is an inevitable and unavoidable event in the lives of all women, healthy or otherwise, who live into their 50s, and all women in industrialized societies who live to average life expectancy will spend a significant proportion of their lifetimes in a postreproductive and somewhat altered hormonal state. Even societies with life expectancy at birth values of less than 50 will have many individuals who live into their seventies, eighties, and nineties.

There are known cases of very old captive nonhuman primate females who have ceased all reproductive capability in the years preceding death. In their clinical profiles this handful of animals share certain similarities with postmenopausal women. However, these few individuals are actually the exceptions that prove the rule from an evolutionary perspective: nothing like the cessation of reproductive function in *all* females *midway through the species life span* has been observed for any species except humans. Reproductive senescence in nonhuman primates is highly individual and is characteristic of extreme old age. Menopause in women is universal and is characteristic of middle age.

Why might natural selection have favored the cessation of reproduction in human females only halfway through the maximum life span of the species? Animals generally reproduce until the end of their life spans, and longevity and reproductive success are both theoretically and empirically correlated. All other things being equal, longer-lived individuals should contribute more offspring to future generations, and empirical investigations of lifetime reproductive success have shown this to be true for a variety of animal species. Throughout the primates there is a progressive increase in longevity from the prosimians through the hominoids, and for all species but *Homo sapiens,* the female egg supply and reproductive function are maintained relative to the species life span.

If the ancestor of early hominids, like chimpanzees, had an oocyte supply that would last 50 years, then in the course of human evolution the hominid life span became longer while the oocyte supply remained the same. The failure of the oocyte supply to keep up is most commonly explained by the adaptive "grandmother hypothesis." Post-menopausal females may enhance their fitness by acting as grandmothers, or surrogate caretakers, for their children's offspring. In this case, selection would favor those women who were not conceiving and were therefore able to help ensure the survival and subsequent reproduction of descendants. An alternative explanation is the nonselectionist pleiotropy perspective, which views menopause as a by-product of some other hominid development. Applying pleiotropy theory to menopause, one would assume that the human female system of egg production and storage is highly adaptive early in life, but that inherent limits in the system (degeneration of the egg supply suspended in anaphase) will express themselves later in life. Possibly a 50-year supply of healthy eggs is a limit in the basic structure of all mammal female reproduction. Few mammals have ever been observed to reproduce much past 50 years. [*See* Primates.]

II. ESTRUS AND THE HORMONAL REGULATION OF SEXUAL BEHAVIOR

Sexual activity in nonhuman primates is constrained by *estrus,* a period of physiological change during which ovulation and mating occur. The external physical indicators of estrus vary widely across the primate order, from extreme redness and swelling of the sexual skin in some species to virtually no physical cues in others, and it is thus best defined behaviorally as the period during which a female shows willingness and/or motivation to mate. In a 20-year life span, most primate females are likely to spend less than 20 weeks during which they engage in sexual activity.

Human females exhibit no obvious period of estrus. They can and do engage in sexual behavior throughout the menstrual cycle, throughout the year, through pregnancy and lactation, and in the postreproductive portion of the life span. Although there may be some subtle changes in behavior and pheromones at different points in the menstrual cycle of women (see the following), there are no obvious external changes to mark ovulation—no physical or behavioral changes directly comparable to the estrus period of most nonhuman primates. There are *cultural* taboos in some societies about interactions with women at certain times during the monthly cycle, or before and after parturition, but human societies are not characterized by regular hormonally based behavioral cycles involving periods of sexual activity alternating with periods of complete sexual dormancy.

Thus sexual activity is less periodic and circumscribed in women than it is in other primate females. Three labels have been applied to this situation in women: human females are said to have lost estrus, evolved concealed ovulation, or become continually sexually receptive. Loss of estrus, as described earlier, refers to the absence of clearly demarcated periods of physiological and behavioral change associated with sexual activity. Human females are generally said to show little or no cyclicity in sexual motivation and undergo no external changes such as swelling and reddening of the skin.

Concealed ovulation relates to loss of estrus in that the external physical and behavioral changes characteristic of estrus indicate internal endocrine events. The degree to which these external changes correlate with internal events (specifically ovulation) is variable, but female nonhuman primates mate only during estrus and get pregnant at this time, so ovulation must occur during estrus. In fact, estrus is widely believed to be a form of advertisement to let males know that a female is ready to conceive. Human females lack external signals that ovulation is occurring: *ovulation in human females is concealed.* Indeed, it is concealed not only from others, but from the woman herself. Great inter- and intraindividual variation in cycle length contributes to the need for rather sophisticated methods of detecting ovulation, attesting to the fact that, for most women, ovulation is indeed concealed.

The term continual receptivity is best understood in historical terms. Until the late 1970s, female seuxality was described and measured in entirely passive terms. Female mammals were *attractive* depending on how many invitations from males they received, and *receptive* depending on how many of these invitations they accepted. Estrus, the period in which females engage in sexual behavior, was conceived of in this passive terminology as the period in which females would *accept* the sexual initiation of males. Estrus was (and often still is) behaviorally defined as that period in which female nonhuman primates are sexually receptive. Though obviously human females do not continuously accept male advances, the situation for human females, who do not have estrus periods, came to be known as continual receptivity.

If nonhuman primate females have estrus periods that circumscribe mating activity and human females do not, then we can safely say that, at some point in the evolution of *Homo sapiens,* estrus disappeared. Most reconstructions of the evolution of the hominid line consider loss of estrus to be one of if not *the* critical adaptive change essential to hominid evolution. The most widespread and popular set of explanations for the evolution of loss of estrus see this change in female sexuality as being at the heart of the male–female pair bond—the "continual" receptivity of the human female is associated with the continuous rather than periodic nature of the human male–female bond. (Critics point out that there are a number of nonhuman primate species with monogamous mating systems involving stable, permanent male–female pair bonds, and none of these have lost estrus.) Another set of explanations for the loss of estrus involve the notion that the presence of females in estrus leads to intense competition and fighting among the males who desired them—the association between females in estrus and increases in male aggression (see Section V) has been cited as evidence in favor of this explanation. Thus it has been argued that the loss of estrus evolved to reduce male competition and aggression and to promote more cooperation and cohesion in the social group. Finally, it has been argued that concealed ovulation evolved because natural selection favored women who were not in tune with their ovulatory cycles because these women had more babies.

These distinctions between human female sexuality and that of the other primates include the directly related notion that human female sexuality is no longer hormonally regulated, whereas the sexuality of the other primates continues to be under rather strict hormonal control. In fact all of these distinctions may be somewhat overblown, and the line between human and nonhuman primate sexuality is considerably more "blurry." First, there is good evidence for hormonal regulation of sexual activity in humans, such as subtle behavioral fluctuations in women that correspond with the hormonal changes in the menstrual cycle. Second, nonhuman primates, despite periodicity in interest, are themselves relatively emancipated from direct hormonal control of sexual behavior. A variety of social factors exert an overriding influence on the sexual activity of female nonhuman primates. Finally, in many nonhuman primate species, ovulation is concealed. Some species show no physical signs of estrus, and in most that do, the correspondence of external signs to internal endocrine events appears to be poor.

III. HOMOSEXUAL BEHAVIOR

In the context of discussions about loss of estrus, it is often argued that men and women engage in sexual activity far more than they need to from a reproductive point of view. Reference to the nonreproductive nature of much of human sexuality is often used as a point of contrast between humans and other animals. Since much of human sexuality is far removed from reproduction, the evolutionary perspective that sheds light on nonhuman sexuality is sometimes felt to be inappropriate to understanding human sexuality. Again, the clarity of the distinction between human and nonhuman primates tends to be exaggerated. Nonhuman primates also engage in sexual activity far more than they need to from a reproductive point of view and thus much of their sexuality is nonreproductive. Nonreproductive sexual behavior occurs in most primate species and includes sexual interactions not close to ovulation, sexual interactions after conception, and sexual interactions with same-sexed others, which will be discussed further here.

For both political and theoretical reasons, homosexual behavior outside of the human species has received much less attention than it should have, given that sexual interactions with same-sexed others appear to be part of the normal behavioral repertoire of most animals. It is important to note that we do not currently possess methods that will allow us to access the cognitive or experiential aspects of any of the behavior of any nonhuman, and therefore, in this context, we must limit our discussion to homosexual *behavior* of nonhuman primates, and not primate homosexuality. Across the order primates, homosexual (or isosexual) behavior has been observed in every major taxa with the exception of prosimians. Specifically it has been reported for several species of New World monkeys, Old World monkeys, lesser apes, and in all four great apes species. These reports come from both captivity and the wild, thus dispelling the notion that this is an abnormality produced by the conditions of captivity. In several species, homosexual behavior has been observed only in free-ranging animals, and not in their captive counterparts.

Sexual interactions between individuals of the same sex appear to be as elaborate and enduring as are those between opposite sex individuals. For example, Japanese monkeys, like many other primate species, exhibit a pattern of sociosexual behavior known as consorts. During the mating season, females who are in estrus form consort relationships with males that are quite unlike pair-bonds involving nonestrus females or pair-bonds occurring outside of the mating season. Male–female consorts involve an exclusive attachment that is sexual in nature but that affects many other social patterns. For a period of time, from a few days to a few weeks, the consort pair travel together, eat together, sleep together, support each other in conflict situations, and engage in repeated copulations. Female–female consorts involve the same kind of intense and enduring bond.

Same-sex sexual interactions appear to be *more varied* in the forms of sexual contact than are opposite-sex ones. Sexual interactions between males and females (i.e., heterosexual behavior) in nonhuman primates are likely to involve primarily genital–genital contact. In the many species in which same-sex sexual behavior occurs in nonhuman primates, it has been observed to include genital–genital, oral–genital, anal–genital, anal–anal, and manual–genital contact. Overall, the sexual repertoires of primates engaging in homosexual interaction appear to be highly flexible and variable.

Homosexual behavior has been reported for both males and females in the nonhuman primates. In most species, male–male and female–female sexual interactions occur with similar frequency, however in some one is more common than the other. For example, in Japanese macaques, female–female sexuality is much more common than is male–male. In mountain gorillas the reverse is true.

A common question raised with respect to homosexual behavior in nonhuman primates is: Do some animals engage in same-sex sexual interactions all of the time, or do all or most animals engage in same-sex interactions some of the time? The answer to this question appears to be that some or most animals engage in same-sex sexual interactions some of the time. Sexual interactions with same-sex others appear to be part of the normal sexual repertoire of all animals, expressed variously over the lifetime of an individual. It ranges from totally absent in some species (howlers and barbary macaques) to levels that match and surpass heterosexual behavior (several species of macaques, langurs, and bonobos). Homosexual behavior occurs rarely to occasionally in most species, but most sexual interaction is of a heterosexual nature.

The evolution of homosexual behavior in humans and other animals can be understood in one of two ways. Some seek to identify direct selective advantages to engaging in this behavior. For example, it has been suggested that female–female homosexuality might have evolved as a form of female–female mate competition in that it serves to prevent the other females from mating with males and conceiving offspring. Others do not attempt to find adaptive advantages to homosexual activity per se, but rather treat homosexual behavior as part of the diverse sexual repertoire of the primates, and focus instead on the need to find evolutionary explanations for the variation, diversity, and plasticity in primate sexual expression.

IV. FEMALE ORGASM

Pleasure has been recognized as a critical element to understanding the motivations of human and nonhuman primates with regard to sexuality. Central to this inquiry is the debate over whether or not female orgasm is an exclusively human experience. The behaviors of male nonhuman primates during ejaculation (muscle rigidity, perianal muscle contractions, a fixed stare) are generally accepted as sufficient evidence that ejaculation is accompanied by a pleasurable sensation for males as we know it usually is in men. Demonstrating the presence of orgasm in nonhuman primate females is less straightforward. There is evidence that female nonhuman primates experience uterine contractions and heart rate increases during sexual interactions that involve genital contact and several researchers have also described characteristic facial expressions and movements observed in females during sexual encounters. But since orgasm is an internal experience and sensation, we cannot ever expect to determine with certainty if our nonlinguistic relatives share it. This evidence—that nonhuman primate females display some of the same physical expressions that are known to be associated with orgasm in women—is the only empirical information available for making the decision about the uniqueness of the human female orgasm.

Perhaps because of the indirect and inconclusive nature of the empirical data, much of the debate over female orgasm is theoretical. One argument for the perspective that female orgasm is uniquely human is that it evolved as part of the continuous receptivity of hominids. On the other side of the debate is the view that orgasm is not unique to women and that it evolved in primates as a proximate mechanism to entice females to engage in behavior that confuses paternity.

An interesting twist on this issue is the possibility that female orgasm during intercourse is actually an ancient primate characteristic that has been partially *lost* in the course of human evolution. This is based on the genital anatomy of monkey, ape, and human females: in monkeys and apes, the clitoris actually lies at or near the base of the vagina, a location that would ensure direct and regular stimulation during intercourse. In human females the clitoris is situated away from the base of the vagina and this results in the indirect and often insufficient stimulation that is widely reported to result in a low frequency of orgasm during intercourse for women. The movement of the clitoris away from the vagina was a by-product of the changes that took place in the skeleton during the evolution of bipedalism. Hominid females had to give birth to increasingly large-brained babies through a birth canal significantly narrowed by the reorganization of the pelvis and sacrum. Selection then favored the movement of the urinary meatus away from the vagina to protect it from trauma during childbirth, and the placement of the clitoris away from the vagina in women is a by-product of this move.

V. SEX AND AGGRESSION

In some aspects of human sexuality, the sex-as-pleasure perspective seems remote, such as when considering the association between sex and aggression. From the violent rape by a stranger, to acquaintance rape, to wife beating, to sadomasochistic sexual practices, to the widespread use of degradation and violence in pornography, it is clear that there is a strong link between two behavior patterns that at many levels seem distinct and incongruous.

The most widely accepted interpretations of sexual violence in humans are those that explain this behavioral connection *in exclusively human terms,* linking the behavior to aspects of human society and culture. The most widely accepted is the feminist perspective, which dissociates sex from violence and interprets sexual aggression as primarily an expression of, or reflection of, male desire to dominate and control women and their sexuality. Rape, as a specific and extreme form of sexual aggression, is not considered to be a sexual act, but a primarily aggressive and violent one that has more in common with nonsexual assaults than it does with consensual sex. The connection between sexuality and aggression in feminist perspectives on rape is seen to be a learned connection and a function of male-dominated patriarchal human society. The most popular alternative to the feminist theories are the social learning theories of rape, which actually overlap considerably with the feminist perspective. Social learning theories essentially see male sexual aggression toward women as stemming from the desensitizing effects of pornography, objectification of women in the mass media, and the acceptance and perpetuation of rape myths (e.g., that women secretly want to be raped). The two have in common the assumption that the connection between sex and aggression is learned and that it is a function of very specific elements of human social life.

If the connection between sex and aggression is a purely learned one that is grounded in patriarchal human society, then a strong connection between the two would not be expected in nonhuman primate society. However, male aggression against females is widespread among primates, and much of it occurs in the context of mating. This has long been recognized by individual researchers, and the primatological literature is filled with reference to male aggression against females and to the substantial increase in aggressive male attacks received by females in estrus. Species in which male sexual aggression against females has been observed include spider monkeys, baboons, macaques, chimpanzees, and mountain gorillas. Forced or attempted forced copulation with an obviously resistant female has not been observed for most nonhuman primates but it is known to occur in orangutans.

The recognition that sexual aggression occurs in several related species of primates has led to the suggestion that an evolutionary perspective is required for a full understanding of the phenomenon. The evolutionary perspectives that have been offered for the connection between sex and aggression are constructed within the framework of Darwinian sexual selection theory and suggest that male sexual aggression is a male reproductive strategy: under some circumstances, some males may be using sexual coercion to increase their access to mates.

VI. THE IMPORTANCE OF THE EVOLUTIONARY PERSPECTIVE

There is strong resistance in the scientific community and among the public to consider any evolved component to sexual aggression and rape more specifically, largely out of fear of the implications of a biologically deterministic view. It is important to realize, however, that the evolutionary perspective is not biologically deterministic, and this includes the evolutionary perspective on male aggression. Although some scientists (and most of the popular media) are guilty of sensationalism and biological reductionism, *biological or genetic determinism is not the basis of an evolutionary perspective.* This misunderstanding of the value of an evolutionary perspective is widespread, despite painstaking efforts to emphasize that the expression of behaviors is obviously highly varied and that environmental factors play a substantial role in most if not all behavioral expression. Critics of the evolutionary perspective on male aggression against women have tended to depict it as an all or none proposition that ignores the learning, situational, and social structural factors. In fact, most evolutionary theorists have emphasized these. A full understanding of the connection between sex and aggression in *human* primates must take into account and attempt to understand the basis of the connection between sex and aggression in primates in general and then the conditions of its specific manifestation in humans. The argument that male sexual aggression may have evolved as a male reproductive strategy includes the prediction that the manifestation of sexual aggression will differ in different groups, as a host of circumstances, such as successful female strategies, are taken into account. Undoubtedly the feminist and social learning theories contain important insights into the wide variation in the expression of this behavior in humans. The cultural and social structural factors to which they point may directly affect whether, and under what circumstances, this primate behavior pattern is likely to manifest itself. And most importantly, no one, and certainly not a knowledgeable evolutionary theorist, would argue that men (and women) are not responsible for their choices and their behavior. [*See* Sexual Behavior.]

Understanding any behavior pattern in the cross-species context provides an important source of information for those interested in the manifestation of the behavior in any one species. As biological animals, human beings have certain evolved potentials, limitations, and tendencies affecting our behavior, and a full understanding of human sexuality cannot be gained from an exclusively cultural point of view. In spite of misunderstanding and misapplication of evolutionary explanations, in this area of human behavior perhaps more than any other, the cross-species and evolutionary perspective is critical.

Biological approaches to understanding behavior have come a long way since the days when behavior was conceived of as *either* instinctive (rigidly programmed, inescapable, or inevitable) *or* purely learned. Even the sociobiological search for adaptive explanations to account for the evolutionary history of a behavior—an exercise that dominated the field of primate behavior from the mid-1970s to the mid-1980s—has been replaced with a recognition of the great range of expression in behavior and the need to understand the immediate physiological and social conditions under which behaviors occur. Animals have become "adaptive decision-makers" with individuals behaving in accordance with a host of complex variables, including the individual personality and choices of the actor. The cross-species evolutionary approach to human behavior is not one that discounts the obvious roles played by language, culture, ideology, and self-reflection in human behavioral expression.

In terms of human sexuality, we have much to gain from an appreciation of our evolved primate tendencies, whether these be shared or uniquely human. Shared mammalian hormonal fluctuations of the menstrual cycle may influence the sexual behavior of individuals. Menopause renders women unable to reproduce halfway through the maximum life span, an evolved constraint that is unique to our species. Exclusive heterosexuality in humans is clearly not an evolved limitation or constraint. The cross-species perspective reveals that sexual expression in primates is flexible and diverse, including same-sex sexuality as part of the normal repertoire of behavior for most animals. Sexual expression beyond the demands of reproduction, including sexual activity during pregnancy and lactation, is also the norm in many primate societies. Our physiological capacity to experience orgasm (both male and female) is probably shared by the monkeys and apes, although the experiential components that are tied to language and conceptual thought are most likely uniquely human. The human female *difficulty* in experiencing orgasm during intercourse may also be evolved—as a by-product of rearrangements in genital anatomy during the evolution of bipedalism. The cross-species capacity for orgasm, extensive nonreproductive sexuality, and masturbation in primates all support the interpretation that sexual pleasure is an evolved feature of primate sexuality. On the darker side, the association between sex and violence that is cause for much concern in human social life is also not uniquely ours. An understanding of the underlying biological connection between sex and aggression will be necessary if we are to fully understand—and influence—the expression of this behavior in humans.

BIBLIOGRAPHY

Abramson, P. R., and Pinkerton, S. D. (eds.) (1995). "Sexual Nature/Sexual Culture." Univ. of Chicago Press, Chicago.

Buss, D. (1994). "The Evolution of Desire: Strategies of Human Mating." Basic Books, New York.

Ellis, L. (1989). "Theories of Rape: Inquiries into the Causes of Sexual Aggression." Hemisphere Publishing, New York.

Pavelka, M. S. M. (1995). Sexual nature: What can we learn from a cross-species perspective? *In* "Sexual Nature/Sexual Culture" (P. R. Abramson and S. D. Pinkerton, eds.), pp. 17–36. Univ. of Chicago Press, Chicago.

Pavelka, M. S. M., and Fedigan, L. M. (1991). Menopause: A comparative life history perspective. *Yearbook Phys. Anthropol.* **34**, 13–38.

Smuts, B. (1992). Male aggression against women: An evolutionary perspective. *Hum. Nature* **3**(1), 1–44.

Smuts, B., and Smuts, R. (1993). Male aggression and sexual coercion of females in nonhuman primates and other mammals: Evidence and theoretical implications. *In* "Advances in the Study of Behavior" (P. J. B. Slater, M. Milinski, J. S. Rosenblatt, and C. T. Snowdon, eds.), Vol. 22, pp. 1–61. Academic Press, San Diego.

Vasey, P. L. (1995). Homosexual behavior in primates: A review of evidence and theory. *Int. J. Primatol.* **16**(2), 173–199.

Wallen, K. (1995). The evolution of female sexual desire. *In* "Sexual Nature/Sexual Culture" (P. R. Abramson and S. D. Pinkerton, eds.), pp. 37–56. Univ. of Chicago Press, Chicago.

Wolfe, L. D. (1991). Evolution and female primate sexual behavior. *In* "Understanding Behavior: What Primate Studies Tell Us About Human Behavior" (J. D. Loy and C. B. Peters, eds.), pp. 121–151. Oxford Univ. Press, New York.

Sexually Transmitted Diseases

WILLARD CATES, JR.
Family Health International

GLOSSARY

AIDS Acquired immunodeficiency syndrome is a condition of immunological deficiency that is associated with infection of the cells of the immune system with the retrovirus HIV. AIDS is usually recognized by the presence of a life-threatening infection (as pneumonia or candidiasis) or of Kaposi's sarcoma in individuals under 60 years of age who have not been subjected to immunosuppressive drugs or an immunosuppressive disease

HIV Human immunodeficiency virus is a retrovirus causing the underlying immunodeficiency in AIDS

NGU Nongonococcal urethritis is not caused by the gonococcus

STD Sexually transmitted diseases are infections spread from person-to-person through intimate human sexual contact

SEXUALLY TRANSMITTED DISEASES ARE INFECTIONS spread from person-to-person through intimate human sexual contact. Few fields of human biology and public health are more dynamic than that of STD control. During the past two decades, this discipline has evolved from one emphasizing the traditional venereal diseases of gonorrhea and syphilis, to one concerned with the bacterial and viral syndromes associated with *Chlamydia trachomatis,* herpes simplex virus (HSV), and human papillomavirus (HPV), to one preoccupied with the fatal systemic infections caused by human immunodeficiency virus (HIV).

This expanded spectrum of infections and new responsibilities has been a two-edged sword. On one hand, it stretched limited resources allocated to prevention/control of STD; however, it also accelerated the acceptance of constructive new approaches to reducing STD transmission among high-risk groups. This article discusses (1) reasons why we are so concerned with STD, (2) general factors affecting STD control, (3) trends in sexually transmitted infections, (4) prevention and control strategies for STD, and (5) future directions in STD.

I. WHY STD ARE SO IMPORTANT NOW

Except for HIV, many STD infections have been around for ages, but they have achieved recent prominence as STD for several reasons:

1. Laboratory diagnostic techniques improved. New diagnostic approaches facilitated epidemic investigations, elucidating the extent, method of transmission, and clinical consequences of STD.
2. The population-at-risk for STD rose. The number of young adults in Western countries increased faster than the total population during the 1960s and 1970s because of the aging of the baby boom generation. Moreover, this cohort was more sexually active than their predecessors.
3. The composition of the STD "core" population changed. During the 1970s, many homosexual men exercised greater sexual liberties than in previous eras. In the 1980s and 1990s, the influence of illicit drugs, especially crack cocaine, has expanded the number of persons exchanging sex-for-drugs.

ENCYCLOPEDIA OF HUMAN BIOLOGY, Second Edition, VOLUME 7. Copyright © 1997 by Academic Press. All rights of reproduction in any form reserved.

4. The incidence of the newer STD increased. For example, in the United States, sexually transmitted viral infections are increasing; estimates of symptomatic genital herpes and genital HPV range up to 1 million new cases annually.

5. A higher proportion of infections with multiple modes of transmission are being transmitted sexually. As sexual behavior has changed, hepatitis A and B viruses, cytomegalovirus, and some enteric pathogens have become frequent sexually transmitted agents, especially among those engaging in anal sex.

6. The STD have been associated with incurable and fatal conditions. The acquired immunodeficiency syndrome (AIDS), HPV-associated genital cancers, and chronic recurrent genital herpes have captured public attention.

7. The key impact of STD on maternal and child health is now apparent. Estimates of the cost of pelvic infection and its sequelae tubal infertility and ectopic pregnancy account for over $2 billion annually in the United States. Those interested in improving reproductive health realize their programs must include activities to prevent and control STD.

8. International travel disseminates STD into a global problem. The rapid spread of HIV and the progressive growth of plasmid-mediated gonococcal resistance are tragic examples of modern STD that have been widely transmitted by world travelers.

II. KEY FACTORS AFFECTING PUBLIC HEALTH

A. Population at Risk

During the past three decades, various sociosexual changes in the United States dramatically influenced the number of those in danger of transmitting or acquiring STD. The size of the sexually active population at risk for STD peaked in 1985, and the absolute numbers declined thereafter. Thus, times are opportune to have an impact on STD trends. In 1970, an estimated 50 million sexually experienced persons were between the ages of 15 and 34 years; in 1985, that number was 69 million; by 1995, it will decline to 64 million.

Several factors have influenced this population at risk: (1) the baby boomers are coming to, and passing through, the most active sexual years; (2) the percentage of young persons who were sexually experienced increased during the 1970s but stabilized in the 1980s; and (3) sexual behaviors of specific high-risk STD "core" populations—primarily homosexual men and those using illicit drugs—have varied. In the 1980s and 1990s, public perceptions of the AIDS and herpes risks apparently led many Americans to change their sexual practices to reduce exposure to STD. These behavioral changes have measurably influenced STD trends among homosexual males, but less impact has been observed in teenage and low-income, inner city, minority heterosexual populations.

B. Governmental Responsibilities for STD Control

National, regional, and local tiers of government have different responsibilities for STD control. At the federal level, in the United States, the Centers for Disease Control and Prevention (CDC) coordinates developing and implementing STD control strategies and undertakes epidemiologic research. The National Institutes of Health (NIH) supports both basic science and applied clinical investigations.

State health departments have the statutory authority for control of communicable diseases, including STD. States (and the largest metropolitan areas) receive federal project grants (number 63 in 1997) from CDC for STD control activities, including disease reporting and program evaluation. Local health departments are charged with providing direct clinical services, which include diagnosis, treatment, patient counseling, and sex partner notification activities.

These differences require federal, state, and local health officials to cooperate closely to ensure an integrated STD program. Crucial factors include (1) identification of local priorities based on well-defined epidemiologic indicators and (2) application of control strategies with the greatest potential for preventing disease. Moreover, each of the three health tiers have increasingly created a matrix of interdepartmental activities that bear on preventing STD—school health programs have required collaboration with education officials, drug abuse programs with law enforcement officials, and so on.

C. Resources for STD Control

Paradoxically, even with recent funding increases, the federal budget allocated to preventing STD has not kept pace with the worsening problem. After adjusting for effects of inflation, the peak year for funding STD control programs was 1947. In that year, over $130 million (in 1997 dollars) was focused on the control of a single STD, syphilis. By 1973, even with the boost of a national gonorrhea control program, *total* federal grant resources for these two diseases was $64 million. By 1997, although this aggregate amount had increased to $110 million, it was spread across a full spectrum of over 20 STD.

Although federal resources directed against AIDS increased markedly from 1985 to 1990, major supplementation of categorical funds for the other STD had much slower growth. Thus, to gain overall leverage on the range of sexually transmitted infections, several simultaneous approaches are necessary: (1) lessons learned in primary prevention of HIV transmission must be rapidly applied to other STD; (2) resources from other governmental programs, namely, education, mental health, family planning, and maternal and child health, must be marshalled to complement STD categorical funds; and (3) the private medical community must see prevention and control of STD as a basic part of primary care.

D. Facilities Providing STD Services

Many perceive STD control efforts as concentrated in the network of the approximately 3500 public health clinics throughout the country; however, an equal number of patients seeking treatment for STD apparently visit the private sector as visit the public health facilities. Approximately 5 million persons are seen each year at public STD clinics for conditions such as nongonococcal urethritis, gonorrhea, genital herpes, trichomoniasis, and genital warts. A similar number of persons visit private physicians for these same complaints. This latter estimate may be low since it does not include visits to hospital-based outpatient facilities.

The ability of STD public sector clinics to provide adequate medical care was severely challenged in the 1980s. The rising number of syphilis and chlamydial infections, accompanied by declining resources available for STD care at the local level, led to long waiting periods and patients going untreated. With the evolving health system in the 1990s, managed care organizations will probably play a greater role in diagnosing and treating STD. Whether they will be able to provide ancillary STD control services remains to be seen.

E. Public Interest in STD

Because the public's perception of the STD problem influences both policy makers and program planners, it has a strong impact on the resources available for STD control. However, the public's interest in STD has varied. It has ignored (or denied) the importance of these infections in some years, and then has overreacted in others. For example, the news media first focused on the problem of genital herpes in 1982, nearly 10 years after the initial rise in number of cases. A media lag of about 1 year occurred with AIDS. Currently, STD, and especially AIDS, make headlines frequently.

Different interest groups also affect policy. One private voluntary organization, the American Social Health Association, has as its mission to increase attention on preventing STD. Other organizations are now recognizing the necessity of controlling STD at both the national level and in their own communities. Recent networking among private health-interest organizations, AIDS-service groups, community-based foundations serving minorities, and the federal/state/local government will help further increase public interest in STD.

III. TRENDS IN SEXUALLY TRANSMITTED INFECTIONS

A. Persistent Viral Infections

1. Human Immunodeficiency Virus

The epidemiology of HIV—and its fatal sequelae AIDS—is well known. In fact, even before the virus was discovered, epidemiologic analysis of those with AIDS allowed development of landmark AIDS prevention guidelines in March 1983, which are still relevant today. Risky behaviors had been identified, a virus was felt to be the causative agent, routes of transmission were understood, and the "core" population of asymptomatic infected persons capable of transmitting the agent was assumed. [*See* Acquired Immune Deficiency Syndrome, Epidemic.]

2. Genital Herpes

Genital herpes, though now less publicized than AIDS, still accounts for sizeable morbidity. It is the main cause of genital ulcers in the United States, accounting for at least 10 times more cases than syphilis. The total number of physician–patient consultations for genital herpes increased 20-fold between 1966 and 1991, from 30,000 to almost 600,000. First office visits—a more likely indicator of first genital infections—also increased 13-fold over this same period, from 18,000 in 1966 to 240,000 in 1991 (Fig. 1). Adults aged 20–29 continued to account for most consultations; women visited physicians' offices more frequently than men for genital herpes.

These data must be interpreted cautiously for several reasons: (1) media attention—especially since 1982—may have increased both physicians' and patients' awareness of the signs and symptoms of genital herpes, thus inflating the numbers of patients seen in recent years; (2) a patient treated for genital herpes by a physician for the first time may not actually represent a newly diagnosed case; (3) asymptomatic infections are not reported; and (4) many of those with symptomatic genital herpes probably did not seek health care at all or did so through public clinics.

Moreover, most symptomatic genital herpes infections are merely a tip of the iceberg—for both disease magnitude and viral transmission. Only one-fourth of those with antibodies to HSV-2 give histories compatible with genital herpes infection. Blacks are more likely to have HSV-2 antibodies than whites. In both races, HSV-2 antibody prevalence was slightly higher in women than in men.

Asymptomatic transmission of primary genital herpes infections is a striking feature. Three-fourths of those who had been the sources of infection for patients with documented primary HSV-2 infections gave no histories of genital lesions at the time of contact. Although all sources had HSV antibodies indicative of prior infection, only one-third were aware that they had had any disease compatible with genital herpes.

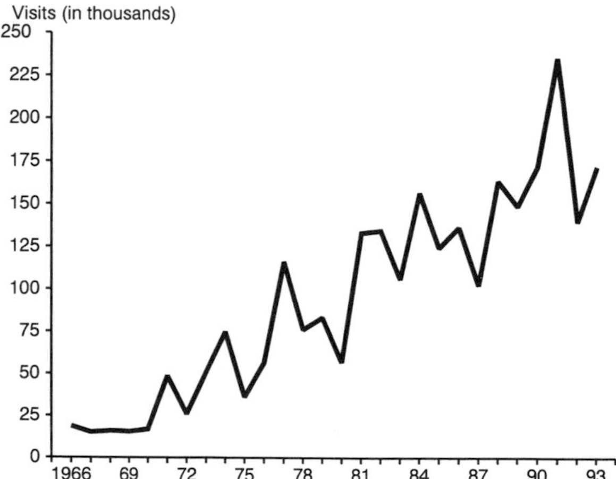

FIGURE 1 Genital herpes simplex virus infections: initial visits to physicians' offices, United States, 1966–1993. [Source: National Disease and Therapeutic Index (IMS American, Ltd.).]

Infection with HSV has been linked to higher risks of HIV transmission in homosexual men. Presumably this is from the virus causing genital ulcers (both recognized and unrecognized), which act as the portal of entry (or egress) for HIV. Because of the magnitude of symptomatic HSV infection in the United States, public health officials are considering potential approaches for HSV control, including HSV vaccine and prophylactic acyclovir in high-risk symptomatic persons.

A serious consequence of genital herpes infection is neonatal herpes. Because it is not a reportable disease, we have no national data with which to calculate incidence. In several regions, the incidence of neonatal herpes reported between 1966 and 1992 has risen more than fourfold. This increase can be attributed to the rise in both symptomatic and asymptomatic genital herpes. Whether these reports represent a true increase in incidence of neonatal herpes or else reflect an improved ability to diagnose the disease is unknown. [*See* Herpesviruses.]

3. Genital Human Papillomavirus Infections

The epidemiology of genital HPV infections is similar to that of genital herpes—except its magnitude of symptomatic infections is nearly threefold higher and the key consequence associated with this infection, cervical neoplasia, is more severe. Using external genital warts as the index for HPV infections, the number of physician–patient consultations for this condition increased over sevenfold between 1966 and 1993, from 179,000 to 1.3 million. First visits also increased nearly sevenfold over the same period, from 54,000 in 1966 to 360,000 in 1987, though the number has declined in recent years (Fig. 2). Persons from 20 to 24 years of age had more frequent genital wart consultations than did patients in other age groups; visits for women outnumbered those for men. As with genital herpes, these data from physician practices have limitations in their interpretation.

Like herpes, genital warts represent only the symptomatic tip of the iceberg of HPV infections. As physician awareness and availability of diagnostic methods increase, subclinical papillomavirus infections of the male and female genital tract are becoming commonly recognized. At present, no serologic test is available, and the virus cannot be recovered through tissue culture. Subclinical infection may be diagnosed by the presence of koilocytes on a cytologic smear or tissue

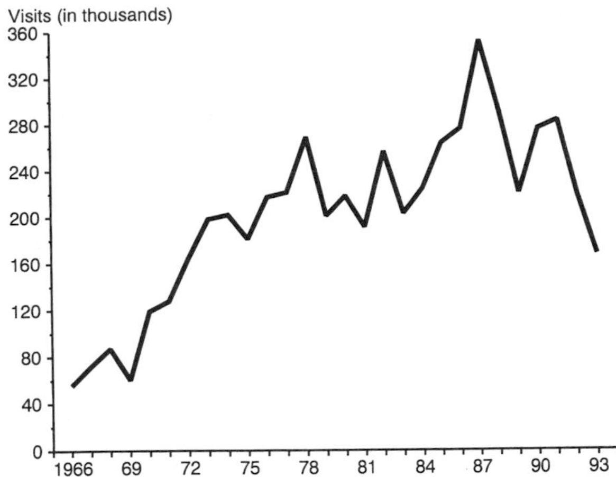

FIGURE 2 Human papillomavirus (genital warts): initial visits to physicians' offices, United States, 1966–1993. [Source: National Disease and Therapeutic Index (IMS American, Ltd.).]

biopsy, HPV DNA sequences detected by hybridization and amplification techniques, HPV antigen detected by immunoperoxidase stains, or certain morphologic features on colposcopy.

Men also have sizeable levels of asymptomatic HPV infection. A large majority of male partners of women with histologic evidence of condyloma acuminata have histologic evidence of penile condyloma. Most have asymptomatic infection, which can be detected colposcopically after application of 5% acetic acid to normal penile epithelium. Thus, HPV infections of the genital tract are probably the most common STD. [*See* Papillomaviruses and Neoplastic Transformation.]

4. Hepatitis B Virus Infections

Nationwide, the incidence of hepatitis B virus (HBV) infections has increased steadily over the last decade in spite of both effective blood-screening programs and the availability of a vaccine. New infections have risen from about 200,000 in 1978 to 300,000 in 1993. Approximately half of those infected suffer symptomatic acute hepatitis. However, the more serious concerns involve the effects of chronic HBV infection. Between 6 and 10% of those infected with HBV become chronic carriers. Chronic active hepatitis develops in more than 25% of carriers and often progresses to cirrhosis and hepatocellular carcinoma. [*See* Liver Cancer and The Role of Hepatitis B Virus.]

Most HBV infections in the United States with known routes of transmission result from sexual exposure. Unfortunately, our vaccination programs have focused primarily on three risk groups—health care workers who are exposed to blood; staff and residents of institutions for the developmentally disabled; and staff and patients in hemodialysis units. These groups, however, account for less than 10% of acute hepatitis B cases. The risk groups that account for most cases—IV drug users, persons acquiring the disease through heterosexual exposure, and homosexual men—are not being reached effectively by current hepatitis B vaccine programs.

The behavior changes of homosexual men to reduce their risk of HIV led to striking decreases in the number of hepatitis B cases (along with the other STD) among this group. However, the number of cases of hepatitis B caused by heterosexual exposure has increased in the 1990s, not surprisingly, primarily among inner city, minority heterosexuals. Of similar concern is the large rise in the proportion of hepatitis B patients with a history of intravenous drug use. Because of the implications of HBV for both chronic hepatitis and hepatocellular carcinoma, the recent trends of sexual transmission carry long-term risks to STD core populations.

B. Bacterial Infections

1. Syphilis

Syphilis remains an important sexually transmmitted organism because of (1) its public health heritage, (2) its association with HIV transmission, (3) its escalating rate among inner city, minority heterosexuals, and (4) its capacity for prevention. Since the introduction of penicillin in the late 1940s, the number of primary/secondary syphilis cases in the United States has declined by 99% (Fig. 3). However, in recent years, infectious syphilis trends have followed a roller coaster course (Fig. 4). In males, the number of reported cases has been affected by homosexual behavior: steadily increasing during the 1970s, but decreasing in the first half of the 1980s. Presumably, this decline reflects behavior changes to reduce the risk of transmitting HIV, which in turn impacts on the other STD.

During the second half of the 1980s, infectious syphilis increased dramatically, to its highest level in 30 years. All of the recent increase in primary/secondary syphilis has occurred in low-income, inner city, minority heterosexual populations. An important contributor to this rise has been the exchange of sex-

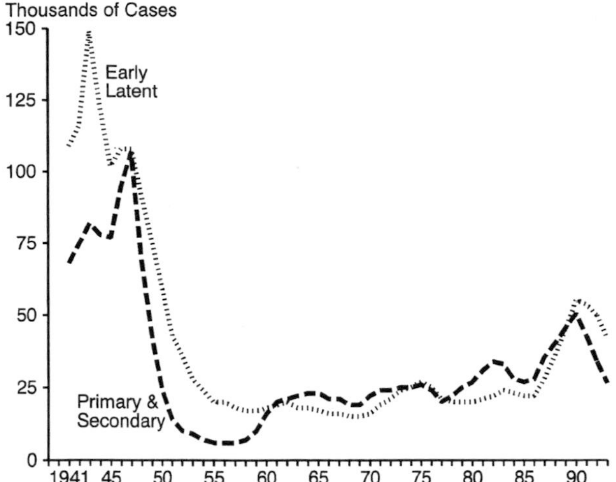

FIGURE 3 Syphilis: reported cases by stage of illness, United States, 1941–1993.

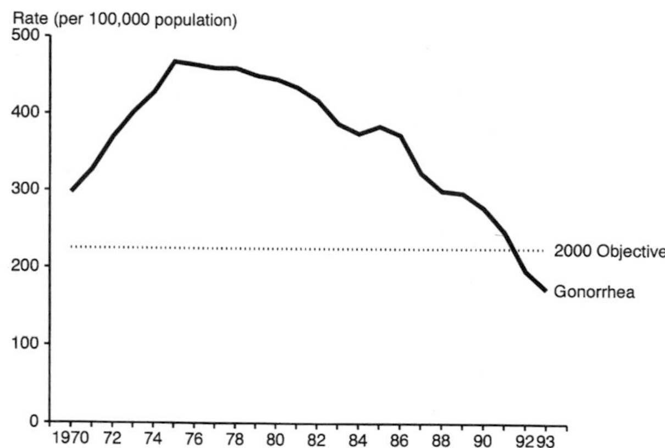

FIGURE 5. Gonorrhea: reported rates, United States, 1970–1993 and the year 2000 objective.

ual services for drugs, especially crack cocaine. The increasing syphilis rate in this traditional STD "core" population had important implications: (1) rises in heterosexual adult syphilis predicted similar trends in congenital syphilis; (2) community health education messages—generated by concerns about HIV—to reduce risky sexual behavior had not permeated the minority, heterosexual community; and (3) because of the association of genital ulcers with HIV transmission, controlling syphilis infections provides an additional opportunity to reduce HIV spread in these communities. However, in the 1990s, rates of syphilis have declined, probably in part due to public health efforts.

Trends in congenital syphilis (CS) reflect recent heterosexual rates. Whereas steady drops in incidence of CS occurred in the 1950s and 1960s, substantial increases have been reported in recent years. Part of the rise observed in 1984 may be attributed to changing surveillance definitions—particularly for stillbirths. However, the increase observed from 1985 on is less attributable to changes in reporting activity. The recent rise in CS incidence suggests that increased vertical transmission may be related to underutilization and inadequacy of prenatal care. With high rates of female syphilis occurring in many areas of the United States, it is particularly important to provide early, high-quality prenatal care to the high-risk populations and to encourage serologic testing in both the first and third trimesters.

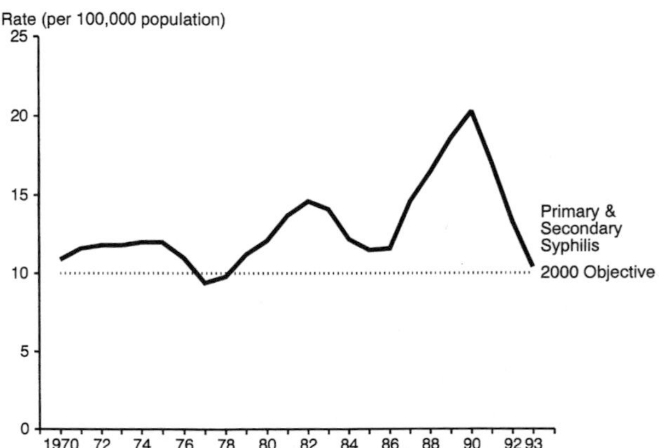

FIGURE 4 Primary and secondary syphilis: reported rates, United States, 1970–1993 and the year 2000 objective.

2. Gonorrhea

An examination of recent gonorrhea trends reveals two major themes: (1) a sustained decrease of penicillin-sensitive organisms and (2) the continued increase in the number and variety of antibiotic-resistant strains. From 1975 to 1993, reported gonorrhea in the United States has declined 55%, with nearly all the decrease occurring since 1981 (Fig. 5). Both males and females have shared in the decline, with the rate of decrease occurring slightly faster in males. Although reported cases account for only an estimated 50% of actual disease incidence, these data have been invaluable in monitoring trends over time. Moreover, trends in gonorrhea visits to private physicians show declines similar to the reported data, beginning even earlier in the 1970s.

The behavioral responses of homosexual men to HIV prevention recommendations have apparently affected gonorrhea trends. In many areas of the country that measure STD by sexual preference, during the 1980s, gonorrhea declined faster in gay males than in other groups. The effect of a decreasing trend in gonococcal infections of homosexual men influences interpretation of the recent heterosexual morbidity trends. Because national gonorrhea data cannot be separated by sexual preference, the sharper decline among the gay population has accelerated the decrease in overall gonorrhea during the 1980s.

Unfortunately, gonorrhea trends in teenagers are disturbing. Teenagers did not share in the gonorrhea decrease over the past two decades. The stable disease rates among the teenage female population mean that gonorrhea control programs designed to lower disease incidence are apparently not yet reaching this key risk group.

Gonococcal antibiotic resistance clouds the generally encouraging trends in reported gonorrhea morbidity. Resistance to antimicrobial agents by the gonococcus has been evolving since the availability of sulfonamides in the 1930s and penicillin in the 1940s. However, the discovery of beta-lactamase-producing *Neisseria gonorrhoeae* (PPNG) in 1976 marked the beginning of an accelerated trend toward greater antibiotic resistance. Since the emergence of PPNG, clinically significant resistance has been described for the three most widely used classes of drugs—the penicillins, tetracyclines, and aminoglycosides (spectinomycin).

The clinical and public health implications of this laboratory-defined resistance remain to be established. Despite the high prevalence of resistant strains in many areas, treatment failure rates have apparently not changed to an appreciable degree. Moreover, the drug spectrum of chromosomal resistance is not restricted to penicillin and tetracycline. Disturbing trends have been also observed on gonococcal susceptibilities to cefoxitin and to spectinomycin.

3. Chlamydia

Genital infections caused by *Chlamydia trachomatis* are the most common bacterial sexually transmitted syndromes in the United States today. Nongonococcal urethritis (NGU) in men is caused by *C. trachomatis* about 40% of the time. In 1972, NGU surpassed gonorrhea as the most common diagnosis for patient visits to private physician offices; the gap has widened in recent years, with NGU now being twice as common as gonococcal urethritis. This pattern of increasing NGU (contrasted with the decreasing gonococcal urethritis) is consistent with trends in other developed countries.

Besides its role in male infections, *C. trachomatis* also plays an important role in causing mucopurulent cervicitis—the female equivalent of NGU. This condition predisposes either to acute pelvic inflammatory disease in nonpregnant women or to infant and puerperal infections in pregnant women, and has ranged from 7 to 12%. Those at highest risk are unwed teenagers living in urban areas, precisely the group at highest risk both for other STD and for adverse pregnancy outcomes.

Based on using *N. gonorrhoeae* as a surrogate, chlamydia is estimated to cause over 4 million infections. Approximately 2.6 million infections occurred in women, 1.8 million in men, and one quarter million in infants. The ratio of *C. trachomatis* to *N. gonorrhoeae* infection was found to be influenced by at least five variables besides gender and age. These included race, pregnancy status, choice of contraception, the proportion of infections without symptoms, and sexual preference. Considerably higher chlamydia-to-gonorrhea ratios were found among whites, pregnant women, oral contraceptive users, and asymptomatic individuals; lower ratios were found among homosexual men.

Efforts to control chlamydia have been hampered by the *relative* difficulties, compared to gonorrhea, of diagnosis and treatment. National chlamydia control guidelines were updated by CDC in 1993. To make maximum use of limited STD funds, recommendations for chlamydia control were primarily based on treating syndromes rather than specific infections. However, broader strategies, where feasible, have proven successful. In areas with the capacity to provide traditional STD strategies such as diagnosis, screening, and partner notification services for chlamydia, apparent declines in sexually transmitted chlamydial infections have occurred.

IV. PREVENTION AND CONTROL STRATEGIES FOR STD

These recent trends in STD have stimulated new directions in prevention and control programs. Previous successful approaches to controlling syphilis and gonorrhea have relied on diagnosis, therapy, and partner notification. However, in the future more emphasis will be needed on primary prevention of STD, through behavioral messages and vaccine development. Many simultaneous activities are necessary to reduce STD. For purposes of simplification, I will group the strategies into six somewhat overlapping categories: health promotion, clinical skills, disease detection, disease treatment, patient counseling, and partner notification.

A. Health Promotion

Health education messages are, in general, an integral part of STD intervention activities. Specific community health education efforts, if presented properly, can supplement other strategies by encouraging primary preventive behaviors in healthy persons at risk. Traditionally, judgmental messages have not been a popular or effective method of changing behavior. In fact, stigmatizing infected individuals by widespread social disapproval may even hinder disease control through delaying care. In today's world of AIDS concerns, public health officials are continually trying to balance their messages between creating excessive fear or excessive reassurance.

To have any chance of success, community health education must stress the benefits of preventive action and promote individual decision making. For example, in our modern society, messages that have emphasized safer sexual practices (postponing sexual initiation, choosing uninfected partners, no receptive anal intercourse, use of condoms) may have already had an impact. In the United States, because of increasing concern with HIV, over half of unmarried persons who believed themselves at risk reported changing their sexual behavior to avoid this disease. As discussed earlier, changing sexual behavior among homosexual men in response to concern about AIDS has also affected other STD.

Health education interventions also affect secondary prevention by reducing risks of complications in those already infected and by limiting the transmission of infection after symptoms appear. Recent experience with use of videotapes to promote actions such as treatment compliance and condom use have been encouraging. These messages are generally given to STD patients as they are being counseled about the appropriate actions they should take both to ensure a cure and also to prevent a reinfection.

School education strategies to increase the students' knowledge of the full spectrum of STD (including AIDS) are also important. Unfortunately, our past heritage of teaching about "VD" in high schools has generally consisted of didactic biomedical lectures concentrating only on syphilis and gonorrhea. To correct this situation, prototype school AIDS and STD curriculum materials for both teachers and students in grades 6 through 12 have been developed, which use a self-instructional format and emphasize behavioral skill-building messages. They have facilitated systematic STD/AIDS education in family life/sex education courses throughout the country.

National hotlines for STD and AIDS provide both general information and specific answers to questions of all callers. In 1988, almost 80,000 calls were answered on the STD hotline. Over 30,000 persons requested referral to confidential medical services in either the public or private sector. Knowing where to turn for STD diagnosis and treatment plays a crucial role in preventing both transmission and complications, and is a major component of community education efforts.

In the future, those concerned with preventing STD need to make better use of the mass media to convey effective health promotion messages. Awareness of herpes and AIDS is widespread largely because of the attention given these conditions by the media. Teenagers are a special audience for the broadcast media; they spend an average of 23 hours a week listening to radio or watching television. Messages are increasingly being aired to encourage condom use in high-risk settings and to publicize hotline numbers for further questions.

B. STD Clinical Care

Medical schools in the United States have been slow to respond to the increasing magnitude of the STD problem. In 1980, less than one in six American medical schools had a specific STD clinic available for STD training. By 1985, only one in five medical schools provided even half its students with STD clinical training. Paradoxically, this occurred just when American physicians need STD training the most; the majority enter specialties requiring knowledge of STD diagnosis and treatment. In 1997, nearly two-thirds of first-year residents chose internal medicine, pediatrics, obstetrics/gynecology, family practice, emergency medicine, and dermatology. Without training in STD problems, the medical community cannot be effectively mobilized to support control programs.

Thus, an integral part of recent STD control strategies in the United States has been to train health care providers in the rapidly changing field of STD. This training must involve all specialties affecting STD control—clinicians, laboratory workers, managers, field investigators, and health educators. For each group, both formal education systems and ongoing inservice training opportunities are essential.

The training initiatives have taken several approaches. To respond to immediate needs, clinical guidelines were developed by CDC to assist STD facilities to improve patient management. National STD treatment recommendations are regularly updated to reflect changing diagnostic capabilities, antibiotic susceptibilities, and pharmacological innovations.

Another training initiative in the United States has involved creating regional STD Prevention/Training Centers. Each center is based on integration of a university medical school with a model public STD clinic for purposes of providing training for midcareer clinicians, as well as for medical, nursing, and paramedical students. By 1997, 10 such multidisciplinary centers were in operation; over 20,000 students have been trained in these facilities since 1979.

Finally, training support has been provided directly to medical schools. To create a cadre of medically qualified clinicians who have a career commitment to the STD discipline, interest in this field must be developed early during clinical training. These individuals could subsequently establish their own STD academic programs, thus multiplying the effect of the training efforts. As of 1997, STD research and/or STD clinical training centers exist in 20 medical schools in the United States.

Physicians interested in STD will need to complement their traditional diagnostic and therapeutic skills with training in additional biosocial sciences—psychology, sociology, and epidemiology. Most clinicians are incapable of taking an adequate sexual history; furthermore, inquiring about the intimate details of their patient's sexual partners does not come naturally. These skills must be taught. Regarding psychosocial training, the increasing emphasis on behavioral factors in STD control is part of the growing awareness in medicine generally of the importance of behavior in both producing and preventing disease. The role of life-style in STD is obvious; the transition from curable bacterial STD to incurable viral STD makes it particularly important that physicians shift their emphasis toward education of individuals to voluntarily modify their behavior.

C. Disease Detection

Early disease detection is crucial to STD intervention strategies. Case finding methods include clinical diagnosis based on symptoms and signs, confirmatory laboratory diagnostic testing in patients with symptoms or signs suggestive of an STD, targeted laboratory diagnostic testing in individuals at high risk for having STD, broad application of diagnostic screening without regard to likelihood of STD, and examination of sex partners of individuals with STD.

Accurate diagnosis is the intervention cornerstone for the early detection strategies of STD control. Whether for making specific diagnoses in those with symptoms, or for screening of persons without symptoms, diagostic tests for STD should ideally be rapid, inexpensive, simply, and accurate. The usual considerations for assessing diagnostic techniques—sensitivity, specificity, and predictive value—have a slightly different interpretation for STD control than for screening of chronic conditions, largely because STD treatment is generally shorter and safer than therapy for other conditions. Moreover, as discussed earlier, curing STD in one individual frequently prevents disease in others. Consequently, achieving high sensitivity by reducing false negatives takes on an increased importance because missed cases (1) place the infected person at continued risk of more serious complications and (2) result in further disease dissemination.

Achieving high specificity by reducing false positives is less important from the public health perspective when treatment is associated with minimal morbidity, cost, and inconvenience. From the perspective of the individual with an STD, especially HIV infection, however, the human and emotional costs of being erroneously stigmatized means that specificity cannot be neglected. In addition, the public health costs of interviewing and notifying large numbers of partners of patients with false-positive tests would drain limited STD resources.

Advances in laboratory techniques have led to many major initiatives in STD control. Serologic capabilities improved syphilis case finding, and selective culture media allowed gonorrhea screening and diagnostic testing. Development of more rapid and less expensive immunodiagnostic tests for chlamydia and HIV have stimulated strategies for controlling these infections. The decision to initiate or abandon a new diagnostic test should be dictated both by the prevalence of the disease in the population being tested and by the sensitivity and specificity of the test being considered. The predictive values of test results are crucial to clinical and public health decisions.

D. Disease Treatment

Once the diagnosis is suspected, treatment should be inexpensive, simple, safe, and effective. Early and adequate treatment of patients and their sexual partners is an effective means of preventing the community spread of STD. To promulgate successful treatment regimens for specific diagnoses, the U.S. Public Health Service has established treatment recommendations as a standard part of its control strategy. Initially, these recommendations covered syphilis and gonorrhea, but in the 1990s they now include 18 other sexually transmitted organisms and syndromes.

Selective prophylactic (preventive or epidemiologic) treatment also has a major role in STD control strategies. In certain instances, waiting to confirm the specific diagnosis prior to initiating therapy is inappropriate. Rather, based on epidemiologic indications, antibiotics should be administered to high-risk individuals when the diagnosis is considered likely, even without clinical signs of infection and before proof of infection by laboratory methods. Thus, selected groups of patients with a high likelihood of infection are identified by epidemiologic analyses—usually because of history of exposure—and treated before confirmation of their infection status. This interrupts the chain of transmission and prevents complications that might occur between the time of testing and treatment, ensures treatment for infected individuals with false-negative laboratory tests, and guarantees treatment for those who might not return when notified of positive tests.

The benefits of this approach are generally thought to outweigh the costs of treating a percentage of uninfected persons, especially of those exposed to gonorrhea, syphilis, or chlamydial infection. Most recently, this approach of selective preventive treatment has effectively limited outbreaks of syphilis, PPNG, and chancroid in metropolitan areas. The same philosophy underlies the recommendation for giving tetracycline concurrently with penicillin to patients with confirmed gonococcal infection, since a relatively high proportion are liable to be harboring coexistent *C. trachomatis*.

E. Patient Counseling

Because of HIV and other incurable STD, counseling (i.e., educating) patients to facilitate changes in their behavior has taken on new importance. STD counselors actively encourage changing patient behaviors to both reduce risks of reinfection or complications and limit spread of STD in the community. The behaviors sought include (1) responding to disease suspicion by promptly seeking appropriate medical evaluation, (2) taking oral medications as directed, (3) returning for follow-up tests when applicable, (4) assuring examination of sexual partners, (5) avoiding future sexual exposure while infectious, and (6) preventing exposure by using barrier protection in high-risk settings.

Risk-reduction counseling to prevent acquisition of STD is increasingly becoming ingrained as a standard part of STD clinical care, whether provided in public STD clinics, other public health facilities, or private physician offices. The expansion of the STD field to include persistent viral infections has lessened the role for prompt treatment and simultaneously raised the need for primary prevention. The concept of "safer sex" has captured worldwide attention because of HIV infection.

Risk-reduction counseling involves much more than emphasizing condom use. Patients need to understand the importance of knowing the risk behaviors of their partners, as well as which sexual practices reduce the potential risk of infection. They should be counseled about the social skills that are essential to negotiating safer behaviors with future partners. Counselors need to adopt nonjudgmental attitudes in discussing potential life-style changes. Unrealistic recommendations will either be ignored or lead to only short-term changes. Patients must find the counseling messages comprehensible, acceptable, and attainable.

Counseling has additional meaning when involved with HIV testing programs. Persons wishing an HIV antibody test should have pretest counseling to establish any risk behaviors for HIV, understand the meaning and implications of test results, allow proper informed consent, and prepare for dealing with the results of the test. Posttest counseling should be delivered by well-trained professionals, one-on-one and face-to-face, not by letter or telephone. Information imparted should build altruistic motivations in seropositive persons and self-protective motivations in seronegative persons. Moreover, HIV-infected individuals should be referred for appropriate medical and psychosocial follow up.

F. Sex Partner Notification

Traditionally, STD control programs in the United States have emphasized active intervention by the health providers to interview the patient, to locate the named sexual partners, and to assure that these individuals are evaluated and treated. The privacy of original patients and contacts is rigorously protected. During the 1970s, in large part due to the expanded spectrum of infections, this process of active intervention was modified in many settings to include a more simplified approach. Instead of relying solely on the health worker, the patient is often encouraged to assume responsibility for locating and referring all of his or her sexual partners.

This self-referral method actively involves patients

in the disease control effort, is inexpensive, is normally acceptable to patients, and reserves scarce staff time for other activities. Potential shortcomings of patient referral methods, however, include (1) limited effectiveness, (2) difficulty in evaluating its outcomes, and (3) nonproductivity with noncompliant patients.

Under most circumstances, patient referral will be the most easily implemented approach to partner notification. Active referral by the health provider is more labor-intensive, time-consuming, and expensive; therefore, active provider referral is restricted to high-yield cases or to high-risk "core" environments. Special situations where this more costly strategy has been useful include (1) introduction of a serious disease (e.g., syphylis or PPNG infection) into a community previously unaffected; (2) men and women with repeated STD infections; (3) female consorts of infectious syphilis cases; (4) STD infections in children; and (5) partners of persons infected with HIV who might not otherwise realize they have been exposed to infection.

V. FUTURE DIRECTIONS

First, reducing the transmission of HIV and controlling its progression to AIDS has become integral to the STD field. Activities will continue to evolve from emphasis on patient and professional education to actual disease intervention through testing and confidential partner notification. Skills will increasingly involve supportive counseling to assist those dealing with consequences of persistent/fatal infections and the need for maintaining safer sexual behaviors.

Second, other facilities beyond STD clinics will increasingly provide STD clinical care to the high-risk populations, for example, drug treatment centers, adolescent, maternal, and child health centers, and family planning clinics. Diagnosis and treatment of STD in these settings will be funded by non-STD public and private resources and will be justified as part of essential medical services provided to these population groups.

Third, health professionals and patients alike will increasingly learn about the medical problems of STD, the need for compliance with therapeutic recommendations, and the responsibility toward sexual partners. The ability to take and provide an accurate sexual history will become part of routine medical examinations.

Fourth, as health reform becomes a reality, the private medical sphere will increasingly assure that modern diagnostic and therapeutic methods are applied to the increasing number of patients served by managed care groups. STD care will be paid for by private insurance and understood by patients and providers alike to be cost-effective. With STD, curative medicine equals preventive medicine.

Firth, acquisition of new computer-based skills will be essential, particularly for modern data collection and their systematic analysis. STD clinical and managerial decisions will be based on these data to obtain the most cost-beneficial results.

BIBLIOGRAPHY

Brandt, A. M. (1985). "No Magic Bullet: A Social History of Venereal Disease in the United States Since 1880." Oxford Univ. Press, New York.

Brunham, R. C. (1991). The concept of core and its relevance to the epidemiology and control of sexually transmitted diseases. *Sexually Transmitted Dis.* **18,** 67–68.

Cates, W., Jr., and Hinman, A. R. (1992). AIDS and absolutism: The demand for perfection in prevention. *N. Engl. J. Med.* **327,** 492–494.

Cates, W., Jr., and Holmes, K. K. (1992). Sexually transmitted diseases. *In* "Maxcy–Rosenau's Public Health and Preventive Medicine" (J. Last and R. B. Wallace, eds.), 13th Ed., pp. 99–114. Appleton & Lange, East Norwalk, Connecticut.

Cates, W., Jr., and Stone, K. M. (1992). Family planning, sexually transmitted diseases, and contraceptive choice: A literature update. *Family Planning Persp.* **24,** 75–84.

Cates, W., Jr., and Wasserheit, J. N. (1991). Genital chlamydial infection: Epidemiology and reproductive sequelae. *Am. J. Obstet. Gynecol.* **164,** 1771–1781.

Centers for Disease Control and Prevention (1997). 1997 sexually transmitted diseases treatment guidelines. *Morbidity and Mortality Weekly Report.* **46**(RR-14), 1–102.

Division of STD Prevention (1997). "Annual Report, 1996." Centers for Disease Control and Prevention, Atlanta.

Holmes, K. K., Mårdh, P.-A., Sparling, P. F., Wiesner, P. J., Cates, W., Jr., Lemon, S. M., and Stamm, W. E. (eds.) (1990). "Sexually Transmitted Diseases," 2nd Ed. McGraw–Hill, New York.

Institute of Medicine (1997). "The Hidden Epidemic: Confronting Sexually Transmitted Diseases," (T. R. Eng and W. T. Butler, eds.). National Academy Press, Washington, D.C.

Kahn, J. G., Walker, C. K., Washington, A. E., Landers, D. V., and Sweet, R. L. (1991). Diagnosing pelvic inflammatory disease. *J. Am. Med. Assoc.* **266,** 2594–2604.

Koutsky, L. A., Stevens, C. E., Holmes, K. K., Ashley, R. L., Kiviat, N. B., Critchlow, C. W., *et al.* (1992). Underdiagnoses of genital herpes by current clinical and viral-isolation procedures. *N. Eng. J. Med.* **326,** 1533–1539.

Wasserheit, J. N. (1989). The significance and scope of reproductive tract infections among third world women. *Int. J. Gynecol. Obstet.* **27** (Suppl. 3), 145–168.